# Multinary Alloys
## Based on II-VI
# Semiconductors

# Multinary Alloys
## Based on II-VI
# Semiconductors

Vasyl Tomashyk

V. Ye. Lashkaryov Institute of Semiconductor
Physics of NAS of Ukraine, Kyiv

CRC Press
Taylor & Francis Group
Boca Raton  London  New York

CRC Press is an imprint of the
Taylor & Francis Group, an **informa** business

CRC Press
Taylor & Francis Group
6000 Broken Sound Parkway NW, Suite 300
Boca Raton, FL 33487-2742

First issued in paperback 2019

© 2016 by Taylor & Francis Group, LLC
CRC Press is an imprint of Taylor & Francis Group, an Informa business

No claim to original U.S. Government works

ISBN-13: 978-1-4822-3651-4 (hbk)
ISBN-13: 978-0-367-37742-7 (pbk

**Visit the Taylor & Francis Web site at**
**http://www.taylorandfrancis.com**

**and the CRC Press Web site at**
**http://www.crcpress.com**

# Contents

# Preface

The II-VI semiconductor compounds have been the subject of extensive research both in fundamental studies and for potential applications in devices. The broad range of band gaps and lattice constants available from these materials, and the unique fundamental phenomena they exhibit, make them attractive for a wide range of applications such as infrared lasers and detectors, blue-green lasers, nonlinear optical materials, magneto-optical devices, photoresistors and transistors, electro-optic modulators, solar cells, components of microwave generators, and solid-state x-ray and γ-ray detectors that can operate at room temperature. Wide band gap II-VI semiconductors are expected to be one of the most vital materials for high-performance optoelectronic devices such as light-emitting diodes and laser diodes operating in the blue or ultraviolet spectral range. Additionally, the high ionicity of these compounds makes them good candidates for high electro-optical and electro-mechanical coupling.

Because of their application in various devices, the search for new semiconductor materials and the improvement of the existing materials are the important fields of study in materials science. Doping by various impurities is the usual method used for modifying and widening physical and chemical semiconductor properties. Multinary compositions allow for the simultaneous adjustment of both the band gap and lattice constant, which allow for an increase in a radiant efficiency at a wider range of wavelengths. Phase diagrams are visual representations of the state of a material as a function of temperature, pressure, and concentration of the constituent components and, therefore, frequently serve as a basic map for alloy choice, processing, development, and understanding.

Though ternary and quaternary phase relations in systems based on zinc, cadmium, and mercury chalcogenides have been published in many journals and collected in handbooks (V. Tomashyk, P. Feychuk, and L. Shcherbak, *Ternary Alloys Based on II-VI Semiconductor Compounds*, London, UK: Taylor & Francis, 2013; and V. Tomashyk, *Quaternary Alloys Based on II-VI Semiconductors*, London, UK: Taylor & Francis, 2013), data pertaining to phase relations in multinary systems based on these semiconductors are preferentially dispersed in the scientific literature. This reference book is intended to describe and illustrate the up-to-date experimental and theoretic information about phase relations based on II-VI semiconductor systems with five or more components. The book contains a critical evaluation of many industrially important systems presented in the form of 2D sections for the condensed phases.

In recent years, much attention has been given to nanoscale II-VI semiconductor compounds. Such materials are obtained from organometallic compounds, which include zinc, cadmium, or mercury and sulfur, selenium, or tellurium. Therefore, the data from the literature on the study of such compounds, published mainly during the last 15–20 years, are summarized in this reference book. In addition to the use of such organometallic compounds as precursors in the preparation of II-VI semiconductor nanoparticles, such compounds have also found wide biological and medicinal applications. They are potential sources of catalysis and could be used for the separation of metal ions, for the deposition of thin films from aqueous solutions, and as templates in multidimensional crystal engineering. Some of them are characterized by good pharmacological properties. The thermolysis of some organo–inorganic compounds provides a new method for the preparation of II-VI semiconductor thin films. The availability of large metal chalcogenide clusters that can be put into solutions may allow for a more rational synthetic approach for a wide variety of solids with semiconducting properties.

All materials are classified according to the periodic table groups of their constituent atoms, that is possible combinations of Zn, Cd, and Hg with chalcogens S, Se, and Te and additional components in the order of their group number. This information is divided into nine chapters.

Every multinary database description contains brief information in the following order: the diagram type, possible phase transformations and physical chemical interaction of the components, methods of the equilibrium investigation, thermodynamic characteristics, and the methods for the sample preparations.

Organometallic compounds are placed in sections according to the increase in the number of atoms of the first element (carbon) in the molecule. At an equal number of carbon atoms in a compound, they are placed according to the increase in the number of atoms of the second (third, fourth, and so on) component.

Most of the figures are presented in their original form, although some have been corrected slightly. If the published data varied essentially, several versions are presented for comparison. The content of system components is usually indicated in mol.%. If the original phase diagram is given in mass.%, the axis with the content in mol.% is also provided in the caption of each such figure.

This book will be helpful for researchers at industrial and national laboratories or universities and graduate students majoring in materials science and engineering. It is also suitable for phase relations researchers, semiconductor chemists, and physicists.

# Author

**Vasyl Tomashyk, Ph.D.** is the executive director and head of the Department of V.Ye. Lashkaryov Institute of Semiconductor Physics of the National Academy of Sciences of Ukraine, Kiev (Ukraine). He graduated from Chernivtsi State University (Ukraine) in 1972 (master of chemistry). He is a doctor of chemical sciences (1992), professor (1999), and author of more than 550 publications in scientific journals and conference proceedings, including 7 books, which are devoted to physical–chemical analysis, chemistry of semiconductors, and chemical treatment of semiconductor surfaces.

Dr. Tomashyk is a international level specialist in the fields of solid-state and semiconductor chemistry, including physical–chemical analysis and technology of semiconductor materials. He was head of research topics within the international program "Copernicus." He has been a member of the Materials Science International Team (Stuttgart, Germany) since 1999, which prepares a series of prestigious reference books under the title *Ternary Alloys* and *Binary Alloys* and he has published 35 articles in *Landolt–Börnstein New Series*. Dr. Tomashyk is actively working with young researchers and graduate students, and under his supervision, 20 Ph.D. theses have been prepared. For many years, he has been a professor at Ivan Franko Zhytomyr State University.

# List of Symbols

| | |
|---|---|
| Ac | Acetyl, $CH_3CO$ |
| Am | Amyl, $C_5H_{11}$ |
| $Am^i$ | $i$-Amyl, $C_5H_{11}$ |
| 2,2′-bipy | 2,2′-Bipyridine, $C_{10}H_8N_2$ |
| 4,4′-bipy | 4,4′-Bipyridine, $C_{10}H_8N_2$ |
| $Bu^i$ | $i$-Butyl, $C_4H_9$ |
| $Bu^n$ | $n$-Butyl, $C_4H_9$ |
| $Bu^s$ | $s$-Butyl, $C_4H_9$ |
| $Bu^t$ | $t$-Butyl, $C_4H_9$ |
| Bz | Benzyl, $C_6H_5CH_2$ |
| CVD | Chemical vapor deposition |
| DMF | Dimethylformamide, $C_3H_7NO$ |
| DMSO | Dimethylsulfoxide, $C_2H_6SO$ |
| DSC | Differential scanning calorimetry |
| DTA | Differential thermal analysis |
| EPMA | Electron probe microanalysis |
| Et | Ethyl, $C_2H_5$ |
| Me | Methyl, $CH_3$ |
| mM | mmol |
| Ph | Phenyl, $C_6H_5$ |
| Phen | 1,10-Phenanthroline, $C_{12}H_8N_2$ |
| $Pr^i$ | $i$-Propyl, $C_3H_7$ |
| $Pr^n$ | $n$-Propyl, $C_3H_7$ |
| Py | Pyridine, $C_5H_5N$ |
| THF | Tetrahydrofuran, $C_4H_8O$ |
| Tol | Tolyl, $C_7H_7$ |
| $Tol^m$ | $m$-Tolyl, $C_7H_7$ |
| $Tol^p$ | $p$-Tolyl, $C_7H_7$ |
| XRD | X-ray diffraction |

# 1 Systems Based on ZnS

## 1.1 Zn–H–Na–Ba–O–Cl–S

**ZnS–NaCl–BaCl$_2$–H$_2$O**. Phase transformations connected with water elimination and BaSO$_4$ and Na$_2$ZnCl$_4$ formation take place in this system at the heat treatment (Yagafarov et al. 1990). This system was investigated by differential thermal analysis (DTA), x-ray diffraction (XRD), and chemical analysis. Heat treatment was realized at 20°C–700°C for 20 min with limited access for oxygen.

## 1.2 Zn–H–Na–C–N–O–S

**Na$_2$[Zn(SCH$_2$CHNH$_2$CO$_2$)$_2$]·4H$_2$O or C$_6$H$_{18}$Na$_2$N$_2$O$_8$S$_2$Zn, sodium bis(L-cysteinato)zincate tetrahydrate**. This multinary compound is formed in this system. To obtain it, a suspension of Zn(SCH$_2$CHNH$_2$CO$_2$)$_2$ (0.01 mol) in 50 mL of H$_2$O was added with stirring 10 mL of 2.0 N NaOH solution (0.02 mol) (Shindo and Brown 1965). Stirring was continued for 1 h and the colloidal precipitate of Zn(OH)$_2$ was filtered off. The filtrate was evaporated to dryness under reduced pressure at room temperature; the white residue was recrystallized from the mixture EtOH/H$_2$O (1:1 volume ratio). The fine white needles were dried in vacuo over P$_2$O$_5$. Na$_2$[Zn(SCH$_2$CHNH$_2$CO$_2$)$_2$]·4H$_2$O was isolated from a solution of 0.02 mol of L-cysteine and 0.01 mol of ZnCl$_2$ in 20 mL of 2 N NaOH (0.04 mol). A white precipitate that at once appears is soon dissolved by stirring. The solution was evaporated to dryness and the residue recrystallized from EtOH/H$_2$O repeatedly.

## 1.3 Zn–H–Na–Sn–O–S

Some compounds are formed in this system.

**[Na$_2$Zn$_{3.5}$Sn$_{3.5}$S$_{13}$]·6H$_2$O**. This compound crystallizes in the cubic structure with the lattice parameter $a = 1786.30 \pm 0.03$ A at 100 K, a calculated density of 2.993 g·cm$^{-3}$, and energy gap 2.9 eV (Wu et al. 2008). Its single crystals were obtained in a reaction containing 0.50 mM of Na$_2$S, 0.80 mM of S, 0.25 mM of Sn, and 0.017 mM of ZnCl$_2$. The starting materials were premixed and grinded and then loaded in a thick wall Pyrex tube. The sample was preheated at 90°C for 1 h and followed by the addition of 0.20 mL of CH$_3$OH and 0.20 mL of H$_2$O. The reaction was heated at 150°C for 7 days. The products were washed with 80% alcohol followed by water. Pale-yellow cubic crystals and crystalline powders were obtained.

**[Na$_{10}$(H$_2$O)$_{32}$][Zn$_5$Sn(μ$_3$-S)$_4$(SnS$_4$)$_4$]·2H$_2$O or Na$_{10}$Zn$_5$Sn$_5$S$_{20}$·34H$_2$O**. This multinary compound is formed in this system. It crystallizes in the triclinic structure with the lattice parameters $a = 1484.57 \pm 0.10$, $b = 1487.72 \pm 0.09$, and $c = 1751.01 \pm 0.11$ pm and $\alpha = 97.402° \pm 0.001°$, $\beta = 100.211° \pm 0.001°$, and $\gamma = 103.956° \pm 0.001°$ pm and a calculated density of 2.197 g·cm$^{-3}$ (Zimmermann et al. 2005). To obtain this compound, $ZnCl_2$ (0.50 mM) was added to a solution of 0.50 mM of $Na_4[SnS_4]$ in 14 mL of $H_2O$. After the mixture was stirred for 4 h, the colorless solution was layered with 10 mL of THF. $Na_{10}Zn_5Sn_5S_{20}·34H_2O$ crystallized as colorless needles over the course of 2 days. Isolation by decanting the mother liquor and the loose precipitate and drying this compound leads to its partial dehydration. All syntheses were performed with strong exclusion of air and external moisture.

## 1.4   Zn–H–Na–N–O–Cl–S

**ZnS–NaCl–NH$_4$NO$_3$.** It was determined that $ZnSO_4$, $ZnO$, $Zn$, $(NH_4)_2SO_4$, and $ZnCl_2$ are formed at the interaction of ZnS, NaCl, and $NH_4NO_3$ at 190°C–380°C (Hîncu 1971).

## 1.5   Zn–H–Na–O–S

**ZnS–NaOH–H$_2$O.** Single crystals of ZnS with sphalerite structure were grown from 1.0 to 10 M solutions of NaOH at 345°C–410°C (Laudise and Ballman 1960).

**ZnS–Na$_2$S–H$_2$O.** Single crystals of ZnS with sphalerite structure were grown from 1.0 to 2.0 M solutions of $Na_2S$ at 250°C–405°C (Laudise and Ballman 1960).

## 1.6   Zn–H–K–C–S

**K$_2$[Zn(S-2,3,5,6-Me$_4$C$_6$H)$_4$] or C$_{40}$H$_{52}$K$_2$S$_4$Zn.** This quinary compound is formed in this system (Gruff and Koch 1989). It could be obtained by the interaction of eight equivalents of $K(S-2,3,5,6-Me_4C_6H)_4$ with $ZnCl_2$ in $CH_3CN$.

## 1.7   Zn–H–K–O–S

**K$_2$[Zn(H$_2$O)$_6$](SO$_4$)$_2$.** This quinary compound is formed in this system. It crystallizes in the monoclinic structure with the lattice parameters $a = 902.9 \pm 0.1$, $b = 1220.4 \pm 0.1$, $c = 614.70 \pm 0.05$ pm, and $\beta = 104.795° \pm 0.009°$ (Euler et al. 2009b). It was prepared by the dissolution of equimolar amounts of $K_2SO_4$ and $ZnSO_4·7H_2O$ in hot distilled $H_2O$ and ensuing evaporation of the solvent. Colorless ellipsoid crystals were obtained.

## 1.8   Zn–H–Rb–O–S

**Rb$_2$[Zn(H$_2$O)$_6$](SO$_4$)$_2$.** This quinary compound is formed in this system. It crystallizes in the monoclinic structure with the lattice parameters $a = 917.9 \pm 0.3$, $b = 1244.1 \pm 0.4$, $c = 623.9 \pm 0.02$ pm, and $\beta = 105.93° \pm 0.03°$ (Euler et al. 2000). It was prepared by the dissolution of equimolar amounts of $Rb_2SO_4$ and $ZnSO_4·7H_2O$ in hot distilled $H_2O$ and ensuing evaporation of the solvent. Colorless sphere-shaped crystals were obtained.

## 1.9　Zn–H–Cs–C–N–O–Re–S

$Cs_2[trans\text{-}[Zn(H_2O)_2][Re_6S_8(CN)_6]$ or $C_6H_4Cs_2N_6O_2Re_6S_8Zn$. This multinary compound is formed in this system. It crystallizes in the orthorhombic structure with the lattice parameters $a = 1848.1 \pm 0.2$, $b = 1065.7 \pm 0.1$, and $c = 1327.0 \pm 0.2$ pm (Beauvais et al. 1998).

## 1.10　Zn–H–Cs–O–S

$Cs_2[Zn(H_2O)_6](SO_4)_2$. This quinary compound is formed in this system. It crystallizes in the monoclinic structure with the lattice parameters $a = 931.4 \pm 0.2$, $b = 1281.7 \pm 0.2$, $c = 636.9 \pm 0.2$ pm, and $\beta = 106.94° \pm 0.02°$ (Euler et al. 2003). It was prepared by the dissolution of equimolar amounts of $Cs_2SO_4$ and $ZnSO_4 \cdot 7H_2O$ in hot distilled $H_2O$ and ensuing evaporation of the solvent.

## 1.11　Zn–H–Cu–C–N–Br–S

Some compounds are formed in this system.

$[Cu(S_2CNBu_2)_2][ZnBr_3]$ or $C_{18}H_{36}CuN_2Br_3S_4Zn$. This compound could be obtained if to a solution of 1 mol of $Zn(S_2CHBu_2)_2$ in chloroform a solution of 2 mol of $Cu(S_2CHBu_2)_2$ in chloroform was added (Cras et al. 1973). After the addition of an equal volume of petroleum ether, crystalline products were formed. They were filtered off and washed with petroleum ether.

Once the composition of the compound is known, a simpler route to it suggested itself. A mixture of 1 mol of $Br_2$ in chloroform and 1 mol of $ZnBr_2$ dissolved in EtOH was added with stirring to a solution of 2 mol of $Cu(S_2CHBu_2)_2$ in chloroform. Upon the addition of petroleum ether and cooling, precipitates of $[Cu(S_2CHBu_2)_2][ZnBr_3]$ were obtained (Cras et al. 1973).

$[Cu_3(S_2CHBu_2)_6][ZnBr_3]_2$ or $C_{54}H_{108}Cu_3N_6Br_6S_{12}Zn_2$. To obtain this compound, a suspension of 1 mol of 3,5-bis($N,N$-dibutyliminium)-1,2,4-trithiolane tetrabromo-di-μ-bromodicuprate(II) in chloroform, a saturated solution of 2 mol of $Zn(S_2CHBu_2)_2$ in the same solvent, was added (Cras et al. 1973). The former compound dissolved slowly, and a crystalline product could be obtained when petroleum ether was added to the solution and cooled to 0°C. The compound was dissolved in chloroform and recrystallized after the addition of an equal amount of petroleum ether.

Once the composition of the compound is known, a simpler route to it suggested itself. A mixture of 1 mol of $Br_2$ in chloroform and 1 mol of $ZnBr_2$ dissolved in EtOH was added with stirring to a solution of 3 mol of $Cu(S_2CHBu_2)_2$ in chloroform. Upon the addition of petroleum ether and cooling, precipitates of $[Cu_3(S_2CHBu_2)_6][ZnBr_3]_2$ were obtained (Cras et al. 1973).

## 1.12　Zn–H–Ba–C–N–O–S

$BaZn(SCN)_4 \cdot 7H_2O$. This multinary compound is formed in this system (Brodersen et al. 1985). It crystallizes in the tetragonal structure with the lattice parameters

$a = 1894.7 \pm 0.6$ and $c = 1032.4 \pm 0.3$ pm and a calculated density of 2.00 g·cm$^{-3}$. The title compound was prepared by a reaction of Ba(SCN)$_2$·3H$_2$O with Zn(NCS)$_2$ in H$_2$O. Its single crystals were obtained by isothermal solvent evaporation.

## 1.13  Zn–H–Cd–C–N–O–S

Some compounds are formed in this system.

**[Zn$_{0.79}$Cd$_{0.21}$\{CH$_2$(COO)$_2$\}\{SC(NH$_2$)$_2$\}$_2$]$_n$** or **C$_5$H$_{10}$Cd$_{0.21}$N$_4$O$_4$S$_2$Zn$_{0.79}$**. This compound crystallizes in the orthorhombic structure with the lattice parameters $a = 948.3 \pm 0.1$, $b = 807.7 \pm 0.5$, and $c = 1476.2 \pm 0.3$ pm at 173 K and a calculated density of 1.936 g·cm$^{-3}$ (Zhang et al. 2000b). To obtain it, to aqueous solutions of ZnSO$_4$·6H$_2$O (0.89 mM) and CdSO$_4$·6H$_2$O (0.11 mM), malonic acid (1.00 mM) and thiourea (2.00 mM) were added. The pH value of the resulting solution was controlled at 5–6 by diluted NaOH and H$_2$SO$_4$. The solution was slowly evaporated at room temperature. Colorless, large needlelike crystals of the title compound were isolated to give the title compound in about 5 days.

**[Zn$_{0.50}$Cd$_{0.50}$\{CH$_2$(COO)$_2$\}\{SC(NH$_2$)$_2$\}$_2$]$_n$** or **C$_5$H$_{10}$Cd$_{0.50}$N$_4$O$_4$S$_2$Zn$_{0.50}$**. This compound crystallizes in the orthorhombic structure with the lattice parameters $a = 948.1 \pm 0.2$, $b = 807.6 \pm 0.1$, and $c = 1471.0 \pm 0.8$ pm at 173 K and a calculated density of 2.024 g·cm$^{-3}$ (Zhang et al. 2000b). To obtain an aqueous solutions of ZnSO$_4$·6H$_2$O (0.67 mM) and CdSO$_4$·6H$_2$O (0.33 mM), malonic acid (1.00 mM) and thiourea (2.00 mM) were added. The pH value of the resulting solution was controlled at 5–6 by diluted NaOH and H$_2$SO$_4$. The solution was slowly evaporated at room temperature. Colorless, large needlelike crystals of the title compound were isolated to give the title compound in about 6 days.

**[Zn$_{0.23}$Cd$_{0.77}$\{CH$_2$(COO)$_2$\}\{SC(NH$_2$)$_2$\}$_2$]$_n$** or **C$_5$H$_{10}$Cd$_{0.77}$N$_4$O$_4$S$_2$Zn$_{0.23}$**. This compound crystallizes in the orthorhombic structure with the lattice parameters $a = 962.9 \pm 0.4$, $b = 812.3 \pm 0.4$, and $c = 1485.1 \pm 0.3$ pm at 173 K and a calculated density of 2.035 g·cm$^{-3}$ (Zhang et al. 2000b). To obtain an aqueous solution of ZnSO$_4$·6H$_2$O (0.50 mM) and CdSO$_4$·6H$_2$O (0.50 mM), malonic acid (1.00 mM) and thiourea (2.00 mM) were added. The pH value of the resulting solution was controlled at 5–6 by diluted NaOH and H$_2$SO$_4$. The solution was slowly evaporated at room temperature. Colorless, large needlelike crystals of the title compound were isolated to give the title compound in about 6 days.

## 1.14  Zn–H–Cd–C–N–S

Some compounds are formed in this system.

**Zn(SCN)$_4$·Cd[CS(NH$_2$)$_2$]$_4$** or **C$_8$H$_{16}$CdN$_{12}$S$_8$Zn**. This compound crystallizes in the tetragonal structure with the lattice parameters $a = 1725.8 \pm 0.2$ and $c = 426.47 \pm 0.07$ pm (Jiang et al. 2000) ($a = 1720.7 \pm 2.0$ and $c = 416 \pm 1$ pm and experimental and calculated densities of 1.89 and 1.97 g·cm$^{-3}$, respectively [Korczyński 1967]). It was synthesized using NH$_4$SCN (KSCN [Jiang et al. 2000]), ZnCl$_2$, CdCl$_2$, and thiourea as raw materials based on the following equations (Bhaskaran et al. 2008):

$$4NH_4SCN + CdCl_2 + ZnCl_2 = ZnCd(SCN)_4 + 4NH_4Cl,$$

$$ZnCd(SCN)_4 + 4CS(NH_2)_2 = Zn(SCN)_4 \cdot Cd[CS(NH_2)_2]_4.$$

In the first step, an aqueous solution of $CdCl_2$ and $NH_4SCN$ was mixed and the resulting solution was slightly warmed with continuous stirring. Then the aqueous solution of $ZnCl_2$ was added to the mixture of $NH_4SCN$ and $CdCl_2$, and $ZnCd(SCN)_4$ got precipitated immediately. It was dissolved in hot $H_2O$ and the coprecipitated $NH_4Cl$ was filtered off. In the second step, the aqueous solution of thiourea was added to $ZnCd(SCN)_4$ solution resulting in white precipitate. After several recrystallization processes, high-purity $Zn(SCN)_4 \cdot Cd[CS(NH_2)_2]_4$ was prepared. Single crystals were grown by slow evaporation of the aqueous solution of the title compound (Bhaskaran et al. 2008).

**$[Zn(NH_2CH_2CH_2NH_2)_3][Cd(S_2CNEt_2)_3]_2$ or $C_{36}H_{84}Cd_2N_{12}S_{12}Zn$.** This compound melts at 135°C and crystallizes in the tetragonal structure with the lattice parameters $a = 2034.8 \pm 0.3$ and $c = 2874.4 \pm 0.9$ pm and a calculated density of 1.52 g·cm$^{-3}$ (Glinskaya et al. 1992, Zemskova et al. 1999). It was prepared by the following procedure. $Zn(S_2CNEt_2)$ (10 mM) and $Cd(S_2CNEt_2)$ (20 mM) were each dissolved in a mixture of 70% ethylenediamine aqueous solution (3 mL) and EtOH (20 mL) at 40°C. Fine-faceted, colorless crystals precipitated when these two solutions were mixed and cooled to room temperature. After 2–3 h, the crystals were filtered, washed with cold EtOH, and dried in air. $Zn_xCd_{1-x}S$ solid solutions are formed at the thermolysis of the title compound (Glinskaya et al. 1992). Single crystals of the title compound were grown by slowly cooling its warm (~40°C) ethanol solutions in the presence of a 10-fold excess of ethylenediamine (Zemskova et al. 1999).

## 1.15  Zn–H–Cd–C–N–I–S

**$[ZnCd(Et_2NCS_2)_2I_2]$ or $C_{10}H_{20}CdN_2I_2S_4Zn$.** This multinary compound is formed in this system. It melts at 184°C with decomposition and crystallizes in the monoclinic structure with the lattice parameters $a = 743.1 \pm 0.2$, $b = 943.7 \pm 0.3$, $c = 1559.3 \pm 0.5$ pm, and $\beta = 93.85° \pm 0.04°$ and a calculated density of 2.216 g·cm$^{-3}$ (Zemskova et al. 1991). To obtain it, to a solution of $Cd(Et_2NCS_2)_2$ in $CH_2Cl_2$ at room temperature, $ZnI_2$ (1.0–1.1:1 molar ratio) was added and the mixture was stirred for 3–5 h. The precipitate that formed was filtered off, washed thoroughly with $CH_2Cl_2$, and dried in air.

## 1.16  Zn–H–Hg–C–N–O–Fe–S

**$[Fe(C_5H_4HgSCN)_2Zn(SCN)_2]\cdot(C_6H_6N_2O)_2$ or $C_{26}H_{20}Hg_2N_8O_2FeS_4Zn$.** This multinary compound is formed in this system, which melts at 185°C with decomposition (Singh and Singh 1986). It was prepared by two methods:

1. A suspension or solution of $[Fe(C_5H_4HgSCN)_2Zn(SCN)_2]$ was prepared in acetone/EtOH mixture, mixed with an ethanol solution of nicotinamide

(1:2 molar ratio), and stirred for 24 h. The precipitate was formed, filtered, washed with EtOH, and dried in vacuum.

2. $Zn(SCN)_2 \cdot 2C_6H_6N_2O$ was dissolved in acetone (Singh and Singh 1986). 1,1'-Bis(thiocyanatomercurio)ferrocene was dissolved in DMSO/acetone mixture. The obtained solutions were mixed in 1:1 molar ratio and stirred for 48 h. The precipitate was formed, filtered, washed in EtOH, and dried in vacuum.

In both cases, dirty white crystals of the title compound were recrystallized from acetone/EtOH mixture.

## 1.17  Zn–H–Hg–C–N–S

$Zn(SCN)_4 \cdot Hg[CS(NH_2)_2]_4$ or $C_8H_{16}HgN_{12}S_8Zn$. This multinary compound is formed in this system. It melts at 180.7°C (Wang et al. 2004) (at 173°C) (Czakis-Sulikowska and Sołoniewicz 1963) and crystallizes in the tetragonal structure with the lattice parameters $a = 1727.71$ and $c = 426.34$ pm (Rajarajan et al. 2006) ($a = 1727.93$ and $c = 426.34$ pm) (Wang et al. 2004). Preparation of the title compound can be easily accomplished by several simple chemical reactions (Wang et al. 2004, Rajarajan et al. 2006):

$$2[CS(NH_2)_2] + ZnX_2 = Zn[CS(NH_2)_2]_2X_2,$$

$$2[CS(NH_2)_2] + HgX_2 = Hg[CS(NH_2)_2]_2X_2,$$

$$4MSCN + Zn[CS(NH_2)_2]_2X_2 + Hg[CS(NH_2)_2]_2X_2 = Zn(SCN)_4 \cdot Hg[CS(NH_2)_2]_4 + 4MX,$$

where $M = K$, Na, $NH_4$; $X = NO_3$, Cl, OAc.

It can also be obtained using a very simple chemical reaction:

$$4[CS(NH_2)_2] + ZnHg(SCN)_4 = Zn(SCN)_4 \cdot Hg[CS(NH_2)_2]_4.$$

Large-sized crystals were grown using the solution of the title compound in a mixed solvent of $EtOH/H_2O$ taken in the volume ratio 3:1 (Rajarajan et al. 2006).

## 1.18  Zn–H–Hg–C–N–Fe–S

Some compounds are formed in this system.

$[Fe(C_5H_4HgSCN)_2Zn(SCN)_2]$ or $C_{14}H_8Hg_2N_4FeS_4Zn$. This compound melts at 178°C with decomposition (Singh and Singh 1986). To obtain it, 1,1'-bis(thiocyanatomercurio)ferrocene was dissolved in a small quantity of DMSO and diluted to 200 mL by acetone. $Zn(SCN)_2$ was similarly dissolved either in acetone or DMSO/acetone mixture separately. The obtained solution were mixed in 1:1 molar ratio and stirred for 24 h. When the quantity of DMSO became more, the precipitate was obtained by addition of EtOH. This precipitate was filtered, washed with EtOH, and dried in vacuum. Light-brown crystals of the title compound were recrystallized from the acetone/EtOH mixture.

$[Fe(C_5H_4HgSCN)_2Zn(SCN)_2]\cdot(C_{10}H_8N_2)$ or $C_{24}H_{16}Hg_2N_6FeS_4Zn$. This compound melts at 162°C with decomposition (Singh and Singh 1986). It was prepared by the next method. A suspension or solution of $[Fe(C_5H_4HgSCN)_2Zn(SCN)_2]$ was prepared in acetone/EtOH mixture, mixed with an ethanol solution of 2,2′-bipyridyl (1:1 molar ratio), and stirred for 24 h. The precipitate was formed, filtered, washed with EtOH, and dried in vacuum. Dirty white crystals of the title compound were recrystallized from acetone/EtOH mixture.

## 1.19    Zn–H–B–C–N–O–S

Some compounds are formed in this system.

$Tt^{t\text{-}Bu}ZnONO_2$ or $C_{21}H_{34}BN_7O_3S_3Zn$ [$Tt^{t\text{-}Bu}$ = tris(1-$t$-butyl)tris(2-thioimidazolyl) hydroborate]. This compound melts at 230°C and crystallizes in the monoclinic structure with the lattice parameters $a = 1104.4 \pm 0.1$, $b = 1589.3 \pm 0.1$, $c = 1603.9 \pm 0.1$ pm, and $\beta = 90.098° \pm 0.002°$ and a calculated density of 1.43 g·cm$^{-3}$ (Tesmer et al. 2001). To obtain it, $Zn(NO_3)_2\cdot4H_2O$ (0.48 mM) in EtOH (10 mL) was added with stirring to $K[Tt^{t\text{-}Bu}]$ (0.48 mM) in EtOH (15 mL). The resulting precipitate was filtered off and the filtrate reduced to 1 mL in vacuo. The precipitated solid was filtered off, washed with 0.5 mL of EtOH, and dried in vacuo, yielding $Tt^{t\text{-}Bu}ZnONO_2$ as colorless crystals.

$\{[Tm^{Ph}]ZnOH\cdot9.5H_2O\}$ or $C_{21}H_{54}BN_6O_{10.5}S_3Zn$ $\{[Tm^{Ph}]$ = tris(2-mercapto-1-phenylimidazolyl)hydroborato ligand$\}$. This compound crystallizes in the trigonal structure with the lattice parameters $a = 1495.0 \pm 0.1$ and $c = 1942.5 \pm 0.2$ pm at 223 K and may be isolated from the reactions of both $Li[Tm^{Ph}]$ with $Zn(ClO_4)_2$ and KOH and $Li[Tm^{Ph}]$ with $Zn(OAc)_2$ and $[(Bu^n)_4N]OH$ (Bridgewater and Parkin 2001).

$[HB(PhC_3N_2H_2)_3ZnSCOMe]$ or $C_{29}H_{25}BN_6OSZn$. This compound crystallizes in the orthorhombic structure with the lattice parameters $a = 1585.4 \pm 0.3$, $b = 1600.1 \pm 0.3$, and $c = 1090.9 \pm 0.2$ pm and experimental and calculated densities 1.41 and 1.40 g·cm$^{-3}$, respectively (Alsfasser et al. 1993). It was obtained by the interaction of $[HB(PhC_3N_2H_2)_3ZnMe]$ with thioacetic acid.

$[HB(OC_6H_4MeC_3N_2H_2)_3ZnSCOMe]$ or $C_{32}H_{31}BN_6O_4SZn$. This compound was obtained by the interaction of $[HB(OC_6H_4MeC_3N_2H_2)_3ZnMe]$ with thioacetic acid (Alsfasser et al. 1993).

$[HB(OC_6H_4MeC_3N_2H_2)_3ZnSEt]$ or $C_{32}H_{33}BN_6O_3SZn$. This compound was obtained by the interaction of $[HB(OC_6H_4MeC_3N_2H_2)_3ZnEt]$ with EtSH (Alsfasser et al. 1993).

$Tt^{Cum}ZnOC_6H_4\text{-}p\text{-}NO_2$ or $C_{42}H_{44}BN_7O_3S_3Zn$ [$Tt^{Cum}$ = tris(1-cumyl)tris(2-thioimidazolyl)hydroborate]. This compound melts at 231°C with decomposition and crystallizes in the triclinic structure with the lattice parameters $a = 1128.7 \pm 0.2$, $b = 1443.4 \pm 0.2$, and $c = 1577.7 \pm 0.2$ pm and $\alpha = 66.776° \pm 0.002°$, $\beta = 72.317° \pm 0.003°$, and $\gamma = 84.854° \pm 0.003°$ and a calculated density of 1.34 g·cm$^{-3}$ (Tesmer et al. 2001). To obtain it, a solution of K[Tt$^{Cum}$] (0.33 mM) in CH$_2$Cl$_2$ (5 mL) was added with stirring to a solution of Zn(ClO$_4$)$_2$·6H$_2$O (0.38 mM) in MeOH (10 mL). After stirring for 2 h, a solution of KOC$_6$H$_4$-$p$-NO$_2$ (0.33 mM) in MeOH (10 ML) was added and the mixture stirred for another 24 h. After evaporation to dryness, the residue was extracted with CH$_2$Cl$_2$ and separated from insoluble parts by filtration. The filtrate was evaporated to dryness again and the residue recrystallized from CH$_2$Cl$_2$/MeCN (1:1 volume ratio) yielding Tt$^{Cum}$ZnOC$_6$H$_4$-$p$-NO$_2$ as yellow crystals.

[{H$_2$B(C$_3$H$_2$N$_2$SC$_9$H$_{11}$)$_2$}ZnNO$_3$]·3C$_6$H$_6$ or C$_{60}$H$_{46}$BN$_5$O$_3$S$_2$Zn. This compound crystallizes in the orthorhombic structure with the lattice parameters $a = 2477.7 \pm 0.5$, $b = 945.1 \pm 0.2$, and $c = 1873.6 \pm 0.4$ pm at 223 K and a calculated density of 1.225 g·cm$^{-3}$ (Kimblin et al. 2000). It was prepared as follows. A mixture of [{H$_2$B(C$_3$H$_2$N$_2$SC$_9$H$_{11}$)$_2$}]$_2$ (0.68 mM) and Zn(NO$_3$)$_2$·6H$_2$O (1.34 mM) in MeOH (ca. 15 mL) was stirred at room temperature for 2 h. The volatile components were removed in vacuo and the residue was extracted into C$_6$H$_6$ (10 mL). The mixture was filtered and the filtrate was concentrated. Colorless microcrystals of the title compound were deposited and isolated by filtration. The volatile components were removed from the filtrate giving the product as a white powder. All manipulations were performed using a combination of glovebox, high-vacuum, or Schlenk techniques.

## 1.20   Zn–H–B–C–N–O–F–S

Some compounds are formed in this system.

[Zn(C$_3$H$_5$NS$_2$)$_2$(BF$_4$)$_2$]·2H$_2$O or C$_6$H$_{14}$B$_2$N$_2$O$_2$F$_8$S$_4$Zn. This compound melts at 127°C (Colombini and Preti 1975). It was prepared by a reaction of Zn(BF$_4$)$_2$ and thiazolidine-2-thione in EtOH. The obtained solid product was washed with ethyl ether.

[Zn(C$_4$H$_7$NOS)$_2$(BF$_4$)$_2$] or C$_8$H$_{14}$B$_2$N$_2$O$_2$F$_8$S$_2$Zn. This compound was obtained by a reaction of Zn(BF$_4$)$_2$ and thiomorpholine-3-one in EtOH (Colombini and Preti 1975). The obtained solid product was washed with ethyl ether.

[Zn(C$_7$H$_9$N$_5$S)(C$_7$H$_8$N$_5$S)·BF$_4$·2H$_2$O] or C$_{14}$H$_{21}$BN$_{10}$O$_2$F$_4$S$_2$Zn. This compound crystallizes in the tetragonal structure with the lattice parameters $a = 1082.0 \pm 0.1$ and $c = 2263.7 \pm 0.3$ pm and a calculated density of 1.448 g·cm$^{-3}$ (Li et al. 2006). To obtain it, Zn(BF$_4$)$_2$·6H$_2$O (0.125 mM) was added at room temperature to acetylpyrazine thiosemicarbazone (0.25 mM) suspended in EtOH (60 mL). After stirring for 30 min, the solid formed was filtered off, washed with EtOH, and dried under vacuum.

[(mepa)$_4$Zn$_3$](BF$_4$)$_2$·2H$_2$O or C$_{32}$H$_{48}$B$_2$N$_8$O$_2$F$_8$S$_4$Zn$_3$ (mepaH = $N$-(2-mercapto-ethyl)picolylamin, C$_8$H$_{12}$N$_2$S). This compound melts at 240°C with decomposition and crystallizes in the tetragonal structure with the lattice parameters $a = 1718.8 \pm 0.2$ and $c = 1523.9 \pm 0.3$ pm and experimental and calculated densities 1.61 and 1.59 g·cm$^{-3}$, respectively (Brand and Vahrenkamp 1995). To obtain it, mepaH (3.34 mM) and NaOH (3.35 mM) were dissolved in H$_2$O (20 mL). A solution of Zn(BF$_4$)$_2$·6H$_2$O (1.67 mM) in H$_2$O (10 mL) was added dropwise with stirring. After being stirred for 30 min, the solution was freeze-dried. The solid residue was picked up in 5 mL of $H_2O$, 10 mL of EtOH was added, and the solution was placed in a desiccator over $CaCl_2$. Within 10 days, $[(mepa)_4Zn_3](BF_4)_2 \cdot 2H_2O$ had precipitated as colorless crystals, which were washed with cold EtOH and dried. Its single crystals were obtained by slow evaporation from an aqueous solution.

## 1.21   Zn–H–B–C–N–O–Cl–S

Some compounds are formed in this system.

$[Tt^{t\text{-}Bu}ZnOClO_3]$ or $C_{21}H_{34}BN_6O_4ClS_3Zn$ [$Tt^{t\text{-}Bu}$ = tris(1-$t$-butyl)tris(2-thioimidazolyl)hydroborate]. This compound melts at 248°C with decomposition and crystallizes in the monoclinic structure with the lattice parameters $a = 1103.1 \pm 0.2$, $b = 1631.4 \pm 0.3$, $c = 1609.3 \pm 0.3$ pm, and $\beta = 90.28° \pm 0.03°$ and a calculated density of 1.47 g·cm$^{-3}$ (Tesmer et al. 2001). To obtain it, a solution of $K[Tt^{t\text{-}Bu}]$ (0.39 mM) in MeOH (15 mL) was added dropwise with stirring within 20 min to a solution of $Zn(ClO_4)_2 \cdot 6H_2O$ (0.39 mM) in MeOH (15 mL). The volume of the solution was reduced to 15 mL in vacuo, the precipitate was filtered off, and the filtrate was evaporated to dryness. The residue was taken up in 10 mL of $CH_2Cl_2$, filtered, and evaporated to dryness again; $Tt^{t\text{-}Bu}ZnOClO_3$ remained as a colorless powder.

$[Tt^{t\text{-}Bu}Zn \cdot EtOH]ClO_4 \cdot EtOH$ or $C_{25}H_{46}BN_6O_6ClS_3Zn$ [$Tt^{t\text{-}Bu}$ = tris(1-$t$-butyl) tris(2-thioimidazolyl)hydroborate]. This compound melts at 126°C with decomposition and crystallizes in the monoclinic structure with the lattice parameters $a = 2534.3 \pm 0.2$, $b = 1962.7 \pm 0.2$, $c = 1851.2 \pm 0.2$ pm, and $\beta = 126.968° \pm 0.002°$ and a calculated density of 1.33 g·cm$^{-3}$ (Tesmer et al. 2001). To obtain it, a solution of $K[Tt^{t\text{-}Bu}]$ (0.39 mM) in EtOH (15 mL) was added dropwise with stirring within 20 min to a solution of $Zn(ClO_4)_2 \cdot 6H_2O$ (0.39 mM) in EtOH (15 mL). The resulting precipitate was filtered off. The addition of 50 mL of hexane to the filtrate produced a precipitate, which was filtered off and dried in vacuo. $[Tt^{t\text{-}Bu}Zn \cdot EtOH]ClO_4 \cdot EtOH$ was obtained as a colorless powder.

$[Tt^{t\text{-}Bu}Zn(PyCHO)]ClO_4$ or $C_{27}H_{39}BN_7O_5ClS_3Zn$ [$Tt^{t\text{-}Bu}$ = tris(1-$t$-butyl)tris (2-thioimidazolyl)hydroborate]. This compound melts at 136°C (Tesmer et al. 2001). To obtain it, a solution of PyCHO (0.31 mM) in chloroform (5 mL) was added while stirring to a suspension of $Tt^{t\text{-}Bu}ZnOClO_3$ (0.31 mM) in chloroform (10 mL). A clear solution resulted. An addition of 20 mL of hexane precipitated $[Tt^{t\text{-}Bu}Zn(PyCHO)]ClO_4$ as a yellow powder, which was filtered off, washed with hexane, and dried in vacuo.

$[Tt^{t\text{-}Bu}Zn(OC_6H_4CHO)] \cdot 0.5CH_2Cl_2$ or $C_{28.5}H_{40}BN_6O_2ClS_3Zn$ [$Tt^{t\text{-}Bu}$ = tris(1-$t$-butyl)tris(2-thioimidazolyl)hydroborate]. This compound melts at 78°C (Tesmer et al. 2001). To obtain it, potassium 2-formylphenolate (0.31 mM) was added to a suspension of $Tt^{t\text{-}Bu}ZnOClO_3$ (0.31 mM) in $CH_2Cl_2$ (10 mL) and the mixture stirred for 1 day. After filtration, the filtrate was evaporated to dryness, leaving $Tt^{t\text{-}Bu}Zn(OC_6H_4CHO) \cdot 0.5CH_2Cl_2$ as a yellow powder.

$[HB(C_{11}H_{12}N_2O_2S)_2(C_{10}H_{10}N_2)Zn(EtOH)]ClO_4$ or $C_{34}H_{42}BN_6O_9ClS_2Zn$. This compound crystallizes in the triclinic structure with the lattice parameters

$a = 1096.2 \pm 0.1$, $b = 1413.1 \pm 0.2$, and $c = 1483.4 \pm 0.2$ pm and $\alpha = 101.011° \pm 0.002°$, $\beta = 109.568° \pm 0.002°$, and $\gamma = 94.234° \pm 0.002°$ at $200 \pm 2$ K (Seebacher et al. 2001). The reaction between two equivalents of thioimidazole ($C_3H_4N_2S$) and one equivalent each of 3-phenyl-5-methylpyrazole ($C_{10}H_{10}N_2$) and $KBH_4$ in boiling toluene leads to very good yields of $KHB(C_{11}H_{12}N_2O_2S)_2(C_{10}H_{10}N_2)$. The reaction between the obtained compound and $Zn(ClO_4)_2 \cdot 6H_2O$ in nondehydrated EtOH produces the raw title compound. Its crystals were obtained by slow evaporation from EtOH.

**[{Tm$^{Mes}$}Zn(MeOH)$_2$](ClO$_4$) or C$_{38}$H$_{48}$BN$_6$O$_6$ClS$_3$Zn {[Tm$^{Mes}$] = tris(2-mercapto-1-mesitylimidazolyl)hydroborato ligand}.** This compound crystallizes in the monoclinic structure with the lattice parameters $a = 1359.20 \pm 0.11$, $b = 1635.98 \pm 0.12$, $c = 2014.84 \pm 0.15$ pm, and $\beta = 92.037° \pm 0.002°$ (Kimblin et al. 1999).

**[(Tt$^{t-Bu}$)$_2$Zn$_3$(PyCH$_2$O)$_2$](ClO$_4$)$_2$ or C$_{54}$H$_{80}$B$_2$N$_{14}$O$_{10}$Cl$_2$S$_6$Zn$_3$ [Tt$^{t-Bu}$ = tris(1-t-butyl)tris(2-thioimidazolyl)hydroborate].** This compound melts at 172°C with decomposition (Tesmer et al. 2001). To obtain it, a solution of PyCH$_2$OH (0.31 mM) in MeOH (15 mL) was added with stirring within 30 min to a solution of Tt$^{t-Bu}$ZnOClO$_3$ (0.31 mM) in MeOH (15 mL). After another 30 min, the solvent was removed in vacuo and the residue was taken up in 20 mL of chloroform and filtered. An addition of 30 mL of hexane to the filtrate precipitated (Tt$^{t-Bu}$)$_2$Zn$_3$(PyCH$_2$O)](ClO$_4$)$_2$ as a colorless powder, which was washed with hexane and dried in vacuo.

**[(Tt$^{t-Bu}$)$_2$Zn$_3$(MePyCH$_2$O)](ClO$_4$)$_2$ or C$_{56}$H$_{84}$B$_2$N$_{14}$O$_{10}$Cl$_2$S$_6$Zn$_3$ [Tt$^{t-Bu}$ = tris(1-t-butyl)tris(2-thioimidazolyl)hydroborate].** This compound melts at 158°C with decomposition and crystallizes in the monoclinic structure with the lattice parameters $a = 1660.4 \pm 0.4$, $b = 2395.0 \pm 0.5$, $c = 2315.3 \pm 0.5$ pm, and $\beta = 108.188° \pm 0.004°$ and a calculated density of 1.47 g·cm$^{-3}$ (Tesmer et al. 2001). It was prepared like (Tt$^{t-Bu}$)$_2$ Zn$_3$(PyCH$_2$O)$_2$](ClO$_4$)$_2$ from MePyCH$_2$OH (0.31 mM) and Tt$^{t-Bu}$ZnOClO$_3$ (0.31 mM). Colorless crystals were formed.

## 1.22   Zn–H–B–C–N–S

Some compounds are formed in this system.

**[H$_2$B(C$_3$H$_2$N$_2$SMe)$_2$]ZnMe or C$_9$H$_{15}$BN$_4$S$_2$Zn.** This compound crystallizes in the monoclinic structure with the lattice parameters $a = 765.2 \pm 0.1$, $b = 1382.2 \pm 0.3$, $c = 1346.9 \pm 0.2$ pm, and $\beta = 102.588° \pm 0.007°$ and a calculated density of 1.527 g·cm$^{-3}$ (Kimblin et al. 2000). To prepare it, a suspension of Tl[H$_2$B(C$_3$H$_2$N$_2$SMe)$_2$] (1.70 mM Tl) in benzene (25 mL) was treated with a solution of ZnMe$_2$ (3.0 mM) in toluene (1.5 mL), thereby resulting in the slow formation of a black precipitate. The mixture was stirred for 2 h and filtered, and the volatile components were removed in vacuo. Crystals for XRD were obtained by recrystallization from benzene. All manipulations were performed using a combination of glovebox, high-vacuum, or Schlenk techniques.

**[Zn{H$_2$B(C$_3$H$_2$N$_2$SMe)$_2$}$_2$] or C$_{16}$H$_{24}$B$_2$N$_8$S$_4$Zn.** This compound melts at 306°C with decomposition (Alvarez et al. 2003) and crystallizes in the monoclinic structure with the lattice parameters $a = 932.7 \pm 0.1$, $b = 1439.3 \pm 0.2$, $c = 1852.0 \pm 0.3$ pm, and

$\beta = 93.811° \pm 0.002°$ at 228 K and a calculated density of 1.456 g·cm$^{-3}$ (Kimblin et al. 2000). To prepare it, a stirred solution of $ZnBr_2$ (0.381 mM) in $H_2O$ (15 mL) was treated with a solution of $Na[H_2B(C_3H_2N_2SMe)_2]$ (0.763 mM) in the same solvent (15 mL), resulting in the formation of a white precipitate (Alvarez et al. 2003). The suspension was stirred for 15 min and the product was isolated by filtration and dried in vacuo for 18 h.

The title compound could also be prepared if a solution of $[H_2B(C_3H_2N_2SMe)_2]$ ZnMe (0.19 mM) in $CHCl_3$ (ca. 5 mL) was allowed to stand at room temperature for 4 days, depositing $[Zn\{H_2B(C_3H_2N_2SMe)_2\}_2]$ as an off-white precipitate, which was separated by filtration, washed with pentane (ca. 5 mL), and dried. All manipulations were performed using a combination of glovebox, high-vacuum, or Schlenk techniques.

$Zn(C_{14}H_{24}BN_4S_2)_2$ **or** $C_{28}H_{48}B_2N_8S_4Zn$. This compound melts at 290°C and crystallizes in the triclinic structure with the lattice parameters $a = 1003.1 \pm 0.4$, $b = 1048.4 \pm 0.4$, and $c = 1855.9 \pm 0.7$ pm and $\alpha = 83.741° \pm 0.007°$, $\beta = 78.966° \pm 0.007°$, and $\gamma = 71.408° \pm 0.009°$ at $243 \pm 2$ K and a calculated density of 1.304 g·cm$^{-3}$ (Alvarez et al. 2003). To prepare it, a stirred solution of $ZnCl_2$ (0.367 mM) in MeOH (15 mL) was treated with a solution of $Na[Bm^{tBu}]$, $[Bm^{tBu}]$ — bis(2-mercapto-1-$t$-butylimidazolyl)borate anion, $C_{14}H_{24}BN_4S_2$, (0.733 mM) in $H_2O$ (15 mL), resulting in the formation of a white precipitate. The suspension was stirred for 20 min and the product was isolated by filtration and dried in vacuo for 1 h.

$[HB(PhC_3N_2H_2)_3ZnSEt]·0.5C_6H_6$ **or** $C_{32}H_{30}BN_6SZn$. This compound crystallizes in the monoclinic structure with the lattice parameters $a = 2613.7 \pm 0.5$, $b = 1761.5 \pm 0.4$, $c = 1405.8 \pm 0.3$ pm, and $\beta = 98.30° \pm 0.03°$ and experimental and calculated densities 1.27 and 1.26 g·cm$^{-3}$, respectively (Alsfasser et al. 1993). It was obtained by the interaction of $[HB(PhC_3N_2H_2)_3ZnBu^t]$ with EtSH.

$[HB(p\text{-}TolC_3N_2H_2)_3ZnSEt]$ **or** $C_{32}H_{33}BN_6SZn$. This compound was obtained by the interaction of $[HB(p\text{-}TolC_3N_2H_2)_3ZnEt]$ with EtSH (Alsfasser et al. 1993).

$[HB(PhC_3N_2H_2)_3ZnSPh]$ **or** $C_{33}H_{27}BN_6SZn$. This compound was obtained by the interaction of $[HB(PhC_3N_2H_2)_3ZnBu^t]$ with PhSH (Alsfasser et al. 1993).

$[Tm^{Ph}]ZnSPh$ **or** $C_{33}H_{27}BN_6S_4Zn$ **{$[Tm^{Ph}]$ = tris(2-mercapto-1-phenylimidazolyl) hydroborato ligand}**. This compound crystallizes in the triclinic structure with the lattice parameters $a = 998.40 \pm 0.10$, $b = 1000.07 \pm 0.10$, and $c = 1772.16 \pm 0.16$ pm and $\alpha = 80.440° \pm 0.002°$, $\beta = 78.784° \pm 0.002°$, and $\gamma = 67.801° \pm 0.002°$ (Bridgewater et al. 2000). It could be obtained by either treatment of $[Tm^{Ph}]ZnI$ with NaSPh or treatment of $[Tm^{Ph}]Li$ with $Zn(SPh)_2$.

$Zn(C_{20}H_{20}BN_4S_2)_2$ **or** $C_{40}H_{40}B_2N_8S_4Zn$. This compound melts at 230°C with decomposition (Alvarez et al. 2003). To prepare it, a stirred solution of $ZnBr_2$ (0.241 mM) in $H_2O$ (15 mL) was treated with a solution of $Na[Bm^{Bz}]$, $[Bm^{Bz}]$—bis(2-mercapto-1-benzylimidazolyl)borate anion, $C_{20}H_{20}BN_4S_2$, (0.483 mM) in a mixture of $H_2O$ (8 mL) and MeOH (3 mL), resulting in the formation of a white precipitate. The suspension was stirred for 15 min and the product was isolated by filtration and dried in vacuo for 2 h.

$Zn(C_{20}H_{20}BN_4S_2)_2$ **or** $C_{40}H_{40}B_2N_8S_4Zn$. This compound melts at 140°C with decomposition (Alvarez et al. 2003). To prepare it, a stirred solution of $ZnCl_2$

(0.271 mM) in $H_2O$ (10 mL) was treated with a solution of Na[Bm$^{p\text{-Tol}}$], [Bm$^{p\text{-Tol}}$]—bis(2-mercapto-1-$p$-tolylimidazolyl)borate anion, $C_{20}H_{20}BN_4S_2$, (0.543 mM) in a MeOH/$H_2O$ (1:1 volume ratio) solvent (20 mL), resulting in the formation of a white precipitate. The suspension was stirred for 20 min and the product was isolated by filtration and dried in vacuo for 2.5 h.

## 1.23   Zn–H–B–C–N–F–S

Some compounds are formed in this system.

[Tt$^{t\text{-Bu}}$ZnF]  or  $C_{21}H_{34}BN_6FS_3Zn$  [Tt$^{t\text{-Bu}}$ = tris(1-$t$-butyl)tris(2-thioimidazolyl) hydroborate]. This compound melts at 238°C and crystallizes in the trigonal structure with the lattice parameters $a = 1151.5 \pm 0.2$ and $c = 3736.2 \pm 0.8$ pm and a calculated density of 1.31 g·cm$^{-3}$ (Tesmer et al. 2001). To obtain it, to a solution of K[Tt$^{t\text{-Bu}}$] (0.19 mM) in MeOH (6 mL) were added solutions of Zn(NO$_3$)$_2$·4H$_2$O (0.19 mM) and KF (0.19 mM) in 4 mL of MeOH each. Upon the addition of CCl$_4$ (30 mL), a precipitate was formed, which was filtered off after 30 min. The filtrate was evaporated to dryness and the residue recrystallized from EtOH/MeCN, yielding Tt$^{t\text{-Bu}}$ZnF as colorless crystals.

$C_{28}H_{48}B_2N_8F_8S_4Zn$, $tetrakis$[1-methyl-3-(2-propyl)-2(3$H$)-imidazolethione]zinc tetrafluoroborate. This compound crystallizes in the monoclinic structure with the lattice parameters $a = 1180.4 \pm 0.2$, $b = 1671.0 \pm 0.3$, $c = 2576.3 \pm 0.5$ pm, and $\beta = 90.14° \pm 0.03°$ at $173 \pm 2$ K and a calculated density of 1.129 g·cm$^{-3}$ (Williams et al. 2006). To obtain it, a mixture of Zn(BF$_4$)$_2$ (4.35 mM) dissolved in EtOH (30 mL) was added with stirring to 20.1 mM of 1-methyl-3-(2-propyl)-2(3$H$)-imidazolethione, which was also dissolved in 30 mL of EtOH. The mixture was stirred for ca. 20 min, and the resulting precipitate was vacuum filtered and washed first with 15 mL of EtOH followed by 30 mL of diethyl ether. The solid was allowed to dry, and the colorless powder was obtained. Colorless crystals were grown by evaporation from a mixture of water and acetonitrile.

$C_{32}H_{56}B_2N_8F_8S_4Zn$, $tetrakis$[1-methyl-3-(1-butyl)-2(3$H$)-imidazolethione]zinc tetrafluoroborate. This compound crystallizes in the tetragonal structure with the lattice parameters $a = 1165.17 \pm 0.16$ and $c = 1682.0 \pm 0.3$ pm at $173 \pm 2$ K and a calculated density of 1.338 g·cm$^{-3}$ (Williams et al. 2006). It was synthesized in a fashion similar to $C_{28}H_{48}B_2N_8F_8S_4Zn$ using 1-methyl-3-(1-butyl)-2(3$H$)-imidazolethione instead of 1-methyl-3-(2-propyl)-2(3$H$)-imidazolethione. Colorless crystals were obtained from MeOH.

## 1.24   Zn–H–B–C–N–Cl–S

Some compounds are formed in this system.

[Zn($C_{12}H_{16}BN_6S_3$)Cl]  or  $C_{12}H_{16}BN_6ClS_3Zn$. This compound crystallizes in the monoclinic structure with the lattice parameters $a = 1272.2 \pm 0.2$, $b = 1262.6 \pm 0.5$, $c = 1196.7 \pm 0.2$ pm, and $\beta = 95.93° \pm 0.02°$ at 123 K (Cassidy et al. 2002). To prepare it, [Tl($C_{12}H_{16}BN_6S_3$)] (0.4 mM) [$C_{12}H_{16}BN_6S_3$ = hydrotris(methimazolyl)borate] was suspended in acetone (30 mL), and a solution of $ZnCl_2$ (0.4 mM) dissolved in the minimum amount of acetone was slowly added. The mixture was stirred overnight at room temperature and the solid formed (TlCl) was filtered off. The filtrate was taken to dryness and the resulting solid dried in vacuo. The crude product could be further purified by dissolving in the minimum amount of $CH_2Cl_2$ and cooling to $-18°C$, whereupon colorless crystals were obtained.

$[Tt^{t\text{-}Bu}ZnCl\cdot MeCN]$ or $C_{23}H_{37}BN_7ClS_3Zn$ $[Tt^{t\text{-}Bu} = tris(1\text{-}t\text{-butyl})tris(2\text{-thioimid-}azolyl)hydroborate]$. This compound melts at $249°C$ with decomposition and crystallizes in the orthorhombic structure with the lattice parameters $a = 3532.2 \pm 0.2$, $b = 3552.3 \pm 0.2$, and $c = 954.0 \pm 0.1$ pm and a calculated density of 1.38 g·cm$^{-3}$ (Tesmer et al. 2001). To obtain it, $ZnCl_2$ (0.48 mM) in MeOH (10 mL) was added to $K[Tt^{t\text{-}Bu}]$ (0.48 mM) in MeOH (4 mL). Within a few minutes, $Tt^{t\text{-}Bu}ZnCl\cdot MeCN$ had precipitated as colorless needles, which were filtered off, washed with MeOH, and dried in vacuo. Its crystals were grown from a mixture EtOH/MeCN.

$[(Tt^{t\text{-}Bu})_2Zn\cdot 3CH_2Cl_2]$ or $C_{45}H_{74}B_2N_{12}Cl_6S_6Zn$ $[Tt^{t\text{-}Bu} = tris(1\text{-}t\text{-butyl})tris(2\text{-thio-}imidazolyl)hydroborate]$. This compound melts at $224°C$ with decomposition and crystallizes in the monoclinic structure with the lattice parameters $a = 4001.7 \pm 0.5$, $b = 1138.7 \pm 0.1$, $c = 2923.0 \pm 0.4$ pm, and $\beta = 110.540° \pm 0.004°$ and a calculated density of 1.36 g·cm$^{-3}$ (Tesmer et al. 2001). To obtain it, $Zn(ClO_4)_2\cdot 6H_2O$ (0.39 mM) in MeOH (5 mL) was added quickly with stirring to $K[Tt^{t\text{-}Bu}]$ (0.78 mM) in MeOH (5 mL). The precipitate was filtered off, the filtrate reduced to 3 mL in vacuo, and the solution was filtered again. The remaining filtrate was evaporated to dryness and the residue recrystallized from $CH_2Cl_2$/hexane (1:2 volume ratio), yielding $(Tt^{t\text{-}Bu})_2Zn\cdot 3CH_2Cl_2$ as colorless crystals.

## 1.25   Zn–H–B–C–N–Cl–F–Ni–S

$[\{(bme\text{-daco})Ni\}_3Zn_2Cl_2][BF_4]_2$ or $C_{30}H_{60}B_2N_6Cl_2F_8Ni_3S_6Zn_2$ $(H_2bme\text{-daco} = N,N'\text{-bis(mercaptoethyl)-1,5-diazacyclootane})$. This multinary compound is formed in this system. It crystallizes in the cubic structure with the lattice parameter $a = 1657.1 \pm 0.3$ (Tuntulani et al. 1992). The reaction of the (bme-daco)Ni(II) complex with $ZnCl_2$ gave the same product regardless of stoichiometry; the following procedure was carried out under a $N_2$ atmosphere. Addition of anhydrous $ZnCl_2$ (0.22 mM) in 10 mL of dry MeOH to a purple solution of (bme-daco)Ni(II) (0.34 mM) in 20 mL of MeOH immediately shifted the solution color to bright red. A pink solid precipitated upon adding 150 mL of anhydrous ether. Ion exchange was required to obtain x-ray-quality crystals. Thus, (bme-daco)Ni(II) and $NaBF_4$ were dissolved in 30 mL of dry MeOH in a test tube and a THF solution of $ZnCl_2$ was carefully layered in by stirring. $[\{(bme\text{-daco})Ni\}_3Zn_2Cl_2][BF_4]_2$ crystallized from this solution as red crystals.

## 1.26   Zn–H–B–C–N–Cl–Br–S

$C_{12}H_{16}BN_6BrS_3Zn\cdot 2.25CHCl_3$. This multinary compound, which crystallizes in the tetragonal structure with the lattice parameters $a = 1329.5 \pm 0.2$ and $c = 1639.4 \pm 0.3$ pm and a calculated density of 1.519 g·cm$^{-3}$, is formed in this system (Garner et al. 1996). Treatment of ZnBr$_2$ with excess of sodium hydrotris(3-methyl-1-imidazolyl-2-thione) borate (C$_{12}$H$_{16}$BN$_6$S$_3$Na) leads to the preferential formation of this compound.

## 1.27   Zn–H–B–C–N–Cl–I–S

[ZnC$_{12}$H$_{16}$BN$_6$S$_3$)I]·0.5CH$_2$Cl$_2$ or C$_{12.5}$H$_{17}$BN$_6$ClIS$_3$Zn. This multinary compound is formed in this system. It crystallizes in the monoclinic structure with the lattice parameters $a = 1116.6 \pm 0.2$, $b = 1525.5 \pm 0.4$, $c = 1374.9 \pm 0.3$ pm, and $\beta = 99.32° \pm 0.02°$ at 123 K (Cassidy et al. 2002). To prepare it, [Tl(C$_{12}$H$_{16}$BN$_6$S$_3$)] (0.4 mM) [C$_{12}$H$_{16}$BN$_6$S$_3$ = hydrotris(methimazolyl)borate] was suspended in acetone (30 mL), and a solution of ZnI$_2$ (0.4 mM) dissolved in the minimum amount of acetone was slowly added. The mixture was stirred overnight at room temperature and the solid formed (TlI) was filtered off. The filtrate was taken to dryness and the resulting solid dried in vacuo. The crude product could be further purified by dissolving in the minimum amount of CH$_2$Cl$_2$ and cooling to −18°C, whereupon colorless crystals were obtained.

## 1.28   Zn–H–B–C–N–Br–S

Some compounds are formed in this system.

[Zn(C$_{12}$H$_{16}$BN$_6$S$_3$)Br] or C$_{12}$H$_{16}$BN$_6$BrS$_3$Zn. To prepare this compound, [Tl(C$_{12}$H$_{16}$BN$_6$S$_3$)] (0.4 mM) [C$_{12}$H$_{16}$BN$_6$S$_3$ = hydrotris(methimazolyl)borate] was suspended in acetone (30 mL), and a solution of ZnBr$_2$ (0.4 mM) dissolved in the minimum amount of acetone was slowly added (Cassidy et al. 2002). The mixture was stirred overnight at room temperature and the solid formed (TlBr) was filtered off. The filtrate was taken to dryness and the resulting solid dried in vacuo. The crude product could be further purified by dissolving in the minimum amount of CH$_2$Cl$_2$ and cooling to −18°C, whereupon colorless crystals were obtained.

[HB(Bu$^t$C$_3$H$_2$N$_2$S)$_2$]ZnBr·MeCN or C$_{23}$H$_{37}$BN$_7$BrS$_3$Zn. This compound melts at 270°C with decomposition and crystallizes in the monoclinic structure with the lattice parameters $a = 969.92 \pm 0.08$, $b = 3303.5 \pm 0.3$, $c = 1028.17 \pm 0.08$ pm, and $\beta = 109.407° \pm 0.002°$ at $238 \pm 2$ K (White et al. 2002). A solution of Na[HB(Bu$^t$C$_3$H$_2$N$_2$S)$_2$] (0.400 mM) in MeOH (6 mL) was added to a stirred solution of ZnBr$_2$ (0.400 mM) in the same solvent (4 mL), resulting in the formation, within 3 min, of a white precipitate. The suspension was stirred for an additional 30 min, treated with H$_2$O (5 mL), and the product was isolated by filtration and dried in vacuo for 1.5 h.

[{Tm$^{Bz}$}ZnBr] or C$_{30}$H$_{28}$BN$_6$BrS$_3$Zn {[Tm$^{Bz}$] = tris(2-mercapto-1-benzylimidazolyl)hydroborato ligand}. This compound melts at 247°C with decomposition and crystallizes in the rhombohedral structure with the lattice parameters $a = 1292.7 \pm 0.2$ and $c = 3195.4 \pm 0.6$ pm at $173 \pm 2$ K and a calculated density of 1.562 g·cm$^{-3}$ (Bakbak et al. 2001). To obtain it, a suspension of ZnBr$_2$ (0.75 mM) and Na[Tm$^{Bz}$] (0.75 mM) in dichloromethane (80 mL) was stirred for 1.5 h at room temperature and filtered.

The colorless filtrate was concentrated under reduced pressure to approximately 10 mL and treated with pentane (15 mL), resulting in the formation of a white solid. The product was isolated by filtration, washed with pentane ($2 \times 20$ mL), and dried in vacuo for 1 h. All reactions were performed under dry oxygen-free $N_2$.

**[{Tm$^{p\text{-Tol}}$}ZnBr or $C_{30}H_{28}BN_6BrS_3Zn$ {[Tm$^{p\text{-Tol}}$] = tris(2-mercapto-1-$p$-tolylimid-azolyl)hydroborato ligand}.** This compound was obtained if a suspension of $ZnBr_2$ (2.00 mM) and Na[Tm$^{p\text{-Tol}}$] (2.00 mM) in dichloromethane (40 mL) was stirred for 2.5 h at room temperature and filtered (Bakbak et al. 2001). The colorless filtrate was concentrated under reduced pressure to approximately 3 mL and treated with pentane (20 mL), resulting in the formation of a white solid. The product was isolated by filtration, washed with pentane ($2 \times 20$ mL), and dried in vacuo for 2 h. All reactions were performed under dry oxygen-free $N_2$.

## 1.29   Zn–H–B–C–N–I–S

Some compounds are formed in this system.

**[{H$_2$B(C$_3$H$_2$N$_2$SMe)$_2$}ZnI] or $C_8H_{12}BN_4IS_2Zn$.** This compound crystallizes in the monoclinic structure with the lattice parameters $a = 1136.2 \pm 0.2$, $b = 993.7 \pm 0.2$, $c = 1348.3 \pm 0.3$ pm, and $\beta = 110.31° \pm 0.02°$ and a calculated density of 2.007 g·cm$^{-3}$ (Kimblin et al. 1997, 2000). To obtain it, a mixture of Tl[H$_2$B(C$_3$H$_2$N$_2$SMe)$_2$] (1.49 mM) and $ZnI_2$ (1.52 mM) in $CH_2Cl_2$ was stirred at room temperature for ca. 3 h resulting in formation of a yellow precipitate. After this period, the mixture was filtered. Removal of the volatile components from the filtrate in vacuo gave [H$_2$B(C$_3$H$_2$N$_2$SMe)$_2$]ZnI as a white solid. Crystals for XRD were obtained by recrystallization from chloroform. All manipulations were performed using a combination of glovebox, high-vacuum, or Schlenk techniques.

**[{HB(C$_3$H$_2$N$_2$SMe)$_2$(C$_3$H$_3$N$_2$H)}ZnI] or $C_{11}H_{14}BN_6IS_2Zn$.** This compound crystallizes in the monoclinic structure with the lattice parameters $a = 1183.3 \pm 0.2$, $b = 1269.9 \pm 0.2$, $c = 1235.3 \pm 0.2$ pm, and $\beta = 99.60° \pm 0.01°$ and a calculated density of 1.805 g·cm$^{-3}$ (Kimblin et al. 1997). To obtain it, a mixture of Tl[HB(C$_3$H$_2$N$_2$SMe)$_2$(C$_3$H$_3$N$_2$H)] (1 mM) and $ZnI_2$ (1.2 mM) in MeOH (20 mL) was stirred at room temperature overnight. After this period, the mixture was filtered and the white precipitate was dried in vacuo and then extracted in $CHCl_3$ (30 mL). The mixture was filtered and the solvent was removed from the filtrate in vacuo, giving [HB(C$_3$H$_2$N$_2$SMe)$_2$(C$_3$H$_3$N$_2$H)]ZnI as a white powder.

This compound could also be prepared if a mixture of [H$_2$B(tim$^{Me}$)$_2$]ZnI (ca. 10 mg) and pyrazole (C$_3$H$_3$N$_2$H) (ca. 5 mg) in $CDCl_3$ (0.7 mL) was heated at 85°C for 2 h (Kimblin et al. 1997).

## 1.30   Zn–H–Ga–C–Sn–N–S

**[Zn$_4$Ga$_{14}$Sn$_2$S$_{35}$](C$_5$H$_{12}$N)$_{12}$ or $C_{60}H_{144}Ga_{14}Sn_2N_{12}S_{35}Zn_4$ (C$_5$H$_{11}$N = piperidine).** This multinary compound is formed in this system. It crystallizes in the cubic structure with the lattice parameter $a = 1825.80 \pm 0.15$ at 150 K and energy gap

3.59 eV (Wu et al. 2010). To obtain it, $SnCl_2$ (0.184 mM), $Ga(NO_3)_3$ (0.988 mM), S (2.233 mM), $Zn(NO_3)_2 \cdot 6H_2O$ (0.259 mM), and piperidine (11.376 mM) were mixed with $H_2O$ (137.8 mM) in a 23 mL Teflon-lined stainless steel autoclave and stirred for 2 h. The vessel was then sealed and heated to 200°C for 10 days. After cooling to room temperature, only pale-yellow block crystals of the title compound were obtained.

## 1.31   Zn–H–Ga–C–Sn–N–Se–S

Some compounds are formed in this system.

$[Zn_4Ga_{13.87}Sn_{2.13}Se_{27.67}S_{5.39}](C_6H_{18}N_2)_5$   or   $C_{30}H_{90}Ga_{13.87}Sn_{2.13}N_{10}Se_{27.67}S_{5.39}Zn_4$ ($C_6H_{16}N_2 = 1,2$-**diamino-2-methylpentan**). This compound crystallizes in the tetragonal structure with the lattice parameters $a = 2370.21 \pm 0.03$ and $c = 4218.58 \pm 0.08$ pm (Wu et al. 2011).

The number of molecules of the organic ligand in the formula is indicated approximately for neutralizing the negative charge of chalcogenide clusters.

$[Zn_4Ga_{14.17}Sn_{1.83}Se_{27.57}S_{5.52}](C_5H_{12}N)_{10}$   or   $C_{50}H_{120}Ga_{14.17}Sn_{1.83}N_{10}Se_{27.57}S_{5.52}Zn_4$ ($C_5H_{11}N =$ **piperidine**). This compound crystallizes in the tetragonal structure with the lattice parameters $a = 2358.49 \pm 0.03$ and $c = 4187.10 \pm 0.17$ pm (Wu et al. 2011).

The number of molecules of the organic ligand in the formula is indicated approximately for neutralizing the negative charge of chalcogenide clusters.

$[Zn_4Ga_{14}Sn_2Se_{31.71}S_{3.29}](C_5H_{12}N)_{12}$ or $C_{60}H_{144}Ga_{14}Sn_2N_{12}Se_{31.71}S_{3.29}Zn_4$ ($C_5H_{11}N =$ **piperidine**). This compound crystallizes in the cubic structure with the lattice parameter $a = 1870.47 \pm 0.10$ (Wu et al. 2011). It was prepared by the next procedure. $Ga(NO_3)_3 \cdot xH_2O$ (ca. 1.0 mM), $Zn(NO_3)_2 \cdot 6H_2O$ (0.25 mM), $SnCl_2$ (0.246 mM), Se (2.24 mM), and piperidine (9.51 mM) were mixed with $H_2O$ (3.0 mL) and DMSO (1 g, 12.80 mM) in a 23 mL Teflon-lined stainless steel autoclave and stirred for 30 min. The vessel was then sealed and heated up to 200°C for 9 days. After cooling to room temperature, a large amount of pale-yellow octahedral crystals were obtained. These raw products were then washed by $H_2O$ and EtOH and dried in air.

The number of molecules of the organic ligand in the formula is indicated approximately for neutralizing the negative charge of chalcogenide clusters.

$[Zn_4Ga_{12.93}Sn_{3.07}Se_{28.13}S_{5.01}](C_{13}H_{28}N_2)_5$   or   $C_{65}H_{140}Ga_{12.93}Sn_{3.07}N_{10}Se_{28.13}S_{5.01}Zn_4$. This compound crystallizes in the tetragonal structure with the lattice parameters $a = 2411.57 \pm 0.02$ and $c = 4195.75 \pm 0.06$ pm (Wu et al. 2011). It was prepared by the following procedure. $Ga(NO_3)_3 \cdot xH_2O$ (ca. 1.0 mM), $Zn(NO_3)_2 \cdot 6H_2O$ (0.25 mM), $SnCl_2$ (0.246 mM), Se (2.24 mM), and 4,4′-trimethylenedipiperidine (9.51 mM) were mixed with $H_2O$ (3.0 mL) and DMSO (2 g, 25.60 mM) in a 23 mL Teflon-lined stainless steel autoclave and stirred for 30 min. The vessel was then sealed and heated up to 200°C for 9 days. After cooling to room temperature, a large amount of pale-yellow octahedral crystals were obtained. These raw products were then washed by $H_2O$ and EtOH and dried in air.

The number of molecules of the organic ligand in the formula is indicated approximately for neutralizing the negative charge of chalcogenide clusters.

$[Zn_4Ga_{13.06}Sn_{2.94}Se_{29.70}S_{3.30}](C_8H_{13}N)_{10}$  or  $C_{80}H_{130}Ga_{13.06}Sn_{2.94}N_{10}Se_{29.70}S_{3.30}Zn_4$ $(C_8H_{12}N = N\text{-ethylcyclohexanamine})$. This compound crystallizes in the tetragonal structure with the lattice parameters $a = 2367.24 \pm 0.02$ and $c = 4226.60 \pm 0.08$ pm (Wu et al. 2011).

The number of molecules of the organic ligand in the formula is indicated approximately for neutralizing the negative charge of chalcogenide clusters.

## 1.32  Zn–H–Ga–C–N–Se–S

Some compounds are formed in this system.

$[Zn_4Ga_{16}Se_{16.89}S_{16.11}](C_6H_{17}N_3)_5$ or $C_{30}H_{85}Ga_{16}N_{15}Se_{16.89}S_{16.11}Zn_4$. This compound crystallizes in the tetragonal structure with the lattice parameters $a = 2289.34 \pm 0.12$ and $c = 4151.6 \pm 0.4$ pm (Wu et al. 2011). To obtain it, $Ga(NO_3)_3 \cdot xH_2O$ (ca. 1.0 mM), $Zn(NO_3)_3 \cdot 6H_2O$ (0.25 mM), Se (2.24 mM), $N$-(2-aminoethyl)piperazine (19.35 mM), and DMSO (25.60 mM) were mixed with $H_2O$ (3.0 mL) in a 23 mL Teflon-lined stainless steel autoclave and stirred for 1 h. The vessel was then sealed and heated up to 200°C for 9 days. After cooling to room temperature, a large amount of pale-yellow octahedral crystals were obtained. These raw products were then washed by $H_2O$ and EtOH and dried in air.

The number of molecules of the organic ligand in the formula is indicated approximately for neutralizing the negative charge of chalcogenide clusters.

$[Zn_4Ga_{16}Se_{20.67}S_{12.33}](C_{11}H_{24}N_2)_5$ or $C_{55}H_{120}Ga_{16}N_{10}Se_{20.67}S_{12.33}Zn_4$ $(C_{11}H_{22}N_2 = \mathbf{dip\text{-}iperidinomethan})$. This compound crystallizes in the tetragonal structure with the lattice parameters $a = 2354.51 \pm 0.02$ and $c = 4122.67 \pm 0.08$ pm (Wu et al. 2011).

The number of molecules of the organic ligand in the formula is indicated approximately for neutralizing the negative charge of chalcogenide clusters.

$[Zn_4Ga_{16}Se_{33-x}S_x](C_{13}H_{28}N_2)_5$  or  $C_{65}H_{140}Ga_{16}N_{10}Se_{33-x}S_xZn_4$. This compound crystallizes in the tetragonal structure with the lattice parameters $a = 4209.3 \pm 0.6$ and $c = 1669.0 \pm 0.3$ pm (Wu et al. 2011). It was prepared by the following procedure. $Ga(NO_3)_3 \cdot xH_2O$ (ca. 1.0 mM), $Zn(NO_3)_3 \cdot 6H_2O$ (0.25 mM), Se (2.24 mM), and 4,4'-trimethylenedipiperidine (9.51 mM) were mixed with $H_2O$ (3.0 mL) and DMSO (2 g, 25.60 mM) in a 23 mL Teflon-lined stainless steel autoclave and stirred for 30 min. The vessel was then sealed and heated up to 200°C for 9 days. After cooling to room temperature, a large amount of pale-yellow octahedral crystals were obtained. These raw products were then washed by $H_2O$ and EtOH and dried in air.

The number of molecules of the organic ligand in the formula is indicated approximately for neutralizing the negative charge of chalcogenide clusters.

## 1.33  Zn–H–In–C–N–O–S

Some compounds are formed in this system.

$[Zn_4In_{16}S_{33}](C_6H_{15}N_3)_5 \cdot xH_2O$   $[C_6H_{15}N_3 \text{—}1\text{-}(2\text{-aminoethyl})\text{piperazine}]$. This compound crystallizes in the tetragonal structure with the lattice parameters $a = 2339.6 \pm 0.8$ and $c = 4344 \pm 2$ pm (Wang et al. 2001a).

**[Zn₄In₁₆S₃₃](C₉H₂₀N₂)₅·xH₂O (C₉H₂₀N₂—4-amino-2,2,6,6-tetramethylenepiperi-dine).** This compound crystallizes in the tetragonal structure with the lattice parameters $a = 2388 \pm 2$ and $c = 4369 \pm 2$ pm (Wang et al. 2001a).

**[Zn₄In₁₆S₃₃](C₁₀H₂₄N₄)₅·xH₂O       [C₁₀H₂₄N₄—1,4-bis(aminopropyl)piperazine].** This compound crystallizes in the tetragonal structure with the lattice parameters $a = 2358.0 \pm 0.8$ and $c = 4392 \pm 1$ pm (Wang et al. 2001a).

**[Zn₄In₁₆S₃₃](C₁₃H₂₆N₂)₅·xH₂O.** This compound crystallizes in the tetragonal structure with the lattice parameters $a = 4181.1 \pm 0.8$ and $c = 1668.5 \pm 0.4$ pm (Wang et al. 2001a). To prepare its crystals, In (85 mg), S (68 mg), $Zn(NO_3)_2 \cdot 6H_2O$ (72 mg), and 4,4′-trimethylenedipiperidine (C₁₃H₂₆N₂) (608 mg) were mixed in a 23 mL Teflon-lined stainless steel autoclave. After the addition of distilled water (4.231 g), the mixture was stirred for 10 min. The vessel was then sealed and heated at 190°C for 10 days. The autoclave was subsequently allowed to cool to room temperature.

## 1.34   Zn–H–In–C–N–S

**[(Zn₁₂In₄₈S₁₇)(Bu₂N)₂₆ or C₂₀₈H₄₆₈In₄₈N₂₆S₉₇Zn₁₂.** This multinary compound is formed in this system. It crystallizes in the cubic structure with the lattice parameter $a = 3498.1 \pm 0.4$ (Bu et al. 2003). To obtain it, indium (123.6 mg), sulfur (121.6 mg), $Zn(NO_3)_2 \cdot 6H_2O$ (125.5 mg), di-$n$-butylamine (2.1036 g), and ethylene glycol (4.2520 g) were mixed in a 23 mL Teflon-lined stainless steel autoclave and stirred for about 20 min. The vessel was then sealed and heated at 150°C for 6 days. The autoclave was subsequently allowed to cool to room temperature. Clear tetrahedral crystals were obtained.

## 1.35   Zn–H–Tl–O–S

**Tl₂[Zn(H₂O)₆](SO₄)₂.** This quinary compound is formed in this system. It crystallizes in the monoclinic structure with the lattice parameters $a = 921.9 \pm 0.2$, $b = 1242.6 \pm 0.2$, $c = 622.6 \pm 0.1$ pm, and $\beta = 106.29° \pm 0.02°$ (Euler et al. 2009a). It was prepared by the dissolution of equimolar amounts of $Tl_2SO_4$ and $ZnSO_4 \cdot 7H_2O$ in hot distilled $H_2O$ and ensuing evaporation of the solvents. Colorless ellipsoid crystals were obtained.

## 1.36   Zn–H–Tb–C–N–O–S

**[TbZn(C₄₄H₂₄N₄O₁₂S₄)H₃O]₂ or C₄₄H₂₇TbN₄O₁₃S₄Zn.** This multinary compound is formed in this system. It crystallizes in the tetragonal structure with the lattice parameters $a = 1542.5 \pm 0.2$ and $c = 983.0 \pm 0.5$ pm at 123.15 K and a calculated density of 1.664 g·cm⁻³ (Chen et al. 2014). It was synthesized by loading $TbCl_3 \cdot 6H_2O$ (0.1 mM), $ZnBr_2$ (0.1 mM), tetra(4-sulfonatophenyl)porphyrin (C₄₄H₂₆N₄O₁₂S₄) (0.1 mM), and 10 mL of distilled $H_2O$ into a 23 mL Teflon-lined stainless steel autoclave and then heating the mixture at 200°C for 1 week. After cooling slowly the mixture down to room temperature at a rate of 6°C/h, red blocklike crystals were obtained. A 3D porous open framework of the title compound is thermally stable up to 336°C.

## 1.37   Zn–H–C–Si–N–O–F–S

[Zn{SiF$_6$(C$_{10}$H$_8$N$_2$S$_2$)$_2$}·3EtOH]$_n$ or C$_{23}$H$_{28}$SiN$_4$O$_3$F$_6$S$_4$Zn. This multinary compound is formed in this system. It is stable up to 100°C and crystallizes in the monoclinic structure with the lattice parameters $a = 1192.4 \pm 0.1$, $b = 1758.7 \pm 0.3$, $c = 1543.6 \pm 0.3$ pm, and $\beta = 101.75° \pm 0.01°$ and a calculated density of 1.560 g·cm$^{-3}$ (Suen and Wang 2006). To obtain it, ZnSiF$_6$·6H$_2$O (1 mM) and 4,4′-dipyridyldisulfide (C$_{10}$H$_8$N$_2$S$_2$) (2 mM) were placed in a flask containing 10 mL MeOH. The mixture was stirred at room temperature for 30 min to yield a white precipitate. The white solid was then filtered off and washed with MeOH. Colorless crystals suitable for XRD were grown by slow diffusion of layered MeOH and H$_2$O solution over 3 days.

## 1.38   Zn–H–C–Si–N–S

Some compounds are formed in this system.

[Zn$_2$(2-SC$_5$H$_3$N-3-SiMe$_3$)$_4$] or C$_{32}$H$_{48}$Si$_4$N$_4$S$_4$Zn$_2$. This compound crystallizes in the monoclinic structure with the lattice parameters $a = 1643.9 \pm 0.3$, $b = 1362.7 \pm 0.2$, $c = 1915.4 \pm 0.3$ pm, and $\beta = 96.69° \pm 0.01°$ and a calculated density of 1.338 g·cm$^{-3}$ (Block et al. 1991). Reaction of Zn(NO)$_3$·6H$_2$O with 2-HSC$_5$H$_3$N-3-SiMe$_3$ in EtOH yields colorless crystals of this compound. It could also be obtained as yellow solids by the electrochemical oxidation of Zn in an acetonitrile solution of 3-trimethylsilyl-pyridine-2-thione (C$_8$H$_{13}$SiNS) (Castro et al. 1993a).

[Zn$_3${N(SiMe$_3$)$_2$}$_2${S(2,4,6-Pr$^i_3$C$_6$H$_2$)}$_4$] or C$_{72}$H$_{128}$Si$_4$N$_2$S$_4$Zn$_3$. This compound melts at 166°C–171°C and crystallizes in the monoclinic structure with the lattice parameters $a = 1878.2 \pm 1.9$, $b = 4174 \pm 4$, $c = 1433.5 \pm 1.6$ pm, and $\beta = 128.28° \pm 0.06°$ and a calculated density of 1.10 g·cm$^{-3}$ (Grützmacher et al. 1992). Addition of 3.47 mM of 2,4,6-Pr$^i_3$C$_6$H$_2$SH to 2.6 mM of Zn[N(SiMe$_3$)$_2$]$_2$ in toluene or CH$_2$Cl$_2$ and stirring of clear solution for 2 h afforded the title compound as colorless crystals after recrystallization at −30°C.

## 1.39   Zn–H–C–Si–N–F–S

Zn$_2$[N(SiMe$_3$)$_2$]{S[2,4,6-(CF$_3$)$_3$C$_6$H$_2$]}$_3$ or C$_{33}$H$_{24}$Si$_2$NF$_{27}$S$_3$Zn$_2$. This multinary compound is formed in this system. It melts at 169°C–174°C and crystallizes in the triclinic structure with the lattice parameters $a = 915.0 \pm 0.4$, $b = 1344.3 \pm 0.6$, and $c = 2046.7 \pm 0.8$ pm and $\alpha = 86.52° \pm 0.03°$, $\beta = 81.58° \pm 0.03°$, and $\gamma = 79.89° \pm 0.03°$ and a calculated density of 1.67 g·cm$^{-3}$ (Grützmacher et al. 1992). An addition of 5.2 mM of 2,6-(SiMe$_3$)$_2$C$_6$H$_3$SH to 2.6 mM of Zn[N(SiMe$_3$)$_2$]$_2$ in toluene or CH$_2$Cl$_2$ and stirring of a clear solution for 2 h afforded the title compound as colorless crystals after recrystallization at −30°C.

## 1.40   Zn–H–C–Si–N–Cl–S

[ZnCl$_2$(2-SC$_5$H$_3$N-6-SiMe$_2$Bu$^t$)$_2$] or C$_{22}$H$_{38}$Si$_2$N$_2$Cl$_2$S$_2$Zn$_2$. This multinary compound is formed in this system. Reaction of ZnCl$_2$ with 2-HSC$_5$H$_3$N-6-SiMe$_2$Bu$^t$ in EtOH yields crystals of the title compound (Block et al. 1991).

## 1.41   Zn–H–C–Si–N–I–S

**[ZnI$_2$(2-SC$_5$H$_3$N-6-SiMe$_2$Bu$^t$)$_2$] or C$_{22}$H$_{38}$Si$_2$N$_2$I$_2$S$_2$Zn$_2$**. This multinary compound is formed in this system. Reaction of ZnI$_2$ with 2-HSC$_5$H$_3$N-6-SiMe$_2$Bu$^t$ in EtOH yields crystals of the title compound (Block et al. 1991).

## 1.42   Zn–H–C–Si–S

Some compounds are formed in this system.

**[CH$_2$SiMe$_3$ZnSCPh$_3$]$_2$ or C$_{46}$H$_{52}$Si$_2$S$_2$Zn$_2$**. This compound melts at 120°C–124°C and crystallizes in the triclinic structure with the lattice parameters $a = 939.6 \pm 0.3$, $b = 1060.9 \pm 0.4$, and $c = 1176.5 \pm 0.4$ pm and $\alpha = 113.88° \pm 0.03°$, $\beta = 94.15° \pm 0.03°$, and $\gamma = 92.89° \pm 0.03°$ at 130 K and a calculated density of 1.33 g·cm$^{-3}$ (Olmstead et al. 1991). To obtain it, triphenylmethanethiol (2 mM) dissolved in 30 mL of a hexane/toluene (1:1 volume ratio) mixture was added dropwise to 2.2 mM of neat zinc dialkyl, and the mixture was stirred for 30 min. The volatile materials were removed under reduced pressure and the residue was redissolved in 5 mL of toluene. Upon the addition of 15 mL of hexane and cooling to −20°C overnight, colorless crystals of the title compound were obtained.

**Zn$_2$\{S[2,6-(SiMe$_3$)$_3$C$_6$H$_3$]\}$_4$ or C$_{48}$H$_{84}$Si$_8$S$_4$Zn$_2$**. This compound melts at 153°C–157°C and crystallizes in the triclinic structure with the lattice parameters $a = 1084.3 \pm 1.0$, $b = 1325.9 \pm 1.3$, and $c = 1342.9 \pm 1.3$ pm and $\alpha = 106.67° \pm 0.08°$, $\beta = 113.06° \pm 0.07°$, and $\gamma = 97.06° \pm 0.08°$ and a calculated density of 1.16 g·cm$^{-3}$ (Grützmacher et al. 1992). Addition of 3.9 mM of 2,4,6-(CF$_3$)$_3$C$_6$H$_2$SH to 2.6 mM of Zn[N(SiMe$_3$)$_2$]$_2$ in toluene or CH$_2$Cl$_2$ and stirring of clear solution for 10 min afforded the title compound as colorless crystals after recrystallization at 0°C.

**[CH$_2$SiMe$_3$ZnS(2,4,6-Pr$^i$$_3$C$_6$H$_2$)]$_3$ or C$_{57}$H$_{102}$Si$_3$S$_3$Zn$_3$**. This compound melts at 110°C–118°C and crystallizes in the monoclinic structure with the lattice parameters $a = 1813.5 \pm 0.8$, $b = 1303.5 \pm 0.5$, $c = 3092.3 \pm 1.0$ pm, and $\beta = 96.34° \pm 0.03°$ at 130 K and a calculated density of 1.12 g·cm$^{-3}$ (Olmstead et al. 1991). It was obtained by the following procedure. Triisopropylthiophenol (3 mM) was added dropwise to an equimolar amount of zinc dialkyl. This afforded a reaction that was accompanied by the evolution of heat and tetramethylsilane and the formation of a white crystalline solid. The resulting solid was redissolved in 2 mL of pentane. Cooling for 2 days in a freezer gave colorless plates of the title compound.

**[CH$_2$SiMe$_3$ZnS(2,4,6-Bu$^t$$_3$C$_6$H$_2$)]$_3$ or C$_{66}$H$_{120}$Si$_3$S$_3$Zn$_3$**. This compound melts at the temperature higher than 200°C with decomposition and crystallizes in the triclinic structure with the lattice parameters $a = 1037.6 \pm 0.3$, $b = 1554.3 \pm 0.6$, and $c = 2647.5 \pm 0.7$ pm and $\alpha = 103.04° \pm 0.03°$, $\beta = 97.35° \pm 0.02°$, and $\gamma = 101.15° \pm 0.01°$ at 130 K and a calculated density of 1.082 g·cm$^{-3}$ (Olmstead et al. 1991). To obtain it, tri-*tert*-butylthiophenol (2 mM) was dissolved in 10 mL of pentane and added dropwise to 2.2 mM of neat bis[(trimethylsilyl)methyl]zinc, and the mixture was stirred for 2 h. The volatile materials were removed, and the residue was redissolved in 4 mL of pentane. Cooling to −20°C afforded colorless needles of the title compound.

## 1.43   Zn–H–C–Ge–N–O–S

**[Ge$_3$S$_6$Zn(H$_2$O)S$_3$Zn(H$_2$O)][(Zn(C$_6$N$_4$H$_{18}$)(H$_2$O)]** or **C$_6$H$_{24}$Ge$_3$N$_4$O$_3$S$_9$Zn$_3$**. This multinary compound is formed in this system. It melts at 153°C–157°C and crystallizes in the trigonal structure with the lattice parameters $a = 960.14 \pm 0.10$ and $c = 837.35 \pm 0.13$ pm and energy gap 3.4 eV (Zheng et al. 2005). To synthesize it, 126.9 mg of GeO$_2$, 172.3 mg of Zn(NO$_3$)$_2$·6H$_2$O, 177.4 mg of sulfur, and 2.0656 g of tris(2-aminoethyl)amine were mixed and stirred for 20 min in a 23 mL Teflon-lined stainless steel autoclave. After sealing the vessel and heating it at 190°C for 5 days, colorless crystals of the title compound were obtained.

## 1.44   Zn–H–C–Ge–N–S

Some compounds are formed in this system.

**(C$_{12}$H$_{25}$NMe$_3$)$_2$[ZnGe$_4$S$_{10}$]** or **C$_{30}$H$_{68}$Ge$_4$N$_2$S$_{10}$Zn**. This compound was prepared at room temperature by the dissolution of C$_{12}$H$_{25}$NMe$_3$Br (1 mM) in EtOH/H$_2$O (200 mL; 1:1 volume ratio) (Wachhold et al. 2000a,b). Na$_4$Ge$_4$S$_{10}$ (0.5 mM) was dissolved in distilled H$_2$O (200 mL). ZnCl$_2$ was dissolved in distilled H$_2$O (20 mL), and the Na$_4$Ge$_4$S$_{10}$ and ZnCl$_2$ solutions were added simultaneously to the solution of C$_{12}$H$_{25}$NMe$_3$Br under constant stirring to form a copious precipitate. The mixture was stirred for 3 h at room temperature and then filtered, washed with water, and vacuum dried overnight. (C$_{12}$H$_{25}$NMe$_3$)$_2$[ZnGe$_4$S$_{10}$] is a wide band gap semiconductor.

**(C$_{14}$H$_{29}$NMe$_3$)$_2$[ZnGe$_4$S$_{10}$]** or **C$_{34}$H$_{76}$Ge$_4$N$_2$S$_{10}$Zn**. This compound was prepared by the same procedure as (C$_{12}$H$_{25}$NMe$_3$)$_2$[ZnGe$_4$S$_{10}$] was obtained using C$_{14}$H$_{29}$NMe$_3$Br instead of C$_{12}$H$_{25}$NMe$_3$Br (Wachhold et al. 2000a,b). It is a wide band gap semiconductor with energy gap 3.13 eV.

**(C$_{16}$H$_{33}$NMe$_3$)$_2$[ZnGe$_4$S$_{10}$]** or **C$_{38}$H$_{84}$Ge$_4$N$_2$S$_{10}$Zn**. This compound was prepared by the same procedure as (C$_{12}$H$_{25}$NMe$_3$)$_2$[ZnGe$_4$S$_{10}$] was obtained using C$_{16}$H$_{33}$NMe$_3$Br instead of C$_{12}$H$_{25}$NMe$_3$Br (Wachhold et al. 2000a,b). It is a wide band gap semiconductor.

**(C$_{18}$H$_{37}$NMe$_3$)$_2$[ZnGe$_4$S$_{10}$]** or **C$_{42}$H$_{92}$Ge$_4$N$_2$S$_{10}$Zn**. This compound was prepared by the same procedure as (C$_{12}$H$_{25}$NMe$_3$)$_2$[ZnGe$_4$S$_{10}$] was obtained using C$_{18}$H$_{37}$NMe$_3$Br instead of C$_{12}$H$_{25}$NMe$_3$Br (Wachhold et al. 2000a,b). It is a wide band gap semiconductor.

## 1.45   Zn–H–C–Sn–N–S

**(CP)$_{1.4}$Zn$_{1.3}$Sn$_2$S$_6$** or **C$_{29.4}$H$_{53.2}$Sn$_2$N$_{1.4}$S$_6$Zn$_{1.3}$** (**CP = cetylpyridinium, C$_{21}$H$_{38}$N$^-$**). This multinary compound is formed in this system. It can be obtained if a solution of 10 mM of CPBr·H$_2$O in 10 mL of formamide was heated at 70°C for a few minutes forming a clear solution (Rangan et al. 2002). Then NH$_3$ was bubbled through the solution to increase the pH ~ 6. To this solution, 1 mM of Na$_2$Sn$_2$S$_6$·14H$_2$O in 10 mL of formamide at 70°C was added to form a clear yellow solution. To the obtained solution, a formamide solution of 1 mM $ZnCl_2$ was added dropwise. A precipitate formed immediately and the mixture was stirred at 70°C for 24 h. The product was isolated by filtration and washed copiously with warm formamide and $H_2O$. The solid was dried under vacuum. $(CP)_{1.4}Zn_{1.3}Sn_2S_6$ is a wide gap semiconductor ($E_g = 3.0$ eV) and is markedly stable in air and shows no appearance of weight loss up to 200°C.

## 1.46   Zn–H–C–N–P–O–S

Some compounds are formed in this system.

$Zn[S_2P(OEt)_2]_2 \cdot (C_2H_8N_2)$ or $C_{10}H_{28}N_2P_2O_4S_4Zn$. This compound is a viscous liquid at room temperature (Harrison et al. 1986a). To obtain it, ethylenediamine dissolved in the minimum amount of EtOH was added to $Zn[S_2P(OEt)_2]_2$ also dissolved in EtOH. After allowing the mixture to stand at room temperature, the colorless crystalline compound was filtered off, recrystallized from hot EtOH, washed with pentane, and dried in air.

$Zn[S_2P(OEt)_2]_2 \cdot (C_4H_4N_2)$ or $C_{12}H_{24}N_2P_2O_4S_4Zn$. This compound melts at 58°C–64°C (Harrison et al. 1986a). To obtain it, pyridazine ($C_4H_4N_2$) dissolved in the minimum amount of EtOH was added to $Zn[S_2P(OEt)_2]_2$ also dissolved in EtOH. After allowing the mixture to stand at room temperature, the colorless crystalline compound was filtered off, recrystallized from hot EtOH, washed with pentane, and dried in air.

$Zn[S_2P(OEt)_2]_2 \cdot (Py)$ or $C_{13}H_{25}NP_2O_4S_4Zn$. This compound melts at 80°C (Harrison et al. 1986a). To obtain it, pyridine dissolved in the minimum amount of EtOH was added to $Zn[S_2P(OEt)_2]_2$ also dissolved in EtOH. After allowing the mixture to stand at room temperature, the colorless crystalline compound was filtered off, recrystallized from hot EtOH, washed with pentane, and dried in air.

$[Zn\{S_2P(OMe)_2\}_2(2,2'-bipy)]$ or $C_{14}H_{20}N_2P_2O_4S_4Zn$. This compound melts at 145°C–147°C and crystallizes in the triclinic structure with the lattice parameters $a = 999.02 \pm 0.07, b = 1014.92 \pm 0.05,$ and $c = 1132.83 \pm 0.08$ pm and $\alpha = 74.947° \pm 0.007°$, $\beta = 80.283° \pm 0.007°$, and $\gamma = 78.538° \pm 0.007°$ and a calculated density of 1.650 g·cm$^{-3}$ (Jeremias et al. 2014). To obtain it, an aqueous solution (25 mL) of $NH_4\{S_2P(OEt)_2\}$ (1.23 mM) was added dropwise to a solution of $Zn(OAc)_2 \cdot 2H_2O$ (0.615 mM) and 2,2'-bipy (0.615 mM) in $H_2O$ (50 mL). After 30 min of stirring, the volume of solution was reduced to 1/3 by evaporation. The precipitate was then filtered off, washed with water, and dried in air at room temperature. The white powder of the title compound was obtained. Single crystals suitable for XRD were obtained by a slow liquid diffusion of hexane into the dichloromethane solution $[Zn\{S_2P(OEt)_2\}_2(2,2'-bipy)]$.

$[Zn\{S_2P(OPr^i)_2\}_2(1,4-NC_4H_4N)_{0.5}]_2$ or $C_{14}H_{30}N_2P_4O_4S_4Zn$. This compound melts at 120°C–121°C and crystallizes in the triclinic structure with the lattice parameters $a = 832.37 \pm 0.14, b = 1126.42 \pm 0.19,$ and $c = 1495.0 \pm 0.4$ pm and $\alpha = 77.797° \pm 0.003°$, $\beta = 73.836° \pm 0.003°$, and $\gamma = 68.317° \pm 0.003°$ at $223 \pm 2$ K and a calculated density of 1.423 g·cm$^{-3}$ (Chen et al. 2006). It was prepared from refluxing (2 h) $Zn\{S_2P(OPr^i)_2\}_2$

(0.2 g) with stoichiometric amount of pyrazine in $CHCl_3$ solution. After the reaction, the solvent was removed in vacuo, and the residue recrystallized by slow evaporation from a $CH_2Cl_2$/hexane (2:1 volume ratio) solution of the title compound.

**$Zn[S_2P(OEt)_2]_2 \cdot (C_6H_{16}N_2)$ or $C_{14}H_{36}N_2P_2O_4S_4Zn$.** This compound melts at 86°C–90°C (Harrison et al. 1986a). To obtain it, $N,N'$-diethylethylenediamine $(C_6H_{16}N_2)$ dissolved in the minimum amount of EtOH was added to $Zn[S_2P(OEt)_2]_2$ also dissolved in EtOH. After allowing the mixture to stand at room temperature, the colorless crystalline compound was filtered off, recrystallized from hot EtOH, washed with pentane, and dried in air.

**$Zn[S_2P(OPr^i)_2]_2 \cdot (C_2H_8N_2)$ or $C_{14}H_{36}N_2P_2O_4S_4Zn$.** This compound melts at 67°C–68°C (Harrison et al. 1986a) and crystallizes in the monoclinic structure with the lattice parameters $a = 1340.2 \pm 0.8$, $b = 1647.0 \pm 0.9$, $c = 1229.4 \pm 0.7$ pm, and $\beta = 99.0° \pm 0.1°$ and experimental and calculated densities 1.34 and 1.36 $g \cdot cm^{-3}$, respectively (Drew et al. 1986). To obtain it, ethylenediamine dissolved in the minimum amount of EtOH was added to $Zn[S_2P(OPr^i)_2]_2$ (1:1 molar ratio) also dissolved in EtOH (Drew et al. 1986, Harrison et al. 1986a). After allowing the mixture to stand at room temperature, the colorless crystalline compound was filtered off, recrystallized from hot EtOH, washed with pentane, and dried in air.

**$Zn[S_2P(OPr^i)_2]_2 \cdot (C_4H_4N_2)$ or $C_{16}H_{32}N_2P_2O_4S_4Zn$.** This compound melts at 78°C–79°C (Harrison et al. 1986a). To obtain it, pyridazine $(C_4H_4N_2)$ dissolved in the minimum amount of EtOH was added to $Zn[S_2P(OPr^i)_2]_2$ also dissolved in EtOH. After allowing the mixture to stand at room temperature, the colorless crystalline compound was filtered off, recrystallized from hot EtOH, washed with pentane, and dried in air.

**$Zn[S_2P(OEt)_2]_2 \cdot (datau)$ or $C_{16}H_{43}N_5P_2O_4S_4Zn$.** This compound melts at 126°C–128°C and crystallizes in the monoclinic structure with the lattice parameters $a = 818.3 \pm 0.2$, $b = 2516.1 \pm 0.2$, $c = 1498.1 \pm 0.2$ pm, and $\beta = 99.72° \pm 0.02°$ and experimental and calculated densities 1.42 and 1.36 $g \cdot cm^{-3}$, respectively (Harrison et al. 1986a). To obtain it, 1,11-diamino-3,6,9-triazaundecane (datau) dissolved in the minimum amount of EtOH was added to $Zn[S_2P(OEt)_2]_2$ also dissolved in EtOH. After allowing the mixture to stand at room temperature, the colorless crystalline compound was filtered off, recrystallized from hot EtOH, washed with pentane, and dried in air.

**$Zn[S_2P(OPr^i)_2]_2 \cdot (Py)$ or $C_{17}H_{33}NP_2O_4S_4Zn$.** This compound melts at 81°C and crystallizes in the monoclinic structure with the lattice parameters $a = 1628.9 \pm 0.3$, $b = 833.1 \pm 0.2$, $c = 2131.6 \pm 0.3$ pm, and $\beta = 99.56° \pm 0.02°$ and experimental and calculated densities 1.34 and 1.33 $g \cdot cm^{-3}$, respectively (Harrison et al. 1986a); and $a = 2128 \pm 1$, $b = 835 \pm 1$, $c = 1627 \pm 1$ pm, and $\beta = 99.5° \pm 0.1°$ and experimental and calculated densities 1.31 and 1.33 $g \cdot cm^{-3}$, respectively (Drew et al. 1986). To obtain it, pyridine dissolved in the minimum amount of EtOH was added to $Zn[S_2P(OPr^i)_2]_2$ also dissolved in EtOH. After allowing the mixture to stand at room temperature, the colorless crystalline compound was filtered off, recrystallized from hot EtOH, washed with pentane, and dried in air.

**$Zn[S_2P(OEt)_2]_2 \cdot (4,4'\text{-bipy})$ or $C_{18}H_{28}N_2P_2O_4S_4Zn$.** This compound crystallizes in the triclinic structure with the lattice parameters $a = 1532.8 \pm 0.7$, $b = 1562.1 \pm 0.6$, and $c = 1242.7 \pm 0.4$ pm and $\alpha = 103.18° \pm 0.03°$, $\beta = 93.98° \pm 0.04°$, and $\gamma = 110.49° \pm 0.03°$ and a calculated density of 1.469 g·cm⁻³ (Zhu et al. 1996). It was prepared by dissolving equimolar quantities of $Zn[S_2P(OEt)_2]_2$ and 4,4'-bipy in EtOH. A white product precipitated immediately from the reaction mixture. Recrystallization from DMF solution gave crystals of $C_{18}H_{28}N_2P_2O_4S_4Zn$.

**$Zn[S_2P(OEt)_2]_2 \cdot (2,2'\text{-bipy})$ or $C_{18}H_{28}N_2P_2O_4S_4Zn$.** This compound melts at 107°C–108°C (Harrison et al. 1986a). To obtain it, 2,2'-bipy dissolved in the minimum amount of EtOH was added to $Zn[S_2P(OEt)_2]_2$ also dissolved in EtOH. After allowing the mixture to stand at room temperature, the colorless crystalline compound was filtered off, recrystallized from hot EtOH, washed with pentane, and dried in air.

**$Zn[S_2P(OPr^i)_2]_2 \cdot (C_6H_{16}N_2)$ or $C_{18}H_{44}N_2P_2O_4S_4Zn$.** This compound melts at 99°C–100°C (Harrison et al. 1986a). To obtain it, $N,N'$-diethylethylenediamine ($C_6H_{16}N_2$) dissolved in the minimum amount of EtOH was added to $Zn[S_2P(OPr^i)_2]_2$ also dissolved in EtOH. After allowing the mixture to stand at room temperature, the colorless crystalline compound was filtered off, recrystallized from hot EtOH, washed with pentane, and dried in air.

**$[Zn\{S_2P(OEt)_2\}_2(Phen)]$ or $C_{20}H_{28}N_2P_2O_4S_4Zn$.** This compound melts at 144°C–146°C (Harrison et al. 1986a, Jeremias et al. 2014). To obtain it, an aqueous solution (25 mL) of $K\{S_2P(OMe)_2\}$ (1.27 mM) was added dropwise to a solution of $Zn(OAc)_2 \cdot 2H_2O$ (0.638 mM) and Phen (0.634 mM) in $H_2O$ (50 mL). After 30 min of stirring, the volume of solution was reduced to 1/3 by evaporation. The precipitate was then filtered off, washed with water, and dried in air at room temperature. The white powder of the title compound was obtained. Single crystals suitable for XRD were obtained by a slow liquid diffusion of hexane into the dichloromethane solution $[Zn\{S_2P(OEt)_2\}_2(Phen)]$.

$Zn[S_2P(OEt)_2]_2 \cdot Phen$ could also be obtained, if Phen dissolved in the minimum amount of EtOH was added to $Zn[S_2P(OEt)_2]_2$ also dissolved in EtOH (Harrison et al. 1986a). After allowing the mixture to stand at room temperature, the colorless crystalline compound was filtered off, recrystallized from hot EtOH, washed with pentane, and dried in air.

**$[Zn\{S_2P(OEt)_2\}_2 \cdot (C_{12}H_{12}N_2)]_n$ or $C_{20}H_{32}N_2P_2O_4S_4Zn$.** This compound melts at 116°C–118°C and crystallizes in the monoclinic structure with the lattice parameters $a = 1168.95 \pm 0.02$, $b = 1695.03 \pm 0.04$, $c = 1469.79 \pm 0.03$ pm, and $\beta = 103.599° \pm 0.001°$ at $120 \pm 2$ K (Tiekink et al. 2007). It was prepared by refluxing $Zn\{S_2P(OEt)_2\}_2$ with 1,2-di-4-pyridylethane. Colorless crystals were isolated by the slow evaporation of an acetonitrile/$CHCl_3$ (1:3 volume ratio) solution of the title compound.

**$Zn[S_2P(OPr^i)_2]_2 \cdot (4,4'\text{-bipy})$ or $C_{22}H_{36}N_2P_2O_4S_4Zn$.** This compound melts at 148°C–150°C and crystallizes in the monoclinic structure with the lattice parameters $a = 1756.9 \pm 0.2$, $b = 2625.3 \pm 0.5$, $c = 2791.5 \pm 0.3$ pm, and $\beta = 100.70° \pm 0.01°$ and a calculated density of 1.361 g·cm⁻³ (Glinskaya et al. 2000c). To obtain this compound,

2.0 mM of Zn[(OPr$^i$)$_2$PS$_2$]$_2$ and 2.0 mM of 4,4′-bipy were dissolved in a minimum quantity of CH$_2$Cl$_2$ at the heating on water bath. The solution was evaporated at room temperature until a white precipitate was appeared and then cooled up to −10°C. The precipitate was filtered off, washed by the cold toluene, and dried in air. Single crystals of Zn(4,4′-bipy)[(OPr$^i$)$_2$PS$_2$]$_2$ were grown by the slow evaporation at room temperature of the solution of Zn(4,4′-bipy)[(OPr$^i$)$_2$PS$_2$]$_2$ in toluene.

**Zn[S$_2$P(OPr$^i$)$_2$]$_2$·(2,2′-bipy)** or **C$_{22}$H$_{36}$N$_2$P$_2$O$_4$S$_4$Zn.** This compound melts at 160°C–161°C and crystallizes in the triclinic structure with the lattice parameters $a = 902.0 \pm 0.2$, $b = 850.2 \pm 0.2$, and $c = 2126.7 \pm 0.2$ pm and $\alpha = 83.12° \pm 0.02°$, $\beta = 97.96° \pm 0.02°$, and $\gamma = 103.46° \pm 0.02°$ and experimental and calculated densities 1.47 and 1.38 g·cm$^{-3}$, respectively (Harrison et al. 1986a). To obtain it, 2,2′-bipy dissolved in the minimum amount of EtOH was added to Zn[S$_2$P(OPr$^i$)$_2$]$_2$ also dissolved in EtOH. After allowing the mixture to stand at room temperature, the colorless crystalline compound was filtered off, recrystallized from hot EtOH, washed with pentane, and dried in air.

**[Zn{S$_2$P(OPr$^i$)$_2$}$_2${4-NC$_5$H$_4$SSC$_5$H$_4$N-4}]$_n$** or **C$_{22}$H$_{36}$N$_2$P$_2$O$_4$S$_6$Zn.** This compound melts at 108°C–109°C and crystallizes in the orthorhombic structure with the lattice parameters $a = 2198.8 \pm 0.5$, $b = 3528.3 \pm 0.7$, and $c = 1756.4 \pm 0.3$ pm at $223 \pm 2$ K and a calculated density of 1.389 g·cm$^{-3}$ (Chen et al. 2006). It was prepared from refluxing (2 h) Zn{S$_2$P(OPr$^i$)$_2$}$_2$ (0.2 g) with stoichiometric amount of 2,2′-dithiopyridine in CHCl$_3$ solution. After the reaction, the solvent was removed in vacuo, and the residue recrystallized by slow evaporation from a chloroform/acetonitrile (3:1 volume ratio) solution of the title compound.

**[Zn{S$_2$P(OPr$^i$)$_2$}$_2${4-NC$_5$H$_4$N(H)C$_5$H$_4$N-4}]$_n$** or **C$_{22}$H$_{37}$N$_3$P$_2$O$_4$S$_4$Zn.** This compound melts at 167°C–169°C and crystallizes in the monoclinic structure with the lattice parameters $a = 1219.69 \pm 0.07$, $b = 838.47 \pm 0.05$, $c = 1557.57 \pm 0.10$ pm, and $\beta = 95.676° \pm 0.002°$ at $223 \pm 2$ K and a calculated density of 1.389 g·cm$^{-3}$ (Chen et al. 2006). It was prepared from refluxing (2 h) Zn{S$_2$P(OPr$^i$)$_2$}$_2$ (0.2 g) with stoichiometric amount of 2,2′-dithiopyridine in CHCl$_3$ solution. After the reaction, the solvent was removed in vacuo, and the residue recrystallized by slow evaporation from a chloroform/acetonitrile (3:1 volume ratio) solution of the title compound.

**Zn[S$_2$P(OEt)$_2$]$_2$·(terpy)** or **C$_{23}$H$_{31}$N$_3$P$_2$O$_4$S$_4$Zn.** This compound melts at 190°C (Harrison et al. 1986a). To obtain it, 2,2′:6,2″-terpyridine (terpy) dissolved in minimum amount of EtOH was added to Zn[S$_2$P(OEt)$_2$]$_2$ also dissolved in EtOH. After allowing the mixture to stand at room temperature, the colorless crystalline compound was filtered off, recrystallized from hot EtOH, washed with pentane, and dried in air.

**[Zn{S$_2$P(OPr$^i$)$_2$}$_2$·(Phen)]** or **C$_{24}$H$_{36}$N$_2$P$_2$O$_4$S$_4$Zn.** This compound melts at 178°C (Harrison et al. 1986a) and crystallizes in the monoclinic structure with the lattice parameters $a = 1926.2 \pm 0.4$, $b = 1039.9 \pm 0.2$, $c = 1655.1 \pm 0.3$ pm, and $\beta = 102.91° \pm 0.03°$ and a calculated density of 1.148 g·cm$^{-3}$ (Glinskaya et al. 2000b) ($a = 1918.3 \pm 0.5$, $b = 1041.8 \pm 0.2$, $c = 1644.4 \pm 0.5$ pm, and $\beta = 103.29° \pm 0.02°$ at 173 K and a calculated density of 1.396 g·cm$^{-3}$ [Tiekink 2001b]; $a = 1931.5 \pm 0.4$,

$b = 1043.8 \pm 0.2$, $c = 1656.7 \pm 0.3$ pm, and $\beta = 102.89° \pm 0.03°$ and a calculated density of 1.371 g·cm$^{-3}$ [Jian et al. 2000]). To obtain this compound, 2 mM of $Zn[(Pr^iO)_2PS_2]_2$ and 2.5 mM of Phen·$H_2O$ were dissolved at the heating in ca. 25 mL of $CHCl_3$ (5 mM of $Zn[(Pr^iO)_2PS_2]_2$ and 5 mM of Phen·$H_2O$ in hot EtOH [50 mL]) (Jian et al. 2000). The obtained solution was filtered off and 20 mL of EtOH was added. Then, approximately half of the solvent volume was evaporated at the heating of the solution on the water bath. After cooling, a white precipitate was obtained, which was filtered off and washed by the cold EtOH. Single crystals of $[Zn\{S_2P(OPr^i)_2\}_2.(Phen)]$ were obtained by the evaporation at room temperature of its saturated (at the heating) solution in the mixture $CHCl_4$/EtOH (1:1 volume ratio) (Glinskaya et al. 2000b). Colorless crystals of the title compound could also be isolated from the slow evaporation of an acetonitrile solution containing equimolar quantities of $Zn\{S_2P(OPr^i)_2\}_2$ and Phen (Tiekink 2001b).

**$[Zn\{S_2P(OPr^i)_2\}_2 \cdot (C_{12}H_{10}N_2)]_n$ or $C_{24}H_{38}N_2P_2O_4S_4Zn$.** This compound melts at 162°C–164°C and crystallizes in the monoclinic structure with the lattice parameters $a = 2100.3 \pm 0.3$, $b = 753.78 \pm 0.08$, $c = 2151.7 \pm 0.3$ pm, and $\beta = 100.289° \pm 0.003°$ at $103 \pm 2$ K (Welte and Tiekink 2007). It was prepared by refluxing $Zn\{S_2P(OPr^i)_2\}_2$ with 1,2-bis(4-pyridyl)ethylene. Colorless crystals were isolated by the slow evaporation of an acetonitrile/$CHCl_3$ (1:1 volume ratio) solution of the title compound.

**$[Zn\{S_2P(OPr^i)_2\}_2 \cdot (C_{12}H_{10}N_4)]_n$ or $C_{24}H_{38}N_4P_2O_4S_4Zn$.** This compound melts at 108°C–111°C and crystallizes in the monoclinic structure with the lattice parameters $a = 1180.9 \pm 0.3$, $b = 1124.3 \pm 0.3$, $c = 1293.3 \pm 0.3$ pm, and $\beta = 108.712° \pm 0.005°$ at $150 \pm 2$ K and a calculated density of 1.434 g·cm$^{-3}$ (Avila and Tiekink 2006). It was prepared by refluxing $Zn\{S_2P(OPr^i)_2\}_2$ with bis(3-pyridylmethylene)hydrazine. Yellow crystals were isolated by the slow evaporation of a $Pr^iOH$ solution of the title compound.

**$[Zn\{S_2P(OPr^i)_2\}_2(C_{12}H_{12}N_2)]_n$ or $C_{24}H_{40}N_2P_2O_4S_4Zn$.** This compound melts at 149°C–150°C and crystallizes in the triclinic structure with the lattice parameters $a = 848.27 \pm 0.06$, $b = 1163.92 \pm 0.07$, and $c = 1749.96 \pm 0.12$ pm and $\alpha = 78.124° \pm 0.003°$, $\beta = 83.169° \pm 0.003°$, and $\gamma = 75.006° \pm 0.003$ at $223 \pm 2$ K and a calculated density of 1.378 g·cm$^{-3}$ (Lai et al. 2004b). It was obtained from refluxing $[Zn\{S_2P(OPr^i)_2\}_2$ (0.2 g) with stoichiometric amount of 1,2-bis(4-pyridyl)ethane ($C_{12}H_{12}N_2$) in $CHCl_3$ solution. After the reaction, the solvent was removed in vacuo, and the residue recrystallized by the slow evaporation of a chloroform/MeOH (3:1 volume ratio) solution of the title compound.

**$[Zn\{S_2P(OPr^i)_2\}_2\{4\text{-}NC_5H_4(CH_2)_3C_5H_4N\text{-}4\}]_n$ or $C_{25}H_{42}N_2P_2O_4S_4Zn$.** This compound melts at 110°C–111°C and crystallizes in the monoclinic structure with the lattice parameters $a = 1856.75 \pm 0.06$, $b = 2399.14 \pm 0.07$, $c = 1614.05 \pm 0.05$ pm, and $\beta = 103.262° \pm 0.001°$ at $223 \pm 2$ K and a calculated density of 1.310 g·cm$^{-3}$ (Chen et al. 2006). It was prepared from refluxing (2 h) $Zn\{S_2P(OPr^i)_2\}_2$ (0.2 g) with stoichiometric amount of bis(4-pyridyl)amine in $CHCl_3$ solution. After the reaction, the solvent was removed in vacuo, and the residue recrystallized by slow evaporation from a chloroform/MeOH (1:1 volume ratio) solution of the title compound.

**Zn[S$_2$P(OPr$^i$)$_2$]$_2$·(terpy)** or **C$_{27}$H$_{39}$N$_3$P$_2$O$_4$S$_4$Zn**. This compound melts at 206°C–208°C and crystallizes in the monoclinic structure with the lattice parameters $a = 884.7 \pm 0.2$, $b = 2521.5 \pm 0.2$, $c = 1574.0 \pm 0.2$ pm, and $\beta = 90.74° \pm 0.02°$ and experimental and calculated densities 1.45 and 1.37 g·cm$^{-3}$, respectively (Harrison et al. 1986a). To obtain it, 2,2′:6,2″-terpyridine (terpy) dissolved in the minimum amount of EtOH was added to Zn[S$_2$P(OPr$^i$)$_2$]$_2$ also dissolved in EtOH. After allowing the mixture to stand at room temperature, the colorless crystalline compound was filtered off, recrystallized from hot EtOH, washed with pentane, and dried in air.

**[Zn{S$_2$P(OBu$^i$)$_2$}$_2$·(C$_{12}$H$_{10}$N$_2$)]$_n$** or **C$_{28}$H$_{46}$N$_2$P$_2$O$_4$S$_4$Zn**. This compound melts at 149°C–153°C and crystallizes in the monoclinic structure with the lattice parameters $a = 937.5 \pm 0.2$, $b = 942.8 \pm 0.2$, $c = 1903.5 \pm 0.5$ pm, and $\beta = 94.797° \pm 0.007°$ at $150 \pm 2$ K and a calculated density of 1.393 g·cm$^{-3}$ (Welte and Tiekink 2006). It was prepared by refluxing Zn{S$_2$P(OBu$^i$)$_2$}$_2$ with 1,2-di-4-pyridylethylene. Colorless crystals were isolated by the slow evaporation of an acetonitrile/CHCl$_3$ (1:1 volume ratio) solution of the title compound.

**[Zn{S$_2$P(OBu$^i$)$_2$}$_2$·(C$_{12}$H$_{12}$N$_2$)]$_n$** or **C$_{28}$H$_{48}$N$_2$P$_2$O$_4$S$_4$Zn**. This compound melts at 147°C–149°C and crystallizes in the monoclinic structure with the lattice parameters $a = 958.74 \pm 0.05$, $b = 944.68 \pm 0.05$, $c = 1940.80 \pm 0.11$ pm, and $\beta = 94.293° \pm 0.001°$ at $223 \pm 2$ K and a calculated density of 1.387 g·cm$^{-3}$ (Lai et al. 2004a). It was prepared by refluxing Zn{S$_2$P(OBu$^i$)$_2$}$_2$ with 1,2-bis(4-pyridyl)ethane. Colorless crystals were isolated by the slow evaporation of a chloroform/acetonitrile (4:1 volume ratio) solution of the title compound.

**[Zn$_2${S$_2$P(OPr$^i$)$_2$}$_4$·(C$_6$H$_{12}$N$_2$)]** or **C$_{30}$H$_{68}$N$_2$P$_4$O$_8$S$_8$Zn$_2$**. This compound melts at 197°C–200°C and crystallizes in the monoclinic structure with the lattice parameters $a = 794.56 \pm 0.18$, $b = 1883.1 \pm 0.4$, $c = 1757.0 \pm 0.4$ pm, and $\beta = 93.624° \pm 0.005°$ at $150 \pm 2$ K and a calculated density of 1.387 g·cm$^{-3}$ (Ellis et al. 2007). It was prepared by refluxing Zn{S$_2$P(OPr$^i$)$_2$}$_2$ with 1,4-diazabicyclo[2.2.2]octane. Colorless crystals were isolated by the slow evaporation of a Pr$^i$OH/CHCl$_3$ (1:1 volume ratio) solution of the title compound.

**[Zn{S$_2$P(OMe)$_2$}$_2$(2,2′-bipy)$_3$]** or **C$_{34}$H$_{36}$N$_6$P$_2$O$_4$S$_4$Zn**. This compound melts at 130°C–132°C and crystallizes in the monoclinic structure with the lattice parameters $a = 1359.67 \pm 0.05$, $b = 1310.08 \pm 0.04$, $c = 2194.94 \pm 0.07$ pm, and $\beta = 103.897° \pm 0.003°$ and a calculated density of 1.484 g·cm$^{-3}$ (Jeremias et al. 2014). The further evaporation of filtrate obtained from the [Zn{S$_2$P(OMe)$_2$}$_2$(2,2′-bipy)$_3$] preparation led to the change of color of the solution from colorless to pink, and subsequently, the formation of pink crystals surrounded by a white powder was noted. These crystals were identified by XRD as the title compound.

**[Zn$_7${(2-C$_5$H$_4$N)CH(OH)PO$_3$}$_6$(H$_2$O)$_6$]SO$_4$·2H$_2$O** or **C$_{36}$H$_{52}$N$_6$P$_6$O$_{36}$SZn$_7$**. This compound crystallizes in the monoclinic structure with the lattice parameters $a = 2269.0 \pm 0.2$, $b = 1667.5 \pm 0.2$, $c = 1815.1 \pm 0.2$ pm, and $\beta = 93.390° \pm 0.002°$ and a calculated density of 1.764 g·cm$^{-3}$ (Cao et al. 2005). It was prepared through hydrothermal treatment of a mixture of ZnSO$_4$·7H$_2$O (0.292 mM) and hydroxyl(2-pyridyl) methylphosphonic acid (0.25 mM) in 8 mL of H$_2$O, adjusted to pH 4.58 with 1 M

NaOH, at 140°C for 24 h. Colorless needlelike crystals were obtained as a monophasic material.

**[Zn{S$_2$P(OC$_6$H$_{11}$)$_2$}$_2${2-NC$_5$H$_4$C(H) = C(H)C$_5$H$_4$N-2}] or C$_{36}$H$_{54}$N$_2$P$_2$O$_4$S$_4$Zn**. This compound melts at 145°C–146°C and crystallizes in the monoclinic structure with the lattice parameters $a = 972.69 \pm 0.08$, $b = 865.85 \pm 0.07$, $c = 2485.3 \pm 0.2$ pm, and $\beta = 100.872° \pm 0.002°$ at $223 \pm 2$ K and a calculated density of 1.348 g·cm$^{-3}$ (Chen et al. 2006). It was prepared from refluxing (2 h) Zn{S$_2$P(OC$_6$H$_{11}$)$_2$}$_2$ (0.2 g) with stoichiometric amount of *trans*-1,2-bis(2-pyridyl)ethylene in CHCl$_3$ solution. After the reaction, the solvent was removed in vacuo, and the residue recrystallized by slow evaporation from a CH$_2$Cl$_2$/hexane (2:1 volume ratio) solution of the title compound.

**[Zn{S$_2$P(OC$_6$H$_{11}$)$_2$}$_2$(C$_{12}$H$_{10}$N$_2$)]$_n$ or C$_{36}$H$_{54}$N$_2$P$_2$O$_4$S$_4$Zn**. This compound melts at 211°C–216°C and crystallizes in the triclinic structure with the lattice parameters $a = 932.37 \pm 0.07$, $b = 1391.38 \pm 0.11$, and $c = 1709.17 \pm 0.12$ pm and $\alpha = 73.994° \pm 0.002°$, $\beta = 77.547° \pm 0.002°$, and $\gamma = 80.795° \pm 0.002$ at $223 \pm 2$ K and a calculated density of 1.339 g·cm$^{-3}$ (Lai et al. 2004b). It was obtained from refluxing Zn{S$_2$P(OC$_6$H$_{11}$)$_2$}$_2$ (0.2 g) with a stoichiometric amount of *trans*-1,2-bis(4-pyridyl)ethylene (C$_{12}$H$_{10}$N$_2$) in CHCl$_3$ solution. After the reaction, the solvent was removed in vacuo, and the residue recrystallized by slow evaporation of a chloroform/acetonitrile (3:1 volume ratio) solution of the title compound.

**[Zn{S$_2$P(OC$_6$H$_{11}$)$_2$}$_2$·(C$_{12}$H$_{10}$N$_4$)]$_n$ or C$_{36}$H$_{54}$N$_4$P$_2$O$_4$S$_4$Zn**. This compound melts at 104°C–106°C and crystallizes in the monoclinic structure with the lattice parameters $a = 1833.40 \pm 0.08$, $b = 1204.04 \pm 0.05$, and $c = 1913.04 \pm 0.08$ pm at $223 \pm 2$ K and a calculated density of 1.356 g·cm$^{-3}$ (Chen et al. 2005). It was prepared by refluxing Zn{S$_2$P(OC$_6$H$_{11}$)$_2$}$_2$ with 4-pyridinealdazine. Colorless crystals were isolated by slow evaporation of a chloroform/acetonitrile (3:1 volume ratio) solution of the title compound.

**[Zn{S$_2$P(OPr$^i$)$_2$}$_2${2-NC$_5$H$_4$C(H) = C(H)C$_5$H$_4$N-2}$_{0.5}$]$_2$ or C$_{36}$H$_{66}$N$_2$P$_4$O$_8$S$_8$Zn$_2$**. This compound melts at 133°C–134°C and crystallizes in the monoclinic structure with the lattice parameters $a = 787.35 \pm 0.08$, $b = 1770.46 \pm 0.15$, $c = 2003.25 \pm 0.19$ pm, and $\beta = 90.638° \pm 0.002°$ at $223 \pm 2$ K and a calculated density of 1.387 g·cm$^{-3}$ (Chen et al. 2006). It was prepared from refluxing (2 h) Zn{S$_2$P(OPr$^i$)$_2$}$_2$ (0.2 g) with a stoichiometric amount of *trans*-1,2-bis(2-pyridyl)ethylene in a CHCl$_3$ solution. After the reaction, the solvent was removed in vacuo, and the residue recrystallized by slow evaporation of a chloroform/acetonitrile (3:1 volume ratio) solution of the title compound.

**[Et$_4$N][Zn{S$_2$P(OC$_6$H$_4$Me)$_2$}$_2$(S$_2$CNMe$_2$)]·2H$_2$O or C$_{39}$H$_{58}$N$_2$P$_2$O$_6$S$_6$Zn**. This compound was obtained by the following procedure (McCleverty et al. 1983). A solution of [Zn{S$_2$P(OC$_6$H$_4$Me-$p$)$_2$}$_2$] (2.3 g) in Pr$^i$OH was added, slowly and with stirring, to a solution of [Et$_4$N][S$_2$CNMe$_2$] (0.8 g) in the same solvent. The complex crystallized as white needles and was collected by filtration.

**[Zn{S$_2$P(OPr$^i$)$_2$}$_2$(C$_{12}$H$_{12}$N$_2$)$_{0.5}$·CH$_3$CN]$_2$ or C$_{40}$H$_{74}$N$_4$P$_4$O$_8$S$_8$Zn$_2$**. This compound melts at 169°C–170°C and crystallizes in the triclinic structure with the lattice parameters $a = 790.94 \pm 0.05$, $b = 1176.29 \pm 0.08$, and $c = 1725.06 \pm 0.11$ pm and

$\alpha = 75.948° \pm 0.001°$, $\beta = 79.099° \pm 0.001°$, and $\gamma = 82.192° \pm 0.002$ at $223 \pm 2$ K and a calculated density of 1.364 g·cm$^{-3}$ (Lai et al. 2004b). It was obtained from refluxing $[Zn\{S_2P(OPr^i)_2\}_2]$ (0.2 g) with a stoichiometric amount of 1,2-bis(4-pyridyl)ethane ($C_{12}H_{12}N_2$) in $CHCl_3$ solution. After the reaction, the solvent was removed in vacuo, and the residue recrystallized by slow evaporation of a chloroform/acetonitrile (4:1 volume ratio) solution of the title compound.

**$[Zn\{S_2P(OEt)_2\}_2(Phen)_3]$ or $C_{44}H_{44}N_6P_2O_4S_4Zn$.** This compound melts at 193°C–195°C and crystallizes in the monoclinic structure with the lattice parameters $a = 1831.62 \pm 0.05$, $b = 1726.68 \pm 0.02$, $c = 1632.12 \pm 0.03$ pm, and $\beta = 121.492° \pm 0.002°$ and a calculated density of 1.473 g·cm$^{-3}$ (Jeremias et al. 2014). The further evaporation of filtrate obtained from $[Zn\{S_2P(OEt)_2\}_2(Phen)]$ preparation led to change of color of the solution from colorless to pink, and subsequently, the formation of pink crystals surrounded by a white powder was noted. These crystals were identified by XRD as the title compound.

**$[Me_4N][Zn\{S_2P(OC_6H_4Me)_2\}_3]$ or $C_{46}H_{54}NP_3O_6S_6Zn$.** This compound crystallizes in the monoclinic structure with the lattice parameters $a = 2931.1 \pm 0.9$, $b = 1103.2 \pm 0.3$, $c = 1751.5 \pm 0.5$ pm, and $\beta = 106.83° \pm 0.02°$ and experimental and calculated densities 1.32 and 1.308 g·cm$^{-3}$, respectively (Kowalski et al. 1981, McCleverty et al. 1983). It was obtained by the following procedures (McCleverty et al. 1983):

1. The complex $[Zn\{S_2P(OC_6H_4Me-p)_2\}_2]$ (3.4 g) was dissolved in Pr$^i$OH and added, slowly with stirring and gentle heating (steam bath), to a suspension of $[Me_4N][S_2P(OC_6H_4Me-p)]$ (1.9 g) in Pr$^i$OH. Upon cooling, flaky crystals of the title compound formed and were filtered off. These could be recrystallized from toluene.

2. A solution of $[Me_4N][OCOC_{17}H_{35}]$ (0.9 g) in Pr$^i$OH was added slowly, and with stirring, to $[Zn\{S_2P(OC_6H_4Me-p)_2\}_2]$ (1.7 g) in Pr$^i$OH. A precipitate, probably basic zinc stearate, formed and was filtered off, leaving a clear solution, which was allowed to stand at room temperature for 24 h. After this time, the colorless crystals of the title compound that had formed were filtered off and washed in the Pr$^i$OH.

**$[Zn\{S_2P(OC_6H_{11})_2\}_2(C_{10}H_8N_2)_{0.5}]_2$ or $C_{58}H_{96}N_2P_4O_8S_8Zn_2$.** This compound melts at 210°C–213°C and crystallizes in the monoclinic structure with the lattice parameters $a = 1329.97 \pm 0.06$, $b = 931.37 \pm 0.05$, $c = 2873.68 \pm 0.13$ pm, and $\beta = 100.794° \pm 0.001°$ at $223 \pm 2$ K and a calculated density of 1.387 g·cm$^{-3}$ (Lai et al. 2004b). It was obtained from refluxing $[Zn\{S_2P(OPr^i)_2\}_2]$ (0.2 g) with stoichiometric amount of 4,4'-bipy in $CHCl_3$ solution. After the reaction, the solvent was removed in vacuo, and the residue recrystallized by slow evaporation of a chloroform/acetonitrile (4:1 volume ratio) solution of the title compound.

**$[Zn_6(m\text{-}O_3SC_6H_4PO_3)_4(4,4'\text{-}bipy)_6(H_2O)_4]\cdot18H_2O$ or $C_{84}H_{108}N_{12}P_4O_{46}S_4Zn_6$.** This compound crystallizes in the monoclinic structure with the lattice parameters $a = 1353.96 \pm 0.01$, $b = 3043.80 \pm 0.05$, $c = 1572.17 \pm 0.03$ pm, and $\beta = 110.535° \pm 0.001°$ and a calculated density of 1.459 g·cm$^{-3}$ (Du et al. 2006). It was prepared by the following procedure. A mixture of $ZnCO_3$ (0.4 mM), $m$-sulfophenylphosphonic acid ($m$-$HO_3S$-$C_6H_4$-$PO_3H_2$) (0.4 mM), and 4,4'-bipy (0.5 mM) in 10 mL of distilled water with an initial pH of about 4.0 was put into a Parr Teflon-lined autoclave (23 mL) and heated at 140°C for 4 days. Colorless plate-shaped crystals of the title compound were collected.

**[$Zn_6$($m$-$O_3SC_6H_4PO_3$)$_4$(Phen)$_8$]$_4$·11$H_2O$ or $C_{120}H_{102}N_{16}P_4O_{35}S_4Zn_6$.** This compound crystallizes in the triclinic structure with the lattice parameters $a = 1399.16 \pm 0.04$, $b = 1406.68 \pm 0.04$, and $c = 1752.88 \pm 0.05$ pm and $\alpha = 97.816° \pm 0.001°$, $\beta = 98.731° \pm 0.001°$, and $\gamma = 114.932° \pm 0.001°$ and a calculated density of 1.638 g·cm$^{-3}$ (Du et al. 2006). To obtain it, a mixture of $ZnCO_3$ (0.3 mM), $m$-sulfophenylphosphonic acid ($m$-$HO_3S$-$C_6H_4$-$PO_3H_2$) (0.32 mM), and Phen (0.32 mM) in 10 mL of distilled water, with its pH value adjusted to about 5.0 via the addition of a 1 M NaOH solution, was put into a Parr Teflon-lined autoclave (23 mL) and heated at 150°C for 4 days. Colorless brick-shaped crystals of the title compound were collected.

**[$Zn$(Phen)$_3$]$_2$[$Zn_4$($m$-$O_3SC_6H_4PO_3$)$_4$(Phen)$_4$]·20$H_2O$ or $C_{144}H_{136}N_{12}P_4O_{46}S_4Zn_6$.** This compound crystallizes in the triclinic structure with the lattice parameters $a = 1409.27 \pm 0.07$, $b = 1460.20 \pm 0.07$, and $c = 1947.40 \pm 0.09$ pm and $\alpha = 95.857° \pm 0.002°$, $\beta = 98.831° \pm 0.001°$, and $\gamma = 109.966° \pm 0.001°$ and a calculated density of 1.581 g·cm$^{-3}$ (Du et al. 2006). It was obtained by the following procedure. A mixture of $ZnCO_3$ (0.3 mM), $m$-sulfophenylphosphonic acid ($m$-$HO_3S$-$C_6H_4$-$PO_3H_2$) (0.3 mM), and Phen (0.5 mM) in 10 mL of distilled water, with its pH value adjusted to about 7.0 via the addition of a 1 M NaOH solution, was put into a Parr Teflon-lined autoclave (23 mL) and heated at 150°C for 4 days. Colorless plate-shaped crystals of the title compound were collected.

## 1.47   Zn–H–C–N–P–O–Cl–S

Some compounds are formed in this system.

**[$Zn${$S_2P$($OC_6H_{11}$)$_2$}$_2$($C_{12}H_{12}N_2$)·0.5CHCl$_3$]$_n$ or $C_{36.5}H_{56.5}N_2P_2O_4Cl_{1.5}S_4Zn$.** This compound melts at 205°C–209°C and crystallizes in the monoclinic structure with the lattice parameters $a = 1003.58 \pm 0.04$, $b = 3027.81 \pm 0.13$, $c = 1512.22 \pm 0.07$ pm, and $\beta = 103.062° \pm 0.001°$ at $223 \pm 2$ K and a calculated density of 1.330 g·cm$^{-3}$ (Lai et al. 2004b). It was obtained from refluxing [$Zn${$S_2P$($OPr^i$)$_2$}$_2$] (0.2 g) with stoichiometric amount of 1,2-bis(4-pyridyl)ethane ($C_{12}H_{12}N_2$) in CHCl$_3$ solution. After the reaction, the solvent was removed in vacuo, and the residue recrystallized by slow evaporation of a chloroform/acetonitrile (3:1 volume ratio) solution of the title compound.

**[$Zn_2${($C_{14}H_{21}N_4S$)$_2Zn$}{$O_2P$($OPh$)$_2$}$_4$]·CH$_2$Cl$_2$     or     $C_{77}H_{84}N_4P_4O_{16}Cl_2S_2Zn_3$.** This compound crystallizes in the triclinic structure with the lattice parameters $a = 1099.6 \pm 0.2$, $b = 1495.9 \pm 0.2$, and $c = 2680.1 \pm 0.3$ pm and $\alpha = 83.64° \pm 0.04°$, $\beta = 84.41° \pm 0.04°$, and $\gamma = 71.78° \pm 0.02°$ (Hammesa and Carrano 2000). To prepare it, a solution of [$MeZn$($C_{14}H_{21}N_4S$)] (0.55 mM) in 20 mL of CH$_2$Cl$_2$ was treated with a CH$_2$Cl$_2$ solution of diphenyl phosphate (0.55 mM). The resulting solution was stirred for 1 h and dried under reduced pressure to give a sticky white solid of the title compound. Crystals were obtained by slow evaporation of a $CH_2Cl_2$/hexane (1.3:1 volume ratio) solution of $[Zn_2\{(C_{14}H_{21}N_4S)_2Zn\}\{O_2P(OPh)_2\}_4]$.

## 1.48  Zn–H–C–N–P–O–I–S

**[(Bu$_4$N)ZnI$_2$\{S$_2$P(OPr$^i$)$_2$\}] or C$_{22}$H$_{50}$NPO$_2$I$_2$S$_2$Zn.** This multinary compound is formed in this system (McCleverty et al. 1983). It was prepared as follows. To a solution of $[Bu^n{}_4N][S_2P(OPr^i)_2]$ (2.3 g) in $Pr^iOH$ was added a solution of $ZnI_2$ (1.6 g) in the same solvent. The reaction mixture was stirred briefly and then allowed to stand for 30 min. The title compound formed slowly as white needle-shaped crystals, which were collected by filtration and washed with $Pr^iOH$.

## 1.49  Zn–H–C–N–P–S

Some compounds are formed in this system.

**[Zn\{N(Pr$^i$$_2$PS)$_2$\}$_2$] or C$_{24}$H$_{56}$N$_2$P$_4$S$_4$Zn.** This compound melts at 144°C and crystallizes in the tetragonal structure with the lattice parameters $a = 1592.8 \pm 0.2$ and $c = 1394.1 \pm 0.2$ pm (Cupertino et al. 1996). To prepare it, $ZnCO_3 \cdot 2Zn(OH)_2 \cdot H_2O$ (0.29 mM) was added to a solution of $HN(Pr^i{}_2PS)_2$ (0.96 mM) in $CH_2Cl_2$ (30 mL), and the mixture was refluxed for 2 h. The cloudy/white mixture was filtered, and the filtrate was reduced by two-thirds and cooled overnight to give the title compound as clear crystals.

**[Zn(2,2′-bipy)(Bu$^i$$_2$PS$_2$)$_2$] or C$_{26}$H$_{44}$N$_2$P$_2$S$_4$Zn.** This compound melts at 127°C–131°C (Shchukin et al. 2000) and crystallizes in the monoclinic structure with the lattice parameters $a = 2421.9 \pm 0.5$, $b = 1138.6 \pm 0.2$, $c = 2491.6 \pm 0.5$ pm, and $\beta = 97.70° \pm 0.03°$ and a calculated density of 1.249 g·cm$^{-3}$ (Klevtsova et al. 2001c). To obtain it, $Zn(Bu^i{}_2PS_2)_2$ (2.1 mM) and 2,2′-bipy (2.2 mM) were dissolved in a minimum amount of toluene at the heating on water bath (Shchukin et al. 2000). The light-yellow solution was evaporated to one-fifth of the initial volume and cooled up to −10°C. White precipitate was filtered with suction, washed with cold toluene, and dried in air. Its single crystals were grown by the slow crystallization from the mixture EtOH/CHCl$_3$ (1:1 volume ratio) (Shchukin et al. 2000, Klevtsova et al. 2001c).

**[Zn(4,4′-bipy)(Bu$^i$$_2$PS$_2$)$_2$] or C$_{26}$H$_{44}$N$_2$P$_2$S$_4$Zn.** This compound crystallizes in the triclinic structure with the lattice parameters $a = 909.0 \pm 0.2$, $b = 946.5 \pm 0.2$, and $c = 1017.9 \pm 0.2$ pm and $\alpha = 93.13° \pm 0.03°$, $\beta = 107.33° \pm 0.03°$, and $\gamma = 103.89° \pm 0.03°$ and a calculated density of 1.322 g·cm$^{-3}$ (Shchukin et al. 2000). To prepare it, $Zn(Bu^i{}_2PS_2)_2$ (2.1 mM) and 4,4′-bipy (2.2 mM) were dissolved in a minimum amount of toluene at the heating on water bath. The light-yellow solution was evaporated to one-fourth of the initial volume and cooled up to −10°C. White precipitate was filtered with suction, washed with cold toluene, and dried in air. Its single crystals were grown by slow crystallization from toluene.

**[Zn(Bu$^i$$_2$PS$_2$)$_2$(Phen)] or C$_{28}$H$_{44}$N$_2$P$_2$S$_4$Zn.** This compound melts at 140°C–142°C (Shchukin et al. 2000) and crystallizes in the monoclinic structure with the lattice parameters $a = 1758.0 \pm 0.4$, $b = 996.9 \pm 0.2$, $c = 1917.5 \pm 0.4$ pm, and $\beta = 90.33° \pm 0.03°$

and a calculated density of 1.313 g·cm$^{-3}$ (Klevtsova et al. 2001c). To obtain it, $Zn(Bu^i_2PS_2)_2$ (2.1 mM) and Phen·H$_2$O (2.6 mM) were dissolved in a minimum amount of CHCl$_3$ (ca. 10 mL) at the heating on water bath (Shchukin et al. 2000). EtOH (30 mL) was added to the obtained yellow solution, and then it was evaporated to one-fourth of the initial volume and cooled up to −10°C. Light-yellow precipitate was filtered with suction, washed with cold EtOH, and dried in air. Its single crystals were grown by the slow crystallization from the mixture EtOH/CHCl$_3$ (1:1 and 4:1 volume ratios) (Shchukin et al. 2000, Klevtsova et al. 2001c).

**[Zn(Py)(S$_2$PPh$_2$)$_2$] or C$_{29}$H$_{25}$NP$_2$S$_4$Zn.** To prepare this compound, a methanolic solution of (Py)(S$_2$PPh$_2$) (1.1 g) was added slowly, with stirring, to a concentrated aqueous solution of Zn(O$_2$CMe)$_2$·4H$_2$O (0.65 g) (McCleverty et al. 1983). The solution that formed was concentrated in vacuo and crystals of the title compound precipitated slowly. These were filtered off and washed with MeOH.

**[Zn$_2$(C$_5$H$_{10}$NS$_2$)$_4$(C$_{10}$H$_{24}$P$_2$) or C$_{30}$H$_{64}$N$_4$P$_2$S$_8$Zn$_2$.** This compound crystallizes in the monoclinic structure with the lattice parameters $a = 1296.0 \pm 0.1$, $b = 1131.1 \pm 0.2$, $c = 2012.9 \pm 0.5$ pm, and $\beta = 96.46° \pm 0.01°$ and a calculated density of 1.254 g·cm$^{-3}$ (Zeng et al. 1994b). The needlelike crystals suitable for XRD were recrystallized from toluene.

**[Me$_4$N][Zn(S$_2$PPh$_2$)$_2$(S$_2$CNMe$_2$)] or C$_{31}$H$_{38}$N$_2$P$_2$S$_6$Zn.** This compound was obtained as follows (McCleverty et al. 1983). A mixture of [Zn(S$_2$PPh$_2$)$_2$] (2.8 g) and [Me$_4$N][S$_2$CNMe$_2$] (1.1 g) was refluxed in acetone (50 mL) for 2 h. The solution was filtered while hot and evaporated to low bulk in vacuo. White needle-shaped crystals of the title compound slowly formed and were collected by filtration and washed with acetone.

**[Et$_4$N][Zn(S$_2$PPh$_2$)$_3$] or C$_{44}$H$_{50}$NP$_3$S$_6$Zn.** This compound crystallizes in the monoclinic structure with the lattice parameters $a = 1116.0 \pm 0.6$, $b = 996.2 \pm 0.5$, $c = 4311 \pm 3$ pm, and $\beta = 104.73° \pm 0.05°$ and experimental and calculated densities 1.35 and 1.354 g·cm$^{-3}$, respectively (Kowalski et al. 1981, McCleverty et al. 1983). To obtain it, a mixture of [Zn(S$_2$PPh$_2$)$_2$] (2.8 g) and [Et$_4$N][S$_2$PPh$_2$] (1.9 g) was refluxed in acetone (50 mL) for 4 h. The solution was then filtered and allowed to stand. The title compound gradually crystallized and was collected by filtration to give white crystals. It was obtained from toluene as colorless eight-faced columnar needles.

[Et$_4$N][Zn(S$_2$PPh$_2$)$_3$] could also be prepared as follows (McCleverty et al. 1983). A mixture of [Zn(S$_2$PPh$_2$)$_2$] (2.8 g) and [Et$_4$N][OCOC$_{17}$H$_{35}$] (2.1 g) in acetone (50 mL) was refluxed for 24 h. A precipitate, probably basic zinc stearate, formed and was filtered off, leaving a clear solution, which was allowed to stand at room temperature for 24 h. After this time, the colorless crystals of the title compound that had formed were filtered off and washed in the Pr$^i$OH.

## 1.50   Zn–H–C–N–P–F–S

**[Zn{S(CH$_2$)$_3$NMe$_3$}$_2$](PF$_6$)$_2$ or C$_{12}$H$_{30}$N$_2$P$_2$F$_{12}$S$_2$Zn.** This multinary compound is formed in this system. It was prepared by the following procedure

(Casals et al. 1991c). 3-Trimethylammonio-1-propanethiol hexafluorophosphate (57.6 mg, corresponding to 0.20 mM of pure thiol) was dissolved in a mixture of acetonitrile (1 mL), MeOH (1 mL), and $H_2O$ (1 mL). Aqueous NaOH (0.32 mM in 2 mL) was added, followed by $Zn(OAc)_2 \cdot 2H_2O$ (0.08 mM) dissolved in 2.0 mL of $H_2O$. The slightly cloudy solution was filtered and kept under nitrogen at room temperature. After several days, the solid product was isolated as a colorless crystalline material.

## 1.51   Zn–H–C–N–P–F–Cl–Pt–S

$[ZnPt_2Cl(2,2'-bipy)(Ph_3P)_4(\mu_3-S)_2][PF_6]$ or $C_{82}H_{68}N_2P_5F_6ClPt_2S_2Zn$. This multinary compound is formed in this system. It crystallizes in the triclinic structure with the lattice parameters $a = 1453.19 \pm 0.02$, $b = 2008.32 \pm 0.02$, and $c = 2922.13 \pm 0.03$ pm and $\alpha = 83.434° \pm 0.001°$, $\beta = 88.522° \pm 0.001°$, and $\gamma = 83.281° \pm 0.001°$ (Li et al. 2000). To obtain it, a suspension of $[Pt_2(Ph_3P)_4(\mu-S)_2]$ (0.1 mM) and $ZnCl_2$ (0.1 mM) was stirred in MeOH (30 mL) for 6 h, and then 2,2'-bipy (0.1 mM) was added. The suspension changed to a clear bright-yellow solution within a few min. The solution was filtered and purified by metathesis with $NH_4PF_6$ to yield the title compound. The product was recrystallized from a $CH_2Cl_2$/MeOH (1:3 volume ratio) mixture to give yellow crystals. Single crystals were grown by slow evaporation of the solution at room temperature in air.

## 1.52   Zn–H–C–N–P–I–S

Some compounds are formed in this system.

$[(Et_4N)\{ZnI_2(S_2PPh_2)\}]$ or $C_{20}H_{30}NPI_2S_2Zn$. This compound was obtained by the next procedure (McCleverty et al. 1983). To a solution of $ZnI_2$ (1.6 g) in acetone was added a hot solution of $[Et_4N][S_2PPh_2]$ (1.9 g) in the same solvent. The mixture was stirred and heated for 15 min, the solvent was then evaporated in vacuo and the remaining oil dissolved in a mixture of acetone and $Pr^iOH$ (1:1 volume ratio). The title compound formed slowly as white crystals and was filtered off and washed with acetone.

$[Ph_3P\{C(NMe_2)S\}\{Zn\{(S_2CNMe_2)I_2\}\}]$ or $C_{24}H_{27}N_2PI_2S_3Zn$. This compound was prepared by the following procedure (McCleverty and Morrison 1976). To a suspension of $[Zn(S_2CNMe_2)_2I_2]$ (2 mM) in $CH_2Cl_2$ (50 mL) was added a solution of $PPh_3$ (1.04 g) in $CH_2Cl_2$ (10 mL). The resulting orange solution had turned yellow after 1 min. The solution was stirred for a further 10 min and then the solvent was evaporated. The solid was shaken with diethyl ether (200 mL) overnight and the mixture was filtered. The residue was extracted with acetone (50 mL), leaving $[ZnI_2(Ph_3P)_2]$, which was recrystallized from $CH_2Cl_2/n$-hexane (0.1 g). Addition of light petroleum to the acetone extract caused precipitation of more $[ZnI_2(Ph_3P)_2]$, which was removed by filtration. Further addition of light petroleum caused precipitation of the title compound, which was collected and recrystallized from acetone/light petroleum. The compound could be obtained in a similar manner from the reaction of $[Zn\{S(SCNMe_2)_2I_2\}]$ (5 mM) with $Ph_3P$ (5 mM).

**[Ph$_3$P{C(NMe$_2$)S}{Zn(Ph$_3$P)I$_3$}] or C$_{39}$H$_{36}$NP$_2$I$_3$SZn**. This compound was prepared by the next procedure (McCleverty and Morrison 1976). To a suspension of
[Zn(S$_2$CNMe$_2$)$_2$I$_2$] (2.8 g) in CHCl$_3$ (30 mL) was added a solution of Ph$_3$P (3.9 g),
and the mixture was shaken. The deep-orange solution had turned yellow after
1 min, and after 30 min, crystals had started to form. The mixture was set aside
for 36 h, and then the yellow crystals of [Ph$_3$P{C(NMe$_2$)S}{Zn(Ph$_3$P)I$_3$}] were filtered off and recrystallized from a acetone/light petroleum mixture. The title compound could be obtained similarly by mixing a suspension of [Zn{S(SCNMe$_2$)$_2$I$_2$}]
(2.64 g) with Ph$_3$P (2.6 g) in CHCl$_3$ (50 mL). It also crystallized from a solution
containing [Ph$_3$P{C(NMe$_2$)S}{Zn{(S$_2$CNMe$_2$)I$_2$}}] (4.0 g) and Ph$_3$P (1.3 g) in CHCl$_3$
(50 mL).

This compound could be prepared by adding a solution of PhP$_3$ (0.65 g) in boiling
MeOH to a solution containing [Ph$_3$P{C(NMe$_2$)S}]I (1.1 g) and ZnI$_2$ (0.8 g) in boiling MeOH (80 mL). The hot solution was filtered and allowed to cool overnight. The
few yellow crystals that formed were washed with CHCl$_3$ and diethyl ether and then
recrystallized from acetone/light petroleum.

## 1.53   Zn–H–C–N–O–S

Some compounds are formed in this system.

**Zn(CH$_4$N$_2$S)$_2$(NO$_3$)$_2$ or C$_2$H$_8$N$_6$O$_6$S$_2$Zn**. This compound crystallizes in the
monoclinic structure with the lattice parameters $a = 1246.6 \pm 0.1$, $b = 750.1 \pm 0.1$,
$c = 1230.3 \pm 0.2$ pm, and $\beta = 90.56° \pm 0.01°$ and a calculated density of 1.973 g·cm$^{-3}$
(Romanenko et al. 2001). To obtain it, a flask equipped with a reflux condenser was
charged with Zn(NO$_3$)$_2$·6H$_2$O (25 mM) and thiourea (50 mM), and then 50 mL of
Pr$^i$OH was added. The mixture was heated on a water bath until complete dissolution
of the solid phases. The resulting solution was refluxed for 0.5 h, filtered, evaporated
on a water bath to about half of the original volume, and then allowed to air crystallization. The precipitated crystals were filtered off with suction, washed with isopropyl alcohol and pentane, and dried in air.

**Zn(CH$_5$N$_3$S)$_2$SO$_4$ or C$_2$H$_{10}$N$_6$O$_4$S$_3$Zn**. This compound decomposes at 265°C
(Mahadevappa and Murthy 1972). It was obtained by simply scratching the sides
of the beaker containing the mixture of aqueous solutions of thiosemicarbazide and
ZnSO$_4$ in the molar ratio 2:1, with a glass rod.

**Zn(CH$_5$N$_3$S)$_2$(NO$_3$)$_2$ or C$_2$H$_{10}$N$_8$O$_6$S$_2$Zn**. This compound decomposes at 171°C
(Mahadevappa and Murthy 1972) and crystallizes in the monoclinic structure
with the lattice parameters $a = 1115.7 \pm 0.2$, $b = 759.3 \pm 0.1$, $c = 1416.1 \pm 0.3$ pm, and
$\beta = 102.08° \pm 0.01°$ and an experimental density of 2.104 g·cm$^{-3}$ (Romanenko et al.
1999). It was obtained by mixing aqueous solutions of thiosemicarbazide and
Zn(NO$_3$)$_2$ in the molar ratio 2:1 (Mahadevappa and Murthy 1972). The resulting solution was slowly evaporated at 60°C on a water bath and then cooled in ice, whereupon crystals of the title compound appeared. Its single crystals were grown from a
mixture of the mother liquor and EtOH (1:1 volume ratio) by the evaporation at room
temperature (Romanenko et al. 1999).

To obtain the title compound another procedure could be used (Babb et al. 2003). Thiosemicarbazide (2.20 mM), dissolved in absolute EtOH (15 mL), was added dropwise to an absolute ethanolic solution of $Zn(NO_3)_2 \cdot 6H_2O$ (1.1 mM in 10 mL) while stirring. After approximately 10 min, a white precipitate of $Zn(CH_5N_3S)_2(NO_3)_2$ was seen to form, and this was collected by filtration.

**$Zn[SC(NH_2)_2]_3SO_4$ or $C_3H_{12}N_6O_4S_4Zn$.** This compound crystallizes in the orthorhombic structure with the lattice parameters $a = 1112.6 \pm 0.5$, $b = 777.3 \pm 0.4$, and $c = 1549.1 \pm 0.5$ pm and experimental and calculated densities 1.923 and 1.926 g·cm$^{-3}$, respectively (Andreetti et al. 1968) ($a = 1111.1$, $b = 779.2$, and $c = 1547.8$ pm [Ushasree et al. 1999]). It was synthesized by stoichiometric incorporation of thiourea and $ZnSO_4 \cdot 7H_2O$ (Ushasree et al. 1999). The required quantities of reactants were estimated from the following equation: $ZnSO_4 \cdot 7H_2O + 3[SC(NH_2)_2] = Zn[SC(NH_2)_2]_3SO_4 + 7H_2O$. The crystals suitable for XRD were grown from an aqueous solution of the title compound by a slow cooling technique.

**$Zn[HCONH_2]_4SO_4$ or $C_4H_{12}N_4O_8SZn$.** This compound was obtained by a crystallization of aqueous solutions of $ZnSO_4$ and an excess of formamide at elevated temperatures (Nardelli and Coghi 1959).

**$Zn(MeCH_4N_3S)_2(NO_3)_2$ or $C_4H_{14}N_8O_6S_2Zn$.** To obtain the title compound, methylthiosemicarbazide (2.20 mM), dissolved in absolute EtOH (15 mL), was added dropwise to an absolute ethanolic solution of $Zn(NO_3)_2 \cdot 6H_2O$ (1.1 mM in 10 mL) while stirring (Babb et al. 2003). After approximately 10 min, a white precipitate of $Zn(MeCH_4N_3S)_2(NO_3)_2$ was seen to form, and this was collected by filtration.

**$Zn[SC(NH_2)_2]_4(NO_3)_2$ or $C_4H_{16}N_{10}O_6S_4Zn$.** This compound crystallizes in the orthorhombic structure with the lattice parameters $a = 2243.4 \pm 2.3$, $b = 946.5 \pm 0.2$, and $c = 885.9 \pm 0.1$ pm and experimental and calculated densities 1.72 and 1.73 g·cm$^{-3}$, respectively (Vega et al. 1978). Its single crystals as colorless prisms were grown by slow evaporation of a mixture of $Zn(NO_3)_2$ and thiourea solutions (1:4 volume ratio).

**$[Zn_2\{S_2CN(CH_2)_4O\}]_4$ or $C_5H_8NOS_2Zn_2$.** This compound was obtained by the next procedure (Ivanov et al. 2003b). To the solution of $ZnCl_2$ (10 mM) in $H_2$ (100 mL) was poured with vigorous stirring a solution of sodium morpholinedithiocarbamate dihydrate, $Na\{S_2CN(CH_2)_4O\} \cdot 2H_2O$, (20 mM) in $H_2O$ (30 mL). The white voluminous precipitate was washed by decantation, separated by filtration, and dried in air. Thermolysis of the title compound leads to the formation of ZnS.

**$[Zn\{CH_2(COO)_2\}\{SC(NH_2)_2\}_2]_n$ or $C_5H_{10}N_4O_4S_2Zn$.** This compound crystallizes in the monoclinic structure with the lattice parameters $a = 756.17 \pm 0.09$, $b = 816.4 \pm 0.1$, $c = 1834.0 \pm 0.2$ pm, and $\beta = 91.40° \pm 0.02°$ (Burrows et al. 2000) ($a = 750.3 \pm 0.2$, $b = 819.9 \pm 0.7$, and $c = 1836 \pm 1$ pm and $\beta = 90.95° \pm 0.03°$ at 173 K and a calculated density of 1.880 g·cm$^{-3}$ [Zhang et al. 2000b]). To obtain it, aqueous solutions of $ZnSO_4 \cdot 6H_2O$ (1.00 mM), malonic acid (1.00 mM), and thiourea (2.00 mM) were mixed together, and the pH value of the resulting solution was controlled at 5–6 by diluted NaOH and $H_2SO_4$ (Zhang et al. 2000b). The solution was slowly evaporated at room temperature. Colorless, large needlelike crystals were isolated to give the title compound in about 1 week. It could also be obtained by adding an aqueous solution of sodium malonate (0.23 mM) to an aqueous solution of [ZnCl$_2${SC(NH$_2$)$_2$}$_4$] (0.23 mM) with no discernible change. After several hours, a colorless crystalline precipitate was observed, which was separated by filtration.

**[{Zn(C$_4$H$_2$O$_4$)[SC(NH$_2$)$_2$](H$_2$O)}$_2$]·4H$_2$O or C$_5$H$_{12}$N$_2$O$_7$SZn.** This compound crystallizes in the triclinic structure with the lattice parameters $a = 908.7 \pm 0.9$, $b = 985.9 \pm 1.0$, and $c = 687 \pm 1$ pm and $\alpha = 108.65° \pm 0.08°$, $\beta = 96.21° \pm 0.05°$, and $\gamma = 97.77° \pm 0.09°$ at 173 K (Zhang et al. 2000a). To obtain it, aqueous solutions (20 mL) of ZnSO$_4$·7H$_2$O (1.0 mM), maleic anhydride (1.0 mM), and thiourea (2.0 mM) were mixed, and the pH value was controlled at 5–6. After stirring for 5 min, the solution was divided into two portions. The first portion was slowly evaporated at room temperature. After 24 h, colorless plate crystals of [{Zn(C$_4$H$_2$O$_4$) [SC(NH$_2$)$_2$](H$_2$O)}$_2$]·4H$_2$O were produced, and then colorless needlelike crystals of [{Zn(C$_4$H$_2$O$_4$)[SC(NH$_2$)$_2$]$_2$}$_n$] were precipitated from the same portion after several days. As the needle crystals of this compound appeared, the plate crystals of the title compound slowly redissolved.

**Zn[Zn(SCH$_2$CHNH$_2$CO$_2$)$_2$] or C$_6$H$_{10}$N$_2$O$_4$S$_2$Zn$_2$, zinc bis(L-cysteinato)zincate.** To obtain this compound, an aqueous solution of ZnCl$_2$ (0.02 mol) was added with stirring to a solution of L-cysteine (0.04 mol) in 100 mL of H$_2$O (Shindo and Brown 1965). The solution was brought to pH 6.0 with dilute NH$_4$OH solution. A white amorphous precipitate appears at pH around 4.0. Stirring was continued for 1 h and the precipitate filtered and washed several times with H$_2$O. The white powder obtained was heated in 100 mL of boiling H$_2$O for 1 h with vigorous stirring, and the hot suspension was filtered. After washing with 50 mL of hot H$_2$O and absolute ethanol, the white powder was dried in vacuo over CaCl$_2$.

Alternatively, Zn[Zn(SCH$_2$CHNH$_2$CO$_2$)$_2$] could be prepared by warming 0.02 mol of L-cysteine in 100 mL of H$_2$O with 0.01 mol of freshly precipitated Zn(OH)$_2$ at 60°C–70°C for 2 h (Shindo and Brown 1965). The white powder obtained is treated in the same way as described earlier.

**[{Zn(C$_4$H$_2$O$_4$)[SC(NH$_2$)$_2$]$_2$}$_n$] or C$_6$H$_{10}$N$_4$O$_4$S$_2$Zn.** This compound crystallizes in the orthorhombic structure with the lattice parameters $a = 1448 \pm 1$, $b = 1564 \pm 1$, and $c = 1095.7 \pm 0.6$ pm at 173 K (Zhang et al. 2000a). To obtain it, aqueous solutions (20 mL) of ZnSO$_4$·7H$_2$O (1.0 mM), maleic anhydride (1.0 mM), and thiourea (2.0 mM) were mixed and the pH value was controlled at 5–6. After stirring for 5 min, the solution was divided into two portions. The first portion was slowly evaporated at room temperature, and the second heated to 100°C for 5 min and then allowed to stand at room temperature. After 24 h, colorless plate crystals of [{Zn(C$_4$H$_2$O$_4$)[SC(NH$_2$)$_2$](H$_2$O)}$_2$]·4H$_2$O were produced from the first portion, and then, colorless needlelike crystals of the title compound were precipitate from the same portion after several days. As the needle crystals of this compound appeared, the plate crystals of [{Zn(C$_4$H$_2$O$_4$)[SC(NH$_2$)$_2$](H$_2$O)}$_2$]·4H$_2$O slowly redissolved. In the second portion, colorless needlelike crystals of [{Zn(C$_4$H$_2$O$_4$)[SC(NH$_2$)$_2$]$_2$}$_n$] were quantitatively isolated after several days.

**[{Zn(C$_4$H$_2$O$_4$)[SC(NH$_2$)$_2$]$_2$}] or C$_6$H$_{10}$N$_4$O$_4$S$_2$Zn.** This compound crystallizes in the orthorhombic structure with the lattice parameters $a = 1483.1 \pm 0.1$, $b = 690.1 \pm 0.1$, and $c = 1137.9 \pm 0.1$ pm (Burrows et al. 2000). To obtain it, an aqueous solution of sodium fumarate (0.23 mM) was added to an aqueous solution of $[ZnCl_2\{SC(NH_2)_2\}_4]$ (0.23 mM) with no discernible change. After several hours, a colorless crystalline precipitate was observed, which was separated by filtration.

**$[Zn(SC(NH_2)_2)_2(CH_2CO_2)_2]_n$ or $C_6H_{12}N_4O_4S_2Zn$.** This compound crystallizes in the orthorhombic structure with the lattice parameters $a = 830.80 \pm 0.10$, $b = 1123.80 \pm 0.10$, and $c = 1353.50 \pm 0.10$ pm (Burrows et al. 2004a). To obtain it, an aqueous solution of sodium succinate (0.23 mM) was added to an aqueous solution of $[Zn(SC(NH_2)_2)_4]Cl_2$ (0.23 mM), with no discernible change. After several days, a colorless crystalline precipitate was observed, which was separated by filtration.

**$[Zn\{C_4H_2O_4\}\{SC(NH_2)_2\}_2]\cdot H_2O$ or $C_6H_{12}N_4O_5S_2Zn$.** This compound crystallizes in the monoclinic structure with the lattice parameters $a = 1275.8 \pm 0.3$, $b = 880.30 \pm 0.10$, $c = 2335.5 \pm 0.4$ pm, and $\beta = 98.07° \pm 0.02°$ (Burrows et al. 2000). To obtain it, an aqueous solution of sodium maleate (0.23 mM) was added to an aqueous solution of $[ZnCl_2\{SC(NH_2)_2\}_4]$ (0.23 mM) with no discernible change. After several hours, a colorless crystalline precipitate was observed, which was separated by filtration.

**$Zn[SC(NH_2)_2]_2(OAc)_2$ or $C_6H_{14}N_4O_4S_2Zn$.** This compound crystallizes in the monoclinic structure with the lattice parameters $a = 693.8 \pm 0.3$, $b = 1767.8 \pm 0.6$, $c = 1179.5 \pm 0.6$ pm, and $\beta = 112°47' \pm 18'$ and experimental and calculated densities 1.61 and 1.671 g·cm$^{-3}$, respectively (Cavalca et al. 1967) ($a = 720$, $b = 1770$, and $c = 1180$ pm and $\beta = 112.7°$) (Nardelli and Chierici 1959). It was obtained by a slow crystallization from the aqueous solution of the title compound.

**$[Zn(CH_4N_2S)_2(CHCO_2)_2]\cdot 2H_2O$ or $C_6H_{14}N_4O_6S_2Zn$.** This compound crystallizes in the triclinic structure with the lattice parameters $a = 779.10 \pm 0.02$, $b = 852.30 \pm 0.02$, and $c = 1094.50 \pm 0.02$ pm and $\alpha = 91.105° \pm 0.001°$, $\beta = 102.602° \pm 0.001°$, and $\gamma = 97.949° \pm 0.001°$ (Burrows et al. 2004a). To obtain it, an aqueous solution of sodium fumarate (0.20 mM) was added to an aqueous solution of $[Zn(CH_4N_2S)_4]$ $(NO_3)_2$ (0.20 mM), with no discernible change. After several hours, a colorless crystalline precipitate was observed, which was separated by filtration.

**$[Zn(CH_5N_3S)_2(H_2O)_2]\cdot(C_4H_2O_4)$ or $C_6H_{16}N_6O_6S_2Zn$.** This compound crystallizes in the triclinic structure with the lattice parameters $a = 640.53 \pm 0.07$, $b = 784.36 \pm 0.08$, and $c = 796.61 \pm 0.09$ pm and $\alpha = 76.182° \pm 0.008°$, $\beta = 78.570° \pm 0.010°$, and $\gamma = 78.530° \pm 0.010°$ (Babb et al. 2003). It was prepared by the following procedure. To an aqueous solution of $Zn(CH_5N_3S)_2(NO_3)_2$ (0.27 mM) was added an aqueous solution of sodium fumarate (0.27 mM). After approximately 24 h, a colorless crystalline material of the title compound was seen to form, which was separated by filtration.

**$Zn(Me_2CH_3N_3S)_2(NO_3)_2$ or $C_6H_{18}N_8O_6S_2Zn$.** To obtain the title compound, dimethylthiosemicarbazide (2.20 mM), dissolved in absolute EtOH (15 mL), was added dropwise to an absolute ethanolic solution of $Zn(NO_3)_2\cdot 6H_2O$ (1.1 mM in 10 mL) while stirring (Babb et al. 2003). After approximately 10 min, a white precipitate of $Zn(Me_2CH_3N_3S)_2(NO_3)_2$ was seen to form, and this was collected by filtration.

**$Zn(EtCH_4N_3S)_2(NO_3)_2$ or $C_6H_{18}N_8O_6S_2Zn$.** To obtain the title compound, ethylthiosemicarbazide (2.20 mM), dissolved in absolute EtOH (15 mL), was added dropwise to an absolute ethanolic solution of $Zn(NO_3)_2 \cdot 6H_2O$ (1.1 mM in 10 mL) while stirring (Babb et al. 2003). After approximately 10 min, a white precipitate of $Zn(EtCH_4N_3S)_2(NO_3)_2$ was seen to form, and this was collected by filtration.

**$Zn[(C_4H_3N)CH(N)(CH_2)_2S] \cdot H_2O$ or $C_7H_{10}N_2OSZn$.** This compound was obtained by the electrochemical oxidation of anodic Zn in acetonitrile solution of the Schiff base derived from 2-pyrrolcarbaldehyde and cysteamine (Castro et al. 1990).

**$\{Zn(CH_4N_2S)_2[CH_2C(CH_2)(CO_2)_2]\}_n$ or $C_7H_{12}N_4O_4S_2Zn$.** This compound crystallizes in the orthorhombic structure with the lattice parameters $a = 832.7 \pm 0.2$, $b = 1174.2 \pm 0.2$, and $c = 1339.6 \pm 0.4$ pm (Burrows et al. 2004a). To obtain it, an aqueous solution of sodium itaconate (0.23 mM) was added to an aqueous solution of $[Zn(CH_4N_2S)_4]Cl_2$ (0.23 mM), with no discernible change. After several days, a colorless crystalline precipitate was observed, which was separated by filtration.

**$[\{Zn(C_5H_4O_4)[SC(NH_2)_2]_2\}]$ or $C_7H_{12}N_4O_4S_2Zn$.** This compound crystallizes in the orthorhombic structure with the lattice parameters $a = 836.0 \pm 0$, $b = 1098.1 \pm 0.2$, and $c = 1471.7 \pm 0.3$ (Burrows et al. 2000). To obtain it, an aqueous solution of sodium citraconate (0.23 mM) was added to an aqueous solution of $[ZnCl_2\{SC(NH_2)_2\}_4]$ (0.23 mM) with no discernible change. After several hours, a colorless crystalline precipitate was observed, which was separated by filtration.

**$\{Zn(CH_4N_2S)_2[EtCH(CO_2)_2]\}_n$ or $C_7H_{14}N_4O_4S_2Zn$.** This compound crystallizes in the orthorhombic structure with the lattice parameters $a = 833.1 \pm 0.2$, $b = 1077.8 \pm 0.2$, and $c = 1520.9 \pm 0.6$ pm (Burrows et al. 2004a). To obtain it, an aqueous solution of sodium ethylmalonate (0.23 mM) was added to an aqueous solution of $[Zn(CH_4N_2S)_4]Cl_2$ (0.23 mM), with no discernible change. After several days, a colorless crystalline precipitate was observed, which was separated by filtration.

**$[Zn(CH_5N_3S)_2(C_5H_4O_4)] \cdot H_2O$ or $C_7H_{16}N_6O_5S_2Zn$.** This compound crystallizes in the monoclinic structure with the lattice parameters $a = 1165.0 \pm 0.3$, $b = 764.00 \pm 0.10$, $c = 1759.0 \pm 0.3$ pm, and $\beta = 106.130° \pm 0.002°$ (Babb et al. 2003). It was prepared by the following procedure. To an aqueous solution of $Zn(CH_5N_3S)_2(NO_3)_2$ (0.27 mM) was added an aqueous solution of sodium citraconate (0.27 mM). After approximately 24 h, a colorless crystalline material of the title compound was seen to form, which was separated by filtration.

**$\{Zn(CH_4N_2S)_2[CH_2CMe(CO_2)_2] \cdot 2H_2O\}_n$ or $C_7H_{17}N_4O_6S_2Zn$.** This compound crystallizes in the triclinic structure with the lattice parameters $a = 713.77 \pm 0.06$, $b = 861.6 \pm 0.1$, and $c = 1265.8 \pm 0.5$ pm and $\alpha = 102.41° \pm 0.01°$, $\beta = 91.176° \pm 0.009°$, and $\gamma = 92.661° \pm 0.008°$ (Burrows et al. 2004a). To obtain it, an aqueous solution of sodium mesaconate (0.23 mM) was added to an aqueous solution of $[Zn(CH_4N_2S)_4]Cl_2$ (0.23 mM), with no discernible change. After several days, a colorless crystalline precipitate was observed, which was separated by filtration.

**$[Zn(HL)_2] \cdot 2H_2O$ or $C_8H_{10}N_4O_6S_2Zn$ ($H_2L$ = thiobarbituric acid, $C_4H_4N_2O_2S$).** To obtain this compound, a solution of Zn salt (10 mM) in EtOH (25 mL) was mixed with a solution of $H_2L$ (20 mM) in the same solvent (50 mL), and the resulting mixture was stirred under reflux for $\approx 2$ h whereupon $C_8H_{10}N_4O_6S_2Zn$ precipitated

(Zaki and Mohamed 2000). It was removed by filtration, washed with EtOH, and dried at 80°C. There is no mass loss for this compound at the heating up to 130°C.

**[Zn(C₂H₃N)₄SO₄] or C₈H₁₂N₄O₄SZn.** This compound crystallizes in the orthorhombic structure with the lattice parameters $a = 2425 \pm 1$, $b = 845.6 \pm 0.3$, and $c = 2074.3 \pm 0.8$ pm (Yang et al. 2004). It was obtained by the following procedure. An aqua-acetonitrile solution of $ZnSO_4 \cdot 6H_2O$ was poured into a Teflon vessel till 30% of its volume was filled. Then the vessel was placed in a stainless steel tank to perform hydrothermal treatment. The crystallization reaction was performed at 160°C under autogenous pressure for 24 h. After the autoclave was cooled and depressurized, colorless slab-shaped crystals were deposited at the bottom of the Teflon vessel. After filtering and washing with aqueous acetonitrile solution for three times, they were dried under $CaCl_2$.

**[Zn{SC(NH₂)(NHMe)}₂(C₄H₂O₄)]ₙ or C₈H₁₄N₄O₄S₂Zn.** This compound crystallizes in the monoclinic structure with the lattice parameters $a = 1481.7 \pm 0.4$, $b = 910.50 \pm 0.10$, $c = 1243.0 \pm 0.3$ pm, and $\beta = 115.84° \pm 0.02°$ and a calculated density of 1.583 g·cm⁻³ (Burke et al. 2003). It was obtained by the following procedure. An aqueous solution of sodium fumarate (0.18 mM) was added to an aqueous solution of $[Zn\{SC(NH_2)(NHMe)\}_4]Cl_2$ (0.18 mM), with no discernable change. After approximately 72 h, a colorless crystalline precipitate was observed, which was separated by filtration.

**[Zn(C₄H₈NOS₂)₂] or C₈H₁₆N₂O₂S₄Zn.** To obtain this compound, *N*-methyl-*N*-ethanoldithiocarbamic acid was prepared by mixing 4 mL of *N*-methyl-*N*-ethanolamine ($C_3H_9NO$) and 2 mL of $CS_2$ diluted with EtOH (20 mL) at 5°C (Thirumaran et al. 1998). To the yellow dithiocarbamic acid solution (20 mM), 10 mM of Zn salt dissolved in $H_2O$ was added with constant stirring. Pale-yellow dithiocarbamate complex separated, which was washed with $H_2O$ and EtOH and then dried.

**[Zn(SCH₂CHNH₂CO₂Me)₂] or C₈H₁₆N₂O₄S₂Zn, bis(methyl-L-cysteinato)zinc.** To obtain this compound, methyl-L-cysteine hydrochloride (0.01 mol) was dissolved in a solution of NaOH (0.02 mol) in 50 mL of $H_2O$ at 0°C; to it was added with vigorous stirring a cold aqueous solution of $ZnCl_2$ (0.005 mol) (Shindo and Brown 1965). The white precipitate formed was quickly filtered, washed three times with cold $H_2O$, and recrystallized from a small amount of hot $H_2O$. The white fine crystals were dried in vacuo over $CaCl_2$.

**[Zn(MeSCH₂CHNH₂CO₂)₂] or C₈H₁₆N₂O₄S₂Zn, bis(S-methyl-L-cysteinato)zinc.** To obtain this compound, *S*-methyl-L-cysteine (0.01 mol) was dissolved in a solution of NaOH (0.01 mol) in 50 mL of $H_2O$; to it was added with stirring an aqueous solution of $ZnCl_2$ (0.005 mol) (Shindo and Brown 1965). The white precipitate began to form soon after the addition. After stirring for 2 h, the precipitate was filtered, washed several times with $H_2O$ until the filtrate gave a negative chloride test, and dried in vacuo over $CaCl_2$.

**[Zn(EtCO₂){CS(NH₂)₂} or C₈H₁₈N₄O₄S₂Zn.** This compound crystallizes in the trigonal structure with the lattice parameters $a = 2505.8 \pm 0.4$ and $c = 1358.5 \pm 0.2$ pm and experimental density 1.471 g·cm$^{-3}$ (Smolander and Ahlgrèn 1994). It was prepared by dissolving 0.01 mM of zinc propionate in 100 mL of $H_2O$ followed by the addition of 0.02 mM of thiourea. The precipitate that formed was filtered off and recrystallized from $H_2O$. The colorless crystals were dried at room temperature.

**[Zn(C$_4$H$_9$N$_3$S)$_2$(NO$_3$)$_2$] or C$_8$H$_{18}$N$_8$O$_6$S$_2$Zn.** This compound was prepared by evaporation of a mixture of ethanolic solutions of stoichiometric quantities of $Zn(NO_3)_2$·6$H_2O$ and acetone thiosemicarbazone ($C_4H_9N_3S$) (1:1 to 1:3 molar ratio) (Ablov and Gerbeleu 1964). A white crystalline precipitate was obtained.

**[Zn(C$_4$H$_{11}$N$_3$S)$_2$(NO$_3$)$_2$] or C$_8$H$_{22}$N$_8$O$_6$S$_2$Zn.** This compound was prepared from the reaction of $Zn(NO_3)_2$·6$H_2O$ with 1,1,4-trimethylthiosemicarbazide in EtOH (Burrows et al. 2003).

**[Zn(C$_4$H$_{11}$N$_3$S)$_2$(H$_2$O)$_2$](NO$_3$)$_2$ or C$_8$H$_{26}$N$_8$O$_8$S$_2$Zn.** This compound crystallizes in the monoclinic structure with the lattice parameters $a = 963.60 \pm 0.03$, $b = 1242.00 \pm 0.04$, $c = 981.70 \pm 0.03$ pm, and $\beta = 118.4370° \pm 0.0016°$ at $170 \pm 2$ K and a calculated density of 1.555 g·cm$^{-3}$ (Burrows et al. 2003). It was prepared by the following procedure. 1,1,4-Trimethylthiosemicarbazide (3.36 mM) in EtOH (10 mL) was added to $Zn(NO_3)_2$·6$H_2O$ (1.68 mM) in EtOH (10 mL), and the solution was stirred for 1 h. The volume of the solution was reduced to 5 mL in vacuo, and the colorless crystals that precipitated overnight were separated by filtration. On recrystallization from $H_2O$, crystals of title compound were obtained.

**[Zn(C$_4$H$_4$NO$_2$CS$_2$)$_2$] or C$_{10}$H$_8$N$_2$O$_4$S$_4$Zn.** This compound melts at 260°C with decomposition (Siddiqi and Nishat 2000). To obtain it, $ZnCl_2$ (10 mM) in EtOH (50 mL) was added to a solution of sodium succinimide dithiocarbamate (20 mM) in hot EtOH (100 mL), and the mixture was stirred for 30 min at room temperature. The colored precipitate thus obtained was filtered, washed with cold EtOH, and dried in vacuo.

**[{Zn(C$_8$H$_4$O$_4$)[SC(NH$_2$)$_2$]$_2$}] or C$_{10}$H$_{12}$N$_4$O$_4$S$_2$Zn.** This compound crystallizes in the orthorhombic structure with the lattice parameters $a = 830.7 \pm 0.2$, $b = 1091.9 \pm 0.2$, and $c = 1589.0 \pm 0.2$ pm (Burrows et al. 2000). To obtain it, an aqueous solution of sodium phthalate (0.23 mM) was added to an aqueous solution of [ZnCl$_2${SC(NH$_2$)$_2$}$_4$] (0.23 mM) with no discernible change. After several hours, a colorless crystalline precipitate was observed, which was separated by filtration.

**[Zn(H$_2$O)(C$_5$H$_5$N$_2$O$_2$S)$_2$] or C$_{10}$H$_{12}$N$_4$O$_5$S$_2$Zn.** This compound crystallizes in the monoclinic structure with the lattice parameters $a = 1248.1 \pm 0.4$, $b = 952.2 \pm 0.3$, $c = 1268.6 \pm 0.5$ pm, and $\beta = 95.800° \pm 0.005°$ (Yang et al. 2009). It was prepared from $ZnCl_2$, 2-aminothiazol-4-acetic acid and deionized $H_2O$ in the molar ratio 1:2:111 by heating at 130°C for 4 days. After preparation, the autoclave was cooled down to room temperature, and the colorless block-shaped single crystals were separated, washed by distilled water, and dried in air.

**[Zn(C$_4$H$_7$NOS)$_2$(SCN)$_2$] or C$_{10}$H$_{14}$N$_4$O$_2$S$_4$Zn.** This compound was obtained by the reaction of $Zn(SCN)_2$ and thiomorpholine-3-one in refluxing chlorobenzene for 2 h (Colombini and Preti 1975). It was separated out by crystallization and washed with ethyl ether.

**[Zn{C₈H₄O₄}{SC(NH₂)₂}₂]·H₂O or C₁₀H₁₄N₄O₅S₂Zn.** This compound crystallizes in the monoclinic structure with the lattice parameters $a = 2025.7 \pm 0.3$, $b = 979.5 \pm 0.1$, $c = 1630.3 \pm 0.3$ pm, and $\beta = 94.50° \pm 0.02°$ (Burrows et al. 2000). To obtain it, an aqueous solution of sodium isophthalate (0.23 mM) was added to an aqueous solution of [ZnCl₂{SC(NH₂)₂}₄] (0.23 mM) with no discernible change. After several hours, a colorless crystalline precipitate was observed, which was separated by filtration.

**[Zn(C₄H₈NCOS)₂] or C₁₀H₁₆N₂O₂S₂Zn.** This compound was prepared by treating 10 mM of *N*-pyrrolidylthiocarbamatosodium (C₄H₈NCOSNa) in H₂O (20 mL) with an equal volume of aqueous solution of Zn(NO₃)₂·6H₂O (5 mM) (McCormick and Greene 1972). The white precipitate that formed was washed with H₂O and diethyl ether and dried in vacuo over anhydrous CaSO₄ for 24 h.

**[Zn(C₃H₅NS₂)₂(Ac)₂] or C₁₀H₁₆N₂O₂S₄Zn.** This compound melts at 92°C (Colombini and Preti 1975). It was prepared by a reaction of ZnCO₃ with thiazolidine-2-thione in AcOH (reflux for 1 h). The title compound, which separated spontaneously during the reaction, was washed with ethyl ether.

**[Zn(C₃H₆N₂S)₂(OAc)₂] or C₁₀H₁₈N₄O₄S₂Zn.** This compound melts at 212°C–220°C and crystallizes in the orthorhombic structure with the lattice parameters $a = 938.54 \pm 0.07$, $b = 1246.47 \pm 0.10$, and $c = 1322.63 \pm 0.11$ pm (in the monoclinic structure with the lattice parameters $a = 910$, $b = 1250$, $c = 1400$ pm, and $\beta = 92.3°$ (Nardelli and Chierici 1959) and a calculated density of 1.665 g·cm⁻³ (Lobana et al. 2008). To obtain it, to a stirred solution of 1,3-imidazolidine-2-thione (C₃H₆N₂S) (0.24 mM) in dry MeOH (8 mL) was added a solution of Zn(OAc)₂ (0.12 mM) in dry MeOH (10 mL). The reaction mixture was stirred for 30 min at room temperature. To this mixture was added CH₂Cl₂ (5 mL) and the slow evaporation at room temperature formed colorless crystals of Zn(C₃H₆N₂S)₂(OAc)₂.

**[Zn{SC(NHMe)₂}₂(C₄H₂O₄)]ₙ or C₁₀H₁₈N₄O₄S₂Zn.** This compound crystallizes in the monoclinic structure with the lattice parameters $a = 779.10 \pm 0.02$, $b = 1462.40 \pm 0.02$, $c = 1510.50 \pm 0.04$ pm, and $\beta = 94.1120° \pm 0.0011°$ at $170 \pm 2$ K and a calculated density of 1.500 g·cm⁻³ (Burke et al. 2003). To obtain it, an aqueous solution of sodium fumarate (0.15 mM) was added to an aqueous solution of [Zn{SC(NHMe)₂}₄]Cl₂ (0.15 mM), with no discernable change. After several hours, a colorless crystalline precipitate was observed, which was separated by filtration.

**[Zn{S₂CN(CH₂CH₂OH)Et}₂]₂ or C₁₀H₂₀N₂O₂S₄Zn.** This compound melts at 149°C–150°C and crystallizes in the monoclinic structure with the lattice parameters $a = 838.21 \pm 0.15$, $b = 1164.32 \pm 0.18$, $c = 1697.8 \pm 0.4$ pm, and $\beta = 102.926° \pm 0.005°$ at 173 K and a calculated density of 1.620 g·cm⁻³ (Benson et al. 2007).

**[Zn{S₂CN(CH₂CH₂OH)₂}₂]₂ or C₁₀H₂₀N₂O₄S₄Zn.** This compound melts at 142°C–145°C (Benson et al. 2007) and crystallizes in the monoclinic structure with the lattice parameters $a = 2901.7 \pm 0.5$, $b = 861.1 \pm 0.4$, $c = 1404.5 \pm 0.2$ pm, and $\beta = 103.17° \pm 0.04°$ and a calculated density of 1.655 g·cm⁻³ (Thirumaran et al. 1998). To obtain it, diethanoldithiocarbamic acid was freshly prepared by mixing 20 mM of diethanolamine and 20 mM of CS₂ (each in 20 mL of MeOH) under ice-cold condition (5°C). $ZnSO_4 \cdot 7H_2O$ (10 mM) in $H_2O$ was added to the obtained solution with continuous stirring to result in a solid complex. Crystals were obtained by recrystallization of the complex from EtOH at room temperature.

According to the date of Benson et al. (2007), the title compound crystallizes in the orthorhombic structure with the lattice parameters $a = 2962.0 \pm 1.5$, $b = 845.8 \pm 0.4$, and $c = 1383.8 \pm 0.7$ pm at 218 K and a calculated density of 1.632 g·cm⁻³.

**[Zn(SOCMe)₂(C₆H₁₆N₂)] or C₁₀H₂₂N₂O₂S₂Zn.** This compound crystallizes in the orthorhombic structure with the lattice parameters $a = 1228.9 \pm 0.2$, $b = 739.8 \pm 0.1$, and $c = 1683.0 \pm 0.1$ pm and a calculated density of 1.440 g·cm⁻³ (Nyman et al. 1996a). To prepare it, $Et_2Zn$ (8.1 mM), 25 mL dry toluene, and 8.1 mM of $N,N,N,N$-tetramethylethylenediamine were combined in a round-bottom flask in an inert-atmosphere box. The flask was removed from the box and placed in a dry-ice acetone bath. Thioacetic acid (16.2 mM) was syringed into the flask while stirring. A white precipitate was formed immediately. The mixture was allowed to warm to room temperature and stirred for several hours. Then it was heated to 70°C to allow most of the precipitate to dissolve, and the hot solution was immediately filtered. The title compound crystallized out of the solution as long colorless blades while standing at room temperature overnight.

**[Zn(NH₂CSNHNH₂)₂(H₂O)₂][C₆H₄(CO₂)₂-1,4]·2H₂O or C₁₀H₂₂N₆O₈S₂Zn.** This compound crystallizes in the triclinic structure with the lattice parameters $a = 616.3 \pm 0.2$, $b = 778.6 \pm 0.2$, and $c = 1031.6 \pm 0.3$ pm and $\alpha = 93.42° \pm 0.02°$, $\beta = 96.17° \pm 0.03°$, and $\gamma = 108.79° \pm 0.02°$ and a calculated density of 1.733 g·cm⁻³ (Burrows et al. 1997). To obtain it, an aqueous solution of $Na_2C_6H_4(CO_2)_2$ (0.54 mM) was added to an aqueous solution of $[Zn(NH_2CSNHNH_2)_2(NO_3)_2]$ (0.54 mM). The solution was allowed to stand for 12 h after which time colorless crystals had grown.

**[Zn(EtMe₂CH₂N₃S)₂(NO₃)₂] or C₁₀H₂₆N₈O₆S₂Zn.** To obtain the title compound, ethyl-dimethyl-thiosemicarbazide (2.20 mM), dissolved in absolute EtOH (15 mL), was added dropwise to an absolute ethanolic solution of $Zn(NO_3)_2 \cdot 6H_2O$ (1.1 mM in 10 mL) while stirring (Babb et al. 2003). After approximately 10 min, a white precipitate of $Zn(EtMe_2CH_2N_3S)_2(NO_3)_2$ was seen to form, and this was collected by filtration.

**[Zn(CH₅N₃S)(C₁₀H₈O₄)] or C₁₁H₁₃N₃O₄S₂Zn.** This compound crystallizes in the monoclinic structure with the lattice parameters $a = 1265.41 \pm 0.10$, $b = 1151.8 \pm 0.2$, $c = 1870.5 \pm 0.3$ pm, and $\beta = 99.516° \pm 0.010°$ (Babb et al. 2003). It was prepared by the following procedure. To an aqueous solution of $Zn(CH_5N_3S)_2(NO_3)_2$ (0.27 mM) was added an aqueous solution of sodium 1,4-phenylenediacetate (0.27 mM). After approximately 24 h, a colorless crystalline material of the title compound was seen to form, which was separated by filtration.

**[Zn{C₉H₆O₄}{SC(NH₂)₂}₂] or C₁₁H₁₄N₄O₄S₂Zn.** This compound crystallizes in the monoclinic structure with the lattice parameters $a = 1676.7 \pm 0.2$, $b = 1134.60 \pm 0.10$, $c = 1704.5 \pm 0.3$ pm, and $\beta = 109.39° \pm 0.01°$ (Burrows et al. 2000). To obtain it, an aqueous solution of sodium homophthalate (0.23 mM) was added to an aqueous solution of $[ZnCl_2\{SC(NH_2)_2\}_4]$ (0.23 mM) with no discernible change.

After several hours, a colorless crystalline precipitate was observed, which was separated by filtration.

**[Zn(C$_5$H$_5$N)(EtOCS$_2$)$_2$] or C$_{11}$H$_{15}$NO$_2$S$_4$Zn**. This compound crystallizes in the monoclinic structure with the lattice parameters $a = 1400.1 \pm 0.5$, $b = 891.1 \pm 0.2$, $c = 1373.1 \pm 0.6$ pm, and $\beta = 104.80° \pm 0.03°$ and experimental and calculated densities $1.56 \pm 0.01$ and $1.55$ g·cm$^{-3}$, respectively (Raston et al. 1976). It was obtained when Zn(EtOCS$_2$)$_2$ (2.5 mM) was dissolved in acetone (30 mL) following the addition of pyridine (5 mL) with molar ratio Zn(EtOCS$_2$)$_2$/C$_5$H$_5$N = 1:1.5 (Leonova et al. 1997). The solution was filtered off and 120 mL of hexane was added to the filtrate. The precipitate was filtered off with suction, washed with hexane, and dried at room temperature in air. After dissolving in acetone (70 mL) with two drops of pyridine, [Zn(C$_5$H$_5$N)(EtOCS$_2$)$_2$] was reprecipitated by adding of hexane (120 mL). The resulting precipitate was filtered off with suction, washed with hexane, and dried at room temperature under vacuum over Mg(ClO$_4$)$_2$.

**[Zn(EtCH$_4$N$_3$S)$_2$(C$_5$H$_4$O$_4$)]·3H$_2$O or C$_{11}$H$_{28}$N$_6$O$_7$S$_2$Zn**. This compound crystallizes in the triclinic structure with the lattice parameters $a = 891.90 \pm 0.06$, $b = 938.60 \pm 0.07$, and $c = 1301.7 \pm 0.1$ pm and $\alpha = 90.099° \pm 0.003°$, $\beta = 88.662° \pm 0.003°$, and $\gamma = 81.684° \pm 0.004°$ (Babb et al. 2003). It was prepared by the following procedure. To an aqueous solution of Zn(EtCH$_4$N$_3$S)$_2$(NO$_3$)$_2$ (0.27 mM) was added an aqueous solution of sodium citraconate (0.27 mM). After approximately 24 h, a colorless crystalline material of the title compound was seen to form, which was separated by filtration.

**[Zn(C$_4$H$_{11}$N$_3$S)(C$_8$H$_4$O$_4$)] or C$_{12}$H$_{15}$N$_3$O$_4$SZn**. This compound crystallizes in the orthorhombic structure with the lattice parameters $a = 1571.6 \pm 0.1$, $b = 1179.2 \pm 0.2$, and $c = 760.3 \pm 0.1$ pm at $150 \pm 2$ K and a calculated density of 1.710 g·cm$^{-3}$ (Burrows et al. 2003). To obtain it, aqueous solutions of sodium isophthalate (0.25 mM) and [Zn(C$_4$H$_{11}$N$_3$S)$_2$(NO$_3$)$_2$] (0.25 mM) were mixed. After approximately 24 h, colorless crystals formed, and these were isolated by filtration.

**[Zn(CH$_4$N$_2$S)$_2${C$_6$H$_4$(CH$_2$CO$_2$)$_2$}]$_n$ or C$_{12}$H$_{16}$N$_4$O$_4$S$_2$Zn**. This compound crystallizes in the monoclinic structure with the lattice parameters $a = 1880.80 \pm 0.03$, $b = 818.00 \pm 0.01$, $c = 2129.60 \pm 0.04$ pm, and $\beta = 98.197° \pm 0.001°$ (Burrows et al. 2004a). To obtain it, an aqueous solution of sodium phenylenediacetate (0.23 mM) was added to an aqueous solution of [Zn(CH$_4$N$_2$S)$_4$]Cl$_2$ (0.23 mM), with no discernible change. After several days, a colorless crystalline precipitate was observed, which was separated by filtration.

**[Zn{SC(NH$_2$)(NHMe)}$_2$(C$_8$H$_4$O$_4$)]$_n$ or C$_{12}$H$_{16}$N$_4$O$_4$S$_2$Zn**. This compound crystallizes in the orthorhombic structure with the lattice parameters $a = 1108.100 \pm 0.010$, $b = 2271.50 \pm 0.03$, and $c = 1384.70 \pm 0.02$ pm at $150 \pm 2$ K and a calculated density of 1.562 g·cm$^{-3}$ (Burke et al. 2003). It was obtained by the following procedure. An aqueous solution of sodium terephthalate (0.18 mM) was added to an aqueous solution of [Zn{SC(NH$_2$)(NHMe)}$_4$]Cl$_2$ (0.18 mM), with no discernable change. After approximately 72 h, a colorless crystalline precipitate was observed, which was separated by filtration.

**[Zn(C$_4$H$_{11}$N$_3$S)(C$_8$H$_4$O$_4$)]·H$_2$O or C$_{12}$H$_{17}$N$_3$O$_5$SZn.** This compound crystallizes in the monoclinic structure with the lattice parameters $a = 848.8 \pm 0.1$, $b = 1444.4 \pm 0.2$, $c = 1367.5 \pm 0.2$ pm, and $\beta = 104.92° \pm 0.01°$ and a calculated density of 1.561 g·cm$^{-3}$ (Burrows et al. 2003). To obtain it, aqueous solutions of sodium isophthalate (0.25 mM) and [Zn(C$_4$H$_{11}$N$_3$S)$_2$(NO$_3$)$_2$] (0.25 mM) were mixed. After approximately 24 h, colorless crystals formed, and these were isolated by filtration.

**[Zn{SC(NH$_2$)(NHMe)}$_2$(C$_8$H$_4$O$_4$)·H$_2$O]$_n$ or C$_{12}$H$_{18}$N$_4$O$_5$S$_2$Zn.** This compound crystallizes in the monoclinic structure with the lattice parameters $a = 1038.30 \pm 0.03$, $b = 1042.60 \pm 0.03$, $c = 1658.70 \pm 0.04$ pm, and $\beta = 94.1980° \pm 0.0010°$ at $150 \pm 2$ K and a calculated density of 1.587 g·cm$^{-3}$ (Burke et al. 2003). It was obtained by the following procedure. An aqueous solution of sodium isophthalate (0.18 mM) was added to an aqueous solution of [Zn{SC(NH$_2$)(NHMe)}$_4$]Cl$_2$ (0.18 mM), with no discernable change. After approximately 72 h, a colorless crystalline precipitate was observed, which was separated by filtration.

**[Zn{(C$_4$H$_{11}$N$_3$S)(H$_2$O)}$_2$(C$_8$H$_4$O$_4$)$_2$]·2H$_2$O or C$_{12}$H$_{19}$N$_3$O$_6$SZn.** This compound crystallizes in the triclinic structure with the lattice parameters $a = 739.7 \pm 0.1$, $b = 938.4 \pm 0.1$, and $c = 1208.5 \pm 0.2$ pm and $\alpha = 98.90° \pm 0.01°$, $\beta = 90.80° \pm 0.02°$, and $\gamma = 99.40° \pm 0.01°$ and a calculated density of 1.621 g·cm$^{-3}$ (Burrows et al. 2003). To obtain it, aqueous solutions of sodium terephthalate (0.22 mM) and [Zn(C$_4$H$_{11}$N$_3$S)$_2$(NO$_3$)$_2$] (0.22 mM) were mixed. After approximately 24 h, colorless crystals formed, and these were isolated by filtration.

**[Zn(C$_5$H$_{10}$NCOS)$_2$] or C$_{12}$H$_{20}$N$_2$O$_2$S$_2$Zn.** This compound was prepared by treating 10 mM of N-piperidylthiocarbamatosodium (C$_5$H$_{10}$NCOSNa) in H$_2$O (20 mL) with an equal volume of aqueous solution of Zn(NO$_3$)$_2$·6H$_2$O (5 mM) (McCormick and Greene 1972). The white precipitate that formed was washed with H$_2$O and diethyl ether and dried in vacuo over anhydrous CaSO$_4$ for 24 h. Recrystallization from benzene solution by the addition of hexane was necessary to obtain pure samples.

**[Zn(C$_4$H$_7$NOS)$_2$(Ac)$_2$] or C$_{12}$H$_{20}$N$_2$O$_4$S$_2$Zn.** This compound melts at 88°C (Colombini and Preti 1975). It was prepared by the reaction of ZnCO$_3$ with thio-morpholine-3-one in AcOH (reflux for 1 h). The title compound, which separated spontaneously during the reaction, was washed with ethyl ether.

**[Zn(C$_4$H$_{11}$N$_3$S)$_2$(H$_2$O)][C$_4$H$_2$O$_4$] or C$_{12}$H$_{26}$N$_6$O$_5$S$_2$Zn.** This compound crystallizes in the triclinic structure with the lattice parameters $a = 812.2 \pm 0.1$, $b = 836.4 \pm 0.1$, and $c = 1624.7 \pm 0.6$ pm and $\alpha = 97.11° \pm 0.02°$, $\beta = 101.98° \pm 0.01°$, and $\gamma = 103.200° \pm 0.009°$ and a calculated density of 1.490 g·cm$^{-3}$ (Burrows et al. 2003). To obtain it, aqueous solutions of sodium fumarate (0.22 mM) and [Zn(C$_4$H$_{11}$N$_3$S)$_2$(NO$_3$)$_2$] (0.22 mM) were mixed. After approximately 24 h, colorless crystals formed, and these were isolated by filtration.

**[NMe$_4$][Zn(S$_2$CNMe$_2$)$_2$(OCOMe)] or C$_{12}$H$_{27}$N$_3$O$_2$S$_4$Zn, tetrametylammonium bis(dimethyldithiocarbamate)acetatozincate.** To obtain this compound, a mixture of [NMe$_4$][OH] (4.8 mL, 25% solution) and acetic acid (1.0 g) in acetone (50 mL) was evaporated to dryness in vacuo at 60°C (McCleverty et al. 1980b). The oil that had formed was redissolved in acetone (100 mL) and [Zn(S$_2$CNMe$_2$)$_2$] (4.0 g) was added. The mixture was refluxed for 2 h, cooled, and filtered, and the filtrate dried over MgSO$_4$. Upon the addition of light petroleum, white crystals precipitated, which were filtered off, washed with cold light petroleum, and dried in vacuo.

**[Zn(C$_4$H$_{11}$N$_3$S)(C$_9$H$_6$O$_4$)]$_2$ or C$_{13}$H$_{17}$N$_3$O$_4$SZn.** This compound crystallizes in the monoclinic structure with the lattice parameters $a = 968.90 \pm 0.02$, $b = 1229.20 \pm 0.03$, $c = 1322.40 \pm 0.03$ pm, and $\beta = 94.835° \pm 0.001°$ at $150 \pm 2$ K and a calculated density of 1.594 g·cm$^{-3}$ (Burrows et al. 2003). To obtain it, aqueous solutions of sodium homophthalate (0.22 mM) and [Zn(C$_4$H$_{11}$N$_3$S)$_2$(NO$_3$)$_2$] (0.22 mM) were mixed. After approximately 24 h, colorless crystals formed, and these were isolated by filtration.

**[Zn(SCN)$_2$(C$_{11}$H$_{17}$N$_3$O)] or C$_{13}$H$_{17}$N$_5$OS$_2$Zn.** This compound crystallizes in the triclinic structure with the lattice parameters $a = 814.3 \pm 0.1$, $b = 917.9 \pm 0.1$, and $c = 1267.7 \pm 0.1$ pm and $\alpha = 74.256° \pm 0.001°$, $\beta = 75.117° \pm 0.002°$, and $\gamma = 71.838° \pm 0.001°$ (Guo 2010). To obtain it, 1-pyridin-2-yl-ethanone (2 mM), NH$_4$SCN (2 mM), Zn(OAc)$_2$ (2 mM), and 2-(2-aminoethylamino)ethanol (2 mM) were dissolved in MeOH (100 mL), yielding a clear yellow solution after stirring in air at room temperature for 3 h. Then yellow block-shaped crystals of the title compound were obtained by slowly evaporating the solvent for about 2 weeks at room temperature. The crystals were filtered off, washed with small amounts of MeOH, and finally dried in vacuum.

**[Zn(C$_5$H$_5$N)(Pr$^i$OCS$_2$)$_2$] or C$_{13}$H$_{19}$NO$_2$S$_4$Zn.** It was obtained by the similar procedure to that described for [Zn(C$_5$H$_5$N)(EtOCS$_2$)$_2$] using Zn(Pr$^i$OCS$_2$)$_2$ instead of Zn(EtOCS$_2$)$_2$ (Leonova et al. 1997).

**[Zn{S$_2$CN(CH$_2$CH$_2$OH)Me}$_2$]$_2$(4,4′-bipy) or C$_{13}$H$_{20}$N$_3$O$_2$S$_4$Zn.** This compound melts at the temperature higher than 320°C and crystallizes in the monoclinic structure with the lattice parameters $a = 865.91 \pm 0.07$, $b = 1400.66 \pm 0.11$, $c = 1559.84 \pm 0.13$ pm, and $\beta = 97.409° \pm 0.002°$ and a calculated density of 1.572 g·cm$^{-3}$ (Benson et al. 2007).

**[NMe$_4$][Zn(S$_2$CNMe$_2$)$_2$(OCOEt)] or C$_{13}$H$_{29}$N$_3$O$_2$S$_4$Zn, tetrametylammonium bis(dimethyldithiocarbamate)propionatozincate.** To obtain this compound, a mixture of [NMe$_4$][OH] (4.8 mL, 25% solution) and propionic acid (1.0 g) in acetone (50 mL) was evaporated to dryness in vacuo at 60°C (McCleverty et al. 1980b). The oil that had formed was redissolved in acetone (100 mL) and [Zn(S$_2$CNMe$_2$)$_2$] (4.0 g) was added. The mixture was refluxed for 2 h, cooled, and filtered, and the filtrate dried over MgSO$_4$. On addition of light petroleum, white crystals precipitated, which were filtered off, washed with cold light petroleum, and dried in vacuo.

**[Zn(C$_7$H$_4$NOS)$_2$]$_n$ or C$_{14}$H$_8$N$_2$O$_2$S$_2$Zn.** To obtain this compound, an ethanol solution (250 mL) of 2-mercaptobenzoxazole (15.1 g) was added KOH (5.6 g) in aqueous EtOH (McCleverty et al. 1980a). The mixture was heated until it became homogeneous, and to it was added an ethanolic solution (100 mL) of [Zn(OCOMe)$_2$]·2H$_2$O (10.9 g). The mixture was allowed to cool, and the pale-yellow precipitate of the title compound was filtered off, washed with ethanol, and dried in vacuo.

**[Zn(C₇H₄NS₂)₂]·H₂O or C₁₄H₁₀N₂OS₄Zn.** This compound was prepared by the electrochemical oxidation of anodic Zn in an acetonitrile solution of benzothiazole-2-thione (Castro et al. 1993b).

**[Zn(C₁₀H₈N₂S₂)(C₄H₂O₄)] or C₁₄H₁₀N₂O₄S₂Zn.** This compound crystallizes in the monoclinic structure with the lattice parameters $a = 2005.0 \pm 0.2$, $b = 2326.5 \pm 0.2$, $c = 1151.5 \pm 0.2$ pm, and $\beta = 118.870° \pm 0.002°$ and a calculated density of 1.129 g·cm⁻³ (Yang et al. 2008). To obtain it, a methanolic solution (10 mL) of 4,4′-bipy (0.1 mM) and fumaric acid (0.1 mM) was slowly diffused into an aqueous solution (10 mL) of Zn(NO₃)₂·6H₂O (0.1 mM) and stirred for 40 min at room temperature. Slow evaporation of the filtrate for 4 days provided colorless block crystals of [Zn(C₁₀H₈N₂S₂)(C₄H₂O₄)].

**[Zn(OAc)₂(C₁₀H₈N₂S₂)] or C₁₄H₁₄N₂O₄S₂Zn.** This compound crystallizes in the monoclinic structure with the lattice parameters $a = 1646.2 \pm 0.4$, $b = 1005.4 \pm 0.2$, $c = 1093.6 \pm 0.3$ pm, and $\beta = 109.556° \pm 0.001°$ at $223 \pm 2$ K and a calculated density of 1.573 g·cm⁻³ (Ng et al. 2004). To obtain it, a solution of Zn(OAc)₂ (0.45 mM) in MeOH (0.5 mL) was added to a solution of 4,4′-bipyridyl disulfide (0.45 mM) in THF (0.5 mL). A clear solution was obtained, and the solution was left for slow evaporation at room temperature. Pale-yellowish rhombic-shaped single crystals of the title compound were formed after a day. The crystals were collected by filtration, washed with Et₂O and dried under vacuum.

**[Zn{SC(NH₂)(NHMe)}₂(C₁₀H₈O₄)]ₙ or C₁₄H₂₀N₄O₄S₂Zn.** This compound crystallizes in the monoclinic structure with the lattice parameters $a = 657.90 \pm 0.02$, $b = 1334.30 \pm 0.05$, $c = 2120.60 \pm 0.09$ pm, and $\beta = 97.9270° \pm 0.0017°$ at $150 \pm 2$ K and a calculated density of 1.577 g·cm⁻³ (Burke et al. 2003). To obtain it, an aqueous solution of sodium 1,3-phenylenediacetate (0.18 mM) was added to an aqueous solution of [Zn{SC(NH₂)(NHMe)}₄]Cl₂ (0.18 mM), with no discernable change. After approximately 72 h, a colorless crystalline precipitate was observed, which was separated by filtration.

**[Zn{SC(NHMe)₂}₂(C₈H₄O₄)]ₙ or C₁₄H₂₀N₄O₄S₂Zn.** This compound crystallizes in the orthorhombic structure with the lattice parameters $a = 1597.84 \pm 0.03$, $b = 1511.77 \pm 0.02$, and $c = 745.66 \pm 0.01$ pm at $150 \pm 2$ K and a calculated density of 1.615 g·cm⁻³ (Burke et al. 2003). To obtain it, an aqueous solution of sodium phthalate (0.15 mM) was added to an aqueous solution of [Zn{SC(NHMe)₂}₄]Cl₂ (0.15 mM), with no discernable change. After several hours, a colorless crystalline precipitate was observed, which was separated by filtration.

**[{Zn(C₄H₃N₂O₂S)₂(H₂O)}·2(OC₃H₆)]ₙ or C₁₄H₂₀N₄O₇S₂Zn.** This compound crystallizes in the tetragonal structure with the lattice parameters $a = 1695.60 \pm 0.10$ and $c = 755.71 \pm 0.08$ pm and a calculated density of 1.485 g·cm⁻³ (Pan et al. 2008). To obtain it, a suspension of ZnSO₄·7H₂O (0.50 mM) and 2-thiobarbituric acid (C₄H₄N₂O₂S) (1.0 mM) in DMF (10 mL) was stirred at room temperature for 10 h. After the mixture was filtered, acetone (10 mL) was layered onto the filtrate; a few weeks later, yellow block crystals of the title compound were isolated.

**[Zn{SC(NHMe)$_2$}$_2$(C$_8$H$_4$O$_4$)·0.5H$_2$O]$_n$** or **C$_{14}$H$_{21}$N$_4$O$_{4.5}$S$_2$Zn**. This compound crystallizes in the triclinic structure with the lattice parameters $a = 855.500 \pm 0.010$, $b = 961.80 \pm 0.02$, and $c = 1303.60 \pm 0.02$ pm and $\alpha = 98.2960° \pm 0.0010°$, $\beta = 106.8340° \pm 0.0010°$, and $\gamma = 100.0610° \pm 0.0010°$ and a calculated density of 1.500 g·cm$^{-3}$ (Burke et al. 2003). To obtain it, an aqueous solution of sodium terephthalate (0.15 mM) was added to an aqueous solution of [Zn{SC(NHMe)$_2$}$_4$]Cl$_2$ (0.15 mM), with no discernable change. After several hours, a colorless crystalline precipitate was observed, which was separated by filtration.

**[Zn{NH(CH$_2$)$_4$O}{S$_2$CN(CH$_2$)$_4$O}$_2$]** or **C$_{14}$H$_{25}$N$_3$O$_3$S$_4$Zn**. This compound was obtained by quantitative absorption of morpholine, NH(CH$_2$)$_4$O, by fine dispersed samples of [Zn$_2${S$_2$CN(CH$_2$)$_4$O}]$_4$ from the gas phase or as a result of their careful wetting (Ivanov et al. 2003b). Thermolysis of the title compound leads to the formation of ZnS.

**[Zn(C$_4$H$_{11}$N$_3$S)(C$_{10}$H$_{14}$O$_4$)]** or **C$_{14}$H$_{25}$N$_3$O$_5$SZn**. This compound crystallizes in the monoclinic structure with the lattice parameters $a = 2662.50 \pm 0.04$, $b = 753.50 \pm 0.01$, $c = 1868.40 \pm 0.04$ pm, and $\beta = 103.7490° \pm 0.0007°$ at $100 \pm 2$ K and a calculated density of 1.448 g·cm$^{-3}$ (Burrows et al. 2003). To obtain it, aqueous solutions of sodium (1R,3S)-(+)-camphorate (0.22 mM) and [Zn(C$_4$H$_{11}$N$_3$S)$_2$(NO$_3$)$_2$] (0.22 mM) were mixed. After approximately 24 h, colorless crystals formed, and these were isolated by filtration.

**[Zn(C$_7$H$_{13}$N$_3$S)$_2$(NO$_3$)$_2$]** or **C$_{14}$H$_{26}$N$_8$O$_6$S$_2$Zn**. This compound was prepared by evaporation of a mixture of ethanolic solutions of stoichiometric quantities of Zn(NO$_3$)$_2$·6H$_2$O and cyclohexanone thiosemicarbazone (C$_7$H$_{13}$N$_3$S) (1:1 to 1:3 molar ratio) (Ablov and Gerbeleu 1964). White crystalline precipitate in the form of hexagonal prisms was obtained.

**[Zn{O(CH$_2$)$_4$NCS$_2$}$_2$(Et$_2$NH)]** or **C$_{14}$H$_{27}$N$_3$O$_2$S$_4$Zn, bis(morpholinedithiocarbamato)diethylaminezinc**. This compound was obtained by quantitative chemisorption of Et$_2$NH vapor for the gas phase on the powder of Zn$_2${O(CH$_2$)$_4$NCS$_2$}$_4$ and by crystallization from toluene solution (Ivanov et al. 2007b).

**[Zn(Me$_2$CH$_3$N$_3$S)$_2$(H$_2$O)]·[C$_8$H$_4$O$_4$]·2H$_2$O** or **C$_{14}$H$_{28}$N$_6$O$_7$S$_2$Zn**. This compound crystallizes in the triclinic structure with the lattice parameters $a = 934.4 \pm 0.2$, $b = 1067.2 \pm 0.2$, and $c = 1275.3 \pm 0.2$ pm and $\alpha = 74.64° \pm 0.02°$, $\beta = 76.87° \pm 0.02°$, and $\gamma = 73.21° \pm 0.02°$ (Babb et al. 2003). It was prepared by the following procedure. To an aqueous solution of Zn(Me$_2$CH$_3$N$_3$S)$_2$(NO$_3$)$_2$ (0.27 mM) was added an aqueous solution of sodium terephthalate (0.27 mM). After approximately 24 h, a colorless crystalline material of the title compound was seen to form, which was separated by filtration.

**[Zn{O(CH$_2$)$_4$NH}{S$_2$CNEt$_2$}$_2$]** or **C$_{14}$H$_{29}$N$_3$OS$_4$Zn**. This compound crystallizes in the monoclinic structure with the lattice parameters $a = 1832.7 \pm 0.1$, $b = 1049.3 \pm 0.1$, $c = 2293.3 \pm 0.1$ pm, and $\beta = 105.52° \pm 0.01°$ at $100 \pm 1$ K and a calculated density of $1.404 \pm 0.001$ g·cm$^{-3}$ (Ivanov et al. 2001c). It was prepared by dissolving [Zn$_2${S$_2$CNEt$_2$}$_4$] in a minute volume of toluene containing morpholine in an amount exceeding the stoichiometric composition by approximately 3%. Both powder and single-crystal samples were isolated by means of slow evaporation of toluene at room temperature. The prepared crystals of the title compound were separated from the mother liquors, washed with a small quantity of toluene, and dried in air to a friable state.

**[Me₄N][Zn(S₂CNMe₂)₂(OCOPrⁿ)] or C₁₄H₃₁N₃O₂S₄Zn, tetrametylammonium bis(dimethyldithiocarbamate)butyratozincate.** To obtain this compound, a mixture of [NMe₄][OH] (4.8 mL, 25% solution) and butyric acid (1.0 g) in acetone (50 mL) was evaporated to dryness in vacuo at 60°C (McCleverty et al. 1980b). The oil that had formed was redissolved in acetone (100 mL), and [Zn(S₂CNMe₂)₂] (4.0 g) was added. The mixture was refluxed for 2 h, cooled, and filtered, and the filtrate dried over $MgSO_4$. Upon the addition of light petroleum, white crystals precipitated, which were filtered off, washed with cold light petroleum, and dried in vacuo.

**[(mepa)Zn][2-(hydroxymethyl)benzenethiolate] or C₁₅H₁₈N₂OS₂Zn (mepaH = N-(2-mercaptoethyl)picolylamin, C₈H₁₂N₂S).** This compound melts at 112°C (Brand and Vahrenkamp 1995). To obtain it, $Zn[N(SiMe_3)_2]_2$ (2.94 mM) was dissolved in toluene (30 mL). A solution of mepaH (2.94 mM) and 2-(hydroxymethyl) benzenethiol (2.94 mM) in toluene (20 mL) was added dropwise within 1 h with stirring. The precipitate was stirred for another 2 h, filtered off, washed in toluene, and dried in vacuo. This compound was purified three times by dissolving in $CH_2Cl_2$ and precipitated by the slow addition of hexane. It resulted as a yellowish powder.

**[Zn(C₇H₈N₅S)₂]·0.5EtOH or C₁₅H₁₉N₁₀O₀.₅S₂Zn.** This compound crystallizes in the tetragonal structure with the lattice parameters $a = 1084.0 \pm 0.2$ and $c = 2207.9 \pm 0.6$ pm and a calculated density of 1.221 g·cm⁻³ (Li et al. 2006). To obtain it, $Zn(OAc)_2 \cdot 4H_2O$ (0.125 mM) was added at room temperature to acetylpyrazine thiosemicarbazone (0.25 mM) suspended in EtOH (60 mL). After stirring for 30 min, the solid formed was filtered off, washed with EtOH, and dried under vacuum.

**[Zn(C₅H₅N)(BuⁿOCS₂)₂] and [Zn(C₅H₅N)(BuⁱOCS₂)₂] or C₁₅H₂₃NO₂S₄Zn.** These were obtained by the similar procedure to that described for [Zn(C₅H₅N)(EtOCS₂)₂] using $Zn(Bu^nOCS_2)_2$ and $Zn(Bu^iOCS_2)_2$ instead of $Zn(EtOCS_2)_2$ (Leonova et al. 1997).

**[NMe₄][Zn(S₂CNMe₂)₂(OCOBuⁿ)] or C₁₅H₃₃N₃O₂S₄Zn, tetrametylammonium bis(dimethyldithiocarbamate)pentanoatozincate.** To obtain this compound, a mixture of [NMe₄][OH] (4.8 mL, 25% solution) and pentanoic acid (1.0 g) in acetone (50 mL) was evaporated to dryness in vacuo at 60°C (McCleverty et al. 1980b). The oil that had formed was redissolved in acetone (100 mL) and [Zn(S₂CNMe₂)₂] (4.0 g) was added. The mixture was refluxed for 2 h, cooled, and filtered, and the filtrate dried over $MgSO_4$. Upon the addition of light petroleum, white crystals precipitated, which were filtered off, washed with cold light petroleum, and dried in vacuo.

**[Zn(C₁₆H₁₅N₆S₂)NO₃] or C₁₆H₁₅N₇O₃S₂Zn.** This compound melts at 260°C (López-Torres et al. 2001). It was prepared as follows. To a suspension of benzilbisthiosemi-carbazone (1.58 mM) in EtOH (40 mM) was added $Zn(NO_3)_2 \cdot 6H_2O$ (1.88 mM) in the same solvent (10 mM). The mixture was stirred under reflux for 4 h. The yellow-orange solid was filtered off, washed with ethanol, and dried in vacuo.

**[Zn(H$_2$O)(C$_{10}$H$_8$N$_2$)(C$_6$H$_2$O$_4$S)]·2H$_2$O or C$_{16}$H$_{16}$N$_2$O$_7$SZn.** This compound crystallizes in the triclinic structure with the lattice parameters $a = 853.56 \pm 0.07$, $b = 953.42 \pm 0.08$, and $c = 1267.0 \pm 0.1$ pm and $\alpha = 69.377° \pm 0.001°$, $\beta = 86.236° \pm 0.001°$, and $\gamma = 67.666° \pm 0.001°$ (Wang et al. 2009). To obtain it, a mixture of Zn(NO$_3$)$_2$·6H$_2$O (0.5 mM), 2,2′-bipy (0.5 mM), 2,5-thiophenedicarboxylic acid (0.5 mM), and H$_2$O (18 mL) was kept in a 30 mL Teflon-lined autoclave under autogenous pressure at 150°C for 5 days. After cooling to room temperature, colorless block-shaped crystals were collected by filtration and washed with distilled water.

**Zn(2,2′-bipy)(EtOCS$_2$) or C$_{16}$H$_{18}$N$_2$O$_2$S$_4$Zn.** This compound crystallizes in the orthorhombic structure with the lattice parameters $a = 984.6 \pm 0.2$, $b = 2059.7 \pm 0.4$, and $c = 2060.2 \pm 0.4$ pm and a calculated density of 1.475 g·cm$^{-3}$ (Glinskaya et al. 2000a). It was obtained when Zn(OAc)$_2$·4H$_2$O (2.5 mM) and 2,2′-bipy (2.5 mM) were dissolved in H$_2$O (100 mL) (Leonova et al. 1997). KS$_2$COEt (5 mM) in 20 mL of H$_2$O was added to this solution. The reaction mixture was stirred for 30 min. The precipitated was filtered off with suction, washed with H$_2$O, and dried at room temperature in air. After dissolving in CH$_2$Cl$_2$ (15 mL), Zn(2,2′-bipy)(EtOCS$_2$) was reprecipitated by adding of hexane (50 mL). The resulting precipitate was filtered off with suction, washed with hexane, and dried at room temperature under vacuum over Mg(ClO$_4$)$_2$.

Single crystals of this compound were grown by the slow evaporation of the diluted solution of Zn(2,2′-bipy)(EtOCS$_2$) in the mixture CH$_2$Cl$_2$/hexane (1:1 volume ratio) at room temperature (Glinskaya et al. 2000a).

**[Zn(PhCOO)$_2${CS(NH$_2$)$_2$}$_2$] or C$_{16}$H$_{18}$N$_4$O$_4$S$_2$Zn.** This compound crystallizes in the orthorhombic structure with the lattice parameters $a = 2466.7 \pm 0.4$, $b = 1057.3 \pm 0.3$, and $c = 1573.6 \pm 0.4$ pm and experimental and calculated densities 1.44 $\pm$ 0.02 and 1.488 g·cm$^{-3}$, respectively (Černák and Adzimová 1995). It was prepared by first adding 1.22 g of solid benzoic acid to a solution of 0.4 g of NaOH in 25 mL of H$_2$O (pH in the range 6.0–6.5). Successive solutions of ZnSO$_4$·7H$_2$O (1.435 g in 10 mL of H$_2$O) and of thiourea (0.762 g in 10 mL of H$_2$O) were then added. After some days, the product separated out in the form of very thin colorless plates. Single crystals suitable for XRD were prepared by recrystallization from EtOH.

**[Zn(C$_3$H$_6$NOS$_2$)$_2$(2,2′-bipy)] or C$_{16}$H$_{20}$N$_4$O$_2$S$_4$Zn.** This compound melts at 189°C and crystallizes in the triclinic structure with the lattice parameters $a = 923.7 \pm 0.2$, $b = 1431.6 \pm 0.2$, and $c = 851.0 \pm 0.3$ pm and $\alpha = 95.45° \pm 0.03°$, $\beta = 111.95° \pm 0.03°$, and $\gamma = 82.64° \pm 0.02°$ (Saravanan et al. 2004). The next procedure was used to obtain it. A hot solution of 2,2′-bipy (2 mM) in EtOH was added to a hot solution of bis(ethanoldithiocarbamato)zinc [Zn(C$_3$H$_6$NOS$_2$)$_2$] (2 mM) in benzene under constant stirring, and the resulting yellow solution was filtered and kept for crystallization. After 2 days, yellow crystals along with some decomposed parent compound were obtained. The yellow crystals were washed with alcohol and filtered out and dried in air.

**[Zn(C₁₄H₂₁N₄S)(OAc)] or C₁₆H₂₄N₄O₂SZn**. This compound crystallizes in the orthorhombic structure with the lattice parameters $a = 1188.9 \pm 0.2$, $b = 1169.4 \pm 0.2$, and $c = 1395.3 \pm 0.3$ pm (Hammesa and Carrano 2000). To prepare it, a solution of [MeZn(C₁₄H₂₁N₄S)] (1.1 mM) in 30 mL of CH₂Cl₂ was treated with HOAc (1.1 mM) as a solution in CH₂Cl₂. The resulting solution was stirred for 1 h, concentrated under reduced pressure, and layered with hexane. Over a 2-day period, the title compound crystallized as colorless blocks, which were filtered off and dried under reduced pressure.

**[Zn(C₄H₆N₂S)₄(NO₃)₂]·H₂O or C₁₆H₂₆N₁₀O₇S₄Zn**. This compound crystallizes in the orthorhombic structure with the lattice parameters $a = 2075.6 \pm 0.8$, $b = 1241.0 \pm 0.5$, and $c = 2205.9 \pm 0.8$ pm and a calculated density of 1.55 g·cm⁻³ (Nowell et al. 1979). Reaction of 1-methyl-2(3*H*)-imidazolinethione with hydrated Zn(NO₃)₂ in anhydrous EtOH gives colorless crystalline material analyzing as the title compound.

**[Zn{O(CH₂)₄NCS₂}₂{NH(CH₂)₆}] or C₁₆H₂₉N₃O₂S₄Zn**. This compound was obtained by quantitative absorption of cyclohexylamine, NH(CH₂)₆, by fine dispersed samples of Zn{O(CH₂)₄NCS₂}₂ from the gas phase, or as a result of their careful wetting (Ivanov et al. 2004b).

**[Zn{O(CH₂)₄NCS₂}₂(Pr^n₂NH)] or C₁₆H₃₁N₃O₂S₄Zn, bis(morpholinedithiocarbamato)di-*n*-propylaminezinc**. This compound was obtained by quantitative chemisorption of Pr^n₂NH vapor for the gas phase on the powder of Zn₂{O(CH₂)₄NCS₂}₄ and by crystallization from toluene solution (Ivanov et al. 2007b).

**[Zn(C₄H₈N₂S)₄(NO₃)₂]·H₂O or C₁₆H₃₄N₁₀O₇S₄Zn**. This compound melts at 145°C (Raper and Nowell 1980). It was prepared by dissolving Zn(NO₃)₂·6H₂O (2 mM) in a small volume (<10 mL) of hot anhydrous EtOH containing 3 mL triethylorthoformate as dehydrating agent. To this solution, the stoichiometric quantity of 1-methylimidazoline-2-thione (C₄H₈N₂S) was added and dissolved in a minimum volume (<5 mL) of hot anhydrous EtOH. The total volume was reduced slightly and then cooled to room temperature. Subsequent slow evaporation of the solvent produced a crystalline colorless product in each instance, which was removed, washed with a small volume of EtOH, and finally vacuum dried at room temperature.

**[Zn(C₁₄H₈O₆S)(DMF)] or C₁₇H₁₅NO₇SZn**. This compound crystallizes in the monoclinic structure with the lattice parameters $a = 2215.1 \pm 0.7$, $b = 1287.2 \pm 0.4$, $c = 999.3 \pm 0.3$ pm, and $\beta = 99.584° \pm 0.005°$ and a calculated density of 1.047 g·cm⁻³ (Zhuang et al. 2007). It was prepared by the following procedure. A mixed solvent of 3 mL MeOH and 3 mL DMF was carefully layered on top of a 2 mL DMF solution of 4,4′-dicarboxybiphenyl sulfone (0.1 mM), to which a solution of Zn(OAc)₂·2H₂O (0.1 mM) in methanol (2 mL) was added. Colorless block crystals were obtained after 2 months at room temperature.

**[MeZn(C₁₄H₂₁N₄S)]·0.28PhMe·0.25H₂O or C₁₇H₂₆.₇N₄O₀.₂₅SZn**. To prepare this compound, a solution of bis(3,5-dimethylpyrazolyl)(1-methyl-1-sulfanylethyl)methane (C₁₄H₂₂N₄S·0.25H₂O) (2.2 mM) in 25 mL of toluene was treated with 1.3 mL of toluene solution of ZnMe₂ (2.6 mM) under argon (Hammesa and Carrano 2000). The reaction mixture containing the precipitated product was stirred for 1 h and subsequent workup was performed in air. The white product was filtered off and washed with ether to yield the title compound.

**[Zn(C$_5$H$_5$N)(Am$^i$OCS$_2$)$_2$] or C$_{17}$H$_{27}$NO$_2$S$_4$Zn.** It was obtained by the similar procedure to that described for [Zn(C$_5$H$_5$N)(EtOCS$_2$)$_2$] using Zn(Am$^i$OCS$_2$)$_2$ instead of Zn(EtOCS$_2$)$_2$ (Leonova et al. 1997).

**[Me$_4$N][Zn(S$_2$CNMe$_2$)$_2$(C$_7$H$_4$NOS)] or C$_{17}$H$_{28}$N$_4$OS$_5$Zn.** This compound was prepared as follows (McCleverty et al. 1980a). To an acetone solution of [Me$_4$N] [C$_7$H$_4$NOS] was added [Zn(S$_2$CNMe$_2$)$_2$], and the mixture was stirred and refluxed for 2 h. The solution was cooled and filtered, and upon the addition of light petroleum, the title compound precipitated. The compound was filtered off, washed with light petroleum, and air-dried.

**[Zn{S$_2$CN(CH$_2$CH$_2$OH)Me}$_2$]$_2$(MeOH) or C$_{17}$H$_{36}$N$_4$O$_5$S$_8$Zn$_2$.** This compound melts at 121°C–124°C and crystallizes in the triclinic structure with the lattice parameters $a = 1047.0 \pm 0.4$, $b = 1234.5 \pm 0.6$, and $c = 1261.5 \pm 0.6$ pm and $\alpha = 95.483° \pm 0.013°$, $\beta = 96.282° \pm 0.010°$, and $\gamma = 100.576° \pm 0.011°$ at 173 K and a calculated density of 1.603 g·cm$^{-3}$ (Benson et al. 2007).

**[Zn(C$_8$H$_4$NO$_2$CS$_2$)$_2$] or C$_{18}$H$_8$N$_2$O$_4$S$_4$Zn.** This compound melts at 165°C with decomposition (Siddiqi and Nishat 2000). To obtain it, ZnCl$_2$ (10 mM) in EtOH (50 mL) was added to a solution of sodium phthalimide dithiocarbamate (20 mM) in hot EtOH (100 mL), and the mixture was stirred for 30 min at room temperature. The colored precipitate thus obtained was filtered, washed with cold EtOH, and dried in vacuo.

**[(mepa)Zn](quinoline-2-carboxylate) or C$_{18}$H$_{17}$N$_3$O$_2$SZn (mepaH = N-(2-mercaptoethyl)picolylamin, C$_8$H$_{12}$N$_2$S).** This compound melts at 260°C with decomposition and crystallizes in the monoclinic structure with the lattice parameters $a = 1363.2 \pm 0.1$, $b = 945.1 \pm 0.1$, $c = 1457.4 \pm 0.1$ pm, and $\beta = 110.14° \pm 0.01°$ and experimental and calculated densities 1.47 and 1.52 g·cm$^{-3}$, respectively (Brand and Vahrenkamp 1995). To obtain it, mepaH (2.38 mM), quinoline-2-carboxylic acid (2.38 mM), and NaOH (4.78 mM) were dissolved in H$_2$O/MeOH (20/40 mL). A solution of Zn(ClO$_4$)$_2$·6H$_2$O (2.39 mM) in H$_2$O/MeOH (10/20 mL) was added dropwise with stirring. After the solution was stirred for 12 h, the volume was reduced to 10 mL in vacuo, during which a precipitate was formed. It was filtered off, washed several times with H$_2$O, and dried in vacuo to yield [(mepa)Zn](quinoline-2-carboxylate) as a brownish-white powder.

**[Zn(Phen)(EtOCS$_2$)] or C$_{18}$H$_{18}$N$_2$O$_2$S$_4$Zn.** This compound crystallizes in the monoclinic structure with the lattice parameters $a = 1167.8 \pm 0.3$, $b = 1921.5 \pm 0.3$, $c = 965.5 \pm 0.1$ pm, and $\beta = 101.23° \pm 0.01°$ and a calculated density of 1.525 g·cm$^{-3}$ (Klevtsova et al. 2006a). To obtain this compound, to a solution of 5 mM of Zn(OAc)$_2$·4H$_2$O and 5 mM of Phen·H$_2$O in 15 mL of H$_2$O was added a solution of 10 mM of KEtOCS$_2$ (10 mM) in 15 mL of H$_2$O through paper filter. The obtained mixture was stirred for 30 min. White precipitate was filtered off through a filter glass with suction, washed with H$_2$O, and dried at room temperature in air. The product was reprecipitated by dissolving in 50 mL of CH$_2$Cl$_2$ and the addition of 100 mL of hexane. The precipitate was filtered off with suction, washed with hexane, and dried at room temperature in air. Single crystals of Zn(Phen)(EtOCS$_2$)$_2$ were grown at slow evaporation of the solution in the mixture CH$_2$Cl$_2$/hexane (1:1 volume ratio) in a refrigerator.

**[Zn(H$_2$O)$_4$(C$_9$H$_5$O$_2$N$_2$S)$_2$] or C$_{18}$H$_{18}$N$_4$O$_8$S$_2$Zn.** This compound crystallizes in the triclinic structure with the lattice parameters $a = 630 \pm 1$, $b = 696 \pm 2$, and $c = 1271 \pm 3$ pm and $\alpha = 103.22° \pm 0.02°$, $\beta = 95.21° \pm 0.02°$, and $\gamma = 106.72° \pm 0.02°$ (Qin et al. 2009). It was synthesized by the reaction of ZnCl$_2$ (0.5 mM), 2-(4-pyridyl)thiazole-4-carboxylate (0.5 mM), and NaOH (0.25 mM) in 7.5 mL of H$_2$O. The reactions were carried out in sealed Teflon-lined stainless steel autoclave reactor at 130°C for 3 days and then slowly cooled down to room temperature. Colorless crystals of the title compound were obtained.

**[Zn(2,2′-bipy)(Pr$^i$OCS$_2$)] or C$_{18}$H$_{22}$N$_2$O$_2$S$_4$Zn.** This compound crystallizes in the triclinic structure with the lattice parameters $a = 1000.2 \pm 0.2$, $b = 1108.0 \pm 0.2$, and $c = 1175.6 \pm 0.2$ pm and $\alpha = 78.46° \pm 0.03°$, $\beta = 75.49° \pm 0.03°$, and $\gamma = 63.50° \pm 0.03°$ and a calculated density of 1.455 g·cm$^{-3}$ (Klevtsova et al. 2002). It was obtained when Zn(OAc)$_2$·4H$_2$O (2.5 mM) and 2,2′-bipy (2.5 mM) were dissolved in H$_2$O (100 mL) (Leonova et al. 1997). KS$_2$COPr$^i$ (5 mM) in 20 mL of H$_2$O was added to this solution. The reaction mixture was stirred for 30 min. The precipitate was filtered off with suction, washed with H$_2$O, and dried at room temperature in air. After dissolving in CH$_2$Cl$_2$ (15 mL), Zn(2,2)(Pr$^i$OCS$_2$) was reprecipitated by adding of hexane (50 mL). The resulting precipitate was filtered off with suction, washed with hexane, and dried at room temperature under vacuum over Mg(ClO$_4$)$_2$.

Its single crystals were grown at slow evaporation of the solution of Zn(2,2′-bipy)(Pr$^i$OCS$_2$) in the mixture CH$_2$Cl$_2$/hexane (1:1 volume ratio) in a refrigerator (Klevtsova et al. 2002).

**[Zn(H$_2$O)$_3$(C$_9$H$_8$NO$_6$S)$_2$] or C$_{18}$H$_{22}$N$_2$O$_{15}$S$_2$Zn.** This compound crystallizes in the orthorhombic structure with the lattice parameters $a = 1109.15 \pm 0.09$, $b = 4206.4 \pm 0.4$, and $c = 509.72 \pm 0.04$ pm (Ma et al. 2007). It was obtained by the following procedure. The mixture of ZnCl$_2$·2H$_2$O (1 mM) and 3-(sulfonyl-glycine)-benzoic acid (2 mM) was stirred into 15 mL aqueous solution at room temperature. Then the pH was adjusted to approximately 7 with 1 M KOH solution. Then 5 mL of ethanolic solution of 4,4′-bipy (1 mM) was added. The reaction mixture was then heated on a water bath for 6 h at 70°C and then filtered. Colorless crystals were obtained from the mother liquor by slow evaporation at room temperature after 6 weeks.

**[Zn(SAc)$_2$(C$_7$H$_9$N)$_2$] or C$_{18}$H$_{24}$N$_2$O$_2$S$_2$Zn.** This compound crystallizes in the triclinic structure with the lattice parameters $a = 814.6 \pm 0.1$, $b = 947.2 \pm 0.1$, and $c = 1400.8 \pm 0.1$ pm and $\alpha = 94.21° \pm 0.01°$, $\beta = 90.99° \pm 0.01°$, and $\gamma = 104.35° \pm 0.01°$ (Nyman et al. 1997). To obtain it, Et$_2$Zn (8.1 mM), 3,5-dimethylpyridine (16.2 mM), and 20 mL of toluene were combined in a round-bottom flask in an inert-atmosphere box. The flask was removed from inert-atmosphere box and placed in an ice water bath. Thioacetic acid (16.2 mM) was added to the solution via a pipet while stirring, and a white precipitate formed immediately. The solution was warmed to room temperature while stirring. After several hours, the solution was heated to 60°C to dissolve most of the reaction products, and the hot solution was filtered immediately. The product was crystallized at room temperature as small colorless blocks.

**[Zn(EtMe$_2$CH$_2$N$_3$S)$_2$(H$_2$O)]·[C$_8$H$_4$O$_4$] or C$_{18}$H$_{32}$N$_6$O$_5$S$_2$Zn.** This compound crystallizes in the monoclinic structure with the lattice parameters $a = 2249.0 \pm 1.2$, $b = 911.8 \pm 0.2$, $c = 1581.9 \pm 0.8$ pm, and $\beta = 129.74° \pm 0.03°$ (Babb et al. 2003). It was prepared by the next procedure. To an aqueous solution of Zn(EtMe$_2$CH$_2$N$_3$S)$_2$(NO$_3$)$_2$ (0.27 mM) was added an aqueous solution of sodium terephthalate (0.27 mM). After approximately 24 h, a colorless crystalline material of the title compound was seen to form, which was separated by filtration.

**[Zn$_2${NH(CH$_2$)$_4$O}{S$_2$CN(CH$_2$)$_4$O}$_2$]·[NH(CH$_2$)$_4$O] or C$_{18}$H$_{34}$N$_4$O$_4$S$_4$Zn.** This compound crystallizes in the monoclinic structure with the lattice parameters $a = 932.6 \pm 0.1$, $b = 1557.5 \pm 0.2$, $c = 1755.8 \pm 0.2$ pm, and $\beta = 96.68° \pm 0.01°$ at $201 \pm 1$ K and a calculated density of $1.474 \pm 0.001$ g·cm$^{-3}$ (Ivanov et al. 2004a). To obtain it, a solution of [Zn$_2${NH(CH$_2$)$_4$O}{S$_2$CN(CH$_2$)$_4$O}$_2$] was washed by morpholine (10 mL), and the reaction mixture was carefully heated (Ivanov et al. 2003b). After full dissolution, the heating was stopped and the solution was left overnight in a refrigerator. The precipitated crystals were collected, filtered off, and dried in air. Thermolysis of the title compound leads to the formation of ZnS.

**[Zn(C$_4$H$_{12}$N$_2$)$_2$(H$_2$O)$_2$](C$_{10}$H$_6$O$_6$S$_2$) or C$_{18}$H$_{34}$N$_4$O$_8$S$_2$Zn.** This compound crystallizes in the monoclinic structure with the lattice parameters $a = 843.27 \pm 0.06$, $b = 1350.61 \pm 0.09$, $c = 1146.13 \pm 0.07$ pm, and $\beta = 108.134° \pm 0.001°$ at $0°C \pm 2°C$ and a calculated density of 1.510 g·cm$^{-3}$ (Lian and Li 2007). It was obtained by the next procedure. $N,N'$-dimethylethylenediamine (2 mM) was added with constant stirring to an aqueous solution (10 mL) of Zn(OAc)$_2$·2H$_2$O (1 mM). The solution was then treated with disodium 2,6-naphthalenedisulfonate (1 mM). Colorless crystals of the title compound were collected after 4 days.

**[Zn(C$_9$H$_{18}$NO$_2$S$_2$)$_2$] or C$_{18}$H$_{36}$N$_2$O$_4$S$_4$Zn.** This compound crystallizes in the tetragonal structure with the lattice parameters $a = 1255.7 \pm 0.2$ and $c = 1660.0 \pm 0.3$ pm and a calculated density of 1.365 g·cm$^{-3}$ (Reck and Becker 2003b). To obtain it, sodium diisobutyldithiocarbamate (one molar equivalent) in acetone was treated with H$_2$O$_2$ (two molar equivalents) in small portions and kept at $0°C–5°C$. A yellow precipitate formed after dilution with water was purified by fractional crystallization from ethanol/$n$-hexane to yield colorless crystals.

**[Zn{O(CH$_2$)$_4$NH}{S$_2$CNEt$_2$}$_2$]·[O(CH$_2$)$_4$NH] or C$_{18}$H$_{38}$N$_4$O$_2$S$_4$Zn.** This compound crystallizes in the monoclinic structure with the lattice parameters $a = 1170.9 \pm 0.1$, $b = 861.7 \pm 0.1$, $c = 1296.3 \pm 0.2$ pm, and $\beta = 98.12° \pm 0.01°$ at $100 \pm 1$ K and a calculated density of $1.373 \pm 0.001$ g·cm$^{-3}$ (Ivanov et al. 2001c). It was prepared by either (1) quantitative absorption of morpholine vapors by polycrystalline sample [Zn{O(CH$_2$)$_4$NH}{S$_2$CNEt$_2$}$_2$] or (2) its cautious wetting with this solvent, keeping it in the solid state. Crystals of the title compound suitable for XRD were isolated by crystallization from solution in morpholine. The resulting crystals were separated from the mother liquors, washed with a small quantity of morpholine, dried in air to a friable state, and placed in a sealed tube.

**[Zn$_2$(CH$_4$N$_2$S)$_3$]·[C$_6$H$_4$(CO$_2$)$_2$-1,4]$_2$·4H$_2$O or C$_{19}$H$_{28}$N$_6$O$_{12}$S$_3$Zn$_2$.** This compound crystallizes in the monoclinic structure with the lattice parameters $a = 1079.8 \pm 0.1$, $b = 2282.1 \pm 0.1$, $c = 1198.8 \pm 0.1$ pm, and $\beta = 103.75° \pm 0.01°$ and a calculated density of 1.758 g·cm$^{-3}$ (Burrows et al. 1997). To obtain it, an aqueous solution of Na$_2$C$_6$H$_4$(CO$_2$)$_2$ (0.41 mM) was added to an aqueous solution of [Zn(CH$_4$N$_2$S)$_4$(NO$_3$)$_2$] (0.20 mM). The solution was allowed to stand for 24 h after which time colorless crystals had grown.

**[Zn(Et$_2$NCS$_2$)$_2$(C$_5$H$_{10}$NH)]·(C$_4$H$_9$NO) or C$_{19}$H$_{40}$N$_4$OS$_4$Zn.** This compound was synthesized via careful wetting of [Zn(Et$_2$NCS$_2$)$_2$(C$_5$H$_{10}$NH)] with morpholine without changing its state of aggregation (Ivanov et al. 2001b). It was kept in sealed tubes.

**[Zn(NO$_3$)$_2$(C$_{10}$H$_8$N$_2$S$_2$)$_2$]$_n$ or C$_{20}$H$_{16}$N$_6$O$_6$S$_4$Zn.** This compound crystallizes in the monoclinic structure with the lattice parameters $a = 1087.9 \pm 0.4$, $b = 1196.1 \pm 0.4$, $c = 1977.4 \pm 0.7$ pm, and $\beta = 103.901° \pm 0.008°$ and a calculated density of 1.675 g·cm$^{-3}$ (Horikoshi and Mikuriya 2005).

**[Zn(NO$_3$)$_2$(C$_{10}$H$_8$N$_2$S$_2$)$_2$·1.5H$_2$O]$_n$ or C$_{20}$H$_{19}$N$_6$O$_{7.5}$S$_4$Zn.** This compound was prepared by the following procedure (Horikoshi and Mikuriya 2005). To a solution of 4,4′-dipyridyl disulfide (C$_{10}$H$_8$N$_2$S$_2$) (0.2 mM) in MeOH (0.5 mL) was added Zn(NO$_3$)$_2$ (0.1 mM) in MeOH (0.5 mL). A white precipitate formed immediately. The reaction mixture was stirred for 30 min, and then the precipitate was filtered off and washed with MeOH (3×0.5 mL). Extraction of the precipitate with MeOH followed by the addition of diethyl ether afforded a white solid of the title compound.

**[Zn(C$_5$H$_5$NS)$_4$(NO$_3$)$_2$] or C$_{20}$H$_{20}$N$_6$O$_6$S$_4$Zn.** To obtain these compounds, an ethanolic solution of Zn(NO$_3$)$_2$·4H$_2$O (2.5 mM) was slowly added to a stirred ethanolic solution of 2-pyridine thiol or 4-pyridine thiol (10 mM) (Kennedy and Lever 1972). The title compound precipitated slowly as the Zn(NO$_3$) was being added.

**[Zn(Phen)(Pr$^i$OCS$_2$)$_2$] or C$_{20}$H$_{22}$N$_2$O$_2$S$_4$Zn.** This compound exists in two modifications, which crystallize in the monoclinic structure with the lattice parameters $a = 1054.3 \pm 0.2$, $b = 1349.4 \pm 0.3$, $c = 1687.5 \pm 0.3$ pm, and $\beta = 102.08° \pm 0.03°$ and a calculated density of 1.460 g·cm$^{-3}$ for α-modification and $a = 1093.1 \pm 0.2$, $b = 1299.6 \pm 0.3$, $c = 1628.8 \pm 0.3$ pm, and $\beta = 92.69° \pm 0.03°$ and a calculated density of 1.483 g·cm$^{-3}$ for β-modification (Klevtsova et al. 2001b). It was obtained by the similar procedure to that described for Zn(2,2′-bipy)(Pr$^i$OCS$_2$) using Phen instead of 2,2′-bipy (Leonova et al. 1997).

Its single crystals were grown at the slow evaporation of the solution of Zn(Phen)(Pr$^i$OCS$_2$)$_2$ in the mixture CH$_2$Cl$_2$/hexane (1:1 volume ratio) at room temperature.

**[Zn(C$_{10}$H$_9$N$_4$O$_2$S)$_2$(NH$_3$)$_2$] or C$_{20}$H$_{24}$N$_{10}$O$_4$S$_2$Zn.** This compound crystallizes in the orthorhombic structure with the lattice parameters $a = 1389.4 \pm 0.1$, $b = 1422.1 \pm 0.1$, and $c = 1260.8 \pm 0.1$ pm and experimental and calculated densities $1.59 \pm 0.01$ and 1.59 g·cm$^{-3}$, respectively (Baenziger et al. 1983). It was synthesized as a fine powder from commercial sulfadiazine (C$_{10}$H$_9$N$_4$O$_2$S) or sodium sulfadiazine plus a zinc salt [Zn(NO$_3$)$_2$ or Zn(OAc)$_2$] in an aqueous solution. Single crystals of zinc sulfadiazine prepared by dissolving zinc sulfadiazine in ammonium hydroxide, driving off ammonia by heating the solution on a water bath until crystals began to appear, and continuing the crystallization in a refrigerator.

**[Zn(C$_{14}$H$_{21}$N$_4$S)(OC$_6$H$_4$NO$_2$)] or C$_{20}$H$_{25}$N$_5$O$_3$SZn.** To prepare this compound, a solution of [MeZn(C$_{14}$H$_{21}$N$_4$S)] (0.46 mM) in 20 mL of CH$_2$Cl$_2$ was treated with a CH$_2$Cl$_2$ solution of *p*-nitrophenol (0.46 mM) (Hammesa and Carrano 2000). The resulting solution was stirred for 24 h and dried under reduced pressure to give the title compound as a light-yellow solid. The product was crystallized by layering a concentrated CH$_2$Cl$_2$ solution of [Zn(C$_{14}$H$_{21}$N$_4$S)(OC$_6$H$_4$NO$_2$)] with hexane.

**[Zn(2,2'-bipy)(Bu$^n$OCS$_2$)$_2$] or C$_{20}$H$_{26}$N$_2$O$_2$S$_4$Zn.** This compound crystallizes in the triclinic structure with the lattice parameters $a = 878.75 \pm 0.03$, $b = 1183.3 \pm 0.1$, and $c = 1334.54 \pm 0.06$ pm and $\alpha = 112.154° \pm 0.002°$, $\beta = 108.503° \pm 0.001°$, and $\gamma = 92.787° \pm 0.002°$ and a calculated density of 1.444 g·cm$^{-3}$ (Klevtsova et al. 2006a). It was obtained when Zn(OAc)$_2$·4H$_2$O (2.5 mM) and 2,2'-bipy (2.5 mM) were dissolved in H$_2$O (100 mL) (Leonova et al. 1997). KS$_2$COBu$^n$ (5 mM) in 20 mL of H$_2$O was added to this solution. The reaction mixture was stirred for 30 min. The precipitate was filtered off with suction, washed with H$_2$O, and dried at room temperature in air. After dissolving in CH$_2$Cl$_2$ (15 mL), Zn(2,2'-bipy)(Bu$^n$OCS$_2$) was reprecipitated by adding of hexane (50 mL). The resulting precipitate was filtered off with suction, washed with hexane, and dried at room temperature under vacuum over Mg(ClO$_4$)$_2$.

Its single crystals were grown at slow evaporation of the solution of Zn(2,2'-bipy)(Bu$^n$OCS$_2$) in the mixture CH$_2$Cl$_2$/hexane (1:1 volume ratio) in refrigerator (Klevtsova et al. 2006a).

**[Zn(2,2'-bipy)(Bu$^i$OCS$_2$)$_2$] or C$_{20}$H$_{26}$N$_2$O$_2$S$_4$Zn.** This compound crystallizes in the triclinic structure with the lattice parameters $a = 876.0 \pm 0.2$, $b = 1252.0 \pm 0.3$, and $c = 1325.2 \pm 0.3$ pm and $\alpha = 63.93° \pm 0.03°$, $\beta = 71.10° \pm 0.03°$, and $\gamma = 88.01° \pm 0.03°$ and a calculated density of 1.410 g·cm$^{-3}$ (Klevtsova et al. 2002). It was obtained by the similar procedure to that described for Zn(2,2'-bipy)(Bu$^n$OCS$_2$) using KS$_2$COBu$^i$ instead of KS$_2$COBu$^n$ (Leonova et al. 1997).

Its single crystals were grown at slow evaporation of the solution of Zn(2,2'-bipy)(Bu$^i$OCS$_2$) in the mixture CH$_2$Cl$_2$/hexane (1:1 volume ratio) in refrigerator (Klevtsova et al. 2002).

**[Zn(4,4'-bipy)(Bu$^i$OCS$_2$)$_2$] or C$_{20}$H$_{26}$N$_2$O$_2$S$_4$Zn.** This compound melts at 127°C with decomposition and crystallizes in the monoclinic structure with the lattice parameters $a = 2628.6 \pm 0.4$, $b = 2226.9 \pm 0.3$, $c = 848.4 \pm 0.1$ pm, and $\beta = 99.04° \pm 0.01°$ and a calculated density of 1.409 g·cm$^{-3}$ (Larionov et al. 1998). To obtain it, a solution of Zn(OAc)$_2$·4H$_2$O (2.5 mM) in H$_2$O (10 mL) was mixed with a solution of 4,4'-bipy (5 mM) in Pr$^i$OH (20 mL). KBu$^i$OCS$_2$ (5 mM) in H$_2$O (10 mL) was added to obtain solution, and the mixture was stirred for 30 min. The separated precipitate was filtered off with suction, washed twice with a mixture Pr$^i$OH/H$_2$O (5:1 volume ratio), and dried in air. The product was reprecipitated by its dissolution in CH$_2$Cl$_2$ (30 mL) following the addition of hexane (70 mL). The precipitate was filtered off with suction and dried at room temperature in vacuo over NaOH. Its single crystals were grown at the storage in a refrigerator of the mother liquor obtained at the reprecipitation.

**[Zn(MeCH$_4$N$_3$S)$_2$(C$_8$H$_5$O$_4$)$_2$]·H$_2$O or C$_{20}$H$_{26}$N$_6$O$_9$S$_2$Zn.** This compound crystallizes in the monoclinic structure with the lattice parameters $a = 792.3 \pm 0.1$, $b = 1437.0 \pm 0.2$, $c = 2339.3 \pm 0.3$ pm, and $\beta = 92.638° \pm 0.06°$ at $150 \pm 2$ K (Babb et al. 2003). It was prepared by the following procedure. To an aqueous solution of Zn(MeCH$_4$N$_3$S)$_2$(NO$_3$)$_2$ (0.27 mM) was added an aqueous solution of sodium phthalate (0.27 mM). After approximately 24 h, a colorless crystalline material of the title compound was seen to form, which was separated by filtration.

**[Zn(NO$_3$)$_2$(C$_{10}$H$_8$N$_2$S$_2$)$_2$(H$_2$O)·4H$_2$O]$_n$ or C$_{20}$H$_{26}$N$_6$O$_{11}$S$_4$Zn.** This compound crystallizes in the orthorhombic structure with the lattice parameters $a = 2058.1 \pm 0.3$, $b = 1066.52 \pm 0.17$, and $c = 2641.0 \pm 0.4$ pm and a calculated density of 1.650 g·cm$^{-3}$ (Horikoshi and Mikuriya 2005). [Zn(NO$_3$)$_2$(C$_{10}$H$_8$N$_2$S$_2$)$_2$·1.5H$_2$O]$_n$ was recrystallized from distilled H$_2$O to give the title compound as transparent crystals.

**[{Zn(NO$_3$)$_2$(C$_{10}$H$_8$N$_2$S$_2$)$_2$(H$_2$O)}(NO$_3$)·4H$_2$O]$_n$ or C$_{20}$H$_{26}$N$_7$O$_{14}$S$_4$Zn.** This compound crystallizes in the orthorhombic structure with the lattice parameters $a = 2043.23 \pm 0.07$, $b = 1062.13 \pm 0.04$, and $c = 2608.12 \pm 0.12$ pm at $110 \pm 2$ K and a calculated density of 1.690 g·cm$^{-3}$ (Seidel et al. 2011). It was prepared by the following procedure. 6 mL of an ethanolic solution of Zn(NO$_3$)$_2$·6H$_2$O (0.114 mM) and a solution of 4,4'-dithiodipyridine (C$_{10}$H$_8$N$_2$S$_2$) (0.229 mM) in 6 mL of EtOH were mixed with stirring. Subsequently, 8 mL of deionized water was added to the stirred mixture. After ca. 3 weeks of standing at room temperature, while the solvent was allowed to evaporate slowly, colorless crystals of [{Zn(NO$_3$)$_2$(C$_{10}$H$_8$N$_2$S$_2$)$_2$(H$_2$O)}(NO$_3$)·4H$_2$O]$_n$ appeared.

**[Zn(C$_8$H$_{10}$N$_5$S)(OAc)]$_2$ or C$_{20}$H$_{26}$N$_{10}$O$_4$S$_2$Zn$_2$.** This compound crystallizes in the monoclinic structure with the lattice parameters $a = 2671.3 \pm 0.3$, $b = 818.3 \pm 0.1$, $c = 1956.8 \pm 0.2$ pm, and $\beta = 131.727° \pm 0.006°$ and a calculated density of 1.384 g·cm$^{-3}$ (García et al. 2002). Its single crystals suitable for analysis by XRD were obtained by slow evaporation of the mother liquor in air at room temperature.

**[(mepa)$_2$Zn$_2$](OAc)$_2$ or C$_{20}$H$_{28}$N$_4$O$_4$S$_2$Zn$_2$ (mepaH = N-(2-mercaptoethyl)picolylamin, C$_8$H$_{12}$N$_2$S).** This compound melts at 95°C and crystallizes in the triclinic structure with the lattice parameters $a = 770.9 \pm 0.2$, $b = 870.8 \pm 0.2$, and $c = 1024.1 \pm 0.2$ pm and $\alpha = 67.16° \pm 0.03°$, $\beta = 67.22° \pm 0.03°$, and $\gamma = 84.06° \pm 0.03°$ and a calculated density of 1.66 g·cm$^{-3}$ (Brand and Vahrenkamp 1995). To obtain it, mepaH (2.26 mM) and Zn(OAc)$_2$·2H$_2$O (2.26 mM) were dissolved with stirring in EtOH (20 mL). After 1 h, the clear solution was reduced in vacuo to 10 mL. Hexane (100 mL) was added and the mixture was kept at −30°C for 24 h. The cold mother liquor was decanted off and the precipitate dried in vacuo at room temperature. It was dissolved in a minimum amount of EtOH, treated with a 10-fold amount of diethyl ether, and kept at −30°C for 24 h again. The precipitate, which was filtered off, washed with a little amount of diethyl ether, and dried in vacuo, consisted of [(mepa)$_2$Zn$_2$](OAc)$_2$, which is very hygroscopic.

**[Zn{(HOC$_2$H$_4$)$_2$NCS$_2$}$_2$(2,2'-bipy)] or C$_{20}$H$_{28}$N$_4$O$_4$S$_4$Zn.** This compound crystallizes in the monoclinic structure with the lattice parameters $a = 721.46 \pm 0.05$, $b = 2617.20 \pm 0.18$, $c = 1594.38 \pm 0.11$ pm, and $\beta = 97.899° \pm 0.002°$ and calculated density 1.297 g·cm⁻³ (Deng et al. 2007). It was prepared from the reaction of $Zn(ClO_4)_2 \cdot 6H_2O$ (0.5 mM), 2,2′-bipy (0.5 mM), and $K(HOC_2H_4)_2NCS_2$ (1 mM) during 6 h in MeOH solution.

**[Zn(C₈H₁₀N₅S)(OAc)]₂·H₂O or C₂₀H₂₈N₁₀O₅S₂Zn₂.** This compound melts at 225°C (García et al. 2002). To obtain it, a solution of 2-pyridineformamide thiosemicarbazone (1.19 mM) in EtOH (20 mL) was added to a solution of $Zn(OAc)_2 \cdot 2H_2O$ (1.19 mM) in 20 mL of 96% EtOH, and the mixture was stirred for several days. The resulting solids were filtered off, washed thoroughly with cold EtOH, and dried over CaCl₂.

**[Zn{S₂CN(CH₂)₄O}₂]₂ or C₂₀H₃₂N₄O₄S₈Zn₂.** To obtain this compound, a solution of $Na\{S_2CN(CH_2)_4O\} \cdot 2H_2O$ (20 mM) in H₂O (30 mL) was added with vigorous stirring to a solution of ZnCl₂ (10 mM) in H₂O (100 mL) (Ivanov et al. 2003a). White voluminous precipitate was washed by decantation, separated by filtration, and dried in air.

**[Zn{O(CH₂)₄NH}{S₂CNEt₂}₂]·(C₆H₆) or C₂₀H₃₅N₃OS₄Zn.** This compound crystallizes in the monoclinic structure with the lattice parameters $a = 1129.9 \pm 0.2$, $b = 860.47 \pm 0.08$, $c = 1366.0 \pm 0.2$ pm, and $\beta = 100.26° \pm 0.02°$ at $190 \pm 1$ K and a calculated density of 1.337 g·cm⁻³ (Ivanov et al. 2001c). It was prepared by either (1) quantitative absorption of benzene vapors by polycrystalline sample [Zn{O(CH₂)₄NH} {S₂CNEt₂}₂] or (2) its cautious wetting with this solvent, keeping it in the solid state. Crystals of the title compound suitable for XRD were isolated by crystallization from the solution in benzene. The resulting crystals were separated from the mother liquors, washed with a small quantity of benzene, dried in air to a friable state, and placed in a sealed tube.

**[Zn{SC(NHMe)₂}₂(C₄H₂O₄)]₂·2H₂O or C₂₀H₄₀N₈O₁₀S₄Zn₂.** This compound crystallizes in the triclinic structure with the lattice parameters $a = 826.4 \pm 0.2$, $b = 964.2 \pm 0.2$, and $c = 1119.9 \pm 0.2$ pm and $\alpha = 78.46° \pm 0.02°$, $\beta = 76.48° \pm 0.02°$, and $\gamma = 83.750° \pm 0.010°$ and a calculated density of 1.589 g·cm⁻³ (Burke et al. 2003). To obtain it, an aqueous solution of sodium maleate (0.15 mM) was added to an aqueous solution of [Zn{SC(NHMe)₂}₄]Cl₂ (0.15 mM), with no discernable change. After several hours, a colorless crystalline precipitate was observed, which was separated by filtration.

**[Zn(C₁₄H₂₁N₄S)(SBz)]·0.25H₂O or C₂₁H₂₈.₅N₄O₀.₂₅S₂Zn.** To prepare this compound, a solution of [MeZn(C₁₄H₂₁N₄S)] (0.60 mM) in 20 mL of CH₂Cl₂ was treated with a CH₂Cl₂ solution of BzSH (0.60 mM) (Hammesa and Carrano 2000). The resulting solution was stirred for 6 h, dried under reduced pressure, and crystallized by layering a CH₂Cl₂ solution of the title compound with hexane.

**[Me₄N][Zn(S₂CNEt₂)₂(C₇H₄NOS)] or C₂₁H₃₆N₄OS₅Zn.** This compound was prepared as follows (McCleverty et al. 1980a). To an acetone solution of [Me₄N] [C₇H₄NOS] was added [Zn(S₂CNEt₂)₂], and the mixture was stirred and refluxed for 2 h. The solution was cooled and filtered, and upon the addition of light petroleum, the title compound precipitated. The compound was filtered off, washed with light petroleum, and air-dried.

**[Zn(SCN)$_2$(C$_{10}$H$_8$N$_2$S$_2$)$_2$·H$_2$O]$_n$** or **C$_{22}$H$_{18}$N$_6$OS$_6$Zn**. This compound was prepared by the following procedure (Horikoshi and Mikuriya 2005). To a solution of 4,4′-bipyridyl disulfide (C$_{10}$H$_8$N$_2$S$_2$) (0.2 mM) in MeOH (0.5 mL) was added Zn(SCN)$_2$ (0.1 mM) in MeOH (0.5 mL). A white precipitate formed immediately. The reaction mixture was stirred for 30 min, and then the precipitate was filtered off and washed with MeOH (3×0.5 mL). Extraction of the precipitate with MeOH followed by the addition of diethyl ether afforded a white solid of the title compound.

**[Zn(C$_6$H$_7$N$_2$S)$_2$(C$_{10}$H$_8$N$_2$)]·0.75H$_2$O** or **C$_{22}$H$_{23.5}$N$_6$O$_{0.75}$S$_2$Zn**. Electrochemical oxidation of Zn in a solution of 4,6-dimethyl-2-thiopyrimidine and 2,2′-bipy in acetonitrile afforded the title compound in the form of a yellow solid (Castro et al. 1994).

**[Zn(C$_{19}$H$_{17}$N$_3$S$_2$)]·DMF** or **C$_{22}$H$_{24}$N$_4$OS$_2$Zn**. This compound melts at 245°C with decomposition and crystallizes in the orthorhombic structure with the lattice parameters $a = 1065.4 \pm 0.2$, $b = 1103.5 \pm 0.7$, and $c = 1912.8 \pm 0.4$ pm and experimental and calculated densities 1.39 and 1.45 g·cm$^{-3}$, respectively (Brand et al. 1996). To obtain it, 2,6-bis[{(2-mercaptophenyl)amino}methyl]pyridine (C$_{19}$H$_{19}$N$_3$S$_2$) (0.16 mM) dissolved in 6 mL of CH$_2$Cl$_2$/MeOH (2:1 volume ratio) was added to a solution of NaOH (33 mol) in 4 mL of H$_2$O/MeOH (1:1 volume ratio). To this solution was added dropwise with stirring a solution of Zn(ClO$_4$)$_2$·6H$_2$O (0.16 mM) in 4 mL of H$_2$O/MeOH (1:1 volume ratio). After 1 h of stirring and pumping off of the CH$_2$Cl$_2$, the mixture was filtered off and the remaining reddish precipitate was washed with H$_2$O and dried in vacuo. Recrystallization from hot acetone/DMF (10:3 volume ratio) yielded C$_{22}$H$_{24}$N$_4$OS$_2$Zn as pale-red crystals.

**[Zn(Phen)(Bu$^n$OCS$_2$)$_2$]** or **C$_{22}$H$_{26}$N$_2$O$_2$S$_4$Zn**. This compound crystallizes in the triclinic structure with the lattice parameters $a = 944.64 \pm 0.03$, $b = 1102.79 \pm 0.04$, and $c = 1365.28 \pm 0.06$ pm and $\alpha = 106.940° \pm 0.001°$, $\beta = 98.382° \pm 0.001°$, and $\gamma = 106.347° \pm 0.001°$ and a calculated density of 1.429 g·cm$^{-3}$ (Klevtsova et al. 2006b). To obtain this compound, to a solution of 2.5 mM of Zn(OAc)$_2$·4H$_2$O and 5 mM of Phen in 10 mL of H$_2$O was added a solution of 5 mM of KBu$^n$OCS$_2$ in 10 mL of H$_2$O through paper filter. The obtained mixture was stirred for 30 min. White precipitate was filtered off and dried in air. The product was reprecipitated by dissolving in 30 mL of CH$_2$Cl$_2$ and the addition of 90 mL of hexane. A light-yellow precipitate was filtered off, washed with ethanol, and dried in air and then in vacuo. Single crystals of Zn(Phen)(Bu$^n$OCS$_2$)$_2$ were grown at slow evaporation of the solution in the mixture CH$_2$Cl$_2$/hexane (1:1 volume ratio) in a refrigerator.

**[Zn(Phen)(Bu$^i$OCS$_2$)$_2$]** or **C$_{22}$H$_{26}$N$_2$O$_2$S$_4$Zn**. This compound crystallizes in the orthorhombic structure with the lattice parameters $a = 2005.3 \pm 0.2$, $b = 1927.2 \pm 0.2$, and $c = 1320.3 \pm 0.1$ pm and a calculated density of 1.416 g·cm$^{-3}$ (Glinskaya et al. 1997, Leonova et al. 1997). To obtain this compound, to a solution of 2.5 mM of Zn(OAc)$_2$·4H$_2$O and 2.5 mM of Phen in 20 mL of H$_2$O was added a solution of 5 mM of KBu$^i$OCS$_2$ in 20 mL of H$_2$O through paper filter. The obtained mixture was stirred for 30 min. The precipitate was filtered off under suction, washed with water, and dried in air at room temperature. The product was reprecipitated by dissolving in 15 mL of CH$_2$Cl$_2$ and the addition of 50 mL of hexane. The precipitate was filtered off under suction, washed with hexane, and dried at room temperature in vacuum over $Mg(ClO_4)_2$. Single crystals of $Zn(Phen)(Bu^iOCS_2)_2$ were grown at slow evaporation of the solution in the mixture $CH_2Cl_2$/hexane (1:1 volume ratio) at room temperature.

**[Zn(S₂COPrⁱ)₂(4,7-Me₂Phen)] or C₂₂H₂₆N₂O₂S₄Zn.** This compound crystallizes in the monoclinic structure with the lattice parameters $a = 1968.34 \pm 0.10$, $b = 980.97 \pm 0.05$, $c = 1281.71 \pm 0.07$ pm, and $\beta = 103.403° \pm 0.001°$ at $183 \pm 2$ K and a calculated density of 1.501 g·cm⁻³ (Soh et al. 2002). Its light-brown crystals were obtained from slow evaporation of an acetone/CHCl₃ (1:1 volume ratio) solution of $Zn(S_2COPr^i)_2(4,7\text{-}Me_2Phen)$ that had been prepared by refluxing equimolar amounts of $Zn(S_2COPr^i)_2$ and 4,7-Me₂Phen in chloroform solution.

**[Zn{(HOC₂H₄)₂NCS₂}₂(Phen)] or C₂₂H₂₈N₄O₄S₄Zn.** This compound was prepared from the reaction of $Zn(ClO_4)_2 \cdot 6H_2O$ (0.5 mM), Phen (0.5 mM), and $K(HOC_2H_4)_2NCS_2$ (1 mM) during 6 h in MeOH solution (Deng et al. 2007).

**[Zn(2,2′-bipy)(AmⁱOCS₂)₂] or C₂₂H₃₀N₂O₂S₄Zn.** This compound was obtained when $Zn(OAc)_2 \cdot 4H_2O$ (2.5 mM) and 2,2′-bipy (2.5 mM) were dissolved in $H_2O$ (100 mL) (Leonova et al. 1997). $KS_2COAm^i$ (5 mM) in 20 mL of $H_2O$ was added to this solution. The reaction mixture was stirred for 30 min. The precipitate was filtered off with suction, washed with $H_2O$, and dried at room temperature in air. After dissolving in $CH_2Cl_2$ (15 mL), $Zn(2,2'\text{-}bipy)(Am^iOCS_2)$ was reprecipitated by adding of hexane (50 mL). The resulting precipitate was filtered off with suction, washed with hexane, and dried at room temperature under vacuum over $Mg(ClO_4)_2$.

**[(ZnC₁₁H₁₅N₂O₂)₂SO₄] or C₂₂H₃₀N₄O₈SZn₂.** This compound crystallizes in the monoclinic structure with the lattice parameters $a = 826.66 \pm 0.06$, $b = 2322.9 \pm 0.2$, $c = 1375.31 \pm 0.09$ pm, and $\beta = 102.329° \pm 0.001°$ (Qiu et al. 2004). To prepare this compound, equimolar amounts of salicylaldehyde (1 mM) and N-2-hydroxyethylaminoethylamine (1 mM) were dissolved in anhydrous alcohol (5 mL) with stirring. To the aforementioned solution was added $ZnSO_4 \cdot 6H_2O$ (1 mM) in $H_2O$ (5 mL). The resulted solution was kept under air feed to evaporate. After about half of the solvents had escaped, colorless block crystals of the title compound were deposited. They were collected by filtration, washed with water and alcohol in turn, and dried in a vacuum desiccator using silica gel.

**{Zn[S₂CN(Me)CH₂CH₂OH]₂}[(3-C₅H₄N)CH₂N(H)C(S)C(S)N(H)CH₂(C₅H₄N-3)] or C₂₂H₃₀N₆O₂S₆Zn.** This compound melts at 132°C–136°C (Poplaukhin and Tiekink 2010). To obtain it, $Zn[S_2CN(Me)CH_2CH_2OH]_2\}_2$ (0.25 mM) and N,N′-bis(pyridin-3-ylmethyl)thiooxalamide (0.5 mM) were separately dissolved in ca. 15 mL of hot CHCl₃/MeCN (3:1 volume ratio) solvent. The solutions were then combined and stirred for 1 h with heating. The mixture was allowed to cool to room temperature and then stirred for another 3 h; during this time, a yellow precipitate of the title compound formed.

**[Zn(EtCH₄N₃S)₂(C₈H₅O₄)₂]·(H₂O) or C₂₂H₃₀N₆O₉S₂Zn.** This compound crystallizes in the monoclinic structure with the lattice parameters $a = 797.00 \pm 0.01$, $b = 1478.10 \pm 0.02$, $c = 2423.20 \pm 0.04$ pm, and $\beta = 92.0270° \pm 0.0009°$ at $170 \pm 2$ K

(Babb et al. 2003). It was prepared by the following procedure. To an aqueous solution of $Zn(EtCH_4N_3S)_2(NO_3)_2$ (0.27 mM) was added an aqueous solution of sodium phthalate (0.27 mM). After approximately 24 h, a colorless crystalline material of the title compound was seen to form, which was separated by filtration.

**$[Zn(C_9H_{12}N_5S)(OAc)]_2$ or $C_{22}H_{30}N_{10}O_4S_2Zn_2$.** This compound melts at the temperature higher than 300°C (Bermejo et al. 2001). It was prepared as follows. A solution of 2-pyridylformamide $N(4)$-dimethylthiosemicarbazone (1 mM) in EtOH (30 mL) was added to a solution of $Zn(OAc)_2$ (1 mM) in 30 mL of 96% EtOH, and the mixture was stirred for several days at room temperature. The resulting yellow solid was filtered off, washed thoroughly with cold EtOH, and stored in a desiccator over Mg and $I_2$.

**$[Zn(C_9H_{12}N_5S)(OAc)]_2$ or $C_{22}H_{30}N_{10}O_4S_2Zn_2$.** This compound crystallizes in the monoclinic structure with the lattice parameters $a = 1030.5 \pm 0.2$, $b = 840.4 \pm 0.5$, $c = 1926.6 \pm 0.3$ pm, and $\beta = 120.81° \pm 0.01°$ and a calculated density of 1.607 g·cm$^{-3}$ (Bermejo et al. 2004a). To obtain it, a solution of $Zn(OAc)_2 \cdot 2H_2O$ (1.12 mM) in EtOH (30 mL) was mixed with a solution of 2-pyridineformamide $N(4)$-ethylthiosemicarbazone ($C_9H_{13}N_5S$) (1.12 mM) in EtOH (30 mL). The resulting solution was stirred at room temperature for 7 days and a yellow precipitate was filtered off. Recrystallization from DMSO gave yellow crystals of the title compound.

**$[Zn(NO_3)_2(C_{10}H_8N_2S_2)_2(H_2O)_2 \cdot (MeOH)_2 \cdot 2H_2O]_n$ or $C_{22}H_{32}N_6O_{12}S_4Zn$.** This compound crystallizes in the monoclinic structure with the lattice parameters $a = 941.7 \pm 0.2$, $b = 1931.2 \pm 0.5$, $c = 975.0 \pm 0.3$ pm, and $\beta = 110.785° \pm 0.004°$ and a calculated density of 1.535 g·cm$^{-3}$ (Horikoshi and Mikuriya 2005). $[Zn(NO_3)_2(C_{10}H_8N_2S_2)_2 \cdot 1.5H_2O]_n$ was recrystallized from MeOH to give the title compound as colorless crystals.

**$[Zn(C_5H_{10}NCOS)_2(C_5N_{10}NH)_2]$ or $C_{22}H_{42}N_4O_2S_2Zn$.** This compound crystallizes in the triclinic structure with the lattice parameters $a = 1027.9 \pm 0.5$, $b = 1269.0 \pm 0.5$, and $c = 1216.1 \pm 0.5$ pm and $\alpha = 107.44° \pm 0.05°$, $\beta = 92.67° \pm 0.05°$, and $\gamma = 117.27° \pm 0.05°$ and experimental and calculated densities $1.31 \pm 0.02$ and 1.32 g·cm$^{-3}$, respectively (Pierpont et al. 1972, Greene et al. 1973). The addition of twofold excess of piperidine to the benzene solution of $[Zn(C_5H_{10}NCOS)_2]$ followed by the addition of hexane until the solution became slightly cloudy provided, upon slow crystallization, white crystals of the title compound (McCormick and Greene 1972). These crystals were washed with a hexane/benzene (3:1 volume ratio) solution and dried in vacuo over anhydrous $CaSO_4$ for 24 h. Colorless crystals were grown from a hexane–benzene solution (Greene et al. 1973).

**$[Zn\{SC(NHMe)_2\}_2(C_5H_4O_4)]_2 \cdot 2H_2O$ or $C_{22}H_{44}N_8O_{10}S_4Zn_2$.** This compound crystallizes in the triclinic structure with the lattice parameters $a = 901.1 \pm 0.2$, $b = 970.1 \pm 0.2$, and $c = 1114.1 \pm 0.2$ pm and $\alpha = 77.79° \pm 0.02°$, $\beta = 77.94° \pm 0.02°$, and $\gamma = 85.63° \pm 0.02°$ and a calculated density of 1.499 g·cm$^{-3}$ (Burke et al. 2003). To obtain it, an aqueous solution of sodium citraconate (0.15 mM) was added to an aqueous solution of $[Zn\{SC(NHMe)_2\}_4]Cl_2$ (0.15 mM), with no discernable change. After several hours, a colorless crystalline precipitate was observed, which was separated by filtration.

**[Zn(C$_6$H$_7$N$_2$S)$_2$(C$_{12}$H$_8$N$_2$)]·1.5H$_2$O or C$_{24}$H$_{25}$N$_6$O$_{1.5}$S$_2$Zn**. Electrochemical oxidation of Zn in a solution of 4,6-dimethyl-2-thiopyrimidine and Phen in acetonitrile afforded the title compound in the form of yellow-orange solid (Castro et al. 1994).

**[Zn(H$_2$O)$_4$(C$_6$H$_6$N$_2$O)$_2$][C$_{12}$H$_8$S$_2$O$_6$] or C$_{24}$H$_{28}$N$_4$O$_{12}$S$_2$Zn**. This compound crystallizes in the monoclinic structure with the lattice parameters $a = 1462.71 \pm 0.03$, $b = 685.58 \pm 0.01$, $c = 1507.43 \pm 0.03$ pm, and $\beta = 111.185° \pm 0.002°$ (Lian et al. 2010). To obtain it, nicotinamide (0.4 mM) was added with constant stirring to an aqueous solution (20 mL) of Zn(NO$_3$)$_2$·6H$_2$O (0.2 mM). The solution was then treated with disodium 4,4′-biphenyldisulfonate (0.2 mM). Colorless crystals of the title compound were collected after 10 days.

**[Zn(Phen)(Am$^i$OCS$_2$)$_2$] or C$_{24}$H$_{30}$N$_2$O$_2$S$_4$Zn**. This compound was obtained by the similar procedure as the one described for Zn(2,2′-bipy)(Am$^i$OCS$_2$) using Phen instead of 2,2′-bipy (Leonova et al. 1997).

**[Zn(C$_6$H$_{12}$N$_2$S$_2$)$_2$(Phen)]·H$_2$O or C$_{24}$H$_{34}$N$_6$OS$_4$Zn**. This compound decomposes at 174°C and crystallizes in the orthorhombic structure with the lattice parameters $a = 1958.8 \pm 0.2$, $b = 1421.0 \pm 0.2$, and $c = 1037.3 \pm 0.3$ pm and a calculated density of 1.454 g·cm$^{-3}$ (Prakasam et al. 2007). It was prepared by adding a hot solution of Phen (1 mM) in benzene to a hot suspension of [Zn(C$_6$H$_{12}$N$_2$S$_2$)$_2$] (1 mM) in benzene/chloroform (1:1 volume ratio) mixture. The resulting pale-yellow solution was filtered and cooled. The yellow compound separated out, which was filtered and recrystallized from benzene/chloroform (1:1 volume ratio) mixture.

**[Zn(SOCCMe$_3$)$_2$(C$_7$H$_9$N)$_2$] or C$_{24}$H$_{36}$N$_2$O$_2$S$_2$Zn**. This compound crystallizes in the triclinic structure with the lattice parameters $a = 985.5 \pm 0.1$, $b = 1071.9 \pm 0.1$, and $c = 1412.5 \pm 0.1$ pm and $\alpha = 91.94° \pm 0.01°$, $\beta = 104.14° \pm 0.01°$, and $\gamma = 105.50° \pm 0.01°$ (Nyman et al. 1997). To obtain it, Et$_2$Zn (8.1 mM), 3,5-dimethylpyridine (16.2 mM), and 20 mL of toluene were combined in a round-bottom flask in an inert-atmosphere box. The flask was removed from inert-atmosphere box and placed in an ice water bath. Thiopivalic acid (16.2 mM) was added to the stirring solution via a pipet and a pale-yellow solution formed. The solution was warmed to room temperature while stirring for several hours and placed in the freezer. Large, block-shaped colorless crystals formed overnight.

Zn(SOCCMe$_3$)$_2$(C$_7$H$_9$N)$_2$ decomposes in toluene to form an amorphous material. Heating this material to 500°C with loss of *surface capping* 3,5-dimethylpyridine results in the formation of sphalerite ZnS with an average crystal size of 5 nm (Nyman et al. 1997).

**[{Zn(C$_{10}$H$_8$N$_2$S$_2$)$_2$(H$_2$O)$_2$}(NO$_3$)$_2$·2EtOH·2H$_2$O]$_n$ or C$_{24}$H$_{36}$N$_6$O$_{12}$S$_4$Zn**. This compound crystallizes in the triclinic structure with the lattice parameters $a = 1102.40 \pm 0.06$, $b = 1159.66 \pm 0.06$, and $c = 1325.20 \pm 0.06$ pm and $\alpha = 80.074° \pm 0.004°$, $\beta = 89.417° \pm 0.004°$, and $\gamma = 89.113° \pm 0.004°$ at $108 \pm 2$ K and a calculated density of 1.581 g·cm$^{-3}$ (Seidel et al. 2011). It was prepared by the following procedure. Zn(NO$_3$)$_2$·6H (0.045 mM) and Phen (0.045 mM) were dissolved in 3 mL of EtOH. A solution of 4,4′-dithiodipyridine (C$_{10}$H$_8$N$_2$S$_2$) (0.045 mM) in 1 mL of EtOH was added with stirring. The mixture was left at room temperature, while the solvent was allowed to evaporate slowly. Colorless crystals of the title compound were found after ca. 3 weeks.

**[Zn{SC(NH$_2$)(NHMe)}$_2$(C$_8$H$_4$O$_4$)]$_2$·4H$_2$O or C$_{24}$H$_{40}$N$_4$O$_{12}$S$_2$Zn.** This compound crystallizes in the monoclinic structure with the lattice parameters $a = 1795.60 \pm 0.03$, $b = 1071.40 \pm 0.02$, $c = 2044.20 \pm 0.04$ pm, and $\beta = 110.6040° \pm 0.0010°$ at $150 \pm 2$ K and a calculated density of 1.593 g·cm$^{-3}$ (Burke et al. 2003). To obtain it, an aqueous solution of sodium phthalate (0.18 mM) was added to an aqueous solution of [Zn{SC(NH$_2$) (NHMe)}$_4$]Cl$_2$ (0.18 mM), with no discernable change. After approximately 72 h, a colorless crystalline precipitate was observed, which was separated by filtration.

**[Zn(C$_4$H$_{11}$N$_3$S)$_2$(H$_2$O)][O$_2$CCH = CC(O)N(Me)C(= NNMe$_2$)S]$_2$·H$_2$O or C$_{24}$H$_{46}$N$_{12}$O$_8$S$_4$Zn.** This compound crystallizes in the triclinic structure with the lattice parameters $a = 962.3 \pm 0.2$, $b = 1184.4 \pm 0.2$, and $c = 1784.8 \pm 0.5$ pm and $\alpha = 102.56° \pm 0.01°$, $\beta = 102.69° \pm 0.01°$, and $\gamma = 92.467° \pm 0.010°$ and a calculated density of 1.420 g·cm$^{-3}$ (Burrows et al. 2003). To obtain it, aqueous solutions of sodium acetylenedicarboxylate (0.22 mM) and [Zn(C$_4$H$_{11}$N$_3$S)$_2$(NO$_3$)$_2$] (0.22 mM) were mixed. After approximately 24 h, yellow crystals formed, and these were isolated by filtration.

**[Bu$^n_4$N][Zn(O$_2$CMe)$_2$(S$_2$CNMe$_2$)] or C$_{24}$H$_{51}$N$_3$O$_2$S$_4$Zn.** To obtain this compound, [Bu$^n_4$N][OH] (6.7 mL, 40% aqueous solution) and acetic acid (0.6 g) were mixed, and the solution was evaporated to dryness in vacuo at 60°C (McCleverty and Morrison 1976). The oil that formed was dissolved in acetone (100 mL) and to this was added [Zn(S$_2$CNMe$_2$)$_2$] (3.0 g). The mixture was shaken for 2 h and then filtered. To the filtrate was added light petroleum, and after gradual evaporation, white crystals of the title compound were formed. These were recrystallized from acetone/light petroleum.

**[Zn$_2$(C$_4$H$_8$NOS$_2$)$_4$(C$_{10}$H$_8$N$_2$)] or C$_{26}$H$_{40}$N$_6$O$_4$S$_8$Zn$_2$.** This compound melts at 175°C and crystallizes in the monoclinic structure with the lattice parameters $a = 860.8 \pm 0.3$, $b = 1419.4 \pm 0.3$, $c = 1567.7 \pm 0.3$ pm, and $\beta = 97.21°$ (Manohar et al. 2001). It was prepared by the following procedure. A hot ethanolic solution of Zn(C$_4$H$_8$NOS$_2$)$_2$ (2 mM) was added to a hot solution of 4,4'-bipy (1 mM) in EtOH. The resulting yellow solution was left to evaporate at room temperature. After 2 days, the yellow crystals were separated out.

**[Me$_4$N][Zn{S$_2$CN(CH$_2$)}$_2$(C$_7$H$_4$NOS)]·C$_3$H$_6$O or C$_{26}$H$_{42}$N$_4$O$_2$S$_5$Zn.** This compound was prepared as follows (McCleverty et al. 1980a). [Zn{S$_2$CN(CH$_2$)}$_2$] was added to an acetone solution of [Me$_4$N][C$_7$H$_4$NOS] and the mixture was stirred and refluxed for 2 h. The solution was cooled and filtered, and upon the addition of light petroleum, the title compound precipitated. The compound was filtered off, washed with light petroleum, and air-dried.

**[Zn(S$_2$COCH$_2$Ph)$_2$(C$_{11}$H$_6$N$_2$O)] or C$_{27}$H$_{20}$N$_2$O$_3$S$_4$Zn.** This compound crystallizes in triclinic structure with the lattice parameters $a = 955.9 \pm 0.2$, $b = 1137.9 \pm 0.3$, and $c = 1309.5 \pm 0.3$ pm and $\alpha = 105.629° \pm 0.003°$, $\beta = 100.745° \pm 0.002°$, and $\gamma = 102.569° \pm 0.002°$ at 98 K and a calculated density of 1.578 g·cm$^{-3}$ (Câmpian et al. 2013). Zn(S$_2$COCH$_2$Ph)$_2$ was obtained from the reaction of Zn(NO$_3$)$_2$·6H$_2$O and potassium $O$-benzyl dithiocarbonate in aqueous solution. The 4.5-diazafluoren-9-one-$N,N'$ ($C_{11}H_6N_2O$) adducts were obtained from stirring at room temperature $Zn(S_2COCH_2Ph)_2$ (0.2 g) with a stoichiometric amount of $C_{11}H_6N_2O$ in EtOH. The solvent was removed in vacuo, and crystals were isolated from slow evaporation of its EtOH/acetone solutions.

**[Zn($C_{14}H_8O_6S$)(2,2'-bipy)($H_2O$)]·Pr$^i$OH or $C_{27}H_{26}N_2O_8SZn$.** This compound crystallizes in the monoclinic structure with the lattice parameters $a = 676.30 \pm 0.09$, $b = 1647.8 \pm 0.2$, $c = 2419.4 \pm 0.3$ pm, and $\beta = 90.674° \pm 0.002°$ and a calculated density of 1.488 g·cm$^{-3}$ (Zhuang et al. 2007). To obtain it, a mixture of $Zn(OAc)_2·2H_2O$ (0.1 mM), 4,4'-dicarboxybiphenyl sulfone (0.1 mM), 2,2'-bipy (0.1 mM), NaOH aqueous solution (0.15 mL, 0.65 M), distilled $H_2O$ (5 mL), Pr$^i$OH (5 mL), and DMF (0.1 mL) was sealed in a Teflon-lined stainless vessel (25 mL) and heated at 180°C for 72 h under autogenous pressure. The vessel was then cooled slowly to room temperature. Colorless block crystals were obtained by filtration, washed with distilled $H_2O$, and dried in air.

**[Zn($C_{10}H_8N_2S_2$)$_2$($H_2O$)$_2$]·2$C_4H_3O_4$ or $C_{28}H_{26}N_4O_{10}S_4Zn$.** This compound crystallizes in the orthorhombic structure with the lattice parameters $a = 1850.5 \pm 2.5$, $b = 1108.78 \pm 0.09$, and $c = 1569.03 \pm 0.13$ pm and a calculated density of 1.523 g·cm$^{-3}$ (Yang et al. 2008). It was obtained by the next procedure. A methanolic solution (10 mL) of 4,4'-bipy (0.1 mM) and maleic acid (0.1 mM) was slowly diffused into an aqueous solution (10 mL) of $Zn(NO_3)_2·6H_2O$ (0.1 mM) and stirring for 40 min at room temperature. Slow evaporation of the filtrate for 1 week provided colorless block crystals of [Zn($C_{10}H_8N_2S_2$)$_2$($H_2O$)$_2$]·2$C_4H_3O_4$.

**[Zn($C_4H_7O_3$)$_2$($C_{10}N_8N_2S_2$)$_2$] or $C_{28}H_{30}N_4O_6S_4Zn$.** This compound crystallizes in the orthorhombic structure with the lattice parameters $a = 3023.0 \pm 0.6$, $b = 872.62 \pm 0.18$, and $c = 1148.8 \pm 0.2$ pm and a calculated density of 1.561 g·cm$^{-3}$ (Carballo et al. 2009). To obtain it, to a suspension of [Zn($C_4H_7O_3$)$_2$($H_2O$)$_2$] (1 mM) in $H_2O$ (20 mL) was slowly added a solution of 4,4'-bipyridyl disulfide (1 mM) in MeOH (10 mL). The mixture was heated under reflux for 2 h, left to cool to room temperature, and stirred for 1 week. Colorless single crystals were obtained after slow evaporation (2 months) of the solution, which results from the isolation by filtration of the white precipitate that corresponds to the insoluble and scarcely reactive precursor. The title compound was also obtained in similar yield by the same synthetic procedure in a 1:2 molar ratio.

**[Zn(SCN)$_2$($C_{10}H_8N_2S_2$)$_2$·(DMF)$_2$]$_n$ or $C_{28}H_{30}N_8O_2S_6Zn$.** This compound crystallizes in the triclinic structure with the lattice parameters $a = 920.8 \pm 0.4$, $b = 1036.4 \pm 0.4$, and $c = 1061.2 \pm 0.4$ pm and $\alpha = 106.197° \pm 0.006°$, $\beta = 96.058° \pm 0.008°$, and $\gamma = 110.384° \pm 0.007°$ (Horikoshi and Mikuriya 2005). [Zn(SCN)$_2$($C_{10}H_8N_2S_2$)$_2$·$H_2O$]$_n$ was recrystallized from DMF to give the title compound as colorless crystals.

**[Zn($C_{12}H_{16}N_5S$)(OAc)]$_2$ or $C_{28}H_{38}N_{10}O_4S_2Zn_2$.** This compound melts at a temperature higher than 300°C and crystallizes in the monoclinic structure with the lattice parameters $a = 2415.11 \pm 0.06$, $b = 1528.6 \pm 0.1$, $c = 991.9 \pm 0.1$ pm, and $\beta = 92.679° \pm 0.005°$ and a calculated density of 1.404 g·cm$^{-3}$ (Castiñeiras et al. 2002). To obtain it, a solution of $C_{12}H_{17}N_5S$ [2-pyridineformamide N(4)-piperidylthiosemicarbazone]

(0.76 mM) in EtOH (30 mL) was added to a solution of $Zn(OAc)_2$ (0.76 mM) in 30 mL of 96% EtOH, and the mixture was stirred for several days at room temperature. The resulting yellow solid was filtered off, washed thoroughly with cold ethanol, and stored in a desiccator over Mg and $I_2$.

**[Zn{SC(NHMe)$_2$}$_2$(C$_8$H$_4$O$_4$)]$_2$·2H$_2$O or C$_{28}$H$_{44}$N$_8$O$_{10}$S$_4$Zn$_2$.** This compound crystallizes in the monoclinic structure with the lattice parameters $a = 1033.90 \pm 0.02$, $b = 1150.00 \pm 0.02$, $c = 1663.30 \pm 0.03$ pm, and $\beta = 90.6080° \pm 0.0007°$ at $150 \pm 2$ K and a calculated density of 1.524 g·cm$^{-3}$ (Burke et al. 2003). To obtain it, an aqueous solution of sodium isophthalate (0.15 mM) was added to an aqueous solution of [Zn{SC(NHMe)$_2$}$_4$]Cl$_2$ (0.15 mM), with no discernable change. After several hours, a colorless crystalline precipitate was observed, which was separated by filtration.

**[Bu$^n_4$N][Zn(S$_2$CNMe$_2$)$_2$(C$_7$H$_4$NOS)] or C$_{29}$H$_{52}$N$_4$OS$_5$Zn.** This compound was prepared as follows (McCleverty et al. 1980a). To an acetone solution (100 mL) of [Bu$^n_4$N][C$_7$H$_4$NOS] (4.1 g) was added [Zn(S$_2$CNMe$_2$)$_2$] (3.1 g), and the mixture was stirred and refluxed for 2 h. The solution was cooled and filtered, and upon the addition of light petroleum, the title compound precipitated. The compound was filtered off, washed with light petroleum, and air-dried.

**[Me$_4$N][Zn(S$_2$CNBu$^n_2$)$_2$(C$_7$H$_4$NOS)]·2H$_2$O or C$_{29}$H$_{56}$N$_4$O$_3$S$_5$Zn.** This compound was prepared as follows (McCleverty et al. 1980a). To an acetone solution of [Me$_4$N][C$_7$H$_4$NOS] was added [Zn(S$_2$CNBu$^n_2$)$_2$], and the mixture was stirred and refluxed for 2 h. The solution was cooled and filtered, and upon the addition of light petroleum, the title compound precipitated. The compound was filtered off, washed with light petroleum, and air-dried.

**[Zn(Phen)$_2$(H$_2$O)$_2$][C$_6$H$_2$(OH)$_2$(SO$_3$)$_2$]·3H$_2$O or C$_{30}$H$_{30}$N$_4$O$_{13}$S$_2$Zn.** This compound crystallizes in the triclinic structure with the lattice parameters $a = 969.68 \pm 0.04$, $b = 1273.92 \pm 0.05$, and $c = 1444.11 \pm 0.06$ pm and $\alpha = 98.224° \pm 0.001°$, $\beta = 97.920° \pm 0.001°$, and $\gamma = 106.409° \pm 0.001°$ (Wang et al. 2005). To obtain it, 4,5-dihydroxy-1,3-benzenedisulfonate monohydrate (1 mM) was added to a solution of $Zn(NO_3)2·H_2O$ (1 mM) in $H_2O$ (20 mL). The colorless solution was mixed with a methanol (5 mL) solution of Phen (2 mM). Upon standing and slow evaporation, light-yellow crystals of the title compound were obtained.

**[Zn{S$_2$CN(Me)CH$_2$CH$_2$OH}$_2$]$_2$·(C$_{14}$H$_{14}$N$_4$S$_2$) or C$_{30}$H$_{46}$N$_8$O$_4$S$_{10}$Zn$_2$.** The title compound was prepared by dissolving bis[N-(2-hydroxyethyl)-N-methyldithiocarbamato]zinc (1.0 mM) and N,N'-bis(pyridin-3-ylmethyl)-thioxalamide (C$_{14}$H$_{14}$N$_4$S$_2$) (0.5 mM) into MeCN (Poplaukhin et al. 2012).

**[Zn{S$_2$CN(Me)CH$_2$CH$_2$OH}$_2$]$_2$·(C$_{14}$H$_{14}$N$_4$S$_2$)·(2S$_8$)　　　or　　　C$_{30}$H$_{46}$N$_8$O$_4$S$_{26}$Zn$_2$.** This compound crystallizes in the triclinic structure with the lattice parameters $a = 941.55 \pm 0.15$, $b = 1098.84 \pm 0.15$, and $c = 1581.0 \pm 0.3$ pm and $\alpha = 78.140° \pm 0.009°$, $\beta = 89.077° \pm 0.009°$, and $\gamma = 65.547° \pm 0.007°$ at 98 K and a calculated density of 1.768 g·cm$^{-3}$ (Poplaukhin et al. 2012). It was obtained when chloroform was diffused into the solution of bis[N-(2-hydroxyethyl)-N-methyldithiocarbamato]zinc (1.0 mM) and N,N'-bis(pyridin-3-ylmethyl)-thioxalamide (C$_{14}$H$_{14}$N$_4$S$_2$) (0.5 mM) in MeCN.

$\{Zn[S_2CN(Me)CH_2CH_2OH]_2\}_2[(3-C_5H_4N)CH_2N(H)C(O)C(O)N(H)$
$CH_2(C_5H_4N-3)]$ **(A) and** $\{Zn[S_2CN(Me)CH_2CH_2OH]_2\}_2[(3-C_5H_4N)CH_2NC(OH)$
$C(OH)NCH_2(C_5H_4N-3)$ **(B) or** $C_{30}H_{46}N_8O_6S_8Zn_2$. Compound A melts at
178°C–180°C and crystallizes in the triclinic structure with the lattice parameters
$a = 846.0 \pm 0.4$, $b = 943.8 \pm 0.4$, and $c = 1330.5 \pm 0.5$ pm and $\alpha = 82.811° \pm 0.017°$,
$\beta = 78.712° \pm 0.013°$, and $\gamma = 89.077° \pm 0.17°$ at 98 K and a calculated density
of 1.610 g·cm$^{-3}$ (Poplaukhin and Tiekink 2010). To obtain it, $\{Zn[S_2CN(Me)$
$CH_2CH_2OH]_2\}_2$ (0.5 mM) and $N,N'$-bis(pyridin-3-ylmethyl)oxalamide (0.5 mM) dissolved in ca. 20 mL of hot MeOH/EtOH (1:1 volume ratio). The mixture was allowed
to cool to room temperature, filtered, and left to evaporate slowly. Colorless blocks
of A appeared in 3 days.

Compound B crystallizes in the monoclinic structure with the lattice parameters
$a = 974.71 \pm 0.14$, $b = 1615.1 \pm 0.2$, $c = 1321.0 \pm 0.2$ pm, and $\beta = 94.743° \pm 0.002°$ at 98 K
and a calculated density of 1.606 g·cm$^{-3}$ (Poplaukhin and Tiekink 2010). The analogous reaction (as for A) of $\{Zn[S_2CN(Me)CH_2CH_2OH]_2\}_2$ with $N,N'$-bis(pyridin-3-
ylmethyl)thiooxalamide in MeCN/CHCl$_3$ (3:1 volume ratio) yielded two products,
and one of them was B. The solution that was filtered off was left to evaporate under
ambient conditions. A few red crystals of B formed over the course of 10 days, along
with further amorphous yellow precipitate.

$[Zn_2(C_5H_{10}NOS_2)_4(C_{10}H_8N_2)]$ **or** $C_{30}H_{48}N_6O_4S_8Zn_2$. This compound melts at 180°C
and crystallizes in the triclinic structure with the lattice parameters $a = 1322.9 \pm 0.3$,
$b = 1559.2 \pm 0.2$, and $c = 1122.5 \pm 0.2$ pm and $\alpha = 105.34° \pm 0.03°$, $\beta = 93.67° \pm 0.03°$,
and $\gamma = 105.58° \pm 0.04°$ and a calculated density of 1.573 g·cm$^{-3}$ (Manohar et al.
2001). It was prepared by the following procedure. $Zn(C_5H_{10}NOS_2)_2$ (2 mM) in EtOH
was added dropwise to a hot solution of 4,4'-bipy (1 mM) in EtOH with continuous
stirring. The resulting yellow solution was left for evaporation at room temperature.
After 2 days, the yellow crystals were separated out.

$[Zn\{S_2CN(CH_2CH_2OH)Et\}_2]_2(4,4'-bipy)$ **or** $C_{30}H_{48}N_6O_4S_8Zn_2$. This compound
melts at a temperature higher than 320°C and crystallizes in the triclinic structure
with the lattice parameters $a = 1107.4 \pm 0.6$, $b = 1324.9 \pm 0.5$, and $c = 1535.0 \pm 0.6$ pm
and $\alpha = 104.546° \pm 0.007°$, $\beta = 107.095° \pm 0.015°$, and $\gamma = 92.740° \pm 0.008°$ at 173 K and
a calculated density of 1.518 g·cm$^{-3}$ (Benson et al. 2007).

$[Zn\{S_2CN(CH_2CH_2OH)_2\}_2]_2(4,4'-bipy)$ **or** $C_{30}H_{48}N_6O_8S_8Zn_2$. This compound melts
at 121°C–124°C and crystallizes in the triclinic structure with the lattice parameters
$a = 1124.9 \pm 0.4$, $b = 1321.4 \pm 0.5$, and $c = 1558.1 \pm 0.6$ pm and $\alpha = 105.335° \pm 0.003°$,
$\beta = 105.092° \pm 0.004°$, and $\gamma = 93.881° \pm 0.005°$ at 218 K and a calculated density of
1.569 g·cm$^{-3}$ (Benson et al. 2007).

$[NBu^n_4][\{Zn(S_2CNMe_2)_2\}_2(OCOMe)]$ **or** $C_{30}H_{63}N_5O_2S_8Zn_2$. This compound crystallizes in the orthorhombic structure with the lattice parameters $a = 1208.1 \pm 1.0$,
$b = 2030.6 \pm 1.6$, and $c = 1909.9 \pm 1.2$ pm and experimental and calculated densities
1.284 and 1.295 g·cm$^{-3}$, respectively (McCleverty et al. 1980b). Its crystals were
obtained from acetone/light petroleum mixtures as colorless needles.

$[Zn_7(\mu_4-O)_2(OAc)_{10}(C_{10}H_8N_2S_2)\cdot0.5MeCN\cdot0.5H_2O$ **or** $C_{31}H_{40.5}N_{2.5}O_{22.5}S_2Zn_7$.
This compound crystallizes in the triclinic structure with the lattice parameters

$a = 1048.0 \pm 0.2$, $b = 1275.5 \pm 0.2$, and $c = 2102.9 \pm 0.3$ pm and $\alpha = 96.962° \pm 0.003°$, $\beta = 99.777° \pm 0.003°$, and $\gamma = 109.041° \pm 0.003°$ at $223 \pm 2$ K and a calculated density of 1.718 g·cm$^{-3}$ (Ng et al. 2003). It was prepared by the next procedure. A methanolic solution (1 mL) of $Zn(OAc)_2$ (0.63 mM) was layered with an MeCN solution (1 mL) of 4,4'-bipyridyldisulfide ($C_{10}H_8N_2S_2$) (0.09 mM). A MeOH buffer layer was introduced between the aforementioned solutions to prevent instant precipitation. Transparent rod-shaped single crystals of the title compound formed after 1 day were washed with $Et_2O$ and dried under vacuum.

**$[Zn(Bu^n_2CNS_2)_2(Phen)]\cdot0.5EtOH$ or $C_{31}H_{47}N_4O_{0.5}S_4Zn$.** This compound melts at 188°C–189°C and crystallizes in the triclinic structure with the lattice parameters $a = 1000.6 \pm 0.4$, $b = 1334.7 \pm 0.2$, and $c = 1553.1 \pm 0.2$ pm and $\alpha = 103.710° \pm 0.010°$, $\beta = 102.78° \pm 0.02°$, and $\gamma = 108.37° \pm 0.02°$ and a calculated density of 1.232 g·cm$^{-3}$ (Reck and Becker 2004). For the preparation of the title compound, a stirred solution of $Zn(Bu^n_2CNS_2)_2$ in EtOH was treated with a solution of Phen (2.2 molar equivalent) in EtOH at ambient temperature. After the addition of $H_2O$, a yellow precipitate was collected and recrystallized from ethanol/water (1:1 volume ratio). Yellow crystals of $[Zn(Bu^n_2CNS_2)_2(Phen)\cdot0.5EtOH$ were obtained.

**$[Bu^n_4N][Zn(S_2CNMe_2)_2(C_7H_4NS_2)]\cdot EtOH$ or $C_{31}H_{58}N_4OS_6Zn$.** This compound crystallizes in the monoclinic structure with the lattice parameters $a = 1679.7 \pm 1.9$, $b = 1534.9 \pm 1.7$, $c = 1667.7 \pm 1.0$ pm, and $\beta = 111.74° \pm 0.04°$ and experimental and calculated densities 1.28 and 1.265 g·cm$^{-3}$, respectively (McCleverty et al. 1980a). It crystallizes from EtOH/light petroleum as regular bricks.

**$[Zn(C_4H_8NOS_2)_2(C_{18}H_{12}N_6)]\cdot1.5C_4H_8O_2$ or $C_{32}H_{40}N_8O_5S_4Zn$.** This compound crystallizes in the triclinic structure with the lattice parameters $a = 1186.3 \pm 1.0$, $b = 1301.9 \pm 1.1$, and $c = 1319.9 \pm 1.1$ pm and $\alpha = 107.214° \pm 0.012°$, $\beta = 105.780° \pm 0.015°$, and $\gamma = 100.892° \pm 0.011°$ at 98 K and a calculated density of 1.502 g·cm$^{-3}$ (Arman et al. 2012). It was prepared by dissolving bis[$N$-(2-hydroxyethyl)-$N$-methyldithiocarbamato-$\kappa S$]zinc(II) (0.5 mM) and 2,4,6-tris(pyridin-2-yl)-1,3,5-triazine (0.5 mM) into a MeOH/EtOH solution. The solution made an abrupt color change from yellow to red. Suitable crystals for XRD were grown by liquid diffusion of dioxane into the MeOH/EtOH solution.

**$[Zn(C_{13}H_{18}N_5S)(OAc)]_2\cdot DMSO$ or $C_{32}H_{48}N_{10}O_5S_3Zn_2$.** This compound crystallizes in the triclinic structure with the lattice parameters $a = 1018.09 \pm 0.08$, $b = 1435.3 \pm 0.1$, and $c = 1507.1 \pm 0.1$ pm and $\alpha = 115.751° \pm 0.001°$, $\beta = 91.868° \pm 0.002°$, and $\gamma = 95.205° \pm 0.002°$ and a calculated density of 1.484 g·cm$^{-3}$ (Bermejo et al. 2004b). To obtain it, a solution of $Zn(OAc)_2\cdot2H_2O$ (0.90 mM) in EtOH (30 mL) was mixed with a solution of 2-pyridineformamide 3-hexamethyleneiminylthiosemicarbazone ($C_{13}H_{19}N_5S$) (0.90 mM) in EtOH (30 mL). The resulting solution was stirred at room temperature for 7 days. The obtained solid was filtered off, washed with anhydrous ether to apparent dryness, and stored in a desiccator over Mg and $I_2$. Recrystallization from DMSO gave orange crystals of the title compound.

**$[(mepa)_4Zn_3](NO_3)_2\cdot2H_2O$ or $C_{32}H_{48}N_{10}O_8S_4Zn_3$ (mepaH = $N$-(2-mercaptoethyl) picolylamin, $C_8H_{12}N_2S$).** This compound melts at 240°C with decomposition

(Brand and Vahrenkamp 1995). To obtain it, mepaH (2.11 mM) and NaOH (2.10 mM) were dissolved in MeOH (20 mL). A solution of $Zn(NO_3)_2 \cdot 4H_2O$ (2.11 mM) in MeOH (10 mL) was added dropwise with stirring. The resulting precipitate was stirred for 30 min and then filtered off, washed with a few mL of MeOH, and dried in vacuo. Recrystallization from $H_2O$/MeOH (1:1 volume ratio) yielded [(mepa)$_4$Zn$_3$] $(NO_3)_2 \cdot 2H_2O$ as colorless crystals.

**[Zn{S$_2$CN(CH$_2$CH$_2$OH)Et}$_2$]$_2$(4,4′-bipy)·2MeOH or C$_{32}$H$_{56}$N$_6$O$_6$S$_8$Zn$_2$.** This compound melts at the temperature higher than 320°C and crystallizes in the triclinic structure with the lattice parameters $a = 954.16 \pm 0.18$, $b = 1130.9 \pm 0.2$, and $c = 1141.8 \pm 0.2$ pm and $\alpha = 75.640° \pm 0.004°$, $\beta = 73.016° \pm 0.004°$, and $\gamma = 85.238° \pm 0.004°$ and a calculated density of 1.466 g·cm$^{-3}$ (Benson et al. 2007).

**[Bu$^n_4$N][Zn(S$_2$CNMe$_2$)(C$_7$H$_4$NOS)$_2$] or C$_{33}$H$_{50}$N$_4$O$_2$S$_4$Zn.** This compound was prepared as follows (McCleverty et al. 1980a). To an acetone solution of [{Zn(C$_7$H$_5$N$_2$S)$_2$}]$_n$ (3.7 g) was added [Bu$^n_4$N][S$_2$CNMe$_2$] (3.8 g), and the mixture was stirred and refluxed for 2 h. The solution was cooled and filtered, and upon the addition of light petroleum, the title compound precipitated. The compound was filtered off, washed with light petroleum, and air-dried.

**[Zn(S-2-PhCONHC$_6$H$_4$)$_2$(1-MeIm)$_2$]·0.5H$_2$O or C$_{34}$H$_{33}$N$_6$O$_{2.5}$S$_2$Zn (1-MeIm = 1-methylimidazole).** This compound could be obtained if an excess of NaBH$_4$ is added in small portions to a THF/EtOH (1:1 volume ratio, 10 mL) solution of bis(2-benzoylaminophenol)disulfide (0.256 mM) (Sun et al. 1999). The mixture was stirred at room temperature to give a clear solution and added dropwise to stirred solution of ZnCl$_2$ (0.256 mM) and 1-MeIm (0.512 mM) in THF (10 mL). The stirring was continued for about 1 h after the addition. The solvents were removed by evaporation, and then toluene (50 mL) was added to the residue and the mixture was stirred overnight at ambient temperature. After filtration, the toluene was removed from the filtrate in vacuo to give crude products. Almost colorless crystals of C$_{34}$H$_{33}$N$_6$O$_{2.5}$S$_2$Zn were obtained by recrystallization from acetonitrile. Single crystals were grown by slow evaporation from recrystallization solution.

**[Zn(H$_2$O)$_4$(C$_6$H$_6$N$_2$O)$_2$][C$_{10}$H$_6$S$_2$O$_6$]·2C$_6$H$_6$N$_2$O·4H$_2$O or C$_{34}$H$_{46}$N$_8$O$_{18}$S$_2$Zn.** This compound crystallizes in the triclinic structure with the lattice parameters $a = 1036.6 \pm 0.6$, $b = 1092.0 \pm 0.6$, and $c = 1094.1 \pm 0.5$ pm and $\alpha = 70.83° \pm 0.05°$, $\beta = 74.81° \pm 0.05°$, and $\gamma = 78.29° \pm 0.05°$ (Zhao et al. 2010). To obtain it, isonicotinamide (0.4 mM) was added with constant stirring to an aqueous solution of $Zn(NO_3)_2 \cdot 6H_2O$ (0.2 mM). The solution was then treated with disodium naphthalene-1,5-disulfonate (0.2 mM). Colorless crystals of the title compound were collected after 10 days.

**[Zn(C$_{17}$H$_{26}$NOS$_2$)$_2$] or C$_{34}$H$_{52}$N$_2$O$_2$S$_4$Zn.** This compound melts at 163°C and crystallizes in the triclinic structure with the lattice parameters $a = 893.6 \pm 0.3$, $b = 1076.2 \pm 0.2$, and $c = 1974.0 \pm 0.3$ pm and $\alpha = 80.63° \pm 0.01°$, $\beta = 88.39° \pm 0.02°$, and $\gamma = 84.11° \pm 0.02°$ and a calculated density of 1.273 g·cm$^{-3}$ (Reck and Becker 2003a). It was prepared by the following procedure. A solution of bis(3,5-di-*tert*-butyl-2-hydroxybenzyl)-*n*-butylamine in EtOH was treated with NaOH solution and excess CS$_2$ and stirred for 2 h at ambient temperature. After the addition of H$_2$O, the yellow oil could be separated with diethyl ether. The obtained solution was stirred for 3 h with aqueous $ZnSO_4$. A precipitate formed after evaporation of organic solvent was recrystallized from $n$-hexane/ethanol to yield colorless crystals of the title compound.

**$[Zn(C_8H_7O_2)_2(C_{10}H_8N_2S_2)_2]\cdot H_2O$ or $C_{36}H_{32}N_4O_5S_4Zn$**. This compound crystallizes in the triclinic structure with the lattice parameters $a = 985.1 \pm 0.2$, $b = 1113.0 \pm 0.2$, and $c = 1831.9 \pm 0.4$ pm and $\alpha = 90.38° \pm 0.03°$, $\beta = 98.88° \pm 0.03°$, and $\gamma = 115.89° \pm 0.03°$ and a calculated density of 1.483 $g\cdot cm^{-3}$ (Zhang and Xu 2010). It was obtained by the next procedure. 0.25 mM $Zn(NO_3)_2\cdot 6H_2O$ and 0.25 mM phenylacetic acid were successively dissolved in a stirred aqueous ethanolic solution consisting of 5 mL EtOH and 10 mL $H_2O$, to which 0.5 mL 1.0 M NaOH was added. The formed white suspension was stirred at 80°C for 30 min and then added was an ethanolic solution of 0.25 mM 4,4′-bipyridyldisulfide in 5 mL EtOH. The final mixture was further stirred at 75°C for 1 h and filtered off. The colorless filtrate (pH = 6.02) was left standing at room temperature for 1 week affording colorless blocklike crystals.

**$[Zn_4(\mu_3\text{-}OH)_2(H_2O)_2(C_8H_4O_7S)_2(4,4'\text{-}bipy)_2]\cdot 3H_2O$ or $C_{36}H_{36}N_4O_{21}S_2Zn_4$**. This compound crystallizes in the triclinic structure with the lattice parameters $a = 1010.6 \pm 0.2$, $b = 1085.1 \pm 0.2$, and $c = 1112.5 \pm 0.2$ pm and $\alpha = 105.66° \pm 0.03°$, $\beta = 112.63° \pm 0.03°$, and $\gamma = 98.18° \pm 0.03°$ and a calculated density of 1.888 $g\cdot cm^{-3}$ (Tao et al. 2004). It was prepared by the following procedure. An aqueous solution (5 mL) of 5-sulfoisophtalic acid ($C_8H_6O_7S$) (0.50 mM) was adjusted to pH = 7 with 1 M NaOH, and transferred into a Parr Teflon-lined stainless steel vessel (23 mL) containing $Zn(NO_3)_2\cdot 6H_2O$ (0.50 mM) and 4,4′-bipy (0.50 mM). An additional 3 mL of water was added to the resultant mixture. The vessel was then sealed and heated to 180°C for 3 days, followed by cooling at 2.5°C per hour to room temperature. Colorless blocklike crystals were collected by hand, washed with distilled water, and dried in air at ambient temperature.

**$[Zn(H_2O)_2(C_{12}H_8N_2)_2][(C_6H_3O_4S)_2]\cdot 7H_2O$ or $C_{36}H_{40}N_4O_{17}S_2Zn$**. This compound crystallizes in the monoclinic structure with the lattice parameters $a = 1669.0 \pm 0.3$, $b = 2290.5 \pm 0.2$, $c = 1294.0 \pm 0.2$ pm, and $\beta = 124.849° \pm 0.006°$ (Liu et al. 2014). To obtain it, a mixture of $Zn(OAc)_2\cdot 2H_2O$ (0.25 mM), Phen (0.25 mM), thiophene-3,4-dicarboxylic acid (0.25 mM), and $H_2O$ (7 mL) was filled in a 23 mL Teflon-lined autoclave under autogenous pressure at 140°C for 5 days. After cooling to room temperature, yellow crystals were collected by filtration and washed with distilled water.

**$[Zn\{S_2CN(Me)CH_2CH_2OH\}_2]_2\cdot (C_{14}H_{14}N_4S_2)\cdot 2(C_3H_7NO)$ or $C_{36}H_{60}N_{10}O_6S_{10}Zn_2$**. This compound crystallizes in the triclinic structure with the lattice parameters $a = 1042.6 \pm 0.3$, $b = 1123.5 \pm 0.3$, and $c = 1310.1 \pm 0.3$ pm and $\alpha = 110.067° \pm 0.003°$, $\beta = 112.286° \pm 0.003°$, and $\gamma = 93.936° \pm 0.003°$ at 98 K and a calculated density of 1.509 $g\cdot cm^{-3}$ (Poplaukhin et al. 2012). It was obtained when DMF was diffused into the solution of bis[$N$-(2-hydroxyethyl)-$N$-methyldithiocarbamato]zinc (1.0 mM) and $N,N'$-bis(pyridin-3-ylmethyl)-thioxalamide ($C_{14}H_{14}N_4S_2$) (0.5 mM) in MeCN.

**$[Zn(Bu^n_2CNOS_2)_2]_2$ or $C_{36}H_{72}N_4O_4S_8Zn_2$**. Bis(zinc-dibutyldithiopercarbamate) crystallizes in two polymorphic forms (Reck et al. 1995). α-Modification crystallizes in the monoclinic structure with the lattice parameters $a = 1840.2 \pm 0.4$, $b = 1378.8 \pm 0.5$, $c = 2044.9 \pm 0.3$ pm, and $\beta = 91.55° \pm 0.01°$ and a calculated density of 1.295 g cm$^{-3}$. β-Modification also crystallizes in the monoclinic structure with the lattice parameters $a = 974.2 \pm 0.2$, $b = 1856.9 \pm 0.3$, $c = 1508.3 \pm 0.4$ pm, and $\beta = 108.62° \pm 0.02°$ and a calculated density of 1.299 g cm$^{-3}$.

**[Bu$^n_4$N][Zn(C$_7$H$_4$NS$_2$)$_3$]·H$_2$O or C$_{37}$H$_{50}$N$_4$OS$_6$Zn.** This compound crystallizes in the monoclinic structure with the lattice parameters $a = 985.4 \pm 0.5$, $b = 1602.7 \pm 1.3$, $c = 2700.8 \pm 2.2$ pm, and $\beta = 94.325° \pm 0.006°$ and experimental and calculated densities 1.28 and 1.286 g·cm$^{-3}$ (Ashworth et al. 1976, McCleverty et al. 1980a). To obtain it, a mixture of Zn(C$_7$H$_4$NS$_2$)$_2$ (2.0 g) and [Bu$^n_4$N(C$_7$H$_4$NS$_2$)] (2.0 g) was shaken in acetone (80 mL) for 1 h. The yellow solution that formed was filtered, and upon the addition of light petroleum followed by partial evaporation in vacuo, pale-yellow needlelike crystals of the title compound formed. These were filtered off and recrystallized from acetone/light petroleum. Crystals suitable for XRD were grown from the solution of the title compound in acetone.

**[Zn{NH(CH$_2$)$_4$O}(S$_2$CNEt$_2$)$_2$]$_2$·CH$_2${N(CH$_2$)$_4$O}$_2$ or C$_{37}$H$_{76}$N$_8$O$_4$S$_8$Zn$_2$.** This compound melts at 50.8°C and crystallizes in the monoclinic structure with the lattice parameters $a = 1299.4 \pm 0.1$, $b = 1059.20 \pm 0.08$, $c = 1980.2 \pm 0.2$ pm, and $\beta = 96.603° \pm 0.001°$ at 30°C and a calculated density of 1.330 g·cm$^{-3}$ (Ivanov et al. 2008). To obtain it, 0.69 mM of binuclear zinc dimethyldithiocarbamate was dissolved in 8 mL of CH$_2$Cl$_2$ with mild heating (40°C). Then, 0.24 mL (2.76 mM) of morpholine was added dropwise to the solution, which was heated to boiling for a short time and left overnight in a closed flask. Crystals of the title compound were obtained at room temperature by slow evaporation of the solvent. The crystals were separated from the mother liquor, dried in air, and kept in sealed ampoules.

**[Zn(C$_{34}$H$_{42}$N$_4$O$_2$)$_2$(DMSO)]·MeCN or C$_{38}$H$_{51}$N$_5$O$_3$SZn.** This compound melts at the temperature higher than 350°C and crystallizes in the monoclinic structure with the lattice parameters $a = 1232.88 \pm 0.07$, $b = 1770.43 \pm 0.09$, $c = 1739.32 \pm 0.09$ pm, and $\beta = 92.4391° \pm 0.0008°$ at 100 K and a calculated density of 1.267 g·cm$^{-3}$ (Aazam et al. 2011). It was prepared by the following procedure. A mixture of diaminomaleonitrile (0.93 mM), 3,5-di-*tert*-butyl-2-hydroxybenzaldehyde (1.86 mM), Zn(OAc)$_2$·2H$_2$O (0.93 mM), and EtOH (5 mL) was placed in a glass Petri dish and capped with a glass cover. The dish was placed in a microwave oven (700 W) and irradiated for 1 min. The reaction mixture was cooled and washed with 15 mL of EtOH. The purple solid was filtered off and washed with EtOH. Recrystallization was by slow evaporation of an acetonitrile/DMSO (9:1 volume ratio) solution, which yielded purple blocks of the title compound.

**[Zn[(C$_7$H$_8$NO$_3$S)$_2$(C$_{13}$H$_{14}$N$_2$)$_2$] or C$_{40}$H$_{44}$N$_6$O$_6$S$_2$Zn.** This compound crystallizes in the orthorhombic structure with the lattice parameters $a = 1682.5 \pm 0.2$, $b = 1302.7 \pm 0.1$, and $c = 1804.1 \pm 0.2$ pm (Li et al. 2014). It was prepared by the following procedure. A mixture of 4-amino-3-methyl-benzenesulfonic acid (0.5 mM), 1,3-bi(4-pyridyl)propane (0.1 mM), Zn(OAc)$_2$·4H$_2$O (0.1 mM), and KOH (0.1 mM) was dissolved in 10 mL of H$_2$O. Then the solution was heated in a 25 mL Teflon-lined autoclave under autogenous pressure at 160°C for 3 days. After cooling to room temperature, colorless block-shaped crystals of the title compound that formed were collected.

**[Zn(C$_8$H$_9$NS$_2$)]$_4$·(3DMF) or C$_{41}$H$_{57}$N$_7$O$_3$S$_8$Zn$_4$.** This compound crystallizes in the monoclinic structure with the lattice parameters $a = 2243.7 \pm 0.7$, $b = 1207.7 \pm 0.5$, $c = 2051.0 \pm 0.5$ pm, and $\beta = 110.96° \pm 0.02°$ and experimental and calculated densities 1.50 and 1.55 g·cm$^{-3}$, respectively (Brand and Vahrenkamp 1996). Colorless crystals of this compound were obtained by slow cooling of hot (60°C) saturated solution of Zn(C$_8$H$_9$NS$_2$) in DMF.

**[Zn$_4$O(C$_7$H$_4$NS$_2$)$_6$] or C$_{42}$H$_{24}$N$_6$OS$_{12}$Zn$_4$.** This compound crystallizes in the trigonal structure with the lattice parameters $a = 1829.27 \pm 0.02$ and $c = 2495.36 \pm 0.06$ pm at 0°C (Jin et al. 2007). The synthesis of the title compound was carried out by the reaction of CeCl$_3$/Zn (0.3 mM), CH$_2$[P(O)(OPr$^i$)$_2$]$_2$ (0.3 mM), and 2-mercaptobenzothiazole (C$_7$H$_5$NS$_2$) (0.3 mM) in EtOH/CH$_2$Cl$_2$ (10 mL, 1:1 volume ratio) solution at room temperature for 6 h. After filtration, the clear yellow solution was obtained and was put at room temperature. Yellow crystals were obtained from the slow evaporation of the solution.

**[Zn(C$_{21}$H$_{16}$N$_3$O$_4$S$_2$)$_2$] or C$_{42}$H$_{32}$N$_6$O$_8$S$_4$Zn.** This compound was prepared by mixing N,N-(iminodiethylene)bisphthalimide dithiocarbamic acid (20 mM) and CS$_2$ (20 mM) in acetonitrile or EtOH under ice-cold conditions (5°C) (Thirumaran et al. 1998, Prakasam et al. 2007). To a yellow dithiocarbamic acid solution, an aqueous solution of ZnSO$_4$ was added with constant stirring. The pale-yellow solid separated from the solution, which was filtered, washed with H$_2$O and alcohol, and then dried in air.

**[Zn(C$_9$H$_7$O$_4$)$_2$(C$_{12}$H$_8$N$_4$S)$_2$]·2H$_2$O or C$_{42}$H$_{34}$N$_8$O$_{10}$S$_2$Zn.** This compound crystallizes in the monoclinic structure with the lattice parameters $a = 3500.4 \pm 0.4$, $b = 814.39 \pm 0.09$, $c = 1577.9 \pm 0.2$ pm, and $\beta = 112.011° \pm 0.001°$ (Li and Chang 2010). Direct mixing of the starting materials in solution leads to immediate fine precipitate formation. In order to obtain crystals, suitable for XRD, the hydrothermal methods were applied. A mixture of 5-methylisophthalic acid (0.1 mM), 2,5-bis(4-pyridyl)-1,3,4-thiadiazole (0.1 mM), Zn(OAc)$_2$·4H$_2$O (0.1 mM), KOH (0.10 mM), and H$_2$O (15 mL) was placed in a Teflon-lined stainless steel vessel, heated to 140°C for 4 days, and then cooled to room temperature over 24 h. Colorless block-shaped crystals of the title compound were obtained.

**[Zn(SBu$^t_3$C$_6$H$_2$-2,4,6)$_2$]$_2$(NC$_5$H$_4$CHO) or C$_{42}$H$_{63}$NOS$_2$Zn.** This compound melts at 164°C–166°C (Bochmann et al. 1994). To obtain it, to a stirred suspension of Zn(SBu$^t_3$C$_6$H$_2$-2,4,6)$_2$]$_2$ (0.3 mM) in 10 mL of petroleum ether at room temperature was added pyridinecarboxaldehyde (0.37 mM). Stirring was continued, and the pale-pink precipitate was filtered off, washed with petroleum ether (3 × 10 mL), and dried in vacuo. All reactions were carried out under an Ar atmosphere using standard vacuum-line techniques.

**[Zn$_{10}$S$_7$(Py)$_9$(SO$_4$)$_3$]·3H$_2$O or C$_{45}$H$_{51}$N$_9$O$_{15}$S$_{10}$Zn$_{10}$.** This compound crystallizes in the monoclinic structure with the lattice parameters $a = 1127.9 \pm 0.6$, $b = 2341.9 \pm 0.6$, $c = 2726 \pm 1$ pm, and $\beta = 107.74° \pm 0.02°$ and a calculated density of 1.87 g·cm$^{-3}$

(Ali et al. 1998). Its preparation was done in an atmosphere of $N_2$ using Schlenk techniques. The precursor for the preparation was a white solid obtained by thermolysis of $[NMe_4][Zn_{10}S_4(SPh)_{16}]$ at 250°C under dynamic vacuum for 1 h. This precursor was treated with the solution (12 mL) of 10 mM $Na_2SO_4$ in 200 mL pyridine at room temperature. Sulfur (0.2 mM of $S_8$) was added with stirring. The mixture was stirred and heated to ca. 70°C for about 2 min until all of the sulfur dissolved. The clear yellow solution was cooled to ambient temperature, layered in air with acetone (2 mL), and allowed to stand at room temperature. Colorless blocky crystals grew during 2 days.

**$[Zn(H_2O)_4(C_{10}H_8N_2)]·[Zn_2(H_2O)_4(C_{10}H_8N_2)_2(C_8H_3SO_7)_2]·4H_2O$ or $C_{46}H_{54}N_6O_{26}S_2Zn$.** This compound crystallizes in the monoclinic structure with the lattice parameters $a = 1941.8 ± 0.3$, $b = 902.4 ± 0.1$, $c = 3169.6 ± 0.4$ pm, and $\beta = 96.708° ± 0.002°$ (Zhao et al. 2009). It was prepared by the following procedure. A mixture of 5-sulfoisophthalic acid monosodium salt (0.1 mM), $Zn(OAc)_2$ (0.1 mM), 2,2'-bipy (0.1 mM), and $H_2O$(10 mL) was stirred for about 30 min. The resulting solution was sealed in a Teflon-lined stainless autoclave and heated to 120°C for 3 days. The bottle was cooled to ambient temperature spontaneously. Colorless single crystals were expected by vacuum filtration and dried in air.

**$[Zn_2(C_{12}H_8N_2)_2](C_6H_4NO_2S)_4·3H_2O$ or $C_{48}H_{38}N_8O_{11}S_4Zn_2$.** This compound crystallizes in the triclinic structure with the lattice parameters $a = 1093.73 ± 0.11$, $b = 1162.01 ± 0.12$, and $c = 1313.71 ± 0.14$ pm and $\alpha = 116.100° ± 0.001°$, $\beta = 97.717° ± 0.002°$, and $\gamma = 108.652° ± 0.002°$ and a calculated density of 1.596 g·cm⁻³ (Gao et al. 2013). To obtain it, an aqueous $Zn(NO_3)_2$ solution (1.0 mM, 10 mL) was mixed with a 2-mercaptonicotinic acid solution (1.0 mM, 10 mL) dissolved by an aqueous KOH solution. Along with stirring, the pH value was adjusted to 7.53 using a HCl solution (1 M), and then the mixture continuously reacted at room temperature. After about 4 h of fully blending, an EtOH solution of Phen (1.0 mM, 10 mL) was added. It was mixed constantly for yet another 4 h; the final pH value of the solution was 7.38, and then it was slowly evaporated at normal temperature. Four weeks later, numerous colorless columnar crystals were obtained from the filtrated concentrated solution.

**$[Zn(C_{12}H_{13}O_4)_2(C_{12}H_8N_4S)]·C_{12}H_8N_4S$ or $C_{48}H_{42}N_8O_8S_2Zn$.** This compound crystallizes in the monoclinic structure with the lattice parameters $a = 720.30 ± 0.09$, $b = 1785.5 ± 0.2$, $c = 3709.2 ± 0.4$ pm, and $\beta = 94.220° ± 0.002°$ (Xu and Tang 2010). To obtain it, a mixture of 5-$t$-butylisophtalic acid (0.2 mM), 2,5-bis(4-pyridyl)-1,3,4-thiadiazole (0.2 mM), $Zn(OAc)_2·2H_2O$ (0.1 mM), KOH (0.1 mM), and $H_2O$ (16 mL) was placed in a Teflon-lined stainless steel vessel, heated to 177°C for 3 days, and then cooled to room temperature over 24 h. Colorless block-shaped crystals were obtained.

**$[Zn_2(C_{12}H_{15}N_2S_2)_4]·2H_2O$ or $C_{48}H_{64}N_8O_2S_8Zn_2$.** This compound loses the lattice $H_2O$ at 44°C and melts at 217°C (Cesur et al. 2001). To obtain it, $Zn(NO_3)_2$ (5 mM) dissolved in distilled $H_2O$ (50 mL) was added with stirring to a solution of potassium benzylpiperazine dithiocarbamate (10 mM) in $H_2O$ (50 mL). The title compound precipitated immediately as white polycrystalline solid and was filtered by suction, washed with $H_2O$ and diethyl ether, and dried in air.

**[Zn$_3$(H$_2$O)$_6$(C$_{10}$H$_8$N$_2$)$_3$(C$_9$H$_6$SNO$_6$)$_2$]·8H$_2$O or C$_{48}$H$_{64}$N$_8$O$_{26}$S$_2$Zn$_3$.** This compound crystallizes in the monoclinic structure with the lattice parameters $a = 2496.2 \pm 0.2$, $b = 1504.4 \pm 0.1$, $c = 1734.2 \pm 0.1$ pm, and $\beta = 117.100° \pm 0.001°$ (Wang 2011). It was synthesized by adding Zn(OAc)$_2$·2H$_2$O (0.25 mm) to 10 mL aqueous solution containing $N$-[(4-carboxyphenyl)sulfonyl]glycine acid (0.5 mM) and 4,4′-bipy (0.25 mM). The mixture was stirred for 8 h at 70°C and then filtered off. The colorless filtrate was allowed to stand at room temperature. Colorless crystals suitable for XRD experiment were obtained after 10 days.

**[Zn(C$_{21}$H$_{16}$N$_3$O$_4$S$_2$)$_2$(Phen)] or C$_{54}$H$_{40}$N$_8$O$_8$S$_4$Zn.** This compound decomposes at 210°C (Prakasam et al. 2007). It was prepared by adding a hot solution of Phen (1 mM) in benzene to a hot suspension of [Zn(C$_{21}$H$_{16}$N$_3$O$_4$S$_2$)$_2$] (1 mM) in benzene/chloroform (1:1 volume ratio) mixture. The resulting pale-yellow solution was filtered and cooled. The yellow compound separated out, which was filtered and recrystallized from benzene/chloroform (1:1 volume ratio) mixture.

**C$_{54}$H$_{44}$N$_4$O$_8$S$_2$Zn$_2$.** This compound crystallizes in the monoclinic structure with the lattice parameters $a = 1329.9 \pm 0.4$, $b = 1201.9 \pm 0.3$, $c = 1532.0 \pm 0.4$ pm, and $\beta = 93.256° \pm 0.004°$ (Shi et al. 2013). To obtain it, a mixture of 2-[3-(carboxymethyl)-4-(phenylthio)phenyl] propanoic acid (1 mM), 2,2′-bipy (1 mM), ZnCl$_2$ (1 mM), H$_2$O (10 mL), EtOH (5 mL), and DMF (5 mL) was placed in a Teflon-lined stainless steel vessel (25 mL). Then an aqueous solution of NaOH was added dropwise until the pH value reached 6.0. The sealed reactor was kept under autogenous pressure at 130°C for 72 h and then slowly cooled to room temperature at a rate of 10°C per hour. Colorless needles of the title complex were obtained.

**[Zn{NH(CH$_2$)$_4$O}{S$_2$CNEt$_2$}]$_4$·NH(CH$_2$)$_4$O·C$_2$H$_4${N(CH$_2$)$_4$O}$_2$ or C$_{70}$H$_{145}$N$_{15}$O$_7$S$_{16}$Zn$_4$.** This compound crystallizes in the triclinic structure with the lattice parameters $a = 1043.00 \pm 0.06$, $b = 1214.73 \pm 0.08$, and $c = 2022.3 \pm 0.1$ pm and $\alpha = 97.443° \pm 0.001°$, $\beta = 96.825° \pm 0.001°$, and $\gamma = 94.907° \pm 0.001°$ and a calculated density of 1.379 g·cm$^{-3}$ (Ivanov et al. 2007a). It was prepared by the following procedure. [Zn$_2$(S$_2$CNEt$_2$)$_4$] (0.69 mM) was dissolved at 40°C to 50°C in 1,2-dichloroethane (8 mL). Morpholine (0.21 mL, 2.45 mM) was added dropwise with stirring and the resulting solution was left overnight. The title compound was crystallized at room temperature by slow evaporation. The crystals obtained were filtered off, dried in air, and kept in sealed glass tubes.

**[(C$_5$H$_4$NCH$_2$O)Zn(SC$_6$H$_2$Pr$^i$$_3$)]$_4$·C$_6$H$_{14}$ or C$_{90}$H$_{130}$N$_4$O$_4$S$_4$Zn$_4$.** This compound melts at 241°C and crystallizes in the monoclinic structure with the lattice parameters $a = 1857.1 \pm 0.4$, $b = 2813.8 \pm 0.6$, $c = 2056.7 \pm 0.9$ pm, and $\beta = 119.14° \pm 0.02°$ and a calculated density of 1.22 g·cm$^{-3}$ (Müller et al. 1999). To obtain it, Zn$_3$(SC$_6$H$_2$Pr$^i$$_3$)$_4$[N(SiMe$_3$)$_2$]$_2$ (0.24 mM) was dissolved in hexane (5 mL). 2-Pyridylmethanol (0.21 mM) was added. Within 12 h, the product had precipitated. The solvent was removed with a syringe, and the precipitate was washed with hexane (1 mL) and dried in vacuo. The title compound remained as colorless crystals. All reactions were carried out in an atmosphere of N$_2$ using carefully dried solvents.

**[BzEt$_3$N]$_2$[Zn$_8$(S)(SBz)$_{16}$]·5MeOH** or **C$_{143}$H$_{176}$N$_2$O$_5$S$_{17}$Zn$_8$**. This compound crystallizes in the triclinic structure with the lattice parameters $a = 1637.9 \pm 0.3$, $b = 1676.2 \pm 0.3$, and $c = 2815.4 \pm 0.6$ pm and $\alpha = 106.82° \pm 0.03°$, $\beta = 100.11° \pm 0.03°$, and $\gamma = 91.14° \pm 0.03°$ and a calculated density of 1.37 g·cm$^{-3}$ (Burth et al. 1998). To obtain it, sodium methoxide (21.3 mL [4.25 mM] of a 0.20 M methanol solution) and benzyl mercaptan (4.25 mM) were combined in MeOH (75 mL). Then Zn(NO$_3$)$_2$·4H$_2$O (1.42 mM) in MeOH (50 mL) was added with vigorous stirring over a period of 2 h. Finally, a solution of [BzEt$_3$N] (1.42 mM) in MeOH (20 mL) was added. The clear solution was slowly reduced to 30 mL in vacuo in a warm water bath, filtered through a fine porosity frit, and left to stand under an inert atmosphere. Over a period of 4 weeks, the solution slowly turned yellow and formed a partly crystalline precipitate of C$_{143}$H$_{176}$N$_2$O$_5$S$_{17}$Zn$_8$.

## 1.54 Zn–H–C–N–O–F–S

Some compounds are formed in this system.

**[(C$_5$H$_4$NCHO)Zn(SC$_6$F$_5$)$_2$]** or **C$_{18}$H$_5$NOF$_{10}$S$_2$Zn**. This compound melts at 110°C with decomposition (Müller et al. 1999). It was prepared by the next procedure. A solution of pyridine-2-carbaldehyde (1.40 mM) in diethyl ether (10 mL) was added dropwise with stirring to a solution of Zn(SC$_6$F$_5$)$_2$ (1.38 mM) in diethyl ether (40 mL). Precipitation started, which was completed by the addition of hexane (75 mL). Filtration, washing with hexane, and drying in vacuo yielded the title compound as a pale-yellow powder. All reactions were carried out in an atmosphere of N$_2$ using carefully dried solvents.

**[(OCHC$_5$H$_3$NCHO)Zn(SC$_6$F$_5$)$_2$]** or **C$_{19}$H$_5$NO$_2$F$_{10}$S$_2$Zn**. This compound melts at 135°C with decomposition (Müller et al. 1999). To obtain it, a solution of pyridine-2,6-dicarbaldehyde (1.33 mM) in diethyl ether (10 mL) was added dropwise with stirring to a solution of Zn(SC$_6$F$_5$)$_2$ (1.3 mM) in diethyl ether (40 mL). Precipitation started, which was completed by the addition of hexane (75 mL). Filtration, washing with hexane, and drying in vacuo yielded the title compound as a yellow powder. All reactions were carried out in an atmosphere of N$_2$ using carefully dried solvents.

**[(MeC$_5$H$_3$NCHO)Zn(SC$_6$F$_5$)$_2$]** or **C$_{19}$H$_7$NOF$_{10}$S$_2$Zn**. This compound melts at 135°C with decomposition and crystallizes in the orthorhombic structure with the lattice parameters $a = 1276.3 \pm 0.8$, $b = 1405.8 \pm 0.7$, and $c = 1160.2 \pm 0.8$ pm and a calculated density of 1.87 g·cm$^{-3}$ (Müller et al. 1999). To obtain it, a solution of picoline-2-carbaldehyde (1.49 mM) in diethyl ether (10 mL) was added dropwise with stirring to a solution of Zn(SC$_6$F$_5$)$_2$ (1.51 mM) in diethyl ether (40 mL). Precipitation started, which was completed by the addition of hexane (75 mL). Filtration, washing with hexane, and drying in vacuo yielded the title compound as a pale-yellow powder. Its crystals were grown from a solution of THF/CH$_2$Cl$_2$ by layering with hexane. All reactions were carried out in an atmosphere of N$_2$ using carefully dried solvents.

**[(OMeC$_5$H$_3$NCHO)Zn(SC$_6$F$_5$)$_2$]** or **C$_{19}$H$_7$NO$_2$F$_{10}$S$_2$Zn**. This compound melts at 127°C and crystallizes in the triclinic structure with the lattice parameters $a = 874.6 \pm 0.2$, $b = 877.7 \pm 0.2$, and $c = 1378.2 \pm 0.3$ pm and $\alpha = 100.22° \pm 0.03°$,

$\beta = 92.17° \pm 0.03°$, and $\gamma = 95.12° \pm 0.03°$ at 173 K and a calculated density of 1.93 g·cm$^{-3}$, (Müller et al. 1999). It was prepared by the following procedure. A solution of 2-methoxypyridine-6-carbaldehyde (0.68 mM) in diethyl ether (10 mL) was added dropwise with stirring to a solution of Zn(SC$_6$F$_5$)$_2$ (0.68 mM) in diethyl ether (40 mL). Precipitation started, which was completed by the addition of hexane (75 mL). Filtration, washing with hexane, and drying in vacuo yielded the title compound as a pale-yellow powder. Its crystals were grown from a solution of CH$_2$Cl$_2$ by layering with hexane. All reactions were carried out in an atmosphere of N$_2$ using carefully dried solvents.

**[(MeC$_5$H$_3$NCH$_2$OH)Zn(SC$_6$F$_5$)$_2$]** or **C$_{19}$H$_9$NOF$_{10}$S$_2$Zn**. This compound melts at 108°C and crystallizes in the monoclinic structure with the lattice parameters $a = 1388.7 \pm 0.3$, $b = 1056.2 \pm 0.2$, $c = 1535.3 \pm 0.3$ pm, and $\beta = 115.46° \pm 0.03°$ and a calculated density of 1.92 g·cm$^{-3}$ (Müller et al. 1999). It was prepared by the next procedure. Zn(SC$_6$F$_5$)$_2$ (0.60 mM) was dissolved in chloroform (10 mL) and 6-picolylmethanol (0.60 mM) was added with stirring. After 30 min, the volume of the solution was reduced to 3 mL in vacuo. The addition of hexane (10 mL) produced a precipitate, which was filtered off, washed with hexane, and dried in vacuo. Its crystals were grown from a solution of CCl$_4$/CH$_2$Cl$_2$ by layering with hexane. All reactions were carried out in an atmosphere of N$_2$ using carefully dried solvents.

**[Zn(C$_{14}$H$_{21}$N$_4$S)(SC$_6$F$_5$)]·H$_2$O** or **C$_{20}$H$_{23}$N$_4$OF$_5$S$_2$Zn**. This compound crystallizes in the monoclinic structure with the lattice parameters $a = 856.3 \pm 0.3$, $b = 3370.1 \pm 0.5$, $c = 945.4 \pm 0.3$ pm, and $\beta = 93.58° \pm 0.02°$ (Hammesa and Carrano 2000). To prepare it, a solution of [MeZn(C$_{14}$H$_{21}$N$_4$S)] (0.48 mM) in 30 mL of CH$_2$Cl$_2$ was treated by one equivalent of HSC$_6$F$_5$ (0.48 mM). The resulting solution was stirred for 1 h, concentrated under reduced pressure, and crystallized by layering the concentrated CH$_2$Cl$_2$ solution with hexane to give the title compound as colorless crystals.

**[(C$_6$H$_4$NMe$_2$CHO)Zn(SC$_6$F$_5$)$_2$]** or **C$_{21}$H$_{11}$NOF$_{10}$S$_2$Zn**. This compound melts at 110°C with decomposition and crystallizes in the monoclinic structure with the lattice parameters $a = 1106.4 \pm 0.4$, $b = 751.0 \pm 0.2$, $c = 1377.9 \pm 0.3$ pm, and $\beta = 90.60° \pm 0.02°$ and a calculated density of 1.78 g·cm$^{-3}$ (Müller et al. 1999). To obtain it, a solution of N-dimethylaniline-2-carbaldehyde (1.21 mM) in diethyl ether (10 mL) was added dropwise with stirring to a solution of Zn(SC$_6$F$_5$)$_2$ (1.21 mM) in diethyl ether (40 mL). A few drops of acetone were added to improve the miscibility of reagents. Precipitation started, which was completed by the addition of hexane (75 mL). Filtration, washing with hexane, and drying in vacuo yielded the title compound as an orange powder. Its crystals were grown from a solution of CH$_2$Cl$_2$ by layering with hexane. All reactions were carried out in an atmosphere of N$_2$ using carefully dried solvents.

**[(C$_6$H$_4$NMe$_2$CH$_2$OH)Zn(SC$_6$F$_5$)$_2$]** or **C$_{21}$H$_{13}$NOF$_{10}$S$_2$Zn**. This compound melts at 133°C (Müller et al. 1999). It was prepared by the next procedure. Zn(SC$_6$F$_5$)$_2$ (2.11 mM) was dissolved in diethyl ether (45 mL) and a few drops of THF. A solution of 2-dimethylaminobenzylalcohole (2.12 mM) in diethyl ether (5 mL) was added dropwise with stirring. Upon the addition of hexane (200 mL), the product was precipitated, which was filtered off, washed with hexane, and dried in vacuo.

The title compound remained as a colorless powder. All reactions were carried out in an atmosphere of $N_2$ using carefully dried solvents.

**[(C₉H₆NCHO)Zn(SC₆F₅)₂] or C₂₂H₇NOF₁₀S₂Zn.** This compound melts at 145°C with decomposition (Müller et al. 1999). To obtain it, a solution of quinoline-2-carbaldehyde (0.46 mM) in diethyl ether (10 mL) was added dropwise with stirring to a solution of $Zn(SC_6F_5)_2$ (0.46 mM) in diethyl ether (40 mL). Precipitation started, which was completed by the addition of hexane (75 mL). Filtration, washing with hexane, and drying in vacuo yielded the title compound as an orange powder. All reactions were carried out in an atmosphere of $N_2$ using carefully dried solvents.

**[(MeC₃H₂N₂CHO)₂Zn(SC₆F₅)₂] or C₂₂H₁₂N₄O₂F₁₀S₂Zn.** This compound melts at 166°C and crystallizes in the triclinic structure with the lattice parameters $a = 771.2 \pm 0.2$, $b = 1170.2 \pm 0.2$, and $c = 1524.7 \pm 0.3$ pm and $\alpha = 81.94° \pm 0.03°$, $\beta = 78.75° \pm 0.03°$, and $\gamma = 73.50° \pm 0.03°$ and a calculated density of 1.76 g·cm⁻³ (Müller et al. 1999). To obtain it, a solution of N-methylimidazole-2-carbaldehyde (1.80 mg) in THF (5 mL) was added dropwise with stirring to a solution of $Zn(SC_6F_5)_2$ (0.90 mM) in diethyl ether (5 mL). After stirring overnight, the precipitate was filtered off, washed with hexane, and dried in vacuo. The title compound remained as a colorless powder. Its crystals were grown from a solution of $CH_2Cl_2$ by layering with hexane. All reactions were carried out in an atmosphere of $N_2$ using carefully dried solvents.

**[Zn(C₈H₄O₂F₃S)₂(C₅H₅N)₂] or C₂₆H₁₈N₂O₄F₆S₂Zn.** This compound crystallizes in the triclinic structure with the lattice parameters $a = 1457.1 \pm 0.3$, $b = 1023.2 \pm 0.3$, and $c = 1278.8 \pm 0.4$ pm and $\alpha = 132.73° \pm 0.21°$, $\beta = 92.82° \pm 0.18°$, and $\gamma = 91.66° \pm 0.19°$ and experimental and calculated densities $1.61 \pm 0.10$ and 1.587 g·cm⁻³, respectively (Pretorius and Boeyens 1978). It was obtained by recrystallization of zinc–thenoyltrifluoroacetone–chelate from Py. These crystals are unstable in the atmosphere.

## 1.55   Zn–H–C–N–O–F–Cl–S

**[(C₉H₈NCH₂O)Zn(SC₆F₅)]₄·2CH₂Cl₂ or C₆₆H₃₆N₄O₄Cl₄S₄Zn₄** multinary compound is formed in this system. It melts at 211°C and crystallizes in the tetragonal structure with the lattice parameters $a = 1380.6 \pm 0.2$ and $c = 3625.4 \pm 0.7$ pm and a calculated density of 1.79 g·cm⁻³ (Müller et al. 1999). To obtain it, a solution of 2-quinolylmethanol (0.43 mM) in MeOH (5 mL) was slowly added to a solution of $Zn(SC_6F_5)_2$ (0.43 mM) in MeOH (10 mL) with stirring. After 2 h, the solvent was removed in vacuo, the residue picked up in a minimum amount of $CH_2Cl_2$, and this was layered with hexane (10 mL). The title compound was precipitated as big yellow crystals. All reactions were carried out in an atmosphere of $N_2$ using carefully dried solvents.

## 1.56   Zn–H–C–N–O–Cl–S

Some compounds are formed in this system.

**[Zn(CH₅N₃S)₂(ClO₄)₂] or C₂H₁₀N₆O₈Cl₂S₂Zn.** This compound decomposes at 270°C (Mahadevappa and Murthy 1972). It was obtained by mixing aqueous solutions of thiosemicarbazide and $Zn(ClO_4)_2$ in the molar ratio 2:1. The resulting solution was slowly evaporated at 60°C on a water bath and then cooled in ice, whereupon crystals of the title compound appeared.

**$[Zn(C_3H_5NOS)_2Cl_2]$ or $C_6H_{10}N_2O_2Cl_2S_2Zn$.** This compound decomposes at 140°C–142°C (Preti et al. 1974). It was obtained by a reaction of $ZnCl_2$ and thiazolidine-2-one in refluxing EtOH for 12 h. The title compound was separated out during crystallization.

**$[Zn(C_2Cl_3O_2)_2(CH_4N_2S)_2]·H_2O$ or $C_6H_{10}N_4O_5Cl_6S_2Zn$.** This compound crystallizes in the monoclinic structure with the lattice parameters $a = 597.2 \pm 0.2$, $b = 2310.3 \pm 1.1$, $c = 1440.0 \pm 0.5$ pm, and $\beta = 102.91° \pm 0.03°$ and a calculated density of 1.922 g·cm$^{-3}$ (Potočňák and Dunaj-Jurčo 1994). It was prepared by mixing a freshly prepared suspension of $Zn(OH)_2$ with an aqueous solution of trichloroacetic acid and an aqueous solution of thiourea in a 1:2:2 molar ratio. After several days, colorless needles of the title compound were filtered off, washed with water and ethanol, and dried in air.

**$[Zn(C_5H_4NC_2H_5N_4S)Cl_2]·H_2O$ or $C_7H_{11}N_5OCl_2SZn$.** This compound melts at a temperature higher than 300°C and crystallizes in the monoclinic structure with the lattice parameters $a = 1033.125 \pm 0.007$, $b = 921.33 \pm 0.04$, $c = 1715.05 \pm 0.06$ pm, and $\beta = 126.239° \pm 0.004°$ and a calculated density of 1.766 g·cm$^{-3}$ (Castiñeiras et al. 2000). To obtain it, a solution of $ZnCl_2$ (2 mM) in EtOH (30 mL) was mixed with a solution of 2-pyridineformamide thiosemicarbazone ($C_5H_4NC_2H_5N_4S$) (2 mM) in the same solvent (30 mL), and the mixture was stirred under reflux for 7 days. Crystals suitable for XRD were grown from the solution of the title compound in 96% EtOH.

**$[ZnCl_2(C_4H_7NOS)_2]$ or $C_8H_{14}N_2O_2Cl_2S_2Zn$.** This compound melts at 116°C–118°C (De Filippo et al. 1971). To obtain it, $ZnCl_2$ was dissolved in an excess of molten thiomorpholine-3-one ($C_4H_7NOS$) (at ca. 105°C; molar ratio $Zn^{2+}/C_4H_7NOS$ ca. 1:2), and the crude product was purified by washing with ethyl ether and recrystallizing from EtOH. The title compound could also be prepared if different amounts of $C_4H_7NOS$ for $Zn^{2+}/C_4H_7NOS$ molar ratios of 1:1, 1:2, 1:3, and 1:4 were added to a solution of $ZnCl_2$ in absolute EtOH, and the mixture was heated under reflux for 4 h. $[ZnCl_2(C_4H_7NOS)_2]$ was separated out by crystallization.

**$[Zn(C_{11}H_{10}N_2S_2)Cl_2]·H_2O$ or $C_{11}H_{12}N_2OCl_2S_2Zn$.** This compound melts at 215°C and crystallizes in the monoclinic structure with the lattice parameters $a = 1010.20 \pm 0.03$, $b = 1099.34 \pm 0.04$, $c = 1420.84 \pm 0.04$ pm, and $\beta = 103.6248° \pm 0.0013°$ (Amoedo-Portela et al. 2002a). To obtain it, a solution of bis(2-pyridylthio)methane (0.3 mM) in 10 mL of an EtOH/$CH_2Cl_2$ (1:1 volume ratio) mixture was added dropwise to a solution of $ZnCl_2$ (0.3 mM) in 10 mL of EtOH, and the mixture was refluxed for 3 h. Upon cooling, a crystalline product formed, which was filtered out and dried in vacuo.

**$[Zn(SC_5H_9NHMe)Cl_2]_2·H_2O$ or $C_{12}H_{18}N_2OCl_4S_2Zn_2$.** This compound, which crystallizes in the monoclinic structure with the lattice parameters $a = 1098.8 \pm 1.2$, $b = 1308.6 \pm 1.2$, and $c = 747.6 \pm 0.9$ pm and $\beta = 91.99° \pm 0.01°$ and experimental and calculated densities 1.69 and 1.71 g·cm$^{-3}$, respectively (Briansó et al. 1981), is formed in this system. To obtain it, a filtered solution of $ZnCl_2$ (52.8 mM) in $H_2O$ (50 mL) was added to a solution of 4-mercapto-$N$-methylpiperidinium chloride (12.5 mM) in $H_2O$ (25 mM). The solution was placed in refrigerator at 5°C for several days. The white crystals were filtered off, washed with cold water, EtOH, and diethyl ether, and dried in vacuo over silica gel.

**[Zn{Zn(SCH₂CHNH₃CO₂)₂}₂][ZnCl₄]·4H₂O** or **C₁₂H₂₄N₄O₈Cl₄S₄Zn₄·4H₂O, tetrakis(L-cysteinato)trizinc tetrachlorozincate**. To obtain this compound, L-cysteine (0.04 mol) was dissolved in 50 mL of warm $H_2O$, and to it was added with stirring a warm solution of $ZnCl_2$ (0.04 mol) in a small amount of $H_2O$ (Shindo and Brown 1965). After 2 h, formation of white crystals was observed. The pH of the solution was around 2.0. Stirring was continued for an additional 3 h; the crystals were filtered and washed with $H_2O$ and absolute EtOH and dried in vacuo over $CaCl_2$. The anhydrous compound was obtained by drying in vacuo at 70°C over $P_2O_5$ under continuous pumping. This compound was converted to $Zn[Zn(SCH_2CHNH_2CO_2)_2]$ by heating in water at 100°C for 1 h.

**[Zn(C₄H₈N₂S)₄(ClO₄)₂]** or **C₁₆H₃₂N₈O₈Cl₂S₄Zn**. This compound melts at 148°C (Raper and Brook 1977). It was prepared by dissolving $Zn(ClO_4)_2·6H_2O$ (2.5 mM) in minimum volume of hot anhydrous EtOH containing 3 mL triethyl orthoformate as dehydrating agent. To this solution, 10.0 mM of 1-methylimidazoline-2-thione ($C_4H_8N_2S$) dissolved in 10 mL of hot anhydrous EtOH was added. The resulting solution refluxed for 3 h. The solvent was allowed to evaporate slowly after cooling to room temperature, and the crystalline product obtained was removed by filtration, washed, and dried at room temperature.

**[Zn(C₉H₁₁N₃O₂S)₂Cl₂]** or **C₁₈H₂₂N₆O₄Cl₂S₂Zn**. This compound was prepared by the slow cooling of a mixture of hot ethanolic solutions of stoichiometric quantities of $ZnCl_2$ and vanillin thiosemicarbazone ($C_9H_{11}N_3O_2S$) (1:2 molar ratio) (Ablov and Gerbeleu 1964). White crystalline precipitate in the form of rhombic plates was obtained, which was filtered off, washed sometimes with EtOH and then with ether, and dried in air.

**[Zn(PySH)₄](ClO₄)₂** or **C₂₀H₂₀N₄O₈Cl₂S₄Zn**. This compound crystallizes in the triclinic structure with the lattice parameters $a = 805.22 \pm 0.01$, $b = 1323.78 \pm 0.02$, and $c = 1331.39 \pm 0.01$ pm and $\alpha = 98.709° \pm 0.001°$, $\beta = 93.555° \pm 0.001°$, and $\gamma = 96.872° \pm 0.001°$ and a calculated density of 1.696 g·cm⁻³ (Anjali et al. 2003). It was prepared by the following procedure. To a solution of 4-pyridine thiol (3.70 mM) in 4 mL of MeOH was added $Zn(ClO_4)_2·4H_2O$ (1.01 mM) in 2 mL of MeOH. The mixture was stirred for 15 min and then left for slow evaporation at room temperature to get bright-yellow crystals, which were separated by filtration, washed with $Et_2O$, and dried in vacuo.

**[Zn(ClO₄)₂(C₁₀H₈N₂S₂)₂·2MeOH]ₙ** or **C₂₂H₂₄N₄O₁₀Cl₂S₄Zn**. This compound was prepared by the following procedure (Horikoshi and Mikuriya 2005). To a solution of 4,4′-bipyridyl disulfide ($C_{10}H_8N_2S_2$) (0.02 mM) in MeOH (0.5 mL) was added $Zn(ClO_4)_2$ (0.01 mM) in MeOH (0.5 mL). A white precipitate formed immediately. The reaction mixture was stirred for 30 min, and then the precipitate was filtered off and washed with MeOH ($3 \times 0.5$ mL). Extraction of the precipitate with MeOH followed by the addition of diethyl ether afforded a white solid of the title compound.

**[Zn(ClO$_4$)$_2$(C$_{10}$H$_8$N$_2$S$_2$)$_2$(DMSO)$_2$]$_n$ or C$_{24}$H$_{28}$N$_4$O$_{10}$Cl$_2$S$_6$Zn.** This compound crystallizes in the tetragonal structure with the lattice parameters $a = 1297.9 \pm 0.6$ and $c = 1081.7 \pm 0.7$ pm and a calculated density of 1.569 g·cm$^{-3}$ (Horikoshi and Mikuriya 2005). [Zn(ClO$_4$)$_2$(C$_{10}$H$_8$N$_2$S$_2$)$_2$·2MeOH]$_n$ was recrystallized from DMSO to give the title compound as transparent crystals.

**[Zn{S$_2$CN(CH$_2$)$_4$}$_2$·(2,9-Me$_2$Phen)]·CHCl$_3$·0.5H$_2$O or C$_{25}$H$_{30}$N$_4$O$_{0.5}$Cl$_3$S$_4$Zn.** This compound melts at 236°C–237°C and crystallizes in the triclinic structure with the lattice parameters $a = 1007.06 \pm 0.08$, $b = 1072.60 \pm 0.09$, and $c = 1409.47 \pm 0.12$ pm and $\alpha = 83.526° \pm 0.002°$, $\beta = 89.515° \pm 0.002°$, and $\gamma = 89.855° \pm 0.002°$ (Lai and Tiekink 2003c). Pale-yellow crystals were isolated from acetonitrile/chloroform (1:2 volume ratio) solution containing equimolar amounts of [Zn{S$_2$CN(CH$_2$)$_4$}$_2$] and (2,9-Me$_2$Phen).

**[(mepa)$_4$Zn$_3$](ClO$_4$)$_2$·2H$_2$O or C$_{32}$H$_{48}$N$_8$O$_{10}$Cl$_2$S$_4$Zn$_3$ (mepaH = $N$-(2-mercapto-ethyl)picolylamin, C$_8$H$_{12}$N$_2$S).** This compound melts at 220°C with decomposition (Brand and Vahrenkamp 1995). To obtain it, mepaH (1.80 mM) and NaOH (1.80 mM) were dissolved in H$_2$O (40 mL). A solution of Zn(ClO$_4$)$_2$·6H$_2$O (1.35 mM) in H$_2$O (15 mL) was added dropwise with stirring. The resulting precipitate was stirred for 30 min and then filtered off, washed with a few mL of H$_2$O, and dried in vacuo. Recrystallization from H$_2$O/MeOH (1:1 volume ratio) yielded [(mepa)$_4$Zn$_3$](ClO$_4$)$_2$·2H$_2$O as colorless crystals.

**[Zn$_2$(C$_8$H$_{18}$N$_2$S$_2$)$_2$Cl$_2$]$_2$·2H$_2$O or C$_{32}$H$_{76}$N$_8$O$_2$Cl$_4$S$_8$Zn$_4$.** This compound crystallizes in the monoclinic structure with the lattice parameters $a = 1187.4 \pm 0.3$, $b = 1026.0 \pm 0.2$, $c = 1426.0 \pm 0.3$ pm, and $\beta = 113.92° \pm 0.01°$ and experimental and calculated densities $1.778 \pm 0.001$ and 1.782 g·cm$^{-3}$, respectively (Hu et al. 1973). To obtain it, $N,N'$-dimethyl-$N,N'$-bis($\beta$-mercaptoethyl)ethylenediamine was synthesized by mercaptoethylation of $N,N'$-dimethylenediamine. To a resulting mixture of 1.5 mol of obtained diamine in 500 mL of anhydrous toluene, 1 mol of ethylene monothiocarbamate was added dropwise over a 15 min period. Refluxing was continued for 15 h under an efficient condenser. The cooled solution was washed with two 100 mL portions of H$_2$O and dried over anhydrous MgSO$_4$. The toluene was removed by distillation under reduced pressure, leaving the crude product.

The reaction of ZnCl$_2$ in a solution of MeOH with $N,N'$-dimethyl-$N,N'$-bis($\beta$-mercaptoethyl)ethylenediamine in ratio of 2:1, followed by recrystallization from H$_2$O, yielded also prismatic crystals of Zn$_2$(C$_8$H$_{18}$N$_2$S$_2$)$_2$·2H$_2$O (Hu et al. 1973).

**[Zn(H$_2$O)$_2$(C$_{12}$H$_8$N$_2$)$_2$](H$_2$NC$_6$H$_3$ClSO$_3$)$_2$·2H$_2$O       or       C$_{36}$H$_{34}$N$_6$O$_{10}$Cl$_2$S$_2$Zn$_4$.** This compound crystallizes in the triclinic structure with the lattice parameters $a = 1082.71 \pm 0.09$, $b = 1248.0 \pm 0.1$, and $c = 1642.7 \pm 0.1$ pm and $\alpha = 75.019° \pm 0.001°$, $\beta = 86.335° \pm 0.001°$, and $\gamma = 68.937° \pm 0.001°$ (Han et al. 2010). It was prepared by the following procedure. A mixture of ZnO (0.5 mM), 1,10-Phen (0.5 mM), and 3-amino-4-chlorobenzenesulphonic acid (1.0 mM) in H$_2$O (20 mL) was sealed into a 25 mL Teflon-lined stainless steel reactor and heated at 170°C for 1 day.

The reactor was then cooled slowly to room temperature, and the solution was filtered. Red block-shaped crystals were obtained from the filtrate after several days at room temperature.

**[Zn(ClO₄)₂(C₁₀H₈N₂S₂)₂(H₂O)₂·(C₁₀H₈N₂S₂)₄]ₙ** or **C₆₀H₅₂N₁₂O₁₀Cl₂S₁₂Zn**. This compound crystallizes in the monoclinic structure with the lattice parameters $a = 3596.2 \pm 0.4$, $b = 1085.46 \pm 0.14$, $c = 2044.9 \pm 0.2$ pm, and $\beta = 115.208° \pm 0.006°$ (Horikoshi and Mikuriya 2005). $[Zn(ClO_4)_2(C_{10}H_8N_2S_2)_2 \cdot 2MeOH]_n$ was recrystallized from distilled $H_2O$ to give the title compound as transparent crystals.

## 1.57 Zn–H–C–N–O–Br–S

Some compounds are formed in this system.

**[Zn(C₃H₅NOS)₂Br₂]** or **C₆H₁₀N₂O₂Br₂S₂Zn**. This compound decomposes at 119°C–121°C (Preti et al. 1974). It was obtained by the reaction of $ZnBr_2$ and thiazolidine-2-one in refluxing EtOH for 12 h. The title compound was separated out during crystallization.

**[ZnBr₂(C₄H₇NOS)₂]** or **C₈H₁₄N₂O₂Br₂S₂Zn**. This compound melts at 122°C–124°C (De Filippo et al. 1971). To obtain it, $ZnBr_2$ was dissolved in an excess of molten thiomorpholine-3-one ($C_4H_7NOS$) (at ca. 105°C; molar ratio $Zn^{2+}/C_4H_7NOS$ ca. 1:2), and the crude product was purified by washing with ethyl ether and recrystallizing from EtOH. The title compound could also be prepared if to a solution of $ZnBr_2$ in absolute EtOH was added different amounts of $C_4H_7NOS$ for $Zn^{2+}/C_4H_7NOS$ molar ratios of 1:1, 1:2, 1:3, and 1:4, and the mixture was heated under reflux for 4 h. $[ZnBr_2(C_4H_7NOS)_2]$ was separated out by crystallization.

**[Zn(C₅H₄NC₂H₅N₄S)Br₂]·(DMSO)** or **C₉H₁₅N₅OBr₂SZn**. This compound crystallizes in the monoclinic structure with the lattice parameters $a = 738.58 \pm 0.06$, $b = 1896.41 \pm 0.11$, $c = 1206.20 \pm 0.07$ pm, and $\beta = 101.460° \pm 0.004°$ and a calculated density of 2.000 g·cm⁻³ (Castiñeiras et al. 2000). It was obtained by the recrystallization of $[Zn(C_5H_4NC_2H_5N_4S)Br_2]$ from DMSO.

**[Zn{Zn(SCH₂CHNH₃CO₂)₂}₂][ZnBr₄]** or **C₁₂H₂₄N₄O₈Br₄S₄Zn₄, tetrakis(L-cysteinato)trizinc tetrabromozincate**. This compound was isolated with some difficulty from a concentrated solution of L-cysteine (0.02 mol) and $ZnBr_2$ (0.02 mol) in 20 mL of $H_2O$ (Shindo and Brown 1965). Stirring was continued for several hours; several drops of 2.0 N NaOH solution was added to promote precipitation (pH 2.0). By allowing the solution to stand overnight, white crystals were deposited; these were filtered, washed several times with $H_2O$ and absolute EtOH, and dried in vacuo over $CaCl_2$.

**[Zn₂(C₆H₁₃NS)₂Br₄]·H₂O** or **C₁₂H₂₈N₂OBr₄S₂Zn₂**. This compound crystallizes in the monoclinic structure with the lattice parameters $a = 1110$, $b = 1314$, $c = 763$ pm, and $\beta = 92.25°$ (Bayon et al. 1982). It was obtained in 10% methanolic solution by adding $ZnBr_2$ to the stoichiometric amount of 1-methyl-4-mercaptopiperidine. The white crystalline product obtained was filtered off; washed with cold water, EtOH, and ethyl ether; and dried in vacuo.

**[Zn(C$_{12}$H$_{17}$N$_5$S)Br$_2$]·(DMSO) or C$_{14}$H$_{23}$N$_5$OBr$_2$S$_2$Zn**. This compound crystallizes in the triclinic structure with the lattice parameters $a = 921.26 \pm 0.11$, $b = 951.76 \pm 0.12$, and $c = 1250.62 \pm 0.15$ pm and $\alpha = 102.756° \pm 0.003°$, $\beta = 93.101° \pm 0.003°$, and $\gamma = 94.958° \pm 0.003°$ and a calculated density of 1.771 g·cm$^{-3}$ (Ketcham et al. 2001). To obtain it, zinc perchlorate, chloride, acetate, or bromide (2 mM) in EtOH (30 mL) was mixed with a solution of 2-pyridineformamide 3-piperidylthiosemicarbazone (2 mM) in EtOH (30 mL), and the mixture was stirred under reflux for 2 h or longer. The resulting yellow crystals were filtered while the solution was warm, washed with anhydrous ether to apparent dryness, and placed on a warm plate at 35°C.

**[Zn(C$_{13}$H$_{19}$N$_5$S)Br$_2$]·EtOH or C$_{15}$H$_{25}$N$_5$OBr$_2$SZn**. This compound could be obtained if a solution of ZnBr$_2$ (0.90 mM) in EtOH (30 mL) was mixed with a solution of 2-pyridineformamide 3-hexamethyleneiminylthiosemicarbazone (C$_{13}$H$_{19}$N$_5$S) (0.90 mM) in EtOH (30 mL) (Bermejo et al. 2004b). The resulting solution was stirred at room temperature for 7 days. The pale-yellow product was filtered off.

**[Zn(C$_9$H$_7$Br$_2$NO$_4$S)(C$_{12}$H$_8$N$_2$)(H$_2$O)]·EtOH or C$_{23}$H$_{23}$N$_3$O$_6$Br$_2$SZn**. This compound crystallizes in the monoclinic structure with the lattice parameters $a = 1870.0 \pm 0.3$, $b = 1971.7 \pm 0.4$, $c = 1572.8 \pm 0.3$ pm, and $\beta = 114.45° \pm 0.01°$ and a calculated density of 1.748 g·cm$^{-3}$ (Zhang et al. 2005). To obtain it, 2-[(E)-3,5-dibromo-2-oxidobenzylideneamino]ethanesulfonic acid (2.0 mM) was dissolved in aqueous EtOH (25 mL). To this solution, Zn(OAc)$_2$·4H$_2$O (2.0 mM) was added, and the mixture was stirred and refluxed at 70°C for 8 h. Phen (2.0 mM) was then added, and the reaction was continued for another 4 h. After cooling to room temperature and filtration, the filtrate was left to stand at room temperature. Red crystals suitable for XRD were obtained.

## 1.58   Zn–H–C–N–O–I–S

Some compounds are formed in this system.

**[Zn(C$_3$H$_5$NOS)$_2$I$_2$] or C$_6$H$_{10}$N$_2$O$_2$I$_2$S$_2$Zn**. This compound decomposes at 108°C–110°C (Preti et al. 1974). It was obtained by the reaction of ZnI$_2$ and thiazolidine-2-one in refluxing EtOH for 12 h. The title compound was separated out during crystallization.

**[ZnI$_2$(C$_4$H$_7$NOS)$_2$] or C$_8$H$_{14}$N$_2$O$_2$I$_2$S$_2$Zn**. This compound melts at 128°C–130°C (De Filippo et al. 1971). To obtain it, ZnI$_2$ was dissolved in an excess of molten thiomorpholine-3-one (C$_4$H$_7$NOS) (at ca. 105°C; molar ratio Zn$^{2+}$/C$_4$H$_7$NOS ca. 1:2), and the crude product was purified by washing with ethyl ether and recrystallizing from EtOH. The title compound could also be prepared if to a solution of ZnI$_2$ in absolute EtOH were added different amounts of C$_4$H$_7$NOS for Zn$^{2+}$/C$_4$H$_7$NOS with molar ratios of 1:1, 1:2, 1:3, and 1:4, and the mixture was heated under reflux for 4 h. [ZnI$_2$(C$_4$H$_7$NOS)$_2$] was separated out by crystallization.

**[Zn$_2$(C$_6$H$_{13}$NS)$_2$I$_4$]·H$_2$O or C$_{12}$H$_{28}$N$_2$OI$_4$S$_2$Zn$_2$**. This compound crystallizes in the monoclinic structure with the lattice parameters $a = 1141$, $b = 1330$, $c = 802$ pm, and

$\beta = 92.39°$ (Bayon et al. 1982). It was obtained in 10% methanolic solution by adding $ZnI_2$ to the stoichiometric amount of 1-methyl-4-mercaptopiperidine. The white crystalline product obtained was filtered off, washed with cold water, EtOH, and ethyl ether, and dried in vacuo.

**[Zn(C$_{13}$H$_{19}$N$_5$S)I$_2$]·0.5EtOH or C$_{14}$H$_{22}$N$_5$O$_{0.5}$I$_2$SZn.** This compound could be obtained if a solution of $ZnI_2$ (0.90 mM) in EtOH (30 mL) was mixed with a solution of 2-pyridineformamide 3-hexamethyleneiminylthiosemicarbazone (C$_{13}$H$_{19}$N$_5$S) (0.90 mM) in EtOH (30 mL) (Bermejo et al. 2004b). The resulting solution was stirred at room temperature for 7 days, and the white product was filtered off.

**[Zn(C$_9$H$_{13}$N$_5$S)$_2$I$_2$]·3H$_2$O or C$_{18}$H$_{32}$N$_{10}$O$_3$I$_2$S$_2$Zn.** To obtain this compound, a solution of $ZnI_2$ (1.12 mM) in EtOH (30 mL) was mixed with a solution of 2-pyridineformamide $N$(4)-ethylthiosemicarbazone (C$_9$H$_{13}$N$_5$S) (0.90 mM) in EtOH (30 mL) (Bermejo et al. 2004a). The resulting solution was stirred at room temperature for 7 days, and after that the yellow precipitate was filtered off.

**[(Me$_4$N)$_2${Zn$_4$(SPh)$_6$I$_4$}]·Me$_2$CO or C$_{47}$H$_{60}$N$_2$OI$_4$S$_6$Zn$_4$.** To obtain this compound, solid iodine (1.2 mM) was added to a stirred solution of (Me$_4$N)$_2$[Zn$_4$(SPh)$_{10}$] (0.6 mM) in 15 mL of acetone (Dean et al. 1987). Immediate decolorization occurred, leaving a clear solution. Acetone was removed under vacuum. The residue was extracted with three 25 mL portions of diethyl ether to remove (PhS)$_2$. The remaining white solid was purified by recrystallization from acetone/cyclohexane. The syntheses were performed under an inert atmosphere.

## 1.59 Zn–H–C–N–O–Re–S

**[Zn(H$_2$O)]$_2$[Re$_6$S$_8$(CN)$_6$]·7H$_2$O or C$_6$H$_{18}$N$_6$O$_9$Re$_6$S$_8$Zn$_2$** multinary compound is formed in this system. It crystallizes in the orthorhombic structure with the lattice parameters $a = 1095.29 \pm 0.05$, $b = 1665.31 \pm 0.08$, and $c = 3298.9 \pm 0.2$ pm at 160 K and a calculated density of 4.024 g·cm$^{-3}$ and undergoes transformation to a new crystalline structure at ca. 200°C (Beauvais et al. 1998). To obtain it, a saturated aqueous solution of $ZnSO_4·7H_2O$ (20 mL) was combined with a 20 mL of aqueous solution of NaCs$_3$Re$_6$S$_8$(CN)$_6$ (0.061 mM), and the mixture was boiled to dryness. The resulting solid was washed with successive aliquots of $H_2O$ (3 × 50 mL) to give [Zn(H$_2$O)]$_2$[Re$_6$S$_8$(CN)$_6$]·7H$_2$O as small orange hexagonal plate crystals.

## 1.60 Zn–H–C–N–S

Some compounds are formed in this system.

**[Zn(N$_2$H$_4$)$_2$(NCS)$_2$] or C$_2$H$_8$N$_6$S$_2$Zn, bis(hydrazine)zinc isothiocyanate.** This compound crystallizes in the monoclinic structure with the lattice parameters $a = 714.1 \pm 0.4$, $b = 1475.6 \pm 0.5$, $c = 421.4 \pm 0.5$ pm, and $\beta = 105°30' \pm 8'$ and experimental and calculated densities 1.914 and 1.906 g·cm$^{-3}$, respectively (Ferrari et al. 1965).

**[MeZnS$_2$CNMe$_2$] or C$_4$H$_9$NS$_2$Zn.** This compound melts at 142°C (Hursthouse et al. 1991). To obtain it, Me$_2$Zn (33.6 mM) and dimethylamine (33.6 mM) were stirred and heated in toluene (30 mL) under a stream of $N_2$ at 70°C for 12 h. To the cooled solution, a cold solution of $CS_2$ (33.6 mM) in toluene (20 mL) was added and the mixture warmed to, and kept at, 40°C for 5 h. Traces of elemental zinc were removed by filtration and the solvent was removed by evaporation. The product was recrystallized from benzene and gave transparent yellow crystals. All manipulations were carried out by using a vacuum line and standard Schlenk techniques.

**[EtZnS₂CNMe₂] or C₅H₁₁NS₂Zn**. This compound melts at 129°C and was obtained in the same way as $MeZnS_2CNMe_2$ was obtained using $Et_2Zn$ instead of $Me_2Zn$ (Hursthouse et al. 1991).

**[Zn(C₂H₃N₃)₂(NCS)₂] or C₆H₆N₈S₂Zn**. To obtain this compound, 5 mM of $Zn(NO_3)_2$ was dissolved in about 15 mL of $H_2O$ (Haasnoot and Groeneveld 1977). 10 mM of 1,2,4-triazole dissolved in about 15 mL of $H_2O$ was added to this solution slowly and under stirring. Next, a solution of 10 mM of $NH_4NCS$ in 15 mL of $H_2O$ was added rapidly. The solution was concentrated in vacuo. Upon standing, crystals separated and were collected by filtration.

[Zn(C₂H₃N₃)₂(NCS)₂] could also be obtained if freshly prepared $Zn(NCS)_2$ (5 mM) was mixed with 50 mL of acetone. 10 mM of 1,2,4-triazole in 20 mL of acetone was added rapidly (Haasnoot and Groeneveld 1977). After stirring, a clear solution was obtained. Crystallization occurred while standing for several days after concentrating the solution in vacuo.

**[Zn(C₃H₄NS)₂]ₙ or C₆H₈N₂S₄Zn**. This compound decomposes at 270°C–272°C (Preti and Tosi 1976). To obtain it, an aqueous solution of $ZnCl_2$ or $Zn(NO_3)_2$ or $Zn(OAc)_2$, maintained at a pH value lower than that of precipitation of the $Zn(OH)_2$, has been treated with an aqueous/ethanol solution of thiazolidine-2-thione ($C_3H_5NS_2$) in the Zn/ligand stoichiometrical ratio of 1:2. It could also be prepared by dissolving of zinc salts in EtOH with the next dropwise addition to the stirred solution of $C_3H_4NS_2$ in a mixture of EtOH and DMSO in the ratio 20:3. The title compound has been purified by means of repeated washing with EtOH and dried over $P_4O_{10}$.

[Zn(C₃H₄NS)₂]ₙ could also be prepared by the next procedure. To an ethanol solution (250 mL) of thiazolidine-2-thione (11.9 g) was added KOH (5.6 g) in aqueous EtOH (McCleverty et al. 1980a). The mixture was heated until it became homogeneous, and to it was added an ethanolic solution (100 mL) of $[Zn(OCOMe)_2]\cdot 2H_2O$ (10.9 g). The mixture was allowed to cool, and the pale-yellow precipitate of the title compound was filtered off, washed with ethanol, and dried in vacuo.

**[Zn{S₂CNMe₂}₂] or C₆H₁₂N₂S₄Zn, zinc dimethyldithiocarbamate**. This compound melts at 250°C (Bell et al. 1989) and crystallizes in the monoclinic structure with the lattice parameters $a = 847.1 \pm 0.1$, $b = 1582.0 \pm 0.2$, $c = 1817.3 \pm 0.2$ pm, and $\beta = 101.87° \pm 0.01°$ (Ramalingam et al. 1998) ($a = 845.5 \pm 0.3$, $b = 1574.7 \pm 0.5$, $c = 1834.5 \pm 0.9$ pm, and $\beta = 104.76° \pm 0.04°$ and experimental and calculated densities 1.66 and 1.72 g·cm⁻³, respectively [Klug 1966]). To obtain it, a solution of $[Bu^n_4N[S_2CNMe_2]$ (0.01 M) in MeOH (100 mL) was mixed with a solution of $ZnI_2$ (0.01 M) in MeOH (100 mL) (McCleverty and Morrison 1976, Ivanov et al. 2003a).

A white precipitate of the title compound formed slowly. It was collected after 30 min and washed with acetone. It could also be prepared if $[Bu^n_4N][Zn(S_2CNMe_2)_3]$ (5 mM) in $CH_2Cl_2$ (25 mL) was allowed to stand for 1 h and then evaporated to dryness. The solid residue with EtOH (15 mL) was shaken overnight and the mixture was then filtered. The residue was washed with acetone (30 mL) and dried in air (McCleverty and Morrison 1976). Crystals were grown by slow cooling of hot saturated solution of $Zn[S_2CNMe_2]_2$ in chloroform (Klug 1966).

**[MeZnS$_2$CNEt$_2$] or C$_6$H$_{13}$NS$_2$Zn.** This compound melts at 142°C and crystallizes in the monoclinic structure with the lattice parameters $a = 677.6 \pm 0.1$, $b = 1034.1 \pm 0.4$, $c = 1447.9 \pm 0.5$ pm, and $\beta = 97.3° \pm 0.3°$ and a calculated density of 1.509 g·cm$^{-3}$ (Hursthouse et al. 1991). It was obtained in the same way as $MeZnS_2CNMe_2$ was obtained using diethylamine instead of dimethylamine.

**[Zn(C$_7$H$_5$N$_2$S)]$_n$ or C$_7$H$_5$N$_2$SZn.** To obtain this compound, to an ethanol solution (250 mL) of 2-mercaptobenzimidazole (15.0 g) was added KOH (5.6 g) in aqueous EtOH (McCleverty et al. 1980a). The mixture was heated until it became homogeneous and to it was added an ethanolic solution (100 mL) of $[Zn(OCOMe)_2]\cdot2H_2O$ (10.9 g). The mixture was allowed to cool, and the pale-yellow precipitate of the title compound was filtered off, washed with ethanol, and dried in vacuo.

**[EtZnS$_2$CNEt$_2$] or C$_7$H$_{15}$NS$_2$Zn.** This compound melts at 115°C and was obtained in the same way as $MeZnS_2CNMe_2$ was obtained using $Et_2Zn$ instead of $Me_2Zn$ and diethylamine instead of dimethylamine (Hursthouse et al. 1991).

**[Zn(C$_4$H$_3$N$_2$S)$_2$] or C$_8$H$_6$N$_4$S$_2$Zn.** This compound was prepared by the electrochemical oxidation of anodic Zn in an acetonitrile solution of pyrimidine-2-thione (Castro et al. 1992). All experiments were carried out under dry $N_2$ atmosphere, which also served to stir the reaction mixture as it bubbled through the solution.

**[Zn(C$_4$H$_4$N$_3$S$_3$)$_2$] or C$_8$H$_8$N$_6$S$_6$Zn.** To obtain this compound, 2-amino-5-methyl-1,3,4-thiadiazolyl carbamate (0.436 mM) in MeOH (10 mL) was added to a solution of $ZnCl_2$ (0.216 mM) in the same solvent (10 mL) (Görgülü and Çukurovali 2002). The title compound precipitated immediately and was filtered off, washed with $H_2O$ several times and dried in vacuo at 60°C.

**[Zn(C$_8$H$_9$NS$_2$)] or C$_8$H$_9$NS$_2$Zn.** This compound melts at 270°C with decomposition (Brand and Vahrenkamp 1996). It was prepared under protective gas ($N_2$) by the following procedures:

1. To a suspension of $Zn(OAc)_2\cdot2H_2O$ (6.36 mM) in EtOH (100 mL) was added dropwise a solution of $N$-(2-mercaptoethyl)-2-mercaptoaniline ($C_8H_{11}NS_2$) (3.18 mM) in the same solvent (Brand and Vahrenkamp 1996). After some time, a precipitate was formed, which was filtered after stirring for 2 h. The white powder was washed with EtOH and vacuum dried.
2. $Zn(ClO_4)_2\cdot6H_2O$ (7.51 mM) was dissolved in MeOH (150 mL) (Brand and Vahrenkamp 1996). To this solution, a solution of $C_8H_{11}NS_2$ (6.01 mM) and NaOH (12.0 mM) in the same solvent (130 mL) was added dropwise.

The precipitate was formed immediately. This mixture was stirred for a further 14 h, filtered off, washed several times with MeOH, and dried in vacuo.

3. $C_8H_{11}NS_2$ (4.57 mM) was dissolved in dry toluene (30 mL), and this solution was mixed with a solution of $Zn[N(SiMe_3)_2]_2$ (4.56 mM) in the same solvent (30 mL) (Brand and Vahrenkamp 1996). After 1 h of stirring, a colorless precipitate was formed, which was washed three times with toluene and vacuum dried.

**[Zn(C₃H₅NS₂)₂(SCN)₂] or C₈H₁₀N₄S₆Zn.** This compound melts at 75°C (Colombini and Preti 1975). It was obtained by the reaction of $Zn(SCN)_2$ and thiazolidine-2-thione in refluxing chlorobenzene for 2 h. The title compound was separated out by crystallization and washed with ethyl ether.

**[ZnS₆(C₄H₆N₂)₂] or C₈H₁₂N₄S₆Zn.** This compound crystallizes in the triclinic structure with the lattice parameters $a = 743.4 \pm 0.4$, $b = 1371.8 \pm 0.3$, and $c = 1633.9 \pm 0.3$ pm and $\alpha = 96.45° \pm 0.02°$, $\beta = 99.66° \pm 0.04°$, and $\gamma = 90.33° \pm 0.04°$ at 198 K and a calculated density of 1.717 g·cm⁻³ (Dev et al. 1990). To obtain it, zinc dust (16.83 mM) was dissolved in a hot (100°C) N-methylimidazole (20 mL) solution of elemental sulfur (101.37 mM). Over the course of 12 h, the reaction proceeded to give a clear, greenish-brown solution. Zn dissolved completely in this reaction. The cooled, filtered solution was layered with toluene (40 mL), and after 24 h at 0°C bright-yellow crystals of the title compound were collected. [ZnS₆(C₄H₆N₂)₂] decomposes at 190°C in vacuum, leaving an off-white residue of ZnS.

**[ZnN{(CH₂)₂NH₂}₃(SCN)](SCN) or C₈H₁₈N₆S₂Zn.** This compound crystallizes in the orthorhombic structure with the lattice parameters $a = 1288.8 \pm 0.2$, $b = 1646.6 \pm 0.3$, and $c = 1363.3 \pm 0.2$ pm and experimental and calculated densities 1.49 and 1.504 g·cm⁻³, respectively (Andreetti et al. 1969) ($a = 1288.6$, $b = 1646.2$, and $c = 1364.3$ pm) (Jain et al. 1968). Its colorless crystals were grown by the recrystallization from $H_2O$.

**[Zn{S(CH₂)₂NMe₂}₂] or C₈H₂₀N₂S₂Zn.** This compound crystallizes in the orthorhombic structure with the lattice parameters $a = 2201.0 \pm 0.5$, $b = 1131.3 \pm 0.3$, and $c = 1017.0 \pm 0.3$ pm and a calculated density of 1.436 g·cm⁻³ (Casals et al. 1990). It was obtained by the electrochemical oxidation of anodic Zn foil into a solution consisting of 2 mL (8.78 mM) of 2-(dimethylamino)ethanethiol disulfide, 6 mL (43 mM) of $Et_3N$, and 0.1 mM of $Et_4NClO_4$ in 50 mL of acetonitrile.

**Zn[SCMe₂CH₂NH₂]₂ or C₈H₂₀N₂S₂Zn, bis(2-amino-1,1-dimethylethanethiolato-N,S)zinc.** This compound crystallizes in the orthorhombic structure with the lattice parameters $a = 557.55 \pm 0.07$, $b = 1058.5 \pm 0.1$, and $c = 2174.1 \pm 0.3$ pm and experimental and calculated densities 1.41 ± 0.01 and 1.417 g·cm⁻³, respectively (Cohen et al. 1978).

**[BuᵗZnS₂CNEt₂] or C₉H₁₉NS₂Zn.** This compound melts at 177°C–179°C (Malik et al. 1992). To obtain it, a solution of (diethyldithiocarbamato)zinc (6.4 mM) in toluene (35 mL) was stirred with $Bu^t_2Zn$ (6.4 mM) at room temperature for 30 min.

The colorless solution, on concentration under vacuum, gave transparent crystals of the title compound.

[**Zn(C₅H₄NS)₂**] or **C₁₀H₈N₂S₂Zn**. This compound was prepared by direct electrochemical oxidation of Zn into a solution of pyridine-2-thione in acetonitrile (Durán et al. 1991). At the end of the electrolysis, the solid was collected, washed with acetonitrile, and dried in vacuo.

[**Zn(S₆)(C₅H₅N)₂**] or **C₁₀H₁₀N₂S₆Zn**, (**hexasulfanediyl**)**di**(**pyridine-*N***)**zinc**. This compound crystallizes in the monoclinic structure with the lattice parameters $a = 1360.8 \pm 0.8$, $b = 899.8 \pm 0.6$, $c = 1533.4 \pm 0.8$ pm, and $\beta = 118.61° \pm 0.04°$ and a calculated density of 1.68 g·cm⁻³ (Li 1994). It was obtained by refluxing a mixture of Zn and S powders in pyridine for 6 h.

[**Zn(C₅H₆N₃S₃)₂**] or **C₁₀H₁₂N₆S₆Zn**. To obtain this compound, 2-amino-5-ethyl-1,3,4-thiadiazolyl carbamate (0.411 mM) in MeOH (10 mL) was added to a solution of ZnCl₂ (0.206 mM) in the same solvent (10 mL) with continuous stirring (Görgülü and Çukurovali 2002). The title compound precipitated immediately and was filtered off, washed with H₂O several times, and dried in vacuo at 60°C.

[**Zn₂(C₅H₈NS₂)₄**] or **C₁₀H₁₆N₂S₄Zn**. This compound crystallizes in the triclinic structure with the lattice parameters $a = 787.07 \pm 0.06$, $b = 798.23 \pm 0.06$, and $c = 1232.46 \pm 0.09$ pm and $\alpha = 74.813° \pm 0.002°$, $\beta = 73.048° \pm 0.002°$, and $\gamma = 88.036° \pm 0.002°$ (Jian et al. 2003) ($a = 775.8 \pm 0.2$, $b = 790.6 \pm 0.2$, and $c = 1226.7 \pm 0.4$ pm and $\alpha = 74.92° \pm 0.09°$, $\beta = 73.36° \pm 0.08°$, and $\gamma = 88.28° \pm 0.11°$ at 100 K and a calculated density of 1.709 g·cm⁻³) (Shahid et al. 2009). It was prepared by the reaction of ZnCl₂ (1.00 mM) in MeOH with 1 or 2 mM of ammonium pyrrolidine dithiocarbamate (NH₄C₅H₉NS₂). The addition of NH₄C₅H₉NS₂ in the colorless Zn²⁺ solution resulted in the formation of white precipitate immediately. After stirring for 30 min, the precipitate was filtered off and dried. The precipitate (0.05 g) was dissolved in DMSO (5 mL) and the resulting solution yielded crystals after a week (Shahid et al. 2009).

The title compound could also be prepared if to a heated aqueous solution of NaC₅H₉NS₂ (4%, 5 mL) was added an EtOH solution of Zn(ClO₄)₂·6H₂O (0.5 mM) with stirring (Jian et al. 2003). The white precipitate was formed. Upon collection by filtration, the deposit was washed with H₂O and dried in air. Colorless prism-like crystals were obtained by recrystallization from the acetonitrile solution. They were collected and dried.

[**ZnS₆(C₅H₈N₂)₂**] or **C₁₀H₁₆N₄S₆Zn**. A solution of sulfur in 1,2-dimethylimidazole, zinc dust, and sulfur afforded greenish-yellow crystals of the title compound (Dev et al. 1990).

[**Zn(S₂CNEt₂)₂**] or **C₁₀H₂₀N₂S₄Zn**. This compound melts at 175°C–176°C (Tiekink 2000) (at 178°C–179°C (Bell et al. 1989) and crystallizes in the monoclinic structure with the lattice parameters $a = 981.4 \pm 0.3$, $b = 1066.7 \pm 0.3$, $c = 1565.5 \pm 0.4$ pm, and $\beta = 104.06° \pm 0.02°$ (Tiekink 2000) ($a = 1001.5 \pm 1.0$, $b = 1066.1 \pm 0.5$, $c = 1635.7 \pm 1.0$ pm, and $\beta = 111.58° \pm 0.05°$) (Bonamico et al. 1965); $a = 1002 \pm 2$, $b = 1080 \pm 5$,

$c = 1600 \pm 2$ pm, and $\beta = 111°$) (Simonsen and Ho 1953) and experimental and calculated densities $1.480 \pm 0.005$ ($1.50$ g·cm$^{-3}$) (Simonsen and Ho 1953) and $1.485$ g·cm$^{-3}$, respectively (Brennan et al. 1990). The title compound was prepared by reacting ZnSCO$_4$ with two molar equivalents of Na(S$_2$CNEt$_2$). The precipitate that formed was filtered off and crystals were obtained from the slow evaporation of a CHCl$_3$ solution.

It could also be prepared by mixing an aqueous solution of Zn(NO$_3$)$_2$·6H$_2$O or Zn(OAc)$_2$·2H$_2$O with an excess of NaEt$_2$NCS$_2$·3H$_2$O (Ivanov et al. 1999b, 2003a, Larionov et al. 1990). The resulting white precipitate was collected, filtered off, washed with H$_2$O, dried in vacuo over P$_2$O$_5$, and recrystallized two times from toluene. Transparent crystals were obtained by recrystallization from chloroform (Simonsen and Ho 1953).

**[Bu$^t$CH$_2$ZnS$_2$CNEt$_2$] or C$_{10}$H$_{21}$NS$_2$Zn.** This compound melts at 158°C–160°C (Malik et al. 1992). To obtain it, a solution of (diethyldithiocarbamato)zinc (6.4 mM) in toluene (35 mL) was stirred with dineopentylzinc (6.4 mM) at room temperature for 30 min. The colorless solution, on concentration under vacuum, gave transparent crystals of the title compound.

**[Zn(Me$_2$NCS$_2$)$_2$(Et$_2$NH)] or C$_{10}$H$_{23}$N$_3$S$_4$Zn, bis(dimethyldithiocarbamato)diethylaminezinc.** This compound was obtained by quantitative chemisorption of Et$_2$NH vapor for the gas phase on the powder of Zn$_2$(Me$_2$NCS$_2$)$_4$ and by crystallization from toluene solution (Ivanov et al. 2007b).

**[Zn(Me$_2$NCS$_2$)$_2$(Bu$^n$NH$_2$)] or C$_{10}$H$_{23}$N$_3$S$_4$Zn.** This compound melts at 132°C–133°C (Higgins and Saville 1963). To obtain it, to a solution of $n$-butylamine (ca. 30 mM) in benzene (20–40 mL) at 50°C was added, in portions, the powdered zinc dimethyldithiocarbamate (20 mM). Upon cooling the resulting solution or partial removal of the solvent or addition of light petroleum (100 $\pm$ 20 mL), the product crystallized. It was collected, washed with light petroleum, and dried in air.

**[Zn(C$_4$H$_3$N)CH(N)(C$_6$H$_4$S] or C$_{11}$H$_8$N$_2$SZn.** This compound was obtained by the electrochemical oxidation of anodic Zn in acetonitrile solution of the Schiff base derived from 2-pyrrolcarbaldehyde and bis(2-aminophenyl)disulfide (Castro et al. 1990).

**[Zn(Me$_2$NCS$_2$)$_2$(C$_5$H$_5$N)] or C$_{11}$H$_{17}$N$_3$S$_4$Zn, bis-($N$,$N$-dimethyldithiocarbamato)pyridinezinc.** This compound crystallizes in the monoclinic structure with the lattice parameters $a = 1293 \pm 3$, $b = 894 \pm 9$, $c = 1415 \pm 3$ pm, and $\beta = 90.0°$ and experimental and calculated densities $1.535$ and $1.563$ g·cm$^{-3}$, respectively (Fraser and Harding 1967). To obtain it, to a solution of pyridine (ca. 30 mM) in benzene (20–40 mL) at 50°C was added, in portions, the powdered zinc dimethyldithiocarbamate (20 mM). Upon cooling the resulting solution or partial removal of the solvent or addition of light petroleum (100 $\pm$ 20 mL), the product crystallized. It was collected, washed with light petroleum, and dried in air (Higgins and Saville 1963). The large transparent plates decomposed slowly in air, losing pyridine (Fraser and Harding 1967).

**[Zn(Me$_2$NCS$_2$)$_2$(C$_5$H$_{10}$NH)]** or **C$_{11}$H$_{23}$N$_3$S$_4$Zn**. This compound melts at 129°C–130°C (Higgins and Saville 1963). To obtain it, to a solution of piperidine (ca. 30 mM) in benzene (20–40 mL) at 50°C was added, in portions, the powdered zinc dimethyldithiocarbamate (20 mM). Upon cooling the resulting solution or partial removal of the solvent or addition of light petroleum (100 ± 20 mL), the product crystallized. It was collected, washed with light petroleum, and dried in air.

**[Zn(C$_5$H$_5$N)$_2$(SCN)$_2$]** or **C$_{12}$H$_{10}$N$_4$S$_2$Zn**. This compound melts at 193°C (Ahuja and Garg 1972). It was obtained by allowing an excess of Py to react with zinc thiocyanate in EtOH.

**[Zn(C$_6$H$_7$N$_2$S)$_2$]** or **C$_{12}$H$_{14}$N$_4$S$_2$Zn**. This compound crystallizes in the orthorhombic structure with the lattice parameters $a = 798.4 \pm 0.1$, $b = 1347.6 \pm 0.2$, and $c = 1370.9 \pm 0.2$ pm and a calculated density of 1.548 g·cm$^{-3}$ (Godino-Salido et al. 1994). To obtain it, a stirred solution of 4,6-dimethyl-2-thiopyrimidine (1 mM) in H$_2$O (20 mL), KOH (1 mM), and ZnCl$_2$ (3 mM) was added subsequently, the final volume reaching ca. 45 mL. From the resulting solution (pH ≈ 7), a powdered precipitate of C$_{12}$H$_{14}$N$_4$S$_2$Zn was collected almost immediately. The mother liquor was then evaporated slowly in a water bath at 40°C for 3 days after which colorless crystals were collected.

The title compound could also be obtained at the electrochemical oxidation of Zn in a solution of 4,6-dimethyl-2-thiopyrimidine in acetonitrile (Castro et al. 1994).

**C$_{12}$H$_{18}$N$_2$S$_2$Zn, [N,N´-ethylenebis-(monothioacetylacetoneiminato)] zinc**. This compound crystallizes in the monoclinic structure with the lattice parameters $a = 1098.8 \pm 0.4$, $b = 1169.2 \pm 0.2$, $c = 1222.7 \pm 0.4$ pm, and $\beta = 116.11° \pm 0.03°$ and experimental density 1.507 g·cm$^{-3}$ (Cini et al. 1980). Its crystals were obtained from the toluene solution.

**[Zn{S$_2$CN(CH$_2$)$_5$}$_2$]** or **C$_{12}$H$_{20}$N$_2$S$_4$Zn**. To obtain this compound, a solution of Na{S$_2$CN(CH$_2$)$_5$}·H$_2$O (20 mM) in H$_2$O (30 mL) was added with vigorous stirring to a solution of ZnCl$_2$ (10 mM) in H$_2$O (100 mL) (Ivanov et al. 2003a). Yellowish voluminous precipitate was washed by decantation, separated by filtration, and dried in air.

**[Zn(C$_6$H$_{12}$N$_2$S$_2$)$_2$]** or **C$_{12}$H$_{24}$N$_4$S$_4$Zn**. This compound was prepared by mixing 4-methylpiperazine-1-thiocarboxylic acid (20 mM) and CS$_2$ (20 mM) in acetonitrile or EtOH under ice-cold conditions (5°C) (Prakasam et al. 2007). To the yellow dithiocarbamic acid solution, an aqueous solution of ZnSO$_4$ was added with constant stirring. The pale-yellow solid separated from the solution, which was filtered, washed with H$_2$O and alcohol, and then dried in air.

**[Zn(Me$_2$NCS$_2$)$_2${NH(CH$_2$)$_6$}]** or **C$_{12}$H$_{25}$N$_3$S$_4$Zn**. This compound melts at 137°C (Higgins and Saville 1963). To obtain it, to a solution of cyclohexylamine (ca. 30 mM) in benzene (20–40 mL) at 50°C was added, in portions, the powdered zinc dimethyldithiocarbamate (20 mM). Upon cooling the resulting solution or partial removal of the solvent or addition of light petroleum (100 ± 20 mL), the product crystallized. It was collected, washed with light petroleum, and dried in air (Higgins and Saville 1963). It could also be prepared by quantitative absorption of cyclohexylamine,

$NH(CH_2)_6$, by fine dispersed samples of $Zn(Me_2NCS_2)_2$ from the gas phase or as a result of their careful wetting (Ivanov et al. 2004b).

**[Zn(Me₂NCS₂)₂(Pr$^n$₂NH)] or C₁₂H₂₇N₃S₄Zn, bis(dimethyldithiocarbamato)di-*n*-propylaminezinc**. This compound was obtained by quantitative chemisorption of $Pr^n_2NH$ vapor for the gas phase on the powder of $Zn_2(Me_2NCS_2)_4$ and by crystallization from toluene solution (Ivanov et al. 2007b).

**[Zn(Et₂NCS₂)₂(Me₂NH)] or C₁₂H₂₇N₃S₄Zn.** This compound melts at 99.5°C–100.5°C (Higgins and Saville 1963). To obtain it, to a solution of dimethylamine (ca. 30 mM) in benzene (20–40 mL) at 50°C was added, in portions, the powdered zinc diethyldithiocarbamate (20 mM). Upon cooling the resulting solution or partial removal of the solvent or addition of light petroleum (100±20 mL), the product crystallized. It was collected, washed with light petroleum, and dried in air.

**[Zn(C₃S₅)(2,2′-bipy)]₂ or C₁₃H₈N₂S₅Zn.** This compound crystallizes in the monoclinic structure with the lattice parameters $a = 1155.6 \pm 0.2$, $b = 914.1 \pm 0.2$, $c = 1502.0 \pm 0.3$ pm, and $\beta = 109.55° \pm 0.03°$ and a calculated density of 1.856 g·cm⁻³ (Liu et al. 2002). Its single crystals were grown in a test tube. Solid [Cu(2,2′-bipy)$(NO_3)_2$]·2H₂O (0.2 mM) was dissolved in 3 mL of $H_2O$ and 5 mL of MeOH, and to this solution was added carefully a solution of $[Et_4N]_2[Zn(C_3S_5)_2]$ (0.2 mM) in acetone (10 mL). The test tube was then allowed to stand undisturbed at room temperature for 3 days, yielding a dark-red blocks of the title compound on the tube wall together with some unknown black amorphous materials at the bottom.

**[Zn(Me₂NCS₂)₂(PhCH₂NH₂)] or C₁₃H₂₁N₃S₄Zn.** This compound melts at 117°C–118°C (Higgins and Saville 1963). To obtain it, to a solution of benzylamine (ca. 30 mM) in benzene (20–40 mL) at 50°C was added, to in portions, the powdered zinc dimethyldithiocarbamate (20 mM). Upon cooling the resulting solution or partial removal of the solvent or addition of light petroleum (100±20 mL), the product crystallized. It was collected, washed with light petroleum, and dried in air.

**[Zn(Et₂NCS₂)₂(C₃H₄N₂)] or C₁₃H₂₄N₄S₄Zn.** This compound crystallizes in the monoclinic structure with the lattice parameters $a = 2049.3 \pm 0.4$, $b = 842.7 \pm 0.1$, $c = 1160.0 \pm 0.2$ pm, and $\beta = 90.53° \pm 0.01°$ and a calculated density of 1.427 g·cm⁻³ (Zemskova et al. 1993b). This compound was prepared by the interaction of $Zn(S_2CNEt_2)_2$ and imidazole (3:4.5 molar ratio) in boiling toluene (Larionov et al. 1990). Needlelike single crystals were grown by slow cooling of the hot saturated solution of the title compound in benzene (Zemskova et al. 1993b). The final product of thermolysis of $[Zn(Et_2NCS_2)_2(C_3H_4N_2)]$ is α-ZnS.

**[Me₄N][Zn(S₂CNMe₂)₃] or C₁₃H₃₀N₄S₆Zn.** This compound was prepared as follows (McCleverty et al. 1980a). To an acetone solution of [Me₄N][S₂CNMe₂] (ca. 1 M) was added [Zn(S₂CNMe₂)₂] (ca. 1 M), and the mixture stirred for 2 h. After filtration, the filtrate was treated with light petroleum, and on partial evaporation, the complex precipitated. The solid was filtered off and recrystallized from acetone/light petroleum. On occasions it proved necessary to reflux the reaction mixture or to use refluxing toluene as the reaction medium. Despite careful control of the stoichiometry reaction, it was often impossible to prevent ligand redistribution.

**[Zn(C$_7$H$_4$NS$_2$)$_2$]$_n$ or C$_{14}$H$_8$N$_2$S$_4$Zn, zinc bis(benzothiazol-2-thiolat).** To obtain this compound, to a solution of benzothiazole-2-thione (16.7 g) in hot EtOH (250 mL) was added KOH (5.7 g) in aqueous solution (30 mL) (McCleverty et al. 1980b). The clear orange-yellow solution that formed was filtered, and a solution of [Zn(OCOMe)$_2$·2H$_2$O (10.9 g) in EtOH was added to the filtrate. After cooling, the pale-yellow precipitate of C$_{14}$H$_8$N$_2$S$_4$Zn was filtered off, washed with EtOH, and dried in vacuo.

**[Zn(C$_6$H$_5$N$_5$)$_2$(SCN)$_2$] or C$_{14}$H$_{10}$N$_{12}$S$_2$Zn.** This compound crystallizes in the monoclinic structure with the lattice parameters $a = 704.2 \pm 0.1$, $b = 1079.7 \pm 0.2$, $c = 1372.5 \pm 0.2$ pm, and $\beta = 95.697° \pm 0.002°$ (Shi et al. 2010). To obtain it, 10 mL of methanol solution of 2-(1H-1,2,4-triazol-1-yl)pyrazine (0.355 mM) was added into 10 mL of water solution containing Zn(ClO$_4$)$_2$·6H$_2$O (0.164 mM) and NaSCN (0.334 mM), and then the mixture was stirred for few minutes. The colorless crystals were obtained after the filtrate had been allowed to stand at room temperature for 2 weeks.

**[Zn(SCN)$_2$(PhNH$_2$)$_2$] or C$_{14}$H$_{14}$N$_4$S$_2$Zn.** This compound crystallizes in the orthorhombic structure with the lattice parameters $a = 1456 \pm 5$, $b = 910 \pm 5$, and $c = 1270 \pm 10$ pm and experimental and calculated densities 1.4 and 1.45 g·cm$^{-3}$, respectively (Shepherd and Woodward 1965). Its needle-shaped single crystals were grown by recrystallization of the title compound from EtOH.

**[Zn(2-MeC$_5$H$_4$N)$_2$(SCN)$_2$], [Zn(3-MeC$_5$H$_4$N)$_2$(SCN)$_2$] and [Zn(4-MeC$_5$H$_4$N)$_2$(SCN)$_2$] or C$_{14}$H$_{14}$N$_4$S$_2$Zn.** [Zn(2-MeC$_5$H$_4$N)$_2$(SCN)$_2$] melts at 154°C, and [Zn(4-MeC$_5$H$_4$N)$_2$(SCN)$_2$] melts at 173°C (Ahuja and Garg 1972). These compounds were obtained by allowing an excess of 2-MePy (3-MePy, 4-MePy) to react with zinc thiocyanate in EtOH.

**[Zn{S$_2$CN(CH$_3$)$_2$}$_2$(C$_5$H$_5$N)]·(0.5C$_6$H$_6$) or C$_{14}$H$_{20}$N$_3$S$_4$Zn.** This compound crystallizes in the monoclinic structure with the lattice parameters $a = 1267 \pm 1$, $b = 796.6 \pm 0.5$, $c = 2013 \pm 2$ pm, and $\beta = 105.27° \pm 0.06°$ and experimental and calculated densities 1.422 and 1.436 g·cm$^{-3}$, respectively (Fraser and Harding 1967). Its fine transparent needles were obtained by dissolving Zn[S$_2$CN(CH$_3$)$_2$]$_2$(C$_5$H$_5$N) in an excess of pyridine and allowing the solution to cool. They lose benzene readily on exposure to air.

**C$_{14}$H$_{22}$N$_2$S$_2$Zn, N,N′-tetramethylenebis(thioacethylacetoniminato)zinc.** This compound melts at 230°C and crystallizes in the monoclinic structure with the lattice parameters $a = 1228.6 \pm 1.0$, $b = 1197.0 \pm 1.0$, $c = 1260.8 \pm 1.0$ pm, and $\beta = 120.67° \pm 0.05°$ and experimental and calculated densities 1.44 and 1.45 g·cm$^{-3}$, respectively (Hewlins 1975). To obtain it, a saturated solution of N,N′-tetramethylenebis (thioacethylacetonimine) (1 g) in MeOH was added with stirring to a filtered saturated solution of Zn(OAc)$_2$·H$_2$O (1 g) in MeOH. Crystals began to grow at once. After 24 h, the product was filtered off and recrystallized from benzene to give the title compound as colorless needles.

**[Zn{S$_2$CN(CH$_2$)$_6$}$_2$] or C$_{14}$H$_{24}$N$_2$S$_4$Zn zinc hexamethylenedithiocarbamate.** This compound crystallizes in the monoclinic structure with the lattice parameters $a = 1094.8 \pm 0.1$, $b = 1047.9 \pm 0.2$, $c = 1740.9 \pm 0.2$ pm, and $\beta = 114°48′ \pm 3′$ and experimental and calculated densities 1.53 and 1.523 $g \cdot cm^{-3}$, respectively (Agre and Shugam 1972) ($a = 1117$, $b = 1067$, $c = 1741$ pm, and $\beta = 117°$ and experimental and calculated densities 1.63 and 1.62 $g \cdot cm^{-3}$, respectively [Agre et al. 1966]). It was prepared by the interaction of $Zn^{2+}$ with appropriate dithiocarbamate anion (Ivanov et al. 2003a). White voluminous precipitate was washed by decantation, separated by filtration, and dried in air. Transparent crystals were obtained by recrystallization from chloroform or DMF.

**[Zn(Et₂NCS₂)₂(Et₂NH)] or C₁₄H₃₁N₃S₄Zn, bis(diethyldithiocarbamato)diethylaminezinc.** This compound melts at 87.5°C (Higgins and Saville 1963). To obtain it, to a solution of diethylamine (ca. 30 mM) in benzene (20–40 mL) at 50°C was added, in portions, the powdered zinc diethyldithiocarbamate (20 mM). Upon cooling the resulting solution or partial removal of the solvent or addition of light petroleum (100 ± 20 mL), the product crystallized. It was collected, washed with light petroleum, and dried in air.

It could also be obtained by quantitative chemisorption of $Et_2NH$ vapor for the gas phase on the powder of $Zn_2(Et_2NCS_2)_4$ and by crystallization from toluene solution (Ivanov et al. 2007b).

**[Zn(Et₂NCS₂)₂(BuⁿNH₂)] or C₁₄H₃₁N₃S₄Zn.** This compound melts at 74.5°C (Higgins and Saville 1963). To obtain it, to a solution of $n$-butylamine (ca. 30 mM) in benzene (20–40 mL) at 50°C was added, in portions, the powdered zinc diethyldithiocarbamate (20 mM). Upon cooling the resulting solution or partial removal of the solvent or addition of light petroleum (100 ± 20 mL), the product crystallized. It was collected, washed with light petroleum, and dried in air.

**[Zn(MePrⁱNCS₂)₂]·(Py) or C₁₅H₂₅N₃S₄Zn.** This compound melts at 175°C and crystallizes in the triclinic structure with the lattice parameters $a = 917.1 ± 1.0$, $b = 1033.5 ± 0.2$, and $c = 1231.8 ± 0.2$ pm and $\alpha = 102.99° ± 0.10°$, $\beta = 78.19° ± 0.10°$, and $\gamma = 114.02° ± 0.10°$ and a calculated density of 1.421 $g \cdot cm^{-3}$ (Malik et al. 1997). It was obtained by the following procedure. Bis($N$-methylisopropyldithiocarbamato) zinc was dissolved in warm Py, and as the excess solvent evaporated, a crude white solid appeared, which was recrystallized from acetone to give colorless, cubic crystals.

**[Zn(S₂CNEt₂)]·(Py) or C₁₅H₂₅N₃S₄Zn.** This compound crystallizes in the orthorhombic structure with the lattice parameters $a = 1021.19 ± 0.08$, $b = 1630.00 ± 0.12$, and $c = 2535.4 ± 0.3$ pm and a calculated density of 1.421 $g \cdot cm^{-3}$ (Ivanov et al. 1999a,b, Ivanov and Antzutkin 2002). It was prepared by dissolving of Py and $Zn(S_2CNEt_2)_2$ with 10% excess of Py in a minimum volume of toluene. Both powder and single-crystal samples were isolated by the slow evaporation of toluene.

The structure of the title compound depends on the pathway of physicochemical conditions during the preparation procedure (Ivanov and Antzutkin 2002). Two isomorphs of the adduct, namely, $\alpha$-[Zn(Py)(S₂CNEt₂)₂] and $\beta$-[Zn(Py)(S₂CNEt₂)₂], could be obtained by recrystallization from toluene of the equimolar solution of [Zn(S₂CNEt₂)₂] and Py. The third isomorph, $\gamma$-[Zn(Py)(S₂CNEt₂)₂], can be obtained by recrystallization from pure pyridine of [Zn(S₂CNEt₂)₂], or by its equimolar absorption of Py, or by desorption of Py from [Zn(Py)(S₂CNEt₂)₂]·Py. $\gamma$-[Zn(Py) (S₂CNEt₂)₂] isomorph recrystallizes from the melt into $\alpha$/$\beta$-isomorphs of the adduct.

**[Zn(S₂CNEt₂)₂]·(C₅H₈N₂) or C₁₅H₂₈N₄S₄Zn.** This compound was prepared by the interaction of Zn(S₂CNEt₂)₂ and 2,3-dimethylpyrazole (5:7.5 molar ratio) in the boiling toluene (Larionov et al. 1990).

**[Zn(Et₂NCS₂)₂(C₅H₁₀NH)] or C₁₅H₃₁N₃S₄Zn.** This compound melts at 115°C–116°C (Higgins and Saville 1963). It was synthesized using the following two methodological procedures (Ivanov et al. 2001b): (1) by dissolving dimeric diethyldithiocarbamatozinc in toluene containing piperidine in an amount exceeding the stoichiometric amount by ~10% and through further crystallization of the adduct at room temperature and (2) by quantitative absorption of piperidine vapors from the gas phase with powdered [Zn₂(Et₂NCS₂)₄]. It could also be obtained if to a solution of piperidine (ca. 30 mM) in benzene (20–40 mL) at 50°C was added, in portions, the powdered zinc diethyldithiocarbamate (20 mM). Upon cooling the resulting solution or partial removal of the solvent or addition of light petroleum (100 ± 20 mL), the product crystallized. It was collected, washed with light petroleum, and dried in air (Higgins and Saville 1963). The title compound was kept in sealed tubes (Ivanov et al. 2001b).

**[Me₄N][Zn(S₂CNEt₂)(S₂CNMe₂)₂] or C₁₅H₃₄N₄S₆Zn.** This compound was prepared as follows (McCleverty et al. 1980a). To an acetone solution of [Me₄N][S₂CNEt₂] (ca. 1 M) was added [Zn(S₂CNMe₂)₂] (ca. 1 M), and the mixture was stirred for 2 h. After filtration, the filtrate was treated with light petroleum, and on partial evaporation, the complex precipitated. The solid was filtered off and recrystallized from acetone/light petroleum. On occasions, it proved necessary to reflux the reaction mixture or to use refluxing toluene as the reaction medium. Despite careful control of the stoichiometry reaction, it was often impossible to prevent ligand redistribution.

**[Zn(C₁₆H₁₄N₆S₂)] or C₁₆H₁₄N₆S₂Zn.** This compound melts at 260°C (López-Torres et al. 2001). An orange solid was formed from the filtrate of the [Zn(C₁₆H₁₅N₆S₂)NO₃] synthesis after cooling for 2 months at –20°C. The solid was filtered off and dried in vacuo.

**[Zn(PhMeNCS₂)₂] or C₁₆H₁₆N₂S₄Zn.** This compound melts at 246°C–248°C (Onwudiwe and Ajibade 2011) and crystallizes in the monoclinic structure with the lattice parameters $a = 1272.76 \pm 0.01$, $b = 669.93 \pm 0.01$, $c = 2225.08 \pm 0.01$ pm, and $\beta = 100.669° \pm 0.001°$ and a calculated density of 1.538 g·cm⁻³ (Baba et al. 2002). The preparation of it was carried out using 25 mL of aqueous solution of Zn(OAc)₂ (1.25 mM), which was added to 25 mL of aqueous solution of N-methyl-N-phenyldithiocarbamate (2.50 mM). A solid precipitate formed immediately, and the mixture was stirred for about 45 min, filtered off, rinsed several times with distilled water, and recrystallized in CH₂Cl₂.

The title compound could also be synthesized from ZnCl₂ (10 mM), methylaniline (20 mM) and CS₂ (20 mM) in ethanol solution (Baba et al. 2002). Crystals were grown by the use of a CH₂Cl₂/MeOH mixture solvent.

**[Zn(SCN)₂(C₇H₈N₄)₂] or C₁₆H₁₆N₁₀S₂Zn.** This compound was prepared by mixing acetone solutions of Zn(SCN)₂ and 5,7-dimethyl[1,2,4]triazolo[1,5-a]pyrimidine (C₇H₈N₄) (Dillen et al. 1983). The title compound was crystallized within several hours and then recrystallized from acetone. White single crystals were grown easily from the solution by cooling slowly to room temperature.

**[Zn(2,4-Me$_2$C$_5$H$_3$N)$_2$(SCN)$_2$] or C$_{16}$H$_{18}$N$_4$S$_2$Zn**. This compound melts at 112°C (Ahuja and Garg 1972). It was obtained by allowing an excess of 2,4-Me$_2$Py to react with zinc thiocyanate in EtOH.

**[Zn(2-EtC$_5$H$_4$N)$_2$(SCN)$_2$],[Zn(3-EtC$_5$H$_4$N)$_2$(SCN)$_2$]and[Zn(4-EtC$_5$H$_4$N)$_2$(SCN)$_2$] or C$_{16}$H$_{18}$N$_4$S$_2$Zn**. These compounds melt at 136°C, 135°C and 159°C, respectively (Ahuja and Garg 1972). They were obtained by allowing an excess of 2-EtPy (3-EtPy, 4-EtPy) to react with zinc thiocyanate in EtOH.

**[Zn(S$_2$CNMe$_2$)$_2$·(2,2′-bipy)] or C$_{16}$H$_{20}$N$_4$S$_4$Zn**. This compound melts at 230°C–232°C (Bell et al. 1989). It was prepared as pale-yellow needles by adding a hot solution of 2,2′-bipy (50 mM) in MeOH to a hot solution of zinc dimethyldithiocarbamate (10 mM) in dichloromethane. The resulting solution was cooled and then added to petroleum ether to produce a yellow precipitate, which was filtered off and recrystallized from dichloromethane.

**[Zn(SCN)$_2$(C$_7$H$_{10}$N$_2$)$_2$] or C$_{16}$H$_{20}$N$_6$S$_2$Zn**. This compound melts 171°C–173°C (Secondo et al. 2000). To obtain it, Zn(NO$_3$)·6H$_2$O (2.0 mM) was dissolved in 25 mL of absolute EtOH, to which a solution of KSCN (4.0 mM) in the same solvent (30 mL) was added with stirring. The precipitated KNO$_3$ was filtered, and to the filtrate was added a solution of 4-(N,N-dimethylamino)pyridine (C$_7$H$_{10}$N$_2$) (4.0 mM) in absolute EtOH (10 mL). The resulting precipitate was isolated by filtration, washed with EtOH and then ether, and dried in vacuo.

**[(mepa)$_2$Zn] or C$_{16}$H$_{22}$N$_4$S$_2$Zn (mepaH = N-(2-mercaptoethyl)picolylamin, C$_8$H$_{12}$N$_2$S)**. This compound melts at 125°C with decomposition and crystallizes in the monoclinic structure with the lattice parameters $a = 1999.8 \pm 0.1$, $b = 879.5 \pm 0.1$, $c = 1179.0 \pm 0.1$ pm, and $\beta = 117.84° \pm 0.01°$ and experimental and calculated densities 1.39 and 1.45 g·cm$^{-3}$, respectively (Brand and Vahrenkamp 1995). To obtain it, a stirred solution of mepaH (10.4 mM) in toluene (40 mL) was treated dropwise with a solution of Zn[N(SiMe$_3$)$_2$]$_2$ (5.2 mM) in toluene (20 mL). The precipitate was filtered off, washed three times with toluene (5 mL), and dried in vacuo yielding (mepa)$_2$Zn as a colorless powder.

**[Zn{S$_2$CN(Me)C$_6$H$_{11}$}$_2$] or C$_{16}$H$_{28}$N$_2$S$_4$Zn**. This compound melts at 212°C and crystallizes in the monoclinic structure with the lattice parameters $a = 1669.90 \pm 0.16$, $b = 1111.19 \pm 0.11$, $c = 2304.9 \pm 0.2$ pm, and $\beta = 107.114° \pm 0.004°$ at 223 ± 2 K (Chan et al. 2004). Pale-yellow crystals of the title compound were grown by the solvent evaporation of an acetonitrile/chloroform (1:3 volume ratio) solution.

**[Zn(Et$_2$NCS$_2$)$_2${NH(CH$_2$)$_6$}] or C$_{16}$H$_{33}$N$_3$S$_4$Zn**. This compound melts at 108°C–109°C (Higgins and Saville 1963). To obtain it, to a solution of cyclohexylamine (ca. 30 mM) in benzene (20–40 mL) at 50°C was added, in portions, the powdered zinc diethyldithiocarbamate (20 mM). On cooling the resulting solution or partial removal of the solvent or addition of light petroleum (100 ± 20 mL), the product crystallized. It was collected, washed with light petroleum, and dried in air.

It could also be obtained by quantitative absorption of cyclohexylamine, $NH(CH_2)_6$, by fine dispersed samples of $Zn(Et_2NCS_2)_2$ from the gas phase or as a result of their careful wetting (Ivanov et al. 2004b).

**[Zn(Et₂NCS₂)₂(Prⁿ₂NH)] or $C_{16}H_{35}N_3S_4Zn$, bis(diethyldithiocarbamato)di-*n*-propylaminezinc.** This compound was obtained by quantitative chemisorption of $Pr^n_2NH$ vapor for the gas phase on the powder of $Zn_2(Et_2NCS_2)_4$ and by crystallization from toluene solution (Ivanov et al. 2007b).

**[(Et₄N){Zn(S₄)₂}] or $C_{16}H_{40}N_2S_8Zn$.** This compound crystallizes in the monoclinic structure with the lattice parameters $a = 1241.0 \pm 0.3$, $b = 1635.0 \pm 0.1$, $c = 1390.6 \pm 0.2$ pm, and $\beta = 95.73° \pm 0.01°$ and experimental and calculated densities $1.38 \pm 0.02$ and $1.37$ g·cm⁻³, respectively (Coucouvanis et al. 1985). It could be obtained if to a solution of $(Et_4N)_2[Zn(SPh)_4]$ (2.0 mM) in 50 mL of $CH_3CN$ was added BzSSSBz (Bz = 3-quinuclidinyl benzilate) (10 mM) in small portions with stirring. After it was stirred for 15 min, the solution was filtered, and dry diethyl ether was added to the filtrate until the first evidence of nucleation became apparent. After the mixture was allowed to stand for 24 h, crystals formed and were isolated, washed with two 10 mL portions each of EtOH and diethyl ether, and dried.

**[Zn{C₂H₄(NH₂)₂}₃][Et₂NCS₂]₂ or $C_{16}H_{44}N_8S_4Zn$.** This compound crystallizes in the triclinic structure with the lattice parameters $a = 1019.5 \pm 0.2$, $b = 1470.9 \pm 0.3$, and $c = 1893.0 \pm 0.3$ pm and $\alpha = 94.58° \pm 0.02°$, $\beta = 98.46° \pm 0.02°$, and $\gamma = 85.59° \pm 0.02°$ and a calculated density of $1.290$ g·cm⁻³ (Glinskaya et al. 1998b). It was prepared by the interaction of $Zn(S_2CNEt_2)_2$ and ethylenediamine (4:60 molar ratio) in the boiling EtOH (Larionov et al. 1990). Its single crystals were grown at slow cooling of the warm ($\approx 40°C$) solution of $[Zn{C_2H_4(NH_2)_2}_3][Et_2NCS_2]_2$ with the addition of ethylenediamine ($[Zn{C_2H_4(NH_2)_2}_3][Et_2NCS_2]_2/C_2H_4(NH_2)_2 = 1:3$ molar ratio) (Glinskaya et al. 1998b, Zemskova et al. 1999).

**[Zn(C₆H₇N₂S)₂(C₅H₅N)] or $C_{17}H_{19}N_5S_2Zn$.** This compound crystallizes in the orthorhombic structure with the lattice parameters $a = 908.6 \pm 0.2$, $b = 1413.7 \pm 0.3$, and $c = 1515.0 \pm 0.3$ pm and a calculated density of $1.443$ g·cm⁻³ (Castro et al. 1994). Electrochemical oxidation of Zn in a solution of 4,6-dimethyl-2-thiopyrimidine and Py in acetonitrile afforded the title compound in the form of yellow well-formed crystals.

**[Zn{NH(CH₂)₅}(S₂CNMe₂)₂] or $C_{17}H_{29}N_7S_4Zn$.** This compound melts at 119.1°C and crystallizes in the monoclinic structure with the lattice parameters $a = 811.72 \pm 0.09$, $b = 2324.9 \pm 0.3$, $c = 1248.01 \pm 0.14$ pm, and $\beta = 108.809° \pm 0.002°$ and a calculated density of $1.397$ g·cm⁻³ (Lutsenko et al. 2010). It was prepared by dissolving binuclear $[Zn_2{S_2CNMe_2}_4]$ (0.80 mM) in benzene (5 mL) containing a 10% excess of piperidine (1.76 mM). The single crystals of the title compound were grown from the obtained solution at room temperature, separated from the mother solution, collected on a filter, and dried in air until loose.

**[Me₄N][Zn(S₂CNMe₂)(S₂CNEt₂)₂] or $C_{17}H_{38}N_4S_6Zn$.** This compound crystallizes in the orthorhombic structure with the lattice parameters $a = 1470.9 \pm 1.4$, $b = 1878.0 \pm 1.3$, and $c = 2013.0 \pm 1.9$ pm and experimental and calculated densities $1.30$

and 1.328 g·cm⁻³, respectively (McCleverty et al. 1980a). It was prepared as follows. To an acetone solution of [Me₄N][S₂CNMe₂] (ca. 1 M) was added [Zn(S₂CNEt₂)₂] (ca. 1 M), and the mixture was stirred for 2 h. After filtration, the filtrate was treated with light petroleum, and on partial evaporation, the complex precipitated. The title compound was filtered off and recrystallized from acetone/light petroleum as color-less needles. On occasions, it proved necessary to reflux the reaction mixture or to use refluxing toluene as the reaction medium. Despite careful control of the stoichi-ometry reaction, it was often impossible to prevent ligand redistribution.

**[Zn(C₉H₆NS)₂] or C₁₈H₁₂N₂S₄Zn, (2-quinolinethiolato)zinc.** This compound crys-tallizes in the orthorhombic structure with the lattice parameters $a = 1300$, $b = 1560$, and $c = 1590$ pm and experimental density 1.57 g·cm⁻³ (Shugam 1958).

**[Zn(C₄H₃N₂S)₂(2,2′-bipy)] or C₁₈H₁₄N₆S₂Zn.** This compound was prepared by the electrochemical oxidation of anodic Zn in an acetonitrile solution of pyrimidine-2-thione in the presence of 2,2′-bipy (Castro et al. 1992). All of the experiments were carried out under a dry N₂ atmosphere, which also served to stir the reaction mixture as it bubbled through the solution.

**[Zn(C₈H₉NS₂)]·(2,2′-bipy) or C₁₈H₁₇N₃S₂Zn.** This compound melts at 160°C (Brand and Vahrenkamp 1996). It was prepared as follows. [Zn(C₈H₉NS₂)] (0.146 mM) was dissolved in 10 mL of acetonitrile. To the obtained suspension, a solution of 2,2′-bipy (0.583 mM) was added. The color of the suspension changed from white to yellow. This mixture was stirred for 16 h before the solid was filtered and vacuum dried.

**[Zn(S₂CNMe₂)₂]·(Phen) or C₁₈H₂₀N₄S₄Zn.** This compound melts at 262°C–263°C (Bell et al. 1989) and crystallizes in the monoclinic structure with the lattice param-eters $a = 1334.0 \pm 0.3$, $b = 1382.7 \pm 0.3$, $c = 2469.8 \pm 0.5$ pm, and $\beta = 102.58° \pm 0.03°$ and a calculated density of 1.452 g·cm⁻³ (Klevtsova et al. 1999b). It was prepared as pale-yellow needles by adding a hot solution of Phen (50 mM) in MeOH to a hot solution of zinc dimethyldithiocarbamate (10 mM) in dichloromethane. The resulting solution was cooled and then added to petroleum ether to produce a yellow precipitate, which was filtered off and recrystallized from MeOH (Bell et al. 1989). Its single crys-tals were grown by the slow cooling and evaporation of a saturated solution in the mixture chloroform/ethanol (2:1 volume ratio). Sublimation of Zn(S₂CNMe₂)₂(Phen) begins at 115°C, and its thermal decomposition leads to the formation of α-ZnS (Klevtsova et al. 1999b).

**[Zn{S₂CN(Et)(C₆H₁₁)}₂] or C₁₈H₃₂N₂S₄Zn.** This compound crystallizes in the monoclinic structure with the lattice parameters $a = 861.8 \pm 0.5$, $b = 1650.0 \pm 0.4$, $c = 1624.2 \pm 0.2$ pm, and $\beta = 97.70° \pm 0.02°$ at 200 K and a calculated density of 1.364 g·cm⁻³ (Cox and Tiekink 1999).

**[Zn(S₂CNBuⁿ₂)₂] or C₁₈H₃₆N₂S₄Zn.** This compound melts at 108°C (Bell et al. 1989) and crystallizes in the monoclinic structure with the lattice parameters $a = 2332.9 \pm 0.3$, $b = 1709.0 \pm 0.2$, $c = 1611.5 \pm 0.2$ pm, and $\beta = 127.560° \pm 0.010°$ and a calculated density of 1.237 g·cm⁻³ (Zhong et al. 2003) ($a = 2330.8 \pm 0.3$, $b = 1712.1 \pm 0.2$, and $c = 1608.0 \pm 0.2$ pm and $\beta = 127.470° \pm 0.002°$ and a calculated density of 1.237 g·cm⁻³ [Ivanov et al. 2003a]). To obtain it, a solution of Na{S₂CNBuⁿ₂}·H₂O (20 mM) in $H_2O$ (100 mL) was added with vigorous stirring to a solution of $ZnCl_2$ (10 mM) in $H_2O$ (100 mL) (Ivanov et al. 2003a). White fine-grained precipitate was formed from the darkened reaction mixture left overnight. The precipitate was separated by filtration, washed with $H_2O$, and dried in air. Single crystals of the title compound were obtained by recrystallization from acetone.

**[Zn($S_2CNEt_2$)$_2$(Bu$^n_2$NH)] or C$_{18}$H$_{39}$N$_3$S$_4$Zn.** This compound melts at 51°C–52°C (Higgins and Saville 1963). To obtain it, to a solution of di-$n$-butylamine (ca. 30 mM) in benzene (20–40 mL) at 50°C was added, in portions, the powdered zinc diethyldithiocarbamate (20 mM). On cooling the resulting solution or partial removal of the solvent or addition of light petroleum (100 ± 20 mL), the product crystallized. It was collected, washed with light petroleum, and dried in air.

**[Zn($C_4H_3N_2S$)$_2$(Phen)] or C$_{20}$H$_{14}$N$_6$S$_2$Zn.** This compound was prepared by the electrochemical oxidation of anodic Zn in an acetonitrile solution of pyrimidine-2-thione in the presence of 2,2′-Phen (Castro et al. 1992). All experiments were carried out under a dry $N_2$ atmosphere, which also served to stir the reaction mixture as it bubbled through the solution.

**[Zn($C_5H_4NS$)$_2$](2,2′-bipy) or C$_{20}$H$_{16}$N$_4$S$_2$Zn.** This compound was prepared by direct electrochemical oxidation of Zn into a solution of pyridine-2-thione with the addition of 2,2′-bipy in acetonitrile (Durán et al. 1991). At the end of the electrolysis, the solid was collected, washed with acetonitrile, and dried in vacuo.

**($C_7H_{10}N$)$_2$[Zn($C_3S_5$)$_2$] or C$_{20}$H$_{20}$N$_2$S$_{10}$Zn.** This compound, which melts at 232°C–233°C and crystallizes in the monoclinic structure with the lattice parameters $a = 967.84 ± 0.08$, $b = 884.67 ± 0.07$, $c = 1662.74 ± 0.14$ pm, and $\beta = 102.250° ± 0.002°$ and a calculated density of 1.610 g·cm$^{-3}$ (Harrison et al. 2000). It was obtained from Na, $CS_2$, and (1,4-Me$_2$Pyridinium)I. Crystals suitable for XRD were grown by recrystallization from a mixture Me$_2$CO/Pr$^i$OH.

**[Zn(SPh)$_2$($C_4H_6N_2$)$_2$] or C$_{20}$H$_{22}$N$_4$S$_2$Zn.** To obtain this compound, zinc dust (0.82 g) dissolved completely in $N$-methylimidazole (15 mL) upon the addition of one equivalent of Ph$_2$S$_2$ (100°C, 12 h). Ether precipitation gave colorless crystals of the title compound (Dev et al. 1990).

**[Zn($S_2CNMe_2$)$_2$·(2,9-Me$_2$Phen)] or C$_{20}$H$_{24}$N$_4$S$_4$Zn.** This compound melts at 210°C–212°C (Bell et al. 1989). It was prepared as yellow needles by adding a hot solution of 2,9Me$_2$Phen (50 mM) in MeOH to a hot solution of zinc dimethyldithiocarbamate (10 mM) in dichloromethane. The resulting solution was cooled and then added to petroleum ether to produce a yellow precipitate, which was filtered off and recrystallized from chloroform.

**[Zn{$S_2CN(CH_2)_4$}$_2$(2,2′-bipy)] or C$_{20}$H$_{24}$N$_4$S$_4$Zn.** This compound melts at 228°C–231°C and crystallizes in the orthorhombic structure with the lattice parameters $a = 2011.5 ± 0.1$, $b = 639.13 ± 0.04$, and $c = 1715.5 ± 0.1$ pm at 223 ± 2 K and a calculated density of 1.548 g·cm$^{-3}$ (Jie and Tiekink 2002). Bright-yellow crystals were obtained from the slow evaporation of a MeCN/CHCl$_3$ (1:1 volume ratio) solution containing equimolar amounts of Zn{$S_2CN(CH_2)_4$}$_2$ and 2,2′-bipy.

**[Zn(S₂CNEt₂)₂·(2,2′-bipy)] or C₂₀H₂₈N₄S₄Zn.** This compound melts at 175°C–176°C (Bell et al. 1989). It was prepared as pale-yellow needles by adding a hot solution of 2,2′-bipy (50 mM) in MeOH to a hot solution of zinc diethyldithiocarbamate (10 mM) in dichloromethane. The resulting solution was cooled and then added to petroleum ether to produce a yellow precipitate, which was filtered off and recrystallized from dichloromethane.

The title compound could also be prepared by the interaction of Zn(S₂CNEt₂)₂ and 2,2′-bipy (2:3 molar ratio) in the boiling toluene (Larionov et al. 1990).

**[MeZnSBuᵗPy]₂ or C₂₀H₃₄N₂S₂Zn₂.** This compound melts at 147°C–155°C with decomposition (Coates and Ridley 1965). It was prepared by the addition of excess pyridine to a solution of methylzinc t-butyl sulfide in benzene followed by removal of all volatile matter and recrystallization from benzene/hexane (1:10 volume ratio).

**[(bme-daco)Zn]₂ or C₂₀H₄₀N₄S₄Zn₂ (H₂bme-daco = N,N′-bis(mercaptoethyl)-1,5-diazacyclooctane).** This compound crystallizes in the monoclinic structure with the lattice parameters $a = 892.6 \pm 0.2$, $b = 1278.5 \pm 0.2$, $c = 1119.9 \pm 0.27$ pm, and $\beta = 101.13° \pm 0.02°$ (Tuntulani et al. 1992). To obtain it, under anaerobic conditions, H₂bme-daco (15 mM) dissolved in dry toluene was added to anhydrous zinc acetylacetonate (15 mM) in 100 mL of toluene without stirring. On standing overnight, the product crystallized as colorless crystals of [(bme-daco)Zn]₂. Crystals suitable for XRD were obtained on recrystallization from a hot mixture MeOH/acetonitrile (50:50 volume ratio).

**Zn(SC₆H₄N)[MeC(C₅H₃N)CMe] or C₂₁H₁₇N₃S₂Zn, bis(2-thiobenzaldimino)-2,6-diacetylpiridinezinc.** This compound crystallizes in the monoclinic structure with the lattice parameters $a = 926.8 \pm 0.8$, $b = 1317.4 \pm 1.3$, $c = 817.2 \pm 0.7$ pm, and $\beta = 105.00° \pm 0.04°$ and experimental and calculated densities $1.50 \pm 0.02$ and $1.519 \pm 0.004$ g·cm⁻³, respectively (Lindoy and Busc 1972, Goedken and Christoph 1973). Reaction of bis-benzothiazoline with Zn(OAc)₂ in DMF/acetone yields red crystals of this compound (Lindoy and Busc 1972).

**[Zn(S₂CNEt₂)₂·Py·C₆H₆] or C₂₁H₃₁N₃S₄Zn.** This compound crystallizes in the monoclinic structure with the lattice parameters $a = 1141.6 \pm 0.3$, $b = 816.8 \pm 0.4$, $c = 1395.7 \pm 0.7$ pm, and $\beta = 101.55° \pm 0.06°$ at $200 \pm 1$ K and a calculated density of $1.535 \pm 0.001$ g·cm⁻³ (Ivanov et al. 1998). It was obtained by the dissolution of Zn(S₂CNEt₂)₂ in benzene following the addition of Py (10% excess) and slow evaporation of the saturated solution at room temperature. Colorless crystals were separated from the mother liquor, washed with a small amount of benzene, and dried in air. They were stored in sealed ampoules.

**[Zn(Et₂NCS₂)₂(C₅H₁₀NH)]·(C₆H₅N) or C₂₁H₃₆N₄S₄Zn.** This compound was synthesized via careful wetting of [Zn(Et₂NCS₂)₂(C₅H₁₀NH)] with pyridine without changing its state of aggregation (Ivanov et al. 2001b). It was kept in sealed tubes.

**[Me₄N][Zn(S₂CNEt₂)₂(C₇H₄NS₂)] or C₂₁H₃₆N₄S₆Zn.** To obtain this compound, a mixture of [Zn(S₂CNEt₂)₂] and [Me₄N][C₇H₄NS₂] was refluxed in acetone for 1 h (McCleverty et al. 1980a). The yellow solution was filtered, and on partial evaporation of the filtrate, yellow crystals of the title compound formed. These were filtered off and recrystallized from ethanol/light petroleum mixture.

**[Zn(Et₂NCS₂)₂(C₅H₁₀NH)]·(C₆H₆) or C₂₁H₃₇N₃S₄Zn**. This compound was synthesized via careful wetting of [Zn(Et₂NCS₂)₂(C₅H₁₀NH)] with benzene without changing its state of aggregation (Ivanov et al. 2001b). It could also be synthesized by dissolving [Zn₂(Et₂NCS₂)₄ in benzene containing piperidine in an amount exceeding the stoichiometric amount by ~10% and through further crystallization of the adduct at room temperature. The title compound was kept in sealed tubes.

**[Zn(C₅H₄NS)₂](Phen) or C₂₂H₁₆N₄S₂Zn**. This compound crystallizes in the monoclinic structure with the lattice parameters $a = 1511.6 \pm 0.2$, $b = 991.8 \pm 0.2$, $c = 1418.1 \pm 0.3$ pm, and $\beta = 108.20° \pm 0.03°$ and a calculated density of 1.532 g·cm⁻³ (Durán et al. 1991). It was prepared by direct electrochemical oxidation of Zn into a solution of pyridine-2-thione with the addition of Phen in acetonitrile. At the end of the electrolysis, the solid was collected, washed with acetonitrile, and dried in vacuo.

**[Zn(C₁₁H₉N₂S)₂] or C₂₂H₁₈N₄S₂Zn, bis(2-[(pyrrol-2-yl)methyleneamino]thiophenolato}zinc**. This compound crystallizes in the orthorhombic structure with the lattice parameters $a = 1376.4 \pm 0.3$, $b = 1781.9 \pm 0.1$, and $c = 1659.3 \pm 0.2$ pm and a calculated density of 1.521 g·cm⁻³ (Castro et al. 1990). It was obtained by the electrochemical oxidation of anodic Zn in acetonitrile solution of the Schiff base derived from 2-pyrrolcarbaldehyde and bis(2-aminophenyl)disulfide.

**[Zn(S₂CNEt₂)₂·(Phen)] or C₂₂H₂₈N₄S₄Zn**. This compound melts at 210°C–211°C (Bell et al. 1989) and exists in two polymorphic modifications (Zemskova et al. 1995). One of them crystallizes in the triclinic structure with the lattice parameters $a = 974.5 \pm 0.2$, $b = 1025.2 \pm 0.2$, and $c = 1433.1 \pm 0.3$ pm and $\alpha = 99.18° \pm 0.02°$, $\beta = 91.01° \pm 0.02°$, and $\gamma = 113.28° \pm 0.02°$ and a calculated density of 1.392 g·cm⁻³. The second modification crystallizes in the monoclinic structure with the lattice parameters $a = 722.0 \pm 0.6$, $b = 1809.5 \pm 0.2$, $c = 1905.0 \pm 0.4$ pm, and $\beta = 95.87° \pm 0.02°$ and a calculated density of 1.454 g·cm⁻³. It was prepared as yellow needles by adding a hot solution of Phen (50 mM) in MeOH to a hot solution of zinc diethyldithiocarbamate (10 mM) in dichloromethane (Bell et al. 1989). The resulting solution was cooled and then added to petroleum ether to produce a yellow precipitate, which was filtered off and recrystallized from MeOH.

The title compound could also be prepared by the interaction of Zn(S₂CNEt₂)₂ and Phen (1:1.05 molar ratio) in the boiling toluene (Larionov et al. 1990).

Sublimation of Zn(S₂CNEt₂)₂·(Phen) begins at 90°C, and its thermal decomposition leads to the formation of α-ZnS (Klevtsova et al. 1999b).

**[Zn(S₂CNMe₂)₂·(3,4,7,8-Me₄Phen)] or C₂₂H₂₈N₄S₄Zn**. This compound melts at 236°C–239°C (Bell et al. 1989). It was prepared as pale-yellow needles by adding a hot solution of 3,4,7,8-Me₄Phen (50 mM) in MeOH to a hot solution of zinc dimethyldithiocarbamate (10 mM) in toluene. The resulting solution was cooled and then added to petroleum ether to produce a yellow precipitate, which was filtered off and recrystallized from chloroform.

**[Zn(C$_5$H$_{10}$NCS$_2$)$_2$]·(2,2′-bipy) or C$_{22}$H$_{28}$N$_4$S$_4$Zn.** This compound crystallizes in the orthorhombic structure with the lattice parameters $a = 3317.7 \pm 0.2$, $b = 999.8 \pm 0.3$, and $c = 1507.5 \pm 0.2$ pm and a calculated density of 1.4402 g·cm$^{-3}$ (Thirumaran et al. 1999). It was prepared by adding a hot solution of 2,2′-bipy (2 mM) in EtOH to a hot solution of (2,2′-bipyridine)bis(piperidinecarbodithioato)zinc (1 mM) in chloroform. The resulting solution was cooled and then added with petroleum ether. The yellow precipitate of the title compound was separated out. Single crystals were obtained by recrystallization from a CHCl$_3$/MeCN mixture solvent.

**[Zn(S$_2$CNEt$_2$)$_2$·(C$_{12}$H$_{10}$N$_2$)]$_2$ or C$_{22}$H$_{30}$N$_4$S$_4$Zn.** This compound melts at 197°C–198°C and crystallizes in the triclinic structure with the lattice parameters $a = 953.11 \pm 0.10$, $b = 1228.20 \pm 0.14$, and $c = 1291.65 \pm 0.10$ pm and $\alpha = 63.369° \pm 0.002°$, $\beta = 80.547° \pm 0.003°$, and $\gamma = 68.465° \pm 0.002°$ at 183 K (Lai and Tiekink 2003b). Bright-yellow crystals were isolated from acetonitrile/chloroform (1:2 volume ratio) solution containing equimolar amounts of Zn(S$_2$CNEt$_2$)$_2$ and *trans*-1,2-bis(4-pyridyl) ethylene (C$_{12}$H$_{10}$N$_2$).

**[Zn(C$_6$H$_{12}$N$_2$S$_2$)$_2$(2,2′-bipy)] or C$_{22}$H$_{32}$N$_6$S$_4$Zn.** This compound decomposes at 169°C (Prakasam et al. 2007). It was prepared by adding a hot solution of 2,2′-bipy (1 mM) in benzene to a hot suspension of [Zn(C$_6$H$_{12}$N$_2$S$_2$)$_2$] (1 mM) in benzene/ chloroform (1:1 volume ratio) mixture. The resulting pale-yellow solution was filtered and cooled. The yellow compound was separated out, which was filtered and recrystallized from benzene/chloroform (1:1 volume ratio) mixture.

**[Zn$_{10}$S$_4$(SEt)$_{12}$(C$_5$H$_3$N)$_4$] or C$_{22}$H$_{36}$N$_2$S$_8$Zn$_5$.** This compound was obtained in the same way as C$_{26}$H$_{48}$N$_2$S$_8$Zn$_5$ was obtained using pyridine instead of 3,5-dimethyl-piridine (Nyman et al. 1996).

**[Et$_4$N]$_2$[Zn(C$_3$S$_5$)$_2$] or C$_{22}$H$_{40}$N$_2$S$_{10}$Zn.** This compound melts at 186°C–188°C and crystallizes in the monoclinic structure with the lattice parameters $a = 957.220 \pm 0.010$, $b = 1786.99 \pm 0.02$, $c = 1941.94 \pm 0.02$ pm, and $\beta = 97.1592° \pm 0.0006°$ at $150 \pm 2$ K and a calculated density of 1.448 g·cm$^{-3}$ (Harrison et al. 2000). It was obtained from Na, CS$_2$, and Et$_4$NBr. Crystals suitable for XRD were grown by recrystallization from a Me$_2$CO/Pr$^i$OH mixture.

**[Me$_4$N][Zn(S$_2$CNEt$_2$)(C$_7$H$_4$NS$_2$)$_2$] or C$_{23}$H$_{30}$N$_4$S$_6$Zn.** To obtain this compound, a mixture of [Zn(C$_7$H$_4$NS$_2$)$_2$]$_n$ and [Me$_4$N][S$_2$CNEt$_2$] was refluxed in acetone for 1 h (McCleverty et al. 1980a). The yellow solution was filtered and partial evaporation of the filtrate afforded yellow crystals of the title compound. These were filtered off and recrystallized from acetone/light petroleum.

**[Me$_4$N][Zn{S$_2$CN(CH$_2$)$_5$}$_2$(C$_7$H$_4$NS$_2$)] or C$_{23}$H$_{36}$N$_4$S$_6$Zn.** To obtain this compound, a mixture of [Zn(S$_2$CN(CH$_2$)$_5$] and [Me$_4$N][C$_7$H$_4$NS$_2$] was refluxed in acetone for 1 h (McCleverty et al. 1980a). The yellow solution was filtered, and on partial evapora-tion of the filtrate, yellow crystals of the title compound formed. These were filtered off and recrystallized from the ethanol/light petroleum mixture.

**[Zn(C$_7$H$_4$NS$_2$)$_2$(2,2′-bipy)] or C$_{24}$H$_{16}$N$_4$S$_4$Zn.** This compound crystallizes in the triclinic structure with the lattice parameters $a = 1127 \pm 1$, $b = 1414 \pm 1$, and

$c = 864 \pm 1$ pm and $\alpha = 105.60° \pm 0.08°$, $\beta = 106.37° \pm 0.09°$, and $\gamma = 102.29° \pm 0.08°$ (Castro et al. 1993b). Yellow block crystals of the title compound were prepared by the electrochemical oxidation of anodic Zn in an acetonitrile solution of benzothiazole-2-thione in the presence of 2,2′-bipy.

**[Zn($C_6H_6S_2$)($C_{18}H_{12}N_2$)] or $C_{24}H_{18}N_2S_2Zn$.** This compound crystallizes in the monoclinic structure with the lattice parameters $a = 1179.9 \pm 0.3$, $b = 1166.5 \pm 0.3$, $c = 1463.0 \pm 0.3$ pm, and $\beta = 93.18° \pm 0.02°$ and a calculated density of 1.53 g·cm$^{-3}$ (Halvorsen et al. 1995). To obtain it, Zn(OAc)$_2$·2H$_2$O (1 mM) was dissolved in 50 mL of hot EtOH. To this solution, 1,2-benzenethiol ($C_6H_6S_2$) dissolved in 25 mL of hot EtOH was added dropwise. A white precipitate formed, which was redissolved by the addition of a small amount of hot DMF, and the resulting solution was brought to reflux. A solution containing 2,2′-biquinoline ($C_{18}H_{12}N_2$) (1 mM) dissolved in 50 mL of hot EtOH was added slowly with vigorous stirring to the aforementioned refluxing solution. The resulting reaction mixture was cooled slowly and allowed to stand for 24 h. A dark purple solid was collected by vacuum filtration, recrystallized from a DMF/EtOH (1:1 volume ratio) solvent, washed twice with EtOH, and dried in vacuo.

**Zn[bis(pyridylmethyl)amine](SPh)$_2$ or $C_{24}H_{23}N_3S_2Zn$.** This compound melts at 158°C with decomposition and crystallizes in the triclinic structure with the lattice parameters $a = 995.1 \pm 0.1$, $b = 1135.4 \pm 0.1$, and $c = 1178.8 \pm 0.1$ pm and $\alpha = 61.800° \pm 0.005°$, $\beta = 77.000° \pm 0.004°$, and $\gamma = 86.150° \pm 0.004°$ and experimental and calculated densities 1.35 and 1.40 g·cm$^{-3}$, respectively (Brand et al. 1996). To obtain it, freshly distilled thiophenol (2.5 mM) in MeOH (30 mL) was treated dropwise under stirring with 10.4 mL of a 0.24 M solution of NaOMe in MeOH. A solution of 1.24 mM of Zn(ClO$_4$)$_2$·6H$_2$O in MeOH (30 mL) and a solution of 1.25 mM of bis(pyridylmethyl)amine in MeOH (20 mL) were added with stirring. The clear, orange reaction mixture was reduced in vacuo to 30 mL. The colorless precipitate that was formed was filtered off, washed with a small amount of MeOH, and dried in vacuo. Recrystallization from acetonitrile yielded $C_{24}H_{23}N_3S_2Zn$ as colorless crystals. All S-containing compounds were handled in a N$_2$ atmosphere.

**[Zn{S$_2$CN(CH$_2$)$_4$}$_2$(4,7-Me$_2$Phen)] or $C_{24}H_{28}N_4S_4Zn$.** This compound melts at 256°C–258°C and crystallizes in the orthorhombic structure with the lattice parameters $a = 1125.58 \pm 0.06$, $b = 1450.78 \pm 0.07$, and $c = 3139.50 \pm 0.17$ pm at $193 \pm 2$ K and a calculated density of 1.467 g·cm$^{-3}$ (Guo et al. 2002). Yellow crystals were obtained from the slow evaporation of a MeCN/CHCl$_3$ (1:1 volume ratio) solution of the title compound that had been prepared by refluxing equimolar amounts of Zn{S$_2$CN(CH$_2$)$_4$}$_2$ and 4,7-Me$_2$Phen in chloroform solution.

**[Zn($C_5H_{10}NCS_2$)$_2$]·(Phen) or $C_{24}H_{28}N_4S_4Zn$.** This compound was prepared by adding a hot solution of Phen (2 mM) in EtOH to a hot solution of (2,2′-bipy)-bis(piperidinecarbodithioato)zinc (1 mM) in chloroform (Thirumaran et al. 1999). The resulting solution was cooled and then added with petroleum ether. A yellow precipitate of the title compound was separated out.

**[Me$_4$N]{Zn(S$_2$CN(CH$_2$)$_5$}(C$_7$H$_4$NS$_2$)$_2$] or C$_{24}$H$_{30}$N$_4$S$_6$Zn.** To obtain this compound, a mixture of [Zn(C$_7$H$_4$NS$_2$)$_2$]$_n$ and [Me$_4$N][S$_2$CN(CH$_2$)$_5$] was refluxed in acetone for 1 h (McCleverty et al. 1980a). The yellow solution was filtered and partial evaporation of the filtrate afforded yellow crystals of the title compound. These were filtered off and recrystallized from acetone/light petroleum.

**[Zn(S$_2$CNEt$_2$)$_2$·(2,9-Me$_2$Phen)] or C$_{24}$H$_{32}$N$_4$S$_4$Zn.** This compound melts at 199°C–202°C (Bell et al. 1989). It was prepared as yellow needles by adding a hot solution of 2,9-Me$_2$Phen (50 mM) in MeOH to a hot solution of zinc diethyldithiocarbamate (10 mM) in dichloromethane. The resulting solution was cooled and then added to petroleum ether to produce a yellow precipitate, which was filtered off and recrystallized from chloroform.

**[Zn(C$_{12}$H$_{16}$N$_5$S)$_2$] or C$_{24}$H$_{32}$N$_{10}$S$_2$Zn.** This compound crystallizes in the orthorhombic structure with the lattice parameters $a = 1384.7 \pm 0.6$, $b = 1496.9 \pm 0.3$, and $c = 2654.5 \pm 0.6$ pm and a calculated density of 1.430 g·cm$^{-3}$ (Castiñeiras et al. 2002). Its single crystals suitable for XRD were obtained by slow evaporation of the mother liquor from preparing [Zn(C$_{12}$H$_{17}$N$_5$S)Cl$_2$] in air at room temperature [C$_{12}$H$_{17}$N$_5$S = 2-pyridineformamide N(4)-piperidylthiosemicarbazone].

**[Zn$_2$(S$_2$CNMe$_2$)$_4$(C$_{12}$H$_{10}$N$_2$)] or C$_{24}$H$_{34}$N$_6$S$_8$Zn$_2$.** This compound melts at 282°C–284°C and crystallizes in the monoclinic structure with the lattice parameters $a = 1306.1 \pm 0.4$, $b = 1590.4 \pm 0.4$, $c = 1765.8 \pm 0.5$ pm, and $\beta = 108.443° \pm 0.004°$ at 98 K and a calculated density of 1.515 g·cm$^{-3}$ (Poplaukhin and Tiekink 2009). It was recrystallized as pale-yellow prisms from the slow evaporation of a chloroform/acetonitrile (3:1 volume ratio) solution.

**[Zn(S$_2$CNPr$^n_2$)$_2$·(2,2'-bipy)] or C$_{24}$H$_{36}$N$_4$S$_4$Zn.** This compound melts at 156°C and crystallizes in the monoclinic structure with the lattice parameters $a = 964.6 \pm 0.2$, $b = 2097.8 \pm 0.4$, $c = 1455.5 \pm 0.3$ pm, and $\beta = 94.95° \pm 0.03°$ and a calculated density of 1.300 g·cm$^{-3}$ (Berus et al. 1998, Glinskaya et al. 1998a). It was prepared by the next procedure (Berus et al. 1998). Zn(S$_2$CNPr$^n_2$) (0.96 mM) and 2,2'-bipy (1.43 mM) were dissolved in toluene (10 mL) at the heating on water bath. The solution was cooled to room temperature and yellow precipitate was filtered off, washed with toluene, and dried in air. To obtain its single crystals, the saturated solution of [Zn(S$_2$CNPr$^n_2$)$_2$·(2,2'-bipy)] in toluene was allowed to stand at ~5°C for 2 days (Glinskaya et al. 1998a).

**[Zn(S$_2$CNPr$^n_2$)$_2$·(4,4'-bipy)] or C$_{24}$H$_{36}$N$_4$S$_4$Zn.** This compound melts at 143°C (Berus et al. 1998) and crystallizes in the monoclinic structure with the lattice parameters $a = 1656.0 \pm 0.2$, $b = 1142.3 \pm 0.1$, $c = 1531.9 \pm 0.2$ pm, and $\beta = 94.27° \pm 0.01°$ and a calculated density of 1.320 g·cm$^{-3}$ (Klevtsova et al. 2001a). To obtain it, Zn(S$_2$CNPr$^n_2$) (1.02 mM) and 4,4'-bipy (1.47 mM) were dissolved in toluene (20 mL) at the heating on water bath (Berus et al. 1998). The solution was cooled to room temperature, and white precipitate was filtered off, washed with toluene, and dried in air. The title compound was recrystallized from toluene. Its single crystals were obtained during synthesis using ethyl acetate instead of toluene (Klevtsova et al. 2001a).

**[Zn$_2$(C$_5$H$_8$NS$_2$)$_4$(4,4′-bipy)]** or **C$_{24}$H$_{40}$N$_6$S$_8$Zn$_2$**. This compound crystallizes in the orthorhombic structure with the lattice parameters $a = 1176.43 \pm 0.02$, $b = 1999.65 \pm 0.03$, and $c = 3298.89 \pm 0.03$ pm and a calculated density of 1.493 g·cm$^{-3}$ (Chen et al. 2000). To obtain it, bis(pyrrolidinedithiocarboxylate)zinc and 4,4′-bipy were dissolved in DMF and refluxed for 1 h. The yellow microcrystals that formed were collected by concentrating the DMF solution. Single crystals suitable for XRD were obtained by recrystallization from MeCN.

**[Zn$_2$(EtPr$^i$CNS$_2$)]** or **C$_{24}$H$_{48}$N$_4$S$_8$Zn$_2$**. This compound crystallizes in the monoclinic structure with the lattice parameters $a = 1086.46 \pm 0.01$, $b = 1505.17 \pm 0.01$, $c = 1129.34 \pm 0.01$ pm, and $\beta = 101.419° \pm 0.001°$ and a calculated density of 1.431 g·cm$^{-3}$ (Baba et al. 2001b). To obtain it, CS$_2$ was added to an ethanol solution of ethylisopropylamine at 4°C followed by an ethanol solution of ZnCl$_2$. The mixture was stirred vigorously and then set aside. The solid that separated was isolated and recrystallized from EtOH to afford the title compound.

**[Bu$^n_4$N][Zn(S$_2$CNMe$_2$)$_3$]** or **C$_{25}$H$_{54}$N$_4$S$_6$Zn**. To obtain this compound, [Zn(S$_2$CNMe$_2$)$_2$] (0.02 M) was added to a solution of [Bu$^n_4$N][S$_2$CNMe$_2$] (0.02 M) in acetone (200 mL) (McCleverty and Morrison 1976). The mixture was shaken until all the solid had dissolved, then light petroleum was added, and the solvent was partially evaporated. The white crystals were collected and recrystallized from acetone/light petroleum.

**[Zn(C$_7$H$_4$NS$_2$)$_2$(Phen)]** or **C$_{26}$H$_{16}$N$_4$S$_4$Zn**. This compound was prepared by the electrochemical oxidation of anodic Zn in an acetonitrile solution of benzothiazole-2-thione in the presence of Phen (Castro et al. 1993b).

**[Zn(Phen)(C$_7$H$_7$S)$_2$]** or **C$_{26}$H$_{22}$N$_2$S$_2$Zn, (Phen)bis(4-toluenethiolato)zinc**. This compound crystallizes in the monoclinic structure with the lattice parameters $a = 1000.9 \pm 0.3$, $b = 2012 \pm 1$, $c = 1214.0 \pm 0.4$ pm, and $\beta = 107.73° \pm 0.01°$ and experimental and calculated densities 1.396 and 1.403 g·cm$^{-3}$, respectively (Cremers et al. 1980). Bright-yellow chunks of [Zn(C$_{12}$H$_8$N$_2$)(C$_7$H$_7$S)$_2$] were prepared by precipitation from an ethanol solution of stoichiometric mixtures of Zn and ligands.

**[Zn(PhS)$_2$(2,9-Me$_2$Phen)]** or **C$_{26}$H$_{22}$N$_2$S$_2$Zn, bis(benzenethiolato)(2,9-Me$_2$Phen) zinc**. This compound melts at 217°C–218°C and is characterized by two polymorphic modifications (Jordan et al. 1991). Phase transition from orthorhombic to monoclinic structure takes place at 136°C–143°C. α-Zn(PhS)$_2$ (2,9-Me$_2$Phen) crystallizes in the orthorhombic structure with the lattice parameters $a = 2729.6 \pm 0.8$, $b = 1394.1 \pm 0.3$, and $c = 1213.9 \pm 0.5$ pm and a calculated density of 1.41 g·cm$^{-3}$. β-Zn(PhS)$_2$(2,9-Me$_2$Phen) crystallizes in the monoclinic structure with the lattice parameters $a = 1541.8 \pm 0.3$, $b = 1315.4 \pm 0.2$, $c = 1155.6 \pm 0.2$ pm, and $\beta = 101.46° \pm 0.01°$ and a calculated density of 1.42 g·cm$^{-3}$.

To obtain this compound, to a solution of Zn(OAc)$_2$·2H$_2$O (1 mM) dissolved in 100 mL of hot EtOH, a solution of redistilled benzenethiol (0.21 mL, 2.1 mM in 10 mL of EtOH) was added dropwise with stirring. As the second millimole was added, a white precipitate formed. After 5 min of continued stirring, 2,9-Me$_2$Phen (1 mM) dissolved in 10 mL of EtOH was added dropwise. The mixture was heated and stirred for 0.5 h as it gradually turned to a deep-yellow solution. It was set aside to cool. After 2–3 days at room temperature, crystals were harvested and washed with EtOH.

Although $Zn(PhS)_2(2,9\text{-}Me_2Phen)$ generally produced monoclinic crystals, recrystallization of a samples at room temperature in an open beaker yielded, adventitiously, a crop of large orthorhombic needles (Jordan et al. 1991).

**$[Zn(C_{13}H_{11}N_4S)_2]$ or $C_{26}H_{22}N_8S_2Zn$, zinc dithizonate.** This compound has two polymorphic modifications. First of them crystallizes in the monoclinic structure with the lattice parameters $a = 788.7 \pm 0.3$, $b = 2240 \pm 1$, $c = 1532.4 \pm 0.6$ pm, and $\beta = 92.82° \pm 0.03°$ and experimental and calculated densities $1.76 \pm 0.01$ and 1.76 g·cm$^{-3}$, respectively (Harrowfield et al. 1983) ($a = 788.7 \pm 0.6$, $b = 2250.1 \pm 1.7$, $c = 1532.7 \pm 1.1$ pm, and $\beta = 92.58° \pm 0.04°$ and experimental and calculated densities 1.40 g·cm$^{-3}$, respectively [Math and Freiser 1971]). Second modification also crystallizes in the monoclinic structure with the lattice parameters $a = 760.7 \pm 0.2$, $b = 1566 \pm 1$, $c = 2276 \pm 1$ pm, and $\beta = 94.97° \pm 0.03°$ and experimental and calculated densities $1.76 \pm 0.01$ and 1.76 g·cm$^{-3}$, respectively.

According to the data of Mawby and Irving (1971, 1972), this compound crystallizes in the monoclinic structure with the lattice parameters $a = 1521 \pm 2$, $b = 2225 \pm 3$, $c = 784 \pm 1$ pm, and $\beta = 91.4° \pm 0.2°$. Fine gray-green needles with a metallic reflex were obtained by recrystallization of a pure sample of $C_{26}H_{22}N_8S_2Zn$ from chloroform.

**$[Zn(Pr^n_2S_2CN)_2 \cdot (Phen)]$ or $C_{26}H_{36}N_4S_4Zn$.** This compound melts at 216°C (Berus et al. 1998) and crystallizes in the orthorhombic structure with the lattice parameters $a = 1862.1 \pm 0.2$, $b = 1470.1 \pm 0.2$, and $c = 1067.6 \pm 0.1$ pm and a calculated density of 1.360 g·cm$^{-3}$ (Glinskaya et al. 1998a). To obtain it, $Zn(S_2CNPr^n_2)$ (1.72 mM) and Phen·H$_2$O (1.92 mM) were dissolved in toluene (100 mL) at the heating on water bath (Berus et al. 1998). The solution was cooled to room temperature, and the yellow precipitate was filtered off, washed with toluene, and dried in air. The title compound was recrystallized from toluene. To obtain its single crystals, the saturated solution of $[Zn(Pr^n_2S_2CN)_2 \cdot (Phen)]$ in toluene was allowed to stand at ~5°C for 2 days (Glinskaya et al. 1998a).

**$[Zn(Pr^i_2S_2CN)_2 \cdot (Phen)]$ or $C_{26}H_{36}N_4S_4Zn$.** This compound crystallizes in the monoclinic structure with the lattice parameters $a = 1874.4 \pm 0.4$, $b = 1095.4 \pm 0.2$, $c = 1454.8 \pm 0.3$ pm, and $\beta = 103.97° \pm 0.03°$ and a calculated density of 1.371 g·cm$^{-3}$ (Klevtsova et al. 1999c, Larionov et al. 1999). To obtain it, $Zn(Pr^i_2S_2CN)_2$ (2.4 mM) and Phen·H$_2$O (3.5 mM) were dissolved in CHCl$_3$ (ca. 20 mL) with heating in a water bath. The resulting yellow solution was filtered under vacuum and an equal amount of EtOH was added. The mixture was cooled to room temperature and then held at −15°C for 2 h. The yellow precipitate was filtered under vacuum, washed with cooled EtOH, and dried in air. Single crystals of the title compound were grown by slow cooling to room temperature a hot saturated solution in a mixture of toluene/chloroform (2:1 volume ratio) and subsequent evaporation of this solution for several days.

**$[Zn(S_2CNEt_2)_2 \cdot (3,4,7,8\text{-}Me_4Phen)]$ or $C_{26}H_{36}N_4S_4Zn$.** This compound melts at 220°C–223°C (Bell et al. 1989). It was prepared as pale-yellow needles by adding a hot solution of 3,4,7,8-Me$_4$Phen (50 mM) in MeOH to a hot solution of zinc diethyldithiocarbamate (10 mM) in dichloromethane. The resulting solution was cooled and then added to petroleum ether to produce the yellow precipitate, which was filtered off and recrystallized from chloroform.

**[Zn{S$_2$CN(C$_6$H$_{11}$)$_2$}$_2$] or C$_{26}$H$_{44}$N$_2$S$_4$Zn.** This compound crystallizes in the trigonal structure with the lattice parameters $a = 1026.4 \pm 1.0$ and $c = 2363.0 \pm 0.9$ pm at 110 K and a calculated density of 1.343 g·cm$^{-3}$ (Cox and Tiekink 1999).

**[Zn$_{10}$S$_4$(SEt)$_{12}${C$_5$H$_3$(Me)$_2$N}$_4$] or C$_{26}$H$_{48}$N$_2$S$_8$Zn$_5$.** This compound crystallizes in the monoclinic structure with the lattice parameters $a = 2579.7 \pm 0.2$, $b = 1777.3 \pm 0.2$, $c = 2165.5 \pm 0.2$ pm, and $\beta = 112.00° \pm 0.01°$ and a calculated density of 1.402 g·cm$^{-3}$ (Nyman et al. 1996). To obtain it, 4.86 mM of diethylzinc and 75 mL of dry toluene were syringed into a 250 mL Schlenk flask. In three equal portions over 2 h, 12.1 mM of S$_8$ was added to the stirring solution of diethyl zinc and toluene at room temperature. The solution was stirred for 15 min more, and a pale-yellow solution remained when the reaction was complete. The volatile components were removed in vacuo, leaving a glassy cream-colored solid. The crude mixture of products was collected and redissolved in 100 mL of 3,5-dimethylpiridine to give a yellow solution, and C$_{26}$H$_{48}$N$_2$S$_8$Zn$_5$ crystallized as a white crystalline solid. Single crystals of this compound were grown by slow evaporation of a dilute 3,5-dimethylpiridine solution to yield small colorless blocks.

**[2{MePr$^i$NCS$_2$}$_2$Zn·Me$_2$N(CH$_2$)$_2$NMe$_2$] or C$_{26}$H$_{56}$N$_6$S$_8$Zn$_2$.** This compound melts at 175°C and crystallizes in the monoclinic structure with the lattice parameters $a = 1131.2 \pm 1.0$, $b = 2561.6 \pm 0.4$, $c = 705.6 \pm 1.0$ pm, and $\beta = 104.02° \pm 0.10°$ and a calculated density of 1.406 g·cm$^{-3}$ (Malik et al. 1997). To obtain this compound, bis(N-methylisopropyldithiocarbamato)zinc was dissolved in N,N,N′,N′-tetramethylethylenediamine, and as the excess solvent evaporated, a crude white solid appeared, which was recrystallized from acetone to give colorless crystals.

**[Me$_4$N][Zn(S$_2$CNBu$^n_2$)(C$_7$H$_4$NS$_2$)$_2$] or C$_{27}$H$_{38}$N$_4$S$_6$Zn.** To obtain this compound, a mixture of [Zn(C$_7$H$_4$NS$_2$)$_2$]$_n$ and [Me$_4$N][S$_2$CNBu$^n_2$] was refluxed in acetone for 1 h (McCleverty et al. 1980a). The yellow solution was filtered and partial evaporation of the filtrate afforded yellow crystals of the title compound. These were filtered off and recrystallized from acetone/light petroleum.

**[Bu$^n_4$N][Zn(S$_2$CNEt$_2$)(S$_2$CNMe$_2$)$_2$] or C$_{27}$H$_{58}$N$_4$S$_6$Zn.** This compound was prepared as follows (McCleverty et al. 1980a). To an acetone solution of [Bu$^n_4$N][S$_2$CNEt$_2$] (ca. 1 M) was added [Zn(S$_2$CNMe$_2$)$_2$] (ca. 1 M), and the mixture stirred for 2 h. After filtration, the filtrate was treated with light petroleum, and on partial evaporation, the complex precipitated. The solid was filtered off and recrystallized from acetone/light petroleum. On occasions it proved necessary to reflux the reaction mixture or to use refluxing toluene as the reaction medium. Despite careful control of the stoichiometry reaction, it was often impossible to prevent ligand redistribution.

**[Zn(S-2,3,5,6-Me$_4$C$_6$H)$_2$(1-Me-imid)$_2$] or C$_{28}$H$_{38}$N$_4$S$_2$Zn.** This compound was obtained by the interaction of Zn(S-2,3,5,6-Me$_4$C$_6$H)$_2$ with two equivalents of 1-methylimidazole in MeCN (Corwin et al. 1987).

**[Zn($C_{14}H_{21}N_4S$)$_2$] or $C_{28}H_{42}N_8S_2Zn$.** This compound crystallizes in the monoclinic structure with the lattice parameters $a = 873.38 \pm 0.02$, $b = 1467.10 \pm 0.05$, $c = 2461.09 \pm 0.06$ pm, and $\beta = 96.225° \pm 0.002°$ (Hammesa and Carrano 2000). To prepare it, a solution of [MeZn($C_{14}H_{21}N_4S$)] (0.48 mM) in 30 mL of $CH_2Cl_2$ was treated with one equivalent of $HSC_6F_5$ (0.48 mM). The resulting solution was stirred for 1 h, concentrated under reduced pressure, and crystallized by layering the concentrated $CH_2Cl_2$ solution with hexane to give the title compound as colorless crystals.

**[Zn($S_2CNBu^n_2$)$_2$·(2,2'-bipy)] or $C_{28}H_{44}N_4S_4Zn$.** This compound melts at 136°C–137°C and crystallizes in the orthorhombic structure with the lattice parameters $a = 2931 \pm 2$, $b = 677 \pm 2$, and $c = 1682 \pm 2$ pm and a calculated density of 1.25 g·cm$^{-3}$ (Bell et al. 1989, Ivanchenko et al. 2000) ($a = 659.7 \pm 0.4$, $b = 2883.1 \pm 0.5$, and $c = 1661.7 \pm 0.3$ pm at 173 K [Tiekink 2001a]). It was prepared as pale-yellow needles by adding a hot solution of 2,2'-bipy (50 mM) in MeOH to a hot solution of zinc di-$n$-butyldithiocarbamate (10 mM) in dichloromethane. The resulting solution was cooled and then added to petroleum ether to produce the yellow precipitate, which was filtered off and recrystallized from dichloromethane (Bell et al. 1989). Pale-yellow crystals of the title compound were obtained from slow evaporation of an acetonitrile solution containing equimolar amounts of [Zn($S_2CNBu^n_2$)$_2$] and 2,2'-bipy.

**[Zn($S_2CNBu^i_2$)$_2$·(2,2'-bipy)] or $C_{28}H_{44}N_4S_4Zn$.** This compound melts at 178°C (Zemskova et al. 1996). To obtain it, Zn($S_2CNBu^i_2$)$_2$ (1.0 mM) and 2,2'-bipy (1.5 mM) were dissolved in toluene (30 mL) at the heating on water bath. At cooling to room temperature, a light-yellow precipitate was formed. The mixture was kept 2–3 h at −15°C and then the precipitate was filtered off, washed with toluene, and dried in air. Thermal decomposition of this compound at 350°C in air leads to the formation of α-ZnS.

**[Zn$_2$($S_2CNEtBu^n$)$_4$] or $C_{28}H_{56}N_4S_8Zn_2$.** This compound crystallizes in the orthorhombic structure with the lattice parameters $a = 1473.01 \pm 0.02$, $b = 1202.54 \pm 0.02$, and $c = 2379.60 \pm 0.01$ and a calculated density of 1.317 g·cm$^{-3}$ (Baba et al. 2001a). To obtain it, $CS_2$ was added to an ethanol solution of $n$-butylethylamine at 4°C followed by an ethanol solution of $ZnCl_2$. The mixture was stirred vigorously and then set aside. The solid that separated was isolated and recrystallized from EtOH to afford the title compound.

**[Zn$_2$($S_2CNPr^n_2$)$_4$] or $C_{28}H_{56}N_4S_8Zn_2$.** This compound was prepared by the interaction of Zn$^{2+}$ with appropriate dithiocarbamate anion (Ivanov et al. 2003a). White voluminous precipitate was washed by decantation, separated by filtration, and dried in air.

**[Zn$_2$($S_2CNPr^i_2$)$_4$] or $C_{28}H_{56}N_4S_8Zn_2$.** This compound crystallizes in the monoclinic structure with the lattice parameters $a = 1685.7 \pm 0.3$, $b = 1116.8 \pm 0.3$, $c = 1140.8 \pm 0.3$ pm, and $\beta = 111.8° \pm 0.2°$ and a calculated density of 1.393 g·cm$^{-3}$ (Miyamae et al. 1979). It was prepared by the interaction of Zn$^{2+}$ with appropriate dithiocarbamate anion (Ivanov et al. 2003a). White voluminous precipitate was washed by decantation, separated by filtration, and dried in air.

**[{Zn$_2$(SEt)$_6$}(Et$_4$N$_2$)] or C$_{28}$H$_{70}$N$_2$S$_6$Zn$_2$.** This compound crystallizes in the monoclinic structure with the lattice parameters $a = 1098.6 \pm 0.2$, $b = 1043.6 \pm 0.2$, $c = 1811.6 \pm 0.4$ pm, and $\beta = 93.56° \pm 0.02°$ and experimental and calculated densities 1.23 and 1.21 g·cm$^{-3}$, respectively (Watson et al. 1985) To obtain it, a solution of 20 mM of ZnCl$_2$ in 50 mL of acetonitrile and 5 mL of DMF was added 60 mM of solid NaSEt with stirring, forming a colorless solution and a precipitate of NaCl. After 50 min, 20 mM of Et$_4$NCl was added and the mixture was stirred for 24 h and filtered. The colorless filtrate was warmed to ~40°C, allowed to cool slowly to −20°C, and stored at this temperature for 24 h. The colorless crystals of C$_{28}$H$_{70}$N$_2$S$_6$Zn$_2$ were collected, washed with 25 mL of ether/acetonitrile (5:1 volume ratio) and 5 mL of ether, and dried in vacuo.

**[Bu$^n_4$N][Zn{S$_2$CN(CH$_2$)$_4$CH$_2$}(C$_7$H$_5$N$_2$S)] or C$_{29}$H$_{51}$N$_4$S$_3$Zn.** This compound was prepared as follows (McCleverty et al. 1980a). To an acetone solution of [Bu$^n_4$N] [C$_7$H$_5$N$_2$S] (4.1 g) was added [Zn{S$_2$CN(CH$_2$)$_4$CH$_2$}] (3.9 g), and the mixture was stirred and refluxed for 2 h. The solution was cooled and filtered, and upon the addition of light petroleum, the title compound precipitated. The compound was filtered off, washed with light petroleum, and air-dried.

**[Me$_4$N][Zn(S$_2$CNBu$^n_2$)$_2$(C$_7$H$_4$NS$_2$)] or C$_{29}$H$_{52}$N$_4$S$_6$Zn.** To obtain this compound, a mixture of [Zn(S$_2$CNBu$^n_2$)$_2$] and [Me$_4$N][C$_7$H$_4$NS$_2$] was refluxed in acetone for 1 h (McCleverty et al. 1980a). The yellow solution was filtered, and on partial evaporation of the filtrate, yellow crystals of the title compound formed. These were filtered off and recrystallized from ethanol/light petroleum mixture.

**[Bu$^n_4$N][Zn(S$_2$CNMe$_2$)$_2$(C$_7$H$_4$NS$_2$)] or C$_{29}$H$_{52}$N$_4$S$_6$Zn.** To obtain this compound, a mixture of [Zn(S$_2$CNMe$_2$)$_2$] (1.8 g) and [Bu$^n_4$N][C$_7$H$_4$NS$_2$] (2.0 g) was refluxed in acetone (80 mL) for 1 h (McCleverty et al. 1980a). The yellow solution was filtered, and on partial evaporation of the filtrate, yellow crystals of the title compound formed. These were filtered off and recrystallized from ethanol/light petroleum mixture.

**[Bu$^n_4$N][Zn(S$_2$CNEt$_2$)$_2$(S$_2$CNMe$_2$)] or C$_{29}$H$_{62}$N$_4$S$_6$Zn.** To obtain this compound, [Zn(S$_2$CNEt$_2$)$_2$] (0.02 M) was added to a solution of [Bu$^n_4$N][S$_2$CNMe$_2$] (0.02 M) in acetone (200 mL) (McCleverty and Morrison 1976, McCleverty et al. 1980a). The mixture was shaken until all the solid had dissolved, then light petroleum was added, and the solvent was partially evaporated. The white crystals were collected and recrystallized from acetone/light petroleum.

**[Zn{S$_2$CN(CH$_2$Ph)$_2$}$_2$] or C$_{30}$H$_{28}$N$_2$S$_4$Zn, zinc dibenzyldithiocarbamate.** This compound melts at 187°C–192°C (Bell et al. 1989) and crystallizes in the orthorhombic structure with the lattice parameters $a = 1621.9 \pm 1.1$, $b = 1900.1 \pm 1.2$, and $c = 937.6 \pm 0.6$ pm and a calculated density of 1.397 g·cm$^{-3}$, (Zhong et al. 2003) ($a = 1609.68 \pm 0.07$, $b = 1893.88 \pm 0.08$, and $c = 928.25 \pm 0.04$ pm at $198 \pm 1$ K [Decken et al. 2004]). Colorless crystals were grown by the solvent evaporation from a CH$_2$Cl$_2$ solution of the title compound layered with diethyl ether at 22°C.

**[Zn(S$_2$CNBu$^n_2$)$_2$·(Phen)] or C$_{30}$H$_{44}$N$_4$S$_4$Zn.** This compound melts at 185°C–186°C and crystallizes in the orthorhombic structure with the lattice parameters

$a = 1538.6 \pm 0.9$, $b = 2248.7 \pm 1.8$, and $c = 992.3 \pm 0.6$ pm and experimental and calculated densities 1.26 and 1.27 g·cm$^{-3}$, respectively (Ivanchenko et al. 2000, Bell et al. 1989). It was prepared as bright-yellow needles by adding a hot solution of Phen (50 mM) in MeOH to a hot solution of zinc di-$n$-butyldithiocarbamate (10 mM) in dichloromethane. The resulting solution was cooled and then added to petroleum ether to produce the yellow precipitate, which was filtered off and recrystallized from MeOH.

**[Zn(S$_2$CNBu$^i_2$)$_2$·(Phen)] or C$_{30}$H$_{44}$N$_4$S$_4$Zn.** This compound melts at 216°C (Zemskova et al. 1996) and crystallizes in the triclinic structure with the lattice parameters $a = 1856.4 \pm 0.4$, $b = 1048.7 \pm 0.1$, and $c = 1750.5 \pm 0.4$ pm and $\alpha = 84.04° \pm 0.02°$, $\beta = 94.88° \pm 0.02°$, and $\gamma = 90.87° \pm 0.02°$ and a calculated density of 1.287 g·cm$^{-3}$ (Klevtsova et al. 1999a). To obtain it, Zn(S$_2$CNBu$^i_2$)$_2$ (1.0 mM) and Phen·H$_2$O (1.5 mM) were dissolved in CHCl$_3$ (25 mL) at the heating on water bath. The obtained yellow solution was cooled, and EtOH (25 mL) was added (Zemskova et al. 1996). A yellow crystalline precipitate was formed, which was filtered off, washed with CHCl$_3$/EtOH (1:1 volume ratio), and dried in air. Thermal decomposition of this compound at 350°C in air leads to the formation of α-ZnS.

Its single crystals of bright-yellow color were grown at a slow cooling of hot saturated Zn(S$_2$CNBu$^i_2$)$_2$ solution in ethyl acetate following evaporation at room temperature (Klevtsova et al. 1999a).

**[Zn(S$_2$CNEt$_2$)$_2$]$_2$·(2,2'-bipy) or C$_{30}$H$_{48}$N$_6$S$_8$Zn$_2$.** This compound crystallizes in the orthorhombic structure with the lattice parameters $a = 1724.6 \pm 0.3$, $b = 679.2 \pm 0.1$, and $c = 2009.3 \pm 0.5$ pm and a calculated density of 1.462 g·cm$^{-3}$ (Zemskova et al. 1993a). Its single crystals were grown by the slow cooling to room temperature the hot saturated solution of this compound in the mixture EtOH/CHCl$_3$ (1:1 volume ratio) following slow evaporation for some days.

**[Zn(S$_2$CNEt$_2$)$_2$]$_2$·(4,4'-bipy) or C$_{30}$H$_{48}$N$_6$S$_8$Zn$_2$.** This compound melts at 200°C and crystallizes in the orthorhombic structure with the lattice parameters $a = 1741.4 \pm 0.4$, $b = 2216.1 \pm 0.7$, and $c = 2153.9 \pm 0.4$ pm and a calculated density of 1.406 g·cm$^{-3}$ (Zemskova et al. 1993a) ($a = 2207.21 \pm 0.09$, $b = 2145.11 \pm 0.09$, and $c = 1730.09 \pm 0.07$ pm at 223 K [Lai and Tiekink 2003a]). To obtain it, Zn(Et$_2$NCS$_2$)$_2$ (1.0 mM) and 4,4'-bipy (1.0 mM) were dissolved in CH$_2$Cl$_2$ (40 mL) at room temperature. Acetone (200 mL) was added to this solution. The yellow precipitate was filtered off, washed with acetone, dried in air, and recrystallized from CH$_2$Cl$_2$. Its single crystals were grown by the slow evaporation of the solution of this compound in CH$_2$Cl$_2$ at room temperature (Zemskova et al. 1993a). Pale-yellow crystals were also isolated from a CH$_2$Cl$_2$/toluene (2:1 volume ratio) solution containing equimolar amounts of [Zn(S$_2$CNEt$_2$)$_2$]$_2$ and 4,4'-bipy (Lai and Tiekink 2003a).

**[Me$_4$N][Zn{S$_2$CN(C$_6$H$_5$)$_2$}(C$_7$H$_4$NS$_2$)$_2$] or C$_{31}$H$_{42}$N$_4$S$_6$Zn.** To obtain this compound, a mixture of [Zn(C$_7$H$_4$NS$_2$)$_2$]$_n$ and [Me$_4$N][S$_2$CN(C$_6$H$_5$)$_2$] was refluxed in acetone for 1 h (McCleverty et al. 1980a). The yellow solution was filtered and partial evaporation of the filtrate afforded yellow crystals of the title compound. These were filtered off and recrystallized from acetone/light petroleum.

**[Bu$^n_4$N][Zn(S$_2$CNBu$^n_2$)(S$_2$CNMe$_2$)$_2$] or C$_{31}$H$_{66}$N$_4$S$_6$Zn**. To obtain this compound, [Zn(S$_2$CNBu$^n_2$)$_2$] (0.02 M) was added to a solution of [Bu$^n_4$N][S$_2$CNMe$_2$] (0.02 M) in acetone (200 mL) (McCleverty and Morrison 1976). The mixture was shaken until all the solid had dissolved, then light petroleum was added, and the solvent was partially evaporated. The white crystals were collected and recrystallized from acetone/light petroleum.

**[Zn{SCN(Ph)NC(Me)CN(Ph)}$_2$] or C$_{32}$H$_{26}$N$_8$S$_2$Zn, bis(l-phenyl-3-methyl-4-phenylhydrazono-pyrazole-5-thionato)zinc**. This compound crystallizes in the triclinic structure with the lattice parameters $a = 992.5 \pm 0.1$, $b = 1262.2 \pm 0.2$, and $c = 1327.4 \pm 0.2$ pm and $\alpha = 68.71° \pm 0.01°$, $\beta = 88.59° \pm 0.01°$, and $\gamma = 80.92° \pm 0.01°$ (Hinsche et al. 1997). It was synthesized by the reaction of Zn(OAc)$_2$ and l-phenyl-3-methyl-4-phenylhydrazono-pyrazole-5-thion in boiling ethanolic solution. Its crystal were obtained by slow cooling of the solution.

**[Zn(S-2,3,5,6-Me$_4$C$_6$H)$_2$(C$_6$H$_5$N)$_2$] or C$_{32}$H$_{36}$N$_2$S$_2$Zn**. This compound was obtained by the interaction of Zn(S-2,3,5,6-Me$_4$C$_6$H)$_2$ with two equivalents of pyridine in MeCN (Corwin et al. 1987).

**[Zn(C$_{12}$H$_{12}$NS)$_2$(C$_4$H$_6$N$_2$)$_2$] or C$_{32}$H$_{36}$N$_6$S$_2$Zn**. This compound melts at 215°C and crystallizes in the monoclinic structure with the lattice parameters $a = 1518.6 \pm 0.3$, $b = 1592.5 \pm 0.3$, $c = 1394.7 \pm 0.3$ pm, and $\beta = 90.88° \pm 0.03°$ and a calculated density of 1.236 g·cm$^{-3}$ (Otto et al. 1999). To obtain it, 2 mM of Et$_3$N, 2 mM of N-(2-thiophenyl)-2,5-dimethylpyrrole, and 2 mM of N-methylimidazole were dissolved in 35 mL of MeOH. Dry ZnCl$_2$ (1 mM) in 10 mL of MeOH was added and big colorless crystals were obtained. These crystals were filtered off, washed with MeOH, and dried in vacuo. The title compound was recrystallized from hot toluene to give colorless needles. All procedures were carried out under an Ar or N$_2$ atmosphere using standard Schlenk and glovebox techniques.

**[Me$_4$N]$_2$[Zn(SPh)$_4$] or C$_{32}$H$_{44}$N$_2$S$_4$Zn**. This compound crystallizes in the monoclinic structure with the lattice parameters $a = 1204.2 \pm 0.3$, $b = 1436.6 \pm 0.1$, $c = 983.7 \pm 0.2$ pm, and $\beta = 90.80° \pm 0.02°$ and a calculated density of 1.269 g·cm$^{-3}$ (Ueyama et al. 1988). To obtain it, to a solution of benzenethiol (0.207 mol) and Me$_4$NCl (0.068 mol) in MeOH (80 mL) at room temperature was added a solution of Zn(NO$_3$)$_2$·6H$_2$O (0.030 mol) in MeOH (40 mL) (Dance et al. 1984). Crystallization was induced in the resulting solution by swirling or seeding and allowed to continue at 0°C. The colorless crystalline product was washed with MeOH and vacuum dried. Recrystallization was from a saturated solution in boiling acetonitrile, as large diamond-shaped crystals. [NMe$_4$]$_2$[Zn(SPh)$_4$] could also be recrystallized from MeOH to give colorless needles (Ueyama et al. 1988).

**[Zn(S$_2$CNBu$^n$)$_2$·(2,9-Me$_2$Phen)] or C$_{32}$H$_{48}$N$_4$S$_4$Zn**. This compound melts at 130°C–132°C (Bell et al. 1989). It was prepared as yellow needles by adding a hot solution of 2,9-Me$_2$Phen (50 mM) in MeOH to a hot solution of zinc di-n-butyldithiocarbamate (10 mM) in dichloromethane. The resulting solution was cooled and then added to petroleum ether to produce the yellow precipitate, which was filtered off and recrystallized from chloroform.

**[Bu$^n_4$N][Zn(S$_2$CNMe$_2$)(C$_7$H$_4$NS$_2$)$_2$] or C$_{33}$H$_{50}$N$_4$S$_6$Zn.** This compound crystallizes in the orthorhombic structure with the lattice parameters $a = 2029.2 \pm 2.6$, $b = 994.7 \pm 1.3$, and $c = 1927.8 \pm 2.4$ pm and experimental and calculated densities 1.29 and 1.298 g·cm$^{-3}$, respectively (McCleverty et al. 1980a). To obtain it, a mixture of [Zn(C$_7$H$_4$NS$_2$)$_2$]$_n$ (1.7 g) and [Bu$^n_4$N][S$_2$CNMe$_2$] (1.5 g) was refluxed in acetone (80 mL) for 1 h. The yellow solution was filtered and partial evaporation of the filtrate afforded yellow crystals of the title compound. These were filtered off and recrystallized from acetone/light petroleum as elongated bricks.

**[Bu$^n_4$N][Zn(S$_2$CNEt$_2$)$_2$(C$_7$H$_4$NS$_2$)] or C$_{33}$H$_{60}$N$_4$S$_6$Zn.** To obtain this compound, a mixture of [Zn(S$_2$CNEt$_2$)$_2$] and [Bu$^n_4$N][C$_7$H$_4$NS$_2$] was refluxed in acetone for 1 h (McCleverty et al. 1980a). The yellow solution was filtered, and on partial evaporation of the filtrate, yellow crystals of the title compound formed. These were filtered off and recrystallized from ethanol/light petroleum mixture.

**[2(C$_5$H$_{10}$NCS$_2$)$_2$·Zn(4,4'-bipy)] or C$_{34}$H$_{48}$N$_6$S$_8$Zn.** This compound crystallizes in the monoclinic structure with the lattice parameters $a = 2202.1 \pm 0.5$, $b = 2221.5 \pm 0.3$, $c = 1753.7 \pm 0.2$ pm, and $\beta = 93.49° \pm 0.02°$ and a calculated density of 1.440 g·cm$^{-3}$ (Liu et al. 2001). To obtain it, zinc $N$-piperidyldithiocarbamate (1 mM) was dissolved in the minimum amount of DMF, and then to the clear DMF solution 4,4'-bipy·2H$_2$O (0.5 mM) was added in small portions over a 30 min period at room temperature. The mixed solution turned yellow, a crude yellow solid appeared, and then the mixture was refluxed for 1 h. Yellow crystals were collected by concentrating the DMF solution of the title compound. Yellow crystals suitable for XRD were obtained by slow evaporation of a MeCN solution at room temperature over 1 week. Thermal decomposition of this compound leads to the formation of ZnS.

**[Zn(S$_2$CNBu$^n_2$)$_2$·(3,4,7,8-Me$_4$Phen)] or C$_{34}$H$_{52}$N$_4$S$_4$Zn.** This compound melts at 170°C–174°C (Bell et al. 1989). It was prepared as pale-yellow needles by adding a hot solution of 3,4,7,8-Me$_4$Phen (50 mM) in MeOH to a hot solution of zinc di-$n$-butyldithiocarbamate (10 mM) in dichloromethane. The resulting solution was cooled and then added to petroleum ether to produce a yellow precipitate, which was filtered off and recrystallized from chloroform.

**[Zn$_2${Ph(SCH$_3$)C=C(S)Ph}$_4$(C$_5$H$_5$N)] or C$_{35}$H$_{31}$NS$_4$Zn.** This compound melts at 65°C with decomposition (Zhang et al. 1991). To obtain it, Zn$_2$[Ph(SCH$_3$)C=C(S) Ph]$_4$ (0.17 mM) was placed into a Schlenk tube of 150 mL capacity and suspended in 10 mL of CH$_2$Cl$_2$. This suspension was added 1.26 mM of pyridine to form a clear solution after stirring for several minutes. This was followed by the addition of 10 mL of $n$-hexane. The colorless crystals that formed overnight were collected and washed with $n$-hexane, affording C$_{37}$H$_{36}$N$_2$S$_4$Zn.

**[Zn{S$_2$CN(CH$_2$Ph)$_2$Py}$_2$] or C$_{35}$H$_{33}$N$_3$S$_4$Zn.** This compound crystallizes in the triclinic structure with the lattice parameters $a = 864.2 \pm 0.6$, $b = 1311.6 \pm 0.9$, and $c = 1662.4 \pm 1.1$ pm and $\alpha = 106.398° \pm 0.001°$, $\beta = 92.633° \pm 0.001°$, and $\gamma = 107.461° \pm 0.001°$ and a calculated density of 1.341 g·cm$^{-3}$ (Zhong et al. 2003).

**[Bu$^n_4$N][Zn(S$_2$CNEt$_2$)(C$_7$H$_4$NS$_2$)$_2$] or C$_{35}$H$_{54}$N$_4$S$_6$Zn.** To obtain this compound, a mixture of [Zn(C$_7$H$_4$NS$_2$)$_2$]$_n$ and [Bu$^n_4$N][S$_2$CNEt$_2$] was refluxed in acetone for 1 h

(McCleverty et al. 1980a). The yellow solution was filtered, and partial evaporation of the filtrate afforded yellow crystals of the title compound. These were filtered off and recrystallized from acetone/light petroleum.

**[Bu$^n_4$N][Zn{S$_2$CN(CH$_2$)$_5$}(C$_7$H$_4$NS$_2$)] or C$_{35}$H$_{60}$N$_4$S$_6$Zn.** To obtain this compound, a mixture of [Zn{S$_2$CN(CH$_2$)$_5$}] and [Bu$^n_4$N][C$_7$H$_4$NS$_2$] was refluxed in acetone for 1 h (McCleverty et al. 1980a). The yellow solution was filtered, and on partial evaporation of the filtrate, yellow crystals of the title compound formed. These were filtered off and recrystallized from ethanol/light petroleum mixture.

**[Zn(C$_6$H$_6$S$_2$)(C$_{12}$H$_8$N$_2$)]$_2$ or C$_{36}$H$_{28}$N$_4$S$_4$Zn$_2$.** This compound crystallizes in the monoclinic structure with the lattice parameters $a = 1098.7 \pm 0.6$, $b = 1076.85 \pm 0.04$, $c = 1334.4 \pm 0.5$ pm, and $\beta = 101.08° \pm 0.04°$ and a calculated density of 1.96 g·cm$^{-3}$ (Halvorsen et al. 1995). To obtain it, Zn(OAc)$_2$·2H$_2$O (1 mM) was dissolved in 50 mL of hot EtOH. To this, a solution of 1,2-benzenethiol (C$_6$H$_6$S$_2$) dissolved in hot EtOH was added dropwise. The resulting precipitate was redissolved with the addition of a small amount of DMF and brought to reflux. To this solution, a solution containing Phen (1 mM) dissolved in 50 mL of hot EtOH was added slowly with vigorous stirring. After the resulting solution stood for 24 h, the solid was collected, then recrystallized from an DMF/EtOH (~2:3 volume ratio) solvent mixture, washed with EtOH, and dried in vacuo.

**[Bu$^n_4$N][Zn{S$_2$CN(CH$_2$)$_5$}(C$_7$H$_4$NS$_2$)$_2$] or C$_{36}$H$_{54}$N$_4$S$_6$Zn.** To obtain this compound, a mixture of [Zn(C$_7$H$_4$NS$_2$)$_2$]$_n$ and [Bu$^n_4$N]{S$_2$CN(CH$_2$)$_5$} was refluxed in acetone for 1 h (McCleverty et al. 1980a). The yellow solution was filtered, and partial evaporation of the filtrate afforded yellow crystals of the title compound. These were filtered off and recrystallized from acetone/light petroleum.

**[Zn(Bu$^n_2$NCS$_2$)$_4$] or C$_{36}$H$_{72}$N$_4$S$_8$Zn$_2$.** This compound crystallizes in the monoclinic structure with the lattice parameters $a = 1603.6 \pm 0.3$, $b = 1660.4 \pm 0.3$, $c = 1848.7 \pm 0.4$ pm, and $\beta = 95.10° \pm 0.03°$ at $180 \pm 2$ K and a calculated density of 1.285 g·cm$^{-3}$ (Paz et al. 2003). It was prepared by the following procedure. CS$_2$ (4.13 mM) was added to an ethanol suspension (ca. 50 mL) containing Bu$^n_2$N (4.13 mM) and freshly prepared Zn(OH)$_2$ (2.07 mM), and the resulting mixture was stirred overnight at ambient temperature. A white precipitate was isolated by vacuum filtration and was air-dried at 60°C. Moderate-quality colorless crystals of the title compound suitable for XRD were obtained by recrystallization from CH$_2$Cl$_2$ over a period of 2 days.

**[Zn$_2${Ph(SCH$_3$)C=C(S)Ph}$_4$(C$_7$H$_{10}$N$_2$)] or C$_{37}$H$_{36}$N$_2$S$_4$Zn.** This compound melts at 110°C with decomposition and crystallizes in the triclinic structure with the lattice parameters $a = 908.87 \pm 0.16$, $b = 967.4 \pm 0.2$, and $c = 1137.7 \pm 0.2$ pm and $\alpha = 81.740° \pm 0.016°$, $\beta = 74.769° \pm 0.015°$, and $\gamma = 69.962° \pm 0.014°$ and experimental and calculated densities 1.36 and 1.359 g·cm$^{-3}$, respectively (Zhang et al. 1991). To obtain this compound, Zn$_2$[Ph(SCH$_3$)C=C(S)Ph]$_4$ (0.17 mM) was placed into a Schlenk tube of 150 mL capacity and suspended in 10 mL of CH$_2$Cl$_2$. To this suspension was added 0.86 mM of 4-(methylamino)pyridine to form a clear solution after stirring for several minutes. This was followed by the addition of 10 mL of *n*-hexane. The colorless precipitate that formed overnight was collected and washed with *n*-hexane, affording $C_{37}H_{36}N_2S_4Zn$.

**[Zn(PhS)$_2$(2,9-Me$_2$-4,7-Ph$_2$Phen)]** or $C_{38}H_{30}N_2S_2Zn$, **bis(benzenethiolato)(2,9-dimethyl-4,7-diphenyl-Phen)zinc**. This compound melts at 181°C–183°C and is characterized by two polymorphic modifications (Jordan et al. 1991). Phase transition from orthorhombic to monoclinic structure takes place at 125°C–130°C.

**[Zn(S-2,4,6-Pr$^i_3$C$_6$H$_2$)$_2$(1-Me-imid)$_2$]** or $C_{38}H_{58}N_4S_2Zn$. This compound was obtained by the interaction of $Zn(S\text{-}2,4,6\text{-}Pr^i_3C_6H_2)_2$ with two equivalents of 1-methylimidazole in MeCN (Corwin et al. 1987).

**[Zn$_2$(S$_2$CNPr$^n_2$)$_4$·(4,4′-bipy)]** or $C_{38}H_{64}N_6S_8Zn_2$. This compound melts at 186°C and crystallizes in the monoclinic structure with the lattice parameters $a = 851.5 \pm 0.1$, $b = 1096.5 \pm 0.1$, $c = 2781.6 \pm 0.3$ pm, and $\beta = 95.860° \pm 0.001°$ and a calculated density of 1.275 g·cm$^{-3}$ (Berus et al. 1998, Klevtsova et al. 2001a). To obtain it, $Zn(S_2CNPr^n_2)$ (1.24 mM) and 4,4′-bipy (0.96 mM) were dissolved in toluene (10 mL) at the heating on water bath (Berus et al. 1998). The solution was cooled to room temperature, and white precipitate was filtered off, washed with toluene, and dried in air. The title compound was recrystallized from toluene. Its single crystals were obtained at the interaction of $Zn(S_2CNPr^n_2)_2$ and 4,4′-bipy in ethyl acetate at molar ratio $Zn(S_2CNPr^n_2)_2/4,4′\text{-bipy} > 1$ or at the recrystallization of $[Zn(S_2CNPr^n_2)_2·(4,4′\text{-bipy})]$ from ethyl acetate (Klevtsova et al. 2001a).

**[Zn$_2$(S$_2$CNPr$^i_2$)$_4$·(4,4′-bipy)]** or $C_{38}H_{64}N_6S_8Zn_2$. This compound was obtained at the heating of $[Zn_2(S_2CNPr^i_2)_4·(4,4′\text{-bipy})]·2(PhMe)$ at 100°C ± 5°C to a constant weight (Larionov et al. 1999).

**[C$_{16}$H$_{36}$N]$_2$[Zn(C$_3$S$_5$)$_2$]** or $C_{38}H_{72}N_2S_{10}Zn$. This compound melts at 163°C–165°C (Comerlato et al. 2002) and crystallizes in the monoclinic structure with the lattice parameters $a = 1858.35 \pm 0.03$, $b = 883.82 \pm 0.01$, $c = 3144.13 \pm 0.04$ pm, and $\beta = 101.967° \pm 0.001°$ (Zhao et al. 2011) ($a = 1925.64 \pm 0.03$, $b = 894.78 \pm 0.02$, $c = 2923.79 \pm 0.05$ pm, and $\beta = 92.9544° \pm 0.0007°$ at $150 \pm 2$ K and a calculated density of 1.245 g·cm$^{-3}$ [Comerlato et al. 2002]). It was obtained by the following procedure. $CS_2$ (24 mL) was added to degassed DMF (48 mL), and the mixture was cooled to 0°C. Na (1.45 g) was added to the solution, and the mixture was vigorously stirred under cooling until the reaction was completed. Several mL of MeOH were slowly added. To this solution, separate solutions of (1) $ZnCl_2·2H_2O$ (2.13 g) dissolved in 25%–28% $NH_3$ (40 mL) and (2) $Bu_4NBr$ (10.12 g) in water (30 mL) were added consecutively with stirring at room temperature. The mixture was stirred overnight. The product was isolated by filtration, washed with $H_2O$ and MeOH, and then solved in acetone. Single crystals of the title compound used for XRD were obtained by evaporation of acetone from the solution held at room temperature.

Red crystals of the title compound were obtained from an acetone/Pr$^i$OH solution (Comerlato et al. 2002).

**[Bu$^n_4$N][Zn(S$_2$CNBu$^n_2$)(C$_7$H$_4$NS$_2$)$_2$]** or $C_{39}H_{62}N_4S_6Zn$. To obtain this compound, a mixture of $[Zn(C_7H_4NS_2)_2]_n$ and $[Bu^n_4N][S_2CNBu^n_2]$ was refluxed in acetone for 1 h

(McCleverty et al. 1980a). The yellow solution was filtered and partial evaporation of the filtrate afforded yellow crystals of the title compound. These were filtered off and recrystallized from acetone/light petroleum.

**[Zn₂{Ph(SCH₃)C=C(S)Ph}₄(C₁₀H₈N₂)] or C₄₀H₃₄N₂S₄Zn.** This compound melts at 110°C with decomposition and crystallizes in the monoclinic structure with the lattice parameters $a = 1385.0 \pm 0.3$, $b = 1761.1 \pm 0.3$, $c = 1662.6 \pm 0.4$ pm, and $\beta = 98.47° \pm 0.02°$ and a calculated density of 1.290 g·cm⁻³ (Zhang et al. 1991). To obtain it, Zn₂[Ph(SCH₃)C=C(S)Ph]₄ (0.17 mM) was placed into a Schlenk tube of 150 mL capacity and suspended in 10 mL of CH₂Cl₂. To this suspension was added 0.74 mM of 2,2'-bipy to form a clear solution after stirring for several minutes. This was followed by the addition of 10 mL of *n*-hexane. The orange crystals that formed overnight were collected and washed with *n*-hexane, affording C₄₀H₃₄N₂S₄Zn.

**[Zn{S₂CN(CH₂Ph)₂}₂·(2,2'-bipy)] or C₄₀H₃₆N₄S₄Zn.** This compound melts at 222°C–223°C (228°C–230°C [Bell et al. 1989]) and crystallizes in the orthorhombic structure with the lattice parameters $a = 3119.7 \pm 0.5$, $b = 702.47 \pm 0.11$, and $c = 846.29 \pm 0.13$ pm at $223 \pm 2$ K (Lai and Tiekink 2004). It was prepared as pale-yellow needles by adding a hot solution of 2,2'-bipy (50 mM) in MeOH to a hot solution of zinc dibenzyldithiocarbamate (10 mM) in dichloromethane. The resulting solution was cooled and then added to petroleum ether to produce the yellow precipitate, which was filtered off and recrystallized from dichloromethane (Bell et al. 1989). Yellow crystals were obtained by the slow evaporation of a chloroform solution containing equimolar amounts of Zn[S₂CN(CH₂Ph)₂]₂ and 2,2'-bipy (Lai and Tiekink 2004).

**[Zn(C₂₀H₂₆N₆S₂)]₂ or C₄₀H₅₂N₁₂S₄Zn₂.** This compound crystallizes in the orthorhombic structure with the lattice parameters $a = 1147.85 \pm 0.06$, $b = 1854.78 \pm 0.07$, and $c = 2011.33 \pm 0.14$ pm and a calculated density of 1.489 g·cm⁻³ (Durán et al. 1999). It was prepared by electrochemical oxidation of Zn in a solution of 1-phenyl-glyoxal bis(3-piperidylthiosemicarbazone) (0.33 mM) in MeCN (30 mL) containing 20 mg of Et₄NClO₄ for 2 h at 10 mA. A loss of 51.2 mg of Zn from the anode resulted. During the reaction, an orange solid was formed at the bottom of the cell. Crystals suitable for XRD were obtained by crystallization from a MeCN/EtOH (1:1 volume ratio) mixture.

**[Buⁿ₄N][Zn(S₂CNMePh)₃] or C₄₀H₆₀N₄S₆Zn.** To obtain this compound, [Zn(S₂CNMePh)₂] (0.02 M) was added to a solution of [Buⁿ₄N][S₂CNMe₂] (0.02 M) in acetone (200 mL) (McCleverty and Morrison 1976). The mixture was shaken until all the solid had dissolved, then light petroleum was added, and the solvent was partially evaporated. The white crystals were collected and recrystallized from acetone/light petroleum.

**[Zn{(SBuᵗ₃C₆H₂-2,4,6)₂}₂(NMeImid)] or C₄₀H₆₄N₂S₂Zn (NMeImid = *N*-methylimidazole).** This compound melts at a temperature higher than 300°C (Bochmann et al. 1994). To obtain it, a suspension of [Zn(SBuᵗ₃C₆H₂-2,4,6)₂]₂ (0.81 mM) was treated with NMeImid (0.79 mM), and the solution was stirred for 2 h. The starting material dissolved, followed by the precipitation of C₄₀H₆₄N₂S₂Zn, which was recrystallized from toluene (0.70 mM). All the reactions were carried out under Ar atmosphere using standard vacuum-line techniques.

**[Zn$_2$(S$_2$CNPr$^i_2$)$_4$(C$_{12}$H$_{10}$N$_2$)] or C$_{40}$H$_{66}$N$_6$S$_8$Zn$_2$**. This compound melts at 240°C–242°C and crystallizes in the triclinic structure with the lattice parameters $a = 826.90 \pm 0.14$, $b = 1116.40 \pm 0.18$, and $c = 1415.6 \pm 0.2$ pm and $\alpha = 80.806° \pm 0.010°$, $\beta = 84.878° \pm 0.009°$, and $\gamma = 72.566° \pm 0.005°$ at 98 K and a calculated density of 1.375 g·cm$^{-3}$ (Arman et al. 2009b). It was prepared by following a standard procedure whereby two equivalents of Zn(S$_2$CNPr$^i_2$)$_2$ were added to *trans*-1,2-bis(4-pyridyl)ethylene. Golden blocks of the title compound were obtained from the slow evaporation of a chloroform/acetonitrile solution (3:1 volume ratio).

**[Zn$_2$(S$_2$CNPr$^n_2$)$_4$(C$_{12}$H$_{12}$N$_2$)] or C$_{40}$H$_{68}$N$_6$S$_8$Zn$_2$**. This compound melts at 178°C–180°C and crystallizes in the monoclinic structure with the lattice parameters $a = 1864.5 \pm 0.5$, $b = 1546.4 \pm 0.5$, $c = 1756.7 \pm 0.4$ pm, and $\beta = 90.756° \pm 0.011°$ at $98 \pm 2$ K and a calculated density of 1.338 g·cm$^{-3}$ (Avila and Tiekink 2008). It was prepared by refluxing Zn(S$_2$CNPr$^n_2$)$_2$ with 1,2-bis(4-pyridyl)ethane. Colorless crystals of the title compound were isolated by the slow evaporation of an EtOH/MeOH (1:1 volume ratio) solution.

**[Zn(SC$_6$H$_2$Bu$^t_3$)$_2$(Bu$^t$CN)] or C$_{41}$H$_{67}$NS$_2$Zn**. This compound decomposed at 200°C. It was obtained if to a stirred suspension of [Zn(2,4,6-Bu$^t_3$C$_6$H$_2$S)$_2$]$_2$ (0.16 mM) in petroleum ether (15 mL) at room temperature was added *t*-butylisocyanide (0.32 mM) (Bochmann et al. 1993). The solid dissolved to give a clear colorless solution from which a white solid precipitated within a few minutes. After stirring the mixture for 2 h, the precipitate was filtered off and recrystallized from hot toluene (5 mL) to give Zn(SC$_6$H$_2$Bu$^t_3$)$_2$(Bu$^t$CN) as colorless crystals.

**[Bu$^n_4$N][Zn(S$_2$CNBu$^n_2$)$_2$(C$_7$H$_4$NS$_2$)] or C$_{41}$H$_{76}$N$_4$S$_6$Zn**. To obtain this compound, a mixture of [Zn(S$_2$CNBu$^n_2$)$_2$] and [Bu$^n_4$N][C$_7$H$_4$NS$_2$] was refluxed in acetone for 1 h (McCleverty et al. 1980a). The yellow solution was filtered, and on partial evaporation of the filtrate, yellow crystals of the title compound formed. These were filtered off and recrystallized from ethanol/light petroleum mixture.

**[Zn{S$_2$CN(PhCH$_2$)$_2$}$_2$·(Phen)] or C$_{42}$H$_{36}$N$_4$S$_4$Zn**. This compound melts at 230°C–231°C (Bell et al. 1989). It was prepared as pale-yellow needles by adding a hot solution of Phen (50 mM) in MeOH to a hot solution of zinc dibenzyldithiocarbamate (10 mM) in dichloromethane. The resulting solution was cooled and then added to petroleum ether to produce the yellow precipitate, which was filtered off and recrystallized from MeOH.

**[Zn(S-2,4,6-Pr$^i_3$C$_6$H$_2$)$_2$(C$_6$H$_5$N)$_2$] or C$_{42}$H$_{56}$N$_2$S$_2$Zn**. This compound was obtained by the interaction of Zn(S-2,4,6-Pr$^i_3$C$_6$H$_2$)$_2$ with two equivalents of pyridine in MeCN (Corwin et al. 1987).

**[(Pr$^n_4$N)Zn(S-2,3,5,6-Me$_4$C$_6$H)$_3$] or C$_{42}$H$_{67}$NS$_3$Zn**. This compound crystallizes in the monoclinic structure with the lattice parameters $a = 1102.9 \pm 0.2$, $b = 1849.9 \pm 0.3$, $c = 2158.8 \pm 0.5$ pm, and $\beta = 96.05° \pm 0.02°$ (Gruff and Koch 1989). The reaction of ZnSO$_4$·7H$_2$O with five equivalents of LiS-2,3,5,6-Me$_4$C$_6$H and one equivalent of $Pr^n_4NBr$ gave a white crystalline $C_{42}H_{67}NS_3Zn$, which was recrystallized from $CH_3CN$.

**$[Bu^n_4N][Zn\{S_2CN(C_6H_{11})_2\}(C_7H_4NS_2)_2]$ or $C_{43}H_{66}N_4S_6Zn$.** To obtain this compound, a mixture of $[Zn(C_7H_4NS_2)_2]_n$ and $[Bu^n_4N][S_2CN(C_6H_{11})_2]$ was refluxed in acetone for 1 h (McCleverty et al. 1980a). The yellow solution was filtered and partial evaporation of the filtrate afforded yellow crystals of the title compound. These were filtered off and recrystallized from acetone/light petroleum.

**$[Zn(SBu^t_3C_6H_2-2,4,6)_2(NC_5H_3Me_2-2,6)]$ or $C_{43}H_{67}NS_2Zn$.** This compound melts at a temperature higher than 300°C and crystallizes in the monoclinic structure with the lattice parameters $a = 1847.5 \pm 0.2$, $b = 2035.2 \pm 0.9$, $c = 1183.5 \pm 0.1$ pm, and $\beta = 95.03° \pm 0.02°$ and a calculated density of 1.090 g·cm$^{-3}$ (Bochmann et al. 1994). To obtain it, to a stirred suspension of $[Zn(SBu^t_3C_6H_2-2,4,6)_2]_2$ (0.24 mM) in 10 mL of petroleum ether at room temperature was added 2,6-lutidine (0.2 mM). The solid dissolved to give a clear colorless solution from which a white solid precipitated during the course of 30 min. The precipitate was filtered off and recrystallized from toluene to give colorless crystals of $C_{43}H_{67}NS_2Zn$. All reactions were carried out under Ar atmosphere using standard vacuum-line techniques.

**$Zn[S_2CN(PhCH_2)_2]_2 \cdot (2,9-Me_2Phen)$ or $C_{44}H_{40}N_4S_4Zn$.** This compound melts at 216°C–217°C (Bell et al. 1989). It was prepared as yellow needles by adding a hot solution of 2,9-Me$_2$Phen (50 mM) in MeOH to a hot solution of zinc dibenzyldithiocarbamate (10 mM) in toluene. The resulting solution was cooled and then added to petroleum ether to produce a yellow precipitate, which was filtered off and recrystallized from chloroform.

**$[Me_4N][Zn(C_{12}H_{12}NS)_3(C_4H_6N_2)]$ or $C_{44}H_{54}N_6S_3Zn$.** This compound decomposes at 210°C and crystallizes in the triclinic structure with the lattice parameters $a = 1185.0 \pm 0.2$, $b = 1323.6 \pm 0.3$, and $c = 1601.2 \pm 0.3$ pm and $\alpha = 89.26° \pm 0.03°$, $\beta = 71.29° \pm 0.03°$, and $\gamma = 64.81° \pm 0.03°$ at $150 \pm 2$ K and a calculated density of 1.285 g·cm$^{-3}$ (Otto et al. 1999). To obtain it, 9 mM of Na was dissolved in 25 mL of MeOH and treated with 9 mM of $N$-(2-thiophenyl)-2,5-dimethylpyrrole and 3 mM of dry ZnCl$_2$. The resulting white precipitate was dissolved through further addition of 25 mL MeOH and 1 h of stirring. Next, 3 mM of $N$-methylimidazole were added. The solution was stirred again for 1 h and Me$_4$NCl (3 mM) was added. The white precipitate was collected, washed with MeOH, and dried in vacuo. Recrystallization with acetone yielded colorless crystals after 1 day. All procedures were carried out under an Ar or N$_2$ atmosphere using standard Schlenk and glovebox techniques.

**$[Zn(SBu^t_3C_6H_2-2,4,6)_2(NMeImid)_2]$ or $C_{44}H_{70}N_4S_2Zn$.** This compound melts at 286°C–288°C and crystallizes in the monoclinic structure with the lattice parameters $a = 1743.8 \pm 0.5$, $b = 1589.4 \pm 0.5$, $c = 1796.4 \pm 0.5$ pm, and $\beta = 109.58° \pm 0.02°$ and a calculated density of 1.108 g·cm$^{-3}$ (Bochmann et al. 1994). To obtain it, $[Zn(SBu^t_3C_6H_2-2,4,6)_2]_2$ (0.81 mM) was treated with NMeImid (8.2 mM), and the solution was stirred for 2 h. The starting material dissolved, followed by the precipitation of $C_{44}H_{70}N_4S_2Zn$, which was recrystallized from toluene (0.76 mM).

All reactions were carried out under Ar atmosphere using standard vacuum-line techniques.

**[Zn₃(C₇H₈S₂)₃(C₁₂H₈N₂)₂] or C₄₅H₄₀N₄S₆Zn₃.** This compound crystallizes in the monoclinic structure with the lattice parameters $a = 1145.4 \pm 0.4$, $b = 2059.0 \pm 0.9$, $c = 1854.8 \pm 0.4$ pm, and $\beta = 95.20° \pm 0.02°$ and a calculated density of 1.55 g·cm⁻³ (Halvorsen et al. 1995). It was prepared in a manner analogous to the preparation described for [Zn(C₆H₆S₂)(C₁₂H₈N₂)]₂ except dimercaptotoluene (C₇H₈S₂) (1 mM) was used instead of 1,2-benzenethiol (C₆H₆S₂). Zn(C₇H₈S₂)₃(C₁₂H₈N₂)₂ formed even in the presence of a large amount of Phen (up to threefold molar excess). Recrystallization was carried out in the same solvent mixture as described for [Zn(C₆H₆S₂)(C₁₂H₈N₂)]₂; however, the percentage of DMF required was much higher due to the lower solubility of this compound in EtOH.

**[Zn{S₂CN(PhCH₂)₂}₂·(3,4,7,8-Me₄Phen)] or C₄₆H₄₆N₄S₄Zn.** This compound melts at 126°C–130°C (Bell et al. 1989). It was prepared as pale-yellow needles by adding a hot solution of 3,4,7,8-Me₄Phen (50 mM) in MeOH to a hot solution of zinc dibenzyl-dithiocarbamate (10 mM) in dichloromethane. The resulting solution was cooled and then added to petroleum ether to produce the yellow precipitate, which was filtered off and recrystallized from chloroform.

**[Zn(SBuᵗ₃C₆H₂-2,4,6)₂(NC₂H₅)₂] or C₄₆H₆₈N₂S₂Zn.** This compound melts at 268°C–270°C (Bochmann et al. 1994). It could be obtained if [Zn(SBuᵗ₃C₆H₂-2,4,6)₂]₂ (0.16 mM) was dissolved in 10 mL of pyridine, and the solution was stirred for 2 h. The solvent was removed in vacuo and white residue recrystallized from toluene with the addition of a small amount of pyridine (0.13 mM). All reactions were carried out under Ar atmosphere using standard vacuum-line techniques.

**[Zn(SC₆H₂Buᵗ₃)₂(BuᵗCN)₂] or C₄₆H₇₆N₂S₂Zn.** This compound decomposed at 192°C. It could be obtained if to a stirred suspension of [Zn(2,4,6-Buᵗ₃C₆H₂S)₂]₂ (0.16 mM) in petroleum ether (15 mL) at room temperature was added t-butylisocyanide (3.19 mM) (Bochmann et al. 1993). The solid dissolved to give a clear colorless solution from which a white solid precipitated within a few minutes. Stirring was continued for 1 h. The precipitate was filtered off and recrystallized from warm toluene (5 mL) to give Zn(SC₆H₂Buᵗ₃)₂(BuᵗCN)₂ as colorless crystals.

**[Zn₂(S₂CNBuⁱ₂)₄·(4,4′-bipy)] or C₄₆H₈₀N₆S₈Zn₂.** This compound melts at 152°C and crystallizes in the triclinic structure with the lattice parameters $a = 950.8 \pm 0.2$, $b = 1289.9 \pm 0.3$, and $c = 1296.9 \pm 0.3$ pm and $\alpha = 100.85° \pm 0.03°$, $\beta = 96.63° \pm 0.03°$, and $\gamma = 108.95° \pm 0.03°$ and a calculated density of 1.265 g·cm⁻³ (Zemskova et al. 1996). To obtain it, Zn(S₂CNBuⁱ₂)₂ (1.0 mM) and 4,4′-bipy (1.5 mM) were dissolved in toluene (30 mL) at the heating on water bath. At cooling to room temperature, a light-yellow precipitate was formed. The mixture was kept 2–3 h at −15°C and then the precipitate was filtered off, washed with toluene, and dried in air. Thermal decomposition of this compound at 350°C in air leads to the formation of α-ZnS.

**[(Prⁿ₄N)Zn(S-2,3,5,6-Me₄C₆H)₃(1-MeImid)(CH₃CN)] or C₄₈H₇₆N₄S₃Zn.** This compound crystallizes in the triclinic structure with the lattice parameters $a = 1139.8 \pm 0.9$, $b = 2269.4 \pm 0.8$, and $c = 1039.6 \pm 0.5$ pm and $\alpha = 93.04° \pm 0.04°$,

$\beta = 107.49° \pm 0.06°$, and $\gamma = 93.86° \pm 0.06°$ (Gruff and Koch 1989). It was obtained at the addition of 1-methylimidazole to the reaction mixture for preparing $[Pr^n_4N]$ $[Zn(S-2,3,5,6-Me_4C_6H)_3]$.

**$[Zn_4(SPh)_8(MeCN)]_n$ or $C_{50}H_{43}NS_8Zn_4$.** This compound could be obtained if to acetonitrile (400 mL) stirred at room temperature, $Zn_4(SPh)_8Bu^n$ was added in small portions only after the previous portion had dissolved (Dance 1980b). At the first indication of clouding and imminent precipitation, the solution was rapidly filtered and allowed to stand for 20 h while the crystallization continued. The mixture was boiled to complete the polymerization and filtered hot, and the product washed with acetonitrile and vacuum dried. All operations were performed in an atmosphere of $N_2$. The composition of this compound is not yet certain.

**$[Zn_2(S_2CNPr^i_2)_4 \cdot (4,4'-bipy)] \cdot 2(PhMe)$ or $C_{52}H_{80}N_6S_8Zn_2$.** This compound crystallizes in the monoclinic structure with the lattice parameters $a = 845.8 \pm 0.2$, $b = 3366.8 \pm 0.7$, $c = 1175.4 \pm 0.2$ pm, and $\beta = 110.68° \pm 0.03°$ and a calculated density of 1.248 g·cm$^{-3}$ (Larionov et al. 1999). To obtain it, a mixture of $Zn(S_2CNPr^i_2)_2$ (2.4 mM) and 4,4'-bipy (2.4 mM) was dissolved in $CH_2Cl_2$ (ca. 15 mL), and the solution obtained of yellow color was evaporated on the water bath to half of the initial volume and evaporated at room temperature to a wet state of the resulting yellow precipitate. It was filtered with suction, washed with cold toluene, and dried in air. The obtained substance was dissolved in a minimum amount of toluene at the heating. A precipitate of light orange color was formed at a rapid cooling of the solution first to room temperature and then to −10°C. This was filtered with suction, washed with cold toluene, and dried in air. Single crystals of the title compound were grown by slow cooling to room temperature a hot saturated solution in $CH_2Cl_2$ and the subsequent evaporation of this solution for several days.

**$[Zn(C_7H_8S_2)(C_{12}H_8N_2)]_2 \cdot 1,4-Ph_2-1,3-C_4H_6$ or $C_{54}H_{46}N_4S_4Zn_2$.** This compound crystallizes in the triclinic structure with the lattice parameters $a = 946.3 \pm 0.4$, $b = 1342.1 \pm 0.5$, and $c = 1346.6 \pm 0.5$ pm and $\alpha = 118.17° \pm 0.03°$, $\beta = 94.44° \pm 0.03°$, and $\gamma = 102.60° \pm 0.03°$ and a calculated density of 1.39 g·cm$^{-3}$ (Halvorsen et al. 1995). To obtain it, $Zn_3(C_7H_8S_2)_3(C_{12}H_8N_2)_2$ (0.039 mM) was dissolved in 50 mL of DMF along with 1,4-diphenyl-1,3-butadiene (100 mM). The solution was slowly cooled and the DMF was allowed to evaporate off slowly over a period of few days. Dark orange crystallographic quality crystals were collected, washed with cold EtOH, and dried in vacuo.

**$[Et_3NH]_2[Zn(C_{12}H_{12}NS)_4] \cdot MeCN$ or $C_{62}H_{83}N_7S_4Zn$.** This compound melts at 183°C and crystallizes in the monoclinic structure with the lattice parameters $a = 2230.2 \pm 0.5$, $b = 1361.1 \pm 0.3$, $c = 2032.6 \pm 0.4$ pm, and $\beta = 91.87° \pm 0.02°$ at $170 \pm 2$ K and a calculated density of 1.205 g·cm$^{-3}$ (Otto et al. 1999). To obtain it, 0.5 mM of dry $ZnCl_2$ dissolved in 5 mL MeOH was added dropwise to a solution of $Et_3N$ (4 mM) and $N$-(2-thiophenyl)-2,5-dimethylpyrrole (3 mM) in 15 mL MeOH. After 30 min of stirring, the white precipitate was collected, washed with MeOH, and dried in vacuo. The title compound was recrystallized from hot acetonitrile to yield colorless prisms. All procedures were carried out under an Ar or $N_2$ atmosphere using standard Schlenk and glovebox techniques.

**[Zn(MeC$_{18}$H$_{18}$N$_4$S)(SBz)$_6$]·2MeCN or C$_{64}$H$_{66}$N$_6$S$_7$Zn$_3$.** This compound crystallizes in the triclinic structure with the lattice parameters $a = 1355.0 \pm 0.2$, $b = 1531.3 \pm 0.2$, and $c = 1549.3 \pm 0.3$ pm and $\alpha = 90.25° \pm 0.02°$, $\beta = 91.87° \pm 0.02°$, and $\gamma = 95.68° \pm 0.02°$ and a calculated density of 1.39 g·cm$^{-3}$ (Burth et al. 1998). To obtain it, a solution of benzyl mercaptan (0.90 mM) in MeOH (150 mL) was treated with sodium methoxide (3.8 mL [0.9 mM] of a 0.24 M methanol solution). Then Zn(NO$_3$)$_2$·4H$_2$O (0.45 mM) in MeOH (50 mL) was added with vigorous stirring over a period of 1 h. MeC$_{18}$H$_{18}$N$_4$S (0.45 mM), dissolved in 20 mL of boiling MeOH, was added, and clear solution was stirred for 1 h. After the volume was reduced to 25 mL in vacuo, the solution was allowed to stand. Within 1 week, a colorless precipitate of C$_{64}$H$_{66}$N$_6$S$_7$Zn$_3$ had formed.

**[Zn{SPh$_2$CCH$_2$}$_2$(C$_5$H$_3$N)]$_2$·MeCN or C$_{68}$H$_{57}$N$_3$S$_4$Zn$_2$.** This compound crystallizes in the orthorhombic structure with the lattice parameters $a = 1796.3 \pm 1.0$, $b = 2743.3 \pm 0.3$, and $c \approx 1146.9 \pm 0.4$ pm and a calculated density of 1.34 g·cm$^{-3}$ (Kaptein et al. 1987).

**[(Me$_4$N)$_2${Zn$_4$(SPh)$_{10}$}] or C$_{68}$H$_{74}$N$_2$S$_{10}$Zn$_4$.** To obtain this compound, a solution of 50 mM of ZnCl$_2$ in 50 mL of MeOH was added to a solution of 125 mM of sodium benzenethiolate (from 2.87 g of Na and 14.2 g of benzenethiol) in 150 mL of MeOH (Hagen et al. 1982). After the solution was stirred for 10 min, NaCl was removed by filtration, and a solution of 22 mM of Me$_4$NBr was added. The resulting precipitate became microcrystalline upon further stirring of the reaction mixture. The product was collected by filtration, washed with MeOH, and dried in vacuo, affording white solid. Pure material was obtained by one recrystallization from acetonitrile.

(Me$_4$N)$_2$[Zn$_4$(SPh)$_{10}$] could also be prepared if a solution of Zn(NO$_3$)$_2$·6H$_2$O (70.6 mM) in MeOH (70 mL) was added to a well-stirred solution of benzenethiol (182 mM) and Et$_3$N (182 mM) in MeOH (40 mL) at room temperature (Dance et al. 1984). A solution of Me$_4$NCl (80 mM) in MeOH (40 mL) was added with brief stirring to dissolve any precipitate, and the mixture was allowed to crystallize at 0°C. The colorless crystalline product was filtered, washed with MeOH, and vacuum dried. Recrystallization was from acetonitrile plus MeOH with the addition of toluene.

**[(Et$_4$N)$_2$Zn(S-2-PhC$_6$H$_4$)$_4$·2MeCN] or C$_{68}$H$_{82}$N$_4$S$_4$Zn.** This compound crystallizes in the tetragonal structure with the lattice parameters $a = 1584.3 \pm 0.4$ and $c = 2475.5 \pm 0.6$ pm (Silver et al. 1993). To obtain it, lithium 2-phenylbenzenethiolate, Li(S-2-PhC$_6$H$_4$), was generated in MeOH from the thiol (3.2 mM) and Li (3.3 mM). The MeOH was removed, and then Zn salt (0.55 mM) was added, followed by the addition of 30 mL of MeCN. After stirring for 3 h, the reaction mixture was filtered, and the filtrate was added to [Et$_4$N]Br (1.1 mM). Large crystals formed after 4 days at −20°C.

**[Zn{S$_2$CN(CH$_2$Ph)$_2$}$_2$]$_2$(4,4′-bipy) or C$_{70}$H$_{64}$N$_6$S$_8$Zn$_2$.** This compound melts at 212°C–213°C and crystallizes in the triclinic structure with the lattice parameters $a = 1068.78 \pm 0.07$, $b = 1262.78 \pm 0.08$, and $c = 1325.86 \pm 0.08$ pm and $\alpha = 93.090° \pm 0.001°$, $\beta = 104.488° \pm 0.001°$, and $\gamma = 108.270° \pm 0.001°$ at 223 K

(Yin et al. 2003). It was prepared by the following procedure. A solution of 4,4′-bipy (1 mM) in THF (10 mL) was added to a solution of zinc dibenzyldithiocarbamate (1 mM) in THF (20 mL). The mixture was stirred for 2 h at room temperature and filtered to give a pale-yellow solution. Yellow crystals were obtained from the slow evaporation of the THF solution of the title compound.

**[(Et$_3$NH)$_2${Zn$_4$(SPh)$_{10}$}] or C$_{72}$H$_{82}$N$_2$S$_{10}$Zn$_4$.** This compound crystallizes in the triclinic structure with the lattice parameters $a = 1281.9 \pm 0.4$, $b = 1367.3 \pm 0.4$, and $c = 2315.8 \pm 0.7$ pm and $\alpha = 108.16° \pm 0.02°$, $\beta = 96.20° \pm 0.03°$, and $\gamma = 98.78° \pm 0.03°$ (Hencher et al. 1981, 1985). It was obtained by the electrochemical oxidation of anodic Zn in an acetonitrile solution of Et$_3$N and PhSH.

**[(Et$_4$N)$_2${Zn$_4$(SPh)$_{10}$}] or C$_{76}$H$_{90}$N$_2$S$_{10}$Zn$_4$.** This compound was prepared by a procedure analogous to that for (Me$_4$N)$_2$[Zn$_4$(SPh)$_{10}$] using Et$_4$NBr instead of Me$_4$NBr (Hagen et al. 1982).

**[Zn$_8${S(CH$_2$)$_3$NMe$_2$}$_{16}$] or C$_{80}$H$_{192}$N$_{16}$S$_{16}$Zn$_8$.** This compound crystallizes in the tetragonal structure with the lattice parameters $a = 2025.3 \pm 0.6$ and $c = 1425.7 \pm 0.6$ pm at 120 K and a calculated density of 1.371 g·cm$^{-3}$ (Casals et al. 1991a). To obtain this compound, an anodic oxidation of Zn in acetonitrile solution containing 3-dimethylamino-1-propanethiol, an excess of triethylamine and tetraethylammonium perchlorate, was carried out over a period of 24 h with a constant potential of 24 V. The addition of diethyl ether and subsequent extraction with $n$-hexane yielded a colorless powder, which recrystallized from dry MeOH to give single crystals.

**[(Me$_4$N)$_2$(c-$\mu_4$-S)($\mu$-SCH$_2$Ph)$_{12}$S$_4$Zn$_8$] or C$_{92}$H$_{108}$N$_2$S$_{17}$Zn$_8$.** This compound crystallizes in the monoclinic structure with the lattice parameters $a = 1452 \pm 2$, $b = 2742 \pm 2$, $c = 2680 \pm 3$ pm, and $\beta = 98.77° \pm 0.09°$ and a calculated density of 1.454 g·cm$^{-3}$ (Guo et al 1999). To obtain it, anhydrous ZnCl$_2$ (67 mM) was added to a flask containing 150 mL of acetonitrile with stirring for 0.5 h. Then, a solution of sodium benzylthiolate (100 mM) in 50 mL of MeOH was added dropwise, and the mixture was stirred for 2 h. Subsequently, Me$_4$NCl (42 mM) was added and agitated for a further 5 h and filtered. The colorless crystals were obtained by diffusion and isolated by filtration, washed with cooled diethyl ether, and dried in vacuo.

**[Zn$_8$S(SPh)$_{14}$·C$_{12}$H$_{10}$N$_2$] or C$_{96}$H$_{80}$N$_2$S$_{15}$Zn$_8$.** This compound crystallizes in the monoclinic structure with the lattice parameters $a = 1460.6 \pm 0.3$, $b = 2583.7 \pm 0.5$, $c = 2690.7 \pm 0.5$ pm, and $\beta = 105.75° \pm 0.03°$ and a calculated density of 1.540 g·cm$^{-3}$ (Xie et al. 2005). To synthesize it, Zn(OAc)$_2$ (5.93 mM), $trans$ 1,2-bis(4-pyridyl) ethylene (0.71 mM), thiourea (1.41 mM), thiophenol (9.17 mM), and distilled H$_2$O (5.93 g) were combined in a 23 mL Teflon-lined stainless steel autoclave and stirred for 20 min. The sealed vessel was heated at 150°C for 7 days. After cooling to room temperature, transparent block crystals were obtained.

**[Zn$_8$S(SPh)$_{14}$·C$_{13}$H$_{14}$N$_2$] or C$_{97}$H$_{84}$N$_2$S$_{15}$Zn$_8$.** This compound crystallizes in the monoclinic structure with the lattice parameters $a = 1345.7 \pm 0.5$, $b = 1768.1 \pm 0.5$, $c = 4083.6 \pm 0.5$ pm, and $\beta = 92.864° \pm 0.005°$ and a calculated density of 1.562 g·cm$^{-3}$ (Xie et al. 2005). To synthesize it, Zn(OAc)$_2$ (3.00 mM), 1,3-bis(4-pyridyl)propane

(1.00 mM), thiourea (0.5 mM), thiophenol (6.04 mM), AgOAc (1.00 mol), and distilled $H_2O$ (5.565 g) were mixed and heated in autoclave (165°C). After 8 days, colorless cube-shaped crystals were recovered.

**[(Me$_4$N)$_4$\{S$_4$Zn$_{10}$(SPh)$_{16}$\}] or C$_{112}$H$_{128}$N$_4$S$_{20}$Zn$_{10}$.** This compound crystallizes in the tetragonal structure with the lattice parameters $a = 1978.3 \pm 0.4$ and $c = 1687.1 \pm 0.5$ pm and experimental and calculated densities $1.41 \pm 0.01$ and $1.42$ g·cm$^{-3}$, respectively (Choy et al. 1982, Dance et al. 1984). To obtain it, to a solution of (Me$_4$N)$_2$[Zn$_4$(SPh)$_{10}$] (6.0 mM) in acetonitrile (150 mL) at room temperature was added finely powdered sulfur (3.0 mM) in one portion, and the flask was immediately sealed. The mixture was stirred only until the sulfur was dissolved (ca. 10 min) and then allowed to stand undisturbed at room temperature, while the product crystallized as long colorless needles, which were filtered, washed with acetonitrile, and vacuum dried.

An alternative preparation of (Me$_4$N)$_4$[S$_4$Zn$_{10}$(SPh)$_{16}$] involves DMF, with a higher S/[Zn$_4$(SPh)$_{10}$]$^{2-}$ ratio: powdered sulfur (8.1 mM) was added in one portion to a stirred solution of (Me$_4$N)$_2$[Zn$_4$(SPh)$_{10}$] (5.1 mM) in DMF (50 mL) at room temperature. The sulfur dissolves quickly to produce a yellow solution, from which the product subsequently precipitated as a colorless fine powder. This mixture was heated to ca. 70°C, with the addition of the minimum volume of DMF necessary to dissolve the precipitate. $H_2O$ (ca. 10 mL) was added to the point of incipient crystallization, and crystallization was allowed to proceed slowly. The colorless crystalline product was filtered with acetone and vacuum dried. The crystals were obtained directly from the preparation in acetonitrile.

### 1.61 Zn–H–C–N–Se–S

**[(Me$_4$N)$_4$\{Se$_4$Zn$_{10}$(SPh)$_{16}$\}] or C$_{112}$H$_{128}$N$_4$Se$_4$S$_{16}$Zn$_{10}$** multinary compound is formed in this system. To obtain it, black Se powder (8.1 mM) was added in one portion to a solution of (Me$_4$N)$_2$[Zn$_4$(SPh)$_{10}$] (5.1 mM) in DMF (50 mL) and stirred at room temperature (Dance et al. 1984). Most of the Se dissolved, generating a yellow-orange solution. After filtration, the solution was allowed to stand at room temperature yielding colorless crystals of the product, which were filtered, washed with acetonitrile, and vacuum dried.

### 1.62 Zn–H–C–N–Se–Cl–S

**[Zn(C$_3$H$_5$NSeS)$_2$Cl$_2$] or C$_6$H$_{10}$N$_2$Se$_2$Cl$_2$S$_2$Zn** multinary compound is formed in this system. It decomposes at 177°C–179°C (Preti et al. 1974). It was obtained by the reaction of ZnCl$_2$ solution in EtOH with thiazolidine-2-selenone dissolved in CHCl$_3$ by refluxing for 12 h. The crude product that separated out during the reaction was purified by washing with chloroform.

### 1.63 Zn–H–C–N–Se–Br–S

**[Zn(C$_3$H$_5$NSeS)$_2$Br$_2$] or C$_6$H$_{10}$N$_2$Se$_2$Br$_2$S$_2$Zn** multinary compound is formed in this system. It decomposes at 169°C–171°C (Preti et al. 1974). It was obtained by reaction of $ZnBr_2$ solution in EtOH with thiazolidine-2-selenone dissolved in $CHCl_3$ by refluxing for 12 h. The crude product that separated out during the reaction was purified by washing with chloroform.

## 1.64   Zn–H–C–N–Se–I–S

$[Zn(C_3H_5NSeS)_2I_2]$ or $C_6H_{10}N_2Se_2I_2S_2Zn$ multinary compound is formed in this system. It decomposes at 188°C–190°C (Preti et al. 1974). It was obtained by the reaction of $ZnI_2$ solution in EtOH with thiazolidine-2-selenone dissolved in $CHCl_3$ by refluxing for 12 h. The crude product that separated out during the reaction were purified by washing with chloroform.

## 1.65   Zn–H–C–N–F–S

Some compounds are formed in this system.

$[Zn(3-CF_3C_5H_3NS)_2]$ or $C_{12}H_6N_2F_6S_2Zn$. Electrochemical oxidation of a Zn anode in a solution of $3-CF_3C_5H_4NS$ (1.492 mM) in acetonitrile (50 mL), containing about 15 mg of $Me_4NClO_4$ as a current carrier, at 20 V and 20 mA for 2 h caused 44.4 mg of Zn to be dissolved (Sousa-Pedrares et al. 2003). During the electrolysis process, $H_2$ was evolved at the cathode, and at the end of the reaction, an insoluble crystalline solid was observed at the bottom of the vessel. The solid was filtered off, washed with acetonitrile and diethyl ether, and dried under vacuum. All manipulations were done under an atmosphere of dry $N_2$.

$[Zn(5-CF_3C_5H_3NS)_2]$ or $C_{12}H_6N_2F_6S_2Zn$. This compound was obtained by the same procedure as for $[Zn(3-CF_3C_5H_3NS)_2]$ using $5-CF_3C_5H_4NS$ instead of $3-CF_3C_5H_4NS$ (45.37 mg of Zn dissolved) (Sousa-Pedrares et al. 2003).

$[Zn(3-CF_3C_5H_3NS)_2(2,2'-bipy)]$ or $C_{22}H_{14}N_4F_6S_2Zn$. This compound crystallizes in the triclinic structure with the lattice parameters $a = 826.1 \pm 0.4$, $b = 1305.0 \pm 0.7$, and $c = 1334.5 \pm 0.7$ pm and $\alpha = 67.986° \pm 0.008°$, $\beta = 85.496° \pm 0.008°$, and $\gamma = 82.258° \pm 0.008°$ and a calculated density of 1.453 g·cm$^{-3}$ (Sousa-Pedrares et al. 2003). It was obtained by the same procedure as for $[Zn(3-CF_3C_5H_3NS)_2]$ by additionally using 2,2'-bipy (1.493 mM) (45.87 mg of Zn dissolved). Its crystals were obtained by crystallization of the initial product from acetonitrile.

$[Zn(5-CF_3C_5H_3NS)_2(2,2'-bipy)]$ or $C_{22}H_{14}N_4F_6S_2Zn$. This compound crystallizes in the monoclinic structure with the lattice parameters $a = 657.55 \pm 0.11$, $b = 838.61 \pm 0.14$, $c = 2044.90 \pm 0.30$ pm, and $\beta = 97.060° \pm 0.003°$ and a calculated density of 1.715 g·cm$^{-3}$ (Sousa-Pedrares et al. 2003). It was obtained by the same procedure as for $[Zn(3-CF_3C_5H_3NS)_2]$ by additionally using 2,2'-bipy (1.493 mM) and $5-CF_3C_5H_4NS$ instead of $3-CF_3C_5H_4NS$ (45.87 mg of Zn dissolved).

$[Zn(3-CF_3C_5H_3NS)_2(Phen)]$ or $C_{24}H_{14}N_4F_6S_2Zn$. This compound crystallizes in the triclinic structure with the lattice parameters $a = 1107.34 \pm 0.17$, $b = 1118.06 \pm 0.17$, and $c = 1185.43 \pm 0.18$ pm and $\alpha = 92.975° \pm 0.002°$, $\beta = 108.253° \pm 0.002°$, and

$\gamma = 102.908° \pm 0.002°$ and a calculated density of 1.585 g·cm$^{-3}$ (Sousa-Pedrares et al. 2003). It was obtained by the same procedure as for [Zn(3-CF$_3$C$_5$H$_3$NS)$_2$] by additionally using Phen (1.63 mM) (46.84 mg of Zn dissolved). Its crystals were obtained by concentration of the solution.

**[Zn(5-CF$_3$C$_5$H$_3$NS)$_2$(Phen)] or C$_{24}$H$_{14}$N$_4$F$_6$S$_2$Zn**. This compound was obtained by the same procedure as for [Zn(3-CF$_3$C$_5$H$_3$NS)$_2$] using additionally Phen (1.638 mM) and 5-CF$_3$C$_5$H$_4$NS instead of 3-CF$_3$C$_5$H$_4$NS (45.86 mg of Zn dissolved) (Sousa-Pedrares et al. 2003).

**[Zn(F$_5$PhS)$_2$(2,9-Me$_2$Phen)] or C$_{26}$H$_{12}$N$_2$F$_{10}$S$_2$Zn, bis(pentafluorobenzenethiolato)(2,9-Me$_2$Phen)zinc**. This compound melts at 206°C–207°C and is characterized by two polymorphic modifications (Jordan et al. 1991). Phase transition from an orthorhombic to monoclinic structure takes place at 180°C–182°C.

**[(Et$_4$N)$_2${Zn(SC$_6$F$_5$)$_4$}] or C$_{40}$H$_{40}$N$_2$F$_{20}$S$_4$Zn**. This compound decomposes at 125°C (Beck et al. 1967). It was prepared by the following procedure. To a solution of C$_6$F$_5$SH (39.3 mM) in 20 mL of 2 N NaOH was added dropwise under constant stirring an aqueous solution (50 mL) of ZnSO$_4$·7H$_2$O (5.2 mM). The obtain mixture was filtered with an excess of Zn(OH)$_2$. An aqueous solution of Et$_4$NBr was added to the filtrate, and the precipitate was thoroughly washed with H$_2$O and dried under high vacuum. The orange crystals were grown by recrystallization of the title compound from mixture solution acetone/diethyl ether.

**[(Bu$_4$N)$_2$(C$_6$F$_5$S)$_4$Zn] or C$_{58}$H$_{72}$N$_2$F$_{20}$S$_4$Zn, bis(butylammonium) tetrakis(pentafluorobenzenethiolato)zincate(II)**. This compound was prepared by a metathetical reaction, with vigorous stirring at room temperature, of the tetrahalozincate(II) (1 mM) dissolved in a dried nonaqueous solvent (usually acetonitrile) with the finally ground C$_6$F$_5$STl or C$_6$F$_5$SAg (4 mM) (Hollebone and Nyholm 1971). The filtered solution was evaporated in vacuo to a small volume, and light-yellow (Bu$_4$N)$_2$(C$_6$F$_5$S)$_4$Zn was isolated by the addition of dried ether or light petroleum. Its crystals were grown by crystallization from acetone with ether. All manipulations were done under N$_2$ atmosphere.

## 1.66   Zn–H–C–N–Cl–S

Some compounds are formed in this system.

**[Zn(CH$_5$N$_3$S)Cl$_2$] or CH$_5$N$_3$Cl$_2$SZn**. This compound crystallizes in the orthorhombic structure with the lattice parameters $a = 1192$, $b = 728$, and $c = 1546$ pm and experimental and calculated densities 2.18 and 2.25 g·cm$^{-3}$, respectively (Cavalca et al. 1960, Nardelli and Chierici 1960). It was obtained at the interaction of ZnCl$_2$ and thiosemicarbazide in the molar ratio of 1:1 in a water solution.

**[Zn{CS(NH$_2$)$_2$}$_2$Cl$_2$] or C$_2$H$_8$N$_4$Cl$_2$S$_2$Zn**. This compound melts at 138°C–140°C (Isab and Wazeer 2005) (155°C $\pm$ 1°C, [Mary and Dhanuskodi 2001]), is stable up to 225°C (Rajasekaran et al. 2001), and crystallizes in the orthorhombic structure with the lattice parameters $a = 1304.0 \pm 0.3$, $b = 1276.7 \pm 0.3$, and $c = 589.3 \pm 0.1$ pm and a calculated density of 1.954 g·cm$^{-3}$ (Bombicz et al. 2007); ($a = 589.8$, $b = 1271.8$, and $c = 1304.7$ pm [Selvakumar et al. 2006]); $a = 590.10 \pm 0.71$, $b = 1274.14 \pm 0.43$, and $c = 1297.78 \pm 1.31$ pm and experimental and calculated densities 1.962 and 1.961 g·cm$^{-3}$, respectively (Mary and Dhanuskodi 2001); $a = 1301.2$, $b = 1276.8$, and $c = 589.0$ pm (Rajasekaran et al. 2001); $a = 1306.5 \pm 0.5$, $b = 1272.2 \pm 0.5$, and $c = 589.0 \pm 0.5$ pm and experimental and calculated densities 1.965 and 1.960 g·cm$^{-3}$, respectively (Kunchur and Truter 1958). To obtain it, granulated Zn (1 mM) was dissolved in 2.5 mL of 1 M HCl, and then 2 mM of thiourea was added and dissolved by warming. After several days at room temperature, well-grown prism-shaped crystals were obtained (Bombicz et al. 2007).

The title compound was also prepared at room temperature by mixing solutions of thiourea with ZnCl$_2$ in a minimum amount of distilled H$_2$O (2–10 mL) in the 2:1 molar ratio and stirring for 1 or 2 h (Kunchur and Truter 1958, Mary and Dhanuskodi 2001, Rajasekaran et al. 2001, 2003, Isab and Wazeer 2005, Selvakumar et al. 2006). The polycrystalline material was synthesized by evaporating the solution to almost dryness at room temperature. Optical quality crystals have been grown at an optimized pH value of 3.0 (Rajasekaran et al. 2003). It was established that the values of the lattice parameters depend from the pH value of the solution, which was used for the crystal growth (Mary and Dhanuskodi 2001, Rajasekaran et al. 2001; 2003). Single crystals have been grown by slow evaporation of saturated aqueous solution at room temperature (Mary and Dhanuskodi 2001) or from low-temperature solution growth method by slow cooling (Rajasekaran et al. 2001).

**[Zn(CH$_5$N$_3$S)$_2$Cl$_2$] or C$_2$H$_{10}$N$_6$Cl$_2$S$_2$Zn.** This compound crystallizes in the orthorhombic structure with the lattice parameters $a = 1671$, $b = 2087$, and $c = 898$ pm (Nardelli and Chierici 1960). It was obtained by the interaction of ZnCl$_2$ and thiosemicarbazide in the molar ratio of 1:2 in a water solution.

**[Zn(C$_4$H$_9$N$_3$S)Cl$_2$] or C$_4$H$_9$N$_3$Cl$_2$SZn.** This compound crystallizes in the monoclinic structure with the lattice parameters $a = 782.3 \pm 0.2$, $b = 1431.0 \pm 0.5$, $c = 869.1 \pm 0.2$ pm, and $\beta = 100.32° \pm 0.02°$ and experimental and calculated densities of 1.85 and 1.855 g·cm$^{-3}$, respectively (Mathew and Palenik 1971). It was prepared by mixing hot ethanolic solutions of ZnCl$_2$ and acetone thiosemicarbazone (C$_4$H$_9$N$_3$S) (1:1 molar ratio) from which the title compound separated upon cooling. Recrystallization from hot EtOH gave long, colorless needles (Ablov and Gerbeleu 1964, Mathew and Palenik 1971).

**[ZnCl$_2${μ-S(CH$_2$)$_3$NH(CH$_3$)$_2$}] or C$_5$H$_{13}$NCl$_2$SZn.** This compound crystallizes in the monoclinic structure with the lattice parameters $a = 1581$, $b = 644$, $c = 1023$ pm, and $\beta = 94.65°$ (Casals et al. 1987). To obtain it, a solution of 3-dimethylamino-1-propanethiol (3.23 mM) in MeOH (10 mL) and H$_2$O (40 mL) at room temperature was added slowly with stirring to a solution of ZnCl$_2$ (3.23 mM) in the same solvent previously acidified with 4.0 M HCl in order to keep the solution clear. A flocculent white precipitate formed, which did not redissolved, and was filtered off just before completion of the addition. The clear solution was then allowed to crystallize at room temperature over a few days. After 48 h, crystalline agglomerates began to appear on the walls and bottom of the baker and also some flakes in the liquid.

Colorless crystals of [$ZnCl_2${$\mu$-$S(CH_2)_3NH(CH_3)_2$}] grew together with a small number of long needles of [{$(CH_3)_2NH(CH_2)_3S$}$_2$][$ZnCl_4$] accompanied by some $Zn(OH)_2$, which could be filtered off owing to its jellied nature. After filtration, it was possible to separate manually crystals of the main product. Attempts to purify the sample by recrystallization were unsuccessful.

**[$ZnCl_2(C_3H_5NS_2)_2$] or $C_6H_{10}N_2Cl_2S_4Zn$.** This compound melts at 143°C (De Filippo et al. 1971). To obtain it, $ZnCl_2$ was dissolved in an excess of molten thiazolidine-2-thione ($C_3H_5NS_2$) (at ca. 105°C; molar ratio $Zn^{2+}/C_3H_5NS_2$ ca. 1:2), and the crude product was purified by washing with ethyl ether and recrystallizing from EtOH. The title compound could also be prepared if to a solution of $ZnCl_2$ in absolute EtOH were added different amounts of $C_3H_5NS_2$ for $Zn^{2+}/C_3H_5NS_2$ molar ratios of 1:1, 1:2, 1:3, and 1:4, and the mixture was heated under reflux for 4 h. [$ZnCl_2(C_3H_5NS_2)_2$] was separated out by crystallization.

**[$Zn${$S(SCNMe_2)_2Cl_2$}] or $C_6H_{12}N_2Cl_2S_3Zn$.** This compound was prepared by the next procedure (McCleverty and Morrison 1976). $ZnCl_2$ was added to a stirred solution of $S(SCNMe_2)_2$ in acetone. After the $ZnCl_2$ had dissolved, most of the acetone was evaporated. When crystals began to form, the mixture was cooled in a bath of acetone/solid $CO_2$. The yellow crystals were collected and washed with diethyl ether.

**[$Zn(C_3H_6N_2S)_2Cl_2$] or $C_6H_{12}N_4Cl_2S_2Zn$, dichlorobis(imydazolidine-2-thione-$S$) zinc.** This compound was prepared by a dropwise addition of ethanol solution of zinc chloride (1 mM in 10 mL) to a stoichiometric amount of imidazoline-2-thione ($C_3H_6N_2S$) in EtOH (2 mM in 50 mL) (in MeOH [Fettouhi et al. 2007]) and refluxing the solution for 30 min on a water bath (Shunmugam and Sathyanarayana 1983). Colorless crystalline product was obtained.

**[$Zn${$Me_2(NH)_2CS$}$_2Cl_2$] or $C_6H_{16}N_4Cl_2S_2Zn$, dichlorobis(1,3-dimethylthio-urea)zinc.** This compound crystallizes in the monoclinic structure with the lattice parameters $a = 1302.30 \pm 0.04$, $b = 894.70 \pm 0.03$, $c = 1243.50 \pm 0.03$ pm, and $\beta = 106.967° \pm 0.002°$ at $150 \pm 2$ K and a calculated density of 1.652 g·cm$^{-3}$ (Burrows et al. 2004b). To obtain it, equimolar aqueous solutions of zinc tetra(1,3-dimethylthiourea) dichloride and sodium salts of succinic, itaconic, or mesaconic acids were allowed to evaporate slowly over a period of 2 weeks in each case resulting in the formation of colorless crystals of [$ZnCl_2(C_3H_6N_2S)_2$].

**[$Zn(C_7H_{13}N_3S)Cl_2$] or $C_7H_{13}N_3Cl_2SZn$.** This compound was prepared by the slow cooling of a mixture of hot ethanolic solutions of stoichiometric quantities of $ZnCl_2$ and cyclohexanone thiosemicarbazone ($C_7H_{13}N_3S$) (Ablov and Gerbeleu 1964). White crystalline precipitate in the form of rhombic plates was obtained.

**[$Zn(C_4H_5N_2SSN_2H_5C_4)Cl_2$] or $C_8H_{10}N_4Cl_2S_2Zn$.** This compound crystallizes in the orthorhombic structure with the lattice parameters $a = 851.86 \pm 0.05$, $b = 1309.05 \pm 0.05$, and $c = 2455.06 \pm 0.10$ pm at 193 K (Matsunaga et al. 2005a). It was obtained from the reaction between $ZnCl_2$ (5.0 mM) and 1-methylimidazoline-2-thione ($C_4H_5N_2S$) (10 mM) in methanol solution (45 mL). After filtration of a white powder and crystals of $[ZnCl_2(C_4H_5N_2S)_2]$, yellow crystals of the title compound were formed by allowing the filtrate to stand at $0°C$ for several months.

$[Zn(C_4H_6N_2S)_2Cl_2]$ or $C_8H_{12}N_4Cl_2S_2Zn$. This compound crystallizes in the orthorhombic structure with the lattice parameters $a = 1345.4 \pm 0.9$, $b = 1367.0 \pm 0.8$, and $c = 783.0 \pm 0.5$ pm and a calculated density of 1.68 g·cm$^{-3}$ (Matsunaga et al. 2005b). It was prepared by the following procedure. To a $CH_3OH$ solution (40 mL) of 1-methyl-imidazoline-2(3$H$)-thione (10 mM), a $CH_3OH$ solution (5 mL) of $ZnCl_2$ (5.0 mM) was added. When the mixture was stirred for 1 h, a white solid appeared. The resulting white powder was collected by filtration and washed with $CH_3OH$. The colorless single crystals for XRD were obtained by cooling the filtrate for a few days in the refrigerator.

$[Zn(C_4H_8N_2S)_2Cl_2]$ or $C_8H_{16}N_4Cl_2S_2Zn$, dichlorobis($N$-methylimydazolidine-2-thione-$S$)zinc. This compound melts at $166°C$ (Raper and Brook 1977) (at $175°C$ with decomposition [Devillanova and Verani 1980]) and crystallizes in the triclinic structure with the lattice parameters $a = 736.75 \pm 0.10$, $b = 771.83 \pm 0.10$, and $c = 1396.44 \pm 0.18$ pm and $\alpha = 94.545° \pm 0.002°$, $\beta = 95.877° \pm 0.002°$, and $\gamma = 108.225° \pm 0.002°$ (Fettouhi et al. 2007). It was prepared by dissolving $ZnCl_2 \cdot 6H_2O$ (1.5 mM) in a small volume of hot anhydrous EtOH (MeOH [Fettouhi et al. 2007]) containing 3 mL triethyl orthoformate as dehydrating agent. To this solution, 3.0 mM of $N$-methylimidazoline-2-thione ($C_4H_8N_2S$) dissolved in 5 mL of hot anhydrous EtOH was added. The resulting solution was reduced in volume by boiling on a water bath and then cooled to room temperature. Subsequent slow evaporation of the solvent was accompanied by the formation of a crystalline colorless product. This was removed by filtration, washed with a small volume of cold anhydrous EtOH and then dry ether, and finally vacuum dried at room temperature (Raper and Brook 1977, Fettouhi et al. 2007). Higher metal/ligand ratios (up to 1:6) produced compound that was identical to that from the aforementioned procedure (Raper and Brook 1977, Shunmugam and Sathyanarayana 1983).

$[Zn(C_4H_8N_2S)_2Cl_2]$ or $C_8H_{16}N_4Cl_2S_2Zn$, dichlorobis(1,3-diazinane-2-thione-$S$) zinc. This compound was prepared by dissolving $ZnCl_2$ in a minimum amount of $H_2O$ and two molar equivalents of 1,3-diazinane-2-thione in MeOH; the mixture was refluxed for 2–3 h (Fettouhi et al. 2007). The solution was allowed to evaporate slowly, and white crystalline products were isolated.

$[(mepa)ZnCl \cdot 0.5CH_2Cl_2]$ or $C_{8.5}H_{12}N_2Cl_2SZn$ (mepaH = $N$-(2-mercaptoethyl) picolylamin, $C_8H_{12}N_2S$). This compound melts at $200°C$ with decomposition and crystallizes in the monoclinic structure with the lattice parameters $a = 2278.9 \pm 0.5$, $b = 727.8 \pm 0.1$, $c = 1772.9 \pm 0.4$ pm, and $\beta = 122.31° \pm 0.03°$ and experimental and calculated densities 1.60 and 1.66 g·cm$^{-3}$, respectively (Brand and Vahrenkamp 1995). To obtain it, a stirred solution of mepaH (10.4 mM) in $CH_2Cl_2$ (40 mL) was treated dropwise with a solution of $Zn[N[SiMe_3]_2]_2$ (5.2 mM) in $CH_2Cl_2$ (20 mL). The precipitate was filtered off, washed three times with $CH_2Cl_2$, and dried in vacuo yielding $[(mepa)ZnCl \cdot 0.5CH_2C_{12}]$ as a colorless crystals.

$[Zn(C_9H_{12}N_4S)Cl_2]$ or $C_9H_{12}N_4Cl_2SZn$. This compound melts at the temperature higher than $300°C$ (Bermejo et al. 1999). It was obtained by treating

2-acetylpyridine 4-methylthiosemicarbazone ($C_9H_{12}N_4S$) with $ZnCl_2$ (1:1 molar ratio) in EtOH. After prolonged stirring (about 1 week) at room temperature, the reaction mixture afforded the title compound, which was filtered off, washed with EtOH, and vacuum dried.

**[Zn($C_9H_{13}N_5S$)$Cl_2$] or $C_9H_{13}N_5Cl_2SZn$.** This compound melts at a temperature higher than 300°C (Bermejo et al. 2001). It was prepared as follows. A solution of 2-pyridylformamide N(4)-dimethylthiosemicarbazone (1 mM) in EtOH (30 mL) was added to a solution of $ZnCl_2$ (1 mM) in 30 mL of 96% EtOH, and the mixture was stirred for several days at room temperature. The resulting yellow solid was filtered off, washed thoroughly with cold EtOH, and stored in a desiccator over Mg and $I_2$.

**[Zn($C_9H_{13}N_5S$)$Cl_2$] or $C_9H_{13}N_5Cl_2SZn$.** To obtain this compound, a solution of $ZnCl_2$ (1.12 mM) in EtOH (30 mL) was mixed with a solution of 2-pyridinefor-mamide N(4)-ethylthiosemicarbazone ($C_9H_{13}N_5S$) (1.12 mM) in EtOH (30 mL) (Bermejo et al. 2004a). The resulting yellow solution was stirred at room temperature for 7 days, and a yellow precipitate was filtered off.

**[Zn($C_5H_4NS$)$_2Cl_2$] or $C_{10}H_8Cl_2N_2S_2Zn$.** This compound crystallizes in the monoclinic structure with the lattice parameters $a = 1231.7 \pm 0.2$, $b = 986.8 \pm 0.2$, $c = 1094.7 \pm 0.2$ pm, and $\beta = 92.37° \pm 0.02°$ at 110 K (Seidel et al. 2012). It was prepared by the following procedure. $ZnCl_2$ (0.05 mM) was dissolved in 2 mL of MeOH, $AgCF_3SO_3$ (0.09 mM) was added, and the mixture was stirred for 20 min in the dark. Subsequently, the precipitate was removed by centrifugation, and a solution of Phen (0.05 mM) in 8 mL of MeOH was added. The resulting solution was carefully layered onto a solution of 4,4'-dithiopdipyridine (0.05 mM) in 10 mL of $CH_2Cl_2$. After standing a couple of days at room temperature, while the solvent was allowed to evaporate slowly, a few colorless crystals of the title compound were found.

**[Zn($C_5H_5NS$)$_2Cl_2$] or $C_{10}H_{10}N_2Cl_2S_2Zn$.** These compounds were prepared by mixing ethanolic solutions of $ZnCl_2$ and 2-pyridine thiol or 4-pyridine thiol (Kennedy and Lever 1972). The resulting precipitate was recrystallized from ethanol by Soxhlet extraction. The title compounds required the addition of HCl to prevent partial hydrolysis.

**[Zn($C_5H_{10}N_2S$)$_2Cl_2$] or $C_{10}H_{20}N_4Cl_2S_2Zn$, dichlorobis[(N,N-dimethylimydazoli-dine-2-thione-S)]zinc.** This compound melts at 212°C (Devillanova and Verani 1980). It was prepared by dissolving $ZnCl_2$ in a minimum amount of $H_2O$ and two molar equivalents of N,N-dimethylimydazolidine-2-thione in MeOH; the mixture was refluxed for 2–3 h (Fettouhi et al. 2007). The solution was allowed to evaporate slowly and white crystalline products were isolated.

**[Zn($C_5H_{10}N_2S$)$_2Cl_2$] or $C_{10}H_{20}N_4Cl_2S_2Zn$, dichlorobis[(N-ethylimydazolidine-2-thione-S)]zinc.** This compound melts at 156°C (Devillanova and Verani 1980). It was prepared by dissolving $ZnCl_2$ in a minimum amount of $H_2O$ and two molar equivalents of N-ethylimydazolidine-2-thione in MeOH; the mixture was refluxed for 2–3 h (Fettouhi et al. 2007). The solution was allowed to evaporate slowly and white crystalline products were isolated.

**[Zn(C$_5$H$_{10}$N$_2$S)$_2$Cl$_2$] or C$_{10}$H$_{20}$N$_4$Cl$_2$S$_2$Zn, dichlorobis[(1,3-diazipane-2-thione-S)]zinc.** This compound was prepared by dissolving ZnCl$_2$ in a minimum amount of H$_2$O and two molar equivalents of 1,3-diazipane-2-thione in MeOH; the mixture was refluxed for 2–3 h (Fettouhi et al. 2007). The solution was allowed to evaporate slowly and white crystalline products were isolated.

**[{(CH$_3$)$_2$NH(CH$_2$)$_3$S}$_2$][ZnCl$_4$] or C$_{10}$H$_{26}$N$_2$Cl$_4$S$_2$Zn.** This compound crystallizes in the orthorhombic structure with the lattice parameters $a = 2334$, $b = 1358$, and $c = 670$ pm (Casals et al. 1987). It was obtained simultaneously with obtaining of [ZnCl$_2$\{μ-S(CH$_2$)$_3$NH(CH$_3$)$_2$\}].

**[Zn(C$_6$H$_7$NS)$_2$Cl$_2$] or C$_{12}$H$_{14}$N$_2$Cl$_2$S$_2$Zn.** This compound was prepared by mixing ethanolic solutions of ZnCl$_2$ and 2-methyl-6-pyridine thiol (Kennedy and Lever 1972). The resulting precipitate was recrystallized from ethanol by Soxhlet extraction. The title compound required the addition of HCl to prevent partial hydrolysis.

**[Zn(dmtp)$_2$Cl$_2$] or C$_{12}$H$_{16}$N$_4$Cl$_2$S$_2$Zn (dmtp = 4,6-dimethyl-2-thiopyrimidine).** This compound crystallizes in the monoclinic structure with the lattice parameters $a = 713.3$, $b = 1631.5$, $c = 740.3$ pm, and $β = 92.48°$ and a calculated density of 1.60 g·cm$^{-3}$ (Lopez-Garzon et al. 1993). It was prepared by dissolving dmtp and ZnCl$_2$ in a 1:3 molar ratio, in the minimum amount of water. The solution was then evaporated on a water bath at 60°C. Pale-yellow crystals precipitated after 2 days.

**[Zn(C$_{12}$H$_{17}$N$_5$S)Cl$_2$] or C$_{12}$H$_{17}$N$_5$Cl$_2$SZn.** This compound melts at the temperature higher than 300°C (Castiñeiras et al. 2002). To obtain it, a solution of C$_{12}$H$_{17}$N$_5$S [2-pyridineformamide N(4)-piperidylthiosemicarbazone (0.76 mM) in EtOH (30 mL) was added to a solution of ZnCl$_2$ (0.76 mM) in 30 mL of 96% EtOH and the mixture was stirred for several days at room temperature. The resulting yellow solid was filtered off, washed thoroughly with cold ethanol, and stored in a desiccator over Mg and I$_2$.

**[Zn(C$_6$H$_{12}$N$_2$S)$_2$Cl$_2$] or C$_{12}$H$_{24}$N$_4$Cl$_2$S$_2$Zn, dichlorobis-[N-methyl(ethyl)imydazolidine-2-thione-S)]zinc.** This compound was prepared by dissolving ZnCl$_2$ in a minimum amount of H$_2$O and two molar equivalents of N-(i-propyl)imydazolidine-2-thione in MeOH; the mixture was refluxed for 2–3 h (Fettouhi et al. 2007). The solution was allowed to evaporate slowly and white crystalline products were isolated.

**[Zn(C$_6$H$_{12}$N$_2$S)$_2$Cl$_2$] or C$_{12}$H$_{24}$N$_4$Cl$_2$S$_2$Zn, dichlorobis-[N-(n-propyl)imydazolidine-2-thione-S)]zinc.** This compound crystallizes in the monoclinic structure with the lattice parameters $a = 1565.76 \pm 0.15$, $b = 876.68 \pm 0.15$, $c = 1492.60 \pm 0.14$ pm, and $β = 104.654° \pm 0.002°$ (Fettouhi et al. 2007). It was prepared by dissolving ZnCl$_2$ in a minimum amount of H$_2$O and two molar equivalents of N-(n-propyl)imydazolidine-2-thione in MeOH; the mixture was refluxed for 2–3 h. The solution was allowed to evaporate slowly and white crystalline products were isolated.

**[Zn(C$_6$H$_{12}$N$_2$S)$_2$Cl$_2$] or C$_{12}$H$_{24}$N$_4$Cl$_2$S$_2$Zn, dichlorobis-N-[(i-propyl)imydazolidine-2-thione-S)]zinc.** This compound was prepared by dissolving ZnCl$_2$ in a minimum amount of H$_2$O and two molar equivalents of N-(i-propyl)imydazolidine-2-thione in MeOH; the mixture was refluxed for 2–3 h (Fettouhi et al. 2007). The solution was allowed to evaporate slowly and white crystalline products were isolated.

**[Zn(C$_6$H$_{12}$N$_2$S)$_2$Cl$_2$] or C$_{12}$H$_{24}$N$_4$Cl$_2$S$_2$Zn, dichlorobis(N-ethyl-1,3-diazinane-2-thione-S)zinc.** This compound was prepared by dissolving ZnCl$_2$ in a minimum amount of H$_2$O and two molar equivalents of N-ethyl-1,3-diazinane-2-thione in MeOH; the mixture was refluxed for 2–3 h (Fettouhi et al. 2007). The solution was allowed to evaporate slowly and white crystalline products were isolated.

**[ZnCl$_2$·C$_{13}$H$_{16}$N$_2$S$_2$] or C$_{13}$H$_{16}$N$_2$Cl$_2$S$_2$Zn.** This compound was obtained at the interaction of ZnCl$_2$ and N-cyclohexyl-2-benzothiazolesulfenamide (C$_{13}$H$_{16}$N$_2$S$_2$) (1:1 molar ratio) (Gudimovich et al. 1984). Its thermal decomposition leads to the formation of β-ZnS.

**[2ZnCl$_2$·C$_{13}$H$_{16}$N$_2$S$_2$] or C$_{13}$H$_{16}$N$_2$Cl$_4$S$_2$Zn$_2$.** This compound was obtained at the interaction of ZnCl$_2$ and N-cyclohexyl-2-benzothiazolesulfenamide (C$_{13}$H$_{16}$N$_2$S$_2$) (2:1 molar ratio) (Gudimovich et al. 1984). Its thermal decomposition leads to the formation of α-ZnS.

**[Zn(C$_{13}$H$_{19}$N$_5$S)Cl$_2$] or C$_{13}$H$_{19}$N$_5$Cl$_2$SZn.** This compound could be obtained if a solution of ZnCl$_2$ (0.90 mM) in EtOH (30 mL) was mixed with a solution of 2-pyridineformamide 3-hexamethyleneiminylthiosemicarbazone (C$_{13}$H$_{19}$N$_5$S) (0.90 mM) in EtOH (30 mL) (Bermejo et al. 2004b). The resulting yellow solution was stirred at room temperature for 7 days and the precipitate was filtered off.

**[ZnCl$_2$(C$_7$H$_{10}$N$_2$S)$_2$] or C$_{14}$H$_{20}$N$_4$Cl$_2$S$_2$Zn (C$_7$H$_{10}$N$_2$S = 1-methyl-3-(prop-2-enyl)-imidazole-2(3H)-thione).** This compound crystallizes in the monoclinic structure with the lattice parameters $a = 1400.82 \pm 0.08$, $b = 1080.3 \pm 0.1$, $c = 1458.1 \pm 0.1$ pm, and $\beta = 116.625° \pm 0.008°$ and a calculated density of 1.50 g·cm$^{-3}$ (Williams et al. 1997). Crystals suitable for XRD were obtained by slow evaporation from the solution of the title compound in CH$_2$Cl$_2$.

**[Zn(C$_{14}$H$_{21}$N$_4$S)Cl] or C$_{14}$H$_{21}$N$_4$ClSZn.** This compound crystallizes in the triclinic structure with the lattice parameters $a = 990.5 \pm 0.2$, $b = 1375.8 \pm 0.2$, and $c = 1478.7 \pm 0.3$ pm and $\alpha = 67.815° \pm 0.012°$, $\beta = 87.213° \pm 0.014°$, and $\gamma = 71.473° \pm 0.014°$ (Hammesa and Carrano 2000). To prepare it, a solution of [MeZn(C$_{14}$H$_{21}$N$_4$S)] (0.86 mM) in 30 mL of CH$_2$Cl$_2$ was treated dropwise with one equivalent of concentrated HCl (0.86 mM as a solution in CH$_2$Cl$_2$). The resulting reaction mixture was stirred for 1 h and concentrated under reduced pressure to yield the title compound. Recrystallization was accomplished by slow evaporation of a CH$_2$Cl$_2$/hexane (3:2 volume ratio) solution.

**[Zn(C$_7$H$_{14}$N$_2$S)$_2$Cl$_2$] or C$_{14}$H$_{28}$N$_4$Cl$_2$S$_2$Zn, dichlorobis[(N,N-diethylimydazolidine-2-thione-S)]zinc.** This compound melts at 95°C (Devillanova and Verani 1980). It was obtained by reacting ZnCl$_2$ with N,N-diethylimydazolidine-2-thione (1:2 molar ratio) in dimethoxypropane.

**[Zn(C$_8$H$_9$N$_3$S)$_2$Cl$_2$] or C$_{16}$H$_{18}$N$_6$Cl$_2$S$_2$Zn.** This compound was prepared by the slow cooling of a mixture of hot ethanolic solutions of stoichiometric quantities of ZnCl$_2$ and benzylidene thiosemicarbazide (C$_8$H$_9$N$_3$S) (1:2 molar ratio) (Ablov and

Gerbeleu 1964). White crystalline precipitate was obtained, which was filtered off, washed sometimes with EtOH and then with ether, and dried in air.

**[Zn₄{S(CH₂)₂NMe₂}₄Cl₄]** or **C₁₆H₄₀N₄Cl₄S₄Zn₄**. To obtain this compound, HS(CH₂)₂NMe₂·HCl (10 mM) was dispersed in acetonitrile (40 mL) (Casals et al. 1991b). Et₃N (2.8 mL) and anhydrous ZnCl₂ (10 mM) were successfully added with stirring under a N₂ atmosphere. Stirring was continued for 24 h. The resulting suspension was filtered in the open atmosphere, washed with copious distilled water, then with MeOH, and with diethyl ether, and vacuum dried.

Alternatively, this compound was obtained by adding an aqueous solution of equimolar amounts of HS(CH₂)₂NMe₂·HCl and NaOH to an aqueous solution of Zn(OAc)₂, the metal/ligand molar ratio being always 1:1 (Casals et al. 1991b). A white powder formed immediately. The mixture was stirred for several hours; the solid was filtered off, washed with cold water and then MeOH, and dried under vacuum. Crystals were obtained by refluxing the compound in DMSO, filtering off the undissolved solid, letting the solution cool to 60°C, and maintaining it at this temperature for 24 h. All these manipulations were performed under a N₂ atmosphere.

**[Zn(S₂CNEt₂)₂(Py)(C₂H₄Cl₂)]** or **C₁₇H₂₉N₃Cl₂S₄Zn**. This compound crystallizes in the monoclinic structure with the lattice parameters $a = 1152.1 \pm 0.1$, $b = 852.5 \pm 0.1$, $c = 2538.4 \pm 0.4$ pm, and $\beta = 96.34° \pm 0.02°$ and a calculated density of $1.477 \pm 0.001$ g·cm⁻³ (Ivanov et al. 1999a). It was prepared by a carefully wetting of the Zn(S₂CNEt₂)₂(Py) samples by a chlorinated solvent.

**[(Buⁿ₄N){Zn(S₂CNMe₂)Cl₂}]** or **C₁₉H₄₂N₂Cl₂S₂Zn**. This compound was prepared by the following method (McCleverty and Morrison 1976). A solution of [Buⁿ₄N][(S₂CNMe₂)] (0.01 M) in acetone (50 mL) was added to a suspension of ZnCl₂ (0.01 M) in acetone (50 mL). The mixture was stirred to dissolve the ZnCl₂ and then light petroleum was added. Partial evaporation of the solvent afforded white crystals (sometimes an oil formed first and then crystallized), which were collected, washed with diethyl ether, and recrystallized from CH₂Cl₂/hexane.

**[Zn(C₂₀H₂₇N₆S₂)Cl]** or **C₂₀H₂₇N₆ClS₂Zn**. This compound crystallizes in the triclinic structure with the lattice parameters $a = 840.6 \pm 0.3$, $b = 1108.6 \pm 0.2$, and $c = 1399.1 \pm 0.3$ pm and $\alpha = 78.702° \pm 0.012°$, $\beta = 86.93° \pm 0.02°$, and $\gamma = 76.465° \pm 0.014°$ (Castiñeiras et al. 1999). It was prepared by mixing 1 mM of ZnCl₂ and 1 mM of 1-phenylglyoxal bis(3-piperidylthiosemicarbazone) in 25 mL of 95% EtOH, with 1 mL of Et₃N added. The mixture was refluxed for 2 h and the resulting orange solid was filtered, while the solution was warm.

**[Zn(SCN)₂(C₇H₁₀N₂)₂]·PhCl** or **C₂₂H₂₅N₆ClS₂Zn**. This compound crystallizes in the monoclinic structure with the lattice parameters $a = 3065.7 \pm 0.3$, $b = 1011.5 \pm 0.2$, $c = 1842.4 \pm 0.2$ pm, and $\beta = 113.089° \pm 0.008°$ and a calculated density of 1.361 g·cm⁻³ (Secondo et al. 2000). Crystals suitable for XRD were grown by recrystallization of [Zn(SCN)₂(C₇H₁₀N₂)₂] from the solution in PhCl.

**[Zn(C₁₁H₁₅N₃S)₂Cl₂]** or **C₂₂H₃₀N₆Cl₂S₂Zn**. This compound was prepared by the slow cooling of a mixture of hot ethanolic solutions of stoichiometric quantities of

$ZnCl_2$ and isopropylbenzaldehyde thiosemicarbazone ($C_{11}H_{15}N_3S$) (1:2 molar ratio) (Ablov and Gerbeleu 1964). White crystalline precipitate in the form of rectangular plates was obtained, which was filtered off, washed sometimes with EtOH and then with ether, and dried in air.

**[Zn(4-ClPhS)$_2$(2,9-Me$_2$Phen)] or $C_{26}H_{21}N_2ClS_2Zn$, bis(4-chlorobenzenethiolato) (2,9-Me$_2$Phen)zinc.** This compound melts at 224°C–225°C and is characterized by two polymorphic modifications (Jordan et al. 1991). Phase transition from orthorhombic to monoclinic structure takes place at 185°C–187°C.

**[Zn$_2$(S$_2$CNEt$_2$)$_4$(C$_{12}$H$_{10}$N$_2$)]·CHCl$_3$ or $C_{33}H_{51}N_6Cl_3S_8Zn_2$.** This compound turned opaque at 76°C–78°C, melts at 128°C–130°C, and crystallizes in the orthorhombic structure with the lattice parameters $a = 1744.3 \pm 0.3$, $b = 1573.9 \pm 0.3$, and $c = 1682.3 \pm 0.3$ pm at 98 K and a calculated density of 1.475 g·cm$^{-3}$ (Arman et al. 2009a). It was prepared by following a standard procedure whereby two equivalents of Zn(S$_2$CNEt$_2$)$_2$ were added to *trans*-1,2-bis(4-pyridyl)ethylene. Crystals of the title compound were obtained by slow evaporation of a chloroform solution.

**[(Et$_4$N)$_2$(μ-SPr$^i$)$_6$(ZnCl)$_4$] or $C_{34}H_{82}N_2Cl_4S_6Zn_4$.** To obtain this compound, a mixture of ZnCl$_2$ (0.98 mM), Et$_4$NCl·H$_2$O (2.0 mM), and Zn(SPr$^i$)$_2$ (3.3 mM) was stirred in CH$_2$Cl$_2$ (20 mL) for 1 h (Dean et al. 1992). The small amount of insoluble material was removed by filtration. The volume of the filtrate was reduced to ca. 4 mL by gentle heating under a flow of N$_2$, and then Et$_2$O (2 mL) was layered on. The mixture was allowed to stand in the refrigerator overnight. The white solid that formed was separated by filtration, washed with Et$_2$O, and dried in vacuo at room temperature.

**[Zn{S$_2$CN(CH$_2$)$_4$}$_2$·(4,7-Ph$_2$Phen)]·CHCl$_3$ or $C_{35}H_{33}N_4Cl_3S_4Zn$.** This compound melts at 205°C–208°C and crystallizes in the triclinic structure with the lattice parameters $a = 1317.03 \pm 0.06$, $b = 1765.86 \pm 0.09$, and $c = 1788.09 \pm 0.09$ pm and $\alpha = 63.013° \pm 0.001°$, $\beta = 86.826° \pm 0.001°$, and $\gamma = 75.315° \pm 0.001°$ (Lai and Tiekink 2003d). Bright-yellow crystals were isolated from acetonitrile/chloroform (1:1 volume ratio) solution containing equimolar amounts of [Zn{S$_2$CN(CH$_2$)$_4$}$_2$] and (4,7-Ph$_2$Phen).

**[(Me$_4$N)$_2${Zn$_4$(SPh)$_6$Cl$_4$}] or $C_{44}H_{54}N_2Cl_4S_6Zn_4$.** To obtain this compound, freshly prepared solid PhICl$_2$ (0.60 mM) was added with stirring to a solution of (Me$_4$N)$_2$[Zn$_4$(SPh)$_{10}$] (0.30 mM) in 15 mL of MeCN (Dean et al. 1987). A rapid reaction occurred with the formation of a precipitate. After 10 min, the solvent was removed under vacuum. The residue was extracted three times with 20 mL portions of diethyl ether to remove (PhS)$_2$ and PhI. The crude residue was purified by recrystallization from a mixture of acetonitrile and diethyl ether. The syntheses were performed under an inert atmosphere.

**[(Me$_4$N)$_2$(μ-SPh)$_6$(ZnSPh)$_2$(ZnCl)$_2$] or $C_{56}H_{64}N_2Cl_2S_8Zn_4$.** This compound crystallizes in the triclinic structure with the lattice parameters $a = 1237.7 \pm 0.2$, $b = 1659.2 \pm 0.3$, and $c = 1696.7 \pm 0.3$ pm and $\alpha = 76.68° \pm 0.01°$, $\beta = 69.92° \pm 0.01°$, and $\gamma = 72.57° \pm 0.01°$ and experimental and calculated densities 1.44 ± 0.02 and 1.46 g·cm$^{-3}$, respectively (Dance 1981). To obtain it, to a stirred solution containing benzenethiol (40 mL),

$Et_3N$ (40 mM), and $Me_4NCl$ (1.8 mM) in MeOH (100 mL) at room temperature was added slowly a solution of $Zn(NO_3)_2 \cdot 6H_2O$ (20 mM) in MeOH (60 mL). During this addition, a white precipitate that formed was redissolved by the addition of minimum volumes of acetone: a total of 35 mL of acetone was required. The flask containing this solution was sealed and allowed to stand undisturbed at 20°C for 10 days. Crystallization commenced within 1 h. The product was filtered, washed with MeOH, and vacuum dried.

Analogous experiments using morpholine (3.5 g) in place of $Et_3N$, with total solvent composition 150 mL of MeOH plus 20 mL of acetonitrile and crystallization at 0°C, or N-methylmorpholine (4.0 g) with total solvent composition 165 mL plus 45 mL of acetonitrile and crystallization at 20°C gave the same crystalline product (Dance 1981).

**$[(Et_4N)_2(\mu\text{-}Bz)_6(ZnCl)_4]$ or $C_{58}H_{82}N_2Cl_4S_6Zn_4$.** To obtain this compound, a mixture of $ZnCl_2$ (0.98 mM), $Et_4NCl \cdot H_2O$ (2.0 mM), and $Zn(SBz)_2$ (3.3 mM) was stirred in $CH_2Cl_2$ (20 mL) for 1 h (Dean et al. 1992). The small amount of insoluble material was removed by filtration. The volume of the filtrate was reduced to ca. 4 mL by gentle heating under a flow of $N_2$, and then pentane (2 mL) was layered on. The mixture was allowed to stand in the refrigerator overnight. The white solid that formed was separated by filtration, washed with $Et_2O$, and dried in vacuo at room temperature.

**$[(Me_4N)_2\{Zn_4(SPh)_9Cl\}]$ or $C_{62}H_{69}N_2ClS_9Zn_4$.** This compound crystallizes in the orthorhombic structure with the lattice parameters $a = 2349.3 \pm 0.3$, $b = 2972.6 \pm 0.4$, and $c = 1783.5 \pm 0.2$ pm and experimental density $1.43 \pm 0.01$ g·cm$^{-3}$ (Dance 1985) [the same crystal structure and the lattice parameters are given for $(Pr_3NH)(Pr_3N)$ $[ClZn_8(SPh)_{16}]$ or $C_{114}H_{123}N_2ClS_{16}Zn_8$ in (Dance 1980a)]. To obtain it, a solution of $Zn(NO_3)_2 \cdot 6H_2O$ (21 mM) in MeOH (100 mL) was added to a well-stirred solution containing benzenethiol (51 mM), $Pr_3N$ (50 mM), and $Me_4NCl$ (16 mM) in MeOH (200 mL) and, after dissolution of all transient precipitate, was sealed under $N_2$ (Dance 1981, 1985). While standing overnight at room temperature, this solution yielded fluid oil, which slowly crystallized on subsequent shaking and heating. Acetone (20 mL) and the acetonitrile (20 mL) were added with heating. The hot solution was filtered from a relatively small proportion of white solid (soluble in acetonitrile alone), and the filtrate allowed to cool slowly. White powder separated overnight, followed by colorless microcrystals during the next 10 days. After precipitation of this crystalline product (A), large colorless blocky crystals (B) began to grow around the upper boundaries of the solution. The mixed solids (A and B) were collected after 6 weeks, washed with EtOH, and vacuum dried. The major component (A) is $(Me_4N)_2[Zn_4(SPh)_9Cl]$.

Similar experiments with decreased Cl$^-$ content invariably led to crystalline $(Me_4N)_2[Zn_4(SPh)_{10}]$, while increased Cl$^-$ proportions led to $(Me_4N)_2[Zn_4(SPh)_9Cl]$ or $(Me_4N)_2[Zn_4(SPh)_8Cl_2]$ (Dance 1985).

**$[(Pr_3NH)(Pr_3N)\{ClZn_8(SPh)_{16}\}]$ or $C_{114}H_{123}N_2ClS_{16}Zn_8$.** This compound crystallizes in the orthorhombic structure with the lattice parameters $a = 2349.3 \pm 0.3$, $b = 2972.6 \pm 0.4$, and $c = 1783.5 \pm 0.2$ pm and experimental density $1.43 \pm 0.01$ g·cm$^{-3}$

(Dance 1980a) (the same crystal structure and the lattice parameters are given for (Me$_4$N)$_2$[Zn$_4$(SPh)$_9$Cl] or C$_{62}$H$_{69}$N$_2$ClS$_9$Zn$_4$ in [Dance 1985]. To obtain it, a mixture of Zn(NO$_3$) 6H$_2$O (21 mM), benzenethiol (51 mM), methanol, and acetone, under N$_2$ atmosphere at room temperature, crystallized two products. After yielding the first microcrystalline product, probably (Me$_4$N)[Zn$_4$(SPh)$_9$Cl], the undisturbed reaction mixture produced well-developed colorless crystals of C$_{114}$H$_{123}$N$_2$ClS$_{16}$Zn$_8$ in a very low yield over a period of 3 weeks. Attempts to develop a reproducible synthesis of this compound have not yet been successful, precluding precise chemical analysis.

## 1.67    Zn–H–C–N–Br–S

Some compounds are formed in this system.

**[Zn(C$_4$H$_9$N$_3$S)Br$_2$] or C$_4$H$_9$N$_3$Br$_2$SZn**. This compound was prepared by mixing hot ethanolic solution (10 mL) of ZnBr$_2$ (10 mM) and hot (70°C) ethanolic solution (15 mL) of acetone thiosemicarbazone (C$_4$H$_9$N$_3$S) (10 mM) (Ablov and Gerbeleu 1964). White crystalline precipitate was obtained at the slow cooling, which was filtered off, washed with EtOH and then with ether, and dried in air.

**[ZnBr$_2${μ-S(CH$_2$)$_3$NH(CH$_3$)$_2$}] or C$_5$H$_{13}$NBr$_2$SZn**. This compound crystallizes in the monoclinic structure with the lattice parameters $a$ = 1591, $b$ = 657, and $c$ = 1020 pm and β = 95.36° (Casals et al. 1987). To obtain it, a solution of 3-dimethylamino-1-propanethiol (2.50 mM) in MeOH (10 mL) and H$_2$O (40 mL) at room temperature was added dropwise with stirring to a solution of ZnBr$_2$ (2.50 mM) in the same solvent that contained sufficient 4.0 M HBr (bromine-free) to keep the solution clear. Soon after completion of the addition, some turbidity developed, which was eliminated by centrifuging and decanting. The clear solution was then allowed to crystallize at room temperature over a few days. After 2 weeks, colorless needles appeared, while the bottom of the baker was encrusted with a white paste. By decanting the liquid, filtering, washing with H$_2$O, and vacuum drying crystals of essentially [ZnBr$_2${μ-S(CH$_2$)$_3$NH(CH$_3$)$_2$}] were separated. The obtained solid includes a mixture of this compound and [{(CH$_3$)$_2$NH(CH$_2$)$_3$S}$_2$][ZnBr$_4$].

**[ZnBr$_2$(C$_3$H$_5$NS)$_2$] or C$_6$H$_{10}$N$_2$Br$_2$S$_4$Zn**. This compound melts at 144°C (De Filippo et al. 1971). To obtain it, ZnBr$_2$ was dissolved in an excess of molten thiazolidine-2-thione (C$_3$H$_5$NS$_2$) (at ca. 105°C; molar ratio Zn$^{2+}$/C$_3$H$_5$NS$_2$ ca. 1:2), and the crude product was purified by washing with ethyl ether and recrystallizing from EtOH. The title compound could also be prepared if to a solution of ZnBr$_2$ in absolute EtOH were added different amounts of C$_3$H$_5$NS$_2$ for Zn$^{2+}$/C$_3$H$_5$NS$_2$ molar ratios of 1:1, 1:2, 1:3, and 1:4, and the mixture was heated under reflux for 4 h. [ZnBr$_2$(C$_3$H$_5$NS$_2$)$_2$] was separated out by crystallization.

**[Zn{S(SCNMe$_2$)$_2$Br$_2$}] or C$_6$H$_{12}$N$_2$Br$_2$S$_3$Zn**. This compound was prepared by the next procedure (McCleverty and Morrison 1976). ZnBr$_2$ was added to a stirred solution of S(SCNMe$_2$)$_2$ in acetone. After the ZnBr$_2$ had dissolved, most of the acetone was evaporated. When crystals began to form, the mixture was cooled in a bath of acetone/solid CO$_2$. The yellow crystals were collected and washed with diethyl ether.

**[ZnBr$_2$·Me$_4$C$_2$N$_2$S$_4$] or C$_6$H$_{12}$N$_2$Br$_2$S$_4$Zn**. This compound was prepared by the following procedure (Brinkhoff et al. 1969). To a suspension of ZnBr$_2$ in CS$_2$, a solution of tetramethylthiuram disulfide in the same solvent was added at room temperature. After ZnBr$_2$ had dissolved, petroleum ether was added, precipitating yellow crystals. The obtained product was filtered and dried.

**[Zn(C$_3$H$_6$N$_2$S)$_2$Br$_2$] or C$_6$H$_{12}$N$_4$Br$_2$S$_2$Zn, dibromobis(imydazolidine-2-thione-$S$) zinc**. This compound melts at 175°C and crystallizes in the monoclinic structure with the lattice parameters $a = 809.84 \pm 0.05$, $b = 1340.94 \pm 0.09$, $c = 1232.30 \pm 0.08$ pm, and $\beta = 100.983° \pm 0.001°$ (Fettouhi et al. 2006). It was prepared by a dropwise addition of ethanol solution of zinc bromide (1 mM in 10 mL) to a stoichiometric amount of imidazoline-2-thione (C$_3$H$_6$N$_2$S) in EtOH (2 mM in 50 mL) (in MeOH [Fettouhi et al. 2006, 2007]) and refluxing the solution for 30 min (3 h [Fettouhi et al. 2006]) on a water bath (Shunmugam and Sathyanarayana 1983). Colorless crystalline product was obtained.

**[Zn(C$_5$H$_4$NC$_2$H$_5$N$_4$S)Br$_2$] or C$_7$H$_9$N$_5$Br$_2$SZn**. This compound melts at 245°C–246°C (Castiñeiras et al. 2000). To obtain it, a solution of ZnBr$_2$ (2 mM) in EtOH (30 mL) was mixed with a solution of 2-pyridineformamide thiosemicarbazone (C$_5$H$_4$NC$_2$H$_5$N$_4$S) (2 mM) in the same solvent (30 mL), and the mixture was stirred under reflux for 7 days.

**[Zn(C$_7$H$_{13}$N$_3$S)Br$_2$] or C$_7$H$_{13}$N$_3$Br$_2$SZn**. This compound was prepared by the slow cooling of a mixture of hot ethanolic solutions of stoichiometric quantities of ZnBr$_2$ and cyclohexanone thiosemicarbazone (C$_7$H$_{13}$N$_3$S) (Ablov and Gerbeleu 1964). White crystalline precipitate in the form of rhombic plates was obtained.

**[(mepa)ZnBr] or C$_8$H$_{11}$N$_2$BrSZn (mepaH = $N$-(2-mercaptoethyl)picolylamin, C$_8$H$_{12}$N$_2$S)**. This compound melts at 225°C with decomposition (Brand and Vahrenkamp 1995). To obtain it, mepaH (2.61 mM) and NaOH (2.61 mM) were dissolved in MeOH (20 mL). A solution of ZnBr$_2$ (2.61 mM) in MeOH (10 mL) was added dropwise with stirring. The precipitate was stirred for 1 h, filtered off, washed with MeOH, and dried in vacuo to yield (mepa)ZnBr as a colorless crystal.

**[Zn(C$_4$H$_6$N$_2$S)$_2$Br$_2$] or C$_8$H$_{12}$N$_4$Br$_2$S$_2$Zn**. This compound crystallizes in the orthorhombic structure with the lattice parameters $a = 1370.6 \pm 0.6$, $b = 1402.1 \pm 0.7$, and $c = 780.9 \pm 0.2$ pm at 220 K and a calculated density of 2.01 g·cm$^{-3}$ (Matsunaga et al. 2005b). It was prepared by the following procedure. To a CH$_3$OH solution (40 mL) of 1-methylimidazoline-2(3$H$)-thione (10 mM), a CH$_3$OH solution (5 mL) of ZnBr$_2$ (5.0 mM) was added. When the mixture was stirred for 1 h, a white solid appeared. The resulting white powder was collected by filtration and washed with CH$_3$OH. The colorless single crystals for XRD were obtained by recrystallization of white powder from a CH$_3$OH/H$_2$O mixed solvent.

**[Zn(C$_4$H$_8$N$_2$S)$_2$Br$_2$] or C$_8$H$_{16}$N$_4$Br$_2$S$_2$Zn, dibromobis($N$-methylimydazolidine-2-thione-$S$)zinc**. This compound melts at 156°C (Raper and Brook 1977) (at 195°C with decomposition [Devillanova and Verani 1980]). It was prepared by dissolving ZnBr$_2$·6H$_2$O (1.5 mM) in a small volume of hot anhydrous EtOH (MeOH,

[Fettouhi et al. 2007]) containing 3 mL triethyl orthoformate as dehydrating agent. To this solution, 3.0 mM of $N$-methylimidazoline-2-thione ($C_4H_8N_2S$) dissolved in 5 mL of hot anhydrous EtOH was added. The resulting solution was reduced in volume by boiling on a water bath and then cooled to room temperature. Subsequent slow evaporation of the solvent was accompanied by the formation of a crystalline colorless product. This was removed by filtration, washed with a small volume of cold anhydrous EtOH and then dry ether, and finally vacuum dried at room temperature (Raper and Brook 1977, Fettouhi et al. 2007). Higher metal/ligand ratios (up to 1:6) produced compound, which were identical to that from the aforementioned procedure (Raper and Brook 1977, Shunmugam and Sathyanarayana 1983).

**[Zn($C_4H_8N_2S$)$_2$Br$_2$] or $C_8H_{16}N_4Br_2S_2Zn$, dibromobis(1,3-diazinane-2-thione-$S$) zinc.** This compound was prepared by dissolving ZnBr$_2$ in a minimum amount of H$_2$O and two molar equivalents of 1,3-diazinane-2-thione in MeOH; the mixture was refluxed for 2–3 h (Fettouhi et al. 2007). The solution was allowed to evaporate slowly and white crystalline products were isolated.

**[Zn($C_9H_{12}N_4S$)Br$_2$] or $C_9H_{12}N_4Br_2SZn$.** This compound melts at 280°C with decomposition (Bermejo et al. 1999). It was obtained by treating 2-acetylpyridine 4-methylthiosemicarbazone ($C_9H_{12}N_4S$) with ZnBr$_2$ (1:1 molar ratio) in EtOH. After prolonged stirring (about 1 week) at room temperature, the reaction mixture afforded the title compound that was filtered off, washed with EtOH, and vacuum dried.

**[Zn($C_9H_{13}N_5S$)Br$_2$] or $C_9H_{13}N_5Br_2SZn$.** This compound melts at the temperature higher than 300°C and crystallizes in the triclinic structure with the lattice parameters $a = 806.5 \pm 0.1$, $b = 849.5 \pm 0.2$, and $c = 1053.4 \pm 0.2$ pm and $\alpha = 87.487° \pm 0.003°$, $\beta = 84.198° \pm 0.003°$, and $\gamma = 85.045° \pm 0.003°$ and a calculated density of 2.084 g·cm$^{-3}$ (Bermejo et al. 2001). It was prepared as follows. A solution of 2-pyridylformamide $N$(4)-dimethylthiosemicarbazone (1 mM) in EtOH (30 mL) was added to a solution of ZnBr$_2$ (1 mM) in 30 mL of 96% EtOH, and the mixture was stirred for several days at room temperature. The resulting yellow solid was filtered off, washed thoroughly with cold EtOH, and stored in a desiccator over Mg and I$_2$. Single crystals suitable for XRD were obtained by slow evaporation of the mother liquor in air at room temperature.

**[Zn($C_9H_{13}N_5S$)Br$_2$] or $C_9H_{13}N_5Br_2SZn$.** To obtain this compound, a solution of ZnBr$_2$ (1.12 mM) in EtOH (30 mL) was mixed with a solution of 2-pyridineformamide $N$(4)-ethylthiosemicarbazone ($C_9H_{13}N_5S$) (1.12 mM) in EtOH (30 mL) (Bermejo et al. 2004a). The resulting solution was stirred at room temperature for 7 days. The white product was filtered off.

**[(ZnBr$_2$){N(CH$_2$CH$_2$SMe)$_3$}] or $C_9H_{21}NBr_2S_3Zn$.** The reaction mixture of ZnBr$_2$ (10 mM) and tris(2-methylthioethyl)amine [N(CH$_2$CH$_2$SMe)$_3$] (12 mM) in Bu$^n$OH (20 mL) separated as sticky oil, which crystallizes on long storage at room temperature (Ciampolini et al. 1968). The solid was ground with petroleum ether containing a few drops of [N(CH$_2$CH$_2$SMe)$_3$], collected, and dried in vacuo at room temperature to give the title compound.

**[Zn($C_5H_5NS$)$_2$Br$_2$] or $C_{10}H_{10}N_2Br_2S_2Zn$.** This compound was prepared by mixing ethanolic solutions of ZnBr$_2$ and 2-pyridine thiol (Kennedy and Lever 1972).

The resulting precipitate was recrystallized from ethanol by Soxhlet extraction. The title compounds required the addition of HBr to prevent partial hydrolysis.

**[Zn(C$_5$H$_{10}$N$_2$S)$_2$Br$_2$] or C$_{10}$H$_{20}$N$_4$Br$_2$S$_2$Zn, dibromobis[(N,N-dimethylimydazolidine-2-thione-S)]zinc.** This compound melts at 115°C (Devillanova and Verani 1980). It was prepared by dissolving ZnBr$_2$ in a minimum amount of H$_2$O and two molar equivalents of N,N-dimethylimydazolidine-2-thione in MeOH; the mixture was refluxed for 2–3 h (Fettouhi et al. 2007). The solution was allowed to evaporate slowly and white crystalline products were isolated.

**[Zn(C$_5$H$_{10}$N$_2$S)$_2$Br$_2$] or C$_{10}$H$_{20}$N$_4$Br$_2$S$_2$Zn, dibromobis[(N-ethylimydazolidine-2-thione-S)]zinc.** This compound melts at 98°C (Devillanova and Verani 1980). It was prepared by dissolving ZnBr$_2$ in a minimum amount of H$_2$O and two molar equivalents of N-ethylimydazolidine-2-thione in MeOH; the mixture was refluxed for 2–3 h (Fettouhi et al. 2007). The solution was allowed to evaporate slowly and white crystalline products were isolated.

**[Zn(C$_5$H$_{10}$N$_2$S)$_2$Br$_2$] or C$_{10}$H$_{20}$N$_4$Br$_2$S$_2$Zn, dibromobis[(1,3-diazipane-2-thione-S)]zinc.** This compound was prepared by dissolving ZnBr$_2$ in a minimum amount of H$_2$O and two molar equivalents of 1,3-diazipane-2-thione in MeOH; the mixture was refluxed for 2–3 h (Fettouhi et al. 2007). The solution was allowed to evaporate slowly and white crystalline products were isolated.

**[{(CH$_3$)$_2$NH(CH$_2$)$_3$S}$_2$][ZnBr$_4$] or C$_{10}$H$_{26}$N$_2$Br$_4$S$_2$Zn.** This compound crystallizes in the orthorhombic structure with the lattice parameters $a = 2346$, $b = 1357$, and $c = 671$ pm (Casals et al. 1987). It was obtained simultaneously with obtaining of [ZnBr$_2${μ-S(CH$_2$)$_3$NH(CH$_3$)$_2$}].

**[Zn(C$_{11}$H$_{10}$N$_2$S$_2$)Br$_2$] or C$_{11}$H$_{10}$N$_2$Br$_2$S$_2$Zn.** This compound melts at 240°C and crystallizes in the orthorhombic structure with the lattice parameters $a = 1263.60 \pm 0.07$, $b = 1417.72 \pm 0.08$, and $c = 1693.21 \pm 0.10$ pm (Amoedo-Portela et al. 2002a). To obtain it, a solution of bis(2-pyridylthio)methane (1.28 mM) in 8 mL of an EtOH/acetonitrile (1:1 volume ratio) mixture was added dropwise to a solution of ZnBr$_2$ (1.29 mM) in 4 mL of the same solvent mixture, and the white suspension that was obtained was refluxed for 6 h. After stirring for 1 day more, the solid was filtered out and dried in vacuo.

**[Zn(EtSCH$_2$)$_2$(C$_5$H$_3$N)Br$_2$] or C$_{11}$H$_{17}$NBr$_2$S$_2$Zn.** This compound crystallizes in the monoclinic structure with the lattice parameters $a = 1262.2 \pm 0.7$, $b = 922.6 \pm 0.3$, $c = 1503.3 \pm 0.4$ pm, and $\beta = 107.85° \pm 0.03°$ (Teixidor et al. 1986). To obtain it, ZnBr$_2$ (1.76 mM) in MeOH (5 mL) was added to a solution of 2,3-bis[(ethylthio)methyl] (pyridine) (1.76 mM) in MeOH (3 mL). After the mixture was allowed to stand at room temperature for 3 days in an open vessel, white crystals were obtained. All operations were performed under N$_2$ atmosphere.

**[Zn(C$_{12}$H$_{17}$N$_5$S)Br$_2$] or C$_{12}$H$_{17}$N$_5$Br$_2$SZn$_2$.** This compound melts at the temperature higher than 300°C (Castiñeiras et al. 2002). To obtain it, a solution of C$_{12}$H$_{17}$N$_5$S [2-pyridineformamide N(4)-piperidylthiosemicarbazone] (0.76 mM) in EtOH (30 mL) was added to a solution of ZnBr$_2$ (0.76 mM) in 30 mL of 96% EtOH and the mixture was stirred for several days at room temperature. The resulting yellow solid was filtered off, washed thoroughly with cold ethanol, and stored in a desiccator over Mg and $I_2$.

**[Zn($C_6H_{12}N_2S)_2Br_2$]** or **$C_{12}H_{24}N_4Br_2S_2Zn$, dibromobis-[$N$-methyl(ethyl)imydazolidine-2-thione-$S$]zinc.** This compound was prepared by dissolving $ZnBr_2$ in a minimum amount of $H_2O$ and two molar equivalents of $N$-(methyl(ethyl))imydazolidine-2-thione in MeOH; the mixture was refluxed for 2–3 h (Fettouhi et al. 2007). The solution was allowed to evaporate slowly and white crystalline products were isolated.

**[Zn($C_6H_{12}N_2S)_2Br_2$]** or **$C_{12}H_{24}N_4Br_2S_2Zn$, dibromobis-[$N$-($n$-propyl)imydazolidine-2-thione-$S$)]zinc.** This compound was prepared by dissolving $ZnBr_2$ in a minimum amount of $H_2O$ and two molar equivalents of $N$-($n$-propyl)imydazolidine-2-thione in MeOH; the mixture was refluxed for 2–3 h. The solution was allowed to evaporate slowly and white crystalline products were isolated.

**[Zn($C_6H_{12}N_2S)_2Br_2$** or **$C_{12}H_{24}N_4Br_2S_2Zn$, dibromobis-$N$-[($i$-propyl)imydazolidine-2-thione-$S$)]zinc.** This compound was prepared by dissolving $ZnBr_2$ in a minimum amount of $H_2O$ and two molar equivalents of $N$-($i$-propyl)imydazolidine-2-thione in MeOH; the mixture was refluxed for 2–3 h (Fettouhi et al. 2007). The solution was allowed to evaporate slowly and white crystalline products were isolated.

**[Zn($C_6H_{12}N_2S)_2Br_2$]** or **$C_{12}H_{24}N_4Br_2S_2Zn$, dibromobis($N$-ethyl-1,3-diazinane-2-thione-$S$)zinc.** This compound was prepared by dissolving $ZnBr_2$ in a minimum amount of $H_2O$ and two molar equivalents of $N$-ethyl-1,3-diazinane-2-thione in MeOH; the mixture was refluxed for 2–3 h (Fettouhi et al. 2007). The solution was allowed to evaporate slowly and white crystalline products were isolated.

**[Zn($C_7H_{12}N_2S)_2Br_2$]** or **$C_{14}H_{24}N_4Br_2S_2Zn$.** This compound melts at 190°C with decomposition and crystallizes in the monoclinic structure with the lattice parameters $a = 1718.7 \pm 0.3$, $b = 899.08 \pm 0.17$, $c = 1556.0 \pm 0.3$ pm, and $\beta = 117.206° \pm 0.003°$ at $243 \pm 2$ K and a calculated density of 1.670 g·cm$^{-3}$ (White et al. 2003). It was obtained by the next procedure. A suspension of $ZnBr_2$ (0.986 mM) and 2-mercapto-1-*tert*-butylimidazole (1.971 mM) in $CH_2Cl_2$ (30 mL) was stirred for 20 min in a warm (ca. 50°C) water bath. The resulting slightly cloudy solution was stirred for an additional 1.5 h and filtered, and the filtrate was concentrated under reduced pressure to ca. 20 mL. The addition of pentane (15 mL) resulted in the formation of a white precipitate, which was isolated by filtration, washed with pentane ($2 \times 20$ mL), and dried in vacuo for 4.5 h. All reactions were performed under argon using a combination of high vacuum and Schlenk techniques.

**[Zn($C_7H_{14}N_2S)_2Br_2$]** or **$C_{14}H_{28}N_4Br_2S_2Zn$, dibromobis[($N,N$-diethylimydazolidine-2-thione-$S$)]zinc.** This compound has been obtained molten (Devillanova and Verani 1980). It was obtained by reacting $ZnBr_2$ with $N,N$-diethylimydazolidine-2-thione (1:2 molar ratio) in dimethoxypropane.

**[Zn($C_8H_9N_3S)_2Br_2$]** or **$C_{16}H_{18}N_6Br_2S_2Zn$.** This compound was prepared by the slow cooling of a mixture of hot ethanolic solutions of stoichiometric quantities of $ZnBr_2$ and benzylidene thiosemicarbazide ($C_8H_9N_3S$) (1:2 molar ratio) (Ablov and

Gerbeleu 1964). White crystalline precipitate was obtained that was filtered off, washed sometimes with EtOH and then with ether, and dried in air.

**[Zn$_4$\{S(CH$_2$)$_2$NMe$_2$\}$_4$Br$_4$] or C$_{16}$H$_{40}$N$_4$Br$_4$S$_4$Zn$_4$.** To obtain this compound, HS(CH$_2$)$_2$NMe$_2$·HBr (5 mM) was dispersed in acetonitrile (40 mL) (Casals et al. 1991b). Et$_3$N (1.4 mL) and anhydrous ZnBr$_2$ (5 mM) were successfully added with stirring under a N$_2$ atmosphere. Stirring was continued for 24 h. The resulting suspension was filtered in an open atmosphere, washed with copious distilled water and then with MeOH and diethyl ether, and vacuum dried.

Alternatively, this compound was obtained by adding an aqueous solution of equimolar amounts of HS(CH$_2$)$_2$NMe$_2$·HBr and NaOH to an aqueous solution of Zn(OAc)$_2$, the metal/ligand molar ratio being always 1:1 (Casals et al. 1991b). A white powder formed immediately. The mixture was stirred for several hours; the solid was filtered off, washed with cold water and then MeOH, and dried under vacuum. Crystals were obtained by refluxing the compound in DMSO, filtering off the undissolved solid, letting the solution cool to 60°C, and maintaining it at this temperature for 24 h. All these manipulations were performed under a N$_2$ atmosphere.

**[ZnBr$_2$·Bu$^n_4$C$_2$N$_2$S$_4$] or C$_{18}$H$_{36}$N$_2$Br$_2$S$_4$Zn.** This compound melts at 138°C–141°C (Brinkhoff et al. 1969). It was prepared as a white precipitate by adding a solution of Br$_2$ (1.7 g) in CS$_2$ (10 mL) to a solution of di-$n$-butyldithiocarbamatozinc (5.0 g) in the same solvent (40 mL) at room temperature. After filtration, the product could be purified by dissolution in chloroform and reprecipitation by petroleum ether. The alternative route includes using ZnBr$_2$ and tetra-$n$-butylthiuram disulfide. To a suspension of ZnBr$_2$ in CS$_2$, a solution of tetra-$n$-butylthiuram disulfide was added at room temperature. After ZnBr$_2$ had dissolved, petroleum ether was added, precipitating the crystals, which was filtered and dried.

**[(Bu$^n_4$N)\{Zn(S$_2$CNMe$_2$)Br$_2$\}] or C$_{19}$H$_{42}$N$_2$Br$_2$S$_2$Zn.** This compound was prepared by the following method (McCleverty and Morrison 1976). A solution of [Bu$^n_4$N] [(S$_2$CNMe$_2$)] (0.01 M) in acetone (50 mL) was added to a suspension of ZnBr$_2$ (0.01 M) in acetone (50 mL). The mixture was stirred to dissolve the ZnBr$_2$ and then light petroleum was added. Partial evaporation of the solvent afforded white crystals (sometimes an oil formed first and then crystallized), which were collected, washed with diethyl ether, and recrystallized from CH$_2$Cl$_2$/hexane.

**[(Et$_4$N)$_2$(μ-SMe)$_6$(ZnBr)$_4$] or C$_{22}$H$_{58}$N$_2$Br$_4$S$_6$Zn$_4$.** To obtain this compound, a mixture of ZnBr$_2$ (0.98 mM), Et$_4$NBr (2.0 mM), and Zn(SMe)$_2$ (3.3 mM) was stirred in CH$_2$Cl$_2$ (20 mL) for 1 h (Dean et al. 1992). The small amount of insoluble material was removed by filtration. The volume of the filtrate was reduced to ca. 4 mL by gentle heating under a flow of N$_2$, and then Et$_2$O (2 mL) was layered on. The mixture was allowed to stand in the refrigerator overnight. The white solid that formed was separated by filtration, washed with Et$_2$O, and dried in vacuo at room temperature.

**[(Et$_4$N)$_2$(μ-SEt)$_6$(ZnBr)$_4$] or C$_{28}$H$_{70}$N$_2$Br$_4$S$_6$Zn$_4$.** To obtain this compound, into a solution of ZnBr$_2$ (4.0 mM) and Et$_4$NBr (2.0 mM) in EtOH (50 mL) at 40°C was stirred EtSH (8.1 mM) followed by Et$_3$N (8.0 mM) (Dean et al. 1992). A white precipitate formed. This dissolved when MeCN (5 mL) was added to the mixture at room temperature. The addition of $Et_2O$ to the resulting solution and cooling to 5°C overnight gave the product as white crystals. However, this compound did not give satisfactory elemental analysis.

**$[(Et_4N)_2(\mu\text{-}SPr^n)_6(ZnBr)_4]$ or $C_{34}H_{82}N_2Br_4S_6Zn_4$.** To obtain this compound, a mixture of $ZnBr_2$ (0.98 mM), $Et_4NBr$ (2.0 mM), and $Zn(SPr^n)_2$ (3.3 mM) was stirred in $CH_2Cl_2$ (20 mL) for 1 h (Dean et al. 1992). The small amount of insoluble material was removed by filtration. The volume of the filtrate was reduced to ca. 4 mL by gentle heating under a flow of $N_2$, and then $Et_2O$ (2 mL) was layered on. The mixture was allowed to stand in the refrigerator overnight. The white solid that formed was separated by filtration, washed with $Et_2O$, and dried in vacuo at room temperature.

**$[(Me_4N)_2Zn_4(SPh)_6Br_4]$ or $C_{44}H_{54}N_2Br_4S_6Zn_4$.** To obtain this compound, $Br_2$ (1.15 mM) in 5 mL of $CCl_4$ was added to a stirred solution containing $(Me_4N)_2[Zn_4(SPh)_{10}]$ (0.575 mM) in 15 mL of acetone (Dean et al. 1987). Decolorization occurred quickly, leaving a clear solution with a small amount of white solid. After 10 min, the solvents were removed under vacuum. The residue was extracted with three 40 mL portions of diethyl ether to remove $(PhS)_2$. The residue was purified by recrystallization from a mixture of acetone and cyclohexane. The syntheses were performed under an inert atmosphere.

**$(Me_4N)_2[(\mu\text{-}SPh)_6(ZnSPh)_2(ZnBr)_2]$ or $C_{56}H_{64}N_2Br_2S_8Zn_4$.** A crystallization solution for obtaining this compound was prepared in the same manner as that for $(Me_4N)_2[(\mu\text{-}SPh)_6(ZnSPh)_2(ZnCl)_2]$ earlier, except that $Me_4NBr$ (18 mM) was used and the total solvent composition was 150 mL of MeOH and 50 mL of acetone (Dance 1981). Crystallization occurred during 3 days at room temperature, as clumps of microcrystals, which were filtered, washed with MeOH, and vacuum dried.

**$[(Et_4N)_2[(\mu\text{-}Bz)_6(ZnBr)_4]$ or $C_{58}H_{82}N_2Br_4S_6Zn_4$.** To obtain this compound, a mixture of $ZnBr_2$ (0.98 mM), $Et_4NBr$ (2.0 mM), and $Zn(SBz)_2$ (3.3 mM) was stirred in $CH_2Cl_2$ (20 mL) for 1 h (Dean et al. 1992). The small amount of insoluble material was removed by filtration. The volume of the filtrate was reduced to ca. 4 mL by gentle heating under a flow of $N_2$, and then pentane (2 mL) was layered on. The mixture was allowed to stand in the refrigerator overnight. The white solid that formed was separated by filtration, washed with $Et_2O$, and dried in vacuo at room temperature.

## 1.68   Zn–H–C–N–I–S

Some compounds are formed in this system.

**$[Zn(CH_4N_2S)_2I_2]$ or $C_2H_8N_4I_2S_2Zn$.** This compound crystallizes in the monoclinic structure with the lattice parameters $a = 1049.90 \pm 0.19$, $b = 747.3 \pm 0.2$, $c = 1487.1 \pm 0.4$ pm, and $\beta = 91.354° \pm 0.018°$ and a calculated density of 2.686 g·cm$^{-3}$ (Albov et al. 2013). In order to prepare it, $ZnI_2$ (3.13 mM) and thiocarbamide (6.26 mM) were dissolved in $H_2O$ (10 mL). The resulting solution was allowed to stand in air up to the precipitation of colorless crystals.

**[Zn(MeCSNH$_2$)$_2$I$_2$] or C$_4$H$_{10}$N$_2$I$_2$S$_2$Zn.** This compound crystallizes in the monoclinic structure with the lattice parameters $a = 755.1 \pm 0.4$, $b = 2351.8 \pm 1.3$, $c = 781.4 \pm 0.5$ pm, and $\beta = 114.03° \pm 0.05°$ and a calculated density of 2.460 g·cm$^{-3}$ (Savinkina et al. 2009). It was obtained by the following procedure. ZnI$_2$ (3.13 mM) and thioacetamide (6.26 mM) were dissolved in MeCN (10 mL). The solution was layered onto a liquid perfluorinated hydrocarbon, 1-methyldecahydronaphthalene, allowing formation of colorless crystals. This compound decomposes within a few hours.

**[Zn(Et$_2$NCS$_2$)I] or C$_5$H$_{10}$NIS$_2$Zn.** This compound melts at 220°C and crystallizes in the monoclinic structure with the lattice parameters $a = 735.5 \pm 0.2$, $b = 943.2 \pm 0.2$, $c = 1543.1 \pm 0.4$ pm, and $\beta = 92.42° \pm 0.04°$ and a calculated density of 2.115 g·cm$^{-3}$ (Zemskova et al. 1991). To obtain it, to a solution of Zn(Et$_2$NCS$_2$)$_2$ in CH$_2$Cl$_2$ at room temperature was added ZnI$_2$ (1.0–1.1:1 molar ratio), and the mixture was stirred for 3–5 h. The precipitate that formed was filtered off, washed thoroughly with CH$_2$Cl$_2$, and dried in air.

**[Zn(C$_3$H$_5$NS$_2$)$_2$I$_2$] or C$_6$H$_{10}$N$_2$I$_2$S$_4$Zn.** This compound melts at 150°C (De Filippo et al. 1971). To obtain it, ZnI$_2$ was dissolved in an excess of molten thiazolidine-2-thione (C$_3$H$_5$NS$_2$) (at ca. 105°C; molar ratio Zn$^{2+}$/C$_3$H$_5$NS$_2$ ca. 1:2), and the crude product was purified by washing with ethyl ether and recrystallizing from EtOH. The title compound could also be prepared if to a solution of ZnI$_2$ in absolute EtOH were added different amounts of C$_3$H$_5$NS$_2$ for Zn$^{2+}$/C$_3$H$_5$NS$_2$ molar ratios of 1:1, 1:2, 1:3, and 1:4, and the mixture was heated under reflux for 4 h. [ZnI$_2$(C$_3$H$_5$NS$_2$)$_2$] was separated out by crystallization.

**[Zn{S(SCNMe$_2$)$_2$I$_2$}] or C$_6$H$_{12}$N$_2$I$_2$S$_3$Zn.** This compound was prepared by the following procedure (McCleverty and Morrison 1976). To a suspension of [Zn(S$_2$CNMe$_2$)$_2$I$_2$] (2 mM) in CH$_2$Cl$_2$ (50 mL) was added a solution of PPh$_3$ (0.53 g) in CH$_2$Cl$_2$ (10 mL). The light-yellow suspension was replaced by a deeper yellow precipitate. The mixture was shaken for 1 h and then the solvent was evaporated. The residue was shaken with diethyl ether (200 mL) for 1 h and this mixture was filtered. The yellow residue, [Zn{S(SCNMe$_2$)$_2$I$_2$}], was recrystallized from a mixture acetone/light petroleum.

The title compound could also be prepared by the alternative procedure (McCleverty and Morrison 1976). ZnI$_2$ (3.2 g) was added to a stirred solution of S(SCNMe$_2$)$_2$ (2.08 g) in acetone (200 mL). After the ZnI$_2$ had dissolved, most of the acetone was evaporated. When crystals began to form, the mixture was cooled in a bath of acetone/solid CO$_2$. The yellow crystals were collected and washed with diethyl ether.

**[Zn(S$_2$CNMe$_2$)$_2$I$_2$] or C$_6$H$_{12}$N$_2$I$_2$S$_4$Zn.** To obtain this compound, a solution of [Bu$^n_4$N] [Zn(S$_2$CNMe$_2$)I$_2$] (0.01 M) in acetone (50 mL) was mixed with a solution of [Bu$^n_4$N] [S$_2$CNMe$_2$] in the same solvent (60 mL) (McCleverty and Morrison 1976). The white precipitate of the title compound that formed slowly was collected after 30 min and washed with acetone. The title compound could also be prepared by the following procedure (McCleverty and Morrison 1976). To a solution of [Zn(S$_2$CNMe$_2$)$_2$]

(0.01 M) in $CH_2Cl_2$ (250 mL) was added a solution of $I_2$ (0.01 M) in the same solvent (250 mL). The solvent was reduced to ca. 100 mL by evaporation and $n$-hexane was added to cause further precipitation of the product. This was collected by filtration, washed with diethyl ether, and recrystallized from $CH_2Cl_2/n$-hexane.

The title compound could also be prepared as follows (Brinkhoff et al. 1969). To a suspension of $ZnI_2$ in $CS_2$, a solution of tetramethylthiuram disulfide in the same solvent was added at room temperature. After the $ZnI_2$ had dissolved, petroleum ether was added, precipitating yellow crystals. The obtained product was filtered and dried.

**[Zn(C$_5$H$_4$NC$_2$H$_5$N$_4$S)I$_2$] or C$_7$H$_9$N$_5$I$_2$SZn.** This compound melts at 285°C–286°C and crystallizes in the monoclinic structure with the lattice parameters $a = 764.34 \pm 0.07$, $b = 1479.24 \pm 0.03$, $c = 1186.20 \pm 0.11$ pm, and $\beta = 90.941° \pm 0.007°$ and a calculated density of 2.548 g·cm$^{-3}$ (Castiñeiras et al. 2000). To obtain it, a solution of $ZnI_2$ (2 mM) in EtOH (30 mL) was mixed with a solution of 2-pyridineformamide thiosemicarbazone (C$_5$H$_4$NC$_2$H$_5$N$_4$S) (2 mM) in the same solvent (30 mL), and the mixture was stirred under reflux for 7 days. Crystals suitable for XRD were grown from the solution of the title compound in anhydrous EtOH.

**[Zn(C$_4$H$_8$N$_2$S)$_2$I$_2$]** or **C$_8$H$_16$N$_4$I$_2$S$_2$Zn,** **diiodobis[($N$-methylimydazolidine-2-thione-$S$)]zinc.** This compound melts at 152°C (Raper and Brook 1977) (at 164°C with decomposition [Devillanova and Verani 1980]). It was prepared by dissolving $ZnI_2$·6H$_2$O (1.5 mM) in a small volume of hot anhydrous EtOH containing 3 mL triethyl orthoformate as dehydrating agent. To this solution 3.0 mM of $N$-methylimidazoline-2-thione (C$_4$H$_8$N$_2$S) dissolved in 5 mL of hot anhydrous EtOH was added. The resulting solution was reduced in volume by boiling on a water bath and then cooled to room temperature. Subsequent slow evaporation of the solvent was accompanied by the formation of a crystalline product. This was removed by filtration, washed with a small volume of cold anhydrous EtOH and then dry ether, and finally vacuum dried at room temperature. Higher metal/ligand ratios (up to 1:6) produced compound, which was identical to that from the aforementioned procedure.

**[Zn(C$_9$H$_12$N$_4$S)I$_2$] or C$_9$H$_12$N$_4$I$_2$SZn.** This compound melts at 287°C (Bermejo et al. 1999). It was obtained by treating 2-acetylpyridine 4-methylthiosemicarbazone (C$_9$H$_12$N$_4$S) with $ZnI_2$ (1:1 molar ratio) in EtOH. After prolonged stirring (about 1 week) at room temperature, the reaction mixture afforded the title compound, which was filtered off, washed with EtOH, and vacuum dried.

**[Zn(C$_9$H$_13$N$_5$S)I$_2$] or C$_9$H$_13$N$_5$I$_2$SZn.** This compound melts at 244°C (Bermejo et al. 2001) and crystallizes in the triclinic structure with the lattice parameters $a = 832.2 \pm 0.1$, $b = 849.1 \pm 0.3$, and $c = 1154.6 \pm 0.2$ pm and $\alpha = 88.07° \pm 0.03°$, $\beta = 82.72° \pm 0.02°$, and $\gamma = 87.58° \pm 0.02°$ and a calculated density of 2.229 g·cm$^{-3}$ (Bermejo et al. 2004a). It was prepared as follows (Bermejo et al. 2001). A solution of 2-pyridylformamide $N$(4)-dimethylthiosemicarbazone (1 mM) in EtOH (30 mL) was added to a solution of $ZnI_2$ (1 mM) in 30 mL of 96% EtOH, and the mixture was stirred for several days at room temperature. The resulting yellow solid was filtered off, washed thoroughly with cold EtOH, and stored in a desiccator over Mg and $I_2$. The title compound could also be prepared upon slow evaporation of the solvent from an aqueous solution of [Zn(C$_9$H$_13$N$_5$S)$_2$I$_2$]·3H$_2$O (Bermejo et al. 2004a).

**(ZnI₂)[N(CH₂CH₂SMe)₃] or C₉H₂₁NI₂S₃Zn.** The reaction mixture of ZnI₂ (10 mM) and tris(2-methylthioethyl)amine [N(CH₂CH₂SMe)₃] (12 mM) in Bu$^n$OH (20 mL) separated as sticky oil, which crystallizes on long storage at room temperature (Ciampolini et al. 1968). The solid was ground with petroleum ether containing a few drops of [N(CH₂CH₂SMe)₃], collected, and dried in vacuo at room temperature to give the title compound.

**[Zn(C₅H₁₀N₂S)₂I₂] or C₁₀H₂₀N₄I₂S₂Zn, diiodobis[(*N,N*-dimethylimydazolidine-2-thione-*S*)]zinc.** This compound melts at 127°C (Devillanova and Verani 1980). It was obtained by reacting ZnI₂ with *N,N*-dimethylimydazolidine-2-thione (1:2 molar ratio) in absolute EtOH.

**[Zn(C₅H₁₀N₂S)₂I₂] or C₁₀H₂₀N₄I₂S₂Zn, diiodobis[(*N*-ethylimydazolidine-2-thione-*S*)]zinc.** This compound melts at 80°C (Devillanova and Verani 1980). It was obtained by reacting ZnI₂ with *N*-ethylimydazolidine-2-thione (1:2 molar ratio) in absolute EtOH.

**[Zn(C₁₁H₁₀N₂S₂)₂I₂] or C₁₁H₁₀N₂I₂S₂Zn.** This compound crystallizes in the monoclinic structure with the lattice parameters $a = 796.90 \pm 0.06$, $b = 1390.93 \pm 0.10$, $c = 1496.54 \pm 0.10$ pm, and $\beta = 94.041 \pm 0.001°$ and a calculated density of 2.222 g·cm$^{-3}$ (Amoedo-Portela et al. 2002b). To obtain it, a solution of (2-pyridylthio)methane (1.32 mM) in 2 mL of dry acetone was added to a solution of ZnI₂ (1.32 mM) in 6 mL of ethanol. The mixture was stirred and refluxed for about 4 h and the white precipitate was filtered off and vacuum dried. Although the isolated product was [Zn(C₁₁H₁₀N₂S₂)₂I₂] 1/2C₂H₆O, by slow evaporation from the mother liquor, yellow crystals of the title compound suitable for XRD were obtained.

**[Zn(C₁₂H₁₇N₅S)I₂] or C₁₂H₁₇N₅I₂SZn₂.** This compound melts at 244°C (Castiñeiras et al. 2002). To obtain it, a solution of C₁₂H₁₇N₅S [2-pyridineformamide *N*(4)-piperidylthiosemicarbazone] (0.76 mM) in EtOH (30 mL) was added to a solution of ZnI₂ (0.76 mM) in 30 mL of 96% EtOH, and the mixture was stirred for several days at room temperature. The resulting yellow solid was filtered off, washed thoroughly with cold ethanol, and stored in a desiccator over Mg and I₂.

**[ZnI(C₄H₆N₂S)₃]I or C₁₂H₁₈N₆I₂S₃Zn.** This compound crystallizes in the orthorhombic structure with the lattice parameters $a = 1196.2 \pm 0.3$, $b = 1128.3 \pm 0.1$, and $c = 1620.3 \pm 0.1$ pm and a calculated density of 2.01 g·cm$^{-3}$ (Matsunaga et al. 2005b). It was prepared by the following procedure. To a CH₃OH solution (40 mL) of 1-methylimidazoline-2(3*H*)-thione (10 mM), a CH₃OH solution (5 mL) of ZnI₂ (5.0 mM) was added. When the mixture was stirred for 1 h, a white solid appeared. The resulting white powder was collected by filtration and washed with CH₃OH. The colorless single crystals for XRD were obtained by cooling the filtrate for a few days in the refrigerator.

**[Zn(S₂CNC₅H₁₀)₂I₂] or C₁₂H₂₀N₂I₂S₄Zn.** This compound was prepared by the following procedure (McCleverty and Morrison 1976). To a solution of [Zn(S₂CNC₅H₁₀)₂] (0.01 M) in CH₂Cl₂ (250 mL) was added a solution of I₂ (0.01 M) in the same solvent (250 mL). The solvent was reduced to ca. 100 mL by evaporation and *n*-hexane was added to cause further precipitation of the product. This was collected by filtration, washed with diethyl ether, and recrystallized from CH₂Cl₂/*n*-hexane.

**[Zn(C$_7$H$_{14}$N$_2$S)$_2$I$_2$]** or **C$_{14}$H$_{28}$N$_4$I$_2$S$_2$Zn, diiodobis[(*N,N*-diethylimydazolidine-2-thione-*S*)]zinc**. This compound melts at 89°C (Devillanova and Verani 1980). It was obtained by reacting ZnI$_2$ with *N,N*-diethylimydazolidine-2-thione (1:2 molar ratio) in dimethoxypropane.

**[ZnI$_2$·Bu$^n$$_4$C$_2$N$_2$S$_4$]** or **C$_{18}$H$_{36}$N$_2$I$_2$S$_4$Zn**. This compound was prepared as a white precipitate by adding a solution of I$_2$ in CS$_2$ to a solution of di-*n*-butyldithiocarbamatozinc in the same solvent at room temperature (Brinkhoff et al. 1969). After filtration, the product could be purified by dissolution in chloroform and reprecipitation by petroleum ether. The alternative route includes using ZnI$_2$ and tetra-*n*-butylthiuram disulfide. To a suspension of ZnI$_2$ in CS$_2$, a solution of tetra-*n*-butylthiuram disulfide was added at room temperature. After ZnI$_2$ had dissolved, petroleum ether was added, precipitating the crystals, which was filtered and dried.

**[(Bu$^n$$_4$N){Zn(S$_2$CNMe$_2$)I$_2$}]** or **C$_{19}$H$_{42}$N$_2$I$_2$S$_2$Zn**. This compound was prepared by the following methods (McCleverty and Morrison 1976).

*Method A*. A solution of [Bu$^n$$_4$N][(S$_2$CNMe$_2$)] (0.01 M) in acetone (50 mL) was added to a suspension of ZnI$_2$ (0.01 M) in acetone (50 mL). The mixture was stirred to dissolve the ZnI$_2$ and then light petroleum was added. Partial evaporation of the solvent afforded white crystals (sometimes an oil formed first and then crystallized), which were collected, washed with diethyl ether, and recrystallized from CH$_2$Cl$_2$/hexane.

*Method B*. A mixture of [Zn(S$_2$CNMe$_2$)$_2$I$_2$] (1 mM), ZnI$_2$ (1 mM), and [Bu$^n$$_4$N]I (2 mM) in acetone (25 mL) was stirred until all the reactants had dissolved. The product was obtained as from (A) earlier.

*Method C*. A solution of [Bu$^n$$_4$N][S$_2$CNMe$_2$] (0.36 g) in acetone (10 mL) was added to a suspension of [Zn{(S$_2$CNMe$_2$)$_2$}I$_2$] (0.56 g) in acetone (15 mL). The addition of diethyl ether (ca. 200 mL) to the yellow solution caused precipitation of the title compound, which was filtered off and recrystallized from CH$_2$Cl$_2$/*n*-hexane.

*Method D*. A solution of [Bu$^n$$_4$N][Zn(S$_2$CNMe$_2$)$_2$] (1 mM) in acetone (20 mL) was mixed with a solution of I$_2$ (1 mM) in the same solvent (50 mL). The yellow solution was evaporated and the residue was shaken with diethyl ether (100 mL) overnight. The mixture was filtered and the residue (the title compound) was recrystallized from acetone/diethyl ether.

**[(Bu$^n$$_4$N){Zn(S$_2$CNEt$_2$)I$_2$}]** or **C$_{21}$H$_{46}$N$_2$I$_2$S$_2$Zn**. This compound was prepared by the following method (McCleverty and Morrison 1976). A mixture of [Zn(S$_2$CNEt$_2$)$_2$I$_2$] (1 mM), ZnI$_2$ (1 mM), and [Bu$^n$$_4$N]I (2 mM) in acetone (25 mL) was stirred until all the reactants had dissolved, and then light petroleum was added. Partial evaporation of the solvent afforded white crystals, which were collected, washed with diethyl ether, and recrystallized from CH$_2$Cl$_2$/hexane.

**[(Et$_4$N)$_2$[(μ-SMe)$_6$(ZnI)$_4$]]** or **C$_{22}$H$_{58}$N$_2$I$_4$S$_6$Zn$_4$**. To obtain this compound, a mixture of ZnI$_2$ (0.98 mM), Et$_4$NBr (2.0 mM), and Zn(SMe)$_2$ (3.3 mM) was stirred in CH$_2$Cl$_2$ (20 mL) for 1 h (Dean et al. 1992). The small amount of insoluble material was removed by filtration. The volume of the filtrate was reduced to ca. 4 mL by gentle heating under a flow of N$_2$, and then Et$_2$O (2 mL) was layered on. The mixture was allowed to stand in the refrigerator overnight. The white solid that formed was separated by filtration, washed with $Et_2O$, and dried in vacuo at room temperature.

**[(Et₄N)₂[(μ-SEt)₆(ZnI)₄]] or C₂₈H₇₀N₂I₄S₆Zn₄.** To obtain this compound, a mixture of $ZnI_2$ (0.98 mM), $Et_4NI$ (2.0 mM), and $Zn(SEt)_2$ (3.3 mM) was stirred in $CH_2Cl_2$ (20 mL) for 1 h (Dean et al. 1992). The initial three-component mixture remained milky and the small amount of insoluble material was removed by filtration. The volume of the filtrate was reduced to ca. 4 mL by gentle heating under a flow of $N_2$, and then $Et_2O$ (2 mL) was layered on. The mixture was allowed to stand in the refrigerator overnight. The white solid that formed was separated by filtration, washed with $Et_2O$, and dried in vacuo at room temperature.

**[(Et₄N)₂[(μ-SPrⁿ)₆(ZnI)₄]] or C₃₄H₈₂N₂I₄S₆Zn₄.** To obtain this compound, a mixture of $ZnI_2$ (0.98 mM), $Et_4NI$ (2.0 mM), and $Zn(SPr^n)_2$ (3.3 mM) was stirred in $CH_2Cl_2$ (20 mL) for 1 h (Dean et al. 1992). The small amount of insoluble material was removed by filtration. The volume of the filtrate was reduced to ca. 4 mL by gentle heating under a flow of $N_2$, and then $Et_2O$ (2 mL) was layered on. The mixture was allowed to stand in the refrigerator overnight. The white solid that formed was separated by filtration, washed with $Et_2O$, and dried in vacuo at room temperature.

**[(Et₄N)₂[(μ-SPrⁱ)₆(ZnI)₄]] or C₃₄H₈₂N₂I₄S₆Zn₄.** This compound was obtained in the same way as $Et_4N)_2[(\mu\text{-}SPr^i)_6(ZnI)_4$ was obtained using $Zn(SPr^i)_2$ instead of $Zn(SPr^n)_2$ (Dean et al. 1992).

**[(Et₄N)₂[(μ-SBuⁿ)₆(ZnI)₄]] or C₄₀H₉₄N₂I₄S₆Zn₄.** To obtain this compound, a mixture of $ZnI_2$ (0.98 mM), $Et_4NI$ (2.0 mM), and $Zn(SBu^n)_2$ (3.3 mM) was stirred in $CH_2Cl_2$ (20 mL) for 1 h (Dean et al. 1992). The small amount of insoluble material was removed by filtration. The volume of the filtrate was reduced to ca. 4 mL by gentle heating under a flow of $N_2$, and then $Et_2O$ (2 mL) was layered on. The mixture was allowed to stand in the refrigerator overnight. The white solid that formed was separated by filtration, washed with $Et_2O$, and dried in vacuo at room temperature.

**[(Et₄N)₂[(μ-Bz)₆(ZnI)₄]] or C₅₈H₈₂N₂I₄S₆Zn₄.** To obtain this compound, a mixture of $ZnI_2$ (0.98 mM), $Et_4NI$ (2.0 mM), and $Zn(SBz)_2$ (3.3 mM) was stirred in $CH_2Cl_2$ (20 mL) for 1 h (Dean et al. 1992). The small amount of insoluble material was removed by filtration. The volume of the filtrate was reduced to ca. 4 mL by gentle heating under a flow of $N_2$, and then pentane (2 mL) was layered on. The mixture was allowed to stand in the refrigerator overnight. The white solid that formed was separated by filtration, washed with $Et_2O$, and dried in vacuo at room temperature.

## 1.69   Zn–H–C–N–Mn–S

**[MnZn₂(S₂CNEt₂)₆]·3(Phen) or C₆₆H₈₄N₁₂S₁₂MnZn₂.** This multinary compound is formed in this system, which is stable up to 200°C. It was prepared by the next procedures (Zemskova et al. 1995).

1. $MnCl_2 \cdot 4H_2O$ (0.5 mM) and $Phen \cdot H_2O$ (1.5 mM) were dissolved with gentle heating in a mixture of (20 mL) $EtOH/H_2O$ (2:1 volume ratio) with the formation of bright-yellow solution. $ZnI_2$ (1.0 mM) and $Na(S_2CNEt_2)$ $3H_2O$

(1.0 mM) were dissolved in the mixture of EtOH (100 mL) and chloroform (25 mL) at the heating on a water bath. The second solution was added in small portions to the first solution with vigorous stirring. A cream-colored precipitate was formed in the mixture. If after complete mixing of the two solutions the mixture was delaminated, a small amount of ethanol was added to the disappearance of delamination. The mixture was stirred for 3–5 min, and the precipitate was filtered through glass filter with suction, washed extensively with EtOH, and dried in air. The title compound was recrystallized from the saturated solution in $CH_2Cl_2$ by the addition of EtOH.

2. $[Mn(S_2CNEt_2)_3]\cdot(Phen)$ (0.3 mM) and $[Zn(S_2CNEt_2)_2]\cdot(Phen)$ (0.3 mM) were dissolved in $CH_2Cl_2$ (ca. 15 mL), and ca. 45 mL of EtOH was added. A cream-colored precipitate was formed in the mixture. The mixture was stirred for 3–5 min, and the precipitate was filtered through a glass filter with suction, washed with EtOH, and dried in air.

Thermolysis of the title compound leads to the formation of $Zn_xMn_{1-x}S$ solid solution with hexagonal structure.

## 1.70   Zn–H–C–N–Co–S

$[Zn_7CoS(SPh)_{14}\cdot C_{13}H_{14}N_2]$ or $C_{97}H_{84}N_2CoS_{15}Zn_7$. This multinary compound is formed in this system. It crystallizes in the monoclinic structure with the lattice parameters $a = 1908.47 \pm 0.15$, $b = 1848.54 \pm 0.14$, $c = 2749.4 \pm 0.2$ pm, and $\beta = 94.347° \pm 0.002°$ and a calculated density of 1.562 g·cm$^{-3}$ (Xie et al. 2005). To synthesize it, $Zn(OAc)_2$ (5.94 mM), 1,3-bis(4-pyridyl)propane (2.11 mM), $Co(NO_3)_2\cdot 6H_2O$ (0.34 mM), thiourea (1.43 mM), thiophenol (9.05 mM), and distilled $H_2O$ (9.528 g) were mixed and then heated in autoclave (150°C). After 10 days, green cubic single crystals were recovered.

## 1.71   Zn–H–C–P–O–S

Some compounds are formed in this system.

$[Zn(Et_2PSO)_2]$ or $C_8H_{20}P_2O_2S_2Zn$. This compound melts at 188°C (Calligaris et al. 1970a). It was prepared by adding stoichiometric quantities of diethylphosphinothionic acid to an ethanol solution of $Zn(OAc)_2$. After heating under reflux for 1–3 h, the white product separated out and was filtered, washed with absolute ethanol, and vacuum dried.

$[Zn\{S_2P(OEt)_2\}_2]$ or $C_8H_{20}P_2O_4S_4Zn$. This compound melts at 74°C–76°C (Harrison et al. 1986a) (81°C–82°C [Harrison et al. 1986b]; 180.0°C–181.0°C [Dickert and Rowe 1967]) and crystallizes in the monoclinic structure with the lattice parameters $a = 1208.4 \pm 0.3$, $b = 1984.0 \pm 0.6$, $c = 846.3 \pm 0.5$ pm, and $\beta = 113.99° \pm 0.03°$, experimental and calculated densities 1.55 and 1.56 g·cm$^{-3}$, respectively (Ito et al. 1969). It was prepared by metathetic reaction using $NH_4S_2P(OEt)_2$ or $NaS_2P(OEt)_2$ and filtered aqueous solution of $ZnCl_2$ in stoichiometric amounts (Dickert and Rowe 1967,

Ivanov et al. 2001a). Solid insoluble $Zn[S_2P(OEt)_2]_2$ were isolated by filtration and purified by recrystallization. Liquid products were extracted from the aqueous mixture with pentane, washed with water, and dried, and the pentane was removed by evaporation under vacuum.

The title compound could also be obtained if ZnO was slowly added to the brown-black solution of $O,O'$-diethyl dithiophosphoric acid yielding a white slurry in a very exothermic reaction (Harrison et al. 1986b). The majority of the white precipitate was dissolved in hot EtOH and the remainder filtered off. From the filtrate was recovered $Zn[S_2P(OEt)_2]_2$.

White powder of this compound was precipitated by combining concentrated aqueous solutions of $ZnSO_4$ and sodium or ammonium diethyldithiophosphate in a molar ratio of 1:2 (Drew et al. 1986, Ito et al. 1969). This was purified by recrystallization from light petroleum (Drew et al. 1986). Crystals were obtained by slow evaporation of an acetone solution of the title compound at room temperature (Ito et al. 1969).

If $Et_2NH$ (0.8 mM) and $Zn[S_2P(OEt)_2]_2$ (4 mM) were refluxed for 1 h in $CCl_4$ (10 mL), after ca. 5 min, the solution became yellow and eventually brown (Harrison et al. 1986b). On standing, white crystals of $Zn[S_2P(OEt)_2]_2 \cdot 2.75Et_2NH$ were obtained.

**$[Zn\{S_2P(OPr^n)_2\}_2]$ or $C_{12}H_{28}P_2O_4S_4Zn$.** This compound could be obtained if $Zn_2[S_2P(OPr^n)_2]_3(OH)$ (1 mM) was added rapidly to Zn dust (1.1 mM), suspended in benzene (500 mL) at 35°C (Bacon and Bor 1962). The reaction was exothermic and the mixture foamed briefly. The mixture was stirred for 2 h at 50°C–70°C and filtered. The benzene was removed by distillation (80°C at 6.7 kPa). The cloudy viscous liquid was filtered and the filtrate was the title compound. It could also be obtained using the filtrate from the $Zn_2[S_2P(OPr^n)_2]_3(OH)$ preparation. The petroleum ether was removed from this filtrate by distillation and further concentrated by heating to 80°C at 6.7 kPa. The slightly cloudy residue was filtered through a sintered glass funnel and the filtrate was $Zn[S_2P(OPr^n)_2]_2$.

The title compound could also be prepared by mixing aqueous solutions of $ZnCl_2$ and $NaS_2P(OPr^n)_2$ (Ivanov et al. 2001a). The liquid compound was separated by extraction into chloroform with subsequent evaporation of the solvent under slight heating.

**$[Zn\{S_2P(OPr^i)_2\}_2]$ or $C_{12}H_{28}P_2O_4S_4Zn$.** This compound melts at 140°C–142°C (Harrison et al. 1986a,b) (at 144.0°C–144.5°C [Dickert and Rowe 1967]; at 147°C–148°C [Wystrach et al. 1956]) and crystallizes in the monoclinic structure with the lattice parameters $a = 1093.4 \pm 0.8$, $b = 1709.8 \pm 0.6$, $c = 2558.7 \pm 1.2$ pm, and $\beta = 99.23° \pm 0.04°$ and experimental and calculated densities $1.37 \pm 0.02$ and $1.384 \pm 0.002$ g·cm$^{-3}$, respectively (Lawton and Kokotailo 1969). It was prepared by metathesis reaction using $NH_4S_2P(OPr^i)_2$ and filtered aqueous solution of $ZnCl_2$ in stoichiometric amounts. Solid insoluble $Zn[S_2P(OPr^i)_2]_2$ was isolated by filtration and purified by recrystallization (Wystrach et al. 1956, Dickert and Rowe 1967, Jian et al. 2000, Chen et al. 2006). Liquid products were extracted from the aqueous mixture with pentane, washed with water, and dried, and the pentane was removed by evaporation under vacuum. Soft, colorless, tabular crystals were obtained by recrystallization of the compound from warm absolute EtOH (Lawton and Kokotailo 1969).

The title compound could also be obtained if ZnO was added to the brown-black solution of $O,O'$-diisopropyl dithiophosphoric acid, and the mixture was stirred for 1 h after the initial exothermic reaction has subsided (Harrison et al. 1986b, McCleverty et al. 1983). From the filtrate was recovered $Zn[S_2P(OPr^i)_2]_2$.

$Zn[S_2P(OPr^i)_2]_2$ could also be prepared by mixing aqueous solutions of $ZnCl_2$ and $NaS_2P(OPr^i)_2$ (Ivanov et al. 2001a). The liquid compound was separated by extraction into chloroform with subsequent evaporation of the solvent under slight heating.

The title compound was also prepared by allowing $Pr^iOH$ (3.9 M) to react with $P_4S_{10}$ (0.375 M) over 30 min (Drew et al. 1986). The resulting mixture was kept at 80°C for a further 30 min. To this warm solution was added $Zn(OAc)_2 \cdot 2H_2O$ (0.60 M) dissolved in EtOH. The mixture was stirred for 15 min and allowed to cool to room temperature during which time white crystals of the product were deposited.

It is possible also to prepare $Zn[S_2P(OPr^i)_2]_2$ if a mixture of ZnO (0.24 g) and $NH_4S_2P(OPr^i)_2$ (1.4 g) was stirred and refluxed in $Pr^iOH$ for 17 h (McCleverty et al. 1983). The clear solution was then filtered and cooled, the colorless crystals of being collected by filtration and washed with $Pr^iOH$.

If $Et_3N$ (29 mM) was added as a solvent to $Zn[S_2P(OPr^i)_2]_2$ (30 mM) dissolved in $CCl_4$ (30 mL), a white precipitate began to form immediately, and after 2 h the compound $Zn_4[S_2P(OPr^i)_2]_6O$ was obtained as a white powder. On allowing to stand, the filtrate became yellow and deposited white crystals of the triethyleneamine adduct $\mathbf{Zn[S_2P(OPr^i)_2]_2 \cdot 1.5Et_3N}$, which melts at 85°C (Harrison et al. 1986b). At the refluxing of $Zn[S_2P(OPr^i)_2]_2$ (0.4 mM) and $Et_3N$ (0.7 mM) for 1 h in $CCl_4$ (10 mL), the adduct $\mathbf{Zn[S_2P(OPr^i)_2]_2 \cdot Et_3N}$ was formed as white solid. When $Zn[S_2P(OPr^i)_2]_2$ (0.4 mM) was dissolved in $Et_3N$ (4 mL), the solution rapidly became yellow, and precipitation of a white solid took place over a period of 30 min. Recrystallization from $Et_3N$ afforded white needle crystals of the compound $\mathbf{Zn[S_2P(OPr^i)_2]_2 \cdot 2.75Et_3N}$, which melts at 104°C–105°C.

$\mathbf{[Zn_4O\{OSP(OMe)_2\}_6]}$ or $\mathbf{C_{12}H_{36}P_6O_{19}S_6Zn_4}$. This compound crystallizes in the rhombohedral structure with the lattice parameters (in hexagonal setting) $a = 1795.6 \pm 0.8$ and $c = 1079 \pm 1$ pm (Menzer et al. 2000). To obtain it, sodium $O,O$-dimethyl monothiophosphate (2.5 g) was treated with dilute HCl (3 M, 200 mL), and the resulting solution extracted into diethyl ether (100 mL). The solvent was removed and the resulting oil dissolved in THF (50 mL). Solid ZnO (0.5 g) was added and heated to reflux for 3 h. The resulting solution was cooled and filtered, and the solvent was removed in vacuo. The title compound was recrystallized from $CH_2Cl_2$/hexane.

$\mathbf{[Zn(Bu^n_2PSO)_2]}$ or $\mathbf{C_{16}H_{36}P_2O_2S_2Zn}$. This compound melts at 80°C (Calligaris et al. 1970a). It was prepared by adding stoichiometric quantities of di-$n$-butylphosphinothionic acid to an ethanol solution of $Zn(OAc)_2$. After heating under reflux for 1–3 h, the white product separated out and was filtered, washed with absolute ethanol, and vacuum dried. Single crystals of the title compound were obtained as platelets by slow evaporation of an EtOH/light petroleum solution. $Zn(Bu^n_2PSO)_2$ may also be obtained in an amorphous form by a dropwise addition of a solution of $ZnSO_4$ to a warm aqueous solution of $Bu^n_2PSONa$.

**[Zn{S$_2$P(OBu$^n$)$_2$}$_2$] or C$_{16}$H$_{36}$P$_2$O$_4$S$_4$Zn.** To obtain this compound, 1 mol of crude *O,O*-di-*n*-butyl phosphorodithioic acid was dissolved in a solution of 1.1 mol of NaOH in 400 mL of H$_2$O (Wystrach et al. 1956). The solution was adjusted to pH = 9.5 with an additional 4 g of NaOH. In order to remove an insoluble, oily contaminant, the solution was treated with 25 g of Darco and 20 g of filter aid and filtered. ZnCl$_2$ (0.5 mol) in 40 mL of H$_2$O was added to the filtered solution, and the resulting mixture was heated at 75°C for about 10 min. The hot product was separated from the supernatant aqueous phase, stripped of residual water, and filtered through a layer of filter aid in a steam-jacketed filter funnel. The final product was semisolid.

Zn[S$_2$P(OBu$^n$)$_2$]$_2$ could also be prepared by mixing aqueous solutions of ZnCl$_2$ and NaS$_2$P(OBu$^n$)$_2$ (Ivanov et al. 2001a). The liquid compound was separated by extraction into chloroform with subsequent evaporation of the solvent under slight heating.

**[Zn{S$_2$P(OBu$^t$)$_2$}$_2$] or C$_{16}$H$_{36}$P$_2$O$_4$S$_4$Zn.** This compound is unstable at room temperature and was prepared by metathetic reaction using NH$_4$S$_2$P(OBu$^t$)$_2$ and filtered aqueous solution of ZnCl$_2$ in stoichiometric amounts (Dickert and Rowe 1967). Solid insoluble Zn[S$_2$P(OBu$^t$)$_2$]$_2$ were isolated by filtration and purified by recrystallization. Liquid products were extracted from the aqueous mixture with pentane, washed with water, and dried, and the pentane was removed by evaporation under vacuum.

**[Zn{S$_2$P(OBu$^i$)$_2$}$_2$] or C$_{16}$H$_{36}$P$_2$O$_4$S$_4$Zn.** This compound could be prepared by mixing aqueous solutions of ZnCl$_2$ and NaS$_2$P(OBu$^i$)$_2$ (Ivanov et al. 2001a). A solid compound was separated by extraction into chloroform with subsequent evaporation of the solvent under slight heating.

**[Zn{S$_2$P(OBu$^s$)$_2$}$_2$] or C$_{16}$H$_{36}$P$_2$O$_4$S$_4$Zn.** This compound could be prepared by mixing aqueous solutions of ZnCl$_2$ and NaS$_2$P(OBu$^s$)$_2$ (Ivanov et al. 2001a). A solid compound was separated by extraction into chloroform with subsequent evaporation of the solvent under slight heating.

**[Zn$_2${S$_2$P(OPr$^n$)$_2$}$_3$(OH)] or C$_{18}$H$_{43}$P$_3$O$_7$S$_6$Zn$_2$, basic zinc double salt of *O,O*-di-*n*-propyl phosphorodithioic acid.** This compound melts at 178°C–179°C (Bacon and Bor 1962). To obtain it, *O,O*-di-*n*-propyl hydrogen phosphorodithioate (1.0 M) was slowly dripped into a slurry of ZnO (1.0 M) and benzene (500 mL) at 35°. The mixture was stirred for 2 h at 50°C, and then the water was removed by refluxing the benzene into a Dean–Stark tube. After all, the water was removed and the solution was cooled and filtered. The benzene was removed by distillation to 90°C at 6.7 kPa. The residue, consisting of solid and liquid, was diluted with petroleum ether (250 mL), and the solid product removed by filtration. The white solid was recrystallized from Pr$^n$OH.

**[Zn$_2${S$_2$P(OPr$^i$)$_2$}$_3$(OH)] or C$_{18}$H$_{43}$P$_3$O$_7$S$_6$Zn$_2$, basic zinc double salt of *O,O*-di-*i*-propyl phosphorodithioic acid.** This compound melts at 74°C–76°C (Wystrach et al. 1956). To obtain it, a solution of potassium di-*i*-propyldithiophosphate (60 mM) in H$_2$O (200 mL) containing 20 mM of NaOH was treated with 2 M solution of ZnCl$_2$ (20 mL). A crystalline solid was obtained. Its crystals were grown at recrystallization from cyclohexane.

**[Zn{S$_2$P(OBu$^t$CH$_2$)$_2$}$_2$]** or **C$_{20}$H$_{44}$P$_2$O$_4$S$_4$Zn**. This compound melts at 223.5°C–224.5°C (Dickert and Rowe 1967). It was prepared by metathetic reaction using NH$_4$S$_2$P(OBu$^t$CH$_2$)$_2$ and filtered aqueous solution of ZnCl$_2$ in stoichiometric amounts. Solid insoluble Zn[S$_2$P(OBu$^t$CH$_2$)$_2$]$_2$ were isolated by filtration and purified by recrystallization. Liquid products were extracted from the aqueous mixture with pentane, washed with water, and dried, and the pentane was removed by evaporation under vacuum.

The title compound could also be prepared by mixing aqueous solutions of ZnCl$_2$ and NaS$_2$P(OBu$^t$CH$_2$)$_2$ (Ivanov et al. 2001a). A solid compound was separated by extraction into chloroform with subsequent evaporation of the solvent under slight heating.

**[Zn(Ph$_2$PSO)$_2$]** or **C$_{24}$H$_{20}$P$_2$O$_2$S$_2$Zn** (**zinc diphenylphosphinothionate**). This compound softens at ca. 150°C. (Calligaris et al. 1970a) It was prepared by adding stoichiometric quantities of diphenylphosphinothionic acid to an ethanol solution of Zn(OAc)$_2$. After heating under reflux for 1–3 h, the white product separated out and was filtered, washed with absolute ethanol, and vacuum dried.

**[Zn{S$_2$P(OC$_6$H$_{11}$)$_2$}$_2$]** or **C$_{24}$H$_{44}$P$_2$O$_4$S$_4$Zn**. This compound crystallizes in the monoclinic structure with the lattice parameters $a = 1222.1 \pm 0.2$, $b = 2048.3 \pm 0.2$, $c = 2611.0 \pm 0.4$ pm, and $\beta = 98.57° \pm 0.02°$ and a calculated density of $1.341 \pm 0.001$ g·cm$^{-3}$ (Ivanov et al. 2001a). It was prepared from the reaction of Zn(NO$_3$)$_2$·6H$_2$O and NH$_4${S$_2$P(OC$_6$H$_{11}$)$_2$} in aqueous solution (Chen et al. 2006). The title compound could also be prepared by mixing aqueous solutions of ZnCl$_2$ and NaS$_2$P(OC$_6$H$_{11}$)$_2$ (Ivanov et al. 2001a). A solid compound was separated by extraction into chloroform with subsequent evaporation of the solvent under slight heating. Its pale-yellow single crystals were grown from absolute EtOH.

**Zn[S$_2$P{OBu$^t$(CHCH$_3$)}$_2$]$_2$** or **C$_{24}$H$_{52}$P$_2$O$_4$S$_4$Zn**. This compound melts at 136.5°C–137.5°C (Dickert and Rowe 1967). It was prepared by metathetic reaction using NH$_4$S$_2$P{OBu$^t$(CHCH$_3$)}$_2$ and filtered aqueous solution of ZnCl$_2$ in stoichiometric amounts. Solid insoluble Zn[S$_2$P{OBu$^t$(CHCH$_3$)}$_2$]$_2$ was isolated by filtration and purified by recrystallization. Liquid products were extracted from the aqueous mixture with pentane, washed with water, and dried, and the pentane was removed by evaporation under vacuum.

**[Zn$_2${S$_2$P(OBu$^n$)$_2$}$_3$(OH)]** or **C$_{24}$H$_{55}$P$_3$O$_7$S$_6$Zn$_2$, basic zinc double salt of *O,O*-di-*n*-butyl phosphorodithioic acid.** This compound melts at 150°C–153°C (Wystrach et al. 1956) (at 149°C–151°C [Bacon and Bor 1962]). It could be obtained as by-product from the preparation of Zn[S$_2$P(OBu$^t$)$_2$]$_2$. The crystalline solid phase from the preparation of C$_{16}$H$_{36}$P$_2$O$_4$S$_4$Zn was centrifuged, washed repeatedly with cold MeOH, and recrystallized from hexane giving large, white, friable crystals.

C$_{24}$H$_{55}$P$_3$O$_7$S$_6$Zn$_2$ could also be prepared if purified *O,O*-di-*n*-butyl phosphorodi-thioic acid (0.1 mol) was added to a solution of 0.133 mol of 97% NaOH in 200 mL of H$_2$O containing 50 g of ice. A very small amount of insoluble oil was removed by filtration of the solution through filter aid. A solution of 0.067 mol of ZnCl$_2$ in H$_2$O (50 mL) was added slowly with manual stirring. The solid crystalline product began to form immediately (Wystrach et al. 1956).

The title compound could be obtained using the following procedure (Bacon and Bor 1962). Crude *O,O*-di-*n*-butyl hydrogen phosphorodithioate (6.0 M), H$_2$O (6.0 M), and benzene (22.3 M) were mixed at 35°C. ZnO (6.0 M) was added slowly at 50°C–60°C over a period of 15 min. The mixture was stirred for 2 h at 60°C. The mater was removed by refluxing the benzene through a Dean–Stark tube. The benzene solution was filtered and the benzene was removed by distillation to 100°C at 4.7 kPa. The residue was cooled and the crude product was filtered. This product was recrystallized from naphtha, and the mother liquors were combined and retreated with ZnO and water as outlined previously. The solid obtained was combined with the previous batch.

**Zn$_2$[S$_2$P(OBu$^i$)$_2$]$_3$(OH) or C$_{24}$H$_{55}$P$_3$O$_7$S$_6$Zn$_2$, basic zinc double salt of *O,O*-di-*i*-butyl phosphorodithioic acid**. This compound melts at 138°C–140°C (Bacon and Bor 1962). To obtain it, crude *O,O*-di-*i*-butyl hydrogen phosphorodithioate (2.0 M), benzene (8.3 M), and H$_2$O (5.5 M) were mixed at 35°C. ZnO (2.0 M) was slowly added at 35°C–50°C over a period of 10 min. The mixture was stirred for 1 h at 50°C and then dried by refluxing the benzene through a Dean–Stark tube. After complete removal of the water (3–4 h), the product was filtered. The benzene was evaporated and the last trace of solvent was removed by heating to 100°C at 4 kPa. The residue was extracted with petroleum ether (500 mL). Upon cooling the petroleum ether, white solid of the title compound was separated.

**[Zn$_4$S{S$_2$P(OEt)$_2$}$_6$] or C$_{24}$H$_{60}$P$_6$O$_{12}$S$_{13}$Zn$_4$**. This compound melts at 202°C–203°C and crystallizes in the trigonal structure with the lattice parameters $a = 2077.6 \pm 0.2$ and $c = 1156.0 \pm 0.2$ and experimental and calculated densities 1.7 and 1.62 g·cm$^{-3}$, respectively (Harrison et al. 1986b). Colorless crystals of the title compound were obtained by the recrystallization of the insoluble material from the preparation of Zn[S$_2$P(OEt)$_2$]$_2$ using hot EtOH/CH$_2$Cl$_2$ (1:1 volume ratio).

**[Zn$_4$O{S$_2$P(OEt)$_2$}$_6$] or C$_{24}$H$_{60}$P$_6$O$_{13}$S$_{12}$Zn$_4$**. This compound melts at 200°C (Harrison et al. 1986b). The addition of tri- or diethyleneamine (0.4 mM) to a solution of Zn[S$_2$P(OEt)$_2$]$_2$ (0.4 mM) in CCl$_4$ (4 mL) resulted in the slow precipitation of the title compound. The same product was also obtained when 1,5-diazabicyclo[4.3.0]non-5-ene (0.1 mM) was added to a solution of Zn[S$_2$P(OEt)$_2$]$_2$ (0.4 mM) in CCl$_4$ (4 mL), and the solution was allowed to stand overnight. Employing toluene as the solvent afforded white crystals of a toluene solvate **[Zn$_4$O{S$_2$P(OEt)$_2$}$_6$]·0.2C$_7$H$_8$**, which melts at 208°C–210°C.

The title compound could also be prepared by the next procedure (Drew et al. 1986). To a solution of [Zn{S$_2$P(OEt)$_2$}$_2$] in MeOH was added a 70% solution of 1,1-dimethylhydroperoxide (4:1 molar ratio). After 1 h of continuous stirring, the product was deposited as a fine white solid. It was recrystallized from light petroleum.

**[Zn{S$_2$P(OC$_6$H$_4$Me)$_2$}$_2$] or C$_{28}$H$_{28}$P$_2$O$_4$S$_4$Zn**. This compound was prepared as follows (McCleverty et al. 1983). To a suspension of P$_4$S$_{10}$ (11.1 g) in toluene (60 mL) was added a solution of *p*-MeC$_6$H$_4$OH (22 g) in toluene (10 mL). The mixture was then refluxed until a clear solution had formed, when the solvent was evaporated in vacuo affording HS$_2$P(OC$_6$H$_4$Me-*p*)$_2$. To a solution of HS$_2$P(OC$_6$H$_4$Me-*p*)$_2$ (31.0 g) in toluene (60 mL) at room temperature was added an aqueous solution of $ZnSO_4 \cdot 7H_2O$ (14.4 g). The mixture was stirred vigorously overnight, the organic layer was then separated, washed with water (2×30 mL), and evaporated in vacuo. The resulting oil was dissolved in diethyl ether/ligroin mixture (1:1 volume ratio) and the solution allowed standing for 24 h during, which time the title compound formed as white microcrystals. These were collected by filtration and washed with ligroin.

**$[Ph_4P][Zn(SC\{O\}Me)_3(H_2O)]$ or $C_{30}H_{31}PO_4S_3Zn$.** This compound crystallizes in the monoclinic structure with the lattice parameters $a = 1161.60 \pm 0.03$, $b = 1921.22 \pm 0.03$, $c = 1439.78 \pm 0.03$ pm, and $\beta = 93.281° \pm 0.001°$ (Sampanthar et al. 1999). It was obtained by the following procedure. Thioacetic acid (28 mM) was added dropwise to a stirred aqueous solution of $Et_3N$ (28 mM) to obtain a solution of $Et_3NHMe\{O\}CS$. To this was added $ZnCl_2 \cdot 1.5H_2O$ (9 mM) in $H_2O$ (5 mL). The resulting solution was stirred for 10 min. When a solution of $Ph_4PCl$ (9 mM) in $H_2O$ was added, a cream-colored precipitate was formed. The mixture was stirred for 1 h, after which $CH_2Cl_2$ (8 mL) was added. The precipitate in the aqueous layer was extracted into the $CH_2Cl_2$ layer, which became yellow. The $CH_2Cl_2$ was separated and washed with 30 mL portions of deionized water three or four times to remove the side product, $Et_3NHCl$. The $Et_2O$ was added to the $CH_2Cl_2$ solution until turbidity was noticed, when the mixture was set aside in a refrigerator overnight, to obtain yellow crystals. They were decanted, washed with $Et_2O$, and dried in vacuo.

**$[Zn\{S_2P(OBu^i)_2\}_2]_2$ or $C_{32}H_{72}P_4O_8S_8Zn_2$.** This compound crystallizes in the triclinic structure with the lattice parameters $a = 1036.52 \pm 0.07$, $b = 1136.68 \pm 0.07$, and $c = 2559.9 \pm 0.2$ pm and $\alpha = 86.473° \pm 0.005°$, $\beta = 86.662° \pm 0.004°$, and $\gamma = 65.324° \pm 0.007°$ at 203 K (Menzer et al. 2000).

**$[Zn_4O\{S_2P(OPr^n)_2\}_6]$ or $C_{36}H_{84}P_6O_{13}S_{12}Zn_4$.** This compound was formed by a spontaneous precipitation from the initial liquid $[Zn\{S_2P(OPr^n)_2\}]$ as a result of oxidation by atmospheric $O_2$ after 6 weeks of its storage in air (Ivanov et al. 2001a). White solid was separated by washing with absolute EtOH and recrystallized from a minute volume of chloroform.

**$[Zn_4O\{S_2P(OPr^i)_2\}_6]$ or $C_{36}H_{84}P_6O_{13}S_{12}Zn_4$.** This compound melts at 198°C–200°C (Harrison et al. 1986b) and crystallizes in the monoclinic structure with the lattice parameters $a = 2493 \pm 8$, $b = 2276 \pm 5$, $c = 1367 \pm 2$ pm, and $\beta = 113°28' \pm 15'$ and a calculated density of 1.47 $g \cdot cm^{-3}$ (Burn and Smith 1965). The addition of tri- or diethyleneamine (0.4 mM) to a solution of $Zn[S_2P(OPr^i)_2]_2$ (0.4 mM) in $CCl_4$ (4 mL) resulted in the slow precipitation of the title compound. It is also produced as a by-product in the synthesis of $Zn[S_2P(OPr^i)_2]_2$. The same product was also obtained when 1,5-diazabicyclo[4.3.0]non-5-ene (0.1 mM) was added to a solution of $Zn[S_2P(OEt)_2]_2$ (0.4 mM) in $CCl_4$ (4 mL), and the solution allowed to stand overnight.

The title compound could also be prepared by the next procedure (Drew et al. 1986). To a solution of $[Zn\{S_2P(OPr^i)_2\}_2]$ in light petroleum was added a 70% solution of 1,1-dimethylhydroperoxide (4:1 molar ratio). After 1 h of continuous stirring the product was deposited as a fine white solid. It was recrystallized from light petroleum.

**[Zn$_2$(S$_2$COPr$^i$)$_4$(Ph$_2$PCH$_2$)$_2$]** or **C$_{42}$H$_{52}$P$_2$O$_4$S$_8$Zn$_2$**. This compound crystallizes in the monoclinic structure with the lattice parameters $a = 1112.05 \pm 0.06$, $b = 1776.59 \pm 0.09$, $c = 2490.02 \pm 0.13$ pm, and $\beta = 92.712° \pm 0.001°$ at $183 \pm 2$ K and a calculated density of 1.446 g·cm$^{-3}$ (Beer et al. 2002). Its colorless crystals were obtained from the slow evaporation of an acetonitrile solution of Zn$_2$(S$_2$COPr$^i$)$_2$ and 1,2-bis(diphenylphosphino)ethane (2:1 molar ratio).

**[(Ph$_4$P){Zn(SOCPh)$_3$}]** or **C$_{45}$H$_{35}$PO$_3$S$_3$Zn**. This compound crystallizes in the triclinic structure with the lattice parameters $a = 1081.9 \pm 0.2$, $b = 1321.9 \pm 0.3$, and $c = 1595.1 \pm 0.3$ pm and $\alpha = 101.75° \pm 0.02°$, $\beta = 97.92° \pm 0.01°$, and $\gamma = 109.18° \pm 0.02°$ and experimental and calculated densities $1.37 \pm 0.05$ and 1.318 g·cm$^{-3}$, respectively (Vittal and Dean 1996). To obtain this compound, Et$_3$NHSCOPh was prepared in situ by mixing Et$_3$N (17.0 mM) in 20 mL of MeOH with thiobenzoic acid (17.0 mM). To this stirred solutions was added Zn(NO$_3$)$_2$·6H$_2$O (5.68 mM) in 15 mL of H$_2$O. The resulting yellow solution with some viscous orange oil at the bottom was treated with Ph$_4$PBr (5.66 mM) in 20 mL of MeOH, and the mixture was heated to boiling, giving a clear yellow solution. This solution was filtered while hot and left in the refrigerator at 5°C for crystallization. The dark-yellow crystals were separated by filtration, washed with MeOH and Et$_2$O, and then dried in a flow of Ar. The washings were combined with the filtrate and allowed to evaporate at room temperature to get a second crop of crystals. All the preparations were carried out under Ar.

**[Zn$_2${MeOC$_6$H$_4$P(OC$_5$H$_9$)S$_2$}$_4$]** or **C$_{48}$H$_{64}$P$_4$O$_8$S$_8$Zn$_2$**. This compound melts at 169°C and crystallizes in the triclinic structure with the lattice parameters $a = 1133.0 \pm 0.2$, $b = 1238.5 \pm 0.2$, $c = 1251.1 \pm 0.2$ pm, and $\beta = 73.984° \pm 0.003°$ (the values of $\alpha$ and $\gamma$ did not indicated in the article) (Karakus et al. 2005). It was prepared by the reaction of MeOC$_6$H$_4$P(OC$_5$H$_9$)(S)(SNH$_4$)$_2$ and ZnSO$_4$·7H$_2$O in water. Colorless crystals of the title compound were obtained from a mixture of chloroform/Pr$^i$OH (3:1 volume ratio).

**[Zn$_4$O{S$_2$P(OBu$^n$)$_2$}$_6$]** or **C$_{48}$H$_{108}$P$_6$O$_{13}$S$_{12}$Zn$_4$**. This compound crystallizes in the trigonal structure with the lattice parameters $a = 2390 \pm 20$ and $c = 1284 \pm 8$ pm and a calculated density of 1.34 g·cm$^{-3}$, the true cell being rhombohedral with $a = 1445$ pm and $\alpha = 111°36'$. (Burn and Smith 1965) (in rhombohedral structure with the lattice parameters $a = 1458 \pm 1$ pm and $\alpha = 112.1° \pm 0.1°$[Drew et al. 1986]). It was formed by a spontaneous precipitation from the initial liquid [Zn{S$_2$P(OBu$^n$)$_2$}] as a result of oxidation by atmospheric O$_2$ after 6 weeks of its storage in air (Ivanov et al. 2001a). Colorless solid was separated by washing with absolute MeOH and recrystallized from a minute volume of chloroform.

**[Zn$_4$O{S$_2$P(OBu$^i$)$_2$}$_6$]** or **C$_{48}$H$_{108}$P$_6$O$_{13}$S$_{12}$Zn$_4$**. This compound crystallizes in the orthorhombic structure with the lattice parameters $a = 1344.7 \pm 0.2$, $b = 2273.4 \pm 0.6$, and $c = 2683.9 \pm 0.6$ pm at 173 K (Menzer et al. 2000).

**[(Ph$_4$P)$_2${Zn(C$_3$S$_5$)$_2$}·DMSO]** or **C$_{56}$H$_{46}$P$_2$OS$_{11}$Zn**. This compound melts at 198°C–201°C crystallizes in the triclinic structure with the lattice parameters $a = 980.970 \pm 0.010$, $b = 1678.11 \pm 0.03$, and $c = 1895.19 \pm 0.03$ pm and $\alpha = 65.5651° \pm 0.0008°$, $\beta = 79.0773° \pm 0.0010°$, and $\gamma = 80.5986° \pm 0.0010°$ at $150 \pm 2$ K and a calculated density of 1.353 g·cm$^{-3}$ (Harrison et al. 2000). It was obtained from Na, CS$_2$, and Ph$_4$PCl. Crystals suitable for XRD were grown by the recrystallization from DMSO.

**[Zn$_4$O{S$_2$P(OPh)$_2$}$_6$] or C$_{72}$H$_{60}$P$_6$O$_{13}$S$_{12}$Zn$_4$.** This compound crystallizes in the cubic structure with the lattice parameter $a = 3252 \pm 2$ pm (Drew et al. 1986). It was prepared as follows. To a hot solution (100°C) of phenol (2.34 M) in xylene (800 mL) was added P$_4$S$_{10}$ (0.292 M). The temperature was raised to 140°C and a clear solution was formed within 15 min. The mixture was allowed to cool to room temperature when gaseous ammonia was passed into it for 30 min. Upon the addition of [Zn(O$_2$CMe)$_2$]·2H$_2$O (0.77 M), a clear solution formed. The mixture was heated to 80°C and quickly poured into cold water (1 L), whereupon a white precipitate of the title compound was deposited.

**[Zn$_4$O{OSP(OPh)$_2$}$_6$]·0.5H$_2$O or C$_{72}$H$_{61}$P$_6$O$_{19.5}$S$_6$Zn$_4$.** This compound crystallizes in the rhombohedral structure with the lattice parameters (in hexagonal setting) $a = 2221.3 \pm 0.7$ and $c = 2802.1 \pm 0.4$ pm (Menzer et al. 2000). To obtain it, sodium $O,O$-diphenyl monothiophosphate (2.5 g) was treated with dilute HCl (3 M, 200 mL) and the resulting solution extracted into diethyl ether (100 mL). The solvent was removed and the resulting oil dissolved in THF (50 mL). Solid ZnO (0.5 g) was added and heated to reflux for 3 h. The resulting solution was cooled and filtered and the solvent was removed in vacuo. The title compound was recrystallized from CH$_2$Cl$_2$/hexane.

## 1.72 Zn–H–C–P–S

Some compounds are formed in this system.

**[Zn(Et$_2$PS$_2$)$_2$] or C$_8$H$_{20}$P$_2$S$_4$Zn, zinc diethyldithiophosphinate.** This compound crystallizes in the monoclinic structure with the lattice parameters $a = 1067 \pm 2$, $b = 3193 \pm 4$, $c = 1250 \pm 2$ pm, and $\beta = 126.7° \pm 0.3°$ and experimental and calculated densities 1.46 and 1.444 g·cm$^{-3}$, respectively (Calligaris et al. 1970b). It was prepared by the next procedure (McCleverty et al. 1983). To an aqueous solution of ZnSO$_4$·7H$_2$O (0.4 g) was added an aqueous solution of Na[S$_2$PEt$_2$] 2H$_2$O (1.0 g) slowly, with stirring and heating. The precipitate of the title compound that formed was filtered off and washed with Pr$^i$OH. Its single crystals as colorless platelike prisms were grown by slow evaporation of an ethanol/benzene solution Calligaris et al. 1970b).

**[Zn{$n$-Pr)$_2$PS$_2$}$_2$] or C$_{12}$H$_{28}$P$_2$S$_4$Zn, bis(di-$n$-propyldithiophosphinato)zinc(II).** This compound crystallizes in the triclinic structure with the lattice parameters $a = 840.9 \pm 0.5$, $b = 977.1 \pm 0.6$, and $c = 1345.1 \pm 0.7$ pm and $\alpha = 90.99° \pm 0.05°$, $\beta = 99.28° \pm 0.02°$, and $\gamma = 105.27° \pm 0.05°$ and a calculated density of 1.35 g·cm$^{-3}$ (Wunderlich 1982).

**[Zn(Bu$^i_2$PS$_2$)$_2$]$_2$ or C$_{16}$H$_{36}$P$_2$S$_4$Zn.** This compound melts at 112°C (at 119°C–121°C [Shchukin et al. 2000]) and crystallizes in the orthorhombic structure with the lattice parameters $a = 4860.2 \pm 0.2$, $b = 1166.5 \pm 0.1$, and $c = 1800.0 \pm 0.1$ pm and calculated density 1.260 g·cm$^{-3}$ (Byrom et al. 2000). To prepare it, ZnSO$_4$·H$_2$O or ZnCl$_2$ (11 mM) in aqueous solution was reacted with Na(Bu$^i_2$PS$_2$) (22 mM) in aqueous solution (Byrom et al. 2000, Shchukin et al. 2000). The white precipitate that formed was filtered under vacuum, washed with distilled H$_2$O, and recrystallized from acetone to give transparent crystals of the title compound. Single crystals of the title compound were grown by recrystallization from a mixture CHCl$_3$/EtOH (1:3 volume ratio) (Shchukin et al. 2000). [Zn(Bu$^i_2$PS$_2$)$_2$]$_2$ could be used as precursor to grow thin films of ZnS (Byrom et al. 2000).

**[Zn(SPh)$_2$·depe] or C$_{22}$H$_{34}$P$_2$S$_2$Zn [depe = 1,2-bis(diethylphosphino)ethane].** This compound melts at 146.8°C–147°C and crystallizes in the orthorhombic structure (Brennan et al. 1990). It could be obtained if to Zn(SPh)$_2$ (0.74 mM) suspended in toluene (10 mL) was added depe (1.6 mM) and pyridine (0.5 mL). The solution was stirred for 12 h, filtered off, concentrated to ca. 15 mL, and cooled (–20°C) to give colorless diamonds. All reactions were performed in an inert atmosphere.

**[Zn(S$_2$PPh$_2$)$_2$] or C$_{24}$H$_{20}$P$_2$S$_4$Zn.** To obtain this compound, a methanolic solution of HS$_2$PPh$_2$ (1.25 g) was added slowly to a gently heated, concentrated aqueous solution of Zn(O$_2$CMe)$_2$·4H$_2$O (0.64 g) (McCleverty et al. 1983). The precipitate of the title compound that formed was collected by filtration and washed with MeOH.

**[Zn(SBu$^t_3$C$_6$H$_2$-2,4,6)$_2$(PMe$_3$)] or C$_{39}$H$_{67}$PS$_2$Zn.** This compound sublimes at 120°C at 1.3 Pa (Bochmann et al. 1994). It could be obtained if to a suspension of [Zn(SBu$^t_3$C$_6$H$_2$-2,4,6)$_2$]$_2$ (0.16 mM) in 10 mL of petroleum ether was added PMe$_3$ (1.31 mM) via syringe. The starting material dissolved within a few seconds, followed by the precipitation of flocculant white product, which was dissolved by adding more petroleum ether and warming on a water bath. Cooling to 0°C gave colorless needles of C$_{39}$H$_{67}$PS$_2$Zn. All reactions were carried out under Ar atmosphere using standard vacuum-line techniques.

**[Zn(SBu$^t_3$C$_6$H$_2$-2,4,6)$_2$(dmpe)] or C$_{42}$H$_{74}$P$_2$S$_2$Zn [dmpe = bis(dimethylphosphino) ethane].** This compound melts at 210°C–212°C (Bochmann et al. 1994). It could be obtained if to a suspension of [Zn(SBu$^t_3$C$_6$H$_2$-2,4,6)$_2$]$_2$ (0.40 mM) in 10 mL of petroleum ether was added dmpe (0.4 mM) via syringe. The starting material dissolved within a few seconds, followed by the precipitation of the product, which was dissolved by adding more petroleum ether and warming on a water bath. After standing, C$_{42}$H$_{74}$P$_2$S$_2$Zn crystallized at room temperature as colorless crystals. All reactions were carried out under Ar atmosphere using standard vacuum-line techniques.

**[(Ph$_4$P)$_2${Zn(S$_4$)$_2$}] or C$_{48}$H$_{40}$P$_2$S$_8$Zn.** To obtain this compound, to a solution of (Ph$_4$P)$_2$[Zn(S$_5$)$_2$] (0.47 mM) in 40 mL of DMF was added with stirring Ph$_3$P (0.94 mM) (Coucouvanis et al. 1985). After it was stirred for ca. 30 min, the solution was filtered, and diethyl ether was added to the filtrate until nucleation was noted. After the mixture was allowed to stand for 24 h, crystals formed and were isolated.

**[(Ph$_4$P)$_2${Zn(S$_5$)$_2$}] or C$_{48}$H$_{40}$P$_2$S$_{10}$Zn.** To obtain this compound, to a solution of (Ph$_4$P)[Zn(SPh)$_4$] (1.02 mM) in 25 mL of CH$_3$CN was added solid BzSSSBz

(Bz = 3-quinuclidinyl benzilate) (10.80 mM) in small portions with stirring (Coucouvanis et al. 1985). After the mixture was stirred for ca. 5 min, diethyl ether was added to incipient nucleation. After several hours, yellow-orange crystals formed and were isolated.

**[(Ph₄P)₂{Zn(S₆)₂}] or C₄₈H₄₀P₂S₁₂Zn.** This compound was obtained by a synthetic procedure identical with the one followed in synthesis of $(Ph_4P)_2[Zn(S_5)_2]$ and using 21.6 mM of BzSSSBz and 1.02 mM of $(Ph_4P)[Zn(SPh)_4]$ in 25 mL of $CH_3CN$ (Coucouvanis et al. 1985). Yellow crystals of $(Ph_4P)_2[Zn(S_6)_2]$ were obtained.

**[Zn(SBuᵗ₃C₆H₂-2,4,6)₂(PMePh₂)] or C₄₉H₇₁PS₂Zn.** This compound melts at 216°C–218°C and crystallizes in the monoclinic structure with the lattice parameters $a = 1697.1 \pm 0.2$, $b = 1588.7 \pm 0.1$, $c = 1950.9 \pm 0.5$ pm, and $\beta = 117.87° \pm 0.01°$ and a calculated density of 1.158 g·cm⁻³ (Bochmann et al. 1994). It could be obtained if to a suspension of $[Zn(SBu^t_3C_6H_2-2,4,6)_2]_2$ (0.40 mM) in 10 mL of petroleum ether was added PMePh₂ (0.4 mM) via syringe. The starting material dissolved within a few seconds, followed by the precipitation of the product, which was dissolved by adding more petroleum ether and warming on a water bath. After standing, C₄₉H₇₁PS₂Zn crystallized at room temperature. All reactions were carried out under Ar atmosphere using standard vacuum-line techniques.

**[(Ph₄P)₂{(PhS)₂Zn(S₄)}] or C₆₀H₅₀P₂S₆Zn.** To obtain this compound, to a solution of $(Ph_4P)[Zn(SPh)_4]$ (1.02 mM) in 25 mL of $CH_3CN$ was added solid BzSSSBz (2.15 mM) in small portions with stirring (Coucouvanis et al. 1985). The color of the solution changed from an initial yellow to orange red. After the mixture was stirred for ca. 5 min, diethyl ether was added to incipient nucleation. After several hours, yellow crystals formed and were isolated.

**[(Ph₄P)₂{(PhS)₂Zn(S₅)}] or C₆₀H₅₀P₂S₇Zn.** This compound was prepared in a manner analogous to the one employed for the synthesis of $(Ph_4P)_2[(PhS)_2Zn(S_4)]$ except that a greater amount of BzSSSBz (5.4 mM) was allowed to react with $(Ph_4P)[Zn(SPh)_4)]$ (1.02 mM) (Coucouvanis et al. 1985).

**[Zn(SBuᵗ₃C₆H₂-2,4,6)₂(PMePh₂)₂] or C₆₂H₈₄P₂S₂Zn.** This compound melts at 192°C–194°C (Bochmann et al. 1994). It could be obtained if to a suspension of $[Zn(SBu^t_3C_6H_2-2,4,6)_2]_2$ (0.19 mM) in 10 mL of petroleum ether was added PMePh₂ (2.0 mM) via syringe. The resulting clear solution was left to stand at room temperature overnight to give colorless crystals. All reactions were carried out under Ar atmosphere using standard vacuum-line techniques.

**[(Ph₄P)₂{Zn(SPh)₄}] or C₇₂H₆₀P₂S₄Zn, bis(tetraphenylphosphonium) tetrakis(benzenethiolato)zincate(II).** This compound crystallizes in the orthorhombic structure with the lattice parameters $a = 1373.5 \pm 0.2$, $b = 1749.9 \pm 0.6$, and $c = 2481.0 \pm 0.5$ pm and a calculated density of 1.31 g·cm⁻³ (Swenson et al. 1978). It was obtained in N₂ atmosphere (Coucouvanis et al. 1982). 3.2 mM of bis(O-ethyl dithiocarbonato)zinc(II), 6.5 mM of [Ph₄P]Cl, and 14 mM of KSPh were added to 20 mL of $CH_3CN$ in a 125 mL Erlenmeyer flask. The suspension was brought to the boiling point of $CH_3CN$ and boiled for 10 min. The mixture was then filtered hot through a medium-porosity sintered-glass filter. The bright-yellow solution was allowed to cool. Upon standing for 30 min, bright-yellow crystals formed, which were removed by filtration and washed with 10 mL portions of absolute EtOH and twice with 10 mL portions of diethyl ether.

**[(Ph$_4$P)$_2${Zn$_2$(SPh)$_6$}]** or **C$_{84}$H$_{70}$P$_2$S$_6$Zn$_2$**. This compound crystallizes in the monoclinic structure with the lattice parameters $a = 1337.9 \pm 0.2$, $b = 1365.4 \pm 0.2$, $c = 2048.0 \pm 0.3$ pm, and $\beta = 106.50° \pm 0.02°$ and a calculated density of 1.356 g·cm$^{-3}$ (Abrahams and Garner 1987). It was initially isolated as one product from a reaction of [Me$_4$N]$_2$[Zn$_4$(SPh)$_{10}$] with [Ph$_4$P][MoS$_4$]; subsequently, a direct, rational synthesis was developed. Zn(NO$_3$)$_2$·6H$_2$O (5 mM) was dissolved in MeOH (10 mL) and added to a solution of [Ph$_4$P][SPh] (15 mM) in MeCN (25 mL). The yellow solution immediately became colorless. Storage at −20°C overnight yielded a mass of colorless hexagonal-shaped plate crystals that were collected by filtration, washed with cold MeCN (5 mL) and diethyl ether, and dried in vacuo.

## 1.73   Zn–H–C–P–I–S

**[(CH$_3$)$_3$PS]$_2$[ZnI$_2$]** or **C$_6$H$_{18}$P$_2$I$_2$Zn**. This quinary compound is formed in this system. It was obtained by the direct reaction of trimethylphosphine sulfide, (CH$_3$)$_3$PS, with ZnI$_2$ in absolute ethanol (Meek and Nicpon 1965).

## 1.74   Zn–H–C–As–O–S

Some compounds are formed in this system.

**[(Ph$_4$As){Zn(SOCPh)$_3$}]** or **C$_{45}$H$_{35}$AsO$_3$S$_3$Zn**. To obtain this compound, Et$_3$NHSCOPh was prepared in situ by mixing Et$_3$N (17.0 mM) in 20 mL of MeOH with thiobenzoic acid (17.1 mM) (Vittal and Dean 1996). To this stirred solutions was added Zn(NO$_3$)$_2$·6H$_2$O (2.1 mM) in 15 mL of H$_2$O. The resulting yellow solution with some viscous orange oil at the bottom was treated with Ph$_4$AsCl·H$_2$O (4.10 mM) in 20 mL of MeOH, and the mixture was heated to boiling, giving a clear yellow solution. This solution was filtered while hot and left in the refrigerator at 5°C for crystallization. The orange needle-shaped crystals were separated by filtration, washed with MeOH and Et$_2$O, and then dried in a flow of Ar. The washings were combined with the filtrate and allowed to evaporate at room temperature to get a second crop of crystals. All the preparations were carried out under Ar.

**[(Ph$_4$As)$_2$(CH$_3$COS)$_4$Zn]** or **C$_{56}$H$_{52}$As$_2$O$_4$S$_4$Zn, bis(tetraphenylarsonium) tetrakis(thioacetato)zincate(II)**. This compound was prepared by a metathetical reaction, with vigorous stirring at room temperature, of the tetrahalozincate(II) (1 mM) dissolved in a dried nonaqueous solvent (usually acetonitrile) with the finally ground thallium(I) or silver thioacetate (4 mM) (Hollebone and Nyholm 1971). The filtered solution was evaporated in vacuo to a small volume and white (Ph$_4$As)$_2$(CH$_3$COS)$_4$Zn was isolated by the addition of dried ether or light petroleum. Its crystals were grown by crystallization from ethanol with ether. All manipulations were done under N$_2$ atmosphere.

## 1.75   Zn–H–C–As–S

Some compounds are formed in this system.

**[SZn$_4$(AsS$_2$Me$_2$)$_6$] or C$_{12}$H$_{36}$As$_6$S$_{13}$Zn$_4$.** This compound crystallizes in the hexagonal structure with the lattice parameters $a = 1542$ and $c = 3465$ pm or in the monoclinic structure with the lattice parameters $a = 1158.6 \pm 0.8$, $b = 3308 \pm 2$, $c = 1199.6 \pm 0.8$ pm, and $\beta = 117.39° \pm 0.01°$ and experimental and calculated densities 2.12 and 2.13 g·cm$^{-3}$, respectively (Johnstone et al. 1972). The reaction of Me$_4$As$_2$S$_2$ with Zn(ClO$_4$)$_2$·6H$_2$O in EtOH or acetone solution yields hexagonal crystals of this compound. Recrystallization of the obtained compound from CH$_2$Cl$_2$ solution yields monoclinic crystals.

**[(C$_{16}$H$_{36}$As)$_2$\{Zn(C$_3$S$_5$)$_2$\}] or C$_{38}$H$_{72}$As$_2$S$_{10}$Zn.** This compound melts at 205°C–208°C and crystallizes in the triclinic structure with the lattice parameters $a = 1034.51 \pm 0.05$, $b = 1505.63 \pm 0.07$, and $c = 1880.83 \pm 0.08$ pm and $\alpha = 73.072° \pm 0.001°$, $\beta = 85.867° \pm 0.001°$, and $\gamma = 72.259° \pm 0.001°$ and a calculated density of 1.524 g·cm$^{-3}$ (Comerlato et al. 2002). Red crystals of the title compound were obtained from a DMSO solution.

## 1.76   Zn–H–C–As–Cl–S

**[Zn(SC$_6$Cl$_5$)$_4$[(C$_6$H$_5$)$_4$As]] or C$_{48}$H$_{20}$AsCl$_{20}$S$_4$Zn.** This multinary compound, which melts with decomposition at 172°C, is formed in this system (Lucas and Peach 1969). It has been formed using penthachlorophenylthio anion as a ligand and was prepared by adding an aqueous solution of ZnSO$_4$·7H$_2$O to an aqueous solution of C$_6$Cl$_5$SH. After filtration, an aqueous solution of tetraphenylarsenium chloride was added to the filtrate to precipitate [Zn(SC$_6$Cl$_5$)$_4$[(C$_6$H$_5$)$_4$As]].

## 1.77   Zn–H–C–O–Se–S

**[Zn(SeC$_6$H$_2$Bu$^t_3$–2,4,6)$_2$(OSC$_4$H$_8$)] or C$_{40}$H$_{66}$OSe$_2$SZn.** This multinary compound, which crystallizes in the monoclinic structure with the lattice parameters $a = 1770.6 \pm 0.6$, $b = 914.1 \pm 0.2$, $c = 2733.8 \pm 0.8$ pm, and $\beta = 104.63° \pm 0.03°$ and a calculated density of 1.204 g·cm$^{-3}$, is formed in this system (Bochmann et al. 1994). This compound could be obtained from the reaction of [Zn(SeBu$^t_3$C$_6$H$_2$–2,4,6)$_2$]$_2$ with excess of tetrahydrothiophene (C$_4$H$_8$S).

## 1.78   Zn–H–C–O–S

Some compounds are formed in this system.

**[Zn(C$_4$H$_4$O$_4$S)·4H$_2$O] or C$_4$H$_{12}$O$_8$SZn, triaquazinc thiodiglycolate monohydrate.** This compound crystallizes in the orthorhombic structure with the lattice parameters $a = 758.9 \pm 0.7$, $b = 928.9 \pm 0.8$, and $c = 1416.8 \pm 1.1$ pm and experimental and calculated densities $1.89 \pm 0.01$ and 1.89 g·cm$^{-3}$, respectively (Drew et al. 1975). Its crystals were prepared by mixing an aqueous solution of thiodiglycolic acid with an excess of Zn(OH)$_2$ and setting aside the filtrate.

**[Zn(S₂COEt)₂] or C₆H₁₀O₂S₄Zn, zinc ethylxanthate**. This compound decomposes at 134°C (Lai et al. 2002) and crystallizes in the monoclinic structure with the lattice parameters $a = 1827.8 \pm 0.4$, $b = 570.0 \pm 0.3$, $c = 1138.1 \pm 1.2$ pm, and $\beta = 101.47° \pm 0.10°$ and a calculated density of 1.758 g·cm⁻³ (Ikeda and Hagihara 1966). It was prepared from the white precipitate obtained on titration of aqueous solutions of potassium ethylxanthate and $ZnCl_2$ or $Zn(NO_3)_2·6H_2O$ (Ikeda and Hagihara 1966, Lai et al. 2002). The precipitate was filtered and dehydrated in a vacuum desiccator. The powder obtained was dissolved in EtOH. Single crystals were collected by the slow evaporation of the solution. The crystals were usually thin rectangular plates in shape. They often showed a needlelike appearance and are stable in air at room temperature.

**[Zn(S₂COPrⁿ)₂] or C₈H₁₄O₂S₄Zn**. This compound decomposes at 140°C and crystallizes in the triclinic structure with the lattice parameters $a = 961.13 \pm 0.06$, $b = 1542.23 \pm 0.11$, and $c = 1586.57 \pm 0.11$ pm and $\alpha = 101.068° \pm 0.001°$ $\beta = 102.869° \pm 0.001°$, and $\gamma = 105.546° \pm 0.001°$ and a calculated density of 1.572 g·cm⁻³ (Lai et al. 2002). It was prepared from the reaction of $Zn(NO_3)_2·6H_2O$ and potassium $n$-propylxanthate in aqueous solution. Crystals were isolated in pure form after recrystallization from an acetonitrile solution of the title compound.

**[Zn(S₂COPrⁱ)₂] or C₈H₁₄O₂S₄Zn**. This compound is stable up to 120°C (Barreca et al. 2005) (decomposes at 152°C [Lai et al. 2002]) and crystallizes in the triclinic structure with the lattice parameters $a = 1091.5 \pm 0.3$, $b = 1316.7 \pm 0.4$, and $c = 1039.3 \pm 0.3$ pm and $\alpha = 100.89° \pm 0.02°$, $\beta = 100.42° \pm 0.02°$, and $\gamma = 101.33° \pm 0.07°$ and experimental and calculated densities 1.58 and 1.59 g·cm⁻³, respectively (Ito 1972). To obtain it, an aqueous solution of potassium ($O$-isopropylxanthate) (15.2 mM) in $H_2O$ (22 mL) was added dropwise to a solution of $Zn(NO_3)_2$ (7.6 mM in 36 mL of $H_2O$), resulting in the precipitation of a white solid (Lai et al. 2002, Barreca et al. 2005). After 1 h stirring, vacuum filtering and drying allowed the recovery of a white powder of the title compound.

The powder of $Zn(S_2COPr^i)_2$ could also be precipitated by combining aqueous solutions of $ZnCl_2$ and potassium isopropylxanthate. Crystals were obtained by recrystallization of the powder from an ethanol solution (Ito 1972). Light exposure during preparation and storage must be avoided, since the resulting compound is slightly photosensitive (Barreca et al. 2005).

**[Zn(C₅H₃OS₂)₂] or C₁₀H₆O₂S₄Zn, bis(dithiofuroato)zinc**. This compound melts at 100°C with decomposition (Fackler et al. 1968). To obtain it, furfural was treated with $(NH_4)_2S$ and sulfur. The reaction mixture was acidified and extracted with ether. The ether layer was extracted with dilute aqueous NaOH and this solution was used to prepare $Zn(C_5H_3OS_2)_2$. The crude product gave orange-brown crystals from xylene.

**[Zn(C₆H₁₂OS₂)₂] or C₁₂H₂₄O₂S₄Zn, bis(O-ethylthioacetothioacetato)zinc**. This compound crystallizes in the monoclinic structure with the lattice parameters $a = 2276.9 \pm 0.3$, $b = 526.15 \pm 0.03$, $c = 1490.7 \pm 0.3$ pm, and $\beta = 106.33° \pm 0.01°$ and experimental and calculated densities 1.49 and 1.503 g·cm⁻³, respectively (Beckett and Hoskins 1972, 1975).

**[Zn(C$_{14}$H$_8$O$_6$S)(H$_2$O)$_2$]** or **C$_{14}$H$_{12}$O$_8$SZn.** This compound crystallizes in the monoclinic structure with the lattice parameters $a = 1331.7 \pm 0.3$, $b = 503.76 \pm 0.13$, $c = 1220.0 \pm 0.3$ pm, and $\beta = 116.295° \pm 0.004°$ and a calculated density of 1.848 g·cm$^{-3}$ (Zhuang et al. 2007). To obtain it, a mixture of Zn(OAc)$_2$·2H$_2$O (0.1 mM), 4,4'-dicarboxybiphenyl sulfone (0.05 mM), NaOH aqueous solution (0.05 mL, 0.65 M), HCl aqueous solution (0.05 mL, 1.20 M), distilled H$_2$O (5 mL), Pr$^i$OH (5 mL), and DMF (0.1 mL) was sealed in a Teflon-lined stainless vessel (25 mL) and heated at 180°C for 72 h under autogenous pressure. The vessel was then cooled slowly to room temperature. Colorless flake crystals were obtained by filtration, washed with distilled water, and dried in air.

The title compound could also be obtained by the diffusion method (Zhuang et al. 2007). A 10 mL aliquot of mixed solvent of distilled H$_2$O and methanol (1:5 volume ratio) was carefully layered on top of a 1 mL DMF solution of 4,4'-dicarboxybiphenyl sulfone (0.1 mM), to which a solution of Zn(OAc)$_2$·2H$_2$O (0.1 mM) in MeOH (2 mL) was added. Colorless flake crystals were obtained after 2 months at room temperature.

**[Zn(S$_2$COBu$^n$)$_2$]·0.75C$_6$H$_6$** or **C$_{14.5}$H$_{22.5}$O$_2$S$_4$Zn.** This compound crystallizes in the triclinic structure with the lattice parameters $a = 1288.5 \pm 0.5$, $b = 1872 \pm 1$, and $c = 856.4 \pm 0.3$ pm and $\alpha = 96.54° \pm 0.04°$, $\beta = 102.48° \pm 0.03°$, and $\gamma = 87.33° \pm 0.04°$ and a calculated density of 1.645 g·cm$^{-3}$ (Cox and Tiekink 1999).

**[Zn(Ph$_2$SOAc)$_2$] 2H$_2$O** or **C$_{16}$H$_{18}$O$_6$S$_2$Zn.** This compound crystallizes in the monoclinic structure with the lattice parameters $a = 3204.8 \pm 1.1$, $b = 531.4 \pm 0.2$, $c = 1072.5 \pm 0.3$ pm, and $\beta = 101.20° \pm 0.02°$ and experimental and calculated densities 1.62 and 1.616 g·cm$^{-3}$, respectively (Mak et al. 1984). It was prepared by reacting (phenylthio)acetic acid in aqueous ethanol with a suspension of excess ZnCO$_3$ (in aqueous ethanol) and digesting for 1 h at 70°C–90°C. The excess carbonate was removed by filtration. The filtrate, on standing at room temperature, yielded crystals, which were recrystallized from absolute ethanol as flat needles or plates.

**[Zn(C$_9$H$_9$OS)$_2$(H$_2$O)$_2$]** or **C$_{18}$H$_{22}$O$_4$S$_2$Zn.** This compound crystallizes in the monoclinic structure with the lattice parameters $a = 3817 \pm 1$, $b = 514.0 \pm 0.1$, $c = 1060.4 \pm 0.3$ pm, and $\beta = 94.30° \pm 0.01°$ and experimental and calculated densities 1.480 and 1.484 g·cm$^{-3}$, respectively (Chan et al. 1987). It was prepared by a reaction of aqueous ethanolic solutions of benzyl thioacetate with a suspension of excess ZnCO$_3$ (in aqueous EtOH) and digesting for ca. 1 h at 70°C–90°C. After removal of excess carbonate by filtration, standing at room temperature resulted in the deposition of crystals, which were recrystallized from absolute EtOH as colorless prisms.

**[Zn$_4$(OH)$_2$[C$_6$H$_4$(COO)$_2$]$_3$(Me$_2$SO)$_4$·2H$_2$O]$_n$** or **C$_{32}$H$_{40}$O$_{20}$S$_4$Zn$_4$.** This compound crystallizes in the orthorhombic structure with the lattice parameters $a = 1446.69 \pm 0.17$, $b = 1704.6 \pm 0.2$, and $c = 1802.0 \pm 0.2$ pm and a calculated density of 1.674 g·cm$^{-3}$ (Zevaco et al. 2007). To obtain it, an aqueous suspension of ZnO and 1,4-benzene dicarboxylic acid (1:1 molar ratio) was stirred under reflux for ca. 4 h. After filtration, the solution was eventually concentrated in vacuo to yield a white amorphous solid of general form [C$_6$H$_4$(COO)$_2$Zn(H$_2$O)$_x$]$_n$ ($x = 1$ or 2). The solid was suspended in hot DMSO (ca. 110°C) and stirred for ca. 4 h. The suspension was then hot filtered and allowed to stand for ca. 4 weeks at room temperature to eventually yield small prismatic single crystals of the title compound.

[$Zn_4(OH)_2(C_8H_4O_4)_3(C_2H_6OS)_4$]·$2H_2O$ or $C_{32}H_{42}O_{20}S_4Zn_4$. This compound crystallizes in the orthorhombic structure with the lattice parameters $a = 1449.88 \pm 0.02$, $b = 1716.59 \pm 0.01$, and $c = 1815.36 \pm 0.05$ pm and a calculated density of 1.671 g·cm$^{-3}$ (Wang et al. 2001b). It was obtained as follows. A mixture of 1,4-benzene dicarboxylic acid (1 mM) and NaOMe (1 mM) in 25 mL of DMSO was stirred for 30 min. To the reaction mixture, solid $Zn(ClO_4)_2$·$6H_2O$ (1 mM) was added. After stirring for 8 h at 50°C, the reaction mixture was filtered to give a colorless solution. Slow diffusion of diethyl ether, containing $Et_3N$ (0.025 mL), into the filtrate and natural evaporation of the filtrate in air yielded a large amount of colorless block crystals of the title compound.

[$Zn(H_2O)_6$]($C_{18}H_{15}O_4SO_3$)$_2$·$4H_2O$ or $C_{36}H_{50}O_{24}S_2Zn$. This compound crystallizes in the triclinic structure with the lattice parameters $a = 724.0 \pm 0.2$, $b = 859.3 \pm 0.2$, and $c = 1227.6 \pm 0.4$ pm and $\alpha = 80.788° \pm 0.004°$, $\beta = 82.971° \pm 0.003°$, and $\gamma = 81.219° \pm 0.003°$ (Wang 2009). To obtain it, formononetin (4 g) was dissolved into NaOH (50 mL, 5%). Diethyl sulfate (6 mL) was added dropwise to the solution with vigorous stirring. The mixture was stirred at room temperature for 3 h and some precipitation appeared. The precipitate was filtered and washed with water until the pH of the filtrate was 7 to obtain monoethoxyformononetin (4′-methoxy-7-ethoxylisoflavone, 3.5 g). Then, monoethoxyformononetin (2 g) was slowly added to the concentrated $H_2SO_4$ (8 mL). The mixture was stirred at room temperature for 1 h and poured into a saturated NaCl solution (60 mL), and a white precipitate formed. This precipitate was collected by filtration and washed with saturated NaCl solution until the pH value of the filtrate was 7. Finally, the precipitate was recrystallized from $H_2O$ to afford sodium monoethoxyformononetin-3′-sulfonate. This compound (1 g) was dissolved in water (10 mL), then mixed with saturated $ZnSO_4$·$7H_2O$ solution (5 mL). The title compound was obtained from water after 24 h. Its crystals were recrystallized from $H_2O$ after 2 weeks at room temperature.

[$ZnOEt_2(SC_6H_2Bu^t_3)_2$] or $C_{40}H_{68}OS_2Zn$. This compound, which became yellow upon heating above 150°C and melts at 202°C–205°C, crystallizes in the monoclinic structure with the lattice parameters $a = 2002.3 \pm 0.7$, $b = 1005.3 \pm 0.6$, $c = 4081.9 \pm 1.2$ pm, and $\beta = 91.12° \pm 0.01°$ (Power and Shoner 1990a,b). To obtain it, under anaerobic and anhydrous conditions, 2,4,6-$Bu^t_3C_6H_2SH$ (4 mM) in pentane (20 mL) was added dropwise to $Zn(CH_2SiMe_3)_2$ (2 mM) in pentane (5 mL). The solution was refluxed for 1 h, and a white precipitate was formed. The volatile products were removed under reduced pressure, and the residue was dissolved in $Et_2O$ (7 mL). Cooling slowly to −25°C resulted in the formation of colorless crystals of $ZnOEt_2(SC_6H_2Bu^t_3)_2$.

[$Zn_2\{C_6H_4(COO)_2\}_4(Me_2SO)_2(H_2O)$·$3Me_2SO$]$_n$ or $C_{42}H_{48}O_{22}S_5Zn_2$. This compound crystallizes in the orthorhombic structure with the lattice parameters $a = 1269.77 \pm 0.16$, $b = 1510.33 \pm 0.19$, and $c = 1890.3 \pm 0.2$ pm and a calculated density of 1.590 g·cm$^{-3}$ (Zevaco et al. 2007). To obtain it, ZnO was added to an aqueous solution of 1,3-benzene dicarboxylic acid (1:1 molar ratio). The resulting cloudy suspension was stirred under reflux for ca. 4 h. The remaining fine particles were then filtered, and the solution was eventually concentrated in vacuo to yield a crystalline solid of general form $[C_6H_4(COO)_2Zn(H_2O)_x]_n$, barely soluble in common solvents. The isolated polycrystalline powder was suspended in hot DMSO (ca. 120°C) and stirred until complete dissolution. This solution was then concentrated and allowed to stand for ca. 4 weeks at room temperature to eventually afford small prismatic single crystals of the title compound.

$[Zn(SC_6H_2Bu^t_3)_2(O=CMe_2)_2]$ or $C_{42}H_{70}O_2S_2Zn$. This compound as colorless crystals was obtained by the interaction of $Zn(SC_6H_2Bu^t_3)_2$ with acetone (Bochmann et al. 1991).

$[Zn_4(SPh)_8(MeOH)]$ or $C_{49}H_{44}OS_8Zn_4$. This compound crystallizes in the orthorhombic structure with the lattice parameters $a = 1652.9 \pm 0.2$, $b = 2212.9 \pm 0.3$, and $c = 1424.1 \pm 0.2$ pm and experimental and calculated densities 1.52 and 1.49 g·cm$^{-3}$, respectively (Dance 1980b). To obtain it, basic zinc carbonate (2.5 g) was added in portions less than 0.5 g to a solution of benzenethiol (5.0 g) in MeOH (80 mL) at a temperature not greater than 30°C. Within 1 h, 2.5 g of basic zinc carbonate had dissolved, but further dissolution was extreme and not attempted. After filtration, the solution was reduced in volume at 30°C to 60 mL and then maintained at 55°C for 5 days while the product crystallized. At higher temperatures, the product separates more rapidly as a microcrystalline precipitate. The colorless crystals were filtered, washed with MeOH, and air-dried. Crystals of $Zn_4(SPh)_8(MeOH)$ of 1 mm dimensions can be obtained if a solution of $Zn(SPh)_2$ in MeOH, prepared as mentioned earlier, was maintained at a temperature close to 20°C in a closed flask such that very slow crystals growth occurred over a period of 4–6 weeks. All operations were performed in an atmosphere of $N_2$.

$[Zn_4(SPh)_8(EtOH)]$ or $C_{50}H_{46}OS_8Zn_4$. This compound crystallizes in the orthorhombic structure (Dance 1980b). To obtain it, basic zinc carbonate (2.5 g) was added in portions less than 0.5 g to a solution of benzenethiol (5.0 g) in absolute EtOH (100 mL) at a temperature less than 30°C, approximately 2.0 g dissolved during 3 h. After filtration, the solution was held at 55°C in a sealed flask while the product crystallized. At reflux temperature, the product separates as a microcrystalline precipitate during 2 h. The isolated product was washed with EtOH and air-dried.

$Zn_4(SPh)_8(MeOH)$ could also be prepared if basic zinc carbonate (2.5 g) was added to a solution of benzenethiol (5.0 g) in acetone (70 mL), and the mixture refluxed for ca. 1 h after which almost complete dissolution had occurred. After filtration, EtOH (100 mL) was added and the acetone removed by distillation. The white crystalline product separated slowly at ambient temperature and rapidly at reflux; it was filtered, washed with EtOH, and vacuum dried. All operations were performed in an atmosphere of $N_2$ (Dance 1980b).

$[Zn_4(SPh)_8(Pr^nOH)]$ or $C_{51}H_{48}OS_8Zn_4$. This compound crystallizes in the orthorhombic structure (Dance 1980b). To obtain it, basic zinc carbonate (2.5 g) was added in portions less than 0.5 g to a solution of benzenethiol (5.0 g) in MeOH (110 mL) at a temperature less than 30°C followed by filtration. An equal volume of $Pr^nOH$ was added, and the MeOH was removed by distillation at less than 10°C. The resulting solution in Pr$^n$OH was warmed to cause crystallization of the product over a period of several hours and finally refluxed to complete precipitation. The microcrystalline, white precipitate was washed with Pr$^n$OH and vacuum dried.

An alternative procedure of $Zn_4(SPh)_8(Pr^nOH)$ obtained includes refluxing acetone as the solvent for reaction of basic zinc carbonate and benzenethiol. The addition of Pr$^n$OH and removal of acetone by distillation at atmospheric pressure yielded a microcrystalline product, which was washed with Pr$^n$OH and vacuum dried. All operations were performed in an atmosphere of $N_2$ (Dance 1980b).

**$[Zn_4(SPh)_8(Bu^nOH)]$** or **$C_{52}H_{50}OS_8Zn_4$**. This compound crystallizes in the orthorhombic structure (Dance 1980b). Both of the procedures described for $Zn_4(SPh)_8(Pr^nOH)$ were used for obtaining this compound, with Bu$^n$OH instead of Pr$^n$OH. The obtained product was fine, microcrystalline, white precipitate, which was vacuum dried.

**$[Zn(SC_6H_2Bu^t_3)_2\{O=CH(2\text{-}MeOC_6H_4)_2\}]$** or **$C_{52}H_{74}O_4S_2Zn$**. Treatment of a suspension of colorless $Zn(SC_6H_2Bu^t_3)_2$ in hexane at room temperature with ca. 10 equivalents of an aromatic aldehyde 2-OMeC$_6$H$_4$CHO leads to an immediate color change to yellow orange and formation of a clear solution, which the complex product precipitate (Bochmann et al. 1991). Recrystallization of the crude product from light petroleum gives $C_{52}H_{74}O_4S_2Zn$ as yellow crystals.

**$[Zn(SC_6H_2Bu^t_3)_2(C_6H_5CHO)]_2$** or **$C_{86}H_{128}O_2S_4Zn_2$**. Treatment of a suspension of colorless $Zn(SC_6H_2Bu^t_3)_2$ in hexane at room temperature with ca. 10 equivalents of an aromatic aldehyde $C_6H_5CHO$ leads to an immediate color change to yellow orange and formation of a clear solution, which the complex product precipitate (Bochmann et al. 1991). Recrystallization of the crude product from toluene gives $C_{86}H_{128}O_2S_4Zn_2$ as yellow crystals.

## 1.79  Zn–H–C–O–F–S

Some compounds are formed in this system.

**$[(p\text{-}TolCHO)Zn(SC_6F_5)_2]_n$** or **$C_{20}H_8OF_{10}S_2Zn$**. This compound melts at the temperature higher than 320°C (Müller et al. 1999). To obtain it, $Zn(SC_6F_5)_2$ (0.39 mM) was dissolved upon warming in 3 mL (8.32 mM) of $p$-tolylaldehyde. After cooling to room temperature, the solution was diluted with CH$_2$Cl$_2$ (2 mL). A dropwise addition of hexane (30 mL) produced a precipitate, which was filtered off, washed with hexane, and dried in vacuo. The title compound was prepared as colorless powder. All reactions were carried out in an atmosphere of $N_2$ using carefully dried solvents.

**$[(C_{10}H_{12}OCHO)Zn(SC_6F_5)_2]_n$** or **$C_{22}H_{12}OF_{10}S_2Zn$**. This compound melts at 144°C and crystallizes in the orthorhombic structure with the lattice parameters $a = 1559.1 \pm 0.4$, $b = 722.7 \pm 0.2$, and $c = 4035.5 \pm 0.5$ pm at 173 K and a calculated density of 1.79 g·cm$^{-3}$ (Müller et al. 1999). To obtain it, $Zn(SC_6F_5)_2$ (0.82 mM) was dissolved upon warming in 10.0 mL (67.8 mM) of mesitylaldehyde. After cooling to room temperature, the solution was diluted with CH$_2$Cl$_2$ (2 mL). A dropwise addition of hexane (30 mL) produced a precipitate, which was filtered off, washed with hexane, and dried in vacuo. The title compound was prepared as colorless powder. Its crystals were grown directly from the reaction solution by layering with hexane. All reactions were carried out in an atmosphere of $N_2$ using carefully dried solvents.

## 1.80   Zn–H–C–O–Cl–S

Some compounds are formed in this system.

**[Zn(ClO₄)₂·4(DMSO)] or C₈H₂₄O₁₂Cl₂S₄Zn.** This compound melts at 73°C–74°C (Selbin et al. 1961). It was obtained by dissolving $Zn(ClO_4)_2 \cdot 6H_2O$ in a minimum amount of acetone and adding a slight excess of DMSO. Upon cooling, white crystals were obtained, which were recrystallized from acetone.

**[Zn(ClO₄)₂·5(DMSO)] or C₁₀H₃₀O₁₃Cl₂S₅Zn.** This compound melts at 182°C–184°C (Ahrland and Björk 1974) (at 192°C–194°C [Currier and Weber 1967]). To prepare it, $Zn(ClO_4)_2 \cdot 6H_2O$ (2 mM) was dissolved in a minimum amount of MeOH. 2,2-dimethoxypropane (20 mM) was added to this solution and to a second flask containing DMSO (12 mM). Both flask were stopped and stirred at 40°C for 2.5 h. The ligand solution was added to the metal ion solution, and the resulting mixture was stirred for an additional 30–60 min. In instances where crystallization did not occur immediately, the solution was concentrated, and approximately 5 mL of anhydrous ethyl ether was added. Crystals formed immediately. They were filtered, pressed between filter paper, and dried in vacuo over $CaCl_2$ for 10–15 h (Currier and Weber 1967).

**[Zn(DMSO)₆(ClO₄)₂] or C₁₂H₃₆O₁₄Cl₂S₆Zn.** This compound melts at 165°C–175°C (Ahrland and Björk 1974). To obtain it, $Zn(ClO_4)_2 \cdot 6H_2O$ (20 mM) was dissolved in a minimum amount of acetone, and DMSO (120 mM) was added. Upon cooling to –20°C, crystals were obtained, which were recrystallized at least twice from dry acetone.

**[(2-MeSC₇H₅O)₂·(ZnCl₂)] or C₁₆H₁₆O₂Cl₂S₂Zn (C₇H₆O = trope).** Yellow crystals of this compound melt at 218°C–220°C (Asao and Kikuchi 1972).

**[Zn(4-ClC₆H₄SCH₂COO)₂]·2H₂O or C₁₆H₁₆O₆Cl₂S₂Zn, zinc di-(4-chlorphenyl-thioacetate) dihydrate.** This compound crystallizes in the triclinic structure with the lattice parameters $a = 523.50 \pm 0.05$, $b = 1072.0 \pm 0.1$, and $c = 1767.2 \pm 0.2$ pm and $\alpha = 88.90° \pm 0.01°$, $\beta = 86.62° \pm 0.01°$, and $\gamma = 86.34° \pm 0.01°$ and a calculated density of 1.684 g·cm⁻³ (Voronkov et al. 2010). It was obtained in the reaction of prota-trane 4-chlorphenylthioacetate 4-ClC₆H₄SCH₂COO₂NH(CH₂CH₂OH)₃ with $ZnCl_2$ in aqueous alcohol.

**[ZnCl₂·2Ph₂SO] or C₂₄H₂₀O₂Cl₂S₂Zn.** This compound was obtained by the next procedure. About 1 g of hydrated zinc chloride and 3 g of Ph₂SO, each dissolved in 25 mL of acetone, were mixed slowly with stirring (Gopalakrishnan and Patel 1967). The solution, on slow evaporation by passing dry air at room temperature, yielded crystals of the title compound in about a day. These were separated, washed several times with petroleum ether to remove any adhering $Ph_2SO$, and recrystallized either from acetone or ethanol.

**[Zn(Pr$^n_2$SO)$_6$](ClO$_4$)$_2$ or C$_{36}$H$_{84}$O$_{14}$Cl$_2$S$_6$Zn**. This compound melts at 74°C–76°C (Currier and Weber 1967). It was prepared in the same way as $Zn(ClO_4)_2 \cdot 5(DMSO)$ was prepared using $Pr^n_2SO$ instead of DMSO.

**[Zn(PhMeSO)$_6$](ClO$_4$)$_2$ or C$_{42}$H$_{48}$O$_{14}$Cl$_2$S$_6$Zn**. This compound melts at 174.5°C–176°C (Currier and Weber 1967). To obtain it, $Zn(ClO_4)_2 \cdot 6H_2O$ (2 mM) and PhMeSO (12 mM) were separately dissolved in a minimum amount of acetone and stirred at 30°C for 30 min. The solution of PhMeSO was added to the metal ion solution and the resulting mixture was stirred at 30°C for 30 min. The solvent was evaporated by means of aspirator until the solution became cloudy. Upon the addition of approximately 5 mL of dry ethyl ether, fine crystals formed. These were filtered, pressed between filter paper, and dried in vacuo over $CaCl_2$ for 10–15 h.

**[Zn(Ph$_2$SO)$_6$](ClO$_4$)$_2$ or C$_{72}$H$_{60}$O$_{14}$Cl$_2$S$_6$Zn**. This compound melts at 176°C–178°C. It was prepared in the same way as [Zn(PhMeSO)$_6$](ClO$_4$)$_2$ was prepared using $Ph_2SO$ instead of PhMeSO (Currier and Weber 1967, Gopalakrishnan and Patel 1967).

## 1.81   Zn–H–C–Se–S

**[Zn(SC$_6$H$_2$Bu$^t_3$–2,4,6)$_2$(C$_4$H$_8$S)] or C$_{40}$H$_{66}$Se$_2$SZn**. This quinary compound, which melts at 210°C, is formed in this system (Bochmann et al. 1994). To obtain this compound, to a suspension of $Zn(SeC_6H_2Bu^t_3–2,4,6)_2$ (0.16 mM) in 10 mL of petroleum ether was added 1.13 mM of tetrahydrothiophene (C$_4$H$_8$S). The starting material dissolved within a few second, followed by the precipitation of the white solid. The mixture was stirred for 1 h, and the precipitate was filtered off and recrystallized from hot toluene with a few drops of tetrahydrothiophene added to give colorless crystals of $C_{40}H_{66}Se_2SZn$.

## 1.82   Zn–H–Ge–N–S

**[Zn$_2$Ge$_4$S$_{10}${Me(CH$_2$)$_{15}$NMe$_3$}$_2$] or C$_{38}$H$_{84}$Ge$_4$N$_2$S$_{10}$Zn$_2$**. This multinary compound is formed in this system (MacLachlan et al. 1999). To obtain this compound, a solution of $[N(CH_3)_4]_4Ge_4S_{10}$ and $[Me(CH_2)_{15}NMe_3]Br$ in formamide was formed at 120°C and cooled to 80°C. A solution of $ZnCl_2$ in formamide was added dropwise over 1 min to the solution of $[N(CH_3)_4]_4Ge_4S_{10}/[Me(CH_2)_{15}NMe_3]Br$, giving an immediate brown precipitate. The mixture was aged at 80°C for 16 h, filtered, washed with copious amounts of formamide (80°C) and $H_2O$ (80°C), and dried under ambient conditions.

## 1.83   Zn–H–N–O–S

**Zn(NH$_4$)$_2$(SO$_4$)$_2 \cdot$6H$_2$O**. This quinary compound is formed in this system, which crystallizes in the monoclinic structure with the lattice parameters $a = 927.9 \pm 1.2$,

$b = 1256.8 \pm 1.5$, $c = 625.3 \pm 0.5$ pm, and $\beta = 106°49' \pm 6'$ (Montgomery and Lingafelter 1964). Its crystals in a rodlike habit elongated in the $c$ axis direction were obtained by quick cooling of a water solution of the title compound.

**ZnS–NH$_4$NO$_3$.** It was determined that $ZnSO_4$, $ZnO$, $Zn$, and $Zn(NH_3)_4(NO_3)_2$ are formed at the interaction of $ZnS$ and $NH_4NO_3$ at 190°C–380°C (Hîncu 1971, Hîncu and Golgoţiu 1972).

## 1.84   Zn–H–V–O–S

**Zn$_7$(OH)$_3$(SO$_4$)(VO$_4$)$_3$.** This quinary compound that crystallizes in the hexagonal structure with the lattice parameters $a = 1281.30 \pm 0.06$ and $c = 514.25 \pm 0.02$ pm is formed in this system (Kato et al. 1998). The title compound was obtained as by-product of a hydrothermal synthesis, in which a suspension of $VO(OH)_2$ powder in $ZnSO_4$ aqueous solution was sealed in a Pyrex glass ampoule and was treated in an autoclave at 280°C for 48 h. The crystals grew on the wall of the ampoule.

## 1.85   Zn–K–Rb–Sn–S

**ZnS–K$_2$S–Rb$_2$S–SnS$_2$.** The phase diagram is not constructed. $K_5RbZn_4Sn_5S_{17}$ and $KRb_5Zn_4Sn_5S_{17}$ quinary compounds are formed in this system. $K_5RbZn_4Sn_5S_{17}$ crystallizes in the tetragonal structure with the lattice parameters $a = 1376.87 \pm 0.10$ and $c = 977.67 \pm 0.06$ pm and energy gap $E_g \approx 2.87$ eV (Manos et al. 2005). $KRb_5Zn_4Sn_5S_{17}$ crystallizes also in the tetragonal structure with the lattice parameters $a = 1383.54 \pm 0.10$ and $c = 988.93 \pm 0.09$ pm.

To obtain $K_5RbZn_4Sn_5S_{17}$, a mixture of Sn, Zn, $K_2S$, $Rb_2S$, and S was sealed under vacuum in a silica tube, heated ($\approx 40$°C/h) to 400°C for 92–96 h, and then cooled to room temperature at a rate of 6°C/h. A typical ion-exchange experimental route for the preparation of $KRb_5Zn_4Sn_5S_{17}$ is as follows (Manos et al. 2005). An excess of solid RbI was added to a suspension of $K_6Zn_4Sn_5S_{17}$ in water. The mixture was stirred for $\approx 12$ h. Then the yellowish-white crystalline material was isolated by filtration, washed several times with water, acetone, and ether (in this order), and dried in air.

## 1.86   Zn–K–Cs–Sn–S

**ZnS–K$_2$S–Cs$_2$S–SnS$_2$.** The phase diagram is not constructed. $K_5CsZn_4Sn_5S_{17}$ quinary compound is formed in this system. It crystallizes in the tetragonal structure with the lattice parameters $a = 1376.84 \pm 0.11$ and $c = 956.30 \pm 0.09$ pm and energy gap $E_g \approx 2.87$ eV (Manos et al. 2005). To obtain this compound, a mixture of Sn, Zn, $K_2S$, $Cs_2S$, and S was sealed under vacuum in a silica tube, heated ($\approx 40$°C/h) to 400°C for 92–96 h, and then cooled to room temperature at a rate of 6°C/h.

## 1.87   Zn–K–Cu–Ba–S

**ZnS–K$_2$S–Cu$_2$S–BaS.** The phase diagram is not constructed. $KBaCu_3ZnS_4$ quinary compound is formed in this system. It crystallizes in the tetragonal structure with the lattice parameters $a = 390.3 \pm 0.4$ and $c = 1296.4 \pm 0.2$ pm (Mouallem-Bahout et al. 2001). This compound was obtained by sulfidation of $Cu_2O+ZnO+BaCO_3+KHCO_3$ mixture in the necessary stoichiometry (Cu/Zn = 3:1) with 50% excess of $KHCO_3$ at $900°C \pm 50°C$ in the $CS_2$ flow.

## 1.88   Zn–Cu–Ag–Hg–Tl–As–Sb–S

**(Cu, Ag, Tl)(Zn, Hg)(As, Sb)S₃**. This (mineral routierite) multinary compound, which crystallizes in the tetragonal structure with the lattice parameters $a = 997.7 \pm 0.2$ and $c = 1229.0 \pm 0.3$ pm and a calculated density of 5.83 g·cm⁻³, is formed in this system (Johan et al. 1974).

## 1.89   Zn–Cu–Ag–Sn–S

**ZnS–Cu₂S–Ag₂S–SnS₂**. The phase diagram is not constructed. CuAgZnSnS₄ quinary compound is formed in this system (Tsuji et al 2010). It could be obtained if an aqueous solution (150 mL) of $Zn(NO_3)_2·6H_2O$ with 10%–15% excess amounts (about 0.029 mol/L) and $SnCl_4·5H_2O$ (0.025 mol/L) was purged with $N_2$ with the next addition of CuCl and AgNO₃ into the mixture solution. The energy gap of CuAgZnSnS₄ is equal to 1.4 eV.

## 1.90   Zn–Cu–Ca–O–S

**ZnS–Cu₂S–CaO**. This system was investigated according to five vertical sections from the CaO corner (Toguzov et al. 1979). Some of these sections are represented in Figure 1.1. Liquidus surface of the ZnS–Na₂S–ZnO quasiternary system includes five fields of primary crystallization (Figure 1.2): ZnS, CaO, CaS, Cu₂S, and zinccalcium oxysulfide, which contains 25.7 at.% (44.65 mass.%) Zn; 21.65 at.% (23.05 mass.%) Ca; 23.25 at.% (19.8 mass.%) S; and 29.4 at.% (12.5 mass.%) O. The fields of CaO and CaS primary crystallization occupy the most part of this liquidus surface.

## 1.91   Zn–Cu–Sr–O–S

**ZnS–Cu₂S–SrO**. The Cu₂Sr₂ZnO₂S₂ quinary compound is formed in this system. It crystallizes in the tetragonal structure with the lattice parameters $a = 400.79$ and $c = 1771.96$ pm (Zhu and Hor 1997). This compound was obtained by the heating of mixtures from stoichiometric quantities of CuO, SrS, and Zn at 920°C for 24 h.

## 1.92   Zn–Cu–Hg–Tl–As–S

**[Hg₀.₇₆(Cu, Zn)₀.₂₄]₁₂Tl₀.₉₆(AsS₃)₈**. This (mineral galkhaite) multinary compound is formed in this system. It crystallizes in the cubic structure with the lattice parameter $a = 1037.9 \pm 0.4$ pm and a calculated density of 5.31 g·cm⁻³ (Divjaković and Nowacki 1975).

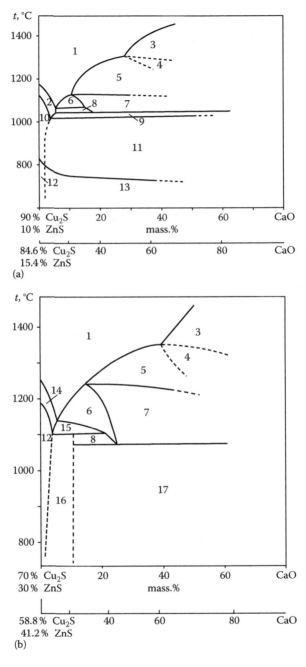

**FIGURE 1.1**  Polythermal sections of the ZnS–Cu₂S–CaO quasiternary system: (a) (84.6 mol.% Cu₂S + 15.4 mol.% ZnS)—CaO; (b) (58.8 mol.% Cu₂S + 41.2 mol.% ZnS)—CaO.

*(Continued)*

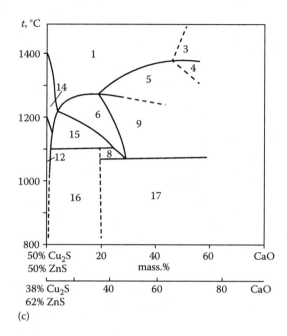

**FIGURE 1.1 (Continued)** Polythermal sections of the ZnS–Cu₂S–CaO quasiternary system: (c) (38.0 mol.% Cu₂S + 62.0 mol.% ZnS)—CaO [1, L; 2, L + (Cu₂S); 3, L + CaO; 4, L + CaO + CaS; 5, L + CaS; 6, L + zinccalcium oxysulfide; 7, L + CaS + zinccalcium oxysulfide; 8, L + (Cu₂S) + zinccalcium oxysulfide; 9, L + (Cu₂S) + CaS(CaO); 10, (Cu₂S); 11, (Cu₂S) + CaS(CaO); 12, (Cu₂S) + ZnS; 13, (Cu₂S) + CaS (CaO + zinccalcium oxysulfide); 14, L + ZnS; 15, L + ZnS + zinccalcium oxysulfide; 16, CaS(CaO) + ZnS + zinccalcium oxysulfide; 17, CaO(CaS) + (Cu₂S) + zinccalcium oxysulfide]. (From Toguzov, M.Z. et al., *Zhurn. neorgan. khimii*, 24(12), 3354, 1979.)

## 1.93 Zn–Cu–In–Mn–S

**ZnS–CuInS₂–MnS**. Isothermal sections of this quasiternary system at 600°C, 850°C, and 1050°C are shown in Figure 1.3 (Sombuthawee and Hummel 1979). The solid solution fields increase with increasing temperature. The limits between the (sphalerite + chalcopyrite) and (wurtzite + chalcopyrite) two-phase fields were not determined. This system was investigated by the XRD.

## 1.94 Zn–Cu–Si–Sn–S

**Cu₂Zn(Si$_x$Sn$_{1-x}$)S₄ (0 ≤ x ≤ 0.5 and 0.8 ≤ x ≤ 1)**. This solid solutions are formed in this system (Hamdi et al. 2014). To obtain them, Cu, Zn, Si, Sn, and S were weighted in stoichiometric amounts, sealed under vacuum in silica tubes, and heated at a temperature ranging from 750°C to 900°C versus the target composition (the higher the Si content the higher the temperature) for 1 week. The as-obtained samples were then ground in an agate mortar and annealed at the synthesis temperature for 96 h

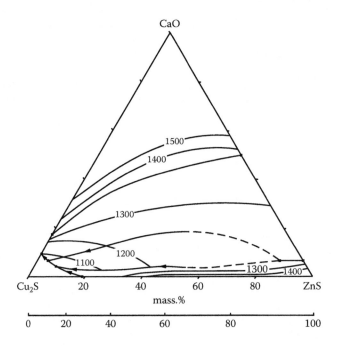

**FIGURE 1.2** Liquidus surface of the ZnS–Cu$_2$S–CaO quasiternary system. (From Toguzov, M.Z. et al., *Zhurn. neorgan. khimii*, 24(12), 3354, 1979.)

and subsequently quenched in an iced bath. Materials crystallize with the kesterite structure type for $x \leq 0.5$, but with the wurtz-stannite (enargite) structure type for $x \geq 0.8$. In between, a miscibility gap exists, and the two solid solutions Cu$_2$ZnSi$_{0.5}$Sn$_{0.5}$S$_4$ and Cu$_2$ZnSi$_{0.8}$Sn$_{0.2}$S$_4$ coexist. The optical band gap increases continuously with the Si content in the whole series.

## 1.95   Zn–Cu–Ge–Fe–S

**Cu$_2$(Fe,Zn)GeS$_4$.** This (mineral briartite) quinary compound, which crystallizes in the tetragonal structure with the lattice parameters $a = 532 \pm 1$ and $c = 1051 \pm 1$ pm, is formed in this system (Francotte et al. 1965).

## 1.96   Zn–Cu–Sn–Se–S

The calculated results revealed that the lattice constants variation of Cu$_2$ZnSn(Se$_x$S$_{1-x}$)$_4$ solid solutions alloys obey Vegard's law (Zhao et al. 2015). The wurtzite-derived alloys have better alloy solubility and component uniform compared with zinc-blende-derived alloys. In the whole range of $x$, the calculated lattice constants and band gaps are nearly linear varying with Se compositions.

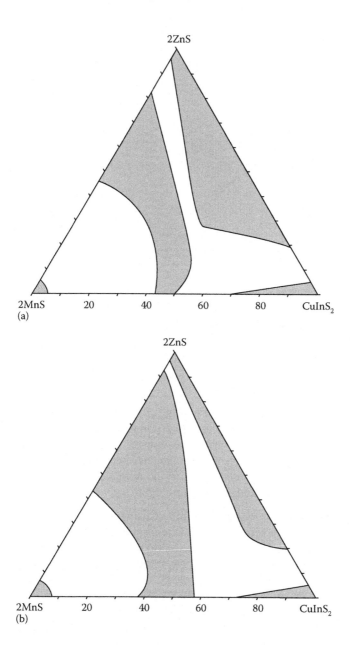

**FIGURE 1.3** Isothermal sections of the $2ZnS$–$CuInS_2$–$2MnS$ quasiternary system at (a) 600°C and (b) 850°C.                                                                        (*Continued*)

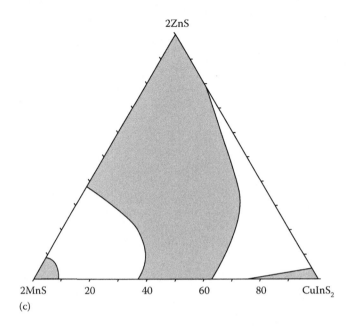

(c)

**FIGURE 1.3** (*Continued*)   Isothermal sections of the 2ZnS–CuInS$_2$–2MnS quasiternary system at (c) 1050°C. (From Sombuthawee, C. and Hummel, F.A., *J. Solid State Chem.*, 30(1), 125, 1979.)

## 1.97   Zn–Cu–Sn–Fe–S

Two minerals, kesterite $Cu_{1.98}Zn_{0.73}Fe_{0.29}Sn_{0.99}S_4$ and stannite $Cu_{1.99}Zn_{0.18}Fe_{0.81}SnS_4$, containing Cd impurity, exist in this quinary system (Hall et al. 1978). The kesterite crystallizes in the tetragonal structure with the lattice parameters $a = 542.7 \pm 0.1$ and $c = 1087.1 \pm 0.5$ pm. Its structure is characterized by a cell that is pseudocubic ($2a \approx c$). The stannite also crystallizes in the tetragonal structure with the lattice parameters $a = 544.9 \pm 0.2$ and $c = 1075.7 \pm 0.3$ pm.

**ZnS–Cu$_2$FeSnS$_4$.** Phase diagram of this quasibinary system is shown in Figure 1.4 (Moh 1975). A complete solid solutions form at elevated temperatures between $\beta$-Cu$_2$SnFeS$_4$ and $\alpha$-ZnS, but only limited solid solutions exist below $\alpha$-$\beta$-Cu$_2$SnFeS$_4$ transformation on both sides. $\alpha$-Cu$_2$SnFeS$_4$ inverts at 706°C ± 5°C into $\beta$-Cu$_2$SnFeS$_4$, but when in stable coexistence with $\alpha$-ZnS, this inversion temperature is lowered to 614°C ± 7°C. At this temperature, $\alpha$-Cu$_2$SnFeS$_4$ takes 10 mass.% $\alpha$-ZnS and $\beta$-Cu$_2$SnFeS$_4$ 13 mass.% $\alpha$-ZnS in the solid solution, whereas the solubility of Cu$_2$SnFeS$_4$ in $\alpha$-ZnS reaches ~3 mass.% and diminishes steadily with lower temperatures, for example 1.5 mass.% at 400°C, but $\beta$-Cu$_2$SnFeS$_4$ narrows only very little with decreasing temperature, at 260°C, it still contains as much as 9 mass.% $\alpha$-ZnS.

Complete $\alpha$-Cu$_2$SnFeS$_4$–$\alpha$-ZnS solid solutions become stable at ~870°C. The peritectic temperature of this system is 1074°C ± 3°C. At this temperature liquid dissolves 8.5 mass.% $\alpha$-ZnS, while the stable $\alpha$-ZnS mixed crystal has a composition of

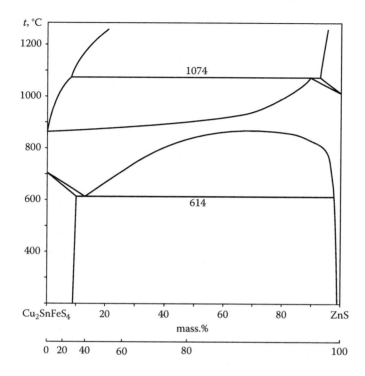

**FIGURE 1.4**   Phase diagram of the ZnS–Cu$_2$SnFeS$_4$ quasibinary system. (From Moh, G.H., *Chem. Erde*, 34(1), 1, 1975.)

91 mass.% α-ZnS and 9 mass.% β-Cu$_2$SnFeS$_4$. Also at 1074°C, β-ZnS takes 5 mass.% β-Cu$_2$SnFeS$_4$ in the solid solution, but the solubility decreases on further heating (Moh 1975).

**ZnS–CuFeS$_2$–Cu$_2$SnFeS$_4$.** The isothermal section of this quasiternary system at 600°C is shown in Figure 1.5 (Moh 1975).

## 1.98   Zn–Cu–Pb–Fe–S

**ZnS–Cu$_{2-x}$S–PbS–FeS.** A part of this quinary system was investigated according to the isoconcentration sections with 5, 10, and 20 mass.% ZnS, and each of these sections were studied according to 6–8 radial sections between ZnS–FeS quasibinary system and ZnS–Cu$_{2-x}$S–PbS quasiternary system (Figure 1.6) (Toguzov et al. 1980).

Liquidus surface of an isoconcentration section with 5 mass.% ZnS (Figure 1.6a) includes the fields of ZnS, PbS, FeS, and Cu$_5$FeS$_4$ primary crystallization. The field of ZnS primary crystallization occupies the most part of this liquidus surface. Temperatures of primary crystallization are minimum (760°C) near the composition of PbS–Cu$_{2-x}$S quasibinary system at the ratio PbS/Cu$_{2-x}$S = 1:1 and maximum (1180°C) near the FeS corner.

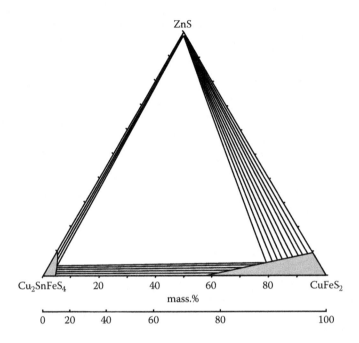

**FIGURE 1.5** The isothermal section of the ZnS–CuFeS$_2$–Cu$_2$SnFeS$_4$ quasiternary system at 600°C. (From Moh, G.H., *Chem. Erde*, 34(1), 1, 1975.)

Liquidus surfaces of isoconcentration sections with 10 and 20 mass.% ZnS (Figure 1.6b,c) include only the fields of ZnS and Cu$_5$FeS$_4$ primary crystallizations. The fields of PbS and FeS primary crystallizations disappear, and the field of Cu$_5$FeS$_4$ decreases noticeably at the ZnS concentration decreasing. Increasing of ZnS and FeS concentrations in this system leads to increasing of melting temperature. Just as in the case of isoconcentration section with 5 mass.% ZnS, minimum melting temperatures (900°C and 1050°C, respectively) was found near the PbS–Cu$_{2-x}$S quasibinary system and maximum melting temperatures (1300°C and 1370°C, respectively) are situated near FeS corner.

This system was investigated by the DTA and metallography (Toguzov et al. 1980).

## 1.99 Zn–Cu–O–Se–S

**ZnS–CuSeO$_3$.** The phase diagram is not constructed. Three exothermic (380°C, 520°C and 730°C) and one endothermic (560°C) effects exist on the thermograms of ZnS + CuSeO$_3$ alloy with an equimolar ratio of starting components (Angelova et al. 1975). Using the results of DTA and XRD, they determined that the next reaction ZnS + 2 CuSeO$_3$ = ZnSO$_4$ + Cu$_2$Se + SeO$_2$ takes place at the heating of such mixture up to 520°C. At temperatures above 560°C, ZnS can interact with ZnSO$_4$ or dissociate forming ZnO and eliminating SO$_2$. Other reactions can also run simultaneously and successively in this system.

This system was investigated by the DTA, XRD, and using thermodynamic simulations (Angelova et al. 1975).

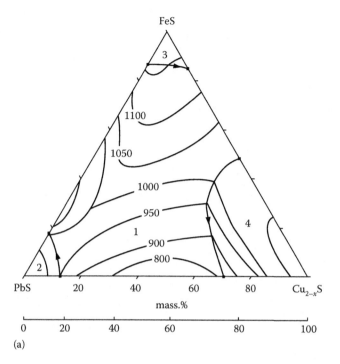

(a)

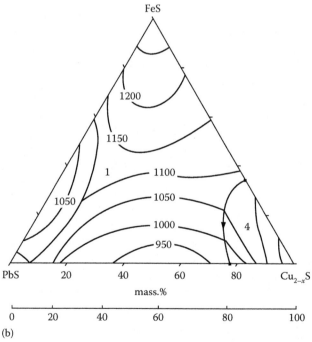

(b)

**FIGURE 1.6** Liquidus surfaces of isoconcentration sections of the ZnS–Cu$_{2-x}$S–PbS–FeS multicomponent system: (a) 5; (b) 10. *(Continued)*

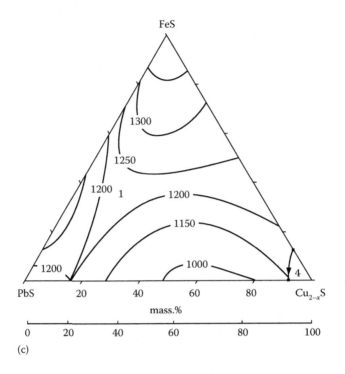

**FIGURE 1.6 (*Continued*)** Liquidus surfaces of isoconcentration sections of the ZnS–Cu$_{2-x}$S–PbS–FeS multicomponent system: (c) 20 mass.% ZnS (1, 2, 3, and 4, the fields of ZnS, PbS, FeS, and Cu$_5$FeS$_4$ primary crystallization, respectively). (From Toguzov, M.Z. et al., *Zhurn. neorgan. khimii*, 25(8), 2237, 1980.)

## 1.100  Zn–Cu–O–Fe–S

**ZnS–Cu$_2$S–FeO**. Liquidus surface of this quasiternary system is shown in Figure 1.7 (Toguzov et al. 1982). It belongs to the systems with four-phase peritectic interaction, which is preceded by one three-phase eutectic and one three-phase peritectic equilibria. New phases are formed in the ZnS–Cu$_2$S–FeO system. The monovariant L$_2$ ↔ L$_1$ + FeO monotectic reaction takes place within the interval of 1345°C–1250°C. Oxidation part consists mainly of FeO with sulfide liquid inclusion. Crystallization of FeO, Cu$_2$S, ZnS, and new phases (bornite solid solution and metallic Cu) takes place at the cooling of melts. Formation of bornite and Cu is a result of the next reaction: 9Cu$_2$S + 2FeO = 2Cu$_5$FeS$_4$ + 8Cu + SO$_2$. Pyrrotine is not formed in the ZnS–Cu$_2$S–FeO system.

This system was investigated by the DTA, metallography, and local XRD (Toguzov et al. 1982).

**ZnS–Cu$_2$S–FeO–FeS**. This system was investigated according to three isoconcentration sections with 5, 10, and 20 mass.% ZnS (Figure 1.8) (Toguzov et al. 1982). All these sections are characterized by the complex phase equilibria and a large immiscibility region.

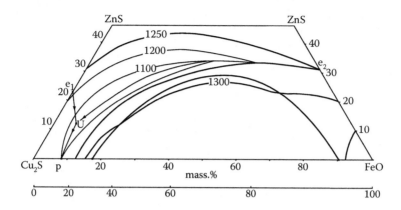

**FIGURE 1.7** A part of the liquidus surface of the ZnS–Cu₂S–FeO quasiternary system. (From Toguzov, M.Z. et al., *Sul'fid. rasplavy tiazh. met.*, M., 122, 1982.)

The liquidus surface of the isoconcentration section with 5 mass.% ZnS (Figure 1.8a) includes the immiscibility region and the fields of ZnS, FeO, FeS, and bornite solid solution primary crystallization. The minimum temperature of primary crystallization corresponds to the composition with 30 mass.% Cu₂S, 25 mass.% FeO, and 45 mass.% FeS. Increasing of ZnS content up to 10 mass.% leads to the small decrease of the immiscibility region, disappearance of the field of pyrrotine primary crystallization, decrease of the field of bornite solid solution primary crystallization, and increase of the field of ZnS primary crystallization (Figure 1.8b).

Isoconcentration section with 20 mass.% ZnS is characterized by the increase of crystallization temperatures, considerable reduction of immiscibility region, further increase of the field of ZnS primary crystallization, and decrease of the field of bornite solid solution primary crystallization (Figure 1.8c).

Minimum crystallization temperature increases up to 1025°C and shifts to the Cu₂S corner with increasing ZnS contents. The FeO + FeS + bornite solid solution ternary eutectic crystallizes in the all isoconcentration sections together with binary eutectics. Complete crystallization in this system takes place at 850°C.

The phases figurative points of which are situated outside of the ZnS–Cu₂S–FeO–FeS system are formed in this system. Therefore, this system must be considered as a part of the Zn–Cu–Fe–O–S multinary system.

This system was investigated by the DTA, metallography, and local XRD (Toguzov et al. 1982).

## 1.101  Zn–Cu–Mn–Fe–S

**$Zn_{0.33}Cu_{0.01}Mn_{0.04}Fe_{0.61}S$.** This (mineral rudashevskyite) quinary compound is formed in this system. It crystallizes in the cubic structure with the lattice parameter $a = 542.6 \pm 0.2$ pm and an experimental density of 3.79 g·cm⁻³ (Britvin et al. 2008). The chemical composition of this mineral varies widely, between 55 and 68 mol.% FeS.

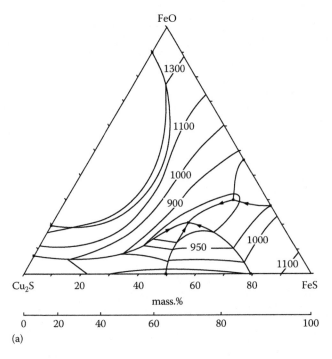

(a)

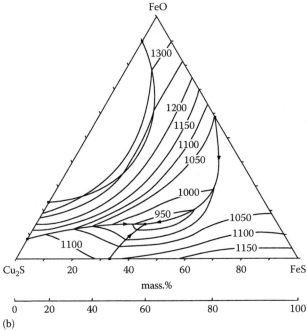

(b)

**FIGURE 1.8** Liquidus surfaces of isoconcentration sections of the ZnS–Cu$_2$S–FeO–FeS multicomponent system: (a) 5; (b), 10.                    (*Continued*)

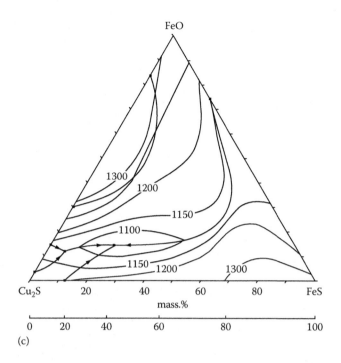

**FIGURE 1.8 (*Continued*)** Liquidus surfaces of isoconcentration sections of the ZnS–Cu₂S–FeO–FeS multicomponent system: (c) 20 mass.% ZnS. (From Toguzov, M.Z. et al., *Sul'fid. rasplavy tiazh. met.*, M., 122, 1982.)

## 1.102  Zn–Mg–Mn–Fe–S

$Zn_{0.32}Mg_{0.04}Mn_{0.16}Fe_{0.46}S$. This (mineral buseckite) quinary compound is formed in this system. It crystallizes in the hexagonal structure with the lattice parameters $a = 383.57$ and $c = 630.02$ pm and a calculated density of 3.697 g·cm⁻³ (Ma 2011, Ma et al. 2012).

## 1.103  Zn–Ca–O–Fe–S

**ZnS–CaO–FeS**. This system was investigated according to six vertical sections from the CaO corner (Toguzov et al. 1981). The phases figurative points of which are situated outside of the ZnS–CaO–FeS system are formed in this system. Therefore, this system must be considered as a part of the Zn–Cu–Fe–O–S multinary system.

As can be seen from Figure 1.9 together with the fields of ZnS, CaO, and FeS primary crystallization, the field of CaS primary crystallization exists on the liquidus surface of the ZnS–CaO–FeS system. Calcium sulfide is formed as the result of CaO interaction with ZnS and FeS. The composition and temperature of ternary eutectic are 84 mass.% FeS + 7.5 mass.% (Zn, Fe)S + 8.5 mass.% CaS and 1065°C.

This system was investigated by the DTA, metallography, and local XRD (Toguzov et al. 1981).

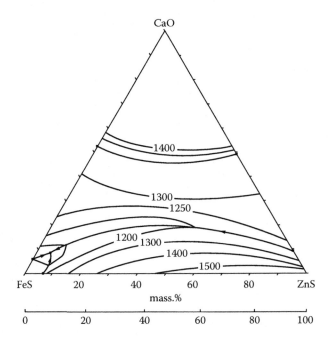

**FIGURE 1.9** Liquidus surface of the ZnS–CaO–FeS system. (From Toguzov, M.Z. et al., *Zhurn. neorgan. khimii*, 26(1), 238, 1981.)

## 1.104 Zn–Sr–Ge–O–S

**Sr₂ZnGe₂OS₆.** This quinary compound is formed in this system (Teske 1985). It crystallizes in the tetragonal structure with the lattice parameters $a = 940.54 \pm 0.07$ and $c = 617.18 \pm 0.07$ pm and experimental and calculated densities 3.61 and 3.62 g·cm⁻³, respectively. This compound was obtained by the interaction of ZnO, SrS, and GeS₂.

## 1.105 Zn–Ba–Ge–O–S

**Ba₂ZnGe₂OS₆.** This quinary compound is formed in this system (Teske 1980). It crystallizes in the tetragonal structure with the lattice parameters $a = 963.59 \pm 0.22$ and $c = 645.06 \pm 0.25$ pm and experimental and calculated densities 3.74 and 3.85 g·cm⁻³, respectively. This compound could be obtained by the interaction according to the two next reactions: $2BaS + ZnS + 2GeS_2 + 1/2O_2 = Ba_2ZnGe_2OS_6$ and $2BaS + ZnO + 2GeS_2 = Ba_2ZnGe_2OS_6$.

## 1.106 Zn–Cd–C–N–S

**ZnCd(SCN)₄.** This quinary compound is formed in this system. It crystallizes in the tetragonal structure with the lattice parameters $a = 1112.61$ and $c = 438.51$ pm (Wang et al. 2003a) ($a = 1110.89 \pm 0.02$ and $c = 444.36 \pm 0.02$ pm and a calculated density of

2.48 g·cm$^{-3}$ [Brodersen et al. 1985]; $a = 1113.5 \pm 0.2$ and $c = 437.6 \pm 0.1$ pm and a calculated density of 2.510 g·cm$^{-3}$ [Tian et al. 1999, Wan et al. 1999]; $a = 1111.41$ and $c = 437.94$ pm [Wang et al. 2003b]; $a = 1110$ and $c = 438$ pm [Joseph et al. 2007]). The title compound was obtained by the interaction of $Zn(SCN)_2$ and $Cd(SCN)_2$ (1:1 molar ratio), or by the interaction of $M_2Cd(SCN)_4$ and $ZnX_2$ (M = NH$_4$, Na, K; X = NO$_3$, Cl, OAc) (1:1 molar ratio), or by the interaction of KSCN, CdCl$_2$, and ZnCl$_2$ (4:1:1 molar ratio) (Tian et al. 1999, Wan et al. 1999, Wang et al. 2003a,b, Joseph et al. 2007). Small single crystals can be crystallized by spontaneous nucleation in a mixing solution made of H$_4$SCN, ZnX$_2$, CdX$_2$ as raw materials with stoichiometric ratios in deionized water. Using the small single crystals obtained as seeds, large single crystals can be obtained by solvent evaporation or temperature lowering of such saturated solutions.

Large single crystals could be grown from the NH$_4$Cl–NH$_4$SCN–H$_2$O mixed solvent by using the solvent evaporation method (Jiang et al. 2001).

Interaction of BaZn(NCS)$_4$·7H$_2$O with CdSO$_4$ in EtOH yields also the title compound after filtration and evaporation of the solvent (Brodersen et al. 1985).

## 1.107   Zn–Cd–Mn–Fe–S

**ZnS–CdS–MnS–FeS.** Cell edges of the (Cd + Mn + Fe)-bearing sphalerites in this system should be linear functions of their compositions, and the following relation should hold: $a(A) = 5.4093 + 0.000456x + 0.00424y + 0.00202z$, where $x$, $y$, and $z$ are the FeS, CdS, and MnS contents in mol.% (Skinner 1961).

It is apparent that functions defining $a$ and $c$ of the solid solutions with the wurtzite structure can be derived by simple addition from the quasibinary system, so $a(A) = 3.8230 + 0.000490x + 0.003124y + 0.001628z$ and $c(A) = 6.2565 + 0.000886x + 0.00455y + 0.002089z$, where $x$, $y$, and $z$ are the FeS, CdS, and MnS content, respectively, in mol.% (Skinner and Bethke 1961). The assumption of additivity for $c$ can only be considered approximate. The $a$ versus composition function is a more suitable one to use in determining the compositions of wurtzite solid solutions from their unit-cell edges.

## 1.108   Zn–Hg–C–N–S

**ZnHg(SCN)$_4$.** This quinary compound is formed in this system. It crystallizes in the tetragonal structure with the lattice parameters $a = 1110.37$ and $c = 445.30$ pm (Wang et al. 2003a) ($a = 1109.12 \pm 0.04$ and $c = 444.14 \pm 0.04$ pm and a calculated density of 3.029 g·cm$^{-3}$ [Xu et al. 1999]). The title compound was obtained by the interaction of $Zn(SCN)_2$ and $Hg(SCN)_2$ (1:1 molar ratio), or by the interaction of $M_2Hg(SCN)_4$ and $ZnX_2$ (M = NH$_4$, Na, K; X = NO$_3$, Cl, OAc) (1:1 molar ratio), or by the interaction of KSCN, HgCl$_2$, and ZnCl$_2$ (4:1:1 molar ratio) (Xu et al. 1999, Wang et al. 2003a). Small single crystals can be crystallized by spontaneous nucleation in a mixing solution made of NH$_4$SCN, ZnX$_2$, HgX$_2$ as raw materials with stoichiometric ratios in deionized water. Using small single crystals obtained as seeds, large single crystals can be obtained by solvent evaporation or temperature lowering of such saturated solutions.

## 1.109   Zn–Ga–La–O–S

**ZnLa$_2$Ga$_2$OS$_6$.** This quinary compound is formed in this system (Teske 1985). It crystallizes in the tetragonal structure with the lattice parameters $a = 936.53 \pm 0.15$ and $c = 609.28 \pm 0.10$ pm and experimental and calculated densities 4.31 and 4.33 g·cm$^{-3}$, respectively. This compound was obtained by the interaction of ZnO, La$_2$S$_3$, and Ga$_2$S$_3$.

## 1.110   Zn–Tl–O–Cl–S

**Tl(ZnSO$_4$)Cl.** This quinary compound is formed in this system. It crystallizes in the monoclinic structure with the lattice parameters $a = 727.8$, $b = 955.1$, $c = 809.2$ pm, and $\beta = 93.97°$ (Bosson 1973).

## 1.111   Zn–C–Pb–N–S

**ZnPb(SCN)$_4$.** This quinary compound is formed in this system (Brodersen et al. 1987). It crystallizes in the orthorhombic structure with the lattice parameters $a = 916.9 \pm 0.4$, $b = 606.1 \pm 0.3$, and $c = 2096.8 \pm 2.0$ pm and experimental and calculated densities 2.97 and 2.95 g·cm$^{-3}$, respectively. The title compound was obtained from the binary Zn(SCN)$_2$ and Pb(NCS)$_2$ in stoichiometric ratio with traces of H$_2$O in an ultrasound bath for about 6 h at room temperature. It could also be obtained by dry milling of Zn(SCN)$_2$ and Pb(SCN)$_2$ or by shaking of Pb(SCN)$_2$ in a concentrated solution of Zn(SCN)$_2$ in H$_2$O (Dost et al. 1968).

## 1.112   Zn–Pb–Sb–O–S

**ZnS–PbO–Sb$_2$O$_3$.** Bulk glasses with starting compositions of 70Sb$_2$O$_3$–(30−$x$) PbO–$x$ZnS ($x = 0.5$–3.0 mol.%) were prepared in this system (Nouadji et al. 2013). The temperature of glass transition ($T_g$) are within the interval from 261°C to 272°C.

## REFERENCES

Aazam E.S., Ng S.W., Tiekink E.R.T. (Dimethyl sulfoxide-κ$O$){4,4′,6,6′-tetra-*tert*-butyl-2,2′-[1,2-dicyanoethene-1,2-diylbis(nitrilomethylidyne)]diphenolato-κ$^4$$O,N,N′,O′$}zinc(II) acetonitrile monosolvate, *Acta Crystallogr.*, E67(3), m314–m315 (2011).

Ablov A.V., Gerbeleu N.V. Metal derivatives thiosemicarbazones [in Russian], *Zhurn. neorgan. khimii*, 9(1), 85–93 (1964).

Abrahams I.L., Garner C.D. Preparation and crystal structures of [PPh$_4$]$_2$[M$_2$(SPh)$_6$] (M = Zn or Cd), *J. Chem. Soc. Dalton Trans.*, (6), 1577–1579 (1987).

Agre V.M., Shugam E.A., Rukhadze E.G. The structure of chelate compounds with metal-sulfur bonds. I. The crystallographic data for certain bivalent metal alkyldithiocarbamates [in Russian], *Zhurn. strukt. khimii*, 7(5), 897–898 (1966).

Agre V.M., Shugam E.A. The structure of chelate compounds with bonds Me–S. 9. Crystal and molecular structure of zinc hexamethylene dithiocarbamate [in Russian], *Zhurn. strukt. khim.*, 13(4), 660–664 (1972).

Ahrland S., Björk N.-O. Metal halide and pseudohalide complexes in dimethyl sulfoxide solution. I. Dimethyl sulfoxide solvates of silver(I), zinc(II), cadmium(II), and mercury(II), *Acta Chem. Scand.*, 28A(8), 823–828 (1974).

Ahuja I.S., Garg A. Complexes of some pyridines and anilines with zinc(II), cadmium(II) and mercury(II) thiocyanates, *J. Inorg. Nucl. Chem.*, 34(6), 1929–1935 (1972).

Albov D.V., Buravlev E.A., Savinkina E.V., Zamilatskov I.A. Synthesis and structure of zinc iodide complex with thiocarbamide, [Zn(CH$_4$N$_2$S)$_2$I$_2$], *Crystallogr. Rep.*, 58(1), 65–67 (2013).

Ali B., Dance I.G., Craig D.C., Scudder M.L. A new type of zinc sulfide cluster: [Zn$_{10}$S$_7$(py)$_9$(SO$_4$)$_3$]·3H$_2$O, *J. Chem. Soc. Dalton Trans.*, (10), 1661–1667 (1998).

Alsfasser R., Powell A.K., Vahrenkamp H., Trofimenko S. Monofunktionelle tetraedrische Zink-Komplexe L$^3$ZnX [L$^3$ = Tris(pyrazolyl)borat], *Chem. Ber.*, 126(3), 685–694 (1993).

Alvarez H.M., Tran T.B., Richter M.A., Alyounes D.M., Rabinovich D., Tanski J.M., Krawiec M. Homoleptic Group 12 metal bis(mercaptoimidazolyl)borate complexes M(Bm$^R$)$_2$ (M = Zn, Cd, Hg), *Inorg. Chem.*, 42(6), 2149–2156 (2003).

Amoedo-Portela A., Carballo R., Casas J.S., García-Martínez E., Gómez-Alonso C., Sánchez-González A., Sordo J., Vázquez-Lopez E.M. The coordination chemistry of the versatile ligand bis(2-pyridylthio)methane, *Z. Anorg. und Allg. Chem.*, 628(5), 939–950 (2002a).

Amoedo-Portela A., Carballo R., Casas J.S., García-Martínez E., Sánchez-Gonzalez A., Sordo J., Vazquez-Lopez E.M. The crystal structure of bis(2-pyridylthio) methanediiodozinc(II), [ZnI$_2$(bpytm)], *Main Group Met. Chem.*, 25(5), 317–318 (2002b).

Andreetti G.D., Cavalca L., Musatti A. The crystal and molecular structure of tris(thiourea) zinc(II) sulphate, *Acta Crystallogr.*, B24(5), 683–690 (1968).

Andreetti G.D., Jain P.C., Lingafelter E.C. The crystal structure of β,β′,β″-triaminotriethyla-mineisothiocyanatozinc(II) thiocyanate, *J. Am. Chem. Soc.*, 91(15), 4112–4115 (1969).

Angelova V., Dimitrov R., Hekimova A. Investigation of interaction of CuSeO$_3$ with ZnS, PbS and FeS [in Bulgarian], *Nauchn. tr. Plovdiv. un-t. Ser. Khimia*, 13(3), 313–322 (1975–1976).

Anjali K.S., Vittal J.J., Dean P.A.W. Syntheses, characterization and thermal properties of [M(Spy)$_2$(SpyH)$_2$] (M = Cd and Hg; Spy$^-$ = pyridine-4-thiolate; SpyH = pyridinium-4-thiolate) and [M(SpyH)$_4$](ClO$_4$)$_2$ (M = Zn, Cd and Hg), *Inorg. Chim. Acta*, 351, 79–88 (2003).

Arman H.D., Poplaukhin P., Tiekink E.R.T. Bis[*N*-(2-hydroxyethyl)-*N*-methyldithiocarbamato-κ$S$][2,4,6-tris(pyridin-2-yl)-1,3,5-triazine-κ$^3$N$^1$,N$^2$,N$^6$]zinc dioxane sesquisolvate, *Acta Crystallogr.*, E68(3), m319–m320 (2012).

Arman H.D., Poplaukhin P., Tiekink E.R.T. (μ-*trans*-1,2-Di-4-pyridylethylene-κ$^2$N:N′) bis[bis(*N*,*N*-diethyldithiocarbamato-κ$^2$S,S′)zinc(II)] chloroform solvate, *Acta Crystallogr.*, E65(11), m1472–m1473 (2009a).

Arman H.D., Poplaukhin P., Tiekink E.R.T. (μ-*trans*-1,2-Di-4-pyridylethylene-κ$^2$N:N′) bis[bis(*N*,*N*-diisopropyldithiocarbamato-κ$^2$S,S′)zinc(II)], *Acta Crystallogr.*, E65(11), m1475 (2009b).

Asao T., Kikuchi (née Ninomiya) Y. The preparation of some tropone complex, *Chem. Lett.*, 1(5), 413–416 (1972).

Ashworth C.C., Bailey N.A., Johnson M., McCleverty J.A., Morrison N., Tabbiner B. X-ray structures of [Zn(S$_2$CNMe$_2$)$^-$ and [Zn(C$_7$H$_4$NS$_2$)$_3$(OH$_2$)]$^-$, possible intermediates in zinc-assisted rubber vulcanization, *J. Chem. Soc. Chem. Commun.*, (18), 743–744 (1976).

Avila V., Tiekink E.R.T. *catena*-Poly[[bis(*O*,*O*′-diisopropyl dithiophosphato-κ$^2$S,S′)zinc(II)]-μ-1,2-bis(3-pyridylmethylene)hydrazine-κ$^2$N:N′], *Acta Crystallogr.*, E62(12), m3530–m3531 (2006).

Avila V., Tiekink E.R.T. μ-1,2-Di-4-pyridylethane-κ$^2$N:N′-bis-[bis(*N*,*N*-diisopropyldithio-carbamato-κ$^2$S,S′)zinc(II)], *Acta Crystallogr.*, E64(5), m680 (2008).

Baba I., Farina Y., Kassim K., Othman A.H., Razak I.A., Fun H.-K., Ng S.W. Bis(μ-*N*-buthyl-*N*-ethyldithiocarbamato-*S*:*S*′)bis-[(*N*-butyl-*N*-ethyldithiocarbamato)zinc(II)], *Acta Crystallogr.*, E57(2), m55–m56 (2001a).

Baba I., Farina Y., Othman A.H., Razak I.A., Fun H.-K., Ng S.W. Bis(μ-*N*-ethyl-*N*-isopropyldithiocarbamato-*S:S′*)bis-[(*N*-ethyl-*N*-isopropyldithiocarbamato-*S,S′*) zinc(II)], *Acta Crystallogr.*, E57(1), m51–m52 (2001b).

Baba I., Lee L.H., Farina Y., Othman A.H., Razak I.A., Usman A., Fun H.-K., Ng S.W. Bis(μ-*N*-methyl-*N*-phenyldithiocarbamato)bis-[(*N*-methyl-*N*-phenyldithiocarbamato) zinc(II)], *Acta Crystallogr.*, E58(12), m744–m745 (2002).

Babb J.E.V., Burrows A.D., Harrington R.W., Mahon M.F. Zinc thiosemicarbazide dicarboxylates: The influence of the anion shape on supramolecular structure, *Polyhedron*, 22(5), 673–686 (2003).

Bacon W.E., Bor J.F. Basic zinc double salts of *O,O*-dialkyl hydrogen phosphorodithioates, *J. Org. Chem.*, 27(4), 1484–1485 (1962).

Baenziger N.C., Modak S.L., Fox Jr. C.L. Diamminebis(2-sulfanilamidopyrimidinato) zinc(II), [Zn(C$_{10}$H$_9$N$_4$O$_2$S)$_2$(NH$_3$)$_2$], *Acta Crystallogr.*, C39(12), 1620–1623 (1983).

Bakbak S., Bhatia V.K., Incarvito C.D., Rheingold A.L., Rabinovich D. Synthesis and characterization of two new bulky tris(mercaptoimidazolyl)borate ligands and their zinc and cadmium complexes, *Polyhedron*, 20(28), 3343–3348 (2001).

Barreca D., Gasparotto A., Maragno C., Seraglia R., Tondello E., Venzo A., Krishnanand V., Bertagnolli H. Synthesis and characterization of zinc bis(*O*-isopropylxanthate) as a single-source chemical vapor deposition precursor for ZnS, *Appl. Organometal. Chem.*, 19(9), 1002–1009 (2005).

Bayon J.C., Casals I., Gaete W., Gonzalez-Duarte P., Ros J. Complexes of 1-methyl-4-mercaptopiperidine with zinc(II), cadmium(II) and mercury(II) halides, *Polyhedron*, 1(2), 157–161 (1982).

Beauvais L.G., Shores M.P., Long J.R. Cyano-bridged Re$_6$Q$_8$ (Q = S, Se) cluster-metal framework solids: A new class of porous materials, *Chem. Mater.*, 10(12), 3783–3786 (1998).

Beckett R., Hoskins B.F. Crystal structure of a tetrahedral zinc(II) complex of a 1,3-dithiolate: Bis(*O*-ethyl thioacetothioacetato)zinc(II), *J. Chem. Soc. Dalton Trans.*, (10), 908–911 (1975).

Beckett R., Hoskins B.F. The crystal and molecular structure of bis(*O*-ethylthioacetothioacetato)zinc(II), *Inorg. Nucl. Chem. Lett.*, 8(8), 679–682 (1972).

Beck W., Stetter K.H., Tadros S., Schwarzhans K.E. Darstellung, IR- und [19]F-KMR-Spektren von Pentafluorphenylmercapto-Metallkomplexen, *Chem. Ber.*, 100(12), 3944–3954 (1967).

Beer S., Lai C.S., Halbach T.S., Tiekink E.R.T. μ-Bis(diphenylphosphino)ethane-*P,P′*-bis[bis(*O*-isopropyldithiocarbonato-*S,S′*)zinc(II)], *Acta Crystallogr.*, E58(5), m194–m196 (2002).

Bell N.A., Johnson E., March L.A., Marsden S.D., Nowell I.W., Walker Y. Complexes of zinc dialkyldithiocarbamates. Part I. Complexes with bidentate nitrogen ligands; crystal structure of 1,10-phenanthroline zinc dibutyldithiocarbamate, *Inorg. Chim. Acta*, 156(2), 205–211 (1989).

Benson R.E., Ellis C.A., Lewis C.E., Tiekink E.R.T. 3D-, 2D- and 1D-supramolecular structures of {Zn[S$_2$CN(CH$_2$CH$_2$OH)R]$_2$}$_2$ and their {Zn[S$_2$CN(CH$_2$CH$_2$OH)R]$_2$}$_2$(4,4′-bipyridine) adducts for R = CH$_2$CH$_2$OH, Me or Et: Polymorphism and pseudo-polymorphism, *CrystEngComm*, 9(10), 930–940 (2007).

Bermejo E., Carballo R., Castiñeiras A., Domínguez R., Liberta A.E., Maichle-Mössmer C., Salberg M.M., West D.X. Synthesis, structural characteristics and biological activities of complexes of Zn$^{II}$, Cd$^{II}$, Hg$^{II}$, Pd$^{II}$, and Pt$^{II}$ with 2-acetylpyridine 4-methylthiosemicarbazone, *Eur. J. Inorg. Chem.*, (6), 965–973 (1999).

Bermejo E., Castiñeiras A., Fostiak L.M., Garcia I., Llamas-Saiz A.L., Swearingen J.K., West D.X. Synthesis, characterization and molecular structure of 2-pyridyl-formamide *N*(4)-dimethylthiosemicarbazone and some five-coordinated zinc(II) and cadmium(II) complexes, *Z. Naturforsch.*, 56B(12), 1297–1305 (2001).

Bermejo E., Castiñeiras A., Fostiak L.M., Santos I.G., Swearingen J.K., West D.X. Spectral and structural studies of Zn and Cd complexes of 2-pyridineformamide N(4)-ethylthiosemicarbazone, *Polyhedron*, 23(14), 2303–2313 (2004a).

Bermejo E., Castiñeiras A., García-Santos I., West D.X. Structural and coordinative variability in zinc(II), cadmium(II), and mercury(II) complexes of 2-pyridineformamide 3-hexamethyleneiminylthiosemicarbazone, *Z. Anorg. und Allg. Chem.*, 630(7), 1096–1109 (2004b).

Berus E.I., Zemskova S.M., Glinskaya L.A., Klevtsova R.F., Vasil'ev A.D., Mazhara A.P., Larionov S.V. Mixed-ligand complexes of dipropyldithiocarbamatozinc(II) with 1,10-phenanthroline, 2,2'-bipyridine and 4,4'-bipyridine [in Russian], *Zhurn. neorgan. khimii*, 43(11), 1847–1851 (1998).

Bhaskaran A., Arjunan S., Raghavan C.M., Kumar R.M., Jayavel R. Investigation on synthesis, growth, structural, optical, thermal and dielectric properties of organometallic non-linear optical tetrathiourea cadmium tetrathiocyanato zincate (TCTZ) single crystals, *J. Cryst. Growth*, 310(21), 4549–4553 (2008).

Block E., Ofori-Okai G., Chen Q., Zubieta J. Complexes of group 12 metals with sterically-hindered thiolate ligands. The crystal and molecular structures of $[Zn_2(2-SC_5H_3N-3-SiMe_3)_4]$ and $[CdI_2(2-SC_5H_3NH-6-SiMe_2Bu^t)_2]$, *Inorg. Chim. Acta*, 189(2), 137–139 (1991).

Bochmann M., Bwembya G.C., Grinter R., Powell A.K., Webb K.J., Hursthouse M.B., Malik K.M.A., Mazid M.A. Synthesis of low-coordinate chalcogenolato complexes of zinc with O, N, S, and P donor ligands. Molecular and crystal structures of $Zn(S-t-Bu_3C_6H_2-2,4,6)_2(L)$ ($L = NC_5H_3Me_2-2,6$, $PMePh_2$), $Zn(Se-t-Bu_3C_6H_2-2,4,6)_2(OSC_4H_8)$ and $Zn(S-t-Bu_3C_6H_2-2,4,6)_2(N$-methylimidazole$)_2$, *Inorg. Chem.*, 33(10), 2290–2296 (1994).

Bochmann M., Bwembya G.C., Powell A.K. Synthesis of isocyanide complexes of zinc. The molecular and crystal structure of $Zn(SeC_6H_2Bu^t_3)_2(CNBu^t)_2$, *Polyhedron*, 12(24), 2929–2932 (1993).

Bochmann M., Webb K., Hursthouse M.B., Mazid M. The first stable aldehyde and ketone complexes of zinc: The structure of $[Zn(SeC_6H_2Bu^t_3)_2(p$-O$=CHC_6H_4OMe)]_2$, *J. Chem. Soc. Chem. Commun.*, (24), 1735–1737 (1991).

Bombicz P., Madarász J., Krunks M., Niinisto L., Pokol G. Multiple secondary interaction arrangement in the crystal structure of dichlorobis(thiourea-S)-zinc(II), *J. Coord. Chem.*, 60(4), 457–464 (2007).

Bonamico M., Mazzone G., Vaciago A., Zambonelli L. Structural studies of metal dithiocarbamates. III. The crystal and molecular structure of zinc diethyldithiocarbamate, *Acta Crystallogr.*, 19(6), 898–909 (1965).

Bosson B. The crystal structure of $Tl(ZnSO_4)Cl$, *Acta Chem. Scand.*, 27(6), 2230–2231 (1973).

Brand U., Burth R., Vahrenkamp H. Design of trigonal-bipyramidal $ZnN_3S_2$ complexes, *Inorg. Chem.*, 35(4), 1083–1086 (1996).

Brand U., Vahrenkamp H. A new tridentate N,N,S ligand and its zinc complexes, *Inorg. Chem.*, 34(12), 3285–3293 (1995).

Brand U., Vahrenkamp H. Zinkkomplexe des chelatisierenden N,S,S-Liganden N-(2-Mercaptoethyl)-2-mercaptoanilin, *Z. Anorg. und Allg. Chem.*, 622(2), 213–218 (1996).

Brennan J.G., Siegrist T., Carroll P.J., Stuczynski S.M., Reynders P., Brus L.E., Steigerwald M.L. Bulk and nanostructure group II-VI compounds from molecular organometallic precursors, *Chem. Mater.*, 2(4), 403–409 (1990).

Briansó M.C., Briansó J.L., Gaete W., Ros J., Suñer C. Synthesis and crystal and molecular structure of di-μ-(N-methylpiperidinium-4-thiolate)-bis[dichlorozinc(II)] monohydrate, *J. Chem. Soc. Dalton Trans.*, (3), 852–854 (1981).

Bridgewater B.M., Fillebeen T., Friesner R.A., Parkin G. A zinc thiolate species which mimics aspects of the chemistry of the Ada repair protein and matrix metalloproteinases: The synthesis, structure and reactivity of the tris(2-mercapto-1-phenylimidazolyl)hydroborato complex $[Tm^{Ph}]ZnSPh$, *J. Chem. Soc. Dalton Trans.*, (24), 4494–4496 (2000).

Bridgewater B.M., Parkin G. A zinc hydroxide complex of relevance to 5-aminolevulinate dehydratase: The synthesis, structure and reactivity of the tris(2-mercapto-1-phenyl-imidazolyl)hydroborato complex [Tm$^{Ph}$]ZnOH, *Inorg. Chem. Commun.*, 4(3), 126–129 (2001).

Brinkhoff H.C., Cras J.A., Steggerda J.J., Willemse J. The oxidation of complexes of nickel, copper and zinc, *Rec. trav. chim. Pays-Bas.*, 88(6), 633–640 (1969).

Britvin S.N., Bogdanova A.N., Boldyreva M.M., Aksenova G.Y. Rudashevskyite, the Fe-dominant analogue of sphalerite, a new mineral: Description and crystal structure, *Am. Mineral.*, 93(4), 902–909 (2008).

Brodersen K., Hummel H.-U., Böhm K., Procher H. Zur Koordination von Thiocyanat in ternären Metallkomplexen: CdZn(NCS)$_4$ und die Kristallstruktur von BaZn(NCS)$_4$·7H$_2$O, *Z. Naturforsch.*, 40B(3), 347–351 (1985).

Brodersen K., Procher H., Hummel H.-U. Zur Darstellung und Kristallstruktur von PbZn(NCS)$_4$, *Z. Naturforsch.*, 42B(6), 679–681 (1987).

Burke N.J., Burrows A.D., Donovan A.S., Harrington R.W., Mahon M.F., Price C.E. Zinc dicarboxylate polymers and dimers: Thiourea substitution as a tool in supramolecular synthesis, *Dalton Trans.*, (20), 3840–3849 (2003).

Burn A.J., Smith G.W. The structure of basic zinc *OO*-dialkyl phosphorodithioates, *Chem. Commun.*, 17, 394–396 (1965).

Burrows A.D., Donovan A.S., Harrington R.W., Mahon M.F. Backbone flexibility and counterion effects on the structure and thermal properties of di(thiourea)zinc dicarboxylate coordination polymers, *Eur. J. Inorg. Chem.*, 23, 4686–4695 (2004a).

Burrows A.D., Harrington R.W., Mahon M.F. Dichlorobis(1,3-dimethylthiourea-κ$S$)zinc(II), *Acta Crystallogr.*, E60(9), m1317–m1318 (2004b).

Burrows A.D., Harrington R.W., Mahon M.F., Price C.E. The influence of hydrogen bonding on the structure of zinc co-ordination polymers, *J. Chem. Soc. Dalton Trans.*, 21, 3845–3854 (2000).

Burrows A.D., Harrington R.W., Mahon M.F., Teat S.J. The influence of functional group orientation on the structure of zinc 1,1,4-trimethylthiosemicarbazide dicarboxylates: Probing the limits of crystal engineering strategies, *Eur. J. Inorg. Chem.*, 4, 766–776 (2003).

Burrows A.D., Menzer S., Mingos D.M.P., White A.J.P., Williams D.J. The influence of the chelate effect on supramolecular structure formation: Synthesis and crystal structures of zinc thiourea and thiosemicarbazide complexes with terephthalate, *J. Chem. Soc. Dalton Trans.*, (22), 4237–4240 (1997).

Burth R., Gelinsky M., Vahrenkamp H. Controlled formation of tri- and octanuclear benzyl-thiolate complexes of zinc, *Inorg. Chem.*, 37(11), 2833–2836 (1998).

Bu X., Zheng N., Li Y., Feng P. Templated assembly of sulfide nanoclusters into cubic-C$_3$N$_4$ type framework, *J. Am. Chem. Soc.*, 125(20), 6024–6025 (2003).

Byrom C., Malik M.A., O'Brien P., White A.J.P., Williams D.J. Synthesis and X-ray single crystal structures of bis(diisobutyldithiophosphinato)cadmium(II) or zinc(II): Potential single-source precursors for II/VI materials, *Polyhedron*, 19(2), 211–215 (2000).

Calligaris M., Ciana A., Meriani S., Nardin G., Randaccio L., Ripamonti A. Structure and polymeric nature of zinc(II) and cobalt(II) phosphino-thionates, *J. Chem. Soc. A: Inorg. Phys. Theor.*, 3386–3392 (1970a).

Calligaris M., Nardin G., Ripamonti A. Crystal and molecular structure of zinc(II) and cobalt(II) diethyldithiophosphinates, *J. Chem. Soc. A: Inorg. Phys. Theor.*, 714–722 (1970b).

Câmpian M.V., Haiduc I., Tiekink E.R.T. Supra- versus intra-molecular π...π interactions in M(S$_2$COCH$_2$Ph)$_2$(dafone) compounds: M=Zn and Cd; dafone=4,5-diazafluoren-9-one-*N*,*N'*, *Z. Kristallogr.*, 228(4), 203–209 (2013).

Cao D.-K., Li Y.-Z., Zheng L.-M. [Zn₇{(2-C₅H₄N)CH(OH)PO₃}₆(H₂O)₆]SO₄·4H₂O: A zinc phosphonate cluster with a drum-like cage structure" *Inorg. Chem.*, 44(9), 2984–2985 (2005).

Carballo R., Covelo B., Fernández-Hermida N., Lago A.B., Vázquez-López E.M. Structural evidence for metalloaromaticity in a new molecular complex with the divergent ligand 4,4'-dipyridyldisulfide, *J. Mol. Struct.*, 936(1–3), 87–91 (2009).

Casals I., Clegg W., González-Duarte P. Preparation and crystal structure of the unprecedented octameric zinc-aminothiolate complex [Zn₈{S(CH₂)₃NMe₂}₁₆], *J. Chem. Soc. Chem. Commun.*, (9), 655–656 (1991a).

Casals I., González-Duarte P., Clegg W., Foces-Foces C., Cano F.H., Martínez-Ripoll M., Gómez M., Solans X. Synthesis and structures of tetranuclear 2-(dimethylamino)ethanethiolato complexes of zinc, cadmium and mercury involving both primary and secondary metal-halogen bonding, *J. Chem. Soc. Dalton Trans.*, (10), 2511–2518 (1991b).

Casals I., González-Duarte P., Clegg W. Preparation of complexes of the 3-trimethylammonio-1-propanethiolato ligand, [M{S(CH₂)₃NMe₃}₂](PF₆)₂, M = Zn, Cd, Hg. Crystal structures of monomeric and polymeric forms of the mercury complex, *Inorg. Chim. Acta*, 184(2), 167–175 (1991c).

Casals I., González-Duarte P., López C., Solans X. Synthesis and structure of amononuclear zinc(II) complex of 2-(dimethylamino)ethanethiol, *Polyhedron*, 9(5), 763–768 (1990).

Casals I., González-Duarte P., Sola J., Font-Bardia M., Solans L., Solans X. Polymeric thiolate complexes of Group 12 metals. Crystal and molecular structures of *catena*-[μ-(3-dimethylammonio-1-propanethiolate)]-dichlorocadmium(II) and bis[3-(dimethylammonio)propyl] disulphide tetrabromocadmate(II), *J. Chem. Soc. Dalton Trans.*, (10), 2391–2395 (1987).

Cassidy I., Garner M., Kennedy A.R., Potts G.B.S., Reglinski R., Slavin P.A., Spicer M.D. The preparation and structures of Group 12 (Zn, Cd, Hg) complexes of the soft tripodal ligand hydrotris(methimazolyl)borate (Tm), *Eur. J. Inorg. Chem.*, (5), 1235–1239 (2002).

Castiñeiras A., Bermejo E., West D.X., Ackerman L.J., Valdés-Martínez J., Hernández-Ortega S. Structural and spectral studies of 1-phenylglyoxal bis(3-piperidylthiosemicarbazone) and its zinc(II), cadmium(II) and platinum(II) complexes, *Polyhedron*, 18(10), 1463–1469 (1999).

Castiñeiras A., García I., Bermejo E., Ketcham K.A., West D.X., El-Sawaf A.K. Coordination of Znᴵᴵ, Cdᴵᴵ, and Hgᴵᴵ by 2-pyridineformamide-3-piperidyl-thiosemicarbazone, *Z. Anorg. und Allg. Chem.*, 628(2), 492–504 (2002).

Castiñeiras A., Garcia I., Bermejo E., West D.X. Structural and spectral studies of 2-pyridineformamide thiosemicarbazone and its complexes prepared with zinc halides, *Z. Naturforsch.*, 55B(6), 511–518 (2000).

Castro J., Romero J., García-Vázquez J.A., Durán M.L., Castiñeiras A., Sousa A., Fenton D.E. Synthesis of nickel(II), copper(II), zinc(II), and cadmium(II) complexes of 2-[(2-mercaptophenyl)iminomethyl]pyrrole and 2-[(2-mercaptoethyl)iminomethyl]pyrrole by the electrochemical cleavage of a disulphide bond: The crystal structure of bis{2-[(pyrrol-2-yl)methyleneamino]thiophenolato-*S,N*}zinc(II), *J. Chem. Soc. Dalton Trans.*, (11), 3255–3258 (1990).

Castro R., Durán M.L., García-Vázquez J.A., Romero J., Sousa A., Castiñeiras A., Hiller W., Strähle J. Direct electrochemical synthesis of pyrimidine-2-thionato complexes of zinc(II) and cadmium(II): The crystal structure of (1,10-phenanthroline)bis(pyrimidine-2-thionato)cadmium(II), *Z. Naturforsch.*, 47B(8), 1067–1074 (1992).

Castro R., García-Vázquez J.A., Romero J., Sousa A., Castiñeiras A., Hiller W., Strähle J. Synthesis of 3-trimethylsilylpyridine-2-thione (3-Me₃SipytH) complexes of nickel, copper, zinc and cadmium: Crystal and molecular structure of [Cd(3-Me₃Sipyt)₂], *Inorg. Chim. Acta*, 211(1), 47–54 (1993a).

Castro R., García-Vázquez J.A., Romero J., Sousa A., McAuliffe C.A., Pritchard R. Electrochemical synthesis of benzothiazole-2-thionato complexes of nickel(II), zinc(II) and cadmium(II): The crystal structure of 2,2'-bipyridine bis(benzothiazole-2-thionato)zinc(II), *Polyhedron*, 12(18), 2241–2247 (1993b).

Castro R., García-Vázquez J.A., Romero J., Sousa A., Hiller W., Strähle J. Electrochemical synthesis of zinc(II) complexes of 4,6-dimethylpyrimidine-2-thione: The crystal structure of pyridine-bis(4,6-dimethylpyrimidine-2-thiolato)zinc(II), *Polyhedron*, 13(2), 273–279 (1994).

Cavalca L, Gasparri G.F., Andreetti G.D., Domiano P. The crystal structure of bisthiourea–zinc acetate, *Acta Crystallogr.*, 22(1), 90–98 (1967).

Cavalca L., Nardelli M., Branchi G. The crystal structure of *mono*-thiosemicarbazide–zinc chloride, *Acta Crystallogr.*, 13(9), 688–693 (1960).

Černák J., Adzimová I. Bis(benzoato-*O*)bis(thiourea-*S*)zinc(II), *Acta Crystallogr.*, C51(3), 392–395 (1995).

Cesur H., Yazicilar T.K., Bati B., Yilmaz V.T. Synthesis, characterization, and spectral and thermal studies of some divalent transition metal complexes of benzylpiperazine dithiocarbamate, *Synth. React. Inorg. Met.-Org. Chem.*, 31(7), 1271–1283 (2001).

Chan M.Y., Lai C.S., Tiekink E.R.T. Bis(*N*-cyclohexyl,*N*-methyldithiocarbamato)zinc(II), *Appl. Organometal. Chem.*, 18(6), 298 (2004).

Chan W.-H., Mak T.C.W., Yip W.-H., Smith G., O'Reilly E.J., Kennard C.H.L. Metal-(phenylthio) alkanoic acid interactions. VII. The crystal and molecular structures of diaquabis[(benzylthio) acetato]-zinc(II), *catena*-[diaqua-tetra[(benzylthio)acetato]-bis[cadmium(II)], *catena*-tetra-μ-{2-methyl-3-(phenylthio)propionato-*O,O'*]-bis[copper(II)]} and tetra-μ-[2-methyl-2-(phenylthio)propionato-*O,O'*]-bis[ethanolcopper(II)], *Polyhedron*, 6(5), 881–889 (1987).

Chen D., Lai C.S., Tiekink E.R.T. *catena*-Poly[[bis(*O,O'*-dicyclohexyldithiophosphato-κ²*S,S'*)zinc(II)]-μ-1,2-bis(3-pyridylmethylene)hydrazine-κ²*N:N'*], *Acta Crystallogr.*, E61(10), m2052–m2054 (2005).

Chen D., Lai C.S., Tiekink E.R.T. Supramolecular aggregation in diimine adducts of zinc(II) dithiophosphates: Controlling the formation of monomeric, dimeric, polymeric (zigzag and helical), and 2-D motifs, *CrystEngComm*, 8(1), 51–58 (2006).

Chen W.-T., Hu R.-H., Wang Y.-F., Zhang X., Liu J. A Tb–Zn tetra(4-sulfonatophenyl)porphyrin hybrid: Preparation, structure, photophysical and electrochemical properties, *J. Solid State Chem.*, 213, 218–223 (2014).

Chen X.-F., Liu S.-H., Zhu X.-H., Vittal J.J., Tan G.-K., You X.-Z. μ-(4,4'-Bipyridine)-*N:N'*-bis[bis-(pyrrolidinedithiocarboxylato-*S,S'*)-zinc(II)], *Acta Crystallogr.*, C56(1), 42–43 (2000).

Choy A., Craig D., Dance I., Scudder M. [S₄M₁₀(SPh)₁₆]⁴⁻ (M=Zn, Cd), a molecular fragment of the sphalerite MS lattice: Structural congruence of metal sulphides and metal thiolates, *J. Chem. Soc. Chem. Commun.*, (21), 1246–1247 (1982).

Ciampolini M., Gelsomini J., Nardi N. Complexes of cobalt(II), nickel(II), copper(II), and zinc(II) with the tripod-like ligands tris (2-methylthioethyl)amine, *Inorg. Chim. Acta*, 2, 343–346 (1968).

Cini R., Cinquantini A., Oriolli P., Sabat M. Crystal and molecular structure of [*N,N'*-ethylenebis-(monothioacetylacetoneiminato)] zinc(II), *Inorg. Chim. Acta*, 41, 151–154 (1980).

Coates G.E., Ridley D. Alkoxy-, thio-, and amino-derivatives of methylzinc, *J. Chem. Soc.*, 1870–1877 (1965).

Cohen B., Mastropaolo D., Potenza J.A., Schugar H.J. Bis(2-amino-1,1-dimethylethanethiolato-N,S)zinc, *Acta Crystallogr.*, B34(9), 2859–2860 (1978).

Colombini G., Preti C. Complexes of thiomorpholin-3-one and thiazolidine-2-thione with zinc(II), cadmium(II) and mercury(II) thiocyanates, acetates and fluoborates, *J. Inorg. Nucl. Chem.*, 37(5), 1159–1165 (1975).

Comerlato N.M., Harrison W.T.A., Howie R.A., Low J.N., Silvino A.C., Wardell J.L., Wardell S.M.S.V. Bis(tetra-*n*-butylammonium) (a redetermination at 150 K) and bis(tetraphenylarsonium) bis(1,3-dithiole-2-thione-4,5-dithiolato)zinc(II) (at 300 K), *Acta Crystallogr.*, C58(2), m105–m108 (2002).

Corwin Jr. D.T., Gruff E.S., Koch S.A. Zinc, cobalt, and cadmium thiolate complexes: Models for the zinc(S-cys)$_2$(his)$_2$ centre in transcription factor IIIA (cys=cysteine; his=histidine), *J. Chem. Soc. Chem. Commun.*, (13), 966–967 (1987).

Coucouvanis D., Murphy C.N., Simhon E., Stremple P., Draganjac M., Neira R. del P. Tetrakis(benzenethiolato)metallate(2–) complexes, [M(SPh)$_4$]$^{2-}$, of manganese, iron, cobalt, zinc, and cadmium and derivatives of the [Fe(SPh)$_4$]$^{2-}$ complex, *Inorg. Synthes.*, 21, 23–28 (1982).

Coucouvanis D., Patil P.R., Kanatzidis M.G., Detering B., Baenziger N.C. Synthesis and reactions of binary metal sulfides. Structural characterization of the [(S$_4$)$_2$Zn]$^{2-}$, [(S$_4$)$_2$Ni]$^{2-}$, [(S$_5$)Mn(S$_6$)]$^{2-}$, and [(CS$_4$)$_2$Ni]$^{2-}$ anions, *Inorg. Chem.*, 24(1), 24–31 (1985).

Cox M.J., Tiekink E.R.T. Structural features of zinc(II) bis(*O*-alkyldithiocarbonate) and zinc(II) bis(*N*,*N*-dialkyldithiocarbamate) compounds, *Z. Kristallogr.*, 214(3), 184–190 (1999).

Cras J.A., Willemse J., Gal A.W., Hummelink-Peters B.G.M.C. Preparation, structure and properties of compounds containing the dipositive tri-copper hexa(*N*,*N*-di-*n*-butyldithiocarbamato) ion, compounds with copper oxidation states II and III, *Rec. trav. chim. Pays-Bas*, 92(6), 641–650 (1973).

Cremers T.L., Bloomquist D.R., Willett R.D., Crosby G.A. Structure of (1,10-phenanthroline) bis(4-toluenethiolato)zinc(II), *Acta Crystallogr.*, B36(12), 3097–3099 (1980).

Cupertino D., Keyte R., Slawin A.M.Z., Williams D.J., Woollins J.D. Preparation and single-crystal characterization of $^i$Pr$_2$P(S)NHP(S)$^i$Pr$_2$ and homoleptic [$^i$Pr$_2$P(S)NP(S)$^i$Pr$_2$]$^-$ complexes of zinc, cadmium, and nickel, *Inorg. Chem.*, 35(9), 2695–2697 (1996).

Currier W.F., Weber J.H. Complexes of sulfoxides. I. Octahedral complexes of manganese(II), iron(II), cobalt(II), nickel(II), and zinc(II), *Inorg. Chem.*, 6(8), 1539–1543 (1967).

Czakis-Sulikowska M., Sołoniewicz R. The crystalline complex compounds containing thiourea and thiocyanates ions. I. Systems Hg$^{++}$–CS(NH$_2$)$_2$–Zn$^{++}$–SCN$^-$ and Hg$^{++}$–CS(NH$_2$)$_2$–Co$^{++}$–SCN$^-$ [in Polish], *Rocz. Chem.*, 37(11), 1405–1410 (1963).

Dance I.G., Choy A., Scudder M.L. Syntheses, properties and molecular and crystal structures of (Me$_4$N)$_4$[E$_4$M$_{10}$(SPh)$_{16}$] (E=S, Se; M=Zn, Cd): Molecular supertetrahedral fragments of the cubic metal chalcogenide lattice, *J. Am. Chem. Soc.*, 116(21), 6285–6295 (1984).

Dance I.G. [ClZn$_8$(SPh)$_{16}$]$^-$, a new, highly symmetrical cage containing metal atoms inside and outside an icosahedron of ligands; X-ray crystallographic study, *J. Chem. Soc. Chem. Commun.*, (17), 818–820 (1980a).

Dance I.G. Formation, crystal structure, and reactions of *catena*-(μ-SPh)[(μ-SPh)$_6$Zn$_4$(CH$_3$OH) (SPh)], a model for zinc thiolate metalloenzymes, *J. Am. Chem. Soc.*, 102(10), 3445–3451 (1980b).

Dance I.G. The prototype *centro*-Cl-*tetrahedro*-Zn$_4$-*icosahedro*-(μ-SPh)$_{12}$-*tetrahedro*-(ZnSPh)$_4$ molecular structure, and topological analysis of surface ligand packing for an ensemble of high symmetry M$_8$(μ-SR)$_{12}$ aggregates and derivatives, *Aust. J. Chem.*, 38(9), 1391–1411 (1985).

Dance I.G. The zinc-benzenethiolate-halide system. Synthesis and structure of the hexakis (μ-(benzenethiolato))-bis(benzenethiolato)dichlorotetrazincate(II) dianion, *Inorg. Chem.*, 20(7), 2155–2160 (1981).

Dean P.A.W., Vittal J.J., Payne N.C. Syntheses of (Me$_4$N)$_2$[(μ-EPh)$_6$(MX)$_4$] (M=Cd, Zn; E=S, Se; X=Cl, Br, I). Crystal and molecular structures of (Me$_4$N)$_2$[(μ-EPh)$_6$(CdBr)$_4$] (E=S, Se) and characterization of [(μ-EPh)$_6$(CdX)$_4$]$^{2-}$ in solution by $^{113}$Cd and $^{77}$Se NMR, *Inorg. Chem.*, 26(11), 1683–1689 (1987).

Dean P.A.W., Vittal J.J., Wu Y. Synthesis, multinuclear magnetic resonance spectra, and chemistry of some complexes [(μ-SR)$_6$(MX)$_4$]$^{2-}$ (R=alkyl; M=Zn or Cd; X=Cl, Br, or I) and the X-ray structural analysis of (Et$_4$N)$_2$[(μ-SPr$^i$)$_6$(CdBr)$_4$, *Can. J. Chem.*, 70(3), 779–791 (1992).

Decken A., Gossage R.A., Chan M.Y., Lai C.S., Tiekink E.R.T. Bis(*N*,*N*-dibenzyldithiocarbamato)zinc(II), *Appl. Organometal. Chem.*, 18(2), 101–102 (2004).

De Filippo D., Devillanova F., Preti C., Verani G. Thiomorpholin-3-one and thiazolidine-2-thione complexes with group IIB metals, *J. Chem. Soc. A: Inorg. Phys. Theor.*, 1465–1468 (1971).

Deng Y.-H., Liu J., Li N., Yang Y.-L., Ma H.-W. Synthesis, characterization and thermostability of IIB Group transition metal complexes with dithiocarbamate [in Chinese], *Acta Chim. Sin.*, 65(24), 2868–2874 (2007).

Devillanova F.A., Verani G. Zn(II) and Cd(II) halide complexes with *N*-mono and *N*,*N*'-disubstituted imidazolines-2-thione, *J. Nucl. Inorg. Chem.*, 42(4), 623–627 (1980).

Dev S., Ramli E., Rauchfuss T.B., Stern C.L. Direct approaches to zinc polychalcogenide chemistry: ZnS$_6$(N-MeIm)$_2$ and ZnSe$_4$(N-MeIm)$_2$, *J. Am. Chem. Soc.*, 112(17), 6385–6386 (1990).

Dickert, Jr. J.J., Rowe C.N. The thermal decomposition of metal *O*,*O*-dialkylphosphorodithioates, *J. Org. Chem.*, 32(3), 647–653 (1967).

Dillen J., Lenstra A.T.H., Haasnoot J.G., Reeddijk J. Transition metal thiocyanates of 5,7-dimethyl[1,2,4]triazolo[1,5-a]-pyrimidine studied by spectroscopic methods. The crystal structures of diaquabis (dimethyltriazolopyrimidine-*N*$^3$)bis(thiocyanato-*N*) cadmium(II) and bis(dimethyltriazolopyrimidine-*N*$^3$)bis(thiocyanato-*S*)-mercury(II), *Polyhedron*, 2(3), 195–201 (1983).

Divjaković V., Nowacki W. Die Kristallstruktur von Galchait [Hg$_{0.76}$(Cu,Zn)$_{0.24}$]$_{12}$Tl$_{0.96}$ (AsS$_3$)$_8$, *Z. Kristallogr.*, 142(3–4), 262–270 (1975).

Dost J., Steinike U., Paudert R. Über die Komplexbildung zwischen Zink- und Blei(II)-thiocyanat, *Krist. und Technol.*, 3(2), 247–254 (1968).

Drew M.G.B., Hasan M., Hobson R.J., Rice D.A. Reactions of [Zn{S$_2$P(OR)$_2$}$_2$] with nitrogen bases and the single-crystal X-ray structures of [Zn{S$_2$P(OPr$^i$)$_2$}$_2$]·H$_2$NCH$_2$CH$_2$NH$_2$ and [Zn{S$_2$P(OPr$^i$)$_2$}$_2$]·NC$_5$H$_5$, *J. Chem. Soc. Dalton Trans.*, (6), 1161–1166 (1986).

Drew M.G.B., Rice D.A., Timewell C.W. Crystal and molecular structure of triaquazinc(II) thiodiglycolate monohydrate, *J. Chem. Soc. Dalton Trans.*, (2), 144–148 (1975).

Durán M.L., Romero J., García-Vázquez J.A., Castro R., Castiñeiras A., Sousa A. Electrochemical synthesis of pyridine-2-thionato complexes of zinc(II) and cadmium(II): The crystal structure of 1,10-phenanthroline-bis(pyridine-2-thionato) zinc(II), *Polyhedron*, 10(2), 197–202 (1991).

Durán M.L., Sousa A., Romero J., Castiñeiras A., Bermejo E., West D.X. Structural study of the electrochemically synthesized binuclear complex bis{1-phenylglyoxal bis(3-piperidylthiosemicarbazone)zinc(II)}, *Inorg. Chim. Acta*, 294(1), 79–82 (1999).

Du Z.-Y., Xu H.-B., Mao J.-G. Three novel zinc(II) sulfonate–phosphonates with tetranuclear or hexanuclear cluster units, *Inorg. Chem.*, 45(16), 6424–6430 (2006).

Ellis C.A., Ng S.W., Tiekink E.R.T. μ-1,4-Diazabicyclo[2.2.2]octane-κ$^2$*N*:*N*'-bis[bis-(*O*,*O*'-diisopropyl dithiophosphato-κ$^2$*S*,*S*')zinc(II)], *Acta Crystallogr.*, E63(1), m108–m110 (2007).

Euler H., Barbier B., Klumpp S., Kirfel A. Crystal structure of Tutton's salts, Rb$_2$[M$^{II}$(H$_2$O)$_6$] (SO$_4$)$_2$, M$^{II}$=Mg, Mn, Fe, Co, Ni, Zn, *Z. Kristallogr. New Cryst. Struct.*, 215(4), 473–476 (2000).

Euler H., Barbier B., Meents A., Kirfel A. Crystal structure of Tutton's salts, Cs$_2$[M$^{II}$(H$_2$O)$_6$] (SO$_4$)$_2$, M$^{II}$=Mg, Mn, Fe, Co, Ni, Zn, *Z. Kristallogr. New Cryst. Struct.*, 218(4), 409–413 (2003).

Euler H., Barbier B., Meents A., Kirfel A. Crystal structures of Tutton's salts $Tl_2[M^{II}(H_2O)_6]$ $(SO_4)_2$, $M^{II}$ = Mg, Mn, Fe, Co, Ni, Zn, Z. *Kristallogr. New Cryst. Struct.*, 224(3), 355–359 (2009a).

Euler H., Barbier B., Meents A., Kirfel A. Refinement of the crystal structure of potassium hexaaquairon(II) sulfate, $K_2[Fe(H_2O)_6](SO_4)_2$ and potassium hexaaquazinc(II) sulfate, $K_2[Zn(H_2O)_6](SO_4)_2$, Z. *Kristallogr. New Cryst. Struct.*, 224(2), 171–173 (2009b).

Fackler J.P., Coucouvanis Jr. D., Fetchin J.A., Seidel W.C Sulfur chelates. VIII. Oxidative addition of sulfur to dithioaryl acid complexes of nickel(II) and zinc(II), *J. Am. Chem. Soc.*, 90(11), 2784–2788 (1968).

Ferrari A., Braibanti A., Bigliardi G., Lanfredi A.M. The crystal structure of bis(hydrazine) zinc isothiocyanate, *Acta Crystallogr.*, 18(3), 367–372 (1965).

Fettouhi M, Wazeer M.I.M., Isab A.A. Crystal structure of dibromo-bis(1,3-imidazolidine-2-thione-*S*)zinc(II), $ZnBr_2(C_3H_6N_2S)_2$, Z. *Kristallogr. New Cryst. Struct.*, 221(2), 221–222 (2006).

Fettouhi M., Wazeer M.I.M., Isab A.A. Zinc halide complexes of imidazolidine-2-thione and its derivatives: X-ray structures, solid state, solution NMR and antimicrobial activity studies, *J. Coord. Chem.*, 60(4), 369–377 (2007).

Francotte J., Moreau J., Ottenburgs R., Lévy C. La briartite, $Cu_2(Fe, Zn)GeS_4$, une nouvelle espèce minérale, *Bull. Soc. fr. minéral. cristallogr.*, 88, 432–437 (1965).

Fraser K.A., Harding M.M. The structure of bis-(*N, N*-dimethyldithiocarbamato)pyridinezine, *Acta Crystallogr.*, 22(1), 75–81 (1967).

Gao E.-J., Liu T.-L., Jiao W., Jiang L.-L., Zhang D., Zhang Y.-J., Xu J., Wu G.-L. Crystal structure of a new transition metal–organic Zn(II) complex, *J. Struct. Chem.*, 54(5), 972–977 (2013).

García I., Bermejo E., El Sawaf A.K., Castiñeiras A., West D.X. Structural studies of metal complexes of 2-pyridineformamide *N*(4)-methylthiosemicarbazone, *Polyhedron*, 21(7), 729–737 (2002).

Garner M., Reglinski J., Cassidy I., Spicer M.D., Kennedy A.R. Hydrotris(methimazolyl) borate, a soft analogue of hydrotris(pyrazolyl)borate. Preparation and crystal structure of a novel zinc complex, *Chem. Commun.*, (16), 1975–1976 (1996).

Glinskaya L.A., Klevtsova R.F., Berus E.I., Zemskova S.M., Larionov S.V. Crystal and molecular structures of di-*n*-propyldithiocarbamate zinc(II) with 2.2'-bipyridyl and 1,10-phenanthroline complexes [in Russian], *Zhurn. strukt. khimii*, 39(4), 688–697 (1998a).

Glinskaya I.A., Klevtsova R.F., Leonova T.G., Larionov S.V. Crystal structure of Zn(2,2'-Bipy)(C_2H_5OCS_2)_2 complex with mono- and bidental ethylxanthate ligands [in Russian], *Zhurn. strukt. khimii*, 41(1), 196–201 (2000a).

Glinskaya I.A., Klevtsova R.F., Shchukin V.G., Larionov S.V. Crystal and molecular structure of mixed-ligand complex $ZnPhen\{(i-C_3H_7O)_2PS_2\}_2$ [in Russian], *Zhurn. strukt. khimii*, 41(1), 191–195 (2000b).

Glinskaya L.A., Leonova T.G., Kirichenko V.N., Klevtsova R.F., Larionov S.V. Synthesis, crystal and molecular structures of complex $Zn(Phen)(S_2COC_4H_9-i)_2$ [in Russian], *Zhurn. strukt. khimii*, 38(1), 142–147 (1997).

Glinskaya L.A., Shchukin V.G., Klevtsova R.F., Mazhara A.P., Larionov S.V. Synthesis, polymeric structure of $[Zn(4,4'-Bipy)\{(i-PrO)_2PS_2\}_2]_n$ complex and thermal properties of $ZnL\{(i-PrO)_2PS_2\}_2$ (L=Phen, 2,2'-Bipy, 4,4'-Bipy) compounds [in Russian], *Zhurn. strukt. khimii*, 41(4), 772–780 (2000c).

Glinskaya L.A., Zemskova S.M, Klevtsova R.F. Crystal structure of tris(ethylenediamine) zinc(II) and tris(ethylenediamine)nickel(II) diethylthiocarbamates [in Russian], *Zhurn. strukt. khimii*, 39(2), 353–360 (1998b).

Glinskaya L.A., Zemskova S.M., Klevtsova R.F., Larionov S.V., Gromilov S.A. The preparation, structures and thermal properties of $[MEn_3][Cd(S_2CNEt_2)_3]_2$ [M = zinc(II), cadmium(II)] complexes, *Polyhedron*, 11(22), 2951–2956 (1992).

Godino-Salido M.L., Gutiérrez-Valero M.D., López-Garzón R., Moreno-Sánchez J.M. Zn(II) complexes with thiopyrimidine derivatives: solution study, synthesis and crystal structure of a zig-zag chain zinc(II) complex with the ligand 4,6-dimethyl-2-thiopyrimidine, *Inorg. Chim. Acta*, 221(1–2), 177–181 (1994).

Goedken V.L., Christoph G.G. Helical coordination. Structure and bonding of the five-coordinate complex bis(2-thiobenzaldimino)-2,6-diacetylpiridinezinc(II), *Inorg. Chem.*, 12(10), 2316–2320 (1973).

Gopalakrishnan J., Patel C.C. Diphenyl sulphoxide complexes of some divalent metal ions, *Inorg. Chim. Acta*, 1, 165–168 (1967).

Görgülü A.O., Çukurovali A. Synthesis and characterization of two 2-amino-alkyl-1,3,4-thiadiazolyl carbamate ligands and their Co(II), Ni(II) and Zn(II) complexes, *Synth. React. Inorg. Met.-Org. Chem.*, 32(6), 1033–1042 (2002).

Greene D.L., McCormick B.J., Pierpont C.G. Amine adducts of thiocarbamate complexes. Crystal and molecular structure of bis(cyclopentamethylenethiocarbamato) bis(piperidine)zinc(II), *Inorg. Chem.*, 12(9), 2148–2152 (1973).

Gruff E.S., Koch S.A. A trigonal planar $[Zn(SR)_3]^{1-}$ complex. A possible new coordination mode for zinc-cysteine centers, *J. Am. Chem. Soc.*, 111(23), 8762–8763 (1989).

Grützmacher H., Steiner M., Pritzkow H., Zsolnai L., Huttner G., Sebald A. Mixed amide thiolate complexes of zinc with low coordination number at the metal atom, *Chem. Ber.*, 125(10), 2199–2207 (1992).

Gudimovich T.F., Prisyazhnyuk A.I., Tikhonenko L.M., Ishkova L.D., Feldman S.V., Kolesnichenko A.B. Complexation of zinc chloride with sulfenamide-C [in Russian], *Zhurn. neorg. khimii*, 29(4), 1038–1041 (1984).

Guo S., Ding E., Chen H., Yin Y., Li X. Synthesis, molecular structure and radical scavenging effect of octanuclear thiolate zinc complex $[NMe_4]_2[(c-\mu_4-S)(\mu-SCH_2Ph)_{12}S_4Zn_8]$, *Polyhedron*, 18(5), 735–740 (1999).

Guo T., Lai C.S., Tan X.J., Teo C.S., Tiekink E.R.T. (4,7-Dimethyl-1,10-phenanthroline) bis(pyrrolinedithiocarbamato)zinc(II), *Acta Crystallogr.*, E58(5), m239–m241 (2002).

Guo Y.-N. Crystal structure of diisothiocyanato-{2-[2-(1-pyridin-2-yl-ethylidene-amino)ethylamino]ethanol}zinc(II), $Zn(NCS)_2(C_{11}H_{17}N_3O)$, *Z. Kristallogr. New Cryst. Struct.*, 225(4), 679–680 (2010).

Haasnoot J.G., Groeneveld W.L. Complexes of transition metal(II) thiocyanates with 1,2,4-triazole, *Z. Naturforsch.*, 32B(5), 533–536 (1977).

Hagen K.S., Stephan D.W., Holm R.H. Metal ion exchange reactions in cage molecules: The systems $[M_{4-n}M'_n(SC_6H_5)_{10}]^{2-}$ (M, M' = Fe(II), Co(II), Zn(II), Cd(II)) with adamantane-like stereochemistry and the structure of $[Fe_4(SC_6H_5)_{10}]^{2-}$, *Inorg. Chem.*, 21(11), 3928–3936 (1982).

Hall S.R., Szymański J.T., Stewart J.M. Kesterite, $Cu_2(Zn, Fe)SnS_4$, and stannite, $Cu_2(Zn, Fe)SnS_4$, structurally similar but distinct minerals, *Can. Mineral.*, 16(2), 131–137 (1978).

Halvorsen K., Crosby G.A., Wacholtz W.F. Synthesis and structural determinations of zinc(II) complexes containing dithiol and *N,N*-heterocyclic ligands. *Inorg. Chem. Acta*, 228(2), 81–88 (1995).

Hamdi M., Lafond A., Guillot-Deudon C., Hlel F., Gargouri M., Jobic S. Crystal chemistry and optical investigations of the $Cu_2Zn(Sn,Si)S_4$ series for photovoltaic applications, *J. Solid State Chem.*, 220, 232–237 (2014).

Hammesa B.S., Carrano C.J. Structure and physical properties of several pseudotetrahedral zinc complexes containing a new alkyl thiolate heteroscorpionate ligand, *J. Chem. Soc. Dalton Trans.*, (19), 3304–3309 (2000).

Han M.-L., Ling X.-L., Wang J.-G., Mi L.-W. Crystal structure of *cis*-diaquabis(1,10-phen-anthroline)zinc(II)bis(3-amino-4-chlorobenzensulphonate)   dihydrate,   [Zn(H$_2$O)$_2$(C$_{12}$H$_8$N$_2$)$_2$](H$_2$NC$_6$H$_3$ClSO$_3$)$_2$·2H$_2$O, *Z. Kristallogr. New Cryst. Struct.*, 225(2), 410–412 (2010).

Harrison P.G., Begley M.J., Kikabhai T., Killer F. Zinc(II) bis(*O,O'*-dialkyl dithiophosphates): Complexation behaviour with pyridine and other multidentate nitrogen donor mole-cules. The crystal and molecular structures of the 1:1 complexes of bis(*O,O'*-di-isopropyl dithiophosphato)zinc(II) with pyridine, 2,2'-bipyridine, and 2,2': 6',2"-terpyridine and of (1,11-diamino-3,6,9-triazaundecane)zinc(II) bis(*O,O'*-diethyl dithiophosphate), *J. Chem. Soc. Dalton Trans.*, (5), 929–938 (1986a).

Harrison P.G., Begley M.J., Kikabhai T., Killer F. Zinc(II) bis(*O,O'*-dialkyl dithiophos-phates): Interaction with small nitrogen bases. The crystal and molecular structure of hexakis(μ-*O,O'*-diethyldithiophosphato)-μ$_4$-thio-tetrazinc, Zn$_4$[S$_2$P(OEt)$_2$]$_6$S, *J. Chem. Soc. Dalton Trans.*, (5), 925–928 (1986b).

Harrison W.T.A., Howie R.A., Wardell J.L., Wardell S.M.S.V., Comerlato N.M., Costa L.A.S., Silvino A.C., de Oliveira A.I., Silva R.M. Crystal structures of three [bis(1,3-dithiole-2-thione-4,5-dithiolato)zincate]$^{2-}$ salts: [Q]$_2$[Zn(dmit)$_2$] (Q = 1,4-Me$_2$-pyridinium or NEt$_4$) and [PPh$_4$]$_2$[Zn(dmit)$_2$]·DMSO. Comparison of the dianion packing arrangements in [Q]$_2$[Zn(dmit)$_2$], *Polyhedron*, 19(7), 821–827 (2000).

Harrowfield J.McB., Pakawatchai C., White A.H. Dimorphism of zinc(II) dithizonate, *Aust. J. Chem.*, 36(4), 825–831 (1983).

Hencher J.L., Khan M., Said F.F., Tuck D.G. Electrochemical preparation and structure of an unusual zinc(II) thiolato anionic cluster, *Inorg. Nucl. Chem. Lett.*, 17(9–12), 287–290 (1981).

Hencher J.L., Khan M.A., Said F.F., Tuck D.G. The direct electrochemical synthesis and crystal structure of salts of [M$_4$(SC$_6$H$_5$)10]$^{2-}$ anions(M = Zn, Cd), *Polyhedron*, 4(7), 1263–1267 (1985).

Hewlins M.J.E. Crystal structure of [*NN'*-tetramethylene bis(thioacetylacetoniminato) (2–)] zinc, *J. Chem. Soc. Dalton Trans.*, (5), 429–432 (1975).

Higgins G.M.C., Saville B. Complexes of amines with zinc dialkyldithiocarbamates, *J. Chem. Soc.*, 2812–2817 (1963).

Hîncu I. Asupra sistemului oxidant-reducator sulfura de metal greu (Cu, Zn, Pb) – azotat de amoniu [in Romanian], *Bul. Inst. politehn. Iasi sec. 2*, 17(3/4), 35–47 (1971).

Hîncu I., Golgoţiu T. Contribuţii la studiul reacţiilor in faza solida a sistemelor ZnS–NH$_4$NO$_3$ şi ZnS–NH$_4$NO$_3$–NaCl. II [in Romanian], *Bul. Inst. politehn. Iaşi sec. 2*, 18(1/2), 15–23 (1972).

Hinsche G., Uhlemann E., Weller F. Crystal structure of bis(1-phenyl-3-methyl-4-phenylhy-drazono-pyrazole-5-thionato)zinc, Zn[SCN(Ph)NC(Me)CN(Ph)]$_2$, *Z. Kristallogr. New Cryst. Struct.*, 212(3), 333–334 (1997).

Hollebone B.R., Nyholm R.S. Pseudo-halide complexes of transition metals. Part I. Synthesis and properties of cobalt(II), nickel(II), copper(II), and zinc(II) derivatives, *J. Chem. Soc. A: Inorg. Phys. Theor.*, 332–337 (1971).

Horikoshi R., Mikuriya M. Self-assembly of repeated rhomboidal coordination polymers from 4,4'-dipyridyl disulfide and ZnX$_2$ salts (X = SCN, NO$_3$, ClO$_4$), *Cryst. Growth Des.*, 5(1), 213–230 (2005).

Hursthouse M.B., Malik M.A., Motevalli M., O'Brien P. Mixed alkyl dialkylthiocarbamates of zinc and cadmium: Potential precursors for II/VI materials. X-ray crystal structure of [MeZnS$_2$CNEt$_2$]$_2$, *Organometallics*, 10(3), 730–732 (1991).

Hu W.J., Barton D., Lippard S.J. Synthesis and structure of a tetranuclear zinc(II) complex of *N,N'*-dimethyl-*N,N'*-bis(β-mercaptoethyl)ethylenediamine, *J. Am. Chem. Soc.*, 95(4), 1170–1173 (1973).

Ikeda T., Hagihara H. The crystal structure of zinc ethylxanthate, *Acta Crystallogr.*, 21(6), 919–927 (1966).

Isab A.A., Wazeer M.I.M. Complexation of Zn(II), Cd(II) and Hg(II) with thiourea and selenourea: A $^1$H, $^{13}$C, $^{15}$N, $^{77}$Se and $^{113}$Cd solution and solid-state NMR study, *J. Coord. Chem.*, 58(6), 529–537 (2005).

Ito T., Igarashi T., Hagihara H. The crystal structure of metal diethyldithiophosphates. I. Zinc diethyldithiophosphate, *Acta Crystallogr.*, B25(11), 2303–2309 (1969).

Ito T. The crystal structure of zinc isopropylxanthate, *Acta Crystallogr.*, B28(6), 1697–1704 (1972).

Ivanchenko A.V., Gromilov S.A., Zemskova S.M., Baidina I.A. Investigation of bis(di-*n*-butyldithiocarbamates) zinc(II) and cadmium(II) complexes with 2,2′-bipyridyl [in Russian], *Zhurn. strukt. khimii*, 41(1), 106–115 (2000).

Ivanov A.V., Antzutkin O.N. Isomorphism of bis(diethyldithiocarbamato)zinc(II) adduct with pyridine, [Zn(Py)(EDtc)$_2$]: Hysteresis in the reaction of the adduct formation, *Polyhedron*, 21(27–28), 2727–2731 (2002).

Ivanov A.V., Antzutkin O.N., Larsson A.-C., Kritikos M., Forsling W. Polycrystalline and surface *O,O′*-dialkyldithiophosphate zinc(II) complexes: Preparation, $^{31}$P CP/MAS NMR and single-crystal x-ray diffraction studies, *Inorg. Chim. Acta*, 315(1), 26–35 (2001a).

Ivanov A.V., Forsling W., Antzutkin O.N., Novikova E.V. Adducts of diethyldithiocarbamate complexes of zinc(II) and copper(II) with piperidine [M(Pip)(Edtc)$_2$] and their solvated forms [M(Pip)(Edtc)$_2$]·L (L=C$_6$H$_6$, C$_5$H$_5$N, C$_4$H$_9$NO): Synthesis, EPR and solid-state ($^{13}$C, $^{15}$N) CP/MAS NMR studies, *Russ. J. Coord. Chem.*, 27(3), 158–166 (2001b).

Ivanov A.V., Ivakhnenko E.V., Gerasimenko A.V. Forsling W. A comparative study of the structural organization of zinc complexes with dialkyl-substituted and cyclic dithiocarbamate ligands: Synthesis, single-crystal x-ray diffraction, and CP/MAS $^{13}$C and $^{15}$N NMR [in Russian], *Zhurn. neorgan. khimii*, 48(1), 52–61 (2003a).

Ivanov A.V., Kritikos M., Antzutkin O.N., Forsling W. The structural reorganisation of bis(diethyldithiocarbamato)morpholine–zinc(II) and –copper(II) in the course of solid-state solvation with morpholine and benzene molecules studied by ESR, solid-state $^{13}$C and $^{15}$N CP/MAS NMR spectroscopy and single-crystal x-ray diffraction, *Inorg. Chim. Acta*, 321(1–2), 63–74 (2001c).

Ivanov A.V., Kritikos M., Antzutkin O.N., Lund A. Structure, EPR, and $^{13}$C and $^{15}$N NMR of clathrates of bis(diethyldithiocarbamato)pyridinezinc(II) and bis(diethyldithiocarbamato)pyridinecopper(II) with 1,2-dichloroethane [in Russian], *Zhurn. neorgan. khimii*, 44(10), 1689–1698 (1999a).

Ivanov A.V., Kritikos M., Lund A., Antsutkin O.N., Rodina T.A. Clathrate formation of bis(dithiocarbamato)pyridinezinc(II) and bis(dithiocarbamato)pyridinecopper(II) with benzene by EPR, high-resolution solid-state $^{13}$C and $^{15}$N NMR, and x-ray crystallography [in Russian], *Zhurn. neorgan. khimii*, 43(9), 1482–1490 (1998).

Ivanov A.V., Leskova S.A., Kritikos M., Forsling W. Adduct formation of the zincmorpholinedithiocarbamate complex with morpholine as studied by CP/MAS $^{15}$N NMR and single-crystal x-ray diffraction [in Russian], *Zhurn. neorg. khimii*, 49(12), 2009–2017 (2004a).

Ivanov A.V., Leskova S.A., Mel'nikova M.A., Rodina T.A., Lund A., Antzutkin O.N., Forsling W. Adducts of zinc and copper(II) morpholine di thio carbamate complexes with morpholine of the composition [M(Mf)(MfDtc)$_2$] and [M(Mf)(MfDtc)$_2$]·Mf: Synthesis, thermal analysis, EPR, and CP/MAS $^{13}$C NMR [in Russian], *Zhurn. neorgan. khimii*, 48(3) 462–468 (2003b).

Ivanov A.V., Lutsenko I.A., Gerasimenko A.V., Merkulov E.B. Synthesis, molecular structure, and thermal properties of the supramolecular complex [Zn{NH(CH$_2$)$_4$O} {S$_2$CN(C$_2$H$_5$)$_2$}$_2$]$_2$·CH$_2${N(CH$_2$)$_4$O}$_2$, *Russ. J. Inorg. Chem.*, 53(2), 293–300 (2008).

Ivanov A.V., Lutsenko I.A., Zaeva A.S., Gerasimenko A.V., Merkulov E.B., Leskova S.A. Formation of supramolecular complexes in reactions of adduct formation of zinc diethyldithiocarbamate with morpholine. Molecular structures and thermal properties, *Russ. J. Coord. Chem.*, 33(11), 815–825 (2007a).

Ivanov A.V., Mitrofanova V.I., Kritikos M., Antzutkin O.N. Rotation isomers of bis(diethyldithiocarbamato)zinc(II) adduct with pyridine, Zn(EDtc)$_2$·Py: ESR, $^{13}$C and $^{15}$N CP/MAS NMR and single-crystal x-ray diffraction studies, *Polyhedron*, 18(15), 2069–2078 (1999b).

Ivanov A.V., Novikova E.V., Leskova S.A., Forsling W. Adduct of zinc and copper(II) dialkyldithiocarbamate complexes with hexamethyleneimine: Synthesis, EPR, and CP/MAS $^{13}$C and $^{15}$N NMR [in Russian], *Zhurn. neorg. khimii*, 49(1), 100–106 (2004b).

Ivanov A.V., Zaeva A.S., Novikova E.V., Rodina T.A., Forsling W. Adducts of zinc and copper(II) dialkyldithiocarbamate complexes with dialkylamines: Synthesis, EPR, and $^{13}$C and $^{15}$N CP/MAS NMR, *Russ. J. Inorg. Chem.*, 52(5), 691–697 (2007b).

Jain P.C., Lingafelter E.C., Paoletti P. Five-coordinate zinc(II) in [Zn(tren)(NCS)](SCN), *J. Am. Chem. Soc.*, 90(2), 519–520 (1968).

Jeremias L., Demo G., Babiak M., Vícha J., Trávníček Z., Novosad J. X-ray structures of heteroleptic zinc(II) complexes involving combinations of *O,O'*-dialkyldithiophosphato and bidentate *N*-donor ligands, *Z. Kristallogr.*, 229(7), 537–542 (2014).

Jian F., Hao Q., Yang X., Lu L., Wang X. Structure of bis(*O,O'*-diisopropyldithiophosphato-*S,S'*)(1,10-phenanthroline-*N*$^1$,*N*$^{10}$)zinc(II) complex, [Zn(phen)(S$_2$P(OiPr)$_2$)$_2$], *J. Chem. Crystallogr.*, 30(7), 469–472 (2000).

Jian F.-F., Xue T., Jiao K., Zhang S.-S. Crystal structures and characterizations of bis(pyrrolidinedithiocarbamato) Cu$^{II}$ and Zn$^{II}$ complexes, *Chin. J. Chem.*, 21(1), 50–55 (2003).

Jiang X.N., Xu D., Yuan D., Lu M., Guo S., Zhang G., Wang X., Fang Q. Growth of zinc cadmium thiocyanate single crystal for laser diode frequency-doubling, *J. Cryst. Growth*, 222(4), 755–759 (2001).

Jiang X.N., Yu W.T., Xu D., Yuan D.R., Lu M.K., Guo S.Y., Meng F.Q., Zhang G.H., Wang X.Q., Jiang M.H. Crystal structure of tetrakis(thiourea)cadmium tetrakis(thiocyanato) zincate, [Cd(NH$_2$CSNH$_2$)$_4$][Zn(SCN)$_4$], *Z. Kristallogr. New Cryst. Struct.*, 215(4), 501–502 (2000).

Jie Q., Tiekink E.R.T. Crystal structure of bis(tetramethylenedithiocarbamato)(2,2'-bipyridine)zinc(II), *Main Group Met. Chem.*, 25(5), 317–318 (2002).

Jin Q.-H., Gao H.-W., Dong J.-C., Yang L., Li P.-Z. Crystal structure of μ$^4$-oxo-hexakis(μ-2-mercaptobenzothiazolato-*N,S'*)tetrazinc(II), Zn$_4$O(C$_7$H$_5$NS$_2$)$_6$, *Z. Kristallogr. New Cryst. Struct.*, 222(3), 233–234 (2007).

Johan Z., Mantienne J., Picot P. La routhiérite, TlHgAsS$_3$, et la laffittite, AgHgAsS$_3$, deux nouvelles espèces minérales, *Bull. soc. fr. minéral. cristallogr.*, 97, 48–53 (1974).

Johnstone D., Fergusson J.E., Robinson W.T. Novel polymetallic cobalt and zinc complexes formed with an arsenic sulphide ligand. The x-ray structure determination of racemic hexa-μ-dithiocacodylato-tetrazincsulphide [SZn$_4${AsS$_2$(CH$_3$)$_2$}$_6$], *Bull. Chem. Soc. Jpn.*, 45(12), 3721 (1972).

Jordan K.J., Wacholtz W.F., Crosby G.A. Structural dependence of the luminescence from bis(substituted benzenethiolato)(2,9-dimethyl-1,10-phenanthroline)zinc(II) complexes, *Inorg. Mater.*, 30(24), 4588–4593 (1991).

Joseph G.P., Rajarajan K., Vimalan M., Thomas P.C., Madhavan J., Kumar S.M.R., Mohamed M.G., Mani G., Sagayaraj P. Growth and characterization of non-linear optical crystal ZnCd(SCN)$_4$, *Cryst. Res. Technol.*, 42(3), 247–252 (2007).

Kaptein B., Wang-Griffin L., Barf G., Kellogg R.M. Metal complexes that model the active site of liver alcohol dehydrogenase, *J. Chem. Soc. Chem. Commun.*, (19), 1457–1459 (1987).

Karakus M., Yilmaz H., Bulak E., Lönnecke P. Bis{μ-[O-cyclopentyl(4-methoxy-phenyl)dithiophosphonato]1κ:S,2κ:S-[O-cyclopentyl(4-methoxyphenyl) dithiophosphonato]1κ²S,S'}dizinc(II), *Appl. Organometal. Chem.*, 19(3), 396–397 (2005).

Kato K., Kanke Y., Oka Y., Yao T. Crystal structure of zinc hydroxide sulfate vanadate(V), Zn$_7$(OH)$_3$(SO$_4$)(VO$_4$)$_3$, *Z. Kristallogr. New Cryst. Struct.*, 213(1), 26 (1998).

Kennedy B.P., Lever A.B.P. Studies of the metal–sulfur bond. Complexes of the pyridine thiols, *Can. J. Chem.*, 50(21), 3488–3507 (1972).

Ketcham K.A., Swearingen J.K., Castiñeiras A., Garcia I., Bermejo E., West D.X. Iron(III), cobalt(II,III), copper(II) and zinc(II) complexes of 2-pyridineformamide 3-piperidyl-thiosemicarbazone, *Polyhedron*, 20(28), 3265–3273 (2001).

Kimblin C., Bridgewater B.M., Churchill D.G., Parkin G. Mononuclear tris(2-mercapto-1-ar-ylimidazolyl)hydroborato complexes of zinc, [Tm$^{Ar}$]ZnX: Structural evidence that a sulfur rich coordination environment promotes the formation of a tetrahedral alcohol complex in a synthetic analogue of LADH, *Chem. Commun.*, (22), 2301–2302 (1999).

Kimblin C., Bridgewater B.M., Hascall T., Parkin G. The synthesis and structural characterization of bis(mercaptoimidazolyl)hydroborato complexes of lithium, thallium and zinc, *J. Chem. Soc. Dalton Trans.*, (6), 891–897 (2000).

Kimblin C., Hascall T., Parkin G. Modeling the catalytic site of liver alcohol dehydrogenase: Synthesis and structural characterization of a [bis(thioimidazolyl)(pyrazolyl)hydrobo-rato]zinc complex, [HB(tim$^{Me}$)$_2$pz]ZnI, *Inorg. Chem.*, 36(25), 5680–5681 (1997).

Klevtsova R.F., Glinskaya L.A., Berus E.I., Larionov S.V. Monodentate and bidentate-bridging function of 4,4'-bipyridine in the crystal structures of [Zn(4,4'-Bipy){(n-C$_3$H$_7$)$_2$NCS$_2$}$_2$] and [Zn$_2$(4,4'-Bipy){(n-C$_3$H$_7$)$_2$NCS$_2$}$_4$] complexes [in Russian], *Zhurn. strukt. khimii*, 42(4), 766–775 (2001a).

Klevtsova R.F., Glinskaya L.A., Leonova T.G., Larionov S.V. Mixed-ligand complexes of Zn(2,2'-Bipy)(ROCS$_2$)$_2$ with monodentate and bidentate ROCS$_2^-$ ligands (R = i-Pr, i-Bu) [in Russian], *Zhurn. strukt. khimii*, 43(1), 132–140 (2002).

Klevtsova R.F., Glinskaya L.A., Leonova T.G., Larionov S.V. Two modifications of the mixed-ligand complex ZnPhen(i-C$_3$H$_7$OCS$_2$)$_2$ with monodentate and bidentate i-C$_3$H$_7$OCS$_2^-$ ligands [in Russian], *Zhurn. strukt. khimii*, 42(2), 293–301 (2001b).

Klevtsova R.F., Glinskaya L.A., Shchukin V.G., Larionov S.V. Molecule packing type in crystal structures of mixed-ligand complexes [ZnPhen{(i-C$_4$H$_9$)$_2$PS$_2$}$_2$] and [Zn(2,2'-Bipy) {(i-C$_4$H$_9$)$_2$PS$_2$}$_2$] [in Russian], *Zhurn. strukt. khimii*, 42(3), 536–544 (2001c).

Klevtsova R.F., Glinskaya L.A., Zemskova S.M., Larionov S.V. Crystal and molecular structure of volatile mixed-ligand Zn(S$_2$CN(i-C$_4$H$_9$)$_2$)$_2$Phen complex [in Russian], *Zhurn. strukt. khimii*, 40(1), 70–76 (1999a).

Klevtsova R.F., Glinskaya L.A., Zemskova S.M., Larionov S.V. Crystal and molecular structure of volatile mixed-ligand Zn[S$_2$CN(CH$_3$)$_2$]Phen complex [in Russian], *Zhurn. strukt. khimii*, 40(1), 77–84 (1999b).

Klevtsova R.F., Glinskaya L.A., Zemskova S.M., Larionov S.V. Synthesis and crystal structure of ZnPhen(i-Pr$_2$NCS$_2$) and molecule packing types in complexes ZnPhen(R$_2$NCS$_2$)$_2$ (R = Me, Et, i-Pr, n-Pr, i-Bu, n-Bu), *Polyhedron*, 18(26), 3559–3565 (1999c).

Klevtsova R.F., Leonova T.G., Glinskaya L.A., Larionov S.V. Crystal and molecular structures of mixed-ligand coordination compounds ZnPhen(C$_2$H$_5$OCS$_2$)$_2$ and Zn(2,2'-Bipy) (n-C$_4$H$_9$OCS$_2$)$_2$ [in Russian], *Zhurn. strukt. khimii*, 47(3), 517–525 (2006a).

Klevtsova R.F., Leonova T.G., Glinskaya L.A., Larionov S.V. Synthesis, crystal and molecular structure of mixed-ligand coordination compound ZnPhen(n-C$_4$H$_9$OCS$_2$)$_2$ [in Russian], *Zhurn. strukt. khimii*, 47(6), 1189–1194 (2006b).

Klug H.P. The crystal structure of zinc dimethyldithiocarbamate, *Acta Crystallogr.*, 21(4), 536–546 (1966).

Korczyński A. The crystal structure of Cd[SC(NH$_2$)$_2$]$_4$Zn(SCN)$_4$ [in Polish], *Rocz. Chemii*, 41(7–8), 1197–1203 (1967).

Kowalski R.S.Z., Bailey N.A., Mulvaney R., Adams H., O'Cleirigh D.A., McCleverty J.A. Anionic tris-dithiophosphato and -dithiophosphinato complexes of zinc(II) and cobalt(II). The structures of [R$_4$N][M(S$_2$PPh$_2$)$_3$] (M=Zn and Co; R=Et or Me) and of [Me$_4$N][Zn{S$_2$P(OC$_6$H$_4$Me-$p$)$_2$}$_3$], *Trans. Met. Chem.*, 6(1), 64–66 (1981).

Kunchur N.R., Truter M.R. The crystal structure of dichlorobisthioureazinc, *J. Chem. Soc.*, 3478–3484 (1958).

Lai C.S., Lim Y.X., Yal T.C., Tiekink E.R.T. Molecular paving with zinc thiolates, *CrystEngChem*, 4(99), 596–600 (2002).

Lai C.S., Liu S., Tiekink E.R.T. *catena*-Poly[[bis($O,O'$-diisobutyldithiophosphato-$\kappa^2 S,S'$) zinc(II)]-μ-1,2-bis(4-pyridyl)ethane-$\kappa^2 N{:}N'$], *Acta Crystallogr.*, E60(7), m1005–m1007 (2004a).

Lai C.S., Liu S., Tiekink E.R.T. Steric control over polymer formation and topology in adducts of zinc dithiophosphates formed with bridging bipyridine ligands, *CrystEngComm*, 6(38), 221–226 (2004b).

Lai C.S., Tiekink E.R.T. (4,4'-Bipyridine)bis[bis($N,N$-diethyldithiocarbamato)zinc(II)], *Appl. Organometal. Chem.*, 17(4), 253–254 (2003a).

Lai C.S., Tiekink E.R.T. (2,2'-Bipyridine)bis($N,N$-dibenzyldithiocarbamato)zinc(II), *Appl. Organometal. Chem.*, 18(4), 197–198 (2004c).

Lai C.S., Tiekink E.R.T. Bis[bis($N,N$-diethyldithiocarbamato)zinc(II)] (*trans*-1,2-bis(4-pyridyl)ethylene)*trans*-1,2-bis(4-pyridyl)ethylene lattice adduct, *Appl. Organometal. Chem.*, 17(4), 251–252 (2003b).

Lai C.S., Tiekink E.R.T. (2,9-Dimethyl-1,10-phenanthroline)bis-($N,N$-pyrrolidinedithiocarbamato)zinc(II) chloroform hemihydrate, *Appl. Organometal. Chem.*, 17(4), 255–256 (2003c).

Lai C.S., Tiekink E.R.T. (4,7-Diphenyl-1,10-phenanthroline) bis(pyrrolinedithiocarbamato) zinc(II) chloroform solvate, *Appl. Organometal. Chem.*, 17(3), 197–198 (2003d).

Larionov S.V., Glinskaya L.A., Leonova T.G., Klevtsova R.F. Polymeric structure of the complex [Zn(4,4'-Bipy)($i$-BuOCS$_2$)$_2$]$_n$ with monodentate isobutyl xanthate ligands [in Russian], *Koord. khim.*, 24(12), 909–914 (1998).

Larionov S.V., Kirichenko V.N., Zemskova C.M., Oglezneva I.M. Synthesis of complexes of diethyldithiocarbamates of zinc(II), cadmium(II), mercury(II) with a nitrogen-containing ligands, and the study of their sublimation [in Russian], *Koord. khim.*, 16(1), 79–84 (1990).

Larionov S.V., Klevtsova R.F., Shchukin V.G., Glinskaya L.A., Zemskova S.M. Heteroligand complexes of Zn(II) diisopropyl dithiocarbamate with 1,10-phenantroline and 4,4'-bipyridine. Crystal and molecular structure of clathrate compound Zn$_2$(4,4'-Bipy)($i$-C$_3$H$_7$)$_2$(NCS$_2$)$_4$·2C$_6$H$_5$CH$_3$ [in Russian], *Koord. khim.*, 25(10), 743–749 (1999).

Laudise R.A., Ballman A.A. Hydrothermal synthesis of zinc oxide and zincsulfide, *J. Phys. Chem.* 64(5), 688–691 (1960).

Lawton S.L., Kokotailo G.T. Crystal and molecular structures of zinc and cadmium $O,O$-diisopropylphosphorodithioates, *Inorg. Chem.*, 8(11), 2410–2421 (1969).

Leonova T.G., Kirichenko V.N., Glinskaya L.A., Klevtsova R.F., Sheludiakova L.A., Batrachenko N.I., Durasov V.B., Lisoyvan V.I., Larionov S.V. Heteroligand coordination compounds of zinc(II) alkylxanthates with nitrogen-containing heterocycles [in Russian], *Koord. khim.*, 23(8), 590–595 (1997).

Lian Z.-X., Li H.-H. *trans*-Diaquabis($N,N'$-dimethylethylenediamine)zinc(II) naphthalene-2,6-disulfonate, *Acta Crystallogr.*, E63(3), m731–m733 (2007).

Lian Z., Zhao N., Liu P. Crystal structure of *trans*-tetraaquabis(nicotinamide)zinc(II) 4,4'-biphenyldisulfonate, [Zn(H$_2$O)$_4$(C$_6$H$_6$N$_2$O)$_2$][C$_{12}$H$_8$S$_2$O$_6$], *Z. Kristallogr. New Cryst. Struct.*, 225(2), 379–380 (2010).

Li G.-F., Chang X.-H. Crystal structure of bis[(hydrogen 5-methylisophthalato)-(2,5-bis(4-pyridyl)-1,3,4-thiadiazole)]zinc(II) dihydrate, Zn(C$_9$H$_7$O$_4$)$_2$(C$_{12}$H$_8$N$_4$S)$_2$·2H$_2$O, *Z. Kristallogr. New Cryst. Struct.*, 225(2), 311–312 (2010).

Li H. (Hexasulfanediyl)di(pyridine-*N*)zinc, *Acta Crystallogr.*, C50(4), 498–500 (1994).

Li M., Sun Q., Bai Y., Duan C., Zhang B., Meng Q. Chiral aggregation and spontaneous resolution of thiosemicarbazone metal complexes, *Dalton Trans.*, (21), 2572–2578 (2006).

Lindoy L.F., Busc D.H. Helical co-ordination: Five-co-ordinate zinc and cadmium complexes formed by metal-ion-induced ligand reactions, *J. Chem. Soc. Chem. Commun.*, (11), 683–684 (1972).

Li S.-H., Zhao Y., Zhang M.-Y., Zhu W.-Q. Crystal structure of poly[bis($\mu_2$–1,3-bi(4-pyridyl)propane-$\kappa^2 N{:}N'$)-bis(4-amino-3-methyl-benzenesulfonate-$\kappa^1 O$)zinc(II)], $C_{40}H_{44}N_6O_6S_2Zn$, *Z. Kristallogr. New Cryst. Struct.*, 229(4), 417–418 (2014).

Liu C.-M., Zhang D.-Q., Song Y.-L., Zhan C.-L., Li Y.-L., Zhu D.-B. Synthesis, crystal structure and third-order nonlinear optical behavior of a novel dimeric mixed-ligand zinc(II) complex of 1,3-dithiole-2-thione-4,5-dithiolate, *Eur. J. Inorg. Chem.*, (7), 1591–1594 (2002).

Liu S.-H., Chen X.-F., Zhu X.-H., Duan C.-Y., You X.-Z. Crystal structure and thermal analysis of 4,4′-bipy-bridged binuclear zinc(II) complex, $2[R_2NCS_2]_2{\cdot}Zn(4,4'\text{-bipy})$ (R=piperidyl), *J. Coord. Chem.*, 53(3), 223–231 (2001).

Liu S.-H., Zhang R.-L., Li E.-X., Xu Q.-Z. Crystal structure of diaqua-bis(1,10-phenanthroline)-zinc(II) bis(thiophene-3-carboxy-4-carboxylate)heptahydrate, $[Zn(H_2O)_2(C_{12}H_8N_2)_2]$ $[(C_6H_3O_4S)_2]{\cdot}7H_2O$, $C_{36}H_{40}N_4O_{17}S_2Zn$, *Z. Kristallogr. New Cryst. Struct.*, 229(2), 141–142 (2014).

Li Z., Loh Z.-H., Fong A.S.-W., Yan Y.-K., Henderson W., Mok K.F., Hor T.S.A. Ligand-stabilization of an unusual square-based pyramidal geometry of Cd(II) and Zn(II) in an heterometallic {MPt$_2$S$_2$} core (M=Cd, Zn), *J. Chem. Soc. Dalton Trans.*, (7), 1027–1031 (2000).

Lobana T.S., Sharma Renu, Sharma Rekha, Sultana R., Butcher R.J. Metal derivatives of 1,3-imidazolidine-2-thione with divalent d$^{10}$ metal ions (Zn–Hg): Synthesis and structures, *Z. Anorg. und Allg. Chem.*, 634(4), 718–723 (2008).

Lopez-Garzon R., Gutierrez-Valero M.D., Godino-Salido M.L., Keppler B.K., Nuber B. Spectroscopic studies of metal-pyrimidine complexes. Crystal structures of 4,6-dimethyl-2-thiopyrimidine complexes with Zn(II) and Cd(II), *J. Coord. Chem.*, 30(2), 111–123 (1993).

López-Torres E., Mendiola M.A., Rodríguez-Procopio J., Sevilla M.T., Colacio E., Moreno J.M., Sobrados I. Synthesis and characterisation of zinc, cadmium and mercury complexes of benzilbisthiosemicarbazone. Structure of cadmium derivative, *Inorg. Chim Acta*, 323(1–2), 130–138 (2001).

Lucas C.R., Peach M.E. Metal derivatives of pentachlorothiophenol, *Inorg. Nucl. Chem. Lett.*, 5(2), 73–76 (1969).

Lutsenko I.A., Ivanov A.V., Zaeva A.S., Gerasimenko A.V. Synthesis, structure, and thermal properties of benzene-solvated forms of bis(dimethyldithiocarbamato)piperidinezinc and -copper(II), *Russ. J. Coord. Chem.*, 36(7), 502–508 (2010).

Ma C., Beckett J.R., Rossman G.R. Buseckite, (Fe, Zn, Mn)S, a new mineral from the Zakłodzie meteorite, *Am. Mineralogist*, 97(7), 1226–1233 (2012).

Ma C. Buseckite, IMA 2011–070. CNMNC Newsletter No. 11, December 2011, page 2891, *Mineral. Mag.*, 75(6), 2887–2893 (2011).

MacLachlan M.J., Coombs N., Ozin G.A. Non-aqueous supramolecular assembly of meso-structured metal germanium sulphides from $(Ge_4S_{10})^{4-}$ clusters, *Nature*, 397(6721), 681–684 (1999).

Mahadevappa D.S., Murthy A.S.A. Some complexes of zinc(II), cadmium(II), and mercury(II) with thiosemicarbazide, *Aust. J. Chem.*, 25(7), 1565–1568 (1972).

Mak T.C.W., Yip W.-H., Smith G., O'Reilly E.J., Kennard C.H.L. Metal–(phenylthio) acetic acid interactions. Part I. The crystal structures of (phenylthio)acetic acid, diaquabis[(phenylthio)acetato]zinc(II), and *catena*{aquabis[(phenylthio)acetato] cadmium(II), *Inorg. Chim. Acta*, 84(1), 57–64 (1984).

Ma L.-F., Wu Y., Lu D.-H. Crystal structure of triaqua-*cis*-bis((2-carboxybenzene)sulfonyl-glycinato)zinc(II), $Zn(H_2O)_3(C_9H_8NO_6S)_2$, *Z. Kristallogr. New Cryst. Struct.*, 222(4), 435–436 (2007).

Malik M.A., Motevalli M., O'Brien P., Walsh J.R. An unusual coordination mode in a bis(dialkyldithiocarbamato)zinc(II) adduct with *N,N,N',N'*-tetramethylethylenediamine: X-ray crystal structures of $2[Me^iPrNCS_2]_2Zn \cdot Me_2N(CH_2)_2NMe_2$ and $[Me^iPrNCS_2]_2Zn \cdot C_5H_5N$, *Inorg. Chem.*, 36(6), 1263–1264 (1997).

Malik M.A., Motevalli M., Walsh J.R., O'Brien P. Neopentyl- or *tert*-butylzinc complexes with diethylthio- or diethylselenocarbamates: Precursors for zinc chalcogens, *Organometallics*, 11(9), 3136–3139 (1992).

Manohar A., Ramalingam K., Bocelli G., Righi L. Synthesis, spectral and cyclic voltammetric studies on (4,4'-bipyridyl)bis(di(2-hydroxyethyl)dithiocarbamato)zinc(II) and (4,4'-bipyridyl)bis(bis(*N*-methyl, *N*-ethanoldithiocarbamato)zinc(II) and their x-ray crystal structures, *Inorg. Chim. Acta*, 314(1–2), 177–183 (2001).

Manos M.J., Iyer R.G., Quarez E., Liao J.H., Kanatzidis M.G. $\{Sn[Zn_4Sn_4S_{17}]^{6-}\}$: A robust open framework based on metal-linked supertetrahedral $[Zn_4Sn_4S_{17}]^{10-}$ clusters possessing ion-exchange properties, *Angew. Chem. Int. Ed. Engl.*, 44(23), 3552–3555 (2005).

Mary P.A.A., Dhanuskodi S. Growth and characterization of a new nonlinear optical crystal: Bis thiourea zinc chloride, *Cryst. Res. Technol.*, 36(11), 1231–1237 (2001).

Mathew M., Palenik G.J. The crystal structure of dichloro-(acetonethiosemicarbazone) zinc(II), *Inorg. Chim Acta*, 5, 349–353 (1971).

Math K.S., Freiser H. Crystal and molecular structure of zinc dithizonate, *Talanta*, 18(4), 435–437 (1971).

Matsunaga Y., Fujisawa K., Amir N., Miyashita Y., Okamoto K.-I. Dichloro[bis(1-methylimidazole-2)disulfide]zinc(II), *Appl. Organometal. Chem.*, 19(1), 208 (2005a).

Matsunaga Y., Fujisawa K., Amir N., Miyashita Y., Okamoto K.-I. Group 12 metal(II) complexes with 1-methylimidazoline-2(3*H*)-thione (mitH): Correlation between crystal structure and physicochemical property, *J. Coord. Chem.*, 58(12), 1047–1061 (2005b).

Mawby A., Irving H.M.N.H. The crystal structure of primary zinc(II) dithizonate and its relevance to the stabilities of its *o*- and *p*-methyl derivatives, *Analyt. Chim. Acta*, 55(1), 269–272 (1971).

Mawby A., Irving H.M.N.H. The crystal structure of primary zinc(II) dithizonate, *J. Inorg. Nucl. Chem.*, 34(1), 109–115 (1972).

McCleverty J.A., Kowalski R.S.Z., Bailey N.A., Mulvaney R., O'Cleirigh D.A. Aspects of the inorganic chemistry of rubber vulcanization. Part 4. Dialkyl- and diaryl-dithiophosphate and -dithiophosphinate complexes of zinc: Phosphorus-31 nuclear magnetic resonance spectral studies and structures of $[NMe_4][Zn\{S_2P(OC_6H_4Me-p)_2\}_3]$ and $[NEt_4]$ $[Zn(S_2PPh_2)_3]$, *J. Chem. Soc. Dalton Trans.*, (4), 627–634 (1983).

McCleverty J.A., Morrison N.J. Metal dithiocarbamates and related species. Part 2. Reaction of zinc thiuram disulphides with triphenylphosphine and other Lewis bases, and anionic dithiocarbamato-complexes of zinc, *J. Chem. Soc. Dalton Trans.*, (21), 2169–2175 (1976).

McCleverty J.A., Morrison N.J., Spencer N., Ashworth C.C., Bailey N.A., Johnson M.R., Smith J.M.A., Tabbiner B.A., Taylor C.R. Aspects of the inorganic chemistry of rubber vulcanisation. Part 2. Anionic mixed-ligand zinc complexes derived from dialkyldithiocarbamates, 2-mercapto-benzothiazole, -benzoxazole, and -benzimidazole, and the crystal and molecular structures of $[NEt_4][Zn(S_2CNMe_2)_3]$, $[NBu^n_4][Zn(C_7H_4NS_2)_3(OH_2)]$, $[NBu^n_4][Zn(S_2CNMe_2)_2(C_7H_4NS_2)] \cdot C_2H_5OH$, and $[NBu^n_4][Zn(S_2CNMe_2)(C_7H_4NS_2)_2]$, *J. Chem. Soc. Dalton Trans.*, (10), 1945–1957 (1980a).

McCleverty J.A., Spencer N., Bailey N.A., Shackleton S.L. Aspects of the inorganic chemistry of rubber vulcanisation. Part 1. Reactions of zinc bis(dithiocarbamates) and bis(benzothiazole-2-thiolates) with carboxylates, and the structure of $[NBu^n_4]$ $[\{Zn(S_2CNMe_2)_2\}_2(\mu\text{-}OCOMe)]$, *J. Chem. Soc. Dalton Trans.*, (10), 1939–1944 (1980b).

McCormick B.J., Greene D.L. Thiocarbamate complexes. II. Derivatives of zinc(II), cadmium(II) and mercury(II), *Inorg. Nucl. Chem. Lett.*, 8(7), 599–703 (1972).

Meek D.W., Nicpon P. Metal complexes of tertiary phosphine sulfides, *J. Am. Chem. Soc.*, 87(21), 4951–4952 (1965).

Menzer S., Phillips J.R., Slawin A.M.Z., Williams D.J., Woollins J.D. Structural characterisation of basic zinc $O,O'$-dialkyl dithiophosphate and two isomeric examples of zinc monothiophosphates, *J. Chem Soc. Dalton Trans.*, (19), 3269–3273 (2000).

Miyamae H., Ito M., Iwasaki H. The structure of zinc(II) $N,N$-diisopropyldithiocarbamate {bis[μ-($N,N$-diisopropyldithiocarbamato-μ-$S,S'$]-bis($N,N$-diisopropyldithiocarbamato) dizinc(II)}, *Acta Crystallogr.*, B35(6), 1480–1482 (1979).

Moh G.H. Tin-containing mineral systems. Part II: Phase relations and mineral assemblages in the Cu–Fe–Zn–Sn–S system, *Chem. Erde*, 34(1), 1–61 (1975).

Montgomery H., Lingafelter E.C. The crystal structure of Tutton's salts. I. Zinc ammonium sulfate hexahydrate, *Acta Crystallogr.*, 17(10), 1295–1299 (1964).

Mouallem-Bahout M., Peña O., Carel C., Ouammou A., Retat M. Obtaining and properties of the $ACuZnS_2$ (A = K, Rb) and $BaKCu_3ZnS_4$ sulfides with low-symmetry structure of the $ThCr_2Si_2$-type [in Russian]. *Zhurn. neorgan. khimii*, 46(5), 742–747 (2001).

Müller B., Schneider A., Tesmer V., Vahrenkamp H. Alcohol and aldehyde adducts of zinc thiolates: Structural modeling of alcoholdehydrogenase, *Inorg. Chem.*, 38(8), 1900–1907 (1999).

Nardelli M., Coghi L. Su alcuni complessi di metalli bivalenti con molecole organiche ossigenate (formamide, acetamide, metilurea, *Ricerca scient.*, 29(1), 134–138 (1959).

Nardelli M., Chierici I. Acetati complessi di metalli bivalenti con molecole solforate (tiourea ed etilentiourea, *Ricerca scient.*, 29(8), 1731–1732 (1959).

Nardelli M., Chierici I. Complessi di cloruri metallici con tiosemicarbazide, *Ricerca scient.*, 30(2), 276–279 (1960).

Ng M.T., Deivaraj T.C., Klooster W.T., McIntyre G.J., Vittal J.J. Hydrogen-bonded polyrotaxane-like structure containing cyclic $(H_2O)_4$ in $[Zn(OAc)_2\cdot(\mu$-bpe)]·$2H_2O$: X-ray and neutron diffraction studies, *Chem. Eur. J.*, 10(22), 5853–5859 (2004).

Ng M.T., Deivaraj T.C., Vittal J.J. Self assembly of heptanuclear zinc(II) clusters linked by angular spacer ligands, *Inorg. Chim. Acta*, 348, 173–178 (2003).

Nouadji M., Ivanjva Z.G., Poulain M., Zavadil J., Attaf A. Glass formation, physicochemical characterization and phototoluminescence properties of new $Sb_2O_3$–PbO–ZnO and $Sb_2O_3$–PbO–ZnS systems, *J. Alloys Compd.*, 549, 158–162 (2013).

Nowell I.W., Cox A.G., Raper E.S. Structure of tetrakis[1-methyl-2(3*H*)-imidazolinethione] zinc(II) nitrate monohydrate, *Acta Crystallogr.*, B35(12), 3047–3050 (1979).

Nyman M.D., Hampden-Smith M.J., Duesler E.N. Low temperature, aerosol-assisted chemical vapor deposition (AACVD) of CdS, ZnS, and $Cd_{1-x}Zn_xS$ using monomeric single-source precursors: $M(SOCCH_3)_2TMEDA$, *Chem. Vap. Depos.*, 2(5), 171–174 (1996).

Nyman M.D., Hampden-Smith M.J., Duesler E.N. Synthesis and characterization of the first neutral zinc–sulfur cluster: $Zn_{10}S_4(SEt)_{12}L_4$, *Inorg. Chem.*, 35(4), 802–803 (1996).

Nyman M.D., Hampden-Smith M.J., Duesler E.N. Synthesis, characterization, and reactivity of group 12 metal thiocarboxylates, $M(SOCR)_2Lut_2$ [M = Cd, Zn; R = $CH_3$, $C(CH_3)_3$; Lut = 3,5-dimethylpyridine (lutidine)], *Inorg. Chem.*, 36(10), 2218–2224 (1997).

Olmstead M.M., Power P.P., Shoner S.C. Synthesis and characterization of novel quasiaromatic zinc–sulfur aggregates and related zinc–oxygen complexes, *J. Am. Chem. Soc.*, 113(9), 3379–3385 (1991).

Onwudiwe D.C., Ajibade P.A. Synthesis, characterization and thermal studies of Zn(II), Cd(II) and Hg(II) complexes of *N*-methyl-*N*-phenyldithiocarbamate: The single crystal structure of $[(C_6H_5)(CH_3)NCS_2]_4Hg_2$, *Int. J. Mol. Sci.*, 12(3), 1964–1978 (2011).

Otto J., Jolk I., Viland T., Wonnemann R., Krebs B. Metal(II) complexes with monodentate S and N ligands as structural models for zinc–sulfur DNA-binding proteins, *Inorg. Chim. Acta*, 285(2), 262–268 (1999).

Pan Z.R., Zhang Y.C., Song Y.L., Zhuo X., Li Y.Z., Zheng H.G. Synthesis, structure and nonlinear optical properties of three dimensional compounds, *J. Coord. Chem.*, 61(20), 3189–3199 (2008).

Paz F.A.A., Neves M.C., Trindade T., Klinowski J. The first dinuclear zinc(II) dithiocarbamate complex with butyl substituent groups, *Acta Crystallogr.*, E59(11), 1067–1069 (2003).

Pierpont C.G., Greene D.L., McCormick B.J. Crystal and molecular structure of the bis-piperidine adduct of bis(pentamethylenethiocarbamato)zinc(II), *J. Chem. Soc. Chem. Commun.*, (16), 960–961 (1972).

Poplaukhin P., Arman H.D., Tiekink E.R.T. Supramolecular isomerism in coordination polymers sustained by hydrogen bonding: Bis[Zn(S₂CN(Me)CH₂CH₂OH)₂]·(N,N'-bis(pyridin-3-ylmethyl)thioxalamide), *Z. Kristallogr.*, 227(6), 363–368 (2012).

Poplaukhin P., Tiekink E.R.T. (μ-1,2-Di-4-pyridylethylene-κ²N:N')bis-[bis(N,N-dimethyl-dithiocarbamato-κ²S,S')zinc(II)], *Acta Crystallogr.*, E65(11), m1474 (2009).

Poplaukhin P., Tiekink E.R.T. Interwoven coordination polymers sustained by tautomeric forms of the bridging ligand, *CrystEngComm*, 12(4), 1302–1306 (2010).

Potočňák I., Dunaj-Jurčo M. Bis(thiourea-κS)bis(trichloroacetato-κO)zinc(II) monohydrate, *Acta Crystallogr.*, C50(12), 1902–1904 (1994).

Power P.P., Shoner S.C. Synthesis and structure of [Et₂OZn(SC₆H₂tBu₃)₂], the first T-shaped zinc complex, *Angew. Chem. Int. Ed. Engl.*, 29(12), 1403–1404 (1990a).

Power P.P., Shoner S.C. Synthese und Struktur von [Et₂OZn(SC₆H₂tBu₃)₂], dem ersten T-förmigen Zink-Komplex, *Angew. Chem.*, 102(12), 1484–1485 (1990b).

Prakasam B.A., Ramalingam K., Bocelli G, Cantoni A. NMR and fluorescence spectral studies on bisdithiocarbamates of divalent Zn, Cd and their nitrogenous adducts: Single crystal x-ray structure of (1,10-phenanthroline)bis(4-methylpiperazinecarbodithioato) zinc(II), *Polyhedron*, 26(15), 4489–4493 (2007).

Preti C., Tosi G., De Filippo D., Verani G. Group IIB metal complexes with thiazolidine-2-selenone and thiazolidine-2-one as ligands, *J. Inorg. Nucl. Chem.*, 36(12), 3725–3729 (1974).

Preti C., Tosi G. Tautomeric equilibrium study of thiazolidine-2-thione. Transition metal complexes of the deprotonated ligand, *Can. J. Chem.*, 54(10), 1558–1562 (1976).

Pretorius J.A., Boeyens J.C.A. Structural aspects of synergistic extraction–IV. Crystal structures of the pyridine adducts of the Co(II), Ni(II), Cu(II) and Zn(II) complexes of thenoyltrifluoroacetone and of trifluoroacetate and their extraction properties, *J. Inorg. Nucl. Chem.*, 40(10), 1745–1763 (1978).

Qin J.-H., Song G., Li W. Crystal structure of tetraaqua-bis(2-(4-pyridyl)thiazole-4-carboxylate)zinc(II), Zn(H₂O)₄(C₉H₅O₂N₂S)₂, *Z. Kristallogr. New Cryst. Struct.*, 224(1), 151–152 (2009).

Qiu X.-Y., Liu Q.-X., Wang Z.-G., Lin Y.-S., Zeng W.-J., Fun H.-K., Zhu H.-L. Crystal structure of sulfatobis{2-[N-(2-hydroxyethylaminoethyl)iminomethyl]phenolatozinc(II)}, [(ZnC₁₁H₁₅N₂O₂)₂SO₄], *Z. Kristallogr. New Cryst. Struct.*, 219(2), 150–152 (2004).

Rajarajan K., Selvakumar S., Joseph G.P., Potheher I.V., Mohamed M.G., Sagayaraj P. Growth, dielectric and photoconducting studies of tetrathiourea mercury(II) tetrathiocyanato zinc(II) NLO single crystals, *J. Cryst. Growth*, 286(2), 470–475 (2006).

Rajasekaran R., Kumar R.M., Jayavel R., Ramasamy P. Influence of pH on the growth and characteristics of nonlinear optical zinc thiourea chloride (ZTC) single crystals, *J. Cryst. Growth*, 252(1–3), 317–327 (2003).

Rajasekaran R., Ushasree P.M., Jayavel R., Ramasamy P. Growth and characterization of zinc thiourea chloride (ZTC): A semiorganic nonlinear optical crystal, *J. Cryst. Growth*, 229(1–4), 563–567 (2001).

Ramalingam K., bin Shawkataly O., Fun H.-K., Razak I.A. Redetermination of the crystal structure of bis{[μ²-(*N,N'*-dimethyldithiocarbamato-*S,S'*)(*N,N'*-dimethyldithiocarbamato-*S,S'*)zinc(II)]}, Zn₂[S₂CN(CH₃)₂]₄, *Z. Kristallogr. New Cryst. Struct.*, 213(2), 371–372 (1998).

Rangan K.K., Trikalitis P.N., Canlas C., Bakas T., Weliky D.P., Kanatzidis M.G. Hexagonal pore organization in meso structured metal tin sulfides built with [Sn₂S₆]4–cluster, *Nano Lett.*, 2(5), 513–517 (2002).

Raper E.S., Brook J.L. Complexes of 1-methylimidazoline-2-thione with Co(II) and Zn(II) halides and perchlorates, *J. Inorg. Nucl. Chem.*, 39(12), 2163–2166 (1977).

Raper E.S., Nowell I.W. Coordination compounds of 1-methylimidazoline-2-(3*H*)-thione and metal nitrates [M(II)=Co, Zn and Cd]. A spectroscopic, thermal analysis and x-ray diffraction study, *Inorg. Chim. Acta*, 43, 165–172 (1980).

Raston C.L., White A.H., Winter G. Crystal structure and electronic spectrum for bis (*O*-ethylxanthato) pyridinezinc(II), *Aust. J. Chem.*, 29(4), 731–738 (1976).

Reck G., Becker R. Bis[*N-n*-butyl-*N*-(3,5-di-*tert*-butyl-2-hydroxybenzyl)dithiocarbamato-κ²*S,S'*]zinc(II)", *Acta Crystallogr.*, E59(5), m234–m235 (2003a).

Reck G., Becker R. (*N,N*-Diisobutyldithiocarbamoylsulfinato)zinc(II), *Acta Crystallogr.*, E59(7), m489–m490 (2003b).

Reck G., Becker R. Phenanthroline complexes of zinc and calcium carbamates, *Acta Crystallogr.*, C60(3), m134–m136 (2004).

Reck G., Becker R., Walther G. Crystal structures of two polymorphic forms of bis-(zinc-dibutyldithiopercarbamate), *Z. Kristallogr.*, 210(10), 769–774 (1995).

Romanenko G.V., Myachina L.I., Larionov S.V. Synthesis, crystal and molecular structure of dinitratobis(thiourea)zinc(II) [Zn(CH₄N₂S)₂(NO₃)₂] [in Russian], *Zhurn. strukt. khimii*, 42(2), 387–391 (2001).

Romanenko G.V., Savel'eva Z.A., Larionov S.V. Crystal structure of zinc(II) nitrate complex with thiosemicarbazide [Zn(NH₂NHC(= S)NH₂)₂](NO₃)₂ [in Russian], *Zhurn. strukt. khimii*, 40(3), 593–596 (1999).

Sampanthar J.T., Deivaraj T.C., Vittal J.J., Philip A.W. Dean P.A.W. Thioacetate complexes of Group 12 metals. Structures of [Ph₄P][Zn(SC{O}Me)₃(H₂O)] and [Ph₄P][Cd(SC{O}Me)₃], *J. Chem. Soc. Dalton Trans.*, (24), 4419–4423 (1999).

Saravanan M., Ramalingam K., Bocelli G., Cantoni A. (2,2'-bipyridyl) bis (ethanoldithiocarbamato) zinc(II), *Appl. Organometal. Chem.*, 18(2), 86–87 (2004).

Savinkina E.V., Buravlev E.A., Zamilatskov I.A., Albov D.V., Kravchenko V.V., Zaitseva M.G., Mavrin B.N. Zinc iodide complexes of propaneamide, benzamide, dimethylurea, and thioacetamide: Syntheses and structures, *Z. Anorg. und Allg. Chem.*, 635(9–10), 1458–1462 (2009).

Secondo P.M., Land J.M., Baughman R.G., Collier H.L. Polymeric octahedral and monomeric tetrahedral Group 12 pseudohalogeno (NCX⁻: X = O, S, Se) complexes of 4-(*N,N*-dimethylamino)pyridine, *Inorg. Chim. Acta*, 309(1–2), 13–22 (2000).

Seebacher J., Shu M., Vahrenkamp H. The best structural model of ADH so far: A pyrazolylbis(thioimidazolyl)borate zinc ethanol complex, *Chem. Commun.*, (11), 1026–1027 (2001).

Seidel R.W., Dietz C., Oppel I.M. Crystal structure of *catena*-[dichlorido-μ₂-4,4'-dithiodipyridine-κ²*N:N'*-zinc(II)], C₁₀H₈Cl₂N₂S₂Zn, *Z. Kristallogr. New Cryst. Struct.*, 227(3), 305–306 (2012).

Seidel R.W., Dietz C., Oppel I.M. Structural insight into repeated-rhomboid coordination polymers from 4,4'-dithiodipyridine and zinc nitrate, *Struct. Chem.*, 22(6), 1225–1232 (2011).

Selbin J., Bull W.E., Holmes Jr. L.H. Metallic complexes of dimethylsulphoxide, *J. Inorg. Nucl. Chem.*, 16(3–4), 219–224 (1961).

Selvakumar S., Rajarajan K., Kumar S.M.R., Potheher I.V., Anand D.P., Ambujam K., Sagayaraj P. Growth and characterization of pure and metal doped bis(thiourea) zinc chloride single crystals, *Cryst. Res. Technol.*, 41(8), 766–770 (2006).

Shahid M., Rüffer T., Lang H., Awan S.A., Ahmad S. Synthesis and crystal structure of a dinuclear zinc(II)-dithiocarbamate complex, *bis*{[(μ²-pyrrolidinedithiocarbamato-*S,S'*)(pyrrolidinedithiocarbamato-*S,S'*)zinc(II)]}, *J. Coord. Chem.*, 62(3), 440–445 (2009).

Shchukin V.G., Glinskaya L.A., Klevtsova R.F., Larionov S.V. Synthesis, structure and thermal properties of heteroligand compounds ZnL{(*i*-C₄H₉)₂PS₂}₂ (L=Phen, 2,2'-Bipy, 4,4'-Bipy. Polymeric structure of [Zn(4,4'-Bipy){(*i*-C₄H₉)₂PS₂}₂]ₙ [in Russian], *Koord. khim.*, 26(5), 354–360 (2000).

Shepherd T.M., Woodward I. An x-ray investigation of the stereochemistry of Zn(NCS)₂(C₆H₅NH₂)₂, *Acta Crystallogr.*, 19(3), 479–482 (1965).

Shindo H., Brown T.L. Infrared spectra of complexes of L-cysteine and related compounds with zinc(II), cadmium(II), mercury(II), and lead(II), *J. Am. Chem. Soc.*, 87(9), 1904–1909 (1965).

Shi Y.-F., Liu C.-Q., Zhao M.-G. Crystal structure of bis[2-(1*H*-1,2,4-triazol-1-yl-κ*N*⁴)-pyrazine-κ*N*⁴]di(thiocyanato-κ*N*)zinc(II), Zn(C₆H₅N₅)₂(NCS)₂, *Z. Kristallogr. New Cryst. Struct.*, 225(1), 81–82 (2010).

Shi Z.-W., Liang P., Cen C.-L., Zhao R.-P., Wei X., Guan N. Crystal structure of bis(2-(3-(carboxylatomethyl-κ*O,O'*)-4-(phenylthio)-phenyl)propanoate-κ*O,O'*)bis(2,2'-dipyridyl-κ*N,N'*)dizinc(II), C₅₄H₄₄N₄O₈S₂Zn₂, *Z. Kristallogr. New Cryst. Struct.*, 228(2), 231–232 (2013).

Shugam E.A. Chelate compounds containing Me–S bond [in Russian], *Kristallografiya*, 3(6), 749–750 (1958).

Shunmugam R., Sathyanarayana D.N. Complexes of imidazoline-2-thione and its 1-methyl analogue with Cu(II), Zn(II), Cd(II) and Hg(II) salts, *J. Coord. Chem.*, 12(3), 151–156 (1983).

Siddiqi K.S., Nishat N. Synthesis and characterization of succinimide and phthalimide dithiocarbamates and their complexes with some transition metal ions, *Synth. React. Inorg. Met.-Org. Chem.*, 30(8), 1505–1518 (2000).

Silver A., Koch S.A., Millar M. X-ray crystal structures of a series of [M^{II}(SR)₄]²⁻ complexes (M=Mn, Fe, Co, Ni, Zn, Cd and Hg) with S₄ crystallographic symmetry, *Inorg. Chim. Acta*, 205(1), 9–14 (1993).

Simonsen S.H., Ho J.W. The unit-cell dimensions and space group of zinc diethyldithiocarbamate, *Acta Crystallogr.*, 6(5), 430 (1953).

Singh P.P., Singh M. 1,1'-Bis(thiocyanatomercurio)ferrocene as ligand towards M(NCS)₂ [M=Mn(II), Co(II), Ni(II), Cu(II), Zn(II)], *Bull. Chem. Soc. Jpn.*, 59(4), 1229–1233 (1986).

Skinner B.J., Bethke P.M. The relationship between unit-cell edges and composition of synthetic wurtzites, *Am. Mineral.*, 46(11–12), 1382–1398 (1961).

Skinner B.J. Unit-cell edges of natural and synthetic sphalerites, *Am. Mineral.*, 46(11–12), 1399–1411 (1961).

Smolander K., Ahlgrèn M. Monomeric (dipropionato-*O*)(dithiourea-*S*)zinc(II), *Acta Crystallogr.*, C50(12), 1900–1902 (1994).

Soh S.F., Lai C.S., Tiekink E.R.T. (4,7-Dimethyl-1,10-phenanthroline)-bis(*O*-isopropyldithiocarbonato)zinc(II), *Acta Crystallogr.*, E58(11), m641–m643 (2002).

Sombuthawee C., Hummel F.A. Subsolidus equilibria in the system ZnS–MnS–CuInS₂, *J. Solid State Chem.*, 30(1), 125–128 (1979).

Sousa-Pedrares A., Romero J., Garcia-Vázquez J., Durán M.L., Casanova I., Sousa A. Electrochemical synthesis and structural characterization of zinc, cadmium and mercury complexes of heterocyclic bidentate ligands (N, S), *Dalton Trans.*, (7), 1379–1388 (2003).

Suen M.-C., Wang J.-C. Syntheses and structural characterization of infinite coordination polymers from bipyridyl ligands and zinc salts, *Struct. Chem.*, 17(3), 315–322 (2006).

Sun W.-Y., Zhang L., Yu K.-B. Structure and characterization of novel zinc(II) and cadmium(II) complexes with 2-(benzoylamino)benzenethiolate and 1-methylimidazole. NH···S hydrogen bonding in metal complexes with a $S_2N_2$ binding site, *J. Chem. Soc. Dalton Trans.*, (5), 795–798 (1999).

Swenson D., Baenziger N.C., Coucouvanis D. Tetrahedral mercaptide complexes. Crystal and molecular structures of $[(C_6H_5)P]_2M(SC_6H_5)_4$ complexes (M = Cd(II), Zn(II), Ni(II), Co(II), and Mn(I1)), *J. Am. Chem. Soc.*, 100(6), 1932–1934 (1978).

Tao J., Yin X., Wei Z.-B., Huang R.-B., Zheng L.-S. Hydrothermal syntheses, crystal structures and photoluminescent properties of three metal-cluster based coordination polymers containing mixed organic ligands, *Eur. J. Inorg. Chem.*, (1), 125–133 (2004).

Teixidor F., Escriche L., Casabó J., Molins E., Miravitlles C. Metal complexes with polydentate sulfur-containing ligands. Crystal structure of (2,6-bis((ethylthio)methyl)pyridine)dibromozinc(II), *Inorg. Chem.*, 25(22), 4060–4062 (1986).

Teske Chr.L. $Ba_2ZnGe_2S_6O$: Ein neues Oxidsulfid mit Tetraedergerüststruktur. *Z. Naturforsch.*, 35B(6), 672–675 (1980).

Teske Chr.L. Über Oxidsulfide mit Åkermanitstruktur $CaLaGa_3S_6O$, $SrLaGa_3S_6O$, $La_2ZnGa_2S_6O$ und $Sr_2ZnGe_2S_6O$, *Z. Anorg. und Allg. Chem.*, 531(12), 52–60 (1985).

Tesmer M., Shu M., Vahrenkamp H. Sulfur-rich zinc chemistry: New tris(thioimidazolyl)hydroborate ligands and their zinc complex chemistry related to the structure and function of alcohol dehydrogenase, *Inorg. Chem.*, 40(16), 4022–4029 (2001).

Thirumaran S., Ramalingam K., Bocelli G., Cantoni A. Electron density distribution studies on $ZnS_4N_2$ chromophore in solution and solid phases: XPS and cyclic voltammetric studies on 1,10-phenanthroline and 2,2'-bipyridine adducts of bis(piperidinecarbodithioato-*S,S'*)zinc(II). Single crystal x-ray structure of (2,2'-bipyridine)bis-(piperidinecarbodithioato-*S,S'*)zinc(II), *Polyhedron*, 18(7), 925–930 (1999).

Thirumaran S., Venkatachalam V., Manohar A., Ramalingam K., Bocelli G., Cantoni A. Synthesis and characterization of bis(*N*-methyl-*N*-ethanol-dithiocarbamato)M(II) (M = Zn, Cd, Hg) and bis(*N,N*-(iminodiethylene)-*bis*phthalimidedithiocarbamato)M(II) (M = Zn, Cd, Hg) complexes. Single crystal x-ray structure of bis(di(2-hydroxyethyl)-dithiocarbamato)zinc(II), *J. Coord. Chem.*, 44(3–4), 281–288 (1998).

Tian Y.-P., Yu W.-T., Fang Q., Wang X.-Q., Yuan D.-R., Xu D., Jiang M.-H. Zinc cadmium thiocyanate (ZCTC), *Acta Crystallogr.*, C55(9), 1393–1395 (1999).

Tiekink E.R.T. Crystal structure of 2,2'-bipyridyl adduct of bis(dibutyldithiocarbamato)zinc(II), $[Zn(S_2CNBu_2)_2(bipy)]$, *Z. Kristallogr. New Cryst. Struct.*, 216(4), 575–576 (2001a).

Tiekink E.R.T. Redetermination of the crystal structure of dimeric bis(N,N-diethyldithiocarbamato)zinc, $[Zn(S_2CNEt_2)_2]_2$, *Z. Kristallogr. New Cryst. Struct.*, 215(3),445–446 (2000).

Tiekink E.R.T. The molecular structure of the 1/1 adduct between bis(isopropyldithiophasphato)zinc(II) and 1,10-phenanthroline, *Main Group Metal Chem.*, 24(1), 63–64 (2001b).

Tiekink E.R.T., Wardell J.L., Welte W.B. *catena*-Poly[[bis(*O,O'*-diethyl dithiophosphato-$\kappa^2 S,S'$)zinc(II)]-$\mu$-1,2-di-4-pyridylethane-$\kappa^2 N:N'$], *Acta Crystallogr.*, E63(3), m790–m792 (2007).

Toguzov M.Z., Kopylov N.I., Sychev A.P. The $Cu_{2-x}S$–PbS–FeS–ZnS system [in Russian], *Zhurn. neorgan. khimii*, 25(8), 2237–2240 (1980).

Toguzov M.Z., Kopylov N.I., Sychev A.P., Yarygin V.I., Minkevich S.M. Phase equilibria in the $Cu_2S$–FeS–ZnS–ZnO system [in Russian], *Sul'fid. rasplavy tiazh. met.*, M., 122–127 (1982).

Toguzov M.Z., Kopylov N.I., Yarygin V.I., Minkevich S.M. The $Cu_2S$–ZnS–CaO system [in Russian], *Zhurn. neorgan. khimii*, 24(12), 3354–3357 (1979).

Toguzov M.Z., Kopylov N.I., Yarygin V.I., Minkevich S.M. The FeS–ZnS–CaO system [in Russian], *Zhurn. neorgan. khimii*, 26(1), 238–241 (1981).

Tsuji I., Shimodaira Y., Kato H., Kobayashi H., Akihiko Kudo A. Novel stannite-type complex sulfide photocatalysts $A^{I}_2$–Zn–$A^{IV}$–$S_4$ ($A^I$ = Cu and Ag; $A^{IV}$ = Sn and Ge) for hydrogen evolution under visible-light irradiation, *Chem. Mater.*, 22(4), 1402–1409 (2010).

Tuntulani T., Reibenspies J.H., Farmer P.J., Darensbourg M.Y. Preparations and structures of a zinc(II) dimer and zinc(II)/nickel(II) pentanuclear derivatives of *N,N'*-bis(mercaptoethyl)-1,5-diazacyclooctane: [(BME-DACO)Zn]$_2$ and {[(BME-DACO)Ni]$_3$[ZnCl]$_2$}{BF$_4$}$_2$, *Inorg. Chem.*, 31(17), 3497–3499 (1992).

Ueyama N., Sugawara T., Sasaki K., Nakamura A., Yamashita S., Wakatsuki Y., Yamazaki H., Yasuoka N. X-ray structures and far-infrared and Raman spectra of tetrahedral thiophenolato and selenophenolato complexes of zinc(II) and cadmium(II), *Inorg. Chem.*, 27(4), 741–747 (1988).

Ushasree P.M., Jayavel R., Subramanian C., Ramasamy P. Growth of zinc thiourea sulfate (ZTS) single crystals: A potential semiorganic NLO material, *J. Cryst. Growth*, 197 (1–2), 216–220 (1999).

Vega R., López-Castro A., Márquez R. Structure of tetrakis(thiourea)zinc(II) nitrate, *Acta Crystallogr.*, B34(7), 2297–2299 (1978).

Vittal J.J., Dean P.A.W. Chemistry of thiobenzoates: Syntheses, structures, and NMR spectra of salts of [M(SOCPh)$_3$]$^-$ (M = Zn, Cd, Hg), *Inorg. Chem.*, 35(11), 3089–3093 (1996).

Voronkov M.G., Fundamensky V.S., Zel'bst E.A., Kashaev A.A., Mirskova A.N., Adamovich S.N. Crystal and molecular structure of dihydrate di(4-chlorophenylthioacetate)zinc [in Russian], *Zhurn. strukt. khimii*, 51(4), 816–818 (2010).

Wachhold M., Rangan K.K., Billinge S.J.L., Petkov V., Heising J., Kanatzidis M.G. Mesostructured non-oxidic solids with adjustable worm-hole shaped pores: M–Ge–Q (Q = S, Se) frameworks based on tetrahedral [Ge$_4$Q$_{10}$]$^{4-}$ clusters, *Adv. Mater.*, 12(2), 85–91 (2000a).

Wachhold M., Rangan K.K., Lei M., Thorpe M.F., Billinge S.J.L., Petkov V., Heising J., Kanatzidis M.G. Mesostructured metal germanium sulfide and selenide materials based on the tetrahedral [Ge$_4$S$_{10}$]$^{4-}$ and [Ge$_4$Se$_{10}$]$^{4-}$ units: Surfactant templated three-dimensional disordered frameworks perforated with worm holes, *J. Solid State Chem.*, 152(1), 21–36 (2000b).

Wang C., Li Y., Bu X., Zheng N., Zivkovic O., Yang C.-S., Feng P. Three-dimensional superlattices built from (M$_4$In$_{16}$S$_{33}$)$^{10-}$ (M = Mn, Co, Zn, Cd) supertetrahedral clusters, *J. Am. Chem. Soc.*, 123(46), 11506–11507 (2001a).

Wang R., Hong M., Liang Y., Cao R. Tris(μ-1,4-benzenedicarboxylate)tetrakis(dimethyl sulfoxide)di-μ$_3$-hydroxo-tetrazinc dihydrate, *Acta Crystallogr.*, E57(7), m277–m279 (2001b).

Wang Q.-W., Li X.-M., Xu Z.-L. Crystal structure of aqua-(2,2'-bipyridyl)-(2,5-thiophenedicarboxylato)zinc(II) dihydrate, Zn(H$_2$O)(C$_{10}$H$_8$N$_2$)(C$_6$H$_2$O$_4$S)·2H$_2$O, *Z. Kristallogr. New Cryst. Struct.*, 224(4), 641–642 (2009).

Wang Q.-Y. Crystal structure of hexaaquazinc(II) bis(monoethoxyformononetin-3'-sulfonate) tetrahydrate, [Zn(H$_2$O)$_6$](C$_{18}$H$_{15}$O$_4$SO$_3$)$_2$·4H$_2$O, *Z. Kristallogr. New Cryst. Struct.*, 224(1), 71–72 (2009).

Wang W.-G., Zhang J., Ju Z.-F., Song L.-J. Diaquabis(1,10-phenanthroline)zinc(II) 4,5-dihydroxy-1,3-benzenedisulfonate trihydrate, *Appl. Organometal. Chem.*, 19(1), 191–192 (2005).

Wang X.Q., Xu D., Lu M.K., Yuan D.R., Huang J., Lu G.W., Zhang G.H., Guo S.Y., Ning H.X., Duan X.L., Chen Y., Zhou Y.Q. A systematic spectroscopic study of four bimetallic thiocyanates of chemical formula AB(SCN)$_4$: ZnCd(SCN)$_4$ and AHg(SCN)$_4$ (A = Zn, Cd, Mn) as UV nonlinear optical crystal materials, *Opt. Mater.*, 23(1–2), 335–341 (2003a).

Wang X., Xu D., Cheng X., Huang J. Preparation and characterization of Hg(N$_2$H$_4$CS)$_4$Zn(SCN)$_4$, *J. Cryst. Growth.*, 271(1–2), 120–127 (2004).

Wang X., Xu D., Lu M., Yuan D., Zhang G., Xu S., Guo S., Jiang X., Liu J., Song C., Ren Q., Huang J., Tian Y. Growth and properties of UV nonlinear optical crystal ZnCd(SCN)$_4$, *Mater. Res. Bul.*, 38(7), 1269–1280 (2003b).

Wang Y.-W. Crystal structure of hexaaquatris(4,4′-bipyridine)bis{*N*-[(4-carboxyphenyl) sulfonyl]glycine}trizinc(II) octahydrate, Zn$_3$(H$_2$O)$_6$(C$_{10}$H$_8$N$_2$)$_3$(C$_9$H$_6$SNO$_6$)$_2$·8H$_2$O, *Z. Kristallogr. New Cryst. Struct.*, 226(3), 368–370 (2011).

Wan X.Q., Xu D., Yuan D.R., Tian Y.P., Yu W.T., Sun S.Y., Yang Z.H., Fang Q., Lu M.K., Yan Y.X., Meng F.Q., Guo S.Y., Zhang G.H., Jiang M.H. Synthesis, structure and properties of a new nonlinear optical material: Zinc cadmium tetrathiocyanate, *Mater Res. Bul.*, 34(12–13), 2003–2011 (1999).

Watson A.D., Rao C.P., Dorfman J.R., Holm R.H. Systematic stereochemistry of metal(II) thiolates: Synthesis and structures of [M$_2$(SC$_2$H$_5$)$_6$]$^{2-}$ (M = Mn(II), Ni(II), Zn(II), Cd(II)), *Inorg. Chem.*, 24(18), 2820–2826 (1985).

Welte W.B., Tiekink E.R.T. *catena*-Poly[[bis(*O,O*′-diisobutyl dithiophosphato-κ$^2$*S,S*′) zinc(II)]-μ-1,2-di-4-pyridylethylene-κ$^2$*N:N*′], *Acta Crystallogr.*, E62(9), m2070–m2072 (2006).

Welte W.B., Tiekink E.R.T. *catena*-Poly[[bis(*O,O*′-diisopropyl dithiophosphato-κ$^2$*S,S*′) zinc(II)]-μ-1,2-di-4-pyridylethylene-κ$^2$*N:N*′], *Acta Crystallogr.*, E63(3), m790–m792 (2007).

White J.L., Tanski J.M., Churchill D.G., Rheingold A.L., Rabinovich D. Synthesis and structural characterization of 2-mercapto-1-*tert*-butylimidazole and its Group 12 metal derivatives (Hmim$^{tBu}$)$_2$MBr$_2$ (M = Zn, Cd, Hg), *J. Chem. Crystallogr.*, 33(5–6), 437–445 (2003).

White J.L., Tanski J.M., Rabinovich D. Bulky tris(mercaptoimidazolyl)borates: Synthesis and molecular structures of the Group 12 metal complexes [Tm$^{tBu}$]MBr (M = Zn, Cd, Hg), *J. Chem. Soc. Dalton Trans.*, (15), 2987–2991 (2002).

Williams D.J., Concepcion J.J., Koether M.C., Arrowood K.A., Carmack A.L., Hamilton T.G., Luck S.M., Ndomo M., Teel C. R., VanDerveer D. The preparation, characterization and x-ray structural analysis of *tetrakis*[1-methyl-3-(2-propyl)-2(3*H*)-imidazolethione] zinc(II) tetrafluoroborate and *tetrakis*[1-methyl-3-(1-butyl)-2(3*H*)-imidazolethione] zinc(II) tetrafluoroborate, *J. Chem. Crystallogr.*, 36(8), 453–457 (2006).

Williams D.J., Ly T.A., Mudge J.W., Pennington W.T., Chimek G.L. Dichlorobis[1-methyl-3-(prop-2-enyl)-imidazole-2(3*H*)-thione-*S*]zinc(II), *Acta Crystallogr.*, C53(4), 415–416 (1997).

Wu M., Emge T.J., Huang X., Li J., Zhang Y. Designing and tuning properties of a three-dimensional porous quaternary chalcogenide built on a bimetallic tetrahedral cluster [M$_4$Sn$_3$S$_{13}$]$^{5-}$ (M = Zn/Sn), *J. Solid State Chem.*, 181(3), 415–422 (2008).

Wunderlich H. Bis(di-*n*-propyldithiophosphinato)zinc(II), *Acta Crystallogr.*, B38(2), 614–617 (1982).

Wu T., Bu X., Zhao X., Khazhakyan R., Feng P. Phase selection and site-selective distribution by tin and sulfur in supertetrahedral zinc gallium selenides, *J. Am. Chem. Soc.*, 133(24), 9616–9625 (2011).

Wu T., Wang L., Bu X., Chau V., Feng P. Largest molecular clusters in the supertetrahedral Tn series, *J Am. Chem. Soc.*, 132(31), 10823–10831 (2010).

Wystrach V.P., Hook E.O., Christopher G.L.M. Basic zinc double salts of *O,O*-dialkyl phos-phorodithioic acids, *J. Org. Chem.*, 21(6), 705–707 (1956).

Xie J., Bu X., Zheng N., Feng P. One-dimensional polymers containing penta-supertetra-hedral sulfide clusters linked by dipyridyl ligands, *Chem. Commun.*, (39), 4916–4918 (2005).

Xu D., Yu W.-T., Wang X.-Q., Yuan D.-R., Lu M.-K., Yang P., Guo S.-Y., Meng F.-Q., Jiang M.-H. Zinc mercury thiocyanate (ZMTC), *Acta Crystallogr.*, C55(8), 1203–1205 (1999).

Xu Y.-Z., Tang W.-P. Crystal structure of *catena*-bis(hydrogen 5-*tert*-butylisophthalato)-[2,5-bis(4-pyridyl)-1,3,4-thiadiazole]zinc(II)— 2,5-bis(4-pyridyl)-1,3,4-thiadiazole (1:1), $Zn(C_{12}H_{13}O_4)_2(C_{12}H_8N_4S) \cdot C_{12}H_8N_4S$, *Z. Kristallogr. New Cryst. Struct.*, 225(3), 444–446 (2010).

Yagafarov Sh.Sh., Gusev V.B., Bamburov V.G., Burmistrov V.A. Phase transformations in the $ZnS–BaCl_2–NaCl–H_2O$ system at heat treatment [in Russian], *Izv. AN SSSR. Neorgan. materialy*, 26(3), 619–621 (1990).

Yang H.-L., Yang F., Zhu H.-L. Crystal structure of tetrakis(acetonitrile)zinc(II) sulfate, $Zn(C_2H_3N)_4SO_4$, *Z. Kristallogr. New Cryst. Struct.*, 219(3), 329–330 (2004).

Yang L., Wang Q., Li S. Crystal structure of monoaqua-bis(2-aminothiazol-4-acetato)zinc(II), $Zn(H_2O)(C_5H_5N_2O_2S)_2$, *Z. Kristallogr. New Cryst. Struct.*, 224(3), 453–454 (2009).

Yang X., Li D., Fu F., Tang L., Yang J., Wang L., Wang Y. Two disparate 2-fold interpen-etrating frameworks constructed from $Zn^{II}$, 4,4'-dipyridyl disulfide and maleic/fumaric acid, *Z. Anorg. und Allg. Chem.*, 634(14), 2634–2638 (2008).

Yin X.,, Zhang W., Fan J., Wei F. X., Lai C.S., Tiekink E.R.T. Bis[bis(*N,N*-dibenzyldithiocarbamato)zinc(II)](4,4'-bipyridine), *Appl. Organomet. Chem.*, 17(11), 889–890 (2003).

Zaki Z.M., Mohamed G.G. Spectral and thermal studies of thiobarbituric acid complexes, *Spectrochim. Acta, Part A*, 56(7), 1245–1250 (2000).

Zemskova S.M., Glinskaya L.A., Durasov V.B., Klevtsova R.F., Larionov S.V. Mixed-ligand complexes of zinc(II) and cadmium(II) diethyldithiocarbamates with 2,2'-bipyridyl and 4,4'-bipyridyl: Preparation, structural and thermal properties [in Russian], *Zhurn. strukt. khimii*, 34(5), 154–156 (1993a).

Zemskova S.M., Glinskaya L.A., Klevtsova R.F., Durasov V.B., Gromilov S.A., Larionov S.V. Volatile mixed-ligand complexes of bis(diisobutyldithiocarbamato)zinc with 1,10-phenanthroline, 2,2'-bipyridyl and 4,4'-bipyridyl. Crystal and molecular structures of binuclear complex $[Zn_2(C_{10}H_8N_2)\{(i\text{-}C_4H_9)_2NCS_2\}_4]$ [in Russian], *Zhurn. strukt. khi-mii*, 37(6), 1114–1121 (1996).

Zemskova S.M., Glinskaya L.A., Klevtsova R.F., Fedotov M.A., Larionov S.V. Structure and properties of mono- and heterometal cadmium, zinc and nickel complexes contain-ing diethyldithiocarbamate ions and ethylenediamine molecules [in Russian], *Zhurn. strukt. khimii*, 40(2), 340–350 (1999).

Zemskova S.M., Glinskaya L.A., Klevtsova R.F., Gromilov S.A., Durasov V.B., Nadolinnyi V.A., Larionov S.V. Volatile complexes $ZnPhen(S_2CNEt_2)_2$, $MnPhen(S_2CNEt_2)_2$ and $MnZn_2Phen_3(S_2CNEt_2)_6$: Thermal properties, crystal and molecular structures of $ZnPhen(S_2CNEt_2)_2$ two modifications [in Russian], *Zhurn. strukt. khimii*, 36(3), 528–540 (1995).

Zemskova S.M., Glinskaya L.A., Klevtsova R.F., Larionov S.V. Preparation, crystal and molecular structure, thermal properties of zinc and cadmium diethyldithiocarbamates complexes with imidazole, $[(C_2H_5)_2NCS_2]_2Zn(C_3H_4N_2)$ and $[(C_2H_5)_2NCS_2]_2Cd(C_3H_4N_2)$ [in Russian], *Zhurn. neorgan. khimii*, 38(3), 466–471 (1993b).

Zemskova S.M., Gromilov S.A., Larionov S.V. Synthesis and x-ray investigation of diethyl-dithiocarbamato-iodide complexes of zinc(II), cadmium(II), mercury(II) [in Russian], *Sib. khim. zhurn.*, (4), 74–77 (1991).

Zeng D., Hampden-Smith M.J., Larson E.M. μ-[1,2-Bis(diethylphosphino)ethane]-*P:P'*-bis[bis(diethyldithiocarbamato-*S,S'*)zinc(II)] ditoluene solvate, *Acta Crystallogr.*, C50(7), 1000–1002 (1994b).

Zevaco T.A., Männle D., Walter O., Dinjus E. An easy way to achieve three-dimensional metal–organic coordination polymers: Synthesis and crystal structure of dizinc diisophthalate bis-dimethylsulfoxide monohydrate: $[Zn_2(ip)_4(DMSO)_2(H_2O)\cdot 3DMSO]_n$, *Appl. Organometal. Chem.*, 21(11), 970–977 (2007).

Zhang C., Chadha R., Reddy H.K., Schrauzer G.N. Pentacoordinate zinc: Synthesis and structures of bis[1-(methylthio)-*cis*-stilbene-2-thiolato]zinc and of its adducts with mono- and bidentate nitrogen bases, *Inorg. Chem.*, 30(20), 3865–3869 (1991).

Zhang J., Xu W. *catena*-Poly[[[(2-phenylacetato-κ*O*)-zinc(II)]bis[μ-4,4'-(disulfanediyl)-dipyridine-κ²*N:N'*]] monohydrate], *Acta Crystallogr.*, E66(7), m788–m789 (2010).

Zhang S.-H., Jiang Y.-M., Yu K.-B. Aqua{2-[(*E*)-3,5-dibromo-2-oxidobenzylideneamino]ethanesulfonato}(1,10-phenanthroline)zinc(II) ethanol solvate, *Acta Crystallogr.*, E61(2), m209–m211 (2005).

Zhang Y., Jianmin L., Nishiura M., Hou H., Deng W., Imamoto T. Thermodynamically controlled products: Novel motif of metallohelicates stabilized by inter- and intra-helix hydrogen bonds, *J. Chem. Soc. Dalton Trans.*, (3), 293–297 (2000a).

Zhang Y., Li J., Chen J., Su Q., Deng W., Nishiura M., Imamoto T., Wu X., Wang Q. A novel α-helix-liked metallohelicate series and their structural adjustments for the isomorphous substitution, *Inorg. Chem.*, 39(11), 2330–2336 (2000b).

Zhao N., Lian Z., Liu P. Crystal structure of tetraaquabis(pyridine-4-carboxamide-*N*)zinc(II) 1,5-naphthalenedisulfonate–isonicotinamide–water (1:2:4), $[Zn(H_2O)_4(C_6H_6N_2O)_2]$ $[C_{10}H_6S_2O_6]\cdot 2C_6H_6N_2O\cdot 4H_2O$, *Z. Kristallogr. New Cryst. Struct.*, 225(3), 461–462 (2010).

Zhao N., Lian Z., Zhang J., Gu Y., Li X., Liu P. Crystal structure of tetraaqua(2,2'-bipyridyl-*N,N'*)zinc(II)–di-μ-aqua-bis[aqua(2,2'-bipyridyl-*N,N'*)-(5-sulfoisophthalato)zinc(II)] tetrahydrate, $Zn(H_2O)_4(C_{10}H_8N_2)\cdot Zn_2(H_2O)_4(C_{10}H_8N_2)_2(C_8H_3SO_7)_2\cdot 4H_2O$, *Z. Kristallogr. New Cryst. Struct.*, 224(2), 255–257 (2009).

Zhao X., Wang Y.L., Zhang B.P., Qin Y.M., Jiang G.C. Refinement of the crystal structure of tetrabutylammonium bis(2-thioxo-1,3-dithiole-4,5-dithiolato)zincate(II), $[C_{16}H_{36}N]_2[Zn(C_3S_5)_2]$, *at room temperature*, *Z. Kristallogr. New Cryst. Struct.*, 226(2), 251–253 (2011).

Zhao Z.-Y., Liu Q.-L., Zhao X. DFT calculations study of structural, electronic, and optical properties of $Cu_2ZnSn(S_{1-x}Se_x)_4$ alloys, *J. Alloys Compd.*, 618, 248–253 (2015).

Zheng N., Bu X., Feng P. Two-dimensional organization of $[ZnGe_3S_9(H_2O)]^{4-}$ supertetrahedral clusters template by a metal complex, *Chem. Commun.*, (22), 2805–2807 (2005).

Zhong Y., Zhang W.-G., Zhang Q.-J., Tan M.-Y., Wang S.-L. Synthesis, structure and thermal stability of zinc complexes with dithiocarbamate [in Chinese], *Acta Chim. Sinica*, 61(11), 1828–1833 (2003).

Zhuang W.-J., Jin L.-P. Syntheses, crystal structures and blue emission of three zinc(II) coordination polymers with a 4,4'-dicarboxybiphenyl sulfone ligand, *Appl. Organometal. Chem.*, 21(2), 76–82 (2007).

Zhu D.-L., Yu Y.-P., Guo G.-C., Zhuang H.-H., Huang J.-S., Liu Q., Xu Z., You X.-Z. *catena*-Poly[bis(*O,O'*-diethyldithiophosphato-*S*)zinc(II)-μ-4,4'-bipyridyl-*N:N'*, *Acta Crystallogr.*, C52(8), 1963–1966 (1996).

Zhu W.J., Hor P.H. Unusual layered transition metal oxysulfides: $Sr_2Cu_2MO_2S_2$ (M = Mn, Zn), *J. Solid State Chem.*, 130(2), 319–321 (1997).

Zimmermann C., Anson C.E., Weigend F., Clérac R., Dehnen S. Unusual syntheses, structures, and electronic properties of compounds containing ternary, T3-type supertetrahedral M/Sn/S anions $[M_5Sn(\mu_3-S)_4(SnS_4)_4]^{10-}$ (M = Zn, Co), *Inorg. Chem.*, 44(16), 5686–5695 (2005).

# 2 Systems Based on ZnSe

## 2.1  Zn–H–Na–Mg–O–Te–Mn–Fe–Se

$(Mg_{0.47}Zn_{0.43}Mn_{0.17}Fe_{1.13})(Te_{2.97}Se_{0.03})O_{9.00}(H_{1.38}Na_{0.22})\cdot3.2H_2O$. This multinary compound (mineral kinichilite) is formed in this system (Hori et al. 1981). It is dark brown in color and crystallizes in a hexagonal structure with the lattice parameters $a=941.9\pm0.5$ and $c=766.6\pm0.5$ pm and a calculated density of 3.96 g·cm⁻³.

## 2.2  Zn–H–Na–C–N–O–Re–Se

$Na_2Zn_3[Re_6Se_8(CN)_6]_2\cdot12H_2O\cdot4MeOH$ or $C_{16}H_{40}Na_2N_{12}O_{16}Re_{12}Se_{16}Zn_3$. This multinary compound is formed in this system. It crystallizes in the trigonal structure with the lattice parameters $a=1708.7\pm0.1$ and $c=4964.3\pm0.6$ pm at 154 K and a calculated density of 3.560 g·cm⁻³ (Bennett et al. 2000). Its single crystals were obtained directly by layering reactant solutions in a narrow-diameter tube. However, the material was best isolated in pure form as follows. A solution of $[Zn(H_2O)_6](ClO_4)_2$ (0.25 mM) in 5 mL of 3.5 M aqueous $NaClO_4$ was added to a solution of $Na_4[Re_6Se_8(CN)_6]$ (1.0 mM) in 10 mL of the same solvent. After the solution was left standing for 30 min, the resulting orange precipitate was collected by centrifugation, washed with methanol, and quickly dried in air to give $Na_2Zn_3[Re_6Se_8(CN)_6]_2\cdot12H_2O\cdot4M$ eOH.

## 2.3  Zn–H–Na–C–O–Se

$[Na(C_{10}H_{20}O_5)]_2[Zn(Se_4)_2]$ or $C_{20}H_{40}Na_2O_{10}Se_8Zn$. This multinary compound is formed in this system (Adel et al. 1988). It crystallizes in the monoclinic structure with the lattice parameters $a=2088.7\pm0.2$, $b=1030.8\pm0.1$, $c=1756.6\pm0.1$ pm, and $\beta=98.43°\pm0.01°$ and a calculated density of 2.10·cm⁻³. To prepare it, a mixture containing 5.70 mM of $Zn(OAc)_2\cdot2H_2O$ and equimolar quantity of $[Na(C_{10}H_{20}O_5)]_2Se_6$ was stirred at room temperature for 12 h and then heated for 1 h at reflux. It was filtered hot of a brown residue and allowed to cool slowly. After 3 days, dark-red single crystals were obtained.

## 2.4  Zn–H–Na–O–Se

$NaZn_2(OH)(SeO_3)_2$. This quinary compound, which crystallizes in the orthorhombic structure with the lattice parameters $a=1333.2\pm0.1$, $b=607.56\pm0.05$, and $c=832.58\pm0.06$ pm and a calculated density of 4.18 g·cm⁻³, is formed in this system (Harrison 1999). Its single crystals were prepared by loading ZnO (10 mM), $Na_2SeO_3$ (10 mM), $SeO_2$ (10 mM), and $H_2O$ (12 mL) into the Teflon cup of a 45 mL capacity hydrothermal bomb. This was sealed and baked for 16 h at 225°C.

## 2.5 Zn–H–K–C–Sn–O–Se

$[K_{10}(H_2O)_{16}(MeOH)_{0.5}][Zn_4(\mu_4\text{-}Se)(SnSe_4)_4]$ or $C_{0.5}H_{34}K_{10}Sn_4O_{16.5}Se_{17}Zn_4$. This multinary compound is formed in this system. It crystallizes in the tetragonal structure with the lattice parameter $a = 1538.4 \pm 0.2$ and $c = 2538.6 \pm 0.5$ pm and energy gap 2.57 eV (Dehnen and Brandmayer 2003). To obtain it, 0.150 mM of $K_4[SnSe_4]\cdot1.5MeOH$ was dissolved in 5 mL of MeOH and added to a solution of 0.150 mM of $ZnCl_2$ in 5 mL of $H_2O$. After stirring for 24 h, traces of a precipitate was removed by filtration. THF at 10 mL was allowed to slowly diffuse into the filtrate. After 1 week the compound was formed.

## 2.6 Zn–H–K–O–Se

$K_2[Zn(H_2O)_6](SeO_4)_2$. This quinary compound is formed in this system. It crystallizes in the monoclinic structure with the lattice parameters $a = 921.2 \pm 0.1$, $b = 1239.1 \pm 0.1$, $c = 627.80 \pm 0.08$ pm, and $\beta = 104.20° \pm 0.01°$ (Euler et al. 2009a). It was prepared by dissolution of equimolar amounts of $K_2SeO_4$ and $ZnSeO_4\cdot6H_2O$ in hot distilled $H_2O$ and ensuing evaporation of the solvents. Colorless ellipsoid crystals were obtained.

## 2.7 Zn–H–Rb–C–O–Se

$[Rb(C_{12}H_{24}O_6)]_2[Zn(Se_4)(Se_6)]$ or $C_{24}H_{48}Rb_2O_{12}Se_{10}Zn$. This multinary compound is formed in this system (Fenske et al. 1991). It crystallizes in the monoclinic structure with the lattice parameters $a = 1963.1 \pm 1.4$, $b = 1125.9 \pm 0.9$, $c = 2060.3 \pm 1.6$ pm, and $\beta = 90.05° \pm 0.03°$ at 223 K and a calculated density of $2.26\cdot cm^{-3}$. It has been prepared by the reaction of $Li_2Se_6$ (5.0 mM) solution in DMF (50 mL) with 18-crown-6 (4.5 mM). Subsequently, $Zn(OAc)_2$ (4 mM) and RbI (4.0 mM) were added. The mixture was stirred for 30 min, then heated to 90°C, filtered by little precipitation, added 25 mL of diethyl ether, and left for several days in a quiet place. The resulting dark-red crystals were filtered, washed with diethyl ether, and vacuum dried.

## 2.8 Zn–H–Rb–Sn–O–Se

$[Rb_{10}(H_2O)_{14.5}][Zn_4(\mu_4\text{-}Se)(SnSe_4)_4]$ or $H_{29}Rb_{10}Sn_4O_{14.5}Se_{17}Zn_4$. This multinary compound is formed in this system (Ruzin and Dehnen 2006). It crystallizes in the triclinic structure with the lattice parameters $a = 1743.1 \pm 0.4$, $b = 1745.9 \pm 0.4$, and $c = 2273.0 \pm 0.5$ pm and $\alpha = 105.82° \pm 0.02°$, $\beta = 99.17° \pm 0.03°$, and $\gamma = 90.06° \pm 0.03°$, a calculated density of $1.646\cdot cm^{-3}$, and energy gap 3.10 eV. To obtain it, $[Rb_4(H_2O)_4][SnSe_4]$ (0.2 mM) was drained in $H_2O$ (5 mL) and added to a solution of $Zn(OAc)_2$ (0.2 mM) in $H_2O$ (5 mL), whereupon a yellow precipitate formed immediately. After stirring for 24 h, the precipitate was removed and the filtrate was layered by THF (10 mL). Yellow rhombuses of $[Rb_{10}(H_2O)_{14.5}][Zn_4(\mu_4\text{-}Se)(SnSe_4)_4]$ crystallized after 3 days. All reactions were performed under an inert atmosphere.

## 2.9 Zn–H–Rb–O–Se

$Rb_2[Zn(H_2O)_6](SeO_4)_2$. This quinary compound is formed in this system. It crystallizes in the monoclinic structure with the lattice parameters $a = 935.2 \pm 0.1$, $b = 1262.6 \pm 0.2$, $c = 636.0 \pm 0.1$ pm, and $\beta = 105.195° \pm 0.008°$ (Euler et al. 2003b). It was prepared by dissolution of equimolar amounts of $Rb_2SeO_4$ and $ZnSeO_4·6H_2O$ in hot distilled $H_2O$ and ensuing evaporation of the solvent.

## 2.10 Zn–H–Rb–O–Cl–Se

$RbZn(HSeO_3)_2Cl$. This multinary compound that crystallizes in the monoclinic structure with the lattice parameters $a = 647.9 \pm 0.1$, $b = 611.0 \pm 0.1$, $c = 1074.1 \pm 0.2$ pm, and $\beta = 103.61° \pm 0.03°$ is formed in this system (Spirovski et al. 2007). Its crystals were synthesized by evaporating a mixture of equal amounts of aqueous solutions of $Rb_2SeO_3$ (0.5 M), $RbCl$ (saturated), and $ZnCl_2$ (0.2 M). The $Rb_2SeO_3$ solution was prepared by dissolving adequate amounts of $Rb_2CO_3$ and $SeO_2$. After 24 h, white crystals with a prismatic shape were formed. The crystals were washed with distilled $H_2O$ and cleaned with acetone.

## 2.11 Zn–H–Rb–O–Br–Se

$RbZn(HSeO_3)_2Br$. This multinary compound that crystallizes in the monoclinic structure with the lattice parameters $a = 652.8 \pm 0.1$, $b = 621.8 \pm 0.1$, $c = 1074.3 \pm 0.2$ pm, and $\beta = 103.57° \pm 0.03°$ is formed in this system (Spirovski et al. 2007). Its crystals were synthesized by the same procedure as for $RbZn(HSeO_3)_2Cl$ using $ZnBr_2$ instead of $ZnCl_2$.

## 2.12 Zn–H–Cs–O–Se

$Cs_2Zn(HSeO_3)_4·2H_2O$ and $Cs_2[Zn(H_2O)_6](SeO_4)_2$. This quinary compounds are formed in this system. The first compound crystallizes in the monoclinic structure and was prepared by crystallization from highly concentrated solutions of $Cs_2SeO_3$, $CsCl$, $ZnCl_2$, and $SeO_2$ at pH = 1 (Spirovski et al. 2006). The second compound also crystallizes in the monoclinic structure with the lattice parameters $a = 948.3 \pm 0.1$, $b = 1300.5 \pm 0.2$, $c = 648.50 \pm 0.08$ pm, and $\beta = 106.16° \pm 0.01°$ (Euler et al. 2003a). It was prepared by dissolution of equimolar amounts of $Cs_2SeO_4$ and $ZnSeO_4·6H_2O$ in hot distilled $H_2O$ and ensuing evaporation of the solvent.

## 2.13 Zn–H–Ag–Sn–N–Se

$\{[Zn(NH_3)_6][Ag_4Zn_4Sn_3Se_{13}]\}_\infty$ or $H_{18}Ag_4Sn_3N_6Se_{13}Zn_5$. This multinary compound is formed in this system. It crystallizes in the cubic structure with the lattice parameter $a = 1896.79 \pm 0.04$ pm and is an $n$-type semiconductor with energy gap 2.09 eV (Xiong et al. 2014). It was synthesized by mixing Ag (0.11 mM), $ZnCl_2$ (0.24 mM), SnSe (0.48 mM), Se (2.43 mM), octanedioic acid ($C_8H_{14}O_4$) (2.54 mM), and hydrazine monohydrate (98%, 3 mL). The mixture was sealed into an autoclave equipped with a Teflon liner (20 mL) and heated at 190°C for 3 days. After cooling to room temperature, the obtained product was filtrated and washed by EtOH. Red block crystals were selected by hand.

{[Zn(NH$_3$)$_6$][Ag$_4$Zn$_4$Sn$_3$Se$_{13}$]}$_\infty$ could also be prepared by the previously mentioned synthetic procedure without the addition of surfactant octanedioic acid, but the quality of single crystals were poor, and the yield of crystals were much lower (Xiong et al. 2014).

## 2.14 Zn–H–Ba–Sn–O–Se

[Ba$_5$(H$_2$O)$_{32}$][Zn$_5$Sn($\mu_3$-Se)$_4$(SnSe$_4$)$_4$ or H$_{64}$Ba$_5$Sn$_5$O$_{32}$Se$_{20}$Zn$_5$. This multinary compound is formed in this system (Ruzin and Dehnen 2006). It crystallizes in the monoclinic structure with the lattice parameters $a = 2523.1 \pm 0.5$, $b = 2477.6 \pm 0.5$, $c = 2539.6 \pm 0.5$ pm, and $\beta = 106.59° \pm 0.03°$, a calculated density of 3.285·cm$^{-3}$, and energy gap 3.15 eV. To obtain it, ZnCl$_2$ (0.1 mM) was added to [Ba$_2$(H$_2$O)$_5$][SnSe$_4$] (0.1 mM) suspended in H$_2$O (10 mL), whereupon a yellow precipitate formed immediately. After stirring for 24 h, the precipitate was removed and the filtrate was layered by THF (10 mL). Yellow parallelepipeds of [Ba$_5$(H$_2$O)$_{32}$][Zn$_5$Sn($\mu_3$-Se)$_4$(SnSe$_4$)$_4$ crystallized after 1 day. All reactions were performed under an inert atmosphere.

## 2.15 Zn–H–B–C–N–O–Se

[HB(OC$_6$H$_4$MeC$_3$N$_2$H$_2$)$_3$ZnSePh] or C$_{36}$H$_{33}$BN$_6$O$_3$SeZn. This multinary compound is formed in this system. It was obtained by the interaction of [HB(OC$_6$H$_4$MeC$_3$N$_2$H$_2$)$_3$ZnEt] with PhSeH (Alsfasser et al. 1993).

## 2.16 Zn–H–B–C–N–S

[HB(PhC$_3$N$_2$H$_2$)$_3$ZnSePh] or C$_{33}$H$_{27}$BN$_6$SeZn. This multinary compound is formed in this system. It was obtained by the interaction of [HB(PhC$_3$N$_2$H$_2$)$_3$ZnBu$^t$] with PhSeH (Alsfasser et al. 1993).

## 2.17 Zn–H–Ga–C–Ge–N–Se

[(Ga$_{2.4}$Ge$_{1.6}$Se$_8$)(Zn$_4$Ga$_{16}$Se$_{33}$)$_4$]·(C$_{12}$H$_{28}$N$_2$)$_x$. This multinary compound is formed in this system. It crystallizes in the tetragonal structure with the lattice parameters $a = 3656.59 \pm 0.03$ and $c = 4661.03 \pm 0.08$ pm and energy gap 1.45 eV (Wu et al. 2009). To obtain it, GeO$_2$ (0.116 mM), Ga$_2$O$_3$ (0.604 mM), Se (2.371 mM), Zn(NO$_3$)$_2$·6H$_2$O (0.277 mM), and 4,4'-trimethylenedipiperidine (C$_{12}$H$_{26}$N$_2$) (9.673 mM) were mixed with H$_2$O (168 mM) in a 23 mL Teflon-lined stainless steel autoclave and stirred for 1 h. The vessel was then sealed and heated to 200°C for 6 days. Pale-yellow column crystals were obtained upon cooling to room temperature.

## 2.18 Zn–H–Ga–C–Sn–N–O–Se

Some compounds are formed in this system.

**[Zn$_4$Ga$_{16-x}$Sn$_x$Se$_{33}$](C$_4$H$_{10}$NO)$_{10}$ or C$_{40}$H$_{100}$Ga$_{16-x}$Sn$_x$N$_{10}$O$_5$Se$_{33}$Zn$_4$ (C$_4$H$_9$NO=morpholine).** This compound crystallizes in the tetragonal structure with the lattice parameters $a = 2407.45 \pm 0.03$ and $c = 4216.53 \pm 0.05$ pm (Wu et al. 2011).

The number of molecules of the organic ligand in the formula is indicated approximately for neutralizing the negative charge of chalcogenide clusters.

**[Zn$_4$Ga$_{16-x}$Sn$_x$Se$_{33}$](C$_6$H$_{14}$NO)$_{10}$ or C$_{60}$H$_{140}$Ga$_{16-x}$Sn$_x$N$_{10}$O$_5$Se$_{33}$Zn$_4$ (C$_6$H$_{13}$NO=2,6-dimethylmorpholine).** This compound crystallizes in the tetragonal structure with the lattice parameters $a = 2413.24 \pm 0.06$ and $c = 4219.76 \pm 0.04$ pm (Wu et al. 2011).

The number of molecules of the organic ligand in the formula is indicated approximately for neutralizing the negative charge of chalcogenide clusters.

**[Zn$_4$Ga$_{16-x}$Sn$_x$Se$_{33}$](C$_9$H$_{20}$N)$_{10}$ or C$_{90}$H$_{190}$Ga$_{16-x}$Sn$_x$N$_{10}$O$_5$Se$_{33}$Zn$_4$ (C$_9$H$_{19}$N=2,2,6,6-tetramethylpiperidine).** This compound crystallizes in the tetragonal structure with the lattice parameters $a = 2416.87 \pm 0.03$ and $c = 4227.24 \pm 0.03$ pm (Wu et al. 2011).

The number of molecules of the organic ligand in the formula is indicated approximately for neutralizing the negative charge of chalcogenide clusters.

## 2.19   Zn–H–Ga–C–Sn–N–Se

Some compounds are formed in this system.

**[Zn$_4$Ga$_{16-x}$Sn$_x$Se$_{33}$](C$_4$H$_{12}$N$_2$)$_5$ or C$_{20}$H$_{60}$Ga$_{16-x}$Sn$_x$N$_{10}$Se$_{33}$Zn$_4$ (C$_4$H$_{10}$N$_2$=piperazine).** This compound crystallizes in the tetragonal structure with the lattice parameters $a = 2412.67 \pm 0.03$ and $c = 4220.13 \pm 0.08$ pm (Wu et al. 2011).

The number of molecules of the organic ligand in the formula is indicated approximately for neutralizing the negative charge of chalcogenide clusters.

**[Zn$_4$Ga$_{13.58}$Sn$_{2.42}$Se$_{33}$](C$_6$H$_{17}$N$_3$)$_5$ or C$_{30}$H$_{85}$Ga$_{13.58}$Sn$_{2.42}$N$_{15}$Se$_{33}$Zn$_4$ (C$_6$H$_{15}$N$_3$=$N$-(2-aminoethyl)piperazine).** This compound crystallizes in the tetragonal structure with the lattice parameters $a = 2307.01 \pm 0.04$ and $c = 4149.75 \pm 0.13$ pm (Wu et al. 2011).

The number of molecules of the organic ligand in the formula is indicated approximately for neutralizing the negative charge of chalcogenide clusters.

**[Zn$_4$Ga$_{14}$Sn$_2$Se$_{33}$](C$_6$H$_{14}$N)$_8$ or C$_{48}$H$_{112}$Ga$_{14}$Sn$_2$N$_8$Se$_{33}$Zn$_4$ (C$_6$H$_{13}$N=2-methylpiperidine).** This compound crystallizes in the tetragonal structure with the lattice parameters $a = 2115.84 \pm 0.12$ and $c = 4193.1 \pm 0.4$ at 150 K (Wu et al. 2010). To obtain it, SnCl$_2$ (0.184 mM), Ga(NO$_3$)$_3$ (0.988 mM), Se (2.233 mM), Zn(NO$_3$)·6H$_2$O (0.259 mM), and 2-methylpiperidine (11.376 mM) were mixed with H$_2$O (137.8 mM) in a 23 mL Teflon-lined stainless steel autoclave and stirred for 2 h. The vessel was then sealed and heated to 200°C for 10 days. The resulting products contain a very small amount of the title compound and a large amount of **[Zn$_4$Ga$_{14}$Sn$_2$Se$_{35}$](C$_6$H$_{14}$N)$_{12}$.**

**[Zn$_4$Ga$_{14}$Sn$_2$Se$_{33}$](C$_6$H$_{14}$N)$_8$ or C$_{48}$H$_{112}$Ga$_{14}$Sn$_2$N$_8$Se$_{33}$Zn$_4$ (C$_6$H$_{13}$N=3-methylpiperidine).** This compound crystallizes in the tetragonal structure with the lattice parameters $a = 2406.93 \pm 0.03$ and $c = 4226.01 \pm 0.10$ at 150 K (Wu et al. 2010). To obtain it, SnCl$_2$ (0.184 mM), Ga(NO$_3$)$_3$ (0.988 mM), Se (2.233 mM), Zn(NO$_3$)$_2$·6H$_2$O (0.259 mM), and 3-methylpiperidine (11.376 mM) were mixed with H$_2$O (137.8 mM) in a 23 mL Teflon-lined stainless steel autoclave and stirred for 2 h. The vessel was then sealed and heated to 200°C for 10 days. After cooling to room temperature, only yellow block crystals of the title compound were obtained.

**[Zn$_4$Ga$_{14}$Sn$_2$Se$_{35}$](C$_5$H$_{12}$N)$_{12}$ or C$_{60}$H$_{144}$Ga$_{14}$Sn$_2$N$_{12}$Se$_{35}$Zn$_4$ (C$_5$H$_{11}$N = piperidine).** This compound crystallizes in the cubic structure with the lattice parameter $a = 1889.51 \pm 0.01$ at 150 K and energy gap 2.71 eV (Wu et al. 2010). To obtain it, SnCl$_2$ (0.184 mM), Ga(NO$_3$)$_3$ (0.988 mM), Se (2.233 mM), Zn(NO$_3$)$_2 \cdot 6$H$_2$O (0.259 mM), and piperidine (11.376 mM) were mixed with H$_2$O (137.8 mM) in a 23 mL Teflon-lined stainless steel autoclave and stirred for 2 h. The vessel was then sealed and heated to 200°C for 10 days. After cooling to room temperature, only yellow block crystals of the title compound were obtained.

**[Zn$_4$Ga$_{14}$Sn$_2$Se$_{35}$](C$_6$H$_{14}$N)$_{12}$ or C$_{72}$H$_{168}$Ga$_{14}$Sn$_2$N$_{12}$Se$_{35}$Zn$_4$ (C$_6$H$_{13}$N = 4-methylpiperidine).** This compound crystallizes in the cubic structure with the lattice parameter $a = 1920.20 \pm 0.03$ at 150 K (Wu et al. 2010). To obtain it, SnCl$_2$ (0.184 mM), Ga(NO$_3$)$_3$ (0.988 mM), Se (2.233 mM), Zn(NO$_3$)$_2 \cdot 6$H$_2$O (0.259 mM), and 4-methylpiperidine (11.376 mM) were mixed with H$_2$O (137.8 mM) in a 23 mL Teflon-lined stainless steel autoclave and stirred for 2 h. The vessel was then sealed and heated to 200°C for 10 days. After cooling to room temperature, only yellow block crystals of the title compound were obtained.

**[Zn$_4$Ga$_{16-x}$Sn$_x$Se$_{33}$](C$_8$H$_{13}$N)$_{10}$ or C$_{80}$H$_{130}$Ga$_{16-x}$Sn$_x$N$_{10}$Se$_{33}$Zn$_4$ (C$_8$H$_{12}$N = N-ethylcyclohexaneamine).** This compound crystallizes in the tetragonal structure with the lattice parameters $a = 2409.85 \pm 0.04$ and $c = 4218.52 \pm 0.06$ pm (Wu et al. 2011). To obtain it, Ga(NO$_3$)$_3 \cdot x$H$_2$O (ca. 1.0 mM), Zn(NO$_3$)$_3 \cdot 6$H$_2$O (0.25 mM), Se (2.24 mM), SnCl$_2$ (0.264 mM), and N-ethylcyclohexaneamine (19.35 mM) were mixed with H$_2$O (3.0 mL) in 23 mL Teflon-lined stainless steel autoclave and stirred for 1 h. The vessel was then sealed and heated up to 200°C for 9 days. After cooling to room temperature, a large amount of pale-yellow octahedral crystals were obtained. These raw products were then washed by H$_2$O and EtOH and dried in air.

The number of molecules of the organic ligand in the formula is indicated approximately for neutralizing the negative charge of chalcogenide clusters.

**[(Ga$_{2.4}$Sn$_{1.6}$Se$_8$)(Zn$_4$Ga$_{16}$Se$_{33}$)$_4$]·(C$_{12}$H$_{28}$N$_2$)$_x$.** This compound crystallizes in the tetragonal structure with the lattice parameters $a = 3601.37 \pm 0.05$ and $c = 4671.90 \pm 0.16$ pm and energy gap 1.54 eV (Wu et al. 2009). The following procedure was used to prepare it. Ga$_2$O$_3$ (0.570 mM), Se (2.047 mM), Zn(NO$_3$)$_2 \cdot 6$H$_2$O (0.247 mM), and 4,4′-trimethylenedipiperidine (C$_{12}$H$_{26}$N$_2$) (9.746 mM) were mixed with H$_2$O (3.122 g) in a 23 mL Teflon-lined stainless steel autoclave and stirred for 2 h. The vessel was then sealed and heated to 200°C for 6 days. Yellow block crystals were obtained upon cooling to room temperature.

## 2.20   Zn–H–Ga–C–Sn–N–Mn–Se

**[(C$_8$H$_{23}$N$_5$)Mn]$_4$[Zn$_2$Ga$_4$Sn$_4$Se$_{20}$] or C$_{32}$H$_{92}$Ga$_4$Sn$_4$N$_{20}$Mn$_4$Se$_{20}$Zn$_2$.** This multinary compound is formed in this system. It is stable up to 260°C and crystallizes in the tetragonal structure with the lattice parameters $a = 1749.51 \pm 0.05$ and $c = 1358.49 \pm 0.08$ pm at 123 K (Xu et al. 2009). It was synthesized by mixing Sn powder (0.33 mM), Ga (1.0 mM), Se (2.0 mM), $Zn(NO_3)_2 \cdot 6H_2O$ (0.33 mM), $MnCl_2 \cdot 4H_2O$ (1.0 mM), and 3.0 mM of tetraethylenepentamine in a 17 mm Teflon-lined stainless steel autoclave. After stirring for 30 min, the vessel was sealed and heated at 190°C for 1 week. The resulting yellow crystals were isolated by washing with EtOH and water. $[(C_8H_{23}N_5)Mn]_4[Zn_2Ga_4Sn_4Se_{20}]$ is a wide band gap semiconductor with an energy gap of 1.99 eV.

## 2.21 Zn–H–Ga–C–N–Se

Some compounds are formed in this system.

$[Zn_4Ga_{16}Se_{33}](C_6H_{17}N_3)_5$ or $C_{30}H_{85}Ga_{16}N_{15}Se_{33}Zn_4$. This compound crystallizes in the tetragonal structure with the lattice parameters $a = 2359.37 \pm 0.08$ and $c = 4262.0 \pm 0.3$ pm (Wu et al. 2011). To obtain it, $Ga(NO_3)_3 \cdot xH_2O$ (ca. 1.0 mM), $Zn(NO_3)_3 \cdot 6H_2O$ (0.25 mM), Se (2.24 mM), and N-(2-aminoethyl)piperazine (19.35 mM) were mixed with $H_2O$ (3.0 mL) in a 23 mL Teflon-lined stainless steel autoclave and stirred for 1 h. The vessel was then sealed and heated up to 200°C for 9 days. After cooling to room temperature, a large amount of pale-yellow octahedral crystals were obtained. These raw products were then washed by $H_2O$ and EtOH and dried in air.

The number of molecules of the organic ligand in the formula is indicated approximately for neutralizing the negative charge of chalcogenide clusters.

$[Zn_4Ga_{16}Se_{33}](C_{13}H_{28}N_2)_5$ or $C_{65}H_{140}Ga_{16}N_{10}Se_{33}Zn_4$. This compound crystallizes in the tetragonal structure with the lattice parameters $a = 4131.1 \pm 0.7$ and $c = 1689.1 \pm 0.4$ pm and energy gap 1.17 eV (Bu et al. 2004). To prepare it Ga (107.5 mg), Se (210.3 mg), $Zn(NO_3)_2 \cdot 6H_2O$ (157.1 mg), and 4.4′-trimethylene-dipiperidine (1.122 g) were mixed with $H_2O$ (4.0212 g) in a 23 mL Teflon-lined stainless steel autoclave and stirred for 30 min. The vessel was sealed and heated to 200°C for 8 days. After cooling to room temperature, pale-yellow crystals were obtained.

## 2.22 Zn–H–Tl–O–Se

$Tl_2[Zn(H_2O)_6](SeO_4)_2$. This quinary compound is formed in this system. It crystallizes in the monoclinic structure with the lattice parameters $a = 938.3 \pm 0.1$, $b = 1261.2 \pm 0.2$, $c = 631.7 \pm 0.1$ pm, and $\beta = 105.59° \pm 0.01°$ (Euler et al. 2009b). It was prepared by dissolution of equimolar amounts of $Tl_2SeO_4$ and $ZnSeO_4 \cdot 6H_2O$ in hot distilled $H_2O$ and ensuing evaporation of the solvents. Colorless ellipsoid crystals were obtained.

## 2.23 Zn–H–C–Si–N–P–Se

Some compounds are formed in this system.

$[Zn\{N(SiMe_3)_2\}\{Bu^t_2P(Se)NPr^i\}]$ or $C_{17}H_{43}Si_2N_2PSeZn$. This compound sublimes at 200° at $6.1 \cdot 10^{-4}$ Pa and melts at temperate higher than 220°C (Bochmann et al. 1995).

The following procedure was used to obtain it. To a suspension of $Bu^t_2P(Se)$ $NHPr_i$ (5.01 mM) in light petroleum (25 mL) was added at room temperature $[Zn\{N(SiMe_3)_2\}_2]$ (5.05 mM) via syringe. The mixture was stirred for 3 h and centrifuged. The supernatant liquid was concentrated and cooled to $-16°C$ to give white crystals.

$[Zn\{N(SiMe_3)_2\}\{Bu^t_2P(Se)N(C_6H_{11})\}]$ or $C_{20}H_{47}Si_2N_2PSeZn$. This compound sublimes at $200°$ at $4.3\cdot10^{-4}$ Pa and melts at a temperature higher than $220°C$ (Bochmann et al. 1995). It was prepared following the procedure for $[Zn\{N(SiMe_3)_2\}\{Bu^t_2P(Se)$ $NPr_i\}]$ from $Bu^t_2P(Se)NH(C_6H_{11})$ (4.52 mM) and $[Zn\{N(SiMe_3)_2\}_2]$ (4.53 mM) as colorless crystals.

## 2.24   Zn–H–C–Ge–N–Se

Some compounds are formed in this system.

$(H_2NC_4H_8NCH_2CH_2NH_2)(H_3NCH_2CH_2NH_2)_3Zn_2Ge_2Se_8$ or $C_{12}H_{43}Ge_2N_9Se_8Zn_2$. This compound is stable up to ca. $105°C$ and crystallizes in the orthorhombic structure with the lattice parameters $a = 1892.8 \pm 0.4$, $b = 3084.9 \pm 0.6$, $c = 639.54 \pm 0.13$ pm, a calculated density of $2.295$ g·cm$^{-3}$ and energy gap $1.8$ eV (Philippidis and Trikalitis 2009). It was prepared by the following procedure. Zn (0.795 mM), Ge (0.399 mM), Se (2.774 mM) and 6 mL of N-(2-aminoethyl)piperazine were placed in a 45 mL Teflon-lined stainless steel autoclave and heated at $200°C$ for 62 h. The product, in the form of transparent, yellow-orange rodlike crystals, was isolated by suction filtration and washed with copious amount of MeOH.

$(C_{12}H_{25}NMe_3)_2[ZnGe_4Se_{10}]$ or $C_{30}H_{68}Ge_4N_2Se_{10}Zn$. To obtain this compound, $(C_{12}H_{25}NMe_3)_4Ge_4Se_{10}$ was dissolved in a warm (ca. $60°C$) EtOH/$H_2O$ mixture (ca. 20 mL; 1:1 volume ratio) (Wachhold et al. 2000a,b). A solution of $ZnCl_2$ in EtOH/ $H_2O$ was then added dropwise to the stirred $(C_{12}H_{25}NMe_3)_4Ge_4Se_{10}$ solution to give instantly a yellow precipitate. The resulting suspension was stirred for another hour, followed by product isolation by centrifugation and subsequent decanting of the solvent. The product was washed with water and acetone to remove all residues of $(C_{12}H_{25}NMe_3)_4Ge_4Se_{10}$ and dried with ether.

This compound could also be obtained if $K_4Ge_4Se_{10}$ and two equivalents of $C_{12}H_{25}NMe_3Br$ were dissolved in $H_2O$/MeOH (ca. 20 mL; 1:1 volume ratio) and heated to $50°C–60°C$ to give a yellow solution (Wachhold et al. 2000a,b). The solution of $ZnCl_2$ in $H_2O$ (100 mL) was slowly added to this mixture under vigorous stirring, which lead to immediate precipitation of the product. The isolation and purification was the same as for the first method. $(C_{12}H_{25}NMe_3)_2[ZnGe_4Se_{10}]$ is a wide band gap semiconductor.

$(C_{14}H_{29}NMe_3)_2[ZnGe_4Se_{10}]$ or $C_{34}H_{76}Ge_4N_2Se_{10}Zn$. This compound was prepared by the same procedure as $(C_{12}H_{25}NMe_3)_2[ZnGe_4Se_{10}]$ was obtained using $(C_{14}H_{29}NMe_3)_4Ge_4Se_{10}$ instead of $(C_{12}H_{25}NMe_3)_4Ge_4Se_{10}$ (first method) and $C_{14}H_{29}$ $NMe_3Br$ instead of $C_{12}H_{25}NMe_3Br$ (second method) (Wachhold et al. 2000a,b). It is a wide band gap semiconductor with energy gap $2.37$ eV.

**$(C_{16}H_{33}NMe_3)_2[ZnGe_4Se_{10}]$** or **$C_{38}H_{84}Ge_4N_2Se_{10}Zn$**. This compound was prepared by the same procedure as $(C_{12}H_{25}NMe_3)_2[ZnGe_4Se_{10}]$ was obtained using $(C_{16}H_{33}NMe_3)_4Ge_4Se_{10}$ instead of $(C_{12}H_{25}NMe_3)_4Ge_4Se_{10}$ (first method) and $C_{16}H_{33}NMe_3Br$ instead of $C_{12}H_{25}NMe_3Br$ (second method) (Wachhold et al. 2000a,b). It is a wide band gap semiconductor.

**$(C_{18}H_{37}NMe_3)_2[ZnGe_4Se_{10}]$** or **$C_{42}H_{92}Ge_4N_2Se_{10}Zn$**. This compound was prepared by the same procedure as $(C_{12}H_{25}NMe_3)_2[ZnGe_4Se_{10}]$ was obtained using $(C_{18}H_{37}NMe_3)_4Ge_4Se_{10}$ instead of $(C_{12}H_{25}NMe_3)_4Ge_4Se_{10}$ (first method) and $C_{18}H_{37}NMe_3Br$ instead of $C_{12}H_{25}NMe_3Br$ (second method) (Wachhold et al. 2000a,b). It is a wide band gap semiconductor.

## 2.25  Zn–H–C–Sn–N–O–Se

**$[Zn(H_2O)_4][Zn_2Sn_3Se_9(MeNH_2)]$** or **$CH_{13}Sn_3NO_4Se_9Zn_3$**. This multinary compound is formed in this system. It crystallizes in the triclinic structure with the lattice parameters $a = 786.32 \pm 0.12$, $b = 861.62 \pm 0.13$, $c = 1021.85 \pm 0.15$ pm and $\alpha = 105.74°$, $\beta = 110.63°$ and $\gamma = 103.17°$, a calculated density of $3.897 \cdot cm^{-3}$ and energy gap 2.08 eV (Manos et al. 2008). To obtain it, Sn (2 mM), Zn (2 mM), Se (6 mM) and methylamine (8 mL, 40% in $H_2O$) were mixed in a 23 mL Teflon-lined stainless steel autoclave. The autoclave was sealed and placed in a temperature-controlled oven operated at 190°C. The autoclave remained undisturbed at this temperature for 9 days. Then, the autoclave was allowed to cool at room temperature. An orange-red crystalline product was isolated, washed several times with $H_2O$, acetone, and ether (in this order) and dried under vacuum.

## 2.26  Zn–H–C–Sn–N–Se

Some compounds are formed in this system.

**$[dbnH][Zn_{0.5}Sn_{0.5}Se_2]$** or **$C_7H_{13}Sn_{0.5}N_2Se_2Zn_{0.5}$** (dbn = 1,5-diazabicyclo[4,3,0] non-5-ene, $C_7H_{12}N_2$). This compound crystallizes in the tetragonal structure with the lattice parameters $a = 1303.3 \pm 0.3$, $c = 654.78 \pm 0.13$ pm, a calculated density of 2.241 g·cm$^{-3}$ and energy gap 2.42 eV (Xiong et al. 2013). To obtain it, a mixture of Sn powder (0.49 mM), $Zn(OAc)_2$ (0.51 mM), Se (1.49 mM), $C_7H_{12}N_2$ (1 mL), and $H_2O$ (0.5 mL) was stirred under ambient conditions. The resulting mixture was sealed into an autoclave equipped with a Teflon liner (23 mL) and heated at 160°C for 9 days. After cooling to room temperature, light-yellow rod crystals of $[dbnH][Zn_{0.5}Sn_{0.5}Se_2]$ were obtained by filtration, washed several times with EtOH, and selected by hand.

**$(Me_4N)_2Zn[Sn_4Se_{10}]$** or **$C_8H_{24}Sn_4N_2Se_{10}Zn$**. This compound crystallizes in the tetragonal structure with the lattice parameters $a = 983.2 \pm 0.3$, $c = 1508.2 \pm 0.7$ pm and energy gap 2.23 eV (Tsamourtzi et al. 2008). To obtain it, $(Me_4N)_4[Sn_4Se_{10}]$ (0.2 mM) was dissolved in 10 mL of deionized water at room temperature forming a clear yellow solution. To this, $ZnCl_2 \cdot 4H_2O$ (0.2 mM) was combined with sodium ethylenediamine tetraacetate in an equimolar amount. All manipulations were carried out inside a $N_2$-filled glovebox.

[(H$_2$NC$_4$H$_8$NCH$_2$CH$_2$NH$_2$)$_2$Zn$_2$Sn$_2$Se$_7$] or C$_{12}$H$_{32}$Sn$_2$N$_6$Se$_7$Zn$_2$. This compound crystallizes in the monoclinic structure with the lattice parameters $a = 1384.7 \pm 0.3$, $b = 1121.6 \pm 0.2$, $c = 1044.7 \pm 0.2$ pm and $\beta = 110.12° \pm 0.03°$ and a calculated density of 2.575 g·cm$^{-3}$ (Philippidis et al. 2009). It can be obtained by reacting stoichiometric amounts of Zn (0.657 mM), Sn (0.657 mM) and Se (2.299 mM) in the presence of N-(2-aminoethyl)piperazine and MeOH in a Teflon-lined stainless steel autoclave at 200°C for 2 days. C$_{12}$H$_{32}$Sn$_2$N$_6$Se$_7$Zn$_2$ is a wide band gap semiconductor with an energy gap of 2.3 eV.

(C$_{21}$H$_{38}$N)$_{4-2x}$Zn$_x$SnSe$_4$ (1.0 < $x$ < 1.3). This compound is a medium band gap semi-conductor with $E_g = 2.5$ eV (Trikalitis et al. 2001). It was prepared by the following procedure. A solution of 9.94 mM of surfactant (cetylpyridinium bromide, C$_{21}$H$_{38}$NBr·H$_2$O) in 20 mL of formamide was heated at 75°C for a few minutes forming a clear solution. To this solution, 1.00 mM of K$_4$SnSe$_4$ was added and the mixture was stirred for 30 min, forming a clear deep-red solution. To this, a solution of 1.00 mM of the zinc salt in 10 mL of formamide was added slowly, using a pipette. A precipitate formed immediately and the mixture was stirred while warm for 20 h. The products were isolated by suction, filtration, and washed with large amounts of warm formamide and H$_2$O. The solids were dried under vacuum overnight. A yellow product was formed. Such preparation gives a cubic phase. The hexagonal phase requires 4.97 mM of surfactant. All reactions were carried out under a N$_2$ atmosphere.

## 2.27   Zn–H–C–N–P–Se

Some compounds are formed in this system.

[Zn{Bu$^t_2$P(Se)NPr$^i$}$_2$] or C$_{22}$H$_{50}$N$_2$P$_2$Se$_2$Zn. This compound sublimes at 200° at $1.3\cdot10^{-4}$·Pa, melts at a temperature higher than 220°C, and crystallizes in the monoclinic structure with the lattice parameters $a = 1581.5 \pm 0.5$, $b = 1184.8 \pm 0.8$, $c = 1604.1 \pm 0.5$ pm and $\beta = 90.56° \pm 0.02°$ and a calculated density of 1.388 g·cm$^{-3}$ (Bochmann et al. 1995). For obtaining it, to a suspension of Bu$^t_2$P(Se)NHPr$_i$ (1.53 mM) in light petroleum (10 mL) was added [Zn{N(SiMe$_3$)$_2$}$_2$] (0.77 mM) at 0°C. The mixture was stirred at this temperature for 5 min and then warm to room temperature and stirred. After 45 min at room temperature, all the suspension had dissolved and a white precipitate had begun to form. After stirring for 3 h the solvent was removed under reduced pressure and the crude product recrystallized from toluene (5 mL) to give colorless crystals after cooling to –16°C.

[Zn{Bu$^t_2$P(Se)N(C$_6$H$_{11}$)}$_2$] or C$_{28}$H$_{58}$N$_2$P$_2$Se$_2$Zn. This compound sublimes at 200° at $2\cdot10^{-4}$ Pa and melts at a temperature higher than 220°C (Bochmann et al. 1995). It was prepared following the procedure for [Zn{Bu$^t_2$P(Se)NPr$_i$}$_2$] from Bu$^t_2$P(Se) NH(C$_6$H$_{11}$) (2.91 mM) and [Zn{N(SiMe$_3$)$_2$}$_2$] (1.45 mM) as colorless crystals.

## 2.28   Zn–H–C–N–Sb–Se

Some compounds are formed in this system.

$(C_2H_9N_2)[Zn(C_2H_8N_2)_3][SbSe_4]$ **or** $C_8H_{33}N_8SbSe_4Zn$. This compound crystallizes in the triclinic structure with the lattice parameters $a = 881.00 \pm 0.02$, $b = 963.24 \pm 0.02$, and $c = 1426.06 \pm 0.04$ pm and $\alpha = 104.963° \pm 0.002°$, $\beta = 92.574° \pm 0.002°$, and $\gamma = 109.713° \pm 0.003°$ at 150 K (Menezes and Fässler 2012). Its single crystals were prepared solvothermally from a mixture of 64 mg Zn, 61 mg Sb, and 240 mg Se. The filling of all starting materials was carried out under inert gas conditions. Ethylenediamine $(C_2H_8N_2)$ (5 mL) was added to the reaction mixture that was then loaded into a Teflon-lined autoclave with an inner volume of 20 mL. The sealed autoclave was heated at 170°C for 7 days before cooling down to room temperature. Air-sensitive yellow needles of the target compound were isolated from the crystalline product, which also contained the unreacted selenium.

$[(C_{14}H_{29}Py)_{5.2}Zn_{4.9}(SbSe_4)_4(SbSe_3)]$ **or** $C_{98.8}H_{176.8}N_{5.2}Sb_5Se_{19}Zn_{4.9}$. This compound crystallizes in the cubic structure with the lattice parameter $a = 8350 \pm 30$ and energy gap 1.79 eV (Ding et al. 2006). To obtain it, $K_3SbSe_4$ (0.5 mM) was dissolved in 3 mL of $H_2O$ upon stirring and heating at 80°C. An orange solution formed quickly. $C_{14}H_{29}PyBr \cdot H_2O$ (5.2 mM) was dissolved in $H_2O$ (20 mL) and $ZnCl_2$ (0.25 mM) was dissolved in 3 mL of $H_2O$ in another flask. These two solutions were added into the $C_{14}H_{29}PyBr/H_2O$ solution simultaneously. A dark-red precipitate formed immediately. However, the reaction mixture was stirred overnight at 80°C. The product was isolated by filtration, washed with warm water, and dried under vacuum.

$[(C_{16}H_{33}Py)_{5.2}Zn_{4.9}(SbSe_4)_2(SbSe_3)_3]$ **or** $C_{109.2}H_{197.6}N_{5.2}Sb_5Se_{17}Zn_{4.9}$. This compound crystallizes in the cubic structure with the lattice parameter $a = 9050 \pm 50$ and energy gap 1.74 eV (Ding et al. 2006). To obtain it, $K_3SbSe_4$ (0.5 mM) was dissolved in 3 mL of $H_2O$ upon stirring and heating at 80°C. An orange solution formed quickly. $C_{16}H_{33}PyBr \cdot H_2O$ (5.2 mM) was dissolved in $H_2O$ (20 mL) and $ZnCl_2$ (0.25 mM) was dissolved in 3 mL of $H_2O$ in another flask. These two solutions were added into the $C_{16}H_{33}PyBr/H_2O$ solution simultaneously. A dark-red precipitate formed immediately. However, the reaction mixture was stirred overnight at 80°C. The product was isolated by filtration, washed with warm water, and dried under vacuum.

## 2.29  Zn–H–C–N–O–Se

Some compounds are formed in this system.

$[Zn_3(SeO_3)_5(CN_3H_6)_4]$ **or** $C_4H_{24}N_{12}O_{15}Se_5Zn_3$. This compound crystallizes in the orthorhombic structure with the lattice parameters $a = 890.07 \pm 0.04$, $b = 1507.71 \pm 0.07$, and $c = 2050.96 \pm 0.09$ pm and a calculated density of 2.586 g·cm$^{-3}$ (Harrison et al. 2000a,b). To obtain it, guanidinium carbonate (10 mM), ZnO (10 mM), $SeO_2$ (20 mM), and $H_2O$ (20 mL) were added to a polytetrafluoroethylene bottle, shaken well, and placed in a 95°C oven. The bottle was vented and recapped after 1 h. Intergrown transparent slabs of $[Zn_3(SeO_3)_5](CN_3H_6)_4$ were recovered by vacuum filtration after 7 days.

$[Zn(C_{12}H_{13}NO_3Se)] \cdot H_2O$ **or** $C_{12}H_{15}NO_4SeZn$. For obtaining this compound, to a warm solution (60°C–70°C) of the D, L-selenomethionine $(C_5H_{11}NO_2Se)$ (5 mM) in

10 mL of $H_2O$, 5 mM of salicylaldehyde ($C_7H_6O_2$) in 10 mL of EtOH was added (Ran et al. 2011). The resulting solution was stirred until D, L-selenomethionine was dissolved. A solution of $Zn(OAc)_2 \cdot H_2O$ (5 mM) dissolved in a minimum quantity of $H_2O$ was added dropwise. The mixture was stirred for 1 h and the colored precipitate obtained was filtered, washed with $H_2O$, EtOH, and $Et_2O$, and dried in vacuo. The resulting solid was recrystallized from either DMF or DMSO.

**$[Zn(C_{16}N_{15}NO_3Se)] \cdot H_2O$ or $C_{16}H_{17}NO_4SeZn$.** The next procedure was used to obtain this compound. To a warm solution (60°C–70°C) of D, L-selenomethionine (5 mM) in 10 mL of $H_2O$, 2-hydroxy-1-naphthaldehyde (5 mM) in 10 mL of MeOH was added (Ran et al. 2010). The resulting solution was stirred until D, L-selenomethionine was completely dissolved. A solution of $Zn(OAc)_2 \cdot H_2O$ (5 mM) dissolved in a minimum quantity of $H_2O$ was added dropwise. The mixture was stirred for 1 h and the colored precipitate was filtered, washed with $H_2O$, EtOH, and ether, and dried in vacuo. The resulting solid was recrystallized from DMSO. ZnSe as the final thermal decomposition of the title compound was observed.

**$[Zn(SeCN)_2(C_7H_{10}N_2)_2]$ or $C_{16}H_{20}N_6Se_2Zn$.** This compound melts at 171°C–172°C (Secondo et al. 2000). To obtain it, $Zn(NO_3) \cdot 6H_2O$ (2.0 mM) was dissolved in 25 ml of absolute EtOH, to which a solution of KSeCN (4.0 mM) in the same solvent (30 mL) was added with stirring. The precipitated $KNO_3$ was filtered and to the filtrate was added a solution of 4-(N,N-dimethylamino)pyridine ($C_7H_{10}N_2$) (4.0 mM) in absolute EtOH (10 mL). The resulting precipitate was isolated by filtration, washed with EtOH and then ether, and dried in vacuo.

**$[Zn_2(C_5H_4NOSe)_4]$ or $C_{20}H_{16}N_4O_4Se_4Zn_2$.** This compound crystallizes in the monoclinic structure with the lattice parameters $a = 851.6 \pm 0.1$, $b = 1011.5 \pm 0.2$, $c = 1401.1 \pm 0.2$ pm, and $\beta = 95.271° \pm 0.001°$ (Ma et al. 2009). It was prepared by the following procedure. 2-Bromopyridine N-oxide hydrochloride dissolved in EtOH was neutralized with a solution of NaOH. The resulting mixture containing the free N-oxide was added to NaHSe solution and pH was adjusted to 2–3 with HOAc. After filtration, it was recrystallized from EtOH to give yellow-green crystals of 2-selenopyridine-N-oxide. This compound and $ZnCl_2$ in a molar ratio of 1:2 were dissolved in $H_2O$ and EtOH (1:1 volume ratio), and dilute solution of NaOH was added dropwise to maintain a pH of 6–7 resulting in the formation of a pale-yellow solid. The precipitate was recovered by filtration, washed with $H_2O$ and EtOH, and dried in vacuo. The colorless square crystals were obtained by recrystallization of the precipitate from DMF.

## 2.30　Zn–H–C–N–O–Re–Se

Some compounds are formed in this system.

**$[Zn(H_2O)]_2[Re_6Se_8(CN)_6] \cdot 8H_2O$ or $C_6H_{20}N_6O_{10}Re_6Se_8Zn_2$.** This compound crystallizes in the monoclinic structure with the lattice parameters $a = 1898.99 \pm 0.02$, $b = 1088.06 \pm 0.03$, $c = 857.88 \pm 0.03$ pm, and $\beta = 108.218° \pm 0.002°$ at 159 K and a calculated density of 4.556 g·cm$^{-3}$ (Bennett et al. 2000). It was prepared by the next procedure. A solution of $[Zn(H_2O)_6](ClO_4)_2$ (0.37 mM) in 10 mL of $H_2O$ was added to a solution of $Na_4[Re_6Se_8(CN)_6]$ (0.050 mM) in 10 mL of $H_2O$. After the solution was left standing for 12 h, large orange-red crystals of product had formed. The crystals were collected by centrifugation, separated from a small amount of orange powder impurity (compound $[Zn(H_2O)_6]Zn_3[Re_6Se_8(CN)_6]_2·18H_2O$) by sonication, washed with $H_2O$, and dried in air to give $[Zn(H_2O)]_2[Re_6Se_8(CN)_6]·8H_2O$.

**$[(H_3O)_2Zn_3\{Re_6Se_8(CN)_6\}_2]·20H_2O$ or $C_{12}H_{46}N_{12}O_{22}Re_{12}Se_{16}Zn_3$.** This compound crystallizes in the trigonal/rhombohedral structure with the lattice parameters $a = 1917.9 \pm 0.2$ pm and $\alpha = 52.97° \pm 0.01°$ and a calculated density of 3.511 $g·cm^{-3}$ (Naumov et al. 2000). To obtain it, to a 0.1 M aqueous solution of $K_4[Re_6Se_8(CN)_6]$, a few drops of acetic acid were added until pH reached 3. A dropwise addition of $ZnCl_2$ (0.05 M aqueous solution) resulted in the precipitation of a fine, insoluble yellow-orange powder. This powder was collected by filtration, washed with $H_2O$, and dried in air.

**$[Zn(H_2O)_6]Zn_3[Re_6Se_8(CN)_6]_2·18H_2O$ or $C_{12}H_{48}N_{12}O_{24}Re_{12}Se_{16}Zn_4$.** This compound crystallizes in the trigonal structure with the lattice parameters $a = 1719.2 \pm 0.1$ and $c = 4936.9 \pm 0.5$ pm at 142 K and a calculated density of 3.551 $g·cm^{-3}$ (Bennett et al. 2000). Its single crystals were obtained by layering a 0.075 M aqueous solution of $[Zn(H_2O)_6](ClO_4)_2$ over a 0.015 M aqueous solution of $Na_4[Re_6Se_8(CN)_6]$ in a narrow-diameter tube. However, the material was best prepared in pure form as follows. A solution of $[Zn(H_2O)_6](ClO_4)_2$ (0.22 mM) in 2 mL of MeOH was added to a solution of $Na_4[Re_6Se_8(CN)_6]$ (0.030 mM) in 2 mL of MeOH. After the solution was left standing for 12 h, the resulting orange precipitate was collected by centrifugation, washed with MeOH, and quickly dried in air.

## 2.31   Zn–H–C–N–Se

Some compounds are formed in this system.

**$[Zn(Me_2NCSe_2)_2]$ or $C_6H_{12}N_2Se_4Zn$, zinc $N,N$-dimethyldiselenocarbamate.** This compound was prepared by the next procedure. A solution of $Na(Me_2NCSe_2)$ was added rapidly to $ZnSO_4$ (6.2 mM) in $H_2O$ (10 mL) at 0°C (Barnard and Woodbridge 1961). The resulting precipitate was filtered off and gave a pink solid with melting temperature of 236°C–239°C. Recrystallization from toluene gave light-brown crystals with melting temperature of 242°C–243°C.

**$[MeZnSe_2CNEt_2]_2$ or $C_6H_{13}NSe_2Zn$.** This compound melts at 155°C (Malik and O'Brien 1991). To obtain it, a solution of ($N,N'$-diethyldiselenocarbamato)zinc (9.1 mM) in toluene (50 mL) was stirred with $Me_2Zn$ (9.1 mM) at room temperature for 30 min. The colorless solution, on concentration under a vacuum, gave transparent crystals of the title compound. All reactions were performed in an inert atmosphere using Schlenk techniques and a vacuum line.

**$[EtZnSe_2CNEt_2]_2$ or $C_7H_{15}NSe_2Zn$.** This compound melts at 131°C (Malik and O'Brien 1991). To obtain it, a solution of ($N,N'$-diethyldiselenocarbamato)zinc (9.1 mM) in toluene (50 mL) was stirred with $Et_2Zn$ (9.1 mM) at room temperature for 30 min. The colorless solution, on concentration under vacuum, gave transparent crystals of the title compound. All reactions were performed in an inert atmosphere using Schlenk techniques and a vacuum line.

**[ZnSe$_4$(C$_4$H$_6$N$_2$)$_2$] or C$_8$H$_{12}$N$_4$Se$_4$Zn.** Interaction of zinc dust, $N$-methylimidazole, and Se at 100°C for 18 h gave a reddish-brown solution, which upon layering with THF afforded the title compound (Dev et al. 1990).

**[Bu$^t$ZnSe$_2$CNEt$_2$] or C$_9$H$_{19}$NSe$_2$Zn.** This compound melts at 163°C (Malik et al. 1992). To obtain it, a solution of (diethyldiselenocarbamato)zinc (6.4 mM) in toluene (35 mL) was stirred with Bu$^t_2$Zn (6.4 mM) at room temperature for 30 min. The colorless solution, on concentration under vacuum, gave transparent crystals of the title compound.

**[Zn(Et$_2$NCSe$_2$)$_2$] or C$_{10}$H$_{20}$N$_2$Se$_4$Zn, zinc $N,N$-diethyldiselenocarbamate.** This compound melts at 154°C–155°C (Barnard and Woodbridge 1961, Furlani et al. 1968) and crystallizes in the monoclinic structure with the lattice parameters $a = 998.5 \pm 1.0$, $b = 1101.7 \pm 1.0$, $c = 1687.3 \pm 1.0$ pm, and $\beta = 110°42' \pm 5'$ and experimental and a calculated density of $2.12 \pm 0.01$ and $2.102$ g·cm$^{-3}$, respectively (Bonamico and Dessy 1971). It was prepared when the resulting solution of NaEt$_2$NCSe$_2$ or KEt$_2$NCSe$_2$ was mixed with an aqueous solution of ZnCl$_2$ in stoichiometric amounts. The raw product cloud could be easily crystallized from several solvents (acetone or DMF) (Furlani et al. 1968).

**[Bu$^t$CH$_2$ZnSe$_2$CNEt$_2$] or C$_{10}$H$_{21}$NSe$_2$Zn.** This compound melts at 135°C (Malik et al. 1992). To obtain it, a solution of (diethyldiselenocarbamato)zinc (6.4 mM) in toluene (35 mL) was stirred with dineopentylzinc (6.4 mM) at room temperature for 30 min. The colorless solution, on concentration under vacuum, gave transparent crystals of the title compound.

**[Zn(Et$_2$NCSe$_2$)$_2$(C$_5$H$_{10}$NH)] or C$_{15}$H$_{31}$N$_3$Se$_4$Zn.** This compound melts at 136°C with decomposition (Higgins and Saville 1963). To obtain it, powdered zinc diethyldiselenocarbamate (20 mM) was added, in portions, to a solution of piperidine (ca. 30 mM) in benzene (20–40 mL) at 50°C. Upon cooling the resultant solution, or partial removal of the solvent or addition of light petroleum (100 ± 20 mL), the product crystallized. It was collected, washed with light petroleum, and dried in air.

**[Zn(Bu$^n_2$NCSe$_2$)$_2$] or C$_{18}$H$_{36}$N$_2$Se$_4$Zn, zinc $N,N$-di-$n$-butyldiselenocarbamate.** This compound was prepared by the next procedure. A solution of Na(Bu$^n_2$NCSe$_2$) was added rapidly to ZnSO$_4$ (6.2 mM) in H$_2$O (10 mL) at 0°C (Barnard and Woodbridge 1961). The resulting precipitate was filtered off and gave an orange solid with melting temperature of 66°C–67°C. Recrystallization from aqueous acetone gave pale-brown crystals with melting temperature of 68°C–68.5°C.

**[(Me$_4$N){Zn(SePh)$_4$}] or C$_{32}$H$_{44}$N$_2$Se$_4$Zn.** This compound crystallizes in the monoclinic structure with the lattice parameters $a = 1208.9 \pm 0.7$, $b = 1458.7 \pm 1.0$, $c = 995.0 \pm 0.7$ pm, and $\beta = 90.84° \pm 0.02°$ and a calculated density of 1.587 g·cm$^{-3}$ (Ueyama et al. 1988). It was synthesized under an Ar atmosphere. Me$_3$SiSePh

(50 mM), NMe$_4$Cl (40 mM), and Bu$^n_3$N (40 mM) were mixed in 30 mL of MeOH. To this solution was added ZnCl$_2$ (7 mM) in 50 mL of MeOH at room temperature. The colorless crystals obtained were collected by filtration, washed with cold MeOH, and recrystallized from hot MeOH.

**[Zn(SeBu$^t_3$C$_6$H$_2$-2,4,6)$_2$(NMeImid)]**   (NMeImid = *N*-methylimidazole) or C$_{40}$H$_{64}$N$_2$Se$_2$Zn. This compound melts at 248°C–250°C (Bochmann et al. 1994). To obtain it, a suspension of Zn(SeBu$^t_3$C$_6$H$_2$-2,4,6)$_2$ (0.81 mM) was treated with NMeImid (0.79 mM), and the solution was stirred for 2 h. The starting material dissolved, followed by the precipitation of C$_{40}$H$_{64}$N$_2$Se$_2$Zn, which was recrystallized from toluene (0.70 mM). All reactions were carried out under an Ar atmosphere using standard vacuum-line techniques.

**C$_{40}$H$_{80}$N$_4$Se$_8$Zn$_2$, *N,N*-diethyldiselenocarbamatozinc.** To obtain this compound, CSe$_2$ was reacted immediately with an excess of (C$_2$H$_5$)$_2$NH in pentane at 0°C to give *N,N*-diethyldiselenocarbamate as the diethylammonium salt (Hursthouse et al. 1992). The selenocarbamate was then reacted with a stoichiometric quantity of an aqueous solution of ZnCl$_2$ to give an insoluble yellow precipitate of C$_{40}$H$_{80}$N$_4$Se$_8$Zn$_2$. The crude product was recrystallized from toluene to give yellow cubic crystals. This compound could be a useful precursor for the deposition of ZnSe films.

**[Zn(SeC$_6$H$_2$Bu$^t_3$)$_2$(Bu$^t$CN)] or C$_{41}$H$_{67}$NSe$_2$Zn.** This compound decomposed at 196°C–198°C. It was obtained if to a stirred suspension of [Zn(2,4,6-Bu$^t_3$C$_6$H$_2$Se)$_2$]$_2$ (0.095 mM) in petroleum ether (10 mL) at room temperature *t*-butylisocyanide (0.18 mM) was added (Bochmann et al. 1993). The solid dissolved to give a clear colorless solution from which a white solid precipitated within a few minutes. After stirring the mixture for 2 h the precipitate was filtered off and recrystallized from hot toluene (5 mL) to give Zn(SeC$_6$H$_2$Bu$^t_3$)$_2$(Bu$^t$CN) as colorless crystals.

**[Zn(SeBu$^t_3$C$_6$H$_2$-2,4,6)$_2$]$_2$(NC$_5$H$_3$Me$_2$-2,6)] or C$_{43}$H$_{67}$NSe$_2$Zn.** This compound melts at 192°C–194°C (Bochmann et al. 1994). To obtain it, to a stirred suspension of [Zn(SeBu$^t_3$C$_6$H$_2$-2,4,6)$_2$]$_2$ (0.24 mM) in 10 mL of petroleum ether at room temperature was added 2,6-lutidine (0.2 mM). The solid dissolved to give a clear colorless solution from which a white solid precipitated during the course of 30 min. The precipitate was filtered off and recrystallized from toluene to give colorless crystals of C$_{43}$H$_{67}$NSe$_2$Zn. All reactions were carried out under an Ar atmosphere using standard vacuum-line techniques.

**[Zn(SeBu$^t_3$C$_6$H$_2$-2,4,6)$_2$]$_2$(NMeImid)$_2$] or C$_{44}$H$_{70}$N$_4$Se$_2$Zn.** This compound melts at 198°C–200°C (Bochmann et al. 1994). To obtain it, [Zn(SeBu$^t_3$C$_6$H$_2$-2,4,6)$_2$]$_2$ (0.81 mM) was treated with NMeImid (8.2 mM), and the solution was stirred for 2 h. The starting material dissolved, followed by the precipitation of C$_{44}$H$_{70}$N$_4$Se$_2$Zn, which was recrystallized from toluene (0.76 mM). All reactions were carried out under an Ar atmosphere using standard vacuum-line techniques.

**[Zn(SeC$_6$H$_2$Bu$^t_3$)$_2$(Bu$^t$CN)$_2$] or C$_{46}$H$_{76}$N$_2$Se$_2$Zn.** This compound decomposed at 190°C. It was obtained if to a stirred suspension of [Zn(2,4,6-Bu$^t_3$C$_6$H$_2$Se)$_2$]$_2$ (0.28 mM) in petroleum ether (15 mL) at room temperature *t*-butylisocyanide

(5.51 mM) was added (Bochmann et al. 1993). The solid dissolved to give a clear colorless solution from which a white solid precipitated within a few minutes. Stirring was continued for 1 h. The precipitate was filtered off and recrystallized from warm toluene (5 mL) to give $Zn(SeC_6H_2Bu^t_3)_2(Bu^tCN)_2$ as colorless crystals.

**$[C_{46}H_{76}N_2Se_2Zn\cdot1.6C_5H_{12}]$ or $C_{54}H_{95.2}N_2Se_2Zn$.** This compound crystallizes in the monoclinic structure with the lattice parameters $a = 957.0 \pm 0.3$, $b = 1862.8 \pm 0.9$, $c = 3144.2 \pm 1.1$ pm, and $\beta = 94.63° \pm 0.03°$ and a calculated density of 1.184 g·cm$^{-3}$ (Bochmann et al. 1993).

**$[(Me_4N)_2\{Zn_4(SePh)_{10}\}]$ or $C_{68}H_{74}N_2Se_{10}Zn_4$.** This compound crystallizes in the triclinic structure with the lattice parameters $a = 1321.4 \pm 0.2$, $b = 2385.9 \pm 0.2$, and $c = 1307.2 \pm 0.1$ pm and $\alpha = 91.134° \pm 0.008°$, $\beta = 113.350° \pm 0.008°$, and $\gamma = 79.865° \pm 0.009°$ and experimental and a calculated density of $1.77 \pm 0.05$ and 1.77 g·cm$^{-3}$, respectively (Vittal et al. 1992). To obtain it, 9.2 mM of benzeneselenol was added to 9.6 mM of sodium metal dissolved in 100 mL of MeOH (Dean et al. 1987). The resultant solution of NaSePh was added to a stirred solution of 3.4 mM of $Zn(NO_3)_2\cdot6H_2O$ in 10 mL of $H_2O$, producing a yellow solution containing some light-yellow precipitate. After the mixture was stirred for about 10 min, a solution of 3.5 mM of $Me_4NCl$ in 10 mL of MeOH was added. The mixture was warmed to 60°C, and acetone and acetonitrile were added until a clear solution was obtained, which was filtered while hot and left at 5°C overnight for crystallization to occur. Pale-yellow transparent crystals formed: these were separated by decantation of the mother liquor; washed with 10 mL of MeOH and, in succession, 20 mL portions of $H_2O$, MeOH, and diethyl ether; and then dried in vacuum (Dean et al. 1987, Vittal et al. 1992). All manipulations were carried out under an atmosphere of Ar.

**$[(Me_4N)_2\{Zn_4(SePh)_{10}\}]\cdot MeCN$ or $C_{70}H_{77}N_3Se_{10}Zn_4$.** This compound crystallizes in the monoclinic structure with the lattice parameters $a = 1424.8 \pm 0.1$, $b = 3972.2 \pm 0.2$, $c = 1340.8 \pm 0.1$ pm, and $\beta = 97.132° \pm 0.005°$ and experimental and a calculated density of $1.78 \pm 0.04$ and 1.74 g·cm$^{-3}$, respectively (Vittal et al. 1992). The omission of the acetone in the procedure described for $(Me_4N)_2[Zn_4(SePH)_{10}]$ gave a comparable yield of this acetonitrile solvate as light-yellow platelike crystals.

## 2.32   Zn–H–C–N–Cl–Se

Some compounds are formed in this system.

**$[Zn\{SeC(NH_2)_2\}_2Cl_2]$ or $C_2H_8N_4Cl_2Se_2Zn$.** This compound melts at 158°C–160°C with decomposition (Isab and Wazeer 2005). It was prepared at room temperature by mixing solutions of selenourea with $ZnCl_2$ in a minimum amount of distilled $H_2O$ (2–10 mL) in the 2:1 molar ratio and stirring for 1 or 2 h. The isolated product was dried thoroughly.

**$[Zn(C_3H_6N_2Se)_2Cl_2]$ or $C_6H_{12}N_4Cl_2Se_2Zn$.** This compound melts at 223°C–224°C with decomposition (Devillanova and Verani 1977). It was obtained by boiling an absolute EtOH solution of $ZnCl_2$ under reflux with ethyleneselenourea (1:2 molar ratio).

$C_{10}H_{16}N_4Cl_2Se_2Zn$, dichlorobis[1,3-dimethyl-2(3$H$)-imidazoleselone]zinc. This compound melts at 197°C–199°C and crystallizes in the monoclinic structure with the lattice parameters $a = 944.37 \pm 0.03$, $b = 1362.11 \pm 0.05$, $c = 1335.22 \pm 0.05$ pm, and $\beta = 107.477° \pm 0.001°$ and a calculated density of 1.972 g·cm$^{-3}$ (Williams et al. 2002). It was obtained if into a 25 mL Erlenmeyer flask with magnetic stir bar was placed 2.3 mM of 1,3-dimethyl-2(3$H$)-imidazoleselone dissolved in 25 mL of boiling acetonitrile. $ZnCl_2$ (1.5 mM) was added with stirring, and the mixture was permitted to continue a slow boil. The solution volume was reduced to approximately 10 mL, and the flask was removed from the heat. Small pale-yellow crystals were noted to form upon cooling of the solution to room temperature. The flask was stoppered and refrigerated. Filtration and washing with small portions of solvents yielded $C_{10}H_{16}N_4Cl_2Se_2Zn$. Its thermal decomposition leads to the formation of ZnSe.

$[(Me_4N)_2\{Zn_4(SePh)_6Cl_4\}]$ or $C_{44}H_{54}N_2Cl_4Se_6Zn_4$. To obtain this compound, freshly prepared solid $PhICl_2$ (0.60 mM) was added with stirring to a solution of $(Me_4N)_2[Zn_4(SePh)_{10}]$ (0.30 mM) in 15 mL of MeCN (Dean et al. 1987). A rapid reaction occurred with the formation of a precipitate. After 10 min the solvent was removed under vacuum. The residue was extracted three times with 20 mL portions of diethyl ether to remove $(PhS)_2$ and PhI. The crude residue was purified by recrystallization from a mixture of acetonitrile and diethyl ether. The syntheses were performed under an inert atmosphere.

$[(Et_4N)_2\{Zn_4Cl_4(SePh)_6\}]$ or $C_{52}H_{70}N_2Cl_4Se_6Zn_4$. This compound crystallizes in the monoclinic structure with the lattice parameters $a = 1212.4 \pm 0.2$, $b = 3726.1 \pm 0.8$, $c = 1379.4 \pm 0.3$ pm, and $\beta = 99.83° \pm 0.03°$ and a calculated density of 1.732 g·m$^{-3}$ (Eichhöfer et al. 1998). To obtain it, a solution of $NEt_4Cl$ (0.75 mM) in MeCN (2 mL) was pippeted to a solution of $ZnCl_2$ (1.5 mM) in THF (30 mL) with the next addition of $PhSeSiMe_3$ (2.23 mM). Colorless crystals of the title compound were obtained after 2–3 days.

$[(Et_4N)_2\{Zn_8Cl_4Se(SePh)_{12}\}]$ or $C_{88}H_{100}N_2Cl_4Se_{13}Zn_8$. This compound crystallizes in the monoclinic structure with the lattice parameters $a = 3848.6 \pm 0.8$, $b = 1784.9 \pm 0.4$, $c = 3432.0 \pm 0.7$ pm, and $\beta = 97.78° \pm 0.03°$ and a calculated density of 1.613 g·m$^{-3}$ (Eichhöfer et al. 1998). To obtain it, a solution of $NEt_4Cl$ (0.55 mM) in MeCN (2 mL) was mixed with a solution of $ZnCl_2$ (2.20 mM) in THF (40 mL). Then a mixture of $PhSeSiMe_3$ (3.30 mM) and $Se(SiMe_3)_2$ (0.30 mM) was added. Colorless crystals of the title compound were obtained after a few days.

## 2.33 Zn–H–C–N–Br–Se

Some compounds are formed in this system.

$[Zn(C_3H_6N_2Se)_2Br_2]$ or $C_6H_{12}N_4Br_2Se_2Zn$. This compound melts at 214°C–215°C with decomposition (Devillanova and Verani 1977). It was obtained by boiling an absolute EtOH solution of $ZnBr_2$ under reflux with ethyleneselenourea (1:2 molar ratio).

**[(Me₄N)₂{Zn₄(SePh)₆Br₄}]** or **C₄₄H₅₄N₂Br₄Se₆Zn₄**. To obtain this compound, $Br_2$ (1.15 mM) in 5 mL of $CCl_4$ was added to a stirred solution containing $(Me_4N)_2[Zn_4(SePh)_{10}]$ (0.575 mM) in 15 mL of acetone (Dean et al. 1987). Decolorization occurred quickly, leaving a clear solution with a small amount of white solid. After 10 min, the solvents were removed under vacuum. The residue was extracted with three 40 mL portions of diethyl ether to remove $(PhS)_2$. The residue was purified by recrystallization from a mixture of acetone and cyclohexane. The syntheses were performed under an inert atmosphere.

## 2.34   Zn–H–C–N–I–Se

Some compounds are formed in this system.

**[Zn(C₃H₆N₂Se)₂I₂]** or **C₆H₁₂N₄I₂Se₂Zn**. This compound melts at 212°C–213°C with decomposition (Devillanova and Verani 1977). It was obtained by boiling an absolute EtOH solution of $ZnI_2$ under reflux with ethyleneselenourea (1:2 molar ratio).

**[(Me₄N)₂{Zn₄(SePh)₆I₄}]** or **C₄₄H₅₄N₂I₄Se₆Zn₄**. To obtain this compound, solid iodine (1.2 mM) was added to a stirred solution of $(Me_4N)_2[Zn_4(SePh)_{10}]$ (0.6 mM) in 15 mL of acetone (Dean et al. 1987). Immediate decolorization occurred, leaving a clear solution. Acetone was removed under vacuum. The residue was extracted with three 25 mL portions of diethyl ether to remove $(PhS)_2$. The remaining white solid was purified by recrystallization from $MeCN/Et_2O$. The syntheses were performed under an inert atmosphere.

## 2.35   Zn–H–C–P–O–Se

Some compounds are formed in this system.

**[Zn{Se₂P(OEt)₂}₂]∞** or **C₈H₂₀P₂O₄Se₄Zn**. This compound crystallizes in the monoclinic structure with the lattice parameters $a = 855.8 \pm 0.3$, $b = 1980.3 \pm 0.7$, $c = 1190.1 \pm 0.4$ pm, and $\beta = 105.272° \pm 0.007°$ and a calculated density of 2.128 g·cm⁻³ (Santra et al. 2003). To obtain it, $Zn(ClO_4)_2 \cdot 6H_2O$ (2 mM) was dissolved in 20 mL of $H_2O$ in a 100 mL Schlenk flask. The water solution was deoxygenated and then transferred by cannula to another 100 mL Schlenk flask containing $NH_4Se_2P(OEt)_2$ (4 mM). The solution mixture was stirred for 2 h at room temperature under a $N_2$ atmosphere. White precipitate was formed during the reaction period. The resulting white precipitate was filtered out and washed with ether. The white residue was then redissolved in $CH_2Cl_2$ and the solution layered with hexane, which afforded a crystalline material of $[Zn\{Se_2P(OEt)_2\}_2]_\infty$. Crystals were grown from $CH_2Cl_2$ layered with hexane.

**[Zn₂{Se₂P(OPrⁱ)₂}₄]** or **C₂₄H₅₆P₄O₈Se₈Zn₂**. This compound crystallizes in the monoclinic structure with the lattice parameters $a = 1123.9 \pm 0.8$, $b = 1739.6 \pm 1.3$,

$c = 2591.5 \pm 1.3$ pm, and $\beta = 99.46° \pm 0.05°$ and a calculated density of 1.806 g·cm$^{-3}$ (Santra et al. 2003). The method for obtaining it is similar to obtaining $[Zn\{Se_2P(OEt)_2\}_2]_\infty$, except that the white precipitate, which formed during the reaction period was redissolved in 10 mL of ether and slow evaporation of the ether solution gave colorless crystals of $[Zn_2\{Se_2P(OPr^i)_2\}_4]$.

**(Ph$_4$P)$_2$\{Zn[Se$_2$C$_2$(OAc)$_2$]$_2$\}** or **C$_{60}$H$_{52}$P$_2$O$_8$Se$_4$Zn**. To obtain this compound, (Ph$_4$P)$_2$[Zn(Se$_4$)$_2$] (0.1 mM) was dissolved in DMF (10 mL), and dimethylacetamide (0.5 mL) was added to the solution (Ansari et al. 1990). This solution was stirred for 0.5 h and then filtered. Toluene (20 mL) was added to the filtrate. Orange crystals formed overnight at room temperature. All the manipulations were carried out under an N$_2$ atmosphere.

## 2.36   Zn–H–C–P–Se

Some compounds are formed in this system.

**[Zn(SePh)$_2$·depe]** or **C$_{22}$H$_{34}$P$_2$Se$_2$Zn [depe = 1,2-bis(diethylphosphino)ethane]**. This compound melts at 137.8°C–138°C and crystallizes in the orthorhombic structure (Brennan et al. 1990). To obtain it, depe (2.5 mM) was added to Zn(SePh)$_2$ (1.7 mM) suspended in toluene (25 mL). The mixture was stirred for 12 h, pyridine (2.8 mL) and heptane (15 mL) were added, and the solution was filtered and cooled (–20°C) to give colorless diamonds. All reactions were performed in an inert atmosphere.

**[Zn(SeBu$^t$$_3$C$_6$H$_2$–2,4,6)$_2$(PMe$_3$)]** or **C$_{39}$H$_{67}$PSe$_2$Zn**. This compound sublimes at 143°C at 1.3 Pa (Bochmann et al. 1994). It could be obtained if to a suspension of [Zn(SeBu$^t$$_3$C$_6$H$_2$–2,4,6)$_2$]$_2$ (0.16 mM) in 10 mL of petroleum ether was added PMe$_3$ (1.31 mM) via syringe. The starting material dissolved within a few seconds, followed by the precipitation of flocculant white product, which was dissolved by adding more petroleum ether and warming on a water bath. Cooling to 0°C gave colorless needles of C$_{39}$H$_{67}$PSe$_2$Zn. All reactions were carried out under an Ar atmosphere using standard vacuum-line techniques.

**[(Ph$_4$P)$_2$\{Zn(Se$_4$)$_2$\}]** or **C$_{48}$H$_{40}$P$_2$Se$_8$Zn**. This compound crystallizes in the monoclinic structure with the lattice parameters $a = 1015.9 \pm 0.5$, $b = 1402.0 \pm 0.4$, $c = 3426 \pm 1$ pm, and $\beta = 92.11° \pm 0.02°$ (Banda et al. 1989) ($a = 1004.8 \pm 1.0$, $b = 1392.3 \pm 1.3$, $c = 3398 \pm 3$ pm, and $\beta = 92.21° \pm 0.03°$ at 110 K and a calculated density of 1.923 g·cm$^{-3}$ [Ansari et al. 1990]). To obtain it, a mixture of Li$_2$Se (2 mM), black Se (6 mM), and Ph$_4$PCl (2 mM) in MeCN (30 mL) and Et$_3$N (5 mL) was stirred for 1 h. A solution of zinc ethylxanthate, Zn(EtCOS$_2$)$_2$, (1 mM) in DMF (10 mL) was added to that mixture. The resulting solution was stirred for 1 h and then filtered. The volume of the filtrate was reduced to 20 mL, and Pr$^i$OH (20 mL) was layered on it. Red crystals were produced in 2 days. All the manipulations were carried out under an N$_2$ atmosphere (Ansari et al. 1990).

Bis-tetraselenide complex [Zn(Se$_4$)$_2$]$^{2-}$ could be readily prepared by the reaction of Se, Na, and ZnCl$_2$ in DMF and crystallized with Ph$_4$P$^+$. The general preparative procedure involves $ZnCl_2$, Se, and Na in a molar ratio of 1:8:4 under a $N_2$ atmosphere, with $Ph_4PBr$ slowly diffused into the filtered reaction solution (Banda et al. 1989).

**[Zn(SeBu$^t_3$C$_6$H$_2$–2,4,6)$_2$(PMePh$_2$)$_2$] or C$_{62}$H$_{84}$P$_2$Se$_2$Zn.** This compound melts at 104°C–105°C (Bochmann et al. 1994). It could be obtained as colorless crystals if to a suspension of [Zn(SeBu$^t_3$C$_6$H$_2$–2,4,6)$_2$]$_2$ (0.35 mM) in 10 mL of petroleum ether was added PMePh$_2$ (3.5 mM) via syringe. All reactions were carried out under an Ar atmosphere using standard vacuum-line techniques.

**[Zn$_8$Se(SePh)$_{14}$(PPr$^n_3$)$_2$] or C$_{102}$H$_{112}$P$_2$Se$_{15}$Zn$_8$.** This compound crystallizes in the orthorhombic structure with lattice parameter $a = 2027.8 \pm 0.4$, $b = 2162.3 \pm 0.4$, and $c = 1668.5 \pm 0.3$ pm and a calculated density of 1.50 g·cm$^{-3}$ (Eichhöfer et al. 1998). To obtain it, $ZnCl_2$ (1.8 mM) and PPr$^n_3$ (1.8 mM) were dissolved in toluene. After an addition to this mixture PhSeSiMe$_3$ (3.6 mM), needlelike crystals of the title compound were grown from the colorless solution.

## 2.37   Zn–H–C–O–Se

Some compounds are formed in this system.

**[Zn(SeC$_6$H$_2$Bu$^t_3$)$_2$(OCMe$_2$)$_2$] or C$_{42}$H$_{70}$O$_2$Se$_2$Zn.** This compound as colorless crystals was obtained by the interaction of Zn(SeC$_6$H$_2$Bu$^t_3$)$_2$ with acetone (Bochmann et al. 1991).

**[Zn(SeC$_6$H$_2$Bu$^t_3$)$_2${OCH(2-MeOC$_6$H$_4$)}$_2$] or C$_{52}$H$_{74}$O$_4$Se$_2$Zn.** Treatment of a suspension of colorless Zn(SeC$_6$H$_2$Bu$^t_3$)$_2$ in hexane at room temperature with ca. 10 equivalents of an aromatic aldehyde 2-OMeC$_6$H$_4$CHO leads to an immediate color change to yellow-orange and formation of a clear solution that the product complex precipitates (Bochmann et al. 1991). Recrystallization of the crude product from light petroleum gives C$_{52}$H$_{74}$O$_4$Se$_2$Zn as orange crystals.

**[Zn(SeC$_6$H$_2$Bu$^t_3$)$_2$(C$_6$H$_5$CHO)]$_2$ or C$_{86}$H$_{128}$O$_2$Se$_4$Zn$_2$.** Treatment of a suspension of colorless Zn(SeC$_6$H$_2$Bu$^t_3$)$_2$ in hexane at room temperature with ca. 10 equivalents of an aromatic aldehyde C$_6$H$_5$CHO leads to an immediate color change to yellow-orange and formation of a clear solution that the product complex precipitates (Bochmann et al. 1991). Recrystallization of the crude product from toluene gives C$_{86}$H$_{128}$O$_2$Se$_4$Zn$_2$ as orange crystals.

**[Zn(SeC$_6$H$_2$Bu$^t_3$)$_2$(p-O=CHC$_6$H$_4$OMe)]$_2$ or C$_{88}$H$_{132}$O$_4$Se$_4$Zn$_2$.** This compound crystallizes in the triclinic structure with the lattice parameters $a = 1131.0 \pm 0.4$, $b = 1292.1 \pm 0.8$, and $c = 1526.7 \pm 0.4$ pm and $\alpha = 93.32° \pm 0.02°$, $\beta = 92.40° \pm 0.01°$, and $\gamma = 100.87° \pm 0.02°$ (Bochmann et al. 1991). Treatment of a suspension of colorless Zn(SeC$_6$H$_2$Bu$^t_3$)$_2$ in hexane at room temperature with ca. 10 equivalents of an aromatic aldehyde C$_6$H$_5$CHO leads to an immediate color change to yellow-orange and formation of a clear solution that the product complex precipitates. Recrystallization of the crude product from toluene gives C$_{86}$H$_{128}$O$_2$Se$_4$Zn$_2$ as orange crystals.

## 2.38 Zn–H–N–O–Se

$(NH_4)_2Zn(SeO_4)_2·6H_2O$. This quinary compound that crystallizes in the monoclinic structure with the lattice parameters $a=638.3±0.1$, $b=1266.6±0.3$, $c=938.7±0.2$ pm, and $\beta=106.24°±0.03°$ is formed in this system (Fleck and Kolitsch 2002). Thick tabular crystals of the title compound up to several mm in diameter were obtained by slow evaporation at room temperature of an acidic aqueous solution of selenic acid, $NH_4OH$, and $ZnCO_3$.

## 2.39 Zn–H–As–O–Se

$Zn_3(AsO_3Se)_2·4H_2O$. This quinary compound is formed in this system that crystallizes in the orthorhombic structure (Gigauri et al. 1999). Thermal decomposition of this compound can be represented by the following scheme: $Zn_3(AsO_3Se)_2·4H_2O$ (120°C) → $Zn_3(AsO_3Se)_2·3H_2O$ (195°C–260°C) → $Zn_3(AsO_3Se)_2·H_2O$ (295°C) → $Zn_3As_2O_4S$.

This compound was obtained by the interaction of $Zn(OAc)_2$ with sodium monothioarsenate at room temperature (Gigauri et al. 1999): $3Zn(OAc)_2+2Na_3AsO_3Se=Zn_3(AsO_3Se)_2·4H_2O+6NaOAc$.

## 2.40 Zn–K–Rb–Sn–Se

$K_3Rb_3Zn_4Sn_3Se_{13}$. This quinary compound that crystallizes in the trigonal structure with the lattice parameters $a=1465.0±0.2$ and $c=1574.4±0.3$, a calculated density of 3.435 g·cm$^{-3}$, and energy gap 2.6 eV is formed in this system (Wu et al. 2005). Single crystals of this compound were grown in a reaction containing 0.5 mM of $K_2Se$, 0.6 mM of Se, 0.25 mM of Sn, 0.5 mM of RbCl, and 0.017 mM of $ZnCl_2$ loaded in thick wall Pyrex tube, followed by the addition of 0.2 mL of $H_2O$ and 0.2 mL of MeOH. The reaction mixture sealed in a tube was heated at 130°C for 6 days. The product was washed with 80% EtOH followed by $H_2O$, and pure yellow crystals were obtained.

## 2.41 Zn–Cu–Mg–Sn–Se

$Cu_{2-x}Mg_xZnSnSe_4$ ($x=0–0.4$) bulks were fabricated by a liquid-phase reactive sintering technique at 600°C for 2 h (Kuo and Wubet 2014). Sintering the Mg-doped $Cu_2ZnSnSe_4$ ingots using $Sb_2S_3$ and Te sintering aids assisted the densification at 600°C and retained the atomic composition of the solid solutions very close to the favored composition design. Mg is a strong promoter of electrical mobility.

## 2.42 Zn–Cu–Al–Sn–Se

$Cu_{1.75}ZnAl_xSn_{1-x}Se_4$ ($x=0–0.6$) bulks were prepared by a liquid-phase reactive sintering method at 600°C with soluble sintering aids of $Sb_2S_3$ and Te (Kuo and Tsega 2013). All ingots are semiconductors and exhibited $p$-type conductivity.

## 2.43   Zn–Cu–Ge–Mn–Se

$Cu_2Zn_{1-x}Mn_xGeSe_4$. Samples in the range $0<x<0.375$ had the tetragonal stannite $\alpha$ structure while for $0.725<x\leq1$ the wurtz-stannite $\delta$ structure (Caldera et al. 2014). The $\alpha$ and $\delta$ fields are separated by a relatively wide three-phase region $(\alpha+\delta+MnSe_2)$. All of the alloys were made by the usual melt and anneal technique. The materials were annealed at 500°C for 1 month.

## 2.44   Zn–Cu–Ge–Fe–Se

At room temperature, a single-phase solid solution $Cu_2Zn_{1-x}Fe_xGeSe_4$ with the tetragonal stannite structure occurs across the whole composition range (Caldera et al. 2008). Undercooling effects occur for samples with $x>0.9$. All alloys were produced by the usual melt and anneal technique.

## 2.45   Zn–Ag–Ba–O–Se

$ZnSe–Ag_2Se–BaO$. $Ag_2Ba_2ZnO_2Se_2$. This quinary compound is formed in this system. It crystallizes in the tetragonal structure with lattice parameter $a=428.49\pm0.04$ and $c=1987.9\pm0.2$ pm at 150 K or in the orthorhombic structure with lattice parameter $a=1990.76\pm0.08$, $b=605.62\pm0.03$, and $c=605.77\pm0.03$ pm at 295 K (Herkelrath et al. 2008). This compound was targeted by a stoichiometric reaction between BaO and elemental Zn, Ag, and Se at 825°C for 48 h in an alumina crucible sealed inside a predried silica tube.

## 2.46   Zn–Cd–C–N–Se

$[ZnCd(SeCN)_4]_n$. This quinary compound is formed in this system that crystallizes in the tetragonal structure with the lattice parameters $a=1133.10\pm0.06$ and $c=463.14\pm0.8$ and a calculated density of 3.338 g·cm$^{-3}$ (Sun et al. 2006). It was obtained by the interaction of $Cd(NO_3)_2\cdot4H_2O$, $ZnCl_2$, and KSeCN in a 4:1:1 molar ratio in $H_2O$. The resulting precipitate was dissolved in a dilute KCl solution. Single crystals suitable for XRD were obtained by slow evaporation of this solution.

## 2.47   Zn–Ga–Si–P–Se

$ZnSe–GaP–Si$. Epitaxy layers of solid solutions $(ZnSe)_{1-x-y}(Si_2)_x(GaP)_y$ ($0\leq x\leq0.03$, $0\leq y\leq0.09$) were grown up from the limited volume of tin solution melting by liquid-phase epitaxy procedure (Saidov et al. 2012). These solid solutions represent the stable phase.

## 2.48   Zn–Ga–Sn–As–Se

$ZnSe–GaAs–Sn$. Isothermal sections of this quasiternary system at 600°C, 650°C, 700°C, and 800°C are shown in Figure 2.1 and some vertical sections are given in

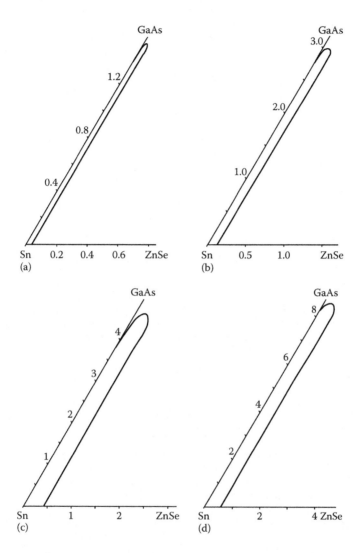

**FIGURE 2.1** Isothermal sections of the ZnSe–GaAs–Sn quasiternary system at (a) 600°C, (b) 650°C, (c) 700°C, and (d) 800°C. (From Novikova, E.M. et al., Phase diagram of the Sn–GaAs–ZnSe quasiternary system from the Sn-rich side [in Russian], *Deposited in VINITI*, No. 3039-74Dep, 1974.)

Figure 2.2 (Novikova et al. 1974). The eutectic temperatures in all vertical sections are degenerated from the Sn-rich side and are equal to 218°C–222°C (the eutectic temperature increases with increasing ZnSe content). $(ZnSe)_{1-x}(GaAs)_x$ solid solutions were obtained by the annealing of thoroughly mixed starting binary compounds at 1050°C for 100 h. These solid solutions were mixed with Sn and annealed at 1050°C for 50 h.

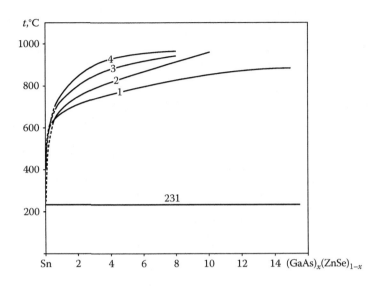

**FIGURE 2.2** The $(ZnSe)_{1-x}(GaAs)_x$–Sn vertical sections at (1) $x = 0.8$, (2) 0.6, (3) 0.4, and (4) 0.2. (From Novikova, E.M. et al., Phase diagram of the Sn–GaAs–ZnSe quasiternary system from the Sn-rich side [in Russian], *Deposited in VINITI*, No. 3039-74Dep, 1974.)

## REFERENCES

Adel J., Weller F., Dehnicke K. Synthese und Kristallstrukturen der spirocyclischen Polyselenido-Komplexe [Na-15-Krone-5]$_2$[M(Se$_4$)$_2$] mit M=Zink, Cadmium und Quecksilber, *Z. Naturforsch.*, 43B(9), 1094–1100 (1988).

Alsfasser R., Powell A.K., Vahrenkamp H., Trofimenko S. Monofunktionelle tetraedrische Zink-Komplexe L$^3$ZnX [L$^3$=Tris(pyrazolyl)borat], *Chem. Ber.*, 126(3), 685–694 (1993).

Ansari M.A., Mahler C.H., Chorghade G.S., Lu Y.-J., Ibers J.A. Synthesis, structure, spectroscopy, and reactivity of M(Se$_4$)$_2^{2-}$ anions, M=Ni, Pd, Zn, Cd, Hg, and Mn, *Inorg. Chem.*, 29(19), 3832–3839 (1990).

Banda R.M.H., Cusick J., Scudder M.L., Craig D.C., Dance I.G. Syntheses and X-ray structures of molecular metal polyselenide complexes [M(Se$_4$)$_2$]$^{2-}$ M=Zn, Cd, Hg, Ni, Pb, *Polyhedron*, 8(15), 1995–1998 (1989).

Barnard D., Woodbridge D.T. Derivatives of *NN*-dialkyldiselenocarbamic acids, *J. Chem. Soc.*, 2922–2926 (1961).

Bennett M.V., Shores M.P., Beauvais L.G., Long J.R. Expansion of the porous solid Na$_2$Zn$_3$[Fe(CN)$_6$]$_2$·9H$_2$O: Enhanced ion-exchange capacity in Na$_2$Zn$_3$[Re$_6$Se$_8$(CN)$_6$]$_2$·24H$_2$O, *J. Am. Chem. Soc.*, 122(28), 6664–6668 (2000).

Bochmann M., Bwembya G.C., Grinter R., Powell A.K., Webb K.J., Hursthouse M.B., Malik K.M.A., Mazid M.A. Synthesis of low-coordinate chalcogenolato complexes of zinc with O, N, S, and P donor ligands. Molecular and crystal structures of Zn(S-t-Bu$_3$C$_6$H$_2$–2,4,6)$_2$(L) (L=NC$_5$H$_3$Me$_2$–2,6, PMePh$_2$), Zn(Se-t-Bu$_3$C$_6$H$_2$–2,4,6)$_2$(OSC$_4$H$_8$) and Zn(S-t-Bu$_3$C$_6$H$_2$–2,4,6)$_2$(N-methylimidazole)$_2$, *Inorg. Chem.*, 33(10), 2290–2296 (1994).

Bochmann M., Bwembya G.C., Hursthouse M.B., Coles S.J. Synthesis of phosphinochalcogenoic amidato complexes of zinc and cadmium. The crystal and molecular structure of [Zn{Bu$^t_2$P(Se)NPr$^i$}$_2$], *J. Chem. Soc., Dalton Trans.*, (17), 2813–2817 (1995).

Bochmann M., Bwembya G.C., Powell A.K. Synthesis of isocyanide complexes of zinc. The molecular and crystal structure of $Zn(SeC_6H_2Bu^t{}_3)_2(CNBu^t)_2$, *Polyhedron*, 12(24), 2929–2932 (1993).

Bochmann M., Webb K., Hursthouse M.B., Mazid M. The first stable aldehyde and ketone complexes of zinc: The structure of $[Zn(SeC_6H_2Bu^t{}_3)_2(p\text{-}O{=}CHC_6H_4OMe)]_2$, *J. Chem. Soc., Chem. Commun.*, (24), 1735–1737 (1991).

Bonamico M., Dessy G. Structural studies of diselenocarbamates. Crystal and molecular structures of nickel(II), copper(II), and zinc(II) diethyldiselenocarbamates, *J. Chem. Soc. A: Inorg. Phys. Theor.*, 264–269 (1971).

Brennan J.G., Siegrist T., Carroll P.J., Stuczynski S.M., Reynders P., Brus L.E., Steigerwald M.L. Bulk and nanostructure group II-VI compounds from molecular organometallic precursors, *Chem. Mater.*, 2(4), 403–409 (1990).

Bu X., Zheng N., Wang X., Wang B., Feng P. Three-dimensional frameworks of gallium selenide supertetrahedral clusters, *Angew. Chem. Int. Ed.*, 43(12), 1502–1505 (2004).

Caldera D., Quintero M., Morocoima M., Moreno E., Quintero E., Grima-Gallardo P., Bocaranda P., Henao J.A., Macías M.A., Briceño J.M., Mora A.E. Lattice parameters values and phase diagram for the $Cu_2Zn_{1-z}Mn_zGeSe_4$ alloy system, *J. Alloys Compd.*, 614, 253–257 (2014).

Caldera D., Quintero M., Morocoima M., Quintero E., Grima P., Marchan N., Moreno E., Bocaranda P., Delgado G.E., Mora A.E., Briceño M., Fernandez J.L. Lattice parameter values and phase diagram for the $Cu_2Zn_{1-z}Fe_zGeSe_4$ alloy system, *J. Alloys Compd.*, 457(1–2), 221–224 (2008).

Dean P.A.W., Vittal J.J., Payne N.C. Syntheses of $(Me_4N)_2[(\mu\text{-}EPh)_6(MX)_4]$ (M = Cd, Zn; E = S, Se; X = Cl, Br, I). Crystal and molecular structures of $(Me_4N)_2[(\mu\text{-}EPh)_6(CdBr)_4]$ (E = S, Se) and characterization of $[(\mu\text{-}EPh)_6(CdX)_4]^{2-}$ in solution by $^{113}Cd$ and $^{77}Se$ NMR, *Inorg. Chem.*, 26(11), 1683–1689 (1987).

Dehnen S., Brandmayer M.K. Reactivity of chalcogenostannate compounds: Syntheses, crystal structures, and electronic properties of novel compounds containing discrete ternary anions $[M^{II}(\mu_4\text{-}Se)(SnSe_4)_4]^{10-}$ (M$^{II}$ = Zn, Mn), *J. Am. Chem. Soc.*, 125(22), 6618–6619 (2003).

Devillanova F.A., Verani G. Complexes of znc(II), cadmium(II) and mercury(II) halides with ethylenselenourea, *Trans. Met. Chem.*, 2(1), 9–11 (1977).

Dev S., Ramli E., Rauchfuss T.B., Stern C.L. Direct approaches to zinc polychalcogenide chemistry: $ZnS_6(N\text{-}MeIm)_2$ and $ZnSe_4(N\text{-}MeIm)_2$, *J. Am. Chem. Soc.*, 112(17), 6385–6386 (1990).

Ding N., Takabayashi Y., Solari P.L., Prassides K., Pcionek R.J., Kanatzidis M.G. Cubic gyroid frameworks in mesostructured metal selenides created from tetrahedral $Zn^{2+}$, $Cd^{2+}$, and $In^{3+}$ ions and the $[SbSe_4]^{3-}$ precursor, *Chem. Mater.*, 18(19), 4690–4699 (2006).

Eichhöfer A., Fenske D., Pfistner H., Wunder M. Zinkselenid- und Zinktelluridcluster mit Phenylselenolat- und Phenyltellurolatliganden. Die Kristallstrukturen von $[NEt_4]_2$ $[Zn_4Cl_4 (SePh)_6]$, $[NEt_4]_2[Zn_8Cl_4Se(SePh)_{12}]$, $[Zn_8Se(SePh)_{14}(P^nPr_3)_2]$, $[HP^nPr_2R]_2[Zn_8 Cl_4Te(TePh)_{12}]$ (R = $^n$Pr, Ph) und $[Zn_{10}Te_4(TePh)_{12}(PR_3)_2]$ (R = $^n$Pr, Ph), *Z. anorg. und allg. Chem.*, 624(11), 1909–1914 (1998).

Euler H., Barbier B., Meents A., Kirfel A. Crystal structure of Tutton's salts, $Cs_2[M^{II}(H_2O)_6]$ $(SO_4)_2$, M$^{II}$ = Mg, Mn, Fe, Co, Ni, Zn, *Z. Kristallogr. New Cryst. Struct.*, 218(4), 409–413 (2003a).

Euler H., Barbier B., Meents A., Kirfel A. Crystal structure of Tutton's salts, $Rb_2[M^{II}(H_2O)_6]$ $(SeO_4)_2$, M$^{II}$ = Mg, Co, Mn, Zn, *Z. Kristallogr. New Cryst. Struct.*, 218(3), 265–268 (2003b).

Euler H., Barbier B., Meents A., Kirfel A. Crystal structures of Tutton's salts $K_2[M^{II}(H_2O)_6]$ $(SeO_4)_2$, M$^{II}$ = Co, Ni, Zn and refinement of the crystal structure of potassium hexaaquamagnesium(II) selenate, $K_2[Mg(H_2O)_6](SeO_4)_2$, *Z. Kristallogr. New Cryst. Struct.*, 224(3), 351–354 (2009a).

Euler H., Barbier B., Meents A., Kirfel A. Crystal structures of Tutton's salts $Tl_2[M^{II}(H_2O)_6]$ $(SeO_4)_2$, $M^{II} = Mg$, Mn, Co, Ni, Cu, Zn, *Z. Kristallogr. New Cryst. Struct.*, 224(3), 360–364 (2009b).

Fenske D., Magull S., Dehnicke K. Synthese und Kristallstruktur von [Rb(18-Krone-6)]$_2$[Zn(Se$_4$)(Se$_6$)], *Z. Naturforsch.*, 46B(8), 1011–1014 (1991).

Fleck M., Kolitsch U. Crystal structure of ammonium hexaaquacobalt(II) selenate, $(NH_4)_2Co(SeO_4)_2\cdot 6H_2O$, and of ammonium hexaaquazinc selenate, $(NH_4)_2Zn(SeO_4)_2\cdot 6H_2O$, hydrogen bonds in two ammonium selenate Tutton's salts, *Z. Kristallogr. New Cryst. Struct.*, 217(4), 471–473 (2002).

Furlani C., Cervone E., Camassei F.D. Transition metal *N,N*-diethyldiselenocarbamates, *Inorg. Chem.*, 7(2), 265–268 (1968).

Gigauri R.D., Machhoshvili R.I., Helashvili G.K., Gigauri R.I., Indzhiya M.A. Ag, Zn, Cd, Hg(II) monothioarsenates [in Russian], *Zhurn. neorgan. khimii*, 44(1), 26–29 (1999).

Harrison W.T.A. Sodium zinc hydroxide selenite, $NaZn_2(OH)(SeO_3)_2$, *Acta Crystallogr.*, C55(9), 1398–1399 (1999).

Harrison W.T.A., Phillips M.L.F., Stanchfield J., Nenoff T.M. $(CN_3H_6)_4[Zn_3(SeO_3)_5]$: The first organically template selenite, *Angew. Chem. Int. Ed.*, 39(21), 3808–3810 (2000a).

Harrison W.T.A., Phillips M.L.F., Stanchfield J., Nenoff T.M. $(CN_3H_6)_4[Zn_3(SeO_3)_5]$: The first organically template selenite, *Angew. Chem.*, 112(21), 3966–3968 (2000b).

Herkelrath S.J.C., Saratovsky I., Hadermann J., Clarke S.J. Fragmentation of an infinite $ZnO_2$ square plane into discrete $[ZnO_2]^{2-}$ linear units in the oxyselenide $Ba_2ZnO_2Ag_2Se_2$, *J. Am. Chem. Soc.*, 130(44), 14426–14427 (2008).

Higgins G.M.C., Saville B. Complexes of amines with zinc dialkyldithiocarbamates, *J. Chem. Soc.*, 2812–2817 (1963).

Hori H., Koyama E., Nagashima K. Kinichilite, a new mineral from the Kawazu mine, Shimoda city, Japan, *Mineral. J.*, 10(7), 333–337 (1981).

Hursthouse M.B., Malik M.A., Majid M., O'Brien P. The crystal and molecular structure of *N,N*-diethyldiselenocarbamatocadmium(II): Cadmium and zinc diethyldiselenocarbamates as precursors for selenides, *Polyhedron*, 11(1), 45–48 (1992).

Isab A.A., Wazeer M.I.M. Complexation of Zn(II), Cd(II) and Hg(II) with thiourea and selenourea: A $^1H$, $^{13}C$, $^{15}N$, $^{77}Se$ and $^{113}Cd$ solution and solid-state NMR study, *J. Coord. Chem.*, 58(6), 529–537 (2005).

Kuo D.-H., Tsega M. Hole mobility enhancement of Cu-deficient $Cu_{1.75}Zn(Sn_{1-x}Al_x)Se_4$ bulks, *J. Solid State Chem.*, 206, 134–138 (2013).

Kuo D.-H., Wubet W. Mg dopant in $Cu_2ZnSnSe_4$: An *n*-type former and a promoter of electrical mobility up to 120 $cm^2\cdot V^{-1}\cdot s^{-1}$, *J. Solid State Chem.*, 215, 122–127 (2014).

Ma D.-L., Zhang H.-J., Shen Z., Niu D.-Z. Crystal structure of zinc(II) 2-selenopyridine-*N*-oxide, $Zn_2(C_5H_4NOSe)_4$, *Z. Kristallogr. New Cryst. Struct.*, 224(2), 283–284 (2009).

Malik M.A., Motevalli M., Walsh J.R., O'Brien P. Neopentyl- or *tert*-butylzinc complexes with diethylthio- or diethylselenocarbamates: Precursors for zinc chalcogens, *Organometallics*, 11(9), 3136–3139 (1992).

Malik M.A., O'Brien P. Mixed methyl- and ethylzinc complexes with diethylselenocarbamate: Novel precursors for ZnSe, *Chem. Mater.*, 3(6), 999–1000 (1991).

Manos M.J., Jang J.I., Ketterson J.B., Kanatzidis M.G. $[Zn(H_2O)_4][Zn_2Sn_2Se_9(MeNH_2)]$: A robust open framework chalcogenides with a large nonlinear optical response, *Chem. Commun.*, (8), 972–974 (2008).

Menezes P.W., Fässler T.F. Crystal structure of 2-aminoethylammonium *tris*-(1,2-ethanediamine) zinc(II)tetraselenoantimonate(V), $(C_2H_9N_2)[Zn(C_2H_8N_2)_3][SbSe_4]$, $C_8H_{33}N_8SbSe_4Zn$, *Z. Kristallogr. New Cryst. Struct.*, 227(4), 439–440 (2012).

Naumov N.G., Virovets A.V., Fedorov V.E. Unusually high porosity in polymeric cluster cyanides: The synthesis and crystal structure of $(H_3O)_2Zn_3[Re_6Se_8(CN)_6]_2\cdot 20H_2O$, *Inorg. Chem. Commun.*, 3(2), 71–72 (2000).

Novikova E.M., Vasil'ev M.G., Evseev V.A., Yershova S.A. Phase diagram of the Sn–GaAs–ZnSe quasiternary system from the Sn-rich side [in Russian], *Deposited in VINITI*, № 3039–74Dep (1974).

Philippidis A., Bakas T., Trikalitis P.N. $(H_2NC_4H_8NCH_2CH_2NH_2)_2Zn_2Sn_2Se_7$: A hybrid ternary semiconductor stabilized by amine molecules acting simultaneously as ligands and counterions, *Chem. Commun.*, (12), 1556–1558 (2009).

Philippidis A., Trikalitis P.N. $(H_2NC_4H_8NCH_2CH_2NH_2)(HNCH_2CH_2NH_2)_3Zn_2Ge_2Se_8$: A new, templated one-dimensional ternary semiconductor stabilized by mixed organic cations, *Polyhedron*, 28(15), 3193–3198 (2009).

Ran X., Wang L., Cao D., Lin Y., Hao J. Synthesis, characterization and in vitro biological activity of cobalt(II), copper(II) and zinc(II) Schiff base complexes derived from salicylaldehyde and D,L-selenomethionine, *Appl. Organometal. Chem.*, 25(1), 9–15 (2011).

Ran X., Wang L., Lin Y., Hao J., Cao D. Syntheses, characterization and biological studies of zinc(II), copper(II) and cobalt(II) complexes with Schiff base ligand derived from 2-hydroxy-1-naphthaldehyde and selenomethionine, *Appl. Organometal. Chem.*, 24(10), 741–747 (2010).

Ruzin E., Dehnen S. Influence of the counterions on the structures of ternary Zn/Sn/Se anions: Synthesis and properties of $[Rb_{10}(H_2O)_{14.5}][Zn_4(\mu_4\text{-}Se)_2(SnSe_4)_4]$ and $[Ba_5(H_2O)_{32}][Zn_5Sn(\mu_3\text{-}Se)_4(SnSe_4)_4]$, *Z. anorg. und allg. Chem.*, 632(5), 749–755 (2006).

Saidov A.S., Usmonov Sn.N., Rakhmonov U.Kh., Kurmantayev A.N., Bahtybayev A.N. Multicomponent solid solutions $(ZnSe)_{1-x-y}(Si_2)_x(GaP)_y$, *J. Mater. Sci. Res.*, 1(2), 150–156 (2012).

Santra B.K., Liaw B.-J., Hung C.-M., Liu C.W., Wang J.-C. Syntheses, solid-state structures, and solution studies by VT $^{31}P$ NMR of $[Zn\{Se_2P(OEt)_2\}_2]_\infty$ and $[Zn_2\{Se_2P(O^iPr)_2\}_4]$, *Inorg. Chem.*, 42(26), 8866–8871 (2003).

Secondo P.M., Land J.M., Baughman R.G., Collier H.L. Polymeric octahedral and monomeric tetrahedral Group 12 pseudohalogeno (NCX⁻: X = O, S, Se) complexes of 4-(N,N-dimethylamino)pyridine, *Inorg. Chim. Acta*, 309(1–2), 13–22 (2000).

Spirovski F., Engelen B., Kang Y., Wagener M. $Cs_2M(HSeO_3)_4·2H_2O$; M(II) = Mn, Co, Ni, Zn, *Z. anorg. und allg. Chem.*, 632(12–13), 2133 (2006).

Spirovski F., Wagener M., Stefov V., Engelen B. Crystal structures of rubidium zinc bis(hydrogenselenate(IV)) chloride, $RbZn(HSeO_3)_2Cl$, and rubidium zinc bis(hydrogenselenate(IV)) bromide, $RbZn(HSeO_3)_2Br$, *Z. Kristallogr. New Cryst. Struct.*, 222(2), 91–92 (2007).

Sun H.-Q., Yu W.-T., Yuan D.-R., Wang X.-Q., Liu L.-Q. Zinc cadmium selenocyanate, *Acta Crystallogr.*, E62(4), i88–i90 (2006).

Trikalitis P.N., Rangan K.K., Bakas T., Kanatzidis M.G. Varied pore organization in mesostructured semiconductors based on the $[SnSe_4]^{4-}$ anion, *Nature*, 410(6829), 671–675 (2001).

Tsamourtzi K., Song J.H., Bakas T., Freeman A.J., Trikalitis P.N., Kanatzidis M.G. Straightforward route to the adamantane clusters $[Sn_4Q_{10}]^{4-}$ (Q = S, Se, Te) and use in the assembly of open-framework chalcogenides $(Me_4N)_2M[Sn_4Se_{10}]$ (M = Mn$^{II}$, Fe$^{II}$, Co$^{II}$, Zn$^{II}$) including the first telluride member $(Me_4N)_2Mn[Ge_4Te_{10}]$, *Inorg. Chem.*, 47(24) 11920–11929 (2008).

Ueyama N., Sugawara T., Sasaki K., Nakamura A., Yamashita S., Wakatsuki Y., Yamazaki H., Yasuoka N. X-ray structures and far-infrared and Raman spectra of tetrahedral thiophenolato and selenophenolato complexes of zinc(II) and cadmium(II), *Inorg. Chem.*, 27(4), 741–747 (1988).

Vittal J.J., Dean P.A.W., Payne N.C. Metal-selenolate chemistry: Stereochemistry of adamantane-type clusters of formula $[(\mu\text{-}SePh)_6(MSePh)_4]^{2-}$ (M = Zn and Cd), *Can. J. Chem.*, 70(3), 792–801 (1992).

Wachhold M., Rangan K.K., Billinge S.J.L., Petkov V., Heising J., Kanatzidis M.G. Mesostructured non-oxidic solids with adjustable worm-hole shaped pores: M–Ge–Q (Q = S, Se) frameworks based on tetrahedral $[Ge_4Q_{10}]^{4-}$ clusters, *Adv. Mater.*, 12(2), 85–91 (2000a).

Wachhold M., Rangan K.K., Lei M., Thorpe M.F., Billinge S.J.L., Petkov V., Heising J., Kanatzidis M.G. Mesostructured metal germanium sulfide and selenide materials based on the tetrahedral $[Ge_4S_{10}]^{4-}$ and $[Ge_4Se_{10}]^{4-}$ units: Surfactant templated three-dimensional disordered frameworks perforated with worm holes, *J. Solid State Chem.*, 152(1), 21–36 (2000b).

Williams D.J., White K.M., VanDerveer D., Wilkinson A.P. Dichlorobis[1,3-dimethyl-2(3*H*)-imidazoleselone]zinc(II): A potential zinc selenide synthon, *Inorg. Chem. Commun.*, 5(2), 124–126 (2002).

Wu M., Su W., Jasutkar N., Huang X., Li J. An open-framework bimetallic chalcogenide structure $K_3Rb_3Zn_4Sn_3Se_{13}$ built on a unique $[Zn_4Sn_3Se_{16}]^{12-}$ cluster: Synthesis, crystal structure, ion exchange and optical properties, *Mater. Res. Bull.*, 40(1), 21–27 (2005).

Wu T., Bu X., Zhao X., Khazhakyan R., Feng P. Phase selection and site-selective distribution by tin and sulfur in supertetrahedral zinc gallium selenides, *J. Am. Chem. Soc.*, 133(24), 9616–9625 (2011).

Wu T., Wang L., Bu X., Chau V., Feng P. Largest molecular clusters in the supertetrahedral T*n* series, *J Am. Chem. Soc.*, 132(31), 10823–10831 (2010).

Wu T., Wang X., Bu X., Zhao X., Wang L., Feng P. Synthetic control of selenide supertetrahedral clusters and three-dimensional co-assembly by charge-complementary metal cations, *Angew. Chem. Int. Ed.*, 48(39), 7204–7207 (2009).

Xiong W.-W., Li P.-Z., Zhou T.-H., Zhao Y., Xu R., Zhang Q. Solvothermal syntheses of three new one-dimensional ternary selenidostannates: $[DBNH][M_{1/2}Sn_{1/2}Se_2]$ (M = Mn, Zn, Hg), *J. Solid State Chem.*, 204, 86–90 (2013).

Xiong W.-W., Miao J., Li P.-Z., Zhao Y., Liu B., Zhang Q. $\{[M(NH_3)_6][Ag_4M_4Sn_3Se_{13}]\}_\infty$ (*M* = Zn, Mn): Three-dimensional chalcogenide frameworks constructed from quaternary metal selenide clusters with two different transition metals, *J. Solid State Chem.*, 218, 146–150 (2014).

Xu G., Guo P., Song S., Zhang H., Wang C. Molecular nanocluster with a $[Sn_4Ga_4Zn_2Se_{20}]^{8-}$ T3 supertetrahedral core, *Inorg. Chem.*, 48(11), 4628–4630 (2009).

# 3 Systems Based on ZnTe

## 3.1 Zn–H–Na–O–Te

$Na_2[Zn_2(TeO_3)_3]\cdot 3H_2O$. This quinary compound, which crystallizes in the hexagonal structure with the lattice parameters $a = 939.5 \pm 0.1$ and $c = 773.3 \pm 0.01$ pm and a calculated density of 4.256 g·cm$^{-3}$, is formed in this system (Miletich 1995b). Its single crystals were grown under hydrothermal conditions. Synthesis experiments were started from the mixture of $ZnO + TeO_2$ with Zn/Te = 2:3 (atomic ratio). A saturated solution of 1–2 mL of NaOH and $H_2O$ was added to set the pH of the aqueous mixture to values ≥9.

## 3.2 Zn–H–Cu–Mg–O–Fe–Te

$(Cu, Mg, Zn)_2(Mg, Fe)TeO_6\cdot 6H_2O$. This multinary compound (mineral leisingite) is formed in this system. It crystallizes in the trigonal structure with lattice parameters $a = 531.6 \pm 0.1$ and $c = 971.9 \pm 0.2$ pm (Margison et al. 1998).

## 3.3 Zn–H–Mg–O–Mn–Fe–Te

$Mg_{0.5}[(Mn, Zn)Fe(TeO_3)_3]\cdot 4.5H_2O$. This multinary compound (mineral kinichilite), which crystallizes in the hexagonal structure with the lattice parameters $a = 945.1 \pm 0.7$ and $c = 768.7 \pm 0.9$ pm and a calculated density of 4.11 g·cm$^{-3}$, is formed in this system (Miletich 1995a). Only a negligibly small amount of $Na_2O$ has been found in this mineral, although before, it was assumed that sodium was a significant part of its composition.

## 3.4 Zn–H–Mg–O–Fe–Te

$Mg_{0.5}[ZnFe(TeO_3)_3]\cdot 4.5H_2O$. This multinary compound (mineral zemannite), which crystallizes in the hexagonal structure with the lattice parameters $a = 940.4 \pm 0.2$ and $942.0 \pm 0.3$ and $c = 763.6 \pm 0.4$ and $765.7 \pm 0.6$ pm and a calculated density of 4.19 and 4.18 g·cm$^{-3}$ for two different species (Miletich 1995a) ($a = 941 \pm 2$ and $c = 764 \pm 2$ pm and a calculated density of $4.36 \pm 0.08$ g·cm$^{-3}$, [Matzat 1967, Mandarino et al. 1976]) is formed in this system. Only a negligibly small amount of $Na_2O$ has been found in this mineral (Miletich 1995a), although before, it was assumed that sodium was a significant part of its composition (Matzat 1967, Mandarino et al. 1976).

## 3.5 Zn–H–C–Si–N–P–Te

Some compounds are formed in this system.

$[Zn\{N(SiMe_3)_2\}\{Bu^t_2P(Te)NPr^i\}]$ or $C_{17}H_{43}Si_2N_2PTeZn$. This compound sublimes at 120°C at $1.5 \cdot 10^{-3}$ Pa and melts at 206°C–208°C (Bochmann et al. 1995a).

The following procedure was used to obtain it. To a yellow suspension of $Bu^t_2P(Te)$ $NHPr_i$ (2.42 mM) in light petroleum (30 mL) was added $[Zn\{N(SiMe_3)_2\}_2]$ (2.46 mM) via syringe at room temperature. The mixture was stirred for 3 h, after which the suspension had dissolved. The solvent was removed under reduced pressure, and the yellow residue recrystallized from hot toluene (10 mL) to give yellow crystals.

**$[Zn\{N(SiMe_3)_2\}\{Bu^t_2P(Te)N(C_6H_{11})\}]$ or $C_{20}H_{47}Si_2N_2PTeZn$.** This compound sublimes at 200° at $4.4 \cdot 10^{-4}$ Pa and melts at the temperate higher than 220°C (Bochmann et al. 1995a). It was prepared following the procedure for $[Zn\{N(SiMe_3)_2\}\{Bu^t_2P(Te)$ $NPr_i\}]$ from $Bu^t_2P(Te)NH(C_6H_{11})$ (3.50 mM) and $[Zn\{N(SiMe_3)_2\}_2]$ (3.63 mM) as yellow crystals.

## 3.6   Zn–H–C–Si–N–Te

Some compounds are formed in this system.

**$[Zn\{SiTe(SiMe_3)_3\}_2](2,2'\text{-}bipy)$ or $C_{28}H_{62}Si_8N_2Te_2Zn$.** To prepare this compound, a solution of 2,2′-Bipy (0.39 mM) in 30 mL of toluene was added to a solution of $[Zn\{SiTe(SiMe_3)_3\}_2]_2$ (0.36 mM) in 30 mL of toluene, resulting in an immediate color change to orange red and precipitation of a fine orange material (Bonasia and Arnold 1992). The toluene was removed under reduced pressure, and the light-orange product was extracted with $CH_2Cl_2$. The resulting dark-red solution was filtered, concentrated from 75 to 45 mL, and allowed to cool to 0°C overnight. Thin orange needles were isolated by filtration. The title compound was also obtained as dark-red diamond-shaped plates by more slowly cooling an analogous solution to 0°C over a period of 24 h. $[Zn\{SiTe(SiMe_3)_3\}_2](2,2'\text{-}Bipy)$ melts to a red-black liquid between 260°C and 270°C.

**$[Zn\{SiTe(SiMe_3)_3\}_2](Py)_2$ or $C_{28}H_{64}Si_8N_2Te_2Zn$.** This compound crystallizes in the triclinic structure with the lattice parameters $a = 1191.91 \pm 0.22$, $b = 1213.97 \pm 0.16$, and $c = 1731.9 \pm 0.3$ pm and $\alpha = 105.328° \pm 0.013°$, $\beta = 99.711° \pm 0.016°$, and $\gamma = 96.278° \pm 0.012°$ at 202 K and a calculated density of 1.38 g·cm$^{-3}$ (Bonasia and Arnold 1992). To obtain it, via a syringe, Py (0.80 mM) was added to a solution of $[Zn\{SiTe(SiMe_3)_3\}_2]_2$ (0.40 mM) in hexane (30 mL). The clear yellow solution quickly turned lighter as a pale-yellow microcrystalline solid precipitated. Volatile components were removed under reduced pressure, and the material was extracted with $CH_2Cl_2$ (25 mL). The clear yellow solution was filtered, concentrated to 10 mL, and cooled to −40°C. After 24 h, clear yellow cubes were observed floating in the $CH_2Cl_2$. Filtration afforded the title compound. The crystals lose Py at 90°C and then slowly decompose to a dark-red-orange material above 150°C.

## 3.7   Zn–H–C–Si–Te

**$[Zn\{SiTe(SiMe_3)_3\}_2]_2$ or $C_{18}H_{54}Si_8Te_2Zn$.** This multinary compound is formed in this system. It melts at 208°C–213°C and crystallizes in the triclinic structure with the lattice parameters $a = 1480.4 \pm 0.5$, $b = 1645.8 \pm 0.5$, and $c = 1823.7 \pm 0.5$ pm and $\alpha = 63.97° \pm 0.02°$, $\beta = 84.61° \pm 0.02°$, and $\gamma = 79.47° \pm 0.02°$ at

168 K and a calculated density of 1.38 g·cm$^{-3}$ (Bonasia and Arnold 1992). To obtain it, a solution of Zn[N(SiMe$_3$)$_2$]$_2$ (0.83 mM) in 25 mL of hexane was added to a solution of HTeSi(SiMe$_3$)$_3$ (1.66 mM) in 25 mL of the same solvent, resulting in the immediate formation of a yellow solution. This mixture was stirred for 30 min, and the solvent was removed under reduced pressure. The dry yellow solid was extracted with hexamethyldisiloxane (40 mL). Then, the solution was filtered, concentrated to 15 mL, and cooled to –40°C for 12 h. Clear yellow cubes of the title compound were isolated by filtration.

## 3.8 Zn–H–C–Sn–N–Te

Some compounds are formed in this system.

**[{Zn(C$_6$H$_{18}$N$_4$)$_2$}$_2$(SnTe$_4$)] or C$_{12}$H$_{36}$SnN$_8$Te$_4$Zn$_2$.** This compound crystallizes in the monoclinic structure with the lattice parameters $a = 842.54 \pm 0.04$, $b = 1130.38 \pm 0.6$, and $c = 1529.73 \pm 0.8$ pm, and $\beta = 100.063° \pm 0.005°$ and energy gap 1.54 eV (Lu et al. 2014). It was prepared by the following procedure. Zn (1.0 mM), Sn (0.50 mM), Te (2.0 mM), KI (1 mM), and triethylenetetramine (C$_6$H$_{18}$N$_4$) (3 mL) were loaded into a polytetrafluoroethylene-lined stainless steel autoclave of 10 mL volume. The sealed autoclave was heated to 150°C for 6 days. After cooling to an ambient temperature, the resulting black plate crystals of the title compound were filtered off, washed with acetone and EtOH, and stored under vacuum.

**[{Zn(C$_6$H$_{18}$N$_4$)$_2$}$_2$(Sn$_2$Te$_6$)] or C$_{12}$H$_{36}$Sn$_2$N$_8$Te$_6$Zn$_2$.** This compound crystallizes in the monoclinic structure with the lattice parameters $a = 835.96 \pm 0.17$, $b = 1300.2 \pm 0.3$, and $c = 1480.9 \pm 0.3$ pm, and $\beta = 95.41° \pm 0.03°$ and energy gap 1.66 eV (Lu et al. 2014). It was prepared using the following procedure. Zn (1.0 mM), Sn (0.50 mM), Te (2.0 mM), KI (1 mM), and triethylenetetramine (C$_6$H$_{18}$N$_4$) (3 mL) were loaded into a polytetrafluoroethylene-lined stainless steel autoclave of 10 mL volume. The sealed autoclave was heated to 190°C for 10 days. After cooling to an ambient temperature, the resulting black polyhedral crystals of the title compound were filtered off, washed with acetone and EtOH, and stored under vacuum.

**[Zn{C$_2$H$_4$(NH$_2$)$_2$}$_2$(Sn$_2$Te$_6$)] or C$_{12}$H$_{48}$Sn$_2$N$_{12}$Te$_6$Zn$_2$.** This compound crystallizes in the monoclinic structure with the lattice parameters $a = 904.8 \pm 0.2$, $b = 2230.0 \pm 0.6$, and $c = 936.0 \pm 0.3$ pm, and $\beta = 103.19° \pm 0.02°$ and a calculated density of 2.699 g·cm$^{-3}$ (Li et al. 1998). Its black column crystals were initially obtained in a reaction attempted in synthesizing a new Cs-containing metal telluride. The reactants included 0.25 mM of Cs$_2$Te, 0.25 mM of ZnCl$_2$, 0.25 mM of SnCl$_2$, and 0.75 mM of Te (Li et al. 1998). The starting materials were weighed and mixed in a glove box under a N$_2$ atmosphere. A thick-walled Pyrex tube was used as the reaction container. Approximately, 5.5 mM of ethylenediamine was added to the mixture. The tube was then sealed under vacuum ($\approx 0.1$ Pa) after the liquid had been condensed by liquid N$_2$. The sample was placed in an oven at 180°C and heated at this temperature for 7 days. After cooling to room temperature, the reaction product was washed with 35% and 95% ethanol and dried with anhydrous diethyl ether. In order to increase the yield, Cs$_2$Te was eliminated in the subsequent reactions.

**[Zn(C₂H₈N₂)₃]₂[Sn₂Te₆]·(C₂H₈N₂)** or **C₁₄H₅₆Sn₂N₁₄Te₆Zn₂**. This compound crystallizes in the monoclinic structure with the lattice parameters $a = 858.0 \pm 0.1$, $b = 1597.0 \pm 0.1$, and $c = 1524.66 \pm 0.09$ pm, and $\beta = 95.076° \pm 0.008°$ and a calculated density of 2.480 g·cm⁻³ (Shreeve-Keyer et al. 1997). The synthesis of the title compound was accomplished by the reaction of Zintl phase K₄SnTe₄ (0/127 mM) with ZnCl₂ (0.127 mM) and Et₄NBr (0.127 mM) in ethylenediamine (1 mL) in a sealed quartz ampoule held at 100°C for 1 day.

**[{Zn(C₄H₁₃N₃)₂}₄(Sn₂Te₆)₁.₇₅(Sn₂Te₈)₀.₂₅]·(C₄H₁₃N₃)** or **C₇₂H₂₃₄Sn₈N₅₄Te₂₅Zn₈**. This compound crystallizes in the monoclinic structure with the lattice parameters $a = 1714.6 \pm 0.3$, $b = 1513.2 \pm 0.3$, and $c = 1715.8 \pm 0.3$ pm, and $\beta = 97.65° \pm 0.03°$ and energy gap 1.98 eV (Lu et al. 2014). To prepare it, Zn (1.0 mM), Sn (0.50 mM), Te (2.0 mM), KI (1 mM), and diethylenetriamine (C₄H₁₃N₃) (3 mL) were loaded into a polytetrafluoroethylene-lined stainless steel autoclave of 10 mL volume. The sealed autoclave was heated to 150°C for 6 days. After cooling to ambient temperature, the resulting black block crystals of the title compound were filtered off, washed with acetone and EtOH, and stored under vacuum.

### 3.9   Zn–H–C–Sn–N–I–Te

**[{Zn(C₆H₁₈N₄)₂}₃(SnTe₄)]I₂** or **C₁₈H₅₄SnN₁₂I₂Te₄Zn₃**. This multinary compound is formed in this system. It crystallizes in the orthorhombic structure with the lattice parameters $a = 1153.6 \pm 0.2$, $b = 1526.6 \pm 0.3$, and $c = 2407.6 \pm 0.5$ pm and energy gap 1.47 eV (Lu et al. 2014). It was prepared using the next procedure. Zn (1.0 mM), Sn (0.50 mM), Te (2.0 mM), KI (3 mM), and triethylenetetramine (C₆H₁₈N₄) (3 mL) were loaded into a polytetrafluoroethylene-lined stainless steel autoclave of 10 mL volume. The sealed autoclave was heated to 150°C for 6 days. After cooling to an ambient temperature, the resulting black crystals of the title compound were filtered off, washed with acetone and EtOH, and stored under vacuum.

### 3.10   Zn–H–C–N–P–Te

Some compounds are formed in this system.

**[Zn{Buᵗ₂P(Te)NPrⁱ}₂]** or **C₂₂H₅₀N₂P₂Te₂Zn**. This compound melts at 135°C with decomposition (Bochmann et al. 1995a). To obtain it, to a yellow solution of Buᵗ₂P(Te)NHPrᵢ (5.41 mM) in light petroleum (20 mL) [Zn{N(SiMe₃)₂}₂] (2.72 mM) was added at –78°C via syringe and the mixture was stirred rapidly. A yellow precipitate formed immediately. The reaction mixture was warmed to room temperature and stirred for 2 h. The precipitate was filtered off and recrystallized from toluene to give yellow crystals.

**[Zn{Buᵗ₂P(Te)N(C₆H₁₁)}₂]** or **C₂₈H₅₈N₂P₂Te₂Zn**. This compound sublimes at 115° at 3.6·10⁻⁴ Pa and melts at a temperate higher than 152°C (Bochmann et al. 1995a). It was prepared following the procedure for [Zn{Buᵗ₂P(Te)NPrᵢ}₂] from Buᵗ₂P(Te)NH(C₆H₁₁) (1.74 mM) and [Zn{N(SiMe₃)₂}₂] (0.91 mM) as yellow crystals.

## 3.11  Zn–H–C–N–V–O–Te

**$Zn_2(2,2'$-bipy$)_2V_4TeO_{14}$ or $C_{20}H_{16}N_4V_4O_{14}TeZn_2$.** This multinary compound is formed in this system. It is stable up to 280°C and crystallizes in the monoclinic structure with the lattice parameters $a = 1071.9 \pm 0.2$, $b = 3763 \pm 1$, and $c = 734.9 \pm 0.2$ pm, and $\beta = 106.841° \pm 0.004°$ and a calculated density of 2.337 g·cm⁻³ (Xie and Mao 2005). This compound was obtained by a hydrothermal reactions of $Zn(OAc)_2$, 2,2'-bipy, $TeO_2$, and $NH_4VO_3$ at 170°C for 5 days.

## 3.12  Zn–H–C–N–O–Re–Te

Some multinary compounds are formed in this system.

**$[\{Zn(H_2O)_2\}\{Zn(H_2O)_4Re_4Te_4(CN)_{12}\}]$ or $C_{12}H_{12}N_{12}O_6Re_4Te_4Zn_2$.** This compound crystallizes in the orthorhombic structure with the lattice parameters $a = 1041.12 \pm 0.10$, $b = 1535.44 \pm 0.14$, and $c = 1884.0 \pm 0.2$ pm, and a calculated density of 3.984 g·cm⁻³ (Mironov et al. 2006). To obtain it, the solution of $K_4[Re_4Te_4(CN)_{12}]\cdot 5H_2O$ (0.006 mM) in $H_2O$ (5 mL) was added to a 25% aqueous ammonia solution (5 mL) containing 0.062 mM of $ZnCl_2$. The obtained mixture was concentrated at room temperature in an open glass for 2 weeks. The resulting dark crystals were separated from the white precipitate by decantation with two or three portions of water, filtered off, and dried in air.

**$[Zn_2(NH_3)_6(\mu\text{-OH})][Zn(NH_3)_4]_{0.5}[Re_4Te_4(CN)_{12}]\cdot 5H_2O$ or $C_{12}H_{35}N_{20}O_6Re_4Te_4Zn_{2.5}$.** This compound crystallizes in the monoclinic structure with the lattice parameters $a = 2323.3 \pm 0.2$, $b = 1459.06 \pm 0.16$, and $c = 1438.25 \pm 0.15$ pm and $\beta = 125.169° \pm 0.001°$ and a calculated density of 3.290 g·cm⁻³ (Mironov et al. 2011). It was obtained by the following procedure. A solution of $ZnCl_2$ (0.12 mM) in aqueous ammonia (5 mL, 25%) was added to a solution containing 0.012 mM of $K_4[Re_4Te_4(CN)_{12}]\cdot 5H_2O$ in 5 mL of $H_2O$. The resuling solution was concentrated at room temperature in a semiclosed vial for 2 days. Deposited dark crystals were collected by filtration and dried in air.

**$[\{Zn(H_2O)(C_2H_8N_2)_2\}\{Zn(C_2H_8N_2)_2\}\{Re_6Te_8(CN)_6\}]\cdot 4H_2O$ or $C_{14}H_{40}N_{14}O_4Re_6Te_8Zn_2$.** This compound crystallizes in the monoclinic structure with the lattice parameters $a = 1076.89 \pm 0.06$, $b = 1655.94 \pm 0.09$, and $c = 2436.31 \pm 0.14$ pm, and $\beta = 92.070° \pm 0.001°$ at 153 K and a calculated density of 4.188 g·cm⁻³ (Brylev et al. 2004). To obtain it, in a narrow-diameter tube, a solution of 0.35 mkM of $Cs_4\{Re_6Te_8(CN)_6\}\cdot 2H_2O$ in 0.5 mL of $H_2O$ was allowed to diffuse into a solution of 0.026 mM of $ZnCl_2$ in 0.5 mL of glycerol to which 0.09 mL (1.35 mM) of ethylenediamine had been added. After 1 week, dark-brown crystals were obtained.

**$[\{Zn(NH_3)_2(C_2H_8N_2)_2\}][\{Zn(NH_3)_2(C_2H_8N_2)\}Re_4Te_4(CN)_{12}]\cdot H_2O$ or $C_{18}H_{38}N_{22}ORe_4Te_4Zn_2$.** This compound crystallizes in the monoclinic structure with lattice parameters $a = 1692.6 \pm 0.3$, $b = 1516.8 \pm 0.3$, and $c = 1633.3 \pm 0.3$ pm and $\beta = 105.63° \pm 0.03°$ and a calculated density of 3.231 g·cm⁻³ (Mironov et al. 2006). To obtain it, to the solution of $K_4[Re_4Te_4(CN)_{12}]\cdot 5H_2O$ (0.003 mM) in $H_2O$ (3 mL) was added to a 25% aqueous ammonia solution (5 mL) containing 0.031 mM of $ZnCl_2$ and 0.37 mM of ethylenediamine. The obtained mixture was concentrated at room temperature for 4 weeks. The resulting crystals were filtered off and dried in air.

**[{Zn$_2$(C$_4$H$_{13}$N$_3$)$_3$}Re$_4$Te$_4$(CN)$_{12}$]·6H$_2$O or C$_{24}$H$_{51}$N$_{21}$O$_6$Re$_4$Te$_4$Zn$_2$.** This compound crystallizes in the monoclinic structure with the lattice parameters $a = 1374.62 \pm 0.03$, $b = 1594.35 \pm 0.05$, and $c = 2273.72 \pm 0.07$ pm, and $\beta = 95.735° \pm 0.001°$ and a calculated density of 2.834 g·cm$^{-3}$ (Mironov et al. 2006). To obtain it, the solution of K$_4$[Re$_4$Te$_4$(CN)$_{12}$]·5H$_2$O (0.003 mM) in H$_2$O (3 mL) was added 25% aqueous ammonia solution (5 mL) containing 0.034 mM of ZnCl$_2$ and 0.25 mM of diethylenetriamine. The obtained mixture was concentrated at room temperature for 4 weeks. The resulting needle-shaped red crystals were filtered off and dried in air.

### 3.13   Zn–H–C–N–Te

Some multinary compounds are formed in this system.

**[Zn(C$_2$H$_8$N$_2$)$_3$][Te$_3$]·0.5(C$_2$H$_8$N$_2$) or C$_7$H$_{28}$N$_7$Te$_3$Zn.** This compound crystallizes in the monoclinic structure with the lattice parameters $a = 850.2 \pm 0.1$, $b = 1555.4 \pm 0.2$, and $c = 1504.6 \pm 0.1$ pm, and $\beta = 102.25° \pm 0.01°$ and a calculated density of 2.249 g·cm$^{-3}$ (Shreeve-Keyer et al. 1997). It was synthesized in a three-compartment, liquid-junction, airtight electrochemical cell equipped with a Ni plate counter electrode in a saturated ethylenediamine solution of Pr$^n_4$NBr, using a cathode of nominal composition, ZnTe$_{10}$. The reaction was run at a constant current using an external supply, which was connected to the leads of the cathode and anode and set to run at the maximum current allowed, 1 mA. The application of such current resulted in the production of a dark-purple stream of anions, which surrounded the cathode and then slowly sank to the bottom of the cathode chamber. After approximately 4 days, small black crystals were found growing throughout the cathode chamber and were isolated.

**[Zn(TeC$_6$H$_2$Me$_3$–2,4,6)$_2$(N-methylimidazole)$_2$] or C$_{26}$H$_{34}$N$_4$Te$_2$Zn.** This compound melts at 152°C–154°C. To obtain it, to a suspension of 0.59 mM of Zn(TeC$_6$H$_2$Me$_3$–2,4,6)$_2$ in 10 mL of petroleum ether was added via syringe an excess of N-methylimidazole (Bochmann et al. 1995b). The starting materials dissolved within a few seconds, followed by the formation of a colorless precipitate. Stirring was continued for 2 h. The product was washed with petroleum ether (3×20 mL) and dried in vacuo. Heating of this compound either without a solvent or in paraffin oil to 270°C–320°C leads to the formation of cubic ZnTe.

**[Zn(TeC$_6$H$_2$Me$_3$–2,4,6)$_2$(Py)$_2$] or C$_{28}$H$_{32}$N$_2$Te$_2$Zn.** This compound melts with decomposition at 114°C–116°C and crystallizes in the monoclinic structure with the lattice parameters $a = 1391.3 \pm 0.4$, $b = 1384.8 \pm 0.4$, and $c = 1579.4 \pm 0.5$ pm, and $\beta = 100.23° \pm 0.02°$ and a calculated density of 1.668 g·cm$^{-3}$ (Bochmann et al. 1995b). It was synthesized in the same way as C$_{26}$H$_{34}$N$_4$Te$_2$Zn was obtained using pyridine instead of N-methylimidazole. Its crystals were grown from pyridine at 10°C as colorless blocks. Heating of this compound either without a solvent or in paraffin oil to 270°C–320°C leads to the formation of cubic ZnTe.

## 3.14   Zn–H–C–P–Te

Some multinary compounds are formed in this system.

**[Zn(TeC$_6$H$_2$Me$_3$–2,4,6)$_2$(PMe$_3$)] or C$_{21}$H$_{31}$PTe$_2$Zn**. This compound melts at 116°C–118°C. To obtain it, to a suspension of 0.59 mM of Zn(TeC$_6$H$_2$Me$_3$–2,4,6)$_2$ in 10 mL of petroleum ether was added via a syringe 0.53 mM of PMe$_3$ (Bochmann et al. 1995b). The starting materials dissolved within a few seconds, followed by the formation of a colorless precipitate. Stirring was continued for 2 h. The product was washed with petroleum ether (3×20 mL) and dried in vacuo. Heating of this compound either without a solvent or in paraffin oil to 270°C–320°C leads to the formation of cubic ZnTe.

**[Zn(TeC$_6$H$_2$Me$_3$–2,4,6)$_2$(PMe$_3$)$_2$] or C$_{24}$H$_{40}$P$_2$Te$_2$Zn**. This compound melts at 106°C–108°C. It was synthesized in the same way as C$_{21}$H$_{31}$PTe$_2$Zn was obtained, while an excess of PMe$_3$ was used (Bochmann et al. 1995b). Heating of this compound either without a solvent or in paraffin oil to 270°C–320°C leads to the formation of cubic ZnTe.

**[Zn(TeC$_6$H$_2$Me$_3$–2,4,6)$_2$(Me$_2$PC$_2$H$_4$PMe$_2$)] or C$_{24}$H$_{44}$P$_2$Te$_2$Zn**. This compound was synthesized in the same way as C$_{21}$H$_{31}$PTe$_2$Zn was obtained, while an excess of Me$_2$PC$_2$H$_4$PMe$_2$ was used (Bochmann et al. 1995b). Heating of this compound either without a solvent or in paraffin oil to 270°C–320°C leads to the formation of cubic ZnTe.

**[(Ph$_4$P)$_2$Zn(Te$_4$)$_2$] or C$_{48}$H$_{40}$P$_2$Te$_8$Zn**. This compound crystallizes in the tetragonal structure with the lattice parameters $a = 2139.6 \pm 0.4$ and $c = 1116.9 \pm 0.3$ and a calculated density of 2.293 g·cm$^{-3}$ (Bollinger et al. 1995). To obtain it, a flask was charged with Li$_2$Te (3.00 mM), Te (6 mM), zinc xanthogenate (1.00 mM), Ph$_4$PBr (2 mM), DMF (30 mL), and Et$_3$P (3.4 mM). The mixture was stirred at 90°C for 1 h and filtered. The solution was layered with Et$_2$O (30 mL) and stored overnight at room temperature to produce crystals of the title compound.

**[Zn$_{10}$Te$_4$(TePh)$_{12}$(PPr$^n_3$)$_2$] or C$_{90}$H$_{102}$P$_2$Te$_{16}$Zn$_{10}$**. This compound crystallizes in the tetragonal structure with the lattice parameters $a = 2566.0 \pm 0.4$ and $c = 2130.1 \pm 0.4$ and a calculated density of 2.007 g·cm$^{-3}$ (Eichhöfer et al. 1998). To prepare it, ZnCl$_2$ (1.5 mM) and PPr$^n_3$ (0.75 mM) were dissolved in THF (30 mL). After mixing with PhTeSiMe$_3$ (3 mM), a colorless solution turned dark yellow. Colorless block crystals of the title compound were obtained after 2 days by layering with Et$_2$O.

**[Zn$_{10}$Te$_4$(TePh)$_{12}$(PPh$_3$)$_2$] or C$_{108}$H$_{90}$P$_2$Te$_{16}$Zn$_{10}$**. This compound crystallizes in the triclinic structure with the lattice parameters $a = 2069.4 \pm 0.4$, $b = 2187.8 \pm 0.4$, and $c = 2351.5 \pm 0.5$ pm and $\alpha = 70.36° \pm 0.03°$, $\beta = 84.62° \pm 0.03°$, and $\gamma = 63.63° \pm 0.03°$ and a calculated density of 1.810 g·cm$^{-3}$ (Eichhöfer et al. 1998). To prepare it, ZnCl$_2$ (1.6 mM) was dissolved in THF (30 mL) with addition of PPr$^n_3$ (1.6 mM). The mixture was cooled to 0°C, and PhTeSiMe$_3$ (3.2 mM) was added dropwise after which the solution turned light yellow. Colorless crystals of the title compound were obtained after 2 days by layering with Et$_2$O at 0°C.

## 3.15    Zn–H–C–P–Cl–Te

Some multinary compounds are formed in this system.

[HPPr$^n_3$]$_2$[Zn$_8$Cl$_4$Te(TePh)$_{12}$] or C$_{90}$H$_{104}$P$_2$Cl$_4$Te$_{13}$Zn$_8$. This compound crystallizes in the monoclinic structure with the lattice parameters $a = 1899.8 \pm 0.4$, $b = 2227.0 \pm 0.5$, and $c = 2939.0 \pm 0.6$ pm, and $\beta = 101.35° \pm 0.03°$ and a calculated density of 1.978 g·cm$^{-3}$ (Eichhöfer et al. 1998). To obtain it, ZnCl$_2$ (2.3 mM) and Pr$^n_3$P (1.15 mM) were dissolved in acetone at 0°C. Addition of PhTeSiMe$_3$ (4.6 mM) leads to the formation of the red solution from which colorless crystals of the title compound were obtained by layering with Et$_2$O at 0°C.

[HPPr$^n_2$Ph]$_2$[Zn$_8$Cl$_4$Te(TePh)$_{12}$] or C$_{96}$H$_{100}$P$_2$Cl$_4$Te$_{13}$Zn$_8$. This compound crystallizes in the monoclinic structure with the lattice parameters $a = 2230.0 \pm 0.8$, $b = 1919.9 \pm 0.4$, and $c = 3139.5 \pm 0.6$ pm, and $\beta = 109.97° \pm 0.04°$ and a calculated density of 1.940 g·cm$^{-3}$ (Eichhöfer et al. 1998). To obtain it, ZnCl$_2$ (2.3 mM) and Pr$^n$PPh (1.15 mM) were dissolved in acetone at 0°C. Addition of PhTeSiMe$_3$ (4.6 mM) leads to the formation of red solution from which at 0°C colorless crystals of the title compound were obtained by layering with Et$_2$O at 0°C.

## 3.16    Zn–H–N–O–Mo–Te

(NH$_4$)$_7$Zn$_{10}$[TeMo$_6$O$_{24}$]$_4$(OH)$_3$·46H$_2$O. This multinary compound is formed in this system (Słoczyński and Śliwa 1978). It decomposes at 350°C–450°C and was precipitated from the solution, containing Zn(NO$_3$)$_2$, telluric acid, and ammonium paramolybdate.

## 3.17    Zn–H–O–Mo–Te

Zn$_3$[TeMo$_6$O$_{24}$]·20H$_2$O. This quinary compound is formed in this system (Słoczyński and Śliwa 1978). It decomposes at 420°C and crystallizes in the hexagonal structure with the lattice parameters $a = 1768 \pm 2$ and $c = 1955 \pm 5$ pm and experimental and calculated densities of 3.11 and 3.09 g·cm$^{-3}$, respectively. This compound was precipitated from the solution, containing Zn(NO$_3$)$_2$, telluric acid, and ammonium paramolybdate, and was isolated by treatment with cold 0.2 M NaOH until all free MoO$_3$ was removed.

## 3.18    Zn–Li–Ag–O–Te

Ag$_6$Li$_2$Zn$_2$Te$_2$O$_{12}$. This quinary compound is formed in this system. It could be synthesized by molten ion exchange of Li$_8$Zn$_2$Te$_2$O$_{12}$ with AgNO$_3$ (Kumar et al. 2012).

## 3.19    Zn–Cu–In–Mn–Te

2ZnTe–CuInTe$_2$–2MnTe. The isothermal section of this system at 600°C is given in Figure 3.1 (Nealt et al. 1989). Four single-phase fields are found to exist, two with

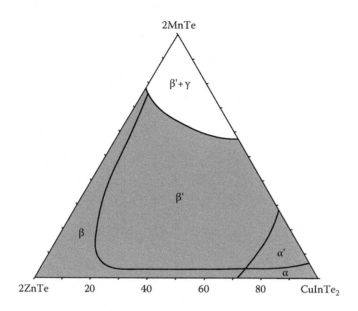

**FIGURE 3.1** Isothermal sections of the 2ZnTe–CuInTe$_2$–2MnTe quasiternary system at 600°C. (From Nealt, C. et al., *J. Phys. D: Appl. Phys.*, 22(9), 1347, 1989.)

normal zincblende (β) and chalcopyrite (α) structures and derived from chalcopyrite (β′) and zincblende (α′) in which Mn atoms show crystallographic ordering on the cation sublattice. Optical absorption measurements were carried out to give values of the room temperature optical energy gap for all single-phase samples (Nealt et al. 1989, Tovar et al. 1990).

Polycrystalline samples were prepared by the usual melt and anneal technique. The alloys were annealed at 600°C for 20–30 days (Nealt et al. 1989, Tovar et al. 1990).

## 3.20   Zn–Cu–O–Cl–Te

**CuZn(TeO$_3$)Cl$_2$.** This quinary compound is formed in this system. It crystallizes in the orthorhombic structure with the lattice parameters $a = 1018.97 \pm 0.08$, $b = 1553.89 \pm 0.12$, and $c = 745.12 \pm 0.06$ pm at 123 K and a calculated density of 4.227 g·cm$^{-3}$ (Johnson and Törnroos 2003). CuZn(TeO$_3$)Cl$_2$ was synthesized from a mixture of CuCl$_2$/ZnO/TeO$_2$ in the molar ratio 1:1:1.

## 3.21   Zn–Ag–In–Mn–Te

**2ZnTe–AgInTe$_2$–2MnTe.** The isothermal section of this system at 600°C is given in Figure 3.2 (Nealt et al. 1989). Four single-phase fields are found to exist, two with normal zincblende (β) and chalcopyrite (α) structures and derived from chalcopyrite (β′) and zincblende (α′) in which Mn atoms show crystallographic ordering on the cation sublattice. There is also a two-phase region (α + β) that penetrates into the

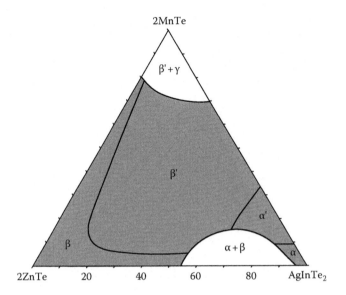

**FIGURE 3.2**  Isothermal sections of the 2ZnTe–AgInTe$_2$–2MnTe quasiternary system at 600°C. (From Nealt, C. et al., *J. Phys. D: Appl. Phys.*, 22(9), 1347, 1989.)

ternary system from the ZnTe–AgInTe$_2$ system. Optical absorption measurements were carried out to give values of the room temperature optical energy gap for all single-phase samples.

Polycrystalline samples were prepared by the usual melt and anneal technique. The alloys were annealed at 600°C during 20–30 days (Nealt et al. 1989).

### 3.22   Zn–Ca–Pb–O–Te

**(Ca,Pb)Zn$_3$Te$_2$O$_{12}$** [**(Zn$_{1.31}$Ca$_{1.32}$Pb$_{0.25}$)$_3$Te$_{1.04}$O$_6$** (Rozhdestvenskaya et al. 1984); **(Zn$_{1.38}$Ca$_{1.36}$Pb$_{0.26}$)$_3$TeO$_6$** (Kim et al. 1982)]. This quinary compound (mineral yafsoanite) is formed in this system. It crystallizes in the cubic structure with lattice parameter $a = 1263.2 \pm 0.2$ pm ($a = 1262.1$ pm [Rozhdestvenskaya et al. 1984]; $a = 631.5 \pm 0.2$ pm [Kim et al. 1982]) and calculated (at Cd/Pb = 6:1) and experimental densities of 5.54 (5.55 [Kim et al. 1982]), and 5.40–5.55 g·cm$^{-3}$, respectively (Jarosh and Zemann 1989).

### 3.23   Zn–Ca–O–Co–Te

**Ca$_3$ZnCo$_2$Te$_2$O$_{12}$.** This quinary compound is formed in this system. It crystallizes in the cubic structure with lattice parameter $a = 1259$ pm (Kasper 1968).

### 3.24   Zn–Ba–O–Cl–Te

**BaZn(TeO$_3$)Cl$_2$.** This quinary compound is formed in this system. It crystallizes in the orthorhombic structure with lattice parameters $a = 1234.5 \pm 0.4$, $b = 564.58 \pm 0.19$, and $c = 1918.6 \pm 0.7$ pm and optical band gap $E_g = 4.48$ eV (Jiang et al. 2006). This compound is stable up to 600°C. Then, it decomposes continuously up to 1200°C, during which $TeO_2$ and $Cl_2$ are released.

This compound was synthesized by a high-temperature reaction of $BaCl_2$, $ZnO$, and $TeO_2$ in an evacuated quartz tube at 720°C for 6 days and then cooled to 300°C at 4°C h$^{-1}$ before switching off the furnace.

## 3.25 Zn–Cd–O–Co–Te

$Zn_2Cd_3CoTe_2O_{12}$. This quinary compound is formed in this system. It crystallizes in the cubic structure with lattice parameter $a = 1255$ pm (Kasper 1968).

## 3.26 Zn–Pb–As–O–Te

$Zn_3Pb_3TeAs_2O_{14}$ (mineral dugganite). This quinary compound is formed in this system. It crystallizes in the trigonal structure with lattice parameters $a = 846.0 \pm 0.2$ and $c = 520.6 \pm 0.2$ pm (Lam et al. 1998).

## 3.27 Zn–O–Br–Co–Te

$ZnCo(TeO_3)Br_2$. This quinary compound, which crystallizes in the orthorhombic structure with lattice parameters $a = 1055.60 \pm 0.03$, $b = 1589.85 \pm 0.04$, and $c = 772.96 \pm 0.02$ pm at 30 K, is formed in this system (Kashi et al. 2008). Its polycrystalline samples were prepared by solid-state reaction: $ZnO$, $TeO_2$, and $CoBr_2$ were mixed in proper molar ratios. Then the mixtures were pressed into pellets, sealed in an evacuated silica tube, and heated at 500°C for 48 h.

## REFERENCES

Bochmann M., Bwembya G.C., Hursthouse M.B., Coles S.J. Synthesis of phosphinochalcogenoic amidato complexes of zinc and cadmium. The crystal and molecular structure of $[Zn\{Bu^t_2P(Se)NPr^i\}_2]$, *J. Chem. Soc., Dalton Trans.*, (17), 2813–2817 (1995a).

Bochmann M., Bwembya G.C., Powell A.K., Song X. Zinc(II) arene tellurolato complexes as precursors to zinc telluride. The crystal and molecular structure of $[Zn(TeC_6H_2Me_3-2,4,6)_2(pyridine)_2]$, *Polyhedron*, 14(23–24), 3495–3500 (1995b).

Bollinger J.C., Roof L.C., Smith D.M., McConnachie J.M., Ibers J.A. Synthesis, X-ray crystal structures, and NMR spectroscopy of $[PPh_4]_2[M(Te_4)_2]$, M = Hg, Cd, Zn, *Inorg. Chem.*, 34(6), 1430–1434 (1995).

Bonasia P.J., Arnold J. Zinc, cadmium, and mercury tellurolates: Hydrocarbon solubility and low coordination numbers enforced by sterically encumbered silyltellurolate ligands, *Inorg. Chem.*, 31(12), 2508–2514 (1992).

Brylev K.A., Mironov Y.V., Naumov N.G., Fedorov V.E., Ibers J.A. New compounds from tellurocyanide rhenium cluster anions and 3d-transition metal cations coordinated with ethylenediamine, *Inorg. Chem.*, 43(16), 4833–4838 (2004).

Eichhöfer A., Fenske D., Pfistner H., Wunder M. Zinkselenid- und Zinktelluridcluster mit Phenylselenolat- und Phenyltellurolatliganden. Die Kristallstrukturen von $[NEt_4]_2[Zn_4Cl_4(SePh)_6]$, $[NEt_4]_2[Zn_8Cl_4Se(SePh)_{12}]$, $[Zn_8Se(SePh)_{14}(P^nPr_3)_2]$, $[HP^nPr_2R]_2$ $[Zn_8Cl_4Te(TePh)_{12}]$ (R = $^n$Pr, Ph) und $[Zn_{10}Te_4(TePh)_{12}(PR_3)_2]$ (R = $^n$Pr, Ph), *Z. anorg. und allg. Chem.*, 624(11), 1909–1914 (1998).

Jarosh D., Zemann J. Yafsoanite: A garnet type calcium–tellurium(VI)–zinc oxide, *Mineral. Petrol.*, 40(2), 111–116 (1989).

Jiang H.-L., Feng M.L., Mao J.-G. Synthesis, crystal structures and characterizations of $BaZn(SeO_3)_2$ and $BaZn(TeO_3)Cl_2$, *J. Solid State Chem.*, 179(6), 1911–1917 (2006).

Johnson M., Törnroos K.W. Synthesis and crystal structures of the layered compound $CuZn(TeO_3)Cl_2$, *Solid State Sci.*, 5(1), 263–266 (2003).

Kashi T., Yasui Y., Moyoshi T., Sato M., Kakurai K., Iikubo S., Igava N. Crystal structure and magnetic properties of $CoZn(TeO_3)Br_2$, *J. Phys. Soc. Jpn.*, 77(8), 084707 (5 p.) (2008).

Kasper H.M. Preparation of $\{A^{2+}_3\}[Te_2](B^{2+}_3)$, *Mater. Res. Bull.*, 3(9), 765–766 (1968).

Kim A.A., Zayakina N.V., Lavrent'yev Yu.G. Yafsoanite $(Zn_{1.38}Ca_{1.36}Pb_{0.26})_3Te_1O_6$—New tellurium mineral [in Russian], *Zap. Vses. Mineral. Obshch.*, 111(1), 118–121 (1982).

Kumar V., Bhardwaj N., Tomar N., Thakral V., Uma S. Novel lithium-containing honeycomb structures, *Inorg. Chem.*, 51(20), 10471–10473 (2012).

Lam A.E., Groat L.A., Ercit T.S. The crystal structure of dugganite, $Pb_3Zn_3Te^{6+}As_2O_{14}$, *Can. Mineral.*, 36(3), 823–830 (1998).

Li J., Chen Z., Emge T.J., Yuen T., Proserpio D.M. Synthesis, structure characterization and magnetic properties of tellurostannates $[M(en)_{32}(Sn_2Te_6)$ (M = Mn, Zn), *Inorg. Chim. Acta*, 273(1–2), 310–315 (1998).

Lu J., Wang F., Shen Y., Tang C., Zhang Y., Jia D. New $\mu$-$SnTe_4$ and $\mu$-$Sn_2Te_6$ ligands to transition metal: Solvothermal syntheses and characterizations of zinc tellurostannates containing polyamine ligands, *J. Solid State Chem.*, 216, 65–72 (2014).

Mandarino J.A., Matzat E., Williams S.J. Zemannite, a zinc tellurite from Moctezuma, Sonora, Mexico, *Can. Mineral.*, 14(3), 387–390 (1976).

Margison S.M., Grice J.D., Groat L.A. The crystal structure of leisingite, $(Cu^{2+}, Mg, Zn)_2(Mg, Fe)Te^{6+}O_6 \cdot 6H_2O$, *Can. Mineral.*, 35(3), 759–763 (1998).

Matzat E. Die Kristallstruktur eines unbenannten zeolithartigen Telluritminerals, $\{(Zn,Fe)_2[TeO_3]_3\}Na_xH_{1-x} \cdot yH_2O$, *Tscherm. Min. Petr. Mitt.*, 12(1), 108–117 (1967).

Miletich R. Crystal chemistry of the microporous tellurite minerals zemannite and kinichilite, $Mg_{0.5}[Me^{2+}Fe^{3+}(TeO_3)_3] \cdot 4.5H_2O$, $(Me^{2+} = Zn; Mn)$, *Eur. J. Mineral.*, 7(3), 509–523 (1995a).

Miletich R. The synthetic microporous tellurites $Na_2[Me_2(TeO_3)_3] \cdot 3H_2O$ (Me = Zn, Co): Crystal structure, de- and rehydration, and ion exchange properties, *Monatsh. Chem.*, 126(4), 417–430 (1995b).

Mironov Y.V., Efremova O.A., Naumov D.Y., Fedorov V.E. Coordination compounds based on the $[Re_4Te_4(CN)_{12}]^{4-}$ cluster anion and $Zn^{2+}$ cation [in Russian], *Izv. RAN. Ser. khim.*,(4), 718–721 (2006).

Mironov Y.V., Shestopalov M.A., Fedorov V.E. Synthesis and crystal structure of tetrahedral complex $[Zn_2(NH_3)_6(\mu\text{-}OH)][Zn(NH_3)_4]_{0.5}[Re_4Te_4(CN)_{12}] \cdot 5H_2O$ [in Russian], *Zhurn. strukt. khimii*, 52(2), 402–405 (2011).

Nealt C., Woolley J.C., Tovar R., Quintero M. $Zn_{2x}(CuIn)_yMn_{2z}Te_2$ and $Zn_{2x}(AgIn)_yMn_{2z}Te_2$ alloys, *J. Phys. D: Appl. Phys.*, 22(9) 1347–1353 (1989).

Rozhdestvenskaya I.V., Zayakina N.V., Kim A.A. Crystal structure of Zn-, Ca-tellurate—yafsoanite [in Russian], *Mineral. Zh.*, 6(2), 75–79 (1984).

Shreeve-Keyer J.L., Warren C.J., Dhingrat S.S., Haushalter R.C. Synthesis and structural characterization of Zintl anions containing the tris(ethylenediamine)zinc cation: $[Zn(N_2C_2H_8)_3][Te_3](0.5en)$ and $[Zn(N_2C_2H_8)_3]_2[Sn_2Te_6](en)$, *Polyhedron*, 16(7), 1193–1199 (1997).

Słoczyński J., Śliwa B. Telluromolybdates of transition metals, *Z. anorg. und allg. Chem.*, 438(1), 295–304 (1978).

Tovar R., Quintero M., Neal C., Woolley J.C. Phase diagram of $Zn_{2x}(CuIn)_yMn_{2z}Te_2$ alloys, *J. Cryst. Growth*, 106(4), 629–634 (1990).

Xie J.-Y., Mao J.-G. $Zn_2(bipy)_2V_4TeO_{14}$: A new type of layered vanadium(V) tellurite with organically covalently bonded zinc(II) complex, *Inorg. Chem. Comm.*, 8(4), 375–378 (2005).

# 4 Systems Based on CdS

## 4.1 Cd–H–Na–C–N–O–S

Some compounds are formed in this system.

**[Cd(C$_{24}$H$_{32}$O$_8$)Na(SCN)$_3$] or C$_{27}$H$_{32}$NaN$_3$O$_8$S$_3$Cd.** This compound crystallizes in the monoclinic structure with the lattice parameters $a = 1611.0 \pm 0.8$, $b = 2038.0 \pm 0.5$, and $c = 1101 \pm 1$ pm and $\beta = 119.87° \pm 0.03°$ and a calculated density of 1.607 g·cm$^{-3}$ (Zhang et al. 2001). It was prepared by the next procedure. With vigorous stirring, a solution of dibenzo-24-crown-8 (C$_{24}$H$_{32}$O$_8$) (1 mM) in acetonitrile (40 mL) was added dropwise to a preformed mixture of CdSO$_4$·3H$_2$O (1 mM) in H$_2$O (10 mL) and NaSCN (3 mM) in H$_2$O (10 mL), over a period of 20 min. White precipitates slowly formed in the process. The solution was stirred for an hour after the completion of the addition. The white precipitate was then collected by filtration and washed several times with methanol. Single crystals suitable for XRD were obtained by recrystallization from aqueous solution of the title compound.

**(Me$_4$N)[Na(Cd{SOCPh}$_3$)$_2$] or C$_{46}$H$_{42}$NaNO$_6$S$_6$Cd$_2$.** This compound crystallizes in the triclinic structure with the lattice parameters $a = 1102.6 \pm 0.1$, $b = 1188.6 \pm 0.2$, and $c = 1090.6 \pm 0.1$ pm and $\alpha = 94.32° \pm 0.01°$, $\beta = 113.25° \pm 0.01°$, and $\gamma = 71.53° \pm 0.01°$ and the experimental and calculated densities 1.528 ± 0.005 and 1.529 g·cm$^{-3}$, respectively (Vittal and Dean 1993). To obtain it, a solution of NaSOPh was prepared by dissolution of Na (15.5 mM) and then PhCOSH (17.0 mM) in 25 mL of MeOH. This solution was added by stirring to a solution of Cd(NO$_3$)$_2$·4H$_2$O (3.83 mM) in 10 mL of H$_2$O, followed by a solution of Me$_4$NCl (7.70 mM) in 25 mL of MeOH. The yellow slurry was heated to 60°C and MeCN was added until most of the solid had dissolved. The remaining solids were removed by filtration. The clear yellow filtrate was refrigerated to 5°C overnight for crystallization to occur. The orange-yellow transparent crystals were separated by decantation, washed with MeOH, and dried in vacuo.

## 4.2 Cd–H–Na–P–O–S

**Na$_{0.5}$Cd$_{0.75}$PS$_3$(H$_2$O)$_{1.88}$.** This multinary compound is formed in this system. It crystallizes in the monoclinic structure with the lattice parameters $a = 657.8$, $b = 1001.3$, and $c = 1287.5$ pm and $\beta = 108.85°$ (Jeevanandam and Vasudevan 1997). This compound was synthesized by the exchange reaction of the K$_{0.57}$Cd$_{0.742}$PS$_3$(H$_2$O)$_{1.2}$ and the aqueous solution, containing N$^+$ ions.

## 4.3   Cd–H–K–C–N–O–S

Some compounds are formed in this system.

**[K(C$_{12}$H$_{24}$O$_6$)][Cd(SCN)$_3$] or C$_{15}$H$_{24}$KN$_3$O$_6$S$_3$Cd**. This compound was prepared according to the following reaction: $CdSO_4 + 3KSCN + C_{12}H_{24}O_6 = [(C_{12}H_{24}O_6)K]$ $[Cd(SCN)_3] + K_2SO_4$, where $C_{12}H_{24}O_6$ – 18-crown-6 (Zhang and Zelmon 2002). Its single crystals can be grown from aqueous solution via two conventional techniques: slow evaporation and temperature-lowering methods.

**[K(C$_{16}$H$_{24}$O$_6$)][Cd(SCN)$_3$] or C$_{19}$H$_{24}$KN$_3$O$_6$S$_3$Cd**. This compound exists in two polymorphic modifications (Pickardt and Dechert 1999). One of them crystallizes in the monoclinic structure with the lattice parameters $a = 1158.31 \pm 0.03$, $b = 1096.55 \pm 0.02$, and $c = 2028.46 \pm 0.04$ pm and $\beta = 99.5261° \pm 0.0002$ and a calculated density of 1.668 g·cm$^{-3}$. The second modification crystallizes in the orthorhombic structure with the lattice parameters $a = 1105.95 \pm 0.03$, $b = 1413.07 \pm 0.04$, and $c = 1617.10 \pm 0.05$ pm and a calculated density of 1.677 g·cm$^{-3}$. Both modifications of the title compound were obtained by a reaction of $K_2[Cd(SCN)_4]$ with benzo-18-crown-6 ($C_{16}H_{24}O_6$) in EtOH.

## 4.4   Cd–H–K–P–O–S

**K$_{0.57}$Cd$_{0.742}$PS$_3$(H$_2$O)$_{1.2}$**. This multinary compound is formed in this system. It crystallizes in the monoclinic structure with the lattice parameters $a = 672.2$, $b = 1033.4$, and $c = 1002.4$ pm and $\beta = 108.76°$ (Jeevanandam and Vasudevan 1997). This compound was synthesized by the ion-exchange intercalation reaction of the layered CdPS$_3$ in the aqueous solution.

## 4.5   Cd–H–Cs–C–N–O–Re–S

Some compounds are formed in this system.

**Cs$_2$[{Cd(H$_2$O)$_2$}{Re$_6$S$_8$(CN)$_6$}] or C$_6$H$_4$Cs$_2$N$_6$O$_2$Re$_6$S$_8$Cd**. This compound crystallizes in the orthorhombic structure with the lattice parameters $a = 1882.1 \pm 0.3$, $b = 1083.0 \pm 0.2$, and $c = 1334.7 \pm 0.3$ pm (Beauvais et al. 1998).

**Cs$_2$[{Cd(H$_2$O)$_2$}$_3${Re$_6$S$_8$(CN)$_6$}]$_2$·$x$H$_2$O or C$_{12}$H$_{12}$Cs$_2$N$_{12}$O$_6$Re$_{12}$S$_{16}$Cd$_3$·$x$H$_2$O**. This compound crystallizes in the trigonal structure with the lattice parameters $a = 2727.7 \pm 0.2$ and $c = 3394.8 \pm 0.4$ pm (Beauvais et al. 1998).

## 4.6   Cd–H–Cs–P–O–S

**Cs$_{0.5}$Cd$_{0.75}$PS$_3$(H$_2$O)$_{1.02}$**. This multinary compound is formed in this system. It crystallizes in the monoclinic structure with the lattice parameters $a = 655.5$, $b = 1081.7$, and $c = 1012.0$ pm and $\beta = 107.47°$ (Jeevanandam and Vasudevan 1997). This compound was synthesized by the ion-exchange intercalation reaction of the layered CdPS$_3$ in the aqueous solution.

## 4.7 Cd–H–Cu–C–N–Br–S

Some compounds are formed in this system.

[Cu(S$_2$CHBu$^n$$_2$)$_2$][CdBr$_3$] or C$_{18}$H$_{36}$CuN$_2$Br$_3$S$_4$Cd. This compound could be obtained if to a solution of 1 mol of Cd(S$_2$CHBu$^n$$_2$)$_2$ in chloroform, a solution of 2 mol of Cu(S$_2$CHBu$^n$$_2$)$_2$ in chloroform was added (Cras et al. 1973). After the addition of an equal volume of petroleum ether, crystalline products were formed. They were filtered off and washed with petroleum ether.

Once the composition of the compound is known, a simpler route to it suggested itself. A mixture of 1 mol of Br$_2$ in chloroform and 1 mol of CdBr$_2$ dissolved in EtOH was added with stirring to a solution of 2 mol of Cu(S$_2$CHBu$_2$)$_2$ in chloroform. Upon the addition of petroleum ether and cooling, precipitates of [Cu(S$_2$CHBu$_2$)$_2$][CdBr$_3$] as dark violet-black needles were obtained (Cras et al. 1973).

[Cu$_3$(S$_2$CHBu$^n$$_2$)$_6$][CdBr$_3$]$_2$ or C$_{54}$H$_{108}$Cu$_3$N$_6$Br$_6$S$_{12}$Cd$_2$. This compound crystallizes in the triclinic structure with the lattice parameters $a = 1262.9 \pm 0.4$, $b = 1370.6 \pm 0.4$, and $c = 1250.6 \pm 0.4$ pm and $\alpha = 95.84° \pm 0.04°$, $\beta = 91.45° \pm 0.04°$, and $\gamma = 86.06° \pm 0.03°$ and an experimental density of 1.6346 g·cm$^{-3}$ (Cras et al. 1973). To obtain it, to a suspension of 1 mol of 3,5-bis($N,N$-dibutyliminium)-1,2,4-trithiolane tetrabromo-di-μ-bromidocuprate(II) in chloroform, a saturated solution of 2 mol of Cd(S$_2$CHBu$^n$$_2$)$_2$ in the same solvent was added. The former compound dissolved slowly and crystalline product could be obtain when petroleum ether was added to the solution and cooled to 0°C. The compound was dissolved in chloroform and recrystallized after the addition of an equal amount of petroleum ether.

Once the composition of the compound is known, a simpler route to it suggested itself. A mixture of 1 mol of Br$_2$ in chloroform and 1 mol of CdBr$_2$ dissolved in EtOH was added with stirring to a solution of 3 mol of Cu(S$_2$CHBu$^n$$_2$)$_2$ in chloroform. Upon the addition of petroleum ether and cooling, precipitates of [Cu$_3$(S$_2$CHBun$_2$)$_6$] [CdBr$_3$]$_2$ were obtained (Cras et al. 1973).

## 4.8 Cd–H–Ba–C–N–O–S

Ba$_2$Cd(SCN)$_6$·7H$_2$O. This multinary compound is formed in this system (Brodersen and Böhm 1986). It crystallizes in the monoclinic structure with the lattice parameters $a = 943.6 \pm 0.3$, $b = 1469.7 \pm 0.6$, and $c = 1827.1 \pm 0.6$ pm and $\beta = 109.02° \pm 0.07°$ and an experimental density of 2.40 g·cm$^{-3}$. The title compound was prepared by the reaction of aqueous solutions of Ba(SCN)$_2$·3H$_2$O and CdSO$_4$·8/3H$_2$O (3:1 molar ratio).

## 4.9 Cd–H–Hg–C–N–O–S

Some compounds are formed in this system.

[CdHg(SCN)$_4$(MeOEtO)] or C$_7$H$_8$HgN$_4$O$_2$S$_4$Cd. This compound is thermally stable up to 115°C and crystallizes in the orthorhombic structure with the lattice parameters $a = 734.7 \pm 0.2$, $b = 1343.6 \pm 0.2$, and $c = 1631.6 \pm 0.1$ pm and a calculated density of 2.523 g·cm$^{-3}$ (Raghavan et al. 2010) [$a = 1643.2 \pm 0.2$, $b = 737.94 \pm 0.09$, and

$c = 1347.1 \pm 0.1$ pm (Zhou et al. 2000)]. Single crystals were grown from a mixed solvent of glycol monomethyl ether/$H_2O$ (1:1 volume ratio) by slow evaporation. The raw materials for synthesis of $CdHg(SCN)_4(MeOC_2Et)$ are $CdCl_2$, $HgCl_2$, $NH_4SCN$, and glycol monoether (MeOEtO). All the reactions were carried out in an aqueous medium. Initially, $CdHg(SCN)_4$ was synthesized at the interaction of $CdCl_2$, $HgCl_2$, and $NH_4SCN$ (Zhou et al. 2000, Raghavan et al. 2010). The formed $CdHg(SCN)_4$ was purified by recrystallization using acetone/$H_2O$ (4:1 volume ratio) as a mixed solvent and used for further synthesis of $CdHg(SCN)_4(MeOC_2Et)$. MeOEtO was diluted with deionized $H_2O$ and added to $CdHg(SCN)_4$ and the obtained solution was stirred for 5 h. The resulting solution was evaporated and the product obtained was subjected to repeated recrystallization using MeOEtO/$H_2O$ (1:1 volume ratio) as a mixed solvent.

**[CdHg(SCN)$_4$(MeNHCHO)$_2$] or C$_8$H$_{10}$HgN$_6$O$_2$S$_4$Cd.** This compound is stable up to 128.5°C (Liu et al. 2012) and crystallizes in the orthorhombic structure with the lattice parameters $a = 1624.4 \pm 0.02$, $b = 776.70 \pm 0.08$, and $c = 1530.43 \pm 0.19$ pm and a calculated density of 2.282 g·cm$^{-3}$ (Wang et al. 2005) [$a = 1626.57$, $b = 777.56$, and $c = 1531.92$ pm (Liu et al. 2012)]. To obtain it, a $N$-methylformamide solution (15 mL) of $CdHg(SCN)_4$ (10 mM) was added slowly to water (25 mL) with stirring at room temperature. After a while, the title compound was precipitated. The crystals were obtained by slow cooling of a solution in $N$-methylformamide/$H_2O$ (1:1 or 2:3 volume ratio) (Wang et al. 2005, Liu et al. 2012).

**[CdHg(SCN)$_4$(DMSO)$_2$]$_n$ or C$_8$H$_{12}$HgN$_4$O$_2$S$_6$Cd.** This compound crystallizes in the orthorhombic structure with the lattice parameters $a = 851.31 \pm 0.03$, $b = 853.61 \pm 0.07$, and $c = 2821.26 \pm 0.01$ pm and a calculated density of 2.121 g·cm$^{-3}$ (Raghavan et al. 2008) [$a = 851.88 \pm 0.06$, $b = 853.98 \pm 0.07$, and $c = 2822.4 \pm 0.6$ pm and an experimental density of 2.270 g·cm$^{-3}$ (Guo et al. 2000a,b, 2001); $a = 851.81 \pm 0.05$, $b = 853.94 \pm 0.07$, and $c = 2823.1 \pm 0.6$ pm and a calculated density of 2.270 g·cm$^{-3}$ (Rajarajan et al. 2006)]. Its single crystals were grown from mixed solution of $CdHg(SCN)_4$ in $H_2O$/DMSO by slow cooling and slow evaporation methods under pH $= 3.5$ (Guo et al. 2000a,b, 2001, Rajarajan et al. 2006, Raghavan et al. 2008).

**[{CdHg(SCN)$_4$(C$_4$H$_9$NO)$_2$}$_2$]$_n$ or C$_{24}$H$_{36}$Hg$_2$N$_{12}$O$_4$S$_8$Cd$_2$.** This compound crystallizes in the monoclinic structure with the lattice parameters $a = 1482.9 \pm 0.3$, $b = 1614.65 \pm 0.19$, and $c = 1928.2 \pm 0.3$ pm and $\beta = 91.27° \pm 0.02°$ and a calculated density of 2.071 g·cm$^{-3}$ (Wang et al. 2002b). To obtain it, NH (40 mM) and $Hg(NO_3)_2 \cdot H_2O$ (10 mM) were dissolved in water (20 mL) with stirring. To the colorless solution, an aqueous solution (10 mL) containing $CdCl_2 \cdot 2.5H_2O$ (10 mM) and $N,N$-dimethylacetamide (10 mL) were added simultaneously. After the colorless solution had been left standing at room temperature for a while, a light-pink title compound precipitated and was separated. The crystals were obtained by slow cooling of a more dilute aqueous solution.

## 4.10    Cd–H–Hg–C–N–S

**CdHg$_2$(SCN)$_6$·C$_6$H$_6$ or C$_{12}$H$_6$Hg$_2$N$_6$S$_6$Cd.** This multinary compound is formed in this system. From a solution containing a $Hg^{2+}$ salt and thiocyanate (molar ratio of about 1:4), a solution of $Cd^{2+}$ salt precipitates, in presence of benzene, colorless microcrystalline title compound (Baur et al. 1962). The obtained compound is thermodynamically unstable.

## 4.11   Cd–H–Hg–C–N–I–S

[CdHg(Et$_2$NCS$_2$)$_2$I$_2$] or C$_{10}$H$_{20}$HgN$_2$I$_2$S$_4$Cd. This multinary compound is formed in this system. It melts at 180°C with decomposition and crystallizes in the monoclinic structure with the lattice parameters $a = 750.2 \pm 0.2$, $b = 941.4 \pm 0.3$, and $c = 1567.1 \pm 0.5$ pm and $\beta = 94.50° \pm 0.03°$ and a calculated density of 2.599 g·cm$^{-3}$ (Zemskova et al. 1991). To obtain it, to a solution of Cd(Et$_2$NCS$_2$)$_2$ in CH$_2$Cl$_2$ at room temperature HgI$_2$ [(1.0–1.1):1 molar ratio] was added, and the mixture was stirred for 3–5 h. The precipitate that formed was filtered off, washed thoroughly with CH$_2$Cl$_2$, and dried in air.

## 4.12   Cd–H–B–C–N–O–S

[Cd(C$_8$H$_{12}$BN$_4$S$_2$)$_2$(DMSO)] or C$_{18}$H$_{30}$B$_2$N$_8$OS$_5$Cd. This multinary compound is formed in this system. It crystallizes in the tetragonal structure with the lattice parameters $a = 1050.3 \pm 0.02$ and $c = 2603.8 \pm 0.5$ pm at $150 \pm 1$ K and a calculated density of 1.546 g·cm$^{-3}$ (Alvarez et al. 2003). It was obtained by recrystallization of Cd(C$_8$H$_{12}$BN$_4$S$_2$)$_2$ from DMSO.

## 4.13   Cd–H–B–C–N–O–F–S

[Cd(C$_4$H$_7$NOS)$_2$(BF$_4$)$_2$] or C$_8$H$_{14}$B$_2$N$_2$O$_2$F$_8$S$_2$Cd. This multinary compound is formed in this system. It melts at 45°C (Colombini and Preti 1975). It was obtained by a reaction of Cd(BF$_4$)$_2$ and thiomorpholine-3-one in EtOH. The obtained solid product was washed with ethyl ether.

## 4.14   Cd–H–B–C–N–S

Some compounds are formed in this system.

[Cd(C$_8$H$_{12}$BN$_4$S$_2$)$_2$] or C$_{16}$H$_{24}$B$_2$N$_8$S$_4$Cd. This compound melts at 273°C (Alvarez et al. 2003). To prepare it, a stirred solution of CdCl$_2$ (0.572 mM) in H$_2$O (15 mL) was treated with a solution of Na[Bm$^{Me}$], [Bm$^{Me}$] – bis(2-mercapto-1-methylimidazolyl) borate anion, C$_8$H$_{12}$BN$_4$S$_2$, (1.144 mM) in the same solvent (15 mL), resulting in the formation of a white precipitate. The suspension was stirred for 15 min and the product was isolated by filtration and dried in vacuo for 18 h.

[HB(3,5-Me$_2$C$_2$N$_2$H)$_3$]Cd[S$_2$CNEt$_2$] or C$_{20}$H$_{32}$BN$_7$S$_2$Cd. This compound decomposes at 245°C–253°C and crystallizes in the monoclinic structure with the lattice parameters $a = 1670.7 \pm 0.7$, $b = 819.7 \pm 0.5$, and $c = 2028.0 \pm 1.1$ pm and $\beta = 111.06° \pm 0.04°$ and a calculated density of 1.434 g·cm$^{-3}$ (Reger et al. 1995). It was obtained by the next procedure. CdCl$_2$ (1.1 mM), K[HB(3,5-Me$_2$C$_2$N$_2$H)$_3$] (1.1 mM), and Na[S$_2$CNEt$_2$]·3H$_2$O (1.1 mM) were charged into a flask along with 25 mL of THF.

The yellow solution was stirred overnight. THF was removed under vacuum and the product extracted with benzene (25 mL). The solution was filtered and the benzene removed under vacuum to leave a yellow solid of the title compound.

**[Cd(C$_{14}$H$_{24}$BN$_4$S$_2$)$_2$] or C$_{28}$H$_{48}$B$_2$N$_8$S$_4$Cd**. This compound melts at 280°C with decomposition and crystallizes in the monoclinic structure with the lattice parameters $a = 1058.46 \pm 0.06$, $b = 3602.1 \pm 0.2$, and $c = 1035.16 \pm 0.06$ pm and $\beta = 110.799° \pm 0.001°$ at $243 \pm 2$ K and a calculated density of 1.366 g·cm$^{-3}$ (Alvarez et al. 2003). To prepare it, a stirred solution of CdCl$_2$ (0.360 mM) in H$_2$O (15 mL) was treated with a solution of Na[Bmt$^{Bu}$], [Bmt$^{Bu}$] – bis(2-mercapto-1-$t$-butylimidazolyl)borate anion, C$_{14}$H$_{24}$BN$_4$S$_2$, (0.22 mM) in the same solvent (15 mL), resulting in the formation of a white precipitate. The suspension was stirred for 30 min and the product was isolated by filtration and dried in vacuo for 1.5 h.

**[{HB(3-Bu$^t$C$_3$H$_2$N$_2$)$_3$}{Cd(SCN)}]·(PhMe) or C$_{29}$H$_{42}$BN$_7$SCd**. This compound melts at 135°C–142°C and crystallizes in the orthorhombic structure with the lattice parameters $a = 1968.4 \pm 0.3$, $b = 1602.9 \pm 0.2$, and $c = 1032.24 \pm 0.15$ pm and a calculated density of 1.313 g·cm$^{-3}$ (Reger et al. 2002). It was prepared by the following procedure. [Cd$_2$(THF)$_5$](BF$_4$)$_4$ (0.50 mM) and TlHB(3-ButC$_3$H$_2$N$_2$)$_3$ (1.0 mM) were charged into a round-bottom Schlenk flask, and THF (35 mL) was added to the flask via syringe. A white precipitate immediately formed, and the mixture was allowed to stir for 2 h at room temperature and then filtered. The volatiles were removed in vacuo leaving a white solid. This solid and KSCN (1.0 mM) were dissolved in THF (20 mL). Immediately, a white precipitate is formed, and this mixture was allowed to stir for 3 h at room temperature and then filtered. The solvent was removed in vacuo to leave a white solid of the title compound. XRD quality crystals were obtained from hot (50°C)/cold (0°C) toluene.

**[HB(3-PhC$_3$N$_2$H$_2$)$_3$]Cd[S$_2$CNEt$_2$] or C$_{32}$H$_{32}$BN$_7$S$_2$Cd**. To obtain this compound, CdCl$_2$ (1.1 mM), K[HB(3-PhC$_3$H$_2$N$_2$)$_3$] (1.1 mM), and Na[S$_2$CNEt$_2$]·3H$_2$O (1.1 mM) were charged into a flask along with 25 mL of THF (Reger et al. 1995). The yellow solution was stirred overnight. THF was removed under vacuum and the product extracted with benzene (25 mL). The solution was filtered and the benzene removed under vacuum to leave a yellow solid of the title compound.

**[HB($p$-TolC$_3$H$_2$N$_2$S)$_3$]CdS·Ph or C$_{36}$H$_{33}$BN$_6$S$_4$Cd**. This compound melts at 232°C with decomposition and crystallizes in the monoclinic structure with the lattice parameters $a = 1501 \pm 4$, $b = 1532 \pm 4$, and $c = 1518 \pm 4$ pm and $\beta = 90.00° \pm 0.05°$ at $173 \pm 2$ K and a calculated density of 1.525 g·cm$^{-3}$ (Bakbak et al. 2002). To prepare it, a yellow suspension of [HB($p$-TolC$_3$H$_2$N$_2$S)$_3$]CdBr (0.32 mM) and Tl(SPh) (0.32 mM) in benzene (20 mL) was stirred for 2 h, and the resulting grayish yellow suspension was filtered. The solvent was removed under reduced pressure from the colorless filtrate to give a white solid, which was washed with pentane (2 × 8 mL) and dried in vacuo for 2 h. All reactions were performed under dry, oxygen-free N$_2$ in a glovebox or under Ar using a combination of high-vacuum and Schlenk techniques.

**[HB(p-TolC$_3$H$_2$N$_2$S)$_3$]CdS·Bz or C$_{37}$H$_{35}$BN$_6$S$_4$Cd**. This compound melts at 160°C with decomposition (Bakbak et al. 2002). It was prepared by the next procedure. A yellow suspension of [HB(p-TolC$_3$H$_2$N$_2$S)$_3$]CdBr (0.30 mM) and Tl(SBz) (0.31 mM) in benzene (25 mL) was stirred for 2 h, and the resulting brownish yellow suspension was filtered. The solvent was removed under reduced pressure from the colorless filtrate to give a white solid, which was washed with pentane (2×5 mL) and dried in vacuo for 1 h. All reactions were performed under dry, oxygen-free N$_2$ in a glovebox or under Ar using a combination of high-vacuum and Schlenk techniques.

**[HB(p-TolC$_3$H$_2$N$_2$S)$_3$]CdS·p-MeC$_6$H$_4$ or C$_{37}$H$_{35}$BN$_6$S$_4$Cd**. This compound melts at 242°C with decomposition (Bakbak et al. 2002). It was prepared using the following procedure. A yellow suspension of [HB(p-TolC$_3$H$_2$N$_2$S)$_3$]CdBr (0.36 mM) and Tl(S-p-MeC$_6$H$_4$) (0.36 mM) in benzene (25 mL) was stirred for 2 h, and the resulting beige suspension was filtered. Concentration of the colorless filtrate under reduced pressure to ca. 3 mL and the addition of diethyl ether (20 mL) resulted in the separation of a white solid, which was isolated by decantation, washed with diethyl ether (2×10 mL), and dried in vacuo for 1 h. All of the reactions were performed under dry, oxygen-free N$_2$ in a glovebox or under Ar using a combination of high-vacuum and Schlenk techniques.

**[Cd(C$_{20}$H$_{20}$BN$_4$S$_2$)$_2$] or C$_{40}$H$_{40}$B$_2$N$_8$S$_4$Cd**. This compound melts at 96°C with decomposition (Alvarez et al. 2003). To prepare it, a stirred solution of CdCl$_2$ (0.241 mM) in H$_2$O (15 mL) was treated with a solution of Na[Bm$^{Bz}$], [Bm$^{Bz}$] – bis(2-mercapto-1-benzylimidazolyl)borate anion, C$_{20}$H$_{20}$BN$_4$S$_2$, (0.483 mM) in a mixture of H$_2$O (10 mL) and MeOH (4 mL), resulting in the formation of a white precipitate. The suspension was stirred for 15 min and the product was isolated by filtration and dried in vacuo for 8 h.

**[Cd(C$_{20}$H$_{20}$BN$_4$S$_2$)$_2$] or C$_{40}$H$_{40}$B$_2$N$_8$S$_4$Cd**. This compound melts at 148°C with decomposition (Alvarez et al. 2003). To prepare it, a stirred solution of CdCl$_2$ (0.284 mM) in H$_2$O (15 mL) was treated with a solution of Na[Bm$^{p-Tol}$], [Bm$^{p-Tol}$] – bis(2-mercapto-1-p-tolylimidazolyl)borate anion, C$_{20}$H$_{20}$BN$_4$S$_2$, (0.567 mM) in a MeOH (15 mL), resulting in the formation of a white precipitate. The suspension was stirred for 15 min and the product was isolated by filtration and dried in vacuo for 3 h.

**[Cd(C$_4$H$_8$N$_2$S)$_4$](Ph$_4$B)$_2$·2MeCN or C$_{68}$H$_{78}$B$_2$N$_{10}$S$_4$Cd**. This compound crystallizes in the triclinic structure with the lattice parameters $a = 1030.0 \pm 0.1$, $b = 1444.4 \pm 0.1$, and $c = 2438.8 \pm 0.2$ pm and $\alpha = 80.53° \pm 0.01°$, $\beta = 79.72° \pm 0.01°$, and $\gamma = 71.38° \pm 0.01°$ and a calculated density of 1.28 g·cm$^{-3}$ (Dean et al. 2002). To prepare it, into MeCN (3 mL) were stirred 1-methylimidazolidine-2-thione (3.0 mM), Cd(CF$_3$SO$_3$)$_2$ (0.49 mM), and NaBPh$_4$ (1.0 mM). Insoluble NaCF$_3$SO$_3$, which formed from the clear solution, was removed by filtration. The filtrate was kept in a refrigerator at 5°C to produce the crystals suitable for XRD.

## 4.15   Cd–H–B–C–N–F–S

Some compounds are formed in this system.

**[Cd(C₃H₅NS₂)₂(BF₄)₂] or C₆H₁₀B₂N₂F₈S₄Cd**. This compound melts at 85°C (Colombini and Preti 1975). It was prepared by a reaction of Cd(BF₄)₂ and thiazolidine-2-thione in EtOH. The obtained solid product was washed with ethyl ether.

**[HB(p-TolC₃H₂N₂S)₃]CdS·C₆F₅ or C₃₆H₂₈BN₆F₅S₄Cd**. This compound melts at 222°C with decomposition (Bakbak et al. 2002). It was prepared using the following procedure. A white suspension of [HB(p-TolC₃H₂N₂S)₃]CdBr (0.26 mM) and Tl(SC₆F₅) (0.27 mM) in benzene (10 mL) was stirred for 1.5 h, and the resulting yellowish suspension was filtered. The solvent was removed from the colorless filtrate under reduced pressure to give a white solid, which was washed with pentane (6 × 10 mL) and dried in vacuo for 4 h. All reactions were performed under dry, oxygen-free N₂ in a glovebox or under Ar using a combination of high-vacuum and Schlenk techniques.

## 4.16   Cd–H–B–C–N–Cl–S

**[Cd(C₁₂H₁₆BN₆S₃)Cl] or C₁₂H₁₆BN₆ClS₃Cd**. This multinary compound is formed in this system. To prepare it, a solution of [Tl(C₁₂H₁₆BN₆S₃)] (0.4 mM) [C₁₂H₁₆BN₆S₃ = hydrotris(methimazolyl)borate] was suspended in acetone (30 mL), and a solution of CdCl₂ (0.4 mM) dissolved in the minimum amount of acetone was slowly added (Cassidy et al. 2002). The mixture was stirred overnight at room temperature and the solid formed (TlCl) was filtered off. The filtrate was taken to dryness and the resulting solid dried in vacuo. The crude product could be further purified by dissolving in the minimum amount of CH₂Cl₂ and cooling to –18°C, whereupon colorless crystals were obtained.

## 4.17   Cd–H–B–C–N–Br–S

Some compounds are formed in this system.

**[Cd(C₁₂H₁₆BN₆S₃)Br] or C₁₂H₁₆BN₆BrS₃Cd**. This compound crystallizes in the monoclinic structure with the lattice parameters $a = 1010.3 \pm 0.2$, $b = 1004.2 \pm 0.1$, and $c = 1895.7 \pm 0.2$ pm and $\beta = 101.39° \pm 0.01°$ at 123 K (Cassidy et al. 2002). To prepare it, [Tl(C₁₂H₁₆BN₆S₃)] (0.4 mM) [C₁₂H₁₆BN₆S₃ means hydrotris(methimazolyl)borate] was suspended in acetone (30 mL), and a solution of CdBr₂ (0.4 mM) dissolved in the minimum amount of acetone was slowly added. The mixture was stirred overnight at room temperature and the solid formed (TlBr) was filtered off. The filtrate was taken to dryness and the resulting solid dried in vacuo. The crude product could be further purified by dissolving in the minimum amount of CH₂Cl₂ and cooling to –18°C, whereupon colorless crystals were obtained.

**[HB(BuᵗC₃H₂N₂S)₂]CdBr·MeCN or C₂₃H₃₇BN₇BrS₃Cd**. This compound melts at 260°C with decomposition and crystallizes in the monoclinic structure with the lattice parameters $a = 982.22 \pm 0.07$, $b = 3299.1 \pm 0.2$, and $c = 1040.59 \pm 0.07$ pm and $\beta = 110.769° \pm 0.001°$ at $243 \pm 2$ K (White et al. 2002). A solution of Na[HBBu*t* $C_3H_2N_2S)_2$](0.390 mM) in MeOH (6 mL) was added to a stirred solution of $CdBr_2$ (0.389 mM) in the same solvent (6 mL). The resulting colorless solution was stirred for 1 h and concentrated under reduced pressure to ca. 4 mL, leading to the formation of a white precipitate. After the addition of $H_2O$ (5 mL), the product was isolated by filtration and dried in vacuo for 16 h.

**[Tm$_{Bz}$]CdBr or C$_{30}$H$_{28}$BN$_6$BrS$_3$Cd {[Tm$_{Bz}$] = tris(2-mercapto-1-benzylimidazolyl) hydroborato ligand}.** This compound melts at 248°C with decomposition and crystallizes in the monoclinic structure with the lattice parameters $a = 1875.3 \pm 0.1$, $b = 1195.1 \pm 0.1$, and $c = 1485.5 \pm 0.1$ pm and $\beta = 111.216° \pm 0.001°$ at $173 \pm 2$ K and a calculated density of 1.652 g·cm$^{-3}$ (Bakbak et al. 2001). To obtain it, a suspension of $CdBr_2$ (0.81 mM) and Na[Tm$^{Bz}$] (0.81 mM) in dichloromethane (40 mL) was stirred for 1.5 h at room temperature and filtered. The solvent was removed under reduced pressure from the colorless filtrate to give a white solid, which was washed with pentane (20 mL) and dried in vacuo for 45 min. All reactions were performed under dry oxygen-free $N_2$.

**[Tm$^{p\text{-Tol}}$]CdBr or C$_{30}$H$_{28}$BN$_6$BrS$_3$Cd {[Tm$^{p\text{-Tol}}$] = tris(2-mercapto-1-$p$-tolylimidazolyl)hydroborato ligand}.** This compound melts at 274°C with decomposition (Bakbak et al. 2001). It was obtain if a suspension of $CdBr_2$ (2.46 mM) and Na[Tm$^{p\text{-Tol}}$] (2.46 mM) in dichloromethane (80 mL) was stirred for 3 h at room temperature and filtered. The solvent was evaporated under reduced pressure from the colorless filtrate to give a white solid, which was washed with pentane ($2 \times 20$ mL) and dried in vacuo for 2.5 h. All reactions were performed under dry oxygen-free $N_2$.

## 4.18  Cd–H–B–C–N–I–S

**[Cd(C$_{12}$H$_{16}$BN$_6$S$_3$)I] or C$_{12}$H$_{16}$BN$_6$IS$_3$Cd.** This multinary compound is formed in this system. To prepare it, [Tl(C$_{12}$H$_{16}$BN$_6$S$_3$)] [0.4 mM) [C$_{12}$H$_{16}$BN$_6$S$_3$ is hydrotris(methimazolyl)borate] was suspended in acetone (30 mL), and a solution of $CdI_2$ (0.4 mM) dissolved in the minimum amount of acetone was slowly added (Cassidy et al. 2002). The mixture was stirred overnight at room temperature and the solid formed (TlI) was filtered off. The filtrate was taken to dryness and the resulting solid dried in vacuo. The crude product could be further purified by dissolving in the minimum amount of $CH_2Cl_2$ and cooling to −18°C, whereupon colorless crystals were obtained.

## 4.19  Cd–H–B–C–F–S

**[(C$_{23}$H$_{23}$N$_3$)Cd(SCN)$_2$][BF$_4$] or C$_{48}$H$_{46}$B$_2$F$_8$S$_2$Cd$_2$.** This multinary compound is formed in this system. It melts at 260°C–262°C with decomposition and crystallizes in the monoclinic structure with the lattice parameters $a = 1488.4 \pm 0.3$, $b = 1146.5 \pm 0.2$, and $c = 1591.1 \pm 0.3$ pm and $\beta = 107.843° \pm 0.004°$ and a calculated density of 1.539 g·cm$^{-3}$ (Reger et al. 2002). It was prepared by the following procedure. [Cd$_2$(THF)$_5$] (BF$_4$)$_4$ (0.25 mM) and 2,6-pyridyl imine (C$_{23}$H$_{23}$N$_3$) (0.50 mM) were charged into a round-bottom Schlenk flask, and THF (20 mL) was added via syringe. The mixture was allowed to stir for 1 h at room temperature and the solvent was removed to leave a yellow solid. This solid (0.15 g) and KSCN (0.50 mM) were dissolved in THF (20 mL) and allowed to stir for 3 h at room temperature. This mixture was filtered and the volatiles were removed in vacuo to leave a yellow solid. Crystals suitable for XRD were grown from a layered solution of $CH_2Cl_2$ and hexane.

## 4.20   Cd–H–In–C–N–O–S

Some compounds are formed in this system.

$[Cd_4In_{16}S_{33}](H_2O)_{20}[(C_{10}H_{28}N_4)_{2.5}]$ or $C_{25}H_{110}In_{16}N_{10}O_{20}S_{33}Cd_4$. This compound crystallizes in the tetragonal structure with the lattice parameters $a = 2358.0 \pm 0.8$ and $c = 4392 \pm 1$ pm and a calculated density of 1.825 $g \cdot cm^{-3}$ (Li et al. 2001). Yellow octahedral crystals of this compound were prepared by reacting $H_2O$ (1.0 mL) and ethylene glycol (0.5 mL) solution mixture of In (0.57 mM), S (1.40 mM), $Cd(OAc)_2 \cdot 2H_2O$ (0.15 mM), and 1,4-bis(3-aminopropyl)piperazine (2.43 mM) at 145°C in a Teflon-lined vessel for 4 days. At 600°C this compound is converted to $CdIn_2S_4$ and $In_2S_3$.

$[Cd_6In_{28}S_{54}] \cdot (Me_4N)_{12}[(HSCH_2COOH)_2]_{3.5}$ or $C_{62}H_{172}In_{28}N_{12}O_{14}S_{61}Cd_6$. This compound crystallizes in the tetragonal structure with the lattice parameters $a = 1962.3 \pm 0.3$ and $c = 2481.8 \pm 0.5$ pm, a calculated density of 2.484 $g \cdot cm^{-3}$, and the energy gap 3.0 eV (Su et al. 2002). Its single crystals were grown under solvothermal conditions. $InCl_3$ (0.5 mM) and $CdCl_2$ (0.5 mM) were added to the solution of $HSCH_2COOH$ (4.5 mM) and $Et_3N$ (4.5 mM) in 10 mL of MeCN. The mixture was stirred for 3 h to give white slurry. This slurry was transferred into a 23 mL Teflon-lined stainless steel vessel, followed by the addition of $Me_4NCl$ (1 mM) and elemental S (2 mM). The vessel was sealed and heated at 145°C for 5 days. After cooling to room temperature, the product was isolated by washing with EtOH and ether. Light-yellow octahedral-shaped crystals were collected.

$[\{Cd_4In_{16}S_{33}\}(C_{13}H_{26}N_2)_5] \cdot xH_2O$ ($x \approx 13$) or $C_{65}H_{130}In_{16}N_{10}S_{33}Cd_4 \cdot xH_2O$. This compound crystallizes in the tetragonal structure with the lattice parameters $a = 4210.5 \pm 0.3$ and $c = 1670.0 \pm 0.2$ pm (Wang et al. 2001). To prepare its crystals, In (85 mg), S (68 mg), $CdCl_2 \cdot 2.5H_2O$ (46 mg), and 4,4′-trimethylenedipiperidine ($C_{13}H_{26}N_2$) (608 mg) were mixed in a 23 mL Teflon-lined stainless steel autoclave. After the addition of distilled water (4.231 g), the mixture was stirred for 10 min. The vessel was then sealed and heated at 190°C for 10 days. The autoclave was subsequently allowed to cool to room temperature. Light-yellow prismatic crystals were obtained.

$[(Cd_{16}In_{64}S_{134})(C_7H_{14}N_2)_{11}(C_8H_{20}N_2O)_{11}(H_2O)_{50}]$ or $C_{165}H_{474}In_{64}N_{44}O_{61}S_{134}Cd_{16}$. This compound crystallizes in the tetragonal structure with the lattice parameters $a = 4329.7 \pm 0.2$ and $c = 2770.67 \pm 0.14$ pm at $158 \pm 2$ K and a calculated density of 2.237 $g \cdot cm^{-3}$ (Li et al. 2003). It was synthesized by combining an aqueous mixture (2.2 mL) of In (0.44 mM), S (2.18 mM), $Cd(OAc)_2 \cdot 2H_2O$ (0.75 mM), 1,5-diazabicyclo[4,3,0]-non-5-ene (3.24 mM), and 4-[2-(dimethylamine)ethyl]morpholine (3.24 mM), which was heated at 138°C in a 23 mL Teflon-lined vessels for 5 days.

## 4.21 Cd–H–In–C–N–S

$[(Cd_{12}In_{48}S_{17})(Bu_2N)_{26}]$ or $C_{208}H_{468}In_{48}N_{26}S_{97}Cd_{12}$. This multinary compound is formed in this system. It crystallizes in the cubic structure with the lattice parameter $a = 3519.4 \pm 0.8$ (Bu et al. 2003). To obtain it, indium, sulfur, $Cd(NO_3)_2 \cdot 4H_2O$, di-$n$-butylamine, and ethylene glycol were mixed in a 23 mL Teflon-lined stainless steel autoclave and stirred for about 20 min. The vessel was then sealed and heated at 150°C for 6 days. The autoclave was subsequently allowed to cool to room temperature. Clear tetrahedral crystals were obtained.

## 4.22 Cd–H–C–Si–N–S

Some compounds are formed in this system.

$[Cd(2\text{-}SC_5H_3N\text{-}3\text{-}SiMe_3)_2]$ or $C_{16}H_{24}Si_2N_2S_2Cd$. This compound crystallizes in the monoclinic structure with the lattice parameters $a = 1181.8 \pm 0.1$, $b = 1364.9 \pm 0.1$, and $c = 1321.8 \pm 0.1$ pm and $\beta = 92.00° \pm 0.01°$ and a calculated density of 1.487 g·cm⁻³ (Castro et al. 1993a). It could be obtained as colorless solids by the electrochemical oxidation of Cd in an acetonitrile solution of 3-trimethylsilylpyridine-2-thione $(C_8H_{13}SiNS)$.

$(Et_4N)_2[Cd(SC_6H_4\text{-}o\text{-}SiMe_3)_4]$ or $C_{52}H_{92}Si_4N_2S_4Cd$. This compound melts at 183°C and crystallizes in the hexagonal structure with the lattice parameters $a = 1750.1 \pm 0.2$ and $c = 3910.0 \pm 0.5$ pm and the experimental and calculated densities 1.07 and 1.03 g·cm⁻³, respectively (Block et al. 1989). To obtain it, to Na (2.1 mM) dissolved in EtOH (5 mL) 2-(trimethylsilyl)benzenethiol (2.1 mM) in MeCN (5 mL) was added. After ca. 5 min of stirring, $CdCl_2$ (0.7 mM) was added with efficient stirring. The mixture was stirred for 1 h, and $Et_4NCl$ (1.4 mM) was added. After the mixture was stirred for 48 h, NaCl was removed by filtration. Ether (0.5 mL) was added to the filtrate, and the solution was maintained at room temperature for several days. The crystals were collected by filtration, washed with ether twice, and dried in air. The product was obtained as colorless, hexagonal-shaped crystals.

## 4.23 Cd–H–C–Si–N–I–S

$[CdI_2(2\text{-}SC_5H_3N\text{-}6\text{-}SiMe_2Bu^t)_2]$ or $C_{22}H_{38}Si_2N_2I_2S_2Cd_2$ multinary compound is formed in this system. It crystallizes in the monoclinic structure with the lattice parameters $a = 2580.4 \pm 0.4$, $b = 1022.8 \pm 0.2$, and $c = 1258.5 \pm 0.3$ pm and $\beta = 104.01° \pm 0.01°$ and a calculated density of 1.682 g·cm⁻³ (Block et al. 1991). Reaction of $CdI_2$ with $2\text{-}HSC_5H_3N\text{-}6\text{-}SiMe_2Bu^t$ in EtOH yields crystals of this compound.

## 4.24 Cd–H–C–Ge–N–S

Some compounds are formed in this system.

$[(Me_4N)_2CdGe_4S_{10}]$ or $C_8H_{24}Ge_4N_2S_{10}Cd$. This compound crystallizes in the tetragonal structure with the lattice parameters $a = 953.87 \pm 0.03$ and $c = 1429.63 \pm 0.08$ pm (Achak et al. 1996). The reaction of $H_2O$, freshly prepared amorphous $GeS_2$, $Me_4NCl$,

$NH_4HCO_3$, and $CdSO_4\cdot8/3H_2O$ in the molar ratio 10:10:5:2:150 for 2 days at 100°C gave white $(Me_4N)_2CdGe_4S_{10}$. This compound that could also be obtained at room temperature was recovered as powder by filtration and washing with $H_2O$ and then dried with ether.

$[(C_{12}H_{25}NMe_3)_2(CdGe_4S_{10})]$ or $C_{30}H_{68}Ge_4N_2S_{10}Cd$. This compound was prepared at room temperature by dissolution of $C_{12}H_{25}NMe_3Br$ (1 mM) in $EtOH/H_2O$ (200 mL; 1:1 volume ratio) (Wachhold et al. 2000a,b). $Na_4Ge_4S_{10}$ (0.5 mM) was dissolved in distilled $H_2O$ (200 mL). $CdCl_2$ was dissolved in distilled $H_2O$ (20 mL) and the $Na_4Ge_4S_{10}$ and $CdCl_2$ solutions were added simultaneously to the solution of $C_{12}H_{25}NMe_3Br$ under constant stirring to form a copious precipitate. The mixture was stirred for 3 h at room temperature and then filtered, washed with water, and vacuum dried overnight. $(C_{12}H_{25}NMe_3)_2[CdGe_4S_{10}]$ is a wide band gap semiconductor.

$[(C_{14}H_{29}NMe_3)_2(CdGe_4S_{10})]$ or $C_{34}H_{76}Ge_4N_2S_{10}Cd$. This compound was prepared by the same procedure as $(C_{12}H_{25}NMe_3)_2[CdGe_4S_{10}]$, and was obtained using $C_{14}H_{29}NMe_3Br$ instead of $C_{12}H_{25}NMe_3Br$ (Wachhold et al. 2000a,b). It is a wide band gap semiconductor with an energy gap of 3.28 eV.

$[(C_{16}H_{33}NMe_3)_2(CdGe_4S_{10})]$ or $C_{38}H_{84}Ge_4N_2S_{10}Cd$. This compound was prepared by the same procedure as $(C_{12}H_{25}NMe_3)_2[CdGe_4S_{10}]$, and was obtained using $C_{16}H_{33}NMe_3Br$ instead of $C_{12}H_{25}NMe_3Br$ (Wachhold et al. 2000a,b). It is a wide band gap semiconductor.

$[(C_{18}H_{37}NMe_3)_2(CdGe_4S_{10})]$ or $C_{42}H_{92}Ge_4N_2S_{10}Cd$. This compound was prepared by the same procedure as $(C_{12}H_{25}NMe_3)_2[CdGe_4S_{10}]$, and was obtained using $C_{18}H_{37}NMe_3Br$ instead of $C_{12}H_{25}NMe_3Br$ (Wachhold et al. 2000a,b). It is a wide band gap semiconductor.

## 4.25    Cd–H–C–Sn–N–S

$(CP)_{1.2}Cd_{1.4}Sn_2S_6$ or $C_{25.2}H_{45.6}Sn_2N_{1.2}S_6Cd_{1.4}$ (CP = cetylpyridinium, $C_{21}H_{38}N^-$). This multinary compound is formed in this system. It was obtained if a solution of 10 mM of $CPBr\cdot H_2O$ in 10 mL of formamide was heated at 70°C for a few minutes forming a clear solution (Rangan et al. 2002). Then $NH_3$ was bubbled through the solution to increase the pH ~ 6. To this solution 1 mM of $Na_2Sn_2S_6\cdot14H_2O$ in 10 mL of formamide at 70°C was added to form a clear yellow solution. To the obtained solution a formamide solution of 1 mM $CdCl_2$ was added dropwise. A precipitate formed immediately and the mixture was stirred at 70°C for 24 h. The product was isolated by filtration and washed copiously with warm formamide and $H_2O$. The solid was dried under vacuum. $(CP)_{1.2}Cd_{1.4}Sn_2S_6$ is a wide-gap semiconductor ($E_g = 2.5$ eV) and is markedly stable in air and show no appearance weight loss up to 200°C.

## 4.26    Cd–H–C–N–P–O–S

Some compounds are formed in this system.

$[Cd\{S_2P(OPr^i)_2\}_2(4,4'-bipy)]_n$ or $C_{22}H_{36}N_2P_2O_4S_4Cd$. This compound crystallizes in the monoclinic structure with the lattice parameters $a = 1986.48 \pm 0.10$,

$b = 1249.97 \pm 0.06$, and $c = 1570.57 \pm 0.08$ pm and $\beta = 123.010° \pm 0.001°$ at $223 \pm 2$ K and a calculated density of 1.412 g·cm⁻³ (Lai and Tiekink 2004). It was obtained from refluxing (2 h) Cd{S₂P(OPr$^i$)₂}₂ (0.2 g) with a stoichiometric amount of 4,4′-bipy in chloroform solution. After reaction, the solvent was removed in vacuo and the residue recrystallized. Crystals were obtained by the slow evaporation of a chloroform/acetonitrile solution (3:1 volume ratio).

**[Cd{S₂P(OPr$^i$)₂}₂(C₁₀H₈N₂S₂)]$_n$ or C₂₂H₃₆N₂P₂O₄S₆Cd.** This compound crystallizes in the triclinic structure with the lattice parameters $a = 988.40 \pm 0.07$, $b = 1202.24 \pm 0.08$, and $c = 1554.25 \pm 0.11$ pm and $\alpha = 73.260° \pm 0.002°$, $\beta = 73.373° \pm 0.001°$, and $\gamma = 80.077° \pm 0.002°$ at $223 \pm 2$ K and a calculated density of 1.495 g·cm⁻³ (Lai and Tiekink 2004). It was obtained from refluxing (2 h) Cd{S₂P(OC₆H₁₁)₂}₂ (0.2 g) with a stoichiometric amount of 4,4′-bipyridylsulfide in chloroform solution. After reaction, the solvent was removed in vacuo and the residue recrystallized. Crystals were obtained by the slow evaporation of a chloroform/acetonitrile solution (3:1 volume ratio).

**[Me₄N][Cd{S₂P(OPr$^i$)₂}₃] or C₂₂H₅₄NP₃O₆S₆Cd.** This compound crystallizes in the rhombohedral structure with the lattice parameters $a = 1541.4 \pm 0.5$ and $\alpha = 50.68° \pm 0.02°$ and the experimental and calculated densities 1.31 and 1.358 g·cm⁻³, respectively (McCleverty et al. 1982). To obtain it, to a solution of [Me₄N][S₂P(OPr$^i$)₂] (10 mM) in Pr$^i$OH a solution of [Cd{S₂P(OPr$^i$)₂}₂]₂ in the same solvent was slowly added. The homogeneous solution was then slowly evaporated in vacuo ultimately affording the title compound as white crystals. This was collected by filtration, washed with Pr$^i$OH, and dried in vacuo. The title compound was obtained from acetone/petroleum ether (1:1 volume ratio) as elongated colorless bricks.

**[Cd{S₂P(OPr$^i$)₂}₂(C₁₂H₁₀N₂)]$_n$ or C₂₄H₃₈N₂P₂O₄S₄Cd.** This compound crystallizes in the orthorhombic structure with the lattice parameters $a = 1418.12 \pm 0.16$, $b = 962.60 \pm 0.11$, and $c = 2307.3 \pm 0.3$ pm at $223 \pm 2$ K and a calculated density of 1.521 g·cm⁻³ (Lai and Tiekink 2004). It was obtained from refluxing (2 h) Cd{S₂P(OPr$^i$)₂}₂ (0.2 g) with a stoichiometric amount of *trans*-1,2-bis(4-pyridyl)ethylene in chloroform solution. After reaction, the solvent was removed in vacuo and the residue recrystallized. Crystals were obtained by the slow evaporation of chloroform.

**[Cd{S₂P(OPr$^i$)₂}₂(C₁₂H₁₀N₂)]$_n$ or C₂₄H₃₈N₂P₂O₄S₄Cd.** This compound crystallizes in the monoclinic structure with the lattice parameters $a = 842.55 \pm 0.06$, $b = 2381.38 \pm 0.17$, and $c = 822.61 \pm 0.06$ pm and $\beta = 105.474° \pm 0.002°$ at $223 \pm 2$ K and a calculated density of 1.506 g·cm⁻³ (Lai and Tiekink 2004). It was obtained from refluxing (2 h) Cd{S₂P(OPr$^i$)₂}₂ (0.2 g) with a stoichiometric amount of *trans*-1,2-bis(2-pyridyl)ethylene in chloroform solution. After the reaction, the solvent was removed in vacuo and the residue recrystallized. Crystals were obtained by the slow evaporation of a CH₂Cl₂/hexane solution (2:1 volume ratio).

**[Cd{S₂P(OPr$^i$)₂}₂{3-NC₅H₄C(H)=NN=C(H)C₅H₄N-3}]$_n$ or C₂₄H₃₈N₄P₂O₄S₄Cd.** This compound melts at 204°C–205°C and crystallizes in the monoclinic structure with the lattice parameters $a = 1221.3 \pm 0.4$, $b = 1109.9 \pm 0.4$, and $c = 1307.6 \pm 0.5$ pm and $\beta = 108.585° \pm 0.008°$ at 223 K and a calculated density of

1.481 g·cm$^{-3}$ (Lai and Tiekink 2006). It was obtained from refluxing (2 h) Cd{S$_2$P(OPr$^i$)$_2$}$_2$ (0.2 g) with a 1:1 stoichiometric amount of the 3-pyridinealdazine in CHCl$_3$ solution. After reaction, the solvent was removed in vacuo and the residue recrystallized by slow evaporation from chloroform/acetonitrile (3:1 volume ratio) solution of the title compound.

[Cd{S$_2$P(OPr$^i$)$_2$}$_2${4-NC$_5$H$_4$C(H) = NN = C(H)C$_5$H$_4$N-4}]$_n$ or C$_{24}$H$_{38}$N$_4$P$_2$O$_4$S$_4$Cd. This compound melts at 181°C–182°C and crystallizes in the monoclinic structure with the lattice parameters $a = 866.43 \pm 0.04$, $b = 1327.22 \pm 0.06$, and $c = 1510.96 \pm 0.07$ pm and $\beta = 98.948° \pm 0.001°$ at 223 K and a calculated density of 1.450 g·cm$^{-3}$ (Lai and Tiekink 2006). It was obtained from refluxing (2 h) Cd{S$_2$P(OPri)$_2$}$_2$ (0.2 g) with a 1:1 stoichiometric amount of the 4-pyridinealdazine in CHCl$_3$ solution. After reaction, the solvent was removed in vacuo and the residue recrystallized by slow evaporation from chloroform/acetonitrile (3:1 volume ratio) solution of the title compound.

[Cd{S$_2$P(OPr$^i$)$_2$}$_2$(C$_{12}$H$_{12}$N$_2$)]$_n$ or C$_{24}$H$_{40}$N$_2$P$_2$O$_4$S$_4$Cd. This compound crystallizes in the orthorhombic structure with the lattice parameters $a = 1414.73 \pm 0.07$, $b = 982.74 \pm 0.05$, and $c = 2307.99 \pm 0.11$ pm at $223 \pm 2$ K and a calculated density of 1.497 g·cm$^{-3}$ (Lai and Tiekink 2004). It was obtained from refluxing (2 h) Cd{S$_2$P(OPr$^i$)$_2$}$_2$ (0.2 g) with a stoichiometric amount of 1,2-bis(4-pyridyl)ethane in chloroform solution. After the reaction, the solvent was removed in vacuo and the residue recrystallized. Crystals were obtained by the slow evaporation of a chloroform/acetonitrile solution (3:1 volume ratio).

[Cd{S$_2$P(OPr$^i$)$_2$}$_2$(C$_{13}$H$_{14}$N$_2$)]$_n$ or C$_{25}$H$_{42}$N$_2$P$_2$O$_4$S$_4$Cd. This compound crystallizes in the tetragonal structure with the lattice parameters $a = 1203.82 \pm 0.14$ and $c = 2434.5 \pm 0.6$ pm at $223 \pm 2$ K and a calculated density of 1.388 g·cm$^{-3}$ (Lai and Tiekink 2004). It was obtained from refluxing (2 h) Cd{S$_2$P(OPr$^i$)$_2$}$_2$ (0.2 g) with a stoichiometric amount of 1,2-bis(4-pyridyl)propane in chloroform solution. After reaction, the solvent was removed in vacuo and the residue recrystallized. Crystals were obtained by the slow evaporation of a chloroform/acetonitrile solution (3:1 volume ratio).

[Cd$_2${S$_2$P(OPr$^i$)$_2$}$_4$(C$_6$H$_{12}$N$_2$)] or C$_{30}$H$_{68}$N$_2$P$_4$O$_8$S$_8$Cd$_2$. This compound melts at 189°C–192°C and crystallizes in the monoclinic structure with the lattice parameters $a = 810.80 \pm 0.15$, $b = 1907.4 \pm 0.3$, and $c = 1721.7 \pm 0.03$ pm and $\beta = 94.141° \pm 0.005°$ at $103 \pm 2$ K and a calculated density of 1.488 g·cm$^{-3}$ (Ellis and Tiekink 2006). It was prepared by refluxing Cd[S$_2$P(OPr$^i$)$_2$]$_2$ with 1,4-diazabicyclo[2.2.2]octane. Colorless crystals were isolated by the slow evaporation of a toluene/CHCl$_3$ (1:1 volume ratio) solution of the title compound.

[Cd{S$_2$P(OC$_6$H$_{11}$)$_2$}$_2$(4,4'-bipy)]$_n$ or C$_{34}$H$_{52}$N$_2$P$_2$O$_4$S$_4$Cd. This compound crystallizes in the monoclinic structure with the lattice parameters $a = 3063.1 \pm 0.2$, $b = 1195.47 \pm 0.08$, and $c = 1115.69 \pm 0.08$ pm and $\beta = 104.983° \pm 0.003°$ at $223 \pm 2$ K and a calculated density of 1.440 g·cm$^{-3}$ (Lai and Tiekink 2004). It was obtained from refluxing (2 h) Cd{S$_2$P(OC$_6$H$_{11}$)$_2$}$_2$ (0.2 g) with a stoichiometric amount of 4,4'-bipy in chloroform solution. After reaction, the solvent was removed in vacuo and the residue recrystallized. Crystals were obtained by the slow evaporation of a chloroform/acetonitrile solution (3:1 volume ratio).

$[Cd\{S_2P(OC_6H_{11})_2\}_2(C_{12}H_{10}N_2)]_n$ or $C_{36}H_{54}N_2P_2O_4S_4Cd$. This compound crystallizes in the triclinic structure with the lattice parameters $a = 1230.15 \pm 0.11$, $b = 1296.51 \pm 0.11$, and $c = 1417.70 \pm 0.12$ pm and $\alpha = 103.761° \pm 0.002°$, $\beta = 90.158° \pm 0.002°$, and $\gamma = 111.701° \pm 0.002°$ at $223 \pm 2$ K and a calculated density of 1.442 g·cm⁻³ (Lai and Tiekink 2004). It was obtained from refluxing (2 h) $Cd\{S_2P(OC_6H_{11})_2\}_2$ (0.2 g) with a stoichiometric amount of *trans*-1,2-bis(4-pyridyl) ethylene in chloroform solution. After reaction, the solvent was removed in vacuo and the residue recrystallized. Crystals were obtained by the slow evaporation of a chloroform/acetonitrile solution (3:1 volume ratio).

$[Cd\{S_2P(OC_6H_{11})_2\}_2(C_{12}H_{10}N_2)]_n$ or $C_{36}H_{54}N_2P_2O_4S_4Cd$. This compound crystallizes in the monoclinic structure with the lattice parameters $a = 2485.32 \pm 0.19$, $b = 869.66 \pm 0.07$, and $c = 1914.22 \pm 0.15$ pm and $\beta = 94.738° \pm 0.002°$ at $223 \pm 2$ K and a calculated density of 1.420 g·cm⁻³ (Lai and Tiekink 2004). It was obtained from refluxing (2 h) $Cd\{S_2P(OC_6H_{11})_2\}_2$ (0.2 g) with a stoichiometric amount of *trans*-1,2-bis(2-pyridyl)ethylene in chloroform solution. After reaction, the solvent was removed in vacuo and the residue recrystallized. Crystals were obtained by the slow evaporation of a chloroform/acetonitrile solution (3:1 volume ratio).

$[Cd\{S_2P(OPr^i)_2\}_2\{2-NC_5H_4C(H)=NN=C(H)C_5H_4N-2\}_{0.5}]_2$ or $C_{36}H_{66}N_4P_4O_8S_8Cd_2$. This compound melts at 147°C–148°C and crystallizes in the triclinic structure with the lattice parameters $a = 880.30 \pm 0.05$, $b = 1197.36 \pm 0.06$, and $c = 1576.17 \pm 0.08$ pm and $\alpha = 67.971° \pm 0.001°$, $\beta = 75.301° \pm 0.001°$, and $\gamma = 76.391° \pm 0.001°$ at 223 K and a calculated density of 1.454 g·cm⁻³ (Lai and Tiekink 2006). It was obtained from refluxing (2 h) $Cd\{S_2P(OPr^i)_2\}_2$ (0.2 g) with a 1:1 stoichiometric amount of the 2-pyridinealdazine in CHCl₃ solution. After reaction, the solvent was removed in vacuo and the residue recrystallized by slow evaporation from CH₂Cl₂/hexane (2:1 volume ratio) solution of the title compound.

$[Cd_3(C_6H_4PO_6S)_2(4,4'-bipy)_3(H_2O)_6]\cdot 4H_2O$ or $C_{42}H_{52}N_6P_2O_{22}S_2Cd_3$. This compound crystallizes in the monoclinic structure with the lattice parameters $a = 2043.31 \pm 0.01$, $b = 1179.79 \pm 0.02$, and $c = 2394.85 \pm 0.03$ pm and $\beta = 112.723° \pm 0.001°$ and a calculated density of 1.824 g·cm⁻³ (Du et al. 2007). To obtain it, a mixture of $Cd(OAc)_2$ (0.4 mM), *m*-phosphonophenylsulfonic acid $(C_6H_7PO_6S)$ (0.3 mM), and 4,4′-bipy (0.42 mM) in 10 mL of distilled water was sealed into a Parr Teflon–lined autoclave (23 mL) and heated at 150°C for 4 days. The initial and final pH values are about 3.8 and 4.3, respectively. Colorless brick-shaped crystals of the title compound were collected.

$(Ph_4P)[Cd(S-2,4,6-Pr^i_3C_6H_2)_3]\cdot(DMF)(Pr^iOH)$ or $C_{75}H_{104}NPO_2S_3Cd$. This compound crystallizes in the triclinic structure with the lattice parameters $a = 1537.9 \pm 0.6$, $b = 2125.5 \pm 0.7$, and $c = 1358.5 \pm 0.6$ pm and $\alpha = 95.04° \pm 0.03°$, $\beta = 114.85° \pm 0.03°$, and $\gamma = 104.40° \pm 0.03°$ (Gruff and Koch 1990). It was prepared by recrystallization of $[Ph_4P][Cd(S-2,4,6-Pr^i_3C_6H_2)_3]$ in a DMF/Pr$^i$OH solvent mixture.

$[Cd_6(C_6H_4PO_6S)_4(Phen)_8]\cdot 14H_2O$ or $C_{120}H_{108}N_{16}P_4O_{38}S_4Cd_6$. This compound crystallizes in the triclinic structure with the lattice parameters $a = 1715.2 \pm 0.4$, $b = 2112.0 \pm 0.5$, and $c = 2120.7 \pm 0.5$ pm and $\alpha = 84.590° \pm 0.009°$. $\beta = 76.168° \pm 0.009°$, and $\gamma = 67.475° \pm 0.006°$ and a calculated density of 1.595 g·cm$^{-3}$ (Du et al. 2007). To obtain it, a mixture of Cd(OAc)$_2$ (0.3 mM), $m$-phosphonophenylsulfonic acid (C$_6$H$_7$PO$_6$S) (0.3 mM), and Phen (0.5 mM) in 10 mL of distilled water was sealed into a Parr Teflon–lined autoclave (23 mL) and heated at 150°C for 4 days. The initial and final pH values did not show much change and was close to 4.0. Colorless prism-shaped crystals of the title compound were collected.

## 4.27   Cd–H–C–N–P–O–Cl–S

Some compounds are formed in this system.

**[Cd{S$_2$P(OC$_6$H$_{11}$)$_2$}$_2$(C$_{13}$H$_{14}$N$_2$)]·CHCl$_3$ or C$_{38}$H$_{59}$N$_2$P$_2$O$_4$Cl$_3$S$_4$Cd.** This compound crystallizes in the triclinic structure with the lattice parameters $a = 1036.81 \pm 0.07$, $b = 1133.86 \pm 0.07$, and $c = 2196.82 \pm 0.13$ pm and $\alpha = 99.897° \pm 0.001°$, $\beta = 92.508° \pm 0.002°$, and $\gamma = 113.129° \pm 0.001°$ at $223 \pm 2$ K and a calculated density of 1.454 g·cm$^{-3}$ (Lai and Tiekink 2004). It was obtained from refluxing (2 h) Cd{S$_2$P(OC$_6$H$_{11}$)$_2$}$_2$ (0.2 g) with a stoichiometric amount of 1,2-bis(4-pyridyl)propane in chloroform solution. After reaction, the solvent was removed in vacuo and the residue recrystallized. Crystals were obtained by the slow evaporation of a chloroform/acetonitrile solution (3:1 volume ratio).

**[Cd$_4$(C$_6$H$_4$PO$_6$S)$_2$(Phen)$_6$(Cl$_2$)(H$_2$O)$_2$]·14H$_2$O or C$_{84}$H$_{88}$N$_{12}$P$_2$O$_{28}$Cl$_2$S$_2$Cd$_4$.** This compound crystallizes in the triclinic structure with the lattice parameters $a = 1167.02 \pm 0.06$, $b = 1391.60 \pm 0.07$, and $c = 1602.47 \pm 0.08$ pm and $\alpha = 69.041° \pm 0.001°$. $\beta = 69.716° \pm 0.001°$, and $\gamma = 84.264° \pm 0.001°$ and a calculated density of 1.720 g·cm$^{-3}$ (Du et al. 2007). It was prepared by the following procedure. A mixture of Cd(OAc)$_2$ (0.3 mM), $m$-phosphonophenylsulfonic acid (C$_6$H$_7$PO$_6$S) (0.3 mM), and Phen (0.6 mM) in 10 mL of distilled water was sealed into a Parr Teflon–lined autoclave (23 mL) and heated at 150°C for 4 days. The initial and final pH values were about 3.5. Colorless prism-shaped crystals of the title compound were collected.

## 4.28   Cd–H–C–N–P–S

Some compounds are formed in this system.

**[Bu$^t_3$PCd(SCN)$_2$] or C$_{14}$H$_{27}$N$_2$PS$_2$Cd.** This compound melts at 205°C (Goel and Ogini 1977). To obtain it, a solution of Bu$^t_3$P (1.20 mM) in 10 mL of CH$_2$Cl$_2$ was added with stirring to a slurry of Cd(SCN)$_2$ (1.01 mM) in 20 mL of CH$_2$Cl$_2$. After 2 h of stirring, a clear solution was obtained. Upon the addition of petroleum ether to the solution, a white precipitate was formed that was filtered off and recrystallized from CH$_2$Cl$_2$. All operations were carried out either in a dry box under an atmosphere of oxygen-free N$_2$ or under vacuum.

**[Cd(SCN)$_2$(Ph$_3$P)] or C$_{20}$H$_{15}$N$_2$PS$_2$Cd.** This compound melts at 150°C (Goel et al. 1981). To obtain it, a solution of Ph$_3$P in EtOH was added dropwise, with stirring, to an alcoholic solution of Cd(SCN)$_2$. The resulting mixture was refluxed for 3 h and then allowed to cool to room temperature. Upon removal of the solvent, a white residue was obtained, which was washed with hexane. The resulting solid was recrystallized from hot EtOH to give the title compound. Attempts to crystallize [Cd(SCN)$_2$(Ph$_3$P)] did not succeed due to its poor solubility.

**[Cd(SCN)$_2$P(C$_6$H$_{11}$)$_3$] or C$_{20}$H$_{33}$N$_2$PS$_2$Cd.** This compound melts at 185°C (Goel et al. 1981). It was obtained by the next procedure. A solution of tricyclohexylphosphine [P(C$_6$H$_{11}$)$_3$] in EtOH was added dropwise, with stirring, to an alcoholic solution of Cd(SCN)$_2$. The resulting mixture was refluxed for 3 h and then allowed to cool to room temperature. Upon removal of the solvent, a white residue was obtained, which was washed with hexane. The resulting solid was recrystallized from hot EtOH to give the title compound. Attempts to crystallize [Cd(SCN)$_2$P(C$_6$H$_{11}$)$_3$] did not succeed due to its poor solubility. P(C$_6$H$_{11}$)$_3$ was handled in an atmosphere of oxygen-free dry N$_2$ with use of glovebox and standard vacuum-line techniques.

**[Cd(SCN)$_2$(Tol$^m$$_3$P)] or C$_{23}$H$_{21}$N$_2$PS$_2$Cd.** This compound melts at 150°C and crystallizes in the monoclinic structure with the lattice parameters $a = 1086.6 \pm 0.7$, $b = 1020.3 \pm 0.3$, and $c = 2207.7 \pm 0.6$ pm and $\beta = 101.14° \pm 0.04°$ and a calculated density of 1.48 g·cm$^{-3}$ (Goel et al. 1981). To obtain it, a solution of tri-$m$-tolylphosphine in EtOH was added dropwise, with stirring, to an alcoholic solution of Cd(SCN)$_2$. The resulting mixture was refluxed for 3 h and then allowed to cool to room temperature. Upon removal of the solvent, a white residue was obtained, which was washed with hexane. The resulting solid was recrystallized from hot EtOH to give the title compound. Attempts to crystallize [Cd(SCN)$_2$(Ph$_3$P)] did not succeed due to its poor solubility.

**[Cd{N(Pr$^i$$_2$PS)$_2$}$_2$] or C$_{24}$H$_{56}$N$_2$P$_4$S$_4$Cd.** This compound melts at 161°C and crystallizes in the triclinic structure with the lattice parameters $a = 928.8 \pm 0.6$, $b = 1292.9 \pm 0.8$, and $c = 1645.2 \pm 1.0$ pm and $\alpha = 78.97° \pm 0.05°$, $\beta = 77.83° \pm 0.05°$, and $\gamma = 69.46° \pm 0.05°$ (Cupertino et al. 1996). To prepare it, CdCO$_3$ (0.58 mM) was added to a solution of HN(Pr$^i$$_2$PS)$_2$ (0.16 mM) in CH$_2$Cl$_2$ (30 mL), and the mixture was refluxed for 2 h. The cloudy/white mixture was filtered, and the filtrate was evaporated to dryness to give a white solid. Colorless crystals of the title compound were obtained from the mixture CH$_2$Cl$_2$/petroleum ether solvent.

**[{Cd(Bu$^i$$_2$PS$_2$)$_2$}·(2,2′-bipy) or C$_{26}$H$_{44}$N$_2$P$_2$S$_4$Cd.** This compound melts at 180°C–183°C (Larionov et al. 2001) and crystallizes in the monoclinic structure with the lattice parameters $a = 1473.7 \pm 0.3$, $b = 2091.8 \pm 0.4$, and $c = 1151.7 \pm 0.2$ pm and $\beta = 105.53° \pm 0.03°$ and a calculated density of 1.334 g·cm$^{-3}$ (Klevtsova et al. 2002). To obtain it, Cd(Bu$^i$$_2$PS$_2$)$_2$ (1.9 mM) and 2,2′-bipy (2 mM) were dissolved in a minimum amount of toluene (ca. 10 mL) when heating on a water bath (Larionov et al. 2001). The obtained solution was cooled to −10°C. The white precipitate was filtered with suction, washed with cooled toluene, and dried in air. Single crystals of the title compound were obtained by the slow crystallization from the mixtures EtOH/toluene (1:2 volume ratio) (Larionov et al. 2001) or EtOH/CHCl$_3$ (4:1 volume ratio) (Klevtsova et al. 2002).

**[{Cd(Bu$^i$$_2$PS$_2$)$_2$}·(4,4′-bipy)] or C$_{26}$H$_{44}$N$_2$P$_2$S$_4$Cd.** This compound melts at 194°C–199°C and crystallizes in the monoclinic structure with the lattice parameters $a = 1467.8 \pm 0.3$, $b = 1186.6 \pm 0.2$, and $c = 1965.5 \pm 0.4$ pm and $\beta = 102.17° \pm 0.03°$ and a calculated density of 1.364 g·cm$^{-3}$ (Larionov et al. 2001). To obtain it, $Cd(Bu^i_2PS_2)_2$ (1.9 mM) and 4,4'-bipy (2 mM) were dissolved separately in the minimum amounts of $CH_2Cl_2$ when heating on a water bath. Then, the solution of 4,4'-bipy was added to the solution of $Cd(Bu^i_2PS_2)_2$ with stirring. The mixture was cooled to $-10°C$. The precipitate was filtered with suction, washed with cooled $CH_2Cl_2$, and dried in air. Single crystals of the title compound were obtained by the slow crystallization from the mixture toluene/CHCl$_3$ (1:1 volume ratio).

**[{Cd(Bu$^i_2$PS$_2$)$_2$}·(Phen)] or C$_{28}$H$_{44}$N$_2$P$_2$S$_4$Cd.** This compound melts at 224°C–226°C (Larionov et al. 2001). To prepare it, $Cd(Bu^i_2PS_2)_2$ (1.9 mM) and Phen·H$_2$O (2.4 mM) were dissolved in a minimum amount of $CH_2Cl_2$ (ca./10 mL) when heating on a water bath. Then, 30 mL of EtOH was added to the obtained yellow solution. The mixture was cooled to $-10°C$. White precipitate was filtered with suction, washed with cooled MeOH, and dried in air. Single crystals of the title compound were obtained by the slow crystallization from the mixture EtOH/CHCl$_3$ (4:1 volume ratio).

**[(Bu$^n_4$N)(Ph$_4$P){Cd(S$_6$)$_2$}] or C$_{40}$H$_{56}$NPS$_{12}$Cd.** This compound melts at 128°C (Banda et al. 1989). To prepare it, a mixture of (Bu$^n_4$N)S$_6$ (6.6 mM) and CdCl$_2$·2.5H$_2$O (3 mM) was dissolved in deoxygenated acetonitrile (30 mL) with ultrasonication for 15 min. After an initial blue coloration, a red-brown solution was obtained. A solution of Ph$_4$PBr (6 mM) in acetonitrile (25 mL) was added and the mixture allowed to stand, while the yellow polycrystalline product separated slowly. The addition of EtOH to the cooled mother liquor yielded more products. The attempts to grow single crystals of the title compound were unsuccessful. All preparations were performed under nitrogen as inert gas.

**[CdPhen(Bu$^i_2$PS$_2$)$_2$] or C$_{96}$H$_{44}$N$_2$P$_2$S$_4$Cd.** This compound crystallizes in the monoclinic structure with the lattice parameters $a = 1564.0 \pm 0.3$, $b = 2079.7 \pm 0.4$, and $c = 1155.9 \pm 0.2$ pm and $\beta = 111.21° \pm 0.03°$ and a calculated density of 1.348 g·cm$^{-3}$ (Klevtsova et al. 2002). Its single crystals were obtained by the slow crystallization from the mixture EtOH/toluene (1:2 volume ratio).

**[Ph$_4$P]$_2$[Cd(C$_{12}$H$_{12}$NS)$_4$] or C$_{96}$H$_{88}$N$_4$P$_2$S$_4$Cd.** This compound decomposes at 207°C and crystallizes in the triclinic structure with the lattice parameters $a = 1308.7 \pm 0.1$, $b = 1347.5 \pm 0.1$, and $c = 2682.1 \pm 0.3$ pm and $\alpha = 95.92° \pm 0.01°$, $\beta = 93.14° \pm 0.01°$, and $\gamma = 116.13° \pm 0.01°$ and a calculated density of 1.266 g·cm$^{-3}$ (Otto et al. 1999). To obtain it, 4.7 mM of $N$-(2-thiophenyl)-2,5-dimethylpyrrole was dissolved in 5 mL of DMF and 45 mL of Pr$^i$OH. Et$_3$N (1 mL), Ph$_4$PCl (1.2 mM), and CdCl$_2$ (1 mM) were added to the solution. After stirring overnight, the white precipitate was collected and dried in vacuo. The title compound was recrystallized from a DMF/Pr$^i$OH solution (1:10 volume ratio) to yield pale-yellow crystals after cooling the solution to 0°C for 24 h. All procedures were carried out under an Ar or N$_2$ atmosphere using standard Schlenk and glovebox techniques.

## 4.29   Cd–H–C–N–P–Se–S

**[Cd{(c-C$_6$H$_{11}$)$_2$P(Se)C(S)NPh}$_2$] or C$_{38}$H$_{54}$N$_2$P$_2$Se$_2$S$_2$Cd.** This multinary compound is formed in this system. It decomposes at 163°C–165°C and crystallizes in the triclinic structure with the lattice parameters $a = 1984.8 \pm 0.4$, $b = 2048.5 \pm 0.3$, and $c = 1096.9 \pm 0.2$ 2 pm and $\alpha = 105.49° \pm 0.01°$, $\beta = 102.37° \pm 0.01°$, and $\gamma = 101.06° \pm 0.01°$ and a calculated density of 1.534 g·cm$^{-3}$ (Kramolowsky et al. 1998). To obtain it, two molar equivalents of the bright-yellow solid $[(c\text{-}C_6H_{11})_2P(Se)C(S)NPh]$ (1.15 mM) were added to a $CH_2Cl_2$/EtOH solution (1:1 volume ratio, 40 mL) of $CdCl_2$ (0.55 mM). After 5 min of stirring, $Et_3N$ was added (2 mL), which resulted in the folding of the solution to a pale-yellow color. After a further 30 min of stirring, the solution was filtered, the solvent removed in vacuo, and the resulting precipitate recrystallized from chloroform solution (10 mL). White crystals of the title compound were obtained.

## 4.30   Cd–H–C–N–P–F–S

Some compounds are formed in this system.

$[Cd\{S(CH_2)_3NMe_3\}_2](PF_6)_2$ or $C_{12}H_{30}N_2P_2F_{12}S_2Cd$. This compound was prepared by the following procedure (Casals et al. 1991b). 3-Trimethylammonio-1-propanethiol hexafluorophosphate (57.6 mg, corresponding to 0.20 mM of pure thiol) was dissolved in a mixture of acetonitrile (1 mL), MeOH (1 mL), and $H_2O$ (1 mL). Aqueous NaOH (0.32 mM in 2 mL) was added, followed by $Cd(OAc)_2 \cdot 2H_2O$ (0.08 mM) dissolved in 2.0 mL of $H_2O$. The slightly cloudy solution was filtered and kept under nitrogen at room temperature. After several days the solid product was isolated as a colorless microcrystalline powder.

$[Cd(C_7H_{12}N_2S)_4(PF_6)_2]$ or $C_{28}H_{48}N_8P_2F_{12}S_4Cd$. This compound crystallizes in the tetragonal structure with the lattice parameters $a = 1247.9 \pm 0.2$ and $c = 2880.6 \pm 0.6$ and a calculated density of 1.521 g·cm$^{-3}$ (Williams et al. 2009). To obtain it, a combination of 7.98 mM of tetrakis[1-methyl-3-(2-propyl)-2(3$H$)-imidazolethion] and of $Cd(OAc)_2 \cdot 2H_2O$ (1.83 mM) was brought up in approximately 100 mL of acetonitrile and mixed well. A stoichiometric quantity of $NH_4PF_6$ was added, and the title compound crystallized out of solution by slow evaporation of the solvent over a period of 1 week to give a batch of analytically pure, crystallographic grade crystals.

## 4.31   Cd–H–C–N–P–F–Cl–Pt–S

$[CdPt_2Cl(2,2'\text{-bipy})(Ph_3P)_4(\mu_3\text{-}S)_2][PF_6]$ or $C_{82}H_{68}N_2P_5F_6ClPt_2S_2Cd$. This multinary compound is formed in this system. It crystallizes in the triclinic structure with the lattice parameters $a = 1452.56 \pm 0.05$, $b = 2011.88 \pm 0.07$, and $c = 2930.02 \pm 0.11$ pm and $\alpha = 83.110° \pm 0.001°$, $\beta = 88.390° \pm 0.001°$, and $\gamma = 82.842° \pm 0.001°$ (Li et al. 2000). To obtain it, a suspension of $[Pt_2(Ph_3P)_4(\mu\text{-}S)_2]$ (0.1 mM) and $CdCl_2$ (0.1 mM) was stirred in MeOH (30 mL) for 6 h, and then 2,2'-bipy (0.1 mM) was added. The suspension changed to a clear bright-yellow solution within a few minutes. The solution was filtered and purified by metathesis with $NH_4PF_6$ to yield the title compound. The product was recrystallized from a $CH_2Cl_2$/MeOH (1:3 volume ratio) mixture to give yellow crystals. Single crystals were grown by the slow evaporation of the solution at room temperature in air.

## 4.32    Cd–H–C–N–P–Br–S

$(Ph_4P)_2[Br_2Cd(\mu-C_3H_4NS_2)_2CdBr_2]$ or $C_{54}H_{48}N_2P_2Br_4S_4Cd_2$. This multinary compound is formed in this system. It crystallizes in the monoclinic structure with the lattice parameters $a = 965.0 \pm 0.2$, $b = 1615.6 \pm 0.3$, and $c = 1836.1 \pm 0.2$ pm and $\beta = 102.78° \pm 0.01°$ (Craig et al. 2005). It was synthesized by the following procedure. A solution of $Cd(NO_3)_2·4H_2O$ (0.25 mM) in 3 mL of MeCN was added with stirring to a mixture of 1,3-thiazolidine-2-thione (0.49 mol) and $Et_3N$ (0.49 m) in 3 mL of MeCN. To the white precipitate was added solid $Ph_4PBr$ (0.491 mM), causing dissolution of the precipitate. Thereafter, diethyl ether was added, and colorless crystals were obtained, which were washed with diethyl ether and dried in vacuo.

## 4.33    Cd–H–C–N–P–I–S

$(Ph_4P)_2[I_2Cd(\mu-C_3H_4NS_2)_2CdI_2]$ or $C_{54}H_{48}N_2P_2I_4S_4Cd_2$. This multinary compound is formed in this system. It crystallizes in the monoclinic structure with the lattice parameters $a = 986.5 \pm 0.1$, $b = 1624.7 \pm 0.2$, and $c = 1852.3 \pm 0.2$ pm and $\beta = 102.292° \pm 0.009°$ and a calculated density of 1.89 g·cm$^{-3}$ (Craig et al. 2005). It was prepared using the next procedure. A solution of 1,3-thiazolidine-2-thione (0.55 mM) and $Et_3N$ (0.53 mM) in $CH_2Cl_2$ (3 mL) was added with stirring to a mixture of $Cd(NO_3)_2·4H_2O$ (0.27 mM) in 10 mL of MeCN, causing formation of a white precipitate. To the mixture was added solid $Ph_4PI$ (0.500 mM), causing the precipitate to dissolve. Diethyl ether was added until turbidity was apparent, at which point the container and contents were placed in a freezer for several days. The colorless crystals that formed were washed with diethyl ether and dried in vacuo. Crystals of the title compound were obtained from a saturated solution in $CH_2Cl_2$.

## 4.34    Cd–H–C–N–P–Fe–S

$[Cd_2Fe(S_2CNEt_2)_4(C_{17}H_{14}P)_2]$ or $C_{54}H_{68}N_4P_2FeS_8Cd_2$. This multinary compound is formed in this system. It melts at 190°C with decomposition and crystallizes in the monoclinic structure with the lattice parameters $a = 1666.41 \pm 0.06$, $b = 1313.96 \pm 0.05$, and $c = 2786.64 \pm 0.11$ pm and $\beta = 98.482° \pm 0.001°$ at $223 \pm 2$ K and a calculated density of 1.510 g·cm$^{-3}$ (Dee and Tiekink 2002a). It was prepared by the following procedure. A $CHCl_3$ (50 mL) mixture of $Cd(S_2CNEt_2)_2$ (200 mg) and an hemimolar amount of 1,1′-bis(diphenylphosphino)ferrocene, $Fe(C_{17}H_{14}P)_2$, was refluxed for 1 h. The solvent was removed in vacuo and the residue recrystallized from a $CHCl_3$/MeCN (3:1 volume ratio) solution to yield yellow blocks of the title compound.

## 4.35    Cd–H–C–N–As–S

Some compounds are formed in this system.

$(Ph_4As)_2[Cd(MeC_6H_3S_2)_2(C_6H_{16}N_2)]$ or $C_{68}H_{68}N_2As_2S_4Cd$. This compound melts at 215°C–217°C (Bustos et al. 1983). To obtain it, 3 mL of $N,N,N'N'$-tetramethylenediamine was added to a solution of 0.3 g of $(Ph_4As)_2[Cd(MeC_6H_3S_2)_2]$

in 10 mL of acetonitrile. The volume of the solution was reduced to approximately 5 mL, yielding the white precipitate of the title compound that was collected, washed with EtOH, and dried in vacuo.

**(Ph₄As)₂[Cd(MeC₆H₃S₂)₂(2,2′-bipy)]** or **C₇₂H₆₀N₂As₂S₄Cd**. This compound was prepared as follows. 2,2′-bipy in EtOH was added to a stirred solution of (Ph₄As)₂[Cd(MeC₆H₃S₂)₂] in acetonitrile; the solution changed color from yellow to orange (Bustos et al. 1983). Diethyl ether was added dropwise until the appearance of cloudiness in the solution, the solid redissolved in acetonitrile, and the resultant solution cooled for approximately 15 h. The final yellow precipitate was filtered off, washed with ethanol, and dried in vacuo.

**(Ph₄As)₂[Cd(MeC₆H₃S₂)₂(Phen)]** or **C₇₄H₆₀N₂As₂S₄Cd**. This compound melts at 256°C–258°C (Bustos et al. 1983). It was prepared by the next procedure. Phen (0.08 g) in 10 mL of EtOH was added to a stirred solution of (Ph₄As)₂[Cd(MeC₆H₃S₂)₂ (0.6 g) in 30 mL of acetonitrile; the solution changed color from yellow to orange. Diethyl ether was added dropwise until the appearance of cloudiness in the solution, the solid redissolved in acetonitrile, and the resultant solution cooled for approximately 15 h. The final yellow precipitate was filtered off, washed with ethanol, and dried in vacuo.

## 4.36 Cd–H–C–N–O–S

Some compounds are formed in this system.

**[Cd{SC(NH₂)₂}SO₄]·2H₂O** or **CH₈N₂O₆S₂Cd, monothiourea-cadmium sulfate dihydrate**. This compound crystallizes in the orthorhombic structure with the lattice parameters $a = 1346.1 \pm 0.5$, $b = 778.3 \pm 0.3$, and $c = 1596.7 \pm 1.2$ pm and the experimental and calculated densities 2.51 and 2.545 g·cm⁻³ (Cavalca et al. 1967).

**[Cd{NH₂NHC(S)N₂}₂(NO₃)₂]** or **C₂H₆N₁₀O₆S₂Cd**. This compound crystallizes in the triclinic structure with the lattice parameters $a = 636.7 \pm 0.1$, $b = 701.9 \pm 0.1$, and $c = 770.1 \pm 0.4$ pm and $\alpha = 115.62° \pm 0.01°$, $\beta = 92.97° \pm 0.01°$, and $\gamma = 94.69° \pm 0.01°$ and a calculated density of 1.6346 g·cm⁻³ (Romanenko et al. 2001).

**[Cd{SC(NH₂)₂}₂(NO₃)₂]** or **C₂H₈N₆O₆S₂Cd, bis(thiourea) cadmium nitrate**. This compound crystallizes in the monoclinic structure with the lattice parameters $a = 426.7 \pm 0.1$, $b = 1195.6 \pm 0.1$, and $c = 1133.0 \pm 0.1$ pm and $\beta = 100.22° \pm 0.06°$ and a calculated density of 2.269 g·cm⁻³ (Petrova et al. 2000) [$a = 409$, $b = 1208$, and $c = 1145$ pm and $\beta = 102°$ and a calculated density of 2.33 g·cm⁻³ (Swaminathan and Natarajan 1967)]. The powder samples of the title compound were synthesized by slow evaporation of an aqueous solution containing Cd(NO₃)₂·4H₂O and thiourea in a 1:2 molar ratio. The appropriate single crystals were obtained after recrystallization from H₂O at room temperature (Petrova et al. 2000).

**[Cd(CH₅N₃S)₂SO₄]** or **C₂H₁₀N₆O₄S₃Cd**. This compound decomposes at 249°C (Mahadevappa and Murthy 1972) and crystallizes in the monoclinic structure with the lattice parameters $a = 2796.2 \pm 1.4$, $b = 656.4 \pm 0.3$, and $c = 1590.8 \pm 0.9$ pm and $\beta = 126.6° \pm 0.6°$ and a calculated density of 2.251 g·cm⁻³ (Larsen and Trinderup 1975). It was precipitated by mixing a 0.05 M aqueous solution of CdSO₄ and a 0.1 M aqueous solution of thiosemicarbazide in a test tube. Crystals developed within 24 h (Mahadevappa and Murthy 1972, Larsen and Trinderup 1975).

**[Cd{SC(NH₂)₂}₃SO₄] or C₃H₁₂N₆O₄S₄Cd.** This compound crystallizes in the triclinic structure with the lattice parameters $a = 877 \pm 2$, $b = 905 \pm 2$, and $c = 983 \pm 1$ pm and $\alpha = 91.3° \pm 0.2°$, $\beta = 111.9° \pm 0.1°$, and $\gamma = 95.5° \pm 0.2°$ and the experimental and calculated densities 2.02 and 2.06 g·cm⁻³, respectively (Cavalca et al. 1970) [$a = 906 \pm 2$, $b = 980 \pm 2$, and $c = 878 \pm 2$ pm and $\alpha = 110.6° \pm 0.5°$, $\beta = 95.1° \pm 0.5°$, and $\gamma = 92.0° \pm 0.5°$ and experimental and calculated densities 2.02 and 2.00 g··cm⁻³, respectively (Corao and Baggio 1969)]. Its crystals were obtained as thick triclinic prisms by slow concentration of an aqueous solution of CdSO₄ and thiourea.

**[Cd(C₂H₈N₂)SCN]·0.5C₂O₄ or C₄H₈N₃O₂SCd.** This compound was prepared by Shvelashvili et al. (1974).

**[Cd(SCN)₂·2(NH₂)₂CO] or C₄H₈N₆O₂S₂Cd.** This compound crystallizes in the monoclinic structure with the lattice parameters $a = 863 \pm 2$, $b = 1349 \pm 2$, and $c = 866 \pm 2$ pm and $\beta = 98°$ and the experimental and calculated densities 2.15 and 2.30 g·cm⁻³, respectively (Tsintsadze et al. 1975).

**[Cd{SC(NH₂)₂}₂(HCOO)₂] or C₄H₁₀N₄O₄S₂Cd, bis(thiourea) cadmium formate.** This compound crystallizes in the orthorhombic structure with the lattice parameters $a = 800.0 \pm 0.9$, $b = 1787.8 \pm 0.5$, and $c = 393.3 \pm 0.7$ pm and the experimental and calculated densities 2.091 and 2.093 g·cm⁻³, respectively (Nardelli et al. 1962, 1965).

**[Cd(CN)₂(CH₄N₂S)₂]·H₂O or C₄H₁₀N₆OS₂Cd.** This compound crystallizes in the monoclinic structure with the lattice parameters $a = 1059.55 \pm 0.06$, $b = 407.82 \pm 0.03$, and $c = 1341.27 \pm 0.08$ pm and $\beta = 98.738° \pm 0.001°$ and a calculated density of 1.940 g·cm⁻³ (Fettouhi et al. 2010). To obtain it, to 1.0 mM of Cd(CN)₂ suspended in 15 mL of H₂O 2 equivalents of thiourea in MeOH was added. Yellow precipitates formed were filtered and the filtrate was kept for crystallization. Crystals were grown by slow evaporation of H₂O/MeOH solution at room temperature.

**[Cd(HCONH₂)₄SO₄] or C₄H₁₂N₄O₈SCd.** This compound was obtained by a crystallization of aqueous solutions of CdSO₄ and an excess of formamide at elevated temperatures (Nardelli and Coghi 1959).

**[Cd{SC(NH₂)₂}₄(NO₃)₂] or C₄H₁₆N₁₀O₆S₄Cd.** This compound exists in two polymorphic modifications, and they are both monoclinic (Petrova et al. 2000). The first of them has the lattice parameters $a = 2138.0 \pm 0.3$, $b = 597.5 \pm 0.1$, and $c = 1753.1 \pm 0.4$ pm and $\beta = 122.56° \pm 0.02°$ and a calculated density of 1.903 g·cm⁻³. The second is characterized by the lattice parameters $a = 1281.1 \pm 0.3$, $b = 915.7 \pm 0.1$, and $c = 1577.0 \pm 0.2$ pm and $\beta = 91.20° \pm 0.01°$ and a calculated density of 1.942 g·cm⁻³. The powder samples of the title compound were synthesized by slow evaporation of an aqueous solution containing Cd(NO₃)₂·4H₂O and thiourea in a 1:4 molar ratio. The appropriate single crystals were obtained after recrystallization from H₂O at room temperature. The second tetrathiourea polymorphic modification crystallizes from water solution of Cd(NO₃)₂·4H₂O and thiourea in stoichiometric amounts by isothermal evaporation at 55°C. They both exist at room temperature, but they apparently do not crystallize together.

**[Cd(OAc)$_2$·CS(NH$_2$)$_2$] or C$_5$H$_{10}$N$_2$O$_4$SCd**. This compound crystallizes in the orthorhombic structure with the lattice parameters $a = 804.7 \pm 0.2$, $b = 1439.1 \pm 0.2$, and $c = 881.9 \pm 0.3$ pm and a calculated density of 1.994 g·cm$^{-3}$ (Tkhanmanivong et al. 1984). It was obtained by the interaction of Cd(OAc)$_2$ and thiourea in an aqueous solution (Surin et al. 1972).

**[Cd{CH$_2$(COO)$_2$}{SC(NH$_2$)$_2$}$_2$]$_n$ or C$_5$H$_{10}$N$_4$O$_4$S$_2$Cd**. This compound crystallizes in the orthorhombic structure with the lattice parameters $a = 963.84 \pm 0.02$, $b = 814.050 \pm 0.010$, and $c = 1487.55 \pm 0.04$ pm and a calculated density of 2.087 g·cm$^{-3}$ (Zhang et al. 2000b). To obtain it, aqueous solutions CdSO$_4$·6H$_2$O (1.00 mM), malonic acid (1.00 mM), and thiourea (2.00 mM) were mixed together, and the pH value of the resulting solution was controlled at 5–6 by diluted NaOH and H$_2$SO$_4$. The solution was slowly evaporated at room temperature. Colorless large needlelike crystals were isolated to give the title compound in about 1 week.

**[Cd{NH$_2$CH(CO$_2$)C(Me$_2$)S}]·H$_2$O or C$_5$H$_{11}$NO$_3$SCd**. This compound crystallizes in the tetragonal structure with the lattice parameters $a = 1050.9 \pm 1.0$ and $c = 709.3 \pm 0.7$ pm and an experimental density of $2.13 \pm 0.03$ g·cm$^{-3}$ (Freeman et al. 1976). It was obtained as hexagonal prisms by the slow evaporation of an aqueous solution containing stoichiometric quantities of CdCl$_2$ and D-penicillamine at pH = 6.

**[Cd(OAc)$_2$·CS(NH$_2$)$_2$·H$_2$O] or C$_5$H$_{12}$N$_2$O$_5$SCd**. This compound was obtained by the interaction of Cd(OAc)$_2$ and thiourea in an aqueous solution (Surin et al. 1972).

**[Cd(DMSO)$_2$(SCN)$_2$] or C$_6$H$_{12}$N$_2$O$_2$S$_4$Cd**. This compound crystallizes in the triclinic structure with the lattice parameters $a = 591.7 \pm 0.2$, $b = 808.8 \pm 0.1$, and $c = 815.4 \pm 0.1$ pm and $\alpha = 114.533° \pm 0.005°$, $\beta = 100.91° \pm 0.01°$, and $\gamma = 95.45° \pm 0.01°$ (Wang et al. 2000). To obtain it, to a crystalline powder of Cd(SCN)$_2$ a mixed solvent of DMSO and water was added. This mixture was heated and stirred until Cd(SCN)$_2$ was dissolved. The colorless solution was left at room temperature. The precipitated crystals were separated.

**[Cd{SC(NH$_2$)$_2$}$_2$(OAc)$_2$] or C$_6$H$_{14}$N$_4$O$_4$S$_2$Cd**. This compound crystallizes in the orthorhombic structure with the lattice parameters $a = 1180.5 \pm 0.2$, $b = 1543.8 \pm 0.3$, and $c = 758.2 \pm 0.2$ pm (Bondar' et al. 1980); $a = 756$, $b = 1186$, and $c = 1568$ pm (Rajesh et al. 2004); $a = 750$, $b = 1530$, and $c = 1180$ pm (Nardelli and Chierici 1959); $a = 756 \pm 2$, $b = 1186 \pm 4$, and $c = 1568 \pm 12$ pm; and experimental and calculated densities 1.79 and 1.81 g·cm$^{-3}$, respectively (Rokabado Merkado et al. 1979). It was synthesized by mixing aqueous solutions of Cd(OAc)$_2$ and thiourea (1:2 molar ratio) (Surin et al. 1972, Rajesh et al. 2004). The obtained mixture had to be stirred well to avoid coprecipitation of multiple phases. The product was purified by repeated recrystallization before it is used for the crystal growth. Crystals of the title compound were grown from an aqueous solution by slow evaporation at room temperature (Nardelli and Chierici 1959, Rokabado Merkado et al. 1979, Rajesh et al. 2004).

**[Cd(C$_5$H$_4$NCO$_2$)(SCN)]$_n$ or C$_7$H$_4$N$_2$O$_2$SCd.** This compound crystallizes in the monoclinic structure with the lattice parameters $a = 1745.4 \pm 0.4$, $b = 587.9 \pm 0.2$, and $c = 1820.0 \pm 0.5$ pm and $\beta = 105.51° \pm 0.03°$ and the experimental and calculated densities $2.17 \pm 0.03$ and $2.160$ g·cm$^{-3}$, respectively (Mautner et al. 1997). It was prepared by mixing 4 mL of an aqueous solution of CdCl$_2$ (1 mM) and picolinic acid (2.5 mM) in 8 mL of EtOH/H$_2$O (1:1 volume ratio). To this clear solution an aqueous solution of KSCN (2 mM) in 3 mL of distilled H$_2$O was added dropwise with thorough shaking. The clear solution was allowed to stand over 10 days to produce colorless crystals.

**[Cd(C$_5$H$_5$NO)(SCN)$_2$] or C$_7$H$_5$N$_3$OS$_2$Cd.** This compound melts at 242°C (Ahuja and Rastogi 1970). It was obtained by allowing an excess of pyridine-$N$-oxide to react with a hot solution of Cd(SCN)$_2$ in EtOH. The title compound that precipitated on stirring the mixture was suction filtered, washed with EtOH, and dried.

**[Cd(SCN)(C$_5$H$_4$NCO$_2$)] or C$_7$H$_6$N$_2$O$_3$SCd.** This compound crystallizes in the monoclinic structure with the lattice parameters $a = 605.2 \pm 0.2$, $b = 1476.3 \pm 0.8$, and $c = 1442.3 \pm 0.5$ pm and $\beta = 94.26° \pm 0.02°$ (Yang et al. 2001b). To obtain it, an aqueous acetonitrile solution (10 mL) of Cd(NO$_3$)$_2$·4H$_2$O (1 mM) and NH$_4$SCN (2 mM) was added to a hot aqueous solution (20 mL) of isonicotinic acid (C$_5$H$_4$NCOOH) (2 mM). The resulting solution was allowed to stand in a desiccator for 2–3 weeks to yield a small amount of pale-yellow prismatic crystals of the title compound.

**[Cd{(C$_4$H$_3$N)CH(N)(CH$_2$)$_2$S}]·2H$_2$O or C$_7$H$_{12}$N$_2$O$_2$SCd.** This compound was obtained by the electrochemical oxidation of anodic Cd in an acetonitrile solution of the Schiff base derived from 2-pyrrolecarbaldehyde and cysteamine (Castro et al. 1990).

**[Cd(C$_5$H$_6$O$_4$)(CH$_4$N$_2$S)$_2$]$_n$ or C$_7$H$_{14}$N$_4$O$_4$S$_2$Cd.** This compound crystallizes in the monoclinic structure with the lattice parameters $a = 768.3 \pm 0.3$, $b = 1156.6 \pm 0.3$, and $c = 1600 \pm 1$ pm and $\beta = 100.32° \pm 0.04°$ and a calculated density of $1.875$ g·cm$^{-3}$ (Ke et al. 2002). To obtain it, CdCO$_3$ (2 mM) and glutaric acid (C$_5$H$_8$O$_4$) (2 mM) were added to 40 mL distilled water. The mixture was heated to a boil for about half an hour. After most of the CdCO$_3$ was dissolved, the solution was filtered and cooled to room temperature. Thiourea of 2 mM was dissolved and the solution was filtered again. The pH value of this solution was about 6. The mixture was allowed to stand at room temperature for several days, and large amounts of colorless crystal were isolated.

**[Cd(C$_4$H$_4$O$_6$){SC(NH$_2$)$_2$}$_3$]$_4$ or C$_7$H$_{16}$N$_6$O$_6$S$_3$Cd.** This compound crystallizes in the monoclinic structure with the lattice parameters $a = 1773.2 \pm 0.2$, $b = 1790.5 \pm 0.2$, and $c = 2069.9 \pm 0.2$ pm and $\beta = 94.111° \pm 0.009°$ (Zhou et al. 2001). It was obtained by the reaction between Cd(OH)$_2$, tartaric acid, and thiourea (1:1:3 molar ratio) in H$_2$O. The crystals used for the XRD were obtained from a more dilute resulting aqueous solution by using temperature-lowering method.

**[Cd(4-CNC$_5$H$_4$NO)(SCN)$_2$] or C$_8$H$_4$N$_2$OS$_2$Cd.** This compound melts at 195°C (Ahuja and Rastogi 1970). It was obtained by allowing an excess of 4-cyanopyridine-$N$-oxide to react with a hot solution of Cd(SCN)$_2$ in EtOH. The title compound that precipitated on stirring the mixture was suction filtered, washed with EtOH, and dried.

**[Cd(2-MeC₅H₄NO)(SCN)₂] or C₈H₇N₃OS₂Cd**. This compound melts at 250°C with decomposition (Ahuja and Rastogi 1970). It was obtained by allowing an excess of 2-methylpyridine-*N*-oxide to react with a hot solution of Cd(SCN)₂ in EtOH. The title compound that precipitated on stirring the mixture was suction filtered, washed with EtOH, and dried.

**[Cd(3-MeC₅H₄NO)(SCN)₂] or C₈H₇N₃OS₂Cd**. This compound melts at 215°C (Ahuja and Rastogi 1970). It was obtained by allowing an excess of 3-methylpyridine-*N*-oxide to react with a hot solution of Cd(SCN)₂ in EtOH. The title compound that precipitated on stirring the mixture was suction filtered, washed with EtOH, and dried.

**[Cd(4-MeC₅H₄NO)(SCN)₂] or C₈H₇N₃OS₂Cd**. This compound melts at 257°C (Ahuja and Rastogi 1970). It was obtained by allowing an excess of 4-methylpyridine-*N*-oxide to react with a hot solution of Cd(SCN)₂ in EtOH. The title compound that precipitated on stirring the mixture was suction filtered, washed with EtOH, and dried.

**[Cd(HL)₂]·2H₂O or C₈H₁₀N₄O₆S₂Cd (H₂L = thiobarbituric acid, C₄H₄N₂O₂S).** This compound crystallizes in the triclinic structure with the lattice parameters $a = 694.33 \pm 0.03$, $b = 722.57 \pm 0.03$, and $c = 740.47 \pm 0.03$ pm and $\alpha = 88.559° \pm 0.002°$, $\beta = 75.346° \pm 0.002°$, and $\gamma = 111.687° \pm 0.001°$ and a calculated density of 2.1604 g·cm⁻³ (Golovnev and Molokeev 2013). To obtain it, to a mixture of Cd(OH)₂ (0.694 mM) and H₂L (1.39 mM) H₂O (10 mL) was added, and the mixture was allowed to stay on a water bath at 50°C–60°C for 8 h until the completion of the interaction. The formed colorless crystalline precipitate was filtered out, washed with acetone, and dried in air.

This compound was also obtained by Zaki and Mohamed (2000), for which a solution of Cd salt (10 mM) in EtOH (25 mL) was mixed with a solution of H₂L (20 mM) in the same solvent (50 mL) and the resulting mixture was stirred under reflux for ≈ 2 h whereupon C₈H₁₀N₄O₆S₂Cd precipitated. It was removed by filtration, washed with EtOH, and dried at 80°C. There is no mass loss for this compound at the heating up to 130°C. The absence of XRD pattern for it does not allow to state that Golovnev and Molokeev (2013) and Zaki and Mohamed (2000) obtained the same phase.

**[Cd(C₄H₈NOS₂)₂] or C₈H₁₆N₂O₂S₄Cd**. To obtain this compound, *N*-methyl-*N*-ethanol-dithiocarbamic acid was prepared by mixing 4 mL of *N*-methyl-*N*-ethanolamine (C₃H₉NO) and 2 mL of CS₂ diluted with EtOH (20 mL) at 5°C (Thirumaran et al. 1998). To the yellow dithiocarbamic acid solution (20 mM), 10 mM of Cd salt dissolved in H₂O was added with constant stirring. Pale-yellow dithiocarbamate complex separated that was washed with H₂O and EtOH and then dried. This compound is stable up to 200°C.

**[Cd(SCN)₂·C₆H₁₂N₄·2H₂O] or C₈H₁₆N₆O₂S₂Cd**. This compound crystallizes in the triclinic structure with the lattice parameters $a = 600.4 \pm 0.2$, $b = 1005.1 \pm 0.3$, and $c = 1283.9 \pm 0.4$ pm and $\alpha = 105.79° \pm 0.09°$, $\beta = 102.63° \pm 0.07°$, and $\gamma = 103.65° \pm 0.03°$ and the experimental and calculated densities 1.88 and 1.946 g·cm⁻³, respectively

(Pickardt and Gong 1993). Its crystals were obtained by diffusion of an aqueous solution of hexamethylenetetramine into an aqueous solution of $Cd(SCN)_2$.

**$[Cd(2,6-Me_2C_5H_3NO)(SCN)_2]$ or $C_9H_9N_3OS_2Cd$.** This compound melts at 197°C (Ahuja and Rastogi 1970). It was obtained by allowing an excess of 2,6-dimethylpyridine-$N$-oxide to react with a hot solution of $Cd(SCN)_2$ in EtOH. The title compound that precipitated on stirring the mixture was suction filtered, washed with EtOH, and dried.

**$[Cd(C_8H_3O_7S)(H_2O)_3]\cdot0.5C_2H_{10}N_2\cdot H_2O$ or $C_9H_{16}NO_{11}SCd$.** This compound crystallizes in the monoclinic structure with the lattice parameters $a = 1447.36 \pm 0.18$, $b = 1747.14 \pm 0.02$, and $c = 1186.50 \pm 0.15$ pm and $\beta = 103.393° \pm 0.001°$ and a calculated density of 2.088 g·cm$^{-3}$ (Liu and Xu 2005b). It was obtained by the following procedure. A methanol solution of 5-sulfoisophthalic acid monosodium salt (0.1 mM) and piperazine (0.1 mM) was carefully layered on top of an aqueous solution of $Cd(NO_3)_2\cdot4H_2O$ (0.1 mM). Colorless crystals of the title compound were obtained after 8 days.

**$[Cd(C_5H_3N_4S)_2]\cdot2H_2O$ or $C_{10}H_{10}N_8O_2S_2Cd$.** This compound crystallizes in the triclinic structure with the lattice parameters $a = 391.8 \pm 0.3$, $b = 937.1 \pm 0.3$, and $c = 1116.2 \pm 0.5$ pm and $\alpha = 67.89° \pm 0.03°$, $\beta = 84.40° \pm 0.04°$, and $\gamma = 76.84° \pm 0.04°$ and the experimental and calculated densities 2.00 and 2.025 g·cm$^{-3}$, respectively (Dubler and Gyr 1988). It was obtained by the next procedure. A 0.2 mM of $CdClO_4\cdot6H_2O$ was added to a solution of 0.35 mM of 6-mercaptopurine monohydrate in 1 L of 0.1 M AcOH/NaOAc buffer (pH = 4.6) at 95°C. The solution was stirred for 3 min and then kept at 60°C. After 2 days, pale-yellow, transparent needles of the title compound could be isolated and were washed with $H_2O$ and dried in a desiccator.

**$[Cd(C_4H_7NOS)_2(SCN)_2]$ or $C_{10}H_{14}N_4O_2S_4Cd$.** This compound melts at 118°C (Colombini and Preti 1975) It was prepared by the reaction of $Cd(SCN)_2$ and thiomorpholine-3-one in refluxing EtOH for 1 h. The title compound obtained by concentrating was washed with ethyl ether.

**$[Cd(C_4H_8NCOS)_2]$ or $C_{10}H_{16}N_2O_2S_2Cd$.** This compound was prepared by treating 10 mM of $N$-pyrrolidylthiocarbamatosodium ($C_4H_8NCOSNa$) in absolute EtOH (20 mL) with an equal volume of EtOH containing $Cd(NO_3)_2\cdot4H_2O$ (5 mM) (McCormick and Greene 1972). The white precipitate that formed was washed twice with 20 mL portions of EtOH and dried in vacuo over anhydrous $CaSO_4$ for 24 h.

**$[Cd(C_3H_5NS_2)_2(Ac)_2]$ or $C_{10}H_{16}N_2O_2S_4Cd$.** This compound melts at 87°C (Colombini and Preti 1975). It was prepared by the reaction of $CdCO_3$ with thiazolidine-2-thione in AcOH (reflux for 1 h). The title compound that separated spontaneously during the reaction was washed with ethyl ether.

**$[Cd(SCN)_2(C_4H_9NO)_2]$ or $C_{10}H_{18}N_4O_2S_2Cd$.** This compound crystallizes in the triclinic structure with the lattice parameters $a = 593.40 \pm 0.06$, $b = 675.65 \pm 0.06$, and $c = 1007.53 \pm 0.13$ pm and $\alpha = 84.277° \pm 0.009°$, $\beta = 82.495° \pm 0.010°$, and $\gamma = 70.349° \pm 0.008°$ and the experimental and calculated densities $1.78 \pm 0.01$ and

1.777 g·cm$^{-3}$, respectively (Moon et al. 2000b). To obtain it, $CdCl_2 \cdot 2.5H_2O$ (5 mM) and KSCN (10 mM) were dissolved in 30 mL of distilled $H_2O$. To this solution, 1.7 mL (20 mM) of neat morpholine was added dropwise with stirring. After the morpholine was added, fine precipitates formed. Then the pH of the solution was adjusted to 9 by adding 2-aminoethanol and citric acid. A small amount of precipitate was filtered off and the aqueous solution was allowed to stand in a refrigerator at 5°C. After a few weeks the crystals were obtained.

**[Cd{SC(NHCH$_2$)$_2$}$_2$(OAc)$_2$] or C$_{10}$H$_{18}$N$_4$O$_4$S$_2$Cd.** This compound crystallizes in the monoclinic structure with the lattice parameters $a = 1400$, $b = 920$, and $c = 1240$ pm and $\beta = 93.1°$ (Nardelli and Chierici 1959). It was obtained by a slow crystallization from the aqueous solution.

**[Cd(C$_4$H$_3$O$_4$){SC(NH$_2$)$_2$}$_2$(H$_2$O)$_2$] or C$_{10}$H$_{18}$N$_4$O$_{10}$S$_2$Cd.** This compound crystallizes in the monoclinic structure with the lattice parameters $a = 580.4 \pm 0.2$, $b = 2497 \pm 2$, and $c = 619.0 \pm 0.5$ pm and $\beta = 92.90° \pm 0.06°$ at 173 K (Zhang et al. 2000a). To obtain it, aqueous solutions (20 mL) of $CdSO_4$ (1.0 mM), maleic anhydride (1.0 mM), and thiourea (2.0 mM) were mixed, and the pH value was controlled at 6. The solution was slowly evaporated at about 10°C. After 2 days, colorless plate crystals of the title compound were produced. Several days later, colorless needlelike crystals of $[\{(C_4H_2O_4)Cd[SC(NH_2)_2]_2\}_2(C_4H_2O_4)Cd[SC(NH_2)_2](H_2O)_2]_n$ were isolated when the title compound gradually disappeared.

**[Cd(SOCMe)$_2$(C$_6$H$_{16}$N$_2$)] or C$_{10}$H$_{22}$N$_2$O$_2$S$_2$Cd.** This compound crystallizes in the orthorhombic structure with the lattice parameters $a = 1261.7 \pm 0.1$, $b = 745.6 \pm 0.1$, and $c = 1691.6 \pm 0.1$ pm and a calculated density of 1.581 g·cm$^{-3}$ (Nyman et al. 1996). To prepare it, $CdCO_3$ (5.8 mM) was placed in a round-bottom flask with 25 mL toluene and 5.8 mM of $N,N,N,N$-tetramethylethylenediamine. While stirring, thioacetic acid was added (11.6 mM). An exothermic reaction took place immediately. The mixture was stirred for ca. 1 h to obtain a toluene solution containing the title compound and a yellowish precipitate containing a mixture of the title compound, CdS, and other unidentified. The toluene solution is placed in the refrigerator and [Cd(SOCMe)$_2$(C$_6$H$_{16}$N$_2$)] crystallizes out overnight as colorless blades.

**[Cd(C$_5$H$_{12}$N$_2$S)$_2$(NO$_3$)$_2$] or C$_{10}$H$_{24}$N$_6$O$_6$S$_2$Cd.** This compound crystallizes in the monoclinic structure with the lattice parameters $a = 1250.7 \pm 0.4$, $b = 727.0 \pm 0.4$, and $c = 2167 \pm 1$ pm and $\beta = 95.42° \pm 0.04°$ and the experimental and calculated densities $1.68 \pm 0.01$ and 1.696 g·cm$^{-3}$, respectively (Griffith et al. 1983). To prepare the title compound, 10 mM of 1,1,3,3-tetramethyl-2-thiourea was dissolved in $H_2O$ (50 mL) and added to a like volume containing 5 mM of $Cd(NO_3)_2 \cdot 4H_2O$. The resulting solution was warm gently (~60°C) for 15 min, allowed to cool, and evaporated slowly at ambient temperature.

**[Cd{NH$_2$(CH$_2$)$_2$SCH$_2$}$_2$(C$_5$H$_3$N)(NO$_3$)$_2$] or C$_{11}$H$_{19}$N$_5$O$_6$S$_2$Cd.** To obtain this compound, $Cd(NO_3)_2 \cdot 4H_2O$ (1.55 mM) suspended in THF (20 mL) was added to a solution of 2,6-bis[(aminoethylthio)methyl](pyridine) (1.55 mM) in THF (20 mL) (Teixidor et al. 1986). The mixture was stirred for 1 h and the resulting solid was filtered out, washed with cold THF, and dried in vacuo. All operations were performed under a $N_2$ atmosphere.

**[{($C_4H_2O_4$)Cd[SC($NH_2$)$_2$]$_2$}$_2$($C_4H_2O_4$)Cd[SC($NH_2$)$_2$]($H_2O$)$_2$]$_n$ or $C_{11}H_{20}N_6O_{10}S_3$ $Cd_2$.** This compound crystallizes in the monoclinic structure with the lattice parameters $a = 823.8 \pm 0.6$, $b = 813.4 \pm 0.6$, and $c = 3457 \pm 5$ pm and $\beta = 92.51° \pm 0.05°$ at 173 K (Zhang et al. 2000a). To obtain it, aqueous solutions (20 mL) of $CdSO_4$ (1.0 mM), maleic anhydride (1.0 mM), and thiourea (2.0 mM) were mixed and the pH value was controlled at 6. The solution was slowly evaporated at about 10°C. After 2 days, colorless plate crystals of [{Cd($C_4H_3O_4$)[SC($NH_2$)$_2$]$_2$($H_2O$)$_2$} were produced. Several days later, colorless needlelike crystals of the title compound were isolated when [{Cd($C_4H_3O_4$)[SC($NH_2$)$_2$]$_2$($H_2O$)$_2$}] gradually disappeared.

**[Cd($C_{10}H_{14}N_4O$)(SCN)$_2$] or $C_{12}H_{14}N_4OS_2Cd$, mono-($N,N$-diethylnicotinamide) cadmium dithiocyanate.** This compound crystallizes in the monoclinic structure with the lattice parameters $a = 973.5 \pm 0.7$, $b = 1620.2 \pm 1.0$, and $c = 1064.0 \pm 0.9$ pm and $\beta = 105.2° \pm 0.1°$ and the experimental and calculated densities 1.643 and 1.668 g·cm$^{-3}$, respectively (Bigoli et al. 1972). Colorless crystals of this compound were obtained by evaporation of an aqueous solution of $CdCl_2$, KSCN, and $N,N$-diethylnicotinamide.

**[Cd{$C_6H_4(SO_2)_2N$}$_2$($H_2O$)$_4$] or $C_{12}H_{16}N_2O_{12}S_4Cd$.** This compound melts at the temperature higher than 330°C and crystallizes in the triclinic structure with the lattice parameters $a = 722.2 \pm 0.3$, $b = 724.0 \pm 0.3$, and $c = 1083.9 \pm 0.4$ pm and $\alpha = 80.25° \pm 0.02°$, $\beta = 88.45° \pm 0.02°$, and $\gamma = 61.08° \pm 0.02°$ and a calculated density of 2.113 g·cm$^{-3}$ (Moers et al. 2002). To obtain it, 1,3,2-benzodithiazole 1,1,3,3-tetraoxide (4.5 mM) was dissolved in $H_2O$ (100 mL) and mixed with $CdCO_3$ (2.6 mM). The obtained suspension was refluxed for 6 h and filtered hot. The filtrate was concentrated at 70°C in a rotary evaporator to approximately 5 mL and left at room temperature for crystallization.

**[Cd($C_8H_3O_7S$)($C_4H_{11}N_2$)($H_2O$)$_2$] or $C_{12}H_{18}N_2O_9SCd$.** This compound crystallizes in the orthorhombic structure with the lattice parameters $a = 1110.96 \pm 0.04$, $b = 1648.60 \pm 0.06$, and $c = 1770.99 \pm 0.06$ pm and a calculated density of 1.961 g·cm$^{-3}$ (Liu and Xu 2005b). It was obtained by the next procedure. A methanol solution of 5-sulfoisophthalic acid monosodium salt (0.1 mM) and diethylamine (0.1 mM) was carefully layered on top of an aqueous solution of Cd($NO_3$)$_2$·$4H_2O$ (0.1 mM). Colorless crystals of the title compound were obtained after 8 days.

**[Cd($C_5H_{10}NCOS$)$_2$] or $C_{12}H_{20}N_2O_2S_2Cd$.** This compound was prepared by treating 10 mM of $N$-piperidylthiocarbamatosodium ($C_5H_{10}NCOSNa$) in absolute EtOH (20 mL) with an equal volume of EtOH containing Cd($NO_3$)$_2$·$4H_2O$ (5 mM) (McCormick and Greene 1972). The white precipitate that formed was washed twice with 20 mL portions of EtOH and dried in vacuo over anhydrous $CaSO_4$ for 24 h.

**[Cd($C_4H_7NOS$)$_2$($Ac$)$_2$] or $C_{12}H_{20}N_2O_4S_2Cd$.** This compound melts at 65°C (Colombini and Preti 1975). It was prepared by a reaction of $CdCO_3$ with thiomorpholine-3-one in AcOH (reflux for 1 h). The title compound that separated spontaneously during the reaction was washed with ethyl ether.

**[{Cd(C₃H₅NS₂)₄}{Cd(NO₃)}₄]** or **C₁₂H₂₀N₈O₁₂S₈Cd₂**. This compound melts at 145.8°C–146.4°C and crystallizes in the tetragonal structure with the lattice parameters $a = 1388.53 \pm 0.11$ and $c = 807.7 \pm 0.2$ pm and a calculated density of 2.025 g·cm$^{-3}$ (Rajalingam et al. 2001). To obtain it, solutions of Cd(NO₃)₂·4H₂O (1.70 mM) in MeCN (8 mL) and thiazolidine-2-thion (C₃H₅NS₂) (6.72 mM) in MeCN (10 mL) were mixed at room temperature. After 30 min, colorless crystals were separated by filtration; the bulk was washed with ice-cold diethyl ether and vacuum dried. The title compound was also formed from mixtures containing Cd(NO₃)₂·4H₂O/ C₃H₅NS₂ = 1:2. All manipulations were carried out under an atmosphere of dry N₂ in a glovebox or with a vacuum–N₂ manifold or under Ar in a glove bag.

**[Cd{Me₂(NH)₂CS}₄(NO₃)₂]** or **C₁₂H₃₂N₁₀O₆S₄Cd**. This compound crystallizes in the orthorhombic structure with the lattice parameters $a = 737.8 \pm 0.2$, $b = 1065.3 \pm 0.4$, and $c = 3516 \pm 1$ pm and the experimental and calculated densities 1.55±0.02 and 1.570 g·cm$^{-3}$, respectively (Rodesiller et al. 1983). To obtain it, 10 mM of dimethyl-thiourea and 5 mM of Cd(NO₃)₂·4H₂O were dissolved in 150 mL of H₂O with gentle heating (ca. 60°C) for approximately 30 min and allowed to evaporate slowly at ambient temperature. The crystals were collected after 6–8 days without washing and air-dried.

**[Cd(H₂O)₂(C₁₂H₈N₂){CH₂(SO₃)₂}]** or **C₁₃H₁₄N₂O₈S₂Cd**. This compound crystallizes in the triclinic structure with the lattice parameters $a = 789.09 \pm 0.05$, $b = 967.12 \pm 0.07$, and $c = 1223.89 \pm 0.07$ pm and $\alpha = 83.706° \pm 0.006°$, $\beta = 72.775° \pm 0.005°$, and $\gamma = 86.733° \pm 0.006°$ (Xia and Liu 2010). The title compound was prepared by the following procedure. A mixture of methanedisulfonic acid (1 mM), Cd(OH)₂ (1 mM), and Phen (1 mM) was sealed in a Teflon-lined reactor (15 mL) with MeOH (5 mL). Then the mixture was heated to 140°C and kept for 3 days. After cooling to room temperature, colorless crystals were obtained and dried at room temperature.

**[Cd(SO₄)(C₃H₆N₂S)(C₁₀H₈N₂)(H₂O)]·H₂O** or **C₁₃H₁₈N₄O₆S₂Cd**. This compound crystallizes in the monoclinic structure with the lattice parameters $a = 637.6 \pm 0.1$, $b = 2572.8 \pm 0.3$, and $c = 1001.1 \pm 0.2$ pm and $\beta = 90.72° \pm 0.02°$ and the experimental and calculated densities 1.91 ± 0.02 and 1.93 g·cm$^{-3}$, respectively (Rodesiler et al. 1987). To obtain it, a solution containing 10 mM of 2,2′-bipy was added slowly with stirring to a solution of 3 mM of 3CdSO₄·8H₂O and 20 mM of 2-imidazolidinethione. The total volume of approximately 150 mL was heated gently (60°C) for 15 min and then allowed to evaporate at ambient temperature. Crystals were collected by gravity filtration after 1 week.

**[Cd(SCN)₂(C₅H₄NCONH₂)₂]** or **C₁₄H₁₂N₆O₂S₂Cd**. This compound crystallizes in the monoclinic structure with the lattice parameters $a = 1178.6 \pm 1.0$, $b = 966.7 \pm 0.4$, and $c = 811.2 \pm 0.4$ pm and $\beta = 107.62° \pm 0.02°$ (Yang et al. 2001b). It was prepared by the following procedure. An aqueous solution of Cd(NO₃)₂·4H₂O (1 mM) and NH₄SCN (2 mM) was added to a hot aqueous solution (20 mL) of isonicotinamide (C₅H₄NCONH₂) (2 mM) with stirring. A white precipitate was formed immediately.

The mixture was slowly heated to nearly boiling, during which time the precipitation dissolved. After cooling to room temperature, the clear solution was allowed to stand for several hours to give colorless crystals of the title compound.

**[Cd(SCN)$_2$(C$_5$H$_4$NCONH$_2$)$_2$]·H$_2$O or C$_{14}$H$_{14}$N$_6$O$_3$S$_2$Cd**. This compound crystallizes in the monoclinic structure with the lattice parameters $a = 2429.0 \pm 1.2$, $b = 986.3 \pm 0.4$, and $c = 761.7 \pm 0.3$ pm and $\beta = 92.67° \pm 0.02°$ (Yang et al. 2001b). It was obtained by the next procedure. An aqueous solution of Cd(NO$_3$)$_2$·4H$_2$O (1 mM) and NH$_4$SCN (2 mM) was added to a hot aqueous solution (20 mL) of nicotinamide (C$_5$H$_4$NCONH$_2$) (2 mM) with stirring. A white precipitate was formed immediately. The mixture was slowly heated to nearly boiling, during which time the precipitation dissolved. After cooling to room temperature, the clear solution was allowed to stand for several hours to give colorless crystals of the title compound.

**[(C$_8$H$_{16}$O$_4$)Cd$_2$(SCN$_6$)] or C$_{14}$H$_{16}$N$_6$O$_4$S$_6$Cd$_2$**. This compound crystallizes in the monoclinic structure with the lattice parameters $a = 1629.7 \pm 0.3$, $b = 2626.7 \pm 0.3$, and $c = 1629.6 \pm 0.2$ pm and $\beta = 90.20° \pm 0.01°$ (Zhang et al. 1997). It was prepared by a dropwise addition, with vigorous stirring, of an aqueous solution of 12-crown-4 (C$_8$H$_{16}$O$_4$) to a mixture of CdSO$_4$ and NH$_4$SCN in H$_2$O (1:4:32 molar ratio). Care must be taken to prevent supersaturation and/or precipitation during the reaction. Colorless square-plate crystals were formed within a few days.

**[Cd(H$_2$O)$_2$(C$_7$H$_8$N$_4$S)$_2$][SO$_4$]·H$_2$O or C$_{14}$H$_{22}$N$_8$O$_7$S$_3$Cd**. This compound crystallizes in the monoclinic structure with the lattice parameters $a = 1368.79 \pm 0.06$, $b = 1861.73 \pm 0.08$, and $c = 1824.70 \pm 0.07$ pm and $\beta = 90.392° \pm 0.001°$ (Zhao et al. 2009). To obtain it, 4-formylpyridine thiosemicarbazone (1 mM) was dissolved in EtOH (15 mL) and refluxed with CdSO$_4$·H$_2$O (1 mM) dissolved in 10 mL of hot H$_2$O. The product that separated out was collected, washed with ether, and dried in vacuo. Single crystals of the title complex were obtained by vapor diffusion of diethyl ether into a DMF solution.

**[Cd$_3$(C$_4$H$_{13}$N$_3$)$_2$(SCN)$_6$]·H$_2$O or C$_{14}$H$_{28}$N$_{12}$OS$_6$Cd$_3$**. This compound crystallizes in the cubic structure with the lattice parameter $a = 1487.1 \pm 0.4$ and the experimental and calculated densities 1.842 and 1.838 g·cm$^{-3}$, respectively (Mondal et al. 1999). To obtain it, diethylenediamine (C$_4$H$_{13}$N$_3$) (2 mM) in MeOH (5 mL) was added dropwise with stirring to Cd(SCN)$_2$ (3 mM) dissolved in MeOH (10 mL). A sticky oily layer separated at the bottom of the container. The supernatant solution was filtered and the filtrate was kept in a CaCl$_2$ desiccator for a few days at ca. 30°C, giving the polymer as shining transparent crystals.

**[NH$_4${Cd(NCS)$_3$}·(C$_{12}$H$_{24}$O$_6$)]$_n$ or C$_{15}$H$_{28}$N$_4$O$_6$S$_3$Cd**. This compound crystallizes in the orthorhombic structure with the lattice parameters $a = 1475.68 \pm 0.06$, $b = 1543.78 \pm 0.06$, and $c = 1063.83 \pm 0.05$ pm and a calculated density of 1.559 g·cm$^{-3}$ (Ramesh et al. 2012). To obtain it, the mixture of 18-crown-6 (C$_{12}$H$_{24}$O$_6$), CdCl$_2$, and NH$_4$SCN (1:1:3 molar ratio) were thoroughly dissolved in double distilled water and stirred for 5 h to obtain a homogeneous mixture. The colorless single crystals were obtained after the filtrate had been allowed to stand at room temperature for 3 weeks.

**[Cd(C$_{16}$H$_{16}$N$_6$S$_2$)(NO$_3$)$_2$] or C$_{16}$H$_{16}$N$_8$O$_6$S$_2$Cd**. This compound melts at 220°C and crystallizes in the orthorhombic structure with the lattice parameters $a = 1452.2 \pm 0.3$, $b = 1116.1 \pm 0.2$, and $c = 2775.4 \pm 0.6$ pm (López-Torres et al. 2001). It was prepared as follows. A solution of Cd(NO$_3$)$_2$·6H$_2$O (1.70 mM) in EtOH (25 mL) was added to a suspension of benzilbisthiosemicarbazone (1.70 mM) in the same solvent (25 mM). The mixture that immediately changed to yellow was stirred at room temperature for 6 h or refluxed with stirring for 6 h. In both procedures, the yellow solid was filtered off and then treated with EtOH with stirring for 15 min. The residual solid was separated by filtration. The filtrate was concentrated until a pale-yellow solid was obtained. Single crystals of the title compound were grown by vapor diffusion of Et$_2$O into a MeOH/acetonitrile (2:1 volume ratio) solution.

**[Cd(SCN)$_2$(C$_7$H$_8$N$_4$)$_2$]·2H$_2$O or C$_{16}$H$_{20}$N$_{10}$O$_2$S$_2$Cd**. This compound crystallizes in the monoclinic structure with the lattice parameters $a = 1582.3 \pm 0.8$, $b = 835.8 \pm 0.7$, and $c = 1765.0 \pm 0.8$ pm and $\beta = 102.82° \pm 0.05°$ and the experimental and calculated densities 1.63 and 1.634 g·cm$^{-3}$, respectively (Dillen et al. 1983). It was prepared by mixing solution of Cd(SCN)$_2$ (2 mM in 20 mL of H$_2$O) with a solution of 5,7-dimethyl[1,2,4]triazolo[1,5-a]pyrimidine (C$_7$H$_8$N$_4$) (4 mM in 20 mL of H$_2$O). The title compound was crystallized within several hours and then recrystallized from H$_2$O. White single crystals were grown easily from the solution by cooling slowly to room temperature.

**[Cd(C$_3$H$_{10}$N$_2$)$_2$]·(C$_{10}$H$_6$O$_6$S$_2$)·2H$_2$O or C$_{16}$H$_{30}$N$_4$O$_8$S$_2$Cd**. This compound crystallizes in the monoclinic structure with the lattice parameters $a = 849.53 \pm 0.11$, $b = 967.32 \pm 0.13$, and $c = 1403.13 \pm 0.18$ pm and $\beta = 90.412° \pm 0.002°$ and a calculated density of 1.668 g·cm$^{-3}$ (Cai et al. 2001). It was prepared as follows. N-methylethylenediamine (2 mM) was added with stirring to an aqueous solution of Cd(OAc)$_2$·2H$_2$O (1 mM). This solution was then treated with sodium 2,6-naphthalenedisulfonate (1 mM). Colorless crystals of the title compound were collected after 4 days.

**[Cd(C$_4$H$_8$N$_2$S)$_4$(NO$_3$)$_2$] or C$_{16}$H$_{32}$N$_{10}$O$_6$S$_4$Cd**. This compound melts at 145°C (Raper et al. 1980). It was prepared by dissolving Cd(NO$_3$)$_2$·4H$_2$O (2 mM) in a small volume (<10 mL) of a hot anhydrous EtOH containing 3 mL triethyl orthoformate as dehydrating agent. To this solution the stoichiometric quantity of 1-methylimidazoline-2-thione (C$_4$H$_8$N$_2$S) was added dissolved in the minimum volume (<5 mL) of hot anhydrous EtOH. The total volume was reduced slightly and then cooled to room temperature. Subsequent slow evaporation of the solvent produced a crystalline colorless product in each instance, which was removed, washed with a small volume of EtOH, and finally vacuum dried at room temperature.

**[Me$_4$N][Cd(OAc)(S$_2$CNEt$_2$)$_2$] or C$_{16}$H$_{35}$N$_3$O$_2$S$_4$Cd**. This compound was obtained by a reaction of [Me$_4$N][OH] (1.8 g, 40% aqueous solution), acetic acid (5 mM), and [Cd(S$_2$CNEt$_2$)$_2$] (5 mM) (McCleverty et al. 1982).

**[Cd(C$_7$H$_4$O$_5$S)(2,2′-bipy)(H$_2$O)]$_n$ or C$_{17}$H$_{14}$N$_2$O$_6$SCd**. This compound crystallizes in the triclinic structure with the lattice parameters $a = 883.50 \pm 0.16$, $b = 947.70 \pm 0.16$, and $c = 1180.19 \pm 0.15$ pm and $\alpha = 99.174° \pm 0.011°$, $\beta = 105.776° \pm 0.013°$, and $\gamma = 113.718° \pm 0.015°$ (Yuan et al. 2001). To obtain it, 0.3 mM of $Cd(OH)_2$, 0.6 mM of 4-sulfobenzoic acid, and 0.1 mM of 2,2′-bipy were placed in a thick Pyrex tube. After the addition of 2 mL of $H_2O$, the tube was frozen by liquid $N_2$, evacuated under vacuum, and sealed with a torch. This tube was then heated at 130°C for 10 h to give colorless crystals of the title compound.

$[Cd(C_7H_4O_6S)(C_5H_5NS)_2] \cdot 2.5H_2O$ or $C_{17}H_{19}N_2O_{8.5}S_3Cd$. This compound crystallizes in the triclinic structure with the lattice parameters $a = 737.30 \pm 0.07$, $b = 992.0 \pm 0.1$, and $c = 1522.8 \pm 0.2$ pm and $\alpha = 79.474° \pm 0.001°$, $\beta = 89.189° \pm 0.001°$, and $\gamma = 81.344° \pm 0.001°$ (Li and Du 2011). It was obtained by the next procedure. $Cd(OAc)_2 \cdot 2H_2O$ (0.5 mM), 5-sulfosalicylic acid (0.5 mM), and pyridine-4-thiol (1.0 mM) were dissolved in 20 mL of $H_2O$. Then the mixed solution was sealed in a 25 mL Teflon-lined reactor and kept under autogenous pressure at 130°C for 5 days. After cooling to room temperature at a rate of 279°C h⁻¹, colorless needlelike crystals of the title compound were obtained and collected.

$C_{17}H_{35}NO_3S_6Cd$, tetraethylammonium tris($O$-ethylxanthato)cadmiumate(II). This compound crystallizes in the monoclinic structure with the lattice parameters $a = 1061.0 \pm 0.1$, $b = 1547.9 \pm 0.1$, and $c = 1833.2 \pm 0.2$ pm and $\beta = 112.83° \pm 0.01°$ and the experimental and calculated densities 1.45 g·cm⁻³ (Hoskins and Kelly 1972).

$[Cd(C_{12}H_8N_2)(EtOCS_2)_2]$ or $C_{18}H_{18}N_2O_2S_4Cd$. This compound crystallizes in the orthorhombic structure with the lattice parameters $a = 1487.0 \pm 0.4$, $b = 1454.9 \pm 0.8$, and $c = 1024.9 \pm 0.1$ pm (Raston and Winter 1976).

$\{[Cd(C_8H_3O_7S)(H-4,4′-bipy)(H_2O)] \cdot 2H_2O\}_n$ or $C_{18}H_{18}N_2O_{10}SCd$. This compound crystallizes in the triclinic structure with the lattice parameters $a = 880.14 \pm 0.06$, $b = 921.40 \pm 0.07$, and $c = 1398.86 \pm 0.10$ pm and $\alpha = 97.4200° \pm 0.0010°$, $\beta = 94.5960° \pm 0.0010°$, and $\gamma = 113.97°$ and a calculated density of 1.851 g·cm⁻³ (Li et al. 2005). It was prepared by the following procedure. A mixture of $NaC_8H_5O_7S$ (monosodium salt of 5-sulfoisophthalic acid) (0.5 mM), $Cd(NO_3)_2 \cdot 4H_2O$ (0.5 mM), 4,4′-bipy·2H_2O (0.5 mM), and $H_2O$ (15 mL) was placed in a 25 mL Teflon-lined stainless autoclave and heated under autogenous pressure at 160°C for 3 days. It was then cooled to room temperature over 24 h. Colorless sheetlike crystals of the title compound were separated from the reaction mixture.

$[Cd(Pr^iOCS_2)_2(2,2′-bipy)]$ or $C_{18}H_{22}N_2O_2S_4Cd$. This compound crystallizes in the orthorhombic structure with the lattice parameters $a = 1814.04 \pm 0.04$, $b = 695.13 \pm 0.01$, and $c = 1758.35 \pm 0.04$ pm and a calculated density of 1.615 g·cm⁻³ (Larionov et al. 2005). To obtain it, a solution of $K(Pr^iOCS_2)$ (5 mM) in $H_2O$ (10 mL) was added through paper filter to a suspension of 2,2′-bipy (2.5 mM) in the solution of $CdCl_2$ (2.5 mM) in $H_2O$ (50 mL). The mixture was stirred for 30 min. The precipitate that formed was filtered with suction, washed with $H_2O$, and dried in air. The title compound was reprecipitated by dissolving in 20 mL of $CH_2Cl_2$, filtration, and the subsequent addition of hexane (80 mL). Single crystals were grown via standing of the mother liquor obtained at the reprecipitation in a refrigerator.

**[Cd(Pr$^i$OCS$_2$)$_2$(4,4′-bipy)] or C$_{18}$H$_{22}$N$_2$O$_2$S$_4$Cd.** This compound crystallizes in the orthorhombic structure with the lattice parameters $a = 1481.2 \pm 0.3$, $b = 1299.3 \pm 0.2$, and $c = 1202.8 \pm 0.2$ pm and the experimental and calculated densities 1.53 and 1.547 g·cm$^{-3}$, respectively (Abrahams et al. 1990). It was prepared by the equimolar addition of 4,4′-bipy to Cd(Me$_2$CHOCS$_2$)$_2$ in CH$_2$Cl$_2$. The resulting solution was evaporated to dryness and the residue was recrystallized from a CH$_2$Cl$_2$/MeOH solution.

**[Cd{S$_2$CO(CH$_2$)$_2$OMe}$_2$(2,2′-bipy)] or C$_{18}$H$_{22}$N$_2$O$_4$S$_4$Cd.** This compound crystallizes in the monoclinic structure with the lattice parameters $a = 2300.1 \pm 0.2$, $b = 850.62 \pm 0.08$, and $c = 1281.76 \pm 0.13$ pm and $\beta = 115.873° \pm 0.002°$ at 183 K and a calculated density of 1.681 g·cm$^{-3}$ (Chen et al. 2003b). It was prepared by refluxing stoichiometric amounts of Cd{S$_2$CO(CH$_2$)$_2$OMe}$_2$ and 2,2′-bipy in acetonitrile/chloroform solution. Crystals were obtained by the slow evaporation of the solution of the title compound.

**[Cd(SAc)$_2$(C$_7$H$_9$N)$_2$] or C$_{18}$H$_{24}$N$_2$O$_2$S$_2$Cd.** This compound crystallizes in the triclinic structure with the lattice parameters $a = 826.7 \pm 0.1$, $b = 946.7 \pm 0.1$, and $c = 1408.7 \pm 0.1$ pm and $\alpha = 94.04° \pm 0.01°$, $\beta = 91.49° \pm 0.01°$, and $\gamma = 104.03° \pm 0.01°$ (Nyman et al. 1997). To obtain it, CdCO$_3$ (5.8 mM), 3,5-dimethylpyridine (11.6 mM), and 20 mL of toluene were combined in a round-bottom flask. Thioacetic acid (11.6 mM) was dropped into the mixture while stirring rapidly, and stirring was continued for 1 h at room temperature. As the reaction proceeded, the solid CdCO$_3$ disappeared, CO$_2$ bubble formation was observed, and the resulting clear solution became yellow. The toluene and volatile by-products of the reaction (H$_2$O) were removed under reduced pressure, and a white crystalline solid and a small amount of yellow material shown to be CdS remained. The solid was redissolved in toluene and filtered to remove CdS. The solution was then placed in the freezer and yielded colorless, crystalline blocks.

Cd(SAc)$_2$(C$_7$H$_9$N)$_2$ decomposes in a solution to form sphalerite, nanocrystalline CdS (Nyman et al. 1997). This reaction has been observed to take place from 25°C to 100°C with increasing rate of CdS formation at elevated temperatures.

**[Cd(C$_4$H$_{12}$N$_2$)$_2$]·(C$_{10}$H$_6$O$_6$S$_2$) or C$_{18}$H$_{30}$N$_4$O$_6$S$_2$Cd.** This compound crystallizes in the monoclinic structure with the lattice parameters $a = 765.48 \pm 0.12$, $b = 1199.68 \pm 0.18$, and $c = 1283.2 \pm 0.2$ pm and $\beta = 103.836° \pm 0.003°$ and a calculated density of 1.669 g·cm$^{-3}$ (Cai et al. 2001). It was prepared as follows. $N,N'$-dimethylethylenediamine (2 mM) was added with stirring to an aqueous solution of Cd(OAc)$_2$·2H$_2$O (1 mM). This solution was then treated with sodium 2,6-naphthalenedisulfonate (1 mM). Colorless crystals of the title compound were collected after 4 days.

**[Cd(C$_4$H$_{12}$N$_2$)$_2$]·(C$_{10}$H$_6$O$_6$S$_2$) or C$_{18}$H$_{30}$N$_4$O$_6$S$_2$Cd.** This compound crystallizes in the triclinic structure with the lattice parameters $a = 843.06 \pm 0.14$, $b = 852.43 \pm 0.14$, and $c = 856.07 \pm 0.14$ pm and $\alpha = 87.652° \pm 0.003°$, $\beta = 67.476° \pm 0.002°$, and $\gamma = 88.910° \pm 0.003°$ and a calculated density of 1.682 g·cm$^{-3}$ (Cai et al. 2001). It was prepared as follows. $N,N'$-dimethylethylenediamine (2 mM) was added with stirring to an aqueous solution of Cd(OAc)$_2$·2H$_2$O (1 mM). This solution was then treated with sodium 1,5-naphthalenedisulfonate (1 mM). Colorless crystals of the title compound were collected after 4 days.

**[Cd(SCN)$_2$(C$_5$H$_4$NCOOH)$_2$]·(C$_5$H$_4$NCOOH)** or **C$_{20}$H$_{15}$N$_5$O$_6$S$_2$Cd**. This compound crystallizes in the triclinic structure with the lattice parameters $a = 730.5 \pm 0.5$, $b = 1162.5 \pm 0.5$, and $c = 1483.7 \pm 0.5$ pm and $\alpha = 109.05° \pm 0.05°$, $\beta = 102.49° \pm 0.05°$, and $\gamma = 94.14° \pm 0.05°$ (Yang et al. 2001b). To prepare it, an aqueous solution of Cd(NO$_3$)$_2$·4H$_2$O (1 mM) and NH$_4$SCN (2 mM) was added to a hot aqueous solution (20 mL, 70°C) of nicotinic acid (2 mM) with stirring. After filtration, the final clear solution was left to stand in a desiccator for several days. Colorless prismatic crystals of the title compound suitable for XRD were formed.

**[Cd(C$_5$H$_5$NS)$_4$(NO$_3$)$_2$]** or **C$_{20}$H$_{20}$N$_6$O$_6$S$_4$Cd**. This compound melts at 170°C–172°C and crystallizes in the tetragonal structure with the lattice parameters $a = 1866.0 \pm 0.3$ and $c = 1521.5 \pm 0.3$ and a calculated density of 1.708 g·cm$^{-3}$ (Rajalingam et al. 2000). The crystals appear to decompose above 158°C but became a yellowish liquid in the range 170°C–172°C. To obtain it, pyridine-2-thione (6.47 mM) was dissolved in MeOH [EtOH (Kennedy and Lever 1972)]. To the clear solution was added at room temperature a solution of Cd(NO$_3$)$_2$·4H$_2$O (1.69 mM) in MeOH [EtOH (Kennedy and Lever 1972)] (1 mL). Light-yellow needle-shaped crystals formed in ca. 10 min. The crystals were collected by filtration, washed with ice-cold MeOH (2 mL), and vacuum dried. All manipulations were carried out under an atmosphere of dry N$_2$ in a glovebox or with vacuum/N$_2$ manifold or under an atmosphere of Ar in a glovebox.

**[Cd(Pr$^i$COS$_2$)$_2$(Phen)]** or **C$_{20}$H$_{22}$N$_2$O$_2$S$_4$Cd**. This compound crystallizes in the triclinic structure with the lattice parameters $a = 1520.2 \pm 0.3$, $b = 1479.3 \pm 0.3$, and $c = 1079.5 \pm 0.3$ pm and $\alpha = 88.97° \pm 0.02°$, $\beta = 99.79° \pm 0.02°$, and $\gamma = 93.02° \pm 0.02°$ and a calculated density of 1.57 g·cm$^{-3}$ (Glinskaya et al. 1990). To obtain it, Cd(Pr$^i$COS$_2$)$_2$ (5.5 mM) and Phen (5.7 mM) were mixed in solid state. To the mixture in small portions with stirring CHCl$_3$ was poured to dissolve the precipitate. The reaction mixture was allowed to stand for 30 min and then poured in a fivefold excess of hexane. After 30 min, the precipitate was filtered off, dried in air, and recrystallized from CHCl$_3$ at a fivefold excess of hexane and dried in vacuo.

**[Cd(Bu$^n$COS$_2$)$_2$(2,2'-bipy)]** or **C$_{20}$H$_{26}$N$_2$O$_2$S$_4$Cd**. This compound crystallizes in the monoclinic structure with the lattice parameters $a = 1055.03 \pm 0.07$, $b = 1705.6 \pm 0.1$, and $c = 1393.58 \pm 0.09$ pm and $\beta = 98.316° \pm 0.001°$ (Jiang et al. 2005); $a = 1047.49 \pm 0.10$, $b = 1685.19 \pm 0.16$, and $c = 1352.28 \pm 0.12$ pm and $\beta = 97.219° \pm 0.002°$ at 183 K and a calculated density of 1.591 g·cm$^{-3}$ (Chen et al. 2003b). It was prepared by stirring or refluxing [Cd(Bu$^n$COS$_2$)$_2$]$_n$ (1 mM) with 2,2'-bipy (1 mM) in acetone or acetonitrile (40 mL) for 1 h at room temperature and filtrated (Chen et al. 2003b, Jiang et al. 2005). Colorless single crystals were obtained by slow evaporation of the solution of the title compound.

**[Cd(Bu$^i$OCS$_2$)$_2$(2,2'-bipy)]** or **C$_{20}$H$_{26}$N$_2$O$_2$S$_4$Cd**. This compound crystallizes in the orthorhombic structure with the lattice parameters $a = 1178.90 \pm 0.03$, $b = 1218.59 \pm 0.03$, and $c = 1753.35 \pm 0.05$ pm and a calculated density of 1.495 g·cm$^{-3}$ (Larionov et al. 2005). To obtain it, a solution of K(Bu$^i$OCS$_2$) (5 mM) in H$_2$O (10 mL) was added through paper filter to a suspension of 2,2'-bipy (2.5 mM) in the solution of $CdCl_2$ (2.5 mM) in $H_2O$ (50 mL). The mixture was stirred for 30 min. The precipitate that formed was filtered with suction, washed with $H_2O$, and dried in air. The title compound was reprecipitated by dissolving in 20 mL of $CH_2Cl_2$, filtration, and subsequent addition of hexane (80 mL). Single crystals were grown via standing of the mother liquor obtained at the reprecipitation in a refrigerator.

**[Cd(Py)$_2$(S$_2$COBu$^n$)$_2$] or C$_{20}$H$_{28}$N$_2$O$_2$S$_4$Cd.** This compound crystallizes in the monoclinic structure with the lattice parameters $a = 1109.4 \pm 1.2$, $b = 615.2 \pm 0.6$, and $c = 1798.5 \pm 1.9$ pm and $\beta = 96.148° \pm 0.002°$ (Jiang et al. 2002). It was synthesized by stirring [Cd(S$_2$COBu$^n$)$_2$] with excess of Py in acetone for 30 min. Pale-yellow columnar single crystals suitable for XRD were obtained by the evaporation of the obtained solution at room temperature for few days.

**[Cd{(HOC$_2$H$_4$)$_2$NCS$_2$}$_2$(2,2′-bipy)] or C$_{20}$H$_{28}$N$_4$O$_4$S$_4$Cd.** This compound crystallizes in the triclinic structure with the lattice parameters $a = 1007.5 \pm 0.2$, $b = 1158.0 \pm 0.2$, and $c = 1177.7 \pm 0.2$ pm and $\alpha = 70.92° \pm 0.03°$, $\beta = 85.71° \pm 0.03°$, and $\gamma = 81.02° \pm 0.03°$ and a calculated density of 1.630 g·cm$^{-3}$ (Deng et al. 2007) [$a = 1007.7 \pm 0.2$, $b = 1156.8 \pm 0.2$, and $c = 1167.6 \pm 0.2$ pm and $\alpha = 70.85° \pm 0.03°$, $\beta = 85.86° \pm 0.03°$, and $\gamma = 81.21° \pm 0.03°$ at 173 K and a calculated density of 1.645 g·cm$^{-3}$ (Song and Tiekink 2009)]. It was prepared from the reaction of Cd(ClO$_4$)$_2$·6H$_2$O (0.5 mM), 2,2′-bipy (0.5 mM), and K(HOC$_2$H$_4$)$_2$NCS$_2$ (1 mM) during 6 h in MeOH solution (Deng et al. 2007). The title compound could also be obtained from the reaction of Cd{(HOC$_2$H$_4$)$_2$NCS$_2$}$_2$ and 2,2′-bipy (Song and Tiekink 2009). Colorless crystals were grown from the slow evaporation of its chloroform/EtOH solution.

**[Cd(C$_4$H$_{12}$N$_2$)$_2$]·(C$_{12}$H$_8$O$_6$S$_2$) or C$_{20}$H$_{32}$N$_4$O$_6$S$_2$Cd.** This compound crystallizes in the triclinic structure with the lattice parameters $a = 760.31 \pm 0.10$, $b = 901.27 \pm 0.12$, and $c = 959.33 \pm 0.13$ pm and $\alpha = 73.237° \pm 0.002°$, $\beta = 84.119° \pm 0.002°$, and $\gamma = 72.192° \pm 0.002°$ and a calculated density of 1.666 g·cm$^{-3}$ (Cai et al. 2001). It was prepared as follows. $N,N'$-dimethylethylenediamine (2 mM) was added with stirring to an aqueous solution of Cd(OAc)$_2$·2H$_2$O (1 mM). This solution was then treated with sodium 4,4′-biphenyldisulfonate (1 mM). Colorless crystals of the title compound were collected after 4 days.

**[Cd(SCN)$_2$(C$_7$H$_{10}$N$_2$)$_2$]·2(DMSO) or C$_{20}$H$_{32}$N$_6$O$_2$S$_4$Cd.** This compound crystallizes in the triclinic structure with the lattice parameters $a = 829.4 \pm 0.1$, $b = 864.4 \pm 0.1$, and $c = 1032.6 \pm 0.2$ pm and $\alpha = 77.85° \pm 0.01°$, $\beta = 76.02° \pm 0.01°$, and $\gamma = 83.49° \pm 0.02°$ and a calculated density of 1.491 g·cm$^{-3}$ (Secondo et al. 2000). Crystals of the title compound suitable for XRD were grown overnight from the saturated [Cd(SCN)$_2$(C$_7$H$_{10}$N$_2$)$_2$] solution in DMSO.

**[Cd(EtOCH$_2$OCS$_2$)$_3$(Et$_4$N)] or C$_{20}$H$_{41}$NO$_6$S$_6$Cd, tetraethylammonium tris(methoxyethylxanthato)cadmium.** This compound crystallizes in the monoclinic structure with the lattice parameters $a = 832.6 \pm 0.2$, $b = 2171.7 \pm 0.4$, and $c = 1769.8 \pm 0.3$ pm and $\beta = 92.28° \pm 0.02°$ and the experimental and calculated densities 1.45 and 1.447 g·cm$^{-3}$, respectively (Abrahams et al. 1988b).

**[Cd{Et$_2$(NH$_2$)$_2$CS}$_4$(SO$_4$)]** or **C$_{20}$H$_{48}$N$_8$O$_4$S$_5$Cd**. This compound melts at 171°C–172°C and crystallizes in the orthorhombic structure with the lattice parameters $a = 1653.58 \pm 0.10$, $b = 1698.77 \pm 0.13$, and $c = 2362.81 \pm 0.17$ pm at $173 \pm 2$ K and a calculated density of 1.476 g·cm$^{-3}$ (Altaf et al. 2011). It was prepared by adding 2 mM methanolic solution of $N,N'$-diethyldithiourea to an aqueous solution of CdSO$_4$ (1 mM). The reaction mixture was stirred for 30 min. The solution was filtered and the filtrate was kept at room temperature for crystallization. As a result, a white crystalline product was obtained that was washed with methanol and dried.

**[Cd(H$_2$NC$_{10}$H$_6$SO$_3$)(C$_{10}$H$_8$N$_2$)(MeOH)(NO$_3$)]$_n$** or **C$_{21}$H$_{20}$N$_4$O$_7$SCd**. This compound crystallizes in the orthorhombic structure with the lattice parameters $a = 1163.23 \pm 0.12$, $b = 1287.05 \pm 0.14$, and $c = 1458.09 \pm 0.16$ pm and a calculated density of 1.780 g·cm$^{-3}$ (Mi et al. 2008). It was prepared by the following procedure. A solution of 4-amino-1-naphthalene sulfonic acid (H$_2$NC$_{10}$H$_6$SO$_3$H) (0.5 mM) and 4,4'-bipy (0.5 mM) in 5 mL of MeOH were added to a solution of Cd(NO$_3$)$_2$ (0.5 mM) in 2 mL of H$_2$O. After adjusting the mixture to alkalescence, the solution was refluxed for 30 min and left to cool to room temperature. Then the mixture was filtered and lay aside for several days. Slow concentration of the resulting red solution afforded transparent yellow single crystals of the title compound. Its yellow crystals were obtained after the filtrate was left at room temperature for several days. By carrying out the reaction of C$_{21}$H$_{20}$N$_4$O$_7$SCd with Na$_2$S in the presence of NaOH, scalelike CdS materials have been obtained.

**{[Cd(C$_{13}$H$_{14}$N$_2$)(C$_8$H$_3$O$_7$S)(H$_2$O)$_3$]·0.5Cd(H$_2$O)$_6$·5H$_2$O}$_n$** or **C$_{21}$H$_{39}$N$_2$O$_{18}$SCd$_{1.5}$**. This compound crystallizes in the monoclinic structure with the lattice parameters $a = 746.13 \pm 0.09$, $b = 2035.43 \pm 0.19$, and $c = 2047.0 \pm 0.2$ pm and $\beta = 95.439° \pm 0.006°$ and a calculated density of 1.735 g·cm$^{-3}$ (Liu and Xu 2005a). It was prepared by the next procedure. A methanol solution of 5-sulfoisophthalic acid monosodium salt (0.1 mM), 1,3-di(4-pyridyl)propane (0.1 mM) and 0.03 mL of Et$_3$N was carefully layered on top of an aqueous solution of Cd(NO$_3$)$_2$·4H$_2$O (0.1 mM). Colorless crystals of the title compound were obtained after 8 days.

**[Cd(C$_{10}$H$_6$N$_6$O)$_2$(SCN)$_2$]$_n$** or **C$_{22}$H$_{12}$N$_{14}$O$_2$S$_2$Cd**. This compound crystallizes in the triclinic structure with the lattice parameters $a = 568.8 \pm 0.2$, $b = 891.6 \pm 0.3$, and $c = 1317.6 \pm 0.5$ pm and $\alpha = 71.365° \pm 0.005°$, $\beta = 82.003° \pm 0.005°$, and $\gamma = 88.528° \pm 0.006°$ and a calculated density of 1.804 g·cm$^{-3}$ (Du et al. 2006). To obtain it, a mixture of 2,5-bis(pyrazinyl)-1,3,4-oxadiazole (0.1 mM) and Cd(ClO$_4$)$_2$·6H$_2$O (0.1 mM) was dissolved in a MeCN/H$_2$O solution (1:1 volume ratio, 15 mL). Then, an excess of NH$_4$SCN (0.3 mM) was added to the aforementioned solution with refluxing for about 30 min. The reaction mixture was filtered and left to stand at ambient environment. Well-shaped yellow block crystals suitable for XRD were obtained after 1 day.

**[Cd(C$_6$H$_7$N$_2$S)$_2$(2,2'-bipy)]·0.5H$_2$O** or **C$_{22}$H$_{23}$N$_6$O$_{0.5}$S$_2$Cd**. This compound was obtained electrochemically (Castro et al. 1994). The cell consists of a tall-form beaker (100 mL) containing an acetonitrile (50 mL) solution of 4,6-dimethylpyrimidine-2-thione, 2,2'-bipy, and a small amount of Et$_4$NClO$_4$ (ca. 10 mg) as the supporting electrolyte; a platinum cathode and the sacrificial anode (Ni or Cd), attached to a platinum wire, served as the electrodes and were connected to a power supply. Hydrogen evolved at the cathode. As the electrolysis proceeded, the color of the solution changed from pale brown to orange. A black product was formed on the cathode but no effort was made to identify it. After electrolysis, the solution was filtered to remove any solid impurities and the solvent evaporated in a Rotavapor to give oil, which was treated with diethyl ether to produce a solid of the title compound.

**[Cd($S_2COBu^n$)$_2$(Phen)] or $C_{22}H_{26}N_2O_2S_4Cd$.** This compound melts at 115°C and crystallizes in the monoclinic structure with the lattice parameters $a = 658.45 \pm 0.02$, $b = 1915.22 \pm 0.06$, and $c = 2070.91 \pm 0.07$ pm and $\beta = 97.106° \pm 0.001°$ and a calculated density of 1.515 g·cm$^{-3}$ (Klevtsova et al. 2007). To obtain it, a solution containing K($S_2COBu^n$) (10 mM) in $H_2O$ (20 mL) was added through the paper filter to a suspension preparing at an addition to a solution of $CdCl_2 \cdot 2.5H_2O$ (5 mM) in 5 mL of $H_2O$ a solution of Phen·$H_2O$ (5 mM) in 10 mL of EtOH. The mixture was stirred for 30 min and standing for 2 h. The precipitate that formed was filtered with suction, washed with $H_2O$, and dried first in air and then in vacuo. The title compound was purified by dissolution in $CH_2Cl_2$ (50 mL) with the next evaporation of the solvent to the volume of 2 mL and addition of 10 mL of hexane. The white precipitate was filtered with suction and dried first in air and then in vacuo over NaOH.

**[Cd($S_2COBu^i$)$_2$(Phen)] or $C_{22}H_{26}N_2O_2S_4Cd$.** This compound melts at 180°C and crystallizes in the orthorhombic structure with the lattice parameters $a = 658.83 \pm 0.03$, $b = 1971.23 \pm 0.10$, and $c = 2019.36 \pm 0.11$ pm and a calculated density of 1.497 g·cm$^{-3}$ (Klevtsova et al. 2007). To obtain it, a solution containing K($S_2COBu^i$) (10 mM) in $H_2O$ (20 mL) was added through the paper filter to a suspension preparing at an addition to a solution of $CdCl_2 \cdot 2.5H_2O$ (5 mM) in 5 mL of $H_2O$ a solution of Phen·$H_2O$ (5 mM) in 10 mL of EtOH. The mixture was stirred for 30 min and standing for 2 h. The precipitate that formed was filtered with suction, washed with $H_2O$, and dried first in air and then in vacuo. The title compound was purified by dissolution in $CH_2Cl_2$ (50 mL) with the next evaporation of the solvent to the volume of 2 mL and addition of 10 mL of hexane. The white precipitate was filtered with suction and dried first in air and then in vacuo over NaOH.

**[Cd{$S_2CO(CH_2)_2OMe$}$_2$(4,7-Me$_2$Phen)] or $C_{22}H_{26}N_2O_4S_4Cd$.** This compound melts at 137°C–138°C and crystallizes in the monoclinic structure with the lattice parameters $a = 897.70 \pm 0.08$, $b = 2338.1 \pm 0.2$, and $c = 1170.06 \pm 0.10$ pm and $\beta = 91.519° \pm 0.002°$ at 183 K (Chen et al. 2003a). It was prepared by the following procedure. To a stirred chloroform/acetonitrile (50 ml) solution of Cd($S_2CO(CH_2)_2OMe$)$_2$ (0.2 g) a stoichiometric amount of 4,7-Me$_2$Phen was added. The mixture was refluxed for 2 h and then filtered and the solvent was removed in vacuo. The precipitate was recrystallized by the slow evaporation of a chloroform solution of the title compound to yield colorless crystals suitable for XRD.

**[Cd{$(HOC_2H_4)_2NCS_2$}$_2$(Phen)] or $C_{22}H_{28}N_4O_4S_4Cd$.** This compound was prepared from the reaction of Cd($ClO_4$)$_2 \cdot 6H_2O$ (0.5 mM), Phen (0.5 mM), and K($HOC_2H_4$)$_2$NCS$_2$ (1 mM) during 6 h in MeOH solution (Deng et al. 2007).

**[Cd(C$_6$H$_6$N$_2$O)$_2$(H$_2$O)$_4$](C$_{10}$H$_8$O$_6$S$_2$) or C$_{22}$H$_{28}$N$_4$O$_{12}$S$_2$Cd.** This compound crystallizes in the monoclinic structure with the lattice parameters $a = 1474.2 \pm 0.8$, $b = 689.9 \pm 0.4$, and $c = 1529.2 \pm 0.8$ pm and $\beta = 110.980° \pm 0.009°$ and a calculated density of 1.695 g·cm$^{-3}$ (Li et al. 2008). To obtain it, a nicotinamide (0.4 mM) was added with constant stirring to an aqueous solution (10 mL) of Cd(OAc)$_2$·2H$_2$O (0.2 mM). The solution was then treated with disodium biphenyl-4,4'-disulfonate (0.2 mM). Colorless crystals of the title compound were collected by slow evaporation at room temperature after 7 days.

**[Cd(C$_9$H$_{12}$N$_5$S)(OAc)]$_2$ or C$_{22}$H$_{30}$N$_{10}$O$_4$S$_2$Cd$_2$.** This compound melts at the temperature higher than 300°C (Bermejo et al. 2001, 2004a). It was prepared as follows. A solution of (2-pyridyl)formamide $N$(4)-dimethylthiosemicarbazone (1 mM) in EtOH (30 mL) was added to a solution of Cd(OAc)$_2$ (1 mM) in 30 mL of 96% EtOH, and the mixture kept stirring for several days at room temperature. The resulting yellow solid was filtered off, washed thoroughly with cold EtOH, and stored in a desiccator over Mg and I$_2$.

**[Cd(C$_8$H$_{16}$O$_4$)$_2$][Cd$_2$(SCN)$_6$] or C$_{22}$H$_{32}$N$_6$O$_8$S$_6$Cd$_3$.** This compound crystallizes in the monoclinic structure with the lattice parameters $a = 1629.7 \pm 0.3$, $b = 2626.7 \pm 0.3$, and $c = 1629.6 \pm 0.2$ pm and $\beta = 90.20° \pm 0.01°$ (Zhang et al. 1999). It was prepared by the following procedure. With vigorous stirring, 18 mL of an aqueous solution of 0.056 M 12-crown-4 (C$_8$H$_{16}$O$_4$) was added dropwise to a preformed mixture of 13.3 mL of 0.3 M aqueous solution of CdSO$_4$ and 32 mL of 1 M aqueous solution of NH$_4$SCN over a period of 30 min. The final C$_8$H$_{16}$O$_4$/CdSO$_4$/NH$_4$SCN molar ratio was 1:4:32. Care must be taken to prevent supersaturation and/or precipitation during the addition. Colorless crystals were grown via slow evaporation.

**[Cd(C$_{10}$H$_8$N$_2$S$_2$)$_2$(H$_2$O)$_2$](NO$_3$)$_2$·2CH$_3$OH·2H$_2$O or C$_{22}$H$_{32}$N$_6$O$_{12}$S$_4$Cd.** This compound crystallizes in the monoclinic structure with the lattice parameters $a = 950.2 \pm 0.1$, $b = 1962.8 \pm 0.2$, and $c = 975.3 \pm 0.1$ pm and $\beta = 110.245° \pm 0.002°$ (Yan 2008). To obtain it, a methanol solution (10 mL) of 4,4'-bipyridyldisulfide (0.1 mM) and maleic acid (0.1 mM) were slowly diffused into an aqueous solution (10 mL) of Cd(NO$_3$)$_2$·4H$_2$O (0.1 mM) with stirring for 40 min at room temperature. Colorless block crystals were obtained in 1 week.

**[Cd(C$_7$H$_4$NS$_2$)$_2$(2,2'-bipy)]·H$_2$O or C$_{24}$H$_{18}$N$_4$OS$_4$Cd.** This compound was prepared by the electrochemical oxidation of anodic Cd in an acetonitrile solution of benzothiazole-2-thione in the presence of 2,2'-bipy (Castro et al. 1993b).

**[Cd(C$_6$N$_7$N$_2$S)$_2$(Phen)]·H$_2$O or C$_{24}$H$_{24}$N$_6$OS$_2$Cd.** This compound was obtained electrochemically (Castro et al. 1994). The cell consists of a tall-form beaker (100 mL) containing an acetonitrile (50 mL) solution of 4,6-dimethylpyrimidine-2-thione, Phen, and a small amount of Et$_4$NClO$_4$ (ca. 10 mg) as the supporting electrolyte; a platinum cathode and the sacrificial anode (Ni or Cd), attached to a platinum wire, served as the electrodes and were connected to a power supply. Hydrogen evolved at the cathode. As the electrolysis proceeded, the color of the solution changed from pale-brown to orange. A black product was formed on the cathode but no effort was made to identify it. After electrolysis, the solution was filtered to remove any solid impurities and the solvent evaporated in a Rotavapor to give oil, which was treated with diethyl ether to produce a solid of the title compound.

**[Cd(C$_8$H$_{16}$O$_4$)$_2$][Cd$_3$(SCN)$_8$] or C$_{24}$H$_{32}$N$_8$O$_8$S$_8$Cd$_4$.** This compound crystallizes in the monoclinic structure with the lattice parameters $a = 1160.4 \pm 0.7$, $b = 1136.6 \pm 0.7$, and $c = 1573.6 \pm 0.4$ pm and $\beta = 92.84° \pm 0.03°$ and a calculated density of 2.029 g·cm$^{-3}$ (Zhang et al. 1999). It was prepared using the next procedure. With vigorous stirring, 18 mL of an aqueous solution of 0.056 M 12-crown-4 (C$_8$H$_{16}$O$_4$) was added dropwise to a preformed mixture of 20 mL of 0.3 M aqueous solution of CdSO$_4$ and 24 mL of 1 M aqueous solution of NH$_4$SCN over a period of 30 min. The final C$_8$H$_{16}$O$_4$/CdSO$_4$/NH$_4$SCN molar ratio was 1:6:24. Care must be taken to prevent supersaturation and/or precipitation during the addition. Colorless crystals were grown via slow evaporation.

**[Cd(SOCCMe$_3$)$_2$(C$_7$H$_9$N)$_2$] or C$_{24}$H$_{36}$N$_2$O$_2$S$_2$Cd.** This compound crystallizes in the monoclinic structure with the lattice parameters $a = 1866.6 \pm 0.1$, $b = 938.0 \pm 0.1$, and $c = 1654.2 \pm 0.1$ pm and $\beta = 94.42° \pm 0.01°$ (Nyman et al. 1997). To obtain it, CdCO$_3$ (5.8 mM), 3,5-dimethylpyridine (11.6 mM), and 20 mL of toluene were combined in a round-bottom flask. Thiopivalic acid (11.8 mM) was pipetted into the rapidly stirring mixture. After approximately 40 min, the solution became clear and yellow, and gas evolution (CO$_2$) was observed over the course of the reaction. After the mixture was stirred for 160 min, the toluene and volatile by-products (H$_2$O) were removed under reduced pressure, and a glassy, white solid remained. The solid was redissolved in toluene, and pentane vapor was introduced slowly. After 4 days, this compound crystallized as long, colorless blades.

Cd(SOCCMe$_3$)$_2$(C$_7$H$_9$N)$_2$ decomposes in solution to form sphalerite, nanocrystalline CdS (Nyman et al. 1997). This reaction has been observed to take place from 25°C to 100°C with increasing rate of CdS formation at elevated temperatures.

**[Cd(C$_{10}$H$_8$N$_2$S$_2$)$_2$(H$_2$O)$_2$(NO$_3$)$_2$]·2EtOH·2H$_2$O or C$_{24}$H$_{36}$N$_6$O$_{12}$S$_4$Cd.** This compound crystallizes in the monoclinic structure with the lattice parameters $a = 897.35 \pm 0.09$, $b = 1983.5 \pm 0.1$, and $c = 1077.8 \pm 0.2$ pm and $\beta = 112.29° \pm 0.01°$ K and a calculated density of 1.574 g·cm$^{-3}$ (Kondo et al. 2000). To obtain it, an aqueous solution (15 mL) of Cd(NO$_3$)$_2$·4H$_2$O (0.78 mM) in H$_2$O was slowly diffused into an ethanol solution (15 mL) of 4,4′-bipyridyldisulfide (2.2 mM). White plate crystals of the title compound were formed in 1 week. One of them was used for XRD, and the residual crystals were collected by filtration, washed with H$_2$O and EtOH, and dried in vacuo. Although the single-crystal structure involves guest EtOH and H$_2$O molecules, solvent molecules were lost during the drying process.

**[Cd(C$_{14}$H$_8$S$_2$O$_4$)(C$_{12}$H$_8$N$_2$)]·0.125H$_2$O or C$_{26}$H$_{16.25}$N$_2$O$_{4.125}$S$_2$Cd.** This compound crystallizes in the triclinic structure with the lattice parameters $a = 1012.3 \pm 0.6$, $b = 1120.3 \pm 0.6$, and $c = 1188 \pm 1$ pm and $\alpha = 96.11° \pm 0.01°$, $\beta = 95.87° \pm 0.01°$, and $\gamma = 116.076° \pm 0.008°$ (Wei and Wang 2009). It was prepared by the hydrothermal method from a mixture of CdCl$_2$·2.5H$_2$O (0.50 mM), Phen (0.50 mM), 2,2′-dithiobis(benzoic acid) (0.50 mM), NaOH(1.0 mM), and H$_2$O (20 mL), heated in a Teflon-lined steel autoclave inside a programmable electric furnace at 150°C for 5 days. After cooling the autoclave to room temperature for 1 day, yellow-brown crystals were obtained.

**[Cd(S$_2$COCH$_2$Ph)$_2$(C$_{11}$H$_6$N$_2$O)] or C$_{27}$H$_{20}$N$_2$O$_3$S$_4$Cd**. This compound crystallizes in the triclinic structure with the lattice parameters $a = 925.92 \pm 0.16$, $b = 1155.7 \pm 0.2$, and $c = 1427.5 \pm 0.3$ pm and $\alpha = 112.993° \pm 0.003°$, $\beta = 91.716° \pm 0.002°$, and $\gamma = 105.280° \pm 0.002°$ at 153 K and a calculated density of 1.637 g·cm$^{-3}$ (Câmpian et al. 2013). Cd(S$_2$COCH$_2$Ph)$_2$ was obtained from the reaction of Cd(NO$_3$)$_2$·4H$_2$O and potassium O-benzyl dithiocarbonate in aqueous solution. The 4.5-diazafluoren-9-one-$N$,$N'$ (C$_{11}$H$_6$N$_2$O) adducts were obtained from stirring at room temperature Cd(S$_2$COCH$_2$Ph)$_2$ (0.2 g) with a stoichiometric amount of C$_{11}$H$_6$N$_2$O in EtOH. The solvent was removed in vacuo and crystals were isolated from slow evaporation of its EtOH/acetone solutions.

**[Cd(C$_7$H$_5$OS)$_3$(C$_6$H$_{16}$N)] or C$_{27}$H$_{31}$NO$_3$S$_3$Cd**. This compound crystallizes in the monoclinic structure with the lattice parameters $a = 1962.19 \pm 0.04$, $b = 1550.67 \pm 0.04$, and $c = 1943.99 \pm 0.05$ pm and $\beta = 103.015° \pm 0.001°$ and a calculated density of 1.443 g·cm$^{-3}$ (Vittal and Dean 1998). To obtain it, thiobenzoic acid (17.0 mM) in 15 mL of MeOH was added to Et$_3$N (17.0 mM). The resultant yellow solution was added with stirring to a solution of Cd(NO$_3$)$_2$·4H$_2$O (5.1 mM) dissolved in 10 mL of H$_2$O to give a milky supernatant and a yellow precipitate. The addition of a solution containing BaCl$_2$ (2.5 mM) in 10 mL of H$_2$O produced a colorless upper layer and a yellow lower layer. The mixture was stirred for 10 min and then the two layers were separated. About 20 mL of Et$_2$O was layered onto the yellow fraction and the mixture was allowed to stand in a refrigerator at 5°C. The yellow crystals that formed were decanted, washed with Et$_2$O, and dried in air. A second crop of crystals was obtained from the mixture of the mother liquor and washings. Single crystals were formed during the preparation. The synthesis was carried out under an Ar atmosphere.

**[Cd(S$_2$O$_3$)$_2$(C$_{14}$H$_{12}$N$_2$)$_2$] or C$_{28}$H$_{24}$N$_4$O$_6$S$_4$Cd$_2$**. This compound crystallizes in the monoclinic structure with the lattice parameters $a = 843.9 \pm 0.1$, $b = 1686.4 \pm 0.2$, and $c = 1101.0 \pm 0.1$ pm and $\beta = 109.56° \pm 0.01°$ and a calculated density of 1.947 g·cm$^{-3}$ (Baggio et al. 1996). The crystalline title compound was obtained by allowing a methanol solution of 2,9-Me$_2$Phen to diffuse into an aqueous solution of Na$_2$S$_2$O$_3$ and Cd(OAc)$_2$ (1:1:1 molar ratio). After several days, small colorless plates appeared.

**[Cd(SCN)$_2$(C$_{13}$H$_{12}$N$_2$S)$_2$]·2H$_2$O or C$_{28}$H$_{28}$N$_6$O$_2$S$_4$Cd**. This compound crystallizes in the monoclinic structure with the lattice parameters $a = 564.3 \pm 0.2$, $b = 1562.5 \pm 0.8$, and $c = 1764.1 \pm 0.8$ pm and $\beta = 89.98° \pm 0.04°$ and a calculated density of 1.540 g·cm$^{-3}$ (Zhu et al. 2000). To obtain it, Cd(NO$_3$)$_2$·4H$_2$O (1 mM), KSCN (2 mM) and diphenylthiourea (2 mM) were dissolved in a small volume of EtOH. The mixture was heated until the white material, which was formed, was completely dissolved. After filtration, the solution was allowed to evaporate slowly. Crystals deposited after several days.

**[Cd(C$_{12}$H$_{16}$N$_5$S)(OAc)]$_2$ or C$_{28}$H$_{38}$N$_{10}$O$_4$S$_2$Cd$_2$**. This compound melts at the temperature higher than 300°C (Castiñeiras et al. 2002). To obtain it, a solution of

$C_{12}H_{17}N_5S$ [2-pyridineformamide $N$(4)-piperidylthiosemicarbazone] (0.76 mM) in EtOH (30 mL) was added to a solution of $Cd(OAc)_2$ (0.76 mmol) in 30 mL of 96% EtOH, and the mixture kept stirring for several days at room temperature. The resulting yellow solid was filtered off, washed thoroughly with cold ethanol, and stored in a desiccator over Mg and $I_2$.

**$[Bu^n_4N][Cd(OAc)(S_2CNEt_2)_2]$ or $C_{28}H_{59}N_3O_2S_4Cd$.** This compound was prepared as follows (McCleverty et al. 1982). To an aqueous solution of $[Bu^n_4N][OH]$ (1.3 g) in acetone (25 mL) HOAc (2 mM) was added, and the mixture was evaporated at 60°C to afford an oil. This oil was dissolved in acetone (50 mL) and $[Cd(S_2CNEt_2)_2]$ (2 mM) was added slowly with vigorous stirring until all the cadmium compound had dissolved. Upon evaporation of the mixture, an oil was formed, which, upon shaking with diethyl ether (30 mL) for 30 min, afforded the title compound as a white powder. This was collected by filtration and dried in vacuo.

**$[Bu^n_4N][Cd(OCOEt)(S_2CNEt_2)_2]$ or $C_{29}H_{61}N_3O_2S_4Cd$.** This compound was prepared by the next procedure (McCleverty et al. 1982). To an aqueous solution of $[Bu^n_4N][OH]$ (1.3 g) in acetone (25 mL) HOCOEt (2 mM) was added, and the mixture was evaporated at 60°C to afford an oil. This oil was dissolved in acetone (50 mL) and $[Cd(S_2CNEt_2)_2]$ (2 mM) was added slowly with vigorous stirring until all the cadmium compound had dissolved. Upon evaporation of the mixture, an oil was formed which, upon shaking with diethyl ether (30 mL) for 30 min, afforded the title compound as a white powder. This was collected by filtration and dried in vacuo.

**$[Cd(H_2O)(C_{10}N_2H_8)_2(C_{10}H_6S_2O_7)]\cdot H_2O$ or $C_{30}H_{26}N_4O_9S_2Cd$.** This compound crystallizes in the monoclinic structure with the lattice parameters $a = 1418.89 \pm 0.02$, $b = 1569.21 \pm 0.01$, and $c = 1420.49 \pm 0.02$ pm and $\beta = 105.449° \pm 0.001°$ (Chunying and Ning 2012). It was prepared by the next procedure. A mixture of 7-hydroxynaphthalene-1,3-disulfonic acid (0.1 mM), $Cd(OAc)_2$ (0.1 mM), 2,2'-bipy (0.1 mM), NaOH (0.2 M, 1.5 mL), and $H_2O$ (10 mL) was stirred for about 30 min. The resulting solution was sealed in a Teflon-lined stainless autoclave and heated to 120°C for 3 days. The bottle was cooled to ambient temperature rapidly. Yellow single crystals were recovered by vacuum filtration and dried in air.

**$[Cd_2(SCN)_4(C_{13}H_{19}N_3O)_2]$ or $C_{30}H_{38}N_{10}O_2S_4Cd_2$.** This compound crystallizes in the monoclinic structure with the lattice parameters $a = 740.31 \pm 0.03$, $b = 2652.6 \pm 0.1$, and $c = 1009.70 \pm 0.04$ pm and $\beta = 108.111° \pm 0.002°$ (Lou and Qiu 2011). To prepare it, a methanol solution of $Cd(OAc)_2 \cdot 2H_2O$ (1 mM) was added to a methanol solution (30 mL) of $NH_4SCN$ (2 mM) and the Schiff base ligand (2-morpholin-4-ylethyl)-(1-pyridin-2-yl-ethylidene)amine (1 mM) with stirring. The resulting solution was allowed to stand at room temperature for a week, yielding colorless block-shaped single crystals.

**$[Cd(C_{13}H_{18}N_5S)(OAc)]_2$ or $C_{30}H_{42}N_{10}O_4S_2Cd_2$.** To obtain this compound, a solution of $Cd(OAc)_2 \cdot 2H_2O$ (0.90 mM) in EtOH (30 mL) was mixed with a solution of 2-pyridineformamide 3-hexamethyleneiminylthiosemicarbazone $(C_{13}H_{19}N_5S)$ (0.90 mM) in EtOH (30 mL) (Bermejo et al. 2004b). The resulting solution was stirred at room temperature for 7 days and the yellow precipitate was filtered off.

**[Me₄N][Cd(6-EtOC₇H₃NS₂)₃] or C₃₁H₃₆N₄O₃S₆Cd**. To obtain this compound, to a solution of [Me₄N][6-EtOC₇H₃NS₂] (0.57 g) in acetone (50 mL) [Cd(6-EtOC₇H₃NS₂)₂] (1.07 g) was added, slowly with shaking (McCleverty et al. 1982). The mixture was shaken vigorously for 2 h. It was then filtered and the filtrate evaporated in vacuo to low bulk. Light petroleum was added to this residual filtrate and the title compound precipitated as a pale-yellow powder. It was filtered off, washed with diethyl ether, and dried in vacuo.

**[Cd{S₂CN(C₇H₇)(C₂H₄OH)}₂(C₁₂H₈N₂)] or C₃₂H₃₂N₄O₂S₄Cd**. This compound crystallizes in the orthorhombic structure with the lattice parameters $a = 726.63 \pm 0.07$, $b = 1860.6 \pm 0.2$, and $c = 2401.0 \pm 0.2$ pm at 223 K (Saravanan et al. 2005). It was prepared by reacting an ethanol solution of Phen with a hot $C_6H_6$/EtOH (2:1 volume ratio) solution containing authenticated Cd[S₂CN(C₇H₇)(CH₂CH₂OH)]₂. Crystals were obtained from the slow evaporation of a benzene/ethanol solution of the title compound.

**[Cd(C₁₀H₈N₂)(H₂O)(C₁₀H₁₁N₂O₅S)₂]·EtOH·2H₂O or C₃₂H₄₂N₆O₁₄S₂Cd**. This compound crystallizes in the monoclinic structure with the lattice parameters $a = 1131.06 \pm 0.08$, $b = 1189.76 \pm 0.08$, and $c = 2905.8 \pm 0.2$ pm and $\beta = 92.325° \pm 0.001°$ (Wang et al. 2008). It was prepared under the hydrothermal conditions. $N$-$p$-acetamidobenzenesulfonyl-glycine acid (1 mM) and 2,2′-bipy in an aqueous solution (10 mL) of NaOH (1 mM) were mixed with an aqueous solution (12 mL) of Cd(ClO₄)₂ (2 mM). After stirring for 5 min on air, Et₃N (0.5 mM) was added and then the mixture was placed into 25 mL Teflon-lined autoclave and heated at 180°C for 96 h. The autoclave was cooled at a rate of 5°C h⁻¹. Colorless prismatic crystals were collected by filtration, washed with water, and dried on air.

**[Cd(H₂O)(C₁₀H₈N₂)₂(C₁₄H₆S₂O₈)]·H₂O or C₃₄H₂₆N₄O₁₀S₂Cd**. This compound crystallizes in the triclinic structure with the lattice parameters $a = 989.7 \pm 0.3$, $b = 1285.6 \pm 0.6$, and $c = 1407.6 \pm 0.4$ pm and $\alpha = 88.31° \pm 0.03°$, $\beta = 80.54° \pm 0.02°$, and $\gamma = 69.54° \pm 0.04°$ (Zhao et al. 2010). To obtain it, a mixture of disodium 9,10-dioxoanthracene-2,6,-disulfonate (0.1 mM), Cd(NO₃)₂·4H₂O (0.1 mM), 2,2′-bipy (0.1 mM), and H₂O 10 mL was stirred for about 30 min and then sealed in a Teflon-lined stainless steel autoclave and heated at 120°C for 3 days. The autoclave was then allowed to cool to an ambient temperature. The title compound was obtained as yellow block-shaped crystals, recovered by vacuum filtration, and dried in air.

**[Cd(S-2-PhCONHC₆H₄)₂(1-MeIm)₂]·0.5H₂O or C₃₄H₃₃N₆O₂.₅S₂Cd (1-MeIm = 1-methylimidazole)**. This compound could be obtained if an excess of NaBH₄ is added in small portions to a THF/EtOH (1:1 volume ratio, 10 mL) solution of bis(2-benzoylaminophenol)disulfide (0.512 mM) (Sun et al. 1999). The mixture was stirred at room temperature to give a clear solution and added dropwise to a stirred solution of Cd(NO₃)₂·4H₂O (0.512 mM) and 1-MeIm (1.024 mM) in THF (10 mL). The stirring was continued for about 1 h after the addition. The solvents were removed by evaporation, then toluene (50 mL) was added to the residue, and the mixture was stirred overnight at ambient temperature. After filtration, the toluene was removed from the filtrate in vacuo to give crude products. Colorless crystals of

$C_{34}H_{33}N_6O_{2.5}S_2Cd$ were obtained by recrystallization from toluene and diethyl ether. Single crystals were grown by slow evaporation from recrystallization solution.

**[Cd{S$_2$CN(CH$_2$)$_2$OHPr$^n$}$_2$(C$_{12}$H$_{10}$N$_4$)$_2$] or C$_{36}$H$_{44}$N$_{10}$O$_2$S$_4$Cd.** This compound crystallizes in the triclinic structure with the lattice parameters $a = 853.2 \pm 0.3$, $b = 1095.1 \pm 0.4$, and $c = 1118.4 \pm 0.5$ pm and $\alpha = 79.59° \pm 0.03°$, $\beta = 88.06° \pm 0.03°$, and $\gamma = 78.23° \pm 0.02°$ at 98 K and a calculated density of 1.468 g·cm$^{-3}$ (Broker and Tiekink 2011). It was prepared from the reaction of Cd[S$_2$CN(CH$_2$)$_2$OHPr$^n$]$_2$ and 4-[(1$E$)-[($E$)-2-(pyridin-4-ylmethylidene)hydrazin-1-ylidene]methyl]pyridine. Yellow plates of the title compound were obtained from the slow evaporation of a chloroform/acetonitrile (3:1 volume ratio) solution.

**[Cd(H$_2$O)$_2$(C$_{12}$H$_8$N$_2$)(C$_7$H$_4$O$_5$S)]$_2$ or C$_{38}$H$_{32}$N$_4$O$_{14}$S$_2$Cd.** This compound crystallizes in the triclinic structure with the lattice parameters $a = 823.84 \pm 0.08$, $b = 1012.99 \pm 0.09$, and $c = 1296.9 \pm 0.1$ pm and $\alpha = 102.000° \pm 0.002°$, $\beta = 101.049° \pm 0.002°$, and $\gamma = 113.453° \pm 0.001°$ (Fan et al. 2005). It was obtained by the following procedure. A mixture of Cd(OAc)$_2$·2H$_2$O (0.25 mM), potassium hydrogen 4-sulfobenzoate (0.26 mM), Phen (0.23 mM), and H$_2$O (15 mL) was sealed in a 30 mL stainless steel reactor with Teflon liner and heated to 150°C for 24 h. After cooling, colorless needle crystals of the title compound were collected by filtration.

**[Cd(PhS)$_2$·(Phen)$_2$]·(DMF) or C$_{39}$H$_{33}$N$_5$OS$_2$Cd.** When the filtrate obtained from the procedure of Cd(PhS)$_2$·0.5CS$_2$·(Phen)$_2$ was stirred for another 48 h and further precipitate was removed, the addition of MeCN to the cold filtrate caused the precipitation of a yellow solid, which is apparently the DMF solvate Cd(PhS)$_2$·(Phen)$_2$·(DMF) (Black et al. 1986).

**[Bu$^n_4$N][Cd(C$_7$H$_4$NS$_2$)$_2$(6-EtOC$_7$H$_3$NS$_2$)] or C$_{39}$H$_{52}$N$_4$OS$_6$Cd.** This compound was prepared as follows (McCleverty et al. 1982). To a solution of [Bu$^n_4$N][OH] (0.65 g, 40% aqueous solution) in acetone (20 mL) 6-ethoxybenzothiazoline-2-thione (0.21 g) was added. The mixture was stirred and then filtered, and to the filtrate was added [Cd(C$_7$H$_4$NS$_2$)$_2$]$_n$ (0.44 g). After shaking the solution that had formed for 15 min, light petroleum was added, and the title compound precipitated as a pale-yellow powder. It was filtered off, washed with diethyl ether, and dried in vacuo.

**[{(C$_6$H$_{11}$)}$_2$C$_3$H$_3$N$_2$)$_2$][Cd$_2$(SCN)$_6$]·C$_3$H$_6$O or C$_{39}$H$_{56}$N$_{10}$OS$_6$Cd$_2$.** This compound crystallizes in the monoclinic structure with the lattice parameters $a = 1330.49 \pm 0.12$, $b = 1755.50 \pm 0.16$, and $c = 2080.12 \pm 0.19$ pm and $\beta = 101.494° \pm 0.002°$ and a calculated density of 1.532 g·cm$^{-3}$ (Chen et al. 2002). It was prepared by the following procedure. To a mixture of 1 mL of 1 M Cd(NO$_3$)$_2$·4H$_2$O and 1 mL of 3 M NH$_4$SCN in H$_2$O [(C$_6$H$_{11}$)$_2$C$_3$H$_3$N$_2$]I ($N,N'$-dicyclohexylimidazolium iodide) (1.0 mM) was added. A white precipitate was immediately formed. Acetone was added to the resulted mixture till all the precipitate dissolved again. Slow evaporation of the pale-yellow solution afforded colorless prisms of the title compound.

**[Cd(C$_8$H$_5$N$_4$O$_2$S)$_2$(C$_{12}$H$_8$N$_2$)$_2$] or C$_{40}$H$_{26}$N$_{12}$O$_4$S$_2$Cd.** This compound crystallizes in the monoclinic structure with the lattice parameters $a = 2341.7 \pm 0.02$, $b = 1095.64 \pm 0.09$, and $c = 1602.4 \pm 0.1$ pm and $\beta = 109.654° \pm 0.001°$ (Hu and Wang 2010). To obtain it, a mixed solution of 1-(4-carboxyphenyl)-5-mercapto-1$H$-tetrazole

(0.05 mM) and Phen (0.05 mM) in the presence of excess 2,6-dimethylpyridine (ca. 0.05 mL for adjusting the pH value to basic condition) in MeOH (10 mL) was carefully layered on top of a $H_2O$ solution (15 mL) of $Cd(ClO_4)_2 \cdot 6H_2O$ (0.1 mM) in a test tube. After ca. 2 weeks at room temperature, colorless single crystals were obtained.

**$[Bu^n{}_4N][Cd(C_7H_4NS_2)(6\text{-}EtOC_7H_3NS_2)_2]$ or $C_{41}H_{56}N_4O_2S_6Cd$.** This compound was prepared by the next procedure (McCleverty et al. 1982). To a solution of $[Bu^n{}_4N][C_7H_4NS_2]$ (0.82 g) in acetone (50 mL) $[Cd(6\text{-}EtOC_7H_3NS_2)_2]_n$ (1.07 g) was added. The mixture was shaken for 2 h during which time a homogeneous solution had been formed. This was filtered and the filtrate was evaporated to low bulk. To the residual solution light petroleum was added, and after a short period of shaking, the title compound formed as a white solid that was filtered off, washed with diethyl ether, and dried in vacuo.

**$[Cd(C_{21}H_{16}N_3O_4S_2)_2]$ or $C_{42}H_{32}N_6O_8S_4Cd$.** This compound was prepared by mixing $N,N$-(iminodiethylene)bisphthalimide dithiocarbamic acid (20 mM) and $CS_2$ (20 mM) in acetonitrile or EtOH under ice-cold conditions (5°C) (Thirumaran et al. 1998, Prakasam et al. 2007). To the yellow dithiocarbamic acid solution aqueous solution of $Cd(NO_3)_2$ was added with constant stirring. Pale-yellow solid separated from the solution, which was filtered, washed with $H_2O$, alcohol, and then dried in air.

**$[Cd_2(C_3H_5OS_2)_4(C_{15}H_{16}N_2O_2S)_2]$ or $C_{42}H_{52}N_4O_8S_{10}Cd_2$.** This compound crystallizes in the triclinic structure with the lattice parameters $a = 943.0 \pm 0.4$, $b = 1182.6 \pm 0.3$, and $c = 1204.4 \pm 0.7$ pm and $\alpha = 92.23° \pm 0.03°$, $\beta = 96.31° \pm 0.04°$, and $\gamma = 91.94° \pm 0.03°$ and a calculated density of 1.60 g·cm$^{-3}$ (Sun et al. 1994). It was synthesized by refluxing $CdBr_2$ with stoichiometric quantities of $CS_2$, sodium ethoxide, and 4-methoxyaniline in EtOH for 2 h. The single crystals of the pale-yellow color were obtained by slow evaporation of the solvent.

**$[(C_{16}H_{36}N)_2\{Cd(C_5O_3S_2)_2\}] \cdot 0.25H_2O$ or $C_{42}H_{72.5}N_2O_{6.25}S_4Cd$.** This compound crystallizes in the triclinic structure with the lattice parameters $a = 982.0 \pm 0.5$, $b = 1500.2 \pm 0.5$, and $c = 1740.6 \pm 0.5$ pm and $\alpha = 74.853° \pm 0.005°$, $\beta = 86.898° \pm 0.005°$, and $\gamma = 87.705° \pm 0.005°$ and a calculated density of 1.272 g·cm$^{-3}$ (Chen et al. 2011). It was prepared by the next procedure. To a solution containing $K_2C_5O_3S_2$ (0.8 mM) in $H_2O$ (20 mL) a solution containing $Cd(NO_3)_2 \cdot 4H_2O$ (0.4 mM) in $H_2O$ (5 mL) was added. The resulting mixture was heated to 70°C for 1 h and then filtered into a solution of $Bu_4NBr$ (0.95 mM) in EtOH (5 mL). The solid product was collected by suction filtration, washed with water, and dried in air. Red block crystals were obtained by recrystallization from acetone.

**$[Bu^n{}_4N][Cd(6\text{-}EtOC_7H_3NS_2)_3]$ or $C_{43}H_{60}N_4O_3S_6Cd$.** To obtain this compound, a solution of $[Bu^n{}_4N][OH]$ (40% aqueous solution) in acetone was added to an appropriate amount of acetic, propionic, or butyric acid (McCleverty et al. 1982). The resulting mixture was evaporated at 60°C to give oil that was redissolved in acetone. To this solution was added $[Cd(6\text{-}EtOC_7H_3NS_2)_3]_n$ and the resulting mixture was shaken vigorously until the cadmium compound had dissolved. The solution was then filtered and diethyl ether was added to the filtrate. After overnight storage in a refrigerator, a white powder was formed. This was collected by filtration, washed with diethyl ether, and dried in vacuo.

**[Cd$_{3.5}$(SC$_6$H$_4$Me-2)$_7$(C$_3$H$_7$NO)] or C$_{52}$H$_{56}$NOS$_7$Cd$_{3.5}$.** This compound crystallizes in the tetragonal structure with the lattice parameters $a = 1402.8 \pm 0.1$ and $c = 5529.5 \pm 0.8$ pm and a calculated density of 1.62 g·cm$^{-3}$ (Dance et al. 1990). To obtain it, a solution of Cd(NO$_3$)$_2$·4H$_2$O (8.6 g) in MeOH (50 mL) was added to a mixture of Et$_3$N (5.6 g) and 2-methylbenzenethiol (6.9 g) in MeOH (50 mL) at 0°C, in the absence of O$_2$. After the slurry was stirred for 1 h, the precipitate was filtered out, washed well with large volumes of MeOH and acetone, and vacuum dried. Crystals were grown by carefully layering a solution in DMF (0.2 g mL$^{-1}$) with EtOH.

**[Cd(C$_{21}$H$_{16}$N$_3$O$_4$S$_2$)$_2$(Phen)] or C$_{54}$H$_{40}$N$_8$O$_8$S$_4$Cd.** This compound decomposes at 198°C (Prakasam et al. 2007). It was prepared by adding a hot solution of Phen (1 mM) in benzene to a hot suspension of [Cd(C$_{21}$H$_{16}$N$_3$O$_4$S$_2$)$_2$] (1 mM) in benzene/chloroform (1:1 volume ratio) mixture. The resulting pale-yellow solution was filtered and cooled. The yellow compound separated out, which was filtered and recrystallized from benzene/chloroform (1:1 volume ratio) mixture.

**{[Cd$_3$(C$_8$H$_3$O$_7$S)$_2$(4,4′-bipy)$_4$(H$_2$O)$_2$]·3H$_2$O}$_n$ or C$_{56}$H$_{48}$N$_8$O$_{19}$S$_2$Cd$_3$.** This compound crystallizes in the monoclinic structure with the lattice parameters $a = 3557.8 \pm 0.4$, $b = 1023.27 \pm 0.10$, and $c = 1621.18 \pm 0.17$ pm and $\beta = 108.913° \pm 0.002°$ and a calculated density of 1.830 g·cm$^{-3}$ (Li et al. 2004, 2005). It was prepared by the following procedure. A mixture of NaC$_8$H$_5$O$_7$S (monosodium salt of 5-sulfoisophthalic acid) (1.0 mM), Cd(NO$_3$)$_2$·4H$_2$O (0.75 mM), 4,4′-bipy·2H$_2$O (1.0 mM), and H$_2$O (10 mL) was placed in a 25 mL Teflon-lined stainless autoclave and heated under autogenous pressure at 160°C for 3 days. It was then slowly cooled to room temperature over 24 h. The light-yellow block-like crystals of the title compound were separated from the reaction mixture.

**[Cd$_4$(μ$_3$-OH)$_2$(H$_2$O)$_2$(C$_8$H$_4$O$_7$S)$_2$(4,4′-bipy)$_4$]·H$_2$O or C$_{56}$H$_{48}$N$_8$O$_{19}$S$_2$Cd$_4$.** This compound crystallizes in the monoclinic structure with the lattice parameters $a = 2560.2 \pm 0.1$, $b = 1657.1 \pm 0.1$, and $c = 1332.2 \pm 0.1$ pm and $\beta = 92.524° \pm 0.001°$ and a calculated density of 1.940 g·cm$^{-3}$ (Tao et al. 2003). It was prepared by the following procedure. An aqueous solution (5 mL) of 5-sulfoisophthalic acid (C$_8$H$_6$O$_7$S) (0.50 mM) was adjusted to pH = 7 with 1 M NaOH and transferred into a Parr Teflon–lined stainless steel vessel (23 mL) containing Cd(NO$_3$)$_2$·4H$_2$O (0.50 mM) and 4,4′-bipy (0.50 mM). An additional 3 mL of water was added to the resultant mixture. The vessel was then sealed and heated to 180°C for 3 days, followed by cooling at 2.5°C h$^{-1}$ to room temperature. Colorless block-like crystals were collected by hand, washed with distilled water, and dried in air at ambient temperature.

**{[Cd$_3$(C$_8$H$_3$O$_7$S)$_2$(2,2′-bipy)$_4$(H$_2$O)$_2$]·6H$_2$O}$_n$ or C$_{56}$H$_{54}$N$_8$O$_{22}$S$_2$Cd$_3$.** This compound crystallizes in the triclinic structure with the lattice parameters $a = 803.33 \pm 0.03$, $b = 1021.11 \pm 0.02$, and $c = 2020.10 \pm 0.07$ pm and $\alpha = 95.68°$, $\beta = 94.43°$, and $\gamma = 109.62°$ and a calculated density of 1.714 g·cm$^{-3}$ (Li et al. 2005). To obtain it, a mixture of NaC$_8$H$_5$O$_7$S (monosodium salt of 5-sulfoisophthalic acid) (0.2 mM), Cd(NO$_3$)$_2$·4H$_2$O

(0.3 mM), 2,2'-bipy·2H$_2$O (0.3 mM), and H$_2$O (1.0 mL) was placed in a 25 mL Teflon-lined stainless autoclave and heated under autogenous pressure at 160°C for 3 days. It was then cooled to room temperature over 24 h. The resulting colorless block-like crystals of the title compound were separated from the reaction mixture.

{[Cd$_2$(C$_8$H$_3$O$_7$S)(4,4'-bipy)$_3$(H$_2$O)$_3$][Cd(C$_8$H$_3$O$_7$S)(4,4'-bipy)(H$_2$O)]·8H$_2$O}$_n$    **or** C$_{56}$H$_{62}$N$_8$O$_{26}$S$_2$Cd$_3$. This compound crystallizes in the triclinic structure with the lattice parameters $a = 1024 \pm 5$, $b = 1797 \pm 10$, and $c = 1874 \pm 12$ pm and $\alpha = 77.71° \pm 0.13°$, $\beta = 87.3° \pm 2°$, and $\gamma = 77.4° \pm 0.2$ and a calculated density of 1.682 g·cm$^{-3}$ (Li et al. 2005). It was prepared by the next procedure. A mixture of NaC$_8$H$_5$O$_7$S (monosodium salt of 5-sulfoisophthalic acid) (0.2 mM), Cd(NO$_3$)$_2$·4H$_2$O (0.3 mM), 4,4'-bipy·2H$_2$O (0.5 mM), EtOH (1.0 mL), and H$_2$O (15 mL) was placed in a 25 mL Teflon-lined stainless autoclave and heated under autogenous pressure at 160°C for 5 days. It was then cooled to room temperature over 24 h. Colorless sheetlike crystals of the title compound were separated from the reaction mixture.

{[Cd$_3${O$_3$SC$_6$H$_3$–1,3-(CO$_2$)$_2$}$_2$(C$_{13}$H$_{14}$N$_2$)$_6$]·4H$_2$O}$_n$    **or**    C$_{94}$H$_{98}$N$_{12}$O$_{18}$S$_2$Cd$_3$. This compound crystallizes in the triclinic structure with the lattice parameters $a = 1022.33 \pm 0.15$, $b = 1284.2 \pm 0.2$, and $c = 1907.3 \pm 0.3$ pm and $\alpha = 101.562° \pm 0.004°$, $\beta = 99.637° \pm 0.005°$, and $\gamma = 104.124° \pm 0.006°$ and a calculated density of 1.495 g·cm$^{-3}$ (Liu et al. 2006). To obtain it, a mixture of Cd(NO$_3$)$_2$·4H$_2$O (0.075 mM), 5-sulfoisophthalic acid monosodium salt (0.05 mM), 1,3-bis(4-pyridyl)propane (0.15 mM), Et$_3$N (0.01 mL), and water (15 mL) was heated at 160°C for 4 days. Colorless rodlike crystals of the title compound were obtained when cooling to room temperature at 5°C h$^{-1}$. The crystals were recovered by filtration, washed with distilled water, and dried in air.

[Cd$_{32}$S$_{14}$(SPh)$_{36}$]·(DMF)$_4$ or C$_{228}$H$_{208}$N$_4$O$_4$S$_{50}$Cd$_{32}$. This compound crystallizes in the cubic structure with the lattice parameter $a = 2185.6 \pm 0.2$ at 203 K (Herron et al. 1993, 2005). Recrystallization of the solid Cd$_{32}$S$_{14}$(SPh)$_{36}$·(DMF)$_4$ from a solution of Py and DMF results in the formation of this compound as pale-yellow cubelike crystals.

## 4.37   Cd–H–C–N–O–Se–S

Some compounds are formed in this system.

[Cd(SeCN)$_2$(C$_7$H$_{10}$N$_2$)$_2$]·2(DMSO) or C$_{20}$H$_{32}$N$_6$O$_2$Se$_2$S$_2$Cd. This compound crystallizes in the triclinic structure with the lattice parameters $a = 828.29 \pm 0.09$, $b = 856.9 \pm 0.2$, and $c = 1058.7 \pm 0.2$ pm and $\alpha = 79.52° \pm 0.01°$, $\beta = 75.68° \pm 0.01°$, and $\gamma = 82.36° \pm 0.01°$ and a calculated density of 1.684 g·cm$^{-3}$ (Secondo et al. 2000). Crystals of the title compound suitable for XRD were grown overnight from the saturated solution of [Cd(SeCN)$_2$(C$_7$H$_{10}$N$_2$)$_2$] in DMSO.

[(Et$_4$N)$_2${SCd$_8$(SePh)$_{16}$}]·(DMF) or C$_{115}$H$_{127}$N$_3$OSe$_{16}$SCd$_8$. This compound crystallizes in the monoclinic structure with the lattice parameters $a = 1967 \pm 1$, $b = 3217.2 \pm 0.9$, and $c = 2004 \pm 1$ pm and $\beta = 90.42°$ and a calculated density of 1.97 g·cm$^{-3}$ (Lee et al. 1990).

## 4.38   Cd–H–C–N–O–F–S

Some compounds are formed in this system.

**[{Cd(4,4'-bipy)(H₂O)₄}(CF₃SO₃)₂]ₙ or C₁₂H₁₆N₂O₁₀F₆S₂Cd.** This compound crystallizes in the orthorhombic structure with the lattice parameters $a = 1160.84 \pm 0.07$, $b = 1237.93 \pm 0.07$, and $c = 1527.89 \pm 0.09$ pm and a calculated density of 1.932 g·cm⁻³ (Xiong et al. 1999). It was obtained as follows. To a 20 mL of ethanol solution containing Cd(CF₃SO₃)₂·6H₂O (1.0 mM) and 4,4'-bipy (2.0 mM) H₂O was added dropwise under stirring until the solution became clear or transparent (ca. 2 mL of H₂O). After refluxing for 3 h and cooling to room temperature, the solution was filtered by suction. The filtrate was allowed to evaporate at ambient temperature for a few weeks, after which time colorless needlelike crystals of the title compound were formed.

**[{Cd(C₃H₅NS₂)₄}{(CF₃SO₃)₂} or C₁₄H₂₀N₄O₆F₆S₁₀Cd.** This compound melts at 145°C–146°C and crystallizes in the orthorhombic structure with the lattice parameters $a = 2013.9 \pm 0.3$, $b = 2333.2 \pm 0.5$, and $c = 1421.4 \pm 0.3$ and a calculated density of 1.765 g·cm⁻³ (Rajalingam et al. 2001). To obtain it, 1,3-thiazolidine-2-thione (10.0 mM) was dissolved in a solution of Cd(CF₃SO₃)₂ (2.48 mM) in 30 mL of MeCN/CH₂Cl₂ (3:1 volume ratio). At this stage, a 5 mL portion of the mixture was diluted with another 10 mL of MeCN/CH₂Cl₂ and then kept in a freezer (−15°C) for the production of single crystals for XRD. The remaining solution was also kept in the freezer, and from it, after a few days, were isolated colorless crystals. These were washed with cold diethyl ether and dried in vacuum. Allowing the filtrate to again stand in the freezer for several days resulted in a further product.

**[Cd(3-CF₃C₅H₃NS)₂(C₃H₇NO)] or C₁₅H₁₃N₃OF₆S₂Cd.** This compound crystallizes in the triclinic structure with the lattice parameters $a = 810.86 \pm 0.17$, $b = 1133.8 \pm 0.2$, and $c = 1246.0 \pm 0.3$ pm and $\alpha = 109.534° \pm 0.005°$, $\beta = 105.748° \pm 0.005°$, and $\gamma = 96.928° \pm 0.005°$ and a calculated density of 1.781 g·cm⁻³ (Sousa-Pedrares et al. 2003). Electrochemical oxidation of a Cd anode in a solution of 3-CF₃C₅H₄NS (1.492 mM) in acetonitrile (50 mL), containing about 15 mg of Me₄NClO₄ as a current carrier, at 20 V and 10 mA for 2 h caused 44.45 mg of Cd to be dissolved. During the electrolysis process, H₂ was evolved at the cathode, and at the end of the reaction, an insoluble crystalline solid was observed at the bottom of the vessel. The solid was filtered off, washed with acetonitrile and diethyl ether, and dried under vacuum. Its crystals were obtained by crystallization of [Cd(3-CF₃C₅H₃NS)₂] from DMF. All manipulations were done under an atmosphere of dry N₂.

**[Cd(C₄H₈N₂S)₄(CF₃SO₃)₂] or C₁₈H₃₂N₈O₆F₆S₆Cd.** This compound crystallizes in the triclinic structure with the lattice parameters $a = 884.54 \pm 0.04$, $b = 1290.65 \pm 0.05$, and $c = 1553.57 \pm 0.06$ pm and $\alpha = 82.571° \pm 0.002°$, $\beta = 78.288° \pm 0.002°$, and $\gamma = 78.584° \pm 0.002°$ and a calculated density of 1.72 g·cm⁻³ (Dean et al. 2002). To prepare it, a solution of 1-methylimidazolidine-2-thione (10 mM) in MeCN (5 mL) was added to a solution of Cd(CF₃SO₃)₂ (2.5 mM) in MeCN (5 mL). To the clear mixture was added Et₂O (ca. 5 mL) until slight turbidity was seen. Thereafter, the mixture was kept in a freezer at −15°C. After 2 weeks, colorless crystals of the title compound were isolated and dried under vacuum. Its crystals were grown by keeping a solution of the salt in MeCN/Et$_2$O (4:1 volume ratio) in the freezer at –15°C.

**[Cd(C$_5$H$_5$NS)$_4$(CF$_3$SO$_3$)$_2$]** or **C$_{22}$H$_{20}$N$_4$O$_6$F$_6$S$_6$Cd**. This compound melts at 144°C–146°C (Rajalingam et al. 2000). To obtain it, pyridine-2-thione (0.67 mM) in CHCl$_3$ (3.0 mL) was added to a solution of Cd(CF$_3$SO$_3$)$_2$ (0.16 mM) in MeCN (1 mL). Diethyl ether was layered onto the clear reaction mixture until slight turbidity was seen. The reaction vial was tightly sealed and kept in a freezer at –15°C. After a week, the first crop of white block-shaped crystals was separated by filtration, washed with cold Et$_2$O, and vacuum dried. All manipulations were carried out under an atmosphere of dry N$_2$ in a glovebox or with vacuum/N$_2$ manifold or under an atmosphere of Ar in a glovebox.

**[Cd(C$_5$H$_5$NS)$_4$(CF$_3$SO$_3$)$_2$]** or **C$_{22}$H$_{20}$N$_4$O$_6$F$_6$S$_6$Cd**. This compound melts at 178°C–180°C (Rajalingam et al. 2000). It was prepared by the following procedure. Pyridine-4-thione (0.67 mM) was dissolved in CHCl$_3$ (8.0 mL). To the clear yellow solution Cd(O$_3$SCF$_3$)$_2$ (0.16 mM) in MeCN (1 mL) was added. The bright-yellow color faded. Diethyl ether was layered onto the clear solution until slight turbidity was seen. The reaction mixture was sealed tightly and kept in a freezer at –15°C. After 2 weeks, a first crop of pale-yellow flaky crystals was separated by filtration, washed with cold Et$_2$O, and vacuum dried.

**[{Cd(4,4′-bipy)(CF$_3$SO$_3$)$_2$(H$_2$O)$_2$}(4,4′-bipy)]$_n$** or **C$_{22}$H$_{20}$N$_4$O$_8$F$_6$S$_2$Cd**. This compound crystallizes in the orthorhombic structure with the lattice parameters $a = 799.08 \pm 0.04$, $b = 1530.13 \pm 0.08$, and $c = 2340.4 \pm 0.1$ pm and a calculated density of 1.764 g·cm$^{-3}$ (Xiong et al. 1999). It was obtained by the following procedure. Samples of Cd(CF$_3$SO$_3$)$_2$·6H$_2$O (0.5 mM), 4,4′-bipy (1 mM), and 4-methoxy-2-nitroaniline (1 mM) were ground, mixed thoroughly in a mortar with pestle, and placed in a thick-walled Pyrex tube. After the addition of 1 mL of 95% EtOH, the tube was frozen with liquid N$_2$, evacuated under vacuum, and sealed with a torch. The tube was heated at 90°C for 12 h to give colorless single crystals of the title compound.

**{[Cd(5-CF$_3$C$_5$H$_3$NS)$_2$]$_2$(C$_3$H$_7$NO)}** or **C$_{27}$H$_{19}$N$_5$OF$_{12}$S$_4$Cd$_2$**. This compound crystallizes in the triclinic structure with the lattice parameters $a = 1141.95 \pm 0.16$, $b = 1249.03 \pm 0.17$, and $c = 1465.1 \pm 0.2$ pm and $\alpha = 64.648° \pm 0.004°$, $\beta = 69.447° \pm 0.004°$, and $\gamma = 77.377° \pm 0.004°$ and a calculated density of 1.904 g·cm$^{-3}$ (Sousa-Pedrares et al. 2003). It was obtained by the same procedure as for [Cd(3-CF$_3$C$_5$H$_3$NS)$_2$(C$_3$H$_7$NO)] using 1.553 mM of 5-CF$_3$C$_5$H$_4$NS instead of 3-CF$_3$C$_5$H$_4$NS (42.6 mg of Cd dissolved). Its crystals were obtained by crystallization of [Cd(5-CF$_3$C$_5$H$_3$NS)$_2$] from DMF.

**[Cd(C$_9$H$_7$NS)$_4$(CF$_3$SO$_3$)$_2$]** or **C$_{38}$H$_{28}$N$_4$O$_6$F$_6$S$_6$Cd**. This compound melts at 192°C–194°C (Rajalingam et al. 2000). It was obtained by the next procedure. Cd(CF$_3$SO$_3$)$_2$ (0.16 mM) in MeCN (1 mL) was added to the intense yellow solution of quinoline-2-thione (0.65 mM) in CHCl$_3$ (3 mL). The reaction mixture quickly turned light yellow. Diethylether (3 mL) was added and the clear solution was kept at room temperature for slow crystallization. After 3 days, the product, as yellow needles, was collected by filtration, washed with Et$_2$O, and dried under vacuum.

[{Cd(4,4'-bipy)$_2$(H$_2$O)$_2$}(CF$_3$SO$_3$)$_2$(4,4'-bipy)(H$_2$O)$_2$(C$_7$H$_8$N$_2$O$_3$)$_2$]$_n$ or C$_{46}$H$_{48}$N$_{10}$O$_{16}$F$_6$S$_2$Cd. This compound crystallizes in the monoclinic structure with the lattice parameters $a = 2050.5 \pm 0.1$, $b = 1181.06 \pm 0.07$, and $c = 2326.6 \pm 0.1$ pm and $\beta = 104.186° \pm 0.001°$ and a calculated density of 1.565 g·cm$^{-3}$ (Xiong et al. 1999). To obtain it, H$_2$O was added dropwise to a 20 mL of EtOH solution containing Cd(CF$_3$SO$_3$)$_2$·6H$_2$O (1 mM), 4,4'-bipy (3 mM), and 4-methoxy-2-nitroaniline (2 mM) under stirring to clear the solution. After refluxing for 3 h and cooling to room temperature, the solution was filtered. The filtrate was allowed to evaporate at ambient temperature for a few weeks, after which time pale-yellow block crystals of the title compound were formed.

## 4.39 Cd–H–C–N–O–Cl–S

Some compounds are formed in this system.

[CdCl$_2$(C$_4$H$_7$NOS)] or C$_4$H$_7$NOCl$_2$SCd. This compound melts at 260°C with decomposition (De Filippo et al. 1971). To obtain it, CdCl$_2$ was dissolved in an excess of molten thiomorpholine-3-one (C$_4$H$_7$NOS) (at ca. 105°C; molar ratio Cd$^{2+}$/C$_4$H$_7$NOS ca. 1:2), and the crude product was purified by washing with ethyl ether and recrystallizing from EtOH. The title compound could also be prepared if to a solution of CdCl$_2$ in absolute EtOH different amounts of C$_4$H$_7$NOS were added for Cd$^{2+}$/C$_4$H$_7$NOS molar ratios of 1:1, 1:2, 1:3, and 1:4, and the mixture was heated under reflux for 4 h. [CdCl$_2$(C$_4$H$_7$NOS)] was separated out by crystallization.

[Cd(SN$_4$C$_5$H$_4$)Cl$_2$H$_2$O]$_2$ or C$_5$H$_6$N$_4$OCl$_2$SCd. This compound crystallizes in the triclinic structure with the lattice parameters $a = 1044.7 \pm 0.1$, $b = 802.6 \pm 0.1$, and $c = 673.5 \pm 0.1$ pm and $\alpha = 96.41° \pm 0.01°$, $\beta = 101.88° \pm 0.01°$, and $\gamma = 74.01° \pm 0.01°$ and the experimental and calculated densities $2.22 \pm 0.02$ and 2.21 g·cm$^{-3}$, respectively (Griffith and Amma 1979). It was prepared by heating 6-mercaptopurine monohydrate (5 mM) and CdCl$_2$·2.5H$_2$O (3 mM) in 100 mL of 0.2 M HCl at 70°C for 3 h. Diffraction quality crystals grew after ca. 2 days of controlled evaporation of the resulting solution.

[Cd(C$_3$H$_5$NOS)$_2$Cl$_2$] or C$_6$H$_{10}$N$_2$O$_2$Cl$_2$S$_2$Cd. This compound melts at 132°C with decomposition (Devillanova and Verani 1978). It was obtained by boiling an absolute ethanol solution of CdCl$_2$ under reflux with 1,3-oxazolidine-2-thione in a 1:2 molar ratio.

[Cd(C$_3$H$_5$NOS)$_2$Cl$_2$] or C$_6$H$_{10}$N$_2$O$_2$Cl$_2$S$_2$Cd. This compound decomposes at 210°C (Preti et al. 1974). It was obtained by a reaction of CdCl$_2$ and thiazolidine-2-one in refluxing EtOH for 12 h. The title compound was separated out during crystallization.

[Cd(C$_5$H$_4$NC$_2$H$_5$N$_4$S)Cl$_2$]·0.25H$_2$O or C$_7$H$_{9.5}$N$_5$O$_{0.25}$Cl$_2$SCd. This compound melts at 284°C (Castiñeiras et al. 2000). To obtain it, a solution of CdCl$_2$ (2 mM) in EtOH (30 mL) was mixed with a solution of 2-pyridineformamide thiosemicarbazone (C$_5$H$_4$NC$_2$H$_5$N$_4$S) (2 mM) in the same solvent (30 mL), and the mixture was stirred under reflux for 2 h. The resulting solid was filtered out, washed with anhydrous ether to apparent dryness, and placed on a warm plate at 35°C.

**[Cd(C$_8$H$_{11}$N$_5$S)Cl$_2$]·H$_2$O or C$_8$H$_{13}$N$_5$OCl$_2$SCd**. This compound melts at 234°C (García et al. 2002). To obtain it, a solution of 2-pyridineformamide thiosemicarbazone (1.19 mM) in EtOH (20 mL) was added to a solution of CdCl$_2$·H$_2$O (1.19 mM) in 20 mL of 96% EtOH, and the mixture was stirred for several days. The resulting solids were filtered off, washed thoroughly with cold EtOH, and dried over CaCl$_2$.

**[Cd(C$_5$H$_4$NC$_2$H$_5$N$_4$S)Cl$_2$]·(DMSO) or C$_9$H$_{15}$N$_5$OCl$_2$S$_2$Cd**. This compound crystallizes in the monoclinic structure with the lattice parameters $a = 1197.0 \pm 0.2$, $b = 1073.7 \pm 0.2$, and $c = 1355.9 \pm 0.2$ pm and $\beta = 108.19° \pm 0.01°$ and a calculated density of 1.832 g·cm$^{-3}$ (Castiñeiras et al. 2000). It was obtain as yellow prisms by the recrystallization of [Cd(C$_5$H$_4$NC$_2$H$_5$N$_4$S)Cl$_2$]·0.25H$_2$O from the mixture DMSO/ EtOH (1:1 volume ratio).

**[Cd(C$_8$H$_{11}$H$_5$S)Cl$_2$]·(DMSO) or C$_{10}$H$_{17}$N$_5$OCl$_2$SCd**. This compound crystallizes in the triclinic structure with the lattice parameters $a = 914.8 \pm 0.3$, $b = 985.6 \pm 0.2$, and $c = 1184.9 \pm 0.2$ pm and $\alpha = 108.56° \pm 0.02°$, $\beta = 93.63° \pm 0.02°$, and $\gamma = 112.46° \pm 0.01°$ and a calculated density of 1.709 g·cm$^{-3}$ (García et al. 2002). Its single crystals suitable for analysis by XRD were obtained by slow evaporation of a DMSO of [Cd(C$_8$H$_{11}$H$_5$S)Cl$_2$] in air at room temperature.

**[Cd{OH(CH$_2$)$_2$SCH$_2$}$_2$(C$_5$H$_3$N)Cl$_2$] or C$_{11}$H$_{17}$NO$_2$Cl$_2$S$_2$Cd**. To obtain this compound, CdCl$_2$·2H$_2$O (1.55 mM) suspended in THF (20 mL) was added to a solution of 2,6-bis[(hydroxyethylthio)methyl](pyridine) (1.55 mM) in THF (20 mL) (Teixidor et al. 1986). The mixture was stirred for 1 h and the resulting solid was filtered out, washed with cold THF, and dried in vacuo. All operations were performed under a N$_2$ atmosphere.

**[Cd(C$_{12}$H$_{17}$N$_4$OS)Cl$_3$]$_2$·H$_2$O or C$_{12}$H$_{19}$N$_4$O$_2$Cl$_3$SCd**. This compound crystallizes in the monoclinic structure with the lattice parameters $a = 699.9 \pm 0.2$, $b = 1281.1 \pm 0.3$, and $c = 2058.7 \pm 0.4$ pm and $\beta = 91.88° \pm 0.02°$ and a calculated density of 1.808 g·cm$^{-3}$ (Casas et al. 1995). It was prepared as follows. Thiamine hydrochloride (C$_{12}$H$_{17}$N$_4$OSCl·HCl) (5 mM) was dissolved in 5 mL of H$_2$O and reacted with 5 mL of 1 M aqueous KOH solution (5 mM). The addition of CdCl$_2$·H$_2$O (5 mM) dissolved in 2 mL of H$_2$O caused almost immediate precipitation of a white solid, which was filtered out. The mother liquor was refrigerated, and after 2 days, a white crystalline solid was formed, which was isolated. Crystals suitable for XRD were obtained by 2 days of refrigeration of the filtered reaction mixture obtained with a C$_{12}$H$_{17}$N$_4$OSCl·HCl/KOH/CdCl$_2$ in a molar ratio of 2:2:1.

**[Cd(C$_{12}$H$_{18}$N$_4$OS)Cl$_4$]·H$_2$O or C$_{12}$H$_{20}$N$_4$O$_2$Cl$_4$SCd**. This compound crystallizes in the monoclinic structure with the lattice parameters $a = 1687.4 \pm 0.2$, $b = 1555.3 \pm 0.2$, and $c = 790.6 \pm 0.2$ pm and $\beta = 97.61° \pm 0.06°$ and the experimental and calculated densities 1.78 and 1.74 g·cm$^{-3}$, respectively (Richardson et al. 1975). It was prepared by mixing aqueous solutions of thiamine hydrochloride and Cd(NO$_3$)$_2$ (2:1 molar ratio) and allowing the solution to evaporate. Well-formed colorless crystals of the title compound were obtained.

**[Cd{Cd(SCH$_2$CHNH$_3$CO$_2$)$_2$}$_2$][CdCl$_4$] or C$_{12}$H$_{24}$N$_4$O$_8$Cl$_4$S$_4$Cd$_4$·4H$_2$O, tetrakis(l-cysteinato)trizinc tetrachlorocadmate**. To obtain this compound, L-cysteine

(0.04 mol) was dissolved in 100 mL of warm $H_2O$, and to it was added with stirring a warm solution of $CdCl_2$ (0.04 mol) in a small amount of $H_2O$ (Shindo and Brown 1965). A white precipitate appeared almost immediately. Stirring was continued for 2 h and the precipitate was filtered, washed several times with $H_2O$ and finally with absolute EtOH, and dried in vacuo over $CaCl_2$. The anhydrous compound was obtained by drying in vacuo at 70°C over $P_2O_5$ under continuous pumping.

**$[Cd(SC_5H_9NHMe)_2(ClO_4)_2]\cdot 2H_2O$ or $C_{12}H_{30}N_2O_{10}Cl_2S_2Cd$.** This compound crystallizes in the tetragonal structure with the lattice parameters $a = 1876 \pm 2$ and $c = 666 \pm 1$ pm and the experimental and calculated densities 1.70 and 1.725 g·cm$^{-3}$, respectively (Bayón et al. 1979). To obtain it, $Cd(ClO_4)_2\cdot 6H_2O$ (7.0 g) dissolved in 50 mL of the mixture MeOH/$H_2O$ (9:1 volume ratio) was added slowly to 1-methyl-4-mercaptopiperidine (4.5 g) in about 50 mL of the same solvent. The reaction mixture did not show any change upon the addition of $Cd^{2+}$ ions into the solution. A stream of $N_2$ was passed through the reaction mixture during the addition, and crystallization took place under a $N_2$ atmosphere. The crystalline white product was isolated by filtration after 48 h, washed with cold $H_2O$ and then with MeOH, and dried in vacuo over silica gel.

**$[Cd(C_9H_{12}N_4S)Cl_2]\cdot(2DMSO)$ or $C_{13}H_{24}N_4O_2Cl_2S_3Cd$.** This compound crystallizes in the monoclinic structure with the lattice parameters $a = 953.5 \pm 0.1$, $b = 1021.7 \pm 0.1$, and $c = 2289.3 \pm 0.2$ pm and $\beta = 90.23° \pm 0.02°$ at $223 \pm 2$ K and a calculated density of 1.632 g·cm$^{-3}$ (Bermejo et al. 1999). Its single crystals were obtained by redissolving $[Cd(C_9H_{12}N_4S)Cl_2]$ in DMSO and slowly evaporating the solvent.

**$[Cd(C_7H_9N_5S)_2(ClO_4)_2]\cdot 2H_2O$ or $C_{14}H_{22}N_{10}O_{10}Cl_2S_2Cd$.** This compound crystallizes in the orthorhombic structure with the lattice parameters $a = 1080.3 \pm 0.2$, $b = 1081.5 \pm 0.2$, and $c = 2330.9 \pm 0.5$ pm and a calculated density of 1.800 g·cm$^{-3}$ (Li et al. 2006). To obtain it, $Cd(ClO_4)_2\cdot 6H_2O$ (0.125 mM) was added at room temperature to acetylpyrazine thiosemicarbazone (0.25 mM) suspended in EtOH (60 mL). After stirring for 30 min, the solid formed was filtered off, washed with EtOH, and dried under vacuum.

**$[Cd(C_{12}H_{17}N_5S)Cl_2]\cdot(DMSO)$ or $C_{14}H_{23}N_5OCl_2SCd$.** This compound crystallizes in the monoclinic structure with the lattice parameters $a = 933.6 \pm 0.1$, $b = 1854.2 \pm 0.2$, and $c = 1291.5 \pm 0.2$ pm and $\beta = 109.608° \pm 0.009°$ and a calculated density of 1.655 g·cm$^{-3}$ (Castiñeiras et al. 2002). Single crystals of the title compound suitable for XRD were obtained by slow evaporation of a DMSO solution of $[Cd(C_{12}H_{17}N_5S)Cl_2]$ in air at room temperature.

**$[Cd(C_{13}H_{19}N_5S)Cl_2]\cdot(DMSO)$ or $C_{15}H_{25}N_5OCl_2S_2Cd$.** This compound crystallizes in the triclinic structure with the lattice parameters $a = 894.2 \pm 0.2$, $b = 1032.8 \pm 0.2$, and $c = 1219.1 \pm 0.3$ pm and $\alpha = 77.258° \pm 0.004°$, $\beta = 79.472° \pm 0.004°$, and $\gamma = 85.003° \pm 0.004°$ and a calculated density of 1.659 g·cm$^{-3}$ (Bermejo et al. 2004b). To obtain it, a solution of $CdCl_2$ (0.90 mM) in EtOH (30 mL) was mixed with a solution of 2-pyridineformamide 3-hexamethyleneiminylthiosemicarbazone ($C_{13}H_{19}N_5S$) (0.90 mM) in EtOH (30 mL). The resulting solution was stirred at room temperature for 7 days and the yellow solid was filtered off. Recrystallization from DMSO gave yellow crystals of the title compound.

**[Cd(SCMe$_2$CH$_2$NH$_2$)$_2$CdCl$_2$]$_2$·2H$_2$O or C$_{16}$H$_{44}$N$_4$O$_2$Cl$_4$S$_4$Cd$_4$.** This compound crystallizes in the monoclinic structure with the lattice parameters $a = 1264.2 \pm 0.8$, $b = 1367.3 \pm 0.7$, and $c = 1155.7 \pm 0.6$ pm and $\beta = 124.64° \pm 0.03°$ and the experimental and calculated densities $2.09 \pm 0.01$ and $2.110$ g·cm$^{-3}$, respectively (Fawcett et al. 1978). An attempt to recrystallize Cd[SCMe$_2$CH$_2$NH$_2$]$_2$ from its mother liquor yielded the title compound as the sole (and unexpected) product. A solution of HSCMe$_2$CH$_2$NH$_2$·HCl (2.0 mM) and of CdCl$_2$·2.5H$_2$O (1.0 mM) in 100 mL of hot (85°C) H$_2$O was made alkaline (pH = 9–10) by the addition of 4 mL of Et$_3$N, filtered through a fine glass frit, and left at 25°C for 1 day. The colorless product consisted entirely of clumps of the platelike Cd[SCMe$_2$CH$_2$NH$_2$]$_2$ compound. An attempt was made to recrystallize this product by adding 100 mL of H$_2$O to the aforementioned mixture, heating it to boiling, and refiltering the resulting colorless solution. After remaining at 25°C for 1 day, the solution deposited pale-yellow rhomboids, which were collected by filtration, washed thoroughly with water, and dried in air.

The title compound may also be prepared directly by the neutralization of equimolar aqueous solutions of CdCl$_2$·2.5H$_2$O and the ligand hydrochloride with 2 equivalents of Et$_3$N.

**[Cd(Cl)(C$_5$H$_2$NO$_4$S)(C$_{12}$H$_8$N$_2$)] or C$_{17}$H$_{10}$N$_3$O$_4$ClSCd.** This compound crystallizes in the monoclinic structure with the lattice parameters $a = 2129.6 \pm 0.3$, $b = 1081.4 \pm 0.1$, and $c = 732.12 \pm 0.09$ pm and $\beta = 91.395° \pm 0.002°$ at 103 K (Liu et al. 2014). The next procedure was used for it to be obtained. A mixture of CdCl$_2$·2.5H$_2$O (0.25 mM), Phen (0.25 mM), 4-nitrothiophene-2-carboxylic acid (0.25 mM), and H$_2$O (8 mL) was placed in a 25 ml Teflon-lined autoclave under autogenous pressure at 160°C for 3 days. After cooling to room temperature, yellow crystals were collected by filtration and washed with distilled H$_2$O.

**[Cd(PySH)$_4$(ClO$_4$)$_2$] or C$_{20}$H$_{20}$N$_4$O$_8$Cl$_2$S$_4$Cd.** This compound crystallizes in the monoclinic structure with the lattice parameters $a = 1191.43 \pm 0.02$, $b = 1294.44 \pm 0.02$, and $c = 1862.27 \pm 0.02$ pm and $\beta = 100.524° \pm 0.001°$ and a calculated density of 1.778 g·cm$^{-3}$ (Anjali et al. 2003). It was prepared by the following procedure. Cd(ClO$_4$)$_2$·6H$_2$O (1.01 mM) in 2 mL of MeOH was added to a solution of 4-pyridine thiol (3.70 mM) in 4 mL of MeOH. The mixture was stirred for 15 min and then left for slow evaporation at room temperature to get bright-yellow crystals, which were separated by filtration, washed with Et$_2$O, and dried in vacuo.

**[CdCl$_2$(DMF)$_2$(C$_8$H$_8$N$_4$O$_2$S)$_2$] or C$_{22}$H$_{30}$N$_{10}$O$_6$Cl$_2$S$_2$Cd.** This compound crystallizes in the triclinic structure with the lattice parameters $a = 882.98 \pm 0.03$, $b = 969.18 \pm 0.03$, and $c = 1057.86 \pm 0.03$ pm and $\alpha = 97.810° \pm 0.001°$, $\beta = 111.192° \pm 0.001°$, and $\gamma = 105.250° \pm 0.001°$ and a calculated density of 1.164 g·cm$^{-3}$ (Tian et al. 1999a). For its preparation, 4-nitrobenzaldehyde thiosemicarbazone (C$_8$H$_8$N$_4$O$_2$S) (10 mM) dissolved in EtOH (50 mL) was added to CdCl$_2$ (10 mM) in EtOH (50 mL). The colorless crystalline solid, [CdCl$_2$(C$_8$H$_8$N$_4$O$_2$S)$_2$], formed after refluxing for 4 h, was isolated and dried in vacuo with P$_2$O$_5$. Single crystals suitable for XRD were obtained by diffusion of ether into a solution of [CdCl$_2$(C$_8$H$_8$N$_4$O$_2$S)$_2$] in DMF.

**C$_{34}$H$_{26}$N$_4$O$_9$Cl$_2$S$_2$Cd.** This compound crystallizes in the triclinic structure with the lattice parameters $a = 1138.95 \pm 0.02$, $b = 1306.68 \pm 0.02$, and $c = 1403.20 \pm 0.04$ pm and $\alpha = 108.454° \pm 0.001°$, $\beta = 107.896° \pm 0.001°$, and $\gamma = 101.988° \pm 0.001°$ (Zhang et al. 2014). To obtain it, at room temperature, di-2-pyridylsulfone (0.1 mM) and cadmium bis(3-chlorobenzoate) (0.1 mM) were dissolved and mixed in 5 mL of acetonitrile with stirring to afford a clear solution. After filtration, the clear solution was left to slowly evaporate in air to give colorless block-shaped crystals of the title compound.

**[{Cd(Phen)(C$_{10}$H$_8$N$_2$S$_2$)$_2$(H$_2$O)}(ClO$_4$)$_2$·2CH$_3$OH·1.5H$_2$O]$_n$ or C$_{34}$H$_{37}$N$_6$O$_{12.5}$Cl$_2$S$_4$Cd.**
This compound crystallizes in the triclinic structure with the lattice parameters $a = 1364.66 \pm 0.05$, $b = 1530.87 \pm 0.09$, and $c = 2151.7 \pm 0.1$ pm and $\alpha = 69.553° \pm 0.005°$, $\beta = 76.379° \pm 0.003°$, and $\gamma = 84.879° \pm 0.004°$ at $108 \pm 2$ K and a calculated density of 1.688 g·cm$^{-3}$ (Seidel et al. 2011). It was prepared by the next procedure. Cd(ClO$_4$)$_2$·6H$_2$O (0.072 mM) and Phen·H$_2$O (0.095 mM) were dissolved in MeOH (5 mL) and 4,4'-dithiopyridine (0.095 mM) was added with stirring. The resulting mixture was left at ambient temperature and the solvent was allowed to evaporate slowly. Colorless crystalline material of the title compound was found after ca. 3 weeks. The material was collected and dried on a filter paper.

## 4.40 Cd–H–C–N–O–Br–S

Some compounds are formed in this system.

**[CdBr$_2$(C$_4$H$_7$NOS)] or C$_4$H$_7$NOBr$_2$SCd.** This compound melts at 230°C with decomposition (De Filippo et al. 1971). To obtain it, CdBr$_2$ was dissolved in an excess of molten thiomorpholine-3-one (C$_4$H$_7$NOS) (at ca. 105°C; molar ratio Cd$^{2+}$/C$_4$H$_7$NOS ca. 1:2), and the crude product was purified by washing with ethyl ether and recrystallizing from EtOH. The title compound could also be prepared if to a solution of CdBr$_2$ in absolute EtOH different amounts of C$_4$H$_7$NOS were added for Cd$^{2+}$/C$_4$H$_7$NOS molar ratios of 1:1, 1:2, 1:3, and 1:4 and the mixture was heated under reflux for 4 h. [CdBr$_2$(C$_4$H$_7$NOS)] was separated out by crystallization.

**[Cd(C$_3$H$_5$NOS)$_2$Br$_2$] or C$_6$H$_{10}$N$_2$O$_2$Br$_2$S$_2$Cd.** This compound melts at 133°C with decomposition (Devillanova and Verani 1978). It was obtained by boiling an absolute ethanol solution of CdBr$_2$ under reflux with 1,3-oxazolidine-2-thione in a 1:2 molar ratio.

**[Cd(C$_3$H$_5$NOS)$_2$Br$_2$] or C$_6$H$_{10}$N$_2$O$_2$Br$_2$S$_2$Cd.** This compound decomposes at 198°C–200°C (Preti et al. 1974). It was obtained by a reaction of CdBr$_2$ and thiazolidine-2-one in refluxing EtOH for 12 h. The title compound was separated out during crystallization.

**[Cd(C$_5$H$_4$NC$_2$H$_5$N$_4$S)Br$_2$]·H$_2$O or C$_7$H$_{11}$N$_5$OBr$_2$SCd.** This compound crystallizes in the triclinic structure with the lattice parameters $a = 752.5 \pm 0.2$, $b = 977.2 \pm 0.4$, and $c = 1039.8 \pm 0.4$ pm and $\alpha = 80.856° \pm 0.017°$, $\beta = 76.391° \pm 0.018°$, and $\gamma = 67.944° \pm 0.011°$ and a calculated density of 2.348 g·cm$^{-3}$ (Castiñeiras et al. 2000). It was obtain as yellow prisms by the recrystallization of [Cd(C$_5$H$_4$NC$_2$H$_5$N$_4$S)Br$_2$] from 96% EtOH.

**[Cd(C₉H₁₃N₅S)Br₂]·(DMSO) or C₁₁H₁₉N₅OBr₂S₂Cd.** This compound crystallizes in the monoclinic structure with the lattice parameters $a = 1354.93 \pm 0.10$, $b = 1298.70 \pm 0.06$, and $c = 1110.62 \pm 0.06$ pm and $\beta = 94.076° \pm 0.006°$ and a calculated density of 1.956 g·cm⁻³ (Bermejo et al. 2004a). Recrystallization of [Cd(C₉H₁₃N₅S)Br₂] from DMSO gave colorless crystals of the title compound.

**[Cd₂(C₆H₁₃NS)₂Br₄]·H₂O or C₁₂H₂₈N₂OBr₄S₂Cd₂.** This compound crystallizes in the monoclinic structure with the lattice parameters $a = 1142$, $b = 1312$, and $c = 781$ pm and $\beta = 92.52°$ (Bayón et al. 1982). It was obtained in 10% methanolic solution by adding CdBr₂ to the stoichiometric amount of 1-methyl-4-mercaptopiperidine. The white crystalline product obtained was filtered off, washed with cold water, EtOH, and ethyl ether, and dried in vacuo.

**[Cd(C₁₃H₁₉N₅S)Br₂]·0.5H₂O or C₁₃H₂₀N₅O₀.₅Br₂SCd.** To obtain this compound, a solution of CdBr₂ (0.90 mM) in EtOH (30 mL) was mixed with a solution of 2-pyridineformamide 3-hexamethyleneiminylthiosemicarbazone (C₁₃H₁₉N₅S) (0.90 mM) in EtOH (30 mL) (Bermejo et al. 2004b). The resulting solution was stirred at room temperature for 7 days and the white solid was filtered off.

**[Cd(C₉H₁₂N₄S)Br₂]·(2DMSO) or C₁₃H₂₄N₄O₂Br₂S₃Cd.** This compound crystallizes in the triclinic structure with the lattice parameters $a = 893.9 \pm 0.4$, $b = 1015.6 \pm 0.5$, and $c = 1359.7 \pm 0.5$ pm and $\alpha = 96.21° \pm 0.03°$, $\beta = 107.74° \pm 0.02°$, and $\gamma = 102.18° \pm 0.03°$ at $213 \pm 2$ K and a calculated density of 1.872 g·cm⁻³ (Bermejo et al. 1999). Its single crystals were obtained by redissolving [Cd(C₉H₁₂N₄S)Br₂] in DMSO and slowly evaporating the solvent.

**[Cd(C₁₂H₁₇N₅S)Br₂]·(DMSO) or C₁₄H₂₃N₅OBr₂SCd.** This compound crystallizes in the triclinic structure with the lattice parameters $a = 912.6 \pm 0.4$, $b = 1002.9 \pm 0.1$, and $c = 1264.0 \pm 0.3$ pm and $\alpha = 77.76° \pm 0.01°$, $\beta = 71.94° \pm 0.03°$, and $\gamma = 79.34° \pm 0.02°$ and a calculated density of 1.912 g·cm⁻³ (Castiñeiras et al. 2002). Single crystals of the title compound suitable for XRD were obtained by slow evaporation of a DMSO solution of [Cd(C₁₂H₁₇N₅S)Br₂] in air at room temperature.

**[Cd(C₁₃H₁₉N₅S)Br₂]·DMSO or C₁₅H₂₅N₅OBr₂S₂Cd.** This compound crystallizes in the triclinic structure with the lattice parameters $a = 916.9 \pm 0.1$, $b = 1021.0 \pm 0.4$, and $c = 1239.4 \pm 0.4$ pm and $\alpha = 77.029° \pm 0.018°$, $\beta = 79.858° \pm 0.016°$, and $\gamma = 83.784° \pm 0.019°$ and a calculated density of 1.878 g·cm⁻³ (Bermejo et al. 2004b). Its yellow crystals were obtained by recrystallization of [Cd(C₁₃H₁₉N₅S)Br₂]·0.5H₂O from DMSO.

**[Cd(C₂₀H₄₆N₆O₄)(SCN)₂Br₂] or C₂₂H₄₆N₈O₄Br₂S₂Cd₂.** This compound crystallizes in the monoclinic structure with the lattice parameters $a = 1290.0 \pm 0.4$, $b = 1199.8 \pm 0.9$, and $c = 1310 \pm 2$ pm and $\beta = 117.41° \pm 0.04°$ and a calculated density of 1.726 g·cm⁻³ (Bazzicalupi et al. 1996). It was prepared as follows. A solution of CdBr₂·4H₂O (0.1 mM) in MeOH (10 mL) was added to a methanolic solution (10 mL) of 1,4,7,16,19,22-hexaza-10,13,25,28-tetraoxacyclotriacontane (C₂₀H₄₆N₆O₄) (0.05 mM). KSCN (0.2 mM) and BuOH (10 mL) were added to the resulting solution. Crystals of the title compound suitable for XRD were obtained by slow evaporation at room temperature of this solution.

[Cd$_8$(SC$_6$H$_4$F)$_{16}$]·(DMF)$_3$ or C$_{105}$H$_{85}$N$_3$O$_3$Br$_{16}$S$_{16}$Cd$_8$. This compound crystallizes in the monoclinic structure with the lattice parameters $a = 2733.1 \pm 1.3$, $b = 1530.3 \pm 0.4$, and $c = 3911.0 \pm 2.0$ pm and $\beta = 126.43° \pm 0.02°$ (Dance et al. 1987b).

## 4.41   Cd–H–C–N–O–I–S

Some compounds are formed in this system.

[Cd(C$_3$H$_5$NOS)$_2$I$_2$] or C$_6$H$_{10}$N$_2$O$_2$I$_2$S$_2$Cd. This compound melts at 133°C with decomposition (Devillanova and Verani 1978). It was obtained by boiling an absolute ethanol solution of CdI$_2$ under reflux with 1,3-oxazolidine-2-thione in a 1:2 molar ratio.

[Cd(C$_3$H$_5$NOS)$_2$I$_2$] or C$_6$H$_{10}$N$_2$O$_2$I$_2$S$_2$Cd. This compound decomposes at 116°C (Preti et al. 1974). It was obtained by a reaction of CdI$_2$ and thiazolidine-2-one in refluxing EtOH for 12 h. The title compound was separated out during crystallization.

[CdI$_2$(C$_4$H$_7$NOS)$_2$] or C$_8$H$_{14}$N$_2$O$_2$I$_2$S$_2$Cd. This compound melts at 115°C (De Filippo et al. 1971). To obtain it, CdI$_2$ was dissolved in an excess of molten thiomorpholine-3-one (C$_4$H$_7$NOS) (at ca. 105°C; molar ratio Cd$^{2+}$/C$_4$H$_7$NOS ca. 1:2), and the crude product was purified by washing with ethyl ether and recrystallizing from EtOH. The title compound could also be prepared if to a solution of CdI$_2$ in absolute EtOH different amounts of C$_4$H$_7$NOS were added for Cd$^{2+}$/C$_4$H$_7$NOS molar ratios of 1:1, 1:2, 1:3, and 1:4 and the mixture was heated under reflux for 4 h. [CdI$_2$(C$_4$H$_7$NOS)$_2$] was separated out by crystallization.

[Cd(C$_9$H$_{13}$N$_5$S)I$_2$]·(DMSO) or C$_{11}$H$_{19}$N$_5$OI$_2$S$_2$Cd. This compound crystallizes in the monoclinic structure with the lattice parameters $a = 1088.6 \pm 0.2$, $b = 1544.5 \pm 0.2$, and $c = 1235.1 \pm 0.4$ pm and $\beta = 100.68° \pm 0.03°$ and a calculated density of 2.173 g·cm$^{-3}$ (Bermejo et al. 2001). Single crystals suitable for XRD were obtained by slow evaporation of a DMSO solution of [Cd(C$_9$H$_{12}$N$_5$S)I$_2$] [C$_9$H$_{12}$N$_5$S – (2-pyridyl) formamide $N$(4)-dimethylthiosemicarbazone] in air at room temperature.

[Cd(C$_9$H$_{13}$N$_5$S)I$_2$]·(DMSO) or C$_{11}$H$_{19}$N$_5$OI$_2$S$_2$Cd. This compound crystallizes in the monoclinic structure with the lattice parameters $a = 1355.57 \pm 0.13$, $b = 1324.59 \pm 0.10$, and $c = 1154.50 \pm 0.14$ pm and $\beta = 94.107° \pm 0.009°$ and a calculated density of 2.145 g·cm$^{-3}$ (Bermejo et al. 2004a). Recrystallization of [Cd(C$_9$H$_{13}$N$_5$S)I$_2$] [C$_9$H$_{12}$N$_5$S – (2-pyridyl)formamide $N$(4)-ethylthiosemicarbazone] from DMSO gave colorless crystals of the title compound.

[Cd$_2$(C$_6$H$_{13}$NS)$_2$I$_4$]·H$_2$O or C$_{12}$H$_{28}$N$_2$OI$_4$S$_2$Cd$_2$. This compound crystallizes in the monoclinic structure with the lattice parameters $a = 1188$, $b = 1339$, and $c = 835$ pm and $\beta = 92.35°$ (Bayón et al. 1982). It was obtained in 10% methanolic solution by adding CdI$_2$ to the stoichiometric amount of 1-methyl-4-mercaptopiperidine. The white crystalline product obtained was filtered off, washed with cold water, EtOH, and ethyl ether, and dried in vacuo.

[Cd(C$_9$H$_{12}$N$_4$S)I$_2$]·(2DMSO) or C$_{13}$H$_{24}$N$_4$O$_2$I$_2$S$_3$Cd. This compound crystallizes in the triclinic structure with the lattice parameters $a = 942.8 \pm 0.3$, $b = 1048.4 \pm 0.5$, and

$c = 1369.9 \pm 0.4$ pm and $\alpha = 93.72° \pm 0.02°$, $\beta = 109.51° \pm 0.03°$, and $\gamma = 104.70° \pm 0.03°$ at 223 K and a calculated density of 1.994 g·cm$^{-3}$ (Bermejo et al. 1999). Its single crystals were obtained by redissolving [Cd(C$_9$H$_{12}$N$_4$S)I$_2$] in DMSO and slowly evaporating the solvent.

**[Cd(C$_{12}$H$_{17}$N$_5$S)I$_2$]·(DMSO) or C$_{14}$H$_{23}$N$_5$OI$_2$SCd.** This compound crystallizes in the triclinic structure with the lattice parameters $a = 967.9 \pm 0.3$, $b = 981.2 \pm 0.3$, and $c = 1278.3 \pm 0.3$ pm and $\alpha = 103.50° \pm 0.04°$, $\beta = 91.55° \pm 0.02°$, and $\gamma = 94.43° \pm 0.03°$ and a calculated density of 1.999 g·cm$^{-3}$ (Castiñeiras et al. 2002). Single crystals of the title compound suitable for XRD were obtained by slow evaporation of a DMSO solution of [Cd(C$_{12}$H$_{17}$N$_5$S)I$_2$] in air at room temperature.

**[Cd(C$_{13}$H$_{19}$N$_5$S)I$_2$]·EtOH or C$_{15}$H$_{25}$N$_5$OI$_2$SCd.** This compound crystallizes in the triclinic structure with the lattice parameters $a = 939.9 \pm 0.1$, $b = 1013.6 \pm 0.1$, and $c = 1197.9 \pm 0.1$ pm and $\alpha = 82.722° \pm 0.005°$, $\beta = 87.055° \pm 0.04°$, and $\gamma = 80.340° \pm 0.03°$ and a calculated density of 2.053 g·cm$^{-3}$ (Bermejo et al. 2004b). Its yellow crystals were obtained by recrystallization of [Cd(C$_{13}$H$_{19}$N$_5$S)I$_2$] from EtOH/acetone mixture.

**[Cd(C$_9$H$_{11}$N$_3$OS)$_2$I$_2$] or C$_{18}$H$_{22}$N$_6$O$_2$I$_2$SCd.** This compound crystallizes in the monoclinic structure with the lattice parameters $a = 1402.68 \pm 0.01$, $b = 1509.88 \pm 0.02$, and $c = 1335.65 \pm 0.02$ pm and $\beta = 111.16° \pm 0.01°$ and a calculated density of 1.976 g·cm$^{-3}$ (Tian et al. 1999b). It was prepared by mixing an ethanol solution of 4-methoxybenzaldehyde thiosemicarbazone (C$_9$H$_{11}$N$_3$OS) and CdI$_2$. The colorless crystalline solid that formed after refluxing for 4 h was isolated and dried under vacuum. Crystals suitable for XRD were obtained by slow evaporation of an ethanol solution of the title compound at room temperature.

## 4.42  Cd–H–C–N–O–Re–S

Some compounds are formed in this system.

**[Cd(ReO$_4$)$_2$·2CS(NH$_2$)$_2$] or C$_2$H$_8$N$_4$O$_8$Re$_2$SCd.** The title compound forms two polymorphs, which crystallize simultaneously from an aqueous solution of stoichiometric amounts of Cd(ReO$_4$)$_2$ and thiourea (Petrova et al. 1996b). Pure phases were obtained from different solvents, α-Cd(ReO$_4$)$_2$·2CS(NH$_2$)$_2$ from cresol, and β-Cd(ReO$_4$)$_2$·2CS(NH$_2$)$_2$ from EtOH. α-Form crystallizes in the monoclinic structure with the lattice parameters $a = 593.5 \pm 0.4$, $b = 1135.8 \pm 0.1$, and $c = 1049.0 \pm 0.2$ pm and $\beta = 94.66° \pm 0.01°$. β-Form crystallizes in the triclinic structure with the lattice parameters $a = 893.1 \pm 0.2$, $b = 906.6 \pm 0.3$, and $c = 1012.5 \pm 0.1$ pm and $\alpha = 76.46° \pm 0.02°$, $\beta = 67.08° \pm 0.01°$, and $\gamma = 66.39° \pm 0.03°$ and a calculated density of 3.688 g·cm$^{-3}$.

**[Cd(ReO$_4$)$_2$·4CS(NH$_2$)$_2$] or C$_4$H$_{16}$N$_8$O$_8$Re$_2$S$_4$Cd.** This compound crystallizes in the monoclinic structure with the lattice parameters $a = 701.6 \pm 0.2$, $b = 1360.3 \pm 0.2$, and $c = 1155.1 \pm 0.2$ pm and $\beta = 107.16° \pm 0.01°$ and a calculated density of 2.892 g·cm$^{-3}$ (Petrova et al. 1996a). A powder sample of the title compound was synthesized by slow evaporation of an aqueous solution containing Cd(ReO$_4$)$_2$ and thiourea in a 1:4 molar ratio at room temperature. Single crystals were obtained after recrystallization from EtOH.

**[Cd{CS(NH₂)₂}₆(ReO₄)₂]·H₂O** or **C₆H₂₆N₁₂O₉Re₂S₆Cd**. This compound crystallizes in the monoclinic structure with the lattice parameters $a = 1554.3 \pm 0.2$, $b = 1402.5 \pm 0.1$, and $c = 1467.6 \pm 0.5$ pm and $\beta = 112.42° \pm 0.02°$ and a calculated density of 2.442 g·cm⁻³ (Petrova et al. 1997). It was synthesized by slow evaporation of an aqueous solution containing Cd(ReO₄)₂ and thiourea in a 1:6 molar ratio at room temperature. Single crystals were obtained after recrystallization from aqueous solution.

**[Cd₂(H₂O)₄][Re₆S₈(CN)₆]·14H₂O** or **C₆H₃₆N₆O₁₈Re₆S₈Cd₂**. To obtain this compound, a solution of CdSO₄·8/3 H₂O (1.1 mM) in 7 mL of H₂O was added to a solution of Na₄[Re₆S₈(CN)₆] (0.17 mM) in 6 mL of H₂O to give an immediate orange precipitate (Shores et al. 1999). The solid was washed with H₂O (3 × 25 mL), collected by centrifugation and dried in air to afford [Cd₂(H₂O)₄][Re₆S₈(CN)₆]·14H₂O as an orange powder. Crystals of [Cd₂(H₂O)₄][Re₆S₈(CN)₆]·17H₂O (C₆H₄₂N₆O₂₁Re₆S₈Cd₂) were grown by carefully layering aqueous solutions of reactants in a narrow-diameter tube. Large orange-red block-shaped crystals formed after 1 day. They crystallize in the orthorhombic structure with the lattice parameters $a = 1326.8 \pm 0.3$, $b = 1523.6 \pm 0.3$, and $c = 1096.3 \pm 0.02$ pm.

## 4.43   Cd–H–C–N–S

Some compounds are formed in this system.

**[Cd(N₂H₄)₂(NCS)₂]** or **C₂H₈N₆S₂Cd, bis(hydrazine)cadmium isothiocyanate**. This compound crystallizes in the monoclinic structure with the lattice parameters $a = 728$, $b = 1494$, and $c = 441$ pm and $\beta = 106°26'$ (Ferrari et al. 1965).

**[Cd(SCN)₂{SC(NH₂)₂}]** or **C₄H₄N₄S₂Cd**. This compound crystallizes in the triclinic structure with the lattice parameters $a = 403.68 \pm 0.03$, $b = 772.37 \pm 0.04$, and $c = 1013.55 \pm 0.05$ pm and $\alpha = 84.607° \pm 0.004°$, $\beta = 80.825° \pm 0.005°$, and $\gamma = 75.318° \pm 0.005°$ and a calculated density of 2.099 g·cm⁻³ (Wang et al. 2002a). To obtain it, the crystalline powder of Cd(SCN)₂ and thiourea were dissolved in H₂O in stoichiometric proportions at about 40°C. The mixture was left to stand at room temperature, producing colorless crystals of [Cd(SCN)₂{SC(NH₂)₂}].

**[Cd(SCN)₂(NH₂CH₂)₂]** or **C₄H₈N₄S₂Cd**. This compound crystallizes in the triclinic structure with the lattice parameters $a = 1246 \pm 2$, $b = 907 \pm 1$, and $c = 752 \pm 1$ pm and $\alpha = 121.8° \pm 0.3°$, $\beta = 123.4° \pm 0.3°$, and $\gamma = 83.9° \pm 0.3°$ and the experimental and calculated densities 1.91 ± 0.02 and 1.89 g·cm⁻³, respectively (Cannas et al. 1977a).

**[Cd{SC(NH₂)₂}₂(SCN)₂]** or **C₄H₈N₆S₄Cd**. This compound crystallizes in the triclinic structure with the lattice parameters $a = 404 \pm 1$, $b = 768 \pm 1$, and $c = 1010 \pm 1$ pm and $\alpha = 90.8° \pm 0.1°$, $\beta = 99.7° \pm 0.1°$, and $\gamma = 105.4° \pm 0.1°$ (Domiano et al. 1969).

**[Cd(SCN)₂(C₂H₃N₃)₂]** or **C₆H₆N₈S₂Cd**. This compound crystallizes in the triclinic structure with the lattice parameters $a = 571.7 \pm 0.2$, $b = 757.3 \pm 0.2$, and $c = 768.3 \pm 0.2$ pm and $\alpha = 79.54° \pm 0.02°$, $\beta = 68.70° \pm 0.02°$, and $\gamma = 89.36° \pm 0.02°$ and the experimental and calculated densities $1.99 \pm 0.01$ and $2.00$ g·cm$^{-3}$, respectively (Haasnoot et al. 1983). It was crystallized from an aqueous solution of stoichiometric amounts of Cd(SCN)$_2$ and 1,2,4-triazole.

**[Cd(C$_3$H$_4$NS$_2$)$_2$] or C$_6$H$_8$N$_2$S$_4$Cd.** This compound melts at 210°C with decomposition (Bell et al. 2004) [decomposes at 180°C–183°C (Preti and Tosi 1976)] and crystallizes in the orthorhombic structure with the lattice parameters $a = 1507.6 \pm 0.4$, $b = 943.6 \pm 0.2$, and $c = 1582.8 \pm 0.6$ pm and a calculated density of 2.06 g·cm$^{-3}$ (Craig et al. 2005). To obtain it, an aqueous solution of CdCl$_2$ or Cd(NO$_3$)$_2$ or Cd(OAc)$_2$, maintained at a pH value lower than that of precipitation of the Cd(OH)$_2$, has been treated with an aqueous/ethanol solution of 1,3-thiazolidine-2-thione (C$_3$H$_4$NS$_2$) in the Cd/ligand molar ratio of 1:2 (Preti and Tosi 1976). It could also be prepared by the dissolving of cadmium salts in EtOH with the next dropwise addition to stirred solution of C$_3$H$_4$NS$_2$ in a mixture of EtOH and DMSO in the ratio 20:3. The compound has been purified by means of repeated washing with EtOH and dried over P$_4$O$_{10}$.

To obtain the title compound, the next procedure could also be used (Bell et al. 2004). Cd(OAc)$_2$·2H$_2$O (5.67 mM), dissolved in boiling EtOH (50 mL), was added to a hot ethanolic solution (50 mL) containing 1,3-thiazolidine-2-thione (11.40 mM) and Et$_3$N (11.50 mM). The white powder that formed immediately was filtered off and dried in vacuo.

Colorless crystals of the title compound were obtained by liquid diffusion at room temperature (Craig et al. 2005): in a vial, a solution of 1,3-thiazolidine-2-thione (1.01 mM) in a mixture of EtOH (8.5 mL) and DMSO (1.5 mL) was allowed to diffuse into a solution of Cd(OAc)$_2$·2H$_2$O (0.51 mM) in 10 mL of water, through an intervening 10 mL layer of EtOH/H$_2$O (1:1 volume ratio) mixture solvent.

**[Cd(SCN)$_2$(C$_4$H$_{12}$N$_2$)] or C$_6$H$_{12}$N$_4$S$_2$Cd.** This compound crystallizes in the orthorhombic structure with the lattice parameters $a = 1420.3 \pm 1.0$, $b = 1044.4 \pm 0.7$, and $c = 1510.9 \pm 1.2$ pm (Mondal et al. 2000). It was prepared by the addition of methanolic solution of $N,N$-dimethylenediamine (1 mM) to a solution of Cd(OAc)$_2$·2H$_2$O (1 mM) in 5 mL of MeOH at 0°C with stirring. To the resulting mixture a methanolic solution of NaSCN (1 mM) was added very slowly. Shiny colorless crystals appeared within a week, which were collected by filtration, washed with MeOH, and dried in air.

**[MeCdS$_2$CNEt$_2$] or C$_6$H$_{13}$NS$_2$Cd.** This compound melts at a temperature higher than 170°C with decomposition (Hursthouse et al. 1991). To obtain it, Me$_2$Cd (33.6 mM) and diethylamine (33.6 mM) were stirred and heated in toluene (30 mL) under a stream of N$_2$ at 70°C for 12 h. To the cooled solution a cold solution of CS$_2$ (33.6 mM) in toluene (20 mL) was added and the mixture warmed to, and kept at, 40°C for 5 h. Traces of elemental Cd were removed by filtration and the solvent was removed by evaporation. The product was recrystallized from benzene and gave transparent yellow crystals. All manipulations were carried out by using a vacuum-line and standard Schlenk techniques.

**[Cd(S$_2$CNHNMe$_2$)$_2$]$_n$ or C$_6$H$_{14}$N$_4$S$_4$Cd.** This compound was prepared by the next procedure (McCleverty et al. 1982). To a solution of CdCl$_2$ (10 mM) in water (30 mL) one drop of dilute HOAc was added. This mixture was added dropwise to a solution of dimethylhydrazinium dimethyldithiocarbazate (20 mM) in water (30 mL). The title compound formed as a thick gelatinous white solid that was filtered off, washed with water, and dried in vacuo for several days.

**[Cd(C$_2$H$_8$N$_2$)$_2$(SCN)$_2$] or C$_6$H$_{16}$N$_6$S$_2$Cd**. This compound crystallizes in the monoclinic structure with the lattice parameters $a = 927 \pm 1$, $b = 1030 \pm 2$, and $c = 815 \pm 1$ pm and $\beta = 123.0° \pm 0.5°$ (Shvelashvili et al. 1974).

**[Cd(SCN)$_3$(Me$_4$N)] or C$_7$H$_{12}$N$_4$S$_3$Cd tetramethylammonium tris(thiocyanate) cadmate**. This compound crystallizes in the orthorhombic structure with the lattice parameters $a = 1050.6 \pm 0.3$, $b = 1379.7 \pm 0.5$, and $c = 945.06 \pm 0.18$ pm the and experimental and calculated densities $1.73 \pm 0.03$ and $1.75$ g·cm$^{-3}$, respectively (Kuniyasu et al. 1987). To obtain it, aqueous solutions of Cd(SCN)$_2$ (2.5 mM) in 20 mL of H$_2$O and Me$_4$N (2.5 mM) in 10 mL of H$_2$O were mixed and left standing overnight in a silica gel desiccator. Crystals of [Cd(SCN)$_3$(Me$_4$N)] were deposited.

**[Cd(SCN)$_2$(PyCN)] or C$_8$H$_4$N$_4$S$_2$Cd**. This compound crystallizes in the monoclinic structure with the lattice parameters $a = 1881.4 \pm 0.5$, $b = 797.3 \pm 0.2$, and $c = 1537.4 \pm 0.4$ pm and $\beta = 108.046° \pm 0.04°$ and a calculated density of $2.016$ g·cm$^{-3}$ (Chen et al. 2002b). To prepare it, Cd(SCN)$_2$ (1 mM) was dissolved in 5 mL of H$_2$O/EtOH (1:4 volume ratio). A solution of isonicotinonitrile (1 mM) in 3 mL of EtOH was added. Upon standing and cooling, colorless crystals were obtained. A small amount of [Cd(SCN)$_2$(PyCN)$_2$] was also formed, which was separated manually under microscope.

**[Cd(C$_4$H$_3$N$_2$S)$_2$] or C$_8$H$_6$N$_4$S$_2$Cd**. This compound was prepared by the electrochemical oxidation of anodic Cd in an acetonitrile solution of pyrimidine-2-thione (Castro et al. 1992). All experiments were carried out under a dry N$_2$ atmosphere, which also served to stir the reaction mixture as it bubbled through the solution.

**[Cd(SCN)$_2$(C$_3$H$_4$N$_2$)$_2$] or C$_8$H$_8$N$_6$S$_2$Cd**. This compound crystallizes in the monoclinic structure with the lattice parameters $a = 780.4 \pm 0.4$, $b = 576.5 \pm 0.2$, and $c = 1416.4 \pm 0.9$ pm and $\beta = 102.37° \pm 0.01°$ and a calculated density of $1.946$ g·cm$^{-3}$ (Chen et al. 1999). To obtain it, a hot EtOH/H$_2$O (1:1 volume ratio) solution (3 mL) of imidazole (2.5 mM) was added to a hot aqueous solution (5 mL) of Cd(SCN)$_2$ (2.5 mM). After cooling to room temperature, the resulting solution was allowed to stand for ca. 1 week to give colorless crystals of [Cd(SCN)$_2$(C$_3$H$_4$N$_2$)$_2$].

**[Cd(C$_4$H$_5$N$_2$S)$_2$] or C$_8$H$_{10}$N$_4$S$_2$Cd**. This compound melts at 300°C with decomposition (Bell et al. 2004). To obtain it, Cd(OAc)$_2$·2H$_2$O (5.0 mM), dissolved in boiling EtOH (50 mL), was added to a hot ethanolic solution (50 mL) containing 1-methylimidazoline-2(3$H$)-thione (10.0 mM) and Et$_3$N (10.0 mM). The white powder that formed immediately was filtered off and dried in vacuo.

**[Cd(C$_3$H$_5$NS$_2$)$_2$(SCN)$_2$] or C$_8$H$_{10}$N$_4$S$_6$Cd**. This compound melts at 137°C–140°C (Colombini and Preti 1975). It was prepared by the reaction of Cd(SCN)$_2$ and thiazolidine-2-thione in refluxing EtOH for 1 h. The title compound obtained by concentrating was washed with ethyl ether.

**[Cd{SC(NHCH₂)₂}₂(SCN)₂] or C₈H₁₂N₆S₄Cd, di-(2-thioimidazolidine)cadmium thiocyanate.** This compound crystallizes in the monoclinic structure with the lattice parameters $a = 1560 \pm 4$, $b = 817 \pm 2$, and $c = 1151 \pm 1$ pm and $\beta = 95°41'$ (Cavalca et al. 1959, 1960). Its crystals were obtained by slow recrystallization of the title compound from alcoholic solution.

**[Cd(SCN)₂(C₆H₁₆N₂)] or C₈H₁₆N₄S₂Cd.** This compound crystallizes in the monoclinic structure with the lattice parameters $a = 1467.2 \pm 0.3$, $b = 852.0 \pm 0.2$, and $c = 1116.4 \pm 0.3$ pm and $\beta = 100.69° \pm 0.02°$ and a calculated density of 1.67 g·cm⁻³ (Zukerman-Schpector et al. 1988). Its crystals were grown from a solution of stoichiometric amounts of Cd(SCN)₂ and $N,N,N',N'$-tetramethylethylenediamine in MeOH by slow evaporation at room temperature.

**[Cd(SCN)₂(C₆H₁₆N₂)] or C₈H₁₆N₄S₂Cd.** This compound crystallizes in the monoclinic structure with the lattice parameters $a = 1321.6 \pm 0.5$, $b = 935.4 \pm 0.5$, and $c = 1075.2 \pm 0.4$ pm and $\beta = 93.01° \pm 0.03°$ (Mondal et al. 2000). It was prepared by the addition of methanolic solution of $N,N$-diethylenediamine (1 mM) to a solution of Cd(OAc)₂·2H₂O (1 mM) in 5 mL of MeOH at 0°C with stirring. To the resulting mixture a methanolic solution of NaSCN (1 mM) was added very slowly. Shiny colorless crystals appeared within a week, which were collected by filtration, washed with MeOH, and dried in air.

**[Cd{Me₂(NH)₂CS}(CN)₂] or C₈H₁₆N₆S₂Cd.** This compound melts at 158°C–159°C and crystallizes in the monoclinic structure with the lattice parameters $a = 981.18 \pm 0.04$, $b = 1551.09 \pm 0.05$, and $c = 970.56 \pm 0.04$ pm and $\beta = 93.183° \pm 0.003°$ at $110 \pm 2$ K and a calculated density of 1.679 g·cm⁻³ (Malik et al. 2013). It was prepared by adding 2 mM of $N,N'$-dimethylthiourea dissolved in 15 mL of MeOH to an aqueous solution (15 mL) of CdCl₂ (1.0 mM) followed by the addition of 2 mM of KCN in H₂O. The mixture was stirred for 15 min at ambient temperature. The colorless solution was filtered and the filtrate was kept at room temperature for crystallization. Crystalline product was washed with MeOH and dried.

**[Cd(C₆H₁₇N₃)(SCN)₂] or C₈H₁₇N₅S₂Cd.** This compound crystallizes in the monoclinic structure with the lattice parameters $a = 777 \pm 1$, $b = 1444 \pm 2$, and $c = 1354 \pm 2$ pm and $\beta = 112.3° \pm 0.3°$ and the experimental and calculated densities 1.68 and 1.70 g·cm⁻³, respectively (Cannas et al. 1977b). It was prepared by the dropwise addition of bis(aminopropyl)amine (C₆H₁₇N₃) to a hot ethanol solution of Cd(SCN)₂. Colorless crystals were obtained by the evaporation of a concentrated methanolic solution of the title compound containing a few drops of H₂O.

**[Cd(SCN)₂(C₇H₉N₃)] or C₉H₉N₅S₂Cd.** This compound crystallizes in the monoclinic structure with the lattice parameters $a = 990.5 \pm 0.2$, $b = 1006.3 \pm 0.2$, and $c = 1361.2 \pm 0.3$ pm and $\beta = 101.774° \pm 0.004°$ at $233 \pm 2$ K and a calculated density of 1.819 g·cm⁻³ (Banerjee et al. 2005). It was prepared by the next procedure. Cd(OAc)₂·2H₂O (5 mM) dissolved in MeOH (20 mL) was added to a methanolic solution (10 mL) of NH₄SCN (10 mM) followed by 1-pyridine-2-yl-ethylidene (C₇H₉N₃) (2.5 mM) dissolved in MeOH (10 mL). The mixture was stirred for few minutes and dried in vacuo. Colorless crystals suitable for XRD were obtained by dissolving the precipitate in hot DMSO and avoiding light for 24 h.

**[Cd(C₅H₄NS)₂]ₙ or C₁₀H₈N₂S₂Cd**. This compound crystallizes in the mono-clinic structure with the lattice parameters $a = 1746.4 \pm 0.2$, $b = 732.3 \pm 0.1$, and $c = 1898.1 \pm 0.2$ pm and $\beta = 110.72° \pm 0.01°$ and a calculated density of 1.95 g·cm⁻³ (Hursthouse et al. 1990). It immediately precipitated on the stoichiometric reaction of an ethanolic solution of pyridinethiol with an aqueous solution of Cd(OAc)₂ (room temperature, 0.01–0.001 M·L⁻¹ in cadmium). The use of Cd(OAc)₂ avoids the precipi-tation of basic complex. Single crystals were obtained by slow recrystallization of the crude compound from hot pyridine. This compound could also be purified by sublimation (160°C and 1.3 Pa).

This compound could also be prepared by direct electrochemical oxidation of Cd into a solution of pyridine-2-thione in acetonitrile (Durán et al. 1991). At the end of the electrolysis, the solid was collected, washed with acetonitrile, and dried in vacuo.

**[Cd(Et₂NCS₂)₂] or C₁₀H₂₀N₂S₄Cd**. This compound melts at 249°C–250°C (Airoldi et al. 1990) and crystallizes in the monoclinic structure with the lat-tice parameters $a = 982.08 \pm 0.03$, $b = 1076.09 \pm 0.03$, and $c = 1599.52 \pm 0.04$ pm and $\beta = 103.581° \pm 0.001°$ at 223 K (Dee and Tiekink 2002b) [$a = 1016.6 \pm 1.0$, $b = 1074.6 \pm 1.0$, and $c = 1671.7 \pm 1.0$ pm and $\beta = 111.53° \pm 0.03°$ (Bonamico et al. 1965, Domenicano et al. 1968); $a = 1003 \pm 1$, $b = 1075 \pm 1$, and $c = 1666 \pm 2$ pm and $\beta = 112°$; and experimental and calculated densities 1.598 and 1.621 g·cm⁻³, respec-tively (Agre et al. 1966, Shugam and Agre 1968)]. To obtain it, an ethanolic solution of NaEt₂NCS₂·3H₂O (0.10 mM) was slowly added to an ethanolic solution of CdCl₂ [Cd(NO₃)₂ (Shugam and Agre 1968)] (0.05 mM) (Airoldi et al. 1990, Ivanov et al. 2005, Larionov et al. 1990). A white precipitate was produced immediately, col-lected by suction, washed with H₂O, and dried in vacuo. The solid was recrystallized from the mixture MeOH/chloroform (1:1 volume ratio). Its colorless crystals were grown by the recrystallization from benzene (Shugam and Agre 1968) or from the slow evaporation of a CHCl₃ solution of the title compound (Dee and Tiekink 2002b).

**[Cd[(C₄H₃N)CH(N)(C₆H₄S)] or C₁₁H₈N₂SCd**. This compound was obtained by the electrochemical oxidation of anodic Cd in acetonitrile solution of the Schiff base derived from 2-pyrrolecarbaldehyde and bis(2-aminophenyl)disulfide (Castro et al. 1990).

**[Et₄N][Cd(SCN)₃] or C₁₁H₂₀N₄S₃Cd**. This compound crystallizes in the ortho-rhombic structure with the lattice parameters $a = 1077.3 \pm 0.2$, $b = 1657.4 \pm 0.4$, and $c = 987.1 \pm 0.2$ pm and the experimental and calculated densities $1.56 \pm 0.03$ and 1.52 g·cm⁻³, respectively (Taniguchi and Ouchi 1989). To prepare it, Cd(SCN)₂ (2.5 mM) was dissolved in MeOH (10 mL), and the acetone solution (5 mL) of Et₄NSCN (2.5 mM) was added. The mixture was stirred and left standing for about 1 h at ambient temperature. The title compound precipitated.

**C₁₁H₂₃N₅S₂Cd, [bis(2-dimethylaminoethyl)methyl-amine]di-isothiocyanatocadmium(II)**. This compound crystallizes in the monoclinic struc-ture with the lattice parameters $a = 1375 \pm 2$, $b = 768 \pm 1$, and $c = 1623 \pm 0.2$ pm and

$\beta = 97.0° \pm 0.3°$ and the experimental and calculated densities 1.52 and 1.57 g·cm$^{-3}$, respectively (Cannas et al. 1976). It was prepared as colorless crystals by evaporating of a concentrated methanolic solution of the title compound, containing a few drops of $H_2O$.

**[Cd(SCN)$_2$(C$_5$H$_5$N)$_2$] or C$_{12}$H$_{10}$N$_4$S$_2$Cd**. This compound melts at 191°C (Ahuja and Garg 1972) and crystallizes in the triclinic structure with the lattice parameters $a = 951.2 \pm 0.2$, $b = 2735.7 \pm 1.0$, and $c = 951.5 \pm 0.2$ pm and $\alpha = 107.00° \pm 0.03°$, $\beta = 112.47° \pm 0.02°$, and $\gamma = 79.19° \pm 0.03°$ (Taniguchi et al. 1987). It was obtained by allowing an excess of Py to react with cadmium thiocyanate in EtOH (Ahuja and Garg 1972).

**[CdS·(C$_{12}$H$_{12}$N$_2$)] or C$_{12}$H$_{12}$N$_2$SCd**. This compound was prepared by the next procedure (Zheng et al. 2006a). A total of 1.690 g of (PPh$_4$Br), 1.940 g of thiourea and 7.652 g of Cd(SPh)$_2$ were mixed together with 100.093 g of acetonitrile and 9.141 g of $H_2O$. After stirring for ca. 30 min, the mixture became a clear solution. To a 25 mL thick-walled glass vial 2.990 g of this clear solution, 0.094 g of 1,2-bis(4-pyridyl) ethane, and 0.245 g of $H_2O$ were added. The vial was sealed and heated at 85°C for 3 days. After cooling to room temperature, the title compound was obtained in the form of pale-yellow crystals.

**[Cd(SCN)$_2$(C$_5$H$_6$N$_2$)$_2$] or C$_{12}$H$_{12}$N$_6$S$_2$Cd**. This compound crystallizes in the orthorhombic structure with the lattice parameters $a = 1408.0 \pm 0.3$, $b = 568.74 \pm 0.13$, and $c = 1866.6 \pm 0.4$ pm at $233 \pm 2$ K and a calculated density of 1.852 g·cm$^{-3}$ (Banerjee et al. 2005). The following procedure was used for it to be obtained. To a methanol solution (5 mL) of Cd(OAc)$_2$·2H$_2$O (2 mM) a methanol solution (10 mL) of NH$_4$SCN (4 mM) was added. The mixture was stirred for 10 min and the resultant colorless solution was then added to 15 mL of 4-amino-pyridine (C$_5$H$_6$N$_2$) solution (2 mM). The filtrate was allowed to stand in air at room temperature for 24 h, yielding colorless crystals.

**[Cd(C$_6$N$_7$N$_2$S)$_2$]$_6$ or C$_{12}$H$_{14}$N$_4$S$_2$Cd**. This compound crystallizes in the rhombohedral structure with the lattice parameters $a = 1878.4 \pm 0.6$ and $c = 3213 \pm 1$ pm (in hexagonal setting) and a calculated density of 1.363 g·cm$^{-3}$ (Castro et al. 1994). It was obtained electrochemically. The cell consists of a tall-form beaker (100 mL) containing an acetonitrile (50 mL) solution of 4,6-dimethylpyrimidine-2-thione and a small amount of Et$_4$NClO$_4$ (ca. 10 mg) as the supporting electrolyte; a platinum cathode and the sacrificial anode (Ni or Cd), attached to a platinum wire, served as the electrodes and were connected to a power supply. Hydrogen evolved at the cathode. Precipitation took place within ca. 15 min of the start of the electrolysis and continued throughout the experiment. The solid was collected, washed with acetonitrile and diethyl ether, and dried in vacuo. A black product was formed on the cathode but no effort was made to identify it. After electrolysis, the solution was filtered to remove any solid impurities and the solvent evaporated in a Rotavapor to give oil, which was treated with diethyl ether to produce a solid of the title compound.

**[Cd(C$_6$H$_{12}$N$_2$S$_2$)$_2$] or C$_{12}$H$_{24}$N$_4$S$_4$Cd**. This compound was prepared by mixing 4-methylpiperazine-1-thiocarboxylic acid (20 mM) and CS$_2$ (20 mM) in acetonitrile or EtOH under ice-cold conditions (5°C) (Prakasam et al. 2007). To the yellow dithiocarbamic acid solution aqueous solution of $Cd(NO_3)_2$ was added with constant stirring. Pale-yellow solid separated from the solution, which was filtered, washed with $H_2O$ and alcohol, and then dried in air.

**$[Cd(SCN)_4(Me_4N)_2]$ or $C_{12}H_{24}N_6S_4Cd$, tetramethylammonium                   tetrakis (thiocyanate)cadmate**. This compound crystallizes in the triclinic structure with the lattice parameters $a = 730.0 \pm 0.2$, $b = 1301.9 \pm 0.4$, and $c = 589.20 \pm 0.09$ pm and $\alpha = 96.32° \pm 0.02°$, $\beta = 93.080° \pm 0.018°$, and $\gamma = 104.63° \pm 0.02°$ and the experimental and calculated densities $1.51 \pm 0.03$ and $1.52$ $g \cdot cm^{-3}$, respectively (Kuniyasu et al. 1987). To obtain it, the solutions of $Cd(SCN)_2$ and $Me_4N$ in the molar ratio of 1:2 [2.5 mM $Cd(SCN)_2$ in 20 mL of acetone and 5.0 mM $Me_4N$ in 10 mL of acetone] were mixed and left standing overnight at ambient temperature. Crystals of $[Cd(SCN)_4(Me_4N)_2]$ were deposited.

**$[Cd(Et_2NCS_2)_2 \cdot \{C_2H_4(NH_2)_2\}]$ or $C_{12}H_{28}N_4S_4Cd$**. This compound was prepared by the interaction of $Cd(S_2CNEt_2)_2$ and ethylenediamine (4:60 molar ratio) in boiling EtOH (Larionov et al. 1990).

**$[Cd(MeC_6H_3S_2)(C_6H_{16}N_2)]$ or $C_{13}H_{12}N_2S_2Cd$**. To obtain this compound, $Cd(MeC_6H_3S_2)$ was dissolved in $N,N,N'N'$-tetramethylenediamine and excess solvent pumped off over several hours (Bustos et al. 1983). The dropwise addition of petroleum ether precipitated a yellow powder, which was dissolved in chloroform, reprecipitated with petroleum ether, and collected and dried in vacuo.

**$[CdS \cdot (C_{13}H_{14}N_2)]$ or $C_{13}H_{14}N_2SCd$**. This compound was prepared by the following procedure (Zheng et al. 2006a). A total of 0.168 g of $(NMe_4)_2[Cd_4(SPh)_{10}]$, 85 mg of 4,4'-trimethylenedipyridine, 0.464 g of a 10% $Na_2S_2O_3 \cdot 5H_2O$ aqueous solution, and 10.119 g of $H_2O$ were mixed in a 23 mL Teflon-lined stainless steel autoclave and stirred for ca. 20 min. The vessel was sealed and heated at 150°C for 3 days. After cooling to room temperature, the filtered solid product was washed with $H_2O$ and EtOH to give crystals of the title compound.

**$[Cd(Et_2NCS_2)_2(C_3H_4N_2)]$ or $C_{13}H_{24}N_4S_4Cd$**. This compound melts at 126°C with decomposition and crystallizes in the monoclinic structure with the lattice parameters $a = 2082.0 \pm 0.5$, $b = 857.6 \pm 0.3$, and $c = 1152.0 \pm 0.3$ pm and $\beta = 90.74° \pm 0.02°$ and a calculated density of 1.541 $g \cdot cm^{-3}$ (Zemskova et al. 1993b). To obtain it, $Cd(Et_2NCS_2)_2$ (1 mM) and imidazole ($C_3H_4N_2$) were dissolved in $CHCl_3$ (15 mL) at the heating on a water bath. The obtained solution was cooled to ca. 20°C and filtered off, and diethyl ether was added to the total volume of 50 mL. To this mixture 200 mL of hexane was added, and a white crystalline precipitate was obtained. The precipitate was filtered off with suction, washed with hexane, and air-dried. Needlelike single crystals were grown by the slow cooling of the hot saturated solution of the title compound in benzene.

**$[Cd(C_7H_4NS_2)_2]_n$ or $C_{14}H_8N_2S_4Cd$**. This compound melts at 250°C with decomposition (Bell et al. 2004) and crystallizes in the monoclinic structure with the lattice parameters $a = 1945.2 \pm 0.3$, $b = 720.3 \pm 0.1$, and $c = 2396.2 \pm 0.3$ pm and $\beta = 110.60° \pm 0.02°$ and a calculated density of 1.60 $g \cdot cm^{-3}$ (Hursthouse et al. 1990). It immediately precipitated on the stoichiometric reaction of an ethanolic solution of benzothiazole-2-thione with an aqueous solution of Cd(OAc)$_2$ (room temperature, 0.01–0.001 M·L$^{-1}$ in cadmium). The use of Cd(OAc)$_2$ avoids the precipitation of basic complex. Single crystals were obtained by slow recrystallization of the crude compound from hot DMF. This compound could also be purified by sublimation (160°C and 1.3 Pa). The title compound could also be prepared by the electrochemical oxidation of anodic Cd in an acetonitrile solution of benzothiazoline-2-thione (Castro et al. 1993b).

To obtain it, the next procedure could also be used (Bell et al. 2004). Cd(OAc)$_2$·2H$_2$O (5.0 mM), dissolved in boiling EtOH (50 mL), was added to a hot ethanolic solution (50 mL) containing 1,3-benzothiazoline-2-thione (10.0 mM) and Et$_3$N (10.0 mM). The pale-yellow crystals that formed immediately was filtered off and dried in vacuo.

[Cd(C$_7$H$_4$NS$_2$)$_2$]$_n$ was also obtained if to a solution of Na(C$_7$H$_4$NS$_2$) (5 mM) dissolved in warm H$_2$O (40 mL) a solution of CdCl$_2$ (2.5 mM) in H$_2$O (5 mL) was added (McCleverty et al. 1982). The title compound was formed as a white precipitate, which was filtered off and dried in vacuo.

**[Cd(SCN)$_2$(PyCN)$_2$] or C$_{14}$H$_8$N$_6$S$_2$Cd.** This compound crystallizes in the monoclinic structure with the lattice parameters $a = 582.71 \pm 0.10$, $b = 1788.6 \pm 0.3$, and $c = 866.27 \pm 0.15$ pm and $\beta = 109.388° \pm 0.003°$ and a calculated density of 1.703 g·cm$^{-3}$ (Chen et al. 2002b). It was obtained by the next procedure. To a solution of Cd(SCN)$_2$ (1 mM) in 10 mL of H$_2$O/EtOH (1:4 volume ratio) isonicotinonitrile (2 mM) was added. Upon standing colorless crystals were obtained.

**[Cd(SCN)$_2$(C$_6$H$_5$N$_5$)$_2$]$_n$ or C$_{14}$H$_{10}$N$_{12}$S$_2$Cd.** This compound crystallizes in the monoclinic structure with the lattice parameters $a = 2581.8 \pm 0.4$, $b = 740.77 \pm 0.10$, and $c = 1102.76 \pm 0.15$ pm and $\beta = 113.843° \pm 0.002°$ and a calculated density of 1.800 g·cm$^{-3}$ (Li and Xie 2009). To obtain it, 6 mL of methanol solution of 2-(1$H$-1,2,4-triazol-1-yl)pyrazine (0.191 mM), 5 mL of Cd(ClO$_4$)$_2$·6H$_2$O (0.193 mM) aqueous solution, and 5 mL NaSCN (0.389 mM) aqueous solution were mixed together and stirred for a few minutes. The colorless single crystals were obtained after the filtrate had been allowed to stand at room temperature for 2 weeks.

**[Cd(SCN)$_2$(2-MeC$_5$H$_4$N)$_2$] or C$_{14}$H$_{14}$N$_4$S$_2$Cd, bis(2-methylpyridine) bis(thiocyanato)cadmium(II).** This compound melts at 137°C (Ahuja and Garg 1972) and crystallizes in the triclinic structure with the lattice parameters $a = 1107.6 \pm 0.3$, $b = 1847.8 \pm 0.8$, and $c \approx 929.9 \pm 0.3$ pm and $\alpha = 104.60° \pm 0.03°$, $\beta = 114.84° \pm 0.02°$, and $\gamma = 81.12° \pm 0.02°$ and the experimental and calculated densities 1.64 $\pm$ 0.03 and 1.65 g·cm$^{-3}$, respectively (Taniguchi et al. 1987). Its crystals were deposited from a mixture of Cd(SCN)$_2$ (2.5 mM) in 15 mL of EtOH/H$_2$O (1:1 volume ratio) and 2-methylpyridine (5.0 mM) in 10 mL of EtOH after being left standing at ambient temperature for several hours.

**[Cd(SCN)$_2$(3-MeC$_5$H$_4$N)$_2$] or C$_{14}$H$_{14}$N$_4$S$_2$Cd, bis(3-methylpyridine) bis(thiocyanato)cadmium(II).** This compound melts at 147°C (Ahuja and Garg 1972) and crystallizes in the triclinic structure with the lattice parameters $a = 876.67 \pm 0.16$, $b = 1665.5 \pm 0.4$, and $c = 584.47 \pm 0.10$ pm and $\alpha = 91.406° \pm 0.018°$, $\beta = 100.571° \pm 0.014°$, and $\gamma = 92.525° \pm 0.018°$ and the experimental and calculated densities 1.65 $\pm$ 0.03 and 1.65 g·cm$^{-3}$, respectively (Taniguchi et al. 1987). Its crystals were obtained from a mixed solution of $Cd(SCN)_2$ (2.5 mM) and 3-methylpyridine (5.0 mM) in 20 and 10 mL of MeOH, respectively, after being left standing for several hours at ambient temperature.

**$[Cd(SCN)_2(4-MeC_5H_4N)_2]$ or $C_{14}H_{14}N_4S_2Cd$, bis(4-methylpyridine) bis(thiocyanato)cadmium(II)**. This compound melts at 190°C (Ahuja and Garg 1972) and crystallizes in the triclinic structure with the lattice parameters $a = 1132.3 \pm 0.4$, $b = 1838.3 \pm 1.0$, and $c = 928.4 \pm 0.5$ pm and $\alpha = 104.62° \pm 0.04°$, $\beta = 114.26° \pm 0.03°$, and $\gamma = 77.71° \pm 0.04°$ and the experimental and calculated densities $1.62 \pm 0.03$ and $1.63$ g·cm$^{-3}$, respectively (Taniguchi et al. 1986). Its crystals were recrystallized from MeOH.

**$[Cd(SCN)_2(PhNH_2)_2]$ or $C_{14}H_{14}N_4S_2Cd$**. This compound melts at 208°C (Ahuja and Garg 1972) and crystallizes in the monoclinic structure with the lattice parameters $a = 1111.32 \pm 0.13$, $b = 585.47 \pm 0.06$, and $c = 1317.06 \pm 0.13$ pm and $\beta = 100.466° \pm 0.009°$ and the experimental and calculated densities $1.64$ and $1.635$ g·cm$^{-3}$, respectively (Moon et al. 2000a). It was obtained by the following procedure. To a 30 mL aqueous solution containing $CdCl_2 \cdot 2.5H_2O$ (5 mM) and KSCN (10 mM) aniline (1.9 mL, 20 mM) was added; the pH of the solution was adjusted to 9 by adding 2-aminoethanol and citric acid. After a small amount of the precipitate had been filtered off, the aqueous solution was allowed to stand in a refrigerator at 5°C. After a few weeks, pale-yellow crystals were obtained (Moon et al. 2000a).

**$[Cd\{S_2CN(CH_2)_6\}_2]$ or $C_{14}H_{24}N_2S_4Cd$, cadmium hexamethylene dithiocarbamate**. This compound crystallizes in the monoclinic structure with the lattice parameters $a = 1117.7 \pm 0.6$, $b = 1062.1 \pm 0.3$, and $c = 1749.4 \pm 0.9$ pm and $\beta = 116.28° \pm 0.05°$ and the experimental and calculated densities $1.63$ and $1.654$ g·cm$^{-3}$, respectively (Agre and Shugam 1968). It was obtained by the interaction of cadmium salt with $Na(CH_2)_6CNS_2 \cdot 3H_2O$ (Ivanov et al. 2005). Its single crystals were grown by the slow recrystallization at room temperature from DMF (Agre and Shugam 1968).

**$[Cd(S_2CNPr^n_2)_2]$ or $C_{14}H_{28}N_2S_4Cd$**. This compound crystallizes in the monoclinic structure with the lattice parameters $a = 825.32 \pm 0.01$, $b = 1945.19 \pm 0.01$, and $c = 1341.63 \pm 0.02$ pm and $\beta = 99.243° \pm 0.001°$ and a calculated density of $1.453$ g·cm$^{-3}$ (Jian et al. 1999b) [$a = 810.7 \pm 0.1$, $b = 1938.0 \pm 0.2$, and $c = 1320.4 \pm 0.2$ pm and $\beta = 98.643° \pm 0.003°$ at $173 \pm 1$ K and a calculated density of $1.506$ g·cm$^{-3}$ (Ivanov et al. 2005)]. It was prepared by the next procedure. To a heated aqueous solution of $NaPr^n_2CNS_2 \cdot H_2O$ (10 mM) an EtOH solution of $Cd(ClO_4)_2$.(54 mM) was added with stirring (Jian et al. 1999b, Ivanov et al. 2005). The white precipitate was collected by filtration, washed with $H_2O$, and dried over $P_2O_5$. The title compound was dissolved in ethyl ether and allowed to evaporate slowly. This resulted in the formation of needlelike crystals that were used for XRD.

**$[Cd(S_2CNPr^i_2)_2]$ or $C_{14}H_{28}N_2S_4Cd$**. This compound crystallizes in the monoclinic structure with the lattice parameters $a = 1161.1 \pm 0.5$, $b = 1122.8 \pm 0.9$, and $c = 1676.7 \pm 0.2$ pm and $\beta = 108.90° \pm 0.02°$ and a calculated density of $1.494$ g·cm$^{-3}$ (Cox and Tiekink 1999) [$a = 1160.17 \pm 0.10$, $b = 1122.68 \pm 0.11$, and $c = 1675.20 \pm 0.15$ pm and $\beta = 108.936° \pm 0.008°$ and a calculated density of $1.497$ g·cm$^{-3}$ (Jian et al. 1999a)].

It was obtained by the interaction of cadmium salt with $NaPr_2^iCNS_2 \cdot 3H_2O$ (Ivanov et al. 2005). For the preparation of crystals of the title compound, bis($N,N$-diisopropyl-dithiocarbamato)cadmium was dissolved in MeCN. Single crystals suitable for XRD were grown by evaporation at room temperature over a period of 2 months (Jian et al. 1999a).

**$[Pr^n_4N][Cd(SCN)_3]$ or $C_{15}H_{28}N_4S_3Cd$.** This compound crystallizes in the triclinic structure with the lattice parameters $a = 1029.7 \pm 0.5$, $b = 1154.2 \pm 0.5$, and $c = 937.4 \pm 0.3$ pm and $\alpha = 90.10° \pm 0.03°$, $\beta = 91.61° \pm 0.04°$, and $\gamma = 80.94° \pm 0.04°$ and the experimental and calculated densities $1.41 \pm 0.03$ and $1.43$ g·cm$^{-3}$, respectively (Taniguchi and Ouchi 1989). To prepare it, $Cd(SCN)_2$ (2.5 mM) was dissolved in $H_2O$ (5 mL), and the acetone solution (7 mL) of $Et_4NSCN$ (2.5 mM) was added. The mixture was stirred and left standing for about 1 h at ambient temperature. The title compound precipitated.

**$[Cd(Et_2NCS_2)_2(C_5H_{10}NH)]$ or $C_{15}H_{31}N_3S_4Cd$.** This compound melts at 114°C–115°C (Higgins and Saville 1963). To obtain it, to a solution of piperidine (ca. 30 mM) in benzene (20–40 mL) at 50°C the powdered cadmium diethyldithio-carbamate (20 mM) was added, in portions. Upon cooling the resultant solution or the partial removal of the solvent or addition of light petroleum (100 ± 20 mL), the product crystallized. It was collected, washed with light petroleum, and dried in air.

**$[Cd(PhMeNCS_2)_2]$ or $C_{16}H_{16}N_2S_4Cd$.** This compound melts at 296°C–298°C (Onwudiwe and Ajibade 2011). The preparation of it was carried out using 25 mL of aqueous solution of $CdCl_2 \cdot 0.5H_2O$ (1.25 mM), which was added to 25 mL of aqueous solution of $N$-methyl-$N$-phenyldithiocarbamate (2.50 mM). A solid precipitate formed immediately and the mixture was stirred for about 45 min, filtered off, rinsed several times with distilled water, and recrystallized.

**$[Cd(C_7H_8N_4)_2(NCS)_2]_n$ or $C_{16}H_{16}N_{10}S_2Cd$.** This compound crystallizes in the monoclinic structure with the lattice parameters $a = 1956.0 \pm 0.5$, $b = 956.2 \pm 0.4$, and $c = 1152.1 \pm 0.4$ pm and $\beta = 94.25° \pm 0.01°$ and the experimental and calculated densities $1.62 \pm 0.01$ and $1.62$ g·cm$^{-3}$, respectively (Cingi et al. 1986). It was obtained by adding [6,8-dimethyl[1,2,4]triazolo[3,4-b]pyridazine] ($C_7H_8N_4$) (4 mM) dissolved in a mixture of EtOH/$H_2O$ (1:1 volume ratio) (10 mL) to a solution of $Cd(NO_3)_2$ (2 mM) and $NH_4SCN$ (4 mM) in $H_2O$ (20 mL).

**$[Cd(2\text{-}MeC_6H_4NH_2)_2(SCN)_2]$, $[Cd(3\text{-}MeC_6H_4NH_2)_2(SCN)_2]$ and $[Cd(4\text{-}MeC_6H_4NH_2)_2(SCN)_2]$ or $C_{16}H_{18}N_4S_2Cd$.** These compounds melt at 154°C, 178°C, and 201°C, respectively (Ahuja and Garg 1972). They were obtained by allowing an excess of 2-methylaniline (3-methylaniline, 4-methylaniline) to react with cadmium thiocyanate in EtOH.

**$[Cd(SCN)_2(BzNH_2)_2]$ or $C_{16}H_{18}N_4S_2Cd$.** This compound exists in two polymorphic modifications (Ouchi and Taniguchi 1988). One of them ($A$) crystallizes in the monoclinic structure with the lattice parameters $a = 2739.0 \pm 1.9$, $b = 577.09 \pm 0.08$, and $c = 587.33 \pm 0.09$ pm and $\beta = 90.35° \pm 0.03°$ and the experimental and calculated densities $1.57 \pm 0.03$ and $1.58$ g·cm$^{-3}$, respectively (Taniguchi and Ouchi 1987a).

The second (B) also crystallizes in the monoclinic structure with the lattice parameters $a = 1395.8 \pm 0.2$, $b = 575.9 \pm 0.1$, and $c = 1146.6 \pm 0.2$ pm and $\beta = 99.24° \pm 0.01°$ and the experimental and calculated densities 1.61 and 1.62 $g \cdot cm^{-3}$, respectively (Ouchi and Taniguchi 1988). To obtain A, $Cd(SCN)_2$ (4 mM) and benzylamine (8.4 mM) were dissolved into 30 and 10 mL of MeOH/EtOH (2:1 volume ratio) mixture solvent, separately. They were warmed to about 80°C and mixed. After the mixture was refluxed about 2 h, it was filtered off while hot in order to remove B (a part of A is also removed). The filtrate was slowly cooled to 6°C, and deposited crystals of A were separated, washed with a little portion of MeOH/EtOH mixture, and dried in open air.

On the other hand, the precipitated crystals were almost pure B, when a little less than the calculated quantity of benzylamine was being added, using MeOH/$H_2O$ (49:1 volume ratio) mixture solvent to increase solubility of A, and the mixed solution was left standing overnight at room temperature (Ouchi and Taniguchi 1988). B could also be obtained by the next procedure (Taniguchi and Ouchi 1987a). $Cd(SCN)_2$ (2.5 mM) was dissolved into 20 mL of MeOH. Benzylamine (5 mM) dissolved into 10 mL of MeOH was added to it and mixed. The mixed solution gradually became turbid and precipitation started soon. After being left standing for about 30 min at ambient temperature, the deposited crystals were filtered off, washed with acetone, and dried in open air.

**$[Cd(2,4-Me_2C_5H_3N)_2(SCN)_2]$ and $[Cd(2,6-Me_2C_5H_3N)_2(SCN)_2]$ or $C_{16}H_{18}N_4S_2Cd$.** The first compound melts at 153°C and the second at 250°C (Ahuja and Garg 1972). It was obtained by allowing an excess of 2,4-$Me_2Py$ and 2,6-$Me_2Py$ to react with cadmium thiocyanate in EtOH.

**$[Cd(2-EtC_5H_4N)_2(SCN)_2]$, $[Cd(3-EtC_5H_4N)_2(SCN)_2]$, and $[Cd(4-EtC_5H_4N)_2(SCN)_2]$ or $C_{16}H_{18}N_4S_2Cd$.** These compounds melt at 160°C, 165°C and 190°C, respectively (Ahuja and Garg 1972). They were obtained by allowing an excess of 2-EtPy (3-EtPy, 4-EtPy) to react with cadmium thiocyanate in EtOH.

**$[(Me_2C_3H_3N_2)_2][Cd_2(SCN)_6]$ or $C_{16}H_{18}N_{10}S_6Cd_2$.** This compound crystallizes in the monoclinic structure with the lattice parameters $a = 1746.8 \pm 0.3$, $b = 772.73 \pm 0.12$, and $c = 1067.50 \pm 0.16$ pm and $\beta = 104.833° \pm 0.002°$ and a calculated density of 1.830 $g \cdot cm^{-3}$ (Chen et al. 2001a). It was prepared by the next procedure. To a mixture of 1 mL of 1 M $Cd(NO_3)_2 \cdot 4H_2O$ and 1 mL of 3 M $NH_4SCN$ in $H_2O$ $[Me_2C_3H_3N_2]I$ (N,N'-dimethylimidazolium iodide) (1.0 mM) was added. A white precipitate was immediately formed. Acetone was added to the resulted mixture till all the precipitate dissolved again. Slow evaporation of the pale-yellow solution afforded colorless prisms of the title compound.

**$[Cd(SCN)_2(C_7H_{10}N_2)_2]_n$ or $C_{16}H_{20}N_6S_2Cd$.** This compound melts at 271°C–273°C and crystallizes in the monoclinic structure with the lattice parameters $a = 606.7 \pm 0.1$, $b = 1101.7 \pm 0.3$, and $c = 1443.4 \pm 0.3$ pm and $\beta = 98.31° \pm 0.02°$ and a calculated density of 1.645 $g \cdot cm^{-3}$ (Secondo et al. 2000). To obtain it, $Cd(NO_3) \cdot 4H_2O$ (2.0 mM) was dissolved in 25 ml of absolute EtOH, to which a solution of KSCN (4.0 mM) in the same solvent (30 mL) was added with stirring. The precipitated $KNO_3$ was filtered and to the filtrate was added a solution of 4-(N,N-dimethylamino)pyridine

$(C_7H_{10}N_2)$ (4.0 mM) in absolute EtOH (10 mL). The resulting precipitate was isolated by filtration, washed with EtOH and then ether, and dried in vacuo. Suitable crystals of the title compound were grown from saturated solution in DMSO over a period of 2 weeks.

**[(3-Bu$^t$C$_3$H$_2$N$_2$)$_2$Cd(SCN)$_2$]$_n$ or C$_{16}$H$_{22}$N$_6$S$_2$Cd.** This compound crystallizes in the monoclinic structure with the lattice parameters $a = 600.16 \pm 0.02$, $b = 1027.59 \pm 0.03$, and $c = 1746.73 \pm 0.05$ pm and $\beta = 92.2898° \pm 0.0016°$ and a calculated density of 1.45 g·cm$^{-3}$ (Reger et al. 2002). X-ray quality crystals were obtained from an attempted crystallization, from a layered solution of acetone and hexane, of the product produced in the [{HB(3-ButC$_3$H$_2$N$_2$)$_3$]Cd(SCN)} preparation.

**[Cd(PhS)$_2$·2CS$_2$·(C$_6$H$_{16}$N$_2$)] or C$_{16}$H$_{34}$N$_2$S$_6$Cd.** When Cd(PhS)$_2$ and excess of $N,N,N',N'$-tetramethylenediamine were dissolved in CS$_2$, a clear yellow solution was obtained (Black et al. 1986). Slow evaporation in vacuo of this solution while it was being cooled in ice produced yellow crystals, which were collected, washed with diethyl ether, and dried in vacuo; this material was shown to be Cd(PhS)$_2$·2CS$_2$(C$_6$H$_{16}$N$_2$).

**[Cd(MeC$_6$H$_3$S$_2$)(2,2′-bipy)] or C$_{17}$H$_{14}$N$_2$S$_2$Cd.** This compound was obtained by the following procedure (Bustos et al. 1983). Cd(MeC$_6$H$_3$S$_2$) was added to a solution of 2,2′-bipy in EtOH. The solution slowly changed color from pale-yellow to deep-yellow on being stirred over a period of 7 and 16 h. The suspended solid was then collected, washed twice with EtOH, and dried in vacuo to give the title compound.

**[Cd(MeC$_6$H$_3$S$_2$)(Py)$_2$] or C$_{17}$H$_{16}$N$_2$S$_2$Cd.** To obtain this compound, Cd(MeC$_6$H$_3$S$_2$) (0.43 g) was dissolved in pyridine (5 mL) with stirring (Bustos et al. 1983). Diethyl ether was added dropwise until the solution became cloudy. A further small quantity of pyridine was added to redissolve the precipitate, and the solution cooled at 0°C for 24 h. The resultant yellow precipitate was collected and dried in vacuo.

**[Cd(C$_4$H$_3$N$_2$S)$_2$(2,2′-bipy)] or C$_{18}$H$_{14}$N$_6$S$_2$Cd.** This compound was prepared by the electrochemical oxidation of anodic Cd in an acetonitrile solution of pyrimidine-2-thione in the presence of 2,2′-bipy (Castro et al. 1992). All experiments were carried out under a dry N$_2$ atmosphere, which also served to stir the reaction mixture as it bubbled through the solution.

**[Cd(C$_9$H$_{12}$N$_5$S)$_2$] or C$_{18}$H$_{24}$N$_{10}$S$_2$Cd.** This compound crystallizes in the monoclinic structure with the lattice parameters $a = 928.4 \pm 0.3$, $b = 1509.6 \pm 0.3$, and $c = 1693.3 \pm 0.5$ pm and $\beta = 95.047° \pm 0.007°$ and a calculated density of 1.565 g·cm$^{-3}$ (Bermejo et al. 2004a). It was prepared by recrystallization of [Cd(C$_9$H$_{12}$N$_5$S)(OAc)]$_2$ from DMSO or upon slow evaporation of the solvent from the aqueous solution of [Cd(C$_9$H$_{13}$N$_5$S)Cl$_2$] as yellow crystals.

**[Cd{S$_2$CN(Et)(C$_6$H$_{11}$)}$_2$] or C$_{18}$H$_{32}$N$_2$S$_4$Cd.** This compound crystallizes in the monoclinic structure with the lattice parameters $a = 854.1 \pm 0.4$, $b = 1968.8 \pm 0.8$, and $c = 1384.7 \pm 0.3$ pm and $\beta = 94.29° \pm 0.03°$ at 200 K and a calculated density of 1.479 g·cm$^{-3}$ (Cox and Tiekink 1999).

**[Cd(S₂CNBuⁿ₂)₂] or C₁₈H₃₆N₂S₄Cd**. This compound was obtained by the interaction of cadmium salt with NaBuⁿ₂CNS₂·H₂O (Ivanov et al. 2005).

**[Cd(S₂CNBuⁱ₂)₂] or C₁₈H₃₆N₂S₄Cd**. This compound crystallizes in the monoclinic structure with the lattice parameters $a = 4997 \pm 5$, $b = 959.7 \pm 0.3$, and $c = 2367.5 \pm 0.8$ pm and $\beta = 116.41° \pm 0.03°$ and the calculated density 1.361 g·cm⁻³ (Cox and Tiekink 1999). It was obtained by the interaction of cadmium salt with NaBuⁱ₂CNS₂·3H₂O (Ivanov et al. 2005).

**Cd₂SCN[(C₈H₁₀N₄S)(C₈H₉N₄S)(SCN)₂] or C₁₉H₁₉N₁₁S₅Cd₂**. This compound crystallizes in the monoclinic structure with the lattice parameters $a = 794.9 \pm 0.2$, $b = 1175.7 \pm 0.3$, and $c = 1610.4 \pm 0.4$ pm and $\beta = 97.403° \pm 0.004°$ and a calculated density of 1.877 g·cm⁻³ (Nfor et al. 2006). It was prepared by the following procedure. To a clear solution of 2-acetylpyridine thiosemicarbazone (0.1 mM) dissolved in MeCN (10 mL) a solution of NH₄NCS (0.1 mM) in H₂O (3 mL) was added. The mixture was stirred at room temperature for 10 min, to which an acetonitrile solution (5 mL) of Cd(OAc)₂·4H₂O (0.1 mM) was added with stirring. The resulting mixture was stirred for another 10 min at room temperature and then filtered. Slow evaporation of the filtrate in air gave colorless block-shaped crystals after 2 weeks.

**[Cd(C₄H₃N₂S)₂(Phen)] or C₂₀H₁₄N₆S₂Cd**. This compound crystallizes in the orthorhombic structure with the lattice parameters $a = 988.2 \pm 0.2$, $b = 1249.1 \pm 0.1$, and $c = 1651.3 \pm 0.2$ pm and a calculated density of 1.678 g·cm⁻³ (Castro et al. 1992). It was prepared by the electrochemical oxidation of anodic Cd in an acetonitrile solution of pyrimidine-2-thione in the presence of Phen. All experiments were carried out under a dry N₂ atmosphere, which also served to stir the reaction mixture as it bubbled through the solution.

**[Cd(PyS)₂(PySH)₂] or C₂₀H₁₈N₄S₄Cd**. This compound crystallizes in the orthorhombic structure with the lattice parameters $a = 917.94 \pm 0.05$, $b = 1022.30 \pm 0.06$, and $c = 1184.65 \pm 0.07$ pm and a calculated density of 1.658 g·cm⁻³ (Anjali et al. 2003). It was obtained by the next procedure. Cd(ClO₄)₂·6H₂O (0.38 mM) in 2 mL of MeOH was added to a solution of 4-pyridine thiol (2.25 mM) in 2 mL of DMSO. The mixture was warmed slightly and stirred for 15 min. The resulting clear solution was cooled to room temperature to get bright greenish yellow crystals. The crystals were separated by filtration, washed with MeOH and then Et₂O, and dried in vacuo. The final product of thermal decomposition of the title compound is CdS.

**[Cd(PhS)₂·2CS₂(2,2′-bipy)] or C₂₀H₂₆N₂S₆Cd**. This compound crystallizes in the monoclinic structure with the lattice parameters $a = 2218.4 \pm 0.3$, $b = 900.1 \pm 0.2$, and $c = 1268.7 \pm 0.6$ pm and $\beta = 99.90° \pm 0.03°$ and a calculated density of 1.565 g·cm⁻³ (Black et al. 1986). When Cd(PhS)₂ (1.38 mM) and 2,2′-bipy (1.41 mM) were added to 30 mL of CS₂, a clear yellow solution was obtained. Slow evaporation in vacuo of this solution while it was being cooled in ice produced yellow crystals, which were collected, washed with diethyl ether, and dried in vacuo; this material was shown to be Cd(PhS)₂·2CS₂(2,2′-bipy).

**[Cd(Et₂NCS₂)₂(2,2′-bipy)] or C₂₀H₂₈N₄S₄Cd**. This compound melts at 204°C–205°C (at 255°C–258°C [Larionov et al. 1990]) and crystallizes in the monoclinic structure with the lattice parameters $a = 1811.8 \pm 0.1$, $b = 839.6 \pm 0.1$, and $c = 1694.1 \pm 0.2$ pm and $\beta = 105.84° \pm 0.01°$ and a calculated density of 1.514 g·cm$^{-3}$ (Airoldi et al. 1990) [$a = 1810.3 \pm 0.3$, $b = 838.1 \pm 0.1$, and $c = 1696.5 \pm 0.3$ pm and $\beta = 105.97° \pm 0.01°$ and the calculated density 1.52 g·cm$^{-3}$ (Glinskaya et al. 1992a)]. To obtain it, an ethanolic solution of 2,2′-bipy (0.63 mM) was slowly added to Cd(Et$_2$NCS$_2$)$_2$ (0.53 mM) dissolved in the mixture EtOH/chloroform (1:1 volume ratio) (Airoldi et al. 1990). A solid appeared immediately, which was left with the solution for 24 h before being collected, dried, and recrystallized from hot chloroform. It could also be prepared by the interaction of Cd(S$_2$CNEt$_2$)$_2$ and 2,2′-bipy (3:3.3 molar ratio) in the boiling mixture CHCl$_3$/toluene (Larionov et al. 1990). Its single crystals were grown by the slow evaporation of the solution of [Et$_2$NCS$_2$]$_2$Cd(2,2′-bipy) in the mixture EtOH/CHCl$_3$ (1:1 volume ratio) at room temperature (Glinskaya et al. 1992a).

**[Cd(Et$_2$NCS$_2$)$_2$(4,4′-bipy)] or C$_{20}$H$_{28}$N$_4$S$_4$Cd.** This compound melts at 180°C (Zemskova et al. 1993a). To obtain it, Cd(Et$_2$NCS$_2$)$_2$ (1.0 mM) and 4,4′-bipy (1.5 mM) were dissolved in CH$_2$Cl$_2$ (40 mL). The resulting solution was evaporated on a water bath until crystallization, then cooled, and kept at –15°C for 2 h. The obtained pale-yellow precipitate was filtered off, washed with cold CH$_2$Cl$_2$, dried in air, and recrystallized from CH$_2$Cl$_2$. Thermal decomposition of this compound at 330°C for 1 h leads to the formation of α-CdS.

**[Cd(SCN)$_2$(C$_{18}$H$_{40}$N$_4$)] or C$_{20}$H$_{40}$N$_6$S$_2$Cd.** This compound was prepared, when the reactants cis-[Cd(C$_{18}$H$_{40}$N$_4$)(NO$_3$)](NO$_3$) (1 mM) and KSCN (2 mM) were taken separately in hot MeOH (25 mL) and the solutions mixed (Roy et al. 2010). The clear and colorless solution was heated on a water bath for about 30 min to allow the reaction to complete. The reaction mixture was kept at room temperature and after 48 h the crystalline white product was separated by filtration. The white product was washed with MeOH followed by diethyl ether and dried in a desiccator over silica gel.

**Cd(SC$_6$H$_4$N)[MeC(C$_5$H$_3$N)CMe] or C$_{21}$H$_{17}$N$_3$S$_2$Cd.** Reaction of bis-benzothiazoline with Cd(OAc)$_2$ in DMF/acetone yields red crystals of this compound (Lindoy and Busc 1972).

**[Cd(PhS)$_2$·CS$_2$·(Phen)] or C$_{21}$H$_{26}$N$_2$S$_4$Cd.** Solution of Phen (1.28 mM) in CS$_2$ (30 mL) readily dissolved Cd(SPh)$_2$ (1.28 mM) (Black et al. 1986). Slow evaporation in vacuo of the resultant yellow solution gave a yellow solid, which was collected, washed with diethyl ether, and dried in vacuo. This product is Cd(PhS)$_2$·CS$_2$·(Phen).

**[Cd(C$_5$H$_4$NS)$_2$(Phen)] or C$_{22}$H$_{16}$N$_4$S$_2$Cd.** This compound was prepared by direct electrochemical oxidation of Cd into a solution of pyridine-2-thione with the addition of Phen in acetonitrile (Durán et al. 1991). At the end of the electrolysis, the solid was collected, washed with acetonitrile, and dried in vacuo.

**C$_{22}$H$_{16}$N$_6$S$_2$Cd, bis(bipyridyl)bis(isothiocyanato)cadmium.** This compound crystallizes in the orthorhombic structure with the lattice parameters $a = 1627.4 \pm 0.6$, $b = 1675 \pm 1$, and $c = 1613.6 \pm 0.9$ pm and the experimental and calculated densities $1.62 \pm 0.02$ and 1.63 g·cm$^{-3}$, respectively (Rodesiler et al. 1989). Bis(pyridyl) cadmium nitrate was prepared in situ dissolving 0.01 M of bipyridyl and 0.005 M of Cd(NO$_3$)$_2$·4H$_2$O in 200 mL of H$_2$O. The resulting solution was warmed gently

(~80°C) for 1 h and 0.01 M of NaSCN dissolved in 50 mL of $H_2O$ was added with stirring. The metathesis reaction was essentially instantaneous and the resultant precipitate was collected, washed, and air-dried. Single crystals were grown from the solution of this compound in DMF.

**[Cd(PhS)₂(2,2′-bipy)] or C₂₂H₁₈N₂S₂Cd.** Electrolysis of Cd into a solution of PhSH (9.7 mM), 2,2′-bipy (5.8 mM), and 40 g $Et_4NClO_4$ in MeCN (50 mL) for 3 h at 15 V and 40 mA led to the dissolution of 0.273 g of Cd and yielded a solution from which an oil precipitated upon slow evaporation of the solvent in vacuo (Black et al. 1986). This oil was triturated with EtOH, whereupon a yellow powder of $Cd(PhS)_2(2,2′$-bipy)$_2$ was formed.

This compound could also be prepared if a solution of $Cd(SPh)_2$ (0.32 mM) and 2,2′-bipy (3.1 mM) in DMF (10 mL) was stirred at room temperature overnight, during which time the colorless mixture turned pink (Black et al. 1986). The resultant solution was cooled to 0°C and MeOH was added, producing a pink oil, which was collected and triturated with cold MeOH and then mixtures of MeOH and diethyl ether at room temperature to yield $Cd(PhS)_2(2,2′$-bipy) as a pink solid.

**[Cd(PhS)₂·2CS₂·(Phen)] or C₂₂H₂₆N₂S₆Cd.** The remaining solution phase after obtaining $Cd(PhS)_2·CS_2·(Phen)$ was evaporated to dryness, and the solid produced was redissolved in 3 mL of $CS_2$; the addition of $n$-pentane caused the precipitation of a yellow solid, which was collected, washed with $n$-pentane, and dried in vacuo (Black et al. 1986). This material was identified as $Cd(PhS)_2·2CS_2·(Phen)$.

**[Cd(Et₂NCS₂)₂(Phen)] or C₂₂H₂₈N₄S₄Cd.** This compound melts at 278°C–280°C [at 266°C–272°C (Larionov et al. 1990)] and crystallizes in the triclinic structure with the lattice parameters $a = 1104.2 \pm 0.2$, $b = 1518.5 \pm 0.2$, and $c = 1697.8 \pm 0.3$ pm and $\alpha = 106.43° \pm 0.01°$, $\beta = 99.84° \pm 0.01°$, and $\gamma = 101.35° \pm 0.01°$ and a calculated density of 1.506 g·cm⁻³ (Airoldi et al. 1990) [$a = 1104.4 \pm 0.2$, $b = 1517.2 \pm 0.4$, and $c = 1696.7 \pm 0.4$ pm and $\alpha = 106.38° \pm 0.01°$, $\beta = 99.79° \pm 0.01°$, and $\gamma = 101.40° \pm 0.01°$ and a calculated density of 1.51 g·cm⁻³ (Glinskaya et al. 1992a)]. To obtain it, an ethanolic solution of Phen·$H_2O$ (0.5 mM) was slowly added to $Cd(Et_2NCS_2)_2$ (0.5 mM) dissolved in the mixture EtOH/chloroform (1:1 volume ratio) (Airoldi et al. 1990). A solid appeared immediately, which was left with the solution for 24 h before being collected, dried, and recrystallized from hot chloroform. Its single crystals were grown by the slow evaporation of the solution of $[Et_2NCS_2]_2Cd(Phen)$ in the mixture EtOH/CHCl₃ (1:1 volume ratio) at room temperature Glinskaya et al. 1992a).

It could also be prepared by the interaction of $Cd(S_2CNEt_2)_2$ and Phen (2:2.2 molar ratio) in the boiling mixture CHCl₃/toluene (Larionov et al. 1990).

**[Cd(CNEt₂)₂(C₆H₅N)₂]ₙ or C₂₂H₃₀N₄S₄Cd.** This compound melts at 198°C–200°C and crystallizes in the monoclinic structure with the lattice parameters $a = 2106.20 \pm 0.17$, $b = 1922.48 \pm 0.17$, and $c = 1516.53 \pm 0.13$ pm and $\beta = 122.439° \pm 0.002°$ at 223 K (Chai et al. 2003). Yellow crystals of the title compound were isolated from the mixture chloroform/toluene (2:1 volume ratio) solution containing equimolar amounts of $Cd(CNEt_2)_2$ and $trans$-1,2-bis(4-pyridyl)ethylene, $(C_6H_5N)_2$.

**[Cd(C$_6$H$_{12}$N$_2$S$_2$)$_2$(2,2′-bipy)]** or **C$_{22}$H$_{32}$N$_6$S$_4$Cd**. This compound decomposes at 184°C (Prakasam et al. 2007). It was prepared by adding a hot solution of 2,2′-bipy (1 mM) in benzene to a hot suspension of [Cd(C$_6$H$_{12}$N$_2$S$_2$)$_2$] (1 mM) in benzene/chloroform (1:1 volume ratio) mixture. The resulting pale-yellow solution was filtered and cooled. The yellow compound separated out, which was filtered and recrystallized from benzene/chloroform (1:1 volume ratio) mixture.

**[Cd$_2$(SCN)$_4$(C$_6$H$_{11}$N$_3$)$_3$]** or **C$_{22}$H$_{33}$N$_{13}$S$_4$Cd$_2$**. This compound crystallizes in the orthorhombic structure with the lattice parameters $a = 1089.8 \pm 0.1$, $b = 1441.3 \pm 0.2$, and $c = 2140.2 \pm 0.2$ pm and the experimental and calculated densities 1.65 and 1.646 g·cm$^{-3}$, respectively (Groeneveld et al. 1982). It was prepared by the rapid addition of a solution of NH$_4$SCN (5 mM) and 4-Bu$^t_4$–1,2,4-triazole (5 mM) in 20 mL of H$_2$O to a boiling solution of Cd(NO$_3$)$_2$ (2.5 mM) in H$_2$O (15 mL). After reduction of the volume to 30 mL, yellow crystals of this compound were formed after several days at ambient temperature.

**[Cd(S-2,4,6-Me$_3$C$_6$H$_2$)$_2$(C$_5$H$_5$N)]** or **C$_{23}$H$_{27}$NS$_2$Cd**. Cd(S-2,4,6-Me$_3$C$_6$H$_2$)$_2$ forms a crystalline pyridine adduct that readily loses Py on heating in vacuo or more slowly on storage at 10°C (Bochmann et al. 1991).

**[Cd(PhS)$_2$(Phen)]** or **C$_{24}$H$_{18}$N$_2$S$_2$Cd**. This compound could be prepared if a solution of equimolar quantities of Cd(SPh)$_2$ and Phen in DMF was stirred at room temperature overnight, during which the precipitation of orange oil occurred, which yielded Cd(PhS)$_2$(Phen) as an orange solid (Black et al. 1986).

**[Cd(Py)$_2$(C$_7$H$_4$NS$_2$)$_2$]** or **C$_{24}$H$_{18}$N$_4$S$_4$Cd**. To prepare this compound, to a solution of Cd(NO$_3$)$_2$ (10 mM) in a mixture (50 mL) of Py/H$_2$O (1:1 volume ratio) benzothiazoline-2-thione (20 mM) dissolved in the mixture (30 mL) Py/EtOH (1:1 volume ratio) was added (McCleverty et al. 1982). The addition of H$_2$O to this mixture caused precipitation of the title compound as a yellow powder.

**[Cd(S$_2$CNEt$_2$)$_2$(2,9-Me$_2$Phen)]** or **C$_{24}$H$_{32}$N$_4$S$_4$Cd**. This compound melts at 259°C–262°C and crystallizes in the monoclinic structure with the lattice parameters $a = 1108.84 \pm 0.07$, $b = 995.70 \pm 0.06$, and $c = 2500.55 \pm 0.15$ pm and $\beta = 93.067° \pm 0.001°$ (Lai and Tiekink 2003). Pale-yellow crystals were isolated from acetonitrile/chloroform (1:2 volume ratio) solution containing equimolar amounts of Cd(S$_2$CNEt$_2$)$_2$ and 2,9-Me$_2$Phen.

**[Cd(C$_6$H$_{12}$N$_2$S$_2$)$_2$(Phen)]** or **C$_{24}$H$_{32}$N$_6$S$_4$Cd**. This compound decomposes at 192°C (Prakasam et al. 2007). It was prepared by adding a hot solution of Phen (1 mM) in benzene to a hot suspension of [Cd(C$_6$H$_{12}$N$_2$S$_2$)$_2$] (1 mM) in benzene/chloroform (1:1 volume ratio) mixture. The resulting pale-yellow solution was filtered and cooled. The yellow compound separated out, which was filtered and recrystallized from benzene/chloroform (1:1 volume ratio) mixture.

**[Cd(C$_{12}$H$_{16}$N$_5$S)$_2$]** or **C$_{24}$H$_{32}$N$_{10}$S$_2$Cd**. This compound crystallizes in the monoclinic structure with the lattice parameters $a = 964.0 \pm 0.4$, $b = 1582.4 \pm 0.3$, and $c = 1766.9 \pm 0.5$ pm and $\beta = 93.619° \pm 0.018°$ and a calculated density of 1.537 g·cm$^{-3}$ (Castiñeiras et al. 2002). Its single crystals suitable for XRD were obtained by slow evaporation of the solution in air at room temperature from the reaction of 2-pyri-dineformamide $N(4)$-piperidylthiosemicarbazone ($C_{12}H_{17}N_5S$) with $Cd(OAc)_2$ in methylcyanide.

**[Bu$^n_4$N][Cd(S$_2$CNMe$_2$)$_3$] or C$_{25}$H$_{54}$N$_4$S$_6$Cd.** This compound was obtained by the next procedure (McCleverty et al. 1982). [Cd(S$_2$CNMe$_2$)] (10 mM) was added in small amounts and with shaking to an acetone solution (20 mL) containing [Bu$^n_4$N] [S$_2$CNMe$_2$] (10 mM). After filtration, the solvent was reduced in vacuo affording the title compound as a very pale-yellow solid. This was collected by filtration and recrystallized from acetone/light petroleum (1:1 volume ratio).

**[Bu$^n_4$N][Cd(S$_2$CNHNMe$_2$)$_3$] or C$_{25}$H$_{57}$N$_7$S$_6$Cd.** This compound was prepared as follows (McCleverty et al. 1982). To a solution of dimethylhydrazinium dimethyldi-thiocarbazate (1.96 g) in H$_2$O (40 mL) a solution of CdCl$_2$ (10 mM) in H$_2$O (10 mL) was added dropwise. To this mixture a warm solution of Bu$_4$NI in aqueous EtOH was added slowly. After stirring, the mixture was filtered off, washed with H$_2$O, EtOH, and diethyl ether, and dried in vacuo.

**[Cd(C$_7$H$_4$NS$_2$)$_2$(Phen)] or C$_{26}$H$_{16}$N$_4$S$_4$Cd.** This compound was prepared by the electrochemical oxidation of anodic Cd in an acetonitrile solution of benzothiazole-2-thione in the presence of Phen (Castro et al. 1993b).

**[Cd(C$_3$H$_4$NS$_2$)$_2$(Py)$_3$]·(Py) or C$_{26}$H$_{28}$N$_2$S$_4$Cd.** This compound crystallizes in the monoclinic structure with the lattice parameters $a = 3117.8 \pm 0.5$, $b = 1192.0 \pm 0.2$, and $c = 1673.8 \pm 0.3$ pm and $\beta = 112.245° \pm 0.005°$ at $250 \pm 2$ K and a calculated density of 1.53 g·cm$^{-3}$ (Craig et al. 2005). Its colorless crystals were produced by recrystalliza-tion of Cd(C$_3$H$_4$NS$_2$)$_2$ from pyridine. This adduct loses pyridine on standing at room temperature, although the loss is not complete even after pumping the adduct under preparative vacuum for 24 h.

**[Cd(S$_2$CNEt$_2$)$_2$(4,7-Me$_2$Phen)]·MeCN or C$_{26}$H$_{35}$N$_5$S$_4$Cd.** This compound crys-tallizes in the monoclinic structure with the lattice parameters $a = 1094.51 \pm 0.07$, $b = 1008.86 \pm 0.06$, and $c = 2843.66 \pm 0.17$ pm and $\beta = 94.198° \pm 0.001°$ at $223 \pm 2$ K and a calculated density of 1.396 g·cm$^{-3}$ (Guo et al. 2002). Yellow crystals were obtained by slow evaporation of a CHCl$_3$/MeCN (1:1 volume ratio) solution of the title compound prepared by refluxing equimolar amounts of Cd(S$_2$CNEt$_2$)$_2$ and 4,7-Me$_2$Phen in chloroform solution. The melting point of the desolvated material was 278°C–282°C.

**[Cd(C$_{13}$H$_{18}$N$_5$S)$_2$] or C$_{26}$H$_{36}$N$_{10}$S$_2$Cd.** This compound crystallizes in the monoclinic structure with the lattice parameters $a = 987.4 \pm 0.1$, $b = 1583.3 \pm 0.2$, and $c = 1851.9 \pm 0.2$ pm and $\beta = 91.562° \pm 0.002°$ and a calculated density of 1.527 g·cm$^{-3}$ (Bermejo et al. 2004b). It was prepared by recrystallization of [Cd(C$_{13}$H$_{19}$N$_5$S)Cl$_2$] from EtOH.

**[Cd(S-2,3,5,6-Me$_4$C$_6$H)$_2$(1-MeImid)$_2$] or C$_{28}$H$_{38}$N$_4$S$_2$Cd.** This compound was obtained by the interaction of Cd(S-2,3,5,6-Me$_4$C$_6$H)$_2$ with two equivalents of 1-methylimidazole in MeCN (Corwin et al. 1987).

**[Cd[Bu$^n_2$NCS$_2$]$_2$(2,2′-bipy)] or C$_{28}$H$_{44}$N$_4$S$_4$Cd.** This compound crystallizes in the orthorhombic structure with the lattice parameters $a = 2871.6 \pm 0.4$, $b = 684.8 \pm 0.6$, and $c = 1718.8 \pm 0.2$ pm and a calculated density of 1.335 g·cm$^{-3}$ (Ivanchenko et al. 2000). To obtain it, the solutions of Cd[Bu$^n_2$NCS$_2$] (1 mM) in toluene (5 mL) and 2,2'-bipy (1.2 mM) in toluene (5 mL) were mixed. After the addition of heptane, a yellow precipitate formed. It was purified by the recrystallization from toluene. Thermolysis of this compound at 350°C leads to the formation of β-CdS. Its single crystals were grown by slow evaporation of the solution in toluene at room temperature.

[Cd[Bu$^i_2$NCS$_2$]$_2$(2,2'-bipy)] or C$_{28}$H$_{44}$N$_4$S$_4$Cd. This compound melts at 188°C–190°C (Zemskova et al. 1998). To obtain it, Cd[Bu$^i_2$NCS$_2$] (10 mM) and 2,2'-bipy (13 mM) were dissolved the mixture of CHCl$_3$ (20 mL) and ethylacetate (15 mL) at 50°C. After cooling and standing the solution at –15°C for 2 h, pale-yellow crystals were formed. The precipitate was filtered with suction in vacuum desiccator over CaCl$_2$. The title compound was recrystallized from the mixture CHCl$_3$/ethylacetate.

[Cd$_2$(SEt)$_6$(Et$_4$N$_2$)] or C$_{28}$H$_{70}$N$_2$S$_6$Cd$_2$. This compound crystallizes in the mono-clinic structure with the lattice parameters $a = 1098.3 \pm 0.4$, $b = 1052.5 \pm 0.4$, and $c = 1844.5 \pm 0.7$ pm and $\beta = 94.56° \pm 0.03°$ and the experimental and calculated densities 1.35 and 1.33 g·cm$^{-3}$, respectively (Watson et al. 1985). To obtain it, to a solution of 19 mM of CdCl$_2$ in ~ 80 mL of acetonitrile 60 mM of solid NaSEt was added with stirring. The mixture was stirred for 4 h, and 20 mM of Et$_4$NCl was added. After the mixture was stirred for a further 4 h, NaCl was removed by fil-tration. The filtrate was condensed to ~ 20 mL and cooled slowly to ~ 0°C. Large colorless rhomblike crystals of C$_{28}$H$_{70}$N$_2$S$_6$Cd$_2$ were collected, washed with ether, and dried in vacuo.

[Et$_4$N][Cd(C$_7$H$_4$NS$_2$)$_3$] or C$_{29}$H$_{32}$N$_4$S$_6$Cd. This compound crystallizes in the cubic structure with the lattice parameter $a = 1859 \pm 1$ and the experimental and calcu-lated densities 1.50 and 1.532 g·cm$^{-3}$, respectively (McCleverty et al. 1982). It was obtained from EtOH as colorless plate-shaped crystals.

[Cd(PhS)$_2$]$_2$[(C$_6$H$_{16}$N$_2$)] or C$_{30}$H$_{26}$N$_2$S$_4$Cd$_2$. At the interaction of Cd(SPh)$_2$ and excess of N,N,N',N'-tetramethylenediamine, the green oil was formed, which proved impossible to crystallize, but the addition of diethyl ether to the supernatant liquid yielded this compound as a colorless solid (Black et al. 1986).

[Cd{S$_2$CN(CH$_2$Ph)$_2$}$_2$]$_2$ or C$_{30}$H$_{28}$N$_2$S$_4$Cd. This compound melts at 194°C–195°C and crystallizes in the monoclinic structure with the lattice parameters $a = 1110.98 \pm 0.04$, $b = 1563.25 \pm 0.05$, and $c = 1666.95 \pm 0.06$ pm and $\beta = 97.9220° \pm 0.0010°$ at 223 K and a calculated density of 1.522 g·cm$^{-3}$ (Fan et al. 2004, Yin et al. 2004). To obtain it, a solution of CdCl$_2$ (0.5 mM) in water (10 mL) was added to a solution of sodium dibenzyldithiocarbamate (1 mM) in EtOH (20 mL). The mixture was stirred for 2 h at room temperature and the precipitated compound was filtered off, washed with ethanol and then with diethyl ether, and dried in vacuum.

[Cd(SCN)$_2${(PhCH$_2$)$_2$NH}$_2$] or C$_{30}$H$_{30}$N$_4$S$_2$Cd. This compound crystallizes in the triclinic structure with the lattice parameters $a = 1128.4 \pm 0.2$, $b = 1140.4 \pm 0.2$, and $c = 597.4 \pm 0.1$ pm and $\alpha = 93.31° \pm 0.02$, $\beta = 98.37° \pm 0.01°$, and $\gamma = 71.84° \pm 0.01°$ and the experimental and calculated densities $1.42 \pm 0.03$ and 1.43 g·cm$^{-3}$, respectively

(Taniguchi and Ouchi 1987b). To obtain it, $Cd(SCN)_2$ (5 mM) was dissolved into 10 mL of MeOH; dibenzylamine (10 mM) dissolved into 10 mL of MeOH was added to it. When the mixed solution was left standing for about 1 h at ambient temperature, colorless needlelike crystals were deposited.

**$[Cd(SCN)_4(C_{13}H_{21}N_3)_2]$ or $C_{30}H_{42}N_{10}S_4Cd_2$.** This compound crystallizes in the triclinic structure with the lattice parameters $a = 733 \pm 3$, $b = 1006 \pm 4$, and $c = 1382 \pm 5$ pm and $\alpha = 73.26° \pm 0.06°$, $\beta = 83.29° \pm 0.07°$, and $\gamma = 74.87° \pm 0.06°$ and a calculated density of 1.581 $g \cdot cm^{-3}$ (Banerjee et al. 2005). To obtain it, a methanolic solution of $Cd(OAc)_2 \cdot 2H_2O$ (2 mM) and $NH_4SCN$ (4 mM) was added to a hot methanolic solution (10 mL) of $N,N$-diethyl-$N'$-(1-pyridine-2-yl-ethylidene)-ethan-1,2-diamine $(C_{13}H_{21}N_3)$ with stirring. The resulting solution was allowed to stand at room temperature for 2 days to yield colorless crystals.

**$[Bu^n_4N][Cd(C_7H_4NS_2)_2]$ or $C_{30}H_{44}N_3S_4Cd$.** This compound was prepared as follows (McCleverty et al. 1982). To a solution containing an excess of $[Bu^n_4N][C_7H_4NS_2]$ in absolute EtOH $[Cd(C_7H_4NS_2)_2]$ was added, slowly and with stirring, until the mixture became homogeneous. On standing for 12 h at room temperature, the title compound formed as pale-yellow needlelike crystals, which were filtered off, washed with diethyl ether, and dried in vacuo.

**$[Cd(Bu^n_2NCS_2)_2(Phen)]$ or $C_{30}H_{44}N_4S_4Cd$.** This compound crystallizes in the orthorhombic structure with the lattice parameters $a = 1559.2 \pm 0.3$, $b = 2272.4 \pm 0.5$, and $c = 992.2 \pm 0.2$ pm and a calculated density of 1.332 $g \cdot cm^{-3}$ (Ivanchenko et al. 1998, 2000). To obtain it, 2 mM of $Cd[Bu^n_2NCS_2]_2$ was dissolved in $CHCl_3$ (5 mL) with next addition to this solution 2.2 mM of $Phen \cdot H_2O$. Then 5 mL of toluene was added to full dissolution and the obtained solution was heated for 10 min. The solution was cooled to room temperature and heptane was added, after which the yellow precipitate was formed. It was purified by recrystallization from toluene. Thermolysis of this compound at 320°C leads to the formation of $\beta$-CdS. Its single crystals were grown by slow evaporation of the solution in toluene at room temperature.

**$[Cd[Bu^i_2NCS_2]_2(Phen)]$ or $C_{30}H_{44}N_4S_4Cd$.** This compound melts at 241°C (Zemskova et al. 1998). To obtain it, $Cd[Bu^i_2NCS_2]$ (10 mM) and $Phen \cdot H_2O$ (13 mM) were dissolved the mixture (35 mL) $CHCl_3$/ethylacetate (6.5:3.5 volume ratio) at 50°C. After cooling and standing the solution at room temperature for 2 h, pale-yellow crystals were formed. The precipitate was filtered with suction in vacuum desiccator over $CaCl_2$. The title compound was recrystallized from the mixture $CHCl_3$/ethylacetate.

**$[Cd(MeC_6H_3S_2)(Phen)_2]$ or $C_{31}H_{22}N_4S_2Cd$.** This compound was obtained by the next procedure (Bustos et al. 1983). To a solution of Phen (0.54 g) in EtOH (40 mL) 0.25 g of $Cd(MeC_6H_3S_2)$ was added. The solution slowly changed color from pale-yellow to orange on being stirred over a period of 6 h. The suspended solid was then collected, washed twice with EtOH, and dried in vacuo to give the title compound.

**$[Me_4N][Cd(S_2CNBu^n_2)_3]$ or $C_{31}H_{66}N_4S_6Cd$.** This compound crystallizes in the monoclinic structure with the lattice parameters $a = 1199.9 \pm 0.7$, $b = 1029.4 \pm 0.7$, and $c = 3659.5 \pm 2.4$ pm and $\beta = 94.510° \pm 0.007°$ and the experimental and calculated densities 1.25 and 1.263 g·cm$^{-3}$, respectively (McCleverty et al. 1982). It was obtained as follows. [Cd(S$_2$CNBu$^n_2$] (10 mM) was added in small amounts and with shaking to an acetone solution (20 mL) containing [Me$_4$N][S$_2$CNBu$^n_2$] (10 mM). After filtration, the solvent was reduced in vacuo affording the title compound as a very pale-yellow solid. This was collected by filtration and recrystallized from acetone/light petroleum (1:1 volume ratio) as elongated colorless bricks.

**[Bu$^n_4$N][Cd(S$_2$CNEt$_2$)$_3$] or C$_{31}$H$_{66}$N$_4$S$_6$Cd.** This compound was obtained by the next procedure (McCleverty et al. 1982). [Cd(S$_2$CNEt$_2$)] (10 mM) was added in small amounts and with shaking to an acetone solution (20 mL) containing [Bu$^n_4$N] [S$_2$CNBu$^n_2$] (10 mM). After filtration, the solvent was reduced in vacuo affording the title compound as a very pale-yellow solid. This was collected by filtration and recrystallized from acetone/light petroleum (1:1 volume ratio). This compound was also obtained by the reaction of [Cd(S$_2$CNEt$_2$)$_3$] with [Bu$^n_4$N][OCOR] (R = Pr$^n$, Am, C$_7$H$_{15}$, and C$_8$H$_{17}$) (McCleverty et al. 1982).

**[Cd$_2$(SCN)$_2$($\mu_2$-SCN)$_2$(PhNHCSNH$_2$)$_2$($\mu_2$-PhNHCSNH$_2$)$_2$]$_n$ or C$_{32}$H$_{32}$N$_{12}$S$_8$Cd$_2$.** This compound crystallizes in the triclinic structure with the lattice parameters $a = 933.6 \pm 0.3$, $b = 1468.6 \pm 0.5$, and $c = 1691.1 \pm 0.5$ pm and $\alpha = 71.36° \pm 0.02°$, $\beta = 84.31° \pm 0.02°$, and $\gamma = 72.470° \pm 0.010°$ and a calculated density of 1.690 g·cm$^{-3}$ (Yang et al. 2001a). To obtain it, phenylthiourea (2 mM), KSCN (2 mM) and Cd(NO$_3$)$_2$·4H$_2$O (1 mM) were dissolved in EtOH with stirring. The resulting mixture was allowed to evaporate to give well-shaped colorless crystals of the title compound within 2 weeks.

**[Cd(S-2,3,5,6-Me$_4$C$_6$H)$_2$(C$_6$H$_5$N)$_2$] or C$_{32}$H$_{36}$N$_2$S$_2$Cd.** This compound crystallizes in the monoclinic structure with the lattice parameters $a = 1654.2 \pm 0.7$, $b = 1589.8 \pm 0.6$, and $c = 1636.4 \pm 1.0$ pm and $\beta = 108.94° \pm 0.13°$ (Corwin et al. 1987). It was obtained by the interaction of Cd(S-2,3,5,6-Me$_4$C$_6$H)$_2$ with two equivalents of pyridine in MeCN.

**[Cd(C$_{12}$H$_{12}$NS)$_2$(C$_4$H$_6$N$_2$)$_2$] or C$_{32}$H$_{36}$N$_6$S$_2$Cd.** This compound crystallizes in the monoclinic structure with the lattice parameters $a = 751.0 \pm 0.1$, $b = 1509.0 \pm 0.1$, and $c = 2848.8 \pm 0.3$ pm and $\beta = 95.68° \pm 0.01°$ and a calculated density of 1.408 g·cm$^{-3}$ (Otto et al. 1999). To obtain it, 2 mM of Et$_3$N, 2 mM of N-(2-thiophenyl)-2,5-dimethylpyrrole, and 2 mM of N-methylimidazole were dissolved in 35 mL of MeOH. Dry CdCl$_2$ (1 mM) in 10 mL of MeOH was added and big colorless crystals were obtained. These crystals were filtered off, washed with MeOH, and dried in vacuo. The title compound was recrystallized from hot acetonitrile to yield colorless needles. All procedures were carried out under an Ar or N$_2$ atmosphere using standard Schlenk and glovebox techniques.

**[Me$_4$N]$_2$[Cd(SPh)$_4$] or C$_{32}$H$_{44}$N$_2$S$_4$Cd.** This compound crystallizes in the monoclinic structure with the lattice parameters $a = 1205.3 \pm 0.2$, $b = 1457.0 \pm 0.2$, and $c = 982.7 \pm 0.5$ pm and $\beta = 90.89° \pm 0.02°$ and a calculated density of 1.343 g·cm$^{-3}$ (Ueyama et al. 1988). To obtain it, a solution of Cd(NO$_3$)$_2$·4H$_2$O (10 mM) in warm propanol (30 mL) was added slowly with stirring to a solution of benzenethiol (70 mM) in Pr$^n$OH [deaerated MeOH (Carson and Dean 1982)] (20 mL) (Dance et al. 1984).

A solution of $Me_4NCl$ (2.2 g) in MeOH (60 mL) was added. The colorless product that crystallized by seeding, if necessary, was filtered, washed with MeOH, and vacuum dried. Recrystallization was from a hot saturated solution in acetonitrile. $[NMe_4]_2[Cd(SPh)_4]$ could also be recrystallized from $Pr^iOH$ and obtained as colorless needles (Ueyama et al. 1988).

**$[\{Cd(PhS)_2\}_2(2,2'\text{-bipy})]$ or $C_{34}H_{28}N_2S_4Cd_2$.** When $CS_2$ (40 mL) was added to a solution of $Cd(PhS)_2$ (1.52 mM) and 2,2'-bipy (5.56 mM) in DMF (10 mL), no precipitate was obtained (Black et al. 1986). After 48 h at room temperature, petroleum ether (30°C–50°C) was added to the mixture, precipitating an oil, which was collected and triturated with EtOH to give a solid, which is $[Cd(PhS)_2](2,2'\text{-bipy})$.

**$[Cd(C_5H_5N)\{S_2CN(PhCH_2)_2\}_2]$ or $C_{35}H_{33}N_3S_4Cd$.** This compound melts at 152.5°– 153.5°C and crystallizes in the monoclinic structure with the lattice parameters $a = 3656.2 \pm 0.2$, $b = 953.95 \pm 0.06$, and $c = 1989.2 \pm 0.1$ pm and $\beta = 103.501° \pm 0.001°$ at 223 K (Wei et al. 2005). To obtain it, pyridine (1 mL) was added to a solution of THF (40 mL) containing bis[bis(N,N-dibenzyldithiocarbamato)cadmium (1 mM). The mixture was stirred for 1 h at room temperature and filtrated. Colorless crystals were obtained by slow evaporation of the solution.

**$[Bu^n_4N][Cd(C_7H_4NS_2)_2(S_2CNEt_2)]$ or $C_{35}H_{54}N_4S_6Cd$.** This compound was prepared if cadmium bis(benzothiazole-2-thiolate) (2.5 mM) was added in small portions to a solution of $\{Bu^n_4N\}[S_2CNEt_2]$ (2.5 mM) in acetone (50 mL) (McCleverty et al. 1982). The mixture was shaken to ensure complete dissolution and then the solvent was removed in vacuo yielding oil. When this oil was shaken with diethyl ether, the title compound was formed as a creamy colorless powder. It was collected by filtration and dried in vacuo.

**$[Cd(PhS)_2(Phen)_2]$ or $C_{36}H_{26}N_4S_2Cd$.** Electrolysis of Cd into a solution of PhSH (9.7 mM), Phen (4.4 mM), and 40 g $Et_4NClO_4$ in MeCN (50 mL) for 2 h at 15 V and 40 mA led to the dissolution of 1.62 mM of Cd (Black et al. 1986). The yellow crystals, which formed in the cell during the experiment, were collected and washed successively with MeCN and n-pentane; during the latter procedure, the crystals disintegrated to a yellow powder, identified as $Cd(PhS)_2(Phen)_2$.

**$[\{Cd(PhS)_2\}_2(Phen)]$ or $C_{36}H_{28}N_2S_4Cd_2$.** This compound could be prepared if a supernatant liquid from the reaction of equimolar quantities of $Cd(SPh)_2$ and Phen in DMF was treated with diethyl ether, which caused the precipitation of the yellow solid (Black et al. 1986).

**$[Cd(Bu^i_2NSC_2)_2]_2$ or $C_{36}H_{72}N_4S_8Cd_2$.** This compound melts at 172°C and crystallizes in the monoclinic structure with the lattice parameters $a = 5008.7 \pm 1.1$, $b = 961.3 \pm 0.2$, and $c = 2370.8 \pm 0.8$ pm and $\beta = 116.51° \pm 0.02°$ and a calculated density of 1.355 g·cm⁻³ (Glinskaya et al. 1999). To obtain it, 0.1 mM of $Bu^i_2NH$ was mixed with 17 ml of MeOH and placed in the flask that was cooled with a mixture of NaCl with ice. A quantity of 0.1 mM of $CS_2$ was added to this solution with vigorous stirring. Then to the reaction mixture a solution of 0.1 mM NaOH in MeOH (100 mL) was added dropwise while cooling and stirring. The resulting clear, slightly yellow solution was poured in portions to a solution of $CdCl_2·2.5H_2O$ (0.05 mM) at the stirring and cooling. White precipitate had appeared, which was filtered off, washed with $H_2O$, and dried in air. It was recrystallized from a mixture $CHCl_3/EtOH$ (1:1 volume ratio). Single crystals were grown at the slow cooling of a hot saturated solution in toluene and the next evaporation at room temperature.

**[Cd(NH$_2$CH$_2$CH$_2$NH$_2$)$_3$][Cd(S$_2$CNEt$_2$)$_3$]$_2$ or C$_{36}$H$_{84}$N$_{12}$S$_{12}$Cd$_3$.** This compound melts at 105°C and crystallizes in the tetragonal structure with the lattice parameters $a = 2040.9 \pm 0.7$ and $c = 2877.7 \pm 0.9$ pm and a calculated density of 1.56 g·cm$^{-3}$ (Glinskaya et al. 1992b, Zemskova et al. 1999). It was prepared by the following procedure. $CdI_2$ (40 mM) was dissolved in a mixture of 70% ethylenediamine aqueous solution (5 mL) and EtOH (20 mL) at 40°C. Fine-faceted, colorless crystals precipitated when the mixture was allowed to cool to room temperature. After 2 h the crystals were filtered, washed with cold EtOH, and dried in air. α-CdS is formed at the thermolysis of the title compound (Glinskaya et al. 1992b). Single crystals of the title compound were grown by slowly cooling warm (~40°C) its ethanol solutions in the presence of a 10-fold excess of ethylenediamine (Zemskova et al. 1999).

**[Cd(PhS)$_2$·0.5CS$_2$·(Phen)$_2$] or C$_{36.5}$H$_{26}$N$_4$S$_3$Cd.** When $CS_2$ (40 mL) was added to a solution of $Cd(PhS)_2$ (1.52 mM) and Phen (5.56 mM) in DMF (10 mL), the solution immediately became deep yellow; after 1 h at room temperature a yellow precipitate forms from the stirred solution (Black et al. 1986). This product is $Cd(PhS)_2 \cdot 0.5CS_2 \cdot (Phen)_2$.

**Cd(S-2,4,6-Pr$^i_3$C$_6$H$_2$)$_2$(1-MeImid)$_2$ or C$_{38}$H$_{58}$N$_4$S$_2$Cd.** This compound crystallizes in the monoclinic structure with the lattice parameters $a = 1654.1 \pm 0.7$, $b = 1588.7 \pm 0.6$, and $c = 1636 \pm 1$ pm and $\beta = 108.91° \pm 0.05$ and a calculated density of 1.22 g·cm$^{-3}$ (Santos et al. 1990). To obtain it, $LiS(2,4,6-Pr^i_3C_6H_2)$ (4.2 mM) and $CdCl_2$ (2.1 mM) were combined in absolute EtOH, and 1-methylimidazole (0.70 mL) was added with a syringe. The resultant white precipitate was filtered and subsequently recrystallized from hot MeCN. The crystals were grown from a slowly cooled acetonitrile solution.

This compound could also be obtained by the interaction of $Cd(S-2,4,6-Pri_3C_6H_2)_2$ with two equivalents of 1-methylimidazole in MeCN (Corwin et al. 1987).

**[Cd{Ph(SMe)CC(S)Ph}$_2$]·(2,2'-bipy) or C$_{40}$H$_{34}$N$_2$S$_4$Cd.** This compound melts at 188°C with decomposition and crystallizes in the triclinic structure with the lattice parameters $a = 1365.2 \pm 0.3$, $b = 1284.3 \pm 0.7$, and $c = 1056.4 \pm 0.2$ pm and $\alpha = 101.39° \pm 0.03°$, $\beta = 94.20° \pm 0.02°$, and $\gamma = 94.36° \pm 0.03°$ and a calculated density of 1.443 g·cm$^{-3}$ (Reddy et al. 1992). To prepare it, $[Cd\{Ph(SMe)CC(S)Ph\}_2] \cdot (DMSO)$ (0.89 mM) and (2,2'-bipy) (2.5 mM) were heated in a sealed glass vial of 10 mL capacity to 100°C for 1 h. When the vial was cooled, a yellow precipitate formed, which was collected, washed several times with $H_2O$, then EtOH, and finally $n$-hexane, and dried. The title compound was recrystallized from the mixture $CH_2Cl_2/n$-hexane (1:1 volume ratio).

**[Cd(S-2,4,6-Pr$^i_3$C$_6$H$_2$)$_2$(2,2'-bipy)] or C$_{40}$H$_{54}$N$_2$S$_2$Cd.** This compound crystallizes in the monoclinic structure with the lattice parameters $a = 1508.1 \pm 0.9$,

$b = 1471.6 \pm 0.9$, and $c = 1772 \pm 1$ pm and $\beta = 100.36° \pm 0.05°$ and a calculated density of 1.27 g·cm$^{-3}$ (Santos et al. 1990). To obtain it, LiS(2,4,6-Pri$_3$C$_6$H$_2$) (6.4 mM) was dissolved in 35 mL of MeOH and CdCl$_2$ (3.2 mM) was added under a N$_2$ atmosphere. Next, 2,2'-bipy (3.2 mM) was added resulting in an immediate color change to lemon yellow. A hot (50°C) filtration, followed by cooling of the solution to –20°C, resulted in precipitation of bright-yellow product. Yellow crystals were grown by slow cooling a hot (60°C) acetonitrile solution to room temperature.

**(Et$_4$N)$_2$[Cd(SPh)$_4$] or C$_{40}$H$_{60}$N$_2$S$_4$Cd.** To obtain this compound, a solution of 25 mM of CdCl$_2$·2.5H$_2$O in 50 mL of MeOH and 10 mL of H$_2$O was added to a solution of 125 mM of sodium benzenethiolate (from 2.87 g of Na and 14.2 g of benzenethiol) in 100 mL of MeOH (Hagen et al. 1982). After the solution was stirred for 30 min, NaCl was removed by filtration and the filtrate was added to 50 mM of solid Et$_4$NBr. The volume of the solution was reduced in vacuum until incipient crystallization; the solution was gently warmed and the slowly cooled to –25°C. The white crystals were collected by filtration, washed with EtOH and ether, and dried in vacuo.

**[Cd(S-2,4,6-Pr$^i$$_3$C$_6$H$_2$)$_2$(1,2-Me$_2$Imid)$_2$] or C$_{40}$H$_{62}$N$_4$S$_2$Cd.** This compound could be prepared by the same procedure as for Cd(S-2,4,6-Pri$_3$C$_6$H$_2$)$_2$(1-MeImid)$_2$ using 1,2-dimethylimidazole instead of 1-methylimidazole (Santos et al. 1990).

**[Cd$_2$(Pr$^n$$_2$NCS$_2$)$_4$(2-C$_5$H$_4$NC(H)=NN=C(H)C$_5$H$_4$N-2)] or C$_{40}$H$_{66}$N$_8$S$_8$Cd$_2$.** This compound melts at 168°C–170°C and crystallizes in the monoclinic structure with the lattice parameters $a = 907.68 \pm 0.16$, $b = 1113.7 \pm 0.2$, and $c = 2538.9 \pm 0.5$ pm and $\beta = 92.216° \pm 0.003°$ at $98 \pm 2$ K and a calculated density of 1.477 g·cm$^{-3}$ (Poplaukhin and Tiekink 2008). Red crystals were grown by the slow evaporation of a MeOH/EtOH (1:1 volume ratio) solution of the title compound.

**[Cd(S-2,4,6-Pr$^i$$_3$C$_6$H$_2$)$_2$(Phen)] or C$_{42}$H$_{54}$N$_2$S$_2$Cd.** This compound crystallizes in the monoclinic structure with the lattice parameters $a = 1012 \pm 2$, $b = 1656 \pm 1$, and $c = 2350 \pm 2$ pm and $\beta = 91.1° \pm 0.1°$ and a calculated density of 1.29 g·cm$^{-3}$ (Santos et al. 1990). To obtain it, LiS(2,4,6-Pr$^i$$_3$C$_6$H$_2$) (4.2 mM), CdCl$_2$ (1.7 mM), and Phen (1.7 mM) were combined in 25 mL of absolute EtOH. The solvent was removed after stirring for 30 min and the crude yellow solid was recrystallized from CH$_2$Cl$_2$/hexane. Crystals were grown by cooling a toluene solution to room temperature.

**[Cd(S-2,4,6-Pr$^i$$_3$C$_6$H$_2$)$_2$(C$_6$H$_5$N)$_2$] or C$_{42}$H$_{56}$N$_2$S$_2$Cd.** This compound was obtained by the interaction of Cd(S-2,4,6-Pr$^i$$_3$C$_6$H$_2$)$_2$ with two equivalents of pyridine in MeCN (Corwin et al. 1987).

**[Cd(SCPh$_3$)$_2$(C$_6$H$_{16}$N$_2$)] or C$_{44}$N$_{46}$N$_2$S$_2$Cd.** This compound melts at 205°C and crystallizes in the orthorhombic structure with the lattice parameters $a = 1776.8 \pm 0.7$, $b = 1952.9 \pm 0.8$, and $c = 2198.4 \pm 0.9$ pm at $153 \pm 2$ K (Edwards et al. 1996). It was prepared under dry oxygen-free Ar using a standard inert atmosphere. $n$-Butyllithium (1.6 M, 3.25 mL) in hexanes (5.0 mM) was added to a solution of Ph$_3$CSH (5 mM) in toluene (10 mL) at 0°C. When the lithiation was complete, [Cd{N(SiMe$_3$)$_2$}] (2.5 mM), tetramethylenediamine (2.5 mM), and THF (5 mL) were added and the reaction mixture stirred at room temperature (5 min). The resulting orange solution was reduced in vacuo to ca. 10 mL, and storage at room temperature (48 h) gave colorless needles of the title compound.

**[Cd{Ph(SMe)CC(S)Ph}₂]·(Me₂N₂Py) or C₄₄H₄₆N₄S₄Cd.** This compound melts at 179°C with decomposition and crystallizes in the monoclinic structure with the lattice parameters $a = 1748.3 \pm 0.4$, $b = 1229.1 \pm 0.3$, and $c = 1996.3 \pm 0.5$ pm and $\beta = 98.998° \pm 0.020°$ and a calculated density of 1.114 g·cm⁻³ (Reddy et al. 1992). To prepare it, [Cd{Ph(SMe)CC(S)Ph}₂]·(DMSO) (0.44 mM) and 4-(dimethylamino) pyridine (2.8 mM) were placed in a glass vial of 10 mL capacity. The vial was sealed and heated to 120°C for 1 h. When the vial was cooled, a yellow precipitate formed, which was collected, washed several times with H₂O, then EtOH, and finally *n*-hexane, and dried. The title compound was recrystallized from the mixture CH₂Cl₂/*n*-hexane (1:1 volume ratio).

**[Cd₂(Buⁱ₂NCS₂)₄(4,4′-bipy)] or C₄₆H₈₀N₆S₈Cd₂.** This compound melts at 170°C and crystallizes in the monoclinic structure with the lattice parameters $a = 1832.4 \pm 0.2$, $b = 1075.9 \pm 0.1$, and $c = 1816.2 \pm 0.2$ pm and $\beta = 106.39° \pm 0.01°$ and a calculated density of 1.356 g·cm⁻³ (Zemskova et al. 1998). To obtain it, Cd[Buⁱ₂NCS₂] (10 mM) and 4,4′-bipy (13 mM) were dissolved in toluene (50 mL) at 50°C. After cooling and standing the solution at room temperature for 2 h, white fine precipitate was formed. The precipitate was filtered with suction, washed with toluene, and dried in air. The title compound was recrystallized from toluene. Its single crystals were grown by the slow cooling of a hot saturated solution in acetonitrile with next evaporation at room temperature within a few days.

**[SCd₈(SBuⁱ)₁₂(CN)₄/₂] or C₅₀H₁₀₈N₂S₁₃Cd₈.** This compound crystallizes in the cubic structure with the lattice parameter $a = 1592.8 \pm 0.2$ (Lee et al. 1988).

**[Cd(C₃₀H₂₈N₂S₄)]₂ or C₆₀H₅₆N₄S₈Cd₂.** This compound crystallizes in the triclinic structure with the lattice parameters $a = 1022.6 \pm 0.2$, $b = 1654.5 \pm 0.3$, and $c = 1788.2 \pm 0.2$ pm and $\alpha = 81.36° \pm 0.02°$, $\beta = 80.45° \pm 0.02°$, and $\gamma = 85.66° \pm 0.02°$ (Saravanan et al. 2004). It was prepared by the next procedure. Dibenzylamine (10 mM) in EtOH (10 mL) and CS₂ (10 mM) in EtOH (10 mL) under cold condition (5°C) were mixed thoroughly and stirred. The resulting yellow dithiocarbamic acid was mixed with Cd(NO₃)₂ (5 mM) in H₂O (5 mL) with constant stirring. The precipitated compound was filtered, washed with H₂O and alcohol, and then dried in air.

**[Cd{SPh₂CCH₂}₂(C₅H₃N)]₂·MeCN or C₆₈H₅₇N₃S₄Cd₂.** This compound crystallizes in the orthorhombic structure with the lattice parameters $a = 1796.3 \pm 1.0$, $b = 2743.3 \pm 0.3$, and $c = 1146.9 \pm 0.4$ pm and a calculated density of 1.34 g·cm⁻³ (Kaptein et al. 1987).

**(Me₄N)₂[Cd₄(SPh)₁₀] or C₆₈H₇₄N₂S₁₀Cd₄.** To obtain this compound, a solution of Cd(NO₃)₂·4H₂O (68 mM) in MeOH (60 mL) was added to a well-stirred solution of benzenethiol (182 mM) and Et₃N (182 mM) in MeOH (60 mL) at room temperature, followed by the addition of a solution of Me₄NCl (77 mM) in MeOH (40 mL) (Choy et al. 1982, Dance et al. 1984). The mixture was stirred only until all precipitate had dissolved and then allowed to stand undisturbed, finally at 0°C, while the product crystallized as colorless flakes. The product was washed well with MeOH and vacuum dried. Recrystallization was from acetonitrile with the addition of toluene.

**[Et₄N]₂[Cd(S-2-PhC₆H₄)₄]·2MeCN or C₆₈H₈₂N₄S₄Cd.** This compound crystallizes in the tetragonal structure with the lattice parameters $a = 1587.6 \pm 0.9$ and $c = 2528.0 \pm 0.6$ pm (Silver et al. 1993). To obtain it, lithium 2-phenylbenzenethiolate, Li(S-2-PhC₆H₄), was generated in MeOH from the thiol (3.2 mM) and Li (3.3 mM). The MeOH was removed, and then Cd salt (0.55 mM) was added, followed by the addition of 30 mL of MeCN. After stirring for 3 h, the reaction mixture was filtered and the filtrate was added to [Et₄N]Br (1.1 mM). Large crystals formed after 4 days at −20°C.

**(Et₃NH)₂[Cd₄(SPh)₁₀] or C₇₂H₈₂N₂S₁₀Cd₄.** This compound crystallizes in the monoclinic structure with the lattice parameters $a = 1375.9 \pm 0.2$, $b = 2050.4 \pm 0.5$, and $c = 1405.6 \pm 0.4$ pm and $\beta = 102.32° \pm 0.02°$ (Hencher et al. 1985). It was obtained by the electrochemical oxidation of anodic Cd in an acetonitrile solution of Et₃N and PhSH.

**(Et₄N)(Et₃NH)[Cd₄(SPh)₁₀] or C₇₄H₈₆N₂S₁₀Cd₄.** This compound crystallizes in the monoclinic structure with the lattice parameters $a = 1384.6 \pm 0.5$, $b = 3478.2 \pm 0.6$, and $c = 1629.1 \pm 0.4$ pm and $\beta = 90.82° \pm 0.02°$ and the experimental and calculated densities 1.50 and 1.502 g·cm⁻³, respectively (Hagen and Holm 1983). Its crystals were prepared by vapor diffusion of ether into MeCN solution of (Et₃NH)₂[Cd₄(SPh)₁₀].

**(Et₄N)₂[Cd₄(SPh)₁₀] or C₇₆H₉₀N₂S₁₀Cd₄.** To obtain this compound, a solution of 40 mM of CdCl₂·2.5H₂O in 100 mL of MeOH and 10 mL of H₂O was added dropwise to a solution of 100 mM Et₃N and 100 mM of benzenethiol in 150 mL of MeOH (Hagen et al. 1982). A solution of 20 mM of Et₄NBr in 50 mL of MeOH was added to the cloudy solution, resulting in the separation of a crystalline solid. Acetone (50 mL) and acetonitrile (10 mL) were added, and the mixture was warmed to 55°C to give a nearly clear solution. This solution was cooled to −25°C, causing separation of a white crystalline solid. This product was collected by filtration, washed with MeOH and ether, and dried in vacuo.

This compound could also be prepared, if a solution of Cd(NO₃)₂·4H₂O (21 g) in 60 mL of MeOH was quickly added to a stirred solution of 20 g of benzenethiol and 18.5 g of Et₃N in 60 mL of MeOH at room temperature in an inert atmosphere glovebox (Farneth et al. 1992). When thoroughly mixed, a solution of 12.7 g of Et₄NCl in 40 mL of MeOH was added all at once. After brief stirring, the solution became clear and was left undisturbed at 0°C, while a viscous oily layer separated in the bottom of the flask. The top layer of MeOH was decanted, and the oily layer was washed and triturated with more MeOH, whereupon it solidified to a low-melting-point white solid.

**(Et₄N)₂[S₄Cd₁₀(SPh)₁₆] or C₁₁₂H₁₀₀N₂S₂₀Cd₁₀.** To obtain this compound, powder of sulfur (0.3 g) was added to a stirred solution of 16 g of (Et₄N)₂[Cd₄(SPh)₁₀] in 50 mL of MeCN at room temperature in an inert atmosphere (Farneth et al. 1992).

After 30 min of stirring, the sulfur had dissolved and the solution was cloudy with a white precipitate. MeCN (150 mL) was added, and the solution was heated close to the boiling point of MeCN in order to dissolve the precipitate and generate a clear solution. Upon cooling overnight, large colorless laths were deposited from MeCN and were collected by filtration.

$(Me_4N)_4[S_4Cd_{10}(SPh)_{16}]$ or $C_{112}H_{128}N_4S_{20}Cd_{10}$. Two crystallographic forms of this compound have been prepared from different solutions and both crystallize in the tetragonal structure (Choy et al. 1982, Dance et al. 1984). When the crystals for XRD were obtained by dissolution of the compound in hot DMF and the addition of $H_2O$, which resulted in the slow growth of well-developed colorless needles, the lattice parameters are $a = 2094.6 \pm 0.2$ and $c = 1477.9 \pm 0.2$ pm and a calculated density of 1.69 g·cm$^{-3}$. If the crystals for XRD were from samples of beautifully formed colorless blocks obtained directly in a preparation using acetonitrile, the lattice parameters are the next: $a = 2014.0 \pm 0.2$ and $c = 1689.6 \pm 0.1$ pm and the experimental and calculated densities $1.593 \pm 0.007$ and 1.597 g·cm$^{-3}$, respectively.

This compound was obtained if finely powdered sulfur (8.9 mM) was added in one portion to a well-stirred solution of $(Me_4N)_2[Cd_4(SPh)_{10}]$ (8.9 mM) in acetonitrile (50 mL) at a temperature not higher than 20°C. The mixture was stirred until the sulfur had reacted and dissolved (ca. 20 min). A white crystalline precipitate of the product formed after or during the dissolution of the sulfur. Acetonitrile was added and the mixture heated to 75°C until the entire product had dissolved; the total volume was ca. 400 mL. The hot solution was allowed to cool slowly, yielding the product as large beautifully formed crystals during a period of 5 days. The product was washed with acetonitrile and vacuum dried.

$(Me_4N)_4[S_4Cd_{10}(SPh)_{16}]$ was prepared also by the following method (Dance et al. 1984). Sulfur (4.5 mM), added in one portion to a solution of $(Me_4N)_2[Cd_4(SPh)_{10}]$ (4.4 mM) in DMF (30 mL) at room temperature, dissolved quickly. The solution was stirred for a further 30 min; then toluene (10 mL) was added and the solution allowed standing undisturbed at 20°C while the product precipitated as colorless microcrystals. After filtration the product was washed with acetonitrile and vacuum dried.

$(Et_3NH)_4[S_4Cd_{10}(SPh)_{16}]$ or $C_{120}H_{144}N_4S_{20}Cd_{10}$. This compound crystallizes in the tetragonal structure with the lattice parameters $a = 2000.7 \pm 0.3$ and $c = 1798.2 \pm 0.3$ pm and a calculated density of 1.57 g·cm$^{-3}$ (Lee et al. 1993). To obtain it, a solution of $Cd(NO_3)_2·4H_2O$ (32.4 mM) dissolved in acetonitrile (20 mL) was added to a solution of PhSH (80.8 mM) and $Et_3N$ (80.8 mM) in acetonitrile (30 mL). A white precipitate formed upon the addition of the $Cd^{2+}$ solution redissolved with stirring. A clear colorless solution remained. Sulfur powder (9.7 mM) was added in one portion and dissolved to give a yellow solution. The white solid that deposited after stirring for 10 min was collected, washed with acetonitrile, and vacuum dried. This crude material (3 g) was redissolved in hot acetonitrile (approximately 20 mL), and the solution was left to cool. Colorless block crystals grew within 10 h at room temperature. They were filtered out, washed with cold acetonitrile, and vacuum dried.

**[Cd$_{17}$S$_4$(SPh)$_{28}$(Me$_4$N)$_2$]** or **C$_{176}$H$_{164}$N$_2$S$_{32}$Cd$_{17}$.** This compound crystallizes in the orthorhombic structure with the lattice parameters $a = 3093.0 \pm 0.5$, $b = 3277.2 \pm 0.5$, and $c = 1999.7 \pm 0.2$ pm and the experimental and calculated densities $1.76 \pm 0.02$ and $1.73$ g·cm$^{-3}$, respectively (Lee et al. 1988). It was obtained by treatment of a solution of PhSH (80 mM) and Et$_3$N (80 mM) in MeCN with Cd(NO$_3$)$_2$ (50 mM) in MeCN and Na$_2$S (20 mM) in MeOH, with alternating additions of the latter two reagents, followed by the addition of Me$_4$NCl (18 mM) in MeOH. The colorless crystalline precipitate was recrystallized from hot MeCN to yield [S$_4$Cd$_{17}$(SPh)$_{28}$(Me$_4$N)$_2$] as colorless block crystals.

**[Cd$_{32}$S$_{14}$(SPh)$_{36}$(C$_{12}$H$_{12}$N$_2$)$_4$]** or **C$_{264}$H$_{228}$N$_8$S$_{50}$Cd$_{32}$ (C$_{12}$H$_{12}$N$_2$ = 1,2-bis(4-pyridyl) ethane).** This compound crystallizes in the monoclinic structure with the lattice parameters $a = 2269.57 \pm 0.04$, $b = 2181.42 \pm 0.04$, and $c = 5766.12 \pm 0.10$ pm and $\beta = 93.245° \pm 0.001°$ (Zheng et al. 2006a).

**[Cd$_{32}$S$_{14}$(SPh)$_{36}$(C$_{13}$H$_{14}$N$_2$)$_4$]** or **C$_{268}$H$_{236}$N$_8$S$_{50}$Cd$_{32}$ (C$_{13}$H$_{14}$N$_2$ = 4,4′-trimethylene-dipyridine).** This compound crystallizes in the monoclinic structure with the lattice parameters $a = 2204.77 \pm 0.02$, $b = 6104.54 \pm 0.07$, and $c = 2228.61 \pm 0.03$ pm and $\beta = 94.011° \pm 0.001°$ (Zheng et al. 2006a).

### 4.44   Cd–H–C–N–Se–S

Some compounds are formed in this system.

**[Cd{SeC(NH$_2$)$_2$}$_2$(SCN)$_2$]** or **C$_4$H$_8$N$_6$Se$_2$S$_2$Cd.** This compound crystallizes in the triclinic structure with the lattice parameters $a = 397 \pm 1$, $b = 793 \pm 1$, and $c = 1009 \pm 1$ pm and $\alpha = 90.3° \pm 0.1°$, $\beta = 99.0° \pm 0.1°$, and $\gamma = 104.0° \pm 0.1°$ (Domiano et al. 1969).

**[Cd(SeCN)$_2$(C$_5$H$_{10}$N$_2$S)]** or **C$_7$H$_{10}$N$_4$Se$_2$SCd.** This compound melts at 156°C–158°C and crystallizes in the orthorhombic structure with the lattice parameters $a = 1495.1 \pm 0.1$, $b = 742.09 \pm 0.06$, and $c = 1192.9 \pm 0.1$ pm and a calculated density of $2.271$ g·cm$^{-3}$ (Fettouhi et al. 2008). To obtain it, $N,N'$-dimethylimidazolidine-2-thione-$S$ (C$_5$H$_{10}$N$_2$S) (20 mM) was dissolved in 15 mL of MeOH and Cd(SeCN)$_2$ (10 mM) was dissolved in 20 mL of MeCN. Both solutions were mixed and refluxed for 2 h, concentrated, and then allowed to crystallize in a refrigerator at 4°C. Subsequently, the solid was collected and dried.

**[(Me$_4$N)$_2${SCd$_8$(SePh)$_{16}$}]** or **C$_{104}$H$_{104}$N$_2$Se$_{16}$SCd$_8$.** To obtain this compound, a deoxygenated solution of NaSePh in EtOH/acetonitrile was treated with solutions of CdI$_2$ in acetonitrile, Na$_2$S in MeOH, and Me$_4$NCl in acetonitrile, yielding a colorless solution (Lee et al. 1990). After stripping of all solvent, the colorless solids were extracted with H$_2$O at 80°C and EtOH at an ambient temperature and then dissolved in acetonitrile at 20°C and filtered, and the title compound crystallized by storage at 0°C.

[(Me$_4$N)$_2${SCd$_8$(SePh)$_{16}$}] is also formed by a reaction of [Cd$_4$(SePh)$_{10}$]$^{2-}$ in acetonitrile with 0.5 equivalent of Na$_2$S or NaSH in MeOH or 1 equivalent of S$_8$ at room temperature and by the reaction of Cd(SePh)$_2$ in DMF with Na$_2$S or NaSH (Lee et al. 1990).

**[(Et$_4$N)$_2${SCd$_8$(SePh)$_{16}$}]** or **C$_{112}$H$_{120}$N$_2$Se$_{16}$SCd$_8$.** To obtain this compound, a deoxygenated solution of NaSePh (15.5 mM) in EtOH (70 mL) + acetonitrile (70 mL) was treated with solutions of $CdI_2$ (6.4 mM) in acetonitrile (20 mL), $Na_2S$ (1.6 mM) in MeOH (20 mL), and $Et_4NCl$ (2.5 mM) in acetonitrile (20 mL), yielding a colorless solution (Lee et al. 1990). After stripping of all solvent, the colorless solids were extracted with $H_2O$ at 80°C and EtOH at ambient temperature and then dissolved in acetonitrile at 20°C and filtered, and the title compound crystallized by storage at 0°C.

**$[(Me_4N)_4\{Se_4Cd_{10}(SPh)_{16}\}]$ or $C_{112}H_{128}N_4Se_4S_{16}Cd_{10}$.** To obtain this compound, black Se powder (8.9 mM) was added in one portion to a solution of $(Me_4N)_2[Cd_4(SPh)_{10}]$ (8.9 mM) in acetonitrile (20 mL) at room temperature, and the mixture was stirred for 3 h, while the Se transformed to a white precipitate. (Dance et al. 1984). The mixture was heated to ca. 75°C and acetonitrile (ca. 150 mL) added until all solid had dissolved. The pale-yellow solution was allowed to cool slowly, yielding the product as well-formed colorless needles, which were washed with acetonitrile and vacuum dried.

**$[(PhCH_2NMe_3)_2\{SCd_8(SePh)_{16}\}]$ or $C_{116}H_{112}N_2Se_{16}SCd_8$.** To obtain this compound, a deoxygenated solution of NaSePh in EtOH/acetonitrile was treated with solutions of $CdI_2$ in acetonitrile, $Na_2S$ in MeOH, and $PhCH_2NMe_3Cl$ in acetonitrile, yielding a colorless solution (Lee et al. 1990). After stripping of all solvent, the colorless solids were extracted with $H_2O$ at 80°C and EtOH at ambient temperature and then dissolved in acetonitrile at 20°C and filtered, and the title compound crystallized by storage at 0°C.

**$[(PhCH_2NEt_3)_2\{SCd_8(SePh)_{16}\}]$ or $C_{122}H_{124}N_2Se_{16}SCd_8$.** To obtain this compound, a deoxygenated solution of NaSePh in EtOH/acetonitrile was treated with solutions of $CdI_2$ in acetonitrile, $Na_2S$ in MeOH, and $PhCH_2NEt_3Cl$ in acetonitrile, yielding a colorless solution (Lee et al. 1990). After stripping of all solvent, the colorless solids were extracted with $H_2O$ at 80°C and EtOH at ambient temperature and then dissolved in acetonitrile at 20°C and filtered, and the title compound crystallized by storage at 0°C.

**$[(Bu_4N)_2\{SCd_8(SePh)_{16}\}]$ or $C_{128}H_{152}N_2Se_{16}SCd_8$.** To obtain this compound, a deoxygenated solution of NaSePh in EtOH/acetonitrile was treated with solutions of $CdI_2$ in acetonitrile, $Na_2S$ in MeOH, and $Bu_4NCl$ in acetonitrile, yielding a colorless solution (Lee et al. 1990). After stripping of all solvent, the colorless solids were extracted with $H_2O$ at 80°C and EtOH at ambient temperature and then dissolved in acetonitrile at 20°C and filtered, and the title compound crystallized by storage at 0°C.

**$Cd_{17}Se_4(SPh)_{26}(C_{13}H_{14}N_2)_2$ or $C_{182}H_{158}N_4Se_4S_{26}Cd_{17}$.** This compound crystallizes in the monoclinic structure with the lattice parameters $a = 3202.8 \pm 0.6$, $b = 2788.9 \pm 0.6$, and $c = 4219.8 \pm 0.9$ pm and $\beta = 99.636° \pm 0.005°$ (Zheng et al. 2006a). It was prepared by the following procedure. To a 23 mL Teflon-lined stainless steel autoclave 0.214 g of 4,4'-trimethylenedipyridine, 0.017 g of Se, 0.168 g of $(NMe_4)_2[Cd_4(SPh)_{10}]$, 2.690 g of acetonitrile, and 0.420 g of $H_2O$ were added. After the mixture was stirred for ca. 20 min, the vessel was sealed and heated at 110°C for 5 days. After cooling to room temperature, a mixture of $Cd_{17}Se_4(SPh)_{26}(C_{13}H_{14}N_2)_2$ and $Cd_{17}S_4(SPh)_{26}$ was obtained.

## 4.45 Cd–H–C–N–Se–Cl–S

[**Cd(C₃H₅NSeS)₂Cl₂**] or **C₆H₁₀N₂Se₂Cl₂S₂Cd**. This multinary compound is formed in this system. It decomposes at 175°C–177°C (Preti et al. 1974). It was obtained by a reaction of CdCl₂ solution in EtOH with thiazolidine-2-selenone dissolved in CHCl₃ by refluxing for 12 h. The crude product that separated out during the reaction was purified by washing with chloroform.

## 4.46 Cd–H–C–N–Se–Br–S

[**Cd(C₃H₅NSeS)₂Br₂**] or **C₆H₁₀N₂Se₂Br₂S₂Cd**. This multinary compound is formed in this system. It decomposes at 158°C–160°C (Preti et al. 1974). It was obtained by a reaction of CdBr₂ solution in EtOH with thiazolidine-2-selenone dissolved in CHCl₃ by refluxing for 12 h. The crude product that separated out during the reaction was purified by washing with chloroform.

## 4.47 Cd–H–C–N–Se–I–S

[**Cd(C₃H₅NSeS)₂I₂**] or **C₆H₁₀N₂Se₂I₂S₂Cd**. This multinary compound is formed in this system. It decomposes at 140°C–142°C (Preti et al. 1974) and was obtained by a reaction of CdI₂ solution in EtOH with thiazolidine-2-selenone dissolved in CHCl₃ by refluxing for 12 h. The crude product that separated out during the reaction was purified by washing with chloroform.

## 4.48 Cd–H–C–N–Te–S

[**(Et₄N)₂{SCd₈(TePh)₁₆}**] or **C₁₁₂H₁₂₀N₂Te₁₆SCd₈**. This multinary compound is formed in this system. To obtain it, a deoxygenated solution of NaTePh in EtOH/acetonitrile was treated with solutions of CdI₂ in acetonitrile, Na₂S in MeOH, and Et₄NCl in acetonitrile, yielding a colorless solution (Lee et al. 1990). After stripping of all solvent, the colorless solids were extracted with H₂O at 80°C and EtOH at ambient temperature and then dissolved in acetonitrile at 20°C and filtered, and the title compound crystallized by storage at 0°C.

## 4.49 Cd–H–C–N–F–S

Some compounds are formed in this system.

[**Cd(C₅H₂N₂F₃S)₂**] or **C₁₀H₄N₄F₆S₂Cd**. This compound crystallizes in the triclinic structure with the lattice parameters $a = 704.19 \pm 0.02$, $b = 1037.10 \pm 0.04$, and $c = 1139.81 \pm 0.03$ pm and $\alpha = 81.768° \pm 0.001°$, $\beta = 78.624° \pm 0.001°$, and $\gamma = 70.736° \pm 0.001°$ and a calculated density of 2.036 g·cm⁻³ (Rodríguez et al. 2005). Electrochemical oxidation of a Cd anode in a solution of 4-(trifluoromethyl)pyrimidine-2-thione (C₅H₃N₂F₃S) (0.83 mM) in acetonitrile (50 mL) at 14 V and 10 mA for 1.5 h caused 36 mg of cadmium to be dissolved. During electrolysis, hydrogen was evolved at the cathode, and at the end of the experiment, a crystalline solid, suitable for XRD, had formed at the bottom of the vessel. The yellow solid was filtered off, washed with acetonitrile and diethyl ether, and dried under vacuum.

**[Cd(C$_5$H$_2$N$_2$F$_3$S)$_2$(2,2'-bipy)]** or **C$_{20}$H$_{12}$N$_6$F$_6$S$_2$Cd**. Electrolysis of an acetonitrile solution (50 mL) containing 4-(trifluoromethyl)pyrimidine-2-thione (C$_5$H$_3$N$_2$F$_3$S) (0.83 mM), 2,2'-bipy (0.42 mM), and a small amount of Et$_4$NClO$_4$ (ca. 10 mg) at 10 mA and 16 V for 1.5 h dissolved 26 mg of Cd (Rodríguez et al. 2005). At the end of the experiment, the white solid of the title compound was filtered off, washed with acetonitrile, and dried in vacuo.

**[Cd(C$_5$H$_2$N$_2$F$_3$S)$_2$(Phen)]** or **C$_{22}$H$_{12}$N$_6$F$_6$S$_2$Cd**. Electrolysis of an acetonitrile solution (50 mL) containing 4-(trifluoromethyl)pyrimidine-2-thione (C$_5$H$_3$N$_2$F$_3$S) (0.83 mM), Phen (0.467 mM), and a small amount of Et$_4$NClO$_4$ (ca. 10 mg) at 10 mA and 12 V for 1.5 h dissolved 27 mg of Cd (Rodríguez et al. 2005). The solid of the title compound was filtered off, washed with cold acetonitrile and diethyl ether, and dried under vacuum.

**[Cd(5-CF$_3$C$_5$H$_3$NS)$_2$(2,2'-bipy)]** or **C$_{22}$H$_{14}$N$_4$F$_6$S$_2$Cd**. Electrochemical oxidation of a Cd anode in a solution of 5-CF$_3$C$_5$H$_4$NS (1.553 mM) and 2,2'-bipy (1.558 mM) in acetonitrile (50 mL) containing about 15 mg of Me$_4$NClO$_4$ as a current carrier, at 20 V and 20 mA for 2 h, caused 83.86 mg of Cd to be dissolved (Sousa-Pedrares et al. 2003). During the electrolysis process, H$_2$ was evolved at the cathode, and at the end of the reaction, an insoluble crystalline solid was observed at the bottom of the vessel. The solid was filtered off, washed with acetonitrile and diethyl ether, and dried under vacuum. All manipulations were done under an atmosphere of dry N$_2$.

**[Cd(C$_6$H$_4$N$_2$F$_3$S)$_2$(2,2'-bipy)]** or **C$_{22}$H$_{16}$N$_6$F$_6$S$_2$Cd**. This compound crystallizes in the orthorhombic structure with the lattice parameters $a = 1591.1 \pm 0.3$, $b = 1793.0 \pm 0.4$, and $c = 909.8 \pm 0.2$ pm and a calculated density of 1.676 g·cm$^{-3}$ (Rodríguez et al. 2005). Electrolysis of an acetonitrile solution (50 mL) containing 4,6-(trifluoromethyl)methylpyrimidine-2-thione (C$_6$H$_5$N$_2$F$_3$S) (1.03 mM), 2,2'-bipy (0.51 mM), and a small amount of Et$_4$NClO$_4$ (ca. 10 mg) at 10 mA and 32 V for 2.75 h dissolved 73 mg of Cd. At the end of the experiment, the yellow solid of the title compound was filtered off, washed with acetonitrile and diethyl ether, and dried in vacuo. Its crystals were obtained by recrystallization from Pr$^i$OH.

**[Cd(3-CF$_3$C$_5$H$_3$NS)$_2$(Phen)]** or **C$_{24}$H$_{14}$N$_4$F$_6$S$_2$Cd**. This compound was obtained by the same procedure as for [Cd(5-CF$_3$C$_5$H$_3$NS)$_2$(2,2'-bipy)] using 3-CF$_3$C$_5$H$_4$NS instead of 5-CF$_3$C$_5$H$_4$NS and Phen (1.711 mM) instead of 2,2'-bipy (100.63 mg of Cd dissolved) (Sousa-Pedrares et al. 2003).

**[Cd(5-CF$_3$C$_5$H$_3$NS)$_2$(Phen)]** or **C$_{24}$H$_{14}$N$_4$F$_6$S$_2$Cd**. This compound was obtained by the same procedure as for [Cd(5-CF$_3$C$_5$H$_3$NS)$_2$(2,2'-bipy)] using Phen (1.711 mM) instead of 2,2'-bipy (90.6 mg of Cd dissolved) (Sousa-Pedrares et al. 2003).

**[Cd(C$_6$H$_4$N$_2$F$_3$S)$_2$(Phen)]** or **C$_{24}$H$_{16}$N$_6$F$_6$S$_2$Cd**. An acetonitrile solution (50 mL) of 4,6-(trifluoromethyl)methylpyrimidine-2-thione (C$_6$H$_5$N$_2$F$_3$S) (1.03 mM), Phen (0.55 mM), and a small amount of Et$_4$NClO$_4$ (ca. 10 mg) was electrolyzed at 10 mA during 2.75 h, and 68 mg of Cd was dissolved from the anode (Rodríguez et al. 2005).

At the end of electrolysis, the solution was concentrated and the resulting green solid was filtered off, washed with cold acetonitrile and diethyl ether, and dried under vacuum.

**[Cd(3-CF$_3$C$_5$H$_3$NS)$_2$(2,2′-bipy)]·MeCN or C$_{24}$H$_{17}$N$_5$F$_6$S$_2$Cd**. This compound crystallizes in the monoclinic structure with the lattice parameters $a = 1368.0 \pm 0.3$, $b = 975.64 \pm 0.18$, and $c = 2041.4 \pm 0.4$ pm and $\beta = 90.684° \pm 0.003°$ and a calculated density of 1.624 g·cm$^{-3}$ (Sousa-Pedrares et al. 2003). It was obtained by the same procedure as for [Cd(5-CF$_3$C$_5$H$_3$NS)$_2$(2,2′-bipy)] (104.0 mg of Cd dissolved). Its crystals were obtained by crystallization of [Cd(3-CF$_3$C$_5$H$_3$NS)$_2$(2,2′-bipy)] from acetonitrile.

**[Cd(C$_{11}$H$_6$N$_2$F$_3$S)$_2$(Phen)] or C$_{34}$H$_{20}$N$_6$F$_6$S$_2$Cd**. This compound crystallizes in the monoclinic structure with the lattice parameters $a = 2333.77 \pm 0.11$, $b = 1363.11 \pm 0.07$, and $c = 1051.63 \pm 0.05$ pm and $\beta = 108.391° \pm 0.001°$ and a calculated density of 1.680 g·cm$^{-3}$ (Rodríguez et al. 2005). An acetonitrile solution (50 mL) of 4,6-(trifluoromethyl)phenylpyrimidine-2-thione (C$_{11}$H$_7$N$_2$F$_3$S) (0.55 mM), Phen (0.19 mM), and a small amount of Et$_4$NClO$_4$ (ca. 10 mg) was electrolyzed at 10 mA and 6 V during 1 h, and 18 mg of Cd was dissolved from the anode. At the end of electrolysis, the solution was concentrated and the crystals were filtered off, washed with cold acetonitrile and diethyl ether, and dried under vacuum.

**[Et$_4$N]$_2$[Cd(SC$_6$F$_5$)$_4$] or C$_{40}$H$_{40}$N$_2$F$_{20}$S$_4$Cd**. This compound decomposes at 125°C (Beck et al. 1967). It was prepared by the next procedure. To a solution of C$_6$F$_5$SH (39.3 mM) in 20 mL of 2 N NaOH was an aqueous solution (50 mL) of CdSO$_4$ (5.2 mM) added dropwise under constant stirring. The obtained mixture was filtered with an excess of Cd(OH)$_2$. An aqueous solution of Et$_4$NBr was added to the filtrate and the precipitate was thoroughly washed with H$_2$O and dried under high vacuum. The colorless needlelike crystals were grown by recrystallization of the title compound from mixture solution acetone/diethyl ether.

## 4.50 Cd–H–C–N–Cl–S

Some compounds are formed in this system.

**[Cd(CH$_5$N$_3$S)Cl$_2$] or CH$_5$N$_3$Cl$_2$SCd**. This compound crystallizes in the monoclinic structure with the lattice parameters $a = 1001$, $b = 1388$, and $c = 689$ pm and $\beta = 123.9°$ (Nardelli and Chierici 1960). It was obtained at the interaction of CdCl$_2$ and thiosemicarbazide in the molar ratio of 1:1 in a water solution.

**[Cd(C$_2$H$_5$NS)Cl$_2$] or C$_2$H$_5$NCl$_2$SCd**. This compound crystallizes in the orthorhombic structure with the lattice parameters $a = 393.8 \pm 0.5$, $b = 1064.6 \pm 0.9$, and $c = 1641.5 \pm 3.2$ pm and the experimental and calculated densities of 2.53 and 2.49 g·cm$^{-3}$, respectively (Rolies and De Ranter 1978). Crystals of the title compound were prepared by heating a solution of CdCl$_2$ in HCl, and a solution of thioacetamide in EtOH was added in proportion as the liquid was evaporated. The final CdCl$_2$/thioacetamide ratio was about 2:1. Upon cooling to room temperature, white needles separated in the course of 24 h. The very fragile crystals were neither well shaped nor could be ground to a sphere.

**[Cd{SC(NH₂)₂}₂Cl₂] or C₂H₈N₄Cl₂S₂Cd.** This compound melts at 205°C–208°C (Isab and Wazeer 2005) [at 180°C (Stoev and Ruseva 1994)] and crystallizes in the orthorhombic structure with the lattice parameters $a = 1314.8 \pm 0.5$, $b = 583.4 \pm 0.1$, and $c = 650.1 \pm 0.2$ pm and the experimental and calculated densities 2.320 and 2.235 g·cm⁻³, respectively (Marcos et al. 1998). It was prepared at room temperature by mixing solutions of thiourea with CdCl₂ in a minimum amount of distilled H₂O (2–10 mL) in the 2:1 molar ratio and stirring for 1 or 2 h. The isolated product was dried thoroughly (Stoev and Ruseva 1994, Isab and Wazeer 2005). Single crystals of the title compound were grown by slow evaporation of aqueous solution containing CdCl₂ and thiourea in the 1:2 molar ratio (Marcos et al. 1998). Thermal dissociation of CdCl₂·2CS(NH₂)₂ allows preparation of CdS from the saturated solutions (Stoev and Ruseva 1994, Isab and Wazeer 2005).

The solubility isotherms of the CdCl₂–Cs(NH₂)₂–MeOH at 25°C has been investigated, and the field of equilibrium existence of CdCl₂·2CS(NH₂)₂ was determined (Stoev and Ruseva 1994).

**[Cd(CH₅N₃S)₂Cl₂] or C₂H₁₀N₆Cl₂S₂Cd.** This compound decomposes at 282°C (Mahadevappa and Murthy 1972). It was obtained by mixing aqueous solutions of thiosemicarbazide and CdCl₂ in the molar ratio 2:1. The resulting solution was slowly evaporated at 60°C on a water bath and then cooled in ice, whereupon crystals of the title compound appeared.

**[Cd{SC(NHNH₂)₂}₂Cl₂] or C₂H₁₂N₈Cl₂S₂Cd, bis(thiocarbohydrazide-*N,S*)cadmium.** This compound crystallizes in the monoclinic structure with the lattice parameters $a = 864 \pm 1$, $b = 578 \pm 1$, and $c = 1378 \pm 1$ pm and $\beta = 119.5° \pm 0.3°$ and the experimental and calculated densities 2.17 and 2.19 g·cm⁻³, respectively (Bigoli et al. 1971). Its crystals have been obtained by evaporation of aqueous solutions of thiocarbohydrazide and CdCl₂ in stoichiometric ratio.

**[Cd(C₃H₅NS₂)Cl₂] or C₃H₅NCl₂S₂Cd.** This compound melts at 224°C–225°C and crystallizes in the monoclinic structure with the lattice parameters $a = 389.61 \pm 0.01$, $b = 2230.76 \pm 0.07$, and $c = 912.98 \pm 0.04$ pm and $\beta = 90.359° \pm 0.001$ and a calculated density of 2.532 g·cm⁻³ (Bell et al. 2004). To obtain it, a warm solution of CdCl₂·2.5H₂O (3.28 mM) in EtOH (25 mL) was added to a hot ethanolic solution (25 mL) of 1,3-thiazolidine-2-thione (3.27 mM) with stirring. The resultant cloudy solution was allowed to cool, resulting in the formation of white powder which was filtered off and dried in vacuo.

**[(C₃H₇N₂S)(CdCl₃) or C₃H₇N₂Cl₃SCd.** This compound crystallizes in the monoclinic structure with the lattice parameters $a = 1443.8 \pm 0.1$, $b = 392.2 \pm 0.2$, and $c = 2095.2 \pm 0.7$ pm and $\beta = 131.72° \pm 0.05°$ and the experimental and calculated densities 2.40 ± 0.03 and 2.41 g·cm⁻³, respectively (Kubiak et al. 1983). Its clear, colorless crystals were obtained from equimolar amounts of Cd(OAc)₂ and (C₃H₆N₂S)·HCl (C₃H₆N₂S = 2-amino-4,5-dihydro-1,3-thiazole) at room temperature.

**[Cd(C₄H₆N₂S)Cl₂] or C₄H₆N₂Cl₂SCd.** This compound melts at 280°C with decomposition (Bell et al. 2004). To obtain it, a warm solution of CdCl₂·2.5H₂O (5.00 mM) in EtOH (25 mL) was added to a hot ethanolic solution (25 mL) of 1-methylimid-azoline-2(3*H*)-thione (5.00 mM) with stirring. The resultant cloudy solution was allowed to cool, resulting in the formation of colorless crystals, which were filtered off and dried in vacuo.

**[Cd{MeNH(NH$_2$)CS}$_2$Cl$_2$] or C$_4$H$_{12}$N$_4$Cl$_2$S$_2$Cd.** This compound melts at 233°C and crystallizes in the triclinic structure with the lattice parameters $a = 697.5 \pm 0.8$, $b = 916.2 \pm 0.8$, and $c = 994.4 \pm 0.6$ pm and $\alpha = 71.97° \pm 0.06°$, $\beta = 76.27° \pm 0.06°$, and $\gamma = 69.75° \pm 0.08°$ at $160 \pm 2$ K and a calculated density of 2.153 g·cm$^{-3}$ (Moloto et al. 2003). To prepare it, a hot solution of *N*-methylthiourea (1.66 mM) in EtOH (10 mL) was added into a heated solution of CdCl$_2$ (0.83 mM) in EtOH (15 mL). The mixture was stirred and refluxed for 2 h. The colorless solution was filtered to remove any traces of unreacted materials while hot and was left in an open beaker at room temperature to crystallize by slow evaporation. Transparent cubic crystals of the title compound were obtained after 24 h. The product was filtered, washed twice with EtOH, and dried under vacuum.

**[Cd{SC(NH$_2$)$_2$}$_4$Cl$_2$] or C$_4$H$_{16}$N$_8$Cl$_2$S$_4$Cd.** This compound crystallizes in the tetragonal structure with the lattice parameters $a = 1380.4 \pm 0.1$ and $c = 926.8 \pm 0.2$ (Jiang et al. 2000). It was prepared by mixing CdCl$_2$ and SC(NH$_2$)$_2$ (1:4 molar ratio) in aqueous solution. Colorless and transparent crystals were formed by slow evaporation of the solvent at room temperature.

**[CdCl$_2${μ-S(CH$_2$)$_3$NH(CH$_3$)$_2$}] or C$_5$H$_{13}$NCl$_2$SCd.** This compound crystallizes in the monoclinic structure with the lattice parameters $a = 1599.5 \pm 0.3$, $b = 653.2 \pm 0.2$, and $c = 1023.7 \pm 0.2$ pm and $\beta = 95.66° \pm 0.02°$ and the experimental and calculated densities 1.68 and 1.660 g·cm$^{-3}$, respectively (Casals et al. 1987). To obtain it, a solution of 3-dimethylamino-1-propanethiol (5 mM) in MeOH (2 mL) and H$_2$O (18 mL) at room temperature was added dropwise with stirring to a solution of CdCl$_2$·2H$_2$O (5 mM) in the same solvent. The white dusty precipitate immediately formed was filtered off and the solution was allowed to crystallize at room temperature. Colorless crystals and microcrystalline solid that appeared after 48 h were both collected after a further 24 h by filtration, washing with H$_2$O, and vacuum drying. It was possible to separate manually the crystals of [CdCl$_2${μ-S(CH$_2$)$_3$NH(CH$_3$)$_2$}]. The obtained solid includes beside this compound the crystals of [{(CH$_3$)$_2$NH(CH$_2$)$_3$S}$_2$][CdCl$_4$].

**[Cd(C$_2$H$_8$N$_2$)$_2$SCNCl] or C$_5$H$_{16}$N$_5$ClSCd.** This compound crystallizes in the orthorhombic structure with the lattice parameters $a = 1240 \pm 2$, $b = 1107 \pm 2$, and $c = 930 \pm 1$ pm (Shvelashvili et al. 1974).

**[CdCl$_2$(C$_3$H$_5$NS$_2$)]$_n$ or C$_6$H$_{10}$N$_2$Cl$_2$S$_4$Cd.** This compound melts at 108°C (De Filippo et al. 1971) and crystallizes in the monoclinic structure with the lattice parameters $a = 393.3 \pm 0.1$, $b = 2241.8 \pm 0.5$, and $c = 993.4 \pm 0.2$ pm and $\beta = 112.81° \pm 0.04°$ and the experimental and calculated densities 2.47 and 2.49 g·cm$^{-3}$, respectively (Kubiak and Głowiak 1985). To obtain it, CdCl$_2$ was dissolved in an excess of molten thiazolidine-2-thione (C$_3$H$_5$NS$_2$) (at ca. 105°C; molar ratio Cd$^{2+}$/C$_3$H$_5$NS$_2$ ca. 1:2), and the crude product was purified by washing with ethyl ether and recrystallizing from EtOH. The title compound could also be prepared if to a solution of $CdCl_2$ in absolute EtOH different amounts of $C_3H_5NS_2$ were added for $Cd^{2+}/C_3H_5NS_2$ molar ratios of 1:1, 1:2, 1:3, and 1:4 and the mixture was heated under reflux for 4 h. $[CdCl_2(C_3H_5NS_2)_2]$ was separated out by crystallization (De Filippo et al. 1971).

**$[Cd(C_3H_6N_2S)_2Cl_2]$ or $C_6H_{12}N_4Cl_2S_2Cd$.** This compound crystallizes in the monoclinic structure with the lattice parameters $a = 625.91 \pm 0.13$, $b = 1461.6 \pm 0.3$, and $c = 1405.4 \pm 0.3$ pm and $\beta = 96.71° \pm 0.03°$ and a calculated density of 2.016 g·cm$^{-3}$ (Al-Arfaj et al. 1998) [$a = 626 \pm 1$, $b = 1454 \pm 2$, and $c = 1459 \pm 1$ pm and $\beta = 108.3° \pm 1.3°$ and the experimental and calculated densities 2.07 and 2.02 g·cm$^{-3}$, respectively (Cavalca et al. 1968)]. It was prepared by a dropwise addition of ethanol solution of cadmium chloride (1 mM in 10 mL) to a stoichiometric amount of imidazoline-2-thione ($C_3H_6N_2S$) in EtOH (2 mM in 50 mL) and refluxing the solution for 2–3 h on a water bath (Shunmugam and Sathyanarayana 1983, Al-Arfaj et al. 1998). The solvents were evaporated at room temperature to low volume and the resulting colorless crystals were collected.

**$[Cd\{Me_2(NH)_2CS\}_2Cl_2]$ or $C_6H_{16}N_4Cl_2S_2Cd$.** This compound melts at 221°C–223°C [at 203°C (Moloto et al. 2003)] and crystallizes in the monoclinic structure with the lattice parameters $a = 1332.3 \pm 0.2$, $b = 901.2 \pm 0.1$, and $c = 1277.9 \pm 0.2$ pm and $\beta = 108.712° \pm 0.002°$ and a calculated density of 1.790 g·cm$^{-3}$ (Malik et al. 2010). It was prepared by adding 2.0 mM methanolic or ethanolic solution of $N,N'$-dimethylthiourea to an aqueous solution of $CdCl_2$ (1.0 mM). The reaction mixture was stirred for 30 min (Moloto et al. 2003, Malik et al. 2010). The colorless solution was then filtered and the filtrate was left at room temperature for crystallization. As a result, a white crystalline product was obtained, which was washed with MeOH or EtOH and dried.

**$[Cd\{EtNH(NH_2)CS\}_2Cl_2]$ or $C_6H_{16}N_4Cl_2S_2Cd$.** This compound melts at 201°C and crystallizes in the triclinic structure with the lattice parameters $a = 854.7 \pm 0.7$, $b = 984.1 \pm 1.0$, and $c = 990.2 \pm 1.0$ pm and $\alpha = 95.50° \pm 0.08°$, $\beta = 114.44° \pm 0.06°$, and $\gamma = 101.36° \pm 0.09°$ and a calculated density of 1.785 g·cm$^{-3}$ (Moloto et al. 2003). To prepare it, a hot solution of $N$-ethylthiourea (1.66 mM) in EtOH (10 mL) was added into a heated solution of $CdCl_2$ (0.83 mM) in the same solvent (15 mL). The mixture was stirred and refluxed for 2 h. The colorless solution was filtered to remove any traces of unreacted materials while hot and was left in an open beaker at room temperature to crystallize by slow evaporation. Transparent cubic crystals of the title compound were obtained after 24 h. The product was filtered, washed twice with EtOH, and dried under vacuum.

**$[Cd(C_7H_5NS_2)Cl_2]$ or $C_7H_5NCl_2S_2Cd$.** This compound melts at 290°C with decomposition (Bell et al. 2004). To obtain it, a warm solution of $CdCl_2·2.5H_2O$ (5.00 mM) in EtOH (25 mL) was added to a hot ethanolic solution (25 mL) of 1,3-benzothiazoline-2-thione (5.00 mM) with stirring. The resultant cloudy solution was allowed to cool, resulting in the formation of pale-yellow powder, which was filtered off and dried in vacuo.

**[CdCl$_2$(C$_4$H$_6$N$_2$S)$_2$] or C$_8$H$_{12}$N$_4$Cl$_2$S$_2$Cd**. This compound crystallizes in the mono-clinic structure with the lattice parameters $a = 1978.5 \pm 0.6$, $b = 771.0 \pm 0.3$, and $c = 961.2 \pm 0.3$ pm and $\beta = 96.77° \pm 0.02°$ and a calculated density of 1.88 g·cm$^{-3}$ (Matsunaga et al. 2005) [$a = 964.64 \pm 0.08$, $b = 762.62 \pm 0.08$, and $c = 1971.51 \pm 0.08$ pm and $\beta = 96.485° \pm 0.006°$ at $150 \pm 2$ K and a calculated density of 1.897 g·cm$^{-3}$ (Beheshti et al. 2005)]. It was prepared by the following procedure. CdCl$_2$·H$_2$O (5.56 mM) was added to a suspension of (NH$_4$)$_2$[WS$_4$] (2.78 mM) in acetone (70 mL) and the mixture was stirred for 1 h. 1-Methylimidazoline-2-thione (6.22 mM) was added to this solution and the mixture was stirred for another 4 h at room temperature. The mixture was centrifuge and the yellow supernatant liquid was decanted and evaporated to dryness in a vacuum. The residue was washed with diethyl ether ($2 \times 5$ mL) and $n$-pentane ($2 \times 5$ mL) to remove any unreacted 1-methylimidazoline-2-thi-one and dried in a vacuum to give an orange-yellow powder. In the solid state, the obtained complex is air stable and can be stored for months in a desiccator, but it decomposed slowly when diethyl ether was diffused slowly into an acetone solution of a product over 3 days at room temperature, resulting in the formation of pale-yellow crystals of [CdCl$_2$(C$_4$H$_6$N$_2$S)$_2$] (Beheshti et al. 2005).

It could also be prepared by the next procedure (Matsunaga et al. 2005). To a CH$_3$OH solution (40 mL) of 1-methylimidazoline-2(3$H$)-thione (10 mM) a CH$_3$OH solution (5 mL) of CdCl$_2$·2.5H$_2$O (5.0 mM) was added. When the mixture was stirred for 1 h, a white solid appeared. The resulting white powder was collected by filtration and washed with CH$_3$OH. The colorless single crystals for XRD were obtained by recrystallization from a CH$_3$OH/H$_2$O mixed solvent.

**[Cd(C$_4$H$_8$N$_2$S)$_2$Cl$_2$] or C$_8$H$_{16}$N$_4$Cl$_2$S$_2$Cd**. This compound melts at 164°C (Devillanova and Verani 1980) and crystallizes in the monoclinic structure with the lattice parameters $a = 1810.57 \pm 0.16$, $b = 902.05 \pm 0.08$, and $c = 1867.17 \pm 0.16$ pm and $\beta = 94.803° \pm 0.001°$ and a calculated density of 1.817 g·cm$^{-3}$ (Wazeer et al. 2007). It was prepared by a dropwise addition of ethanol solution of cadmium chloride (1 mM in 10 mL) to a stoichiometric amount of $N$-methylimidazoline-2-thione (C$_4$H$_8$N$_2$S) in EtOH (2 mM in 50 mL) and refluxing the solution for 30 min on a water bath (Shunmugam and Sathyanarayana 1983, Wazeer et al. 2007). Colorless crystalline product was obtained.

**[Cd(C$_9$H$_{12}$N$_4$S)Cl$_2$] or C$_9$H$_{12}$N$_4$Cl$_2$SCd**. This compound melts at 296°C (Bermejo et al. 1999). It was obtained by treating 2-acetylpyridine 4-methylthiosemicarbazone (C$_9$H$_{12}$N$_4$S) with CdCl$_2$ (1:1 molar ratio) in EtOH. After prolonged stirring (about 1 week) at room temperature, the reaction mixture afforded the title compound, which was filtered off, washed with EtOH, and vacuum dried.

**[Cd(C$_9$H$_{13}$N$_5$S)Cl$_2$] or C$_9$H$_{13}$N$_5$Cl$_2$SCd**. This compound melts at 255°C (Bermejo et al. 2001). It was prepared as follows. A solution of (2-pyridyl)formamide $N$(4)-dimethylthiosemicarbazone (1 mM) in EtOH (30 mL) was added to a solution of CdCl$_2$ (1 mM) in 30 mL of 96% EtOH, and the mixture kept stirring for several days at room temperature. The resulting yellow solid was filtered off, washed thoroughly with cold EtOH, and stored in a desiccator over Mg and I$_2$.

**[Cd(C$_9$H$_{13}$N$_5$S)Cl$_2$]** or **C$_9$H$_{13}$N$_5$Cl$_2$SCd**. To obtain this compound, a solution of CdCl$_2$ (1.12 mM) in EtOH (30 mL) was mixed with a solution of 2-pyridineformamide N(4)-ethylthiosemicarbazone (C$_9$H$_{13}$N$_5$S) (1.12 mM) in EtOH (30 mL) (Bermejo et al. 2004a). The resultant solution was stirred at room temperature for 7 days and a white solid was separated by filtration.

**[Cd(C$_5$H$_5$NS)$_2$Cl$_2$]** or **C$_{10}$H$_{10}$N$_2$Cl$_2$S$_2$Cd**. This compound was prepared by mixing ethanolic solutions of CdCl$_2$ and 2-pyridine thiol (Kennedy and Lever 1972). It could not be recrystallized.

**[Cd(C$_5$H$_{10}$N$_2$S)$_2$Cl$_2$]** or **C$_{10}$H$_{20}$N$_4$Cl$_2$S$_2$Cd**. This compound melts at 281°C with decomposition (Devillanova and Verani 1980). It was obtained by reacting CdCl$_2$ with N,N-dimethylimidazolidine-2-thione (1:2 molar ratio) in absolute EtOH.

**[Cd(C$_5$H$_{10}$N$_2$S)$_2$Cl$_2$]** or **C$_{10}$H$_{20}$N$_4$Cl$_2$S$_2$Cd**. This compound melts at 148°C (Devillanova and Verani 1980) and crystallizes in the monoclinic structure with the lattice parameters $a = 1279.64 \pm 0.14$, $b = 1039.97 \pm 0.11$, and $c = 1332.43 \pm 0.14$ pm and $\beta = 99.588° \pm 0.002°$ and a calculated density of 1.686 g·cm$^{-3}$ (Wazeer et al. 2007). It was obtained by reacting CdCl$_2$ with N-ethylimidazolidine-2-thione (1:2 molar ratio) in absolute EtOH.

**[Cd{(NHEt$_2$)CS}$_2$Cl$_2$]** or **C$_{10}$H$_{24}$N$_4$Cl$_2$S$_2$Cd**. This compound melts at 204°C (Moloto et al. 2003). To prepare it, a hot solution of N,N'-diethylthiourea (1.66 mM) in EtOH (10 mL) was added into a heated solution of CdCl$_2$ (0.83 mM) in the same solvent (15 mL). The mixture was stirred and refluxed for 2 h. The colorless solution was filtered to remove any traces of unreacted materials while hot and was left in an open beaker at room temperature to crystallize by slow evaporation. Transparent cubic crystals of the title compound were obtained after 24 h. The product was filtered, washed twice with EtOH, and dried under vacuum.

**[{(CH$_3$)$_2$NH(CH$_2$)$_3$S}$_2$][CdCl$_4$]** or **C$_{10}$H$_{26}$N$_2$Cl$_4$S$_2$Cd**. This compound crystallizes in the orthorhombic structure with the lattice parameters $a = 2344$, $b = 1356$, and $c = 670$ pm (Casals et al. 1987). It was obtained simultaneously with obtaining of [CdCl$_2${μ-S(CH$_2$)$_3$NH(CH$_3$)$_2$}].

**[Cd(dmtp)$_2$Cl$_2$]** or **C$_{12}$H$_{16}$N$_4$Cl$_2$S$_2$Cd (dmtp = 4,6-dimethyl-2-thiopyrimidine)**. This compound crystallizes in the monoclinic structure with the lattice parameters $a = 1319.7$, $b = 843.8$, and $c = 1586.2$ pm and $\beta = 97.30°$ and a calculated density of 1.75 g·cm$^{-3}$ (Lopez-Garzon et al. 1993). It was prepared by dissolving dmtp and CdCl$_2$ in a 1:3 molar ratio, in the minimum amount of water. The solution was then evaporated on a water bath at 60°C. Pale-yellow crystals precipitated after 2 days.

**[Cd(C$_{12}$H$_{17}$N$_5$S)Cl$_2$]** or **C$_{12}$H$_{17}$N$_5$Cl$_2$SCd**. This compound melts at 255°C (Castiñeiras et al. 2002). To obtain it, a solution of C$_{12}$H$_{17}$N$_5$S [2-pyridineformamide N(4)-piperidylthiosemicarbazone] (0.76 mM) in EtOH (30 mL) was added to a solution of CdCl$_2$ (0.76 mM) in 30 mL of 96% EtOH, and the mixture kept stirring for several days at room temperature. The resulting yellow solid was filtered off, washed thoroughly with cold ethanol, and stored in a desiccator over Mg and I$_2$.

**Cd(SCN)$_2$(ClC$_6$H$_4$NH$_2$)$_2$]** or **C$_{14}$H$_{12}$N$_4$Cl$_2$S$_2$Cd**. This compound melts at 160°C (Ahuja and Garg 1972). It was obtained by allowing an excess of chloraniline to react with cadmium thiocyanate in EtOH.

**[Cd{PhNH(NH$_2$)CS}$_2$Cl$_2$]** or **C$_{14}$H$_{16}$N$_4$Cl$_2$S$_2$Cd**. This compound melts at 210°C (Moloto et al. 2003). To prepare it, a hot solution of *N*-phenylthiourea (1.66 mM) in EtOH (10 mL) was added into a heated solution of CdCl$_2$ (0.83 mM) in EtOH (15 mL). The mixture was stirred and refluxed for 2 h. The colorless solution was filtered to remove any traces of unreacted materials while hot and was left in an open beaker at room temperature to crystallize by slow evaporation. Transparent cubic crystals of the title compound were obtained after 24 h. The product was filtered, washed twice with EtOH, and dried under vacuum.

**[Cd(C$_7$H$_{14}$N$_2$S)$_2$Cl$_2$]** or **C$_{14}$H$_{28}$N$_4$Cl$_2$S$_2$Cd, dichlorobis[(*N*,*N*-diethylimidazolidine-2-thione-*S*)]zinc**. This compound melts at 195°C with decomposition (Devillanova and Verani 1980). It was obtained by reacting CdCl$_2$ with *N*,*N*-diethylimidazolidine-2-thione (1:2 molar ratio) in dimethoxypropane.

**[Cd$_4${S(CH$_2$)$_2$NMe$_2$}$_4$Cl$_4$]** or **C$_{16}$H$_{40}$N$_4$Cl$_4$S$_4$Cd$_4$**. To obtain this compound, HS(CH$_2$)$_2$NMe$_2$·HCl (10 mM) was dispersed in acetonitrile (40 mL). Et$_3$N (2.8 mL) and anhydrous CdCl$_2$ (10 mM) were successfully added with stirring under a N$_2$ atmosphere (Casals et al. 1991a). Stirring was continued for 24 h. The resulting suspension was filtered in the open atmosphere, washed with copious distilled water and then with MeOH and diethyl ether, and vacuum dried.

Alternatively, this compound was obtained by adding an aqueous solution of equimolar amounts of HS(CH$_2$)$_2$NMe$_2$·HCl and NaOH to an aqueous solution of Cd(OAc)$_2$, the metal/ligand molar ratio being always 1:1 (Casals et al. 1991a). A white powder formed immediately. The mixture was stirred for several hours; the solid was filtered off, washed with cold water and then MeOH, and dried under vacuum. Crystals were obtained by refluxing the compound in DMSO, filtering off the undissolved solid, letting the solution cool to 60°C, and maintaining it at this temperature for 24 h. All these manipulations were performed under a N$_2$ atmosphere.

**[Cd(C$_{10}$H$_8$N$_2$S$_2$)$_2$Cl$_2$]$_n$** or **C$_{20}$H$_{16}$N$_4$Cl$_2$S$_4$Cd**. This compound is stable up to 302°C and crystallizes in the monoclinic structure with the lattice parameters $a = 761.46 \pm 0.03$, $b = 1826.06 \pm 0.08$, and $c = 932.03 \pm 0.02$ pm and $\beta = 100.26° \pm 0.01°$ and a calculated density of 1.625 g·cm$^{-3}$ (Luo et al. 2003). To obtain it, a colorless solution of CdCl$_2$·2.5H$_2$O (0.5 mM) in MeOH/H$_2$O (5:1 volume ratio, 5 mL) was carefully layered onto a solution of 4,4′-bipyridyl disulfide (1 mM) in acetonitrile (5 mL). Diffusion between the two phases over a period of 5 days produced light-yellow crystals of the title compound.

**[Cd(C$_5$H$_4$N$_4$S)$_4$Cl$_2$]** or **C$_{20}$H$_{16}$N$_{16}$Cl$_2$S$_4$Cd**. This compound crystallizes in the triclinic structure with the lattice parameters $a = 808.3 \pm 0.2$, $b = 1190.4 \pm 0.3$, and $c = 737.4 \pm 0.2$ pm and $\alpha = 99.04° \pm 0.02°$, $\beta = 101.30° \pm 0.02°$, and $\gamma = 91.53° \pm 0.02°$ and the experimental and calculated densities 1.92 and 1.917 g·cm$^{-3}$, respectively (Dubler and Gyr 1988). It was obtained by the next procedure. A 3 mM of CdCl$_2$·2.5H$_2$O was added to a suspension of 5 mM of 6-mercaptopurine monohydrate in 100 mL of 0.2 M HCl, and the mixture was stirred for 3 h at 80°C. After the solution was cooled to room temperature, yellow transparent plates of the title compound formed within a few hours. The crystals were washed with water and dried in a desiccator. The compound is thermally stable until 280°C. The subsequent decomposition up to 800°C finally leads to CdO.

**[Cd($C_{20}H_{27}N_6S_2$)Cl] or $C_{20}H_{27}N_6ClS_2Cd$.** This compound crystallizes in the triclinic structure with the lattice parameters $a = 869.19 \pm 0.08$, $b = 1127.27 \pm 0.08$, and $c = 1452.0 \pm 1.0$ pm and $\alpha = 77.466° \pm 0.008°$, $\beta = 87.949° \pm 0.009°$, and $\gamma = 75.026° \pm 0.007°$ (Castiñeiras et al. 1999). It was prepared by mixing 1 mM of CdCl$_2$ and 1 mM of 1-phenylglyoxal bis(3-piperidylthiosemicarbazone in 25 mL of 95% EtOH, with 1 mL of Et$_3$N added. The mixture was refluxed for 2 h and the resulting orange solid was filtered while the solution was warm.

**[Cd$_4${S(CH$_2$)$_2$NMe$_2$}$_6$Cl$_2$] or $C_{24}H_{60}N_6Cl_2S_6Cd_4$.** This compound crystallizes in the monoclinic structure with the lattice parameters $a = 1284.6 \pm 0.1$, $b = 1310.9 \pm 0.1$, and $c = 1345.6 \pm 0.1$ pm and $\beta = 109.130° \pm 0.007°$ and a calculated density of 1.777 g·cm$^{-3}$ (Casals et al. 1991a). For its preparation, a solution of NaS(CH$_2$)$_2$NMe$_2$ (freshly prepared by mixing HCl with sodium methoxide in a 1:2 molar ratio in MeOH) was added to a suspension of an equimolar amount of [Cd{S(CH$_2$)$_2$NMe$_2$} Cl] in acetonitrile under an inert atmosphere. The reaction mixture was stirred under N$_2$ overnight. The filtered solution was concentrated until a colorless solid separated out. This was filtered off, washed with acetonitrile, dried under vacuum, and then recrystallized from MeOH. To increase the overall yield, the filtrate from the reaction was evaporated to dryness and the solid residue recrystallized from MeOH; several crops of the crystals could be obtained in this way. Crystals of this compound were obtained by slow evaporation of methanol solution under an inert atmosphere.

**[Cd{(NHPh$_2$)CS}$_2$Cl$_2$] or $C_{26}H_{24}N_4Cl_2S_2Cd$.** This compound melts at 198°C (Moloto et al. 2003). To prepare it, a hot solution of $N,N'$-diphenylthiourea (1.66 mM) in EtOH (10 mL) was added into a heated solution of CdCl$_2$ (0.83 mM) in the same solvent (15 mL). The mixture was stirred and refluxed for 2 h. The colorless solution was filtered to remove any traces of unreacted materials while hot and was left in an open beaker at room temperature to crystallize by slow evaporation. Transparent cubic crystals of the title compound were obtained after 24 h. The product was filtered, washed twice with EtOH, and dried under vacuum.

**[Cd{S$_2$CN(C$_6$H$_{11}$)$_2$}$_2$]·CH$_2$Cl$_2$ or $C_{27}H_{46}N_2Cl_2S_4Cd$.** This compound crystallizes in the triclinic structure with the lattice parameters $a = 1355.7 \pm 0.7$, $b = 1449 \pm 2$, and $c = 960.4 \pm 0.2$ pm and $\alpha = 100.70° \pm 0.06°$, $\beta = 91.55° \pm 0.03°$, and $\gamma = 115.58° \pm 0.06°$ and a calculated density of 1.421 g·cm$^{-3}$ (Cox and Tiekink 1999).

**[Cd(PhNHCSNH$_2$)$_4$Cl$_2$] or $C_{28}H_{32}N_8Cl_2S_4Cd$.** This compound crystallizes in the monoclinic structure with the lattice parameters $a = 2705.7 \pm 1.3$, $b = 810.8 \pm 0.3$, and $c = 1675.1 \pm 0.8$ pm and $\beta = 114.46°$ and a calculated density of 1.573 g·cm$^{-3}$ (Yang et al. 2001a). To obtain it, phenylthiourea (2 mM) was dissolved in 30 mL of warm H$_2$O and the pH value was adjusted to ca. 4 by dilute HCl solution. An aqueous solution of CdCl$_2$·5H$_2$O (1 mM) was added to the aforementioned solution and a white precipitate formed immediately. The mixture was heated to near boiling with stirring until all of the solids were dissolved. The resulting solution was slowly cooled to room temperature to give pale-yellow crystals of the title compound.

**(Et$_4$N)$_2$[(μ-SEt)$_6$(CdCl)$_4$] or C$_{28}$H$_{70}$N$_2$Cl$_4$S$_6$Cd$_4$.** To obtain this compound, a mixture of CdCl$_2$ (0.98 mM), Et$_4$NCl·H$_2$O (2.0 mM), and Cd(SEt)$_2$ (3.3 mM) was stirred in CH$_2$Cl$_2$ (20 mL) for 1 h (Dean et al. 1992). The initial three-component mixture remained milky and the small amount of insoluble material was removed by filtration. The volume of the filtrate was reduced to ca. 4 mL by gentle heating under a flow of N$_2$, and then Et$_2$O (2 mL) was layered on. The mixture was allowed to stand in the refrigerator overnight. The white solid that formed was separated by filtration, washed with Et$_2$O, and dried in vacuo at room temperature.

**(Et$_4$N)$_2$[(μ-SPr$^i$)$_6$(CdCl)$_4$] or C$_{34}$H$_{82}$N$_2$Cl$_4$S$_6$Cd$_4$.** To obtain this compound, a mixture of CdCl$_2$ (0.98 mM), Et$_4$NCl·H$_2$O (2.0 mM), and Cd(SPr$^i$)$_2$ (3.3 mM) was stirred in CH$_2$Cl$_2$ (20 mL) for 1 h (Dean et al. 1992). The small amount of insoluble material was removed by filtration. The volume of the filtrate was reduced to ca. 4 mL by gentle heating under a flow of N$_2$, and then Et$_2$O (2 mL) was layered on. The mixture was allowed to stand in the refrigerator overnight. The white solid that formed was separated by filtration, washed with Et$_2$O, and dried in vacuo at room temperature.

**(Me$_4$N)$_2$[Cd$_4$(SPh)$_6$Cl$_4$] or C$_{44}$H$_{54}$N$_2$Cl$_4$S$_6$Cd$_4$.** To obtain this compound, freshly prepared solid PhICl$_2$ (0.60 mM) was added with stirring to a solution of (Me$_4$N)$_2$[Cd$_4$(SPh)$_{10}$] (0.30 mM) in 15 mL of MeCN (Dean et al. 1987). A rapid reaction occurred with the formation of a precipitate. After 10 min the solvent was removed under vacuum. The residue was extracted three times with 20 mL portions of diethyl ether to remove (PhS)$_2$ and PhI. The crude residue was purified by recrystallization from a mixture of acetonitrile and diethyl ether. The syntheses were performed under an inert atmosphere.

**[(Et$_4$N)$_2$[(μ-SBz)$_6$(CdCl)$_4$] or C$_{58}$H$_{82}$N$_2$Cl$_4$S$_6$Cd$_4$.** To obtain this compound, a mixture of CdCl$_2$ (0.98 mM), Et$_4$NCl·H$_2$O (2.0 mM), and Cd(SBz)$_2$ (3.3 mM) was stirred in CH$_2$Cl$_2$ (20 mL) for 1 h (Dean et al. 1992). The small amount of insoluble material was removed by filtration. The volume of the filtrate was reduced to ca. 4 mL by gentle heating under a flow of N$_2$, and then Et$_2$O (2 mL) was layered on. The mixture was allowed to stand in the refrigerator overnight and oily white solids were obtained. These solids were separated by decantation, washed several times with pentane, and dried in vacuo to give white foamy solids.

## 4.51   Cd–H–C–N–Cl–Br–S

**[Cd(C$_8$H$_8$N$_3$ClS)$_2$Br$_2$] or C$_{16}$H$_{16}$N$_6$Cl$_2$Br$_2$S$_2$Cd.** This multinary compound is formed in this system. It melts at 253°C and crystallizes in the monoclinic structure with the lattice parameters $a = 817.5 \pm 0.1$, $b = 1417.6 \pm 0.1$, and $c = 2107.3 \pm 0.1$ pm and $\beta = 94.02° \pm 0.01°$ and a calculated density of 1.910 g·cm$^{-3}$ (Tian et al. 1997). To obtain it, an ethanol solution of chlorobenzaldehyde thiosemicarbazone (C$_8$H$_8$N$_3$ClS) (2 mM) and CdBr$_2$ (2 mM) were mixed. The colorless crystalline solid formed after refluxing for 4 h was isolated, washed with EtOH, and dried in vacuo over $P_2O_5$. Crystals were obtained by slowly evaporating of dichloromethane solution in air.

## 4.52   Cd–H–C–N–Cl–I–S

$[Cd(C_8H_8N_3ClS)_2I_2]$ or $C_{16}H_{16}N_6Cl_2I_2S_2Cd$. This multinary compound is formed in this system. It melts at 223°C and crystallizes in the triclinic structure with the lattice parameters $a = 786.2 \pm 0.2$, $b = 1054.7 \pm 0.2$, and $c = 1609.9 \pm 0.3$ pm and $\alpha = 83.86° \pm 0.02°$, $\beta = 94.02° \pm 0.01°$, and $\gamma = 70.54° \pm 0.02°$ and a calculated density of 2.132 g·cm$^{-3}$ (Tian et al. 1997). To obtain it, an ethanol solution of chlorobenzaldehyde thiosemicarbazone ($C_8H_8N_3ClS$) (2 mM) and $CdI_2$ (2 mM) were mixed. The colorless crystalline prisms after refluxing for 4 h were isolated, washed with EtOH, and dried in vacuo over $P_2O_5$. Crystals were obtained by slowly evaporating dichloromethane solution into the air.

## 4.53   Cd–H–C–N–Br–S

Some compounds are formed in this system.

$[CdBr_2 \cdot 2CS(NH_2)_2]$ or $C_2H_8N_4Br_2S_2Cd$. This compound melts at 180°C (Stoev and Ruseva 1994) and crystallizes in the orthorhombic structure with the lattice parameters $a = 1305.2 \pm 1.8$, $b = 589.92 \pm 1.8$, and $c = 1354.2 \pm 0.3$ pm and a calculated density of 2.663 g·cm$^{-3}$ (Marcos et al. 1998). It could be isolated from the saturated solutions in MeOH by isothermal evaporation of the solvent (Stoev and Ruseva 1994). Single crystals of the title compound were grown by slow evaporation of aqueous solution containing $CdBr_2$ and thiourea in the 1:2 molar ratio (Marcos et al. 1998). Thermal dissociation of $CdBr_2 \cdot 2CS(NH_2)_2$ allows preparation of CdS from the saturated solutions (Stoev and Ruseva 1994).

The solubility isotherms of the $CdBr_2$–$CS(NH_2)_2$–MeOH at 25°C have been investigated, and the field of equilibrium existence of $CdBr_2 \cdot 2CS(NH_2)_2$ was determined (Stoev and Ruseva 1994).

$[Cd(C_3H_5NS_2)Br_2]$ or $C_3H_5NBr_2S_2Cd$. This compound melts at 185°C–186°C (Bell et al. 2004). To obtain it, a warm solution of $CdBr_2 \cdot 4H_2O$ (5.23 mM) in EtOH (25 mL) was added to a hot ethanolic solution (25 mL) of 1,3-thiazolidine-2-thione (5.21 mM) with stirring. The resultant cloudy solution was allowed to cool, resulting in the formation of colorless crystals which were filtered off and dried in vacuo.

$[Cd(C_4H_6N_2S)Br_2]$ or $C_4H_6N_2Br_2SCd$. This compound melts at 217°C–220°C (Bell et al. 2004). To obtain it, a warm solution of $CdBr_2 \cdot 4H_2O$ (5.00 mM) in EtOH (25 mL) was added to a hot ethanolic solution (25 mL) of 1-methylimidazoline-2(3H)-thione (5.00 mM) with stirring. The resultant cloudy solution was allowed to cool, resulting in the formation of colorless crystals, which were filtered off and dried in vacuo.

$[CdBr_2\{\mu\text{-}S(CH_2)_3NH(CH_3)_2\}]$ or $C_5H_{13}NBr_2SCd$. This compound crystallizes in the monoclinic structure with the lattice parameters $a = 1594$, $b = 656$, and $c = 1021$ pm and $\beta = 95.14°$ (Casals et al. 1987). To obtain it, a solution of 3-dimethylamino-1-propanethiol (2.50 mM) in MeOH (10 mL) and $H_2O$ (40 mL) at room temperature was added slowly with stirring to a solution of $CdBr_2 \cdot 4H_2O$ (2.50 mM) in the same solvent. Soon after completion of the addition, some white flakes developed and eventually settled at the bottom of the beaker while small colorless crystals appeared on the walls. After 48 h the flakes were separated by filtration and the crystals were washed with $H_2O$ and vacuum dried. It was then possible to separate manually a very few crystals of $[\{(CH_3)_2NH(CH_2)_3S\}_2][CdBr_4]$. The mixture of crystalline solids indicate the presence of a mixture of this compound and $[CdBr_2\{\mu\text{-}S(CH_2)_3NH(CH_3)_2\}]$.

**$[Cd(C_3H_5NS_2)_2Br_2]$ or $C_6H_{10}N_2Br_2S_4Cd$.** This compound melts at 127°C–128°C (Bell et al. 2004) [at 125°C (De Filippo et al. 1971)]. It was prepared by the next two procedures:

1. A warm solution of $CdBr_2 \cdot 4H_2O$ (3.90 mM) in EtOH (25 mL) was added to a hot ethanolic solution (25 mL) of 1,3-thiazolidine-2-thione (7.80 mM) with stirring (Bell et al. 2004). The resultant cloudy solution was allowed to cool, resulting in the formation of colorless crystals, which were filtered off and dried in vacuo.

2. $CdBr_2$ was dissolved in an excess of molten thiazolidine-2-thione ($C_3H_5NS_2$) (at ca. 105°C; molar ratio $Cd^{2+}/C_3H_5NS_2$ ca. 1:2), and the crude product was purified by washing with ethyl ether and recrystallizing from EtOH (De Filippo et al. 1971). The title compound could also be prepared if to a solution of $CdBr_2$ in absolute EtOH different amounts of $C_3H_5NS_2$ were added for $Cd^{2+}/C_3H_5NS_2$ in molar ratios of 1:1, 1:2, 1:3, and 1:4 and the mixture was heated under reflux for 4 h. The title compound was separated out by crystallization.

**$[Cd(C_3H_6N_2S)_2Br_2]$ or $C_6H_{12}N_4Br_2S_2Cd$.** This compound melts at 190°C–195°C and crystallizes in the monoclinic structure with the lattice parameters $a = 633.15 \pm 0.09$, $b = 1481.7 \pm 0.2$, and $c = 1384.43 \pm 0.19$ pm and $\beta = 96.282° \pm 0.002°$ and a calculated density of 2.452 g·cm$^{-3}$ (Lobana et al. 2008). To obtain it, to a stirred solution of imidazoline-2-thione ($C_3H_6N_2S$) (0.24 mM) in dry EtOH (8 mL) a solution of $CdBr_2$ (0.12 mM) in dry EtOH (8 mL) was added. White precipitates were formed. To these precipitates MeOH (5 mL) was added and slow evaporation at room temperature formed yellow needles of $[Cd(C_3H_6N_2S)_2Br_2]$.

**$[Cd\{Me_2(NH)_2CS\}_2Br_2]$ or $C_6H_{16}N_4Br_2S_2Cd$.** This compound melts at 186°C–188°C and crystallizes in the monoclinic structure with the lattice parameters $a = 1372.33 \pm 0.15$, $b = 897.37 \pm 0.07$, and $c = 1300.00 \pm 0.14$ pm and $\beta = 108.562° \pm 0.007°$ at $172 \pm 2$ K and a calculated density of 2.103 g·cm$^{-3}$ (Ahmad et al. 2011). It was prepared by adding 2 mM methanolic solution (15 mL) of $N,N'$-dimethylthiourea to an aqueous solution (5 mL) of $CdBr_2$ (1.0 mM) and stirring the mixture for 30 min. White precipitate was formed on mixing. The colorless solution was filtered and the filtrate was kept at room temperature for crystallization. As a result, white crystalline products were obtained, which were washed with MeOH and dried.

**[Cd(C₇H₅NS₂)Br₂] or C₇H₅NBr₂S₂Cd.** This compound melts at 265°C with decomposition (Bell et al. 2004). To obtain it, a warm solution of $CdBr_2 \cdot 4H_2O$ (4.50 mM) in EtOH (25 mL) was added to a hot ethanolic solution (25 mL) of 1,3-benzothiazoline-2-thione (4.50 mM) with stirring. The resultant cloudy solution was allowed to cool, resulting in the formation of pale-yellow powder, which was filtered off and dried in vacuo.

**[Cd(C₅H₄NC₂H₅N₄S)Br₂] or C₇H₉N₅Br₂SCd.** This compound melts at 240°C (Castiñeiras et al. 2000). To obtain it, a solution of $CdBr_2$ (2 mM) in EtOH (30 mL) was mixed with a solution of 2-pyridineformamide thiosemicarbazone ($C_5H_4NC_2H_5N_4S$) (2 mM) in the same solvent (30 mL), and the mixture was stirred under reflux for 2 h. The resulting solid was filtered out, washed with anhydrous ether to apparent dryness, and placed on a warm plate at 35°C.

**[Cd(C₄H₆N₂S)₂Br₂] or C₈H₁₂N₄Br₂S₂Cd.** This compound melts at 175°C–177°C and crystallizes in the orthorhombic structure with the lattice parameters $a = 976.90 \pm 0.07$, $b = 3486.8 \pm 0.3$, and $c = 2649.14 \pm 0.19$ pm and a calculated density of 2.211 $g \cdot cm^{-3}$ (Bell et al. 2004) [$a = 2665.8 \pm 0.9$, $b = 1164.2 \pm 1.0$, and $c = 976.9 \pm 0.4$ pm at 219 K and calculated density 2.19 $g \cdot cm^{-3}$ (Matsunaga et al. 2005)]. To obtain it, a warm solution of $CdBr_2$ (5.00 mM) in EtOH (25 mL) was added to a hot ethanolic solution (25 mL) of 1-methylimidazoline-2(3$H$)-thione (10.00 mM) with stirring (Bell et al. 2004). The resultant cloudy solution was allowed to cool, resulting in the formation of colorless crystals, which were filtered off and dried in vacuo.

The title compound could also be prepared by the following procedure (Matsunaga et al. 2005). To a $CH_3OH$ solution (40 mL) of 1-methylimidazoline-2(3$H$)-thione (10 mM) a $CH_3OH$ solution (5 mL) of $CdBr_2 \cdot 4H_2O$ (5.0 mM) was added. When the mixture was stirred for 1 h, a white solid appeared. The resulting white powder was collected by filtration and washed with $CH_3OH$. The colorless single crystals for XRD were obtained by recrystallization of white powder from a $CH_3OH/H_2O$ mixed solvent.

**[Cd(C₄H₈N₂S)₂Br₂] or C₈H₁₆N₄Br₂S₂Cd, dibromobis[(N-methylimidazolidine-2-thione-S)]cadmium.** This compound melts at 173°C (Devillanova and Verani 1980). It was prepared by a dropwise addition of ethanol solution of $CdBr_2$ (1 mM in 10 mL) to a stoichiometric amount of N-methylimidazoline-2-thione ($C_4H_8N_2S$) in EtOH (2 mM in 50 mL) and refluxing the solution for 30 min on a water bath (Shunmugam and Sathyanarayana 1983). Colorless crystalline product was obtained.

**[Cd(C₉H₁₂N₄S)Br₂] or C₉H₁₂N₄Br₂SCd.** This compound melts at 291°C (Bermejo et al. 1999). It was obtained by treating 2-acetylpyridine 4-methylthiosemicarbazone ($C_9H_{12}N_4S$) with $CdBr_2$ (1:1 molar ratio) in EtOH. After prolonged stirring (about 1 week) at room temperature, the reaction mixture afforded the title compound that was filtered off, washed with EtOH, and vacuum dried.

**[Cd(C₉H₁₃N₅S)Br₂] or C₉H₁₃N₅Br₂SCd.** This compound melts at 252°C (Bermejo et al. 2001). It was prepared as follows. A solution of (2-pyridyl)formamide N(4)-dimethylthiosemicarbazone (1 mM) in EtOH (30 mL) was added to a solution of

CdBr$_2$ (1 mM) in 30 mL of 96% EtOH, and the mixture kept stirring for several days at room temperature. The resulting yellow solid was filtered off, washed thoroughly with cold EtOH, and stored in a desiccator over Mg and I$_2$.

**[Cd(C$_9$H$_{13}$N$_5$S)Br$_2$] or C$_9$H$_{13}$N$_5$Br$_2$SCd.** To obtain this compound, a solution of CdBr$_2$ (1.12 mM) in EtOH (30 mL) was mixed with a solution of 2-pyridineformamide $N$(4)-ethylthiosemicarbazone (C$_9$H$_{13}$N$_5$S) (1.12 mM) in EtOH (30 mL) (Bermejo et al. 2004a). The resultant solution was stirred at room temperature for 7 days and a yellow formed solid was filtered off.

**[Cd(C$_5$H$_5$NS)$_2$Br$_2$] or C$_{10}$H$_{10}$N$_2$Br$_2$S$_2$Cd.** This compound was prepared by mixing ethanolic solutions of CdBr$_2$ and 2-pyridine thiol (Kennedy and Lever 1972). It could not be recrystallized.

**[(Et$_2$NCS)$_2$S][CdBr$_4$] or C$_{10}$H$_{20}$N$_2$Br$_4$S$_3$Cd.** This compound was obtained as follows (McCleverty et al. 1982). To a suspension of [Cd(S$_2$CNEt$_2$)$_2$] (10 mM) in CS$_2$ (75 mL) bromine (10 mM) in CS$_2$ (20 mL) was added, and the mixture was shaken for 15 min. The title compound was formed as a pale-yellow solid, which was filtered off, washed with diethyl ether, and dried in vacuo.

**[Cd(C$_5$H$_{10}$N$_2$S)$_2$Br$_2$] or C$_{10}$H$_{20}$N$_4$Br$_2$S$_2$Cd, dibromobis[($N$,$N$-dimethylimidazolidine-2-thione-$S$)]cadmium.** This compound melts at 194°C (Devillanova and Verani 1980). It was obtained by reacting CdBr$_2$ with $N$,$N$-dimethylimidazolidine-2-thione (1:2 molar ratio) in absolute EtOH.

**[Cd(C$_5$H$_{10}$N$_2$S)$_2$Br$_2$] or C$_{10}$H$_{20}$N$_4$Br$_2$S$_2$Cd, dibromobis[($N$-ethylimidazolidine-2-thione-$S$)]cadmium.** This compound melts at 114°C (Devillanova and Verani 1980). It was obtained by reacting CdBr$_2$ with $N$-ethylimidazolidine-2-thione (1:2 molar ratio) in absolute EtOH.

**[CdBr$_2$(C$_5$H$_{12}$N$_2$S)$_2$] or C$_{10}$H$_{24}$N$_4$Br$_2$S$_2$Cd.** This compound crystallizes in the monoclinic structure with the lattice parameters $a = 1861.33 \pm 0.17$, $b = 1006.90 \pm 0.09$, and $c = 1346.00 \pm 0.12$ pm and $\beta = 130.834° \pm 0.001°$ and a calculated density of 1.868 g·cm$^{-3}$ (Nawaz et al. 2010a). To obtain it, 1.0 mM of CdBr$_2$·4H$_2$O dissolved in 10 mL of H$_2$O was added to two equivalents of tetramethylthiourea in 15 mL MeOH. A white precipitate formed and was filtered off. The filtrate was kept for crystallization. As a result, an off-white crystalline product was obtained.

**[{(CH$_3$)$_2$NH(CH$_2$)$_3$S}$_2$][CdBr$_4$] or C$_{10}$H$_{26}$N$_2$Br$_4$S$_2$Cd.** This compound crystallizes in the orthorhombic structure with the lattice parameters $a = 2340.6 \pm 0.3$, $b = 1353.1 \pm 0.2$, and $c = 673.8 \pm 0.1$ pm and the experimental and calculated densities 2.04 and 2.080 g·cm$^{-3}$, respectively (Casals et al. 1987). It was obtained simultaneously with obtaining of [CdBr$_2${μ-S(CH$_2$)$_3$NH(CH$_3$)$_2$}].

**[Cd(C$_{12}$H$_{17}$N$_5$S)Br$_2$] or C$_{12}$H$_{17}$N$_5$Br$_2$SCd.** This compound melts at 252°C (Castiñeiras et al. 2002). To obtain it, a solution of C$_{12}$H$_{17}$N$_5$S [2-pyridineformamide $N$(4)-piperidylthiosemicarbazone] (0.76 mM) in EtOH (30 mL) was added to a solution of CdBr$_2$ (0.76 mM) in 30 mL of 96% EtOH, and the mixture kept stirring for several days at room temperature. The resulting yellow solid was filtered off, washed thoroughly with cold ethanol, and stored in a desiccator over Mg and $I_2$.

$[Cd(C_7H_5NS_2)_2Br_2]$ or $C_{14}H_{10}N_2Br_2S_4Cd$. This compound melts at 220°C with decomposition (Bell et al. 2004). To obtain it, a warm solution of $CdBr_2 \cdot 4H_2O$ (3.00 mM) in EtOH (25 mL) was added to a hot ethanolic solution (25 mL) of 1,3-benzothiazoline-2-thione (6.00 mM) with stirring. The resultant cloudy solution was allowed to cool, resulting in the formation of pale-yellow powder, which was filtered off and dried in vacuo.

$[Cd(C_7H_{12}N_2S)_2Br_2]$ or $C_{14}H_{24}N_4Br_2S_2Cd$. This compound melts at 148°C with decomposition and crystallizes in the triclinic structure with the lattice parameters $a = 746.25 \pm 0.06$, $b = 961.49 \pm 0.09$, and $c = 3102.0 \pm 0.3$ pm and $\alpha = 93.485° \pm 0.002°$, $\beta = 94.579° \pm 0.002°$, and $\gamma = 103.872° \pm 0.002°$ at $238 \pm 2$ K and a calculated density of 1.809 $g \cdot cm^{-3}$ (White et al. 2003). It was obtained by the next procedure. A suspension of $CdBr_2$ (0.845 mM) and 2-mercapto-1-*tert*-butylimidazole (1.689 mM) in $CH_2Cl_2$ (30 mL) was stirred for 30 min in a warm (ca. 50°C) water bath. The resulting slightly cloudy solution was stirred for an additional 1.5 h and filtered, and the filtrate was concentrated under reduced pressure to ca. 10 mL. The addition of pentane (15 mL) resulted in the formation of a white precipitate, which was isolated by filtration, washed with pentane ($2 \times 20$ mL), and dried in vacuo for 4 h. All of the reactions were performed under argon using a combination of high-vacuum and Schlenk techniques.

$[Cd(C_7H_{14}N_2S)_2Br_2]$ or $C_{14}H_{28}N_4Br_2S_2Cd$, dibromobis[(*N,N*-diethylimidazolidine-2-thione-*S*)]zinc. This compound melts at 190°C (Devillanova and Verani 1980). It was obtained by reacting $CdBr_2$ with *N,N*-diethylimidazolidine-2-thione (1:2 molar ratio) in dimethoxypropane.

$[Cd_4\{S(CH_2)_2NMe_2\}_4Br_4]$ or $C_{16}H_{40}N_4Br_4S_4Cd_4$. This compound crystallizes in the monoclinic structure with the lattice parameters $a = 939.77 \pm 0.04$, $b = 1349.93 \pm 0.11$, and $c = 1390.05 \pm 0.12$ pm and $\beta = 108.675 \pm 0.004°$ and a calculated density of 2.357 $g \cdot cm^{-3}$ (Casals et al. 1991a). To obtain it, $HS(CH_2)_2NMe_2 \cdot HBr$ (5 mM) was dispersed in acetonitrile (40 mL). $Et_3N$ (1.4 mL) and anhydrous $CdBr_2$ (5 mM) were successfully added with stirring under a $N_2$ atmosphere. Stirring was continued for 24 h. The resulting suspension was filtered in the open atmosphere, washed with copious distilled water and then with MeOH and diethyl ether, and vacuum dried.

Alternatively, this compound was obtained by adding an aqueous solution of equimolar amounts of $HS(CH_2)_2NMe_2 \cdot HBr$ and NaOH to an aqueous solution of $Cd(OAc)_2$, the metal/ligand molar ratio being always 1:1 (Casals et al. 1991a). A white powder formed immediately. The mixture was stirred for several hours; the solid was filtered off, washed with cold water and then MeOH, and dried under vacuum. Crystals were obtained by refluxing the compound in DMSO, filtering off the undissolved solid, letting the solution cool to 60°C, and maintaining it at this temperature for 24 h. All these manipulations were performed under a $N_2$ atmosphere.

$[CdBr_2 \cdot Bu^n_4C_2N_2S_4]$ or $C_{18}H_{36}N_2Br_2S_4Cd$. This was prepared as a white precipitate by adding a solution of $Br_2$ in $CS_2$ to a solution of cadmium di-*n*-butyldithiocarbamate in the same solvent at room temperature (Brinkhoff et al. 1969). After filtration the product could be purified by dissolution in chloroform and reprecipitation by petroleum ether. The alternative route includes using $CdBr_2$ and tetra-$n$-butylthiuram disulfide. To a suspension of $CdBr_2$ in $CS_2$ a solution of tetra-$n$-butylthiuram disulfide was added at room temperature. After $CdBr_2$ had dissolved, petroleum ether was added, precipitating the crystals, which was filtered and dried.

**$[Cd_4\{S(CH_2)_2NMe_2\}_6Br_2]$ or $C_{24}H_{60}N_6Br_2S_6Cd_4$.** For the preparation of this compound, a solution of $NaS(CH_2)_2NMe_2$ (freshly prepared by mixing HBr with sodium methoxide in a 1:2 molar ratio in MeOH) was added to a suspension of an equimolar amount of $[Cd\{S(CH_2)_2NMe_2\}Br]$ in acetonitrile under an inert atmosphere (Casals et al. 1991a). The reaction mixture was stirred under $N_2$ overnight. The filtered solution was concentrated until a colorless solid separated out. This was filtered off, washed with acetonitrile, dried under vacuum, and then recrystallized from MeOH. To increase the overall yield, the filtrate from the reaction was evaporated to dryness and the solid residue recrystallized from MeOH; several crops of the crystals could be obtained in this way. Crystals of this compound were obtained by slow evaporation of methanol solution under an inert atmosphere.

**$(Et_4N)_2[(\mu\text{-}SEt)_6(CdBr)_4]$ or $C_{28}H_{70}N_2Br_4S_6Cd_4$.** To obtain this compound, a mixture of $CdBr_2$ (0.98 mM), $Et_4NBr$ (2.0 mM), and $Cd(SEt)_2$ (3.3 mM) was stirred in $CH_2Cl_2$ (20 mL) for 1 h (Dean et al. 1992). The initial three-component mixture remained milky and the small amount of insoluble material was removed by filtration. The volume of the filtrate was reduced to ca. 4 mL by gentle heating under a flow of $N_2$, and then $Et_2O$ (2 mL) was layered on. The mixture was allowed to stand in the refrigerator overnight. The white solid that formed was separated by filtration, washed with $Et_2O$, and dried in vacuo at room temperature.

The title compound could also be obtained if into a solution of $CdBr_2$ (4.0 mM) and $Et_4NBr$ (2.0 mM) in EtOH (50 mL) at 40°C EtSH (8.1 mM) was stirred, followed by $Et_3N$ (8.0 mM) (Dean et al. 1992). A white precipitate formed. This dissolved when MeCN (5 mL) was added to the mixture at room temperature. Cooling the resulting solution to 5°C overnight gave the product as white crystals.

**$(Et_4N)_2[(\mu\text{-}SPr^i)_6(CdBr)_4]$ or $C_{34}H_{82}N_2Br_4S_6Cd_4$.** This compound crystallizes in the monoclinic structure with the lattice parameters $a = 2407.9 \pm 0.3$, $b = 1136.5 \pm 0.2$, and $c = 2256.1 \pm 0.3$ pm and $\beta = 113.89° \pm 0.01°$ and the experimental and calculated densities $1.75 \pm 0.05$ and $1.743$ g·cm$^{-3}$, respectively (Dean et al. 1992). To obtain it, a mixture of $CdBr_2$ (0.98 mM), $Et_4NBr$ (2.0 mM), and $Cd(SPr^i)_2$ (3.3 mM) was stirred in $CH_2Cl_2$ (20 mL) for 1 h. The small amount of insoluble material was removed by filtration. The volume of the filtrate was reduced to ca. 4 mL by gentle heating under a flow of $N_2$, and then $Et_2O$ (2 mL) was layered on. The mixture was allowed to stand in the refrigerator overnight. The white solid that formed was separated by filtration, washed with $Et_2O$, and dried in vacuo at room temperature. Single crystals of this compound were grown by diffusion of diethyl ether into acetone solution of the compound at 5°C.

**(Me₄N)₂[Cd₄(SPh)₆Br₄] or C₄₄H₅₄N₂Br₄S₆Cd₄.** This compound crystallizes in the cubic structure with the lattice parameter $a = 1786.9 \pm 0.2$ and the experimental and calculated densities 1.86 ± 0.02 and 1.83 g·cm⁻³, respectively (Dean et al. 1987). To obtain it, Br₂ (1.15 mM) in 5 mL of CCl₄ was added to a stirred solution containing (Me₄N)₂[Cd₄(SPh)₁₀] (0.575 mM) in 15 mL of acetone. Decolorization occurred quickly, leaving a clear solution with a small amount of white solid. After 10 min, the solvents were removed under vacuum. The residue was extracted with three 40 mL portions of diethyl ether to remove (PhS)₂. The residue was purified by recrystallization from a mixture of acetone and cyclohexane. The syntheses were performed under an inert atmosphere. Colorless air-stable crystals were grown by diffusion of cyclohexane into acetone at room temperature (Dean et al. 1987) or by recrystallization from C₆H₁₂/Me₂CO after removal of Ph₂S₂ by extraction with Et₂O (Dean and Vittal 1985).

**(Et₄N)₂[(μ-SBz)₆(CdBr)₄] or C₅₈H₈₂N₂Br₄S₆Cd₄.** To obtain this compound, a mixture of CdBr₂ (0.98 mM), Et₄NBr (2.0 mM), and Cd(SBz)₂ (3.3 mM) was stirred in CH₂Cl₂ (20 mL) for 1 h (Dean et al. 1992). The small amount of insoluble material was removed by filtration. The volume of the filtrate was reduced to ca. 4 mL by gentle heating under a flow of N₂, and then Et₂O (2 mL) was layered on. The mixture was allowed to stand in the refrigerator overnight and oily white solids were obtained. These solids were separated by decantation, washed several times with pentane, and dried in vacuo to give white foamy solids.

## 4.54   Cd–H–C–N–I–S

Some compounds are formed in this system.

**[CdI₂·2CS(NH₂)₂] or C₂H₈N₄I₂S₂Cd.** This compound melts at 180°C (Stoev and Ruseva 1994) and crystallizes in the monoclinic structure with the lattice parameters $a = 1048.1 \pm 0.2$, $b = 764.0 \pm 0.2$, and $c = 1513.3 \pm 0.4$ pm and $\beta = 91.00° \pm 0.02°$ and the calculated density 2.842 g·cm⁻³ (Marcos et al. 1998) ($a = 1050 \pm 3$, $b = 761 \pm 1$, and $c = 1516 \pm 2$ pm and $\beta = 90°30' \pm 30'$ and the experimental and calculated densities 2.85 g·cm⁻³ [Durski et al. 1975]). Its colorless needlelike crystals were prepared by the evaporation at room temperature of aqueous solutions of arbitrary concentrations containing CdI₂ and CS(NH₂)₂ at the molar ratio 1:2. It could also be isolated from the saturated solutions in MeOH by isothermal evaporation of the solvent (Stoev and Ruseva 1994, Marcos et al. 1998). During the thermal decomposition of this compound, CdS is formed (Durski et al. 1975, Stoev and Ruseva 1994).

The solubility isotherms of CdBr₂–Cs(NH₂)₂–MeOH at 25°C have been investigated and the field of equilibrium existence of CdBr₂·2CS(NH₂)₂ was determined (Stoev and Ruseva 1994).

**[CdI₂(MeCSNH₂)₂] or C₄H₁₀N₄I₂S₂Cd.** This compound crystallizes in the orthorhombic structure with the lattice parameters $a = 1036.9 \pm 0.9$, $b = 980.9 \pm 0.8$, and $c = 1279.2 \pm 0.05$ pm and a calculated density of 2.637 g·cm⁻³ (Zamilatskov et al. 2007). To obtain it, CdI₂ (1.0 g) and thioacetamide (0.4 g) were dissolved in H₂O

(6 mL) at room temperature. The solution was layered on a liquid hydrocarbon 1-methyldecahydronaphthalene. In 30 min, colorless crystals formed and felt in 1-methyldecahydronaphthalene, which served as protective environment.

**[Cd(Et₂NCS₂)I] or C₅H₁₀NIS₂Cd.** This compound melts at 210°C with decomposition and crystallizes in the monoclinic structure with the lattice parameters $a = 746.6 \pm 0.9$, $b = 943.5 \pm 0.2$, and $c = 1566.7 \pm 0.3$ pm and $\beta = 94.68° \pm 0.01°$ and a calculated density of 2.340 g·cm⁻³ (Zemskova et al. 1991). To obtain it, to a solution of Cd(Et₂NCS₂)₂ in CH₂Cl₂ at room temperature CdI₂ ([1.0–1.1]:1 molar ratio) was added, and the mixture was stirred for 3–5 h. The precipitate that formed was filtered off, washed thoroughly with CH₂Cl₂, and dried in air.

**[Cd(C₃H₅NS₂)₂I₂] or C₆H₁₀N₂I₂S₄Cd.** This compound melts at 148°C–150°C [at 147°C (De Filippo et al. 1971)] and crystallizes in the monoclinic structure with the lattice parameters $a = 1477.99 \pm 0.04$, $b = 952.91 \pm 0.02$, and $c = 1076.68 \pm 0.04$ pm and $\beta = 98.124° \pm 0.001°$ and a calculated density of 2.675 g·cm⁻³ (Bell et al. 2004). To obtain it, a warm solution of CdI₂ (4.40 mM) in EtOH (25 mL) was added to a hot ethanolic solution (25 mL) of 1,3-thiazolidine-2-thione (8.90 mM) with stirring. The resultant cloudy solution was allowed to cool, resulting in the formation of colorless prisms, which were filtered off and dried in vacuo.

The title compound could also be obtained by the next procedure (De Filippo et al. 1971). CdI₂ was dissolved in an excess of molten 1,3-thiazolidine-2-thione (C₃H₅NS₂) (at ca. 105°C; molar ratio Cd²⁺/C₃H₅NS₂ ca. 1:2), and the crude product was purified by washing with ethyl ether and recrystallizing from EtOH. The title compound could also be prepared if to a solution of CdI₂ in absolute EtOH different amounts of C₃H₅NS₂ were added for Cd²⁺/C₃H₅NS₂ in molar ratios of 1:1, 1:2, 1:3, and 1:4 and the mixture was heated under reflux for 4 h. [CdI₂(C₃H₅NS₂)₂] was separated out by crystallization.

**[Cd(C₃H₆N₂S)₂I₂] or C₆H₁₂N₄I₂S₂Cd.** This compound melts at 155°C–160°C and crystallizes in the orthorhombic structure with the lattice parameters $a = 1384.87 \pm 0.10$, $b = 1442.32 \pm 0.11$, and $c = 706.59 \pm 0.05$ pm and a calculated density of 2.685 g·cm⁻³ (Lobana et al. 2008). To obtain it, to a stirred solution of imidazoline-2-thione (C₃H₆N₂S) (0.24 mM) in dry MeOH (8 mL) a solution of CdI₂ (0.12 mM) in dry MeOH (8 mL) was added, followed by stirring for 2 h, and slow evaporation at room temperature formed colorless crystals of [Cd(C₃H₆N₂S)₂I₂].

**[Cd{Me₂(NH)₂CS}₂I₂] or C₆H₁₆N₄I₂S₂Cd.** This compound melts at 108°C–110°C and crystallizes in the monoclinic structure with the lattice parameters $a = 1419.7 \pm 0.2$, $b = 814.43 \pm 0.08$, and $c = 1474.4 \pm 0.2$ pm and $\beta = 109.566° \pm 0.010°$ at $172 \pm 2$ K and a calculated density of 2.376 g·cm⁻³ (Ahmad et al. 2011). It was prepared by adding 2 mM methanolic solution (15 mL) of N,N′-dimethylthiourea to an aqueous solution (5 mL) of CdI₂ (1.0 mM) and stirring the mixture for 30 min. The colorless clear solution was filtered and the filtrate was kept at room temperature for crystallization. As a result, white crystalline products were obtained, which were washed with MeOH and dried.

**[Cd(C$_5$H$_4$NC$_2$H$_5$N$_4$S)I$_2$] or C$_7$H$_9$N$_5$I$_2$SCd**. This compound melts at 210°C and crystallizes in the monoclinic structure with the lattice parameters $a = 762.7 \pm 0.5$, $b = 1498.0 \pm 0.4$, and $c = 1435.6 \pm 0.3$ pm and $\beta = 121.978° \pm 0.001°$ and a calculated density of 2.680 g·cm$^{-3}$ (Castiñeiras et al. 2000). To obtain it, a solution of CdI$_2$ (2 mM) in EtOH (30 mL) was mixed with a solution of 2-pyridineformamide thiosemicarbazone (C$_5$H$_4$NC$_2$H$_5$N$_4$S) (2 mM) in the same solvent (30 mL), and the mixture was stirred under reflux for 2 h. The resulting solid was filtered out, washed with anhydrous ether to apparent dryness, and placed on a warm plate at 35°C. Crystals suitable for XRD were grown by the recrystallization from 96% EtOH.

**[Cd(C$_4$H$_6$N$_2$S)$_2$I$_2$] or C$_8$H$_{12}$N$_4$I$_2$S$_2$Cd**. This compound melts at 186°C–188°C (Bell et al. 2004) and crystallizes in the orthorhombic structure with the lattice parameters $a = 1154.2 \pm 0.6$, $b = 1413.8 \pm 0.5$, and $c = 992.6 \pm 0.3$ pm and a calculated density of 2.44 g·cm$^{-3}$ (Matsunaga et al. 2005). It was prepared by the following procedures:

1. To a CH$_3$OH solution (40 mL) of 1-methylimidazoline-2(3$H$)-thione (10 mM) a CH$_3$OH solution (5 mL) of CdI$_2$ (5.0 mM) was added (Matsunaga et al. 2005). When the mixture was stirred for 1 h, a white solid appeared. The resulting white powder was collected by filtration and washed with CH$_3$OH. The colorless single crystals for XRD were obtained by recrystallization of white powder from a CH$_3$OH/H$_2$O mixed solvent.

2. A warm solution of CdI$_2$ (5.00 mM) in EtOH (25 mL) was added to a hot ethanolic solution (25 mL) of 1-methylimidazoline-2(3$H$)-thione (10.00 mM) with stirring (Bell et al. 2004). The resultant cloudy solution was allowed to cool, resulting in the formation of pale-yellow crystals, which were filtered off and dried in vacuo.

**[Cd(C$_4$H$_8$N$_2$S)$_2$I$_2$] or C$_8$H$_{16}$N$_4$I$_2$S$_2$Cd, diiodobis[($N$-methylimidazolidine-2-thione-$S$)]cadmium**. This compound melts at 140°C with decomposition (Devillanova and Verani 1980). It was obtained by reacting CdI$_2$ with $N$-methylimidazolidine-2-thione (1:2 molar ratio) in absolute EtOH.

**[Cd(C$_9$H$_{12}$N$_4$S)I$_2$] or C$_9$H$_{12}$N$_4$I$_2$SCd**. This compound melts at 270°C with decomposition (Bermejo et al. 1999). It was obtained by treating 2-acetylpyridine 4-methylthiosemicarbazone (C$_9$H$_{12}$N$_4$S) with CdI$_2$ (1:1 molar ratio) in EtOH. After prolonged stirring (about 1 week) at room temperature, the reaction mixture afforded the title compound, which was filtered off, washed with EtOH, and vacuum dried.

**[Cd(C$_9$H$_{13}$N$_5$S)I$_2$] or C$_9$H$_{13}$N$_5$I$_2$SCd**. This compound melts at 233°C (Bermejo et al. 2001). It was prepared as follows. A solution of (2-pyridyl)formamide $N$(4)-dimethylthiosemicarbazone (1 mM) in EtOH (30 mL) was added to a solution of CdI$_2$ (1 mM) in 30 mL of 96% EtOH, and the mixture kept stirring for several days at room temperature. The resulting yellow solid was filtered off, washed thoroughly with cold EtOH, and stored in a desiccator over Mg and I$_2$.

**[Cd(C$_9$H$_{13}$N$_5$S)I$_2$] or C$_9$H$_{13}$N$_5$I$_2$SCd**. To obtain this compound, a solution of CdI$_2$ (1.12 mM) in EtOH (30 mL) was mixed with a solution of 2-pyridineformamide

$N$(4)-ethylthiosemicarbazone ($C_9H_{13}N_5S$) (1.12 mM) in EtOH (30 mL) (Bermejo et al. 2004a). The resultant solution was stirred at room temperature for 7 days and the yellow precipitate was filtered off.

**[CdI$_2${S(SCNEt$_2$)$_2$}] or C$_{10}$H$_{20}$N$_2$I$_2$S$_3$Cd**. This compound was prepared as follows (McCleverty et al. 1982). To a solution of [CdI$_2${(S$_2$CNEt$_2$)$_2$}] (0.66 g) in chloroform (25 mL) a solution of Ph$_3$P (0.26 g) in the same solvent (10 mL) was added. A pale-yellow solution was formed, which was stirred for 30 min. Evaporation of the solvent yielded an oil that was shaken with diethyl ether giving the title compound as a yellow powder.

**[CdI$_2${(S$_2$CNEt$_2$)$_2$}] or C$_{10}$H$_{20}$N$_2$I$_2$S$_4$Cd**. This compound was prepared by the next procedure (McCleverty et al. 1982). To a solution of [Cd(S$_2$CNEt$_2$)$_2$] (10 mM) in chloroform (100 mL) a solution of iodine (10 mM) was added, dropwise with vigorous stirring. The resulting solution was evaporated to low bulk and $n$-pentane was added. After standing, cream-colored needles of the title compound separated and were collected by filtration, washed with diethyl ether, and dried in vacuo.

**[Cd(C$_5$H$_{10}$N$_2$S)$_2$I$_2$] or C$_{10}$H$_{20}$N$_4$I$_2$S$_2$Cd, diiodobis[($N,N$-dimethylimidazolidine-2-thione-$S$)]cadmium**. This compound melts at 120°C (Devillanova and Verani 1980). It was obtained by reacting CdI$_2$ with $N,N$-dimethylimidazolidine-2-thione (1:2 molar ratio) in absolute EtOH.

**[Cd(C$_5$H$_{10}$N$_2$S)$_2$I$_2$] or C$_{10}$H$_{20}$N$_4$I$_2$S$_2$Cd, diiodobis[($N$-ethylimidazolidine-2-thione-$S$)]cadmium**. This compound melts at 80°C (Devillanova and Verani 1980). It was obtained by reacting CdI$_2$ with $N$-ethylimidazolidine-2-thione (1:2 molar ratio) in absolute EtOH.

**[CdI$_2$(C$_5$H$_{12}$N$_2$S)$_2$] or C$_{10}$H$_{24}$N$_4$I$_2$S$_2$Cd**. This compound crystallizes in the monoclinic structure with the lattice parameters $a = 1898.5 \pm 0.5$, $b = 1039.5 \pm 0.3$, and $c = 1371.9 \pm 0.4$ pm and $\beta = 130.740° \pm 0.004°$ and a calculated density of 2.042 g·cm$^{-3}$ (Nawaz et al. 2010b). For its preparation, to 1.0 mM of CdI$_2$ in 10 mL of H$_2$O two equivalents of tetramethylthiourea in 15 mL of MeOH was added. A clear solution was obtained that was stirred for 30 min. The colorless solution was filtered and the filtrate was kept at room temperature for crystallization. As a result, a white crystalline product was obtained, which was washed with MeOH and dried.

**[Cd(C$_{12}$H$_{17}$N$_5$S)I$_2$] or C$_{12}$H$_{17}$N$_5$I$_2$SCd**. This compound melts at 233°C (Castiñeiras et al. 2002). To obtain it, a solution of C$_{12}$H$_{17}$N$_5$S [2-pyridineformamide $N$(4)-piperidylthiosemicarbazone] (0.76 mM) in EtOH (30 mL) was added to a solution of CdI$_2$ (0.76 mM) in 30 mL of 96% EtOH, and the mixture kept stirring for several days at room temperature. The resulting yellow solid was filtered off, washed thoroughly with cold ethanol, and stored in a desiccator over Mg and I$_2$.

**[Cd(C$_{13}$H$_{19}$N$_5$S)I$_2$] or C$_{13}$H$_{19}$N$_5$I$_2$SCd**. To obtain this compound, a solution of CdI$_2$ (0.90 mM) in EtOH (30 mL) was mixed with a solution of 2-pyridineformamide 3-hexamethyleneiminylthiosemicarbazone (C$_{13}$H$_{19}$N$_5$S) (0.90 mM) in EtOH (30 mL) (Bermejo et al. 2004b). The resulting solution was stirred at room temperature for 7 days and the white precipitate was filtered off.

**[Cd(C₇H₅NS₂)₂I₂] or C₁₄H₁₀N₂I₂S₄Cd.**

[Cd(C$_7$H$_5$NS$_2$)$_2$I$_2$] or C$_{14}$H$_{10}$N$_2$I$_2$S$_4$Cd. This compound melts at 203°C–205°C (Bell et al. 2004). To obtain it, a warm solution of CdI$_2$ (4.25 mM) in EtOH (25 mL) was added to a hot ethanolic solution (25 mL) of 1,3-benzothiazoline-2-thione (8.51 mM) with stirring. The resultant cloudy solution was allowed to cool, resulting in the formation of yellow powder, which was filtered off and dried in vacuo.

**[Cd(C$_7$H$_{14}$N$_2$S)$_2$I$_2$] or C$_{14}$H$_{28}$N$_4$I$_2$S$_2$Cd, diiodobis[(N,N-diethylimidazolidine-2-thione-S)]zinc.** This compound melts at 173°C with decomposition (Devillanova and Verani 1980). It was obtained by reacting CdI$_2$ with N,N-diethylimidazolidine-2-thione (1:2 molar ratio) in dimethoxypropane.

**[CdI$_2$·Bu$^n$$_4$C$_2$N$_2$S$_4$] or C$_{18}$H$_{36}$N$_2$I$_2$S$_4$Cd.** This compound was prepared as a white precipitate by adding a solution of I$_2$ in CS$_2$ to a solution of cadmium di-$n$-butyldithiocarbamate in the same solvent at room temperature (Brinkhoff et al. 1969). After filtration, the product could be purified by dissolution in chloroform and reprecipitation by petroleum ether. The alternative route includes using CdI$_2$ and tetra-$n$-butylthiuram disulfide. To a suspension of CdI$_2$ in CS$_2$ a solution of tetra-$n$-butylthiuram disulfide was added at room temperature. After CdI$_2$ had dissolved, petroleum ether was added, precipitating the crystals, which were filtered and dried.

**[Bu$^n$$_4$N][CdI$_2${(S$_2$CNEt$_2$)$_2$}] or C$_{21}$H$_{46}$N$_2$I$_2$S$_2$Cd.** This compound was prepared by the next procedure (McCleverty et al. 1982). To a solution of [CdI$_2${(S$_2$CNEt$_2$)$_2$}] (1 mM) in acetone (25 mL) either [Bu$^n$$_4$N][S$_2$CNEt$_2$] (1 mM) or [Bu$^n$$_4$N][C$_7$H$_4$NS$_2$] (0.1 mM) in the same solvent (25 mL) was added. Upon evaporation of the mixture to low bulk followed by the addition of diethyl ether, the title compound was precipitated as a white powder. It was collected by filtration, washed with diethyl ether, and dried in vacuo.

**(Et$_4$N)$_2$[(μ-SEt)$_6$(CdI)$_4$] or C$_{28}$H$_{70}$N$_2$I$_4$S$_6$Cd$_4$.** To obtain this compound, a mixture of CdI$_2$ (0.98 mM), Et$_4$NI (2.0 mM), and Cd(SEt)$_2$ (3.3 mM) was stirred in CH$_2$Cl$_2$ (20 mL) for 1 h (Dean et al. 1992). The initial three-component mixture remained milky and the small amount of insoluble material was removed by filtration. The volume of the filtrate was reduced to ca. 4 mL by gentle heating under a flow of N$_2$, and then Et$_2$O (2 mL) was layered on. The mixture was allowed to stand in the refrigerator overnight. The white solid that formed was separated by filtration, washed with Et$_2$O, and dried in vacuo at room temperature.

The title compound could also be obtained if into a solution of CdI$_2$ (4.0 mM) and Et$_4$NI (2.0 mM) in EtOH (50 mL) at 40°C EtSH (8.1 mM) followed by Et$_3$N (8.0 mM) was stirred (Dean et al. 1992). A white precipitate formed. This dissolved when MeCN (5 mL) was added to the mixture at room temperature. Cooling the resulting solution to 5°C overnight gave the product as white crystals.

**[(Et$_4$N)$_2$(μ-SPr$^n$)$_6$(CdI)$_4$] or C$_{34}$H$_{82}$N$_2$I$_4$S$_6$Cd$_4$.** To obtain this compound, a stoichiometric mixture of CdI$_2$, Et$_4$NI, and Cd(SPr$^n$$_2$) in CH$_2$Cl$_2$ was stirred at room temperature (Dean and Manivannan 1990). The colorless product obtained by refrigeration of the reaction mixture overnight was separated by filtration, washed with Et$_2$O, and dried in vacuo. All synthesis and preparation of samples were carried out under an Ar or N$_2$ atmosphere.

**[(Et$_4$N)$_2$(µ-SPr$^i$)$_6$(CdI)$_4$] or C$_{34}$H$_{82}$N$_2$I$_4$S$_6$Cd$_4$.** To obtain this compound, a mixture of CdI$_2$ (0.98 mM), Et$_4$NI (2.0 mM), and Cd(SPr$^i$)$_2$ (3.3 mM) was stirred in CH$_2$Cl$_2$ (20 mL) for 1 h (Dean et al. 1992). The small amount of insoluble material was removed by filtration. The volume of the filtrate was reduced to ca. 4 mL by gentle heating under a flow of N$_2$, and then Et$_2$O (2 mL) was layered on. The mixture was allowed to stand in the refrigerator overnight. The white solid that formed was separated by filtration, washed with Et$_2$O, and dried in vacuo at room temperature.

**(Et$_4$N)$_2$[(µ-SBu$^n$)$_6$(CdI)$_4$] or C$_{40}$H$_{94}$N$_2$I$_4$S$_6$Cd$_4$.** To obtain this compound, a mixture of CdI$_2$ (0.98 mM), Et$_4$NI (2.0 mM), and Cd(SBu$^n$)$_2$ (3.3 mM) was stirred in CH$_2$Cl$_2$ (20 mL) for 1 h (Dean et al. 1992). The small amount of insoluble material was removed by filtration. The volume of the filtrate was reduced to ca. 4 mL by gentle heating under a flow of N$_2$, and then Et$_2$O (2 mL) was layered on. The mixture was allowed to stand in the refrigerator overnight. The white solid that formed was separated by filtration, washed with Et$_2$O, and dried in vacuo at room temperature.

**(Me$_4$N)$_2$[Cd$_4$(SPh)$_6$I$_4$] or C$_{44}$H$_{54}$N$_2$I$_4$S$_6$Cd$_4$.** To obtain this compound, solid iodine (1.2 mM) was added to a stirred solution of (Me$_4$N)$_2$[Cd$_4$(SPh)$_{10}$] (0.6 mM) in 15 mL of acetone (Dean et al. 1987). Immediate decolorization occurred, leaving a clear solution. Acetone was removed under vacuum. The residue was extracted with three 25 mL portions of diethyl ether to remove (PhS)$_2$. The remaining white solid was purified by recrystallization from MeCN/Et$_2$O. The syntheses were performed under an inert atmosphere. Colorless crystals were obtained by recrystallization from C$_6$H$_{12}$/Me$_2$CO after removal of (PhS)$_2$ by extraction with Et$_2$O (Dean and Vittal 1985).

**(Et$_4$N)$_2$[(µ-SBz)$_6$(CdI)$_4$] or C$_{58}$H$_{82}$N$_2$I$_4$S$_6$Cd$_4$.** To obtain this compound, a mixture of CdI$_2$ (0.98 mM), Et$_4$NI (2.0 mM), and Cd(SBz)$_2$ (3.3 mM) was stirred in CH$_2$Cl$_2$ (20 mL) for 1 h (Dean et al. 1992). The small amount of insoluble material was removed by filtration. The volume of the filtrate was reduced to ca. 4 mL by gentle heating under a flow of N$_2$, and then Et$_2$O (2 mL) was layered on. The mixture was allowed to stand in the refrigerator overnight and oily white solids were obtained. These solids were separated by decantation, washed several times with pentane, and dried in vacuo to give white foamy solids.

## 4.55 Cd–H–C–N–Fe–S

Some compounds are formed in this system.

**[Cd$_{32}$S$_{14}$(SPh)$_{38}$Fe(2,2'-bipy)$_3$] or C$_{258}$H$_{214}$N$_6$FeS$_{52}$Cd$_{32}$.** This compound crystallizes in the orthorhombic structure with the lattice parameters $a = 4637.4 \pm 0.2$, $b = 5158.1 \pm 0.2$, and $c = 5340.7 \pm 0.2$ pm at 90–150 K (Zheng et al. 2006b).

**[Cd$_{32}$S$_{14}$(SPh)$_{38}$Fe(Phen)$_3$] or C$_{264}$H$_{214}$N$_6$FeS$_{52}$Cd$_{32}$.** This compound crystallizes in the monoclinic structure with the lattice parameters $a = 4825.0 \pm 0.2$, $b = 4865.0 \pm 0.2$, and $c = 3565.2 \pm 0.2$ pm and $\beta = 131.386° \pm 0.002°$ at 90–150 K (Zheng et al. 2006b). To obtain it, a solution containing 67 mg of FeCl$_2$, 283 mg of Phen, and 25.122 g of CH$_3$CN was prepared. 2.557 g of this solution, 85 mg of Cd(SPh)$_2$, 41 mg of thiourea, and 0.508 g of water was placed in a 23 mL Teflon-lined stainless steel autoclave and stirred for ca. 20 min. The vessel was then sealed and heated at 130°C for 5 days. After cooling to room temperature, red crystals of the title compound were obtained.

[$Cd_{32}S_{14}(SPh)_{40}\{Fe(Phen)_3\}_2$] or $C_{312}H_{248}N_{12}Fe_2S_{54}Cd_{32}$. This compound crystallizes in the triclinic structure with the lattice parameters $a = 2093.25 \pm 0.04$, $b = 4634.39 \pm 0.09$, and $c = 3421.72 \pm 0.07$ pm and $\alpha = 76.220° \pm 0.030°$, $\beta = 92.061° \pm 0.001°$, and $\gamma = 84.380° \pm 0.030°$ at 90–150 K (Zheng et al. 2006b).

[$Cd_{32}S_{14}(SPh)_{40}\{Fe(3,4,7,8-Me_4Phen)_3\}_2$] or $C_{336}H_{296}N_{12}Fe_2S_{54}Cd_{32}$. This compound crystallizes in the monoclinic structure with the lattice parameters $a = 2158.8 \pm 0.4$, $b = 2232.3 \pm 0.4$, and $c = 4384.4 \pm 0.9$ pm and $\beta = 78.61° \pm 0.03°$ at 90–150 K (Zheng et al. 2006b).

## 4.56   Cd–H–C–N–Ru–S

[$Cd_{32}S_{14}(SPh)_{38}Ru(Phen)_3$] or $C_{264}H_{214}N_6RuS_{52}Cd_{32}$. This multinary compound is formed in this system. It crystallizes in the monoclinic structure with the lattice parameters $a = 4844.8 \pm 0.6$, $b = 4889.9 \pm 0.7$, and $c = 3585.1 \pm 0.5$ pm and $\beta = 131.502° \pm 0.003°$ at 90–150 K (Zheng et al. 2006b).

## 4.57   Cd–H–C–P–O–S

Some compounds are formed in this system.

[$Cd\{S_2P(OMe)_2\}_2$] or $C_4H_{12}P_2O_4S_4Cd$. This compound crystallizes in the orthorhombic structure with the lattice parameters $a = 928.2 \pm 0.4$, $b = 1783.7 \pm 0.5$, and $c = 848.8 \pm 0.4$ pm and a calculated density of 2.017 g·cm$^{-3}$ (Ito and Otake 1996). To obtain it, powdered $P_2S_5$ (50 mM) was added to 70 mL of MeOH and heated to 50°C–60°C. After ca. 1 h, when the generation of $H_2S$ vapor ceased, powdered $CdSO_4$ (50 mM) was added to the solution to give a white precipitate of the title compound. Its colorless crystals were obtained by recrystallization of the precipitate from an acetone solution.

[$Cd\{S_2P(OPr^n)_2\}_2$] or $C_{12}H_{28}P_2O_4S_4Cd$. This compound was prepared by the following procedure. To a solution of $Cd(ClO_4)\cdot6H_2O$ (0.50 mM) in 5 mL of $H_2O$ di-$n$-propyldithiophosphate (1 mM) in $H_2O$ (25 mL) was added dropwise under vigorous stirring (Yin et al. 2003). The addition lasted for 20 min. After the time period, white precipitate was collected and washed by water several times. The solid was dried in vacuo and then redissolved in MeOH to recrystallize. Three days or so later, needle crystals were harvested.

[$Cd\{S_2P(OPr^i)_2\}_2$] or $C_{12}H_{28}P_2O_4S_4Cd$. This compound crystallizes in the monoclinic structure with the lattice parameters $a = 1096.4 \pm 0.6$, $b = 1690.6 \pm 0.8$, and $c = 2649.0 \pm 0.8$ pm and $\beta = 99.91° \pm 0.02°$ and the experimental and calculated densities $1.46 \pm 0.02$ and $1.480 \pm 0.002$ g·cm$^{-3}$, respectively (Lawton and Kokotailo 1969). It was obtained in the same way as $Cd[S_2P(OPr^n)_2]_2$ was prepared using di-$i$-propyldithiophosphate instead of di-$n$-propyldithiophosphate (Yin et al. 2003).

Soft, colorless, tabular crystals were obtained by recrystallization of the compound from warm absolute EtOH (Lawton and Kokotailo 1969).

The title compound could also be obtained from the reaction of $CdSO_4 \cdot H_2O$ and $NH_4[S_2P(OPr^i)_2]$ in aqueous solution (Lai and Tiekink 2006) or $Cd(NO_3)_2 \cdot 6H_2O$ and $HS_2P(OPr^i)_2$ in $Pr^iOH$ (McCleverty et al. 1982).

**[Cd{S$_2$P(OBu$^n$)$_2$}$_2$] or C$_{16}$H$_{36}$P$_2$O$_4$S$_4$Cd.** This compound was prepared by the following procedure. To a solution of $Cd(ClO_4)_2 \cdot 6H_2O$ (0.50 mM) in 5 mL of $H_2O$ di-$n$-butyldithiophosphate (1 mM) in $H_2O$ (25 mL) was added dropwise under vigorous stirring (Yin et al. 2003). The addition lasted for 20 min. After the period, white precipitate was collected and washed by water several times. The solid was dried in vacuo and then redissolved in MeOH to recrystallize. Three days or so later, needle crystals were harvested.

**[Cd{S$_2$P(OBu$^i$)$_2$}$_2$] or C$_{16}$H$_{36}$P$_2$O$_4$S$_4$Cd.** This compound crystallizes in the monoclinic structure with the lattice parameters $a = 3003.3 \pm 0.5$, $b = 1031.9 \pm 0.2$, and $c = 1922.2 \pm 0.3$ pm and $\beta = 111.570° \pm 0.11°$ (Yin et al. 2003). It was obtained in the same way as $Cd[S_2P(OBu^n)_2]_2$ was prepared using di-$i$-butyldithiophosphate instead of di-$n$-butyldithiophosphate.

**[Cd{MeOC$_6$H$_4$P(OC$_5$H$_9$)S$_2$}$_2$] or C$_{24}$H$_{32}$P$_2$O$_4$S$_4$Cd.** This compound melts at 200°C–201°C and crystallizes in the monoclinic structure with the lattice parameters $a = 1118.1 \pm 0.4$, $b = 2480.4 \pm 0.3$, and $c = 1150.1 \pm 0.2$ pm and $\beta = 114.60° \pm 0.02°$ (Karakus et al. 2004). It was prepared by the interaction of $Cd(OAc)_2 \cdot 2H_2O$ and $MeOC_6H_4P(OC_5H_9)(S)(SNH_4)$ in dry EtOH. Colorless crystals were obtained from a mixture of chloroform/MeOH (1:1 volume ratio).

**[Cd(S$_2$COPr$^i$)$_2$PPh$_3$] or C$_{26}$H$_{29}$PO$_2$S$_4$Cd.** This compound crystallizes in the monoclinic structure with the lattice parameters $a = 1016.5 \pm 0.2$, $b = 1822.5 \pm 0.4$, and $c = 1632.4 \pm 0.2$ pm and $\beta = 103.39° \pm 0.01°$ and the experimental and calculated densities 1.42 and 1.456 g·cm$^{-3}$, respectively (Abrahams et al. 1986).

**[Cd(S$_2$COPr$^i$)$_2$P(c-C$_6$H$_{11}$)$_3$] or C$_{26}$H$_{47}$PO$_2$S$_4$Cd.** Cadmium bis(isopropylxanthate) $[Cd(S_2COPr^i)_2]$ reacts with tricyclohexylphosphine $[P(c-C_6H_{11})_3]$ to give the title compound (Abrahams et al. 1986).

**[Ph$_4$P][Cd(SC{O}Me)$_3$] or C$_{30}$H$_{29}$PO$_3$S$_3$Cd.** This compound crystallizes in the monoclinic structure with the lattice parameters $a = 1071.95 \pm 0.01$, $b = 1305.61 \pm 0.02$, and $c = 1121.99 \pm 0.02$ pm and $\beta = 100.701° \pm 0.001°$ (Sampanthar et al. 1999). It was obtained by the following procedure. Thioacetic acid (28 mM) was added dropwise to a stirred aqueous solution of $Et_3N$ (28 mM) to obtain a solution of $Et_3NHMe\{O\}CS$. To this was added $CdCl_2 \cdot H_2O$ (9 mM) in $H_2O$ (5 mL). The resulting solution was stirred for 10 min. When a solution of $Ph_4PCl$ (9 mM) in $H_2O$ was added, a cream-colored precipitate was formed. The mixture was stirred for 1 h, after which $CH_2Cl_2$ (8 mL) was added. The precipitate in the aqueous layer was extracted into the $CH_2Cl_2$ layer, which became yellow. The $CH_2Cl_2$ was separated and washed with 30 mL portions of deionized water 3 or 4 times to remove the side product, $Et_3NHCl$. The $Et_2O$ was added to the $CH_2Cl_2$ solution until turbidity was noticed, when the mixture was set aside in a refrigerator overnight, to obtain yellow crystals. They were decanted, washed with $Et_2O$, and dried in vacuo.

**(Ph$_4$P)[Cd(SOCPh)$_3$]** or **C$_{45}$H$_{35}$PO$_3$S$_3$Cd**. This compound has two polymorphic modifications. α-(Ph$_4$P)[Cd(SOCPh)$_3$] crystallizes in the monoclinic structure with the lattice parameters $a = 1334.4 \pm 0.2$, $b = 1477.6 \pm 0.2$, and $c = 2073.1 \pm 0.3$ pm and $\beta = 92.57° \pm 0.01°$ and the experimental and calculated densities $1.40 \pm 0.05$ and 1.40 g·cm$^{-3}$, respectively (Dean et al. 1998). β-(Ph$_4$P)[Cd(SOCPh)$_3$] crystallizes in the rhombohedral structure with the lattice parameters $a = 3555.5 \pm 0.6$ and $c = 1964.3 \pm 0.6$ (in hexagonal setting) and a calculated density of 1.40 g·cm$^{-3}$. The crystalline orange-yellow product obtaining during the literature synthesis of this compound proved to be β-form. A paler-yellow α-form was obtained in crystalline form by layering Et$_2$O onto a solution of β-form in CH$_2$Cl$_2$/MeOH and allowing diffusion to occur at room temperature.

To obtain this compound, solid PhCOSH (17.0 mM) was added to Et$_3$N (17.0 mM) in 20 mL of MeOH (Vittal and Dean 1996). To this mixture was added a solution of Cd(NO$_3$)$_2$·4H$_2$O (5.67 mM) in 15 mL of H$_2$O, producing a pale-yellow solution with some yellow oil on the bottom. A light-yellowish precipitate was obtained upon the addition of Ph$_4$PBr (5.67 mM) in 20 mL of MeOH. The addition of MeCN (20 mL) plus 5 mL of CH$_2$Cl$_2$ produced a clear yellow solution, which was left at 5°C overnight. The bright-yellow crystals so produced were isolated by decantation and then washed and dried. The washings were combined with the filtrate and allowed to evaporate at room temperature to get a second crop of crystals. All preparations were carried out under an Ar atmosphere.

**[Ph$_4$P]$_2$[Cd(1,2-S$_2$C$_6$H$_{10}$)$_2$]·4H$_2$O** or **C$_{60}$H$_{68}$P$_2$O$_4$S$_4$Cd**. This compound crystallizes in the triclinic structure with the lattice parameters $a = 1371.6 \pm 0.1$, $b = 1830.6 \pm 0.1$, and $c = 1333.0 \pm 0.1$ pm and $\alpha = 110.511° \pm 0.006°$, $\beta = 111.133° \pm 0.006°$, and $\gamma = 69.290° \pm 0.005°$ (Govindaswamy et al. 1992). The reaction of CdCl$_2$ with 3 equivalents of the lithium salt of 1,2-*trans*-cyclohexanedithiolate and 2 equivalents of Ph$_4$PBr in H$_2$O gives crystals of [Ph$_4$P]$_2$[Cd(1,2-C$_6$H$_{10}$)$_2$]·4H$_2$O.

## 4.58   Cd-H-C-P-O-Se-F-S

**[Cd{SeP(c-C$_6$H$_{11}$)$_3$}$_4$(O$_3$SCF$_3$)$_2$]** or **C$_{74}$H$_{132}$P$_4$O$_6$Se$_4$F$_6$S$_2$Cd**. This multinary compound is formed in this system. To obtain it, to a solution of (c-C$_6$H$_{11}$)$_3$PSe (0.700 mM) in 3 mL of the mixture Me$_2$CO/CH$_2$Cl$_2$ (1:1 volume ratio) solid Cd(O$_3$SCF$_3$)$_2$ (0,161 mM) was added, with stirring (Rajalingam et al. 2000). When the cadmium salt had dissolved, Et$_2$O (ca. 4 mL) was layered onto the mixture, which was then refrigerated (–15°C) overnight. The white crystals of the title compound were separated by filtration, washed with ice-cold Et$_2$O, and vacuum dried. All manipulations were carried out under an atmosphere of dry N$_2$ in a glovebox or with vacuum/N$_2$ manifold or under an atmosphere of Ar in a glovebox.

## 4.59   Cd-H-C-P-Se-Cl-S

Some compounds are formed in this system.

**[Cd$_4$(SPr$^j$)$_6$(PPh$_3$)$_2$(ClO$_4$)$_2$]** or **C$_{54}$H$_{72}$P$_3$O$_8$Cl$_2$S$_6$Cd$_4$**. To prepare this compound, a mixture of Cd(PPh$_3$)$_2$(ClO$_4$)$_2$ (0.5 mM), Cd(SPr$^j$)$_2$ (1.5 mM), and PPh$_3$ (2.0 mM) in $CH_2Cl_2$ (10 mL) was stirred for 20 min (Dean et al. 1993). A small amount of gelatinous insoluble material was then removed by filtration. The addition of $Et_2O$ (2.5 mL) to the filtrate resulted in a white crystalline product after the $CH_2Cl_2/Et_2O$ mixture was refrigerated overnight. The product was separated by filtration, washed with $Et_2O$, and dried in vacuo.

**$[Cd_4(SPr^i)_6(PPh_3)_2(ClO_4)_2]\cdot EtOH$ or $C_{56}H_{78}P_3O_8Cl_2S_6Cd_4$.** This compound crystallizes in the monoclinic structure with the lattice parameters $a = 1500.6 \pm 0.2$, $b = 1625.0 \pm 0.2$, and $c = 2822.7 \pm 0.3$ pm and $\beta = 98.79° \pm 0.01°$ at $233 \pm 2$ K and a calculated density of 1.625 $g\cdot cm^{-3}$ (Dean et al. 1993). It was obtained by the recrystallization of $[Cd_4(SPr^i)_6(PPh_3)_2(ClO_4)_2]$ from ethanolic solution.

**$[Cd_4(SPr^n)_6(PPh_3)_3(ClO_4)_2]$ or $C_{72}H_{87}P_3O_8Cl_2S_6Cd_4$.** To prepare this compound, a mixture of $Cd(PPh_3)_2(ClO_4)_2$ (0.5 mM), $Cd(SPr^n)_2$ (1.5 mM), and $PPh_3$ (2.0 mM) in $CH_2Cl_2$ (10 mL) was stirred for 20 min (Dean et al. 1993). A small amount of gelatinous insoluble material was then removed by filtration. The addition of $Et_2O$ (2.5 mL) to the filtrate resulted in a white crystalline product after the $CH_2Cl_2/Et_2O$ mixture was refrigerated overnight. The product was separated by filtration, washed with $Et_2O$, and dried in vacuo.

**$[Cd_4(SC_6H_{11})_6(PPh_3)_2(ClO_4)_2]$ or $C_{72}H_{96}P_2O_8Cl_2S_6Cd_4$.** To prepare this compound, a mixture of $Cd(PPh_3)_2(ClO_4)_2$ (0.5 mM) and $Cd(SC_6H_{11})_2$ (1.5 mM) dissolved completely in $CH_2Cl_2$ (5 mL) when stirred for 1 h (Dean et al. 1993). The addition of $Et_2O$ (3 mL) and refrigeration of the $CH_2Cl_2/Et_2O$ mixture overnight gave the title compound in a white microcrystalline form.

**$[Cd_4(SPh)_6(PPh_3)_4(ClO_4)_2]$ or $C_{108}H_{90}P_4O_8Cl_2S_6Cd_4$.** To prepare this compound, a mixture of $Cd(PPh_3)_2(ClO_4)_2$ (0.5 mM), $Cd(SPh)_2$ (1.5 mM), and $PPh_3$ (1.0 mM) in $CH_2Cl_2$ (10 mL) was stirred for 20 min (Dean et al. 1993). A small amount of gelatinous insoluble material was then removed by filtration. The addition of $Et_2O$ (6 mL) to the filtrate resulted in a white crystalline product after the $CH_2Cl_2/Et_2O$ mixture was refrigerated overnight. The product was separated by filtration, washed with $Et_2O$, and dried in vacuo.

## 4.60   Cd–H–C–P–S

Some compounds are formed in this system.

**$[Cd(Et_2PS_2)_2]$ or $C_8H_{20}P_2S_4Cd$.** This compound crystallizes in the orthorhombic structure with the lattice parameters $a = 1206.6 \pm 0.1$, $b = 2032.0 \pm 0.2$, and $c = 1381.3 \pm 0.2$ pm and the experimental and calculated densities 1.6 and 1.64 $g\cdot cm^{-3}$, respectively (Wunderlich 1986) ($a = 1195.9 \pm 0.4$, $b = 2023.5 \pm 1.3$, and $c = 1374.7 \pm 0.5$ pm at 173 K and a calculated density of $1.673 \pm 0.001$ $g\cdot cm^{-3}$ [Swensson and Albertsson 1991]). The crystallization of the title compound from a solution in chloroform yielded large crystals containing cavities and inclusions of the solvent. Recrystallization from $Bu^nOH$ gave well-shaped transparent crystals. Its single crystals were obtained by slow evaporation from a dioxane solution, yielding colorless crystals with an approximately spherical habit (Swensson and Albertsson 1991).

**[Cd(Bu$^i_2$PS$_2$)$_2$]$_2$ or C$_{16}$H$_{36}$P$_2$S$_4$Cd**. This compound melts at 125°C (at 83°C–86°C [Larionov et al. 2001]) and crystallizes in the monoclinic structure with the lattice parameters $a = 1114.0 \pm 0.2$, $b = 1207.0 \pm 0.2$, and $c = 1883.6 \pm 0.3$ pm and $\beta = 92.54° \pm 0.01°$ and a calculated density of 1.394 g·cm$^{-3}$ (Byrom et al. 2000). To prepare it, CdSO$_4$·3/8H$_2$O (11 mM) in aqueous solution was reacted with Na(Bu$^i_2$PS$_2$) (22 mM) in aqueous solution. The white precipitate that formed was filtered under vacuum, washed with distilled H$_2$O, and recrystallized from acetone to give transparent cubic crystals of the title compound.

The title compound could also be prepared as follows. To a solution of CdCl$_2$·2H$_2$O (30 mM) in H$_2$O (170 mL) a mixture of 50% aqueous solution (24 mL) of Na(Bu$^i_2$PS$_2$) (6 mM) in H$_2$O (70 mL) was added (Larionov et al. 2001). A pH value of 3–4 was maintained by the dropwise addition of HCl. White precipitate was filtered with suction, washed with H$_2$O, and dried in air. [Cd(Bu$^i_2$PS$_2$)$_2$]$_2$ was recrystallized from the mixture CHCl$_3$/MeOH (1:5 volume ratio).

[Cd(Bu$^i_2$PS$_2$)$_2$]$_2$ could be used as precursor to grow thin films of CdS (Byrom et al. 2000).

**[Cd(SPh)$_2$·depe] or C$_{22}$H$_{34}$P$_2$S$_2$Cd [depe = 1,2-bis(diethylphosphino)ethane]**. This compound melts at 129.9°C–131.3°C and crystallizes in the orthorhombic structure (Brennan et al. 1990). To obtain it, depe (0.97 mM) was added to Cd(SPh)$_2$ (0.39 mM) suspended in toluene (20 mL). The colorless solution was saturated with heptane (5 mL) and cooled (–10°C) to give colorless diamonds. All reactions were performed in an inert atmosphere.

**[Cd(Ph$_2$PS$_2$)$_2$] or C$_{24}$H$_{20}$P$_2$S$_4$Cd**. This compound melts at 170°C with decomposition and crystallizes in the triclinic with the lattice parameters $a = 927.3 \pm 0.1$, $b = 1042.6 \pm 0.1$, and $c = 1384.5 \pm 0.3$ pm and $\alpha = 94.38° \pm 0.01°$, $\beta = 103.16° \pm 0.01°$, and $\gamma = 104.63° \pm 0.01°$ and a calculated density of 1.625 g·cm$^{-3}$ (Casas et al. 1994). To obtain it, a solution of Cd(ClO$_4$)$_2$·6H$_2$O (1.3 mM) in EtOH (5 mL) was slowly added, with stirring, to NH$_4$S$_2$PPh$_2$ (2.6 mM) dissolved in the same solvent (10 mL). The white solid formed after stirring was filtered out, washed with EtOH, and dried under vacuum. Single crystals suitable for XRD studies were grown by recrystallization from DMSO.

**[(Ph$_4$P)(CdSPh$_3$)]$_2$ or C$_{42}$H$_{35}$PS$_3$Cd**. This compound crystallizes in two polymorphic modifications (Ali et al. 2002). One of them is triclinic with the lattice parameters $a = 1099.6 \pm 0.4$, $b = 1290.3 \pm 0.5$, and $c = 1427.6 \pm 0.6$ pm and $\alpha = 73.33° \pm 0.03°$, $\beta = 72.54° \pm 0.03°$, and $\gamma = 88.86° \pm 0.02°$ and a calculated density of 1.40 g·cm$^{-3}$. To prepare this polymorph, solid Ph$_4$PBr (5.3 mM) was added to a stirred solution of PhSH (5.3 mM) and Et$_3$N (5.3 mM) in 20 mL of MeCN. To the resulting clear solution was added in one portion Cd(NO$_3$)$_2$·4H$_2$O (2 mM) in 40 mL of EtOH, whereupon the solution turned to pale yellow. When the solution was cooled (0°C), colorless crystals were obtained, filtered, washed well with small portions of EtOH and finally with diethyl ether, and then dried in vacuo. It was found that many different crystallizations yield the title compound as the monoclinic modification.

For the obtaining of the monoclinic modification of the title compound, the following three procedures were used (Ali et al. 2002):

1. A solution of $Cd(NO_3)_2 \cdot 4H_2O$ (34 mM) in 30 mL of MeOH was added to a well-stirred solution of PhSH (91.1 mM) and $Et_3N$ (91.1 mM) in 30 mL of MeOH, yielding a white solid during the addition. With stirring the solid dissolved, forming a clear colorless solution, which was treated with a 30 mL MeOH solution of $Ph_4PBr$ (38.2 mM). After completion of the addition, a small amount of white solid had precipitated. The reaction mixture was stirred for 20 min and the solution became clear and colorless. It was allowed to stand undisturbed at room temperature, yielding colorless crystals that were filtered, washed with MeOH, and dried under vacuum.

2. A solution containing PhSH (4.5 mM), $Et_3N$ (4.5 mM), and $Ph_4PBr$ (6.2 mM) in 40 mL of acetone was prepared. $Cd(NO_3)_2 \cdot 4H_2O$ (1.0 mM) in 8 mL of EtOH was added to this colorless solution, yielding a white solid. An additional 30 mL of EtOH was added and the mixture heated to 60°C–70°C for 3 min. The reaction mixture was filtered and the solid was redissolved in warm acetonitrile and then kept at 0°C. After 15 min colorless crystals were obtained. The product was washed with portions of EtOH and dried under vacuum.

3. A well-stirred solution of PhSH (3 mM) and $Et_3N$ (3 mM) in 20 mL of MeOH was treated with solid $CdCl_2$ (0.91 mM), followed by the addition of $Ph_4PBr$ (1.0 mM). The mixture was then stirred with warming until a clear colorless solution was obtained. It was allowed to stand undisturbed to cool at room temperature and then stored at <0°C. Colorless crystals formed within 2 days and were filtered, washed with small portions of cold acetonitrile, and dried in vacuo.

**$(Ph_4P)_2[Cd(S_5)_2]$ or $C_{48}H_{40}P_2S_{10}Cd$.** To obtain this compound, to a solution of $(Ph_4P)$ $[Cd(SPh)_4]$ (1.02 mM) in 25 mL of $CH_3CN$ solid BzSSSBz (Bz = 3-quinuclidinyl benzilate) (10.80 mM) in small portions was added with stirring (Coucouvanis et al. 1985). After the mixture was stirred for ca. 5 min, diethyl ether was added to incipient nucleation. After several hours yellow crystals formed and were isolated.

**$[(Ph_4P)_2\{Cd(S_6)_2\}]$ or $C_{48}H_{40}P_2S_{12}Cd$.** This compound melts at 188°C–190°C and crystallizes in the triclinic structure with the lattice parameters $a = 1184.9 \pm 0.6$, $b = 1199.0 \pm 0.6$, and $c = 1985.6 \pm 1.0$ pm and $\alpha = 92.05° \pm 0.03°$, $\beta = 92.88° \pm 0.03°$, and $\gamma = 111.88° \pm 0.02°$ and a calculated density of 1.50 g·cm$^{-3}$ (Banda et al. 1989). To prepare it, a mixture of $Na_2S \cdot 9H_2O$ (17.4 mM), $S_8$ (157.5 mM of S), and DMF (40 mL) was stirred at 70°C for 10 min under an atmosphere of $N_2$. The color changed from green to dark red after 30 s. To the red solution was added CdS (11.8 mM), and the mixture was stirred for a further 30 min until all CdS had dissolved. After the mixture was cooled to room temperature, a solution of $Ph_4PBr$ (24 mM) in acetonitrile (40 mL) was added. The large orange crystals that grew within 12 h were collected, washed with acetonitrile, and dried under $N_2$.

**$[(Ph_4P)_2\{Cd(S_6)(S_7)\}]$ or $C_{48}H_{40}P_2S_{13}Cd$.** This compound melts at 180°C and crystallizes in the triclinic structure with the lattice parameters $a = 1191.3 \pm 0.5$, $b = 1201.0 \pm 0.5$, and $c = 1978.5 \pm 0.8$ pm and $\alpha = 92.01° \pm 0.03°$, $\beta = 93.05° \pm 0.03°$, and $\gamma = 111.65° \pm 0.02°$ and a calculated density of 1.49 g·cm$^{-3}$ (Banda et al. 1989).

To prepare it, $Na_2S \cdot 9H_2O$ (52.1 mM) was dehydrated by heating under a stream of $N_2$ to obtain a white solid. CdS (34.7 mM) and $S_8$ (469 mM of S) were added and the mixture was treated with dry DMF (175 mL) and stirred at 90°C for 2 h, generating a very dark-red solution. The small amount of solid that remained was dissolved by heating the mixture briefly to 135°C. After the mixture was cooled, the small amount of residue was removed by filtration, and a solution of $Ph_4PBr$ (70 mM) in acetonitrile (250 mL) was added. The first rapidly precipitated product was separated and washed with acetonitrile. The mother liquor yielded dark-red crystals of the title compound during 2 days.

$(Ph_4P)_2[(PhS)_2Cd(S_5)]$ or $C_{60}H_{50}P_2S_7Cd$. For obtaining this compound, to a solution of $(Ph_4P)[Cd(SPh)_4]$ (1.02 mM) in 25 mL of $CH_3CN$ solid BzSSSBz was added (2.15 mM) in small portions with stirring (Coucouvanis et al. 1985). The color of the solution changed from an initial yellow to orange red. After the mixture was stirred for ca. 5 min, diethyl ether was added to incipient nucleation. After several hours yellow crystals formed and were isolated.

$[Cd(PhS)_2]_4PPh_3$ or $C_{66}H_{55}PS_8Cd_4$. The reaction of $Cd(SPh)_2$ (0.90 mM) and $Ph_3P$ (1.86 mM) in DMF (10 mL) at room temperature gave a clear solution (Black et al. 1986). The addition of MeOH gave a white precipitate of this compound, which was collected and dried in vacuo

$[Ph_4P][Cd(S-2,4,6-Pr^i_3C_6H_2)_3]$ or $C_{69}H_{89}PS_3Cd$. To obtain this compound, LiS-$2,4,6-Pr^i_3C_6H_2$ (4.7 mM), $CdCl_2$ (1.0 mM), and $Ph_4PBr$ (1.9 mM) were combined in a mixture of 5 mL of DMF and 50 mL of $Pr^iOH$, and the pale-yellow mixture was stirred for 2 h (Gruff and Koch 1990). This mixture was then cooled to −20°C overnight and the resultant crystalline white solid was filtered, washed with $Pr^iOH$, and dried.

$[Ph_4P]_2[Cd(SPh)_4]$ or $C_{72}H_{60}P_2S_4Cd$, bis(tetraphenylphosphonium) tetrakis(benzenethiolato)cadmate(II). This compound crystallizes in the orthorhombic structure with the lattice parameters $a = 1398.3 \pm 0.2$, $b = 1757.2 \pm 0.4$, and $c = 2490.6 \pm 0.4$ pm and a calculated density of 1.33 g·cm$^{-3}$ (Swenson et al. 1978). It was obtained in a $N_2$ atmosphere (Coucouvanis et al. 1982). To 20 mL of $CH_3CN$ in a 125 mL Erlenmeyer flask were added 3.2 mM of bis(O-ethyl dithiocarbonato) cadmium(II), 6.5 mM of $[Ph_4P]Cl$, and 14 mM of KSPh. The suspension was brought to the boiling point of $CH_3CN$ and boiled for 10 min. The mixture was then filtered hot through a medium-porosity sintered glass filter. The bright-yellow solution was allowed to cool. Upon standing for 30 min, bright-yellow crystals formed, which was removed by filtration and washed with 10 mL portions of absolute EtOH and twice with 10 mL portions of diethyl ether.

$[Ph_4P]_2[Cd_2(SPh)_6]$ or $C_{84}H_{70}P_2S_6Cd_2$. This compound crystallizes in the monoclinic structure with the lattice parameters $a = 1359.9 \pm 0.1$, $b = 1358.4 \pm 0.1$, and $c = 2065.5 \pm 0.2$ pm and $\beta = 106.06° \pm 0.01°$ and a calculated density of 1.411 g·cm$^{-3}$ (Abrahams and Garner 1987). To obtain this compound, a solution of $Cd(NO_3)_2 \cdot 4H_2O$ (3 mM) in MeOH (10 mL) was added to a solution of $[Ph_4P][SPh]$ (10 mM) in MeCN (10 mL), and an immediate white precipitate was obtained, which was redissolved on stirring and heating to ca. 70°C. The hot filtered solution was cooled to −20°C, to yield a colorless microcrystalline precipitate, which was filtered off, washed with MeOH (5 mL) and diethyl ether (30 mL), and dried in vacuo.

Alternatively, [Ph$_4$P][SPh] (5 mM) in MeCN (20 mL) was added to a suspension of Cd(SPh)$_2$ (5 mM) in MeCN (20 mL). Cd(SPh)$_2$ dissolved over a period of ca. 10 min. The resultant colorless solution was heated to 60°C and concentrated to a volume of ca. 10 mL under a reduced pressure. Slow cooling to room temperature produced colorless hexagonal-shaped crystals, which were washed with MeCN ($3 \times 5$ mL at $-20$°C) and diethyl ether (30 mL) and dried in vacuo.

## 4.61 Cd–H–C–P–Se–S

[(Ph$_4$P)$_2${SCd$_8$(SePh)$_{16}$}] or C$_{144}$H$_{120}$P$_2$Se$_{16}$SCd$_8$. This multinary compound is formed in this system. To obtain it, a deoxygenated solution of NaSePh in EtOH/acetonitrile was treated with solutions of CdI$_2$ in acetonitrile, Na$_2$S in MeOH, and Ph$_4$PCl in acetonitrile, yielding a colorless solution (Lee et al. 1990). After stripping of all solvent, the colorless solids were extracted with H$_2$O at 80°C and EtOH at ambient temperature and then dissolved in acetonitrile at 20°C and filtered, and the title compound crystallized by storage at 0°C.

## 4.62 Cd–H–C–P–I–S

[(CH$_3$)$_3$PS]$_2$[CdI$_2$] or C$_6$H$_{18}$P$_2$I$_2$S$_2$Cd. This quinary compound is formed in this system. It was obtained by the direct reaction of trimethylphosphine sulfide, (CH$_3$)$_3$PS, with CdI$_2$ in absolute ethanol (Meek and Nicpon 1965).

## 4.63 Cd–H–C–As–O–S

Some compounds are formed in this system.

(Ph$_4$As)[Cd(SOCPh)$_3$] or C$_{45}$H$_{35}$AsO$_3$S$_3$Cd. This compound crystallizes in the triclinic structure with the lattice parameters $a = 1074.1 \pm 0.2$, $b = 1316.2 \pm 0.2$, and $c = 1580.9 \pm 0.2$ pm and $\alpha = 101.00° \pm 0.01°$, $\beta = 97.65° \pm 0.01°$, and $\gamma = 109.88° \pm 0.01°$ and the experimental and calculated densities $1.51 \pm 0.05$ and $1.502$ g·cm$^{-3}$, respectively (Vittal and Dean 1996). To obtain it, solid PhCOSH (1.63 mM) was added to Et$_3$N (16.3 mM) in 20 mL of MeOH. To this mixture was added a solution of Cd(NO$_3$)$_2$·4H$_2$O (2.0 mM) in 15 mL of H$_2$O, producing a pale-yellow solution with some yellow oil on the bottom. A light-yellowish precipitate was obtained upon the addition of Ph$_4$AsCl·H$_2$O (3.94 mM) in 20 mL of MeOH. The addition of MeCN (20 mL) plus 5 mL of CH$_2$Cl$_2$ produced a clear yellow solution, which was left at 5°C overnight. The bright-yellow crystals so produced were isolated by decantation and then washed and dried. The washings were combined with the filtrate and allowed to evaporate at room temperature to get a second crop of crystals. All preparations were carried out under an Ar atmosphere.

(Ph$_4$As)$_2$[Cd(MeC$_6$H$_3$S$_2$)$_2$]·2EtOH or C$_{66}$H$_{64}$As$_2$O$_2$S$_4$Cd. This compound melts at 230°C–232°C and crystallizes in the triclinic structure with the lattice parameters $a = 2303.2 \pm 0.4$, $b = 1239.6 \pm 0.4$, and $c = 2252.9 \pm 0.5$ and $\beta = 111.106° \pm 0.002°$

and the experimental and calculated densities 1.409 and 1.410 g·cm$^{-3}$, respectively (Bustos et al. 1983). It was prepared as follows. NaOH (0.8 g) dissolved in 30 mL of EtOH was mixed with a stirred solution of MeC$_6$H$_3$(SH)$_2$ in 15 mL of EtOH at room temperature, Cd(MeC$_6$H$_3$S$_2$) (1.03 g) added slowly, and the mixture stirred until complete dissolution took place (ca. 16 h). An ethanol solution of Ph$_4$As was then added, and after 5 min a yellow powder slowly precipitated. This material was collected, washed twice with EtOH, and redissolved in EtOH/MeCN, and the solution cooled overnight at 0°C. The resultant yellow crystals of the title compound were collected and dried in vacuo.

## 4.64    Cd–H–C–As–Se–S

[(Ph$_4$As)$_2${SCd$_8$(SePh)$_{16}$}] or C$_{144}$H$_{120}$As$_2$Se$_{16}$SCd$_8$. This multinary compound is formed in this system. To obtain it, a deoxygenated solution of NaSePh in EtOH/acetonitrile was treated with solutions of CdI$_2$ in acetonitrile, Na$_2$S in MeOH, and Ph$_4$AsCl in acetonitrile, yielding a colorless solution (Lee et al. 1990). After stripping of all solvent, the colorless solids were extracted with H$_2$O at 80°C and EtOH at ambient temperature and then dissolved in acetonitrile at 20°C and filtered, and the title compound crystallized by storage at 0°C.

## 4.65    Cd–H–C–O–S

Some compounds are formed in this system.

[Cd(S$_2$COMe)$_2$] or C$_4$H$_6$O$_2$S$_4$Cd. This compound crystallizes in the monoclinic structure with the lattice parameters $a = 1900.6 \pm 0.2$, $b = 397.28 \pm 0.03$, and $c = 1251.26 \pm 0.14$ pm and $\beta = 107.993° \pm 0.004°$ at $123 \pm 2$ K and a calculated density of 2.415 g·cm$^{-3}$ (Young and Tiekink 2002). Pale-yellow crystals were obtained from the slow evaporation of an acetone solution of the title compound.

[Cd{S(CH$_2$COO)$_2$}]·H$_2$O or C$_4$H$_6$O$_5$SCd. This compound crystallizes in the monoclinic structure with the lattice parameters $a = 800.9 \pm 0.1$, $b = 535.2 \pm 0.1$, and $c = 914.3 \pm 0.2$ pm and $\beta = 116.03° \pm 0.01°$ and a calculated density of 1.31 g·cm$^{-3}$ (Whitlow 1975). Cadmium thiodiacetate hydrate, prepared from stoichiometric mixture of Cd(NO$_3$)$_2$ and thiodiacetic acid in aqueous solution, crystallizes as pale-yellow prisms.

[Cd(EtOCS$_2$)$_2$] or C$_6$H$_{10}$O$_2$S$_4$Cd, cadmium ethylxanthate. This compound crystallizes in the monoclinic structure with the lattice parameters $a = 1128.9 \pm 1.0$, $b = 587.2 \pm 0.4$, and $c = 905.7 \pm 0.5$ pm and $\beta = 90.28° \pm 0.05°$ and the experimental and calculated densities 1.98 and 2.02 g·cm$^{-3}$, respectively (Iimura et al. 1972). The powder of the title compound was precipitated by mixing aqueous solutions of Cd(OAc)$_2$ and potassium ethylxanthate slowly in a molar ratio of 1: 2. The precipitate was filtered and dried in a vacuum desiccator. Crystals were obtained by slow evaporation of a methanol solution of the precipitate at room temperature.

Cd(EtOCS$_2$)$_2$ could also be prepared by the next procedure (Barreca et al. 2005). KOH (25.0 mM) was ground to a fine powder and placed in a flask where a slight excess of EtOH was added. The mixture was stirred until the solid was completely dissolved (ca. 2 h). After the addition of $CS_2$ (25 mM), the mixture was stirred for 1 h to give a pale-yellow solution. Deionized $H_2O$ (10 mL) was added, leading to the final formation of light-yellow needles of $K(S_2COEt)$. An aqueous solution of the obtained compound (22 mM in 30 mL of $H_2O$) was added to a solution of $CdCl_2$ (11 mM in 50 mL of $H_2O$). A white flocculent precipitate formed immediately. The mixture was stirred for 1 h, subsequently vacuum filtered, and finally dried to obtain a white powder. Because the compound obtained is slightly photosensitive, precautions were taken during preparation and storage to avoid light exposure. The title compound is a potential single-source molecular precursor for the CVD of CdS thin films.

**[Cd(Pr$^i$OCS$_2$)$_2$] or C$_8$H$_{14}$O$_2$S$_4$Cd.** This compound crystallizes in the orthorhombic structure with the lattice parameters $a = 2065.2 \pm 0.2$, $b = 889.34 \pm 0.06$, and $c = 759.17 \pm 0.05$ pm and a calculated density of 1.822 g·cm$^{-3}$ (Tomlin et al. 1999). It was obtained by the following procedure (Tomlin et al. 1999, Barreca et al. 2005). KOH (25.0 mM) was ground to a fine powder and placed in a flask where a slight excess of Pr$^i$OH was added. The mixture was stirred until the solid was completely dissolved (ca. 2 h). After the addition of $CS_2$ (25 mM), the mixture was stirred for 1 h to give a pale-yellow solution. Deionized $H_2O$ (10 mL) was added, leading to the final formation of light-yellow needles of $K(S_2COPr^i)$. An aqueous solution of the obtained compound (22 mM in 30 mL of $H_2O$) was added to a solution of $CdCl_2$ (11 mM in 50 mL of $H_2O$). A white flocculent precipitate formed immediately. The mixture was stirred for 1 h, subsequently vacuum filtered, and finally dried to obtain a white powder. Crystals were grown from the solution in THF (Tomlin et al. 1999). Because the compound obtained is slightly photosensitive, precautions were taken during preparation and storage to avoid light exposure. The title compound is potential single-source molecular precursor for the CVD of CdS thin films (Barreca et al. 2005).

**[Cd(SCH$_2$COOEt)$_2$] or C$_8$H$_{14}$O$_4$S$_2$Cd.** This compound melts at 172°C–173°C and crystallizes in the tetragonal structure with the lattice parameters $a = 1937.1 \pm 0.3$ and $c = 1376.4 \pm 0.3$ pm and the experimental and calculated densities 1.81 $\pm$ 0.02 and 1.80 g·cm$^{-3}$, respectively (Dance et al. 1983). To obtain it, a solution of Cd(NO$_3$)$_2$·4H$_2$O (19 mM) in warm absolute EtOH (30 mL) was added slowly to a stirred solution of ethyl-2-mercaptoacetate (40 mM) and Et$_3$N (40 mM) in absolute EtOH (50 mL). The mixture was sealed and allowed to crystallize during 3 days. The product was filtered, washed with MeOH, dried under vacuum, and recrystallized from CHCl$_3$/MeOH.

**[Cd(S$_2$COCH$_2$CH$_2$OMe)$_2$] or C$_8$H$_{14}$O$_4$S$_4$Cd, cadmium (methoxyethyl)xanthate.** This compound crystallizes in the monoclinic structure with the lattice parameters $a = 1271.2 \pm 0.2$, $b = 421.5 \pm 0.1$, and $c = 1340.0 \pm 0.2$ pm and $\beta = 104.60° \pm 0.01°$ and the experimental and calculated densities 2.01 and 1.983 g·cm$^{-3}$, respectively (Abrahams et al. 1988a). It was prepared from potassium (methoxyethyl)xanthate and CdCl$_2$. The crystals were obtained by the slow evaporation of an acetone solution of the title compound.

**C$_8$H$_{20}$O$_4$S$_4$Cd$_2$, cadmium bisthioglycolate.** This compound melts at 172°C and crystallizes in the monoclinic structure with the lattice parameters $a = 1006$, $b = 1981$, and $c = 874.4$ pm and $\beta = 107.37°$ (Bürgi 1974). To obtain it, $Cd(OAc)_2$ (11.3 mM) was dissolved in 30 mL of $H_2O$ and mixed with thioglycol (20 mM) dissolved in another 30 mL of $H_2O$. The mixture was stored in desiccator together with a beaker of 1% aqueous ammonia solution. At pH ~ 8 transparent, colorless crystals, which are stable in air, started to crystallize.

**$[Cd(MeC_6H_3S_2)(DMSO)]$ or $C_9H_{12}OS_3Cd$.** To prepare this compound, $Cd(MeC_6H_3S_2)$ was dissolved in 7 mL of DMSO, and the excess of the solvent removed under vacuum over several hours (Bustos et al. 1983). Brown heavy oil was collected and redissolved in chloroform. A dropwise addition of diethyl ether precipitated the title compound as a white powder.

**$[Cd(Bu^nOCS_2)_2]$ or $C_{10}H_{18}O_2S_4Cd$.** This compound crystallizes in the monoclinic structure with the lattice parameters $a = 1159 \pm 17$, $b = 584 \pm 9$, and $c = 2570 \pm 40$ pm and $\beta = 101°44' \pm 12'$ and the experimental and calculated densities $1.66 \pm 0.02$ and $1.61 \pm 0.6$ g·cm$^{-3}$, respectively (Rietveld and Maslen 1965). Its crystals had been grown by recrystallization from hexanol.

**$[Cd(PhCOS)_2 \cdot EtOH]_2$ or $C_{16}H_{16}O_3S_2Cd$.** This compound crystallizes in the monoclinic structure with the lattice parameters $a = 2687 \pm 2$, $b = 576.3 \pm 0.5$, and $c = 1100 \pm 1$ pm and $\beta = 101.03° \pm 0.09°$ and a calculated density of 1.72 g·cm$^{-3}$ (Musaev et al. 1978).

**$[Cd(Ph_2SOAc)_2] \cdot H_2O$ or $C_{16}H_{16}O_5S_2Cd$.** This compound crystallizes in the orthorhombic structure with the lattice parameters $a = 3382.6 \pm 0.8$, $b = 511.9 \pm 0.1$, and $c = 987.2 \pm 0.3$ pm and the experimental and calculated densities 1.87 and 1.806 g·cm$^{-3}$, respectively (Mak et al. 1984). It was prepared by reacting phenylthioacetic acid in aqueous ethanol with a suspension of excess $CdCO_3$ (in aqueous ethanol) and digesting for 1 h at 70°C–90°C. The excess carbonate was removed by filtration. The filtrate, on standing at room temperature, yielded crystals that were recrystallized from absolute ethanol as flat needles or plates.

**$[Cd_5(\mu_2\text{-}OH)_2(\mu_3\text{-}OH)_2\{O_3SC_6H_3\text{-}1,3\text{-}(CO_2)_2\}_2(H_2O)_5]_n$ or $C_{16}H_{20}O_{23}S_2Cd_5$.** This compound crystallizes in the triclinic structure with the lattice parameters $a = 816.570 \pm 0.010$, $b = 1204.52 \pm 0.02$, and $c = 1523.66 \pm 0.03$ pm and $\alpha = 78.749° \pm 0.005°$, $\beta = 85.824° \pm 0.006°$, and $\gamma = 72.229° \pm 0.005°$ and a calculated density of 2.863 g·cm$^{-3}$ (Liu et al. 2006). To obtain it, a mixture of $Cd(NO_3)_2 \cdot 4H_2O$ (0.2 mM), 5-sulfoisophthalic acid monosodium salt (0.2 mM), $Et_3N$ (0.05 mL), and water (15 mL) (pH = 7) was heated at 160°C for 4 days. Colorless prism-like crystals of the title compound were recovered by filtration, washed with distilled water, and dried in air.

**$[Cd_2(\mu\text{-}H_2O)_2\{HO_3SC_6H_3\text{-}1,3\text{-}(CO_2)_2\}_2(H_2O)_6]$ or $C_{16}H_{24}O_{22}S_2Cd_2$.** This compound crystallizes in the monoclinic structure with the lattice parameters $a = 2310.4 \pm 0.6$, $b = 687.56 \pm 0.11$, and $c = 873.15 \pm 0.18$ pm and $\beta = 107.007° \pm 0.014°$ and a calculated density of 2.146 g·cm$^{-3}$ (Liu et al. 2006). To obtain it, a mixture of $Cd(NO_3)_2 \cdot 4H_2O$ (0.2 mM), 5-sulfoisophthalic acid monosodium salt (0.2 mM), $Et_3N$ (0.01 mL), and water (15 mL) (pH = 2) was heated at 160°C for 4 d. Colorless plate-like crystals of the title compound were obtained when cooling to room temperature at 5°C h$^{-1}$. The crystals were recovered by filtration, washed with distilled water, and dried in air.

**[Cd$_5$(SC$_2$H$_4$OH)$_8$SO$_4$]·2H$_2$O or C$_{16}$H$_{44}$O$_{14}$S$_9$Cd$_5$.** This compound crystallizes in the monoclinic structure with the lattice parameters $a = 2521$, $b = 1227$, and $c = 2552$ pm and $\beta = 96.5°$ (Strickler 1969).

**[Cd(C$_{24}$H$_{31}$O$_4$S)OAc]·2H$_2$O or C$_{26}$H$_{38}$O$_8$SCd.** This compound melts at 245°C with decomposition (Pouskouleli et al. 1977). To prepare it, equimolar quantities of spironolactone (208.3 mg) and Cd(OAc)$_2$·2H$_2$O (133.2 mg) were refluxed overnight in absolute EtOH. A white amorphous precipitate was collected and washed with absolute EtOH.

**[Cd{Ph(SMe)CC(S)Ph}$_2$]·(DMSO) or C$_{32}$H$_{32}$OS$_5$Cd.** This compound melts at 165°C with decomposition and crystallizes in the triclinic structure with the lattice parameters $a = 1152.9 \pm 0.4$, $b = 1299.5 \pm 0.5$, and $c = 2280.4 \pm 0.8$ pm and $\alpha = 106.480° \pm 0.020°$, $\beta = 92.590° \pm 0.020°$, and $\gamma = 94.680° \pm 0.020$ pm and a calculated density of 1.443 g·cm$^{-3}$ (Reddy et al. 1992). From a hot solution of [Cd{Ph(SMe) CC(S)Ph}$_2$] in DMSO a crystalline title compound was precipitated upon cooling.

**[Cd$_2$(C$_9$H$_9$OS)$_4$(H$_2$O)$_2$]$_n$ or C$_{36}$H$_{40}$O$_8$S$_2$Cd$_2$.** This compound crystallizes in the triclinic structure with the lattice parameters $a = 896.8 \pm 0.2$, $b = 1159.6 \pm 0.3$, and $c = 1887.5 \pm 0.5$ pm and $\alpha = 95.67° \pm 0.02°$, $\beta = 91.33° \pm 0.02°$, and $\gamma = 90.25° \pm 0.02$ (Chan et al. 1987). It was prepared by the reaction of aqueous ethanolic solutions of benzyl thioacetate with a suspension of excess CdCO$_3$ (in aqueous EtOH) and digesting for ca. 1 h at 70°C–90°C. After removal of excess carbonate by filtration, standing at room temperature resulted in deposition of crystals which were recrystallized from absolute EtOH as colorless prisms.

**[Cd$_{17}$S$_4$(SCH$_2$CH$_2$OH)$_{26}$] or C$_{52}$H$_{130}$O$_{26}$S$_{30}$Cd$_{17}$.** This compound crystallizes in the monoclinic structure with the lattice parameters $a = 2514.7 \pm 0.4$, $b = 4009.6 \pm 0.5$, and $c = 2535.8 \pm 0.5$ pm and $\beta = 94.0° \pm 0.1°$ (Vossmeyer et al. 1995a). To obtain it, Cd(ClO$_4$)$_2$ and 2-mercaptoethanol were dissolved in H$_2$O, and H$_2$S was injected under vigorous stirring in alkaline solution (molar ratio 1.0:2.4:0.2). After it was stirred for several hours at room temperature, the solution was dialyzed exhaustively against H$_2$O; then crystals began to grow in the dialysis tube as small colorless needles, while the solution reached a pH of 9. These crystals decompose above 160°C and are very instable.

**[Cd$_{32}$S$_{14}${SCH$_2$CH(OH)Me}$_{36}$]·4H$_2$O or C$_{108}$H$_{260}$O$_{40}$S$_{50}$Cd$_{32}$.** This compound crystallizes in the rhombohedral structure with the lattice parameters $a = 2153.6 \pm 0.3$ and $c = 9585.7 \pm 1.0$ pm and a calculated density of 1.915 g·cm$^{-3}$ (Vossmeyer et al. 1995b). It was prepared by the following procedure. A solution of 4.70 mM of Cd(ClO$_4$)$_2$·6H$_2$O and 11.39 mM of 1-mercapto-2-propanol in 250 mL of H$_2$O was adjusted to pH 11.2 with 1 M NaOH before the addition of 2.04 mM of H$_2$S with vigorous stirring. The transparent solution was heated to 100°C. After cooling to room temperature, the crude cluster solution was dialyzed exhaustively against deionized H$_2$O. The crystals were found to crack during drying.

$[Cd_{17}S_4(SPh)_{24}(CH_3OCS_2)_{4/2}]_n \cdot nCH_3OH$ or $C_{149}H_{130}O_3S_{32}Cd_{17}$. This compound melts at the temperature higher than 250°C and crystallizes in the orthorhombic structure with the lattice parameters $a = 3720.0 \pm 1.5$, $b = 4154 \pm 2$, and $c = 4564.8 \pm 1.5$ pm and a calculated density of 1.848 g·cm$^{-3}$ (Jin et al. 1996). A yellow solution resulted when $[Cd_4(SPh)_{24}(SPh)_{4/2}]_n$ was dissolved in DMF/CS$_2$ (3:1 volume ratio). Some colorless crystals of the title compound were formed by slow diffusion of MeOH into that yellow solution after several weeks. They are air stable at room temperature. All reactions were carried out under a N$_2$ atmosphere.

## 4.66   Cd–H–C–O–Cl–S

Some compounds are formed in this system.

$[CdCl_2 \cdot (DMSO)]$ or $C_2H_6OCl_2SCd$. This compound decomposes at the heating (Selbin et al. 1961). To prepare it, CdCl$_2$·2H$_2$O was dissolved in hot DMSO and the solution was cooled. The addition of ether promoted crystallization. The white product was filtered and washed with ether.

$[(2\text{-MeSC}_7H_5O) \cdot (CdCl_2)]$ or $C_8H_8OCl_2SCd$ ($C_7H_6O$ = tropone). This compound crystallizes as pale-yellow needles and melts at the temperature higher than 300°C (Asao and Kikuchi 1972).

$[CdCl_2 \cdot Ph_2SO]$ or $C_{12}H_{10}OCl_2SCd$. This compound was obtained by the next procedure. Anhydrous CdCl$_2$ and Ph$_2$SO were each dissolved in EtOH, and the obtained solutions were mixed slowly with stirring (Gopalakrishnan and Patel 1967). The mixture, upon slow evaporation by passing dry air at room temperature, yielded crystals of the title compound in about a day. These were separated, washed several times with petroleum ether to remove any adhering Ph$_2$SO, and recrystallized from ethanol.

$[Cd(C_{12}H_{24}S_4)(ClO_4)_2]$ or $C_{12}H_{24}O_8Cl_2S_4Cd$. This compound crystallizes in the triclinic structure with the lattice parameters $a = 834.6 \pm 0.2$, $b = 834.9 \pm 0.3$, and $c = 868.8 \pm 0.2$ pm and $\alpha = 69.87° \pm 0.02°$, $\beta = 68.97° \pm 0.02°$, and $\gamma = 82.33° \pm 0.02$ and a calculated density of 1.903 g·cm$^{-3}$ (Setzer et al. 1991). To obtain it, a solution of 1,5,9,13-tetrathiacyclohexadecane (0.134 mM) in 8 mL of anhydrous nitromethane was added to a solution of Cd(ClO$_4$)$_2$·6H$_2$O (0.0956 mM) in 2 mL of anhydrous nitromethane and 4 drops of acetic anhydride. Colorless crystals were grown from the reaction mixture by solvent diffusion with anhydrous diethyl ether. The crystalline solid was dried overnight under high vacuum to give the title compound as a colorless crystalline solid.

$[Cd(DMSO)_6(ClO_4)_2]$ or $C_{12}H_{36}O_{14}Cl_2S_6Cd$. This compound melts at 188°C–190°C (Ahrland and Björk 1974) and crystallizes in the orthorhombic structure with the lattice parameters $a = 1254 \pm 1$, $b = 2023 \pm 1$, and $c = 2553 \pm 1$ pm and the experimental and calculated densities of $1.58 \pm 0.01$ and 1.60 g·cm$^{-3}$, respectively (Sandström 1978). To obtain it, Cd(ClO$_4$)$_2$·6H$_2$O (20 mM) was dissolved in a minimum amount (12 mL) of acetone (Ahrland and Björk 1974). After the addition of 2,2-dimethyl-ethoxypropane (120 mM), the solution was shaken for 2 h. DMSO (120 mM) was then added and the resulting mixture was shaken for another 30 min. Upon cooling to $-20°C$ crystals were formed, which were recrystallized from acetone. This compound could also be recrystallized from DMSO (Sandström 1978). Upon slow evaporation, colorless prismatic crystals were formed from a saturated 0.70 M solution at room temperature.

**$[ClCd_8(SCH_2CH_2OH)_{12}Cl_3]$ or $C_{24}H_{60}O_{12}Cl_4S_{12}Cd_8$.** This compound crystallizes in the rhombohedral structure with the lattice parameters $a = 1170.0 \pm 0.1$ and $\alpha = 103.57° \pm 0.01$ pm and the experimental and calculated densities $2.30 \pm 0.01$ and $2.27$ g·cm$^{-3}$, respectively (Dance et al. 1987a). It was prepared by the following procedure. A solution of 2-hydroxyethanethiol (52.9 mM) in $H_2O$ (20 mL) was added dropwise to a solution of $Cd(MeCO_2)_2(H_2O)_2$ (29.3 mM) in $H_2O$ (30 mL) and stirred until all precipitate had dissolved. A solution of NaCl (58.7 mM) in $H_2O$ (10 mL) was added, yielding a turbid solution that settled to a sticky white precipitate. This mixture was allowed to stand at room temperature for 50 h, while colorless crystals of the title compound grew: these crystals were separated manually from the remaining fine precipitate, washed with water, and vacuum dried. This compound decomposes when dissolved in $H_2O$. All preparations were performed under an atmosphere of $N_2$.

**$[Cd_{10}(SCH_2CH_2OH)_{16}Cl_4]$ or $C_{32}H_{80}O_{16}Cl_4S_{16}Cd_{10}$.** This compound crystallizes in the orthorhombic structure with the lattice parameters $a = 1480.2 \pm 0.3$, $b = 2495.0 \pm 0.5$, and $c = 2089.1 \pm 0.4$ pm (Dance et al. 1986). To obtain it, $Cd_{10}(SCH_2CH_2OH)_{16}(ClO_4)_2]\cdot5.5H_2O$ (0.21 mM) and $Bu_4NCl\cdot H_2O$ (0.85 mM) were dissolved together in DMF (3 mL) at room temperature. This solution in a 1 cm inner diameter tube was layered with acetone (12 mL). Colorless crystals of the title compound began to grow within 1 day and crystallization continued for 3 days. The mother liquor was withdrawn and the crystals were washed first with DMF/acetone (1:5 molar ratio) and then briefly with acetone and were dried in a stream of $N_2$. The obtained crystals decomposed rapidly under irradiation.

This compound has been prepared in higher yield by the following procedure (Dance et al. 1986). $Cd_{10}(SCH_2CH_2OH)_{16}(ClO_4)_2]\cdot5.5H_2O$ (0.382 mM) and $Bu_4NCl\cdot H_2O$ (1.53 mM) were dissolved in DMF (5.45 mL), treated with acetone (25 mL) with stirring, and allowed to stand for 12 h, and then the white crystalline precipitate was filtered, washed with acetone, and vacuum dried.

**$[Cd_{10}(SCH_2CH_2OH)_{16}(ClO_4)_2]\cdot5.5H_2O$ or $C_{32}H_{91}O_{37.5}Cl_4S_{16}Cd_{10}$.** This compound was prepared if a solution of 2-hydroxyethanethiol (159 mM) in $H_2O$ was added slowly to a solution of $Cd(OAc)_2\cdot2H_2O$ (87.9 mM) in $H_2O$ (100 mL), under $N_2$ at ambient temperature, allowing time for any precipitate to redissolved (Dance et al. 1986). A solution of $NaClO_4$ (60 g) in $H_2O$ (80 mL) was added with stirring, and the resulting precipitate was separated and dissolved in the minimum quantity of hot $H_2O$, filtered, and cooled. The colorless crystals that formed were separated, washed with cold water, and dried in vacuum.

**$[Cd_{10}(SCH_2CH_2OH)_{16}(ClO_4)_4]\cdot8H_2O$ or $C_{32}H_{96}O_{40}Cl_4S_{16}Cd_{10}$.** This compound crystallizes in the monoclinic structure with the lattice parameters $a = 3207.4 \pm 0.6$, $b = 1314.17 \pm 0.06$, and $c = 2516.2 \pm 0.4$ pm and $\beta = 126.12° \pm 0.01°$ at 143 K and a calculated density of 2.25 g·cm$^{-3}$ (Lacelle et al. 1984).

**[Cd(Ph₂SO)₆(ClO₄)₂]** or **C₇₂H₆₀O₁₄Cl₂S₆Cd**. This compound was obtained by the next procedure. Hydrated $CdCl_2$ and $Ph_2SO$ were each dissolved in acetone, and the obtained solutions were mixed slowly with stirring (Gopalakrishnan and Patel 1967). The mixture, upon slow evaporation by passing dry air at room temperature, yielded crystals of the title compound in about a day. These were separated, washed several times with petroleum ether to remove any adhering $Ph_2SO$, and recrystallized either from acetone or ethanol.

## 4.67  Cd–H–C–O–Br–S

Some compounds are formed in this system.

**[CdBr₂·(DMSO)]** or **C₂H₆OBr₂SCd**. This compound melts at 173°C (Selbin et al. 1961). It was prepared if $CdBr_2 \cdot 2H_2O$ was dissolved in hot DMSO and the solution was cooled. The addition of ether promoted crystallization. The white product was filtered and washed with ether.

**[BrCd₈(SCH₂CH₂OH)₁₂Br₃]** or **C₂₄H₆₀O₁₂Br₄S₁₂Cd₈**. This compound crystallizes in the rhombohedral structure with the lattice parameters $a = 1168.1 \pm 0.1$ and $\alpha = 102.38° \pm 0.01$ pm and the experimental and calculated densities $2.44 \pm 0.01$ and $2.43$ g·cm⁻³, respectively (Dance et al. 1987a). It was obtained by the next procedure. A solution of 2-hydroxyethanethiol (52.9 mM) in $H_2O$ (20 mL) was added dropwise to a solution of $Cd(MeCO_2)_2(H_2O)_2$ (29.3 mM) in $H_2O$ (30 mL) and stirred until all precipitate had dissolved. A solution of NaBr (235 mM) in $H_2O$ (10 mL) was added, yielding a turbid solution that settled to a sticky white precipitate. This mixture was allowed to stand at room temperature for 50 h, while colorless crystals of the title compound grew: these crystals were separated manually from the remaining fine precipitate, washed with water, and vacuum dried. This compound decomposes when dissolved in $H_2O$. All preparations were performed under an atmosphere of $N_2$.

## 4.68  Cd–H–C–O–I–S

**[Cd₂(SCH₂CH₂OH)₃I]₄·H₂O** or **C₂₄H₆₂O₁₃I₄S₁₂Cd₈**. This multinary compound is formed in this system. It crystallizes in the triclinic structure with the lattice parameters $a = 2787$, $b = 1077$, and $c = 1294$ pm and $\alpha = 73.1°$, $\beta = 116.1°$, and $\gamma = 120.0°$ and the experimental and calculated densities 2.51 and 2.59 g·cm⁻³, respectively (Bürgi et al. 1976). To obtain it, $Cd(OAc)_2$ (22 mM) was dissolved in 60 mL of $H_2O$ and 2-hydroxyethanethiol (33 mM) was added. NaI (40 mM) was dissolved in 10 mL of $H_2O$. On mixing the solutions, a precipitate was formed, which is best recrystallized from NaI solution. The obtained solid was dissolved in MeCN (30 mL) and repre-cipitated by the addition of $H_2O$ (Dance et al. 1987a).

## 4.69  Cd–H–C–Se–S

**[Cd₁₇Se₄(SPh)₂₆]** or **C₁₅₆H₁₃₀Se₄S₂₆Cd₁₇**. This multinary compound is formed in this system. It crystallizes in the orthorhombic structure with the lattice parameters $a = 3797.7 \pm 0.2$, $b = 4105.0 \pm 0.2$, and $c = 4521.8 \pm 0.2$ pm at 150 K (Zheng et al. 2006a). It was prepared as a by-product upon obtaining $Cd_{17}Se_4(SPh)_{26}(C_{13}H_{14}N_2)_2$.

## 4.70 Cd–H–C–F–S

$Cd_4(SC_6H_4F)_8$ or $C_{48}H_{32}F_8S_8Cd_4$. This multinary compound is formed in this system. It crystallizes in the tetragonal structure with the lattice parameters $a = 1579.3 \pm 0.2$ and $c = 2112.1 \pm 0.5$ pm (Dance et al. 1987b).

## 4.71 Cd–H–C–Cl–S

$[(C_7H_6S) \cdot (CdCl_2)]$ or $C_7H_6Cl_2SCd$ ($C_7H_6S$ = thiotropone). This multinary compound is formed in this system. Its orange crystals melt at 140°C with decomposition (Asao and Kikuchi 1972).

## 4.72 Cd–H–N–O–S

$Cd(NH_4)_2(SO_4)_2 \cdot 6H_2O$. This quinary compound is formed in this system that crystallizes in the monoclinic structure with the lattice parameters $a = 943.3 \pm 1.0$, $b = 1282.3 \pm 1.5$, and $c = 628.6 \pm 0.6$ pm and $\beta = 106°52' \pm 6'$ (Montgomery and Lingafelter 1966). It was obtained from neutral aqueous solutions containing $CdCl_2$ and thiosemicarbazide.

## 4.73 Cd–H–As–O–S

$Cd_3(AsO_3S)_2 \cdot 2H_2O$. This quinary compound is formed in this system that crystallizes in the orthorhombic structure (Gigauri et al. 1999). Thermal decomposition of this compound can be represented by the next scheme: $Cd_3(AsO_3S)_2 \cdot 2H_2O$ (60°C–190°C) $\rightarrow Cd_3(AsO_3S)_2$ (340°C–775°C) $\rightarrow Cd_3As_2O_4S + $ "O" (775°C–790°C) $\rightarrow Cd_3As_2O_5S$. It can be obtained by the interaction of $Cd(OAc)_2$ with sodium monothioarsenate at room temperature: $3Cd(OAc)_2 + 2Na_3AsO_3S = Cd_3(AsO_3S)_2 \cdot 2H_2O + 6NaOAc$.

## 4.74 Cd–Rb–C–N–S

$RbCd(SCN)_3$. This quinary compound is formed in this system (Thiele and Messer 1980). It crystallizes in the monoclinic structure with the lattice parameters $a = 571.20 \pm 0.12$, $b = 1319.44 \pm 0.22$, and $c = 1268.28 \pm 0.20$ pm and $\beta = 94.20° \pm 0.015°$ and the experimental and calculated densities 2.60 and 2.59 g·cm⁻³, respectively. This compound was obtained at the interaction of stoichiometric amounts of RbSCN and $Cd(SCN)_2$ in the solution of $EtOH/H_2O$ (1:2 volume ratio).

## 4.75 Cd–Cs–C–N–S

$CsCd(SCN)_3$ quinary compound is formed in this system (Thiele and Messer 1980). It crystallizes in the monoclinic structure with the lattice parameters $a = 1081.08 \pm 0.25$, $b = 722.37 \pm 0.18$, and $c = 1239.01 \pm 0.28$ pm and $\beta = 90.37° \pm 0.06°$ and the experimental and calculated densities 2.92 and 2.88 g·cm⁻³, respectively. This compound was obtained at the interaction of stoichiometric amounts of CsSCN and $Cd(SCN)_2$ in the solution of $EtOH/H_2O$ (1:2 volume ratio).

## 4.76  Cd–Cu–Ga–In–S

**2CdS–CuGaS₂–CuInS₂.** Liquidus surface of this quasiternary system (Figure 4.1) consists of three fields of primary crystallization of the solid solutions based on the system components (Marushko et al. 2010). The primary crystallization fields are separated with three monovariant lines and four invariant points, one of which is a ternary transition point (U, 1099°C, 38 mol.% $2CdS + 59$ mol.% $CuGaS_2 + 3$ mol.% $CuInS_2$).

The isothermal section of the $2CdS$–$CuGaS_2$–$CuInS_2$ quasiternary system at 600°C is shown in Figure 4.2 (Marushko et al. 2010). Three single-phase fields exist in this system at this temperature. The first is the continuous solid solution between $CuGaS_2$ and $CuInS_2$ with the chalcopyrite structure. The second field belongs to the solid solution based on CdS. The third single-phase field is the solid solution with the sphalerite structure based on the one of the high-temperature modifications of $CuInS_2$. Moreover, there are three two-phase fields and one three-phase field in this system.

This system was investigated by the DTA and XRD, and the alloys were annealed at 600°C for 500 h and quenched in cold water (Marushko et al. 2010).

## 4.77  Cd–Cu–Ga–In–Se–S

**2CdS–CuGaSe₂–CuInSe₂.** Projection of the liquidus surface of the $2CdS$–$CuGaSe_2$–$CuInSe_2$ quasiternary system (Figure 4.3) has been constructed only in the $CuInSe_2$-rich part relevant for the single crystal growth of the solid solutions

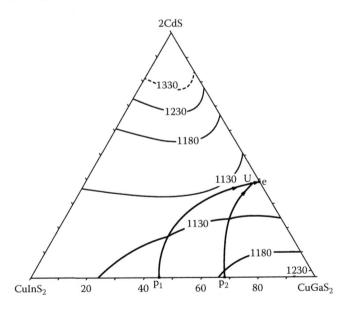

**FIGURE 4.1**  Liquidus surface of the $2CdS$–$CuGaS_2$–$CuInS_2$ quasiternary system. (From Marushko, L.P. et al., *J. Alloys Compd.*, 492(1–2), 184, 2010.)

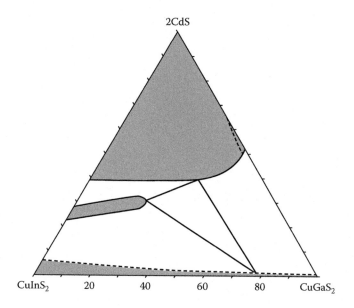

**FIGURE 4.2** The isothermal section of the 2CdS–CuGaS$_2$–CuInS$_2$ quasiternary system at 600°C. (From Marushko, L.P. et al., *J. Alloys Compd.*, 492(1–2), 184, 2010.)

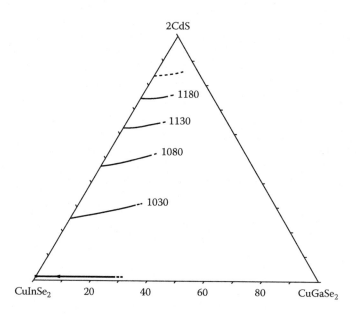

**FIGURE 4.3** The part of the liquidus surface of the 2CdS–CuGaSe$_2$–CuInSe$_2$ quasiternary system. (From Parasyuk, O.V. et al., *J. Cryst. Growth*, 318(1), 332, 2011.)

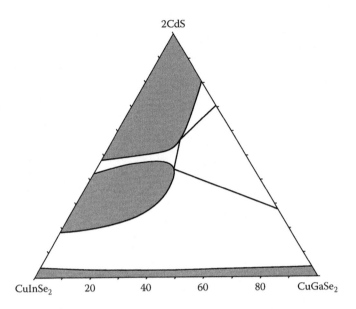

**FIGURE 4.4** The isothermal section of the 2CdS–CuGaSe$_2$–CuInSe$_2$ quasiternary system at 600°C. (From Parasyuk, O.V. et al., *J. Cryst. Growth*, 318(1), 332, 2011.)

based on this compound (Parasyuk et al. 2011). The liquidus surface consists of two fields of the primary crystallization of the solid solutions based on CdS and CuInSe$_2$, and the field of the solid solutions based on CdS occupies the largest area on the concentration triangle.

The isothermal section of this system at 600°C is shown in Figure 4.4 (Parasyuk et al. 2011). There are three single-phase fields at this temperature. The first field is the continuous range of the solid solutions with the chalcopyrite structure between CuGaSe$_2$ and α-CuInSe$_2$. The second field belongs to the solid solutions based on CdS with the wurtzite structure. The third single-phase field is the solid solutions based on β-CuInSe$_2$ stabilized by CdS to the annealing temperature. There are also three two-phase fields in this system as well as one three-phase field.

This system was investigated by the DTA and XRD (Parasyuk et al. 2011). The alloys were annealed at 600°C for 500 h. Single crystals of the solid solutions based on β-CuInSe$_2$ with the sphalerite structure have been grown by the Bridgman method. The obtained crystals are of the *p*-type conductivity, and their band gap energy varies from 1.16 to 1.32 eV.

## 4.78    Cd–Cu–Ga–Se–S

**2CdS + CuGaSe$_2$ ⇔ 2CdSe + CuGaS$_2$.** The isothermal section of this ternary mutual system at 600°C is shown in Figure 4.5 (Piskach et al. 2008). Three single-phase fields based on the solid solutions CdS$_x$Se$_{1-x}$ and CuGa(S$_x$Se$_{1-x}$)$_2$ and CuCd$_2$GaSe$_4$ quaternary compound exist in this system. These single-phase fields are divided by

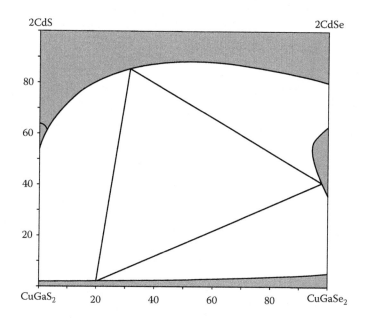

**FIGURE 4.5** The isothermal section of the 2CdS + CuGaSe$_2$ ⇔ 2CdSe + CuGaS$_2$ ternary mutual system at 600°C. (From Piskach, L.V. et al., *Nauk. visnyk Volyns'k. nats. un-tu im. Lesi Ukrainky. Khim. Nauky*, (16), 47, 2008.)

the three two-phase fields and the biggest part of the concentration quadrangle occupies the three-phase field. The alloys were annealed at 600°C for 600 h with the next quenching in the 25% NaCl aqueous solution.

## 4.79  Cd–Cu–In–Se–S

**2CdS–CuInSe$_2$.** This section is a nonquasibinary section of the 2CdS + CuInSe$_2$ ⇔ 2CdSe + CuInSe$_2$ ternary mutual system (Figure 4.6) (Parasyuk et al. 2006). The limiting solubility of CdS in CuInSe$_2$ after different treatments has been determined to correspond to $(0.025 \pm 0.005) \leq x \leq (0.040 \pm 0.005)$ depending on the heat treatment conditions of the solid solution $(CdS)_{2x}(CuInSe_2)_{1-x}$ (Rogacheva et al. 2001). Within the interval $0.005 \leq x \leq 0.01$, anomalies have been revealed in concentration dependences of microhardness and XRD line width. The CdS impurity introduction into CuInSe$_2$ results in the conductivity type change from *p* to *n*. Obtained solid solutions crystallize in the chalcopyrite structure.

**2CdSe–CuInS$_2$.** This section is also a nonquasibinary section of the 2CdS + CuInSe$_2$ ⇔ 2CdSe + CuInS$_2$ ternary mutual system (Parasyuk et al. 2006). Its liquidus and solidus are represented by the lines of the beginning and the end of the primary crystallization of the solid solutions based on CdSe.

**2CdS ± CuInSe$_2$ ⇔ 2CdSe ± CuInS$_2$.** The liquidus surface of this system is shown in Figure 4.7 (Parasyuk et al. 2006). It consists of two fields of primary crystallization

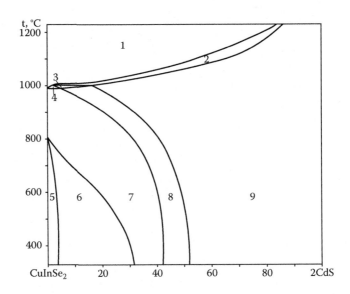

**FIGURE 4.6** Phase relations in the 2CdS–CuInSe$_2$ system: 1, L; 2, L+(CdS); 3, L+(β-CuInSe$_2$)+(CdS); 4, L+(β-CuInSe$_2$); 5, (α-CuInSe$_2$); 6, (α-CuInSe$_2$)+(β-CuInSe$_2$); 7, (β-CuInSe$_2$); 8, (β-CuInSe$_2$)+(CdS); and 9, (CdS). (From Parasyuk, O.V. et al., *J. Solid State Chem.*, 179(10), 2998, 2006.)

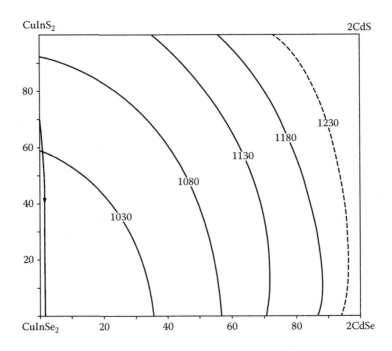

**FIGURE 4.7** Liquidus surface of the 2CdS+CuInSe$_2$ ⇔ 2CdSe+CuInS$_2$ ternary mutual system. (From Parasyuk, O.V. et al., *J. Solid State Chem.*, 179(10), 2998, 2006.)

that correspond to the solid solution based on $CdSe_xS_{1-x}$ and $CuIn(Se_xS_{1-x})_2$, which are separated by a monovariant line. The field of the solid solution based on $CdSe_xS_{1-x}$ occupies the largest part of the concentration quadrangle.

The isothermal section of this system at 600°C and 450°C is shown in Figure 4.8 (Parasyuk et al. 2006). They contain three single-phase fields of solid solutions, two of which are stretched along the side systems of the triangle that form the continuous solid solution series, and the third one ($\gamma$-phase) is located between them. The decreasing temperature leads to narrowing of the homogeneity regions of the solid solutions.

The $\gamma$-phase crystallizes in the cubic structure of the sphalerite type and can be considered as the solid solution of the high-temperature modification of $CuInS(Se)_2$, stabilized at room temperature by the addition of $CdS(Se)$ (Romanyuk et al. 2008).

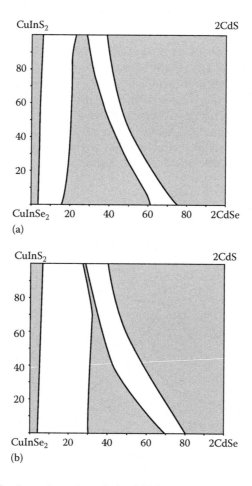

(a)

(b)

**FIGURE 4.8** The isothermal section of the $2CdS + CuInSe_2 \Leftrightarrow 2CdSe + CuInS_2$ ternary mutual system at (a) 600°C and (b) 450°C. (From Parasyuk, O.V. et al., *J. Solid State Chem.*, 179(10), 2998, 2006.)

The alloys were annealed at 600°C for 500 h (250 h [Romanyuk et al. 2008]) and at 450°C during 1440 h with the next quenching into cold water (Parasyuk et al. 2006). Single crystals of the γ-phase were grown by a modified Bridgman method (Romanyuk et al. 2008).

## 4.80 Cd–Cu–Ge–Sn–S

**CdS–Cu₂GeS₃–Cu₂SnS₃.** The liquidus surface of this quasiternary system is shown in Figure 4.9 (Marushko et al. 2009a). It consists of five fields of the primary crystallization of the solid solutions based on CdS (β), $Cu_2Cd_3GeS_6$ (ε), $Cu_2CdGeS_4$ (γ), and $Cu_2CdSnS_4$ (δ) quaternary compounds and $Cu_2Ge_xSn_{1-x}S_3$ (α) solid solutions. The field of the solid solutions based on CdS occupies the largest part of the concentration triangle. These fields of the primary crystallization are separated by seven monovariant lines and eight invariant points. The types, the temperatures, and the coordinates of the ternary invariant points are presented in Table 4.1.

The isothermal section of the CdS–Cu₂GeS₃–Cu₂SnS₃ quasiternary system at 400°C is shown in Figure 4.10 (Marushko et al. 2009a). The existence of four single-phase regions is established at this temperature. They are the continuous solution series of Cu₂GeS₃ and Cu₂SnS₃, solid solutions of the quaternary compounds stretched along the Cu₂CdGeS₄–Cu₂CdSnS₄ section, solid solution range of Cu₂CdGeS₄ with the orthorhombic structure that extends to 9 mol.% Cu₂CdSnS₄, solid solution range of Cu₂CdSnS₄ with the stannite structure that extends to 14 mol.% Cu₂CdGeS₄, and solid solution range of CdS with the wurtzite structure.

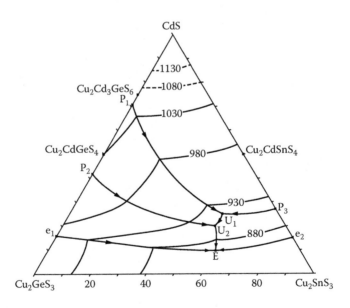

**FIGURE 4.9** Liquidus surface of the CdS–Cu₂GeS₃–Cu₂SnS₃ quasiternary system. (From Marushko, L.P. et al., *J. Alloys Compd.*, 484(1–2), 147, 2009a.)

**TABLE 4.1**

**Temperatures and Compositions of the Ternary Invariant Points in the CdS–Cu₂GeS₃–Cu₂SnS₃ Quasiternary System**

| Invariant Points | Reaction | Composition, mol. % | | | t, °C |
|---|---|---|---|---|---|
| | | CdS | Cu₂GeS₃ | Cu₂SnS₃ | |
| E | $L \Leftrightarrow \alpha + \gamma + \delta$ | 10 | 30 | 60 | 844 |
| $U_1$ | $L + \beta \Leftrightarrow \delta + \varepsilon$ | 25 | 20 | 55 | 914 |
| $U_2$ | $L + \varepsilon \Leftrightarrow \gamma + \delta$ | 20 | 25 | 55 | 903 |

*Source:* Marushko, L.P. et al., *J. Alloys Compd.*, 484(1–2), 147, 2009a.

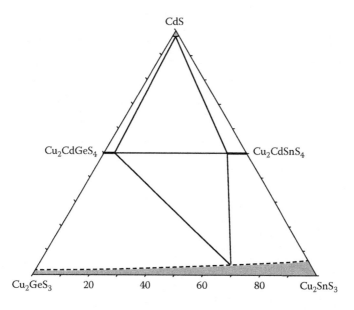

**FIGURE 4.10** The isothermal section of the CdS–Cu₂GeS₃–Cu₂SnS₃ quasiternary system at 400°C. (From Marushko, L.P. et al., *J. Alloys Compd.*, 484(1–2), 147, 2009a.)

This system was investigated by the DTA and XRD, and the samples were annealed at 400°C for 500 h, followed by quenching into cold water (Marushko et al. 2009a).

## 4.81 Cd–Cu–Ge–Se–S

**3CdS + Cu₂GeSe₃ $\Leftrightarrow$ 3CdSe ± Cu₂GeS₃.** Liquidus surface of this ternary mutual system is shown in Figure 4.11 (Marushko et al. 2009b). It consists of five fields of the primary crystallization that correspond to the CdSe$_x$S$_{1-x}$, Cu₂Cd₃GeSe$_{6(1-x)}$S$_{6x}$ and Cu₂CdGeSe$_{4(1-x)}$S$_{4x}$ solid solutions and solid solution based on the Cu₂GeS₃ and Cu₂GeSe₃ ternary compounds. The fields of the primary crystallization are

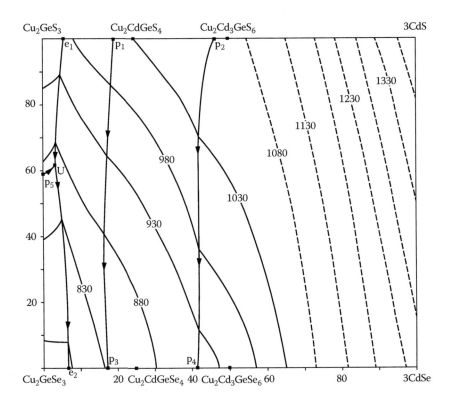

**FIGURE 4.11** Liquidus surface of the $3CdS + Cu_2GeSe_3 \Leftrightarrow 3CdSe + Cu_2GeS_3$ ternary mutual system. (From Marushko, L.P. et al., *J. Alloys Compd.*, 473(1–2), 94, 2009b.)

separated by five monovariant lines and eight invariant points. Ternary transition point $U$ takes place at 852°C. Neither of the sections that form the quadrangle diagonals is a quasibinary one.

Five single-phase regions exist in the $3CdS + Cu_2GeSe_3 \Leftrightarrow 3CdSe + Cu_2GeS_3$ ternary mutual system at 400°C (Figure 4.12): the limited solid solution ranges of $Cu_2GeS_3$ and $Cu_2GeSe_3$, the limited solid solutions of the quaternary compounds (solid solution range of $Cu_2CdGeS_4$ with the orthorhombic structure and a small range of solid solution of low-temperature modification of $Cu_2CdGeSe_4$ with the stannite structure), and the continuous solid solution series of CdS and CdSe with the wurtzite structure (Marushko et al. 2009b).

The samples were annealed at 400°C for 500 h and then quenched into cold water (Marushko et al. 2009b).

## 4.82   Cd–Cu–Ge–Fe–S

**CdS–Cu₂S–GeS₂–FeS**. The phase diagram is not constructed. **Cu₂(Cd, Fe)GeS₄** quinary compound (mineral barquillite) is formed in this system. It crystallizes in the tetragonal structure with the lattice parameters $a = 545 \pm 4$ and $c = 1060 \pm 10$ pm and a calculated density of 4.53 g·cm⁻³ (Jambor and Roberts 1999, Mandarino 1999,

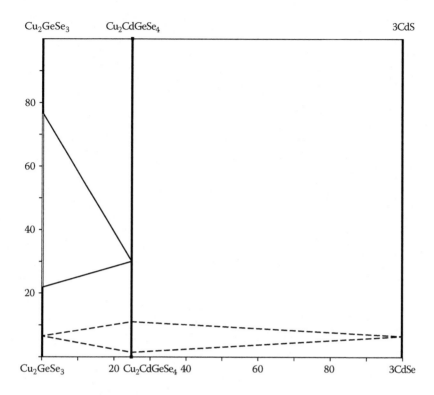

**FIGURE 4.12** The isothermal section of the 3CdS + Cu₂GeSe₃ ⇔ 3CdSe + Cu₂GeS₃ ternary mutual system at 400°C. (From Marushko, L.P. et al., *J. Alloys Compd.*, 473(1–2), 94, 2009b.)

Murciego et al. 1999). This mineral contains 2.1 at.% (2.2 mass.%) Fe and its composition shows extensive Cd-for-Fe substitution and extends to at least Cd: Fe = 1: 1 ratio.

## 4.83  Cd–Ag–Ba–Sn–S

**CdS–Ag₂S–BaS–SnS₂.** The phase diagram is not constructed. **Ag₂Ba₆CdSn₄S₁₆** quinary compound is formed in this system. It crystallizes in the cubic structure with lattice parameter $a = 1472.53 \pm 0.07$ pm (Teske 1985).

## 4.84  Cd–Ag–Ga–Se–S

**2CdS–AgGaSe₂.** This section is a nonquasibinary section of the 2CdS + AgGaSe₂ ⇔ 2CdSe + AgGaS₂ ternary mutual system (Figure 4.13) (Olekseyuk et al. 2004).

**2CdSe–AgGaS₂.** This section is also a nonquasibinary section of the 2CdS + AgGaSe₂ ⇔ 2CdSe + AgGaS₂ ternary mutual system (Olekseyuk et al. 2004). The liquidus consists of two fields of primary crystallization of the AgGa(S$_x$Se$_{1-x}$)₂ (α-phase) and CdS$_x$Se$_{1-x}$ (β-phase) solid solutions.

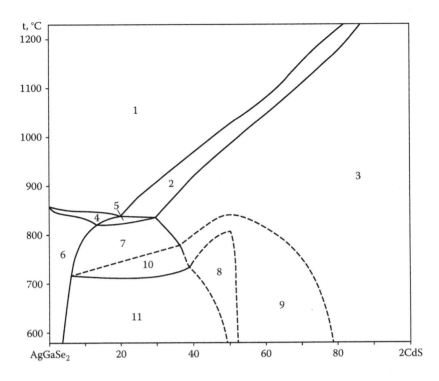

**FIGURE 4.13** Phase relations in the 2CdS–AgGaSe$_2$ section: 1, L; 2, L+(CdS); 3, (CdS); 4, L+(AgGaSe$_2$); 5, L+(CdS)+(AgGaSe$_2$); 6, (AgGaSe$_2$); 7, (CdS)+(AgGaSe$_2$); 8, AgCd$_2$GaS$_{4(1-x)}$Se$_{4x}$; 9, (CdS)+AgCd$_2$GaS$_{4(1-x)}$Se$_{4x}$; 10, (AgGaSe$_2$)+(CdS)+AgCd$_2$GaS$_{4(1-x)}$Se$_{4x}$; and 11, (AgGaSe$_2$)+AgCd$_2$GaS$_{4(1-x)}$Se$_{4x}$. (From Olekseyuk, I.D. et al., *J. Alloys Compd.*, 367(1–2), 25, 2004.)

**2CdS + AgGaSe$_2$ ⇔ 2CdSe + AgGaS$_2$.** The liquidus surface of this ternary mutual system (Figure 4.14) consists of three fields of primary crystallization corresponding to the AgGaSe$_{2(1-x)}$S$_x$ (α-phase), CdSe$_{1-x}$S$_x$ (β-phase), and AgCd$_2$Ga(S$_x$Se$_{1-x}$)$_4$ (γ-phase) solid solutions (Olekseyuk et al. 2004). The β-phase field occupies the largest part of the concentration quadrangle and the γ-solid solution range is small, which is mainly caused by the solid-state formation of AgCd$_2$GaSe$_4$. The fields of primary crystallization are divided by three monovariant lines and four invariant points, one of which is a ternary transition point (*U*, 885°C, 22 mol.% 2CdS + 18 mol.% AgGaSe$_2$ + 60 mol.% AgGaS$_2$).

The isothermal section of this system at 600°C (Figure 4.15) (Olekseyuk et al. 2004) includes three single-phase solid solution ranges, two of which are located along the quadrangle sides where continuous solid solutions occur (α- and β-solid solutions), while the third one is localized between the quaternary intermediate phases (γ-solid solution).

This system was investigated by the DTA, XRD, and metallography, and the alloys were annealed at 600°C for 500 h with subsequent quenching in cold water (Olekseyuk et al. 2004).

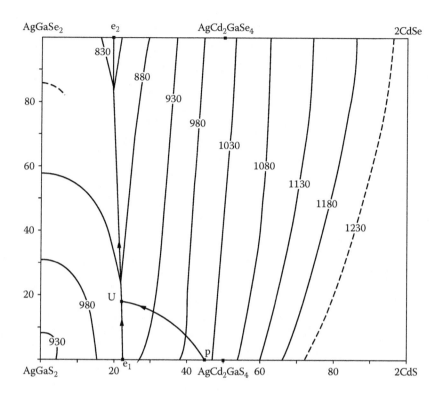

**FIGURE 4.14** Liquidus surface of the 2CdS + AgGaSe$_2$ ⇔ 2CdSe + AgGaS$_2$ ternary mutual system. (From Olekseyuk, I.D. et al., *J. Alloys Compd.*, 367(1–2), 25, 2004.)

## 4.85 Cd–Hg–C–N–S

**CdHg(SCN)$_4$.** This quinary compound is formed in this system. It crystallizes in the tetragonal structure with the lattice parameters $a = 1148.7 \pm 0.3$ and $c = 421.8 \pm 0.1$ pm and a calculation density of 3.254 g·cm$^{-3}$ (Yuan et al. 1997, Raghavan et al. 2009) ($a = 1148.18$ and $c = 421.96$ pm (Wang et al. 2003a,b); $a = 1148 \pm 2$ and $c = 433 \pm 2$ pm; and experimental and calculation densities 3.062 and 3.079 g·cm$^{-3}$, respectively [Iizuka and Sudo 1968]). The title compound was obtained by the interaction of Cd(SCN)$_2$ and Hg(SCN)$_2$ (1:1 molar ratio), or by the interaction of M$_2$Hg(SCN)$_4$ and CdX$_2$ (M = NH$_4$, Na, K; X = NO$_3$, Cl, OAc) (1:1 molar ratio), or by the interaction of KSCN, HgCl$_2$, and CdCl$_2$ (4:1:1 molar ratio) (Iizuka and Sudo 1968; Wang et al. 2003a,b; Raghavan et al. 2008, 2009). Small single crystals can be crystallized by spontaneous nucleation in a mixing solution made of H$_4$SCN, CdX$_2$, and HgX$_2$ as raw materials with stoichiometric ratios in deionized water. Using the small single crystals obtained as seeds, large single crystals can be obtained by solvent evaporation or temperature lowering of such saturated solutions Wang et al. 2003a,b).

Single crystals of the title compound could also be grown from acetone/H$_2$O (4:1 volume ratio) by the slow evaporation solution technique (Raghavan et al. 2008, 2009),

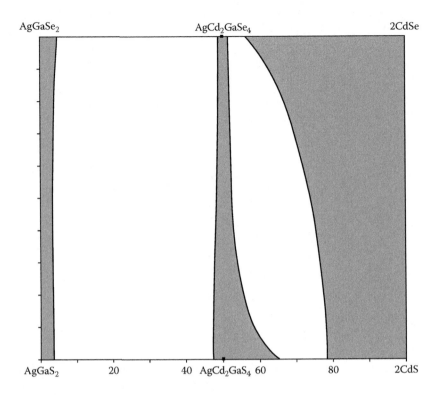

**FIGURE 4.15**   The isothermal section of the $2CdS + AgGaSe_2 \Leftrightarrow 2CdSe + AgGaS_2$ ternary mutual system at 600°C. (From Olekseyuk, I.D. et al., *J. Alloys Compd.*, 367(1–2), 25, 2004.)

or by using the temperature-lowering method from water/NaCl and water/KCl mixture solvents (Yuan et al. 1998).

## 4.86   Cd–Hg–C–N–Cl–S

**CdHg₃Cl₂(SCN)₆.** This multinary compound is formed in this system. It crystallizes in the rhombohedral structure with the lattice parameters (in the hexagonal setting) $a = 1120.66 \pm 0.02$ and $c = 5992.0 \pm 0.1$ pm and a calculation density of 3.47 g·cm⁻³ (Mosset et al. 2002). To prepare it, mixtures of $CdCl_2$ (1 mM) + $Hg(SCN)_2$ (2 mM) and $CdCl_2$ (1 mM) + $Hg(SCN)_2$ (3 mM) were, respectively, dissolved in 30 mL of MeOH. The resulting clear solutions were slowly evaporated at room temperature and yielded millimetric crystalline platelets and rhombohedra of the title compound. Crystals were filtered off, washed with MeOH, and dried at room temperature.

## 4.87   Cd–Hg–C–N–Br–S

**CdHg₄Br₄(SCN)₆.** This multinary compound is formed in this system. It crystallizes in the orthorhombic structure with the lattice parameters $a = 2232.2 \pm 0.1$,

$b = 1846.4 \pm 0.1$, and $c = 636.78 \pm 0.03$ pm and a calculation density of 4.006 g·cm⁻³ (Mosset et al. 2002). To prepare it, a mixture of $CdBr_2 \cdot 4H_2O$ (1 mM) + $Hg(SCN)_2$ (3 mM) was dissolved in 30 mL of MeOH. The resulting clear solution were slowly evaporated at room temperature and yielded millimetric crystalline platelets and rhombohedra of the title compound. Crystals were filtered off, washed with MeOH, and dried at room temperature.

## REFERENCES

Abrahams B.F., Corbett M., Dakternieks D., Gable R.W., Hoskins B.F., Tiekink E.R.T., Winter G.N.M.R. Studies of phosphine adducts of mercury and cadmium xanthates: Crystal and molecular structures of $Cd(S_2COPr^i)_2PPh_3$, $Hg(S_2COPr^i)_2PPh_3$ and $Hg(S_2COPr^i)_2P(c\text{-}C_6H_{11})_3$, *Aust. J. Chem.*, 39(12), 1993–2001 (1986).

Abrahams B.F., Hoskins B.F., Tiekink E.R.T., Winter G. Investigation of a new xanthate ligand. The crystal and molecular structures of nickel and cadmium (methoxyethyl) xanthates, *Aust. J. Chem.*, 41(7), 1117–1122 (1988a).

Abrahams B.F., Hoskins B.F., Winter G. The structure of cadmium bis(isopropylxanthate)-4,4'-bipyridine, *Aust. J. Chem.*, 43(10), 1759–1765 (1990).

Abrahams B.F., Hoskins B.F., Winter G., Tiekink E.R.T. The structure of the cadmium tris(methoxyethylxanthato) anion, $Cd(CH_3OCH_2CH_2OCS_2)^-_3$, as its tetraethylammonium salt, Inorg. *Chim. Acta*, 150(2), 147–148 (1988b).

Abrahams I.L., Garner C.D. Preparation and crystal structures of $[PPh_4]_2[M_2(SPh)_6]$ (M = Zn or Cd), *J. Chem. Soc., Dalton Trans.*, (6), 1577–1579 (1987).

Achak O., Pivan J.Y., Maunaye M., Loüer M., Loüer D. Structure refinement by the Rietveld method of the thiogermanates $[(CH_3)_4N]_2MGe_4S_{10}$ (M = Fe, Cd), *J. Solid State Chem.*, 121(2), 473–478 (1996).

Agre V.M., Shugam E.A. The structure of chelate compounds with bonds Me–S. 8. X-ray diffraction study cadmium hexamethylene dithiocarbamate [in Russian], *Kristallografiya*, 17(2), 303–308 (1968).

Agre V.M., Shugam E.A., Rukhadze E.G. The structure of chelate compounds with metal-sulfur bonds. I. The crystallographic data for certain bivalent metal alkyldithiocarbamates [in Russian], *Zhurn. Strukt. Khim.*, 7(5), 897–898 (1966).

Ahmad S., Altaf M., Stoeckli-Evans H., Isab A.A., Malik M.R., Ali S., Shuja S. Synthesis and structural characterization of dibromidobis(*N,N'*-dimethylthiourea-κS)cadmium(II) and diiodidobis(*N,N'*-dimethylthiourea-κS)cadmium(II), *J. Chem. Crystallogr.*, 41(8), 1099–1104 (2011).

Ahrland S., Björk N.-O. Metal halide and pseudohalide complexes in dimethylsulfoxide solution. I. Dimethyl sulfoxide solvates of silver(I), zinc(II), cadmium(II), and mercury(II), *Acta Chem. Scand.*, 28A(8), 823–828 (1974).

Ahuja I.S., Garg A. Complexes of some pyridines and anilines with zinc(II), cadmium(II) and mercury(II) thiocyanates, *J. Inorg. Nucl. Chem.*, 34(6), 1929–1935 (1972).

Ahuja I.S., Rastogi P. Complexes of pyridine N-oxides with cadmium(II) thiocyanate, *J. Inorg. Nucl. Chem.*, 32(4), 1381–1383 (1970).

Airoldi C., De Oliveira S.F., Ruggiero S.G., Lechat J.R. Adducts of cadmium diethyldithiocarbamate with bidentate nitrogen ligands, *Inorg. Chim. Acta*, 174(1), 103–108 (1990).

Al-Arfaj A.R., Reibenspies J.H., Isab A.A., Hussain M.S. Dichlorobis(1,3-imidazolidine-2-thione-S)cadmium(II). *Acta Crystallogr.*, C54(1), 51–53 (1998).

Ali B., Dance I., Scudder M., Craig D. Dimorphs of $(Ph_4P)_2[Cd_2(SPh)_6]$: Crystal packing analyses and the interplay of intermolecular and intramolecular energies, *Crystal Growth Des.*, 2(6), 601–607 (2002).

Altaf M., Stoeckli-Evans H., Murtaza G., Isab A.A., Ahmad S., Shaheen M.A. Structural characterization of a cadmium(II)-sulfato complex, [Cd(N,N'-diethyl thiourea)₄(SO₄)] [in Russian], Zhurn. Strukt. Khim., 52(3), 640–644 (2011).

Alvarez H.M., Tran T.B., Richter M.A., Alyounes D.M., Rabinovich D., Tanski J.M., Krawiec M. Homoleptic Group 12 metal bis(mercaptoimidazolyl)borate complexes M(Bmᴿ)₂ (M = Zn, Cd, Hg), Inorg. Chem., 42(6), 2149–2156 (2003).

Anjali K.S., Vittal J.J., Dean P.A.W. Syntheses, characterization and thermal properties of [M(Spy)₂(SpyH)₂] (M = Cd and Hg; Spy⁻ = pyridine-4-thiolate; SpyH = pyridinium-4-thiolate) and [M(SpyH)₄](ClO₄)₂ (M = Zn, Cd and Hg), Inorg. Chim. Acta, 351, 79–88 (2003).

Asao T., Kikuchi (née Ninomiya) Y. The preparation of some tropone complex, Chem. Lett., 1(5), 413–416 (1972).

Baggio R., Baggio S., Pardo M.I., Garland M.T. Bis(μ-thiosulfato)-1κO,1:2κ²S;2κO,2κ²S-bis[(2,9-dimethyl-1,10-phenanthroline-N,N')cadmium(II)], Acta Crystallogr., C52(8), 1939–1942 (1996).

Bakbak S., Bhatia V.K., Incarvito C.D., Rheingold A.L., Rabinovich D. Synthesis and characterization of two new bulky tris(mercaptoimidazolyl)borate ligands and their zinc and cadmium complexes, Polyhedron, 20(28), 3343–3348 (2001).

Bakbak S., Incarvito C.D., Rheingold A.L., Rabinovich D. Synthesis and characterization of novel mononuclear cadmium thiolate complexes in a sulfur-rich environment, Inorg. Chem., 41(4), 998–1001 (2002).

Banda R.M.H., Dance I.G., Bailey T.D., Craig D.C., Scudder M.L. Cadmium polysulfide complexes, [Cd(Sₓ)(Sᵧ)]²⁻: Syntheses, crystal and molecular structures, and ¹¹³Cd NMR studies, Inorg. Chem., 28(10), 1862–1871 (1989).

Banerjee S., Wu B., Lassahn P.-G., Janiak C., Ghosh A. Synthesis, structure and bonding of cadmium(II) thiocyanate systems featuring nitrogen based ligands of different denticity, Inorg. Chim. Acta, 358(3), 535–544 (2005).

Barreca D., Gasparotto A., Maragno C., Seraglia R., Tondello E., Venzo A., Krishnanand V., Bertagnolli H. Cadmium O-alkylxanthates as CVD precursors of CdS: A chemical characterization, Appl. Organomet. Chem., 19(1), 59–67 (2005).

Baur R., Schellenberg M., Schwarzenbach G. Aromatenkomplexe von Quecksilber, Helv. Chim. Acta, 45(3), 775–783 (1962).

Bayón J.C., Briansó M.C., Briansó J.L., Duarte P.G. Synthesis and crystal and molecular structure of catena-bis[μ-(N-methylpiperidinium-4-thiolato)]-cadmium(II) perchlorate dihydrate, Inorg. Chem., 18(12), 3478–3482 (1979).

Bayon J.C., Casals I., Gaete W., Gonzalez-Duarte P., Ros J. Complexes of 1-methyl-4-mercaptopiperidine with zinc(II), cadmium(II) and mercury(II) halides, Polyhedron, 1(2), 157–161 (1982).

Bazzicalupi C., Bencini A., Bianchi A., Giorgi C., Fusi V., Paoletti P., Valtancoli B. Binuclear metal assemblies inside an oxa-aza macrocyclic receptor, Inorg. Chim. Acta, 246(1–2), 125–131 (1996).

Beauvais L.G., Shores M.P., Long J.R. Cyano-bridged Re₆Q₈ (Q = S, Se) cluster-metal framework solids: A new class of porous materials, Chem. Mater., 10(12), 3783–3786 (1998).

Beck W., Stetter K.H., Tadros S., Schwarzhans K.E. Darstellung, IR- und ¹⁹F-KMR-Spektren von Pentafluorphenylmercapto-Metallkomplexen, Chem. Ber., 100(12), 3944–3954 (1967).

Beheshti A., Brooks N.R., Clegg W., Hyvadi R. Dichlorobis[1-methylimidazoline-2(3H)-thione]cadmium(II), Acta. Crystallogr., E62(7), m1383–m1385 (2005).

Bell N.A., Clegg W., Coles S.J., Constable C.P., Harrington R.W., Hursthouse M.B., Light M.E., Raper E.S., Chris S., Walker M.R. Complexes of heterocyclic thiones and group 12 metals: Part VI. Preparation and characterisation of complexes of cadmium(II) halides with

1-methylimidazoline-2(3H)-thione, 1,3-thiazolidine-2-thione and 1,3-benzothiazoline-2-thione. Crystal structures of polymeric (1,3-thiazolidine-2-thione)cadmium(II) chloride, bis(1,3-thiazolidine-2-thione)cadmium(II) iodide and monomeric bis(1-methylimidazoline-2(3H)-thione)cadmium(II) bromide, *Inorg. Chim. Acta*, 357(7), 2091–2099 (2004).

Bermejo E., Carballo R., Castiñeiras A., Domínguez R., Liberta A.E., Maichle-Mössmer C., Salberg M.M., West D.X. Synthesis, structural characteristics and biological activities of complexes of Zn$^{II}$, Cd$^{II}$, Hg$^{II}$, Pd$^{II}$, and Pt$^{II}$ with 2-acetylpyridine 4-methylthiosemicarbazone, *Eur. J. Inorg. Chem.*, (6), 965–973 (1999).

Bermejo E., Castiñeiras A., Fostiak L.M., Garcia I., Llamas-Saiz A.L., Swearingen J.K., West D.X. Synthesis, characterization and molecular structure of 2-pyridyl-formamide $N$(4)-dimethylthiosemicarbazone and some five-coordinated zinc(II) and cadmium(II) complexes, *Z. Naturforsch.*, 56B(12), 1297–1305 (2001).

Bermejo E., Castiñeiras A., Fostiak L.M., Santos I.G., Swearingen J.K., West D.X. Spectral and structural studies of Zn and Cd complexes of 2-pyridineformamide $N$(4)-ethylthiosemicarbazone, *Polyhedron*, 23(14), 2303–2313 (2004a).

Bermejo E., Castiñeiras A., García-Santos I., West D.X. Structural and coordinative variability in zinc(II), cadmium(II), and mercury(II) complexes of 2-pyridineformamide 3-hexamethyleneiminylthiosemicarbazone, *Z. anorg. und allg. Chem.*, 630(7), 1096–1109 (2004b).

Bigoli F., Braibanti A., Lanfredi A.M.M., Tiripicchio A., Camellini M.T. The crystal and molecular structure of bis(thiocarbohydrazide-$N$,$S$)cadmium dichloride, *Inorg. Chim. Acta*, 5, 392–396 (1971).

Bigoli F., Braibanti A., Pellinghelli M.A., Tiripicchio A. The crystal and molecular structure of mono-($N$,$N$-diethylnicotinamide)cadmium dithiocyanate, *Acta Crystallogr.*, B28(3), 962–966 (1972).

Black S.J., Einstein F.W.B., Hayes P.C., Kumar R., Tuck D.G. Reactions of Cd(SR)$_2$ (R=n-butyl, phenyl) and the molecular structure of the 2,2′-bipyridine adduct of bis (n-butyl thioxanthato)cadmium(II), *Inorg. Chem.*, 25(23), 4181–4184 (1986).

Block E., Gernon M., Kang H., Ofori-Okai G., Zubieta J. Coordination chemistry of sterically hindered thiolate ligands. Preparation and structural characterization of the oligomeric homoleptic complexes [Cu(SC$_6$H$_4$-o-SiMe$_3$)]$_{12}$ and [{Ag(SC$_6$H$_4$-o-SiMe$_3$)}$_4$]$_2$ and a comparison to the structure of the mononuclear species (Et$_4$N)$_2$[Cd(SC$_6$H$_4$-o-SiMe$_3$)$_4$], *Inorg. Chem.*, 28(7), 1263–1271 (1989).

Block E., Ofori-Okai G., Chen Q., Zubieta J. Complexes of group 12 metals with sterically-hindered thiolate ligands. The crystal and molecular structures of [Zn$_2$(2-SC$_5$H$_3$N-3-SiMe$_3$)$_4$] and [CdI$_2$(2-SC$_5$H$_3$NH-6-SiMe$_2$Bu$^t$)$_2$], *Inorg. Chim. Acta*, 189(2), 137–139 (1991).

Bochmann M., Webb K., Hursthouse M.B., Mazid M. Sterically hindered chalcogenolato complexes. Mono- and di-meric thiolates and selenolates of zinc and cadmium; structure of [{Cd(SeC$_6$H$_2$Bu$^t_3$-2,4,6)$_2$}$_2$], the first three-co-ordinate cadmium–selenium complex, *J. Chem. Soc., Dalton Trans.*, (9), 2317–2323 (1991).

Bonamico M., Mazzone G., Vaciago A., Zambonelli L. Structural studies of metal dithiocarbamates. III. The crystal and molecular structure of zinc diethyldithiocarbamate, *Acta Crystallogr.*, 19(6), 898–909 (1965).

Bondar' V.I., Rau V.G., Struchkov Y.T., Akimov V.M., Molodkin A.K., Ilyukhin V.V., Belov N.V. Crystal structure of Cd(CH$_3$COO)$_2$·2CSN$_2$H$_4$ [in Russian], *Dokl. Akad. Nauk SSSR*, 250(4), 852–854 (1980).

Brennan J.G., Siegrist T., Carroll P.J., Stuczynski S.M., Reynders P., Brus L.E., Steigerwald M.L. Bulk and nanostructure group II-VI compounds from molecular organometallic precursors, *Chem. Mater.*, 2(4), 403–409 (1990).

Brinkhoff H.C., Cras J.A., Steggerda J.J., Willemse J. The oxidation of dithiocarbamato complexes of nickel, copper and zinc, *Rec. Trav. Chim. Pays-Bas*, 88(6), 633–640 (1969).

Brodersen K., Böhm K. Zur Koordination von Thiocyanat in ternären Metallkomplexen. Die Kristallstruktur von $Ba_2Cd(SCN)_6 \cdot 7H_2O$, *Z. Naturforsch.*, 41B(4), 439–443 (1986).

Broker G.A., Tiekink E.R.T. Bis[*N*-(2-hydroxyethyl)-*N*-propyldithiocarbamato-$\kappa^2 S,S'$]bis (4-{[(pyridin-4-yl-methylidene)hydrazinylidene]methyl}pyridine-$\kappa N^1$)cadmium, *Acta Crystallogr.*, E67(3), m320–m321 (2011).

Bu X., Zheng N., Li Y., Feng P. Templated assembly of sulfide nanoclusters into cubic-$C_3N_4$ type framework, *J. Am. Chem. Soc.*, 125(20), 6024–6025 (2003).

Bürgi H.-B. Stereochemistry of polynuclear cadmium(II) thioglycolates: Crystal structure of cadmium(II) bisthioglycolate, *Helv. Chim. Acta*, 57(3), 513–519 (1974).

Bürgi H.B., Gehrer H., Strickler P., Winkler F.K. Stereochemistry of polynuclear cadmium(II) thioglycolates: Crystal structure of $[ICd_8(SCH_2CH_2OH)_{12}]^{3+} \cdot 3I^- \cdot H_2O$, *Helv. Chim. Acta*, 59(7), 2558–2565 (1976).

Bustos L., Khan M.A., Tuck D.G. Neutral and anionic derivatives of toluene-3,4-dithiolatocadmium(II); the crystal structure of di(tetraphenylarsonium) bis(toluene-3,4-dithiolato)cadmate(II), *Can. J. Chem.*, 61(6), 1146–1152 (1983).

Byrom C., Malik M.A., O'Brien P., White A.J.P., Williams D.J. Synthesis and X-ray single crystal structures of bis(diisobutyldithiophosphinato)cadmium(II) or zinc(II): Potential single-source precursors for II/VI materials, *Polyhedron*, 19(2), 211–215 (2000).

Cai J., Chen C.-H., Feng X.-L., Liao C.-Z., Chen X.-M. A novel supramolecular synthon for H-bonded coordination networks: Syntheses and structures of extended 2-dimensional cadmium(II) arenedisulfonates, *J. Chem. Soc., Dalton Trans.*, (16), 2370–2375 (2001).

Câmpian M.V., Haiduc I., Tiekink E.R.T. Supra-versus intra-molecular $\pi \ldots \pi$ interactions in $M(S_2COCH_2Ph)_2$(dafone) compounds: M=Zn and Cd; dafone=4,5-diazafluoren-9-one-*N,N'*, *Z. Kristallogr.*, 228(4), 203–209 (2013).

Cannas M., Carta G., Cristini A., Marongiu G. Five-co-ordinate cadmium(II) in [bis (2-dimethylaminoethyl)methyl-amine]di-isothiocyanatocadmium(II), *J. Chem. Soc., Dalton Trans.*, (3), 210–212 (1976).

Cannas M., Carta G., Cristini A., Marongiu G. Unusual mode of coordination of bis (2-aminoethyl)amine in *catena*-bis[μ-bis(2-aminoethyl)amine]-bis[μ-thiocyanato) bis(isothiocyanato)dicadmium(II), *Inorg. Chem.*, 16(2), 228–230 (1977a).

Cannas M., Cristini A., Marongiu G. The crystal and molecular structure of di-(3-aminopropyl) aminedi-isothiocyanatecadmium(II), *Inorg. Chim. Acta*, 22, 233–237 (1977b).

Carson G.K., Dean P.A.W. The metal NMR spectra of thiolate and phenyl selenolate complexes of zinc(II) and mercury(II), *Inorg. Chim. Acta*, 66, 157–161 (1982).

Casals I., González-Duarte P., Clegg W., Foces-Foces C., Cano F.H., Martínez-Ripoll M., Gómez M., Solans X. Synthesis and structures of tetranuclear 2-(dimethylamino)eth-anethiolato complexes of zinc, cadmium and mercury involving both primary and secondary metal-halogen bonding, *J. Chem. Soc., Dalton Trans.*, (10), 2511–2518 (1991a).

Casals I., González-Duarte P., Clegg W. Preparation of complexes of the 3-trimethylam-monio-1-propanethiolato ligand, $[M\{S(CH_2)_3NMe_3\}_2](PF_6)_2$, M=Zn, Cd, Hg. Crystal structures of monomeric and polymeric forms of the mercury complex, *Inorg. Chim. Acta*, 184(2), 167–175 (1991b).

Casals I., González-Duarte P., Sola J., Font-Bardia M., Solans L., Solans X. Polymeric thiolate complexes of Group 12 metals. Crystal and molecular structures of *catena*-[μ-(3-dimethylammonio-1-propanethiolate)]-dichlorocadmium(II) and bis [3-(dimethylammonio)propyl] disulphide tetrabromocadmate(II), *J. Chem. Soc., Dalton Trans.*, (10), 2391–2395 (1987).

Casas J.S., Castellano E.E., Couce M.D., Sánchez A., Sordo J., Varela J.M., Zukerman-Schpector J. Vitamin B1: Chemical interaction with $CdCl_2$ and in vivo effects on cadmium toxicity in rats. Crystal structure of $[Cd(thiamine)Cl_3]_2 \cdot 2H_2O$, a complex containing pyrimidine and cadmium–hydroxyethyl bonds, *Inorg. Chem.*, 34(9), 2430–2437 (1995).

Casas J.S., García-Tasende M.S., Sánchez A., Sordo J., Vázquez-López E.M., Castellano E.E., Zukerman-Schpector J. Synthesis, crystal structure and spectroscopic properties of bis(diphenyldithiophosphinato)cadmium(II), *Inorg. Chim. Acta*, 219(1–2), 115–119 (1994).

Cassidy I., Garner M., Kennedy A.R., Potts G.B.S., Reglinski R., Slavin P.A., Spicer M.D. The preparation and structures of Group 12 (Zn, Cd, Hg) complexes of the soft tripodal ligand hydrotris(methimazolyl)borate (Tm), *Eur. J. Inorg. Chem.*, (5), 1235–1239 (2002).

Castiñeiras A., Bermejo E., West D.X., Ackerman L.J., Valdés-Martínez J., Hernández-Ortega S. Structural and spectral studies of 1-phenylglyoxal bis(3-piperidylthiosemicarbazone) and its zinc(II), cadmium(II) and platinum(II) complexes, *Polyhedron*, 18(10), 1463–1469 (1999).

Castiñeiras A., García I., Bermejo E., Ketcham K.A., West D.X., El-Sawaf A.K. Coordination of $Zn^{II}$, $Cd^{II}$, and $Hg^{II}$ by 2-pyridineformamide-3-piperidyl-thiosemicarbazone, *Z. anorg. und allg. Chem.*, 628(2), 492–504 (2002).

Castiñeiras A., Garcia I., Bermejo E., West D.X. Structures of 2-pyridineformamide thiosemicarbazone and its complexes with cadmium halides, *Polyhedron*, 19(15), 1873–1880 (2000).

Castro J., Romero J., García-Vázquez J.A., Durán M.L., Castiñeiras A., Sousa A., Fenton D.E. Synthesis of nickel(II), copper(II), zinc(II), and cadmium(II) complexes of 2-[(2-mercaptophenyl)iminomethyl]pyrrole and 2-[(2-mercaptoethyl)iminomethyl] pyrrole by the electrochemical cleavage of a disulphide bond: the crystal structure of bis{2-[(pyrrol-2-yl)methyleneamino]thiophenolato-S,N}zinc(II), *J. Chem. Soc., Dalton Trans.*, (11), 3255–3258 (1990).

Castro R., Durán M.L., García-Vázquez J.A., Romero J., Sousa A., Castiñeiras A., Hiller W., Strähle J. Direct electrochemical synthesis of pyrimidine-2-thionato complexes of zinc(II) and cadmium(II): The crystal structure of (1,10-phenanthroline)bis(pyrimidine-2-thionato)cadmium(II), *Z. Naturforsch.*, 47B(8), 1067–1074 (1992).

Castro R., García-Vázquez J.A., Romero J., Sousa A., Castiñeiras A., Hiller W., Strähle J. Synthesis of 3-trimethylsilylpyridine-2-thione (3-Me₃SipytH) complexes of nickel, copper, zinc and cadmium: Crystal and molecular structure of [Cd(3-Me₃Sipyt)₂], *Inorg. Chim. Acta*, 211(1), 47–54 (1993a).

Castro R., García-Vázquez J.A., Romero J., Sousa A., McAuliffe C.A., Pritchard R. Electrochemical synthesis of benzothiazole-2-thionato complexes of nickel(II), zinc(II) and cadmium(II): the crystal structure of 2,2'-bipyridine bis(benzothiazole-2-thionato) zinc(II), *Polyhedron*, 12(18), 2241–2247 (1993b).

Castro R., García-Vázquez J.A., Romero J., Sousa A., Pritchard R., McAuliffe C.A. 4,6-dimethylpyrimidine-2-thionato (dmpymt⁻) complexes of nickel(II) and cadmium(II). Crystal structure of [Cd(dmpymt)₂]: A compound with a calixarene-like structure, *J. Chem. Soc., Dalton Trans.*, (7), 1115–1120 (1994).

Cavalca L., Domiano P., Gasparri G.F., Boldrini P. The crystal structure of monothiourea-cadmium sulphate dihydrate, *Acta Crystallogr.*, 22(6), 878–885 (1967).

Cavalca L., Domiano P., Musatti A., Sgarabotto P. The crystal structure of bis(imidazoline-2-thione)cadmium chloride, *Chem. Commun.*, (18), 1136–1137 (1968).

Cavalca L., Nardelli M., Fava G. The crystal structure of bis-ethylenethiourea-cadmium thiocyanate, *Acta Crystallogr.*, 13(2), 125–130 (1960).

Cavalca L., Nardelli M., Fava G. The crystal structure of di-(2-thioimidazolidine)cadmium thiocyanate, *Proc. Chem. Soc.*, 159–160 (May 1959).

Cavalca L., Villa A.C., Mangia A., Palmieri C. The crystal structure of tris(thiourea)cadmium sulphate, *Inorg. Chim. Acta*, 4, 463–470 (1970).

Chai J., Lai C.S., Yan J., Tiekink E.R.T. Polymeric [bis(N,N-diethyldithiocarbamato) (trans-1,2-bis(4-pyridyl)ethylene)cadmium(II)], *Appl. Organomet. Chem.*, 17(4), 249–250 (2003).

Chan W.-H., Mak T.C.W., Yip W.-H., Smith G., O'Reilly E.J., Kennard C.H.L. Metal-(phenylthio)alkanoic acid interactions. VII. The crystal and molecular structures of diaquabis[(benzylthio)acetato]-zinc(II), *catena*-[diaqua-tetra[(benzylthio) acetato]-bis[cadmium(II)], *catena*-tetra-μ-{2-methyl-3-(phenylthio)propionato-O,O']-bis[copper(II)]} and tetra-μ-[2-methyl-2-(phenylthio)propionato-O,O']-bis [ethanolcopper(II)], *Polyhedron*, 6(5), 881–889 (1987).

Chen D., Lai C.S., Tiekink E.R.T. Bis(*O*-methoxyethyldithiocarbonato)(4,7-dimethyl-1, 10-phenanthroline)cadmium(II), *Appl. Organomet. Chem.*, 17(4), 247–248 (2003a).

Chen D., Lai C.S., Tiekink E.R.T. Crystal structures of 2,2'-bipyridine adducts of two cadmium O-alkyl dithiocarbonates: rationalisation of disparate coordination geometries based on different crystal packing environments, *Z. Kristallogr.*, 218(11), 747–752 (2003b).

Chen H.-J., Yang G., Chen X.-M. *catena*-Poly[[bis(imidazole-N$^3$)cadmium(II)]bis (μ-thiocyanato)-S:N;N:S], *Acta Crystallogr.*, C55(12), 2012–2014 (1999).

Chen H.-Y., Xia G.-M., Zhang Z.-W., Li P. Bis(tetrabutylammonium) bis(3,4,5-trioxocyclopent-1-ene-1,2-dithiolato-κ$^2$S,S')cadmate(II) 0.25-hydrate, *Acta Crystallogr.*, E67(1), m12–m13 (2011).

Chen W., Liu F., You X. Inorganic-organic hybrid materials: Synthesis and X-ray structure of N,N'-dimethylimidazolium salts [(Me$_2$Im)$_2$][Cd$_2$(SCN)$_6$] and N,N'-dicyclohexylimidazolium [(Cy$_2$Im)$_2$][Cd$_2$(SCN)$_6$]·C$_3$H$_6$O, *J. Solid State Chem.*, 167(1), 119–125 (2002a).

Chen W., Liu F., You X. Reactions of [Cd(SCN)$_2$] and isonicotinonitrile, synthesis and X-ray characterization of 1- and 3D coordination polymers, [Cd(SCN)$_2$(pyCN) $_n$] (pyCN = isonicotinonitrile, n = 1, 2), *Bull. Chem. Soc. Jpn.*, 75(7), 1559–1560 (2002b).

Choy A., Craig D., Dance I., Scudder M. [S$_4$M$_{10}$(SPh)$_{16}$]$^{4-}$ (M = Zn, Cd), a molecular fragment of the sphalerite MS lattice: Structural congruence of metal sulphides and metal thiolates, *J. Chem. Soc., Chem. Commun.*, (21), 1246–1247 (1982).

Chunying L., Ning Z. Crystal structure of aqua-bis-(2,2'-bipyridyl-N,N)(7-hydroxynaphthalene-1,3-disulfonato) cadmium(II)monohydrate, [Cd(H$_2$O)(C$_{10}$N$_2$H$_8$)$_2$](C$_{10}$H$_6$S$_2$O$_7$)·H$_2$O, C$_{30}$H$_{26}$CdN$_4$O$_9$S$_2$, *Z. Kristallogr New Cryst. Struct.*, 227(4), 538–540 (2012).

Cingi M.B., Lanfredi A.M.M., Tiripicchio A., Haasnoot J.G., Reedijk J. Single-strand polymer in *catena*-poly{[trans-bis(6,8-dimethyl[1,2,4]triazolo[3,4-b]-pyridazine-N$^2$) cadmium]-μ-(thiocyanato-N:S)-μ-(thiocyanato-S:N)}, *Acta Crystallogr.*, C42(11), 1509–1512 (1986).

Colombini G., Preti C. Complexes of thiomorpholin-3-one and thiazolidine-2-thione with zinc(II), cadmium(II) and mercury(II) thiocyanates, acetates and fluoborates, *J. Inorg. Nucl. Chem.*, 37(5), 1159–1165 (1975).

Corao E., Baggio S. The crystal structure of a five-coordinated cadmium(II) complex: Tristhiourea-cadmium sulphate, *Inorg. Chim. Acta*, 3, 617–622 (1969).

Corwin, Jr. D.T., Gruff E.S., Koch S.A. Zinc, cobalt, and cadmium thiolate complexes: Models for the zinc(S-cys)$_2$(his)$_2$ centre in transcription factor IIIA (cys = cysteine; his = histidine), *J. Chem. Soc., Chem. Commun.*, (13), 966–967 (1987).

Coucouvanis D., Murphy C.N., Simhon E., Stremple P., Draganjac M., del Neira R.P. Tetrakis(benzenethiolato)metallate(2–) complexes, [M(SPh)$_4$]$^{2-}$, of manganese, iron, cobalt, zinc, and cadmium and derivatives of the [Fe(SPh)$_4$]$^{2-}$ complex, *Inorg. Synth.*, 21, 23–28 (1982).

Coucouvanis D., Patil P.R., Kanatzidis M.G., Detering B., Baenziger N.C. Synthesis and reactions of binary metal sulfides. Structural characterization of the [(S$_4$)$_2$Zn]$^{2-}$, [(S$_4$)$_2$Ni]$^{2-}$, [(S$_5$)Mn(S$_6$)]$^{2-}$, and [(CS$_4$)$_2$Ni]$^{2-}$ anions, *Inorg. Chem.*, 24(1), 24–31 (1985).

Cox M.J., Tiekink E.R.T. Structural characteristics of cadmium(II) bis(N,N-dialkyldithiocarbamate) compounds, *Z. Kristallogr.*, 214(10), 670–676 (1999).

Craig D.C., Dance I.G., Dean P.A.W., Hook J.M., Jenkins H.A., Kirby C.W., Scudder M.L., Rajalingam U. A synthetic, structural and $^{113}$Cd NMR study of cadmium complexes of 1,3-thiazolidine-2-thionate, including the structures of Cd(C$_3$H$_4$NS$_2$)$_2$, (Cd(C$_3$H$_4$NS$_2$)$_2$(C$_5$H$_5$N)$_3$·C$_5$H$_5$N, and (Ph$_4$P)$_2$[I$_2$Cd(μ-C$_3$H$_4$NS$_2$)$_2$CdI$_2$], *Can. J. Chem.*, 83(2), 174–184 (2005).

Cras J.A., Willemse J., Gal A.W., Hummelink-Peters B.G.M.C. Preparation, structure and properties of compounds containing the dipositive tri-copper hexa(*N,N*-di-n-butyldithiocarbamato) ion, compounds with copper oxidation states II and III, *Rec. trav. chim. Pays-Bas*, 92(6), 641–650 (1973).

Cupertino D., Keyte R., Slawin A.M.Z., Williams D.J., Woollins J.D. Preparation and single-crystal characterization of $^i$Pr$_2$P(S)NHP(S)$^i$Pr$_2$ and homoleptic [$^i$Pr$_2$P(S)NP(S)$^i$Pr$_2$]$^-$ complexes of zinc, cadmium, and nickel, *Inorg. Chem.*, 35(9), 2695–2697 (1996).

Dance I.G., Choy A., Scudder M.L. Syntheses, properties and molecular and crystal structures of (Me$_4$N)$_4$[E$_4$M$_{10}$(SPh)$_{16}$] (E = S, Se; M = Zn, Cd): Molecular supertetrahedral fragments of the cubic metal chalcogenide lattice, *J. Am. Chem. Soc.*, 116(21), 6285–6295 (1984).

Dance I.G., Garbutt R.G., Craig D.C. Applications of cadmium NMR to polycadmium compounds. 5. Diagonally coordinated halide within an octametal cage. Crystal and solution structures of [XCd$_8$(SCH$_2$CH$_2$OH)$_{12}$X$_3$], X = Cl, Br, I, *Inorg. Chem.*, 26(22), 3732–3740 (1987a).

Dance I.G., Garbutt R.G. Craig D.C. Applications of Cd N.M.R. to polycadmium complexes. IV. Reactions of [Cd$_{10}$(SCH$_2$CH$_2$OH)$_{16}$]$^{4+}$ with chloride ion, and the crystal and molecular structure of hexadeca(2-hydroxyethanethiolato)-tetrachloro-decacadmium [Cd$_{10}$(SCH$_2$CH$_2$OH)$_{16}$Cl$_4$], *Aust. J. Chem.*, 39(9), 1449–1463 (1986).

Dance I.G., Garbutt R.G., Craig D.C, Scudder M.L., Bailey T.D. New non-molecular polyadamantanoid crystal structures of Cd(SAr)$_2$. Zeolitic Cd$_4$(SC$_6$H$_4$Me-4)$_8$, and solvated Cd$_8$(SC$_6$H$_4$Br-4)$_{16}$(DMF)$_3$ (DMF = dimethylformamide), in relation to the molecular structures of aggregates in solution, *J. Chem. Soc., Chem. Commun.*, (15), 1164–1167 (1987b).

Dance I.G., Garbutt R.G., Scudder M.L. A new three-dimensionally nonmolecular Cd–S lattice in Cd$_7$(SC$_6$H$_4$CH$_3$-2)$_{14}$(DMF)$_2$, *Inorg. Chem.*, 29(8), 1571–1575 (1990).

Dance I.G., Scudder M.L., Secomb R. Cadmium thiolates. Tetrahedral CdS$_4$ and dodecahedral CdS$_4$O$_4$ coordination in *catena*-bis(carbethoxymethanethiolato)cadmium(II), *Inorg. Chem.*, 22(12), 1794–1797 (1983).

Dean P.A.W., Jennings M., Rajalingam U., Craig D.C., Scudder M.L., Dance I.G. The crystal packing of [Cd(C$_4$H$_8$N$_2$S)$_4$](BPh$_4$)$_2$·2MeCN and [Cd(C$_4$H$_8$N$_2$S)$_4$](CF$_3$SO$_3$)$_2$. Tetraphenylborate forming an aryl box, *CrystEngComm*, 4(9), 46–50 (2002).

Dean P.A.W., Manivannan V. Synthesis and multinuclear ($^{13}$C, $^{77}$Se, $^{125}$Te, $^{199}$Hg) magnetic resonance spectra of adamantane-like anions of mercury(II), [(μ-ER)$_6$(HgX)$_4$]$^{2-}$ (E = S, Se, Te; X = Cl, Br, I), *Inorg. Chem.*, 29(16), 2997–3002 (1990).

Dean P.A.W., Payne N.C., Vittal J.J., Wu Y. Synthesis and NMR spectra ($^{31}$P, $^{111/113}$Cd, $^{77}$Se) of adamantane-like phosphine complexes with the (μ-ER)$_6$Cd core and the crystal and molecular structure of [(μ-SPr$^i$)$_6$(CdPPh$_3$)$_2$(CdOClO$_3$)$_2$]·EtOH, *Inorg. Chim.*, 32(21), 4632–4639 (1993).

Dean P.A.W., Vittal J.J. Simple synthesis and $^{113}$Cd NMR spectroscopic characterization of the fully terminally substituted clusters [(μ-EPh)$_6$(CdX)$_4$]$^{2-}$ (E = S or Se; X = Br or I), *Inorg. Chem.*, 24(23), 3722–3723 (1985).

Dean P.A.W., Vittal J.J., Craig D.C., Scudder M.L. Polytopal isomerism of the [Cd(S{O}CPh)$_3$]$^-$ anion, *Inorg. Chem.*, 37(7), 1661–1664 (1998).

Dean P.A.W., Vittal J.J., Payne N.C. Syntheses of (Me$_4$N)$_2$[(μ-EPh)$_6$(MX)$_4$] (M = Cd, Zn; E = S, Se; X = Cl, Br, I). Crystal and molecular structures of (Me$_4$N)$_2$[(μ-EPh)$_6$(CdBr)$_4$] (E = S, Se) and characterization of [(μ-EPh)$_6$(CdX)$_4$]$^{2-}$ in solution by $^{113}$Cd and $^{77}$Se NMR, *Inorg. Chem.*, 26(11), 1683–1689 (1987).

Dean P.A.W., Vittal J.J., Wu Y. Synthesis, multinuclear magnetic resonance spectra, and chemistry of some complexes $[(\mu\text{-SR})_6(MX)_4]^{2-}$ (R = alkyl; M = Zn or Cd; X = Cl, Br, or I) and the X-ray structural analysis of $(Et_4N)_2[(\mu\text{-SPr}^i)_6(CdBr)_4$, *Can. J. Chem.*, 70(3), 779–791 (1992).

Dee C.M., Tiekink E.R.T. [μ-1,1′-Bis(diphenylphosphino)ferrocene-P,P′]bis[bis(*N,N*-diethyldithiocarbamato-S,S′)cadmium(II)], *Acta Crystallogr.*, E58(4), m136–m138 (2002a).

Dee C.M., Tiekink E.R.T. Refinement of the crystal structure of dimeric bis(*N,N*-diethyldithiocarbamato)cadmium(II), $[Cd(S_2CNEt_2)_2]_2$, *Z. Kristallogr New Cryst. Struct.*, 217(1), 85–86 (2002b).

De Filippo D., Devillanova F., Preti C., Verani G. Thiomorpholin-3-one and thiazolidine-2-thione complexes with group IIB metals, *J. Chem. Soc. A: Inorg., Phys., Theor.*, 1465–1468 (1971).

Deng Y.-H., Liu J., Li N., Yang Y.-L., Ma H.-W. Synthesis, characterization and thermostability of IIB Group transition metal complexes with dithiocarbamate [in Chinese], *Acta Chim. Sinica.*, 65(24), 2868–2874 (2007).

Devillanova F.A., Verani G. Reactions of 1,3-oxazolidine-2-thione with zinc(II), cadmium(II) and mercury(II) halides, *J. Coord. Chem.*, 7(3), 177–180 (1978).

Devillanova F.A., Verani G. Zn(II) and Cd(II) halide complexes with *N*-mono and *N,N*′-disubstituted imidazolines-2-thione, *J. Nucl. Inorg. Chem.*, 42(4), 623–627 (1980).

Dillen J., Lenstra A.T.H., Haasnoot J.G., Reeddijk J. Transition metal thiocyanates of 5,7-dimethyl[1,2,4]triazolo[1,5-a]-pyrimidine studied by spectroscopic methods. The crystal structures of diaquabis (dimethyltriazolopyrimidine-$N^3$)bis(thiocyanato-N) cadmium(II) and bis(dimethyltriazolopyrimidine-$N^3$)bis(thiocyanato-S)-mercury(II), *Polyhedron*, 2(3), 195–201 (1983).

Domenicano A., Torelli L., Vaciago A., Zambonelli L. Structural studies of metal dithiocarbamates. Part IV. The crystal and molecular structure of cadmium(II) *NN*-diethyldithiocarbamate, *J. Chem. Soc. A: Inorg., Phys., Theor.*, 1351–1361 (1968).

Domiano P., Manfredotti A.G., Grossoni G., Nardelli M., Tani M.E.V. X-ray powder data, infrared spectra and crystal structures of some bis(selenourea)metal(II) thiocyanates, *Acta Crystallogr.*, B25(3), 591–595 (1969).

Du M., Li C.-P., Guo J.-H. Distinct $Cd^{II}$ and $Co^{II}$ thiocyanate coordination complexes with 2,5-bis(pyrazinyl)-1,3,4-oxadiazole: Metal-directed assembly of a 1-D polymeric chain and a 3-D supramolecular network, *Inorg. Chim Acta*, 359(8), 2575–2582 (2006).

Du Z.-Y., Li X.-L., Liu Q.-Y., Mao J.-G. Novel cadmium(II) phosphonatophenylsulfonate cluster compounds: Syntheses, structures, and luminescent properties, *Cryst. Growth Des.*, 7(8), 1501–1507 (2007).

Dubler E., Gyr E. New metal complexes of the antitumor drug 6-mercaptopurine. Syntheses and X-ray structural characterizations of dichloro(6-mercaptopurinium)copper(I), dichlorotetrakis(6-mercaptopurine)cadmium(II), and bis(6-mercaptopurinato) cadmium(II) dehydrate, *Inorg. Chem.*, 27(8), 1466–1473 (1988).

Durán M.L., Romero J., García-Vázquez J.A., Castro R., Castiñeiras A., Sousa A. Electrochemical synthesis of pyridine-2-thionato complexes of zinc(II) and cadmium(II): the crystal structure of 1,10-phenanthroline-bis(pyridine-2-thionato) zinc(II), *Polyhedron*, 10(2), 197–202 (1991).

Durski Z., Boniuk H., Majorowski S. Lattice constant and space groups examination of $CdI_2 \cdot 2CO(NH_2)_2$ and $CdI_2 \cdot 2CS(NH_2)_2$ crystals, *Rocz. Chem.*, 49(12), 2101–2105 (1975).

Edwards A.J., Fallaize A., Raithby P.R., Rennie M.-A., Steiner A., Verhorevoort K.L., Wright D.S. A halide-free route to groups 12 and 13 organometallic and metalloorganic complexes, *J. Chem. Soc., Dalton Trans.*, (1), 133–137 (1996).

Ellis C.A., Tiekink E.R.T. μ-1,4-Diazabicyclo[2.2.2]octane-κ²$N$:$N'$-bis[bis(O,O'-diisopropyl dithiophosphato-κ²S,S')cadmium(II)], *Acta Crystallogr.*, E62(11), m3049–m3051 (2006).

Fan J., Yin X., Zhang W.-G., Zhang Q.-J., Lai C.-S., Tiekink E.R.T., Fan Y., Nuang M.Y. Synthesis, structure and thermal stability of metal complexes with $N,N'$-dibenzyl dithiocarbamate [in Chinese], *Acta Chim. Sinica*, 62(17), 1626–1634 (2004).

Fan S.-R., Zhang L.-P., Xiao H.-P., Cai G.-Q., Zhu L.-G. Crystal structure of bis[diaqua(1,10-phenanthroline)(μ-4-sulfobenzoato)cadmium(II)], [Cd(H₂O)₂(C₁₂H₈N₂)(C₇H₄O₅S)]₂, Z. *Kristallogr. New Cryst. Struct.*, 220(1), 61–64 (2005).

Farneth W.E., Herron N., Wang Y. Bulk semiconductors from molecular solids: A mechanistic investigation, *Chem. Mater.*, 4(4), 916–922 (1992).

Fawcett T.G., Ou C.-C., Potenza J.A., Schugar H.J. Crystal and molecular structure of the tetrameric Cd(II) complex (Cd[SC(CH₃)₂CH₂NH₂]₂CdCl₂)₂·2H₂O, *J. Am. Chem. Soc.*, 100(7), 2058–2062 (1978).

Ferrari A., Braibanti A., Bigliardi G., Lanfredi A.M. The crystal structure of bis(hydrazine) zinc isothiocyanate, *Acta Crystallogr.*, 18(3), 367–372 (1965).

Fettouhi M., Malik M.R., Ali S., Isab A.A., Ahmad S. Dicyanidobis(thiourea-κS)cadmium(II) monohydrate, *Acta Crystallogr.*, E66(8), m997 (2010).

Fettouhi M., Wazeer M.I.M., Isab A.A. A novel polymeric Cd[SSe₂N₂] central core five-coordinate complex: Synthesis, X-ray structure and ¹¹³Cd, ⁷⁷Se CP MAS NMR characterization of *catena*(bis(μ₂-selenocyanato-N,Se)-(N,N'-dimethylimidazolidine-2-thione-S)-cadmium(II)), *Inorg. Chem. Commun.*, 11(3), 252–255 (2008).

Freeman H.C., Huq F., Stevens G.N. Metal binding by D-penicillamine: Crystal structure of D-penicillaminatocadmium(II), *J. Chem. Soc., Chem. Commun.*, (3), 90–91 (1976).

García I., Bermejo E., El Sawaf A.K., Castiñeiras A., West D.X. Structural studies of metal complexes of 2-pyridineformamide N(4)-methylthiosemicarbazone, *Polyhedron*, 21(7), 729–737 (2002).

Gigauri R.D., Machhoshvili R.I., Helashvili G.K., Gigauri R.I., Indzhiya M.A. Ag, Zn, Cd, Hg(II) monothioarsenates [in Russian], *Zhurn. Neorgan. Khim.*, 44(1), 26–29 (1999).

Glinskaya L.A., Klevtsova R.F., Zemskova S.M. Crystal and molecular structures of volatile complexes of bis(diethyldithiocarbamate)cadmium with 1,10-phenanthroline and 2,2'-bipyridyl [in Russian], *Zhurn. Strukt. Khim.*, 33(1), 106–114 (1992a).

Glinskaya L.A., L'vov P.E., Klevtsova R.F., Larionov S.V. Synthesis, crystal and molecular structure of (1,10-phenanthroline)bis(isopropylxanthogenato)cadmium (lead) [in Russian], *Zhurn. Neorgan. Khim.*, 35(4), 911–917 (1990).

Glinskaya L.A., Zemskova S.M., Klevtsova R.F. Crystal and molecular structure of diisobutyldithiocarbamato cadmium(II)){Cd[(i-C₄H₉)₂NSC₂]₂}₂ [in Russian], *Zhurn. Strukt. Khim.*, 40(6), 1206–1212 (1999).

Glinskaya L.A., Zemskova S.M., Klevtsova R.F., Larionov S.V., Gromilov S.A. The preparation, structures and thermal properties of [MEn₃][Cd(S₂CNEt₂)₃]₂ [M=zinc(II), cadmium(II)] complexes, *Polyhedron*, 11(22), 2951–2956 (1992b).

Goel R.G., Henry W.P., Olivier M.J., Beauchamp A.L. Preparation and spectral characterization of cadmium(II) thiocyanate complexes of tricyclohexyl-, triphenyl- and tri-m-tolylphosphines and the crystal and molecular structure of the tri-m-tolylphosphine complex, *Inorg. Chem.*, 20(11), 3924–3928 (1981).

Goel R.M., Ogini W.O. Preparation, characterization, and spectral studies of neutral tri-*tert*-butylphosphine complexes of zinc(II) and cadmium(II), *Inorg. Chem.*, 16(8), 1968–1972 (1977).

Golovnev N.N., Molokeev M.S. Crystal structure of *catena*-bis(2-thiobarburato-O,S) diaquacadmium, *Rus. J. Inorg. Chem.*, 58(10), 1193–1196 (2013).

Gopalakrishnan J., Patel C.C. Diphenyl sulphoxide complexes of some divalent metal ions, *Inorg. Chim. Acta*, 1, 165–168 (1967).

Govindaswamy N., Moy J., Millar M., Koch S.A. A distorted $[Hg(SR)_4]^{2-}$ complex with alkanethiolate ligands: The fictile coordination sphere of monomeric $[Hg(SR)_x]$ complexes, *Inorg. Chem.*, 31(26), 5343–5344 (1992).

Griffith E.A.H., Amma E.L. Crystal structure and $_{48}{}^{113}Cd$ N.M.R. spectrum of di-μ-chloro-dichlorobis-(6-mercaptopurine)diaquodicadmium(II), *J. Chem. Soc., Chem. Commun.*, 1013–1014 (1979).

Griffith E.A.H., Charles N.G., Rodesiler P.F., Amma E.L. A tetrahedral $CdS_2O_2$ stereochemistry: the structure of (dinitrato-O)bis(1,1,3,3-tetramethyl-2-thiourea-S)cadmium(II), $Cd(C_5H_{12}N_2S)_2(NO_3)_2$, and its $^{113}Cd$ NMR spectra, *Acta Crystallogr.*, C39(3), 331–333 (1983).

Groeneveld L.R., Vos G., Verschoor G.C., Reedijk J. Crystal structure of *catena*-poly-(thiocyanato-N)cadmium-di-μ-(thiocyanato-N)-(thiocyanato-N)cadmium-tris-μ-(4-t-butyl-1,2,4-triazole-$N^1$:$N^2$). A new, alternating zig-zag chain containing N-bonded, bridging thiocyanate, *J. Chem. Soc., Chem. Commun.*, (11), 620–621 (1982).

Gruff E.S., Koch S.A. Trigonal-planar $[M(SR)_3]^{1-}$ complexes of cadmium and mercury. Structural similarities between mercury–cysteine and cadmium–cysteine coordination centers, *J. Am. Chem. Soc.*, 112(3), 1245–1247 (1990).

Guo S.Y., Xu D., Lv M., Yuan D., Yang Z., Zhang G., Sun S., Wang X., Zhou M., Jiang X. A novel organometallic nonlinear optical complex crystal: Cadmium mercury thiocyanate dimethyl-sulphoxide, *Progr. Cryst. Growth Charact. Mater.*, 40(1–4), 111–114 (2000a).

Guo S.Y., Xu D., Yang P., Yu W.T., Lv M.K., Yuan D.R., Yang Z.H. et al. A novel nonlinear optical complex crystal with an organic ligand coordinated through an O atom: Tetrathiocyanatocadmuimmercury–dimethyl sulfoxide, *Cryst. Res. Technol.*, 36(6), 609–614 (2001).

Guo S., Yuan D., Xu D., Zhang G., Sun S., Meng F., Wang X., Jiang X., Jiang M. Growth of cadmium mercury thiocyanate dimethyl-sulphoxide single crystal for laser frequency doubling, *Progr. Cryst. Growth Charact. Mater.*, 40(1–4), 75–79 (2000b).

Guo T., Lai C.S., Tan X.J., Teo C.S., Tiekink E.R.T. Bis(diethyldithiocarbamato)(4,7-dimethyl-1,10-phenanthroline)cadmium(II) acetonitrile solvate, *Acta Crystallogr.*, E58(8), m439–m440 (2002).

Haasnoot J.G., De Keyzer G.C.M., Verschoor G.C. *catena*-Di-μ-thiocyanato-N,S-bis(1H-1,2,4-triazole-$N^4$)cadmium, $[Cd(C_2H_3N_3)_2(NCS)_2]$; a third structure type of composition $M(NCS)_2(1,2,4-triazole)_2$, *Acta Crystallogr.*, 39(9), 1207–1209 (1983).

Hagen K.S., Holm R.H. Stereochemistry of $[Cd_4(SC_6H_5)_{10}]^{2-}$, a cage complex related to the cadmium-cysteinate aggregates in metallothioneins, *Inorg. Chem.*, 22(21), 3171–3174 (1983).

Hagen K.S., Stephan D.W., Holm R.H. Metal ion exchange reactions in cage molecules: The systems $[M_{4-n}M'_n(SC_6H_5)_{10}]^{2-}$ (M, M' =Fe(II), Co(II), Zn(II), Cd(II)) with adamantane-like stereochemistry and the structure of $[Fe_4(SC_6H_5)_{10}]^{2-}$, *Inorg. Chem.*, 21(11), 3928–3936 (1982).

Hencher J.L., Khan M.A., Said F.F., Tuck D.G. The direct electrochemical synthesis and crystal structure of salts of $[M_4(SC_6H_5)_{10}]^{2-}$ anions (M=Zn,Cd), *Polyhedron*, 4(7), 1263–1267 (1985).

Herron N.J., Calabrese C., Farneth W.E., Wang Y. Crystal structure and optical properties of $Cd_{32}S_{14}(SC_6H_5)_{36}·DMF_4$, a cluster with a 15 angstrom CdS core, *Science*, 259(5100), 1426–1428 (1993).

Herron N.J., Calabrese C., Farneth W.E., Wang Y. Crystal structure and optical properties of $Cd_{32}S_{14}(SC_6H_5)_{36}·DMF_4$, a cluster with a 15 angstrom CdS core, SPIE Milestone Series, MS180, 61–63 (2005).

Higgins G.M.C., Saville B. Complexes of amines with zinc dialkyldithiocarbamates, *J. Chem. Soc.*, 2812–2817 (1963).

Hoskins B.F., Kelly B.P. A novel five-covalent cadmium(II) complex: The crystal structure of tetraethylammonium tris(O-ethylxanthato)cadmiumate(II), *Inorg. Nucl. Chem. Lett.*, 8(10), 875–878 (1972).

Hu M., Wang R. Crystal structure of bis[(1-(4-carboxyphenyl)-5-mercapto-1H-tetrazole) (1,10-phenanthroline)]cadmium(II), $Cd(C_8H_5N_4O_2S)_2(C_{12}H_8N_2)_2$, *Z. Kristallogr. New Cryst. Struct.*, 225(1), 185–187 (2010).

Hursthouse M.B., Khan O.F.Z., Mazid M., Motevalli M., O'Brien P. The X-ray crystal structures of the cadmium complexes of pyridine-1-thiol and mercaptobenzothiazole, $[Cd(C_5H_4NS)_2]_n$ and $[Cd(C_7H_4N_2S_2)_2]_n$: Two unusual volatile polymeric complexes, *Polyhedron*, 9(4), 541–544 (1990).

Hursthouse M.B., Malik M.A., Motevalli M., O'Brien P. Mixed alkyl dialkylthiocarbamates of zinc and cadmium: Potential precursors for II/VI materials. X-ray crystal structure of $[MeZnS_2CNEt_2]_2$, *Organometallics*, 10(3), 730–732 (1991).

Iimura Y., Ito T., Hagihara H. The crystal structure of cadmium ethylxanthate, *Acta Crystallogr.*, B28(7), 2271–2279 (1972).

Iizuka M., Sudo T. Crystal structure of cadmium mercuric thiocyanate, $CdHg(SCN)_4$, *Z. Kristallogr.*, 126(5–6), 376–378 (1968).

Isab A.A., Wazeer M.I.M. Complexation of Zn(II), Cd(II) and Hg(II) with thiourea and sele-nourea: A $^1H$, $^{13}C$, $^{15}N$, $^{77}Se$ and $^{113}Cd$ solution and solid-state NMR study, *J. Coord. Chem.*, 58(6), 529–537 (2005).

Ito T., Otake M. Linear-chain structure of bis(O,O′-dimethyldithiophosphato)cadmium(II), *Acta Crystallogr.*, C52(12), 3024–3025 (1996).

Ivanchenko A.V., Gromilov S.A., Zemskova S.M., Baidina I.A. Investigation of bis(di-n-butyldithiocarbamates) zinc(II) and cadmium(II) complexes with 2,2′-bipyridyl [in Russian], *Zhurn. Strukt. Khim.*, 41(1), 106–115 (2000).

Ivanchenko A.V., Gromilov S.A., Zewmskova S.M., Baidina I.A., Glinskaya L.A. Synthesis and crystal structure of bis(di-n-butyldithiocarbamato)(1,10-phenanthroline) cadmium(II) [in Russian], *Zhurn. Strukt. Khim.*, 39(4), 682–687 (1998).

Ivanov A.V., Konzelko A.A., Gerasimenko A.V., Ivanov M.A., Antsutkin O.N., Forsling W. Structural organization of cadmium(II) and copper(II) dithiocarbamate complexes with dialkyl-substituted and cyclic ligands: synthesis, single-crystal X-ray diffraction, EPR, and CP/MAS $^{13}C$, $^{15}N$, and $^{113}Cd$ NMR [in Russian], *Zhurn. Neorgan. Khim.*, 50(11), 1827–1844 (2005).

Jambor J.L., Roberts A.C. New mineral names, *Am. Mineral.*, 84(9), 1464–1468 (1999).

Jeevanandam P., Vasudevan S. Preparation and characterization of $Cd_{0.75}PS_3A_{0.5}(H_2O)_y$ [A = Na, K and Cs], Solid State Ionics, 104(1–2), 45–55 (1997).

Jian F.-F., Wang Z.-X., Fun H.-K., Bai Z.-P., You X.-Z. A binuclear cadmium(II) complex: Bis[bis(N,N-diisopropyldithiocarbamato)-cadmium(II)], *Acta Crystallogr.*, C55(2), 174–176 (1999a).

Jian F., Wang Z., Bai Z., You X., Fun H.-K., Chinnakali K. Structure of bis(dipropyldithiocarbamate) cadmium(II), $[Cd_2(n-Pr_2dtc)_4]$ (dtc = dithiocarbamate), *J. Chem. Crystallogr.*, 29(2), 227–231 (1999b).

Jiang X.H., Zhang W.G., Zhong Y., Wang S.L. Synthesis and structure of the cadmium (II) complex: $[Cd(C_5H_5N)_2(S_2CO-n-C_4H_9)_2]$, *Molecules*, 7(7), 549–553 (2002).

Jiang X.-H., Zhang W.-G., Zhong Y., Wang S.-L. Crystal structure of (2,2′-bipyridine)bis(n-butyldithiocarbonato-S,S′)-cadmium(II), $Cd(C_4H_9COS_2)_2(C_{10}H_8N_2)$, *Z. Kristallogr. New Cryst. Struct.*, 220(4), 569–570 (2005).

Jiang X.N., Yu W.T., Yuan D.R., Xu D., Lu M.K., Wang X.Q., Guo S.Y., Jiang M.H. Redetermination of the crystal structure of dichlorotetrakisthioureacadmium, $C_4H_{16}CdCl_2N_8S_4$, *Z. Kristallogr. New Cryst. Struct.*, 215(4), 499–500 (2000).

Jin X., Tang K., Jia S., Tang Y. Synthesis and crystal structure of a polymeric complex [$S_4Cd_{17}(SPh)_{24}(CH_3OCS_2)_{4/2}]_n·nCH_3OH$, *Polyhedron*, 15(15), 2617–2622 (1996).

Kaptein B., Wang-Griffin L., Barf G., Kellogg R.M. Metal complexes that model the active site of liver alcohol dehydrogenase, *J. Chem. Soc., Chem. Commun.*, (19), 1457–1459 (1987).

Karakus M., Yilmaz H., Ozcan Y., Ide S. Bis{μ-[O-cyclopentyl(4-methoxyphenyl) dithiophosphonato]1κ:S,2κ:S-[O-cyclopentyl(4-methoxyphenyl)dithio-phosphonato]1κ²S,S'}dicadmium(II), *Appl. Organometal. Chem.*, 18(3), 141–142 (2004).

Ke Y., Li J., Zhang Y. Synthesis and crystal structure analysis of one dimensional supramo-lecular [$Cd(Glu)(Tu)_2]_n$, *Cryst. Res. Technol.*, 37(5), 501–508 (2002).

Kennedy B.P., Lever A.B.P. Studies of the metal–sulfur bond. Complexes of the pyridine thiols, *Can. J. Chem.*, 50(21), 3488–3507 (1972).

Klevtsova R.F., Glinskaya L.A., Shchukin V.G., Larionov S.V. Peculiarities of molecule packings in crystal structures of mixed-ligand complexes [$CdPhen\{(i-C_4H_9)_2PS_2\}_2$] and [$Cd(2,2'-Bipy)\{(i-C_4H_9)_2PS_2\}_2$] [in Russian], *Zhurn. Strukt. Khim.*, 43(1), 141–150 (2002).

Klevtsova R.F., Leonova T.G., Glinskaya L.A., Larionov S.V. Synthesis, crystal and molecu-lar structure of mixed-ligand complexes (iso-$C_4H_9OCS_2)_2$ and CdPhen(n-$C_4H_9OCS_2)_2$ [in Russian], *Zhurn. Strukt. Khim.*, 48(2), 274–281 (2007).

Kondo M., Shimamura M., Noro S.-I., Kimura Y., Uemura K., Kitagawa S. Synthesis and structures of coordination polymers with 4,4'-dipyridyldisulfide, *J. Solid State Chem.*, 152(1), 113–119 (2000).

Kramolowsky R., Sawluk J., Siasios G., Tiekink R.T. Synthesis and crystal structure of [$Cd\{(c-C_6H_{11})_2P(Se)C(S)NPh\}_2$], *Inorg. Chim. Acta*, 269(2), 317–321 (1998).

Kubiak M., Głowiak T. Structure of *catena*-poly[(1,3-thiazolidine-2-thionecadmium)-μ-chloro-$μ_3$-chloro], *Acta Crystallogr.*, C41(11), 1580–1582 (1985).

Kubiak M., Głowiak T., Kozlowski H. Structure of 2-amino-4,5-dihydro-3H⁺-1,3-thiazolium trichlorocadmate(II), $C_3H_7N_2S^+·CdCl_3^-$, *Acta Crystallogr.*, C39(12), 1637–1639 (1983).

Kuniyasu Y., Suzuki Y., Taniguchi M., Ouchi A. The synthesis and the X-ray crystal structure of tetramethylammonium tetrakis- and tris(thiocyanato)cadmates(II), [$(CH_3)_4N]_2[Cd(SCN)_4$] and [$(CH_3)_4N][Cd(SCN)_3$], *Bull. Chem. Soc. Jpn.*, 60(1), 179–183 (1987).

Lacelle S., Stevens W.C., Kurtz, Jr. D.M., Richardson, Jr. J.W., Jacobson R.A. Crystal and molecular structure of [$Cd_{10}(SCH_2CH_2OH)_{16}](ClO_4)_4·8H_2O$. Correlations with ¹¹³Cd NMR spectra of the solid and implications for cadmium-thiolate ligation in proteins, *Inorg. Chem.*, 23(7), 930–935 (1984).

Lai C.S., Tiekink E.R.T. Bis(*N,N*-diethyldithiocarbamato)(2,9-dimethyl-1,10-phenanthroline) cadmium(II), *Appl. Organometal. Chem.*, 17(2), 139–140 (2003).

Lai C.S., Tiekink E.R.T. Engineering polymers with variable topology—Bipyridine adducts of cadmium dithiophosphates, *CrystEngComm*, 6(97), 593–605 (2004).

Lai C.S., Tiekink E.R.T. Polymeric topologies in cadmium(II) dithiophosphate adducts of the isomeric n-pyridinealdazines, n = 2, 3 and 4, *Z. Kristallogr.*, 221(4), 288–293 (2006).

Larionov S.V., Glinskaya L.A., Leonova T.G., Klevtsova R.F. Synthesis, crystal and molecu-lar structures of mixed-ligand $Cd(2,2'-Bipy)(i-PrOCS_2)_2$ and $Cd(2,2'-Bipy)(i-BuOCS_2)_2$ complexes [in Russian], *Zhurn. Strukt. Khim.*, 46(6), 1064–1071 (2005).

Larionov S.V., Kirichenko V.N., Zemskova C.M., Oglezneva I.M. Synthesis of complexes of diethyldithiocarbamates of zinc(II), cadmium(II), mercury(II) with a nitrogen-containing ligands, and the study of their sublimation [in Russian], *Koord. Khim.*, 16(1), 79–84 (1990).

Larionov S.V., Shchukin V.G., Glinskaya L.A., Klevtsova R.F., Mazhara A.P. Mixed-ligand complexes of coordination compounds $CdL\{i-C_4H_9\}_2PS_2\}$ (L=Phen, 2,2′-Bipy, 4,4′-Bipy): Synthesis, structure, and thermal properties. Polymeric structure of $[Cd(4,4′-Bipy)\{i-C_4H_9\}_2PS_2\}]_n$ [in Russian], *Koord. Khim.*, 27(7), 498–503 (2001).

Larsen E., Trinderup P. The crystal and molecular structure of bis(thiosemicarbazide)-cadmium(II) sulfate, *Acta Chem. Scand. A*, 29(5), 481–484 (1975).

Lawton S.L., Kokotailo G.T. Crystal and molecular structures of zinc and cadmium O,O-diisopropylphosphorodithioates, *Inorg. Chem.*, 8(11), 2410–2421 (1969).

Lee G.S.H., Craig D.C., Ma I., Scudder M.L., Bailey T.D., Dance I.G. $[S_4Cd_{17}(SPh)_{28}]^{2-}$, the first member of a third series of tetrahedral $[S_WM_X(SR)_y]^{z-}$ clusters, *J. Am. Chem. Soc.*, 110(14), 4863–4864 (1988).

Lee G.S.H., Fisher K.J., Craig D.C., Scudder M.L., Dance I.G. $[ECd_8(E′Ph)_{16}]^{2-}$cluster chemistry (E,E′=sulfur, selenium, tellurium), *J. Am. Chem. Soc.*, 112(17), 6435–6437 (1990).

Lee G.S.H., Fisher K.J., Vassallo A.M., Hanna J.V., Dance I.G. Solid state [113]Cd NMR of three structural isomers of $[S_4Cd_{10}(SPh)_{16}]^{4-}$, *Inorg. Chem.*, 32(1), 66–72 (1993).

Li C., Chen M., Shao C. trans-Tetraaquabis(nicotinamide-κN)cadmium(II) biphenyl-4,4′-disulfonate, *Acta Crystallogr.*, E64(2), m424 (2008).

Li H., Kim J., Groy T.L., O'Keeffe M., Yaghi O.M. 20 Å $Cd_4In_{16}S_{35}^{14-}$ supertetrahedral T4 clusters as building units in decorated cristobalite frameworks, *J. Am. Chem. Soc.*, 123(20), 4867–4868 (2001).

Li H., Kim J., O'Keeffe M., Yaghi O.M. $[Cd_{16}In_{64}S_{134}]^{44-}$: 31-Å tetrahedron with a large cavity, *Angew. Chem. Int. Ed.*, 42(16), 1819–1821 (2003).

Li H., Xie L.M. *catena*-Poly[[bis[2-(1H-1,2,4-triazol-1-yl-κN⁴)pyrazine]cadmium(II)]-di-μ-thiocyanato-κ²S:N;κ²N:S], *Acta Crysallogr.*, E65(8), m935 (2009).

Li J.-X., Du Z.-X. Crystal structure of *catena*-[(5-sulfosalicylato-κ²O,O′)-bis(μ₂-4-thiolatopyridinium-κ²S:S)cadmium(II)] hydrate, $Cd(C_7H_4O_6S)(C_5H_5NS)_2·2.5H_2O$, *Z. Kristallogr. New Cryst. Struct.*, 226(3), 331–332 (2011).

Li M., Sun Q., Bai Y., Duan C., Zhang B., Meng Q. Chiral aggregation and spontaneous resolution of thiosemicarbazone metal complexes, *Dalton Trans.*, (21), 2572–2578 (2006).

Li X., Cao R., Bi W., Yuan D., Sun D. Self-assembly of 1D to 3D cadmium complexes: structural characterization and properties, *Eur. J. Inorg. Chem.*, (15), 3156–3166 (2005).

Li X., Cao R., Sun D., Bi W., Yuan D. A novel 3-D self-penetrating topological network assembled by mixed bridging ligands, *Eur. J. Inorg. Chem.*, (11), 2228–2231 (2004).

Li Z., Loh Z.-H., Fong A.S.-W., Yan Y.-K., Henderson W., Mok K.F., Hor T.S.A. Ligand-stabilization of an unusual square-based pyramidal geometry of Cd(II) and Zn(II) in an heterometallic $\{MPt_2S_2\}$ core (M=Cd, Zn), *J. Chem. Soc., Dalton Trans.*, (7), 1027–1031 (2000).

Lindoy L.F., Busc D.H. Helical co-ordination: five-co-ordinate zinc and cadmium complexes formed by metal-ion-induced ligand reactions, *J. Chem. Soc., Chem. Commun.*, (11), 683–684 (1972).

Liu Q.-Y., Wang Y.-L., Xu L. Synthesis, crystal structures and photoluminescent properties of three novel cadmium(II) compounds constructed from 5-sulfoisophthalic acid ($H_3SIP$), *Eur. J. Inorg. Chem.*, (23), 4843–4851 (2006).

Liu Q.-Y., Xu L. $(H_2O)_{12}$-containing infinite chain encapsulated in supramolecular open framework built of cadmium(II), 1,3-di(4-pyridyl)propane and 5-sulfoisophthalic acid monosodium salt, *CrystEngComm*, 7(12), 87–89 (2005a).

Liu Q.-Y., Xu L. Synthesis, crystal structures and photophysical properties of two supramolecular complexes of cadmium(II), *Inorg. Chem. Commun.*, 8(4), 401–405 (2005b).

Liu X.T., Wang X.Q., Lin X.J., Sun G.H., Zhang G.H., Xu D. Growth, characterization and theoretical study of a novel organometallic nonlinear optical crystal:
CdHg(SCN)$_4$(C$_2$H$_5$NO)$_2$, *Appl. Phys. A*, 107(4), 949–957 (2012).

Liu Y., Chen K., Zhang C.-C. Crystal structure of poly[chlorido-(1,10-phenanthroline)-4-
nitro-thiophene-2-carboxylato-κO:O')cadmium(II)],        Cd(Cl)(C$_5$H$_2$NO$_4$S)(C$_{12}$H$_8$N$_2$),
C$_{17}$H$_{10}$CdClN$_3$O$_4$S, *Z. Kristallogr. New Cryst. Struct.*, 229(3), 237–238 (2014).

Lobana T.S., Sharma Renu, Sharma Rekha, Sultana R., Butcher R.J. Metal derivatives of
1,3-imidazolidine-2-thione with divalent d$^{10}$ metal ions (Zn–Hg): Synthesis and structures, *Z. anorg. und allg. Chem.*, 634(4), 718–723 (2008).

Lopez-Garzon R., Gutierrez-Valero M.D., Godino-Salido M.L., Keppler B.K., Nuber
B. Spectroscopic studies of metal-pyrimidine complexes. Crystal structures of
4,6-dimethyl-2-thiopyrimidine complexes with Zn(II) and Cd(II), *J. Coord. Chem.*,
30(2), 111–123 (1993).

López-Torres E., Mendiola M.A., Rodríguez-Procopio J., Sevilla M.T., Colacio E., Moreno
J.M., Sobrados I. Synthesis and characterisation of zinc, cadmium and mercury com-
plexes of benzilbisthiosemicarbazone. Structure of cadmium derivative, *Inorg. Chim
Acta*, 323(1–2), 130–138 (2001).

Lou S.-F., Qiu X.-Y. Crystal structure of bis(μ$_2$-thiocyanato-N,S)-bis(isothiocyanato)-bis
[(2-morpholin-4-ylethyl)-(1-pyridin-2-ylethylidene)amine-N,N',N'']-dicadmium(II),
Cd$_2$(NCS)$_4$(C$_{13}$H$_{19}$N$_3$O)$_2$, *Z. Kristallogr. New Cryst. Struct.*, 226(2), 213–214 (2011).

Luo J., Hong M., Wang R., Yuan D., Cao R., Han L., Xu Y., Lin Z. Self-assembly of three
Cd$^{II}$- and Cu$^{II}$-containing coordination polymers from 4,4'-dipyridyl disulfide, *Eur. J.
Inorg. Chem.*, (19), 3623–3632 (2003).

Mahadevappa D.S., Murthy A.S.A. Some complexes of zinc(II), cadmium(II), and mercury(II)
with thiosemicarbazide, *Aust. J. Chem.*, 25(7), 1565–1568 (1972).

Mak T.C.W., Yip W.-H., Smith G., O'Reilly E.J., Kennard C.H.L. Metal–(phenylthio)
acetic acid interactions. Part I. The crystal structures of (phenylthio)acetic acid,
diaquabis[(phenylthio)acetato]zinc(II),        and        catena{aquabis[(phenylthio)acetato]
cadmium(II), *Inorg. Chim. Acta*, 84(1), 57–64 (1984).

Malik M.R., Ali S., Fettouhi M., Isab A.A., Ahmad S. Structural characterization of
dichlorobis(N,N'-dimethylthiourea-S)cadmium(II), J. Struct. Chem., 51(5), 976–979 (2010).

Malik M.R., Rüffer T., Lang H., Isab A.A., Ali S., Ahmad S., Stoeckli-Evans H. Structural
characterization of dicyanidobis(N,N'-dimethylthiourea-kS)cadmium(II), *Zhurn.
Strukt. Khim.*, 54(4), 763–767 (2013).

Mandarino J.A. Barquillite Cu$_2$(Cd, Fe)GeSe$_4$, *Mineral. Rec.*, 30(5), 999 (1999).

Marcos C., Alía J.M., Adovasio V., Prieto M., García-Granda S. Bis(thiourea) cadmium
halides, *Acta Crystallogr.*, C54(9), 1225–1229 (1998).

Marushko L.P., Piskach L.V., Parasyuk O.V., Ivashchenko I.A., Olekseyuk I.D. Quasiternary
system Cu$_2$GeS$_3$–Cu$_2$SnS$_3$–CdS, *J. Alloys Compd.*, 484(1–2), 147–153 (2009a).

Marushko L.P., Piskach L.V., Parasyuk O.V., Olekseyuk I.D., Volkov S.V., Pekhnyo V.I. The
reciprocal system Cu$_2$GeS$_3$+3CdSe ⇔ Cu$_2$GeSe$_3$+3 CdS, *J. Alloys Compd.*, 473(1–2),
94–99 (2009b).

Marushko L.P., Piskach L.V., Romanyuk Y.E., Parasyuk O.V., Olekseyuk I.D., Volkov S.V.,
Pekhnyo V.I. Quasiternary system CuGaS$_2$–CuInS$_2$–2CdS, *J. Alloys Compd.*, 492
(1–2), 184–189 (2010).

Matsunaga Y., Fujisawa K., Amir N., Miyashita Y., Okamoto K.-I. Group 12 metal(II) complexes with 1-methylimidazoline-2(3H)-thione (mitH): correlation between crystal
structure and physicochemical property, *J. Coord. Chem.*, 58(12), 1047–1061 (2005).

Mautner F.A., Abu-Youssef M.A.M., Goher M.A.S. Polymeric complexes of cadmium(II)
bridged simultaneously by tetradentate picolinato and μ(1,1)-azido or μ(N,S)-
thiocyanato anions. Synthesis and structural characterization of [Cd(picolinato)(N$_3$)$_n$
and [Cd(picolinato)(NCS)]$_n$, *Polyhedron*, 16(2), 235–242 (1997).

McCleverty J.A., Gill S., Kowalski R.S.Z., Bailey N.A., Adams H., Lumbard K.W., Murphy M.A. Aspects of the inorganic chemistry of rubber vulcanization. Part 3. Anionic cadmium complexes derived from dialkyldithiocarbamates, 2-mercaptobenzothiazole and its derivatives, and dialkyl dithiophosphates, and the crystal and molecular structures of $[NBu^n_4][Cd(S_2CNEt_2)_3]$, $[NEt_4][Cd(C_7H_4NS_2)_3]$, and $[NMe_4][Cd\{S_2P(OPr^i)_2\}_3$, *J. Chem. Soc., Dalton Trans.*, (3), 493–503 (1982).

McCormick B.J., Greene D.L. Thiocarbamate complexes. II. Derivatives of zinc(II), cadmium(II) and mercury(II), *Inorg. Nucl. Chem. Lett.*, 8(7), 599–703 (1972).

Meek D.W., Nicpon P. Metal complexes of tertiary phosphine sulfides, *J. Am. Chem. Soc.*, 87(21), 4951–4952 (1965).

Mi L.-W., Han M.-L., Li D.-P., Chen W.-H., Zheng Z. A novel two-dimensional cadmium polymeric aminonaphthalene sulfonate and its application in the synthesis of CdS materials, *Z. anorg. und allg. Chem.*, 634(2), 373–376 (2008).

Moers O., Henschel D., Blaschette A., Jones P.G. Lamellare Schichten auf der Grundlage von Wasserstoffbrücken und π-Stapelung: Kristallstrukturen der Metall(II)-Komplexe $[Cd\{C_6H_4(SO_2)_2N\}_2(H_2O)_4]$ und $[Cu\{C_6H_4(SO_2)_2N\}_2(H_2O)_4]_2$ $H_2O$, *Z. anorg. und allg. Chem.*, 628(2), 505–511 (2002).

Moloto M.J., Malik M.A., O'Brien P., Motevalli M., Kolawole G.A. Synthesis and characterisation of some N-alkyl/aryl and N,N'-dialkyl/aryl thiourea cadmium(II) complexes: the single crystal X-ray structures of $[CdCl_2(CS(NH_2)NHCH_3)_2]_n$ and $[CdCl_2(CS(NH_2)NHCH_2CH_3)_2]$, *Polyhedron*, 22(4), 595–603 (2003).

Mondal A., Mostafa G., Ghosh A., Laskar I.R., Chaudhuri N.R. Construction of a unique three-dimensional array with cadmium(II), *J. Chem. Soc., Dalton Trans.*, (1), 9–10 (1999).

Mondal N., Saha M.K., Mitra S., Gramlich V. Synthesis and crystal structure of thiocyanato or azido bridged one- or two-dimensional polymeric complexes of cadmium(II), *J. Chem. Soc., Dalton Trans.*, (18), 3218–3221 (2000).

Montgomery H., Lingafelter E.C. The crystal structure of Tutton's salts. IV. Cadmium ammonium sulfate hexahydrate, *Acta Crystallogr.*, 20(6), 728–730 (1966).

Moon H.-S., Kim C.-H., Lee S.-G. One-dimensional polymeric chain structure of bis(aniline) dithiocyanato-cadmium(II), *Acta Crystallogr.*, C56(4), 425–426 (2000a).

Moon H.-S., Kim C.-H., Lee S.-G. Preparation and one-dimensional coordination structure of the bis(morpholine)dithiocyanatocadmium(II) complex, $[Cd(SCN)_2(C_4H_9NO)_2]$, *Bull. Korean Chem. Soc.*, 21(3), 339–341 (2000b).

Mosset A., Bagieu-Beucher M., Lecchi A., Masse R., Zaccaro J. Crystal engineering strategy of thiocyanates for quadratic nonlinear optics. $Hg_3CdCl_2(SCN)_6$ and $Hg_4CdBr_4(SCN)_6$, *Solid State Sci.*, 4(6). 827–834 (2002).

Murciego A., Pascua M.J., Babkine J., Dusausoy Y., Medenbach O., Bernhardt H.-J. Barquillite $Cu_2(Cd, Fe)GeSe_4$, a new mineral from the Barquilla Deposit, Salamanca, Spain, *Eur. J. Min.*, 11(1), 111–117 (1999).

Musaev F.N., Mamedov Kh.S., Movsumov E.M., Shnulin A.N., Amiraslanov I.R. Crystal and molecular structure of di(monothiobenzoato)cadmium(II)monoethanol, *J. Struct. Chem.*, 19(3), 440–445 (1978).

Nardelli M., Chierici I. Acetati complessi di metalli bivalenti con molecole solforate (tiourea ed etilentiourea, Ricerca scient., 29(8), 1731–1732 (1959).

Nardelli M., Chierici I. Complessi di cloruri metallici con tiosemicarbazide, *Ricerca Sci.*, 30(2), 276–279 (1960).

Nardelli M., Coghi L. Su alcuni complessi di metalli bivalenti con molecole organiche ossigenate (formamide, acetamide, metilurea, *Ricerca Sci.*, 29(1), 134–138 (1959)

Nardelli M., Fava G., Boldrini P. Complessi a ponti di solfo: la struttura del formato di ditiourea–cadmio, *Gazz. Chim. Ital.*, 92(12), 1392–1400 (1962).

Nardelli M., Gasparri G.F., Boldrini P. The crystal and molecular structure of bisthiourea–cadmium formate, *Acta Crystallogr.*, 18(4), 618–623 (1965).

Nawaz S., Sadaf S., Fettouhi M., Fazal A., Ahmad S. Dibromidobis($N,N,N',N'$-tetramethylthiourea-κS)cadmium(II), *Acta Crystallogr.*, E66(8), m950 (2010a).

Nawaz S., Sadaf S., Fettouhi M., Fazal A., Ahmad S. Diiodidobis($N,N,N',N'$-tetramethylthiourea-κS)cadmium(II), *Acta Crystallogr.*, E66(8), m951 (2010b).

Nfor E.N., Liu W., Zuo J.-L., You X.-Z., Offiong O.E., Eno E.A. Synthesis, crystal structure and luminescent property of the novel μ-thiocyanato-bis[(2-acetylpyridinethiosemicabazonato)(thiocyanate)] cadmium(II) complex, *Trans. Met. Chem.*, 31(7), 837–841 (2006).

Nyman M.D., Hampden-Smith M.J., Duesler E.N. Low temperature, aerosol-assisted chemical vapor deposition (AACVD) of CdS, ZnS, and $Cd_{1-x}Zn_xS$ using monomeric single-source precursors: $M(SOCCH_3)_2TMEDA$, *Chem. Vap. Depos.*, 2(5), 171–174 (1996).

Nyman M.D., Hampden-Smith M.J., Duesler E.N. Synthesis, characterization, and reactivity of group 12 metal thiocarboxylates, $M(SOCR)_2Lut_2$ [M = Cd, Zn; R = CH$_3$, C(CH$_3$)$_3$; Lut = 3,5-dimethylpyridine (lutidine)], *Inorg. Chem.*, 36(10), 2218–2224 (1997).

Olekseyuk I.D., Husak O.A., Gulay L.D., Parasyuk O.V. The $AgGaS_2 + 2CdSe$ ⇔ $AgGaSe_2 + 2CdS$ system, *J. Alloys Compd.*, 367(1–2), 25–35 (2004).

Onwudiwe D.C., Ajibade P.A. Synthesis, characterization and thermal studies of Zn(II), Cd(II) and Hg(II) complexes of $N$-methyl-$N$-phenyldithiocarbamate: The single crystal structure of $[(C_6H_5)(CH_3)NCS_2]_4Hg_2$, *Int. J. Mol. Sci.*, 12(3), 1964–1978 (2011).

Otto J., Jolk I., Viland T., Wonnemann R., Krebs B. Metal(II) complexes with monodentate S and N ligands as structural models for zinc–sulfur DNA-binding proteins, *Inorg. Chim. Acta*, 285(2), 262–268 (1999).

Ouchi A., Taniguchi M. Synthesis, and crystal and molecular structure of bis(benzylamine) bis(thiocyanato)cadmium(II), $[Cd(SCN)_2(C_6H_5CH_2NH_2)_2]$, in planar polymeric form: allotrope of the linear polymeric complex, *Bull. Chem. Soc. Jpn.*, 61(9), 3347–3349 (1988).

Parasyuk O.V., Atuchin V.V., Romanyuk Y.E., Marushko L.P., Piskach L.V., Olekseyuk I.D., Volkov S.V., Pekhnyo V.I. The $CuGaSe_2–CuInSe_2–2CdS$ system and single crystal growth of the γ-phase, *J. Cryst. Growth*, 318(1), 332–336 (2011).

Parasyuk O.V., Olekseyuk I.D., Zaremba V.I., Dzham O.A., Lavrynyuk Z.V., Piskach L.V., Yanko O.G., Volkov S,V., Pekhnyo V.I. The reciprocal $CuInS_2 + 2CdSe$ ⇔ $CuInSe_2 + 2CdS$ system. Pt. 2. Liquid–solid equilibria in the system, *J. Solid State Chem.*, 179(10), 2998–3006 (2006).

Petrova R., Angelova O., Bakardjieva S., Macíček J. Molecular adducts of inorganic salts. VII. $Cd(ReO)_4$·4tu (tu = thiourea), *Acta Crystallogr.*, C52(10), 2432–2434 (1996a).

Petrova R., Angelova O., Macíček J. Molecular adducts of inorganic salts. VI. The dimorphism of $Cd(ReO)_4$·2tu (tu = thiourea), *Acta Crystallogr.*, C52(8), 1935–1939 (1996b).

Petrova R., Angelova O., Macíček J. Molecular adducts of inorganic salts. VII. Cadmium tetraoxorhenium hexakis(thiourea) hydrate, *Acta Crystallogr.*, C53(5), 565–568 (1997).

Petrova R., Bakardjieva S., Todorov T. Structure of molecular adducts of inorganic salts. VIII. Thiourea complexes of cadmium nitrate, *Z. Kristallogr.*, 215(2), 118–121 (2000).

Pickardt J., Dechert S. Kristallstrukturen "supramolekularer" (Benzo-18-krone-6) kaliumtetrathiocyanatometallate: Ein dimerer Komplex {[K(Benzo-18-krone-6)]$_2$[Hg(SCN)$_4$]}$_2$ und zwei isomere Komplexe [K(Benzo-18-krone-6)][Cd(SCN)$_3$] mit kettenförmigen Trithiocyanatocadmat-Anionen, *Z. anorg. und allg. Chem.*, 625(1), 153–159 (1999).

Pickardt J., Gong G.-T. Kristallstruktur des Cadmiumthiocyanat-Hexamethylentetramin-Addukts $Cd(SCN)_2$·$C_6H_{12}N_4$·$2H_2O$, *Z. Naturforsch.*, 48B(1), 23–26 (1993).

Piskach L.V., Lavrynyuk Z.V., Parasyuk O.V., Zmiy O.F., Kadykalo E.M., Pekhnyo V.I., Volkov S.V. Isothermal section of the $CuGaSe_2 + 2CdS \Leftrightarrow CuGaSe_2 + 2CdSe$ mutual system at 870 K [in Ukrainian], Nauk. visnyk Volyns'k. nats. un-tu im. Lesi Ukrainky. *Khim. Nauky*, (16), 47–51 (2008).

Poplaukhin P., Tiekink E.R.T. ($\mu$-2-Pyridinealdazine-$\kappa^4 N,N':N'',N'''$)bis[bis($N,N$-di-n-propyldithiocarbamato-$\kappa^2 S,S'$)cadmium(II)], *Acta Crystallogr.*, E64(9), m1176 (2008).

Pouskouleli G., Kourounakis P., Theophanides T. Mercury and cadmium thiosteroid complexes, *Inorg. Chim. Acta*, 24, 45–51 (1977).

Prakasam B.A., Ramalingam K., Bocelli G, Cantoni A. NMR and fluorescence spectral studies on bisdithiocarbamates of divalent Zn, Cd and their nitrogenous adducts: Single crystal X-ray structure of (1,10-phenanthroline)bis(4-methylpiperazinecarbodithioato) zinc(II), *Polyhedron*, 26(15), 4489–4493 (2007).

Preti C., Tosi G., De Filippo D., Verani G. Group IIB metal complexes with thiazolidine-2-selenone and thiazolidine-2-one as ligands, *J. Inorg. Nucl. Chem.*, 36(12), 3725–3729 (1974).

Preti C., Tosi G. Tautomeric equilibrium study of thiazolidine-2-thione. Transition metal complexes of the deprotonated ligand, *Can. J. Chem.*, 54(10), 1558–1562 (1976).

Raghavan C.M., Bhaskaran A., Sankar R., Jayavel R. Studies on the growth, structural, optical, thermal and electrical properties of nonlinear optical cadmium mercury thiocyanate glycol monomethyl ether single crystal, *Curr. Appl. Phys.*, 10(2), 479–483 (2010).

Raghavan C.M., Pradeepkumar R., Bhagavannarayana G., Jayavel R. Growth of cadmium mercury thiocyanate single crystals using acetone–water mixed solvent and their characterization studies, *J. Cryst. Growth*, 311(11), 3174–3178 (2009).

Raghavan C.M., Sankar R., Kumar R.M., Jayavel R. Growth and characterization of non-linear optical bis-(dimethylsulfoxide) cadmium mercury thiocyanate single crystal, *J. Cryst. Growth*, 310(21), 4570–4575 (2008).

Rajalingam U., Dean P.A.W, Jenkins H.A. Solution multinuclear ($^{31}P$, $^{111}Cd$, $^{77}Se$) magnetic resonance studies of cadmium complexes of heterocyclic aromatic thiones and the structure of [tetrakis(2(1H)-pyridinethione)cadmium] nitrate, $[Cd(C_5H_5NS)_4](NO_3)_2$, *Can. J. Chem.*, 78(5), 590–597 (2000).

Rajalingam U., Dean P.A.W., Jenkins H.A., Jennings M., Hook J.M. Cadmium complexes of thiones. Part II. A synthetic, solution, and solid-state MAS $^{111/113}Cd$ NMR study of cadmium complexes of 1,3-thiazolidine-2-thione, and the structures of [tetrakis(1,3-thiazolidine-2-thione)cadmium] trifluoromethanesulfonate ($[Cd(C_3H_5NS_2)_4](CF_3SO_3)_2$) and [tetrakis(1,3-thiazolidine-2-thione)cadmium][tetrakis(nitrato-O,O')cadmate] ($[Cd(C_3H_5NS_2)_4][Cd(O_2NO)_4]$), *Can. J. Chem.*, 79(9), 1330–1337 (2001).

Rajarajan K., Selvakumar S., Joseph G.P., Samikkannu S., Potheher I.V., Sagayaraj P. Optical, dielectric and photoconductivity studies of bis(dimethylsulfoxide) tetrathiocyanato-cadmium(II) mercury(II) NLO single crystals, *Opt. Mater.*, 28(10), 1187–1191 (2006).

Rajesh N.P., Kannan V., Ashok M., Sivaji K., Raghavan P.S., Ramasamy P. A new nonlinear optical semi-organic material: cadmium thiourea acetate, *J. Cryst. Growth*, 262(1–4), 561–566 (2004).

Ramesh V., Rajarajan K., Kumar K.S., Subashini A., NizamMohideen M. *catena*-Poly[ammonium (cadmium-tri-$\mu$-thiocyanato-$\kappa^4 S$:N;$\kappa^2 N$:S)-1,4,10,13,16-hexaoxacyclooctadecane (1/1)], *Acta Crystallogr.*, E68(3), m335–m336 (2012).

Rangan K.K., Trikalitis P.N., Canlas C., Bakas T., Weliky D.P., Kanatzidis M.G. Hexagonal pore organization in mesostructured metal tin sulfides built with $[Sn_2S_6]^{4-}$ cluster, *Nano Lett.*, 2(5), 513–517 (2002).

Raper E.S., Nowell I.W. Coordination compounds of l-methylimidazoline-2-(3H)-thione and metal nitrates [M(II)=Co, Zn and Cd]. A spectroscopic, thermal analysis and X-ray diffraction study, *Inorg. Chim. Acta*, 43, 165–172 (1980).

Raston C.L., Winter G. Crystal structure of bis(O-ethylxanthato)-1,10-phenanthroline-cadmium(II), *Aust. J. Chem.*, 29(4), 739–742 (1976).

Reddy H.K., Zhang C., Schlemper E.O., Schrauzer G.N. Synthesis and structures of the bis(cis-1-methylthiostilbene-2-thiolate) of cadmium and is adducts with dimethyl sulfoxide, 4-(dimethylamino)pyridine, and 2,2'-bipyridyl: hexacoordination of cadmium vs pentacoordination in the corresponding zinc derivatives, *Inorg. Chem.*, 31(9), 1673–1677 (1992).

Reger D.L., Myers S.M., Mason S.S., Rheingold A.L., Haggerty B.S., Ellis P.D. Syntheses and cadmium-113 NMR studies of five-coordinate complexes with $CdN_5$, $CdN_3O_2$, and $CdN_3S_2$ central cores. Solid state structures of $[HB(3-Phpz)_3]Cd[H_2B(3,5-Me_2pz)_2]$ and $[HB(3,5-Me_2pz)_3]Cd[S_2CNEt_2]$ (pz = pyrazolyl), *Inorg. Chem.*, 34(20), 4996–5002 (1995).

Reger D.L., Wright T.D., Smith M.D., Rheingold A.L., Kassel S., Concolino T., Rhagitan B. Syntheses and structures of mono-thiocyanate complexes of cadmium(II) and lead(II) containing bulky nitrogen based polydentate ligands, *Polyhedron*, 21(18), 1795–1807 (2002).

Richardson M.F., Franklin K., Thompson D.M. Reactions of metals with vitamins. I. Crystal and molecular structure of thiaminium tetrachlorocadmate monohydrate, *J. Am. Chem. Soc.*, 97(11), 3204–3209 (1975).

Rietveld H.M., Maslen E.N. The crystal structure of cadmium n-butyl xanthate, *Acta Crystallogr.*, 18(3), 429–436 (1965).

Rodesiller P.F., Charles N.G., Griffith E.A.H., Amma E.L. An approximately tetrahedral $CdS_4$ moiety. The structure of tetrakis(N,N'-dimethylthiourea-S)cadmium(II) nitrate, $C_{12}H_{32}CdN_8S_4 \cdot 2NO_3$, and its [113]Cd CP/MAS NMR, *Acta Crystallogr.*, C39(10), 1350–1352 (1983).

Rodesiler P.F, Charles N.G., Griffith E.A.H., Amma E.L. Structure of catena-aqua(2,2'-bipyridine)(2-imidazolidinethione-S)-μ-[sulfato(2–)-O:O']-cadmium(II)    monohydrate and its solid-state [113]Cd NMR spectrum, *Acta Crystallogr.*, C43(6), 1058–1061 (1987).

Rodesiler P.F., Turner R.W., Charles N.G., Griffith E.A.H., Amma E.L. Solution and solid-state [113]Cd NMR of $Cd(\alpha,\alpha'-bpy)_2X_2$ (X = Cl[-], Br[-], $NCS^-$, $NO_3^-$, $H_2O$) and crystal structures of the nitrate (monohydrate) and the isothiocyanate derivatives, *Inorg. Chem.*, 23(8), 999–1004 (1989).

Rodríguez A., Sousa-Pedrares A., García-Vázquez J.A., Romero J., Sousa A. Electrochemical synthesis and structural characterisation of cadmium and mercury complexes containing pyrimidine-2-thionate ligands, *Eur. J. Inorg. Chem.*, (11), 2242–2254 (2005).

Rogacheva E.I., Tavrina T.V., Galkin S.N. CdS effect on $CuInSe_2$ structure and properties, *Funct. Mater.*, 8(4), 635–641 (2001).

Rokabado Merkado T.R., Akimov V.M., Molodkin A.K. Synthesis, crystal morphology and lattice parameters of cadmium acetate dithiourea [in Russian], *Zhurn. Neorgan. Khim.*, 24(12), 3381–3383 (1979).

Rolies M.M., De Ranter C.J. The crystal and molecular structure of di-μ-chloro-bis[thioacetamide(chloro)cadmium(II)], *Acta Crystallogr.*, B34(11), 3216–3218 (1978).

Romanenko G.V., Savel'eva Z.A., Larionov S.V. Crystal and molecular structure of cadmium(II) nitrate complexes with thiosemicarbazide [Cd{NH$_2$NHC(S)NH$_2$}$_2$(NO$_3$)$_2$] [in Russian], *Zhurn. Strukt. Khimi.* 42(5), 1032–1035 (2001).

Romanyuk Y.E., Yu K.M., Walukewicz Yu.W., Lavrynyuk Z.V., Pekhnyo V.I., Parasyuk O.V. Single crystal growth and properties of γ-phase in the $CuInSe_2 + 2CdS \Leftrightarrow CuInS_2 + 2CdSe$ reciprocal system, *Sol. Energy Mater. Sol. Cells*, 92(11), 1495–1499 (2008).

Roy T.G., Hazari S.K.S., Barua K.K., Anwar N., Zukerman-Schpector J., Tiekink E.R.T. Synthesis, characterization and anti-microbial studies of cadmium(II) compounds containing 3,10-C-meso-Me$_8$[14]ane$_C$. Crystal and molecular structure of cis-[CdL$_C$(NO$_3$)] (NO$_3$), *Appl. Organometal. Chem.*, 24(12), 878–887 (2010).

Sampanthar J.T., Deivaraj T.C., Vittal J.J., Philip A.W. Dean P.A.W. Thioacetate complexes of Group 12 metals. Structures of [Ph$_4$P][Zn(SC{O}Me)$_3$(H$_2$O)] and [Ph$_4$P][Cd(SC{O}Me)$_3$], *J. Chem. Soc., Dalton Trans.*, (24), 4419–4423 (1999).

Sandström M. Crystal and molecular structure of hexakis(dimethylsulfoxide)cadmium(II) perchlorate, [Cd((CH$_3$)$_2$SO)$_6$](ClO$_4$)$_2$, *Acta Chem. Scand.*, 32A(6), 519–525 (1978).

Santos R.A., Gruff E.S., Koch S.A., Harbison G.S. Single-crystal, solid-state, and solution $^{113}$Cd NMR studies of [Cd(SR)$_2$(N-donor)$_2$] complexes. Structural and spectroscopic analog for biologically occurring [M(S-Cys)$_2$(His)$_2$] centers, *J. Am. Chem. Soc.*, 112(25), 9257–9263 (1990).

Saravanan M., Ramalingam R., Arulprakasam B., Bocelli G., Cantoni A., Tiekink E.R.T. Crystal structure of bis[(N-benzyl,N-hydroxyethyl)-dithiocarbamato](1,10-phenanthroline)cadmium(II), Cd[S$_2$CN(C$_7$H$_7$)(C$_2$H$_2$OH)]$_2$(C$_{12}$H$_8$N$_2$), *Z. Kristallogr. New Cryst. Struct.*, 220(3), 477–478 (2005).

Saravanan M., Ramalingam K., Bocelli G., Olla R. A new polymorph for bis[bis(N,N-dibenzyldithiocarbamato)cadmium(II)], *Appl. Organometal. Chem.*, 18(2), 103 (2004).

Secondo P.M., Land J.M., Baughman R.G., Collier H.L. Polymeric octahedral and monomeric tetrahedral Group 12 pseudohalogeno (NCX$^-$: X = O, S, Se) complexes of 4-(N,N-dimethylamino)pyridine, *Inorg. Chim. Acta*, 309(1–2), 13–22 (2000).

Selbin J., Bull W.E., Holmes, Jr. L.H. Metallic complexes of dimethylsulphoxide, *J. Inorg. Nucl. Chem.*, 16(3–4), 219–224 (1961).

Seidel R.W., Dietz C., Oppel I.M. A polymeric arched chain and a chair-like M$_2$L$_2$-metallamacrocycle *crystal engineered* of [M(phen)]$^{2+}$ (M = Cu, Cd; phen = 1,10-phenanthroline) corner units and the bent ligand 4,4'-dithiodipyridine, *Z. anorg. und allg. Chem.*, 637(1), 94–101 (2011).

Setzer W.N., Tang Y., Grant G.J., VanDerveer D.G. Synthesis and X-ray crystal structures of heavy-metal complexes of 1,5,9,13-tetrathiacyclohexadecane, *Inorg. Chem.*, 30(19), 3652–3656 (1991).

Shindo H., Brown T.L. Infrared spectra of complexes of L-cysteine and related compounds with zinc(II), cadmium(II), mercury(II), and lead(II), *J. Am. Chem. Soc.*, 87(9), 1904–1909 (1965).

Shores M.P., Beauvais L.G., Long J.R. [Cd$_2$(H$_2$O)$_4$][Re$_6$S$_8$(CN)$_6$]·14H$_2$O: A cyano-bridged cluster–cluster framework solid with accessible cubelike cavities, *Inorg. Chem.*, 38(8), 1648–1649 (1999).

Shugam E.A., Agre V.M. The structure of chelate compounds with bonds Me–S. 5 The crystal and molecular structure of cadmium dimethyldithiocarbamate [in Russian], *Kristallografiya*, 13(2), 253–257 (1968).

Shunmugam R., Sathyanarayana D.N. Complexes of imidazoline-2-thione and its 1-methyl analogue with Cu(II), Zn(II), Cd(II) and Hg(II) salts, *J. Coord. Chem.*, 12(3), 151–156 (1983).

Shvelashvili A.E., Porai-Koshits M.A., Kvitashvili A.I., Shchedrin B.M., Sarishvili L.P. Crystal structures of rhodanide ethylenediamine compounds of cadmium(II), *J. Struct. Chem.*, 15(2), 293–295 (1974).

Silver A., Koch S.A., Millar M. X-ray crystal structures of a series of [M$^{II}$(SR)$_4$]$^{2-}$ complexes (M = Mn, Fe, Co, Ni, Zn, Cd and Hg) with S$_4$ crystallographic symmetry, *Inorg. Chim. Acta*, 205(1), 9–14 (1993).

Song J.C., Tiekink E.R.T. (2,2'-Bipyridyl)bis[N,N-bis(2-hydroxyethyl)dithiocarbamato-κ$^2$S,S']-cadmium(II), *Acta Crystallogr.*, E65(12), m1667–m1668 (2009).

Sousa-Pedrares A., Romero J., Garcia-Vázquez J., Durán M.L., Casanova I., Sousa A. Electrochemical synthesis and structural characterization of zinc, cadmium and mercury complexes of heterocyclic bidentate ligands (N, S), *Dalton Trans.*, (7), 1379–1388 (2003).

Stoev M., Ruseva S. Preparation of cadmium sulfide from $CdCI_2(CdBr_2, CdI_2)–CS(NH_2)_2–CH_3OH$ systems, *Monatsh. Chem.*, 125(6–7), 599–606 (1994).

Strickler P. The structure of a novel polynuclear complex related to the sphalerite lattice, *J. Chem. Soc., Chem. Commun.*, (12), 655–656 (1969).

Su W., Huang X., Li J., Fu H. Crystal of semiconducting quantum dots built on covalently bonded T5 $[In_{28}Cd_6S_{54}]^{-12}$: the largest supertetrahedral cluster in solid state, *J. Am. Chem. Soc.*, 124(44), 12944–12945 (2002).

Sun W.-Y., Zhang L., Yu K.-B. Structure and characterization of novel zinc(II) and cadmium(II) complexes with 2-(benzoylamino)benzenethiolate and 1-methylimidazole. NHS hydrogen bonding in metal complexes with a $S_2N_2$ binding site, *J. Chem. Soc., Dalton Trans.*, (5), 795–798 (1999).

Sun Z.-F., Duan C.-Y., You X.-Z. A mixed-ligand cadmium(II) complex of xanthic acid and $N,N'$-bis(4-methoxy-phenyl)thiourea, *Acta Crystallogr.*, C50(7), 1012–1014 (1994).

Surin Yu.V., Molodkin A.K., Druzhinin I.G., Linichenko M.P. Thiourea–cadmium acetate–water system at 20 and 30°C [in Russian], *Zhurn. Neorgan. Khim.*, 17(10), 2776–2779 (1972).

Swaminathan S., Natarajan S. Crystal structure of bisthiourea cadmium nitrate, *Curr. Sci.*, 36(19), 513 (1967).

Swensson G., Albertsson J. On the crystal structure of dithiophosphorus compounds. The crystal structure of $Pb[S_2P(C_2H_5)_2]_2$ at 293 K and $Cd[S_2P(C_2H_5)_2]_2$ at 173 K, *Acta Chem. Scand.*, 45, 820–827 (1991).

Swenson D., Baenziger N.C., Coucouvanis D. Tetrahedral mercaptide complexes. Crystal and molecular structures of $[(C_6H_5)P]_2M(SC_6H_5)_4$ complexes (M=Cd(II), Zn(II), Ni(II), Co(II), and Mn(II)), *J. Am. Chem. Soc.*, 100(6), 1932–1934 (1978).

Taniguchi M., Ouchi A. Crystal and molecular structure of bis(benzylamine)bis(thiocyanato) cadmium(II), $[Cd(SCN)_2(C_6H_5CH_2NH_2)_2]$, *Bull. Chem. Soc. Jpn.*, 60(11), 4172–4174 (1987a).

Taniguchi M., Ouchi A. The crystal and molecular structure of bis(dibenzylamine) bis(thiocyanato)cadmium(II), $Cd(SCN)_2\{(C_6H_5CH_2)_2NH_2\}_2$, *Bull. Chem. Soc. Jpn.*, 60(3), 1192–1194 (1987b).

Taniguchi M., Ouchi A. Synthesis, and crystal and molecular structures of tetraethylammonium and tetrapropylammonium tris(thiocyanato)cadmates(II), $[R_4N][Cd(SCN)_3]$, $(R=C_2H_5, C_3H_7)$, *Bull. Chem. Soc. Jpn.*, 62(2), 424–428 (1989).

Taniguchi M., Shimoi M., Ouchi A. The crystal and molecular structures of bis(4-methylpyridine)bis(thiocyanato)cadmium(II) in polymeric form $[Cd(SCN)_2(CH_3C_5H_4N)_2]_n$, *Bull. Chem. Soc. Jpn.*, 59(7), 2299–2302 (1986).

Taniguchi M., Sugita Y., Ouchi A. The crystal and molecular structures of bis(2-methylpyridine)-, bis(3-methylpyridine)bis(thiocyanato)cadmium(II) in polymeric forms $[Cd(SCN)_2(CH_3C_5H_4N)_2]_n$, *Bull. Chem. Soc. Jpn.*, 60(4), 1321–1326 (1987).

Tao J., Yin X., Wei Z.-B., Huang R.-B., Zheng L.-S. Hydrothermal syntheses, crystal structures and photoluminescent properties of three metal-cluster based coordination polymers containing mixed organic ligands, *Eur. J. Inorg. Chem.*, (1), 125–133 (2003).

Teixidor F., Escriche L., Casabó J., Molins E., Miravitlles C. Metal complexes with polydentate sulfur-containing ligands. Crystal structure of (2,6-bis((ethylthio)methyl)pyridine) dibromozinc(II), *Inorg. Chem.*, 25(22), 4060–4062 (1986).

Teske Chr.L. Darstellung und Kristallstruktur von $Ba_3CdSn_2S_8$ mit einer Anmerkung über $Ba_6CdAg_2Sn_4S_{16}$, *Z. anorg. und allg. Chem.*, 522(3), 122–130 (1985).

Thiele G., Messer D. S-Thiocyanato- und N-lsothiocyanato-Bindungsisomerie in den Kristallstrukturen von RbCd(SCN)$_3$ und CsCd(SCN)$_3$, *Z. anorg. und allg. Chem.*, 464(1), 255–267 (1980).

Thirumaran S., Venkatachalam V., Manohar A., Ramalingam K., Bocelli G., Cantoni A. Synthesis and characterization of bis(*N*-methyl-*N*-ethanol-dithiocarbamato)M(II) (M = Zn, Cd, Hg) and bis(*N,N*-(iminodiethylene)-bisphtalimidedithiocarbamato)M(II) (M = Zn, Cd, Hg) complexes. Single crystal X-ray structure of bis(di(2-hydroxyethyl)-dithiocarbamato)zinc(II), *J. Coord. Chem.*, 44(3–4), 281–288 (1998).

Tian Y.-P., Duan C.-Y., Zhao C.-Y., You X.-Z., Mak T.C.W., Zhang Z.-Y. Synthesis, crystal structure, and second-order optical nonlinearity of bis(2-chlorobenzaldehyde thiosemicarbazone)cadmium halides (CdL$_2$X$_2$; X = Br, I), *Inorg. Chem.*, 36(6), 1247–1252 (1997).

Tian Y.-P., Wu J.-W., Xie F.-X., Shanmuga Sundara Raj S, Yang P., Fun H.-K. Dichlorobis(dimethylformamide-O)bis-(4-nitrobenzaldehyde thiosemicarbazone-S)-cadmium(II), *Acta Crystallogr.*, C55(10), 1641–1644 (1999a).

Tian Y.-P., Yu W.-T., Jiang M.-H., Raj S.S.S., Yang P., Fun H.-K. Diiodobis(4-methoxybenzaldehyde thio-semicarbazone-S)cadmium(II), *Acta Crystallogr.*, C55(10), 1639–1641 (1999b).

Tkhanmanivong T., Akimov V.M., Andrianov V.G., Struchkov Yu.T., Molodkin A.K. Crystal structure of cadmium monothiocarbamide acetate, Cd(CH$_3$COO)$_2$·CS(NH$_2$)$_2$ [in Russian], *Zhurn. Neorgan. Khim.*, 29(4), 1033–1037 (1984).

Tomlin D.W., Cooper T.M., Zelmon D.E., Gebeyehu Z., Hughes J.M. Cadmium isopropylxanthate, *Acta Crystallogr.*, C55(5), 717–719 (1999).

Tsintsadze O.V., Tsivtsivadze T.I., Orbeladze V.F. Structure of crystals of Cd(NCS)$_2$·2ur and Ni(NCS)$_2$·4DMAA, *J. Struct. Chem.*, 16(2), 304 (1975).

Ueyama N., Sugawara T., Sasaki K., Nakamura A., Yamashita S., Wakatsuki Y., Yamazaki H., Yasuoka N. X-ray structures and far-infrared and Raman spectra of tetrahedral thiophenolato and selenophenolato complexes of zinc(II) and cadmium(II), *Inorg. Chem.*, 27(4), 741–747 (1988).

Vittal J.J., Dean P.A.W. Chemistry of thiobenzoates: syntheses, structures, and NMR spectra of salts of [M(SOCPh)$_3$]$^-$ (M = Zn, Cd, Hg), *Inorg. Chem.*, 35(11), 3089–3093 (1996).

Vittal J.J., Dean P.A.W. Synthesis, structure, and nuclear magnetic resonance spectra ($^{111}$Cd, $^{23}$Na) of tetramethylammonium hexakis(thiobenzoato)(dicadmiumsodium)ate. A trinuclear anion with cadmium in nearly trigonal planar coordination and sodium in octahedral coordination, *Inorg. Chem.*, 32(6), 791–794 (1993).

Vittal J.J., Dean P.A.W. Triethylammonium tris(thiobenzoato-O,S)-cadmate(II), *Acta Crystallogr.*, C54(3), 319–321 (1998).

Vossmeyer T., Reck G., Katsikas L., Haupt E.T.K., Schulz B., Weller H. A "double-diamond superlattice" built up of Cd$_{17}$S$_4$(SCH$_2$CH$_2$OH)$_{26}$ clusters, *Science*, 267(5203), 1476–1479 (1995a).

Vossmeyer T., Reck G., Schulz B., Katsikas L., Weller H. Double-layer superlattice structure built up of Cd$_{32}$S$_{14}$(SCH$_2$CH(OH)CH$_3$)$_{36}$·4H$_2$O clusters, *J. Am. Chem. Soc.*, 117(51), 12881–12882 (1995b).

Wachhold M., Rangan K.K., Billinge S.J.L., Petkov V., Heising J., Kanatzidis M.G. Mesostructured non-oxidic solids with adjustable worm-hole shaped pores: M–Ge–Q (Q = S, Se) frameworks based on tetrahedral [Ge$_4$Q$_{10}$]$^{4-}$ clusters, *Adv. Mater.*, 12(2), 85–91 (2000a).

Wachhold M., Rangan K.K., Lei M., Thorpe M.F., Billinge S.J.L., Petkov V., Heising J., Kanatzidis M.G. Mesostructured metal germanium sulfide and selenide materials based on the tetrahedral [Ge$_4$S$_{10}$]$^{4-}$ and [Ge$_4$Se$_{10}$]$^{4-}$ units: surfactant templated three-dimensional disordered frameworks perforated with worm holes, *J. Solid State Chem.*, 152(1), 21–36 (2000b).

Wang C., Li Y., Bu X., Zheng N., Zivkovic O., Yang C.-S., Feng P. Three-dimensional super-lattices built from $(M_4In_{16}S_{33})^{10-}$ (M = Mn, Co, Zn, Cd) supertetrahedral clusters, *J. Am. Chem. Soc.*, 123(46), 11506–11507 (2001).

Wang T.-X., Song G., You H.-W. Crystal structure of monoaqua-bis(*N-p*-acetamido-benzenesulfonyl glycinato)(2,2′-bipyridine)cadmium(II)—ethanol—water (1:1:2), $[Cd(C_{10}H_8N_2)(H_2O)(C_{10}H_{11}N_2O_5S)_2]·C_2H_6O·2H_2O$, *Z. Kristallogr. New Cryst. Struct.*, 223(2), 160–162 (2008).

Wang X.Q., Xu D., Lu M.K., Yuan D.R., Huang J., Li S.G., Lu G.W. et al. Physicochemical behavior of nonlinear optical crystal $CdHg(SCN_4$, *J. Cryst. Growth*, 247(3–4), 432–437 (2003a).

Wang X.Q., Xu D., Lu M.K., Yuan D.R., Huang J., Lu G.W., Zhang G.H. et al. A systematic spectroscopic study of four bimetallic thiocyanates of chemical formula $AB(SCN)_4$: $ZnCd(SCN)_4$ and $AHg(SCN)_4$ (A = Zn, Cd, Mn) as UV nonlinear optical crystal materi-als, *Opt. Mater.*, 23(1–2). 335–341 (2003b).

Wang X.Q., Yu W.T., Xu D., Lu M.K., Yuan D.R. *catena*-Poly[[bis(thiocyanato-κN) cadmium(II)]-di-μ-thiourea-κ⁴S:S], *Acta Crystallogr.*, C58(6), m336–m337 (2002a).

Wang X.Q., Yu W.T., Xu D., Lu M.K., Yuan D.R., Lu G.T. Polymeric bis(*N,N*-dimethyl-acetamide)tetrakis(thiocyanato)cadmium(II)mercury(II), *Acta Crystallogr.*, C58(6), m341–m343 (2002b).

Wang X.-Q., Yu W.-T., Xu D., Yuan D.-R., Lu M.-K., Tian Y.-P., Yang P., Meng F.-Q., Guo S.-Y., Jiang M.-H. Crystal structure of *catena*-bis(dimethyl sulfoxide)bis(μ²-thiocyanato)-cadmium(II), $C_6H_{12}CdN_2O_2S_4$, *Z. Kristallogr. New Cryst. Struct.*, 215(1), 91–92 (2000).

Wang X.-Q., Yu W.-T., Xu D., Zhang G.-H. Polymeric bis(*N*-methylformamide) tetrathiocyanatocadmium(II)mercury(II), *Acta Crystallogr.*, E61(6), m1147–m1149 (2005).

Watson A.D., Rao C.P., Dorfman J.R., Holm R.H. Systematic stereochemistry of metal(II) thiolates: synthesis and structures of $[M_2(SC_2H_5)_6]^{2-}$ (M = Mn(II), Ni(II), Zn(II), Cd(II)), *Inorg. Chem.*, 24(18), 2820–2826 (1985).

Wazeer M.I.M., Isab A.A., Fettouhi M. New cadmium chloride complexes with imidazoli-dine-2-thione and its derivatives: X-ray structures, solid state and solution NMR and antimicrobial activity studies, *Polyhedron*, 26(8), 1725–1730 (2007).

Wei L.-H., Wang Z.-L. Crystal structure of (1,10-phenanthroline)-[2,2′-dithiobis(benzoato)] cadmium(II)-water (1:0.125), $Cd(C_{14}H_8S_2O_4)(C_{12}H_8N_2)·0.125H_2O$, *Z. Kristallogr. New Cryst. Struct.*, 224(3), 433–434 (2009).

Wei F.-X., Yin X., Zhang W.-G., Fan J., Jiang X-.H., Wang S.-L. Crystal structure of (pyridine-*N*)-bis(*N,N*-dibenzyldithiocarbamato)cadmium(II), $Cd(C_5H_5N)[S_2CN(C_6H_5CH_2)_2]_2$, *Z. Kristallogr. New Cryst. Struct.*, 220(3), 417–419 (2005).

White J.L., Tanski J.M., Churchill D.G., Rheingold A.L., Rabinovich D. Synthesis and structural characterization of 2-mercapto-1-*tert*-butylimidazole and its Group 12 metal derivatives $(Hmim^{tBu})_2MBr_2$ (M = Zn, Cd, Hg), *J. Chem. Crystallogr.*, 33(5–6), 437–445 (2003).

White J.L., Tanski J.M., Rabinovich D. Bulky tris(mercaptoimidazolyl)borates: synthesis and molecular structures of the Group 12 metal complexes $[Tm^{tBu}]MBr$ (M = Zn, Cd, Hg), *J. Chem. Soc., Dalton Trans.*, (15), 2987–2991 (2002).

Whitlow S.H. Cadmium(II) thiodiacetate hydrate, *Acta Crystallogr.*, 31(10), 2531–2533 (1975).

Williams D.J., Gulla D., Arrowood K.A., Bloodworth L.M., Carmack A.L., Evers T.J., Wilson M.S. et al.. The preparation, characterization and X-ray structural analysis of tetrakis[1-methyl-3-(2-propyl)-2(3H)-imidazolethione] cadmium(II) hexafluorophos-phate, *J. Chem. Crystallogr.*, 39(8), 581–584 (2009).

Wunderlich H. Structure of bis(diethyldithiophosphinato)cadmium(II), *Acta Crystallogr.*, C42(5), 631–32 (1986).

Xia D.-C., Liu Y.-Y, Crystal structure of diaqua(1,10-phenanthroline)cadmium(II) methane-disulfonate, Cd(H₂O)₂(C₁₂H₈N₂)[CH₂(SO₃)₂], Z. *Kristallogr. New Cryst. Struct.*, 225(3), 527–528 (2010).

Xiong R.-G., Liu C.-M., Zuo J.-L., You X.-Z. Guest-induced dimension change. A novel network intercalation complex: {[Cd(4,4′-bipy)₂(H₂O)₂](CF₃SO₃)₂(4,4′-bipy) (H₂O)₂(C₇H₈N₂O₃)₂}∞, *Inorg. Chem. Commun.*, 2(7), 292–297 (1999).

Yang G., Liu G.-F., Zheng S.-L., Chen X.-M. Synthesis and structures of dichlorotetrakis(phenylthiourea)cadmium(II) and *catena*-bis(thiocyanate) bis(phenylthiourea)cadmium(II), *J. Coord. Chem.*, 53(3), 269–279 (2001a).

Yang G., Zhu H.-G., Liang B.-H., Chen X.-M. Syntheses and crystal structures of four metal–organic co-ordination networks constructed from cadmium(II) thiocyanate and nicotinic acid derivatives with hydrogen bonds, *J. Chem. Soc., Dalton Trans.*, (5), 580–585 (2001b).

Yan L. Crystal structure of diaqua-bis(4,4′-dipyridyldisulfide)cadmium(II)dinitrate-methanol-water(1:2:2), [Cd(C₁₀H₈N₂S₂)₂(H₂O)₂][NO₃]₂·2CH₃OH·2H₂O, Z. *Kristallogr. New Cryst. Struct.*, 223(4), 533–534 (2008).

Yin X., Zhang W., Zhang Q., Fan J., Lai C.S., Tiekink E.R.T. Bis[bis(N,N-dibenzyldithiocarbamato) cadmium(II)], *Appl. Organometal. Chem.*, 18(3), 139–140 (2004).

Yin Y.-G., Forsling W., Boström D., Antzutkin O., Lindberg M., Ivanov A. Polymeric structure and solid NMR spectra of cadmium (II) di-alkyldithiophosphates (alkyl = propyl, butyl, isopropyl and isobutyl), *Chin. J. Chem.*, 21(3), 291–295 (2003).

Young, Jr V.G., Tiekink E.R.T. Bis(O-methyldithiocarbonato)cadmium(II), *Acta Crystallogr.*, E58(10), m537–m539 (2002).

Yuan D., Xu D., Liu M., Qi F., Yu W., Hou W., Bing Y., Sun S., Jiang M. Structure and properties of a complex crystal for laser diode frequency doubling: Cadmium mercury thiocyanate, *Appl. Phys. Lett.*, 70(5), 544–546 (1997).

Yuan D., Zhong Z., Liu M., Xu D., Fang Q., Bing Y., Sun S., Jiang M. Growth of cadmium mercury thiocyanate single crystal for laser diode frequency doubling, *J. Cryst. Growth*, 186(1–2), 240–244 (1998).

Yuan R.-X., Xiong R.-G., Xie Y.-L., You X.-Z., Peng S.-M., Lee G.-H. A novel 1D Cd^II coordination polymer with 4-sulfobenzoate and 2,2′-bipyridine containing an approximate rectangular molecular box unit with blue fluorescent emission, *Inorg. Chem. Commun.*, 4(8), 384–387 (2001).

Zaki Z.M., Mohamed G.G. Spectral and thermal studies of thiobarbituric acid complexes, *Spectrochim. Acta, Part A*, 56(7), 1245–1250 (2000).

Zamilatskov I.A., Buravlev E.A., Savinkina E.V., Roukk N.S., Albov D.V. Diiodidobis(thioacetamide-κS)cadmium(II), *Acta Crystallogr.*, E63(11), m2669 (2007).

Zemskova S.M., Glinskaya L.A., Durasov V.B., Klevtsova R.F., Larionov S.V. Mixed-ligand complexes of zinc(II) and cadmium(II) diethyldithiocarbamates with 2,2′-bipyridyl and 4,4′-bipyridyl: Preparation, structural and thermal properties [in Russian], *Zhurn. Strukt. Khim.*, 34(5), 154–156 (1993a).

Zemskova S.M., Glinskaya L.A., Klevtsova R.F., Fedotov M.A., Larionov S.V. Structure and properties ov mono- and heterometal cadmium, zinc and nickel complexes containing diethyldithiocarbamate ions and ethylenediamine molecules [in Russian], *Zhurn. Strukt. Khim.*, 40(2), 340–350 (1999).

Zemskova S.M., Glinskaya L.A., Klevtsova R.F., Larionov S.V. Preparation, crystal and molecular structure, thermal properties of zinc and cadmium diethyldithiocarbamates complexes with imidazole, [(C₂H₅)₂NCS₂]₂Zn(C₃H₄N₂) and [(C₂H₅)₂NCS₂]₂Cd(C₃H₄N₂) [in Russian], *Zhurn. Neorgan. Khim.*, 38(3), 466–471 (1993b).

Zemskova S.M., Gromilov S.A., Larionov S.V. Synthesis and X-ray investigation of diethyldithiocarbamato-iodide complexes of zinc(II), cadmium(II), mercury(II) [in Russian], *Sib. Khim. Zhurn.*, (4), 74–77 (1991).

Zemskova S.M., Prashad G., Glinskaya L.A., Klevtsova R.F., Durasov V.B., Tkachev S.V., Gromilov S.A., Larionov S.V. Volatile mixed-ligand complexes of bis(diisobutyldithiocarbamato)cadmium with 1,10-phenanthroline, 2,2'-bipyridyl and 4,4'-bipyridyl. Crystal and molecular structure of the $Cd[(CN(i-C_4H_9)_2]_2(C_{12}H_8N_2)$ complex [in Russian], *Zhurn. Neorgan. Khim.*, 43(10), 1644–1650 (1998).

Zhang H., Wang X., Zelmon D.E., Teo B.K. Synthesis and structure of [(DB24C8)Na] [Cd(SCN)$_3$]. Formation of a novel linear Cd···Cd···Cd chain with a mer-$CdN_3S_3$ coordination configuration and a new coiled [(DB24C8)Na]$^+$ cation, *Inorg. Chem.*, 40(7),1501–1507 (2001).

Zhang H., Wang X., Zhu H., Xiao W., Teo B.K. Synthesis and structure of [(12C4)$_2$Cd] [Cd$_2$(SCN)$_6$]. Formation of a novel tetragonal net of anionic layered structure of $[Cd_2(SCN)_6{}^{2-}]_\infty$ templated by the square-shaped sandwich [(12C4)$_2$Cd]$^{2+}$ cation, *J. Am. Chem. Soc.*, 119(23), 5463–5464 (1997).

Zhang H., Wang X., Zhu H., Xiao W., Zhang K., Teo B.K. Anisotropic templating effect in the formation of two-dimensional anionic cadmium-thiocyanate coordination solids [(12C4)$_2$Cd][Cd$_2$(SCN)$_6$] and [(12C4)$_2$Cd][Cd$_3$(SCN)$_8$] with checkerboard and herringbone patterns, respectively, *Inorg. Chem.*, 38(5), 886–892 (1999).

Zhang H., Zelmon D.E. Crystal growth of a new hybrid nonlinear optical compound [(18C6) K][Cd(SCN)$_3$] from aqueous solution, *J. Cryst. Growth*, 234(2–3), 529–532 (2002).

Zhang Y., Jianmin L., Nishiura M., Hou H., Deng W., Imamoto T. Thermodynamically controlled products: Novel motif of metallohelicates stabilized by inter- and intra-helix hydrogen bonds, *J. Chem. Soc., Dalton Trans.*, (3), 293–297 (2000a).

Zhang Y., Li J., Chen J., Su Q., Deng W., Nishiura M., Imamoto T., Wu X., Wang Q. A novel α-helix-liked metallohelicate series and their structural adjustments for the isomorphous substitution, *Inorg. Chem.*, 39(11), 2330–2336 (2000b).

Zhang Y., Sun X.-Z., Wan C.-Q. Crystal structure of poly[aqua($\mu_2$-di-4-pyridylsulfone-$\kappa^2 N{:}N'$) (3-chlorobenzoato-$\kappa^2 O,O'$-3-chlorobenzoato-$\kappa^1 O$) cadmium(II)] – di-4-pyridylsulfone (1:1), $C_{34}H_{26}CdCl_2N_4O_9S_2$, *Z. Kristallogr. New Cryst. Struct.*, 229(4), 417–418 (2014).

Zhao J.-Y., Tang G.-D., Li R.-Q., Zhang Z.-C. Crystal structure of *catena*-poly[diaquabis (4-formylpyridine-thiosemicarbazone-N,S)cadmium(II)] sulfate hydrate, [$Cd(H_2O)_2$ $(C_7H_8N_3S)_2][SO_4]\cdot H_2O$, *Z. Kristallogr. New Cryst. Struct.*, 224(2), 205–207 (2009).

Zhao N., Lian Z., Liu P. Crystal structure of aquabis(2,2'-bipyridyl-*N,N'*)-(9,10-dioxoanthracene-2,6-disulfonato)cadmium(II) monohydrate, $Cd(H_2O)(C_{10}H_8N_2)_2(C_{14}H_6S_2O_8)\cdot H_2O$, *Z. Kristallogr. New Cryst. Struct.*, 225(3), 454–456 (2010).

Zheng N., Bu X., Lauda J., Feng P. Zero- and two-dimensional organization of tetrahedral cadmium chalcogenide clusters with bifunctional covalent linkers, *Chem. Mater.*, 18(18), 4307–4311 (2006a).

Zheng N., Lu H., Bu X., Feng P. Metal-chelate dye-controlled organization of $Cd_{32}S_{14}(SPh)_{40}{}^{4-}$ nanoclusters into three-dimensional molecular and covalent open architecture, *J. Am. Chem. Soc.*, 128(14), 4528–4529 (2006b).

Zhou M., Yu W.T., Xu D., Guo S.Y., Lu M.K., Yuan D.R. Crystal structure of cadmium mercury tetrathiocyanate (glycol monomethyl ether), $C_7H_8CdHgN_4O_2S_4$, *Z. Kristallogr. New Cryst. Struct.*, 215(3), 425–426 (2000).

Zhou M., Yu W.T., Xu D., Lu M.K., Yuan D.R., Wang X.Q., Gou S.Y., Meng F.Q. Crystal structure of (DL)-trithioureatartrato-$O^1,O^2,O^3$-cadmium, $C_{28}H_{64}Cd_4N_{24}O_{24}S_{12}$, *Z. Kristallogr. New Cryst. Struct.*, 216(2), 201–204 (2001).

Zhu H.-G., Yang G., Chen X.-M., Ng S.W. *catena*-Poly[[[bis(*N,N'*-diphenylthiourea) cadmium(II)]-di-$\mu$-thiocyanato] dihydrate], *Acta Crystallogr.*, C56(10), e430–e431 (2000).

Zukerman-Schpector J., Castellano E.E., Mauro A.E., Muraoko T.K. *catena*-(*N,N,N',N'*-tetramethylethylenediamine)-di-$\mu$-thiocyanato-N,S-cadmium(II), *Acta Crystallogr.*, C44(7), 1207–1209 (1988).

# 5 Systems Based on CdSe

## 5.1 Cd–H–Li–C–O–Se

[Li$_3$(C$_8$H$_{16}$O$_4$)$_3$O$_2$CMe][Cd(Se$_4$)$_2$] or C$_{26}$H$_{51}$Li$_3$O$_{14}$Se$_8$Cd. This multinary compound is formed in this system (Kräuter et al. 1989). It crystallizes in the orthorhombic structure with the lattice parameters $a = 1981.5 \pm 0.6$, $b = 1400.3 \pm 0.1$, and $c = 3268.9 \pm 0.9$ pm and a calculated density of 1.98 g·cm$^{-3}$. The dark-red crystals of the title compound have been prepared by the reaction of [Li(C$_8$H$_{16}$O$_4$)]$_2$Se$_6$ with Cd(OAc)$_2$ in ethanolic solution.

## 5.2 Cd–H–Na–C–O–Se

[Na(C$_{10}$H$_{20}$O$_5$)]$_2$[Cd(Se$_4$)$_2$] or C$_{20}$H$_{40}$Na$_2$O$_{10}$Se$_8$Cd. This multinary compound is formed in this system (Adel et al. 1988). It crystallizes in the monoclinic structure with the lattice parameters $a = 2099.7 \pm 0.7$, $b = 1046.6 \pm 0.2$, and $c = 1750.2 \pm 0.6$ pm, and $\beta = 99.08° \pm 0.03°$, and a calculated density of 2.15·cm$^{-3}$. To prepare it, a mixture containing 5.70 mM of Cd(OAc)$_2$·2H$_2$O and an equimolar quantity of [Na(C$_{10}$H$_{20}$O$_5$)]$_2$Se$_6$ was stirred at room temperature for 12 h and then heated for 1 h at reflux. It was filtered hot of a brown residue and allowed to cool slowly. After 3 days, dark-red single crystals were obtained.

## 5.3 Cd–H–K–C–Sn–O–Se

[K$_{10}$(H$_2$O)$_{16}$(MeOH)$_{0.5}$][Cd$_4$($\mu_4$-Se)(SnSe$_4$)$_4$] or C$_{0.5}$H$_{34}$K$_{10}$Sn$_4$O$_{16.5}$Cd$_4$Se$_{17}$. This multinary compound is formed in this system. It crystallizes in the tetragonal structure with the lattice parameters $a = 1562.3 \pm 0.2$ and $c = 2543.4 \pm 0.5$ at 203 K, a calculated density of 3.169 g·cm$^{-3}$, and energy gap 2.68 eV (Brandmayer et al. 2004). To obtain it, K$_4$[SnSe$_4$]·1.5MeOH (0.300 mM) was suspended in MeOH (5 mL) and added to a solution of CdCl$_2$ (0.240 mM) in H$_2$O (5 mL). After stirring for 24 h, the solution was removed from the precipitate by decanting, and the yellow solid was redissolved in H$_2$O (10 mL) and layered with THF (10 mL). Over 1 week, yellow crystals formed selectively, while layering of the decanted reaction solution only yielded crystals of the coproduct [K$_4$(MeOH)$_4$][Sn$_2$Se$_6$].

## 5.4 Cd–H–K–Sn–Se

K$_{14-x}$H$_x$Cd$_{15}$Sn$_{12}$Se$_{46}$ ($x \approx 7$). This quinary compound, which crystallizes in the cubic structure with the lattice parameter $a = 2308.97 \pm 0.05$ pm and a calculated density of 3.790 g·cm$^{-3}$ (for K$_{7.92}$H$_{6.08}$Cd$_{15}$Sn$_{12}$Se$_{46}$), is formed in this system (Ding and Kanatzidis 2006). It was obtained after stirring K$_6$Cd$_4$Sn$_3$Se$_{13}$ in a concentrated aqueous solution of HI (pH = 1.1–1.8).

## 5.5   Cd–H–Ba–C–N–O–Se

**[Ba(C$_{12}$H$_{24}$O$_6$)(DMF)$_4$][Cd(Se$_4$)$_2$] or C$_{24}$H$_{52}$BaN$_4$O$_{10}$Se$_8$Cd**. This multinary compound is formed in this system. It crystallizes in the monoclinic structure with the lattice parameters $a = 2021.9 \pm 0.12$, $b = 1019.8 \pm 0.6$, and $c = 2270.8 \pm 1.4$ pm, and $\beta = 106.98° \pm 0.04°$ at 200 K and a calculated density of 2.12 g·cm$^{-3}$ (Magull et al. 1992). It has been prepared by the reaction of a DMF solution (50 mL) of lithium polyselenide (5 mM) with BaSe$_2$ (4.17 mM) and Cd(OAc)$_2$·2H$_2$O (4.0 mM) in the presence of 18-crown-6 (C$_{12}$H$_{24}$O$_6$) (1.04 g). The obtained solution was heated slowly to 90°C. After 3 h, the solution was cooled, filtered from the Se powder, and concentrated in vacuo to 30 mL. To this 20 mL of diethyl ether was added and the mixture was left to stand quiet for 2 days. Black crystals were formed, which were filtered off, washed with diethyl ether, and dried in a stream of Ar.

## 5.6   Cd–H–C–Ge–N–Se

Some compounds are formed in this system.

**(C$_{12}$H$_{25}$NMe$_3$)$_2$[CdGe$_4$Se$_{10}$] or C$_{30}$H$_{68}$Ge$_4$N$_2$Se$_{10}$Cd**. To obtain this compound, (C$_{12}$H$_{25}$NMe$_3$)$_4$Ge$_4$Se$_{10}$ was dissolved in a warm (ca. 60°C) EtOH/H$_2$O mixture (ca. 20 mL; 1:1 volume ratio) (Wachhold et al. 2000a,b). A solution of CdCl$_2$ in EtOH/H$_2$O was then added dropwise to the stirred (C$_{12}$H$_{25}$NMe$_3$)$_4$Ge$_4$Se$_{10}$ solution to give instantly a yellow precipitate. The resulting suspension was stirred for another hour, followed by product isolation by centrifugation and subsequent decanting of the solvent. The product was washed with water and acetone to remove all residues of (C$_{12}$H$_{25}$NMe$_3$)$_4$Ge$_4$Se$_{10}$ and dried with ether.

This compound could also be obtained if K$_4$Ge$_4$Se$_{10}$ and two equivalents of C$_{12}$H$_{25}$NMe$_3$Br were dissolved in H$_2$O/MeOH (ca. 20 mL; 1:1 volume ratio) and heated to 50°C–60°C to give a yellow solution (Wachhold et al. 2000a,b). The solution of CdCl$_2$ in H$_2$O (100 mL) was slowly added to this mixture under vigorous stirring, which led to immediate precipitation of the product. The isolation and purification were the same as for the first method. (C$_{12}$H$_{25}$NMe$_3$)$_2$[CdGe$_4$Se$_{10}$] is a wide band gap semiconductor.

**(C$_{14}$H$_{29}$NMe$_3$)$_2$[CdGe$_4$Se$_{10}$] or C$_{34}$H$_{76}$Ge$_4$N$_2$Se$_{10}$Cd**. This compound was prepared by the same procedure as (C$_{12}$H$_{25}$NMe$_3$)$_2$[CdGe$_4$Se$_{10}$] was obtained using (C$_{14}$H$_{29}$NMe$_3$)$_4$Ge$_4$Se$_{10}$ instead of (C$_{12}$H$_{25}$NMe$_3$)$_4$Ge$_4$Se$_{10}$ (first method) and C$_{14}$H$_{29}$NMe$_3$Br instead of C$_{12}$H$_{25}$NMe$_3$Br (second method) (Wachhold et al. 2000a,b). It is a wide band gap semiconductor with energy gap 2.37 eV.

**(C$_{16}$H$_{33}$NMe$_3$)$_2$[CdGe$_4$Se$_{10}$] or C$_{38}$H$_{84}$Ge$_4$N$_2$Se$_{10}$Cd**. This compound was prepared by the same procedure as (C$_{12}$H$_{25}$NMe$_3$)$_2$[CdGe$_4$Se$_{10}$] was obtained using (C$_{16}$H$_{33}$NMe$_3$)$_4$Ge$_4$Se$_{10}$ instead of (C$_{12}$H$_{25}$NMe$_3$)$_4$Ge$_4$Se$_{10}$ (first method) and C$_{16}$H$_{33}$NMe$_3$Br instead of C$_{12}$H$_{25}$NMe$_3$Br (second method) (Wachhold et al. 2000a,b). It is a wide band gap semiconductor.

**(C$_{18}$H$_{37}$NMe$_3$)$_2$[CdGe$_4$Se$_{10}$] or C$_{42}$H$_{92}$Ge$_4$N$_2$Se$_{10}$Cd**. This compound was prepared by the same procedure as (C$_{12}$H$_{25}$NMe$_3$)$_2$[CdGe$_4$Se$_{10}$] was obtained using

$(C_{18}H_{37}NMe_3)_4Ge_4Se_{10}$ instead of $(C_{12}H_{25}NMe_3)_4Ge_4Se_{10}$ (first method) and $C_{18}H_{37}NMe_3Br$ instead of $C_{12}H_{25}NMe_3Br$ (second method) (Wachhold et al. 2000a,b). It is a wide band gap semiconductor.

## 5.7  Cd–H–C–Sn–N–Se

$[(C_{21}H_{38}N)_{4-2x}Cd_xSnSe_4]$ **(1.0 < x < 1.3)**. This multinary compound is formed in this system. It is a medium band gap semiconductor with $E_g = 2.4$ eV (Trikalitis et al. 2001) and was prepared by the following procedure. A solution of 9.94 mM of surfactant (cetylpyridinium bromide, $C_{21}H_{38}NBr·H_2O$) in 20 mL of formamide was heated at 75°C for a few minutes forming a clear solution. To this solution, 1.00 mM of $K_4SnSe_4$ was added and the mixture was stirred for 30 min, forming a clear deep-red solution. To this a solution of 1.00 mM of the cadmium salt in 10 mL of formamide was added slowly, using a pipet. A precipitate formed immediately and the mixture was stirred while warmed for 20 h. The products were isolated by suction and filtration and washed with large amounts of warm formamide and $H_2O$. The solids were dried under vacuum overnight. A yellow product was formed. All reactions were carried out under a $N_2$ atmosphere.

## 5.8  Cd–H–C–N–P–Se

Some compounds are formed in this system.

$[Cd\{Bu^t_2P(Se)NPr^i\}_2]$ **or** $C_{22}H_{50}N_2P_2Se_2Cd$. This compound sublimes at 200° at $2·10^{-4}$ Pa and melts at a temperature higher than 220°C (Bochmann et al. 1995). To obtain it, to a suspension of $Bu^t_2P(Se)NHPr_i$ (1.49 mM) in light petroleum (10 mL) was added $[Cd\{N(SiMe_3)_2\}_2]$ (0.745 mM) at 0°C. The mixture was stirred at this temperature for 5 min and then warmed to room temperature and stirred. After 45 min at room temperature, all the suspension had dissolved and a white precipitate had begun to form. After stirring for 3 h, the solvent was removed under reduced pressure and the crude product recrystallized from toluene (5 mL) to give colorless crystals.

$[Cd\{Bu^t_2P(Se)N(C_6H_{11})\}_2]$ **or** $C_{28}H_{58}N_2P_2Se_2Cd$. This compound sublimes at 200° at $1.7·10^{-4}$ Pa and melts at a temperature higher than 220°C (Bochmann et al. 1995). It was prepared following the procedure for $[Cd\{Bu^t_2P(Se)NPr_i\}_2]$ from $Bu^t_2P(Se)$ $NH(C_6H_{11})$ (1.48 mM) and $[Cd\{N(SiMe_3)_2\}_2]$ (0.74 mM) as colorless crystals.

## 5.9  Cd–H–C–N–P–F–Se

Some compounds are formed in this system.

$C_{20}H_{32}N_8P_2F_{12}Se_4Cd$, **tetrakis[1,3-dimethyl-2(3H)-imidazoleselone] cadmium(II) hexafluorophosphate**. This multinary compound is formed in this system. It melts at 230°C with decomposition and crystallizes in the orthorhombic structure with the lattice parameters $a = 1278.3 \pm 0.3$, $b = 2220.6 \pm 0.4$, and $c = 1315.3 \pm 0.3$ pm at $158 \pm 2$ K and a calculated density of 1.962 g·cm$^{-3}$ (Williams et al. 2007). It was synthesized by the addition of 5.07 mM of

1,3-dimethyl-2(3*H*)-imidazoleselone and 1.03 mM of $Cd(OAc)_2 \cdot 2H_2O$ in warm acetonitrile. The solution was left to cool to room temperature. In 40 mL of EtOH, 2.00 mM of ammonium hexafluorophosphate was dissolved and added to the cadmium solution, which was then boiled down to approximately 25 mL and left to cool to room temperature. Once cooled to room temperature, yellow-orange crystals formed.

## 5.10   Cd–H–C–N–Sb–Se

Some compounds are formed in this system.

**[(C₁₄H₂₉Py)₅.₈Cd₄.₆(SbSe₄)₅]** or $C_{110.2}H_{197.2}N_{5.8}Sb_5Se_{20}Cd_{4.6}$. This compound crystallizes in the cubic structure with the lattice parameter $a = 8310 \pm 40$ and energy gap 1.88 eV (Ding et al. 2006). To obtain it, $K_3SbSe_4$ (0.5 mM) was dissolved in 3 mL of $H_2O$ upon stirring and heating at 80°C. An orange solution formed quickly. $C_{14}H_{29}PyBr \cdot H_2O$ (5.2 mM) was dissolved in $H_2O$ (20 mL) and $CdCl_2$ (0.25 mM) was dissolved in 3 mL of $H_2O$ in another flask. These two solutions were added to the $C_{14}H_{29}PyBr/H_2O$ solution simultaneously. A dark-red precipitate formed immediately. However, the reaction mixture was stirred overnight at 80°C. The product was isolated by filtration, washed with warm water, and dried under vacuum.

**[(C₁₆H₃₃Py)₆.₂Cd₄.₄(SbSe₄)₅]** or $C_{130.2}H_{235.6}N_{6.2}Sb_5Se_{20}Cd_{4.4}$. This compound crystallizes in the cubic structure with the lattice parameter $a = 9240 \pm 40$ and energy gap 1.79 eV (Ding et al. 2006). To obtain it, $K_3SbSe_4$ (0.5 mM) was dissolved in 3 mL of $H_2O$ upon stirring and heating at 80°C. An orange solution formed quickly. $C_{16}H_{33}PyBr \cdot H_2O$ (5.2 mM) was dissolved in $H_2O$ (20 mL) and $CdI_2$ (0.25 mM) was dissolved in 3 mL of $H_2O$ in another flask. These two solutions were added into the $C_{16}H_{33}PyBr/H_2O$ solution simultaneously. A dark-red precipitate formed immediately. However, the reaction mixture was stirred overnight at 80°C. The product was isolated by filtration, washed with warm water, and dried under vacuum.

## 5.11   Cd–H–C–N–O–Se

Some compounds are formed in this system.

**[Cd(SeCN)₂(C₇H₁₀N₂)₂]ₙ** or $C_{16}H_{20}N_6Se_2Cd$. This compound melts at 236°C–238°C and crystallizes in the monoclinic structure with the lattice parameters $a = 617.1 \pm 0.1$, $b = 1109.3 \pm 0.2$, and $c = 1459.9 \pm 0.3$ pm and $\beta = 98.89° \pm 0.02°$ and a calculated density of 1.906 g·cm⁻³ (Secondo et al. 2000). To obtain it, $Cd(NO_3) \cdot 4H_2O$ (2.0 mM) was dissolved in 25 ml of absolute EtOH, to which a solution of KSeCN (4.0 mM) in the same solvent (30 mL) was added with stirring. The precipitated $KNO_3$ was filtered, and to the filtrate was added a solution of 4-(*N,N*-dimethylamino)pyridine (C₇H₁₀N₂) (4.0 mM) in absolute EtOH (10 mL). The resulting precipitate was isolated by filtration, washed with EtOH and then ether, and dried in vacuo. Suitable crystals of the title compound were grown from the saturated solution in DMSO over a period of 2 weeks.

**[Cd₈Se(SePh)₁₄(DMF)₃]** or $C_{93}H_{91}N_3O_3Se_{15}Cd_8$. This compound crystallizes in the orthorhombic structure with the lattice parameters $a = 2653.8 \pm 0.5$,

$b = 1359.2 \pm 0.3$, and $c = 3202.5 \pm 0.6$ pm (Behrens and Fenske 1997). The dissolution of $[Cd_8Se(SePh)_{14}(PPh_3)_2]$ (0.02 mM) in DMF (4 mL) and subsequently the layering with EtOH resulted in crystals of $[Cd_8Se(SePh)_{14}(DMF)_3]$. The synthesis was performed under a $N_2$ atmosphere.

## 5.12   Cd–H–C–N–O–Cl–Se

Some compounds are formed in this system.

$[(Pr^n_4N)_2\{Cd_4(SePh)_6Cl_4\}] \cdot C_4H_8O$ or $C_{64}H_{94}N_2OCl_4Se_6Cd_4$. This compound crystallizes in the monoclinic structure with the lattice parameters $a = 2281.0 \pm 0.5$, $b = 1315.0 \pm 0.3$, and $c = 2767.0 \pm 0.6$ pm and $\beta = 113.48° \pm 0.03°$ at 200 K and a calculated density of 1.714 $g \cdot cm^{-3}$ (Soloviev et al. 2001). It was prepared by the following procedure. First, 1.15 mM of $CdCl_2$ together with 0.57 mM of $Pr^n_4NCl$ were dissolved in 60 mL of $CH_2Cl_2$. Then 2.29 mM of $PhSeSiMe_3$ was added and the resulting clear solution stirred overnight. Addition of small amounts of diethyl ether led to the formation of colorless crystals of the title compound. Standard Schlenk techniques were employed throughout the syntheses using a double-manifold vacuum line with high-purity dry nitrogen.

$[Cd(DMF)_6][Cd_8Se(SePh)_{12}Cl_4]$ or $C_{90}H_{102}N_6O_6Cl_4Se_{13}Cd_9$. This compound crystallizes in the cubic structure with the lattice parameter $a = 2797.4 \pm 0.3$ pm (Behrens and Fenske 1997). It formed after some days if $C_{288}H_{240}P_4Cl_4Se_{71}Cd_{25}$ (0.007 mM) is dissolved in DMF (2 mL). The synthesis was performed under a $N_2$ atmosphere and absolute THF was used.

## 5.13   Cd–H–C–N–Se

Some compounds are formed in this system.

$[Me_4N][Cd(SeCN)_3]$ or $C_7H_{12}N_4Se_3Cd$. This compound was prepared as follows (Zhang et al. 2000). With vigorous stirring, 20 mL of an aqueous solution of 1 M $Me_4NCl$ was added dropwise to a preformed mixture of 50 mL of 0.35 M aqueous solution of $CdSO_4$ and 50 mL of a 1 M aqueous solution of KSeCN, respectively, over a period of 30 min. The solution mixture was then cooled with an ice bath. Fine crystals that formed after 2 days were collected by filtration and dried in a vacuum oven.

$[(C_5H_9N_2)Cd(SeCN)_3]$ or $C_8H_9N_5Se_3Cd$. This compound crystallizes in the orthorhombic structure with the lattice parameters $a = 965.86 \pm 0.17$, $b = 1069.75 \pm 0.19$, and $c = 1362.7 \pm 0.3$ pm at 123 K and a calculated density of 2.474 $g \cdot cm^{-3}$ (Liu and Li 2003a). $N,N'$-dimethylimidazolium iodide, $(C_5H_9N_2)I$ (1.0 mM), was added to the premixed solution of 1 M $Cd(NO_3)_2 \cdot 4H_2O$ and 3 mL of 1 M KSeCN. Colorless crystals were obtained by slow evaporation of the solution at room temperature.

$[Cd(CH_3NHCH_2)_2(SeCN)_2]$ or $C_{10}H_{24}N_6Se_2Cd$. This compound crystallizes in the orthorhombic structure with the lattice parameters $a = 1585.4 \pm 1.1$, $b = 1585.9 \pm 0.9$, and $c = 2929 \pm 2$ pm and a calculated density of 1.799 $g \cdot cm^{-3}$ (Choudhury et al. 2003).

The title compound and $[Cd_2(CH_3NHCH_2)_2(SeCN)_4]_n$ were synthesized by the addition of a methanolic solution of N,N-dimethyl ethylenediamine (1 mM) to a solution of $Cd(OAc)_2$ (1 mM) in 10 mL of MeOH at 0°C with constant stirring. To the resulting mixture, a methanolic solution of KSeCN (1 mM) was added very slowly. The resulting solution was kept at room temperature. Shiny colorless crystals of two different shapes appeared after 10 days, and they were separated in two batches by their different shapes.

$(EtMe_3N)_2[Cd(Se_4)]$ or $C_{10}H_{28}N_2Se_8Cd$. This compound crystallizes in the monoclinic structure with the lattice parameters $a = 1251.25 \pm 0.02$, $b = 1132.73 \pm 0.03$, and $c = 1672.90 \pm 0.03$ pm, and $\beta = 95.174° \pm 0.001°$ and a calculated density of 2.589 g·cm$^{-3}$ (Kim and Kim 2011). All experiments were performed under an atmosphere of dry Ar or $N_2$ using either a glove box or a Schlenk line. To a 50 mL DMF solution of 1.5 mM $K_2Se_4$ and 1.5 mM $EtMe_3NI$, a 10 mL DMF solution of 0.75 mM $CdI_2$ was added dropwise over a 20 min period. Ether at 60 mL was slowly layered over the filtrate solution, after removing undissolved precipitates by filtration. Upon standing at room temperature for 3 days, dark-purple crystals were obtained. These crystals were isolated and washed with ether several times. More crystals were obtained upon layering an additional 50 mL ether over the solution after isolation of the first crop of crystals.

$(Et_4N)[Cd(SeCN)_3]$ or $C_{11}H_{20}N_4Se_3Cd$. This compound crystallizes in the orthorhombic structure with the lattice parameters $a = 993.8 \pm 0.1$, $b = 1686.8 \pm 0.2$, and $c = 1105.4 \pm 0.1$ pm and a calculated density of 1.998 g·cm$^{-3}$ (Zhang et al. 2000). It was prepared as follows. With vigorous stirring, 20 mL of an aqueous solution of 1 M $Et_4NCl$ was added dropwise to a preformed mixture of 50 mL of 0.35 M aqueous solution of $CdSO_4$ and 50 mL of 1 M aqueous solution of KSeCN, respectively, over a period of 30 min. The solution mixture was then cooled with an ice bath. Fine crystals that formed after 2 days were collected by filtration and dried in a vacuum oven.

$[Cd_2(CH_3NHCH_2)_2(SeCN)_4]_n$ or $C_{12}H_{24}N_8Se_4Cd_2$. This compound crystallizes in the monoclinic structure with the lattice parameters $a = 1281.3 \pm 0.7$, $b = 859.2 \pm 0.5$, and $c = 2322.7 \pm 1.2$ pm, and $\beta = 104.38° \pm 0.04°$ and a calculated density of 2.202 g·cm$^{-3}$ (Choudhury et al. 2003). The title compound and $[Cd(CH_3NHCH_2)_2(SeCN)_2]$ were synthesized by the addition of a methanolic solution of N,N-dimethyl ethylenediamine (1 mM) to a solution of $Cd(OAc)_2$ (1 mM) in 10 mL of MeOH at 0°C with constant stirring. To the resulting mixture, a methanolic solution of KSeCN (1 mM) was added very slowly. The resulting solution was kept at room temperature. Shiny colorless crystals of two different shapes appeared after 10 days, and they were separated in two batches by their different shapes.

$[Cd(SeCN)_2(C_6H_4N_2)_2]$ or $C_{14}H_8N_6Se_2Cd$. This compound crystallizes in the monoclinic structure with the lattice parameters $a = 597.02 \pm 0.10$, $b = 1774.0 \pm 0.3$, and $c = 874.25 \pm 0.15$ pm, and $\beta = 109.351° \pm 0.003°$ at 123 K and a calculated density of 2.017 g·cm$^{-3}$ (Li and Liu 2003). To a solution of $Cd(NO_3)_2$ (1 mM) in $H_2O$ (10 mL) was added KSeCN (2 mM). The resulting colorless solution was treated with an ethanolic solution of isonicotinonitrile ($C_6H_4N_2$) (2 mM). Upon standing and slow evaporation, colorless crystals of the title compound were obtained.

**[XylIm$_2$][Cd(SeCN)$_4$]** or **C$_{20}$H$_{20}$N$_8$Se$_4$Cd** **[XylIm$_2$=$p$-bis($N$-methylimidazolyl) xylylene]**. This compound crystallizes in the monoclinic structure with the lattice parameters $a=1213.4\pm0.3$, $b=452.46\pm0.10$, and $c=2379.9\pm0.5$ pm, and $\beta=90.898°\pm0.004°$ (Liu and Li 2003b). To obtain it, XylIm$_2$Cl$_2$ was added to the premixed solution of 1 mL of Cd(NO$_3$)$_2$·4H$_2$O (1 M) and 3 mL of KSeCN (1 M). Colorless crystals were obtained by slow evaporation on the filtrate at room temperature.

**[Me$_4$N][Cd(SePh)$_4$]** or **C$_{32}$H$_{44}$N$_2$Se$_4$Cd**. This compound crystallizes in the orthorhombic structure with the lattice parameters $a=2015.7\pm1.5$, $b=1458.3\pm0.6$, and $c=1221.4\pm0.8$ pm and a calculated density of 1.637 g·cm$^{-3}$ (Ueyama et al. 1988). It was synthesized under an Ar atmosphere. Me$_3$SiSePh (50 mM), NMe$_4$Cl (40 mM), and Bu$^n_3$N (40 mM) were mixed in 30 mL of MeOH. CdCl$_2$ (7 mM) in 50 mL of MeOH was added to this solution at room temperature. The colorless needles obtained were collected by filtration, washed with cold MeOH, and recrystallized from hot MeOH.

**C$_{40}$H$_{80}$N$_4$Se$_8$Cd$_2$,** ***N,N*-diethyldiselenocarbamatocadmium**. To obtain this compound, CSe$_2$ was reacted immediately with an excess of (C$_2$H$_5$)$_2$NH in pentane at 0°C to give $N,N$-diethyldiselenocarbamate as the diethylammonium salt (Hursthouse et al. 1992). The selenocarbamate was then reacted with a stoichiometric quantity of an aqueous solution of CdCl$_2$ to give an insoluble yellow precipitate of C$_{40}$H$_{80}$N$_4$Se$_8$Cd$_2$. The crude product was recrystallized from toluene to give yellow cubic crystals. This compound could be a useful precursor for the deposition of CdSe films.

**(Me$_4$N)$_2$[Cd$_4$(SePh)$_{10}$]** or **C$_{68}$H$_{74}$N$_2$Se$_{10}$Cd$_4$**. This compound crystallizes in the monoclinic structure with the lattice parameters $a=2083.0\pm0.2$, $b=1428.2\pm0.1$, and $c=2587.2\pm0.1$ pm, and $\beta=99.626°\pm0.006°$ and experimental and calculated densities $1.91\pm0.05$ and 1.89 g·cm$^{-3}$, respectively (Vittal et al. 1992). To obtain it, a 9 mM portion of PhSeH was added to an equimolar amount of sodium metal dissolved in 10 mL of MeOH (Dean and Vittal 1986). The resultant solution of NaSePh was added to a stirred solution of Cd(NO$_3$)$_2$·4H$_2$O in 20 mL of MeOH, producing a white precipitate. Following the addition, with stirring of a solution of 6 mM of Me$_4$NCl in MeOH and then 100 mL of H$_2$O, the mixture was warmed to 80°C and acetonitrile was added in small portions until the precipitate dissolved. After hot filtration, the clear solution was left at 5°C for crystallization to occur. The colorless crystals were separated by filtration, washed with MeOH, H$_2$O, and the MeOH again, and finally dried under vacuum. The syntheses were carried out under an Ar atmosphere.

**[(Me$_4$N)$_2${SeCd$_8$(SePh)$_{16}$}]** or **C$_{104}$H$_{104}$N$_2$Se$_{17}$Cd$_8$**. To obtain this compound, a deoxygenated solution of NaSePh in EtOH/acetonitrile was treated with solutions of CdI$_2$ in acetonitrile, Na$_2$Se in MeOH, and Me$_4$NCl in acetonitrile, yielding a colorless solution (Lee et al. 1990). After stripping of all solvent, the colorless solids were extracted with H$_2$O at 80°C and EtOH at ambient temperature, then dissolved in acetonitrile at 20°C, and filtered, and the title compound crystallized by storage at 0°C.

**[(Et₄N)₂{SeCd₈(SePh)₁₆}]** or **C₁₁₂H₁₂₀N₂Se₁₇Cd₈.** To obtain this compound, a deoxygenated solution of NaSePh in EtOH/acetonitrile was treated with solutions of $CdI_2$ in acetonitrile, $Na_2Se$ in MeOH, and $Et_4NCl$ in acetonitrile, yielding a colorless solution (Lee et al. 1990). After stripping of all solvent, the colorless solids were extracted with $H_2O$ at 80°C and EtOH at ambient temperature, then dissolved in acetonitrile at 20°C, and filtered, and the title compound crystallized by storage at 0°C.

## 5.14   Cd–H–C–N–Te–Se

**[(Et₄N)₂{TeCd₈(SePh)₁₆}]** or **C₁₁₂H₁₂₀N₂TeSe₁₆Cd₈.** This multinary compound is formed in this system. To obtain it, a deoxygenated solution of NaSePh in EtOH/ acetonitrile was treated with solutions of $CdI_2$ in acetonitrile, $Na_2Te$ in MeOH, and $Et_4NCl$ in acetonitrile, yielding a colorless solution (Lee et al. 1990). After stripping of all solvent, the colorless solids were extracted with $H_2O$ at 80°C and EtOH at ambient temperature, then dissolved in acetonitrile at 20°C, and filtered, and the title compound crystallized by storage at 0°C.

## 5.15   Cd–H–C–N–Cl–Se

Some compounds are formed in this system.

**[Cd{SeC(NH₂)₂}₂Cl₂]** or **C₂H₈N₄Cl₂Se₂Cd.** This compound melts at 220°C–222°C with decomposition (Isab and Wazeer 2005). It was prepared at room temperature by mixing solutions of selenourea with $CdCl_2$ in a minimum amount of distilled $H_2O$ (2–10 mL) in the 2:1 molar ratio and stirring for 1 or 2 h. The isolated product was dried thoroughly.

**[Cd(C₃H₆N₂Se)₂Cl₂]** or **C₆H₁₂N₄Cl₂Se₂Cd.** This compound melts at 231°C–232°C with decomposition (Devillanova and Verani 1977). It was obtained by boiling an absolute EtOH solution of $CdCl_2$ under reflux with ethyleneselenourea (1:2 molar ratio).

**[Cd(C₅H₈N₂Se)₂Cl₂]** or **C₁₀H₁₆N₄Cl₂Se₂Cd.** This compound melts at 233°C–235°C (Williams et al. 2009). To obtain it, a 10.0 mM charge of 1,3-dimethyl-2(3H)-imidazoleselone was added to $CdCl_2 \cdot 2.5H_2O$ (5.0 mM) in 150 mL of boiling water. The water was boiled down to half the volume and allowed to cool. Fine-white crystals were obtained upon cooling the mixture in a refrigerator.

**(Me₄N)₂[Cd₄(SePh)₆Cl₄]** or **C₄₄H₅₄N₂Cl₄Se₆Cd₄.** To obtain this compound, freshly prepared solid $PhICl_2$ (0.60 mM) was added with stirring to a solution of $(Me_4N)_2[Cd_4(SePh)_{10}]$ (0.30 mM) in 15 mL of MeCN (Dean et al. 1987). A rapid reaction occurred with the formation of a precipitate. After 10 min the solvent was removed under vacuum. The residue was extracted three times with 20 mL portions of diethyl ether to remove $(PhS)_2$ and PhI. The crude residue was purified by recrystallization from a mixture of acetonitrile and diethyl ether. The syntheses were performed under an inert atmosphere.

## 5.16 Cd–H–C–N–Br–Se

Some compounds are formed in this system.

**[Cd(C₃H₆N₂Se)₂Br₂] or C₆H₁₂N₄₂Br₂Se₂Cd**. This compound melts at 224°C–225°C with decomposition (Devillanova and Verani 1977). It was obtained by boiling an absolute EtOH solution of $CdBr_2$ under reflux with ethyleneselenourea (1:2 molar ratio).

**(Me₄N)₂[Cd₄(SePh)₆Br₄] or C₄₄H₅₄N₂Br₄Se₆Cd₄**. This compound crystallizes in the cubic structure with the lattice parameter $a = 1806.2 \pm 0.2$ and experimental and calculated densities $2.11 \pm 0.02$ and 2.09 g·cm⁻³, respectively (Dean et al. 1987). To obtain it, $Br_2$ (1.15 mM) in 5 mL of $CCl_4$ was added to a stirred solution containing (Me₄N)₂[Cd₄(SePh)₁₀] (0.575 mM) in 15 mL of acetone. Decolorization occurred quickly, leaving a clear solution with a small amount of white solid. After 10 min, the solvents were removed under vacuum. The residue was extracted with three 40 mL portions of diethyl ether to remove (PhS)₂. The residue was purified by recrystallization from a mixture of acetone and cyclohexane. The syntheses were performed under an inert atmosphere. Colorless air-stable crystals were grown by diffusion of cyclohexane into acetone at room temperature (Dean et al. 1987) or by recrystallization from $C_6H_{12}$/Me₂CO after removal of (PhS)₂ by extraction with Et₂O (Dean and Vittal 1985).

## 5.17 Cd–H–C–N–I–Se

Some compounds are formed in this system.

**[Cd(C₃H₆N₂Se)₂I₂] or C₆H₁₂N₄I₂Se₂Cd**. This compound melts at 202°C–203°C with decomposition (Devillanova and Verani 1977). It was obtained by boiling an absolute EtOH solution of $CdI_2$ under reflux with ethyleneselenourea (1:2 molar ratio).

**(Me₄N)₂[Cd₄(SePh)₆I₄] or C₄₄H₅₄N₂I₄Se₆Cd₄**. To obtain this compound, solid iodine (1.2 mM) was added to a stirred solution of (Me₄N)₂[Cd₄(SePh)₁₀] (0.6 mM) in 15 mL of acetone (Dean et al. 1987). Immediate decolorization occurred, leaving a clear solution. Acetone was removed under vacuum. The residue was extracted with three 25 mL portions of diethyl ether to remove (PhS)₂. The remaining white solid was purified by recrystallization from MeCN/Et₂O. The syntheses were performed under an inert atmosphere. Colorless crystals were obtained by recrystallization from $C_6H_{12}$/Me₂CO after removal of Ph₂S₂ by extraction with Et₂O (Dean and Vittal 1985).

## 5.18 Cd–H–C–P–O–Cl–Se

Some compounds are formed in this system.

**[Cd₄(SePh)₆(PPh₃)₄(ClO₄)₂] or C₁₀₈H₉₀P₄O₈Cl₂Se₆Cd₄**. To prepare this compound, a mixture of Cd(PPh₃)₂(ClO₄)₂ (0.5 mM), Cd(SePh)₂ (1.5 mM), and PPh₃ (1.0 mM) in $CH_2Cl_2$ (5 mL) was stirred for 20 min (Dean et al. 1993). A small amount of gelatinous insoluble material was then removed by filtration. The addition of Et₂O (5 mL) to the filtrate gave an oil. After refrigeration to 0°C overnight, the supernatant liquid was removed. The remaining oil was shaken with a further 5 mL portion of $Et_2O$ and refrigerated for several days to give a white solid product. The product was separated by filtration, washed with $Et_2O$, and dried in vacuo.

**[{$Cd_{17}Se_4(SePh)_{24}(Ph_2Pr^nP)_4$}{$Cd_8Se(SePh)_{12}Cl_4$}] or $C_{300}H_{296}P_4O_6Cl_4Se_{41}Cd_{25}$.**
This compound crystallizes in the cubic structure with the lattice parameter $a = 3266.4 \pm 0.4$ at 200 K and a calculated density of 1.853 g·cm$^{-3}$ (Soloviev et al. 2001). It was prepared by the next procedure. First, 1.58 mM of $CdCl_2$ was dissolved in 40 mL of THF under the addition of 3.16 mM of $Ph_2Pr^nP$. Then 2.28 mM of $PhSeSiMe_3$ was added and the resulting clear solution was stirred overnight. The addition of 0.32 mM of $Se(SiMe_3)_2$ led to the formation of a clear solution from which the title compound could be crystallized by overlayering with heptane. Standard Schlenk techniques were employed throughout the syntheses using a double-manifold vacuum line with high-purity dry nitrogen.

## 5.19   Cd–H–C–P–Se

Some compounds are formed in this system.

**[$Cd(SePh)_2$]$_2$[$Et_2PCH_2CH_2PEt_2$] or $C_{17}H_{22}PSe_2Cd$.** This compound melts at 158.6°C–159°C and crystallizes in the monoclinic structure with the lattice parameters $a = 2186.7 \pm 0.8$, $b = 2226.1 \pm 0.8$, and $c = 813.1 \pm 0.2$ pm, and $\beta = 100.08° \pm 0.05°$ and a calculated density of 1.733 g·cm$^{-3}$ (Brennan et al. 1989). To obtain it, a solution of $CdMe_2$ (11 mM) and $Et_2PCH_2CH_2PEt_2$ (9 mM) in toluene (10 mL) was added HSePH (22 mM) in a dropwise fashion. After a period of minutes to hours, a white precipitate formed. Recrystallization from pyridine/heptane (1:8 volume ratio) gave colorless needles of $C_{17}H_{22}PSe_2Cd$. Thermolysis of this compound leads to the formation zincblende nanometer-sized clusters of CdSe.

**[$Cd(SePh)_2$·depe] or $C_{22}H_{34}P_2Se_2Cd$ [depe = 1,2-bis(diethylphosphino)ethane].**
This compound melts at 162°C–164°C and crystallizes in the orthorhombic structure with the lattice parameters $a = 1616.3 \pm 1.3$, $b = 1637.7 \pm 1.0$, and $c = 1952.3 \pm 1.4$ pm (Brennan et al. 1990). To obtain it, depe (0.77 mM) was added to $Cd(SePh)_2$ (0.34 mM) in a mixture of acetonitrile (20 mL) and pyridine (0.50 mL). The solution was stirred for 2 h, filtered, and cooled (−10°C) to give colorless diamonds. All reactions were performed in an inert atmosphere.

**[{$Cd(SePh)_2$}$_2$·dmpe] or $C_{30}H_{36}P_4Se_4Cd_2$ [dmpe = 1,2-bis(dimethylphosphino)ethane].** This compound melts at 162°C–164°C (Brennan et al. 1990). To obtain it, dmpe (0.33 mM) was added to $Cd(SePh)_2$ (0.36 mM) dissolved in pyridine (5 mL). The solution was saturated in heptane (13 mL) upon which colorless needles immediately began to precipitate. The needles were collected and dried under vacuum. All reactions were performed in an inert atmosphere.

**($Ph_4P$)$_2$[$Cd(Se_4)_2$] or $C_{48}H_{40}P_2Se_8Cd$.** This compound crystallizes in the monoclinic structure with the lattice parameters $a = 1018.5 \pm 0.5$, $b = 1413.7 \pm 0.4$, and $c = 3439 \pm 2$ pm, and $\beta = 92.22° \pm 0.02°$ (Banda et al. 1989). To obtain it, a mixture of

Li$_2$Se (2 mM), black Se (6 mM), and Ph$_4$PCl (2 mM) in MeCN (30 mL) and Et$_3$N (5 mL) was stirred for 1 h. A solution of cadmium ethylxanthate, Cd(EtCOS$_2$)$_2$, (1 mM) in DMF (10 mL) was added to that mixture. The resulting solution was stirred for 1 h and then filtered. The volume of the filtrate was reduced to 20 mL, and Pr$^i$OH (20 mL) was layered on it. Crystals were produced in 2 days. All the manipulations were carried out under an N$_2$ atmosphere (Ansari et al. 1990).

Bis-tetraselenide complex [Cd(Se$_4$)$_2$]$^{2-}$ could be readily prepared by the reaction of Se, Na, and Cd(NO$_3$)$_2$·4H$_2$O in DMF and crystallized with Ph$_4$P$^+$. The general preparative procedure involves Cd(NO$_3$)$_2$·4H$_2$O, Se, and Na in a molar ratio of 1:8:4 under a N$_2$ atmosphere, with Ph$_4$PBr slowly diffused into the filtered reaction solution (Banda et al. 1989).

**[Cd$_{10}$Se$_4$(SePh)$_{12}$(Pr$^n_3$P)$_4$] or C$_{108}$H$_{144}$P$_4$Se$_{16}$Cd$_{10}$.** This compound crystallizes in the tetragonal structure with the lattice parameters $a = 2544.1 \pm 0.4$ and $c = 2030.8 \pm 0.4$ pm at 200 K and a calculated density of 1.998 g·cm$^{-3}$ (Soloviev et al. 2001). It was prepared by the following procedure. First, 1.04 mM of CdCl$_2$ was dissolved in 40 mL of toluene under the addition of 2.07 mM of Pr$^n_3$P. Then 1.35 m of PhSeSiMe$_3$ was added and the resulting clear solution was stirred overnight. The addition of 0.36 mM of Se(SiMe$_3$)$_2$ led to the formation of a pale-yellow solution from which the title compound could be crystallized by overlayering with heptane. Standard Schlenk techniques were employed throughout the syntheses using a double-manifold vacuum line with high-purity dry nitrogen.

**[Cd$_8$Se(SePh)$_{14}$(PPh$_3$)$_2$] or C$_{120}$H$_{100}$P$_2$Se$_{15}$Cd$_8$.** This compound crystallizes in the orthorhombic structure with the lattice parameters $a = 1829.3 \pm 0.4$, $b = 2639.9 \pm 0.5$, and $c = 2975.9 \pm 0.6$ pm (Behrens and Fenske 1997). To obtain it, PhSeSiMe$_3$ (0.88 mM) was added dropwise to a suspension of [CdCl$_2$(PPh$_3$)$_2$] (0.44 mM) in dichloroethane (35 mL). A colorless solution was formed from which colorless cuboids of this compound crystallized after some days. The reaction of [CdCl$_2$(PPh$_3$)$_2$] with PhSeSiMe$_3$ in dimethyl ether and dichloromethane also yields crystals of [Cd$_8$Se(SePh)$_{14}$(PPh$_3$)$_2$].

**[Cd$_{32}$Se$_{14}$(SePh)$_{36}$(PPh$_3$)$_4$] or C$_{288}$H$_{240}$P$_4$Se$_{50}$Hg$_{32}$.** This compound crystallizes in the monoclinic structure with the lattice parameters $a = 2534.05 \pm 0.76$, $b = 6813.53 \pm 0.94$, and $c = 2540.53 \pm 0.36$ pm, and $\beta = 113.519° \pm 0.020°$ at 200 K (Behrens et al. 1996a,b). It was prepared by the following procedure. First, 1.47 mM of CdCl$_2$ was dissolved in 40 mL of acetone under the addition of 5.89 mM of Ph$_3$P (Behrens et al. 1996a,b, Soloviev et al. 2001). Then 2.94 mM of PhSeSiMe$_3$ was added and the resulting clear solution was stirred overnight. The addition of 0.29 mM of Se(SiMe$_3$)$_2$ at −30°C led to the formation of a pale-yellow solution. The title compound crystallizes from this solution after several days. Standard Schlenk techniques were employed throughout the syntheses using a double-manifold vacuum line with high-purity dry nitrogen.

## 5.20   Cd–H–C–P–Cl–Se

**[Cd$_{17}$Se$_{34}$(SePh)$_{24}$(PPh$_3$)$_4$][Cd$_8$Se(SePh)$_{12}$Cl$_4$] or C$_{288}$H$_{240}$P$_4$Cl$_4$Se$_{71}$Cd$_{25}$.** This multinary compound is formed in this system. It crystallizes in the cubic structure with the lattice parameter $a = 3337.6 \pm 0.4$ pm (Behrens and Fenske 1997). To obtain it, a suspension of $[CdCl_2(PPh_3)_2]$ (0.83 mM) in THF (50 mL) was prepared. The addition of $PhSeSiMe_3$ (3.33 mM) to this suspension yields a colorless solution. After one night cubic crystals of this compound were formed. The synthesis was performed under a $N_2$ atmosphere and absolute THF was used.

## 5.21   Cd–H–C–P–I–Se

**$CdI_2 \cdot 2Bu^n_3PSe$ or $C_{24}H_{54}P_2I_2Se_2Cd$**. This multinary compound is formed in this system. Its colorless crystals were obtained by the interaction of $Bu^n_3PSe$ and $CdI_2$ in EtOH or MeOH (Nicpon and Meek 1966).

## 5.22   Cd–H–C–As–I–Se

**$CdI_2 \cdot Ph_3AsSe$ or $C_{18}H_{15}AsI_2SeCd$**. This multinary compound is formed in this system. It was prepared by the interaction of $Ph_3AsSe$ and $CdI_2$ in EtOH or MeOH (Nicpon and Meek 1966).

## 5.23   Cd–H–V–O–Se

**$Cd(VO_2)_4(SeO_3)_3 \cdot H_2O$**. This quinary compound is formed in this system. It crystallizes in the triclinic structure with the lattice parameters $a = 629.99 \pm 0.04$, $b = 740.78 \pm 0.04$, and $c = 1575.25 \pm 0.09$ pm and $\alpha = 100.881° \pm 0.001°$, $\beta = 98.796° \pm 0.001°$, and $\gamma = 93.690° \pm 0.001°$ at $173 \pm 2$ K and a calculated density of 3.943 $g \cdot cm^{-3}$ (Kim et al. 2001). To obtain it, $V_2O_5$, $SeO_2$, and $Cd(NO_3)_2 \cdot H_2O$ in a molar ratio of 2:20:5 were placed in a Teflon container of 20 mL. Distilled $H_2O$ (10 mL) was added to make the degree of fill about 80%. The solution pH reached 0.42 immediately after preparing the $V_2O_5$-suspended solution. The container was placed in an autoclave, which was placed in a 230°C oven. After the hydrothermal reaction for 3 days, the autoclave was cooled to room temperature, and the brown crystalline products were recovered by filtration, washing with distilled $H_2O$, and air-drying.

## 5.24   Cd–H–O–Br–Se

**$[Cd_{10}(SeO_3)_8Br_4] \cdot HBrH_2O$**. This quinary compound is formed in this system. It is stable up to 200°C and crystallizes in the orthorhombic structure with the lattice parameters $a = 1088.2 \pm 0.3$, $b = 1627.5 \pm 0.5$, and $c = 1872.8 \pm 0.6$ pm and a calculated density of 5.123 $g \cdot cm^{-3}$, and energy gap 1.65 eV (Chen et al. 2013). This compound was prepared from the reaction of $CdBr_2 \cdot 4H_2O$ and Se. The starting materials were loaded into a silica tube, which was flame sealed under a 0.1 Pa atmosphere, heated to 200°C in 6 h at room temperature and kept for 24 h, then heated to 450°C, and hold for 6 days, followed by cooling to 100°C at a rate of $6°C \cdot h^{-1}$ and then power off. $[Cd_{10}(SeO_3)_8Br_4] \cdot HBr \cdot H_2O$ is very stable in air and water.

## 5.25   Cd–K–Cu–O–Cl–Se

**$KCdCu_7O_2(SeO_3)Cl_9$**. This multinary compound (mineral burnsite) is formed in this system. It crystallizes in the hexagonal structure with the lattice parameters

$a = 878.05 \pm 0.08$ and $c = 1552.1 \pm 0.2$ pm and a calculated density of 3.85 g·cm$^{-3}$ (Burns et al. 2002, Krivovichev et al. 2002, Grew et al. 2003).

## 5.26 Cd–K–Mg–Sn–Se

$K_{14-2x}Mg_xCd_{15}Sn_{12}Se_{46}$. These quinary compounds are formed in this system (Ding and Kanatzidis 2006). To obtain them, $K_2Se$ (0.6 mM), Cd (1.3 mM), Sn (1.0 mM), and Se (3.3 mM) were combined in an evacuated and flame-sealed fused-silica tube and melted at 900°C. The product was loaded into a 9 mm Pyrex tube along with $MgCl_2 \cdot 6H_2O$ (0.7 mM) and deionized water (0.2 mL). The tube was then evacuated to 0.4 Pa and flame sealed. The tube was kept at 115°C for 18–24 h. The products, dark-orange tetrahedral or polyhedral crystals, were collected by filtration and washed with water, ethanol, and diethyl ether. The obtained crystals are stable in air for months.

## 5.27 Cd–K–Ca–Sn–Se

$K_{14-2x}Ca_xCd_{15}Sn_{12}Se_{46}$. This quinary compounds are formed in this system (Ding and Kanatzidis 2006). It crystallizes (for $x \approx 1.5$) in the cubic structure with the lattice parameter $a = 2328.07 \pm 0.04$ pm and a calculated density of 3.8807 g·cm$^{-3}$. They were prepared by the similar experiment described for $K_{14-2x}Mg_xCd_{15}Sn_{12}Se_{46}$ using $CaCl_2$ (0.6 mM) instead of $MgCl_2 \cdot 6H_2O$.

## 5.28 Cd–Cu–Ga–In–Se

$2CdSe–CuGaSe_2–CuInSe_2$. Figure 5.1 presents the isothermal section of this quasiternary system at 600°C (Marushko et al. 2010). There are three single-phase regions in this system. The first is the continuous α-solid solution between $CuGaSe_2$ and $CuInSe_2$ with the chalcopyrite structure type. The second single-phase region belongs to the β-solid solution based on CdSe with the wurtzite structure type. The third region is the γ-solid solution with the sphalerite structure type between the high-temperature modification of $CuInSe_2$, which is stabilized at the annealing temperature by CdSe and $CuCd_2GaSe_4$.

The samples were annealed at 600°C for 500 h and then quenched into cold water (Marushko et al. 2010). Single crystals of the γ-solid solutions were grown by a horizontal variation of the Bridgman technique.

## 5.29 Cd–Cu–In–Te–Se

$2CdTe–CuInSe_2$. This section is a nonquasibinary section of the $2CdSe + CuInTe_2$ $\Leftrightarrow 2CdTe + CuInSe_2$ ternary mutual system (Figure 5.2) (Lavrynyuk et al. 2009, Parasyuk et al. 2009). The liquidus and the solidus are represented by the lines of the beginning and the end of the primary crystallization of the continuous solid solutions (γ-phase). The subsolidus region shows the decomposition of the γ-solid solutions due to the polymorphous transition of $CuInSe_2$ at 807°C and the formation

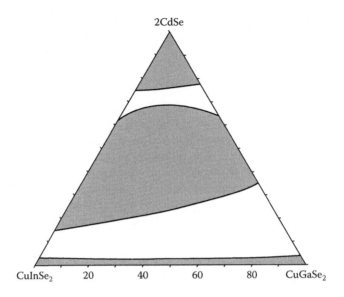

**FIGURE 5.1** The isothermal section of the 2CdSe–CuGaSe$_2$–CuInSe$_2$ quasiternary system at 600°C. (From Marushko, L.P. et al., *J. Alloys Compd.*, 505(1), 101, 2010b.)

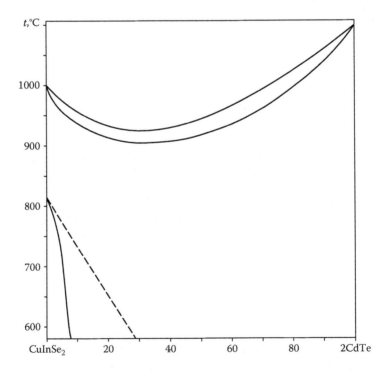

**FIGURE 5.2** Phase relations in the 2CdTe–CuInSe$_2$ section. (From Lavrynyuk, Z.F. et al., *Pol. J. Chem.*, 83(1), 7, 2009; Parasyuk, O.V. et al., *J. Cryst. Growth*, 311(8), 2381, 2009.)

of the α-phase. The α-solid solutions range up to 6 mol.% 2CdTe at 600°C and γ-solid solutions extend from 26 to 100 mol.% 2CdTe at this temperature.

Single crystals of the γ-solid solutions were grown by a modified Bridgman method (Parasyuk et al. 2009). They crystallize in the cubic structure of sphalerite type and possess intrinsic $p$-type conductivity.

**2CdSe–CuInTe₂.** This section is also a nonquasibinary section of the 2CdSe + CuInTe₂ ⇔ 2CdTe + CuInSe₂ ternary mutual system (Figure 5.3) (Lavrynyuk et al. 2009). Its liquidus consists of two parts: the primary crystallization of the β- and γ-solid solutions. The subliquidus part has a three-phase field that belongs to the volume of the secondary crystallization and corresponds to the only monovariant process in the system. The subsolidus region shows a partial decomposition of the γ-solid solutions with the formation of α-solid solutions. The limits of the two-phase region (β + γ) range from 72 to 84 mol.% 2CdSe at 600°C.

**2CdSe + CuInTe₂ ⇔ 2CdTe + CuInSe₂.** The liquidus surface of the 2CdSe + CuInTe₂ ⇔ 2CdTe + CuInSe₂ ternary mutual system (Figure 5.4) consists of the primary crystallization of the β- and γ-solid solutions (Lavrynyuk et al. 2009).

The existence of three homogeneity regions at 600°C was established in this ternary mutual system (Figure 5.5) (Lavrynyuk et al. 2009). The α-solid solutions with the chalcopyrite structure are stretched along the CuInTe₂–CuInSe₂ boundary side and only slightly penetrate into the quadrangle. The β-solid solutions possess the

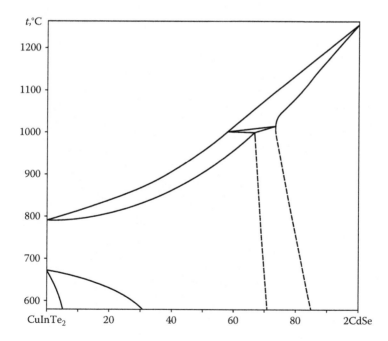

**FIGURE 5.3** Phase relations in the 2CdSe–CuInTe₂ section. (From Lavrynyuk, Z.F. et al., *Pol. J. Chem.*, 83(1), 7, 2009.)

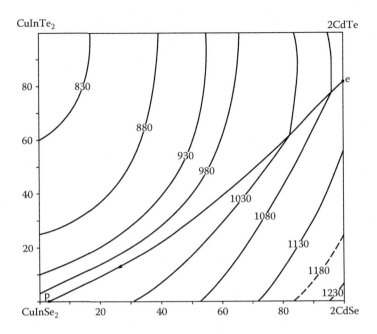

**FIGURE 5.4** Liquidus surface of the $2CdSe + CuInTe_2 \Leftrightarrow 2CdTe + CuInSe_2$ ternary mutual system. (From Lavrynyuk, Z.F. et al., *Pol. J. Chem.*, 83(1), 7, 2009.)

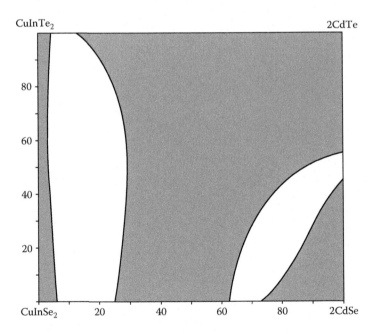

**FIGURE 5.5** The isothermal section of the $2CdSe + CuInTe_2 \Leftrightarrow 2CdTe + CuInSe_2$ ternary mutual system at 600°C. (From Lavrynyuk, Z.F. et al., *Pol. J. Chem.*, 83(1), 7, 2009.)

hexagonal wurtzite structure and the γ-solid solutions crystallize in the cubic sphalerite structure.

This system was investigated by the DTA and XRD, and the alloys were annealed at 600°C for 600 h and then quenched into saturated saline solution (Lavrynyuk et al. 2009).

## 5.30 Cd–Cu–Ge–Mn–Se

**CdSe–Cu$_2$Se–GeSe$_2$–MnSe. Cu$_2$Cd$_{1-x}$Mn$_x$GeSe$_4$** solid solution is formed in this system (Quintero et al. 2002, 2007). In the composition range $0.05 < x < 0.2$, the alloys were found to be of two phases α + δ (the tetragonal stannite and orthorhombic wurtz-stannite structures), while that for the range $0.2 \leq x < 0.7$, the samples were found to be single phase with the wurtz-stannite structure. For $0.7 \leq x \leq 1.0$, the samples were found to be of two phases δ + MnSe (Quintero et al. 2002, 2007). The values of the crystal parameters $a$ and $c$ in the wurtz-stannite δ composition range decrease linearly with the composition $x$, while the values of $b$ decrease nonlinearly (Quintero et al. 2007). The experimental values for $a$ and $c$ were least squared fitted to a linear equation, but the $b$ versus $x$ curve was fitted to a quadratic form, and the obtained results are given by $a(x) = 0.8068 - 0.0086x$ (nm), $b(x) = 0.6869 - 0.0010x - 0.0060x^2$ (nm), and $c(x) = 0.6602 - 0.0057x$ (nm).

The structure of the **Cu$_2$Cd$_{0.5}$Mn$_{0.5}$GeSe$_4$** alloy was refined from the XRD pattern (Delgado et al. 2004). It crystallizes in the orthorhombic structure with the lattice parameters $a = 802.53 \pm 0.02$, $b = 685.91 \pm 0.02$, and $c = 657.34 \pm 0.02$ pm and a calculated density of 5.52 g·cm$^{-3}$.

All of the alloy samples were produced by the usual melt and anneal techniques (Quintero et al. 2002, 2007).

## 5.31 Cd–Cu–Ge–Fe–Se

**Cu$_2$Cd$_{1-x}$Fe$_x$GeSe$_4$** solid solution is formed in this system (Quintero et al. 2007). There are only two single solid fields, namely, the tetragonal stannite α and the wurtz-stannite δ fields. The crystallographic parameter values appear, within the limits of experimental error, to follow the usual linear Vegard behavior. The experimental values of $a$ and $c$ were least squared fitted to a linear equation, and the obtained results are given by $a(x) = 0.5746 - 0.0148x$ (nm) and $c(x) = 1.1057 - 0.5213 \cdot 10^{-4}x$ (nm). All of the alloy samples were produced by the usual melt and anneal techniques (Quintero et al. 2007).

## 5.32 Cd–Cu–Sn–Mn–Se

**CdSe–Cu$_2$Se–SnSe$_2$–MnSe. Cu$_2$Cd$_{1-x}$Mn$_x$SnSe$_4$** solid solution is formed in this system (Sachanyuk et al. 2001, Moreno et al. 2009). Only two single solid phase fields, the tetragonal stannite α and the wurtz-stannite δ structures, were found to occur. In addition to the α phase, extra XRD lines due to MnSe were observed as grown samples in the range $0.7 < x < 1.0$ (Moreno et al. 2009). However, it was found that the amount of the extra phase decreased for the compressed samples.

The value of the crystal parameter $a$ decreases linearly with the composition $x$, while the values of $c$ decreases nonlinearly. The experimental values for $a$ were least squared fitted to a linear equation, but the $c$ versus $x$ curve was fitted to a quadratic form, and the obtained results are given by $a(x) = 0.58306 - 0.0106x$ (nm) and $c(x) = 1.13998 - 0.004264x$ (nm). All of the alloy samples were produced by the usual melt and anneal techniques.

According to the data of Sachanyuk et al. (2001), at 400°C there is a continuous series of solid solutions of substitution between the quaternary chalcogenides. A $Cu_2Cd_{0.5}Mn_{0.5}SnSe_4$ solid solution crystallizes in the tetragonal structure with the lattice parameters $a = 579.35 \pm 0.02$ and $c = 1140.36 \pm 0.05$ and a calculated density of $5.5951 \pm 0.0007$ g·cm$^{-3}$.

## 5.33   Cd–Cu–Sn–Fe–Se

A $Cu_2Cd_{1-x}Fe_xSnSe_4$ solid solution is formed in this system (Moreno et al. 2009). Only two single solid phase fields, the tetragonal stannite α and the wurtz-stannite δ structures, were found to occur. In addition to the α phase, extra XRD lines due to $FeSe_2$ were observed as grown samples in the range $0.7 < x < 1.0$. However, it was found that the amount of the extra phase decreased for the compressed samples. The crystallographic parameter values appear, within the limits of experimental error, to follow the usual linear Vegard behavior: $a(x) = 0.58306 - 0.005297x$ (nm) and $c(x) = 1.13998 - 0.002575x$ (nm). Solid solutions $Cu_2Cd_{0.8}Fe_{0.2}SnSe_4$ and $Cu_2Cd_{0.2}Fe_{0.8}SnSe_4$ crystallize in the tetragonal structure with the lattice parameters $a = 580.50 \pm 0.02$ and $c = 1139.43 \pm 0.04$ and $a = 573.54 \pm 0.01$ and $c = 1131.62 \pm 0.03$ pm, respectively. All of the alloy samples were produced by the usual melt and anneal techniques.

## 5.34   Cd–Ag–As–Te–Se

$CdTe–Ag_2Te–As_2Se_3$. The glass forming region in this system is shown in Figure 5.6 (Vassilev et al. 2008). It is extended to the area rich in $As_2Se_3$ and partially lies on the faces $Ag_2Te–As_2Se_3$ (50–100 mol.% $As_2Se_3$) and $CdTe–As_2Se_3$ (0–15 mol.% CdTe) faces of the concentration triangle. The introduction of CdTe and $Ag_2Te$ decreases, in an almost equal degree, the glass transition temperature ($T_g$), which changes within the interval from 138°C to 189°C.

## 5.35   Cd–Ag–As–I–Se

$CdSe–AgI–As_2Se_3$. The glass-forming region in this quasiternary system is shown in Figure 5.7 (Kassem et al. 2011). Its range for $(CdSe)_x(AgI)_{0.5-x/2}(As_2Se_3)_{0.5-x/2}$ compositions was found to be limited to $x \leq 0.20$ without any significant change in the glass transition temperature but with two subregions with decreased ($x \leq 0.05$) and increased ($x \geq 0.05$) crystallization ability (Kassem et al. 2010). $CdSe + 2AgI \Leftrightarrow CdI_2 + Ag_2Se$ exchange reaction takes place in the glass-forming melt. The maximum CdSe solubility in the glass ternary system is only 20 mol.%. The glass transition temperature ($T_g$) changes within the interval from 116°C to 159°C (Kassem et al. 2011) [from 122°C to 142°C (Kassem et al. 2010)].

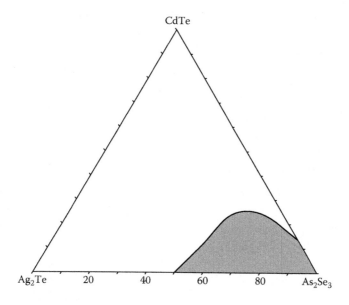

**FIGURE 5.6** Glass formation region in the CdTe–Ag$_2$Te–As$_2$Se$_3$ system. (From Vassilev, V. et al., *J. Phys. Chem. Solids*, 69(7), 1835, 2008.)

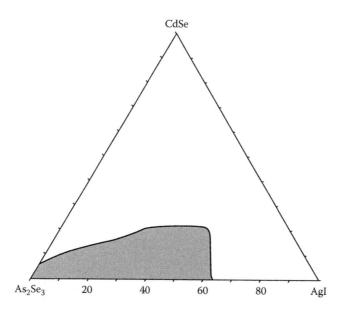

**FIGURE 5.7** Glass formation region in the CdSe–AgI–As$_2$Se$_3$ system. (From Kassem, F. et al., *Mater. Res. Bull.*, 46(2), 210, 2011.)

The samples were prepared using the appropriate proportion of previously synthesized CdSe, AgI, and $Ag_2Se_3$ (Kassem et al. 2010). The mixtures were heated slowly in a rocking furnace to 850°C at a rate of 5°C·min$^{-1}$, maintaining at this temperature for 24 h, and cooled down to 650°C before quenching in cold salt/water mixture.

The glass-forming region in this system was investigated through XRD and DSC (Kassem et al. 2010, 2011).

## 5.36  Cd–Ge–Sb–Te–Se

**CdTe–GeSe$_2$–Sb$_2$Se$_3$**. The glass-forming region in this quasiternary system is shown in Figure 5.8 (Vassilev et al. 2002). The glasses have been obtained in the GeSe$_2$-rich region. The maximum CdTe solubility in the glass ternary system is only 15 mol.%. The glass transition temperature ($T_g$) changes within the interval from 185°C to 250°C. The glass-forming region in this system was investigated visually and through XRD, DTA, and EPMA.

## 5.37  Cd–As–Sb–Te–Se

**CdTe–As$_2$Se$_3$–Sb$_2$Se$_3$**. The glass-forming region in this quasiternary system is shown in Figure 5.9 (Vassilev et al. 2002). The glasses have been obtained in the

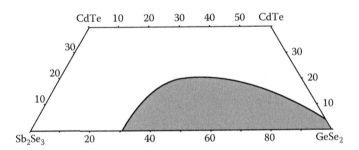

**FIGURE 5.8**  Glass formation region in the CdTe–GeSe$_2$–Sb$_2$Se$_3$ system. (From Vassilev, V.S. et al., *Mater. Lett.*, 52(1–2), 126, 2002.)

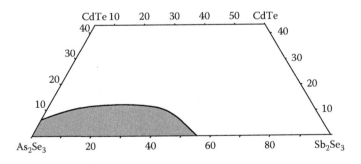

**FIGURE 5.9**  Glass formation region in the CdTe–As$_2$Se$_3$–Sb$_2$Se$_3$ system. (From Vassilev, V.S. et al., *Mater. Lett.*, 52(1–2), 126, 2002.)

As$_2$Se$_3$-rich region. The maximum CdTe solubility in the glass ternary system is only 12 mol.%. The glass transition temperature ($T_g$) changes within the interval from 80°C to 165°C. The glass-forming region in this system was investigated visually and through XRD, DTA, and EPMA.

## REFERENCES

Adel J., Weller F., Dehnicke K. Synthese und Kristallstrukturen der spirocyclischen Polyselenido-Komplexe [Na-15-Krone-5]$_2$[M(Se$_4$)$_2$] mit M=Zink, Cadmium und Quecksilber, *Z. Naturforsch.*, 43B(9), 1094–1100 (1988).

Ansari M.A., Mahler C.H., Chorghade G.S., Lu Y.-J., Ibers J.A. Synthesis, structure, spectroscopy, and reactivity of M(Se$_4$)$_2$$^{2-}$ anions, M=Ni, Pd, Zn, Cd, Hg, and Mn, *Inorg. Chem.*, 29(19), 3832–3839 (1990).

Banda R.M.H., Cusick J., Scudder M.L., Craig D.C., Dance I.G. Syntheses and x-ray structures of molecular metal polyselenide complexes [M(Se$_4$)$_2$]$^{2-}$ M=Zn, Cd, Hg, Ni, Pb, *Polyhedron*, 8(15), 1995–1998 (1989).

Behrens S., Bettenhausen M., Deveson A.C., Eichhöfer, A., Fenske D., Lohde A., Woggon U. Synthese und Struktur der Nanocluster [Hg$_{32}$Se$_{14}$(SePh)$_{36}$], [Cd$_{32}$Se$_{14}$(SePh)$_{36}$(PPh$_3$)$_4$], [Et$_2$P(Ph)C$_4$H$_8$OSiMe$_3$]$_5$[Cd$_{18}$I$_{17}$(PSiMe$_3$)$_{12}$] und [Et$_3$NC$_4$H$_8$OSiMe$_3$]$_5$[Cd$_{18}$I$_{17}$(PSiMe$_3$)$_{12}$, *Angew. Chem.*, 108(19), 2360–2363 (1996a).

Behrens S., Bettenhausen M., Deveson A.C., Eichhöfer, A., Fenske D., Lohde A., Woggon U. Synthesis and structure of the nanoclusters [Hg$_{32}$Se$_{14}$(SePh)$_{36}$], [Cd$_{32}$Se$_{14}$(SePh)$_{36}$(PPh$_3$)$_4$], [P(Et)$_2$(Ph)C$_4$H$_8$OSiMe$_3$]$_5$[Cd$_{18}$I$_{17}$(PSiMe$_3$)$_{12}$], and [N(Et)$_3$C$_4$H$_8$OSiMe$_3$]$_5$[Cd$_{18}$I$_{17}$(PSiMe$_3$)$_{12}$], *Angew. Chem. Int. Ed. Engl.*, 35(19), 2215–2218 (1996b).

Behrens S., Fenske D. Cadmium nanoclusters with phenylselenolato- and phenyltellurato ligands: synthesis and structural characterization of [Cd$_{17}$Se$_4$(SePh)$_{24}$(PPh$_3$)$_4$] [Cd$_8$Se(SePh)$_{12}$Cl$_4$], [Cd(DMF)$_6$] [Cd$_8$Se(SePh)$_{12}$Cl$_4$], [Cd$_8$Se(SePh)$_{14}$(PPh$_3$)$_2$], [Cd$_8$Se(SePh)$_{14}$(DMF)$_3$] and [Cd$_8$Te(TePh)$_{14}$(PEt$_3$)$_3$], *Ber. Bunsenges. Phys. Chem.*, 101(11), 1588–1592 (1997).

Bochmann M., Bwembya G.C., Hursthouse M.B., Coles S.J. Synthesis of phosphinochalcogenoic amidato complexes of zinc and cadmium. The crystal and molecular structure of [Zn{Bu$^t$$_2$P(Se)NPr$^i$}$_2$], *J. Chem. Soc., Dalton Trans.*, (17), 2813–2817 (1995).

Brandmayer M.K., Clérac R., Weigend F., Dehnen S. Ortho-chalcogenostannates as ligands: Syntheses, crystal structures, electronic properties, and magnetism of novel compounds containing ternary anionic substructures [M$_4$($\mu_4$-Se)(SnSe$_4$)$_4$]$^{10-}$ (M=Mn, Zn, Cd, Hg), {[Hg$_4$($\mu_4$-Se)(SnSe$_4$)$_3$]$^{6-}$}, or {[HgSnSe$_4$]$^{2-}$}, *Chem. Eur. J.*, 10(20), 5147–5157 (2004).

Brennan J.G., Siegrist T., Carroll P.J., Stuczynski S.M., Brus L.E., Steigerwald M.L. The preparation of large semiconductor clusters via the pyrolysis of a molecular precursor, *J. Am. Chem. Soc.*, 111(11), 4141–4143 (1989).

Brennan J.G., Siegrist T., Carroll P.J., Stuczynski S.M., Reynders P., Brus L.E., Steigerwald M.L. Bulk and nanostructure group II-VI compounds from molecular organometallic precursors, *Chem. Mater.*, 2(4), 403–409 (1990).

Burns P.C., Krivovichev S.V., Filatov S.K. New Cu$^{2+}$ coordination polyhedra in the crystal structure of burnsite, KCdCu$_7$O$_2$(SeO$_3$)$_2$Cl$_9$, *Can. Mineral.*, 40(6), 1587–1595 (2002).

Chen W.-T., Wang M.-S., Wang G.-E, Chen H.-F., Guo G.-C. Solid-state synthesis, structure and properties of a novel open-framework cadmium selenite bromide: [Cd$_{10}$(SeO$_3$)$_8$Br$_4$]·HBr·H$_2$O, *J. Solid State Chem.*, 204, 153–158 (2013).

Choudhury C.R., Dey S.K., Mondal N., Mitra S., Gramlich V. Synthesis and structural studies of selenocyanato-bridged polymeric and molecular complexes of cadmium(II): Presence of both end-to-end and end-on coordination, *Inorg. Chim. Acta*, 353, 217–222 (2003).

Dean P.A.W., Payne N.C., Vittal J.J., Wu Y. Synthesis and NMR spectra ($^{31}$P, $^{111/113}$Cd, $^{77}$Se) of adamantane-like phosphine complexes with the ($\mu$-ER)$_6$Cd core and the crystal and molecular structure of [($\mu$-SPr$^i$)$_6$(CdPPh$_3$)$_2$(CdOClO$_3$)$_2$]·EtOH, *Inorg. Chim.*, 32(21), 4632–4639 (1993).

Dean P.A.W., Vittal J.J. Cadmium-113 nuclear magnetic resonance spectroscopic study of mixed-ligand tetranuclear clusters of the type [Cd$_4$(EPh)$_x$(E'R)$_{10-x}$]$^{2-}$ and of the mixed-metal clusters [Cd$_x$Zn$_{4-x}$(SPh)$_{10}$]$^{2-}$, *Inorg. Chem.*, 25(4), 514–519 (1986).

Dean P.A.W., Vittal J.J., Payne N.C. Syntheses of (Me$_4$N)$_2$[($\mu$-EPh)$_6$(MX)$_4$] (M = Cd, Zn; E = S, Se; X = Cl, Br, I). Crystal and molecular structures of (Me$_4$N)$_2$[($\mu$-EPh)$_6$(CdBr)$_4$] (E = S, Se) and characterization of [($\mu$-EPh)$_6$(CdX)$_4$]$^{2-}$ in solution by $^{113}$Cd and $^{77}$Se NMR, *Inorg. Chem.*, 26(11), 1683–1689 (1987).

Dean P.A.W., Vittal J.J. Simple synthesis and $^{113}$Cd NMR spectroscopic characterization of the fully terminally substituted clusters [($\mu$-EPh)$_6$(CdX)$_4$]$^{2-}$ (E = S or Se; X = Br or I), *Inorg. Chem.*, 24(23), 3722–3723 (1985).

Delgado G.E., Quintero E., Tovar R., Quintero M. X-ray powder diffraction study of the semiconducting alloy Cu$_2$Cd$_{0.5}$Mn$_{0.5}$GeSe$_4$, *Cryst. Res. Technol.*, 39(9), 807–810 (2004).

Devillanova F.A., Verani G. Complexes of zinc(II), cadmium(II) and mercury(II) halides with ethylenselenourea, *Trans. Met. Chem.*, 2(1), 9–11 (1977).

Ding N., Kanatzidis M.G. Acid-induced conversions in open-framework semiconductors: From [Cd$_4$Sn$_3$Se$_{13}$]$^{6-}$ to [Cd$_{15}$Sn$_{12}$Se$_{46}$]$^{14-}$, a remarkable disassembly/reassembly process, *Angew. Chem. Int. Ed.*, 45(9), 1397–1401 (2006).

Ding N., Takabayashi Y., Solari P.L., Prassides K., Pcionek R.J., Kanatzidis M.G. Cubic gyroid frameworks in mesostructured metal selenides created from tetrahedral Zn$^{2+}$, Cd$^{2+}$, and In$^{3+}$ ions and the [SbSe$_4$]$^{3-}$ precursor, *Chem. Mater.*, 18(19), 4690–4699 (2006).

Grew E.S., Pertsev N.N., Roberts A.C. New mineral names, *Am. Mineral.*, 88(7), 1175–1180 (2003).

Hursthouse M.B., Malik M.A., Majid M., O'Brien P. The crystal and molecular structure of *N,N*-diethyldiselenocarbamatocadmium(II): Cadmium and zinc diethyldiselenocarbamates as precursors for selenides, *Polyhedron*, 11(1), 45–48 (1992).

Isab A.A., Wazeer M.I.M. Complexation of Zn(II), Cd(II) and Hg(II) with thiourea and selenourea: A $^1$H, $^{13}$C, $^{15}$N, $^{77}$Se and $^{113}$Cd solution and solid-state NMR study, *J. Coord. Chem.*, 58(6), 529–537 (2005).

Kassem M., Le Coq D., Bokova M., Bychkov E. Chemical and structural origin of conductivity changes in CdSe–AgI–As$_2$Se$_3$ glasses, *Solid State Ionics*, 181(11–12), 466–472 (2010).

Kassem M., Le Coq D., Fourmentin M., Hindle F., Bokova M., Cuisset A., Masseline P., Bychkov E. Synthesis and properties of new CdSe–AgI–As$_2$Se$_3$ chalcogenide glasses, *Mater. Res. Bull.*, 46(2), 210–215 (2011).

Kim J., Kim K.-W. Bis(*N,N,N*-trimethylethanaminium) bis(1,4-tetraselenido-$\kappa^2$*Se$^1$,Se$^4$*)cadmate, *Acta Crystallogr.*, E67(4), m394 (2011).

Kim Y.-T., Kim Y.-H., Park K., Kwon Y.-U., Young, Jr. V.G. Cd(VO$_2$)$_4$(SeO$_3$)$_3$·H$_2$O: A new bimetallic vanadium selenite compound with heptacoordinated cadmium ion, *J. Solid State Chem.*, 161(1), 23–30 (2001).

Kräuter G., Weller F., Dehnicke K. Synthesen und Kristallstrukturen der Kronenether-Komplexe [Li$_3$(12-Krone-4)$_3$O$_2$CCH$_3$][Cd(Se$_4$)$_2$], {[K(18-Krone-6)]$_2$[Hg(Se$_4$)$_2$]}$_2$ und [Na(15-Krone-5)]NO$_3$, *Z. Naturforsch.*, 44B(4), 444–454 (1989).

Krivovichev S.V., Vergasova L.P., Starova G.L., Filatov S.K., Britvin S.N., Roberts A.C., Steele I.M. Birnsite, KCdCu$_7$O$_2$(SeO$_3$)Cl$_9$, a new mineral species from Tolbachik volcano, Kamchatka peninsula, Russia *Can. Mineral.*, 40(4), 1171–1175 (2002).

Lavrynyuk Z.F., Zmiy O.F., Parasyuk O.V., Olekseyuk I.D., Pekhnyo V.I. The reciprocal CuInSe$_2$ + 2CdTe $\Leftrightarrow$ CuInTe$_2$ + 2CdSe system, *Pol. J. Chem.*, 83(1), 7–18 (2009).

Lee G.S.H., Fisher K.J., Craig D.C., Scudder M.L., Dance I.G. [ECd$_8$(E'Ph)$_{16}$]$^{2-}$ cluster chemistry (E, E' = sulfur, selenium, tellurium), *J. Am. Chem. Soc.*, 112(17), 6435–6437 (1990).

Li D., Liu D. Bis(isonicotinonitrile)cadmium diselencyanate, [Cd(SeCN)$_2$(pyCN)$_2$] (pyCN = isonicotinonitrile), *Appl. Organomet. Chem.*, 17(5), 321–322 (2003).

Liu D., Li D. *N,N'*-Dimethylimidazolium tris(selenocyanate) cadmium(II), [Me$_2$Im] [Cd(SeCN)$_3$] (Me$_2$Im = *N,N'*-dimethylimidazolium), *Appl. Organomet. Chem.*, 17(10), 809–810 (2003a).

Liu D., Li D. *p*-Bis(*N*-methylimidazolyl)xylylene tetra(selenocyanate)cadmium(II), [XylIm$_2$] [Cd(SeCN)$_4$], *Appl. Organomet. Chem.*, 17(10), 811–812 (2003b).

Magull S., Dehnicke K., Fenske D. Synthese und Kristallstruktur von [Ba(18-Krone-6) (DMF)$_4$][Cd(Se$_4$)$_2$], *Z. anorg. und allg. Chem.*, 608(2), 17–22 (1992).

Marushko L.P., Romanyuk Y.E., Piskach L.V., Parasyuk O.V., Olekseyuk I.D., Volkov S.V., Pekhnyo V.I. The CuInSe$_2$–CuGaS$_2$–CdSe system and crystal growth of the γ-solid solutions, *J. Alloys Compd.*, 505(1), 101–107 (2010).

Moreno E., Quintero M., Morocoima M., Quintero E., Grima P., Tovar R., Bocaranda P. et al. Lattice parameter values and phase transitions for the Cu$_2$Cd$_{1-z}$Mn$_z$SnSe$_4$ and Cu$_2$Cd$_{1-z}$Fe$_z$SnSe$_4$ alloys, *J. Alloys Compd.*, 486(1–2), 212–218 (2009).

Nicpon P., Meek D.W. Metal complexes of tertiary arsine sulphides and phosphine selenides, *Chem. Commun.*, (13), 398–399 (1966).

Parasyuk O.V., Lavrynyuk Z.F., Zmiy O.F., Romanyuk Y.E. Single crystal growth and properties of the γ-phase in the CuInSe$_2$–2CdTe system, *J. Cryst. Growth*, 311(8), 2381–2384 (2009).

Quintero E., Tovar R., Quintero M., Delgado G.E., Morocoima M., Caldera D., Ruiz J., Mora A.E., Briceño M., Fernandez J.L. Lattice parameter values and phase transitions for the Cu$_2$Cd$_{1-z}$Mn$_z$GeSe$_4$ and Cu$_2$Cd$_{1-z}$Fe$_z$GeSe$_4$ alloys, *J. Alloys Compd.*, 432(1–2), 142–148 (2007).

Quintero E., Tovar R., Quintero M., Morocoima M., Ruiz J., Delgado G., Broto J.M., Rakoto H. Structural characterization and magnetic properties for the semiconducting semimagnetic system Cu$_2$Cd$_{1-z}$Mn$_z$GeSe$_4$ alloys, *Phys. B*, 320(1–4), 384–387 (2002).

Sachanyuk V.P., Olekseyuk I.D., Parasyuk O.V. X-ray powder diffraction study of the Cu$_2$Cd$_{1-x}$Mn$_x$SnSe$_4$ alloys, *Phys. Status Solidi (a)*, 203(3), 459–465 (2001).

Secondo P.M., Land J.M., Baughman R.G., Collier H.L. Polymeric octahedral and monomeric tetrahedral Group 12 pseudohalogeno (NCX$^-$: X = O, S, Se) complexes of 4-(*N,N*-dimethylamino)pyridine, *Inorg. Chim. Acta*, 309(1–2), 13–22 (2000).

Soloviev V.N., Eichhöfer A., Fenske D., Banin U. Size-dependent optical spectroscopy of a homologous series of CdSe cluster molecules, *J. Am. Chem. Soc.*, 123(10), 2354–2364 (2001).

Trikalitis P.N., Rangan K.K., Bakas T., Kanatzidis M.G. Varied pore organization in mesostructured semiconductors based on the [SnSe$_4$]$^{4-}$ anion, *Nature*, 410(6829), 671–675 (2001).

Ueyama N., Sugawara T., Sasaki K., Nakamura A., Yamashita S., Wakatsuki Y., Yamazaki H., Yasuoka N. X-ray structures and far-infrared and Raman spectra of tetrahedral thiophenolato and selenophenolato complexes of zinc(II) and cadmium(II), *Inorg. Chem.*, 27(4), 741–747 (1988).

Vassilev V., Karadashka I., Parvanov S. New chalcogenide glasses in the Ag$_2$Te–As$_2$Se$_3$–CdTe system, *J. Phys. Chem. Solids*, 69(7), 1835–1840 (2008).

Vassilev V.S., Boycheva S.V., Petkov P. Glass formation in the GeSe$_2$(As$_2$Se$_3$)–Sb$_2$Se$_3$–CdTe, *Mater. Lett.*, 52(1–2), 126–129 (2002).

Vittal J.J., Dean P.A.W., Payne N.C. Metal-selenolate chemistry: Stereochemistry of adamantane-type clusters of formula [(μ-SePh)$_6$(MSePh)$_4$]$^{2-}$ (M = Zn and Cd), *Can. J. Chem.*, 70(3), 792–801 (1992).

Wachhold M., Rangan K.K., Billinge S.J.L., Petkov V., Heising J., Kanatzidis M.G. Mesostructured non-oxidic solids with adjustable worm-hole shaped pores: M–Ge–Q (Q = S, Se) frameworks based on tetrahedral $[Ge_4Q_{10}]^{4-}$ clusters, *Adv. Mater.*, 12(2), 85–91 (2000a).

Wachhold M., Rangan K.K., Lei M., Thorpe M.F., Billinge S.J.L., Petkov V., Heising J., Kanatzidis M.G. Mesostructured metal germanium sulfide and selenide materials based on the tetrahedral $[Ge_4S_{10}]^{4-}$ and $[Ge_4Se_{10}]^{4-}$ units: Surfactant templated three-dimensional disordered frameworks perforated with worm holes, *J. Solid State Chem.*, 152(1), 21–36 (2000b).

Williams D.J., Gulla D., Arrowood K.A., Bloodworth L.M., Carmack A.L., Evers T.J., Wilson M.S. et al. The preparation, characterization and x-ray structural analysis of *tetrakis*[1-methyl-3-(2-propyl)-2(3H)-imidazolethione] cadmium(II) hexafluorophosphate, *J. Chem. Crystallogr.*, 39(8), 581–584 (2009).

Williams D.J., McKinney B.J., Baker B., Gwaltney K.P., VanDerveer D. The preparation, characterization and x-ray structural analysis of tetrakis[1,3-dimethyl-2(3H)-imidazoleselone]cadmium(II) hexafluorophosphate, *J. Chem. Crystallogr.*, 37(10), 691–694 (2007).

Zhang H., Zelmon D.E., Price G.E., Teo B.K. Wide spectral range nonlinear optical crystals of one-dimensional coordination solids $[Et_4N][Cd(SCN)_3]$ and $[Et_4N][Cd(SeCN)_3]$ and the general design criteria for $[R_4N][Cd(XCN)_3]$ (where R = alkyl and X = S, Se, Te) as NLO crystals, *Inorg. Chem.*, 39(9), 1868–1873 (2000).

# 6 Systems Based on CdTe

## 6.1 Cd–H–Na–C–N–O–Te

[Na(C$_{10}$H$_{20}$O$_5$)$_4$][Cd$_4$Te$_{12}$]·8DMF or C$_{64}$H$_{136}$NaN$_8$O$_{28}$Te$_{12}$Cd$_4$. This multinary compound is formed in this system (Schreiner et al. 1993). It crystallizes in the monoclinic structure with the lattice parameters $a = 1622.9 \pm 0.6$, $b = 2038.2 \pm 0.5$, $c = 1739.8 \pm 0.2$ pm, and $\beta = 92.26° \pm 0.02°$ at 223 K and a calculated density of 2.04 g·cm$^{-3}$. This compound was prepared by the reaction of Na$_2$Te$_3$ with Cd(OAc)$_2$·2H$_2$O in a DMF solution in the presence of 15-crown-5 (1,4,7,10,13-pentaoxacyclopentadecane, C$_{10}$H$_{20}$O$_5$), forming black crystal needles.

## 6.2 Cd–H–C–Si–N–Te

[Cd{SiTe(SiMe$_3$)$_3$}$_2$](2,2′-bipy) or C$_{28}$H$_{62}$Si$_8$N$_2$Te$_2$Cd. This multinary compound is formed in this system. It melts at 241°C–243°C (Bonasia and Arnold 1992). To prepare it, a solution of 2,2′-bipy (0.30 mM) in 10 mL of toluene was added to a solution of [Cd{SiTe(SiMe$_3$)$_3$}$_2$]$_2$ (0.35 mM) in the same solvent (25 mL). After being stirred for several minutes, the clear red solution became orange yellow and a fine yellow powder precipitated. The toluene was removed under reduced pressure and the residue was extracted with CH$_2$Cl$_2$ (40 mL). The dark-orange-red solution was filtered, concentrated to 15 mL, and cooled to −40°C for 12 h. Thin yellow needles of the title compound were collected by filtration.

## 6.3 Cd–H–C–Si–P–Te

[Cd{SiTe(SiMe$_3$)$_3$}$_2$](C$_6$H$_{16}$P$_2$) or C$_{24}$H$_{70}$Si$_8$P$_2$Te$_2$Cd. This multinary compound is formed in this system. It crystallizes in the orthorhombic structure with the lattice parameters $a = 1314.98 \pm 0.21$, $b = 2555.25 \pm 0.34$, and $c = 2853.49 \pm 0.36$ pm at 156 K and a calculated density of 1.40 g·cm$^{-3}$ (Bonasia and Arnold 1992). To obtain it, via syringe 1,2-bis(dimethylphosphino)ethane (0.47 mM) was added to a solution of [Cd{SiTe(SiMe$_3$)$_3$}$_2$]$_2$ (0.468 mM) in 50 mL of hexane. After being stirred for several minutes, the yellow solution became cloudy and pale-yellow needles precipitated. The hexane was removed under reduced pressure and the residue was extracted with CH$_2$Cl$_2$ (40 mL). The filtrate was concentrated to 10 mL and cooled to −40°C for 24 h to afford the title compound as large clear yellow crystals. [Cd{SiTe(SiMe$_3$)$_3$}$_2$] (C$_6$H$_{16}$P$_2$) loses 1,2-bis(dimethylphosphino)ethane by 200°C and slowly decomposes from 280°C to 300°C.

## 6.4 Cd–H–C–Si–Te

[Cd{SiTe(SiMe$_3$)$_3$}$_2$]$_2$ or C$_{18}$H$_{54}$Si$_8$Te$_2$Cd. This multinary compound is formed in this system. It melts at 210°C–220°C (Bonasia and Arnold 1992). To obtain it, a solution of Cd[N(SiMe₃)₂]₂ (1.41 mM) in 25 mL of hexane was combined with HTeSi(SiMe₃)₃ (2.82 mM) dissolved in 25 mL of hexane, resulting in the immediate formation of a yellow solution. This mixture was stirred for 30 min and the solvent was removed under reduced pressure. The dry yellow solid was extracted with hexamethyldisiloxane (40 mL). Then the solution was filtered, concentrated to 15 mL, and recrystallized from hexane (15 mL) at −40°C to give a yellow powder of the title compound.

## 6.5 Cd–H–C–Sn–N–Mn–Te

**[Mn(C₂H₈N₂)₃]CdSnTe₄** or **C₆H₂₄SnN₆MnTe₄Cd**. This multinary compound is formed in this system (Chen et al. 2000). It crystallizes in the triclinic structure with the lattice parameters $a = 913.4 \pm 0.2$, $b = 1008.5 \pm 0.3$, and $c = 1269.1 \pm 0.3$ pm and $\alpha = 73.52° \pm 0.02°$, $\beta = 86.05° \pm 0.02°$, and $\gamma = 76.43° \pm 0.02°$, a calculated density of 2.977 g·cm⁻³, and an energy gap of 1.75 eV. Its single crystals were obtained from a solvothermal reaction containing 0.25 mM of SnTe, 0.75 mM of Te, 0.25 mM of MnCl₂, and 0.25 mM of CdCl₂. The reagents were mixed under an inert atmosphere in a glove box. The mixture was then transferred to a Pyrex tube, and approximately 0.4 mL of ethylenediamine (C₂H₈N₂) was added to the sample. After the liquid was condensed by liquid N₂, the tube was sealed under vacuum and heated at 180°C for 7 days. After being cooled to room temperature, the mixture was washed with 30% and 95% EtOH followed by drying with anhydrous diethyl ether. Dark-red prism-like crystals of [Mn(C₂H₈N₂)₃]CdSnTe₄ were isolated from the final product.

## 6.6 Cd–H–C–N–P–Te

Some compounds are formed in this system.

**[Cd{Buᵗ₂P(Te)NPrⁱ}₂]** or **C₂₂H₅₀N₂P₂Te₂Cd**. This compound sublimes at 115° at 2·10⁻⁴ Pa and melts at 188°C (Bochmann et al. 1995). To obtain it, a yellow solution of Buᵗ₂P(Te)NHPrᵢ (4.64 mM) in light petroleum (20 mL) [Cd{N(SiMe₃)₂}₂] (2.31 mM) was added via a syringe at 0°C, and the reaction mixture was stirred vigorously. A yellow precipitate formed instantaneously. The reaction mixture was warmed to room temperature and stirred for 2 h. The precipitate was filtered off and recrystallized from toluene to give yellow crystals.

**[Cd{Buᵗ₂P(Te)N(C₆H₁₁)}₂]** or **C₂₈H₅₈N₂P₂Te₂Cd**. This compound sublimes at 180° at·10⁻³ Pa and melts at 213°C–216°C (Bochmann et al. 1995). It was prepared following the procedure for [Cd{Buᵗ₂P(Te)NPrᵢ}₂] from Buᵗ₂P(Te)NH(C₆H₁₁) (1.21 mM) and [Cd{N(SiMe₃)₂}₂] (0.58 mM) as yellow crystals.

## 6.7 Cd–H–C–N–O–Re–Te

Some compounds are formed in this system.

**Cd₂[Re₄Te₄(CN)₁₂]·6H₂O** or **C₁₂H₁₂N₁₂O₆Re₄Te₄Cd₂**. This compound crystallizes in the orthorhombic structure with the lattice parameters $a = 1068.4 \pm 0.1$,

$b = 1552.8 \pm 0.2$, and $c = 1911.8 \pm 0.3$ pm and a calculated density of 3.980 g·cm⁻³ (Mironov et al. 2001b). Its single crystals were prepared by slow diffusion in the opposite direction of an aqueous solution of $K_4[Re_4Te_4(CN)_{12}]$ (0.1 mM·L⁻¹) and $CdSO_4$ (0.5 mM·L⁻¹) through the silica gel for 4 weeks. The crystals were separated manually from the silica gel.

**$[Cd(NH_3)_5][Cd(NH_3)_3][Re_4Te_4(CN)_{12}]\cdot 4H_2O$ or $C_{12}H_{32}N_{20}O_4Re_4Te_4Cd_2$.** This compound crystallizes in the triclinic structure with the lattice parameters $a = 1272.6 \pm 0.2$, $b = 1340.8 \pm 0.2$, and $c = 1391.8 \pm 0.2$ pm and $\alpha = 111.31° \pm 0.02°$, $\beta = 116.77° \pm 0.02°$, and $\gamma = 90.17° \pm 0.02°$ pm and a calculated density of 3.433 g·cm⁻³ (Mironov et al. 2001a). To obtain this compound, a solution of $CdSO_4 \cdot 8H_2O$ (20 mg) in concentrated aqueous ammonia (3 mL) was mixed with a solution of $K_4[Re_4Te_4(CN)_{12}]\cdot 5H_2O$ (10 mg) in $H_2O$ (2 mL). At ambient temperature, the resulting solution was transferred to a crystallizing dish, which was then covered by a watch glass. After 2 days, black crystals of $Cd(NH_3)_5][Cd(NH_3)_3][Re_4Te_4(CN)_{12}]\cdot 4H_2O$ had formed, which were collected by filtration and dried on filter paper.

## 6.8 Cd–H–C–N–Te

**$(Et_4N)_4[Cd_4Te_{12}]$ or $C_{32}H_{80}N_4Te_{12}Cd_4$.** This multinary compound is formed in this system. It crystallizes in the orthorhombic structure with the lattice parameters $a = 1548.2 \pm 0.5$, $b = 2256.3 \pm 0.4$, and $c = 1802.7 \pm 0.4$ pm and a calculated density of 2.639 g·cm⁻³ (Kim and Kanatzidis 1994). It was prepared by the reaction of $CdI_2$, $K_2Te_2$ and $Et_4NBr$ in a 2:3:2 molar ratio in DMF. To a 40 mL DMF solution of 1.1 mM $K_2Te_2$ and 0.75 mM $Et_4NBr$, a 35 mL DMF solution of 0.74 mM $CdI_2$ was added dropwise over 20 min. When all the $CdI_2$ solution was added, the reaction solution turned brownish black in color. After removal of the undissolved black precipitate by filtration, the solution was placed into several long test tubes and ether was layered over it. Upon allowing the solution to stand for a week, black polyhedral chunky crystals were obtained.

## 6.9 Cd–H–C–P–Te

Some compounds are formed in this system.

**$Cd(TePh)_2\cdot depe$ or $C_{22}H_{34}P_2Te_2Cd$ [depe = 1,2-bis(diethylphosphino)ethane].** This compound melts at 87.5°C–89°C (Brennan et al. 1990). To obtain it, depe (1.90 mM) was added to $Cd(TePh)_2$ (0.58 mM) suspended in MeCN and pyridine (5 mL). The yellow solution was concentrated to ca. 3 mL, saturated with toluene/heptane (5:1 volume ratio), and cooled (–20°C) to give yellow diamonds. All reactions were performed in an inert atmosphere.

**$[(Ph_4P)_2Cd(Te_4)_2]$ or $C_{48}H_{40}P_2Te_8Cd$.** This compound crystallizes in the tetragonal structure with the lattice parameters $a = 2144.2 \pm 0.3$ and $c = 1125.9 \pm 0.2$ and a calculated density of 2.325 g·cm⁻³ (Bollinger et al. 1995). To obtain it, a flask was charged with $Li_2Te$ (3.00 mM), Te (6 mM), cadmium xanthogenate (1.00 mM), $Ph_4PBr$ (2 mM), DMF (30 mL), and $Et_3P$ (3.4 mM). The mixture was stirred at 90°C for 1 h and filtered. The solution was layered with $Et_2O$ (30 mL) and stored overnight at room temperature to produce crystals of the title compound.

**$[Cd_8Te(TePh)_{14}(PEt_3)_3]$ or $C_{102}H_{115}P_3Te_{15}Cd_8$.** This compound crystallizes in the orthorhombic structure with the lattice parameters $a = 1469.1 \pm 0.3$, $b = 2826.9 \pm 0.6$, and $c = 3501.8 \pm 0.7$ pm (Behrens and Fenske 1997). To obtain it, $CdCl_2$ (0.60 mM) was suspended in THF (25 mL). The addition to this suspension of $PEt_3$ (1.2 mM) yields a colorless solution. With the next addition of $PhTeSiMe_3$ (2.4 mM), an orange-brown solution was formed. The solution was layered with $n$-heptane and kept at $-30°C$. Pale-yellow crystals of $[Cd_8Te(TePh)_{14}(PEt_3)_3]$ were formed. The synthesis was performed under a $N_2$ atmosphere and absolute THF and $n$-heptane were used.

## 6.10   Cd–Cu–Ag–In–Te

**$CdTe–CuInTe_2–AgInTe_2$.** The phase diagram is not constructed. A wide range of solid solutions for both the chalcopyrite ($\alpha$-phase) and the sphalerite ($\beta$-phase) exists in this system (Guerrero et al. 1988). Over the whole composition range, the $\beta$-phase is bounded on the higher temperature by a two-phase field. Polycrystalline samples were prepared by the melt and anneal technique. This system was investigated by DTA, XRD, and metallography, and the alloys were annealed at 600°C at least 20–30 days.

## 6.11   Cd–Cu–Ag–In–Mn–Te

DTA and XRD for the **$(AgCd_2In)_x(CuIn)_{2y}Mn_{4z}Te_4$** alloys show that a wide range of solid solutions occurs for both the zincblende and chalcopyrite structures and that, for both structures, an ordered form, attributed to the Mn ordering on the cation sublattice, exists at approximately below 400°C for $z > 0.05$ $(x + y + z = 1)$ (Quintero et al. 1988). The results show that various sections cannot be treated as quasibinary, although for temperatures less than 600°C, this condition is satisfied to a good approximation.

The alloys were produced by the usual melt and anneal technique. The ingots were annealed at 600°C for 20–30 days (Quintero et al. 1988).

## 6.12   Cd–Cu–Ga–Mn–Te

**$2CdTe–CuGaTe_2–2MnTe$.** Three vertical sections of this quasiternary system were constructed (Goudreault et al. 1993). Using the obtained results, the isothermal sections at 25°C, 650°C, 800°C, 900°C, 950°, and 1000°C were estimated (Figure 6.1a,b,c,d,e). The ordering transition $\beta–\beta'$ involving the Mn atoms was observed at about 200°C. At higher temperatures, the diagram in the vicinity of $CuGaTe_2$ is complicated by the occurrence of a two-phase field $(\beta + \eta)$, where both phases lie outside the triangular section. At higher temperatures, three-phase fields occur, which probably involve a phase composition close to $MnGa_2Te_4$.

The temperature versus composition diagram of the alloy system $Cd_{2x}(CuGa)_yMn_{2z}Te_2$ $(x + y + z = 1)$ was investigated in the $0 < z < 0.8$ range by DTA and XRD (Goudreault et al. 1993).

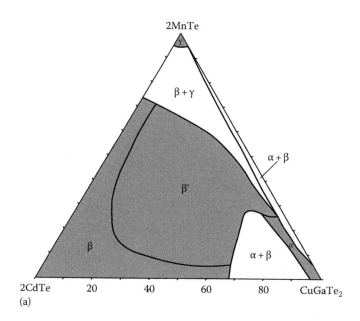

(a)

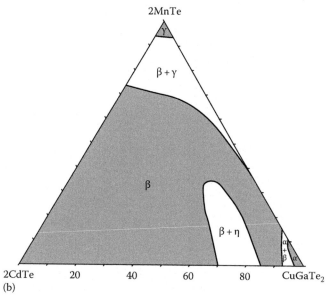

(b)

**FIGURE 6.1** Isothermal sections of the 2CdTe–CuGaTe$_2$–2MnTe system at (a) 25°C and (b) 650°C. *(Continued)*

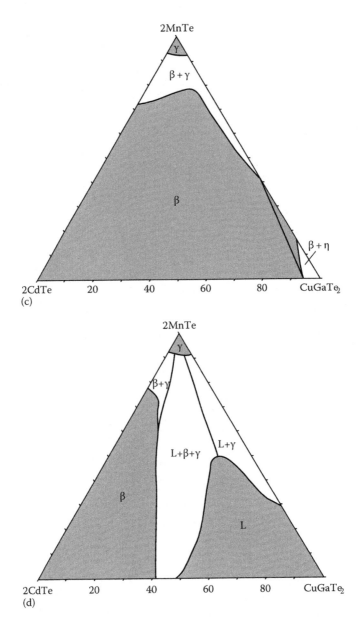

**FIGURE 6.1 (Continued)**  Isothermal sections of the 2CdTe–CuGaTe$_2$–2MnTe system at (c) 800°C, and (d) 900°C.                    (*Continued*)

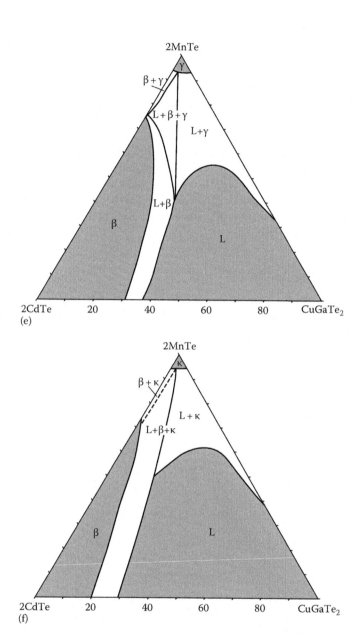

**FIGURE 6.1 (*Continued*)** Isothermal sections of the 2CdTe–CuGaTe₂–2MnTe system at (e) 950°C and (f) 1000°C. (From Goudreault, R. et al., *J. Solid State Chem.*, 107(1), 264, 1993.)

## 6.13 Cd–Cu–In–Mn–Te

**2CdTe–CuInTe₂–2MnTe**. The isothermal section of this system at 600°C is given in Figure 6.2 (Quintero et al. 1986). A wide range of solid solutions occurs for both the chalcopyrite (*ch*) and sphalerite (*s*) structures (Quintero et al. 1990). For both structures, an ordered form, attributed to the Mn ions ordering on the cation sublattice, occurs below approximately 550°C for $z < 0.5$ (Quintero et al. 1990). Variation of lattice parameter $a$ versus $z$ of the zincblende structure based on $AgInCd_2Te_4$ for the section at $x = 3y$ could be expressed by the following equation $a = 0.6368 - 0.00374z$ (nm) (Quintero et al. 1987).

Polycrystalline samples of $Cd_{2x}(CuIn)_y Mn_{2z}Te_2$ $(x + y + z = 1)$ alloys were prepared by a melt and anneal technique. It was found that with the addition of Mn to the zincblende and chalcopyrite structures, a partial cubic structure was obtained, plus a two-phase field at higher $z$ values. It was determined that the variation of $a$ is practically linear with the composition. The ingots were annealed at 600°C for 20–30 days (Quintero et al. 1986, 1990).

## 6.14 Cd–Ag–Hg–In–Te

**CdTe–AgInTe₂–HgTe**. The solid solution region in this quasiternary system is shown in Figure 6.3 (Rodot 1961). This system was investigated by DTA, metallography, and XRD.

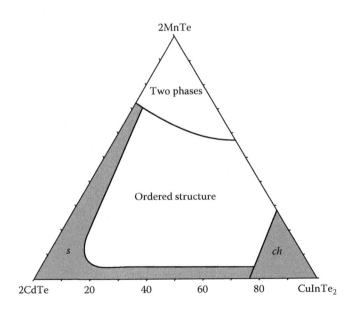

**FIGURE 6.2** The isothermal section of the 2CdTe–CuInTe₂–2MnTe system at 600°C. (From Quintero, M. et al., *J. Solid State Chem.*, 63(1), 110, 1986.)

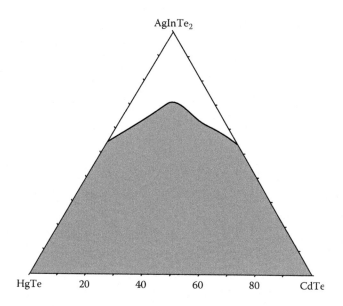

**FIGURE 6.3** Solid solution region in the CdTe–AgInTe₂–HgTe quasiternary system. (From Rodot, H., Solutions solides entre composés semiconducteurs binaires et ternaires, in: *Proceedings of the International Conference on Semiconductor Physics*, Prague, Czech Republic, 1961, pp. 1010–1014.)

## 6.15  Cd–Ag–In–Mn–Te

**2CdTe–AgInTe₂–2MnTe**. The isothermal section of this system showing the phase boundaries at 600°C is given in Figure 6.4 (Quintero and Woolley 1985). A wide range of solid solutions occurs for both the sphalerite (*s*) and chalcopyrite (*ch*) structures, and for both structures an ordered form, attributed to the Mn ordering on the cation sublattice, exists (Quintero and Woolley 1985, Quintero et al. 1989).

In general, none of the investigated sections of this system are quasibinary in character and the 2CdTe–AgInTe₂–2MnTe diagram cannot be quasiternary (Quintero et al. 1989). For temperatures below 600°C, these quasibinary and quasiternary conditions appear to be satisfied to a good approximation.

Polycrystalline samples of $Cd_{2x}(AgIn)_yMn_{2z}Te_2$ ($x+y+z=1$) alloys were prepared by a melt and anneal technique. It was determined that the variation of *a* is practically linear with composition. The ingots were annealed at 600°C for 20–30 days (Quintero and Woolley 1985, Quintero et al. 1989).

## 6.16  Cd–Ag–As–I–Te

**CdTe–AgI–As₂Te₃**. The glass-forming region in this system is presented in Figure 6.5 (Kassem et al. 2012). This domain is located around the binary AgI–As₂Te₃ line, in the As₂Te₃ side where glassy compositions are observed from pure As₂Te₃ until 65 mol.% AgI. Whatever the binary composition in the AgI–As₂Te₃ system, the solubility of CdTe remains below 16 mol.%.

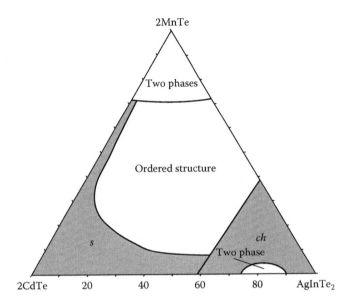

**FIGURE 6.4** The isothermal section of the 2CdTe–AgInTe₂–2MnTe system at 600°C. (From Quintero, M. and Woolley, J.C., *Phys. Status Solidi (a)*, 92(2), 449, 1985.)

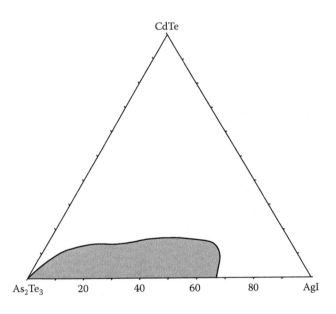

**FIGURE 6.5** Glass-forming region in the CdTe–AgI–As₂Te₃ system. (From Kassem, M. et al., *Mater. Res. Bull.*, 47(2), 198, 2012.)

The glass transition temperature $(T_g)$ changes within the interval from 109°C to 119°C (Kassem et al. 2012). The CdTe increment leads to a gradual decrease of glass transition temperature and to a decrease of glass-forming ability. The density almost linearly changes with the increae of CdTe and AgI contents.

## REFERENCES

Behrens S., Fenske D. Cadmium nanoclusters with phenylselenolato-and phenyltelluro-lato ligands: Synthesis and structural characterization of [Cd$_{17}$Se$_4$(SePh)$_{24}$(PPh$_3$)$_4$] [Cd$_8$Se(SePh)$_{12}$Cl$_4$], [Cd(DMF)$_6$] [Cd$_8$Se(SePh)$_{12}$Cl$_4$], [Cd$_8$Se(SePh)$_{14}$(PPh$_3$)$_2$], [Cd$_8$Se (SePh)$_{14}$(DMF)$_3$] and [Cd$_8$Te(TePh)$_{14}$(PEt$_3$)$_3$], *Ber. Bunsenges. Phys. Chem.*, 101(11), 1588–1592 (1997).

Bochmann M., Bwembya G.C., Hursthouse M.B., Coles S.J. Synthesis of phosphinochalcoge-noic amidato complexes of zinc and cadmium. The crystal and molecular structure of [Zn{Bu$^t_2$P(Se)NPr$^i$}$_2$], *J. Chem. Soc., Dalton Trans.*, (17), 2813–2817 (1995).

Bollinger J.C., Roof L.C., Smith D.M., McConnachie J.M., Ibers J.A. Synthesis, X-ray crystal structures, and NMR spectroscopy of [PPh$_4$]$_2$[M(Te$_4$)$_2$], M = Hg, Cd, Zn, *Inorg. Chem.*, 34(6), 1430–1434 (1995).

Bonasia P.J., Arnold J. Zinc, cadmium, and mercury tellurolates: Hydrocarbon solubility and low coordination numbers enforced by sterically encumbered silyltellurolate ligands, *Inorg. Chem.*, 31(12), 2508–2514 (1992).

Brennan J.G., Siegrist T., Carroll P.J., Stuczynski S.M., Reynders P., Brus L.E., Steigerwald M.L. Bulk and nanostructure group II–VI compounds from molecular organometallic precursors, *Chem. Mater.*, 2(4), 403–409 (1990).

Chen Z., Wang R.-J., Li J. [Mn(en)$_3$CdSnTe$_4$ and [Mn(en)$_3$]Ag$_6$Sn$_2$Te$_8$: New intermetallic tellurides synthesized in superheated organic medium, *Chem. Mater.*, 12(3), 762–766 (2000).

Goudreault R., Woolley J.C., Quintero M., Tovar R. Phase diagram, lattice parameter, and optical energy gap values for the Cd$_{2x}$(CuGa)$_y$Mn$_{2z}$Te$_2$ alloys, *J. Solid State Chem.*, 107(1), 264–272 (1993).

Guerrero E., Quintero M., Wooley J.C. Phase relations in (CuIn)$_x$(AgIn)$_y$Cd$_{2z}$Te$_2$ alloys, *J. Cryst. Growth*, 92(1–2), 150–154 (1988).

Kassem M., Le Coq D., Boidin R., Bychkov E. New chalcogenide glasses in the CdTe–AgI–As$_2$Te$_3$ system, *Mater. Res. Bull.*, 47(2), 193–198 (2012).

Kim K.-W., Kanatzidis M.G. Synthesis, structure, and properties of the polychalcogenides [M$_4$Te$_{12}$]$^{4-}$ (M = Cd, Hg), *Inorg. Chim. Acta*, 224(1–2), 163–169 (1994).

Mironov Y.V., Oeckler O., Simon A., Fedorov V.E. New types of complexes based on Re$_4$ chal-cocyanide clusters—Syntheses and crystal structures of [Ni(NH$_3$)$_5$]$_2$[Re$_4$Te$_4$(CN)$_{12}$]·3.4H$_2$O and [Cd(NH$_3$)$_5$][Cd(NH$_3$)$_3$][Re$_4$Te$_4$(CN)$_{12}$]·4H$_2$O, *Eur. J. Inorg. Chem.*, (11), 2751–2753 (2001a).

Mironov Y.V., Virovets A.V., Sheldrick W.S., Fedorov V.E. Novel inorganic polymeric com-pounds based on the Re$_4$ chalcocyanide cluster complexes: Synthesis and crystal struc-tures of Mn$_2$[Re$_4$Se$_4$(CN)$_{12}$]·6H$_2$O, Cd$_2$[Re$_4$Te$_4$(CN)$_{12}$]·6H$_2$O, Cu$_2$[Re$_4$Te$_4$(CN)$_{12}$]·4H$_2$O and K$_4$Re$_4$Se$_4$(CN)$_{12}$·6H$_2$O, *Polyhedron*, 20(9–10), 969–974 (2001b).

Quintero M., Dierker L., Woolley J.C. Crystallography and optical energy gap values for Cd$_{2x}$(CuIn)$_y$Mn$_{2z}$Te$_2$ alloys, *J. Solid State Chem.*, 63(1), 110–117 (1986).

Quintero M., Grima P., Guerrero G., Tovar R., Woolley J.C. Phase relation and the effects of ordering in (AgCd$_2$In)$_p$(CuIn)$_{2y}$Mn$_{4z}$Te$_4$ ($p + y + z = 1$) alloys, *J. Cryst. Growth*, 89(2–3), 301–307 (1988).

Quintero M., Guerrero E., Grima P., Woolley J.C. Phase relations and the effects of ordering in Cd$_{2x}$(AgIn)$_y$Mn$_{2z}$Te$_2$ ($x + y + z = 1$) alloys, *J. Electrochem. Soc.*, 136(4), 1220–1223 (1989).

Quintero M., Guerrero E., Tovar R., Pérez G.S., Woolley J.C. Phase diagram of $Cd_{2x}(CuIn)_yMn_{2z}Te_2$ $(x+y+z=1)$ alloys, *J. Solid State Chem.*, 87(2), 456–462 (1990).

Quintero M., Sagredo V., Tovar R., Grima P., Pérez G.S. Crystallographic and magnetic properties of the system $(AgInCd_2)_x(CuIn)_{2y}Mn_{4z}Te_4$ $(x+y+z=1)$ with $x=3y$, *Solid State Commun.*, 64(4), 407–410 (1987).

Quintero M., Woolley J.C. Crystallography and optical energy gap values for $Cd_{2x}(AgIn)_y$ $Mn_{2z}Te_2$ alloys, *Phys. Status Solidi (a)*, 92(2), 449–456 (1985).

Rodot H. Solutions solides entre composés semiconducteurs binaires et ternaires. In: *Proceedings of the International Conference on Semiconductor Physics*, Prague, Czech Republic, pp. 1010–1014 (1961).

Schreiner B., Dehnicke K., Fenske D. Synthese und Kristallstruktur von [Na(15-Krone-5)]$_4$[Cd$_4$Te$_{12}$]·8 DMF, *Z. anorg. und allg. Chem.*, 619(6), 1127–1131 (1993).

# 7 Systems Based on HgS

## 7.1 Hg–H–Na–O–S

**HgS–Na$_2$S–H$_2$O**. The phase diagram is not constructed. **Na$_2$Hg$_3$S$_4$·(H$_2$O)$_2$** multinary compound is formed in this system, which crystallizes in the orthorhombic structure with the lattice parameters $a = 658.23 \pm 0.06$, $b = 1092.2 \pm 0.1$, and $c = 1437.9 \pm 0.2$ pm and a calculated density of 5.22 g·cm$^{-3}$ (Banda et al. 1991). To obtain it, Na$_2$S·9H$_2$O (3.0 mM) was partially dehydrated in a stream of N$_2$ at 80°C, and black HgS (1.0 mM) was added. Deoxygenated and dried DMF (5 mL) was introduced, the vessel was sealed under N$_2$, and the mixture was subjected to low-power ultrasonication at room temperature. After 24 h, the solid was yellow and the solution pale green: after 40 h, the solid was white, and after 52 h, the volume of the white solid had grown to about 4 cm$^3$. When the mixture was heated briefly to 110°C, the pale-yellow mother liquor changed to green and returned to a pale yellow upon cooling, but there was no evidence that the polycrystalline solid was soluble or would crystallize. Upon standing at room temperature for several months, large buff-colored prismatic crystals grew in the solution. The mother liquor was removed, and the crystals dried in a stream of N$_2$. The crystals were hygroscopic and on brief exposure to the atmosphere appeared to undergo a morphological change without losing their single crystallinity.

## 7.2 Hg–H–K–C–N–O–S

**[K(C$_{16}$H$_{24}$O$_6$)]$_2$[Hg(SCN)$_4$]** or **C$_{36}$H$_{48}$K$_2$N$_4$O$_{12}$S$_4$Hg**. This multinary compound is formed in this system. It crystallizes in the monoclinic structure with the lattice parameters $a = 1737.35 \pm 0.02$, $b = 1377.16 \pm 0.02$, and $c = 1984.12 \pm 0.03$ pm and $\beta = 100.6370° \pm 0.0006$ and a calculated density of 1.617 g·cm$^{-3}$ (Pickardt and Dechert 1999). The title compound was obtained by the reaction of K$_2$[Hg(SCN)$_4$] with benzo-18-crown-6 (C$_{16}$H$_{24}$O$_6$) in EtOH.

## 7.3 Hg–H–Cu–Au–C–N–Br–S

**[Cu$_2$Au(Bu$^n_2$NCS$_2$)$_6$][HgBr$_3$]$_2$** or **C$_{54}$H$_{108}$Cu$_2$AuN$_6$Br$_6$S$_{12}$Hg$_2$**. This multinary compound is formed in this system. It crystallizes in the triclinic structure with the lattice parameters $a = 1261.4 \pm 0.4$, $b = 1369.1 \pm 0.4$, and $c = 1250.2 \pm 0.4$ pm and $\alpha = 96.37° \pm 0.04°$, $\beta = 91.47° \pm 0.04$, and $\gamma = 85.43° \pm 0.04$ and a calculated density of 1.887 g·cm$^{-3}$ (Gal et al. 1976). It was prepared by the following procedure. To a solution of 1 mol of [Au(Bu$^n_2$NCS$_2$)]$_2$ in chloroform, a solution of 1 mol of Bu$^n_4$(NCS$_2$)$_2$ and a solution of 4 mol of [CuAu(Bu$^n_2$NCS$_2$)$_2$][HgBr$_2$] in the same solvent were added in succession. After the addition of an equal volume of petroleum ether (40°C–60°C) to the dark-brown solution, in a few days, dark-green crystals were obtained.

## 7.4   Hg–H–Cu–C–N–O–Fe–S

$[Fe(C_5H_4HgSCN)_2Cu(SCN)_2]\cdot(C_6H_6N_2O)_2$ or $C_{26}H_{20}CuN_8O_2FeS_4Hg_2$. This multinary compound is formed in this system, which melts at 190°C with decomposition (Singh and Singh 1986). It was prepared by the next method. A suspension or solution of $[Fe(C_5H_4HgSCN)_2Cu(SCN)_2]$ was prepared in acetone/EtOH mixture, mixed with an ethanol solution of nicotinamide (1:2 molar ratio), and stirred for 24 h. The precipitate was formed, filtered, washed with EtOH, and dried in vacuum. Green crystals of the title compound were recrystallized from acetone/EtOH mixture.

## 7.5   Hg–H–Cu–C–N–S

$[Hg(SCN)_4][Cu(C_4H_8N_2)_2]$ or $C_8H_{16}CuN_8S_4Hg$. This multinary compound is formed in this system. It crystallizes in the monoclinic structure with the lattice parameters $a = 751$, $b = 1803 \pm 2$, and $c = 1420 \pm 1$ pm and $\beta = 96.32° \pm 0.01°$ and an experimental density of $2.12 \pm 0.07$ g·cm$^{-3}$ (Scouloudi 1953). Its deep-purple crystals were crystallized from acetone and H$_2$O in well-developed monoclinic prismatic plates.

## 7.6   Hg–H–Cu–C–N–Br–S

Some compounds are formed in this system.

$[Cu(S_2CHBu_2)_2][HgBr_3]$ or $C_{18}H_{36}CuN_2Br_3S_4Hg$. This compound could be obtained if to a solution of 1 mol of Hg(S$_2$CHBu$_2$)$_2$ in chloroform a solution of 2 mol of Cu(S$_2$CHBu$_2$)$_2$ in chloroform was added (Cras et al. 1973). After the addition of an equal volume of petroleum ether, crystalline products were formed. They were filtered off and washed with petroleum ether.

Once the composition of the compound is known, a simpler route to it suggested itself. A mixture of 1 mol of Br$_2$ in chloroform and 1 mol of HgBr$_2$ dissolved in EtOH was added by stirring to a solution of 2 mol of Cu(S$_2$CHBu$_2$)$_2$ in chloroform. Upon the addition of petroleum ether and cooling, precipitates of $[Cu(S_2CHBu_2)_2][HgBr_3]$ were obtained (Cras et al. 1973).

$[Cu_3(S_2CHBu_2)_6][HgBr_3]_2$ or $C_{54}H_{108}Cu_3N_6Br_6S_{12}Hg_2$. To obtain this compound, to a suspension of 1 mol of 3,5-bis($N,N$-dibutyliminium)-1,2,4-trithiolane tetrabromo-di-µ-bromodicuprate(II) in chloroform a saturated solution of 2 mol of Hg(S$_2$CHBu$_2$)$_2$ in the same solvent was added (Cras et al. 1973). The former compound dissolved slowly, and crystalline product could be obtained when petroleum ether was added to the solution and cooled to 0°C. The compound was dissolved in chloroform and recrystallized after the addition of an equal amount of petroleum ether.

Once the composition of the compound is known, a simpler route to it suggested itself. A mixture of 1 mol of Br$_2$ in chloroform and 1 mol of HgBr$_2$ dissolved in EtOH was added by stirring to a solution of 3 mol of Cu(S$_2$CHBu$_2$)$_2$ in chloroform. Upon the addition of petroleum ether and cooling, precipitates of $[Cu_3(S_2CHBu_2)_6]$ $[HgBr_3]_2$ were obtained (Cras et al. 1973).

## 7.7   Hg–H–Cu–C–N–Fe–S

Some compounds are formed in this system.

**[Fe(C$_5$H$_4$HgSCN)$_2$Cu(SCN)$_2$] or C$_{14}$H$_8$CuN$_4$FeS$_4$Hg$_2$.** This compound melts at 185°C with decomposition (Singh and Singh 1986). To obtain it, 1,1′-bis(thiocyanatomercurio)ferrocene was dissolved in a small quantity of DMSO and diluted to 200 mL by acetone. Cu(SCN)$_2$ was similarly dissolved in either acetone or DMSO/acetone mixture separately. The obtained solutions were mixed in a 1:1 molar ratio and stirred for 24 h. When the quantity of DMSO became more, the precipitate was obtained by the addition of EtOH. This precipitate was filtered, washed with EtOH, and dried in vacuum. Light-green crystals of the title compound were recrystallized from acetone/EtOH mixture.

**[Fe(C$_5$H$_4$HgSCN)$_2$Cu(SCN)$_2$]·(C$_{10}$H$_8$N$_2$) or C$_{24}$H$_{16}$CuN$_6$FeS$_4$Hg$_2$.** This compound melts at 175°C with decomposition (Singh and Singh 1986). It was prepared by the next method. A suspension or solution of [Fe(C$_5$H$_4$HgSCN)$_2$Cu(SCN)$_2$] was prepared in acetone/EtOH mixture, mixed with an ethanol solution of 2,2′-bipyridyl (1:1 molar ratio), and stirred for 24 h. The precipitate was formed, filtered, washed with EtOH, and dried in vacuum. Mustard crystals of the title compound were recrystallized from acetone/EtOH mixture.

## 7.8   Hg–H–Mg–C–N–O–S

**MgHg(SCN)$_4$·2H$_2$O.** This multinary compound is formed in this system (Brodersen von and Humnel 1982). It crystallizes in the monoclinic structure with the lattice parameters $a = 1335.1 \pm 0.6$, $b = 531.6 \pm 0.5$, and $c = 1867.0 \pm 1.4$ and $\beta = 92.3° \pm 0.1°$ and the experimental and calculated densities 2.37 and 2.49 g·cm$^{-3}$, respectively. It was obtained by crystallization of an aqueous solution containing Mg(SCN)$_2$·4H$_2$O and Hg(SCN)$_2$. The resulting product can be recrystallized from EtOH.

## 7.9   Hg–H–B–C–N–O–F–S

**[Hg(C$_4$H$_7$NOS)$_2$(BF$_4$)$_2$] or C$_8$H$_{14}$B$_2$N$_2$O$_2$F$_8$S$_2$Hg.** This multinary compound is formed in this system. To prepare it, HgO was dissolved in a cold solution of HBF$_4$ and thiomorpholine-3-one. The obtained solid product was washed with ethyl ether (Colombini and Preti 1975).

## 7.10   Hg–H–B–C–N–O–Cl–S

**[Hg(C$_{12}$H$_{16}$BN$_6$S$_3$)Cl]·0.25Et$_2$O or C$_{13}$H$_{18.5}$BN$_6$O$_{0.25}$ClS$_3$Cd.** This multinary compound is formed in this system. To prepare it, HgCl$_2$ (2 mM) and [Na(C$_{12}$H$_{16}$BN$_6$S$_3$)] [C$_{12}$H$_{16}$BN$_6$S$_3$ = hydrotris(methimazolyl)borate] (2 mM) were placed in a flask, and acetone (25 mL) was added (Cassidy et al. 2002). The mixture was stirred at room temperature for 20 h. The precipitate that formed (NaCl) was removed by filtration, and the solvent was removed in vacuo to yield a crude product, which was recrystallized by slow diffusion of *n*-hexane into a saturated CH$_2$Cl$_2$ solution.

## 7.11   Hg–H–B–C–N–S

Some compounds are formed in this system.

**[Hg(C$_8$H$_{12}$BN$_4$S$_2$)$_2$] or C$_{16}$H$_{24}$B$_2$N$_8$S$_4$Hg**. This compound melts at 150°C (Alvarez et al. 2003). To prepare it, in the absence of light, a stirred solution of HgI$_2$ (0.381 mM) in MeOH (10 mL) was treated with a solution of Na[Bm$^{Me}$], [Bm$^{Me}$]—bis(2-mer-capto-1-methylimidazolyl)borate anion, C$_8$H$_{12}$BN$_4$S$_2$, (0.763 mM) in the same solvent (15 mL), resulting in the formation of a white precipitate. The suspension was stirred for 10 min, and the product was isolated by filtration and dried in vacuo for 4 h.

**[Hg(C$_{14}$H$_{24}$BN$_4$S$_2$)$_2$] or C$_{28}$H$_{48}$B$_2$N$_8$S$_4$Hg**. This compound melts at 200°C with decomposition and crystallizes in the triclinic structure with the lattice parameters $a = 1000.07 \pm 0.10$, $b = 1065.38 \pm 0.11$, and $c = 1895.10 \pm 0.18$ pm and $\alpha = 81.699° \pm 0.002°$, $\beta = 76.592° \pm 0.002°$, and $\gamma = 71.855° \pm 0.002°$ at $243 \pm 2$ K and a calculated density of 1.512 g·cm$^{-3}$ (Alvarez et al. 2003). To prepare it, in the absence of light, a stirred solution of HgI$_2$ (0.361 mM) in H$_2$O (15 mL) was treated with a solution of Na[Bm$^{tBu}$], [Bm$^{tBu}$]–bis(2-mercapto-1-$t$-butylimidazolyl)borate anion, C$_{14}$H$_{24}$BN$_4$S$_2$, (0.722 mM) in the same solvent (15 mL), resulting in the formation of a white precipitate. The suspension was stirred for 20 min, and the product was isolated by filtration and dried in vacuo for 1.5 h.

**[Hg(C$_{20}$H$_{20}$BN$_4$S$_2$)$_2$] or C$_{40}$H$_{40}$B$_2$N$_8$S$_4$Hg**. This compound melts at 169°C with decomposition (Alvarez et al. 2003). To prepare it, in the absence of light, a stirred solution of Na[Bm$^{Bz}$], [Bm$^{Bz}$]—bis(2-mercapto-1-benzylimidazolyl)borate anion, C$_{20}$H$_{20}$BN$_4$S$_2$, (0.362 mM) in MeOH (8 mL) was treated with a solution of HgI$_2$ (0.181 mM) in the same solvent (5 mL), resulting in the formation of a white precipi-tate. The suspension was stirred for 10 min, and the product was isolated by filtration and dried in vacuo for 2 h.

**[Hg(C$_{20}$H$_{20}$BN$_4$S$_2$)$_2$] or C$_{40}$H$_{40}$B$_2$N$_8$S$_4$Hg**. This compound melts at 132°C with decom-position (Alvarez et al. 2003). To prepare it, a stirred solution of HgCl$_2$ (0.214 mM) in MeOH (10 mL) was treated with a solution of Na[Bm$^{p\text{-}Tol}$], [Bm$^{p\text{-}Tol}$]—bis(2-mer-capto-1-$p$-tolylimidazolyl)borate anion, C$_{20}$H$_{20}$BN$_4$S$_2$, (0.427 mM) in the same solvent (15 mL), resulting in the formation of a white precipitate. The suspension was stirred for 15 min, and the product was isolated by filtration and dried in vacuo for 2 h.

## 7.12   Hg–H–B–C–N–F–S

**[Hg(C$_3$H$_5$NS$_2$)$_2$(BF$_4$)$_2$] or C$_6$H$_{10}$B$_2$N$_2$F$_8$S$_4$Hg**. This multinary compound is formed in this system. It melts at 190°C (Colombini and Preti 1975). To prepare this com-pound, HgO was dissolved in a cold solution of HBF$_4$ and thiazolidine-2-thione. The obtained solid product was washed with ethyl ether.

## 7.13   Hg–H–B–C–N–Br–S

Some compounds are formed in this system.

**[Hg(C$_{12}$H$_{16}$BN$_6$S$_3$)Br] or C$_{12}$H$_{16}$BN$_6$BrS$_3$Hg**. This compound crystallizes in the monoclinic structure with the lattice parameters $a = 1007.99 \pm 0.02$, $b = 1005.28 \pm 0.2$, and $c = 1888.40 \pm 0.04$ pm and $\beta = 101.2261° \pm 0.0003°$ at 123 K (Cassidy et al. 2002). To prepare it, $HgBr_2$ (2 mM) and $[Na(C_{12}H_{16}BN_6S_3)]$ $[C_{12}H_{16}BN_6S_3 = $ hydrotris(methimazolyl)borate] (2 mM) were placed in a flask, and acetone (25 mL) was added. The mixture was stirred at room temperature for 20 h. The precipitate that was formed (NaBr) was removed by filtration, and the solvent was removed in vacuo to yield crude product, which was recrystallized by slow diffusion of $n$-hexane into a saturated $CH_2Cl_2$ solution.

**$[HB(Bu^tC_3H_2N_2S)_2]HgBr \cdot C_6H_6$ or $C_{27}H_{30}BN_6BrS_3Hg$.** This compound melts at 160°C with decomposition and crystallizes in the monoclinic structure with the lattice parameters $a = 1770.24 \pm 0.10$, $b = 1088.38 \pm 0.06$, and $c = 3501.3 \pm 0.2$ pm and $\beta = 96.7900° \pm 0.0010°$ at $243 \pm 2$ K (White et al. 2002). A solution of $Na[HB(Bu^tC_3H_2N_2S)_2]$ (0.400 mM) in MeOH (6 mL) was added to a stirred solution of $HgBr_2$ (0.400 mM) in the same solvent (6 mL), resulting in the immediate formation of an ivory precipitate, which dissolved within 3 min to give a clear pale-yellow solution. After stirring for 45 min, this solution was concentrated under reduced pressure to ca. 4 mL, leading to the formation of an ivory precipitate. After the addition of $H_2O$ (5 mL), the product was isolated by filtration and dried in vacuo for 16 h.

## 7.14   Hg–H–B–C–N–I–S

**$[Hg(C_{12}H_{16}BN_6S_3)I]$ or $C_{12}H_{16}BN_6IS_3Hg$.** This multinary compound is formed in this system. To prepare it, $HgI_2$ (2 mM) and $[Na(C_{12}H_{16}BN_6S_3)]$ $[C_{12}H_{16}BN_6S_3 = $ hydrotris(methimazolyl)borate] (2 mM) were placed in a flask, and acetone (25 mL) was added (Cassidy et al. 2002). The mixture was stirred at room temperature for 20 h. The precipitate that was formed (NaI) was removed by filtration, and the solvent was removed in vacuo to yield crude product, which was recrystallized by slow diffusion of $n$-hexane into a saturated $CH_2Cl_2$ solution.

## 7.15   Hg–H–C–Si–N–S

Some compounds are formed in this system.

**$[Hg\{(3-Bu^tSiMe_2)(C_5H_3NS)\}_2]$ or $C_{22}H_{36}Si_2N_2S_2Hg$.** This compound has been obtained by an electrochemical oxidation of Hg in an acetonitrile solution (50 mL) of 3-(*tert*-butyldimethylsilyl)pyridine-2-thione with a small amount of $(Et_4N)ClO_4$ (Tallon et al. 1995). As electrolysis proceeded, the color of the solution changed. At the end of the process, the solution was filtered to remove any solid impurities, and the filtrate was slowly concentrated at room temperature to give the solid title compound.

**$[Hg\{(6-Bu^tSiMe_2)(C_5H_3NS)\}_2]$ or $C_{22}H_{36}Si_2N_2S_2Hg$.** This compound crystallizes in the monoclinic structure with the lattice parameters $a = 1413.4 \pm 0.3$, $b = 1360.4 \pm 0.3$, and $c = 709.2 \pm 0.1$ and $\beta = 92.77° \pm 0.03°$ and a calculated density of 1.583 g·cm$^{-3}$ (Tallon et al. 1995). It has been obtained by an electrochemical oxidation of Hg in an acetonitrile solution (50 mL) of 6-(*tert*-butyldimethylsilyl)pyridine-2-thione with a small amount of $(Et_4N)ClO_4$. As electrolysis proceeded, the color of the solution changed. At the end of the process, the solution was filtered to remove any solid impurities, and the filtrate was slowly concentrated at room temperature to give the title compound as light-yellow blocks.

**$[Hg\{(3,6-Bu^tSiMe_2)_2(C_5H_2NS)\}_2]$ or $C_{34}H_{64}Si_4N_2S_2Hg$.** This compound has been obtained by an electrochemical oxidation of Hg in an acetonitrile solution (50 mL) of 3,6-bis(*tert*-butyldimethylsilyl)pyridine-2-thione with a small amount of $(Et_4N)ClO_4$ (Tallon et al. 1995). As electrolysis proceeded, the color of the solution changed. At the end of the process, the solution was filtered to remove any solid impurities, and the filtrate was slowly concentrated at room temperature to give the solid title compound.

## 7.16 Hg–H–C–Si–O–S

**$[\{(Bu^tO)_3SiS\}_2Hg]$ or $C_{24}H_{54}Si_2O_6S_2Hg$.** This multinary compound is formed in this system. It melts at 55°C–57°C and crystallizes in the triclinic structure with the lattice parameters $a = 930.8 \pm 0.3$, $b = 944.0 \pm 0.1$, and $c = 1204.8 \pm 0.3$ pm and $\alpha = 100.34° \pm 0.05°$, $\beta = 97.64° \pm 0.03$, and $\gamma = 116.50° \pm 0.03$ and a calculated density of 1.395 g·cm$^{-3}$ (Wojnowski et al. 1985). It was obtained as colorless triclinic plates by the reaction of HgO with Bu$^t$OSiSH in *n*-hexane.

## 7.17 Hg–H–C–Si–S

Some compounds are formed in this system.

**$[Hg(Me_3SiSCH_2)_2]$ or $C_8H_{22}Si_2S_2Hg$.** This compound was prepared using the next procedure (Block et al. 1990). A solution containing $HgCl_2$ (1.0 mM), $Et_3N$ (2.0 mM), and (trimethylsilyl)methanethiol (1.0 mM) in MeOH (15 mL) was stirred at room temperature for 1 h. The solid product, which formed immediately upon mixing of the reagents, was collected by filtration and washed with MeOH and distilled water. After solvent removal in vacuo, a white solid of the title compound was obtained.

**$[MeHg(SC_6H_4-2-SiMe_3)]$ or $C_{10}H_{16}SiSHg$.** To obtain this compound, a solution containing $MeHgNO_3$ (3.0 mM), 2-(trimethylsilyl)benzenethiol (3.1 mM), and $Et_3N$ (3.1 mM) in MeOH (30 mL) was stirred for 1 h at room temperature (Block et al. 1990). The resultant solution was filtered, and the filtrate was kept at −20°C for several days. A white precipitate was collected and washed with distilled $H_2O$. After solvent removal at 13.3 Pa, a white powder of the title compound was obtained.

**$[MeHg(SC_6H_4-2-SiEt_3)]$ or $C_{13}H_{22}SiSHg$.** To obtain this compound, a solution containing $MeHgNO_3$ (0.3 mM), 2-(trimethylsilyl)benzenethiol (0.3 mM), and $Et_3N$ (0.3 mM) in MeOH (5 mL) was stirred for 3 h at room temperature (Block et al. 1990). The resultant solution was concentrated to 3 mL at −20°C for 3 days. Colorless block crystals were collected by filtration and washed with distilled $H_2O$. Solvent removal at 13.3 Pa yielded the colorless crystals of the title compound.

**[MeHg{SC$_6$H$_3$-2,6-(SiMe$_3$)$_2$}]** or **C$_{13}$H$_{24}$Si$_2$SHg**. To obtain this compound, a solution containing MeHgCl (3.86 mM), 2,6-bis(trimethylsilyl)benzenethiol (3.86 mM), and Et$_3$N (3.86 mM) in MeOH (20 mL) was stirred for 1 h at room temperature (Block et al. 1990). MeOH was removed in vacuo, and the solid product was washed with distilled H$_2$O and a small amount of MeOH. Solvent removal at 13.3 Pa yielded a white powder of the title compound.

**[Hg{(SC$_6$H$_4$)$_2$SiMe$_2$}]** or **C$_{14}$H$_{14}$SiS$_2$Hg**. This compound crystallizes in the triclinic structure with the lattice parameters $a = 868.0 \pm 0.2$, $b = 1270.5 \pm 0.3$, and $c = 1505.3 \pm 0.4$ pm and $\alpha = 73.94° \pm 0.02°$, $\beta = 77.56° \pm 0.02°$, and $\gamma = 70.85° \pm 0.02°$ at 236 K and a calculated density of 2.12 g·cm$^{-3}$ (Block et al. 1990). It was prepared by the following procedure. A solution containing HgCl$_2$ (2.0 mM) and bis(2-mercaptophenyl)dimethylsilane (4.0 mM, two fold excess) in acetonitrile/MeOH (1:1 volume ratio) mixture solvent (50 mL) was stirred at room temperature for 5 min, and then Et$_3$N (8.2 mM) was added. The solution was stirred for 24 h and then concentrated in vacuo to a yellow oil. The oil was dissolved in 10 mL of acetonitrile to which 30 mL of MeOH was carefully added. After several days, colorless crystals of the title compound were collected by filtration and dried in air.

**[MeHg(SC$_6$H$_3$-4-Bu$^t$-2-SiMe$_3$)]** or **C$_{14}$H$_{24}$SiSHg**. This compound was obtained by the following procedure. A solution containing MeHgCl (5.57 mM) and 4-(*tert*-butyl)-2-(trimethylsilyl)benzenethiol (5.57 mM) in mixture solvent MeOH/acetonitrile (1:1 volume ratio, 20 mL) was stirred for 10 min at room temperature (Block et al. 1990). Upon the addition of Et$_3$N (5.60 mM), a white precipitate formed immediately. After 10 min of stirring at room temperature, the solid was collected by filtration and washed with distilled H$_2$O and MeOH. Solvent removal at 13.3 Pa yielded a white powder of the title compound.

**[Hg{(Me$_3$Si)$_2$SCH}$_2$]** or **C$_{14}$H$_{38}$Si$_4$S$_2$Hg**. This compound was prepared using the next procedure (Block et al. 1990). A solution containing HgCl$_2$ (1.0 mM), Et$_3$N (2.0 mM), and bis(trimethylsilyl)methanethiol (1.0 mM) in MeOH (15 mL) was stirred at room temperature for 1 h. The solid product, which formed immediately upon mixing of the reagents, was collected by filtration and washed with MeOH and distilled water. After solvent removal in vacuo, a white solid of the title compound was obtained.

**[Hg(SC$_6$H$_4$-2-SiMe$_3$)$_2$]** or **C$_{18}$H$_{26}$Si$_2$S$_2$Hg**. This compound crystallizes in the monoclinic structure with the lattice parameters $a = 690.0 \pm 0.1$, $b = 1285.6 \pm 0.2$, and $c = 2597.4 \pm 0.5$ pm and $\beta = 104.20° \pm 0.01°$ and a calculated density of 1.67 g·cm$^{-3}$ (Block et al. 1990). It was prepared by the following procedure. To a solution containing HgCl$_2$ (2.0 mM) and Et$_3$N (2.0 mM) in MeOH (5 mL) 2-(trimethylsilyl)benzenethiol (4.0 mM) in MeOH (10 mL) was added. White solid was produced immediately, and the solution was stirred at room temperature for 1 h. The solid was collected by filtration and washed with MeOH and distilled H$_2$O to yield, after solvent removal at 13.3 Pa, a white powder of the title compound. Recrystallization from warm CH$_2$Cl$_2$/MeCN afforded needlelike colorless crystals, which were collected by filtration and dried in air.

**[Hg(SC$_6$H$_5$-2-SiEt$_3$)$_2$] or C$_{24}$H$_{40}$Si$_2$S$_2$Hg.** This compound was prepared by the next procedure (Block et al. 1990). To a solution containing HgCl$_2$ (2.0 mM) and Et$_3$N (2.0 mM) in MeOH (5 mL) 2-(trimethylsilyl)-benzenethiol (10.0 mM) in MeOH (10 mL) was added. White solid was produced immediately, and the solution was stirred at room temperature for 1 h. The solid was collected by filtration and washed with MeOH and distilled H$_2$O to yield, after solvent removal at 13.3 Pa, a white powder of the title compound.

**[MeHg(SC$_6$H$_4$-2-SiPh$_3$)] or C$_{25}$H$_{22}$SiSHg.** To obtain it, a solution containing MeHgCl (2.43 mM), 2-(triphenylsilyl)benzenethiol (2.43 mM) and Et$_3$N (2.65 mM) in a mixture of solvent MeOH/acetonitrile (1:1 volume ratio, 20 mL) were stirred for 20 h at room temperature (Block et al. 1990). The white precipitate wew collected and washed with distilled H$_2$O and MeOH. Solvent removal at 13.3 Pa yielded a white powder of the title compound.

**[Hg(SC$_6$H$_3$-4-Bu$^t$-2-SiMe$_3$)$_2$] or C$_{26}$H$_{42}$Si$_2$S$_2$Hg.** This compound was prepared by the next procedure (Block et al. 1990). To a solution containing HgCl$_2$ (2.0 mM) and Et$_3$N (2.0 mM) in MeOH (5 mL) 2-(trimethylsilyl)-benzenethiol (2.0 mM) in MeOH (10 mL) was added. White solid was produced immediately, and the solution was stirred at room temperature for 1 h. The solid was collected by filtration and washed with MeOH and distilled H$_2$O to yield, after solvent removal at 13.3 Pa, a white powder of the title compound.

## 7.18  Hg–H–C–Ge–N–S

Some compounds are formed in this system.

**(C$_{12}$H$_{25}$NMe$_3$)$_2$[HgGe$_4$S$_{10}$] or C$_{30}$H$_{68}$Ge$_4$N$_2$S$_{10}$Hg.** This compound was prepared at room temperature by dissolution of C$_{12}$H$_{25}$NMe$_3$Br (1 mM) in EtOH/H$_2$O (200 mL; 1:1 volume ratio) (Wachhold et al. 2000a,b). Na$_4$Ge$_4$S$_{10}$ (0.5 mM) was dissolved in distilled H$_2$O (200 mL). HgCl$_2$ was also dissolved in distilled H$_2$O (20 mL), and Na$_4$Ge$_4$S$_{10}$ and HgCl$_2$ solutions were added simultaneously to the solution of C$_{12}$H$_{25}$NMe$_3$Br under constant stirring to form a copious precipitate. The mixture was stirred for 3 h at room temperature, and then filtered, washed with water, and vacuum-dried overnight. (C$_{12}$H$_{25}$NMe$_3$)$_2$[HgGe$_4$S$_{10}$] is a wide band gap semiconductor.

**(C$_{14}$H$_{29}$NMe$_3$)$_2$[HgGe$_4$S$_{10}$] or C$_{34}$H$_{76}$Ge$_4$N$_2$S$_{10}$Hg.** This compound was prepared by the same procedure as (C$_{12}$H$_{25}$NMe$_3$)$_2$[HgGe$_4$S$_{10}$] was obtained using C$_{14}$H$_{29}$NMe$_3$Br instead of C$_{12}$H$_{25}$NMe$_3$Br (Wachhold et al. 2000a,b). It is a wide band gap semiconductor with an energy gap of 2.89 eV.

**(C$_{16}$H$_{33}$NMe$_3$)$_2$[HgGe$_4$S$_{10}$] or C$_{38}$H$_{84}$Ge$_4$N$_2$S$_{10}$Hg.** This compound was prepared by the same procedure as (C$_{12}$H$_{25}$NMe$_3$)$_2$[HgGe$_4$S$_{10}$] was obtained using C$_{16}$H$_{33}$NMe$_3$Br instead of C$_{12}$H$_{25}$NMe$_3$Br (Wachhold et al. 2000a,b). It is a wide band gap semiconductor.

**(C$_{18}$H$_{37}$NMe$_3$)$_2$[HgGe$_4$S$_{10}$] or C$_{42}$H$_{92}$Ge$_4$N$_2$S$_{10}$Hg.** This compound was prepared by the same procedure as (C$_{12}$H$_{25}$NMe$_3$)$_2$[HgGe$_4$S$_{10}$] was obtained using

$C_{18}H_{37}NMe_3Br$ instead of $C_{12}H_{25}NMe_3Br$ (Wachhold et al. 2000a,b). It is a wide band gap semiconductor.

## 7.19 Hg–H–C–N–Si–S

Some compounds are formed in this system.

**[MeHg(2-SC$_5$H$_3$N-6-SiBu$^t$Me$_2$)] or C$_{12}$H$_{21}$NSiSHg.** To obtain this compound, a solution containing MeHgCl (1.0 mM), 6-(*tert*-butyldimethylsilyl)-2-mercaptopyridine (1.0 mM), and Et$_3$N (1.0 mM) in MeOH (20 mL) was stirred for 1 h at room temperature (Block et al. 1990). MeOH was removed in vacuo, and the solid product was washed with distilled H$_2$O and a small amount of MeOH. Solvent removal at 13.3 Pa yielded a white powder of the title compound.

**[MeHg(2-SC$_5$H$_3$N-3-SiPhMe$_2$)] or C$_{14}$H$_{17}$NSiSHg.** To obtain this compound, a solution containing MeHgCl (1.0 mM), 3-(phenyldimethylsilyl)-2-mercaptopyridine (1.0 mM), and Et$_3$N (1.0 mM) in MeOH (20 mL) was stirred for 1 h at room temperature (Block et al. 1990). MeOH was removed in vacuo, and the solid product was washed with distilled H$_2$O and a small amount of MeOH. Solvent removal at 13.3 Pa yielded a white powder of the title compound.

**[Hg(2-SC$_5$H$_3$N-3-SiMe$_3$)$_2$] or C$_{16}$H$_{24}$N$_2$Si$_2$S$_2$Hg.** This compound crystallizes in the triclinic structure with the lattice parameters $a = 905.8 \pm 0.2$, $b = 998.2 \pm 0.2$, and $c = 1263.9 \pm 0.3$ pm and $\alpha = 82.15° \pm 0.01°$, $\beta = 71.65° \pm 0.02°$, and $\gamma = 79.98° \pm 0.02°$ at 263 K and a calculated density of 1.77 g·cm$^{-3}$ (Block et al. 1990). It was prepared by the next procedure. To a solution containing HgCl$_2$ (2.0 mM) and Et$_3$N (2.0 mM) in MeOH (5 mL) 3-(trimethylsilyl)-2-mercaptopyridine (2.0 mM) in MeOH (10 mL) was added. A white solid was produced immediately, and the solution was stirred at room temperature for 1 h. The solid was collected by filtration and washed with MeOH and distilled H$_2$O to yield, after solvent removal at 13.3 Pa, a white powder of the title compound.

**[MeHg{2-SC$_5$H$_2$N-3,6-(SiBu$^t$Me$_2$)$_2$}] or C$_{18}$H$_{35}$NSi$_2$SHg.** To obtain this compound, a solution containing MeHgCl (3.41 mM), 3,6-bis(*tert*-butyldimethylsilyl)-2-mercaptopyridine (3.41 mM), and Et$_3$N (3.41 mM) in MeOH (20 mL) was stirred for 1 h at room temperature (Block et al. 1990). MeOH was removed in vacuo, and the solid product was washed with distilled H$_2$O and a small amount of MeOH. Solvent removal at 13.3 Pa yielded a white powder of the title compound.

**[Hg(2-SC$_5$H$_3$N-3-SiEt$_3$)$_2$] or C$_{22}$H$_{36}$N$_2$Si$_2$S$_2$Hg.** This compound was prepared by the next procedure (Block et al. 1990). To a solution containing HgCl$_2$ (2.0 mM) and Et$_3$N (2.0 mM) in MeOH (5 mL) 3-(trimethylsilyl)-2-mercaptopyridine (2.0 mM) in MeOH (10 mL) was added. White solid was produced immediately, and the solution was stirred at room temperature for 1 h. The solid was collected by filtration and washed with MeOH and distilled H$_2$O to yield, after solvent removal at 13.3 Pa, a white powder of the title compound.

**[Hg(2-SC$_5$H$_3$N-3-Bu$^t$Me$_2$)$_2$] or C$_{22}$H$_{36}$N$_2$Si$_2$S$_2$Hg.** This compound was obtained by the following procedure (Block et al. 1990). To a solution containing HgCl$_2$ (2.0 mM) and Et$_3$N (2.0 mM) in MeOH (5 mL) 6-(*tert*-butyldimethylsilyl)-2-mercaptopyridine (2.0 mM) in MeOH (10 mL) was added. White solid was produced immediately, and the solution was stirred at room temperature for 1 h. The solid was collected by filtration and washed with MeOH and distilled H$_2$O to yield, after solvent removal at 13.3 Pa, a white powder of the title compound.

**[Hg(2-SC$_5$H$_3$N-3-SiPhMe$_2$)$_2$] or C$_{26}$H$_{28}$N$_2$Si$_2$S$_2$Hg.** This compound was prepared by the following procedure (Block et al. 1990). To a solution containing HgCl$_2$ (2.0 mM) and Et$_3$N (2.0 mM) in MeOH (5 mL) 3-(triphenyldimethylsilyl)-2-mercaptopyridine (2.0 mM) in MeOH (10 mL) was added. White solid was produced immediately, and the solution was stirred at room temperature for 1 h. The solid was collected by filtration and washed with MeOH and distilled H$_2$O to yield, after solvent removal at 13.3 Pa, a white powder of the title compound.

**[Hg{(2-SC$_5$H$_2$N-3,6-(SiBu$^t$Me$_2$)$_2$}$_2$] or C$_{34}$H$_{64}$N$_2$Si$_2$S$_2$Hg.** This compound was prepared using the next procedure (Block et al. 1990). A solution containing HgCl$_2$ (2.0 mM) and Et$_3$N (2.0 mM) in MeOH (5 mL) was added 3,6-bis(*tert*-butyldimethylsilyl)-2-mercaptopyridine (2.0 mM) in MeOH (10 mL). White solid was produced immediately, and the solution was stirred at room temperature for 1 h. The solid was collected by filtration and washed with MeOH and distilled H$_2$O to yield, after solvent removal at 13.3 Pa, a white powder of the title compound.

## 7.20   Hg–H–C–N–P–S

Some compounds are formed in this system.

**[Hg(SCN)$_2$·Ph$_3$P] or C$_{20}$H$_{15}$N$_2$PS$_2$Hg, dithiocyanato(triphenylphosphine) mercury(II).** This compound melts at 110°C–111°C (Jain and Rivest 1970). β-Form of the title compound crystallizes in the monoclinic structure with the lattice parameters $a = 2107.1 \pm 0.5$, $b = 1771.6 \pm 0.5$, and $c = 1128.5 \pm 0.3$ pm and $\beta = 101.66° \pm 0.05°$ and a calculated density of 1.864 g·cm$^{-3}$ (Makhija et al. 1979). Another modifications of Hg(SCN)$_2$·Ph$_3$P also crystallizes in the monoclinic structure with the lattice parameters $a = 1007.5 \pm 0.6$, $b = 2113.6 \pm 1.1$, and $c = 1074.5 \pm 0.4$ pm and $\beta = 112.37° \pm 0.04°$ and a calculated density of 1.817 g·cm$^{-3}$ (Gagnon and Beauchamp 1979).

To obtain this compound, Hg(SCN)$_2$ and PhP$_3$ (1:1 molar ratio) were mixed in acetone or in THF. Its crystals were grown by slow evaporation of the Hg(SCN)$_2$·Ph$_3$P solution at room temperature (Gagnon and Beauchamp 1979, Makhija et al. 1979). They were recrystallized from THF, filtered, washed with small portions of cold THF, and dried in vacuo at room temperature (Jain and Rivest 1970).

**[Hg(SCN)$_2${P(C$_6$H$_{11}$)$_3$}] or C$_{20}$H$_{33}$N$_2$PS$_2$Hg.** This compound melts at 214°C–215°C and crystallizes in the monoclinic structure with the lattice parameters $a = 1085.0 \pm 0.1$, $b = 982.5 \pm 0.2$, and $c = 2197.7 \pm 0.3$ pm and $\beta = 94.34° \pm 0.01°$ and the experimental and a calculated densities 1.70 g·cm$^{-3}$ (Alyea et al. 1977). The colorless crystals were recrystallized from Pr$^i$OH.

**[Hg(CH$_2$PSPh$_2$)$_2$] or C$_{26}$H$_{24}$P$_2$S$_2$Hg.** This compound melts at the temperature higher than 230°C and crystallizes in the monoclinic structure with the lattice parameters $a = 2509.0 \pm 0.8$, $b = 1088.6 \pm 0.5$, and $c = 918.9 \pm 0.4$ pm and $\beta = 91.83° \pm 0.03°$ and a calculated density of 1.76 g·cm$^{-3}$ (Wang and Fackler 1989). To prepare it, a solution of LiCH$_2$PSPh$_2$ (4.0 mM) in 20 mL of THF was cooled to −78°C. Then, 1.8 mM of HgCl$_2$ was added to this solution. After the mixture was slowly warmed up to 0°C, the white solid of the title compound formed and precipitated from the solution. The solution was stirred for 1 h at 0°C. After filtration, the white solid was washed with cold THF, then ethyl alcohol, and finally diethyl ether. Colorless single crystals were grown from CH$_2$Cl$_2$/diethyl ether solution by slow diffusion of the solvent at 22°C.

**[Ph$_4$P][Hg(SCN)$_3$] or C$_{27}$H$_{20}$N$_3$PS$_3$Hg, tetraphenylphosphonium trithiocyanatomercurate(II).** This compound crystallizes in the monoclinic structure with the lattice parameters $a = 1157.4 \pm 0.4$, $b = 2001.4 \pm 0.8$, and $c = 1214.4 \pm 0.4$ pm and $\beta = 103.56° \pm 0.03°$ and the experimental and calculated densities 1.68 ± 0.01 and 1.66 g·cm$^{-3}$, respectively (Sakhri and Beauchamp 1975a). It was obtained by evaporation of a 1:1 molar ratio solution of Hg(SCN)$_2$ and [Ph$_4$P]SCN in EtOH.

**[Hg(SCN)$_2$(Ph$_3$P)$_2$] or C$_{38}$H$_{30}$N$_2$P$_2$S$_2$Hg, dithiocyanatobis(triphenylphosphine)mercury(II).** This compound melts at 202°C–203°C (Davis et al. 1970) (at 190°C–191°C [Jain and Rivest 1970]) and crystallizes in the monoclinic structure with the lattice parameters $a = 1738.5 \pm 0.9$, $b = 1058.1 \pm 0.4$, and $c = 1930.4 \pm 0.5$ pm and $\beta = 91.41° \pm 0.06°$ and the experimental and calculated densities 1.58 ± 0.01 and 1.585 g·cm$^{-3}$, respectively (Makhija et al. 1973). Ph$_3$P and Hg(SCN)$_2$ were allowed to react in the boiling ether (in warm THF [Jain and Rivest 1970]) (2:1 molar ratio). After boiling for several minutes, the reaction mixture was cooled. The colorless crystals that appeared were filtered off and washed thoroughly with absolute ether to remove any unreacted Ph$_3$P. The title compound is stable in air (Davis et al. 1970). Twice recrystallizations from acetone (from THF [Jain and Rivest 1970]) and slow evaporation of the solvent yield diamond-shaped crystals (Makhija et al. 1973).

**[Hg(SCN)$_2$(PhPC$_{12}$H$_8$)]$_2$ or C$_{40}$H$_{26}$N$_4$P$_2$S$_4$Hg$_2$.** This compound crystallizes in the monoclinic structure with the lattice parameters $a = 1489.9 \pm 0.2$, $b = 1781.5 \pm 0.2$, and $c = 1613.1 \pm 0.4$ pm and $\beta = 114.400° \pm 0.010°$ and a calculated density of 1.966 g·cm$^{-3}$ (Gallagher et al. 1999). It was synthesized from Hg(SCN)$_2$ and 1-phenyldibenzophosphole in aqueous ethanol solution. Recrystallization from EtOH/CH$_2$Cl$_2$ solution afforded colorless crystals.

**[Hg(SCN)$_2$(Ph$_3$P)]$_2$ or C$_{40}$H$_{30}$N$_4$P$_2$S$_4$Hg$_2$.** This compound melts at 70°C (Davis et al. 1970). Ph$_3$P and Hg(SCN)$_2$ were allowed to react in the boiling ether (1:1 molar ratio). After boiling for several minutes, the reaction mixture was cooled. The colorless crystals that appeared were filtered off and washed thoroughly with absolute ether to remove any unreacted Ph$_3$P. The title compound is stable in air.

**[(Ph$_3$P)$_2$N][Hg(SBu$^t$)$_3$] or C$_{48}$H$_{57}$NP$_2$S$_3$Hg.** This compound melts at 215°C–225°C (Bowmaker et al. 1984). It was prepared by the next procedure. 2-Methylpropane-2-thiol (3.9 mM) was dissolved in a solution, of NaOH in absolute EtOH (10 mL). To the resulting solution [N(Ph$_3$P)$_2$]Cl (0.78 mM) and Hg(SBu$^t$)$_2$ (0.79 mM) were added. The resulting mixture was heated at 60°C, and DMF (4.5 mL) was added.

The clear solution was filtered, and the white crystals of the title compound formed upon cooling. These were collected by filtration and dried under vacuum.

**[Ph$_4$P]$_2$[Hg(SCN)$_4$] or C$_{52}$H$_{40}$N$_4$P$_2$S$_4$Hg**. This compound crystallizes in the triclinic structure with the lattice parameters $a = 1285.9 \pm 0.6$, $b = 1349.4 \pm 0.7$, and $c = 1658.3 \pm 0.7$ pm and $\alpha = 89.68° \pm 0.03°$, $\beta = 76.78° \pm 0.03°$, and $\gamma = 61.59° \pm 0.03°$ and a calculated density of 1.509 g·cm$^{-3}$ (Sakhri and Beauchamp 1975b). It was prepared by evaporation of the ethanolic solution of Hg(SCN)$_2$ and [Ph$_4$P]SCN (1:2 molar ratio).

## 7.21   Hg–H–C–N–P–F–S

Some compounds are formed in this system.

**[Hg{S(CH$_2$)$_3$NMe$_3$}$_2$](PF$_6$)$_2$ or C$_{12}$H$_{30}$N$_2$P$_2$F$_{12}$S$_2$Hg**. This compound crystallizes in the triclinic structure with the lattice parameters $a = 1509.7 \pm 0.4$, $b = 1896.8 \pm 0.5$, and $c = 2036.5 \pm 0.6$ pm and $\alpha = 71.74° \pm 0.01°$, $\beta = 68.50° \pm 0.01°$, and $\gamma = 75.46° \pm 0.02°$ and a calculated density of 1.975 g·cm$^{-3}$ (Casals et al. 1991b). It was prepared by the next procedure. 3-Trimethylammonio-1-propanethiol hexafluorophosphate (57.6 mg, corresponding to 0.20 mM of pure thiol) was dissolved in a mixture of acetonitrile (1 mL), MeOH (1 mL), and H$_2$O (1 mL). Aqueous NaOH (0.32 mM in 2 mL) was added, followed by Hg(OAc)$_2$·2H$_2$O (0.08 mM) dissolved in 2.0 mL of H$_2$O. The slightly cloudy solution was filtered and kept under nitrogen at room temperature. After several days, the solid product was isolated. Some preparations give pale-yellow crystals of the title compound.

**[Hg{S(CH$_2$)$_3$NMe$_3$}$_2$](PF$_6$)$_2$·0.5[S(CH$_2$)$_3$NMe$_3$] or C$_{15}$H$_{37.5}$N$_{2.5}$P$_2$F$_{12}$S$_{2.5}$Hg**. This compound crystallizes in the tetragonal structure with the lattice parameters $a = 1976.8 \pm 0.1$ and $c = 701.61 \pm 0.04$ and a calculated density of 1.9991 g·cm$^{-3}$ (Casals et al. 1991b). To obtain it, 3-trimethylammonio-1-propanethiol hexafluorophosphate (57.6 mg, corresponding to 0.20 mM of pure thiol) was dissolved in a mixture of acetonitrile (1 mL), MeOH (1 mL), and H$_2$O (1 mL). Aqueous NaOH (0.32 mM in 2 mL) was added, followed by Hg(OAc)$_2$·2H$_2$O (0.08 mM) dissolved in 2.0 mL of H$_2$O. The slightly cloudy solution was filtered and kept under nitrogen at room temperature. After several days, the solid product was isolated. Some preparations give a colorless microcrystalline solid, which upon recrystallization from concentrated aqueous solutions always yielded colorless crystals of the title compound.

## 7.22   Hg–H–C–N–P–Co–S

**[HgCo(SCN)$_4$(Ph$_3$P)$_2$]$_2$ or C$_{40}$H$_{30}$N$_4$P$_2$CoS$_4$Hg**. This multinary compound is formed in this system. It crystallizes in the triclinic structure with the lattice parameters $a = 1193.4 \pm 0.2$, $b = 1464.2 \pm 0.9$, and $c = 1180.1 \pm 0.6$ pm and $\alpha = 90.41° \pm 0.05°$, $\beta = 94.04° \pm 0.02°$, and $\gamma = 83.82° \pm 0.02°$ (Kinoshita et al. 1985b). It was prepared by the next procedure. HgCo(SCN)$_4$ was finely powdered (4.1 mM) and then poured into 20 mL of acetonitrile solution containing Ph$_3$P (9.2 mM), after which the mixture was stirred for 6 h at an ambient temperature. The addition reaction occurred rapidly, and the solid product thus obtained was filtered off, washed with acetonitrile, and dried on silica gel in a vacuum desiccator. By the gradual evaporation of the filtrate, some additional crop was obtained in larger crystal grains.

## 7.23  Hg–H–C–N–As–S

Some compounds are formed in this system.

**(Me$_4$N)[HgAs$_3$S$_6$] or C$_4$H$_{12}$NAs$_3$S$_6$Hg.** This compound crystallizes in the monoclinic structure with the lattice parameters $a = 1860.7 \pm 0.7$, $b = 712.6 \pm 0.1$, and $c = 2652.4 \pm 0.6$ pm and $\beta = 91.87° \pm 0.02°$ and a calculated density of 2.615 g·cm$^{-3}$ and an energy gap of 2.8 eV (Chou and Kanatzidis 1995). To obtain it, a mixture of HgCl$_2$ (0.5 mM), K$_3$AsS$_3$ (0.5 mM), and Me$_4$NCl (1 mM) was sealed under vacuum with 0.3–0.5 mL of H$_2$O in a Pyrex tube (ca. 4 mL). The reaction was run at 130°C for 1 week. Large pale-yellow transparent crystals were isolated and washed with H$_2$O, MeOH, and anhydrous ether.

**[Hg(SCN)$_2$Ph$_3$As] or C$_{20}$H$_{15}$N$_2$AsS$_2$Hg.** This compound crystallizes in the monoclinic structure with the lattice parameters $a = 1029.0 \pm 0.7$, $b = 2119.9 \pm 2.3$, and $c = 1071.9 \pm 0.7$ pm and $\beta = 112.00° \pm 0.02°$ and the experimental and calculated densities 1.90 ± 0.01 and 1.909 g·cm$^{-3}$, respectively (Hubert et al. 1975) ($a = 1076$, $b = 2124$, and $c = 1070$ pm and $\beta = 111°54'$ and the experimental and calculated densities 1.90 and 1.917 g·cm$^{-3}$, respectively [Makhija et al. 1972]). It was prepared by the reaction of Hg(SCN)$_2$ (1.9 mM) and Ph$_3$As (2.1 mM) in EtOH (125 mL) (Hubert et al. 1975). By slow evaporation of the solution, well-shaped crystals were obtained. Its single crystals were grown from an acetone solution (Makhija et al. 1972).

**[Hg(SCN)$_2$(Ph$_3$As)$_2$] or C$_{38}$H$_{30}$N$_2$As$_2$S$_2$Hg, dithiocyanatobis(triphenylarsine) mercury(II).** This compound melts at 122°C–124°C (Davis et al. 1970). Ph$_3$As and Hg(SCN)$_2$ were allowed to react in the boiling ether (2:1) molar ratio). After boiling for several minutes, the reaction mixture was cooled. The colorless crystals that appeared were filtered off and washed thoroughly with absolute ether to remove any unreacted Ph$_3$As. The title compound is stable in air.

**[Hg(SCN)$_2$(Ph$_3$As)]$_2$ or C$_{40}$H$_{30}$N$_4$As$_2$S$_4$Hg$_2$.** This compound melts at 129°C–131°C (Davis et al. 1970). Ph$_3$As and Hg(SCN)$_2$ were allowed to react in the boiling ether (1:1 molar ratio). After boiling for several minutes, the reaction mixture was cooled. The colorless crystals that appeared were filtered off and washed thoroughly with absolute ether to remove any unreacted Ph$_3$As. The title compound is stable in air.

## 7.24  Hg–H–C–N–O–S

Some compounds are formed in this system.

**[MeHg{SC(NH$_2$)$_2$}]NO$_3$ or C$_2$H$_7$N$_3$O$_3$SHg.** To obtain this compound, hot deoxygenated aqueous solution of MeHgNO$_3$ (5.0 mM) was added to a refluxing solution of a stoichiometric amount of thiourea under N$_2$ (Carty et al. 1979). After ca. 30 min, the solution was filtered and slowly evaporated to low volume. Failure to adequately protect the solution from air leads to extensive decomposition. Colorless crystals of the title compound were filtered off, dried in a desiccator, and stored in the dark.

**[MeHg(C₃H₅NS₂)(NO₃)] or C₄H₈N₂O₃S₂Hg.** This compound crystallizes in the monoclinic structure with the lattice parameters $a = 715.8 \pm 1.4$, $b = 1015.6 \pm 0.7$, and $c = 1347.2 \pm 1.2$ pm and $\beta = 108.21° \pm 0.04°$ and a calculated density of 2.83 g·cm⁻³ (Norris et al. 1990).

**[Hg(MeC₃H₆NO₂S)]·H₂O or C₄H₁₁NO₃SHg, L-cysteinato(methyl)mercury(II).** This compound crystallizes in the orthorhombic structure with the lattice parameters $a = 638.6 \pm 0.6$, $b = 2602.6 \pm 1.3$, and $c = 528.2 \pm 0.4$ pm and the experimental and calculated densities 2.65 and 2.676 g·cm⁻³, respectively (Taylor et al. 1975). It was prepared by the following procedure. MeHgOH (2.16 mM) and L-cysteine (C₃H₇NO₂S) (2.16 mM) were mixed in 50% EtOH (150 mL) and stirred for 3 h. The volume of the filtered solution was reduced to ca. 20 mL, and the almost saturated solution set aside until prismatic crystals were obtained. The same compound was prepared by the reaction of MeHgCl with L-cysteine in basic aqueous ethanol (pH ca. 10) solution. Crystals were grown from aqueous ethanol solution.

**[HgMe(C₄H₄N₃OS)]·H₂O or C₅H₉N₃O₂SHg, (4-amino-2-mercapto-6-pyrimidi-nonato)methylmercury(II).** This compound crystallizes in the monoclinic structure with the lattice parameters $a = 731.2 \pm 0.4$, $b = 870.2 \pm 0.4$, and $c = 1491.1 \pm 0.7$ pm and $\beta = 94.43° \pm 0.05°$ and the experimental and calculated densities 2.59 and 2.63 g·cm⁻³, respectively (Stuart et al. 1980). To obtain it, an aqueous solution of MeHgOAc (1 mM) was added to an aqueous solution of a mixture of 4-amino-2-mercapto-6-pyrimidinol (C₄H₅N₃OS) (1 mM) and NaHCO₃ (1 mM). The resulting precipitate was filtered and recrystallized from hot acetone. Crystals were grown from a solution of this compound in a mixture of DMSO, acetone, and H₂O.

**[MeHg(C₄H₆N₂S)](NO₃) or C₅H₉N₃O₃SHg (C₄H₆N₂S = 1-methylimidazoline-2-thione).** This compound crystallizes in the triclinic structure with the lattice parameters $a = 714.6 \pm 0.2$, $b = 778.3 \pm 0.3$, and $c = 1018.4 \pm 0.5$ pm and $\alpha = 72.95° \pm 0.03°$, $\beta = 71.09° \pm 0.03°$, and $\gamma = 81.83° \pm 0.02°$ and the experimental and calculated densities 2.52 and 2.543 g·cm⁻³, respectively (Norris et al. 1983).

**[(MeHg)₂(C₃H₄NS₂)(NO₃)] or C₅H₁₀N₂O₃S₂Hg₂.** This compound crystallizes in the monoclinic structure with the lattice parameters $a = 2520.0 \pm 1.0$, $b = 702.9 \pm 0.6$, and $c = 1794.6 \pm 0.8$ pm and $\beta = 128.99° \pm 0.03°$ and a calculated density of 3.9 g·cm⁻³ (Norris et al. 1990).

**[MeHgC₅H₄OS] or C₆H₇NOSHg.** This compound melts at 117°C (Sytsma and Kline 1973). To obtain it, MeHgOAc was dissolved in water and added to 2-mer-captopyridine-N-oxide dissolved in MeOH or methanolic KOH solution. The crude products were purified by the recrystallization in H₂O/EtOH.

**[MeHg(C₅H₄NS)](NO₃) or C₆H₇N₂O₃SHg.** This compound melts at 90°C–91°C (Canty and Marker 1976). Reaction of pyridine-2(1H)-thione (1.35 mM) in acetone (10 mL) with MeHgNO₃ (1.35 mM) gave colorless crystals of the title compound when left to stand for 24 h. Nitric acid (0.1 mM, 5 M) added to [MeHg(C₅H₄NS)] (0.104 g) in MeOH (5 mL) also gave colorless crystals of [MeHg(C₅H₅NS)](NO₃) after being left to stand for 3 days.

**[Hg(CN)₂·2(DMSO)] or C₆H₁₂N₂O₂S₂Hg**. This compound melts at 96°C–97°C (Jain and Rivest 1969b). It was prepared by dissolving Hg(CN)₂ in a minimum amount of warm DMSO forming a clear solution. The title compound crystallized out from the reaction mixture on standing overnight at room temperature. It was filtered out under suction and the excess DMSO was removed under vacuum.

**[(MeSHg)₂(O₂CMe)(C₃H₃N₂) or C₆H₁₂N₂O₂S₂Hg₂**. This compound melts at 124°C–129°C (Canty and Tyson 1978). It was prepared by the next procedure. Imidazole (1.29 mM) in H₂O (2 mL) was added to a solution of MeSHgO₂CMe (1.29 mM) in H₂O (36 mL) and HOAc (5 M, 2 mL). The solution was filtered and evaporated at ambient temperature to low volume. Addition of ethanol resulted in slow precipitation of a white powder, which was collected, washed with EtOH, and dried in a vacuum over P₂O₅. A similar preparation using MeSHgO₂CMe/imidazole of 2:1 molar ratio gave the same product.

The title compound could also be prepared if MeSHgO₂CMe (0.67 mM) was added to a solution of imidazole (0.67 mM) and the suspension was stirred for 24 h. The white powder was collected, washed in H₂O, and dried in a vacuum over P₂O₅.

**[Hg(SCN)₂·2(DMSO)] or C₆H₁₂N₂O₂S₄Hg**. This compound melts at 88°C–89°C (Jain and Rivest 1969b). It was prepared by dissolving Hg(SCN)₂ in a minimum amount of warm DMSO forming a turbid solution, which had to be filtered warm. The title compound crystallized out from the reaction mixture on standing overnight at room temperature. It was filtered out under suction and the excess DMSO was removed under vacuum.

**[(MeHg)(C₅H₁₀NO₂S)]·H₂O or C₆H₁₅NO₃SHg**. This compound crystallizes in the monoclinic structure with the lattice parameters $a = 2315.4 \pm 3.3$, $b = 981.9 \pm 1.4$, and $c = 1932.1 \pm 2.7$ pm and $\beta = 106.4° \pm 0.1°$ and the experimental and calculated densities 2.40 and 2.14 g·cm⁻³, respectively (Wong et al. 1977) ($a = 2333.1$, $b = 992.1$, and $c = 1949.6$ pm and $\beta = 106.5°$ and the experimental and calculated densities 2.40 and 2.43 g·cm⁻³, respectively [Wong et al. 1973a]). To obtain it, HgCl₂ (1.5 g) was dissolved in absolute EtOH (20 mL), and 3 mL of 2 M NaOH was added with stirring. The solution was heated to boiling, allowed to cool to room temperature, and finally cooled in an ice bath. The cold solution was rapidly filtered through sintered glass funnel to remove the precipitated NaCl. An equivalent amount of DL-penicillamine (C₅H₁₁NO₂S) (0.89 g) in 50% aqueous EtOH (20 mL) was added to the filtered solution, and the mixture was stirred for 12 h at 25°C. The volume of the solution was reduced to ca. 15 mL by evaporation at room temperature. Colorless crystals were obtained by the slow evaporation of the solution at room temperature (Wong et al. 1977). Colorless prisms of the title compound were also obtained from aqueous ethanolic solutions containing equimolar quantities of DL-penicillamine and MeHgOH or MeHgCl in basic solution (Wong et al. 1973a,b).

**[Hg(C₅H₄NS)(OAc)]ₙ or C₇H₇NO₂SHg**. This compound melts at 175°C and crystallizes in the monoclinic structure with the lattice parameters $a = 869.3 \pm 0.5$, $b = 1299.1 \pm 0.5$, and $c = 811.0 \pm 0.4$ pm and $\beta = 103.45° \pm 0.04°$ and a calculated density of 2.76 g·cm⁻³ (Wang and Fackler 1989). It was synthesized by the following procedure. First, 0.94 mM of Hg(OAc)₂ was dissolved in 10 mL of CH₂Cl₂ at room temperature. Then, 0.90 mM of mercaptopyridine ($C_5H_5NS$) was added to the solution. The yellow color of $C_5H_5NS$ disappeared rapidly. A clear colorless solution was obtained within a few minutes. After 1 h, the solution was concentrated to 5 mL and excess diethyl ether was added. Colorless crystals of the title compound were precipitated. Single crystals were grown from $CH_2Cl_2$/diethyl ether solution by slow diffusion of the solvent at 22°C. All reactions were performed by using standard Schlenk techniques under an atmosphere of dry $N_2$.

**[(MeHg)$_2$(C$_5$H$_{10}$NO$_2$S)] or C$_7$H$_{16}$NO$_2$SHg.** This compound crystallizes in the orthorhombic structure with lattice parameters $a = 1969.5 \pm 2.3$, $b = 1047.8 \pm 1.3$, and $c = 1197.1 \pm 2.6$ pm and the experimental and calculated densities 3.06 and 3.11 g·cm$^{-3}$, respectively (Wong et al. 1977) ($a = 1970$, $b = 1048$, and $c = 1197$ pm and the experimental and calculated densities 3.06 and 3.114 g·cm$^{-3}$, respectively [Wong et al. 1973b]). To obtain it, $HgCl_2$ (0.99 g) was dissolved in absolute EtOH (20 mL), and 4 mL of 2 M NaOH was added with stirring. The solution was heated to boiling, allowed to cool to room temperature, and finally cooled in an ice bath. The cold solution was rapidly filtered through sintered glass funnel to remove the precipitated NaCl. DL-penicillamine ($C_5H_{11}NO_2S$) (0.30 g) in 50% aqueous EtOH (20 mL) was added to the filtered solution, and the mixture was stirred for 12 h at 25°C. The volume of the solution was reduced to 10 mL. The precipitate obtained by slow evaporation of the solution was recrystallized from aqueous EtOH giving colorless needles of the title compound (Wong et al. 1977). Interaction of MeHgCl with DL-penicillamine ($C_5H_{11}NO_2S$) in EtOH also gives [(MeHg)$_2$(C$_5$H$_{10}$NO$_2$S)].

**(MeSHgOAc)(Py) or C$_8$H$_{11}$NO$_2$SHg.** This compound crystallizes in the monoclinic structure with the lattice parameters $a = 1009.3 \pm 0.3$, $b = 702.6 \pm 0.2$, and $c = 1535.0 \pm 0.5$ pm and $\beta = 105.01° \pm 0.03°$ and a calculated density of 2.44 g·cm$^{-3}$ (Canty et al. 1978b). To obtain it, MeSHgOAc was dissolved in Py to slowly deposit colorless crystals (Canty et al. 1978a). The precipitate was collected, washed with EtOH, and dried over $P_2O_5$ in a vacuum. The obtained crystals slowly lose pyridine at ambient temperature and pressure but may be stored at ca. −20°C in a sealed tube.

**[(MeSHg)$_2$(O$_2$CMe)(C$_4$H$_5$N$_2$) or C$_8$H$_{14}$N$_2$O$_2$S$_2$Hg$_2$.** This compound melts at 121°C–123°C (Canty and Tyson 1978). It was prepared using a 1:1 molar ratio of mercurial and methylimidazole in ethanol.

**[Hg{S(CH$_2$)$_2$CONHMe}$_2$] or C$_8$H$_{16}$N$_2$O$_2$S$_2$Hg.** A mixture of two equal volumes of aqueous solution of 0.02 M methylcarbamoylethanethiol and 0.01 M $HgCl_2$ causes the immediate formation of a white precipitate of this compound (Perchard et al. 1981). Using $D_2O$ instead of $H_2O$ leads to the formation of the deuterated compound [Hg{S(CH$_2$)$_2$CONDMe}$_2$]. The crystals were grown by slow evaporation of the filtrate at room temperature for some days.

**[Hg(C$_4$H$_8$NOS$_2$)$_2$] or C$_8$H$_{16}$N$_2$O$_2$S$_4$Hg$_2$.** To obtain this compound, N-methyl-N-ethanoldithiocarbamic acid was prepared by mixing 4 mL of N-methyl-N-ethanolamine ($C_3H_9NO$) and 2 mL of $CS_2$ diluted with EtOH (20 mL) at 5°C (Thirumaran et al. 1998). To the yellow dithiocarbamic acid solution (20 mM),

10 mM of Hg salt dissolved in $H_2O$ was added with constant stirring. Pale-yellow dithiocarbamate complex separated, which was washed with $H_2O$ and EtOH and then dried. This compound is stable up to 200°C.

**[($Et_2S)_2$·Hg(NO$_3)_2$] or $C_8H_{20}N_2O_6S_2Hg$.** This compound melts at 63°C (Faragher et al. 1929). It was isolated as follows. $Et_2S$ was dissolved in hexane and treated with 8% aqueous solution of Hg(NO$_3)_2$ by shaking for 5 min. After standing for a few hours, the hexane and the liquid reaction product were decanted from the mercury and the liquid product was separated from the hexane. The syrupy reaction product was filtered twice to remove all traces of metallic mercury. The filtrate was placed under a vacuum of 270 Pa, and after a few hours, long white needles appeared. The needles were removed and kept under vacuum again for 15 h. The crystals are very hygroscopic.

**[(EtSHgOAc)(Py)] or $C_9H_{13}NO_2SHg$.** To obtain it, EtSHgOAc was dissolved in Py to slowly deposit colorless crystals (Canty et al. 1978a). The precipitate was collected, washed with EtOH, and dried over $P_2O_5$ in a vacuum. The obtained crystals slowly lose pyridine at an ambient temperature and pressure but may be stored at ca. −20°C in a sealed tube.

**[(MeSHgOAc)(MePy)] or $C_9H_{13}NO_2SHg$.** This compound crystallizes in the monoclinic structure with the lattice parameters $a = 1661.7 \pm 0.9$, $b = 727.1 \pm 0.3$, and $c = 2047.6 \pm 0.2$ pm and $\beta = 113.31° \pm 0.15°$ and a calculated density of 2.34 g·cm$^{-3}$ (Canty et al. 1979b).

**[Hg{NO$_3$HgC$_4$H$_2$SCOO}$_2$] or $C_{10}H_4N_2O_{10}S_2Hg_3$, bis[(5-nitratomercurio)thiophen-2-carboxylato]mercury(II).** It was prepared by the slow addition of an aqueous solution (ca. 50 mL) of Hg(NO$_3)_2$·H$_2O$ (1.46 mM; with some drops of 20% HNO$_3$) to an aqueous solution (ca. 30 mL) of thiophene-2-carboxylic acid (1.46 mM) (Popović et al. 2000b). The mercurated product was obtained immediately upon mixing the solutions. The reaction mixture was left to stand for a few hours and then filtered off. The precipitate was washed with $H_2O$ and EtOH and dried.

**[Bu$^t$HgSC$_6$H$_4$NO$_2$] or $C_{10}H_{13}NO_2SHg$, t-butylmercury p-nitrophenylmercaptide.** This compound melts at 135°C–137°C with decomposition (Bach and Weibel 1976). A solution of t-butylmercury chloride (1 mM) in 20 mL of MeOH containing p-nitrophenylmercaptan (0.17 g of 80% purity) and 0.7 mL of 3N NaOH in 20 mL of MeOH rapidly afforded a light-colored precipitate. After the solution was refrigerated overnight, it was filtered to yield a crude solid, which on recrystallization from CH$_2$Cl$_2$/pentane (1:1 volume ratio) gave yellow platelets of Bu$^t$HgSC$_6$H$_4$NO$_2$.

**[Hg(C$_4$H$_7$NOS)$_2$(CN)$_2$] or $C_{10}H_{14}N_4O_2S_2Hg$.** This compound melts at 72°C–75°C (Colombini and Preti 1975). It was prepared by adding thiomorpholine-3-one dissolved in EtOH to a hot solution of Hg(SCN)$_2$ in EtOH. The title compound obtained by concentrating the solution was washed with ethyl ether.

**[Hg(C$_5$H$_7$N$_2$OS)$_2$] or $C_{10}H_{14}N_4O_2S_2Hg$.** To obtain this compound, an equivalent of 0.1 M NaOH was added to Hg(C$_5$H$_8$N$_2$OS)XY (X, Y = Cl, Br, I) dissolved in the least amount of EtOH ($C_5H_8N_2OS$—5,5-dimethylimidazolidine-2-thione-4-one) (Bellon et al. 1988). The title compound precipitated immediately as a white powder. Except for the chlorocompound, this reaction was also carried out in $H_2O$ to yield the same product. This compound was obtained by reacting $Hg(ClO_4)_2$ and $C_5H_8N_2OS$ (1:2 molar ratio) in $H_2O$ (Cristiani et al. 1990).

**[Hg($C_4H_7NOS$)$_2$(SCN)$_2$] or $C_{10}H_{14}N_4O_2S_4Hg$.** This compound melts at 96°C–98°C (Colombini and Preti 1975). It was obtained by heating a suspension of $Hg(SCN)_2$ with thiomorpholine-3-one in EtOH and by filtering the resultant mixture while still hot. The title compound crystallized out, and upon cooling was washed with ethyl ether.

**[(EtSHgOAc)(MePy)] or $C_{10}H_{15}NO_2SHg$.** This compound crystallizes in the monoclinic structure with the lattice parameters $a = 934.4 \pm 0.2$, $b = 1791.7 \pm 0.9$, and $c = 717.9 \pm 0.2$ pm and $\beta = 106.24° \pm 0.03°$ and a calculated density of 2.30 g·cm$^{-3}$ (Canty et al. 1979b).

**[Hg($C_4H_8NCOS$)$_2$] or $C_{10}H_{16}N_2O_2S_2Hg$.** This compound was prepared by treating 10 mM of sodium N-pyrrolidylthiocarbamate ($C_4H_8NCOSNa$) in $H_2O$ (20 mL) with an equal volume of aqueous solution of $Hg(NO_3)_2 \cdot H_2O$ (5 mM) (McCormick and Greene 1972). The white precipitate that formed was washed twice with 20 mL portions of EtOH and dried in vacuo over anhydrous $CaSO_4$ for 24 h.

**[Hg{$S_2CN(CH_2CH_2OH)_2$}$_2$] or $C_{10}H_{20}N_2O_4S_4Hg$.** This compound melts at 95°C–97°C and crystallizes in the orthorhombic structure with the lattice parameters $a = 1381.73 \pm 0.07$, $b = 453.07 \pm 0.01$, and $c = 2602.61 \pm 0.12$ pm and a calculated density of 2.287 g·cm$^{-3}$ (Howie et al. 2009). Its pale-yellow crystals were grown by the slow evaporation of an ethanol solution of the title compound.

**[($C_5H_{12}N_2S$)$_2$Hg(CN)$_2$]·2H$_2$O or $C_{12}H_{28}N_6O_2S_2Hg$.** This compound crystallizes in the monoclinic structure with the lattice parameters $a = 1016.10 \pm 0.15$, $b = 797.55 \pm 0.08$, and $c = 1512.2 \pm 0.2$ pm and $\beta = 92.480° \pm 0.013°$ (Altaf et al. 2010a). It was prepared by mixing methanolic solution of tetramethylthiourea ($C_5H_{12}N_2S$) and $Hg(CN)_2$ (2:1 molar ratio), and the crystals were obtained by adding 5 mL $H_2O$ to the resulting solution.

**[Hg(SCN)($C_5N_4NCOO$)($C_5N_4NCOOH$)] or $C_{13}N_9N_3O_4SHg$.** To obtain this compound, a mixture of an ethanolic solution of $Hg(SCN)_2$ (0.95 mM in 75 mL) and pyridine-2-carboxylic acid (2.03 mM in 20 mL) was left to concentrate at room temperature (Popović et al. 1999a). When the volume was reduced to ca. 50 mL, white needles started to crystallize. Crystals were filtered off, washed with EtOH, and dried.

**[Hg($C_{12}H_{12}N_2O_2$)(SCN)$_2$] or $C_{14}H_{12}N_4O_2S_2Hg$.** This compound crystallizes in the monoclinic structure with the lattice parameters $a = 909.9 \pm 0.1$, $b = 1416.7 \pm 0.1$, and $c = 1319.2 \pm 0.2$ pm and $\beta = 101.035° \pm 0.009°$ (Esmhosseini et al. 2011). It was obtained by the following procedure. A solution of di(2-pyridyl)ketone (1.10 mM) in MeOH (20 mL) was added to a solution of $Hg(SCN)_2$ (1.10 mM) in MeOH (20 mL), and the resulting colorless solution was stirred for 20 min at 40°C. This solution was left to evaporate slowly at room temperature. After 1 week, colorless block-shaped crystals of the title compound were isolated.

**[Hg{PhN·C(OMe)S}$_2$] or C$_{16}$N$_{16}$N$_2$O$_2$S$_2$Hg.** This compound melts at 97°C–98°C (Davies and Peddle 1965) and crystallizes in the monoclinic structure with the lattice parameters $a = 1412$, $b = 619$, and $c = 2110$ pm and $\beta = 106°42'$ and a calculated density of 2.00 g·cm$^{-3}$ (McEwen and Sim 1967a). An interaction of mercury dimethoxide Hg(OMe)$_2$ with phenyl isothiocyanate PhSCN gives Hg[PhN·C(OME)S]$_2$ as white crystals (Davies and Peddle 1965).

**[Hg(C$_{16}$H$_{16}$N$_6$S$_2$)(NO$_3$)$_2$] or C$_{16}$H$_{16}$N$_8$O$_6$S$_2$Hg.** This compound melts at 180°C with decomposition (López-Torres et al. 2001). It was prepared as follows. To a suspension of benzylbisthiosemicarbazone (1.11 mM) in EtOH (30 mM), a suspension of Hg(NO$_3$)$_2$·H$_2$O (1.11 mM) was added in the same solvent (40 mM). The mixture was stirred for 12 h at room temperature. The yellow solid that formed was filtered off and then partially dissolved in EtOH (filtering again to remove the insoluble fraction) and concentrated in vacuo until a yellow solid was obtained.

**[Hg(C$_7$H$_6$N$_2$S)$_2$(SCN)$_2$]·EtOH or C$_{18}$H$_{18}$N$_6$OS$_4$Hg.** This compound crystallizes in the triclinic structure with the lattice parameters $a = 780.37 \pm 0.14$, $b = 1028.5 \pm 0.3$, and $c = 1520.2 \pm 0.3$ pm and $\alpha = 90.02° \pm 0.2°$, $\beta = 98.18° \pm 0.02°$, and $\gamma = 96.96° \pm 0.02°$ and a calculated density of 1.838 g·cm$^{-3}$ (Popović et al. 2002a). To prepare it, a solution of benzo-1,3-imidazole-2-thione (2.80 mM) in 20 mL of 96% EtOH was added dropwise to a solution of Hg(SCN)$_2$ (1.26 mM) in 120 mL of 96% EtOH. The reaction mixture was left in the dark due to the photosensitivity of the compound. After a few weeks, the pale-yellow crystals were filtered off, washed with cold EtOH, and dried in air.

**[(PhHg)$_2$(C$_3$H$_6$OS$_2$)]·(Py) or C$_{20}$H$_{21}$NOS$_2$Hg$_2$.** This compound melts at 74°C–75°C (Canty and Kishimoto 1977). Its colorless crystals were obtained by dissolving (PhHg)$_2$(C$_3$H$_6$OS$_2$) in Py during several days. These were collected and washed with Py.

**[Hg(C$_{18}$H$_{36}$N$_2$O$_6$)][{Hg(SCN)$_3$}$_2$] or C$_{24}$H$_{36}$N$_8$O$_6$S$_6$Hg$_3$.** This compound crystallizes in the monoclinic structure with the lattice parameters $a = 2478.3 \pm 1.2$, $b = 1540.4 \pm 0.4$, and $c = 1049.4 \pm 0.3$ pm and $\beta = 108.77° \pm 0.03°$ and a calculated density of 2.32 g·cm$^{-3}$ (Pickardt et al. 1995a). The reaction of 2,2,2-cryptand with Hg(SCN)$_2$ in MeOH yields crystals of the title compound.

**[(C$_4$H$_{12}$N)Hg(C$_7$H$_5$OS)$_3$] or C$_{25}$H$_{27}$NO$_3$S$_3$Hg.** This compound crystallizes in the monoclinic structure with the lattice parameters $a = 1148.0 \pm 0.2$, $b = 1584.6 \pm 0.3$, and $c = 1514.0 \pm 0.3$ pm and $\beta = 91.50° \pm 0.03°$ and the experimental and calculated densities 1.67 ± 0.05 and 1.656 g·cm$^{-3}$, respectively (Vittal and Dean 1997). The synthesis was carried out under an Ar atmosphere. Thiobenzoic acid (17.85 mM) in 15 mL of MeOH was added to NaOMe prepared in situ by dissolving Na (16.06 mM) in 10 mL of MeOH. The resulting yellow solution was added by stirring to a solution containing HgCl$_2$ (5.35 mM) dissolved in 20 mL of MeOH producing a yellow precipitate. This turned into a white precipitate upon the addition of a solution containing Me$_4$NCl (5.35 mM) in 20 mL of MeOH. The mixture was stirred for 10 min, and then 10 mL of $H_2O$ was added, followed by 25 mL of MeCN. Warming the mixture to 40°C for 30 min gave a colorless clear solution, which was filtered hot and left to crystallize at 5°C. Colorless crystals were separated by decantation, washed with MeOH and diethyl ether, and dried in a stream of Ar. A second crop of crystals was obtained from the mixture of the mother liquor and washings. Single crystals were grown from a mixture of acetone and diethyl ether by diffusion.

**[Hg(CN)$_2$(C$_5$H$_{12}$N$_2$S)]$_4$·4H$_2$O or C$_{28}$H$_{56}$N$_{16}$O$_4$S$_4$Hg$_4$.** This compound crystallizes in the monoclinic structure with the lattice parameters $a = 1143.69 \pm 0.18$, $b = 805.71 \pm 0.09$, and $c = 1053.38 \pm 0.18$ pm and $\beta = 109.088° \pm 0.011°$ (Altaf et al. 2010a). To prepare it, tetramethylthiourea (C$_5$H$_{12}$N$_2$S) was dissolved in $H_2O$ keeping the C$_5$H$_{12}$N$_2$S/Hg(CN)$_2$ molar ratio of 2:1.

**[Hg$_2$(C$_{15}$H$_{19}$N$_4$O$_2$)$_2$(SCN)$_4$] or C$_{34}$H$_{38}$N$_{12}$O$_4$S$_4$Hg$_2$.** This compound crystallizes in the triclinic structure with the lattice parameters $a = 707.67 \pm 0.05$, $b = 1207.22 \pm 0.08$, and $c = 1268.14 \pm 0.08$ pm and $\alpha = 81.813° \pm 0.001°$, $\beta = 89.790° \pm 0.001°$, and $\gamma = 76.450° \pm 0.001°$ (Lou et al. 2012). It was synthesized by adding KSCN (1 mM) to 20 mL of an ethanol solution containing HgCl$_2$ (1 mM) and 4,4,5,5-tetramethyl-2-(1-methyl-1$H$-benzimidazol-2-yl)-4,5-dihydroimidazole-1-oxyl-3-oxide (C$_{15}$H$_{19}$N$_4$O$_2$) (1 mM). The mixture was stirred for 4 h at room temperature and then filtered off. The black filtrate was allowed to stand at room temperature. Black crystals were obtained after 2 weeks.

**[Hg(C$_{21}$H$_{16}$N$_3$O$_4$S$_2$)$_2$] or C$_{42}$H$_{32}$N$_6$O$_8$S$_4$Hg.** To obtain this compound, $N,N$-(iminodiethylene)bisphthalimide in acetonitrile was added to CS$_2$ in acetonitrile in equimolar ratio (Thirumaran et al. 1998). After an hour, fine yellow crystals separated out. Hg(C$_{21}$H$_{16}$N$_3$O$_4$S$_2$)$_2$ was prepared by mixing Hg salt and obtained earlier solution in a 1:2 ratio.

**[Bu$^n_4$N]$_2$[Hg(C$_5$O$_3$S$_2$)$_2$]·0.28H$_2$O or C$_{42}$H$_{72.56}$N$_2$O$_{6.28}$S$_4$Hg.** This compound crystallizes in the triclinic structure with the lattice parameters $a = 978.7 \pm 0.05$, $b = 1499.9 \pm 0.5$, and $c = 1734.2 \pm 0.5$ pm and $\alpha = 74.973° \pm 0.005°$, $\beta = 87.124° \pm 0.005°$, and $\gamma = 87.609° \pm 0.005°$ (Zhang et al. 2012). It was prepared by the following procedure. To a solution containing K$_2$(C$_5$O$_3$S$_2$) (0.8 mM) in $H_2O$ (20 mL), a solution containing Hg(OAc)$_2$ (0.4 mM) in $H_2O$ (5mL) was added. The resulting mixture was kept at 70°C for 1 h, and then filtered into a solution of Bu$_4$NBr (0.95 mM) in EtOH (5 mL). The solid product was collected by suction filtration, washed with water, and dried in air. Red crystals were obtained by recrystallization from acetone.

**(Bu$_4$N)[HgPh$_4$(C$_4$S$_3$)$_2$COOEt] or C$_{51}$H$_{61}$NO$_2$S$_6$Hg.** This compound melts at 143°C–145°C and crystallizes in the monoclinic structure with the lattice parameters $a = 1864.6 \pm 0.3$, $b = 1504.3 \pm 0.3$, and $c = 2015.1 \pm 0.3$ pm and $\beta = 116.090° \pm 0.010°$ and a calculated density of 1.456 g·cm$^{-3}$ (Kim et al. 2001). It was prepared by the next procedure. To a 30 mL of ethanol suspension of thieno[2,3-$d$]-1,3-dithiol-2-one (C$_5$H$_2$OS$_3$) (1.2 mM) KOH (2.3 mM) dissolved in 20 mL of ethanol under an N$_2$ atmosphere was added, and the mixture was stirred for 30 min at room temperature. After treating with, HgCl$_2$ (0.5 mM) and $n$-Bu$_4$NBr (1.0 mM) dissolved separately in 20 mL of ethanol, the yellow precipitate was filtered, washed with $H_2O$, EtOH, and diethyl ether, and dried under vacuum. The needle-shaped yellow crystals of the title compound were obtained by recrystallization from acetone/ethyl acetate.

**[Pr$^n_4$N][Hg(S-2,4,6-Pr$^i_3$C$_6$H$_2$)$_3$]·EtOH or C$_{58}$H$_{99}$NOS$_3$Hg.** This compound crystallizes in the monoclinic structure with the lattice parameters $a = 1429.8 \pm 1.0$, $b = 1871.0 \pm 0.5$, and $c = 2454.3 \pm 0.8$ pm and $\beta = 106.14° \pm 0.04°$ (Gruff and Koch 1990). It was prepared by recrystallization of [Pr$^n_4$N][Hg(S-2,4,6-Pr$^i_3$C$_6$H$_2$)$_3$] in EtOH.

## 7.25   Hg–H–C–N–O–F–S

**[{CF$_3$SHg(NO$_3$)}·0.5H$_2$O]$_2$ or C$_2$H$_2$N$_2$O$_7$F$_6$S$_2$Hg$_2$.** This multinary compound is formed in this system. Upon cooling, a solution of Hg(NO$_3$)$_2$·H$_2$O (2.57 g) and bis(trifluoromethylthio)mercury (3.02 g) in 10% HNO$_3$ (10–15 mL) yielded crystals of the title compound, which were filtered off, washed with water and benzene, and dried in vacuo over KOH (Downs et al. 1962). Anhydrous trifluoromethylmercuric nitrate could not be obtained by vacuum sublimation at temperatures up to 140°C.

## 7.26   Hg–H–C–N–O–Cl–S

Some compounds are formed in this system.

**[Hg{SC(NH$_2$)$_2$}Cl]Cl·0.5H$_2$O or CH$_5$N$_2$O$_{0.5}$Cl$_2$SHg.** This compound melts at 234°C with decomposition (Aucken and Drago 1960). To obtain it, equimolar proportions of HgCl$_2$ (27.15 g) and thiourea (7.61 g) in aqueous solution were heated together on a water bath for a short time. The cooled white precipitate of needle-shaped crystals may be washed with small quantities of cold H$_2$O and dried on filter paper or with alcohol followed by drying in vacuo.

**[Hg{SCH$_2$CH(NH$_3$)COOH}Cl$_2$] or C$_3$H$_7$NO$_2$Cl$_2$SHg.** This compound crystallizes in the orthorhombic structure with the lattice parameters $a = 1469.9 \pm 0.3$, $b = 801.7 \pm 0.2$, and $c = 702.5 \pm 0.1$ pm and the experimental and calculated densities 3.16 and 3.150 g·cm$^{-3}$, respectively (Taylor and Carty 1977). The reaction of HgCl$_2$ (1.0 g) in EtOH (20 mL) with L-cysteine (C$_3$H$_7$NO$_2$S) (0.45 g) in H$_2$O (20 mL) gave a white precipitate. Dissolution in a minimum quantity of HCl and crystallization afforded clear prisms of the title compound. From very dilute solutions, this compound can also be obtained, without the addition of acid, by crystallization of the mother liquor after the removal of the initial precipitate.

**[HgCl$_2$·(MeNHCSOEt)] or C$_4$H$_9$NOCl$_2$SHg.** This compound exists in two forms *A* and *B* (Faraglia et al. 1981). **Form *A*** (melting temperature 94°C–95°C). To obtain it, a suspension of HgCl$_2$ (1.3 mM) in a benzene solution (10 mL) of MeNHCSOEt (1.3 mM) was heated up to 50°C. The undissolved residue was separated; by cooling, the filtrate gave the white flocculent solid. By adding *n*-pentane to the residual solutions, small fractions of Hg(MeNHCSOEt)$_2$Cl$_2$ were obtained. **Form *B*** (melting temperature 117°C) precipitated by the addition of MeNHCSOEt (1.5 mM) to an acetone solution (2 mL) of HgCl$_2$ (1.5 mL). It was filtered and washed with an acetone/*n*-pentane solution mixture. By adding *n*-pentane to the filtrate, a mixture of the two forms, richer in $A$, was obtained. Either in solution or more slowly in the solid state, form $A$ changes to form $B$.

**[Hg(C$_5$H$_7$N$_2$OS)Cl] or C$_5$H$_7$N$_2$OClSHg**. To obtain this compound, an equivalent of 0.1 M NaOH was added to Hg(C$_5$H$_8$N$_2$OS)Cl$_2$ dissolved in the least amount of EtOH (C$_5$H$_8$N$_2$OS—5,5-dimethylimidazolidine-2-thione-4-one) (Bellon et al. 1988). The title compound precipitated immediately as a white powder. The same compound was also obtained by reacting Hg(C$_5$H$_8$N$_2$OS)Cl$_2$ in absolute EtOH with an equivalent of both ethoxy anion and alcoholic KOH.

**[Hg(C$_5$H$_8$N$_2$OS)Cl$_2$] or C$_5$H$_8$N$_2$OCl$_2$SHg**. This compound crystallizes in the monoclinic structure with the lattice parameters $a = 999.8 \pm 0.2$, $b = 1376.9 \pm 0.3$, and $c = 779.8 \pm 0.1$ pm and $\beta = 100.61° \pm 0.04°$ (Bellon et al. 1988). To prepare it, a suspension of Hg(C$_5$H$_7$N$_2$OS)Cl in EtOH was treated with an equivalent amount of HCl with stirring. The complex dissolved almost immediately, and after some time, the title compound was isolated by slow evaporation of the solvent.

**[Hg(C$_5$H$_8$N$_2$OS)(ClO$_4$)$_2$]·$n$H$_2$O or C$_5$H$_8$N$_2$O$_9$Cl$_2$SHg·$n$H$_2$O**. This compound was obtained by the reaction of Hg(ClO$_4$)$_2$ and 5,5-dimethylimidazolidine-2-thione-4-one (C$_5$H$_8$N$_2$OS) (1:1 molar ratio) in H$_2$O (Cristiani et al. 1990). C$_5$H$_8$N$_2$OS dissolved in the least amount of H$_2$O was added to a solution of Hg(ClO$_4$)$_2$. No precipitation was observed, and hygroscopic title compound was obtained by evaporating the solvent. Every analyzed sample showed a variable amount of H$_2$O. However, if the aforementioned solution is left at room temperature for a few days, **[Hg(C$_5$H$_7$N$_2$OS) (ClO$_4$)]·2H$_2$O (C$_5$H$_{11}$N$_2$O$_7$ClSHg)** is formed.

The reaction in diethyl ether was carried out by dissolving Hg(ClO$_4$)$_2$ and C$_5$H$_8$N$_2$OS separately in the least amount of solvent (Cristiani et al. 1990). C$_5$H$_8$N$_2$OS is then added to the perchlorate dropwise in a 1:1 molar ratio. The title compound was precipitated immediately or after some time by evaporation of the solvent.

**[Hg(C$_5$H$_{11}$NOS)Cl$_2$] or C$_5$H$_{11}$NOCl$_2$SHg**. This compound crystallizes in the monoclinic structure with the lattice parameters $a = 1073 \pm 1$, $b = 1341 \pm 1$, and $c = 771 \pm 1$ pm and $\beta = 96.90° \pm 0.05°$ and a calculated density of 2.44 g·cm$^{-3}$ (Faraglia et al. 1982). HgCl$_2$ dissolved easily in a benzene solution of $N,N$-dimethyl $O$-ethylthiocarbamate (ca. 1:3 molar ratio). By the addition of $n$-hexane, the product separated as oil, which crystallized slowly as white solid of the title compound by washing with $n$-hexane.

**[Hg(C$_5$H$_7$N$_2$OS)(ClO$_4$)$_2$]·8H$_2$O or C$_5$H$_{23}$N$_2$O$_{17}$Cl$_2$SHg**. This compound was obtained by reacting Hg(ClO$_4$)$_2$ and 5,5-dimethylimidazolidine-2-thione-4-one (C$_5$H$_8$N$_2$OS) (1:1 molar ratio) in H$_2$O (Cristiani et al. 1990). It was precipitated after 1 day by evaporation of the solvent.

**[Hg(C$_3$H$_5$NOS)$_2$Cl$_2$] or C$_6$H$_{10}$N$_2$O$_2$Cl$_2$S$_2$Hg**. This compound melts at 295°C with decomposition (Devillanova and Verani 1978). It was obtained by boiling an absolute ethanol solution of HgCl$_2$ under reflux with 1,3-oxazolidine-2-thione in a 1:2 molar ratio.

**[Hg(C$_3$H$_5$NOS)$_2$Cl$_2$] or C$_6$H$_{10}$N$_2$O$_2$Cl$_2$S$_2$Hg**. This compound decomposes at 187°C (Preti et al. 1974). It was obtained by the interaction of HgCl$_2$ and thiazolidine-2-one by refluxing EtOH for 12 h. The title compound was separated out spontaneously during the reaction.

**[Hg(C₆H₁₂NS)](ClO₄)₂ or C₆H₁₂NO₈Cl₂SHg**. To prepare this compound, a methanolic solution of 1-methyl-4-mercaptopiperidine (7.5 mM in 50 mL) was slowly added to a stirred clean methanolic solution of Hg(ClO₄)₂·6H₂O (6.7 mM in 200 mL) (Barrera et al. 1982). The title compound precipitated slowly as the ligand was being added.

**[Hg(EtOCSNH₂)₂Cl₂] or C₆H₁₄N₂O₂Cl₂S₂Hg**. This compound melts at 105°C–107°C and crystallizes in the monoclinic structure with the lattice parameters $a = 1099.4 \pm 1.2$, $b = 1243.7 \pm 0.5$, and $c = 1056.8 \pm 0.6$ pm and $\beta = 101.59° \pm 0.07°$ and the experimental and calculated densities 2.24 and 2.26 g·cm⁻³, respectively (Bandoli et al. 1975). It was obtained as follows. Ethyl thiocarbamate (EtOCSNH₂) (20 mM) was dissolved in EtOH (5 mL). HgCl₂ (10 mM) in EtOH (15 mL) was gradually added with stirring. After 10 h, the solution was filtered to remove a slight quantity of an initial yellowish-white precipitate. The crystals, which appeared after a few days, were washed with EtOH, recrystallized from acetone/light petroleum (1:1 volume ratio), and dried in vacuo over P₂O₅.

**[Hg{SCH₂CH(NH₃)COO}}SCH₂CH(NH₃)COOH}Cl]·2.5H₂O or C₆H₁₈N₂O₆.₅ClS₂Hg**. This compound crystallizes in the monoclinic structure with the lattice parameters $a = 2418.1 \pm 0.8$, $b = 509.3 \pm 0.3$, and $c = 1200.6 \pm 0.6$ pm and $\beta = 118.83° \pm 0.03°$ and the experimental and calculated densities 2.50 and 2.494 g·cm⁻³, respectively (Taylor and Carty 1977). It can be synthesized from HgCl₂ (1.0 g) and L-cysteine (C₃H₇NO₂S) (0.9 g) in a mixture (EtOH/H₂) (1:1 volume ratio) as needlelike crystals.

**[ClHgC₇H₈NO₂S] or C₇H₈NO₂ClSHg, 3-(5-chloromercurio-2-thienyl)alanine**. It was prepared by slow addition of an aqueous solution (ca. 50 mL) of HgCl₂ (1.84 mM) and NaOAc·3H₂O (3.68 mM) to an aqueous solution (ca. 30 mL) of 3-(2-thienyl)alanine (1.84 mM) (Popović et al. 2000b). The reaction mixture was left overnight, and then the precipitated product was filtered off, washed with water, and dried in vacuo.

**[Hg(C₇H₁₄NS)](ClO₄)₂·H₂O or C₇H₁₆NO₉Cl₂SHg**. While a methanolic solution of 1-methyl-3(mercaptomethyl)piperidine (7.5 mM in 50 mL) was slowly added to a stirred clean methanolic solution of Hg(ClO₄)₂·6H₂O (6.7 mM in 200 mL), a yellow oil formed (Barrera et al. 1982). It was left in the refrigerator overnight and then washed with absolute EtOH several times. The oil became a white solid by staying under vacuum.

**[Hg(C₈H₁₁N₂O₃S)(Cl)]ₙ or C₈H₁₁N₂O₃ClSHg**. This compound crystallizes in the monoclinic structure with the lattice parameters $a = 692.32 \pm 0.07$, $b = 2408.9 \pm 0.2$, and $c = 750.97 \pm 0.07$ pm and $\beta = 110.609° \pm 0.001°$ (Du and Li 2008). To obtain it, 2-(2-pyridylmethylamino)ethanesulfonic acid (1 mM) was dissolved in 15 mL of H₂O. To this solution, HgCl₂ (1 mM) was added, and the resulting mixture was stirred at 60°C for 3 h and then cooled to room temperature. After filtration, the filtrate was left to stand at room temperature. Colorless claviform crystals were obtained about 1 week later.

**[HgCl₂(C₄H₇NOS)₂]** or **C₈H₁₄N₂O₂Cl₂S₂Hg**. This compound melts at 118°C (De Filippo et al. 1971). To obtain it, $HgCl_2$ was dissolved in an excess of molten thiomorpholine-3-one ($C_4H_7NOS$) (at ca. 105°C; $Hg^{2+}/C_4H_7NOS$ ca. 1:2 molar ratio), and the crude product was purified by washing with ethyl ether and recrystallizing from EtOH. The title compound could also be prepared if to a solution of $HgCl_2$ in acetone/EtOH mixture different amounts of $C_4H_7NOS$ were added for $Hg^{2+}/C_4H_7NOS$ molar ratios of 1:1, 1:2, 1:3, and 1:4, and the mixture was heated under reflux for 4 h. [HgCl₂(C₄H₇NOS)₂] was separated out spontaneously during the reaction.

**[Hg(OHC₄H₇N₂S)₂Cl₂]** or **C₈H₁₆N₄O₂Cl₂S₂Hg**. This compound melts at 164°C (Isab and Perzanowski 1996). It was prepared by a dropwise addition of ethanol solution of $HgCl_2$ to a stoichiometric amount of 5(OH)-1,3-diazinane-2-thione in EtOH.

**[Hg(C₈H₁₆NS)](ClO₄)₂·0.5H₂O** or **C₈H₁₇NO₈.₅Cl₂SHg**. While slowly adding an aqueous solution of 1-methyl-2(2-mercaptoethyl)piperidine in 0.1 M $HClO_4$ to a stirred red solution of $Hg(ClO_4)_2·6H_2O$ in the same medium (1:1 molar ratio), an oil formed, which became a white solid through continuous stirring for several hours (Barrera et al. 1982). This solid was collected, washed with 0.1 M $HClO_4$ solution and $H_2O$, and dried under vacuum over $P_2O_5$.

**[Hg(MeNHCSOEt)₂Cl₂]** or **C₈H₁₈N₂O₂Cl₂S₂Hg**. This compound melts at 97°C–98°C (Faraglia et al. 1981). It was obtained by dissolving $HgCl_2$ in a benzene solution of MeNHCSOEt (molar ratio from 1:3 to 1:4). The white title compound was isolated by the addition of $n$-pentane to this solution.

**[Hg(C₅H₈N₂OS)₂(ClO₄)]·3H₂O** or **C₁₀H₂₂N₄O₉ClS₂Hg**. The experiments in aqueous $HClO_4$ were performed by dissolving 1.78 g of $Hg(ClO_4)_2·6H_2O$ (3.5 mM) in 30 g of 25% $HClO_4$ and by adding the proper amount of $C_5H_8N_2OS$ (Cristiani et al. 1990). The title compound was precipitated after 1 day by evaporation of the solvent.

[Hg(C₅H₈N₂OS)₂(ClO₄)]·3H₂O could also be obtained by reacting $Hg(ClO_4)_2$ and $C_5H_8N_2OS$ [1:(3–5) molar ratios] (Cristiani et al. 1990). Its crystals appeared immediately.

**[Hg(C₅H₈N₂OS)₂(ClO₄)₂]·3H₂O** or **C₁₀H₂₂N₄O₁₃Cl₂S₂Hg**. This compound crystallizes in the monoclinic structure with the lattice parameters $a = 750.5 ± 0.3$, $b = 2437.2 ± 0.6$, and $c = 1311.0 ± 0.6$ pm and $\beta = 99.33° ± 0.03°$ and a calculated density of 2.082 g·cm⁻³ (Cristiani et al. 1990) The experiments in aqueous $HClO_4$ were performed by dissolving 1.78 g of $Hg(ClO_4)_2·6H_2O$ (3.5 mM) in 30 g of $HClO_4$ at various concentrations (5%, 10%, 15%, and 20%) and by adding the proper amount of $C_5H_8N_2OS$. The title compound was precipitated after a month (5% and 10% $HClO_4$) or after a day (15% and 20% $HClO_4$) by evaporation of the solvent. The crystal used for the structural analysis was obtained from these solutions.

[Hg(C₅H₈N₂OS)₂(ClO₄)₂]·3H₂O could also be obtained by reacting $Hg(ClO_4)_2$ and $C_5H_8N_2OS$ (1:2 molar ratio) (Cristiani et al. 1990). Its crystals, in the presence of some Hg, appeared when the solution was left to stand for 1 month.

**[HgCl₂{PhCOCH₂(Me₂NCS₂)}]** or **C₁₁H₁₃NOCl₂S₂Hg**. This compound melts at 134°C–138°C (Brinkhoff and Dautzenberg 1972). It was prepared by the addition of $HgCl_2$ in EtOH to a solution of $PhCOCH_2(Me_2NCS_2)$ (1:1 molar ratio). The title compound was recrystallized from EtOH.

**$[Hg(C_6H_{12}NS)_2](ClO_4)_2$ or $C_{12}H_{24}N_2O_8Cl_2S_2Hg$.** This compound crystallizes in the orthorhombic structure with the lattice parameters $a = 1316.1 \pm 0.3$, $b = 658.9 \pm 0.2$, and $c = 2474.0 \pm 0.4$ pm and the experimental and calculated densities 2.03 and 2.05 g·cm$^{-3}$, respectively (Barrera et al. 1982). A white crystalline solid was formed while adding slowly 15 mM of 1-methyl-4-mercaptopiperidine dissolved in 75 mL of 0.1 M $HClO_4$ aqueous solution to 75 mL of the same solvent where $Hg(ClO_4)_2 \cdot 6H_2O$ (8 mM) had been dissolved. Crystals suitable for x-ray diffraction (XRD) were grown in 0.1 M $HClO_4$.

**$[HgCl_2\{PhCOCH_2(Et_2NCS_2)\}]$ or $C_{13}H_{17}NOCl_2S_2Hg$.** This compound melts at 120°C–122°C (Brinkhoff and Dautzenberg 1972). It was prepared by the addition of $HgCl_2$ in EtOH to a solution of $PhCOCH_2(Et_2NCS_2)$ (1:1 molar ratio). The title compound was recrystallized from EtOH.

**$[Hg(C_9H_{12}N_4S)Cl_2] \cdot (2DMSO)$ or $C_{13}H_{24}N_4O_2Cl_2S_3Hg$.** This compound crystallizes in the monoclinic structure with the lattice parameters $a = 860.5 \pm 0.1$, $b = 1010.9 \pm 0.1$, and $c = 1337.5 \pm 0.1$ pm and $\alpha = 97.07° \pm 0.01°$, $\beta = 104.16° \pm 0.01°$, and $\gamma = 100.61° \pm 0.01°$ at $223 \pm 2$ K and a calculated density of 1.935 g·cm$^{-3}$ (Bermejo et al. 1999). Its single crystals were obtained by redissolving $[Hg(C_9H_{12}N_4S)Cl_2]$ in DMSO and slowly evaporating the solvent.

**$[Hg(C_7H_{14}NS)_2](ClO_4)_2$ or $C_{14}H_{28}N_2O_8Cl_2S_2Hg$.** An oil formed when $NaClO_4$ concentrated solution was added to 60 mL of a solution 0.1 M $HClO_4$ containing 1.58 mM of $Hg(ClO_4)_2 \cdot 6H_2O$ and 3.27 mM of 1-methyl-3(mercaptomethyl)piperidine (Barrera et al. 1982). This oil was placed in the refrigerator where it turned to a white solid of the title compound.

**$[Hg(C_5H_8N_2OS)_3(ClO_4)_2]$ or $C_{15}H_{24}N_6O_{11}Cl_2S_3Hg$.** The reaction in diethyl ether was carried out by dissolving $Hg(ClO_4)_2$ and $C_5H_8N_2OS$ separately in the least amount of solvent (Cristiani et al. 1990). $C_5H_8N_2OS$ is then added to the perchlorate dropwise in 1:2 or 1:3 molar ratios. The title compound was precipitated immediately or after some time by evaporation of the solvent.

**$[Hg(C_{13}H_{19}N_5S)Cl_2] \cdot DMSO$ or $C_{15}H_{25}N_5OCl_2S_2Hg$.** This compound crystallizes in the triclinic structure with the lattice parameters $a = 898.3 \pm 0.3$, $b = 1027.5 \pm 0.3$, and $c = 1213.1 \pm 0.4$ pm and $\alpha = 77.258° \pm 0.005°$, $\beta = 78.944° \pm 0.006°$, and $\gamma = 85.141° \pm 0.005°$ and a calculated density of 1.945 g·cm$^{-3}$ (Bermejo et al. 2004). To obtain it, a solution of $HgCl_2$ (0.90 mM) in EtOH (30 mL) was mixed with a solution of 2-pyridineformamide 3-hexamethyleneiminylthiosemicarbazone ($C_{13}H_{19}N_5S$) (0.90 mM) in EtOH (30 mL). The resulting solution was stirred at room temperature for 7 days and the pale-yellow solid was filtered off. Recrystallization from DMSO gave yellow crystals of the title compound.

**$[Hg(C_8H_{16}NS)_2](ClO_4)_2$ or $C_{16}H_{32}N_2O_8Cl_2S_2Hg$.** While slowly adding an aqueous solution of 1-methyl-2(2-mercaptoethyl)piperidine in 0.1 M $HClO_4$ to a stirred red solution of $Hg(ClO_4)_2 \cdot 6H_2O$ in the same medium (2:1 mole ratio), an oil is formed (Barrera et al. 1982). This oil was treated with MeOH several times and finally dried under vacuum over $P_2O_5$. The white solid obtained is highly hygroscopic.

**$[Hg(PySH)_4](ClO_4)_2$ or $C_{20}H_{20}N_4O_8Cl_2S_4Hg$.** This compound crystallizes in the monoclinic structure with the lattice parameters $a = 1187.41 \pm 0.07$, $b = 1292.26 \pm 0.08$, and $c = 1858.47 \pm 0.11$ pm and $\beta = 100.590° \pm 0.001°$ and a calculated density of 2.000 $g \cdot cm^{-3}$ (Anjali et al. 2003). It was prepared by the following procedure. To a solution of 4-pyridine thiol (3.70 mM) in 4 mL of MeOH $Hg(ClO_4)_2 \cdot 6H_2O$ (1.01 mM) in 2 mL of MeOH was added. The mixture was stirred for 15 min and then left for slow evaporation at room temperature to get bright-yellow needlelike crystals that were separated by filtration, washed with $Et_2O$, and dried in vacuo.

**$[HgCl_2(C_{20}H_{24}N_2O_4S_2)]$ or $C_{20}H_{24}N_2O_4Cl_2S_2Hg$.** This compound crystallizes in the monoclinic structure with the lattice parameters $a = 1538.7 \pm 0.5$, $b = 2234.9 \pm 0.5$, and $c = 1457.9 \pm 0.5$ pm and $\beta = 95.337° \pm 0.005°$ and a calculated density of 1.842 $g \cdot cm^{-3}$ (Kubo et al. 2000). Its single crystals were obtained by crystallization of an equimolar mixture of $C_{20}H_{24}N_2O_4S_2$ and $HgCl_2$ from $CH_3CN$.

**$C_{20}H_{24}N_4O_4Cl_4S_2Hg_2$, bis{(μ-chloro)-chloro-[N-benzoyl-N'-(2-hydroxyethyl) thiourea]mercury(II)}.** This compound crystallizes in the triclinic structure with the lattice parameters $a = 734.97$, $b = 803.56$, and $c = 1185.56$ pm and $\alpha = 77.812°$, $\beta = 81.932°$, and $\gamma = 83.145°$ pm and a calculated density of 2.440 $g \cdot cm^{-3}$ (Zhang et al. 2005). To obtain it, an EtOH (30 mL) solution of N-benzoyl-N(2-hydroxyethyl)thiourea (2 mM) $HgCl_2$ (2 mM) was added under constant stirring. After stirring the reaction mixture at room temperature for 2 h, the mixture was filtered to obtain white solid, which was dried in air and recrystallized from EtOH. Single crystals of the title compound were obtained after 1 week by slow evaporation of the ethanol solution.

**$[Hg(C_5H_8N_2OS)_4(ClO_4)_2]$ or $C_{20}H_{32}N_8O_{12}Cl_2S_4Hg$.** The reaction in diethyl ether was carried out by dissolving $Hg(ClO_4)_2$ and $C_5H_8N_2OS$ separately in the least amount of solvent (Cristiani et al. 1990). $C_5H_8N_2OS$ is then added to the perchlorate dropwise in a 1:4 molar ratio. The title compound was precipitated immediately or after some time by evaporation of the solvent.

**$[Hg(C_5H_8N_2OS)_5(ClO_4)_2]$ or $C_{25}H_{40}N_{10}O_{13}Cl_2S_5Hg$.** The reaction in diethyl ether was carried out by dissolving $Hg(ClO_4)_2$ and $C_5H_8N_2OS$ separately in the least amount of solvent (Cristiani et al. 1990). $C_5H_8N_2OS$ is then added to the perchlorate dropwise in 1:5–10 molar ratios. The title compound was precipitated immediately or after some time by evaporation of the solvent.

**$[PhC(OMe)NC(S)NEt_2HgCl_2]_2$ or $C_{26}H_{36}N_4O_2Cl_4S_2Hg_2$.** This compound melts at 177°C and crystallizes in the triclinic structure with the lattice parameters $a = 933.0 \pm 0.1$, $b = 953.0 \pm 0.1$, and $c = 1146 \pm 1$ pm and $\alpha = 71.99° \pm 0.01°$, $\beta = 88.86° \pm 0.01$, and $\gamma = 65.01° \pm 0.1°$ pm and a calculated density of 1.989 $g \cdot cm^{-3}$ (Leßmann et al. 2000). To obtain it, 0.2 mM of N-(diethylaminothiocarbonyl)benzimido-O-methyl ester was dissolved in 10 mL of $CHCl_2$. To this solution, 0.2 mM of $HgCl_2$ dissolved in 15 mL of MeOH was added. The title compound that was crystallized overnight as colorless prisms was filtered, dried, and washed with MeOH.

[{Hg(C$_{14}$H$_{24}$N$_2$S$_2$)Cl$_2$}$_2$]·EtOH or C$_{30}$H$_{54}$N$_4$OCl$_4$S$_4$Hg$_2$. This compound crystallizes in the monoclinic structure with the lattice parameters $a = 856.3 \pm 0.1$, $b = 2141.4 \pm 0.3$, and $c = 1164.9 \pm 0.2$ pm and $\beta = 103.58° \pm 0.02°$ and a calculated density of 1.85 g·cm$^{-3}$ (Baggio et al. 1995). It was synthesized from *N,N'*-dicyclohexyldithiooxamide (2.5 mM) and HgCl$_2$ (1.25 mM) in acetone/EtOH (1:1 volume ratio) solution (100 mL) at room temperature. The reaction mixture was stirred for 24 h at 25°C and then filtered. The filtrate was kept at 10°C for 3 days. Light-green-yellow prismatic crystals of the title compound were collected and dried under vacuum. Its crystals were obtained by slow diffusion of EtOH in a diluted solution of [{Hg(C$_{14}$H$_{24}$N$_2$S$_2$)Cl$_2$}$_2$]·EtOH. These crystals are stable when kept in the dark in a dry atmosphere at room temperature.

[Hg$_5$(C$_5$H$_{10}$NS$_2$)$_8$](ClO$_4$)$_2$ or C$_{40}$H$_{80}$N$_8$O$_8$Cl$_2$S$_{16}$Hg$_5$. This compound crystallizes in the triclinic structure with the lattice parameters $a = 1276.6 \pm 0.2$, $b = 1405.0 \pm 0.2$, and $c = 1082.0 \pm 0.2$ pm and $\alpha = 101.97° \pm 0.02°$, $\beta = 105.86° \pm 0.02$, and $\gamma = 84.03° \pm 0.1°$ pm and the experimental and calculated densities 2.16 and 2.17 g·cm$^{-3}$, respectively (Bond et al. 1987). Oxidative electrolysis of Hg(C$_5$H$_{10}$NOS$_2$)$_2$ in CH$_2$Cl$_2$ (0.1 M Bu$^n_4$NClO$_4$) was carried out at a Hg pool electrode. With the passage of the expected charge (10°C), a green solution was produced. Slow evaporation of the solution to dryness gave yellow crystals of the title compound and a white solid [Bu$^n_4$NClO$_4$ and Hg(ClO$_4$)$_2$]. The residue was taken up in EtOH, in which the former is only sparingly soluble, and filtered to give yellow crystals of [Hg$_5$(C$_5$H$_{10}$NOS$_2$)$_8$](ClO$_4$)$_2$. Oxidative electrolysis of Hg(C$_5$H$_{10}$NOS$_2$)$_2$ in CH$_2$Cl$_2$ at a Pt electrode also gave a green solution, which upon evaporation produced yellow crystals of this compound.

Addition of solid Hg$_2$(ClO$_4$)$_2$·2H$_2$O to a dichloromethane solution of Hg(C$_5$H$_{10}$NOS$_2$)$_2$ (2.5 mM) (1:1 molar ratio) with stirring for 1–2 h gave a color change (yellow to yellow/green), and the appearance of elemental mercury in the reaction vessel was noted. The solution was filtered, reduced by boiling to about one-fifth of its original volume, and then cooled to 0°C, which yielded yellow crystals of the title compound (Bond et al. 1987).

## 7.27   Hg–H–C–N–O–Br–S

Some compounds are formed in this system.

[Hg{SCH$_2$CH(NH$_3$)COOH}Br$_2$] or C$_3$H$_7$NO$_2$Br$_2$SHg. The reaction of HgBr$_2$ (1.0 g) in EtOH (20 mL) with ʟ-cysteine (C$_3$H$_7$NO$_2$S) (0.45 g) in H$_2$O (20 mL) gave a white precipitate (Taylor and Carty 1977). Dissolution in a minimum quantity of HBr and crystallization afforded clear prisms of the title compound. From very dilute solutions, this compound can also be obtained, without the addition of acid, by crystallization of the mother liquor after removal of the initial precipitate.

[HgBr$_2$·(MeNHCSOEt)] or C$_4$H$_9$NOBr$_2$SHg. This compound exists in two forms $A$ and $B$ (Faraglia et al. 1981). By adding *n*-pentane to an acetone solution of HgBr$_2$ and MeNHCSOEt (1:1 molar ratio), a small amount of an oily pink product precipitated, which was discarded. The progressive addition of several small fractions of *n*-pentane allowed separating first the form $A$ (melting temperature 69°C–71°C) and then the form $B$ (melting temperature 73°C–76°C).

**[Hg(C₅H₈N₂OS)Br₂]** or **C₅H₈N₂OBr₂SHg**. This compound crystallizes in the monoclinic structure with the lattice parameters $a = 1038.9 \pm 0.9$, $b = 1385.4 \pm 0.4$, and $c = 781.7 \pm 0.7$ pm and $\beta = 102.91° \pm 0.07°$ and a calculated density of 3.056 g·cm⁻³ (Bellon et al. 1988). To prepare it, a suspension of Hg(C₅H₇N₂OS)Br in EtOH was treated with an equivalent amount of HBr with stirring. The complex dissolved almost immediately, and after some time, the title compound was isolated by slow evaporation of the solvent.

**[Hg(C₅H₁₁NOS)Br₂]** or **C₅H₁₁NOBr₂SHg**. HgBr₂ dissolved easily in a benzene solution of N,N-dimethyl O-ethylthiocarbamate (ca. 1:3 molar ratio) (Faraglia et al. 1982). By the addition of n-hexane, the product separated as oil, which crystallized slowly as white solid of the title compound by washing with n-hexane.

**[Hg(C₃H₅NOS)₂Br₂]** or **C₆H₁₀N₂O₂Br₂S₂Hg**. This compound melts at 103°C with decomposition (Devillanova and Verani 1978). It was obtained by boiling an absolute ethanol solution of HgBr₂ under reflux with 1,3-oxazolidine-2-thione in a 1:2 molar ratio.

**[Hg(C₃H₅NOS)₂Br₂]** or **C₆H₁₀N₂O₂Br₂S₂Hg**. This compound decomposes at 168°C–170°C (Preti et al. 1974). It was obtained by the interaction of HgBr₂ and thiazolidine-2-one in refluxing EtOH for 12 h. The title compound was separated out spontaneously during the reaction.

**[HgBr₂(C₄H₇NOS)₂]** or **C₈H₁₄N₂O₂Br₂S₂Hg**. This compound melts at 114°C (De Filippo et al. 1971). To obtain it, HgBr₂ was dissolved in an excess of molten thiomorpholine-3-one (C₄H₇NOS) (at ca. 105°C; Hg²⁺/C₄H₇NOS ca. 1:2 molar ratio), and the crude product was purified by washing with ethyl ether and recrystallizing from EtOH. The title compound could also be prepared if to a solution of HgBr₂ in acetone/EtOH mixture different amounts of C₄H₇NOS were added for Hg²⁺/C₄H₇NOS molar ratios of 1:1, 1:2, 1:3, and 1:4, and the mixture was heated under reflux for 4 h. [HgBr₂(C₄H₇NOS)₂] was separated out spontaneously during the reaction.

**[Hg(OHC₄H₇N₂S)₂Br₂]** or **C₈H₁₆N₄O₂Br₂S₂Hg**. This compound melts at 152°C (Isab and Perzanowski 1996). It was prepared by a dropwise addition of ethanol solution of HgBr₂ to a stoichiometric amount of 5(OH)-1,3-diazinane-2-thione in EtOH.

**[Hg(MeNHCSOEt)₂Br₂]** or **C₈H₁₈N₂O₂Br₂S₂Hg**. This compound melts at 94°C–96°C (Faraglia et al. 1981). It was obtained by dissolving HgBr₂ in a benzene solution of MeNHCSOEt (molar ratio from 1:3 to 1:4). White title compound was isolated by the addition of n-pentane to this solution.

**[HgBr₂{PhCOCH₂(Me₂NCS₂)}]** or **C₁₁H₁₃NOBr₂S₂Hg**. This compound melts at 130°C–134°C (Brinkhoff and Dautzenberg 1972). It was prepared by the addition of HgBr₂ in EtOH to a solution of PhCOCH₂(Me₂NCS₂) (1:1 molar ratio). The title compound was recrystallized from EtOH.

**[Hg(C₉H₁₃N₅S)Br₂]·(DMSO)** or **C₁₁H₁₉N₅OBr₂SHg**. This compound crystallizes in the monoclinic structure with the lattice parameters $a = 1061.0 \pm 0.1$, $b = 1504.4 \pm 0.3$, and $c = 1233.9 \pm 0.3$ pm and $\beta = 102.87° \pm 0.02°$ and a calculated density of 2.289 g·cm⁻³ (Bermejo et al. 2003). Its single crystals suitable for XRD were obtained by slow evaporation from an EtOH/DMSO (1:1 volume ratio) mixture.

[Hg₂(C₆H₁₃NS)₂Br₄]·H₂O or C₁₂H₂₈N₂OBr₄S₂Hg₂. This compound crystallizes in the monoclinic structure with the lattice parameters $a = 1156$, $b = 1313$, and $c = 790$ pm and $\beta = 92.85°$ (Bayon et al. 1982). It was obtained in 10% methanolic solution by adding HgBr₂ to the stoichiometric amount of 1-methyl-4-mercaptopiperidine. The white crystalline product obtained was filtered off, washed with cold water, EtOH, and ethyl ether, and dried in vacuo.

[HgBr₂{PhCOCH₂(Et₂NCS₂)}] or C₁₃H₁₇NOBr₂S₂Hg. This compound melts at 150°C–152°C (Brinkhoff and Dautzenberg 1972). It was prepared by the addition of HgBr₂ in EtOH to a solution of PhCOCH₂(Et₂NCS₂) (1:1 molar ratio). The title compound was recrystallized from EtOH.

[Hg(C₁₂H₁₇N₅S)Br₂]·(DMSO) or C₁₄H₂₃N₅OBr₂SHg. This compound crystallizes in the triclinic structure with the lattice parameters $a = 909.94 \pm 0.15$, $b = 996.56 \pm 0.16$, and $c = 1270.7 \pm 0.2$ pm and $\alpha = 78.201° \pm 0.003°$, $\beta = 71.678° \pm 0.003°$, and $\gamma = 79.976° \pm 0.003°$ and a calculated density of 2.192 g·cm⁻³ (Castiñeiras et al. 2002). Single crystals of the title compound suitable for XRD were obtained by slow evaporation of a DMSO solution of [Cd(C₁₂H₁₇N₅S)Br₂] in air at room temperature.

[Hg₃(C₁₃H₁₉N₅S)(C₁₃H₁₈N₅S)Br₅]·H₂O or C₂₆H₃₉N₁₀OBr₅S₂Hg₃. This compound crystallizes in the triclinic structure with the lattice parameters $a = 932.2 \pm 0.2$, $b = 955.2 \pm 0.4$, and $c = 2371.6 \pm 0.7$ pm and $\alpha = 87.860° \pm 0.014°$, $\beta = 88.946° \pm 0.017°$, and $\gamma = 80.06° \pm 0.02°$ and a calculated density of 2.559 g·cm⁻³ (Bermejo et al. 2004). Its yellow crystals were obtained by recrystallization of [Hg(C₁₃H₁₉N₅S)Br₂] from DMSO/EtOH.

## 7.28  Hg–H–C–N–O–I–S

Some compounds are formed in this system.

[HgI₂·(MeNHCSOEt)] or C₄H₉NOI₂SHg. This compound exists in two forms A and B (Faraglia et al. 1981). It was obtained as follows. HgI₂ (1.3 mM) and MeNHCSOEt (1.3 mM) were dissolved in acetone. By adding *n*-pentane, an oily product was obtained, which give an orange solid (form A, melting temperature 65°C–67°C). On standing, the solution gave pale-green crystals of the form B (melting temperature 67°C–68°C). The orange form turns to green if heated at about 50°C.

[HgI₂{Ac(Me₂NCS₂)}] or C₅H₉NOI₂S₂Hg. This compound was prepared by the addition of an equivalent amounts of HgI₂ (1:1 molar ratio) to a solution of Ac(Me₂NCS₂) in ether, acetone, chloroform, or CS₂ (Brinkhoff and Dautzenberg 1972). Precipitation could, if necessary, be facilitated by the addition of petroleum ether. The title compound was recrystallized from acetone.

[Hg(C₅H₁₁NOS)I₂] or C₅H₁₁NOI₂SHg. HgI₂ dissolved easily in a benzene solution of N,N-dimethyl O-ethylthiocarbamate (ca. 1:3 molar ratio) (Faraglia et al. 1982).

By the addition of *n*-hexane, the product separated as oil, which crystallized slowly as pale-yellow needles of the title compound by washing with *n*-hexane.

**[HgI$_2$(C$_4$H$_7$NOS)$_2$] or C$_8$H$_{14}$N$_2$O$_2$I$_2$S$_2$Hg**. This compound melts at 110°C (De Filippo et al. 1971). To obtain it, HgI$_2$ was dissolved in an excess of molten thio-morpholine-3-one (C$_4$H$_7$NOS) (at ca. 105°C; Hg$^{2+}$/C$_4$H$_7$NOS ca. 1:2 molar ratio), and the crude product was purified by washing with ethyl ether and recrystallizing from EtOH. The title compound could also be prepared if to a solution of HgI$_2$ in acetone/EtOH mixture different amounts of C$_4$H$_7$NOS were added for Hg$^{2+}$/ C$_4$H$_7$NOS molar ratios of 1:1, 1:2, 1:3, and 1:4, and the mixture was heated under reflux for 4 h. [HgI$_2$(C$_4$H$_7$NOS)$_2$] was separated out spontaneously during the reaction.

**[Hg(MeNHCSOEt)$_2$I$_2$] or C$_8$H$_{18}$N$_2$O$_2$I$_2$S$_2$Hg**. This compound melts at 58°C (Faraglia et al. 1981). It was obtained by dissolving HgI$_2$ in a benzene solution of MeNHCSOEt (molar ratio from 1:3 to 1:4). White title compound was isolated by the addition of *n*-pentane to this solution.

**[HgI$_2${PhCOCH$_2$(Me$_2$NCS$_2$)}] or C$_{11}$H$_{13}$NOI$_2$S$_2$Hg**. This compound melts at 107°C–110°C (Brinkhoff and Dautzenberg 1972). It was prepared by the addition of an equivalent amounts of HgI$_2$ (1:1 molar ratio) to a solution of PhCOCH$_2$(Me$_2$NCS$_2$) in ether, acetone, chloroform, or CS$_2$. Precipitation could, if necessary, be facilitated by the addition of petroleum ether. The title compound was recrystallized from CS$_2$.

**[Hg$_2$(C$_6$H$_{13}$NS)$_2$I$_4$]·H$_2$O or C$_{12}$H$_{28}$N$_2$OI$_4$S$_2$Hg$_2$**. This compound crystallizes in the monoclinic structure with the lattice parameters *a* = 1172, *b* = 1322, and *c* = 822 pm and β = 92.61° (Bayon et al. 1982). It was obtained in 10% methanolic solution by adding HgI$_2$ to the stoichiometric amount of 1-methyl-4-mercaptopiperidine. A slight excess of NaI was used in order to make HgI$_2$ soluble. The slightly yellow crystalline product obtained was filtered off; washed with cold water, EtOH, and ethyl ether; and dried in vacuo.

**[HgI$_2${PhCOCH$_2$(Et$_2$NCS$_2$)}] or C$_{13}$H$_{17}$NOI$_2$S$_2$Hg**. This compound melts at 134°C–135°C (Brinkhoff and Dautzenberg 1972). It was prepared by the addition of an equivalent amount of HgI$_2$ (1:1 molar ratio) to a solution of PhCOCH$_2$(Et$_2$NCS$_2$) in ether, acetone, chloroform, or CS$_2$. Precipitation could, if necessary, be facili-tated by the addition of petroleum ether. The title compound was recrystallized from acetone.

**[HgI$_2$(CH$_4$N$_2$S)$_2$]·2C$_{11}$H$_6$N$_2$O or C$_{24}$H$_{20}$N$_8$O$_2$I$_2$S$_2$Hg, (diiodobisthiourea) mercury(II)-bis(diazafluoren-9-one)**. This compound crystallizes in the triclinic structure with the lattice parameters *a* = 734.73 ± 0.19, *b* = 1319.1 ± 0.3, and *c* = 1624.8 ± 0.4 pm and α = 88.984° ± 0.005°, β = 77.062° ± 0.005°, and γ = 80.452° ± 0.005° and a calculated density of 2.131 g·cm$^{-3}$ (Wu et al. 2004). To obtain it, an aqueous solution (15 mL) containing 4,5-diazafluoren-9-one (0.5 mM), HgI$_2$ (0.25 mM), and thiourea (0.5 mM) was refluxed for 2 h. The solution was fil-tered, and the filtrate was kept in thermostat at 60°C. Yellow single crystals were obtained after 2 days.

## 7.29   Hg–H–C–N–O–I–Cl–S

[HgI$_2$(C$_{14}$H$_{11}$N$_2$OClS)$_2$] or C$_{28}$H$_{22}$N$_4$O$_2$Cl$_2$I$_2$S$_2$Hg. This multinary compound is formed in this system. It crystallizes in the monoclinic structure with the lattice parameters $a = 2090.0 \pm 0.4$, $b = 885.0 \pm 0.2$, and $c = 1984.5 \pm 0.4$ pm and $\beta = 118.26° \pm 0.03°$ and a calculated density of 2.128 g·cm$^{-3}$ (Yusof et al. 2004). It was prepared by the following procedure. A solution of o-chlorophenylbenzoylthiourea (4.3 mM) in EtOH (50 mL) was added dropwise to 50 mL of an ethanol solution containing an equimolar amount of HgI$_2$ in a two-neck round-bottomed flask. The solution was refluxed for about 2 h. The light-yellow solution was filtered, and light-yellow crystals were obtained from the filtrate after evaporation for 3 days.

## 7.30   Hg–H–C–N–O–Mn–S

Some compounds are formed in this system.

[HgMn(SCN)$_4$(C$_3$H$_8$O$_2$)]$_n$ or C$_7$H$_8$N$_4$O$_2$MnS$_4$Hg. This compound crystallizes in the orthorhombic structure with the lattice parameters $a = 1620.46 \pm 0.16$, $b = 729.74 \pm 0.05$, and $c = 1350.90 \pm 0.14$ pm and a calculated density of 2.345 g·cm$^{-3}$ (Wang et al. 2000b). It was prepared by the following procedure. To a crystalline powder of HgMn(SCN)$_4$ (16.0 mM) about 12 mL of a mixed solvent glycol monomethyl ether/H$_2$O (5:7 volume ratio) was added. This mixture was heated and stirred until the HgMn(SCN)$_4$ dissolved. The solution, which was slightly pale red, was left standing at about 20°C. The deposited crystals were separated. The crystals used for the XRD were obtained from a more dilute solution of the mixed solvent, which had been allowed to stand overnight.

[HgMn(SCN)$_4$(C$_2$H$_5$NO)$_2$]$_n$ or C$_8$H$_{10}$N$_6$O$_2$MnS$_4$Hg. This compound crystallizes in the orthorhombic structure with the lattice parameters $a = 1612.03 \pm 0.15$, $b = 773.73 \pm 0.07$, and $c = 1521.35 \pm 0.18$ pm and a calculated density of 2.121 g·cm$^{-3}$ (Wang et al. 2005b). To obtain it, N-methylformamide solution (5 mL) of HgMn(SCN)$_4$ (9 mM) was added slowly to H$_2$O (25 mL) with stirring at room temperature. After a while, the title compound precipitated. The crystals used for the XRD were obtained from a mixed solvent of N-methylformamide/H$_2$O (1:3 volume ratio) using a temperature-lowering method.

[HgMn(SCN)$_4$(DMSO)$_2$]$_n$ or C$_8$H$_{12}$N$_4$O$_2$MnS$_6$Hg. This compound crystallizes in the orthorhombic structure with the lattice parameters $a = 849.37 \pm 0.06$, $b = 852.26 \pm 0.04$, and $c = 2788.4 \pm 0.4$ pm and a calculated density of 2.120 g·cm$^{-3}$ (Wang et al. 2000c). To obtain it, a crystalline powder of HgMn(SCN)$_4$ was added to a mixed solvent of DMSO and H$_2$O. This mixture was heated and stirred until HgMn(SCN)$_4$ had dissolved. The colorless solution was left at room temperature until crystals of the title compound were formed.

[{HgMn(SCN)$_4$(H$_2$O)$_2$}·2C$_4$H$_9$NO]$_n$ or C$_{12}$H$_{22}$N$_6$O$_4$MnS$_4$Hg, diaquatetra-μ-thiocyanato-manganese(II)mercury(II) bis(N,N-di-methylacetamide). This compound crystallizes in the tetragonal structure with the lattice parameters

$a = 1224.36 \pm 0.06$ and $c = 807.08 \pm 0.06$ pm and a calculated density of 1.916 g·cm$^{-3}$ (Wang et al. 2000a). To obtain it, to a crystalline powder of HgMn(SCN)$_4$ (41 mM) about 100 mL of mixed solvent of H$_2$O and $N,N$-dimethylacetamide (1:1 volume ratio) was added; the pH of the solution was adjusted to 3 by the addition of HCl. This mixture was heated and stirred until HgMn(SCN)$_4$ dissolved. The slightly pale-red solution was allowed to stand in an oven at 30°C. After 1 day, the colorless crystals were obtained from this solution.

**[HgMn(SCN)$_4$(H$_2$O)$_2$·2(C$_3$H$_6$CONMe] or C$_{14}$H$_{22}$N$_6$O$_4$MnS$_4$Hg**. This compound crystallizes in the tetragonal structure with the lattice parameters $a = 1212.94$ and $c = 822.38$ pm (Wang et al. 2002a). To prepare it, 304.5 g of NH$_4$SCN and 342.6 g of Hg(NO$_3$)$_2$·H$_2$O were dissolved in 300 mL of H$_2$O with stirring. A 358 g of Mn(NO$_3$)$_2$ aqueous solution (ca. 50%) and 200 mL of $N$-methyl-2-pyrrolidone were added to the colorless solution at the same time. After the pale-red solution was left standing at room temperature for a while, white precipitate of the title compound was separated and dried. The product was purified by repeated recrystallizations, typically twice from a mixed hot solvent of water and $N$-methyl-2-pyrrolidone. Single crystals of this compound can be obtained by the slow temperature-lowering technique.

## 7.31   Hg–H–C–N–O–Mn–Fe–S

**[Fe(C$_5$H$_4$HgSCN)$_2$Mn(SCN)$_2$]·(C$_6$H$_6$N$_2$O)$_2$ or C$_{26}$H$_{20}$N$_8$O$_2$MnFeS$_4$Hg$_2$**. This multinary compound is formed in this system, which melts at 170°C with decomposition (Singh and Singh 1986). It was prepared by two methods.

1. A suspension or solution of [Fe(C$_5$H$_4$HgSCN)$_2$Mn(SCN)$_2$] was prepared in acetone/EtOH mixture, mixed with an ethanol solution of nicotinamide (1:2 molar ratio), and stirred for 24 h. The precipitate was formed, filtered, washed with EtOH, and dried in vacuum.
2. Mn(SCN)$_2$·2C$_6$H$_6$N$_2$O was dissolved in ethyl acetate. 1,1′-Bis(thiocyanato-mercurio)ferrocene was dissolved in DMSO/acetone mixture. The obtained solutions were mixed in 1:1 molar ratio and stirred for 48 h. The precipitate was formed, filtered, washed in EtOH, and dried in vacuum.

In both cases, gray crystals of the title compound were recrystallized from the acetone/EtOH mixture.

## 7.32   Hg–H–C–N–O–Fe–Co–S

**[Fe(C$_5$H$_4$HgSCN)$_2$Co(SCN)$_2$]·(C$_6$H$_6$N$_2$O)$_2$ or C$_{26}$H$_{20}$N$_8$O$_2$FeCoS$_4$Hg$_2$**. This multinary compound is formed in this system, which melts at 195°C with decomposition (Singh and Singh 1986). It was prepared by two methods.

1. A suspension or solution of [Fe(C$_5$H$_4$HgSCN)$_2$Co(SCN)$_2$] was prepared in acetone/EtOH mixture, mixed with an ethanol solution of nicotinamide

(1:2 molar ratio), and stirred for 24 h. The precipitate was formed, filtered, washed with EtOH, and dried in vacuum.

2. $Co(SCN)_2 \cdot 2(C_6H_6N_2O)$ was dissolved in acetone. 1,1'-Bis(thiocyanatomercurio) ferrocene was dissolved in DMSO/acetone mixture. The obtained solutions were mixed in 1:1 molar ratio and stirred for 48 h. The precipitate was formed, filtered, washed in EtOH, and dried in vacuum.

In both cases, dirty-pink crystals of the title compound were recrystallized from the acetone/EtOH mixture.

## 7.33 Hg–H–C–N–O–Fe–Ni–S

$[Fe(C_5H_4HgSCN)_2Ni(SCN)_2] \cdot (C_6H_6N_2O)_2$ or $C_{26}H_{20}N_8O_2FeNiS_4Hg_2$. This multinary compound is formed in this system, which melts at 200°C with decomposition (Singh and Singh 1986). It was prepared by two methods.

1. A suspension or solution of $[Fe(C_5H_4HgSCN)_2Ni(SCN)_2]$ was prepared in acetone/EtOH mixture, mixed with an ethanol solution of nicotinamide (1:2 molar ratio), and stirred for 24 h. The precipitate was formed, filtered, washed with EtOH, and dried in vacuum.

2. $Ni(SCN)_2 \cdot 2C_6H_6N_2O$ was dissolved in acetone. 1,1'-Bis(thiocyanatomercurio) ferrocene was dissolved in DMSO/acetone mixture. The obtained solutions were mixed in 1:1 molar ratio and stirred for 48 h. The precipitate was formed, filtered, washed in EtOH, and dried in vacuum.

In both cases, brown crystals of the title compound were recrystallized from the acetone/EtOH mixture.

## 7.34 Hg–H–C–N–O–Co–S

Some compounds are formed in this system.

$[HgCo(SCN)_4(NCONHMe)_2]$ or $C_8H_{10}N_6O_2CoS_4Hg$. This compound crystallizes in the orthorhombic structure with the lattice parameters $a = 1495.3 \pm 0.4$, $b = 1595.5 \pm 0.4$, and $c = 768.38 \pm 0.09$ pm and the experimental and calculated densities $2.21 \pm 0.03$ and $2.21$ g·cm$^{-3}$, respectively (Kinoshita and Ouchi 1986). A crystalline powder of $HgCo(SCN)_4$ (2.0 mM) was dissolved into $N$-methylformamide (20 mL), and 3 mL of $H_2O$ was added to the solution. It was left standing for several days, and the crystals of the title compound were deposited. The red-violet crystals were filtered off and dried over silica gel.

$[Hg(SCN)_4Co(DMF)_2]$ or $C_{10}H_{14}N_6O_2CoS_4Hg$. This compound crystallizes in the monoclinic structure with the lattice parameters $a = 916.3 \pm 0.2$, $b = 1405.7 \pm 0.3$, and $c = 1627.6 \pm 0.3$ pm and $\beta = 92.88° \pm 0.03°$ and the experimental and calculated densities $2.02 \pm 0.2$ and $2.023$ g·cm$^{-3}$, respectively (Udupa and Krebs 1980). To obtain it, to a hot solution of $HgCl_2$ (3.4 g) in $H_2O$ (50 mL) 10 mL of an aqueous solution containing 3.5 g of $CoSO_4 \cdot 7H_2O$ and 3.8 g of $NH_4SCN$ was added. The resultant solution was stirred, and the precipitated blue compound was filtered. The dried $HgCo(SCN)_4$ was dissolved in a minimum amount of hot DMF when a pink solution was obtained, which upon cooling gave a pink crystalline product of the title compound.

[$HgCo(SCN)_4(H_2O)_2 \cdot 2(MeCONMe_2)$] or $C_{12}H_{22}N_6O_4CoS_4Hg$. This compound crystallizes in the tetragonal structure with the lattice parameters $a = 1212.4 \pm 0.5$ and $c = 798.5 \pm 0.2$ pm and the experimental and calculated densities $1.99 \pm 0.03$ and 1.99 g·cm$^{-3}$, respectively (Kinoshita et al. 1985a). To obtain it, to a finely powdered $HgCo(SCN)_4$ (2 mM) about 10 mL of mixed solvent of $N,N$-dimethylacetamide/ $H_2O$ (1:1 volume ratio) was added. The solution, which was reddish violet in color, was left standing at about 15°C overnight. The deposited crystals were separated, washed with mixed solvent and $H_2O$, and dried by pressing between filter paper sheets.

[$HgCo(SCN)_4(H_2O)_2 \cdot 2(C_3H_6CONMe)$] or $C_{14}H_{22}N_6O_4CoS_4Hg$. This compound crystallizes in the tetragonal structure with the lattice parameters $a = 1207.45$ and $c = 821.58$ pm (Potheher et al. 2008) ($a = 1208.2 \pm 0.3$ and $c = 809.7 \pm 0.2$ pm and a calculated density of 2.04 g·cm$^{-3}$ [Kinoshita et al. 1986]). To obtain it, $HgCo(SCN)_4$ was dissolved into 20 mL of $N$-methyl-2-pyrrolidone (Kinoshita et al. 1986). To the obtained blue solution, an equal volume of $H_2O$ was added; the color of the solution turned red. After the solution was left standing overnight, precipitated title compound was separated. Its single crystals were grown from a mixture solvent of $N$-methyl-2-pyrrolidone/$H_2O$ (3:1 volume ratio). Within 40–50 days, deep-purple transparent crystals were obtained by slow evaporation at room temperature (Potheher et al. 2008).

[$HgCo(SCN)_4(EtC_4H_6NO)_3$] or $C_{22}H_{33}N_7O_3CoS_4Hg$. This compound crystallizes in the monoclinic structure with the lattice parameters $a = 1826.4 \pm 0.7$, $b = 1622.3 \pm 0.7$, and $c = 1062.3 \pm 0.5$ pm and $\beta = 92.77° \pm 0.05°$ and the experimental and calculated densities $1.75 \pm 0.03$ and 1.76 g·cm$^{-3}$, respectively (Kinoshita and Ouchi 1988). A crystalline powder of $HgCo(SCN)_4$ (2.0 mM) was dissolved into $N$-ethyl-2-pyrrolidone (10 mL), and 10 mL of THF was added to the solution. The mixture was then left standing for several days; thereby, red-purple crystals of the title compound were deposited. They were filtered off and dried over silica gel.

## 7.35 Hg–H–C–N–O–Ni–S

$HgNi(SCN)_4 \cdot H_2O$. This multinary compound is formed in this system (Porai-Koshits 1963) and crystallizes in the monoclinic structure with the lattice parameters $a = 1323$, $b = 527$, and $c = 1835$ pm and $\beta = 92.9°$.

## 7.36 Hg–H–C–N–S

Some compounds are formed in this system.

[$MeHgSCN$] or $C_2H_3NSHg$. This compound melts at 125°C (Relf et al. 1972). To obtain it, MeHgI (15 mM) and AgSCN (15 mM) were shaken in EtOH (25 mL) for 3 h. AgI and excess of AgSCN were filtered off, and the filtrate was evaporated to dryness under a vacuum. The product prepared was recrystallized from benzene, and white platelets were obtained.

**[Hg(S₂CNH₂)₂] or C₂H₄N₂S₄Hg, mercury(II) dithiocarbamate**. This compound melts at 165°C–166°C and crystallizes in the orthorhombic structure with the lattice parameters $a = 785.1 \pm 0.3$, $b = 1756.5 \pm 0.7$, and $c = 1205.1 \pm 0.3$ pm and the experimental and calculated densities $3.05 \pm 0.04$ and 3.076 g·cm⁻³, respectively (Chieh and Cheung 1981). To obtain it, to 15 mL aqueous/ethanol (1:1 volume ratio) solution of ammonium dithiocarbamate (4 mM), 10 mL ethanol solution of HgCl₂ (2 mM) was added slowly at 0°C. A white precipitate (**A**) that formed instantly was filtered off. The clear filtrate was allowed to evaporate slowly, and colorless crystals (**B**) were obtained among a black material (HgS) resulting from decomposition. In an attempt to recrystallize white precipitate **A**, it was suspended in an aqueous/ethanol solution, and the white slurry turned black after a few hours. The filtrate from this yielded crystals similar to **B** when dried.

**[Hg(SCN)₂·NH₄SCN] or C₃H₄N₄S₃Hg**. This compound crystallizes in the monoclinic structure with the lattice parameters $a = 1118.5$, $b = 408$, and $c = 1093.5$ pm and $\beta = 114°45'$ and the experimental and calculated densities 3.05 and 2.86 g·cm⁻³ (Zhdanov and Sanadze 1952). It was prepared by the slow mixing of a diluted aqueous solution of HgSO₄ and a concentrated solution of NH₄SCN to obtain insoluble precipitate by heating this mixture. After cooling the solution, yellowish needlelike crystals were obtained.

**[MeHg(C₃H₄NS₂)] or C₄H₇NS₂Hg**. This compound crystallizes in the hexagonal structure with the lattice parameters $a = 1350.2 \pm 0.8$ and $c = 698.4 \pm 0.7$ pm and a calculated density of 3.02 g·cm⁻³ (Norris et al. 1990).

**[Hg(CH₄N₂S)₂(CN)₂] or C₄H₈N₆S₂Hg**. This compound melts at 178°C (Isab et al. 2011). It was prepared by adding two equivalents of thiourea in 15 mL of MeOH to a solution of Hg(CN)₂ (1.0 mM) in 10 mL of MeOH, and the solution was stirred for 30 min. White crystalline product was obtained from the resulting colorless solution, which was washed with MeOH.

**[Hg{SC(NH₂)₂}₂(SCN)₂] or C₄H₈N₆S₄Hg**. This compound melts at 123°C (Czakis-Sulikowska and Sołoniewicz 1963) and crystallizes in the orthorhombic structure with the lattice parameters $a = 863 \pm 1$, $b = 935 \pm 1$, and $c = 1592 \pm 1$ pm and the experimental and calculated densities 2.535 and 2.567 g·cm⁻³, respectively (Korczyński 1963b, 1966). It was prepared by the addition of 0.5 M aqueous solution of thiourea to 0.5 M aqueous solution of (NH₄)₂[Hg(SCN)₄] containing 2 mL of concentrated HNO₃ up to dissolution of formed precipitate (Czakis-Sulikowska and Sołoniewicz 1963). After cooling the solution, needlelike colorless crystal were obtained.

**[Hg(C₃H₄N₂S)(SCN)₂] or C₅H₄N₄S₃Hg**. This compound melts at 180°C (Popović et al. 1999b). It was obtained by adding an ethanolic solution of 1,3-imidazole-2-thione to an ethanolic solution of Hg(SCN)₂ (1:1 molar ratio).

**C₅H₆N₂SHg, methyl-2-mercaptopyrimidinatomercury(II)**. This compound crystallizes in the orthorhombic structure with the lattice parameters $a = 406.3 \pm 0.2$, $b = 990.1 \pm 0.4$, and $c = 1880.8 \pm 1.5$ pm and the experimental and calculated densities $2.88 \pm 0.01$ and $2.867$ g·cm$^{-3}$, respectively (Chieh 1978c). When two aqueous solutions containing (not completely dissolved) 0.3 g of 2-mercaptopyrimidine and 0.62 g of methylmercuric hydroxide were mixed, a fine white precipitate was obtained leaving a clear colorless filtrate. Recrystallization by slow evaporation of an ethanol solution gave crystals of suitable size for XRD. Crystals of the same form were also obtained by slow evaporation of the filtrate.

**[MeHg(C₄H₅N₂S)] or C₅H₈N₂SHg (C₄H₆N₂S = 1-methylimidazoline-2-thione)**. This compound crystallizes in the monoclinic structure with the lattice parameters $a = 773.4 \pm 0.1$, $b = 1712.5 \pm 0.3$, and $c = 606.1 \pm 0.3$ pm and $\beta = 99.66° \pm 0.03$ and the experimental and calculated densities $2.74 \pm 0.02$ and $2.759$ g·cm$^{-3}$, respectively (Norris et al. 1983).

**[Hg(C₃H₃N₂S)₂] or C₆H₆N₄S₂Hg**. This compound melts at 174°C (Popović et al. 1999b). It was obtained by adding an ethanolic solution of 1,3-imidazole-2-thione to an ethanolic solution of Hg(OAc)₂ (2:1 molar ratio).

**[MeHg(C₅H₄NS)] or C₆H₇NSHg**. This compound melts at 54°C–55°C (at 52°C [Canty and Marker 1976]; at 53°C [Sytsma and Kline 1973]) and crystallizes in the monoclinic structure with the lattice parameters $a = 1083.4 \pm 0.5$, $b = 420.6 \pm 0.3$, and $c = 1714.4 \pm 0.2$ pm and $\beta = 101.91° \pm 0.01$ at 190 K and a calculated density of $2.834$ g·cm$^{-3}$ (Castiñeiras et al. 1986). To obtain it, a solution of MeOH was prepared by stirring HgMeCl (2.9 mM) and freshly precipitated Ag₂O (0.36 g) for 24 h in H₂O (ca. 100 mL) (Sytsma and Kline 1973, Castiñeiras et al. 1986). To this solution, pyridine-2(1H)-thione (3.0 mM) in MeOH (ca. 30 mL) was added. On addition of mercaptan, a white precipitate was formed immediately. The solid was recrystallized from MeOH by slow evaporation of the solvent.

The title compound could also be prepared as follows. Pyridine-2(1H)-thione (1.41 mM) in acetone (10 mL) was added to MeHgNO₃ (1.42 mM) in acetone (10 mL). Na₂CO₃ (1.42 mM) and H₂O (1 mL) were added, and the solution was stirred until all reagents dissolved (Canty and Marker 1976). Upon standing, a white crystalline [MeHg(C₅H₄NS)] was formed. Another procedure could also be used to obtain it. Na₂CO₃ (0.39 mM) was added to [MeHg(C₅H₅NS)](NO₃) (0.38 mM) in acetone (10 mL) and H₂O (1 mL). The solution was stirred for 30 min and filtered to remove undissolved Na₂CO₃: colorless crystals were allowed to form over 2 days.

**[Hg(C₅H₈NS₂)(CN)] or C₆H₈N₂S₂Hg**. For the preparation of this compound, Hg(CN)₂ was prepared first by the reaction of 1 mM of HgCl₂ in MeOH with 2 mM of KCN in H₂O (Altaf et al. 2010b). Then, 1 mM of Hg(CN)₂ in MeOH (15 mL) was mixed with a solution of 1 mM of pyrrolidine dithiocarbamate (C₅H₉NS₂) in MeOH (20 mL), resulting in the formation of white precipitate immediately. The mixture was stirred for further 30 min. After stirring, the precipitate was filtered off, washed with methanol, and dried in air.

**[Hg(C₅H₈NS₂)(SCN)]** or **C₆H₈N₂S₃Hg**. This compound was prepared by mixing 1 mM of $HgCl_2$ in 15 mL of MeOH with 2 mM of the solution of KSCN in MeOH (20 mL) (Altaf et al. 2010b). After stirring the mixture for 15 min, one equivalent of pyrrolidine dithiocarbamate ($C_5H_9NS_2$) in MeOH (20 mL) was added. The mixture was stirred for further 30 min. After stirring, the precipitate was filtered off, washed with methanol, and dried in air.

**[Hg(C₃H₄NS₂)₂]** or **C₆H₈N₂S₄Hg**. This compound melts at 72°C–74°C (Bell et al. 2001b) and decomposes at 160°C–162°C (Preti and Tosi 1976). To obtain it, an ethanolic solution (50 mL) of $Hg(OAc)_2$ (6.0 mM) was added with stirring to a warm ethanolic solution (50 mL) of 1,3-thiazolidine-2-thione (12.0 mM) in which $Et_3N$ (12.0 mM) had been dissolved. A pale-yellow precipitate formed immediately, which was then filtered off, washed with EtOH and diethyl ether, and dried in vacuo (Bell et al. 2001b).

The title compound could also be prepared by the next procedure (Preti and Tosi 1976). An aqueous solution of $HgCl_2$ or $Hg(NO_3)_2$ or $Hg(OAc)_2$, maintained at a pH value lower than that of precipitation of $Hg(OH)_2$, has been treated with an aqueous/ethanol solution of 1,3-thiazolidine-2-thione ($C_3H_4NS_2$) in the Hg/ligand stoichiometrical ratio of 1:2. The compound has been purified by means of repeated washing with EtOH and dried over $P_4O_{10}$.

**[Hg(C₄H₈N₂S)(CN)₂]** or **C₆H₈N₄SHg**. This compound melts at 203°C with decomposition (Popović et al. 2001). It was prepared by adding dropwise a methanol solution of 3,4,5,6-tetrahydropyrimidine-2-thione to an equimolar methanol solution of the appropriate of $Hg(CN)_2$. The crystalline solids, which were formed after standing for several days, were filtered off, washed with MeOH, and dried. The isolated compound is crystalline and yellow substance.

**[Hg(C₄H₈N₂S)(SCN)₂]** or **C₆H₈N₄S₃Hg**. This compound melts at 122°C (Popović et al. 2001). It was prepared by adding dropwise a methanol solution of 3,4,5,6-tetrahydropyrimidine-2-thione to an equimolar methanol solution of $Hg(SCN)_2$. The crystalline solids, which were formed after standing for several days, were filtered off, washed with MeOH, and dried. The isolated compound is crystalline and a colorless substance. It is photosensitive and should be kept in the dark.

**[HgMe(C₅H₆N₃S)]** or **C₆H₉N₃SHg, 4-amino-5-methyl-2-pyrimidinethiolato) methylmercury(II)**. This compound crystallizes in the monoclinic structure with the lattice parameters $a = 1193.1 \pm 0.6$, $b = 582.4 \pm 0.3$, and $c = 1289.5 \pm 0.6$ pm and $\beta = 99.75° \pm 0.06°$ and the experimental and calculated densities 2.64 and 2.67 g·cm⁻³, respectively (Stuart et al. 1980). To obtain it, an aqueous solution of MeHgOAc (1 mM) was added to an aqueous solution of a mixture of 4-amino-5-methyl-2-pyrimidinethiol ($C_5H_7N_3S$) (1 mM) and $NaHCO_3$ (1 mM). The resulting precipitate was filtered and recrystallized from hot acetone. Crystals were grown from a solution of this compound in a mixture of DMSO, acetone, and $H_2O$.

**[Hg(S₂CNMe₂)₂]** or **C₆H₁₂N₂S₄Hg**. This compound crystallizes in the monoclinic structure with the lattice parameters $a = 1730.8 \pm 0.1$, $b = 755.8 \pm 0.3$, and

$c = 997.8 \pm 0.2$ pm and $\beta = 113.24° \pm 0.01°$ and a calculated density of 2.442 g·cm$^{-3}$ (Cox and Tiekink 1997). It was synthesized as follows. A solution of Na(S$_2$CNMe$_2$)·$x$H$_2$O (5 mM) in H$_2$O (50 mL) was added to a vigorously stirred solution of Hg(NO$_3$)$_2$·H$_2$O (10 mM) in H$_2$O (20 mL). The resulting voluminous yellow precipitate was washed by decanting, filtered off, and dried in air (Ivanov et al. 2008). Flattened prismatic, transparent yellow crystals of the title compound for XRD were grown in chloroform.

**[Hg(MeCH$_3$N$_2$S)$_2$(CN)$_2$] or C$_6$H$_{12}$N$_6$S$_2$Hg.** This compound melts at 141°C–142°C and crystallizes in the monoclinic structure with the lattice parameters $a = 1080.5 \pm 0.4$, $b = 426.18 \pm 1.5$, and $c = 2829.9 \pm 0.9$ pm and $\beta = 96.533° \pm 0.007°$ and a calculated density of 2.221 g·cm$^{-3}$ (Isab et al. 2011). It was prepared by adding two equivalents of methylthiourea in 15 mL of MeOH to a solution of Hg(CN)$_2$ (1.0 mM) in 10 mL of MeOH, and the solution was stirred for 30 min. A white crystalline product was obtained from the resulting colorless solution, which was washed with MeOH.

**[MeHg(S$_2$CNEt$_2$)] or C$_6$H$_{13}$NS$_2$Hg, methyl($N,N$-diethyldithiocarbamato) mercury(II).** This compound melts at 141°C and crystallizes in the monoclinic structure with the lattice parameters $a = 715.5 \pm 0.5$, $b = 775.7 \pm 0.4$, and $c = 1824.8 \pm 1.7$ pm and $\beta = 96.1° \pm 0.1°$ and the experimental and calculated densities 2.39 and 2.399 g·cm$^{-3}$, respectively (Chieh and Leung 1976). To obtain it, equal molar amounts of tetraethylthiuram disulfide and MeHgCl dissolved in chloroform and EtOH, respectively, were mixed at room temperature. No immediate change was observable, and the reaction mixture was left in a cold room (0°C) partly covered. After 2 weeks, two kinds of crystals appeared in the same beaker. The brown crystals proved to be the α-form of bis($N,N$-diethyldithiocarbamato)mercury(II).

**[PhHgSCN] or C$_7$H$_5$NSHg.** This compound melts at 232°C (Dehnicke 1967). It was prepared from the reaction of equimolar amounts of C$_6$H$_5$Hg and (SCN)$_2$ in the benzene solution: after a few minutes, the first crystals were formed.

**[MeHg(S$_2$CC$_5$H$_6$NH$_2$-2)] or C$_7$H$_{11}$NS$_2$Hg.** This compound exists in two polymorphic modifications. Yellow-orange, flattened octahedral crystals of the title compound melt at 198°C–199°C and bright-yellow needles at 201°C–202°C (Lai and Tiekink 2003d). α-MeHg(S$_2$CC$_5$H$_6$NH$_2$-2) crystallizes in the monoclinic structure with the lattice parameters $a = 748.10 \pm 0.06$, $b = 1224.34 \pm 0.10$, and $c = 1088.63 \pm 0.09$ pm and $\beta = 108.526° \pm 0.002°$ at 223 K and a calculated density of 2.627 g·cm$^{-3}$. β-MeHg(S$_2$CC$_5$H$_6$NH$_2$-2) also crystallizes in the monoclinic structure with the lattice parameters $a = 1231.86 \pm 0.14$, $b = 1277.07 \pm 0.15$, and $c = 1234.38 \pm 0.13$ pm and $\beta = 91.394° \pm 0.002°$ at 223 K and a calculated density of 2.558 g·cm$^{-3}$. It was prepared from the reaction of stoichiometric amounts of MeHgCl in CH$_2$Cl$_2$ solution (50 mL), and 2-amino-cyclopent-1-ene-1-carbodithioate was added as a solid. The original clear solution turned orange with time, and after 2 h of stirring, the solution was filtered, and the filtrate allowed to stand. Crystals appeared after 2 days. These were generally air and light sensitive, turning black after days.

**[Hg(SCN)$_2$·(C$_6$H$_4$N$_2$)] or C$_8$H$_4$N$_4$S$_2$Hg.** This compound melts at 137°C–138°C (Jain and Rivest 1969b). It was prepared by mixing Hg(SCN)$_2$ and 3-cyanopyridine (C$_6$H$_4$N$_2$) (approximately 1:1 molar ratio) in warm THF. A crystalline solid was obtained from the reaction mixture by partially evaporating the THF. The colorless title compound that was obtained was recrystallized from THF solution, filtered, and dried in vacuo at room temperature.

**[Hg(C$_4$H$_3$N$_2$S)$_2$] or C$_8$H$_6$N$_4$S$_2$Hg**. This compound has been obtained by an electrochemical oxidation of Hg in an acetonitrile solution (50 mL) of pyrimidine-2-thione with a small amount of (Et$_4$N)ClO$_4$ (Tallon et al. 1995). As electrolysis proceeded, the color of the solution changed. At the end of the process, the solution was filtered to remove any solid impurities, and the filtrate was slowly concentrated at room temperature to give the solid title compound.

**[MeHg(C$_7$H$_4$NS$_2$)] or C$_8$H$_7$NS$_2$Hg**. This compound crystallizes in the triclinic structure with the lattice parameters $a = 800.9 \pm 0.4$, $b = 1004.2 \pm 0.4$, and $c = 1307.4 \pm 0.3$ pm and $\alpha = 101.25° \pm 0.02°$, $\beta = 102.61° \pm 0.03°$, and $\gamma = 101.42° \pm 0.03°$ pm and the experimental and calculated densities 2.65 and 2.745 g·cm$^{-3}$ (Bravo et al. 1985). To obtain it, a solution of MeHgOH was obtained by stirring MeHgCl (0.008 mM) with an excess of freshly precipitated Ag$_2$O for 48 h in deionized H$_2$O. The resulting solution, once filtered, was mixed with 0.005 mM of 2-mercaptobenzothiazole in MeOH. The yellow solid formed was removed by filtration, washed with H$_2$O, and recrystallized from EtOH under slow evaporation until crystals suitable for XRD were obtained.

**[Hg(C$_3$H$_4$N$_2$S)$_2$(SCN)$_2$] or C$_8$H$_8$N$_6$S$_4$Hg**. This was obtained by adding an ethanolic solution of 1,3-imidazole-2-thione to an ethanolic solution of HgCl$_2$ (1:1 molar ratio) (Popović et al. 1999b).

**[Hg(C$_4$H$_5$N$_2$S)$_2$] or C$_8$H$_{10}$N$_4$S$_2$Hg**. This compound melts at 241°C–243°C (Bell et al. 2000b) (at 217°C with decomposition [Popović et al. 1999b]) and crystallizes in the monoclinic structure with the lattice parameters $a = 1121.7 \pm 0.2$, $b = 1154.30 \pm 0.10$, and $c = 882.81 \pm 0.08$ pm and $\beta = 94.047° \pm 0.014°$ and a calculated density of 2.487 g·cm$^{-3}$ (Popović et al. 1999b). To obtain it, Hg(OAc)$_2$ (2.35 mM) was dissolved in 50 mL of H$_2$O (Popović et al. 1999b, Bell et al. 2000b). This solution was added to methyl-1,3-imidazole-2-thione (0.500 mM), also dissolved in 50 mL of H$_2$O, with continuous stirring. Et$_3$N (5.44 mM) was rapidly added to the clear solution of reactants, also with continuous stirring. A colorless precipitate immediately formed and was recovered by filtration and dried in vacuo. The title compound was also obtained as by-product by the electrochemical oxidation of Hg(C$_4$H$_5$N$_2$S)$_2$.

Hg(C$_4$H$_5$N$_2$S)$_2$ could also be obtained by an electrochemical oxidation of Hg in an acetonitrile solution (50 mL) of methyl-1,3-imidazole-2-thione with a small amount of (Et$_4$N)ClO$_4$ (Tallon et al. 1995). As electrolysis proceeded, the color of the solution changed. At the end of the process, the solution was filtered to remove any solid impurities, and the filtrate was slowly concentrated at room temperature to give the solid title compound.

**[Hg(C$_6$H$_{12}$N$_2$)(SCN)$_2$] or C$_8$H$_{12}$N$_4$S$_2$Hg**. This compound crystallizes in the orthorhombic structure with the lattice parameters $a = 740.5 \pm 0.3$, $b = 609.1 \pm 0.4$, and $c = 1220.0 \pm 0.8$ pm and a calculated density of 2.59 g·cm$^{-3}$ (Pickardt et al. 1995b). Reaction of 1,4-diazabicyclo[2.2.2]octane (C$_6$H$_{12}$N$_2$) with Hg(SCN)$_2$ in DMF yields crystals of the title compound.

**[Hg(SCN)₂·(C₆H₁₆N₂)] or C₈H₁₆N₄S₂Hg**. This compound melts at 110°C–111°C (Jain and Rivest 1969b). It was prepared by mixing $Hg(SCN)_2$ and $N,N,N',N'$-tetramethylenediamine ($C_6H_{16}N_2$) (approximately 1:1 molar ratio) in warm THF. A crystalline solid is separated out upon allowing the reaction mixture to stay overnight at room temperature. The colorless title compound that was obtained was recrystallized from THF solution, filtered, and dried in vacuo at room temperature.

**[Hg(EtNHC(S)NH₂)₂(CN)₂] or C₈H₁₆N₆S₂Hg**. This compound crystallizes in the monoclinic structure with the lattice parameters $a = 1486.91 \pm 0.06$, $b = 1246.32 \pm 0.05$, and $c = 838.29 \pm 0.04$ pm and $\beta = 94.0660° \pm 0.0010°$ and a calculated density of 1.976 g·cm⁻³ (Sadaf et al. 2012). To prepare it, 1 mM of $Hg(CN)_2$ was dissolved in 15 mL of MeOH and mixed with two equivalents of ethylthiourea in 15 mL of MeOH. After stirring for 15 min, the resulting mixture was kept at room temperature. The title compound is light and moisture sensitive and should be stored in the dark.

**[Hg{Me₂(NH)₂CS}₂(CN)₂] or C₈H₁₆N₆S₂Hg**. This compound melts at 158°C–160°C (Isab et al. 2011) and crystallizes in the monoclinic structure with the lattice parameters $a = 1811.61 \pm 0.11$, $b = 775.33 \pm 0.05$, and $c = 1405.3 \pm 0.08$ pm and $\beta = 128.533° \pm 0.003°$ and a calculated density of 1.983 g·cm⁻³ (Malik et al. 2010). To obtain it, to 1.0 mM of $Hg(CN)_2$ in 10 mL of MeOH two equivalents of $N,N'$-dimethylthiourea in MeOH were added. On mixing, a clear solution was obtained. It was stirred for 30 min after which it was filtered, and the filtrate was kept at room temperature for crystallization by slow evaporation of the solvent. As a result, colorless block-like crystals were obtained (Malik et al. 2010, Isab et al. 2011).

**[Hg(C₇H₆N₂S)(SCN)₂] or C₉H₆N₄S₃Hg**. To prepare this compound, a solution of benzo-1,3-imidazole-2-thione (1.40 mM) in 20 mL of 96% EtOH was added dropwise to a solution of $Hg(SCN)_2$ (1.26 mM) in 120 mL of 96% EtOH (Popović et al. 2002a). The reaction mixture was left in the dark due to the photosensitivity of the compound. After a few weeks, the yellow crystalline product was filtered off, washed with cold EtOH, and dried in air.

**[Hg(C₅H₄NS)₂] or C₁₀H₈N₂S₂Hg**. This compound melts at 191°C and crystallizes in the monoclinic structure with the lattice parameters $a = 1107.8 \pm 0.2$, $b = 408.73 \pm 0.06$, and $c = 1253.3 \pm 0.2$ pm and $\beta = 101.253° \pm 0.007°$ and a calculated density of 2.51 g·cm⁻³ (Wang and Fackler 1989). To obtain it, two procedures could be used.

1. In 20 mL of $CH_2Cl_2$, 1.57 mM of $Hg(OAc)_2$ was dissolved. Then, 3.14 mM of 2-pyridine thiol ($C_5H_5NS$) was added to this solution. After the mixture was stirred for 1 h at room temperature, a colorless solution was obtained. This solution was then concentrated to 10 mL. Excess diethyl ether was added. Colorless fiber-like crystals of the title compound were obtained. Single crystals were grown from $CH_2Cl_2$/diethyl ether solution by slow diffusion of the solvent at 22°C.
2. In 5 mL of $CH_2Cl_2$, 0.149 mM of [Hg(C₅H₄NS)(OAc)] was dissolved. Then, 0.144 mM of $C_5H_5NS$ was added to this solution. The mixture was stirred for 1 h, and the solution was then concentrated to 1 mL. Diethyl ether was

added, and the colorless crystals of the title compound were obtained. All reactions were performed by using standard Schlenk techniques under an atmosphere of dry $N_2$.

[Hg($C_5H_4NS$)$_2$] could also be prepared by the next procedure (Kennedy and Lever 1972). Sodium salt of 2-pyridine thiol (Na$C_5H_4NS$) in EtOH was added to $HgCl_2$ in the same solvent. The white precipitate was recrystallized by Soxhlet extraction with MeOH, giving needlelike crystals.

**[MeHgS$C_9H_6N$] or $C_{10}H_9NSHg$.** This compound melts at 163°C (Sytsma and Kline 1973). To obtain it, MeHgOAc was dissolved in water and added to 8-mercapto-quinoline dissolved in MeOH or methanolic KOH solution. The crude products were purified by the recrystallization in $H_2O$/acetone.

**[MeHgS$C_9H_6N$] or $C_{10}H_9NSHg$.** This compound melts at 132°C (Sytsma and Kline 1973). To obtain it, MeHgOAc was dissolved in water and added to 2-mercapto-quinoline dissolved in MeOH or methanolic KOH solution. The crude products were purified by the recrystallization in hexane.

**[Hg{$S_2$CN($CH_2$)$_4$}$_2$] or $C_{10}H_{16}N_2S_4Hg$.** This compound melts at 235°C–239°C and crystallizes in the monoclinic structure with the lattice parameters $a = 1855.33 \pm 0.16$, $b = 833.22 \pm 0.07$, and $c = 1116.92 \pm 0.10$ pm and $\beta = 122.542° \pm 0.001°$ at 183 K (Lai and Tiekink 2003a). Its crystals were isolated from an acetonitrile/chloroform (1:1 volume ratio) solution containing equimolar amounts of Hg($S_2$CNEt$_2$)$_2$ and [Zn{$S_2$CN($CH_2$)$_4$}$_2$].

**[Hg($C_4H_8N_2S$)$_2$(CN)$_2$] or $C_{10}H_{16}N_6S_2Hg$.** This compound melts at 147°C and crystallizes in the monoclinic structure with the lattice parameters $a = 1250.82 \pm 0.07$, $b = 903.02 \pm 0.05$, and $c = 1408.57 \pm 0.10$ pm and $\beta = 95.583° \pm 0.006°$ and a calculated density of 2.034 g·cm$^{-3}$ (Popović et al. 2000a). It was prepared by reacting one equiv-alent of Hg(CN)$_2$ and two equivalents of 3,4,5,6-tetrahydropyrimidine-2-thione in MeOH. The crystalline solids, which formed upon standing for several days, were filtered off, washed with MeOH, and dried. The isolated compound is crystalline and a colorless substance.

**[Hg($C_4H_8N_2S$)$_2$(SCN)$_2$] or $C_{10}H_{16}N_6S_4Hg$.** This compound melts at 133°C and crystallizes in the triclinic structure with the lattice parameters $a = 765.08 \pm 0.07$, $b = 849.76 \pm 0.07$, and $c = 1417.5 \pm 0.2$ pm and $\alpha = 105.968° \pm 0.010°$, $\beta = 89.941° \pm 0.009°$, and $\gamma = 92.003° \pm 0.010°$ and a calculated density of 2.060 g·cm$^{-3}$ (Popović et al. 2000a). It was prepared by reacting one equivalent of Hg(SCN)$_2$ and two equiva-lents of 3,4,5,6-tetrahydropyrimidine-2-thione in MeOH. The crystalline solids, which formed on standing for several days, were filtered off, washed with MeOH, and dried. The isolated compound is crystalline and a colorless substance. It is pho-tosensitive and should be kept in the dark.

**[Hg(Et$_2$CNS$_2$)] or $C_{10}H_{20}N_2S_4Hg$.** This compound crystallizes in the mono-clinic structure with the lattice parameters $a = 981.2 \pm 0.4$, $b = 1096.1 \pm 0.7$, and $c = 1586.5 \pm 0.5$ pm and $\beta = 103.008° \pm 0.003°$ at $100 \pm 2$ K and a calculated density of 1.986 g·cm$^{-3}$ (Mendoza et al. 1997). To obtain it, to a solution of tetraethylthiuram disulfide (0.3 g) in acetone (50 mL) $Hg(ClO_4)_2$ (0.23 g) dissolved in $EtOH/H_2O$ (9:1 volume ratio) (50 mL) was added. The solution was shaken for a few minutes at room temperature. After standing for several days, yellow crystals were obtained, separated, washed with acetone, and air-dried.

**$[Hg(S_2CNEt_2)_2]$ or $C_{10}H_{20}N_2S_4Hg$, bis($N,N$-diethyldithiocarbamato)mercury(II).** This compound exists in two polymorphic modifications. The first polymorph crystallizes in the monoclinic structure with the lattice parameters $a = 1186.1 \pm 0.8$, $b = 1502 \pm 1$, and $c = 475.1 \pm 0.4$ pm and $\beta = 106.30° \pm 0.06°$ and the experimental and calculated densities $1.98 \pm 0.03$ and $2.04$ g·cm$^{-3}$, respectively (Healy and White 1973) ($a = 1180.7 \pm 0.3$, $b = 1493.0 \pm 0.5$, and $c = 472.3 \pm 0.4$ pm and $\beta = 106.26° \pm 0.04°$ and the experimental and calculated densities $2.06 \pm 0.03$ and $2.064$ g·cm$^{-3}$, respectively [Iwasaki et al. 1973]). The second polymorph also crystallizes in the monoclinic structure with the lattice parameters $a = 1666.2 \pm 0.8$, $b = 1095.4 \pm 0.6$, and $c = 986.1 \pm 0.6$ pm and $\beta = 111.8° \pm 0.2°$ and a calculated density of $1.976$ g·cm$^{-3}$, respectively (Iwasaki et al. 1973) ($a = 960.48 \pm 0.07$, $b = 1091.19 \pm 0.08$, and $c = 1579.8 \pm 0.1$ pm and $\beta = 102.375° \pm 0.002°$ at 183 K [Lai and Tiekink 2002]). The title compound was prepared by the addition of a stoichiometric quantity of $NaEt_2CNS$ in $H_2O$ to an aqueous solution of $Hg(NO_3)_2·0.5H_2O$ and was recrystallized from chloroform. Crystals were grown as large, irregular, interlocked, yellow leaves by the slow evaporation of a chloroform (acetonitrile/chloroform [1:1 volume ratio] [Lai and Tiekink 2002]) solution (Healy and White 1973, Iwasaki et al. 1973, Larionov et al. 1990, Ivanov et al. 2008).

**$[Hg(Pr^nNHC(S)NH_2)_2(CN)_2]$ or $C_{10}H_{20}N_6S_2Hg$.** This compound crystallizes in the triclinic structure with the lattice parameters $a = 823.51 \pm 0.07$, $b = 895.14 \pm 0.07$, and $c = 1289.24 \pm 0.11$ pm and $\alpha = 89.471° \pm 0.004$, $\beta = 81.451° \pm 0.004°$, and $\gamma = 75.023° \pm 0.004°$ and a calculated density of $1.790$ g·cm$^{-3}$ (Sadaf et al. 2012). To prepare it, 1 mM of $Hg(CN)_2$ was dissolved in 15 mL of MeOH and mixed with two equivalents of $n$-propylthiourea in 15 mL of MeOH. After stirring for 15 min, the resulting mixture was kept at room temperature. The title compound is light and moisture sensitive and should be stored in the dark.

**$[PhHg(C_5H_4NS)]$ or $C_{11}H_9NSHg$.** This compound melts at 82°C–83°C (Castiñeiras et al. 1986). When to a solution of phenylmercury(II) acetate (4.7 mM) in MeOH (50 mL) a solution of pyridine-2(1$H$)-thione (4.7 mM) in MeOH (25 mL) was added, a white precipitate was formed. Recrystallization from MeOH yielded a crystalline solid.

**$[PhHg\{S_2CN(CH_2)_4\}]$ or $C_{11}H_{13}NS_2Hg$.** This compound melts at 154°C–156°C and crystallizes in the monoclinic structure with the lattice parameters $a = 1337.11 \pm 0.10$, $b = 636.41 \pm 0.05$, and $c = 1458.26 \pm 0.11$ pm and $\beta = 95.535° \pm 0.002°$ at $183 \pm 2$ K and a calculated density of $2.280$ g·cm$^{-3}$ (Lai et al. 2002). It was prepared by the next procedure. To a stirred solution of PhHgCl (0.1939 g) in $CH_2Cl_2$ (30 mL) a stoichiometric amount of $NH_4S_2CN(CH_2)_4$ in $H_2O$ (20 mL) was added. After 3 h, the organic layer was separated, dried over $MgSO_4$, and evaporated to dryness. Colorless crystals were obtained from the recrystallization of the precipitate from a $CH_2Cl_2/CH_3OH$ (1:1 volume ratio) solution.

**[PhHg(S₂CNEt₂)] or $C_{11}H_{15}NS_2Hg$**. This compound melts at 104°C–105°C and crystallizes in the triclinic structure with the lattice parameters $a = 995.9 \pm 0.2$, $b = 1235.9 \pm 0.4$, and $c = 1309.8 \pm 0.2$ pm and $\alpha = 65.53° \pm 0.02°$, $\beta = 65.81° \pm 0.02°$, and $\gamma = 81.26° \pm 0.02°$ and a calculated density of 2.114 g·cm⁻³ (Tiekink 1987b). It was prepared by the following procedure. To a stirred solution of PhHgCl (0.5 g, 50 mL of $CH_2Cl_2$) the stoichiometric quantity of $KS_2CNEt_2$ (20 mL of $H_2O$) was added. After 1 h of stirring, the organic layer was separated and dried over $Na_2SO_4$. Crystals were obtained by the slow evaporation of a mixture benzene/petroleum spirit solution of the title compound.

**[Et₄N][Hg(SMe)₃]₂ or $C_{11}H_{29}NS_3Hg$**. This compound melts at 77°C–78°C and crystallizes in the triclinic structure with the lattice parameters $a = 865.6 \pm 0.5$, $b = 913.0 \pm 0.6$, and $c = 1236.8 \pm 0.8$ pm and $\alpha = 102.16° \pm 0.04°$, $\beta = 105.51° \pm 0.04°$, and $\gamma = 105.29° \pm 0.04°$ and a calculated density of 1.81 g·cm⁻³ (Bowmaker et al. 1984). To obtain it, methanethiol (30 mM) was added to a suspension of bis(methanethiolato) mercury(II) (10 mM) in a mixture of 25°C aqueous Et₄NOH (20 mM) and absolute EtOH (10 mL). The solution became clear on warming to 60°C and was filtered and reduced in volume to give white crystals of the title compound.

**[Hg(SCN)₂·(2,2′-bipy)] or $C_{12}H_8N_4S_2Hg$**. This compound melts at 183°C–185°C (Jain and Rivest 1970). It was obtained by slowly adding a warm solution of 2,2′-bipy in THF to a warm solution of Hg(SCN)₂ in the same solvent. The molar ratio of reactants was approximately 1:1. The colorless crystalline product that was precipitated immediately was filtered under reduced pressure on a sintered glass crucible, washed three times with a little amount of THF, and dried in vacuo for several hours.

**[Hg(C₅H₅N)₂(SCN)₂] or $C_{12}H_{10}N_4S_2Hg$**. This compound melts at 79°C (Ahuja and Garg 1972). It was obtained by heating a suspension of Hg(SCN)₂ in hot EtOH with an excess of Py and filtering the resultant mixture while still hot.

**[Hg(C₅H₅NS)₂(SCN)₂] or $C_{12}H_{10}N_4S_4Hg$**. This compound exists in two polymorphic modifications. The first polymorph (α-form) crystallizes in the monoclinic structure with the lattice parameters $a = 1124.8 \pm 0.1$, $b = 866.9 \pm 0.2$, and $c = 1781.9 \pm 0.2$ pm and $\beta = 106.84° \pm 0.01°$ and a calculated density of 2.153 g·cm⁻³ (Popović et al. 1999a). The second polymorphic modification (β-form) crystallizes in the triclinic structure with the lattice parameters $a = 706.8 \pm 0.2$, $b = 774.4 \pm 0.2$, and $c = 1628.2 \pm 0.3$ pm and $\alpha = 94.47° \pm 0.02°$, $\beta = 99.95° \pm 0.02°$, and $\gamma = 107.83° \pm 0.02°$ pm and a calculated density of 2.164 g·cm⁻³. [Hg(C₅H₅NS)₂(SCN)₂] was prepared by the following procedure. An ethanolic solution of pyridine-2-thione (1.35 mM in 20 mL) was added to an ethanolic solution of Hg(SCN)₂ (0.63 mM in 60 mL). The reaction mixture was left to evaporate until colorless crystals of α-modification formed. The product was collected by filtration, washed with EtOH, and dried in air. Recrystallization from EtOH gave β-modification.

**[PhHg(S₂CC₅H₆NH₂-2)] or $C_{12}H_{13}NS_2Hg$**. This compound melts at 166°C–167°C and crystallizes in the orthorhombic structure with the lattice parameters $a = 1112.94 \pm 0.07$, $b = 893.65 \pm 0.06$, and $c = 2577.99 \pm 0.16$ pm at 183 K and a calculated density of 2.259 g·cm⁻³ (Lai and Tiekink 2003d). It was prepared from the reaction of stoichiometric amounts of PhHgCl in $CH_2Cl_2$ solution (50 mL) and 2-amino-cyclopent-1-ene-1-carbodithioate was added as a solid. The original clear solution turned orange with time, and after 2 h of stirring, the solution was filtered, and the filtrate allowed to stand. Crystals appeared after 2 days. These were generally air and light sensitive, turning black after days.

**$[Hg(C_6H_7N_2S)_2]_n$ or $C_{12}H_{14}N_4S_2Hg$.** This compound crystallizes in the monoclinic structure with the lattice parameters $a = 1180.8 \pm 0.2$, $b = 919.8 \pm 0.2$, and $c = 1470.2 \pm 0.3$ pm and $\beta = 112.19° \pm 0.03°$ and the experimental and calculated densities 2.148 and 2.152 g·cm$^{-3}$, respectively (Das and Seth 1997). Interaction of 4,4-dimethylpyrimidine-2-thiol with $HgCl_2$ by refluxing in an aqueous solution of NaOAc leads to the formation of the title compound as a white crystalline solid. The product was recrystallized as colorless blocks by slow evaporation of its solution in an equal volume mixture of acetone and EtOH.

It could also be obtained by an electrochemical oxidation of Hg in an acetonitrile solution (50 mL) of 4,6-dimethylpyrimidine-2-thiol with a small amount of $(Et_4N)ClO_4$ (Tallon et al. 1995). As electrolysis proceeded, the color of the solution changed. At the end of the process, the solution was filtered to remove any solid impurities, and the filtrate was slowly concentrated at room temperature to give the solid title compound.

**$[Hg(C_5H_8N_2)_2(SCN)_2]$ or $C_{12}H_{16}N_6S_2Hg$.** crystallizes in the triclinic structure with the lattice parameters $a = 773.2 \pm 0.2$, $b = 1025.2 \pm 0.2$, and $c = 1180.2 \pm 0.2$ pm and $\alpha = 65.57° \pm 0.03°$, $\beta = 72.19° \pm 0.03°$, and $\gamma = 88.13° \pm 0.02°$ pm (Mahjoub et al. 2003). It was prepared by dissolving $Hg(SCN)_2$ (1 mM) in distilled $H_2O$ and adding an alcoholic solution of 2,2′-bis(4,5-dimethyleimidazole) (2 mM). The resulting solution was stirred for 5 h at room temperature, and then it was allowed to stand for 2–3 days in a refrigerator (ca. 6°C). Black crystals of the title compound precipitated, which were filtered off, washed with acetone, and ether and air dried.

**$[Hg(C_4H_5N_2S)_2(C_4H_6N_2S)]$ or $C_{12}H_{16}N_6S_3Hg$.** This compound melts and crystallizes in the monoclinic structure with the lattice parameters $a = 965.25 \pm 0.05$, $b = 1566.76 \pm 0.07$, and $c = 1130.01 \pm 0.05$ pm and $\beta = 94.554° \pm 0.002°$ at 160 K and a calculated density of 2.110 g·cm$^{-3}$ (Bell et al. 2000b). The electrochemical synthesis of the title compound was performed in a slim-form 50 mL beaker. Saturated solution of 1-methylimidazoline-2(3H)-thione was prepared in acetonitrile. A small quantity (10 mg) of the supporting electrolyte, $Bu^n_4N(BF_4)$, was added to the saturated ligand solution. The reaction mixture was electrolyzed for 4 h using a potential of 6 V and a current of 10 mA. At the completion of the electrolysis, both a grayish-white material of indeterminate composition and a small quantity of black HgS fell from the anode surface. These were removed, and the electrolyzed solution was covered and left overnight at room temperature. Pale-yellow microcrystals and colorless well-formed crystals formed overnight and were deposited on the base and sides of the reaction vessel, together with a small quantity of unreacted ligand. Chemical analysis of the pale-yellow crystalline product (m.p. 239°C–244°C) indicated a formal 1:2 (metal/ligand) composition corresponding to $C_8H_{10}N_4S_2Hg$. The colorless crystals of the title compound were removed, washed with a little cold acetonitrile, dried, and stored in vacuo.

**[PhHg(C_7H_4NS_2)] or C_13H_9NS_2Hg**. To obtain this compound, solutions of PhHgOAc (0.005 mM) in EtOH and 2-mercaptobenzothiazole in the same solvent were mixed, and a white precipitate formed (Bravo et al. 1985). The solvent was removed by filtration, and the solid was redissolved in hot EtOH. Crystals were obtained by slow evaporation.

**[PhHg(S_2CNPr$^n$_2)] or C_13H_19NS_2Hg**. This compound melts at 56°C–58°C and crystallizes in the monoclinic structure with the lattice parameters $a = 994.59 \pm 0.06$, $b = 1350.69 \pm 0.08$, and $c = 1212.73 \pm 0.07$ pm and $\beta = 109.732° \pm 0.001°$ at 183 K (Lai and Tiekink 2003c). It was obtained by the next procedure. To a stirred dichloromethane solution of PhHgCl (64 mM) a stoichiometric amount of KS_2CNPr$^n$_2 dissolved in H_2O (20 mL) was added. After stirring the mixture for 2 h, the organic layer was separated and dried over MgSO_4. The crude product was recrystallized as colorless crystals from CH_2Cl_2/MeOH (1:1 volume ratio) solution.

**[Hg(C_7H_4NS_2)_2] or C_14H_8N_2S_4Hg**. This compound melts at 184°C–185°C (Bell et al. 2001b) and crystallizes in the monoclinic structure with the lattice parameters $a = 1188.6 \pm 0.3$, $b = 603.6 \pm 0.2$, and $c = 3181.6 \pm 0.7$ pm and $\beta = 96.36° \pm 0.04°$ and a calculated density of 2.341 g·cm$^{-3}$ (Popović et al. 2002b) ($a = 1181.7 \pm 0.2$, $b = 598.30 \pm 0.08$, and $c = 3168.9 \pm 1.0$ pm and $\beta = 97.34° \pm 0.04°$ at $150 \pm 2$ K and a calculated density of 2.390 g·cm$^{-3}$ [Bell et al. 2001b]). It was prepared by the next procedure (Bell et al. 2001b, Popović et al. 2002b). HgCl_2 or Hg(OAc)_2 (6 mM) dissolved in warm EtOH (50 mL) was added by stirring to a warm solution of benzo-1,3-thiazoline-2-thione (12 mM) dissolved in warm EtOH (50 mL) containing Et_3N (12 mM). The yellow crystalline precipitate that formed overnight was filtered off and then recrystallized from boiling acetone as shiny sheet-like yellow crystals, which were filtered off and dried in vacuo.

The title compound could also be obtained by an electrochemical oxidation of Hg in an acetonitrile solution (50 mL) of 1,3-benzothiazole-2-thiol with a small amount of (Et_4N)ClO_4 (Tallon et al. 1995). As electrolysis proceeded, the color of the solution changed. At the end of the process, the solution was filtered to remove any solid impurities, and the filtrate was slowly concentrated at room temperature to give the solid title compound.

**[Hg(SCN)_2·(Phen)] or C_14H_8N_4S_2Hg**. This compound melts at 205°C–206°C (Jain and Rivest 1970). It was obtained by slowly adding a warm solution of Phen in THF to a warm solution of Hg(SCN)_2 in the same solvent. The molar ratio of the reactants was approximately 1:1. The colorless crystalline product that was precipitated immediately was filtered under reduced pressure on a sintered glass crucible, washed three times with a little amount of THF, and dried in vacuo for several hours.

**[Hg(C_7H_5N_2S)_2] or C_14H_10N_4S_2Hg**. To prepare this compound, a solution of benzo-1,3-imidazole-2-thione (2.73 mM) in 30 mL of MeOH was added dropwise to a solution of Hg(OAc)_2 (1.26 mM) in 30 mL of MeOH (Popović et al. 2002a). The colorless crystalline product was filtered off, washed with cold EtOH, and dried in air.

**[MeHgC_13H_11N_4S] or C_14H_14N_4SHg**. This compound melts at 144°C with decomposition (Sytsma and Kline 1973). To obtain it, MeHgOAc was dissolved in water and added to dithizone (diphenylthiocarbazone) dissolved in MeOH or methanolic KOH solution. The crude products were purified by the recrystallization in MeOH.

**[Hg(2-MeC$_5$H$_4$N)$_2$(SCN)$_2$] or C$_{14}$H$_{14}$N$_4$S$_2$Hg**. This compound melts at 81°C (Ahuja and Garg 1972). It was obtained by heating a suspension of Hg(SCN)$_2$ in hot EtOH with an excess of 2-MePy and by filtering the resultant mixture while still hot.

**[Hg{S$_2$CN(CH$_2$)$_6$}$_2$] or C$_{14}$H$_{24}$N$_2$S$_4$Hg**. This compound crystallizes in the monoclinic structure with the lattice parameters $a = 1100.82 \pm 0.08$, $b = 1073.37 \pm 0.08$, and $c = 1589.21 \pm 0.01$ pm and $\beta = 101.465° \pm 0.002°$ and a calculated density of 1.982 g·cm$^{-3}$ (Ivanov et al. 2008). It was synthesized as follows. A solution of Na{S$_2$CN(CH$_2$)$_6$}·2H$_2$O (5 mM) in H$_2$O (50 mL) was added to a vigorously stirred solution of Hg(NO$_3$)$_2$·H$_2$O (10 mM) in H$_2$O (20 mL). The resulting voluminous yellow precipitate was washed by decanting, filtered off, and dried in air. Flattened prismatic, transparent yellow crystals of the title compound for XRD were grown in chloroform.

**[Hg(C$_7$H$_{12}$NS$_2$)$_2$] or C$_{14}$H$_{24}$N$_2$S$_4$Hg, mercury(II) 4-methyl-piperidine dithio-carbamate**. This compound crystallizes in the monoclinic structure with the lattice parameters $a = 869.7 \pm 0.2$, $b = 1915.6 \pm 0.3$, and $c = 1209.8 \pm 0.2$ pm and $\beta = 108.14° \pm 0.01°$ and the experimental and calculated densities 1.88 and 1.90 g·cm$^{-3}$, respectively (Benedetti et al. 1988). It was prepared by adding a H$_2$O/EtOH (1:1 volume ratio) solution of HgCl$_2$ to an aqueous solution of 4-methylpiperidine dithiocarbamate in 1:2 metal/ligand molar ratio. The powder precipitated was collected by filtration and recrystallized from chloroform.

**[Hg{Pr$^i_2$CNS$_2$}] or C$_{14}$H$_{28}$N$_2$S$_4$Hg, bis(N,N-diisopropyldithiocarbamato) mercury(II)**. This compound exists in two polymorphic modifications, and both crystallize in the monoclinic structure (Iwasaki et al. 1978, Ito and Iwasaki 1979). α-Hg[Pr$^i_2$CNS$_2$] is characterized by the lattice parameters $a = 1946.8 \pm 0.1$, $b = 801.7 \pm 0.1$, and $c = 1483.2 \pm 0.1$ pm and $\beta = 119.18° \pm 0.03°$ and a calculated density of 1.818 g·cm$^{-3}$, and β-Hg[Pr$^i_2$CNS$_2$] has the lattice parameters $a = 3506 \pm 2$, $b = 980.6 \pm 0.3$, and $c = 1978 \pm 2$ pm and $\beta = 115.7° \pm 0.1°$ and a calculated density of 1.80 g·cm$^{-3}$. Hg[Pr$^i_2$CNS$_2$] was prepared by adding HgCl$_2$ to an aqueous solution of sodium N,N-diisopropyldithiocarbamate. Recrystallization from acetone solution gave two types of crystals (α + β), both being platelike and yellowish.

**[Hg(MeC$_{15}$H$_{10}$N$_3$S)] or C$_{16}$H$_{13}$N$_3$SHg**. This compound melts at 140°C and crystallizes in the orthorhombic structure with the lattice parameters $a = 599.23 \pm 0.10$, $b = 1402.88 \pm 0.03$, and $c = 1858.10 \pm 0.03$ pm and a calculated density of 2.041 g·cm$^{-3}$ (López-Torres et al. 2006). To obtain it, over a solution of 5-methoxy-5,6-diphenyl-4,5-dihydro-2H-[1,2,4]triazine-3-thione (1.1 mM) and LiOH·H$_2$O (1.1 mM) in 50 mL of MeOH was added dropwise a solution of HgMeCl (1.1 mM) in the same solvent. The solution was stirred at room temperature for 6 h. Then, it was evaporated until a yellow solid appeared, which was filtered off and vacuum dried. Single crystals suitable for XRD were obtained by slow evaporation of a solution of the title compound in DMF.

**[Hg(SCN)$_2$(C$_7$H$_8$N$_4$)$_2$] or C$_{16}$H$_{16}$N$_{10}$S$_2$Hg**. This compound crystallizes in the triclinic structure with the lattice parameters $a = 1034.3 \pm 0.7$, $b = 1322.5 \pm 0.8$, and

$c = 806.0 \pm 0.6$ pm and $\alpha = 97.15° \pm 0.04°$, $\beta = 103.43° \pm 0.03°$, and $\gamma = 79.01° \pm 0.04°$ and the experimental and calculated densities 1.94 and 1.941 g·cm$^{-3}$, respectively (Dillen et al. 1983). It was prepared by mixing acetone solutions of Hg(SCN)$_2$ and 5,7-dimethyl[1,2,4]triazolo[1,5-a]pyrimidine (C$_7$H$_8$N$_4$). The title compound was crystallized within several hours and then recrystallized from acetone. White single crystals were grown easily from the solution by cooling slowly to room temperature.

**[Hg(2,6-Me$_2$C$_5$H$_3$N)$_2$(SCN)$_2$] or C$_{16}$H$_{18}$N$_4$S$_2$Hg.** This compound melts at 112°C (Ahuja and Garg 1972). It was obtained by heating a suspension of Hg(SCN)$_2$ in hot EtOH with an excess of 2,6-Me$_2$Py and by filtering the resultant mixture while still hot.

**[Hg(2-EtC$_5$H$_4$N)$_2$(SCN)$_2$] and [Hg(3-EtC$_5$H$_4$N)$_2$(SCN)$_2$] or C$_{16}$H$_{18}$N$_4$S$_2$Hg.** These compounds melt at 58°C and 60°C, respectively (Ahuja and Garg 1972). They were obtained by heating a suspension of Hg(SCN)$_2$ in hot EtOH with an excess of 2-EtPy (3-EtPy) and by filtering the resultant mixture while still hot.

**[Hg(Me$_2$NC$_6$H$_4$S)$_2$] or C$_{16}$H$_{20}$N$_2$S$_2$Hg.** This compound changes color depending on the temperature (Lecher 1915).

**[Hg(C$_{14}$H$_{24}$N$_2$S$_2$)(SCN)$_2$] or C$_{16}$H$_{24}$N$_4$S$_4$Hg.** This compound crystallizes in the monoclinic structure with the lattice parameters $a = 1131.8 \pm 0.1$, $b = 1772.7 \pm 0.2$, and $c = 1153.6 \pm 0.2$ pm and $\beta = 109.08° \pm 0.02°$ and a calculated density of 1.83 g·cm$^{-3}$ (Baggio et al. 1995). To obtain it, $N,N'$-dicyclohexyldithiooxamide (2.50 mM) was dissolved in dry acetone (100 mL) to which Hg(SCN)$_2$ (1.25 mM) was added in small portions. The reaction mixture was stirred at 35°C for 4 h. The yellow precipitate that immediately formed was collected by filtration, washed with acetone, and dried in vacuo. The product was recrystallized by dissolving a very small portion in warm acetone and by allowing the solution to cool slowly. The yellow needle crystals of the title compound were isolated by filtration and dried in vacuo.

**[Hg(S$_2$CNPr$^i_2$)$_2$] or C$_{16}$H$_{28}$N$_2$S$_4$Hg.** This compound was synthesized as follows. A solution of Na(S$_2$CNPr$^i_2$)·3H$_2$O (5 mM) in H$_2$O (50 mL) was added to a vigorously stirred solution of Hg(NO$_3$)$_2$·H$_2$O (10 mM) in H$_2$O (20 mL). The resulting voluminous yellow precipitate was washed by decanting, filtered off, and dried in air (Ivanov et al. 2008).

**[Hg$_{1.5}$(SCN)$_3$(C$_7$H$_{10}$N$_2$)$_2$] or C$_{17}$H$_{20}$N$_7$S$_2$Hg$_{1.5}$.** This compound melts at 124°C–126°C (Secondo et al. 2000). To obtain it, an ethanolic solution of 4-(N,N-dimethylamino) pyridine (C$_7$H$_{10}$N$_2$) (4 mM) was added to a solution of mercury(III) thiocyanate (2 mM) in the same solvent, and the solution was stirred for 2 h. The solvent was evaporated at room temperature, upon which precipitation of the title compound occurred. It was filtered, washed with ether, and dried in vacuo.

**[Hg{S$_2$CNEt(C$_6$H$_{11}$)}$_2$] or C$_{18}$H$_{32}$N$_2$S$_4$Hg.** This compound melts at 160°C–162°C and crystallizes in the monoclinic structure with the lattice parameters $a = 845.9 \pm 0.6$, $b = 1950.6 \pm 0.9$, and $c = 1395.5 \pm 0.6$ pm and $\beta = 92.89° \pm 0.04°$ and a calculated density of 1.748 g·cm$^{-3}$ (Cox and Tiekink 1999a). It was recrystallized from acetone/ Pr$^i$OH (1:1 volume ratio), and yellow crystals were obtained.

**[Hg(S₂CNBuⁿ₂)] or C₁₈H₃₆N₂S₄Hg.** This compound melts at 59°C–61°C and crystallizes in the monoclinic structure with the lattice parameters $a = 2321.0 \pm 0.6$, $b = 1648 \pm 1$, and $c = 1628.5 \pm 0.6$ pm and $\beta = 126.81° \pm 0.02°$ and a calculated density of 1.623 g·cm⁻³ (Cox and Tiekink 1999a). It was recrystallized from acetonitrile, and yellow crystals were obtained.

**[Hg(S₂CNBuⁱ₂)] or C₁₈H₃₆N₂S₄Hg.** This compound melts at 87°C–88°C and crystallizes in the monoclinic structure with the lattice parameters $a = 2627 \pm 2$, $b = 1365.4 \pm 0.8$, and $c = 1386.3 \pm 0.5$ pm and $\beta = 96.64° \pm 0.04°$ and a calculated density of 1.639 g·cm⁻³ (Cox and Tiekink 1999a). It was recrystallized from DMF, and pale-yellow crystals were obtained.

**[Hg(C₁₀H₆NS₅)₂)] or C₂₀H₁₂N₂S₁₀Hg.** This compound melts at 162°C–163°C with decomposition and crystallizes in the triclinic structure with the lattice parameters $a = 817.9 \pm 0.2$, $b = 912.70 \pm 0.10$, and $c = 926.50 \pm 0.10$ pm and $\alpha = 96.860° \pm 0.010°$, $\beta = 109.970° \pm 0.010°$, and $\gamma = 90.370° \pm 0.010°$ and a calculated density of 2.065 g·cm⁻³ (Nam et al. 2004). To prepare it, after obtaining and filtration of Hg(C₁₀H₇NS₅)X₂ (X = Cl, Br, I), the filtrate was concentrated slowly in air. The yellow crystals collected from the filtrate were washed with MeOH and dried in vacuo. The chemical composition of the compounds obtained from the filtrate of Hg(C₁₀H₇NS₅)X₂ was identical regardless of which halide ion was used.

**[Hg(C₁₈H₁₂N₂S₂)(SCN)₂] or C₂₀H₁₂N₄S₄Hg.** This compound crystallizes in the monoclinic structure with the lattice parameters $a = 1316.7 \pm 0.3$, $b = 730.32 \pm 0.15$, and $c = 2247.4 \pm 0.8$ pm and $\beta = 92.99° \pm 0.02°$ and a calculated density of 1.961 g·cm⁻³ (Mahjoub and Morsali 2003). It was obtained by reacting 2,2′-diphenyl-4,4′-bithiazole (2 mM) with Hg(SCN)₂ (1 mM) in MeOH (5 mL) at room temperature, with stirring, for 48 h. The white solid formed was filtered out and dried under vacuum. The crude product was dissolved in CH₃CN (10 mL), and Et₂O was diffused into it, forming a mixture of white precipitate and light-yellow crystals.

**[Hg(PyS)₂(PySH)₂] or C₂₀H₁₈N₄S₄Hg.** This compound crystallizes in the orthorhombic structure with the lattice parameters $a = 905.74 \pm 0.04$, $b = 1030.75 \pm 0.04$, and $c = 1182.03 \pm 0.04$ pm and a calculated density of 1.936 g·cm⁻³ (Anjali et al. 2003). It was obtained by the next procedure. To a solution of 4-pyridine thiol (2.25 mM) in 2 mL of DMSO was added Hg(ClO₄)₂·6H₂O (0.38 mM) in 2 mL of MeOH. The mixture was warmed slightly and stirred for 15 min. The resulting clear solution was cooled to room temperature to get bright-greenish-yellow crystals. The crystals were separated by filtration, washed with MeOH and Et₂O, and dried in vacuo. The final product of thermal decomposition of the title compound is HgS.

**[(C₇H₁₀N)₂{Hg(C₃S₅)₂}] or C₂₀H₂₀N₂S₁₀Hg.** This compound crystallizes in the monoclinic structure with the lattice parameters $a = 2028.6 \pm 0.4$, $b = 1459.8 \pm 0.3$, and $c = 1000.7 \pm 0.2$ pm and $\beta = 104.964° \pm 0.003°$ (Wang et al. 2006). Its single crystals were obtained by slow evaporation of an acetone solution held at approximately 4°C.

**[Hg(C₅H₈NS₂)₂]₂ or C₂₀H₃₂N₄S₈Hg₂.** This compound crystallizes in the triclinic structure with the lattice parameters $a = 748.09 \pm 0.10$, $b = 795.2 \pm 1.0$, and $c = 1249.06 \pm 0.13$ pm and $\alpha = 101.060° \pm 0.012°$, $\beta = 100.416° \pm 0.012°$, and $\gamma = 91.536° \pm 0.008°$ at $173 \pm 2$ K and a calculated density of 2.288 g·cm$^{-3}$ (Altaf et al. 2010b). It was prepared by mixing 1 mM of HgCl$_2$ in 15 mL of MeOH and pyrrolidine dithiocarbamate (C$_5$H$_9$NS$_2$) in MeOH (20 mL) (1:2 mole ratio). The addition of C$_5$H$_9$NS$_2$ in the colorless metal ion solution resulted in the formation of yellow precipitate immediately. After stirring for 30 min, the precipitate was filtered off and dried. Yellow crystals of the title compound were prepared by dissolving [Hg(C$_5$H$_8$NS$_2$)$_2$]$_2$ in DMSO on heating and then cooling the resulting solution.

**[Hg{S$_2$CNPr$^i$(C$_6$H$_{11}$)}$_2$] or C$_{20}$H$_{36}$N$_2$S$_4$Hg.** This compound melts at 172°C–173°C and crystallizes in the triclinic structure with the lattice parameters $a = 1229.1 \pm 0.9$, $b = 1153.4 \pm 0.6$, and $c = 955.9 \pm 0.4$ pm and $\alpha = 69.82° \pm 0.04°$, $\beta = 88.94° \pm 0.04°$, and $\gamma = 72.00° \pm 0.05°$ and a calculated density of 1.747 g·cm$^{-3}$ (Cox and Tiekink 1999a). It was recrystallized from acetone/Pr$^i$OH (1:1 volume ratio), and yellow crystals were obtained.

**[Hg(CN)$_2$(C$_9$H$_{20}$N$_2$S)$_2$] or C$_{20}$H$_{40}$N$_6$S$_2$Hg.** This compound crystallizes in the monoclinic structure with the lattice parameters $a = 1746.92 \pm 0.03$, $b = 959.28 \pm 0.02$, and $c = 1746.99 \pm 0.04$ pm and $\beta = 115.540° \pm 0.001°$ and a calculated density of 1.535 g·cm$^{-3}$ (Ahmad et al. 2009). To obtain it, 1 mM of Hg(CN)$_2$ dissolved in 15 mL of MeOH was mixed with two equivalents of $N,N'$-dibutylthiourea in 15 mL of MeOH. After stirring for 15 min, the resulting mixture was filtered, and the filtrate was kept at room temperature. After 24 h, white crystals were obtained.

**[Et$_4$N][Hg(SBu$^t$)$_3$] or C$_{20}$H$_{47}$NS$_3$Hg.** This compound melts at 170°C–175°C with decomposition (Bowmaker et al. 1984) and crystallizes in the monoclinic structure with the lattice parameters $a = 1153.9 \pm 0.1$, $b = 1104.6 \pm 0.1$, and $c = 2102.7 \pm 0.2$ pm and $\beta = 90.65° \pm 0.01°$ and a calculated density of 1.48 g·cm$^{-3}$ (Watton et al. 1990). It was prepared by the next procedure (Bowmaker et al. 1984). 2-Methylpropane-2-thiol (60 mM) was dissolved in a mixture of 25% aqueous Et$_4$NOH (30 mM) and absolute EtOH (30 mL). HgO (10 mM) was added to the resulting solution, and the mixture was warmed to 55°C and stirred until all HgO had dissolved. The resulting solution was filtered and reduced to about half of its original volume. The white crystalline product was collected and washed with absolute EtOH.

The title compound was also prepared by the following method (Bowmaker et al. 1984). 2-Methylpropane-2-thiol (13 mM) and bis(2-methylpropane-2-thiolato) mercury(II) (7 mM) were added to a mixture of 25% aqueous Et$_4$NOH (13 mM) and EtOH (10 mL). The solution became clear on warming to 60°C and was filtered and reduced in volume to give white crystals of [Et$_4$N][Hg(SBu$^t$)$_3$]. Single crystals of [Et$_4$N][Hg(SBu$^t$)$_3$] were obtained by slow evaporation of an EtOH/H$_2$O mixture (Watton et al. 1990).

**[Hg(Et$_2$CNS$_2$)$_2$(Phen)] or C$_{22}$H$_{28}$N$_4$S$_4$Hg.** This compound crystallizes in the triclinic structure with the lattice parameters $a = 1030.8 \pm 0.2$, $b = 1117.1 \pm 0.3$, and $c = 1155.2 \pm 0.2$ pm and $\alpha = 94.16° \pm 0.02°$, $\beta = 96.66° \pm 0.01°$, and $\gamma = 105.17° \pm 0.02°$ pm and a calculated density of 1.774 g·cm$^{-3}$ (Klevtsova et al. 2002). It was prepared by the interaction of Hg(Et$_2$CNS$_2$)$_2$ and Phen (2:2.4 molar ratio) in the boiling toluene (Larionov et al. 1990). Its single crystals were obtained by the interaction of $[Hg(Et_2CNS_2)_2]$ with the excess of Phen by heating in toluene with the next solution cooling (Klevtsova et al. 2002).

**$[Me_4N]_2[Hg(MeC_6H_3S_2)$ or $C_{22}H_{36}N_2S_4Hg$.** This compound was synthesized under an $N_2$ atmosphere. Toluene-3,4-dithiol (32 mM), $Bu^n_3N$ (24 mM), $Me_4NCl$ (20 mM), and $Hg(NO_3)_2$ (3.4 mM) were mixed in deaerated MeOH (Carson and Dean 1982). Deaerated $Bu^nOH$ was added to lower the solubility of the colorless compound crystallized upon cooling.

**$[Me_4N][Hg(SC_6H_{11})_3]$ or $C_{22}H_{45}NS_3Hg$.** To obtain this compound, a solution of $HgCl_2$ (1 mM) in acetonitrile (15 mL) was slowly added with stirring to acetonitrile (5 mL) containing $NaSC_6H_{11}$ (3 mM) and EtOH (ca. 1 mL) (Alsina et al. 1992b). After precipitation of NaCl and addition of EtOH (10 mL) containing 1 mM $Me_4NCl$, the resulting mixture was stirred for several hours. NaCl was filtered off, and the filtrate was evaporated to dryness. Extraction of the residue with hot acetonitrile, filtration to remove some insoluble material, and addition of diethyl ether to the filtrate caused separation of a white crystalline solid. This material was recrystallized from acetonitrile/ether at $-20°C$ to give the title compound as colorless crystals, which were filtered off at a low temperature, washed with acetonitrile/ether (2:1 volume ratio), and dried under vacuum.

**$[Hg(C_{18}H_{12}N_2)(SCN)_2]·0.5C_6H_6$ or $C_{23}H_{15}N_4S_2Hg$.** This compound melts at 210°C and crystallizes in the monoclinic structure with the lattice parameters $a = 1118.6 ± 0.5$, $b = 1513.7 ± 0.7$, and $c = 1235.5 ± 0.6$ pm and $\beta = 100.28° ± 0.01°$ at 110 K (Ramazani et al. 2004). It was obtained by reacting 2,2'-biquinoline (2 mM) with $Hg(SCN)_2$ (1 mM) in MeOH (5 mL) at room temperature with stirring for 48 h. The white solid formed was filtered out and dried under vacuum.

**$[Hg(C_7H_4NS_2)_2]·(2,2'-bipy)]$ or $C_{24}H_{16}N_4S_4Hg$.** This compound crystallizes in the triclinic structure with the lattice parameters $a = 955.3 ± 0.2$, $b = 1162.0 ± 0.2$, and $c = 1166.0 ± 0.2$ pm and $\alpha = 72.13° ± 0.02°$, $\beta = 74.41° ± 0.02°$, and $\gamma = 85.94° ± 0.02°$ and a calculated density of 1.929 $g·cm^{-3}$ (Popović et al. 2002b). It crystallized from a DMF/MeOH (1:1 volume ratio) mixture solvent of equimolar amount of $[Hg(C_7H_4NS_2)_2]$ and 2,2'-bipy.

**$[(C_4H_3S)_2Hg·(3,4,7,8-Me_4Phen)]$ or $C_{24}H_{22}N_2S_2Hg$.** This compound melts at 180°C–210°C (Bell et al. 2004). The reaction of 2-thienylmercury(II) chloride with 3,4,7,8-tetramethyl-1,10-phenanthroline in ethanolic solution produced the title compound as colorless prism.

**$[Hg(SCN)_2(Phen)_2]$ or $C_{26}H_{16}N_6S_2Hg$.** This compound melts at 244°C–245°C (Jain and Rivest 1970) and crystallizes in the triclinic structure with the lattice parameters $a = 1325.2 ± 0.5$, $b = 1107.7 ± 0.4$, and $c = 844.3 ± 0.3$ pm and $\alpha = 105.20° ± 0.03°$, $\beta = 83.25° ± 0.03°$, and $\gamma = 90.92° ± 0.03°$ and the experimental and calculated densities 1.95 and 1.99 $g·cm^{-3}$, respectively (Beauchamp et al. 1971, 1974). It was obtained by slowly adding under constant stirring a warm solution of Phen in THF to a warm solution of $Hg(SCN)_2$ in the same solvent. The molar ratio of the reactants was approximately

1:3 (Jain and Rivest 1970). The colorless crystalline product that was precipitated immediately was filtered under reduced pressure on a sintered glass crucible, washed three times with a little amount of THF, and dried in vacuo for several hours.

**$Hg[S_2CN(C_6H_{11})_2]_2$ or $C_{26}H_{44}N_2S_4Hg$, bis(dicyclohexyldithiocarbamato) mercury(II).** This compound melts at 195°C–198°C and crystallizes in the trigonal structure with the lattice parameters $a = 1037.0 \pm 0.7$ and $c = 2395.3 \pm 0.9$ pm at 173 K and a calculated density of 1.593 g·cm$^{-3}$ (Cox and Tiekink 2000). Its pale-yellow-green crystals were obtained from an acetone solution of the compound.

**$[\{(C_5H_{12}N_2S)_2Hg(CN)_2\}_2 \cdot Hg(CN)_2]$ or $C_{26}H_{48}N_{14}S_4Hg_3$.** This compound melts at 126°C–128°C and crystallizes in the monoclinic structure with the lattice parameters $a = 1043.27 \pm 0.09$, $b = 1365.35 \pm 0.09$, and $c = 1441.83 \pm 0.13$ pm and $\beta = 94.759° \pm 0.011°$ at $173 \pm 2$ K and a calculated density of 2.084 g·cm$^{-3}$, (Altaf et al. 2010a). It was prepared by adding 1.75 equivalents of tetramethylthiourea ($C_5H_{12}N_2S$) in $H_2O$ (10 mL) to a solution of $Hg(CN)_2$ (1 mM) in MeOH (15 mL) and stirring the solution for 0.5 h. White crystalline product was obtained from the resulting colorless solution, which was washed with MeOH.

**$[Et_4N][Hg(SC_6H_{11})_3]$ or $C_{26}H_{53}NS_3Hg$.** This compound crystallizes in the triclinic structure with the lattice parameters $a = 1072.4 \pm 0.54$, $b = 1244.0 \pm 0.5$, and $c = 1264.3 \pm 0.5$ pm and $\alpha = 72.40° \pm 0.02°$, $\beta = 79.36° \pm 0.02°$, and $\gamma = 73.33° \pm 0.02°$ pm and a calculated density of 1.467 g·cm$^{-3}$ (Alsina et al. 1992b). To obtain it, a solution of $HgCl_2$ (1 mM) in acetonitrile (15 mL) was slowly added with stirring to acetonitrile (5 mL) containing $NaSC_6H_{11}$ (3 mM) and EtOH (ca. 1 mL). After precipitation of NaCl and addition of MeCN (10 mL) containing 1 mM $Et_4NCl$, the resulting mixture was stirred for several hours. NaCl was filtered off, and the filtrate was evaporated to dryness. Extraction of the residue with hot acetonitrile, filtration to remove some insoluble material, and addition of diethyl ether to the filtrate caused separation of a white crystalline solid. This material was recrystallized from acetonitrile/ether at –20°C to give the title compound as colorless crystals, which were filtered off at a low temperature, washed with acetonitrile/ether (2: 1 volume ratio), and dried under vacuum.

**$[Bu^n_4N][Hg(SBu^t)_3]$ or $C_{28}H_{63}NS_3Hg$.** This compound melts at 89°C–93°C (Bowmaker et al. 1984). It was prepared by the next procedure. 2-Methylpropane-2-thiol (29.6 mM) was added to a mixture of 40% aqueous $Bu^n_4NOH$ (21.1 mM) and absolute EtOH (15 mL) to produce a cloudy solution. To this the solution of $Hg(SBu^t)_2$ (7 mM) was added, and the resulting mixture was stirred with heating until all of the solid had dissolved. The clear solution was filtered, and the volume of the filtrate was reduced to about 10 mL, whereupon white crystals of the title compound formed. These were collected by filtration and dried under vacuum.

**$[Hg(C_{15}H_{10}N_3S)_2]$ or $C_{30}H_{20}N_6S_2Hg$.** This compound melts at 218°C and crystallizes in the monoclinic structure with the lattice parameters $a = 1035.74 \pm 0.02$, $b = 1154.51 \pm 0.02$, and $c = 1150.16 \pm 0.03$ pm and $\beta = 98.9730° \pm 0.0010°$ and a calculated density of 1.783 g·cm$^{-3}$ (López-Torres et al. 2006). Single crystals of the title compound suitable for XRD were grown by slow evaporation of its solution in acetonitrile.

**[Hg{S$_2$CN(CH$_2$Ph)$_2$}$_2$] or C$_{30}$H$_{28}$N$_2$S$_4$Hg.** This compound melts at 167°C–168°C (209°C–210°C [Lai et al. 2004]) and crystallizes in the orthorhombic structure with the lattice parameters $a = 1647.38 \pm 0.01$, $b = 1864.18 \pm 0.14$, and $c = 940.00 \pm 0.06$ pm ($a = 1638.51 \pm 0.05$, $b = 1857.74 \pm 0.06$, and $c = 931.90 \pm 0.03$ pm at 223 K [Lai et al. 2004]) and a calculated density of 1.715 g·cm$^{-3}$ (Fan et al. 2004). Colorless crystals were obtained by slow evaporation of a chloroform solution of the title compound (Lai et al. 2004).

**[Hg$_2$(PhMeNCS$_2$)$_4$] or C$_{32}$H$_{32}$N$_4$S$_8$Hg$_2$.** This compound melts at 235°C–236°C and crystallizes in the monoclinic structure with the lattice parameters $a = 1271.68 \pm 0.10$, $b = 651.98 \pm 0.06$, and $c = 2226.12 \pm 0.09$ pm and $\beta = 98.341° \pm 0.003°$ at $100 \pm 2$ K and a calculated density of 2.056 g·cm$^{-3}$ (Onwudiwe and Ajibade 2011). The preparation of it was carried out using 25 mL of aqueous solution of HgCl$_2$ (1.25 mM), which was added to 25 mL of aqueous solution of $N$-methyl-$N$-phenyldithiocarbamate (2.50 mM). A solid precipitate formed immediately, and the mixture was stirred for about 45 min, filtered off, rinsed several times with distilled water, and recrystallized from CH$_2$Cl$_2$/ethyl acetate (3:1 volume ratio) mixture.

**[Me$_4$N]$_2$[Hg(SPh)$_4$] or C$_{32}$H$_{44}$N$_2$S$_4$Hg.** This compound was synthesized under N$_2$ atmosphere. PhSH (32 mM), Bu$^n{}_3$N (24 mM), Me$_4$NCl (20 mM), and Hg(NO$_3$)$_2$ (3.4 mM) were mixed in deaerated MeOH (Carson and Dean 1982). Deaerated Bu$^n$OH was added to lower the solubility of the colorless compound crystallized upon cooling.

**[Bu$^n{}_4$N][Hg(SPh)$_3$] or C$_{34}$H$_{51}$NS$_3$Hg.** This compound melts at 125°C–127°C (Bowmaker et al. 1996) and crystallizes in the monoclinic structure with the lattice parameters $a = 2066.3 \pm 0.7$, $b = 1681.2 \pm 0.6$, and $c = 975.7 \pm 0.3$ pm and $\beta = 95.52° \pm 0.02°$ (Christou et al. 1984). To obtain it, Na (80 mM) was dissolved in MeOH (100 mL), and benzenethiol (80 mM) was added, followed by solid HgO (20 mM). The solid was slowly dissolved with stirring and slight warming to produce a pale-yellow solution. After 30 min, Bu$^n{}_4$NBr (40 mM) in EtOH (80 mL) was added to yield a copious precipitate of small platelike crystals. The flask was left at 0°C overnight and the solid collected by filtration, washed copiously with EtOH, and dried in vacuo. The crude material was dissolved in warm MeCN (45 mL, ~ 50°C) and filtered, and EtOH was added at this temperature. Slow cooling to ambient temperature and then 0°C overnight produced large yellow prisms, which were collected by filtration, washed copiously with EtOH, and dried in vacuo. All manipulations were performed under an N$_2$ atmosphere.

The title compound could also be prepared by the reaction of NaSPh, HgO, and [NBu$^n{}_4$]Br in a 5:1:2 molar ratio in MeOH (Bowmaker et al. 1996).

**[Hg(NHPhNHCSN$_2$Ph)$_2$·2C$_5$H$_5$N] or C$_{36}$H$_{32}$N$_{10}$S$_2$Hg.** This compound crystallizes in the orthorhombic structure with the lattice parameters $a = 3530 \pm 3$, $b = 3630 \pm 3$, and $c = 540 \pm 2$ pm and a calculated density of 1.66 g·cm$^{-3}$ (Harding 1958). Deep-red needles of Hg(NHPhNHCSN$_2$Ph)$_2$·2C$_5$H$_5$N were obtained by crystallization from aqueous pyridine.

**[Bu$^n{}_4$N]$_2$[Hg(C$_3$S$_5$)$_2$] or C$_{38}$H$_{72}$N$_2$S$_{10}$Hg.** This compound crystallizes in the monoclinic structure with the lattice parameters $a = 1015.1 \pm 0.1$, $b = 2865.9 \pm 0.4$, and

$c = 928.6 \pm 0.2$ pm and $\beta = 110.18° \pm 0.01°$ (Wang et al. 2005a). Two methods were used to prepare the title compound. (1) To degassed DMF (48 mL), $CS_2$ (24 mL) was added, and the mixture was cooled to 0°C. Sodium (1.45 g) was added to the solution, and the mixture was vigorously stirred with cooling until the reaction was completed. An adequate volume of MeOH was slowly added. To this solution, separate solutions of $HgCl_2$ (4.24 g) dissolved in acetone (10 mL) and $Bu^n_4NBr$ (10.12 g) in $H_2O$ (30 mL) were added consecutively with stirring at room temperature. The mixture was stirred overnight, and then the product was isolated by filtration and washed with $H_2O$ and MeOH, to afford the crystals of the title compound. (2) $[Bu^n_4N]_2[Hg(C_3S_5)_2]$ could also be obtained by the metathetical reaction of $[Bu^n_4N]_2[Zn(C_3S_5)_2]$ (9.44 g) and $HgCl_2$ (2.72 g) in acetone (55 mL). The dark-red single crystals used for XRD were obtained by slow evaporation of an acetone solution.

**$[Hg\{S_2CNMe(EtPh)\}_2]_2$ or $C_{40}H_{52}N_4S_8Hg_2$.** This compound was prepared by dissolving 6.1 mM of $Hg(OAc)_2$ in 150 mL of EtOH followed by filtration (Green et al. 2004). This a mixture of $CS_2$ (12 mM) and *N*-methylphenethylamine ($C_9H_{13}N$) (12 mM) in 50 mL of EtOH was added dropwise, producing an immediate white precipitate. After 4 h stirring, the precipitate was filtered, air-dried, and redissolved in 50 mL of $CH_2Cl_2$. Storage at 4°C resulted in yellow needlelike crystals, which are indefinitely stable at room temperature. The title compound has been used as a room temperature precursor to a metastable phase of HgS not normally accessible via low-temperature reaction pathways.

**$[Hg(SC_6H_2Bu^t_3)_2(C_5H_5N)]$ or $C_{41}H_{63}NS_2Hg$.** This compound crystallizes in the triclinic structure with the lattice parameters $a = 1054.1 \pm 0.2$, $b = 1123.4 \pm 0.3$, and $c = 2010.5 \pm 0.6$ pm and $\alpha = 87.16° \pm 0.02°$, $\beta = 79.31° \pm 0.02°$, and $\gamma = 63.32° \pm 0.02°$ pm and a calculated density of 1.327 g·cm⁻³ (Bochmann et al. 1992). Recrystallization of $Hg(SC_6H_2Bu^t_3)_2$ from dry Py at room temperature gave colorless crystals of this compound, which released Py above 140°C and decomposed above 200°C.

**$[Me_4N]_2[Hg_2(SPh)_6]$ or $C_{44}H_{54}N_2S_6Hg_2$.** This compound melts at 116°C–119°C and crystallizes in the monoclinic structure with the lattice parameters $a = 1984 \pm 1$, $b = 1566.6 \pm 0.3$, and $c = 1628.4 \pm 0.9$ pm and $\beta = 110.91° \pm 0.03°$ and a calculated density of 1.69 g·cm⁻³ (Bowmaker et al. 1996). It was prepared by the following procedure. Benzenethiol (30 mM) was added to HgO (10 mM) suspended in EtOH (5 mL). $Me_4NOH$ (10 mM) was added to this mixture, and benzenethiol and $Me_4NOH$ containers were each rinsed with ethanol (5 mL), which was then added to the reaction mixture. The mixture was heated to about 90°C and stirred for 1.25 h, after which time most of the solid material had dissolved. Upon cooling to room temperature, an oil is separated, and this solidified upon further cooling to 0°C. The supernatant liquid was decanted off and the solid redissolved in EtOH (100 mL) at 90°C. The solution was filtered while hot and a large amount of white cloudy oil separated from the filtrate upon cooling. More EtOH (20 mL) was added, and the mixture was heated until a clear yellow solution was obtained. At first, an oil separated again upon cooling, but this quickly solidified, and the remainder of the product separated upon further cooling as a white crystalline solid. All reactions were carried out in Schlenk tubes under an atmosphere of oxygen-free nitrogen.

**[Et$_4$N]$_2$[Hg(S-2-PhC$_6$H$_4$)$_4$]·2MeCN or C$_{68}$H$_{82}$N$_4$S$_4$Hg.** This compound crystallizes in the tetragonal structure with the lattice parameters $a = 1585.7 \pm 0.8$ and $c = 2529 \pm 1$ pm (Silver et al. 1993). To obtain it, lithium 2-phenylbenzenethiolate, Li(S-2-PhC$_6$H$_4$), was generated in MeOH from thiol (3.2 mM) and Li (3.3 mM). The MeOH was removed, and then Hg salt (0.55 mM) was added, followed by the addition of 30 mL of MeCN. After stirring for 3 h, the reaction mixture was filtered, and the filtrate was added to [Et$_4$N]Br (1.1 mM). Large crystals formed after 4 days at –20°C.

**[{Bu$^n$$_4$N}$_2$[Hg$_2$(Ph$_2$C$_4$S$_4$)$_3$] or C$_{80}$H$_{102}$N$_2$S$_{12}$Hg$_2$.** This compound melts at 180°C–181°C and crystallizes in the triclinic structure with the lattice parameters $a = 1214.4 \pm 0.2$, $b = 1221.9 \pm 0.2$, and $c = 2958.8 \pm 0.8$ pm and $\alpha = 96.82° \pm 0.04°$, $\beta = 91.326° \pm 0.009°$, and $\gamma = 99.380° \pm 0.005°$ pm and a calculated density of 1.451 g·cm$^{-3}$ (Noh et al. 1997). To prepare it, 1,3-dithiole-2-one was treated with KOH dissolved in MeOH under N$_2$. To this solution, separate solutions of HgCl$_2$ and [Bu$^n$$_4$N] Br in MeOH were added. The yellow precipitate was filtered off and washed with MeOH. The precipitate was recrystallized from acetone/ethyl acetate in a refrigerator to yield the title compound.

## 7.37  Hg–H–C–N–Se–Cl–S

**[Hg(C$_3$H$_5$NSeS)$_2$Cl$_2$] or C$_6$H$_{10}$N$_2$Se$_2$Cl$_2$S$_2$Hg.** This multinary compound is formed in this system. It decomposes at 187°C–189°C (Preti et al. 1974) and was obtained by the reaction of HgCl$_2$ solution in EtOH with thiazolidine-2-selenone dissolved in CHCl$_3$ by refluxing for 12 h. The crude product that separated out during the reaction was purified by washing with chloroform.

## 7.38  Hg–H–C–N–Se–Br–S

**[Hg(C$_3$H$_5$NSeS)$_2$Br$_2$] or C$_6$H$_{10}$N$_2$Se$_2$Br$_2$S$_2$Hg.** This multinary compound is formed in this system. It decomposes at 180°C–182°C (Preti et al. 1974) and was obtained by the reaction of HgBr$_2$ solution in EtOH with thiazolidine-2-selenone dissolved in CHCl$_3$ by refluxing for 12 h. The crude product that separated out during the reaction was purified by washing with chloroform.

## 7.39  Hg–H–C–N–Se–I–S

**[Hg(C$_3$H$_5$NSeS)$_2$I$_2$] or C$_6$H$_{10}$N$_2$Se$_2$I$_2$S$_2$Hg.** This multinary compound is formed in this system. It decomposes at 197°C–199°C (Preti et al. 1974) and was obtained by the reaction of HgI$_2$ solution in EtOH with thiazolidine-2-selenone dissolved in CHCl$_3$ by refluxing for 12 h. The crude product that separated out during the reaction was purified by washing with chloroform.

## 7.40  Hg–H–C–N–F–S

Some compounds are formed in this system.

**[Hg(SCF$_3$)$_2$·(HSCNMe$_2$)] or C$_5$H$_7$NF$_6$S$_3$Hg.** This compound melts at 91°C–92°C (Man et al. 1959).

**[Hg(C$_5$H$_2$N$_2$F$_3$S)$_2$] or C$_{10}$H$_4$N$_4$F$_6$S$_2$Hg.** A solution of acetonitrile (50 mL) containing 4-(trifluoromethyl)pyrimidine-2-thione (C$_5$H$_3$N$_2$F$_3$S) (0.83 mM) was electrolyzed at 10 mA with Hg anode during 2 h (Rodríguez et al. 2005). No precipitation occurred initially, but within several hours, a white solid had formed, and this was filtered off, washed with acetonitrile, and dried.

**[Hg(3-CF$_3$C$_5$H$_3$NS)$_2$] or C$_{12}$H$_6$N$_2$F$_6$S$_2$Hg.** This compound crystallizes in the orthorhombic structure with the lattice parameters $a = 757.88 \pm 0.15$, $b = 1880.0 \pm 0.4$, and $c = 2147.9 \pm 0.4$ pm and a calculated density of 2.417 g·cm$^{-3}$ (Sousa-Pedrares et al. 2003). To obtain it, a solution of 3-CF$_3$C$_5$H$_4$NS (1.117 mM) and NEt$_3$ (1.680 mL) in MeOH (10 mL) was heated under reflux, and a solution of Hg(OAc)$_2$ (0.558 mM) in MeOH (10 mL) was added. A white solid was formed immediately, and the reaction mixture was stirred for 4 h. The solid was filtered off, washed with EtOH and diethyl ether, and dried under vacuum. Crystals were obtained from the mother liquor. All manipulations were done under the atmosphere of dry N$_2$.

**[Hg(5-CF$_3$C$_5$H$_3$NS)$_2$] or C$_{12}$H$_6$N$_2$F$_6$S$_2$Hg.** This compound crystallizes in the monoclinic structure with the lattice parameters $a = 1463.9 \pm 0.6$, $b = 1507.5 \pm 0.6$, and $c = 693.4 \pm 0.6$ pm and $\beta = 92.085° \pm 0.008°$ and a calculated density of 2.419 g·cm$^{-3}$ (Sousa-Pedrares et al. 2003). It was prepared by the similar procedure to that described for [Hg(3-CF$_3$C$_5$H$_3$NS)$_2$] using 5-CF$_3$C$_5$H$_4$NS instead of 3-CF$_3$C$_5$H$_4$NS. The solid was filtered off, washed with MeOH and diethyl ether, and dried under vacuum. Crystals were obtained by crystallization of this compound from diethyl ether. All manipulations were done under the atmosphere of dry N$_2$.

**[Hg(C$_6$H$_4$N$_2$F$_3$S)$_2$] or C$_{12}$H$_8$N$_4$F$_6$S$_2$Hg.** This compound crystallizes in the orthorhombic structure with the lattice parameters $a = 763.2 \pm 0.2$, $b = 3185.5 \pm 0.6$, and $c = 696.9 \pm 0.1$ pm and a calculated density of 2.301 g·cm$^{-3}$ (Rodríguez et al. 2005). A solution of acetonitrile (50 mL) containing 4,6-(trifluoromethyl)methylpyrimidine-2-thione (C$_6$H$_5$N$_2$F$_3$S) (1.03 mM) was electrolyzed at 10 mA with Hg anode during 2 h. At the end of the experiment, a microcrystalline white solid of the title compound had formed, and this was filtered off, washed with acetonitrile, and dried. Crystallization from EtOH/acetone gave single crystals suitable for XRD.

**[Et$_4$N]$_2$[Hg(SC$_6$F$_5$)$_4$] or C$_{40}$H$_{40}$N$_2$F$_{20}$S$_4$Hg.** This compound decomposes at 172°C (Beck et al. 1967). It was prepared by the next procedure. To a solution of C$_6$F$_5$SH (39.3 mM) in 20 mL of 2 N NaOH an aqueous solution (50 mL) of HgCl$_2$ was added dropwise up to saturation under constant stirring. The obtained mixture was filtered, and the filtrate was precipitated by Et$_4$N. The precipitate was washed with H$_2$O and dried under high vacuum. The colorless needlelike crystals were grown by recrystallization of the title compound from a mixture solution of acetone/diethyl ether.

## 7.41   Hg–H–C–N–F–Cl–S

**Me$_4$N[Hg(SCF$_3$)$_2$Cl] or C$_6$H$_{12}$NF$_6$ClS$_2$Hg.** This multinary compound is formed in this system. To obtain it, powdered bis(trifluoromethylthio)mercury and tetramethylammonium chloride were mixed in a molar ratio of 1.1:1.0 (Jellinek and Lagowski 1960). The mixture initially formed a paste, which was homogenized. After some time, the mixture solidified; it was heated at 50°C, crushed, and homogenized several times. The product was cooled to room temperature, quickly washed three times with cold ether (which removes the excess of bis(trifluoromethylthio)mercury), and dried at 50°C. The preparation was carried out in dry $N_2$, because of the hygroscopicity of the starting material. The title compound decomposes at 131°C forming $HgCl_2$ and a liquid, which decomposes slowly on further heating to form mercuric sulfide, $Me_4NF$, and volatile C–F–S compounds.

## 7.42    Hg–H–C–N–F–I–S

**$Me_4N[Hg(SCF_3)_2I]$ or $C_6H_{12}NF_6IS_2Hg$**. This multinary compound is formed in this system. It melts at 105°C (Jellinek and Lagowski 1960). On further heating, $HgI_2$ separates from the melt, which in turn decomposes slowly. To obtain it, powdered bis(trifluoromethylthio)mercury and tetramethylammonium iodide were mixed in a molar ratio of 1.1:1.0. The mixture initially formed a paste, which was homogenized. After some time, the mixture solidified; it was heated at 50°C, crushed, and homogenized several times. The product was cooled to room temperature, quickly washed three times with cold ether (which removes the excess of bis(trifluoromethylthio) mercury), and dried at 50°C. The title compound decomposes at 131°C forming $HgI_2$ and a liquid, which decomposes slowly on further heating to form mercuric sulfide, $Me_4NF$, and volatile C–F–S compounds.

## 7.43    Hg–H–C–N–Cl–S

Some compounds are formed in this system.

**$(NH_4)[HgCl_2(SCN)]$ or $CH_4N_2Cl_2SHg$**. This compound decomposes at 150°C and crystallizes in the monoclinic structure with the lattice parameters $a = 929.7 \pm 0.1$, $b = 417.1 \pm 0.1$, and $c = 919.8 \pm 0.1$ pm and $\beta = 92.827° \pm 0.005°$ and a calculated density of 3.24 g·cm$^{-3}$ (Gacemi et al. 2003) ($a = 929.4 \pm 0.3$, $b = 417.1 \pm 0.3$, and $c = 919.0 \pm 0.3$ pm and $\beta = 92.84° \pm 0.02$ and a calculated density of 3.24 g·cm$^{-3}$ [Dupont et al. 1983]). To prepare the title compound, an equimolar mixture of $HgCl_2$ and $NH_4SCN$ (17 mM of each) was dissolved in EtOH (20 mL). The resulting clear solution was slowly evaporated at room temperature and yielded "millimetric," transparent needles of $NH_4[HgCl_2(SCN)]$. Crystals were filtered off, washed with methanol, and dried at room temperature.

The crystals of this compound could also be obtained by slow evaporation at room temperature of the solution containing 0.5 g of urea, 13.5 g of $HgCl_2$, and 3.8 g of $NH_4SCN$ in 20 mL of MeOH (Dupont et al. 1983). The experiments showed that the title compound could not be crystallized from the aqueous solution.

**$[HgCl_2\{SC(NH_2)_2\}]·0.5HgCl_2$ or $CH_4N_2Cl_3SHg_{1.5}$**. This compound crystallizes in the monoclinic structure with the lattice parameters $a = 800 \pm 1$, $b = 1499 \pm 2$, and $c = 721 \pm 1$ pm and $\beta = 93.5° \pm 0.5°$ and a calculated density of 3.72 g·cm$^{-3}$ (Brotherton and White 1973b). During the preparation of $HgCl_2[SC(NH_2)_2]$, slow evaporation over a period of weeks yielded in addition a minimum quantity of the title compound as colorless needles.

**[Hg(CH₅N₃S)Cl₂] or CH₅N₃Cl₂SHg.** This compound melts at 182°C and crystallizes in the monoclinic structure with the lattice parameters $a = 896.6 \pm 1.1$, $b = 681.7 \pm 0.9$, and $c = 1209.2 \pm 1.4$ pm and $\beta = 100.3° \pm 0.1°$ and the experimental and calculated densities 3.31 and 3.313 g·cm⁻³, respectively (Chieh and Cowell 1977) ($a = 892$, $b = 675$, and $c = 1197$ pm and $\beta = 100.0°$ [Nardelli and Chierici 1960]). It was prepared by the next procedure. An EtOH/H₂O solution containing 2 mM of thiosemicarbazide (2 mM) of HgCl₂ was added. Needlelike crystals started to form, and the crystals were visible in 10 min. The solution was covered, and the reaction proceeded slowly at room temperature overnight. These crystals were collected by filtration. There was a trace of black material due to decomposition and the sample appeared silvery (Chieh and Cowell 1977).

**(NH₄)[HgCl(SCN)₂] or C₂H₄N₃ClS₂Hg.** This compound decomposes at 134°C and crystallizes in the monoclinic structure with the lattice parameters $a = 708.8 \pm 0.1$, $b = 1998.6 \pm 0.2$, and $c = 595.8 \pm 0.1$ pm and $\beta = 100.718° \pm 0.005°$ and a calculated density of 2.97 g·cm⁻³ (Gacemi et al. 2003). To prepare the title compound, an equimolar mixture of Hg(SCN)₂ (17 mM) dissolved in EtOH (30 mL) and NH₄Cl (17 mM) was dissolved in EtOH (10 mL). The resulting clear solution was slowly evaporated at room temperature and yielded very large, transparent platelets of NH₄[HgCl(SCN)₂]. Crystals were filtered off, washed with methanol, and dried at room temperature.

**[Hg{SC(NH₂)₂}₂Cl₂] or C₂H₈N₄Cl₂S₂Hg.** This compound melts at 248°C–252°C with decomposition (Aucken and Drago 1960, Isab and Wazeer 2005) and crystallizes in the orthorhombic structure with the lattice parameters $a = 644 \pm 1$, $b = 1276 \pm 2$, and $c = 591 \pm 1$ pm ($a = 1279$, $b = 589$, and $c = 644$ pm [Cheung et al. 1965]) and the experimental and calculated densities $2.76 \pm 0.01$ and 2.76 g·cm⁻³, respectively (Brotherton et al. 1973). It was prepared by the slow evaporation at room temperature of a solution containing a stoichiometric (1:2) molar ratio of HgCl₂ and thiourea, the solution having been filtered to remove a slight quantity of an initial black precipitate (presumably Hg), which formed during the first day. Crystals were deposited as thin square plates (Aucken and Drago 1960, Cheung et al. 1965, Brotherton et al. 1973, Isab and Wazeer 2005).

**[Hg(CH₅N₃S)₂Cl₂] or C₂H₁₀N₆Cl₂S₂Hg.** This compound decomposes at 218°C (Mahadevappa and Murthy 1972) and crystallizes in the orthorhombic structure with the lattice parameters $a = 867.5 \pm 0.7$, $b = 812.3 \pm 0.6$, and $c = 1578.6 \pm 1.1$ pm and the experimental and calculated densities 2.70 and 2.709 g·cm⁻³, respectively (Chieh 1977b) ($a = 863$, $b = 810$, and $c = 1570$ pm [Nardelli and Chierici 1960]). The mixed solvent from equal volumes of absolute EtOH and deionized H₂O was used. To 40 mL of the solution containing 4 mM of thiosemicarbazide, 2 mM of solid HgCl₂ was added. No stirring was applied to this natural heterogeneous mixture. The reaction was left to proceed slowly at room temperature. Some long, fine needles formed very quickly around the crystals of HgCl₂. The reaction vessel was then covered. On the next day, the crystals of HgCl₂ had disappeared, and the long needles were no longer there. In their place were colorless prismatic crystals. These prisms were separated by filtration. The filtrate was allowed to evaporate in a partially covered beaker at room temperature and more colorless prismatic crystals of the title compound were obtained (Nardelli and Chierici 1960, Mahadevappa and Murthy 1972, Chieh 1977b).

**[Hg(C₃H₄N₂S)Cl₂] or C₃H₄N₂Cl₂SHg**. This compound melts at 205°C (Popović et al. 1999b). It was obtained by adding an ethanolic solution of 1,3-imidazole-2-thione to an ethanolic solution of HgCl₂ (1:1 molar ratio).

**[Hg(C₃H₅NS₂)Cl₂] or C₃H₅NCl₂S₂Hg**. This compound melts at 184°C–188°C (Bell et al. 2001a). To obtain it, a solution of 1,3-thiazolidine-2-thione (12 mM), dissolved in warm EtOH (25 mL), was added to a warm solution of HgCl₂ (12 mM) in EtOH (25 mL). A white precipitate formed immediately, which, after cooling, was filtered off under suction, washed with EtOH and then diethyl ether, and dried in vacuo. The white powder was recrystallized as small colorless needles from hot acetone.

**[Hg{SC(NH₂)₂}₃Cl₂] or C₃H₁₂N₆Cl₂S₃Hg**. This compound crystallizes in the orthorhombic structure with the lattice parameters $a = 811 \pm 1$, $b = 1987 \pm 2$, and $c = 864 \pm 1$ pm and the experimental and calculated densities 2.35 and 2.39 g·cm⁻³, respectively (Korczyński 1968). It was obtained by the slow evaporation of an acidified by HCl solution (1 L) containing HgCl₂ (0.13 mol) and thiourea (0.8 mol). The dried crystals sinter at about 174°C and begin to decompose at about 179°C (Aucken and Drago 1960).

**[Hg(C₄H₆N₂S)Cl₂] or C₄H₆N₂Cl₂SHg**. This compound melts at 166°C–167°C and was obtained by the following procedure (Raper et al. 1998). Both HgCl₂ (12 mM) and 1-methylimidazoline-2-thione (C₄H₆N₂S) (12 mM) were separately dissolved in 50 mL of aqueous EtOH and heated until they dissolved. The warm solution of (C₄H₆N₂S) was added dropwise to that of the HgCl₂ solution with continuous stirring. An initial white precipitate subsequently redissolved. A white crystalline product was eventually obtained after the complete addition of the ligand. The initial product was recrystallized from hot acetone as white microcrystalline needles.

**[Hg(C₄H₈N₂S)Cl₂] or C₄H₈N₂Cl₂SHg**. This compound melts at 185°C and crystallizes in the triclinic structure with the lattice parameters $a = 1004.51 \pm 0.15$, $b = 1353.0 \pm 0.2$, and $c = 1416.20 \pm 0.7$ pm and $\alpha = 103.008° \pm 0.013°$, $\beta = 91.997° \pm 0.015°$, and $\gamma = 90.34° \pm 0.02°$ pm and a calculated density of 2.748 g·cm⁻³ (Popović et al. 2001). It was prepared by adding dropwise a methanol solution of 3,4,5,6-tetrahydropyrimidine-2-thione to an equimolar methanol solution of HgCl₂. The crystalline solids, which were formed after standing for several days, were filtered off, washed with MeOH, and dried. The isolated compound is crystalline and a colorless substance.

**[HgCl₂{Me(Me₂NCS₂)}] or C₄H₉NCl₂S₂Hg**. This compound melts at 121°C–125°C (Brinkhoff and Dautzenberg 1972). It was prepared by the addition of HgCl₂ in EtOH to a solution of Me(Me₂NCS₂) (1:1 molar ratio). The title compound was recrystallized from EtOH.

**[Hg{SC(NH₂)₂}₄Cl₂] or C₄H₁₆N₈Cl₂S₄Hg, β-tetrakis(thiourea)mercury(II) chloride**. This compound crystallizes in the monoclinic structure with the lattice parameters $a = 3345 \pm 5$, $b = 847 \pm 2$, and $c = 609 \pm 1$ pm and $\beta = 92.3° \pm 0.3$ and the experimental and calculated densities $2.20 \pm 0.02$ and $2.22$ g·cm$^{-3}$, respectively (Brotherton and White 1973a). To obtain it, fairly concentrated aqueous solutions containing 1 M HgCl$_2$ to 4 M (or more) of thiourea were mixed, the white precipitate redissolving in the excess of thiourea (Aucken and Drago 1960). The clear solution was evaporated down to half bulk and filtered hot. Large, colorless, well-formed, prismatic crystals of the tetraderivative slowly grow out of the cooling solution. They may be quickly washed two or three times with small quantities of cold H$_2$O, redissolved in hot H$_2$O, and recrystallized quickly by shaking and cooling the hot concentrated solution, whereupon microscopic hexagons crystallize out and may be washed as before and dried. They sinter at 140°C and decompose at 182°C.

**[Cl(SH)Hg{SC(NH$_2$)$_2$}$_4$] or C$_4$H$_{17}$N$_8$ClS$_5$Hg.** This compound crystallizes in the monoclinic structure with the lattice parameters $a = 607.6 \pm 0.4$, $b = 849.1 \pm 0.6$, and $c = 3338.9 \pm 2.5$ pm and $\beta = 91.75° \pm 0.05°$ and the experimental and calculated densities 2.28 and 2.31 g·cm$^{-3}$, respectively (Criado et al. 1975). Its crystals were obtained by adding aqueous solutions of thiourea and HgCl$_2$ (4:1 molar ratio) at 70°C.

**[Hg(C$_5$H$_5$NS)Cl$_2$] or C$_5$H$_5$NCl$_2$SCd.** This compound was prepared by mixing ethanolic solutions of HgCl$_2$ and 4-pyridine thiol (Kennedy and Lever 1972). It could not be recrystallized.

**[Hg(C$_5$H$_8$NS$_2$)Cl] or C$_5$H$_8$NClS$_2$Hg.** This compound was prepared by mixing 1 mM of HgCl$_2$ in 15 mL of MeOH and pyrrolidine dithiocarbamate (C$_5$H$_9$NS$_2$) in MeOH (20 mL) (1:1 molar ratio) (Altaf et al. 2010b). The addition of C$_5$H$_9$NS$_2$ in the colorless metal ion solution resulted in the formation of white precipitate immediately. After stirring for 30 min, the precipitate was filtered off and dried.

**[Hg(Me$_2$C$_3$H$_4$N$_2$S)Cl$_2$] or C$_5$H$_{10}$N$_2$Cl$_2$SHg.** This compound melts at 213°C with decomposition (Cannas et al. 1981). It was obtained by adding $N,N'$-dimethylimidazolidine-2-thione to HgCl$_2$ (2:1 molar ratio), using MeOH as a solvent.

**[HgCl$_2${Et(Me$_2$NCS$_2$)}] or C$_5$H$_{11}$NCl$_2$S$_2$Hg.** This compound melts at 111°C–115°C (Brinkhoff and Dautzenberg 1972). It was prepared by the addition of HgCl$_2$ in EtOH to a solution of Et(Me$_2$NCS$_2$) (1:1 molar ratio). The title compound was recrystallized from EtOH.

**[HgCl$_2${$\mu$-S(CH$_2$)$_3$NH(CH$_3$)$_2$}] or C$_5$H$_{13}$NCl$_2$SHg.** This compound crystallizes in the monoclinic structure with the lattice parameters $a = 1013.6 \pm 0.2$, $b = 651.9 \pm 0.1$, and $c = 1594.0 \pm 0.6$ pm and $\beta = 97.20° \pm 0.03°$ and a calculated density of 2.48 g·cm$^{-3}$ (Casals et al. 1988) ($a = 1589$, $b = 658$, and $c = 1021$ pm and $\beta = 95.57°$ (Casals et al. 1987]). To obtain it, a solution of 3-dimethylamino-1-propanethiol (5 mM) in H$_2$O (12.5 mL) and EtOH (12.5 mL) at room temperature was added slowly with stirring to 25 mL of a solution of HgCl$_2$ (5 mM) in the same solvent at the same temperature. Just before completion of the addition, very small colorless crystals started to develop. After 1 week, crystals of regular size were filtered off, washed with H$_2$O, and vacuum dried. These crystals include a mixture of [HgCl$_2${$\mu$-S(CH$_2$)$_3$NH(CH$_3$)$_2$}] and [{(CH$_3$)$_2$NH(CH$_2$)$_3$S}$_2$][HgCl$_4$] (Casals et al. 1987).

The title compound could also be prepared using as a solvent the mixture of MeOH (8 mL) and DMSO (2 mL) at 50°C. The resultant solution was allowed to cool to room temperature and was stored in the dark (Casals et al. 1988). After 2 weeks, pink, prismatic-shaped crystals were separated. They were filtered off, washed with cold $H_2O$, and dried in vacuo.

**[Hg(C$_6$H$_7$NS)Cl$_2$] or C$_6$H$_7$NCl$_2$SHg**. This compound was prepared by mixing ethanolic solutions of $HgCl_2$ and 2-methyl-6-pyridine thiol (Kennedy and Lever 1972). It could not be recrystallized.

**[Hg(C$_3$H$_4$N$_2$S)$_2$Cl$_2$] or C$_6$H$_8$N$_4$Cl$_2$S$_2$Hg**. This compound melts at 205°C (Popović et al. 1999b) and crystallizes in the orthorhombic structure with the lattice parameters $a = 752.96 \pm 0.11$, $b = 1372.09 \pm 0.19$, and $c = 1175.87 \pm 0.16$ pm and a calculated density of 2.579 g·cm$^{-3}$ (Pavlović et al. 2000a). It was obtained by adding an ethanolic solution of 1,3-imidazole-2-thione to an ethanolic solution of $HgCl_2$ (1:1 molar ratio) (Popović et al. 1999b). Its crystals were formed from a dilute ethanolic solution of $HgCl_2$ and 1,2-imidazole-2-thione in a 1:2 molar ratio at room temperature after standing for several days (Pavlović et al. 2000a).

**[Hg(C$_3$H$_5$NS$_2$)$_2$Cl$_2$] or C$_6$H$_{10}$N$_2$Cl$_2$S$_4$Hg**. This compound melts at 188°C–191°C (Bell et al. 2001b) (at 179°C [De Filippo et al. 1971]). To obtain it, a warm ethanolic solution (25 mL) of $HgCl_2$ (6.0 mM) was added to a hot solution of 1,3-thiazolidine-2-thione (C$_3$H$_5$NS$_2$) (12.0 mM) in EtOH (25 mL) with stirring. The white precipitate that formed immediately was filtered off under suction and was recrystallized from boiling acetone solution as fine white crystals and dried in vacuo (Bell et al. 2001b).

The title compound could also be obtained by the next procedure (De Filippo et al. 1971). $HgCl_2$ was dissolved in an excess of molten 1,3-thiazolidine-2-thione (at ca. 105°C; Hg$^{2+}$/C$_3$H$_5$NS$_2$ ca. 1:2 molar ratio), and the crude product was purified by washing with ethyl ether and recrystallizing from EtOH.

**HgCl$_2$[MeSCSN(CH$_2$)$_4$] or C$_6$H$_{11}$NCl$_2$S$_2$Hg**. This compound crystallizes in the triclinic structure with the lattice parameters $a = 861.8 \pm 0.1$, $b = 863.7 \pm 0.1$, and $c = 854.4 \pm 0.1$ pm and $\alpha = 90.72° \pm 0.01°$, $\beta = 101.67° \pm 0.01°$, and $\gamma = 113.73° \pm 0.01°$ pm and the experimental and calculated densities 2.50 and 2.53 g·cm$^{-3}$, respectively (Brotherton et al. 1974).

**[Hg(C$_3$H$_6$N$_2$S)$_2$Cl$_2$] or C$_6$H$_{12}$N$_4$Cl$_2$S$_2$Hg**. This compound melts at 250°C–251°C (Isab and Perzanowski 1990). It was prepared by a dropwise addition of ethanol solution of mercury chloride (1 mM in 10 mL) to a stoichiometric amount of imidazoline-2-thione (C$_3$H$_6$N$_2$S) in EtOH (2 mM in 50 mL) and refluxing the solution for 30 min on a water bath (Shunmugam and Sathyanarayana 1983, Isab and Perzanowski 1990). Colorless crystalline product was obtained.

**[HgCl$_2$(Me$_2$NHCS)$_2$] or C$_6$H$_{14}$N$_2$Cl$_2$S$_2$Hg**. This compound crystallizes in the monoclinic structure with the lattice parameters $a = 919.3 \pm 0.2$, $b = 1509.4 \pm 0.2$, and $c = 1012.1 \pm 0.2$ pm and $\beta = 105.10° \pm 0.01°$ and the experimental and calculated densities 2.22 and 2.203 g·cm$^{-3}$, respectively (Stålhandske et al. 1997). Its colorless crystals were obtained by cooling saturated N,N-dimethylthioformamide solution of $HgCl_2$.

**[Hg(C₇H₅NS₂)Cl₂]** or **C₇H₅NCl₂S₂Hg**. This compound melts at 230°C–238°C (Bell et al. 2001a,b). To prepare it, $HgCl_2$ (5.0 mM) dissolved in warm EtOH or MeOH (50 mL) was added to a stirred solution of 1,3-benzo-2-thiazoline-2-thione (10.0 mM), dissolved in warm EtOH (50 mL), resulting in the immediate deposition of fine, bright-yellow crystals (Bell et al. 2001a,b; Popović et al. 2002b). These were filtered off under suction and dried in vacuo.

**[Hg(C₇H₆N₂S)Cl₂]** or **C₇H₆N₂Cl₂SHg**. To prepare this compound, a solution of benzo-1,3-imidazole-2-thione (1.66 mM) in 20 mL of MeOH was added dropwise to a solution of $HgCl_2$ (1.47 mM) in 10 mL of MeOH (Popović et al. 2002a). A pale-pink crystalline product was obtained the following day, which was filtered off, washed with cold MeOH, and dried in air.

**[Hg(Et₂C₃H₄N₂S)Cl₂]** or **C₇H₁₄N₂Cl₂SHg**. This compound melts at 103°C with decomposition and crystallizes in the orthorhombic structure with the lattice parameters $a = 1478 \pm 2$, $b = 1244 \pm 2$, and $c = 661 \pm 1$ pm and the experimental and calculated densities 2.30 and 2.35 g·cm⁻³, respectively (Cannas et al. 1981). It was obtained by adding $N,N'$-diethylimidazolidine-2-thione to $HgCl_2$ (2:1 molar ratio), using MeOH as a solvent.

**[Hg(C₄H₆N₂S)₂Cl₂]** or **C₈H₁₂N₄Cl₂S₂Hg**. This compound melts at 185°C–187°C (at 182°C [Popović et al. 1999b]) and crystallizes in the monoclinic structure with the lattice parameters $a = 1410.2 \pm 0.2$, $b = 833.8 \pm 0.2$, and $c = 1252.7 \pm 0.3$ pm and $\beta = 97.869° \pm 0.011°$ and a calculated density of 2.275 g·cm⁻³ (Bell et al. 2000a). Both $HgCl_2$ (6 mM) and methyl-1,3-imidazole-2-thione (12 mM) were separately dissolved in 50 mL of aqueous EtOH and heated until they dissolved (Shunmugam and Sathyanarayana 1983, Popović et al. 1999b, Bell et al. 2000a). The warm solution of $HgCl_2$ was added, dropwise, to that of the solution of the ligand with continuous stirring. A colorless crystalline compound was precipitated after 2 h from a pale-yellow solution. The initial product was recrystallized from hot acetone as white microcrystalline needles.

**[Hg(C₄H₈N₂S)₂Cl₂]** or **C₈H₁₆N₄Cl₂S₂Hg**. This compound melts at 179°C and crystallizes in the orthorhombic structure with the lattice parameters $a = 1483.40 \pm 0.07$, $b = 1283.40 \pm 0.03$, and $c = 1685.30 \pm 0.03$ pm and a calculated density of 2.086 g·cm⁻³ (Popović et al. 2000a). It was prepared by reacting one equivalent of $HgCl_2$ and two equivalents of 3,4,5,6-tetrahydropyrimidine-2-thione in MeOH. The crystalline solids, which formed upon standing for several days, were filtered off, washed with MeOH, and dried. The isolated compound is a crystalline and a colorless substance.

**[Hg(MeC₃H₅N₂S)₂Cl₂]** or **C₈H₁₆N₄Cl₂S₂Hg**. This compound melts at 152°C–153°C (Isab and Perzanowski 1990) (at 170°C [Cannas et al. 1981]). To obtain it, a solution of $HgCl_2$ (2 mM) in 20 mL of EtOH or MeOH was added dropwise to a solution of methylimidazoline-2-thione ($C_4H_8N_2S$) (4 mM) in 100 mL of EtOH or MeOH (Cannas et al. 1981, Isab and Perzanowski 1990). The reaction mixture was heated on a water bath for 0.5 h, after which the white crystalline solid was filtered off, washed with ethanol, and dried.

**[Hg(C₄H₈N₂S)₂Cl₂] or C₈H₁₆N₄Cl₂S₂Hg**. This compound melts at 178°C (Isab and Perzanowski 1996). It was prepared by a dropwise addition of an ethanol solution of HgCl₂ to a stoichiometric amount of 1,3-diazinane-2-thione in EtOH.

**[Hg(C₉H₁₂N₄S)Cl₂] or C₉H₁₂N₄Cl₂SHg**. This compound melts at 165°C (Bermejo et al. 1999). It was obtained by treating 2-acetylpyridine 4-methylthiosemicarbazone (C₉H₁₂N₄S) with HgCl₂ (1:1 molar ratio) in EtOH. After prolonged stirring (about 1 week) at room temperature, the reaction mixture afforded the title compound, which was filtered off, washed with EtOH, and vacuum dried.

**[Hg(C₉H₁₃N₅S)Cl₂] or C₉H₁₃N₅Cl₂SHg**. This compound melts at 230°C with decomposition (Bermejo et al. 2003). It was prepared as follows. A solution of 2-pyridineformamide N(4)-dimethylthiosemicarbazone (1.12 mM) in EtOH (20 mL) was added to a solution of HgCl₂ (1.12 mM) in 20 mL of 96% ethyl alcohol, and the mixture was stirred for several days at room temperature. The resulting solids were filtered out, washed thoroughly with cold ethanol, and stored in a desiccator over CaSO₄.

**[Hg(C₁₀H₇NS₅)Cl₂] or C₁₀H₇NCl₂S₅Hg**. This compound melts at 140°C–141°C with decomposition (Nam et al. 2004). To obtain it, to a 20 mL CH₂Cl₂ solution of 5-pyridine-2-yl-5,6-dihydro-[1,3]dithio[4,5-b]dithiin-2-thione (C₁₀H₇NS₅) HgCl₂ (0.50 mM) dissolved in MeCN with stirring at room temperature was added. The red crystals formed from the mixture after standing for 2 days were filtered and washed with MeOH and then dried in vacuo.

**[Hg(C₅H₄N₄S)₂Cl₂] or C₁₀H₈N₈Cl₂S₂Hg, dichlorobis(6-mercaptopurine) mercury(II)**. This compound crystallizes in the orthorhombic structure with the lattice parameters $a = 954.3 \pm 0.9$, $b = 799.6 \pm 0.8$, and $c = 2110 \pm 2$ pm and the experimental and calculated densities $2.38 \pm 0.1$ and 2.374 g·cm⁻³, respectively (Lavertue et al. 1976). To prepare it, HgCl₂ was added to a suspension of 6-mercaptopurine in H₂O (1:2 molar ratio), and the mixture was stirred at room temperature for 1 h. The insoluble colorless powder of C₁₀H₈N₈Cl₂S₂Hg was obtained. It was dissolved (0.30 g) in boiling 0.58 N aqueous HCl (300 mL). The hot mixture was filtered and left to crystallize; yellow needles formed overnight.

**[Hg(C₅H₅NS)₂Cl₂] or C₁₀H₁₀N₂Cl₂S₂Cd**. This compound was prepared by mixing ethanolic solutions of HgCl₂ and 2-pyridine thiol (Kennedy and Lever 1972). It could not be recrystallized.

**[HgCl₂{Bz(Me₂NCS₂)}] or C₁₀H₁₃NCl₂S₂Hg**. This compound melts at 126°C–128°C (Brinkhoff and Dautzenberg 1972). It was prepared by the addition of HgCl₂ in EtOH to a solution of Bz(Me₂NCS₂) (1:1 molar ratio). The title compound was recrystallized from EtOH.

**[Hg(Me₂C₃H₄N₂S)₂Cl₂] or C₁₀H₂₀N₄Cl₂S₂Hg**. This compound melts at 68°C–69°C (Isab and Perzanowski 1990). To obtain it, a solution of HgCI₂ (2 mM) in 20 mL of EtOH was added dropwise to a solution of N,N'-dimethylimidazoline-2-thione (C₅H₁₀N₂S) (4 mM) in 100 mL of EtOH. The reaction mixture was heated on a water bath for 0.5 h, after which the white powder was filtered off, washed with ethanol, and dried.

**[Hg(EtC$_3$H$_5$N$_2$S)$_2$Cl$_2$] or C$_{10}$H$_{20}$N$_4$Cl$_2$S$_2$Hg.** This compound melts at 161°C–162°C (Cannas et al. 1981, Isab and Perzanowski 1990) and crystallizes in the triclinic structure with the lattice parameters $a = 1366 \pm 3$, $b = 843 \pm 2$, and $c = 813 \pm 2$ pm and $\alpha = 105.3° \pm 0.5°$, $\beta = 96.1° \pm 0.5°$, and $\gamma = 103.2° \pm 0.5°$ pm and the experimental and calculated densities 2.01 and 2.04 g·cm$^{-3}$, respectively (Cannas et al. 1981). To obtain it, a solution of HgCl$_2$ (2 mM) in 20 mL of EtOH or MeOH was added dropwise to a solution of ethylimidazoline-2-thione (C$_5$H$_{10}$N$_2$S) (4 mM) in 100 mL of EtOH or MeOH Cannas et al. 1981, Isab and Perzanowski 1990). The reaction mixture was heated on a water bath for 0.5 h, after which the pale-yellow powder was filtered off, washed with ethanol, and dried.

**[Hg(C$_5$H$_{10}$N$_2$S)$_2$Cl$_2$] or C$_{10}$H$_{20}$N$_4$Cl$_2$S$_2$Hg.** This compound melts at 172°C (Isab and Perzanowski 1996). It was prepared by a dropwise addition of ethanol solution of HgCl$_2$ to a stoichiometric amount of 1,3-diazipane-2-thione in EtOH.

**[HgCl$_2$(C$_5$H$_{12}$N$_2$S)$_2$] or C$_{10}$H$_{24}$N$_4$Cl$_2$S$_2$Hg.** This compound crystallizes in the monoclinic structure with the lattice parameters $a = 1874.18 \pm 0.12$, $b = 959.20 \pm 0.06$, and $c = 1351.77 \pm 0.09$ pm and $\beta = 130.834° \pm 0.001°$ and a calculated density of 1.936 g·cm$^{-3}$ (Nawaz et al. 2010). To obtain it, to 1.0 mM of HgCl$_2$ in 10 mL of MeOH two equivalents of tetramethylthiourea in 15 mL of MeOH were added. A clear solution was obtained which was stirred for 30 min. The colorless solution was filtered, and the filtrate was kept at room temperature for crystallization. As a result, a white crystalline product was obtained, which was finally washed with MeOH and dried.

**[HgCl$_2$(C$_5$H$_{12}$N$_2$S)$_2$] or C$_{10}$H$_{24}$N$_4$Cl$_2$S$_2$Hg.** This compound crystallizes in the monoclinic structure with the lattice parameters $a = 797.13 \pm 0.02$, $b = 1723.21 \pm 0.05$, and $c = 2751.43 \pm 0.07$ pm and $\beta = 94.870° \pm 0.001°$ and a calculated density of 1.891 g·cm$^{-3}$ (Mufakkar et al. 2010). To obtain it, to 1.0 mM of HgCl$_2$ in 10 mL of MeOH two equivalents of N,N'-diethylthiourea in 15 mL of MeOH were added. On mixing, a clear solution was obtained which was stirred for 30 min. The colorless solution was filtered, and the filtrate was kept at room temperature for crystallization. A white crystalline product was obtained, which was washed with MeOH and dried.

**[{(CH$_3$)$_2$NH(CH$_2$)$_3$S}$_2$][HgCl$_4$] or C$_{10}$H$_{26}$N$_2$Cl$_4$S$_2$Hg.** This compound crystallizes in the orthorhombic structure with the lattice parameters $a = 2349$, $b = 1362$, and $c = 667$ pm (Casals et al. 1987). It was obtained simultaneously by obtaining [HgCl$_2$\{μ-S(CH$_2$)$_3$NH(CH$_3$)$_2$\}].

**[Hg$_2$(C$_{11}$H$_{10}$N$_2$S$_2$)Cl$_4$] or C$_{11}$H$_{10}$N$_2$Cl$_4$S$_2$Hg$_2$.** This compound melts at 159°C and crystallizes in the monoclinic structure with the lattice parameters $a = 781.31 \pm 0.02$, $b = 1644.09 \pm 0.02$, and $c = 1436.45 \pm 0.04$ pm and $\beta = 101.0710° \pm 0.0005°$ (Amoedo-Portela et al. 2002). To obtain it, a solution of bis(2-pyridylthio)methane (0.43 mM) in 5 mL of an EtOH/CH$_2$Cl$_2$ (2:1 volume ratio) mixture was added dropwise to a solution of HgCl$_2$ (0.43 mM) in 5 mL of EtOH. The white suspension that was obtained was refluxed for 3 h. After stirring for 3 days, the solid was filtered out and dried in vacuo. Crystals suitable for XRD were obtained from a THF solution.

**[Hg$_2$Cl$_4$(C$_6$H$_7$NS)$_2$] or C$_{12}$H$_{14}$N$_2$Cl$_4$S$_2$Hg$_2$.** This compound crystallizes in the ortho-rhombic structure with the lattice parameters $a = 1778.6 \pm 0.2$, $b = 1450.4 \pm 0.1$, and

$c = 743.7 \pm 0.3$ pm and a calculated density of 2.747 g·cm⁻³ (Davidović et al. 1998). To obtain it, $HgCl_2(1.84$ mM) was dissolved in EtOH (60 mL). Into this solution, an ethanol solution of benzothiazole (20 mL) was added with stirring. Slow evaporation of the solvent at room temperature yielded crystals of the title compound.

**[Hg(dmtp)₂Cl₂] or C₁₂H₁₆N₄Cl₂S₂Hg (dmtp = 4,6-dimethyl-2-thiopyrimidine).** This compound was prepared by dissolving dmtp and $HgCl_2$ in a 1:2 molar ratio, in the minimum amount of water (Lopez-Garzon et al. 1993). The solution was then evaporated on a water bath at 60°C. White crystals precipitated after 2 days.

**[Hg(C₁₂H₁₇N₅S)Cl₂] or C₁₂H₁₇N₅Cl₂SHg.** This compound melts at 296°C (Castiñeiras et al. 2002). To obtain it, a solution of $C_{12}H_{17}N_5S$ [2-pyridineformamide N(4)-piperidylthiosemicarbazone] (0.76 mM) in EtOH (30 mL) was added to a solution of $HgCl_2$ (0.76 mM) in 30 mL of 96% EtOH, and the mixture was kept stirring for several days at room temperature. The resulting yellow solid was filtered off, washed thoroughly with cold ethanol, and stored in a desiccator over Mg and $I_2$.

**[Hg(C₆H₁₂N₂S)₂Cl₂] or C₁₂H₂₄N₄Cl₂S₂Hg.** This compound melts at 122°C–123°C (Isab and Perzanowski 1990). To obtain it, a solution of $HgCI_2$ (2 mM) in 20 mL of EtOH was added dropwise to a solution of N-n-propylimidazoline-2-thione $(C_6H_{12}N_2S)$ (4 mM) in 100 mL of EtOH. The reaction mixture was heated on a water bath for 0.5 h, after which the white powder was filtered off, washed with ethanol, and dried.

**[Hg(C₆H₁₂N₂S)₂Cl₂] or C₁₂H₂₄N₄Cl₂S₂Hg.** This compound melts at 150°C–151°C (Isab and Perzanowski 1990). To obtain it, a solution of $HgCI_2$ (2 mM) in 20 mL of EtOH was added dropwise to a solution of N-i-propylimidazoline-2-thione $(C_6H_{12}N_2S)$ (4 mM) in 100 mL of EtOH. The reaction mixture was heated on a water bath for 0.5 h, after which the white crystalline solid was filtered off, washed with ethanol, and dried.

**[C₁₃H₁₀N₄S·HgCl₂] or C₁₃H₁₀N₄Cl₂SHg.** This compound crystallizes in the monoclinic structure with the lattice parameters $a = 2571.7 \pm 1.0$, $b = 647.6 \pm 1.0$, and $c = 1147.6 \pm 0.8$ pm and $\beta = 102.79° \pm 0.03°$ and the experimental and calculated densities $1.95 \pm 0.05$ and $1.89$ g·cm⁻³, respectively (Kozarek and Fernando 1972, 1973). It was obtained by mixing equimolar amounts of $HgCl_2$ in $H_2O$ and dehydrodithizone in EtOH. Pale-yellow needle-shaped crystals were obtained by crystallization from acetone.

**[(BuˡSHgCl)₂(Py)] or C₁₃H₂₃NCl₂S₂Hg₂.** This compound crystallizes in the triclinic structure with the lattice parameters $a = 1439.9 \pm 0.6$, $b = 989.3 \pm 0.6$, and $c = 959.7 \pm 0.5$ pm and $\alpha = 121.82° \pm 0.01°$, $\beta = 102.52° \pm 0.04°$, and $\gamma = 102.94° \pm 0.04°$ pm and the experimental and calculated densities $2.36 \pm 0.01$ and $2.34$ g·cm⁻³, respectively (Canty et al. 1978b). To obtain it, BuˡSHgCl was dissolved in Py to slowly deposit colorless crystals (Canty et al. 1978a). The precipitate was collected, washed with EtOH, and dried over $P_2O_5$ in a vacuum.

**[Hg(C₇H₅NS₂)₂Cl₂] or C₁₄H₁₀N₂Cl₂S₄Hg.** This compound was prepared as yellow crystals by the mixing of methanol solutions of $HgCl_2$ and 1,3-benzo-2-thiazoline-2-thione (Popović et al. 2002b).

**[Hg(C$_7$H$_6$N$_2$S)$_2$Cl$_2$] or C$_{14}$H$_{12}$N$_4$Cl$_2$S$_2$Hg**. To prepare this compound, a solution of benzo-1,3-imidazole-2-thione (3.26 mM) in 40 mL of MeOH was added dropwise to a solution of HgCl$_2$ (1.47 mM) in 10 mL of MeOH (Popović et al. 2002a). A pale-pink crystalline product was obtained the following day, which was filtered off, washed with cold MeOH, and dried in air.

**[Hg(C$_{14}$H$_{24}$N$_2$S$_2$)Cl$_2$] or C$_{14}$H$_{24}$N$_2$Cl$_2$S$_2$Hg**. To obtain this compound, *N,N'*-dicyclohexyldithiooxamide (2.50 mM) was dissolved in dry acetone (100 mL) to which HgCl$_2$ (1.25 mM) was added with stirring in small portions (Baggio et al. 1995). The deep-yellow reaction mixture was stirred at 25°C for 4 h. It was then filtered, and the filtrate was stored in a refrigerator for a few days whereupon the major product deposited as bright-yellow needles, which were collected by filtration. The crystals cracked within a few hours presumably due to loss of acetone molecules of crystallization. Total removal of acetone under vacuum yielded the powdery yellow solid of the title compound.

**[Hg(C$_7$H$_{14}$N$_2$S)$_2$Cl$_2$] or C$_{14}$H$_{28}$N$_4$Cl$_2$S$_2$Hg**. This compound melts at 297°C–298°C (Isab and Perzanowski 1990). To obtain it, a solution of HgCI$_2$ (2 mM) in 20 mL of EtOH was added dropwise to a solution of *N,N'*-diethylimidazoline-2-thione (C$_7$H$_{14}$N$_2$S) (4 mM) in 100 mL of EtOH. The reaction mixture was heated on a water bath for 0.5 h, after which the dark-brown powder was filtered off, washed with ethanol, and dried.

**[N(CH$_2$CH$_2$SCHMe$_2$)$_3$HgCl]$_2$(Hg$_2$Cl$_6$) or C$_{15}$H$_{33}$NCl$_4$S$_3$Hg$_2$**. This compound crystallizes in the triclinic structure with the lattice parameters $a = 1011.4 \pm 0.2$, $b = 1123.0 \pm 0.5$, and $c = 1219.6 \pm 0.7$ pm and $\alpha = 86.11° \pm 0.04°$, $\beta = 73.47° \pm 0.04°$, and $\gamma = 74.45° \pm 0.04°$ pm and a calculated density of 2.250 g·cm$^{-3}$ (Cecconi et al. 1998). To prepare it, a solution of N(CH$_2$CH$_2$SCHMe$_2$) (1 mM) in CH$_2$Cl$_2$ (20 mL) was added to a solution of HgCl$_2$ (2 mM) in MeOH (40 mM). Upon slow evaporation of the solvent in air, white crystals precipitated, which were filtered off, washed with MeOH, and dried under vacuum.

**[Hg$_4${S(CH$_2$)$_2$NMe$_2$}$_4$Cl$_4$] or C$_{16}$H$_{40}$N$_4$Cl$_4$S$_4$Hg$_4$**. To obtain this compound, HS(CH$_2$)$_2$NMe$_2$·HCl (10 mM) was dispersed in acetonitrile (40 mL) (Casals et al. 1991a). Et$_3$N (2.8 mL) and anhydrous HgCl$_2$ (10 mM) were successively added with stirring under an N$_2$ atmosphere. Stirring was continued for 24 h. The resulting suspension was filtered in the open atmosphere, washed with copious distilled water, then with MeOH, and with diethyl ether, and vacuum dried.

Alternatively, this compound was obtained by adding an aqueous solution of equimolar amount of HS(CH$_2$)$_2$NMe$_2$·HCl and NaOH to an aqueous solution of Hg(OAc)$_2$, the metal/ligand molar ratio being always 1:1 (Casals et al. 1991a). A white powder formed immediately. The mixture was stirred for several hours; the solid was filtered off, washed with cold water and then MeOH, and dried under vacuum. Crystals were obtained by refluxing the compound in DMSO, filtering off the undissolved solid, letting the solution cool to 60°C, and maintaining it at this temperature for 24 h. All these manipulations were performed under an N$_2$ atmosphere.

**[C$_4$H$_3$SHgCl·(2,9-Me$_2$Phen)]** or **C$_{18}$H$_{15}$N$_2$ClSHg$_2$.** This compound melts at 159°C–163°C (Bell et al. 2004). The reaction of 2-thienylmercury(II) chloride with 2,9-dimethyl-1,10-phenanthroline in ethanolic solution produced the title compound as white powder.

**[Hg(C$_9$H$_{12}$N$_5$S)Cl]$_2$** or **C$_{18}$H$_{24}$N$_{10}$Cl$_2$S$_2$Hg$_2$.** This compound crystallizes in the monoclinic structure with the lattice parameters $a = 771.4 \pm 0.2$, $b = 2100.5 \pm 0.3$, and $c = 833.7 \pm 0.1$ pm and $\beta = 109.31° \pm 0.02°$ and a calculated density of 2.388 g·cm$^{-3}$ (Bermejo et al. 2003). Its single crystals suitable for XRD were obtained by slow evaporation of an ethanolic solution in air at room temperature.

**[HgCl$_2$·Bu$^n$$_4$(CNS)$_2$)$_2$]** or **C$_{18}$H$_{36}$N$_2$Cl$_2$S$_4$Hg.** This compound melts at 105°C–112°C (Brinkhoff et al. 1970). It was precipitated as colorless needles by the addition of an ethanolic solution of HgCl$_2$ (1 M) to an ethanolic solution of Bu$^n$$_4$C$_2$N$_2$S$_4$ (1 M). The product was purified by the addition of petroleum ether to a concentrated solution in CHCl$_3$.

**[Hg(C$_9$H$_{18}$N$_2$S)$_2$Cl$_2$]** or **C$_{18}$H$_{36}$N$_4$Cl$_2$S$_2$Hg.** This compound melts at 144°C–145°C (Isab and Perzanowski 1990). To obtain it, a solution of HgCl$_2$ (2 mM) in 20 mL of EtOH was added dropwise to a solution of N,N′-di-i-propylimidazoline-2-thione (C$_9$H$_{18}$N$_2$S) (4 mM) in 100 mL of EtOH. The reaction mixture was heated on a water bath for 0.5 h, after which the pale-yellow powder was filtered off, washed with ethanol, and dried.

**[Hg(S$_2$CNEt$_2$)(4,7-Me$_2$Phen)Cl]·0.5CHCl$_3$** or **C$_{19.5}$H$_{22.5}$N$_3$Cl$_{2.5}$S$_2$Hg.** This compound melts at 127°C–129°C and crystallizes in the monoclinic structure with the lattice parameters $a = 1971.45 \pm 0.15$, $b = 1058.93 \pm 0.08$, and $c = 2403.61 \pm 0.08$ pm and $\beta = 114.076° \pm 0.002°$ (Lai and Tiekink 2003b). Bright-yellow crystals of the title compound were obtained from the slow evaporation of an acetonitrile/chloroform solution of a solid that had precipitated from the refluxing (1 h) of equimolar amounts of Hg(S$_2$CNEt$_2$)$_2$ and 4,7-Me$_2$Phen in chloroform solution.

**[C$_4$H$_3$SHgCl·(3,4,7,8-Me$_4$Phen)]** or **C$_{20}$H$_{19}$N$_2$ClSHg$_2$.** This compound melts at 220°C–221°C (Bell et al. 2004). The reaction of 2-thienylmercury(II) chloride with 3,4,7,8-tetramethyl-1,10-phenanthroline in ethanolic solution produced the title compound as white powder.

**[Hg$_3$Cl$_2$(S$_2$CNEt$_2$)$_4$]** or **C$_{20}$H$_{40}$N$_4$Cl$_2$S$_8$Hg$_3$.** This compound crystallizes in the triclinic structure with the lattice parameters $a = 1853 \pm 1$, $b = 1017.4 \pm 0.6$, and $c = 1101.6 \pm 0.6$ pm and $\alpha = 112.82° \pm 0.04°$, $\beta = 97.17° \pm 0.05°$, and $\gamma = 91.15° \pm 0.06°$ pm and the experimental and calculated densities 2.22 and 2.219 g·cm$^{-3}$, respectively (Book and Chieh 1980) ($a = 1856.2 \pm 0.2$, $b = 1015.7 \pm 0.8$, and $c = 1101.2 \pm 0.4$ pm and $\alpha = 113.06° \pm 0.05°$, $\beta = 97.06° \pm 0.02°$, and $\gamma = 91.24° \pm 0.03°$ pm and the experimental and calculated densities 2.2 and 2.223 g·cm$^{-3}$, respectively [Iwasaki 1972, 1973]). It was prepared from HgCl$_2$ and sodium diethyldithiocarbamate. When HgCl$_2$ crystals were added to an ethanolic solution of Na(S$_2$CNEt$_2$), a black precipitate formed immediately, indicating disproportionation of mercurous chloride. The filtrate from this reaction gave a mixture of α-[Hg(S$_2$CNEt$_2$)$_2$]$_2$ and Hg$_3$Cl$_2$(S$_2$CNEt$_2$)$_4$ crystals, when the solvent had evaporated at room temperature. The former are deep-yellow rhomboids and the later are pale-yellow needles elongated along the $c$-axis (Iwasaki 1972, 1973, Book and Chieh 1980).

[{Hg(S$_2$CNBu$^n$$_2$)Cl}(Phen)] or C$_{21}$H$_{26}$N$_3$ClS$_2$Hg. This compound crystallizes in the orthorhombic structure with the lattice parameters $a = 1952.2 \pm 0.8$, $b = 2193.2 \pm 0.4$, and $c = 1045.0 \pm 0.2$ pm at 173 K (Tiekink 2001). It was prepared from the reaction of Hg(S$_2$CNBu$^n$$_2$)Cl with Phen (1:1 molar ratio).

[(C$_4$H$_3$SHgCl)$_4$·(C$_{10}$H$_8$N$_2$S$_2$)] or C$_{26}$H$_{20}$N$_2$Cl$_4$S$_6$Hg$_4$. This compound melts at 175°C–180°C (Bell et al. 2004). The reaction of 2-thienylmercury(II) chloride with 2,2′-bipyridyl disulfide in ethanolic solution produced the title compound as white powder.

[(Bu$^t$SHgCl)$_4$(MePy)$_2$] or C$_{28}$H$_{50}$N$_2$Cl$_4$S$_4$Hg$_4$. This compound crystallizes in the monoclinic structure with the lattice parameters $a = 1233.4 \pm 0.7$, $b = 1746.8 \pm 0.9$, and $c = 999.9 \pm 0.5$ pm and $\beta = 91.18° \pm 0.04°$ and a calculated density of 2.29 g·cm$^{-3}$ (Canty et al. 1979b). It crystallized from a solution of Bu$^t$SHgCl in 4-methylpyridine.

(Me$_4$N)$_2$[Hg(SC$_6$H$_4$Cl)$_4$] or C$_{32}$H$_{49}$N$_2$Cl$_4$S$_4$Hg. This compound crystallizes in the monoclinic structure with the lattice parameters $a = 1220.9 \pm 0.3$, $b = 1451.5 \pm 0.1$, and $c = 1149.5 \pm 0.2$ pm and $\beta = 111.83° \pm 0.01°$ and the experimental and calculated densities 1.62 ± 0.02 and 1.62 g·cm$^{-3}$, respectively (Choudhury et al. 1983). It was obtained by the following procedure. A solution of Hg(NO$_3$)$_2$·H$_2$O (10 mM) in MeOH (30 mL) was added to a solution containing 4-chlorobenzenethiol (40 mM), Et$_3$N (40 mM), and Me$_4$NCl (27 mM) in MeOH (70 mL) at room temperature. Flocculent precipitation of Hg(SC$_6$H$_4$Cl)$_2$ during the addition quickly redissolved. Pr$^n$OH (70 mL) was added and MeOH was removed by distillation at room temperature until crystallization began. The mixture was heated to dissolve the product, sealed, and allowed to crystallize slowly, first at room temperature and then at 0°C. The large well-formed crystals, almost colorless, were filtered, washed with Pr$^n$OH, and vacuum dried. All solutions were deoxygenated with nitrogen.

[Hg(μ-Cl)Cl(Ph$_2$C$_5$H$_2$S$_5$)]$_2$ or C$_{38}$H$_{24}$N$_2$Cl$_4$S$_{10}$Hg$_2$. This compound melts at 136°C and crystallizes in the monoclinic structure with the lattice parameters $a = 1690.5 \pm 0.2$, $b = 726.2 \pm 0.3$, and $c = 1879.8 \pm 0.4$ pm and $\beta = 98.10° \pm 0.02°$ and a calculated density of 1.995 g·cm$^{-3}$ (Lee et al. 1999). To obtain it, to a methylene chloride solution (3 mL) of 4,5-(1′,2′-diphenylethylenedithio)-1,3-dithiole-2-thione (Ph$_2$C$_5$H$_2$S$_5$) (0.25 mM) an acetonitrile solution (5 mL) of HgCl$_2$ (0.5 mM) with stirring at room temperature was added. Reddish-orange precipitates were formed immediately. The mixture was heated until the precipitate was redissolved to form orange crystals. The precipitate was collected from filtrate. Diamond-shaped orange crystals suitable for XRD were obtained from the filtrate after a few minutes standing.

## 7.44   Hg–H–C–N–Br–S

Some compounds are formed in this system.

[MeHg{SC(NH$_2$)$_2$}]Br or C$_2$H$_7$N$_2$BrSHg. To obtain this compound, hot deoxygenated aqueous solution of MeHgBr (5.0 mM) was added to a refluxing solution of a stoichiometric amount of thiourea under $N_2$ (Carty et al. 1979). After ca. 30 min, the solution was filtered and slowly evaporated to low volume. Failure to adequately protect the solution from air leads to extensive decomposition. Colorless crystals of the title compound were filtered off, dried in a desiccator, and stored in the dark.

**[Hg(C$_3$H$_4$N$_2$S)Br$_2$] or C$_3$H$_4$N$_2$Br$_2$SHg.** This compound melts at 173°C (Popović et al. 1999b). It was obtained by adding an ethanolic solution of 1,3-imidazole-2-thione to an ethanolic solution of HgBr$_2$ (1:1 molar ratio).

**[Hg(C$_3$H$_5$NS$_2$)Br$_2$] or C$_3$H$_5$NBr$_2$S$_2$Hg.** This compound melts at 149°C–152°C and crystallizes in the triclinic structure with the lattice parameters $a = 774.28 \pm 0.11$, $b = 796.88 \pm 0.11$, and $c = 948.89 \pm 0.12$ pm and $\alpha = 66.718° \pm 0.003°$, $\beta = 69.588° \pm 0.003°$, and $\gamma = 73.733° \pm 0.003°$ pm at 160 K and a calculated density of 3.582 g·cm$^{-3}$ (Bell et al. 2001a). To obtain it, a solution of 1,3-thiazolidine-2-thione (10 mM), dissolved in warm EtOH (25 mL), was added to a warm solution of HgBr$_2$ (10 mM) in EtOH (25 mL). A white precipitate was formed immediately, which, after cooling, was filtered off under suction, washed with EtOH and then diethyl ether, and dried in vacuo. The very fine white crystals were recrystallized from boiling acetone. Crystals of the title compound were grown by slow evaporation of acetone solution.

**[Hg(C$_4$H$_6$N$_2$S)Br$_2$] or C$_4$H$_6$N$_2$Br$_2$SHg.** This compound melts at 137°C–139°C and was obtained by the following procedure (Raper et al. 1998). Both HgBr$_2$ (10 mM) and 1-methylimidazoline-2-thione (C$_4$H$_6$N$_2$S) (10 mM) were separately dissolved in 50 mL of aqueous EtOH and heated until they dissolved. The warm solution of (C$_4$H$_6$N$_2$S) was added dropwise to that of the HgBr$_2$ solution with continuous stirring. An initial white precipitate subsequently redissolved. A white crystalline product was eventually obtained after complete addition of the ligand. The initial product was recrystallized from hot acetone as white microcrystalline needles.

**[Hg(C$_4$H$_8$N$_2$S)Br$_2$] or C$_4$H$_8$N$_2$Br$_2$SHg.** This compound melts at 142°C and crystallizes in the triclinic structure with the lattice parameters $a = 1043.9 \pm 0.3$, $b = 1374.9 \pm 0.2$, and $c = 1450.0 \pm 0.3$ pm and $\alpha = 102.6970° \pm 0.0017°$, $\beta = 92.082° \pm 0.002°$, and $\gamma = 90.1747° \pm 0.0019°$ pm and a calculated density of 3.121 g·cm$^{-3}$ (Popović et al. 2001). It was prepared by adding dropwise a methanol solution of 3,4,5,6-tetrahydropyrimidine-2-thione to an equimolar methanol solution of HgBr$_2$. The crystalline solids, which were formed after standing for several days, were filtered off, washed with MeOH, and dried. The isolated compound is crystalline and a colorless substance.

**[Hg(C$_2$H$_4$N$_4$S)$_2$Br$_2$] or C$_4$H$_8$N$_8$Br$_2$S$_2$Hg.** This compound crystallizes in the orthorhombic structure with the lattice parameters $a = 970.7 \pm 0.2$, $b = 860.9 \pm 0.1$, and $c = 1612.8 \pm 0.2$ pm and a calculated density of 2.92 g·cm$^{-3}$ (Baraldi et al. 1996). To obtain it, HgBr$_2$ (0.1 mM) dissolved in H$_2$O (2 mL with 2 drops of 48% HBr) was added to 3-amino-2-mercapto-1,2,4-triazole (0.3 mM) dissolved in H$_2$O (15 mL with 2 drops of 48% HBr). Colorless crystals formed after a few days of slow evaporation.

**[HgBr$_2${Me(Me$_2$NCS$_2$)}] or C$_4$H$_9$NBr$_2$S$_2$Hg.** This compound melts at 114°C–116°C (Brinkhoff and Dautzenberg 1972). It was prepared by the addition of HgBr$_2$ in EtOH to a solution of Me(Me$_2$NCS$_2$) (1:1 molar ratio). The title compound was recrystallized from EtOH.

**[Hg($C_5H_5NS$)$Br_2$] or $C_5H_5NBr_2SHg$.** This compound was prepared by mixing ethanolic solutions of $HgBr_2$ and 4-pyridine thiol (Kennedy and Lever 1972). It could not be recrystallized.

**[Hg($Me_2C_3H_4N_2S$)$Br_2$] or $C_5H_{10}N_2Br_2SHg$.** This compound melts at 201°C with decomposition (Cannas et al. 1981). It was obtained by adding *N,N′*-dimethylimidazolidine-2-thione to $HgBr_2$ (2:1 molar ratio), using MeOH as a solvent.

**[$HgBr_2${μ-S($CH_2$)$_3$NH($CH_3$)$_2$}] or $C_5H_{13}NBr_2SHg$.** This compound crystallizes in the monoclinic structure with the lattice parameters $a=1585$, $b=660$, and $c=1020$ pm and $\beta=95.53°$ (Casals et al. 1987). To obtain it, a solution of 3-dimethylamino-1-propanethiol (5 mM) in $H_2O$ (18 mL) and EtOH (2 mL) at room temperature was added very slowly to a solution of $HgBr_2$ (5 mM) in 20 mL of the same solvent. During the first minutes of the addition, a white precipitate was developed, and it eventually became a white mass that adhered to the walls of the beaker. Meanwhile, a white microcrystalline solid started to appear. These crystals were separated by decanting and then filtering, washing with $H_2O$, and vacuum drying. These crystals are a mixture of [$HgBr_2${μ-S($CH_2$)$_3$NH($CH_3$)$_2$}] and [{($CH_3$)$_2$NH($CH_2$)$_3$S}$_2$][$HgBr_4$].

**[Hg($C_6H_7NS$)$Br_2$] or $C_6H_7NBr_2SHg$.** This compound was prepared by mixing ethanolic solutions of $HgBr_2$ and 2-methyl-6-pyridine thiol (Kennedy and Lever 1972). It could not be recrystallized.

**[Hg($C_3H_4N_2S$)$_2$$Br_2$] or $C_6H_8N_4Br_2S_2Hg$.** This compound melts at 180°C with decomposition and crystallizes in the monoclinic structure with the lattice parameters $a=863.26\pm0.08$, $b=998.85\pm0.06$, and $c=1461.62\pm0.08$ pm and $\beta=96.188°\pm0.006°$ and a calculated density of 2.972 g·cm$^{-3}$ (Popović et al. 1999b). It was obtained by adding an ethanolic solution of 1,3-imidazole-2-thione to an ethanolic solution of $HgBr_2$ (2:1 molar ratio).

**[Hg($C_3H_5NS_2$)$_2$$Br_2$] or $C_6H_{10}N_2Br_2S_4Hg$.** This compound melts at 133°C–136°C (at 146°C [De Filippo et al. 1971]) and crystallizes in the triclinic structure with the lattice parameters $a=762.30\pm0.10$, $b=895.10\pm0.10$, and $c=1091.4\pm0.2$ pm and $\alpha=82.760°\pm0.010°$, $\beta=73.330°\pm0.006°$, and $\gamma=80.54°\pm0.02°$ at $150\pm2$ K and a calculated density of 2.836 g·cm$^{-3}$ (Bell et al. 2001b). To obtain it, a warm ethanolic solution (25 mL) of $HgBr_2$ (6.0 mM) was added to a hot solution of 1,3-thiazolidine-2-thione ($C_3H_5NS_2$) (12.0 mM) also in EtOH (25 mL) with stirring. The white precipitate that formed immediately was filtered off under suction and was recrystallized from boiling acetone solution as fine white crystals and dried in vacuo (Bell et al. 2001b).

The title compound could also be obtained by the next procedure (De Filippo et al. 1971). $HgBr_2$ was dissolved in an excess of molten 1,3-thiazolidine-2-thione ($C_3H_5NS_2$) (at ca. 105°C; molar ratio $Hg^{2+}$/$C_3H_5NS_2$ ca. 1:2), and the crude product was purified by washing with ethyl ether and recrystallizing from EtOH.

**[Hg($C_3H_6N_2S$)$_2$$Br_2$] or $C_6H_{12}N_4Br_2S_2Hg$.** This compound melts at 236°C (Isab and Perzanowski 1996). It was prepared by a dropwise addition of ethanol solution of $HgBr_2$ (1 mM in 10 mL) to a stoichiometric amount of imidazoline-2-thione ($C_3H_6N_2S$) in EtOH (2 mM in 50 mL) and refluxing the solution for 30 min on a water bath (Shunmugam and Sathyanarayana 1983). Colorless crystalline product was obtained.

**[HgBr₂(Me₂NHCS)₂] or C₆H₁₄N₂Br₂S₂Hg.** This compound crystallizes in the monoclinic structure with the lattice parameters $a = 940.6 \pm 0.2$, $b = 1520.8 \pm 0.3$, and $c = 1022.6 \pm 0.2$ pm and $\beta = 104.66° \pm 0.01°$ and the experimental and calculated densities 2.54 and 2.528 g·cm⁻³, respectively (Stålhandske et al. 1997). Its colorless crystals were obtained by cooling saturated $N,N$-dimethylthioformamide solution of HgBr₂.

**[Hg(C₇H₅NS₂)Br₂] or C₇H₅NBr₂S₂Hg.** This compound melts at 211°C–214°C (Bell et al. 2001a). It was prepared by the next procedure (Bell et al. 2001a,b; Popović et al. 2002b). HgBr₂ (5.0 mM) dissolved in warm EtOH or MeOH (50 mL) was added to a stirred solution of 1,3-benzo-2-thiazoline-2-thione (10.0 mM), dissolved in warm EtOH (50 mL), resulting in the immediate deposition of fine needlelike pale-yellow crystals. These were filtered off under suction and dried in vacuo.

**[Hg(C₇H₆N₂S)Br₂] or C₇H₆N₂Br₂SHg.** To prepare this compound, a solution of benzo-1,3-imidazole-2-thione (1.26 mM) in 20 mL of MeOH was added dropwise to a solution of HgBr₂ (1.11 mM) in 10 mL of MeOH (Popović et al. 2002a). A pale-yellow crystalline product was obtained the following day, which was filtered off, washed with cold MeOH, and dried in air.

**[Hg(Et₂C₃H₄N₂S)Br₂] or C₇H₁₄N₂Br₂SHg.** This compound melts at 150°C with decomposition (Cannas et al. 1981). It was obtained by adding $N,N'$-diethylimidazolidine-2-thione to HgBr₂ (2:1 molar ratio), using MeOH as a solvent.

**[Hg(C₄H₆N₂S)₂Br₂] or C₈H₁₂N₄Br₂S₂Hg.** This compound melts at 185°C–186°C (at 160°C with decomposition [Popović et al. 1999b]) and crystallizes in the orthorhombic structure with the lattice parameters $a = 2646.8 \pm 0.3$, $b = 1159.31 \pm 0.14$, and $c = 976.82 \pm 0.12$ pm at $160 \pm 2$ K and a calculated density of 2.609 g·cm⁻³ (Bell et al. 2000a). Both HgBr₂ (5 mM) and methyl-1,3-imidazole-2-thione (10 mM) were separately dissolved in 50 mL of aqueous EtOH and heated until they dissolved. The warm solution of HgBr₂ was added, dropwise, to that of the solution of the ligand with continuous stirring. A colorless crystalline compound was precipitated after 2 h from a pale-yellow solution. The initial product was recrystallized from hot EtOH as pale-cream-colored microcrystalline needles (Shunmugam and Sathyanarayana 1983, Popović et al. 1999b, Bell et al. 2000a).

**[Hg(MeC₃H₅N₂S)₂Br₂] or C₈H₁₆N₄Br₂S₂Hg.** This compound melts at 137°C (Isab and Perzanowski 1996) (at 175°C [Cannas et al. 1981]). It was prepared by a dropwise addition of ethanol solution of HgBr₂ to a stoichiometric amount of methylimidazoline-2-thione in EtOH or MeOH (Cannas et al. 1981, Isab and Perzanowski 1996).

**[Hg(C₄H₈N₂S)₂Br₂] or C₈H₁₆N₄Br₂S₂Hg.** This compound melts at 99°C (Isab and Perzanowski 1996). It was prepared by a dropwise addition of ethanol solution of HgBr₂ to a stoichiometric amount of 1,3-diazinane-2-thione in EtOH.

**[Hg(C₄H₈N₂S)₂Br₂] or C₈H₁₆N₄Br₂S₂Hg.** This compound melts at 148°C and crystallizes in the monoclinic structure with the lattice parameters $a = 786.98 \pm 0.07$, $b = 1564.4 \pm 0.2$, and $c = 1288.21 \pm 0.10$ pm and $\beta = 98.79° \pm 0.02°$ and a calculated density of 2.512 g·cm$^{-3}$ (Popović et al. 2000a). It was prepared by reacting one equivalent of $HgBr_2$ and two equivalents of 3,4,5,6-tetrahydropyrimidine-2-thione in MeOH. The crystalline solids, which formed on standing for several days, were filtered off, washed with MeOH, and dried. The isolated compound is crystalline and a colorless substance.

**[Hg(C$_9$H$_{12}$N$_4$S)Br$_2$] or C$_9$H$_{12}$N$_4$Br$_2$SHg**. This compound melts at the temperature higher than 300°C (Bermejo et al. 1999). It was obtained by treating 2-acetylpyridine 4-methylthiosemicarbazone (C$_9$H$_{12}$N$_4$S) with $HgBr_2$ (1:1 molar ratio) in EtOH. After prolonged stirring (about 1 week) at room temperature, the reaction mixture afforded the title compound, which was filtered off, washed with EtOH, and vacuum dried.

**[Hg(C$_9$H$_{13}$N$_5$S)Br$_2$] or C$_9$H$_{13}$N$_5$Br$_2$SHg**. This compound melts at 294°C (Bermejo et al. 2003). It was prepared as follows. A solution of 2-pyridineformamide $N(4)$-dimethylthiosemicarbazone (1.12 mM) in EtOH (20 mL) was added to a solution of $HgBr_2$ (1.12 mM) in 20 mL of 96% ethyl alcohol, and the mixture was stirred for several days at room temperature. The resulting solids were filtered out, washed thoroughly with cold ethanol, and stored in a desiccator over $CaSO_4$.

**[(Bu$^t$SHgBr)(Py)] or C$_9$H$_{14}$NBrSHg**. To obtain this compound, Bu$^t$SHgBr was dissolved in Py to slowly deposit colorless crystals (Canty et al. 1978a). The precipitate was collected, washed with EtOH, and dried over $P_2O_5$ in a vacuum. The obtained crystals slowly lose pyridine at an ambient temperature and pressure but may be stored at ca. −20°C in a sealed tube.

**[Hg(C$_{10}$H$_7$NS$_5$)Br$_2$] or C$_{10}$H$_7$NBr$_2$S$_5$Hg**. This compound melts at 125°C–126°C with decomposition and crystallizes in the triclinic structure with the lattice parameters $a = 785.4 \pm 0.2$, $b = 873.4 \pm 0.2$, and $c = 1359.4 \pm 0.5$ pm and $\alpha = 90.03° \pm 0.03°$, $\beta = 102.12° \pm 0.02°$, and $\gamma = 116.61° \pm 0.02°$ and a calculated density of 2.713 g·cm$^{-3}$ (Nam et al. 2004). To obtain it, to a 20 mL $CH_2Cl_2$ solution of 5-pyridine-2-yl-5,6-dihydro-[1,3] dithio[4,5-b]dithiin-2-thione (C$_{10}$H$_7$NS$_5$) $HgBr_2$ (0.50 mM) dissolved in MeCN with stirring at room temperature was added. The red crystals formed from the mixture after standing for 2 days were filtered and washed with MeOH and then dried in vacuo.

**[Hg(C$_5$H$_5$NS)$_2$Br$_2$] or C$_{10}$H$_{10}$N$_2$Br$_2$S$_2$Cd**. This compound was prepared by mixing ethanolic solutions of $HgBr_2$ and 2-pyridine thiol (Kennedy and Lever 1972). It could not be recrystallized.

**[HgBr$_2${Bz(Me$_2$NCS$_2$)}] or C$_{10}$H$_{13}$NBr$_2$S$_2$Hg**. This compound melts at 112°C–116°C (Brinkhoff and Dautzenberg 1972). It was prepared by the addition of $HgBr_2$ in EtOH to a solution of Bz(Me$_2$NCS$_2$) (1:1 molar ratio). The title compound was recrystallized from EtOH.

**[HgBr$_2$·{Et$_4$(NCS$_2$)$_2$}] or C$_{10}$H$_{20}$N$_2$Br$_2$S$_4$Hg, dibromotetraethylthiuramdisulfide mercury(II)**. This compound crystallizes in the orthorhombic structure with the lattice parameters $a = 1305.0 \pm 0.5$, $b = 1359.3 \pm 0.7$, and $c = 2237.3 \pm 1.2$ pm and the experimental and calculated densities $2.20 \pm 0.05$ and 2.198 g·cm$^{-3}$, respectively (Chieh 1977c, 1978a). The obtained crystals are colorless plates.

**[HgBr$_2$·Hg{(Et$_2$NCS$_2$)$_2$}] or C$_{10}$H$_{20}$N$_2$Br$_2$S$_4$Hg$_2$.** This compound crystallizes in the monoclinic structure with the lattice parameters $a = 1042.1 \pm 0.7$, $b = 1524.4 \pm 1.3$, and $c = 1464.9 \pm 0.8$ pm and $\beta = 113.3° \pm 0.1°$ and the experimental and calculated densities 2.65 and 2.664 g·cm$^{-3}$, respectively (Chieh 1978b). When two ethanolic solutions of HgBr$_2$ and tetraethylthiuram disulfide were mixed at 25°C, a white precipitate was obtained. Recrystallization of the precipitate in EtOH/chloroform (1:1 volume ratio) gave two kinds of crystals. One was colorless and the other pale yellow. The latter was the title compound.

**[Hg(Me$_2$C$_3$H$_4$N$_2$S)$_2$Br$_2$] or C$_{10}$H$_{20}$N$_4$Br$_2$S$_2$Hg.** This compound melts at 141°C (Isab and Perzanowski 1996). It was prepared by a dropwise addition of ethanol solution of HgBr$_2$ to a stoichiometric amount of dimethylimidazoline-2-thione in EtOH.

**[Hg(EtC$_3$H$_5$N$_2$S)$_2$Br$_2$] or C$_{10}$H$_{20}$N$_4$Br$_2$S$_2$Hg.** This compound melts at 127°C (Isab and Perzanowski 1996) (at 125°C [Cannas et al. 1981]). It was prepared by a dropwise addition of ethanol solution of HgBr$_2$ to a stoichiometric amount of ethylimidazoline-2-thione in EtOH or MeOH (Cannas et al. 1981, Isab and Perzanowski 1996).

**[Hg(C$_5$H$_{10}$N$_2$S)$_2$Br$_2$] or C$_{10}$H$_{20}$N$_4$Br$_2$S$_2$Hg.** This compound melts at 136°C (Isab and Perzanowski 1996). It was prepared by a dropwise addition of ethanol solution of HgBr$_2$ to a stoichiometric amount of 1,3-diazipane-2-thione in EtOH.

**[{Me$_2$NH(CH$_2$)$_3$S}$_2$][HgBr$_4$] or C$_{10}$H$_{26}$N$_2$Br$_4$S$_2$Hg.** This compound crystallizes in the orthorhombic structure with the lattice parameters $a = 2348$, $b = 1363$, and $c = 669$ pm (Casals et al. 1987). It was obtained simultaneously by obtaining [HgBr$_2${μ-S(CH$_2$)$_3$NH(CH$_3$)$_2$}].

**[Hg(C$_{11}$H$_{10}$N$_2$S$_2$)Br$_2$] or C$_{11}$H$_{10}$N$_2$Br$_2$S$_2$Hg.** This compound melts at 124°C and crystallizes in the monoclinic structure with the lattice parameters $a = 833.13 \pm 0.18$, $b = 2400.1 \pm 0.5$, and $c = 757.03 \pm 0.16$ pm and $\beta = 94.245° \pm 0.004°$ (Amoedo-Portela et al. 2002). It was prepared by the following procedure. A solution of bis(2-pyridylthio)methane (0.79 mM) in 5 mL of an EtOAc/acetonitrile (2:3 volume ratio) mixture was added dropwise to a solution of HgBr$_2$ (0.79 mM) in 5 mL of EtOAc/acetonitrile (3:2 volume ratio). The white suspension that was obtained was refluxed for 5 h. After stirring for 2 days, the solid was filtered out and dried in vacuo. Crystals suitable for XRD were obtained from a THF solution.

**[Hg$_2$(C$_{11}$H$_{10}$N$_2$S$_2$)Br$_4$] or C$_{11}$H$_{10}$N$_2$Br$_4$S$_2$Hg$_2$.** This compound melts at 116°C (Amoedo-Portela et al. 2002). To obtain it, a solution of bis(2-pyridylthio)methane (0.43 mM) in 2.4 mL of an EtOAc/acetonitrile (1:1 volume ratio) mixture was added dropwise to a solution of HgBr$_2$ (0.86 mM) in 5 mL of the same solvent mixture. After stirring for 3 days without reflux, a white solid appeared. Stirring was continued for 12 days more, and the solid was then filtered out and dried in vacuo.

**[Hg(C$_{12}$H$_{17}$N$_5$S)Br$_2$] or C$_{12}$H$_{17}$N$_5$Br$_2$SHg.** This compound melts at 294°C (Castiñeiras et al. 2002). To obtain it, a solution of C$_{12}$H$_{17}$N$_5$S [2-pyridineformamide N(4)-piperidylthiosemicarbazone] (0.76 mM) in EtOH (30 mL) was added to a solution of HgBr$_2$ (0.76 mM) in 30 mL of 96% EtOH, and the mixture was kept stirring for several days at room temperature. The resulting yellow solid was filtered off, washed thoroughly with cold ethanol, and stored in a desiccator over Mg and I$_2$.

**[Hg(Pr$^n$C$_3$H$_5$N$_2$S)$_2$Br$_2$]** and **[Hg(Pr$^i$C$_3$H$_5$N$_2$S)$_2$Br$_2$]** or **C$_{12}$H$_{24}$N$_4$Br$_2$S$_2$Hg**. These compounds melt at 98°C and 137°C, respectively (Isab and Perzanowski 1996). They were prepared by a dropwise addition of ethanol solution of HgBr$_2$ to a stoichiometric amount of $n$-propylimidazoline-2-thione or $i$-propylimidazoline-2-thione in EtOH.

**[Hg(C$_4$H$_8$N$_2$S)$_3$Br$_2$]** or **C$_{12}$H$_{24}$N$_6$Br$_2$S$_3$Hg**. This compound melts at 125°C, decomposes at 206°C, and crystallizes in the trigonal structure with the lattice parameters $a = 1158.0$ and $c = 2893.9$ pm and a calculated density of 2.10 g·cm$^{-3}$ (Hou et al. 1993).

**[Hg(C$_{13}$H$_{19}$N$_5$S)Br$_2$]** or **C$_{13}$H$_{19}$N$_5$Br$_2$SHg**. This compound was obtained if a solution of HgBr$_2$ (0.90 mM) in EtOH (30 mL) was mixed with a solution of 2-pyridine-formamide 3-hexamethyleneiminylthiosemicarbazone (C$_{13}$H$_{19}$N$_5$S) (0.90 mM) in EtOH (30 mL), and the resulting solution was stirred at room temperature for 4 days (Bermejo et al. 2004). The obtained yellow solid was filtered off.

**[Hg(C$_7$H$_5$NS$_2$)$_2$Br$_2$]** or **C$_{14}$H$_{10}$N$_2$Br$_2$S$_4$Hg**. This compound crystallizes in the monoclinic structure with the lattice parameters $a = 679.67 \pm 0.11$, $b = 3172.0 \pm 0.4$, and $c = 894.43 \pm 0.11$ pm and $\beta = 96.205° \pm 0.013°$ and a calculated density of 2.408 g·cm$^{-3}$ (Popović et al. 2002b). It was prepared as yellow crystals by the mixing of methanol solutions of HgBr$_2$ and 1,3-benzo-2-thiazoline-2-thione in the appropriate molar ratio.

**[Hg(C$_7$H$_6$N$_2$S)$_2$Br$_2$]** or **C$_{14}$H$_{12}$N$_4$Br$_2$S$_2$Hg**. To prepare this compound, a solution of benzo-1,3-imidazole-2-thione (2.46 mM) in 30 mL of MeOH was added dropwise to a solution of HgBr$_2$ (1.11 mM) in 10 mL of MeOH (Popović et al. 2002a). After a few days, the pale-pink crystalline product was filtered off, washed with cold MeOH, and dried in air.

**[Hg(C$_7$H$_{12}$N$_2$S)$_2$Br$_2$]** or **C$_{14}$H$_{24}$N$_4$Br$_2$S$_2$Hg**. This compound melts at 140°C with decomposition and crystallizes in the monoclinic structure with the lattice parameters $a = 689.08 \pm 0.06$, $b = 1023.97 \pm 0.09$, and $c = 2985.9 \pm 0.3$ pm and $\beta = 94.364° \pm 0.002°$ at $243 \pm 2$ K and a calculated density of 2.128 g·cm$^{-3}$ (White et al. 2003). It was obtained by the next procedure. A suspension of HgBr$_2$ (0.583 mM) and 2-mercapto-1-$tert$-butylimidazole (1.165 mM) in CH$_2$Cl$_2$ (25 mL) was stirred for 25 min in a warm (ca. 50°C) water bath. The resulting slightly cloudy solution was stirred for an additional 1.5 h and filtered, and the filtrate was concentrated under reduced pressure to ca. 20 mL. Addition of pentane (15 mL) resulted in the formation of a white precipitate, which was isolated by filtration, washed with pentane (2 × 20 mL), and dried in vacuo for 18 h. All reactions were performed under argon using a combination of high-vacuum and Schlenk techniques.

**[Hg(Et$_2$C$_3$H$_4$N$_2$S)$_2$Br$_2$]** or **C$_{14}$H$_{28}$N$_4$Br$_2$S$_2$Hg**. This compound melts at 94°C (Isab and Perzanowski 1996). It was prepared by a dropwise addition of ethanol solution of HgBr$_2$ to a stoichiometric amount of diethylimidazoline-2-thione in EtOH.

**[N(CH$_2$CH$_2$SCHMe$_2$)$_3$HgBr]$_2$(Hg$_2$Br$_6$)** or **C$_{15}$H$_{33}$NBr$_4$S$_3$Hg$_2$**. To prepare this compound, a solution of N(CH$_2$CH$_2$SCHMe$_2$) (1 mM) in CH$_2$Cl$_2$ (20 mL) was added to a solution of HgBr$_2$ (2 mM) in MeOH (40 mM) (Cecconi et al. 1998). Upon slow evaporation of the solvent in air, white crystals precipitated, which were filtered off, washed with MeOH, and dried under vacuum.

**[Hg$_4${S(CH$_2$)$_2$NMe$_2$}$_4$Br$_4$]** or **C$_{16}$H$_{40}$N$_4$Br$_4$S$_4$Hg$_4$**. This compound crystallizes in the monoclinic structure with the lattice parameters $a = 1396.0 \pm 0.3$, $b = 1310.7 \pm 0.3$, and $c = 956.3 \pm 0.2$ pm and $\beta = 110.60° \pm 0.02°$ and a calculated density of 2.759 g·cm$^{-3}$ (Casals et al. 1991a). To obtain it, HS(CH$_2$)$_2$NMe$_2$·HBr (5 mM) was dissolved in acetonitrile (40 mL). Et$_3$N (1.4 mL) and anhydrous HgBr$_2$ (5 mM) were successfully added with stirring under an N$_2$ atmosphere. Stirring was continued for 24 h. The resulting suspension was filtered in the open atmosphere; washed with copious distilled water, then with MeOH, and with diethyl ether; and vacuum dried.

Alternatively, this compound was obtained by adding an aqueous solution of equimolar amount of HS(CH$_2$)$_2$NMe$_2$·HBr and NaOH to an aqueous solution of Hg(OAc)$_2$, the metal/ligand molar ratio being always 1:1 (Casals et al. 1991a). A white powder formed immediately. The mixture was stirred for several hours; the solid was filtered off, washed with cold water and then MeOH, and dried under vacuum. Crystals were obtained by refluxing the compound in DMSO, filtering off the undissolved solid, letting the solution cool to 60°C, and maintaining it at this temperature for 24 h. All these manipulations were performed under an N$_2$ atmosphere.

**[Hg(C$_9$H$_{12}$N$_5$S)$_2$Br]$_2$** or **C$_{18}$H$_{24}$N$_{10}$Br$_2$S$_2$Hg$_2$**. This compound crystallizes in the monoclinic structure with the lattice parameters $a = 786.2 \pm 0.2$, $b = 2123.4 \pm 0.6$, and $c = 849.5 \pm 0.3$ pm and $\beta = 109.53° \pm 0.01°$ and a calculated density of 2.507 g·cm$^{-3}$ (Bermejo et al. 2003). Its single crystals suitable for XRD were obtained by slow evaporation of an ethanolic solution in the air at room temperature.

**[HgBr$_2$·Bu$^n_4$(CNS$_2$)$_2$]** or **C$_{18}$H$_{36}$N$_2$Br$_2$S$_4$Hg**. This compound melts at 132.5°C–135.5°C (Brinkhoff et al. 1970). It was precipitated as colorless needles by the addition of an ethanolic solution of HgBr$_2$ (1 M) to an ethanolic solution of Bu$^n_4$C$_2$N$_2$S$_4$ (1 M). The product was purified by the addition of petroleum ether to a concentrated solution in CHCl$_3$. It could also be prepared as a white precipitate by adding a solution of Br$_2$ in CS$_2$ to a solution of di-$n$-butyldithiocarbamatomercury in the same solvent at room temperature (Brinkhoff et al. 1969). After filtration, the product could be purified by dissolution in chloroform and reprecipitation by petroleum ether. The alternative route includes using HgBr$_2$ and Bu$^n_4$C$_2$N$_2$S$_4$. To a suspension of HgBr$_2$ in CS$_2$ a solution of Bu$^n_4$C$_2$N$_2$S$_4$ was added at room temperature. After HgBr$_2$ had dissolved, petroleum ether was added, precipitating the crystals, that were filtered and dried.

**[Hg(Pr$^i_2$C$_3$H$_4$N$_2$S)$_2$Br$_2$]** or **C$_{18}$H$_{36}$N$_4$Br$_2$S$_2$Hg**. This compound melts at 145°C (Isab and Perzanowski 1996). It was prepared by a dropwise addition of ethanol solution of HgBr$_2$ to a stoichiometric amount of diisopropylimidazoline-2-thione in EtOH.

**[Hg(C$_{13}$H$_{18}$N$_5$S)$_2$Br]$_2$** or **C$_{26}$H$_{36}$N$_{10}$Br$_2$S$_2$Hg$_2$**. This compound crystallizes in the monoclinic structure with the lattice parameters $a = 1278.7 \pm 0.3$, $b = 745.0 \pm 0.2$, and $c = 1802.9 \pm 0.3$ pm and $\beta = 106.314° \pm 0.013°$ and a calculated density of 2.244 g·cm$^{-3}$ (Bermejo et al. 2004). Its yellow crystals were obtained by recrystallization of [Hg(C$_{13}$H$_{19}$N$_5$S)Br$_2$] from EtOH.

## 7.45   Hg–H–C–N–I–S

Some compounds are formed in this system.

**[Hg(C$_3$H$_4$N$_2$S)I$_2$] or C$_3$H$_4$N$_2$I$_2$SHg**. This compound melts at 188°C with decomposition and crystallizes in the orthorhombic structure with the lattice parameters $a = 762.26 \pm 0.07$, $b = 777.7 \pm 0.2$, and $c = 1616.93 \pm 0.13$ pm and a calculated density of 3.843 g·cm$^{-3}$ (Popović et al. 1999b). It was obtained by adding an ethanolic solution of 1,3-imidazole-2-thione to an ethanolic solution of HgI$_2$ (1:1 molar ratio).

**[Hg(C$_3$H$_5$NS$_2$)I$_2$] or C$_3$H$_5$NI$_2$S$_2$Hg**. This compound melts at 96°C–97°C (Bell et al. 2001a). To obtain it, 1,3-thiazolidine-2-thione (3.5 mM), dissolved in hot EtOH, was added to a warm ethanolic solution of HgI$_2$ (3.5 mM). The obtained pale-yellow solution was allowed to evaporate slowly, resulting in the formation of pale-yellow microcrystals that were filtered off under suction and dried in vacuo.

**[Hg(C$_4$H$_6$N$_2$S)I$_2$] or C$_4$H$_6$N$_2$I$_2$SHg**. This compound melts at 110°C–112°C and crystallizes in the monoclinic structure with the lattice parameters $a = 770.63 \pm 0.13$, $b = 1447.1 \pm 0.3$, and $c = 2570.5 \pm 0.4$ pm and $\beta = 95.114° \pm 0.009°$ and a calculated density of 3.312 g·cm$^{-3}$ (Raper et al. 1998). It was obtained by the following procedure. Both HgI$_2$ (18.8 mM) and 1-methylimidazoline-2-thione (C$_4$H$_6$N$_2$S) (18.8 mM) were separately dissolved in 50 mL of aqueous EtOH and heated until they dissolved. The warm solution of (C$_4$H$_6$N$_2$S) was added dropwise to that of the HgI$_2$ solution with continuous stirring. An initial white precipitate subsequently redissolved. A white crystalline product was eventually obtained after complete addition of the ligand. The initial product was recrystallized from hot acetone as white microcrystalline needles.

**[Hg(C$_4$H$_8$N$_2$S)I$_2$] or C$_4$H$_8$N$_2$I$_2$SHg**. This compound melts at 128°C (Popović et al. 2001) and crystallizes in the orthorhombic structure with the lattice parameters $a = 1282.6 \pm 0.2$, $b = 739.20 \pm 0.10$, and $c = 1144.7 \pm 0.2$ pm at $200 \pm 2$ K and a calculated density of 3.492 g·cm$^{-3}$ (Matković-Čalogović et al. 2001). It was prepared by adding dropwise a methanol solution of 3,4,5,6-tetrahydropyrimidine-2-thione to an equimolar methanol solution of HgI$_2$ (Popović et al. 2001). The crystalline solids, which were formed after standing for several days, were filtered off, washed with MeOH, and dried. The isolated compound is crystalline and yellow substance.

**[HgI$_2${Me(Me$_2$NCS$_2$)}] or C$_4$H$_9$NI$_2$S$_2$Hg**. This compound melts at 106°C–108°C (Brinkhoff and Dautzenberg 1972). It was prepared by the addition of an equivalent amounts of HgI$_2$ (1:1 molar ratio) to a solution of Me(Me$_2$NCS$_2$) in ether, acetone, chloroform, or CS$_2$. Precipitation could, if necessary, be facilitated by the addition of petroleum ether. The title compound was recrystallized from acetone/petroleum ether.

**[HgI$_2${NCCH$_2$(Me$_2$NCS$_2$)}] or C$_5$H$_8$N$_2$I$_2$S$_2$Hg**. This compound was prepared by the addition of an equivalent amounts of HgI$_2$ (1:1 molar ratio) to a solution of

NCCH$_2$(Me$_2$NCS$_2$) in ether, acetone, chloroform, or CS$_2$ (Brinkhoff and Dautzenberg 1972). Precipitation could, if necessary, be facilitated by the addition of petroleum ether. The title compound was recrystallized from chloroform.

**[(Et$_2$NCS$_2$)(HgI)] or C$_5$H$_{10}$NIS$_2$Hg.** This compound melts at 136°C (at 164°C [Zemskova et al. 1991]) and crystallizes in the monoclinic structure with the lattice parameters $a = 764.5 \pm 0.3$, $b = 787.7 \pm 0.4$, and $c = 1800.5 \pm 0.7$ pm and $\beta = 102.08° \pm 0.04°$ and the experimental and calculated densities $3.00 \pm 0.01$ and 3.014 g·cm$^{-3}$, respectively (Chieh 1977a). It was obtained by the following procedure. To 20 mL of ethanolic solution of 0.235 g HgI$_2$, 20 mL of chloroform solution of 0.154 g of tetraethylthiuram disulfide was added. There was a small amount of small deep-yellow square plates formed when the solutions were mixed. The mixture was covered and left at room temperature for 3 days. Then, it was placed in a refrigerator (0°C) and partly covered only to allow slow evaporation. After a week, the solution was reduced to about 0.5 mL. Yellow hexagonal prismatic crystals were obtained after filtration.

The title compound was also prepared as follows (Zemskova et al. 1991). To a solution of Hg(Et$_2$NCS$_2$)$_2$ in CH$_2$Cl$_2$ at room temperature HgI$_2$ ([1.0–1.1]:1 molar ratio) was added, and the mixture was stirred for 3–5 h. The precipitate that formed was filtered off, washed thoroughly with CH$_2$Cl$_2$, and dried in air.

**[Hg(Me$_2$C$_3$H$_4$N$_2$S)I$_2$] or C$_5$H$_{10}$N$_2$I$_2$SHg.** This compound melts at 104°C with decomposition (Cannas et al. 1981). It was obtained by adding $N,N'$-dimethylimidazolidine-2-thione to HgI$_2$ (2:1 molar ratio), using MeOH as a solvent.

**[HgI$_2${Et(Me$_2$NCS$_2$)}] or C$_5$H$_{11}$NI$_2$S$_2$Hg.** This compound melts at 83°C–87°C (Brinkhoff and Dautzenberg 1972). It was prepared by the addition an equivalent amounts of HgI$_2$ (1:1 molar ratio) to a solution of Et(Me$_2$NCS$_2$) in ether, acetone, chloroform, or CS$_2$. Precipitation could, if necessary, be facilitated by the addition of petroleum ether. The title compound was recrystallized from acetone/petroleum ether.

**[Hg(C$_3$H$_4$N$_2$S)$_2$I$_2$] or C$_6$H$_8$N$_4$I$_2$S$_2$Hg.** This compound melts at 176°C (Popović et al. 1999b). It was obtained by adding an ethanolic solution of 1,3-imidazole-2-thione to an ethanolic solution of HgI$_2$ (2:1 molar ratio).

**[Hg(C$_3$H$_5$NS$_2$)$_2$I$_2$] or C$_6$H$_{10}$N$_2$I$_2$S$_4$Hg.** This compound melts at 130°C–132°C (Bell et al. 2001b) (at 126°C [De Filippo et al. 1971]). To obtain it, HgI$_2$ (3.0 mM) dissolved in warm EtOH (25 mL) was added to 1,3-thiazolidine-2-thione (6.0 mM) and also dissolved in hot EtOH (25 mL). The clear solution was allowed to evaporate slowly at room temperature until yellow crystals were formed, which were filtered off under suction. Recrystallization from boiling acetone yielded pale-yellow crystals, which were filtered off and dried in vacuo.

The title compound could also be prepared by the next procedure (De Filippo et al. 1971). HgI$_2$ was dissolved in an excess of molten 1,3-thiazolidine-2-thione (C$_3$H$_5$NS$_2$) (at ca. 105°C; Hg$^{2+}$/C$_3$H$_5$NS$_2$ ca. 1:2 molar ratio), and the crude product was purified by washing with ethyl ether and recrystallizing from EtOH.

**[HgI$_2$(Me$_4$C$_2$N$_2$S$_3$)$_2$]** or **C$_6$H$_{12}$N$_2$I$_2$S$_3$Hg**. This compound melts at 119°C–120°C (Brinkhoff et al. 1970) and crystallizes in the orthorhombic structure with the lattice parameters $a = 1291.0 \pm 0.3$, $b = 1244.1 \pm 0.3$, and $c = 986.3 \pm 0.2$ pm and the experimental and calculated density of $2.79 \pm 0.01$ and 2.78 g·cm$^{-3}$, respectively (Skelton and White 1977). It was prepared as yellow plates by adding HgI$_2$ (1 M) to a concentration solution of tetramethylthiuram monosulfide (Me$_4$C$_2$N$_2$S$_3$) (1 M) in CS$_2$. The mixture was stirred for 1 h. The reaction product was filtered off and purified by dissolving in CHCl$_3$ and reprecipitating by adding petroleum ether.

**[HgI$_2$(S$_2$CNMe$_2$)$_2$]** or **C$_6$H$_{12}$N$_2$I$_2$S$_4$Hg**. This compound crystallizes in the monoclinic structure with the lattice parameters $a = 795.7 \pm 0.4$, $b = 2263.9 \pm 0.7$, and $c = 995.6 \pm 0.6$ pm and $\beta = 112.14° \pm 0.05°$ and the experimental and calculated densities $2.76 \pm 0.03$ and $2.775 \pm 0.004$ g·cm$^{-3}$, respectively (Beurskens et al. 1971).

**[Hg(C$_3$H$_6$N$_2$S)$_2$I$_2$]** or **C$_6$H$_{12}$N$_4$I$_2$S$_2$Hg**. This compound melts at 164°C–165°C and crystallizes in the monoclinic structure with the lattice parameters $a = 602.6 \pm 0.2$, $b = 1452.9 \pm 0.5$, and $c = 813.3 \pm 0.3$ pm and $\beta = 92.365° \pm 0.006°$ and a calculated density of 3.075 g·cm$^{-3}$ (Lobana et al. 2008). To obtain it, to a stirred solution of imidazoline-2-thione (C$_3$H$_6$N$_2$S) (0.24 mM) in dry MeOH (10 mL) was added a solution of HgI$_2$ (0.12 mM) in dry MeOH (10 mL). The reaction mixture was stirred for 2 h and slow evaporation at room temperature formed yellow crystals of [Hg(C$_3$H$_6$N$_2$S)$_2$I$_2$].

**[HgI$_2${Me(Et$_2$NCS$_2$)}]** or **C$_6$H$_{13}$NI$_2$S$_2$Hg**. This compound melts at 97°C–99°C (Brinkhoff and Dautzenberg 1972). It was prepared by the addition an equivalent amounts of HgI$_2$ (1:1 molar ratio) to a solution of Me(Et$_2$NCS$_2$) in ether, acetone, chloroform, or CS$_2$. Precipitation could, if necessary, be facilitated by the addition of petroleum ether. The title compound was recrystallized from acetone/petroleum ether.

**[HgI$_2$(Me$_2$NHCS)$_2$]** or **C$_6$H$_{14}$N$_2$I$_2$S$_2$Hg**. This compound crystallizes in the monoclinic structure with the lattice parameters $a = 975.7 \pm 0.1$, $b = 1554.6 \pm 0.3$, and $c = 1041.6 \pm 0.2$ pm and $\beta = 104.47° \pm 0.01°$ and the experimental and calculated densities 2.78 and 2.747 g·cm$^{-3}$, respectively (Stålhandske et al. 1997). Its colorless crystals that show tendency for twin formation were obtained by cooling saturated $N,N$-dimethylthioformamide solution of HgI$_2$.

**[Hg(C$_7$H$_5$NS$_2$)I$_2$]** or **C$_7$H$_5$NI$_2$S$_2$Hg**. This compound melts at 179°C–182° (Bell et al. 2001a) and crystallizes in the triclinic structure with the lattice parameters $a = 755.37 \pm 0.16$, $b = 887.51 \pm 0.17$, and $c = 966.53 \pm 0.18$ pm and $\alpha = 71.013° \pm 0.015°$, $\beta = 78.113° \pm 0.016°$, and $\gamma = 88.795° \pm 0.016°$ and a calculated density of 3.488 g·cm$^{-3}$. It was prepared by the following procedure (Bell et al. 2001a,b; Popović et al. 2002b). HgI$_2$ (5.0 mM) dissolved in warm EtOH or MeOH (50 mL) was added to a stirred solution of benzo-1,3-thiazoline-2-thione (10.0 mM), dissolved in warm EtOH (50 mL), resulting in the immediate deposition of fine crystals. These were filtered off under suction and dried in vacuo.

**[Hg(C$_7$H$_6$N$_2$S)I$_2$]** or **C$_7$H$_6$N$_2$I$_2$SHg**. This compound crystallizes in the monoclinic structure with the lattice parameters $a = 1741.7 \pm 0.3$, $b = 671.40 \pm 0.10$, and

$c = 1108.0 \pm 0.2$ pm and $\beta = 114.60° \pm 0.03°$ at 200 K and a calculated density of 3.409 g·cm⁻³ (Popović et al. 2002a). To prepare it, a solution of benzo-1,3-imidazole-2-thione (0.99 mM) in 10 mL of MeOH was added dropwise to a solution of HgI₂ (0.88 mM) in 20 mL of MeOH. After a few days, the yellow crystals were filtered off, washed with cold MeOH, and dried in air.

**[Hg(Et₂C₃H₄N₂S)I₂] or C₇H₁₄N₂I₂SHg.** This compound melts at 147°C with decomposition (Cannas et al. 1981). It was obtained by adding *N,N'*-diethylimidazolidine-2-thione to HgI₂ (2:1 molar ratio), using MeOH as a solvent.

**[Hg(C₄H₆N₂S)₂I₂ or C₈H₁₂N₄I₂S₂Hg.** This compound melts at 178°C–180°C (Bell et al. 2000a) (at 175°C [Popović et al. 1999b]) and crystallizes in the orthorhombic structure with the lattice parameters $a = 1154.3 \pm 0.2$, $b = 1399.6 \pm 0.5$, and $c = 991.80 \pm 0.07$ pm and a calculated density of 2.830 g·cm⁻³ (Pavlović et al. 2000b) ($a = 1151.2 \pm 0.2$, $b = 1385.1 \pm 0.2$, and $c = 984.57 \pm 0.10$ pm at $160 \pm 2$ K and a calculated density of 2.89 g·cm⁻³ [Bell et al. 2000a]). Both HgI₂ (4.4 mM) and methyl-1,3-imidazole-2-thione (8.8 mM) were separately dissolved in ca. 200 mL of aqueous EtOH and heated until they dissolved (Popović et al. 1999b, Bell et al. 2000a). The warm solution of HgI₂ was added, dropwise, to that of the solution of the ligand with continuous stirring. The crude product precipitated from solution after 3 days as dendritic pale-green crystals. The title compound was recrystallized from hot EtOH as pale-green microcrystalline needles.

Its crystals were formed from a very dilute methanol solution of the title compound at room temperature on standing for several days (Pavlović et al. 2000b).

**[Hg(C₄H₈N₂S)₂I₂] or C₈H₁₆N₄I₂S₂Hg.** This compound melts at 135°C and crystallizes in the monoclinic structure with the lattice parameters $a = 1473.71 \pm 0.17$, $b = 785.50 \pm 0.05$, and $c = 1507.97 \pm 0.11$ pm and $\beta = 96.345° \pm 0.008°$ and a calculated density of 2.629 g·cm⁻³ (Popović et al. 2000a). It was prepared by reacting one equivalent of HgI₂ and two equivalents of 3,4,5,6-tetrahydropyrimidine-2-thione in MeOH. The crystalline solids, which formed on standing for several days, were filtered off, washed with MeOH, and dried. The isolated compound is crystalline and yellow substance.

**[Hg(MeC₃H₅N₂S)₂I₂] or C₈H₁₆N₄I₂S₂Hg.** This compound melts at 122°C (Cannas et al. 1981). It was obtained by adding *N*-methylimidazolidine-2-thione to HgI₂ (2:1 molar ratio), using MeOH as a solvent.

**[Hg(C₉H₁₂N₄S)I₂] or C₉H₁₂N₄I₂SHg.** This compound melts at 206°C with decomposition (Bermejo et al. 1999). It was obtained by treating 2-acetylpyridine 4-methylthiosemicarbazone (C₉H₁₂N₄S) with HgI₂ (1:1 molar ratio) in EtOH. After prolonged stirring (about 1 week) at room temperature, the reaction mixture afforded the title compound, which was filtered off, washed with EtOH, and vacuum dried.

**[Hg(C₉H₁₃N₅S)I₂] or C₉H₁₃N₅I₂SHg.** This compound melts at 179°C (Bermejo et al. 2003). It was prepared as follows. A solution of 2-pyridineformamide *N*(4)-dimethylthiosemicarbazone (1.12 mM) in EtOH (20 mL) was added to a solution of HgI₂ (1.12 mM) in 20 mL of 96% ethyl alcohol, and the mixture was stirred for several days at room temperature. The resulting solids were filtered out, washed thoroughly with cold ethanol, and stored in a desiccator over $CaSO_4$.

**[Hg(C$_{10}$H$_7$NS$_5$)I$_2$] or C$_{10}$H$_7$NI$_2$S$_5$Hg.** This compound melts at 125°C–126°C with decomposition and crystallizes in the triclinic structure with the lattice parameters $a = 821.9 \pm 0.2$, $b = 878.5 \pm 0.3$, and $c = 1382.6 \pm 0.3$ pm and $\alpha = 79.16° \pm 0.03°$, $\beta = 77.66° \pm 0.03°$, and $\gamma = 62.61° \pm 0.03°$ and a calculated density of 2.915 g·cm$^{-3}$ (Nam et al. 2004). To obtain it, to a 20 mL $CH_2Cl_2$ solution of 5-pyridine-2-yl-5,6-dihydro-[1,3]dithio[4,5-$b$]dithiin-2-thione (C$_{10}$H$_7$NS$_5$) HgI$_2$ (0.50 mM) was added and dissolved in MeCN with stirring at room temperature. The red crystals formed from the mixture after standing for 2 days were filtered and washed with MeOH and then dried in vacuo.

**[HgI$_2${Bz(Me$_2$NCS$_2$)}] or C$_{10}$H$_{13}$NI$_2$S$_2$Hg.** This compound melts at 104°C–106°C (Brinkhoff and Dautzenberg 1972). It was prepared by the addition an equivalent amounts of HgI$_2$ (1:1 molar ratio) to a solution of Bz(Me$_2$NCS$_2$) in ether, acetone, chloroform, or $CS_2$. Precipitation could, if necessary, be facilitated by the addition of petroleum ether. The title compound was recrystallized from acetone/petroleum ether.

**[HgI$_2$·{Et$_4$(NCS$_2$)$_2$}] or C$_{10}$H$_{20}$N$_2$I$_2$S$_4$Hg, diiodotetraethylthiuramdisulfidemercury(II).** This compound melts at 132.5°C–136°C (Brinkhoff et al. 1970) and crystallizes in the orthorhombic structure with the lattice parameters $a = 1310.1 \pm 0.9$, $b = 1399.9 \pm 0.9$, and $c = 2274.2 \pm 1.5$ pm and the experimental and calculated densities $2.40 \pm 0.01$ and 2.391 g·cm$^{-3}$, respectively (Chieh 1977c). It was prepared by the following procedure. To 50 mL solution of tetraethylthiuram disulfide (0.21 g) in chloroform/ethanol (1:1 volume ratio), red crystalline HgI$_2$ (0.32 g) was added. At room temperature, the reaction proceeded rather quickly. In about 10 min, yellow square plates were seen under a 30× microscope. Both the red HgI$_2$ and square plate crystals dissolved slowly resulting in a pale-yellow solution in a few hours. The solution was then partially covered to allow slow evaporation. When the solution was reduced to 1 mL, both square plates (the title compound) and hexagonal prisms (Et$_2$NCS$_2$HgI)$_2$ were obtained after filtration (Chieh 1977c). The title compound could also be obtained by the interaction of Hg(Et$_2$CNS$_2$)$_2$ (1 M) with I$_2$ (1 M) in CHCl$_3$.

**[Hg(EtC$_3$H$_5$N$_2$S)I$_2$] or C$_{10}$H$_{20}$N$_4$I$_2$S$_2$Hg.** This compound melts at 76°C with decomposition (Cannas et al. 1981). It was obtained by adding $N$-ethylimidazolidine-2-thione to HgI$_2$ (2:1 molar ratio), using MeOH as a solvent.

**[Hg(C$_{12}$H$_{17}$N$_5$S)I$_2$] or C$_{12}$H$_{17}$N$_5$I$_2$SHg.** This compound melts at 205°C (Castiñeiras et al. 2002). To obtain it, a solution of C$_{12}$H$_{17}$N$_5$S [2-pyridineformamide $N$(4)-piperidylthiosemicarbazone] (0.76 mM) in EtOH (30 mL) was added to a solution of HgI$_2$ (0.76 mM) in 30 mL of 96% EtOH, and the mixture was kept stirring for several days at room temperature. The resulting yellow solid was filtered off, washed thoroughly with cold ethanol, and stored in a desiccator over Mg and I$_2$.

**[Hg(C$_7$H$_5$NS$_2$)$_2$I$_2$] or C$_{14}$H$_{10}$N$_2$I$_2$S$_4$Hg.** This compound was prepared as yellow crystals by the mixing of methanol solutions of HgI$_2$ and 1,3-benzo-2-thiazoline-2-thione in the appropriate molar ratio (Popović et al. 2002b).

**[Hg(C₇H₆N₂S)₂I₂] or C₁₄H₁₂N₄I₂S₂Hg.** This compound crystallizes in the triclinic structure with the lattice parameters $a = 817.11 \pm 0.07$, $b = 862.17 \pm 0.10$, and $c = 1474.94 \pm 0.10$ pm and $\alpha = 76.246° \pm 0.009°$, $\beta = 81.199° \pm 0.009°$, and $\gamma = 86.001° \pm 0.011°$ and a calculated density of 2.515 g·cm⁻³ (Popović et al. 2002a). To prepare it, a solution of benzo-1,3-imidazole-2-thione (1.93 mM) in 20 mL of MeOH was added dropwise to a solution of HgI₂ (0.88 mM) in 20 mL of MeOH. After a few days, the yellow crystals were filtered off, washed with cold MeOH, and dried in air.

**[HgI₂·Buⁿ₄(NCS₂)₂] or C₁₈H₃₆N₂I₂S₄Hg.** This compound melts at 129°C–131.5°C (Brinkhoff et al. 1969). It was prepared as a white precipitate by adding a solution of I₂ in CS₂ to a solution of di-*n*-butyldithiocarbamatomercury in the same solvent at room temperature. After filtration, the product could be purified by dissolving in chloroform and reprecipitation by petroleum ether. The alternative route includes using HgI₂ and tetra-*n*-butylthiuram disulfide. To a suspension of HgI₂ (2.7 g) in CS₂ (10 mL), a solution of tetra-*n*-butylthiuram disulfide (2.0 g) was added at room temperature. After the red HgI₂ had dissolved, petroleum ether was added, precipitating the yellow crystals, which was filtered and dried.

**[Hg(C₁₃H₁₈N₅S)I] or C₂₆H₃₆N₁₀I₂S₂Hg₂.** This compound crystallizes in the triclinic structure with the lattice parameters $a = 881.7 \pm 0.3$, $b = 878.3 \pm 0.3$, and $c = 1285.5 \pm 0.2$ pm and $\alpha = 90.675° \pm 0.015°$, $\beta = 105.587° \pm 0.015°$, and $\gamma = 114.41° \pm 0.02°$ and a calculated density of 2.320 g·cm⁻³ (Bermejo et al. 2004). To obtain it, a solution of HgI₂ (0.90 mM) in EtOH (30 mL) was mixed with a solution of 2-pyridineformamide 3-hexamethyleneiminylthiosemicarbazone (C₁₃H₁₉N₅S) (0.90 mM) in EtOH (30 mL), and the resulting solution was stirred at room temperature for 4 days. The bright-yellow precipitate was filtered off. Recrystallization from MeOH gave yellow crystals of the title compound.

**[(Et₄N)₂(μ-SPrⁿ)₆(HgI)₄] or C₃₄H₈₂N₂I₄S₆Hg₄.** To obtain this compound, a stoichiometric mixture of HgI₂, Et₄NI, and Hg(SPrⁿ₂) in CH₂Cl₂ was stirred at room temperature (Dean and Manivannan 1990b). The colorless plates obtained by refrigeration of the reaction mixture overnight were separated by filtration, washed with Et₂O, and dried in vacuo. All syntheses and preparation of samples were carried out under an Ar or N₂ atmosphere.

## 7.46   Hg–H–C–N–Mn–S

**Hg[SC(NH₂)₂]₄Mn(SCN)₄ or C₈H₁₆N₁₂MnS₈Hg.** This multinary compound is formed in this system. It crystallizes in the tetragonal structure with the lattice parameters $a = 1742.97 \pm 0.08$ and $c = 417.54 \pm 0.03$ pm and a calculated density of 2.074 g·cm⁻³ (Yu et al. 2001). To obtain it, to an aqueous solution (20 mL) containing MnHg(SCN)₄ (10.7 mM), thiourea (13.8 mM) was added. The pH of the solution was adjusted to 3 by adding HCl. This mixture was heated and stirred until the MnHg(SCN)₄ had dissolved. The aqueous solution was then allowed to stand at room temperature. After a few hours, colorless crystals of the title compound suitable for XRD were obtained.

## 7.47 Hg–H–C–N–Mn–Fe–S

Some compounds are formed in this system.

**[Fe(C₅H₄HgSCN)₂Mn(SCN)₂]** or **C₁₄H₈N₄MnFeS₄Hg₂**. This compound melts at 164°C with decomposition (Singh and Singh 1986). To obtain it, 1,1′-bis(thiocyanatomercurio)ferrocene was dissolved in a small quantity of DMSO and diluted to 200 mL by acetone. Mn(SCN)₂ was similarly dissolved in ethyl acetate. The obtained solution was mixed in 1:1 molar ratio and stirred for 24 h. When the quantity of DMSO became more, the precipitate was obtained by the addition of EtOH. This precipitate was filtered, washed with EtOH, and dried in vacuum. Gray crystals of the title compound were recrystallized from acetone/EtOH mixture.

**[Fe(C₅H₄HgSCN)₂Mn(SCN)₂]·(C₁₀H₈N₂)** or **C₂₄H₁₆N₆MnFeS₄Hg₂**. This compound melts at 180°C with decomposition (Singh and Singh 1986). It was prepared by the next method. A suspension or solution of [Fe(C₅H₄HgSCN)₂Mn(SCN)₂] was prepared in acetone/EtOH mixture, mixed with an ethanol solution of 2,2′-bipy (1:1 molar ratio), and stirred for 24 h. The precipitate was formed, filtered, washed with EtOH, and dried in vacuum. Light-brown crystals of the title compound were recrystallized from acetone/EtOH mixture.

## 7.48 Hg–H–C–N–Fe–Co–S

Some compounds are formed in this system.

**[Fe(C₅H₄HgSCN)₂Co(SCN)₂]** or **C₁₄H₈N₄FeCoS₄Hg₂**. This compound melts at 230°C with decomposition (Singh and Singh 1986). To obtain it, 1,1′-bis(thiocyanatomercurio) ferrocene was dissolved in a small quantity of DMSO and diluted to 200 mL by acetone. Co(SCN)₂ was similarly dissolved either in acetone or in DMSO/acetone mixture separately. The obtained solution was mixed in 1:1 molar ratio and stirred for 24 h. When the quantity of DMSO became more, the precipitate was obtained by the addition of EtOH. This precipitate was filtered, washed with EtOH, and dried in vacuum. Blue crystals of the title compound were recrystallized from acetone/EtOH mixture.

**[Fe(C₅H₄HgSCN)₂Co(SCN)₂]·(C₁₀H₈N₂)** or **C₂₄H₁₆N₆FeCoS₄Hg₂**. This compound melts at 190°C with decomposition (Singh and Singh 1986). It was prepared by the next method. A suspension or solution of [Fe(C₅H₄HgSCN)₂Co(SCN)₂] was prepared in acetone/EtOH mixture, mixed with an ethanol solution of 2,2′-bipy (1:1 molar ratio), and stirred for 24 h. The precipitate was formed, filtered, washed with EtOH, and dried in vacuum. Dirty-pink crystals of the title compound were recrystallized from acetone/EtOH mixture.

## 7.49 Hg–H–C–N–Fe–Ni–S

Some compounds are formed in this system.

**[Fe(C₅H₄HgSCN)₂Ni(SCN)₂]** or **C₁₄H₈N₄FeNiS₄Hg₂**. This compound melts at 240°C with decomposition (Singh and Singh 1986). To obtain it, 1,1′-bis(thiocyanatomercurio) ferrocene was dissolved in a small quantity of DMSO and diluted to 200 mL by acetone. Ni(SCN)$_2$ was similarly dissolved either in acetone or in DMSO/acetone mixture separately. The obtained solution was mixed in 1:1 molar ratio and stirred for 24 h. When the quantity of DMSO became more, the precipitate was obtained by the addition of EtOH. This precipitate was filtered, washed with EtOH, and dried in vacuum. Light-green crystals of the title compound were recrystallized from acetone/EtOH mixture.

[Fe(C$_5$H$_4$HgSCN)$_2$Ni(SCN)$_2$]·(C$_{10}$H$_8$N$_2$) or C$_{24}$H$_{16}$N$_6$FeNiS$_4$Hg$_2$. This compound melts at 210°C with decomposition (Singh and Singh 1986). It was prepared by the next method. A suspension or solution of [Fe(C$_5$H$_4$HgSCN)$_2$Ni(SCN)$_2$] was prepared in acetone/EtOH mixture, mixed with an ethanol solution of 2,2'-bipy (1:1 molar ratio), and stirred for 24 h. The precipitate was formed, filtered, washed with EtOH, and dried in vacuum. Mustard crystals of the title compound were recrystallized from acetone/EtOH mixture.

## 7.50   Hg–H–C–N–Co–S

Some compounds are formed in this system.

[Hg{SC(NH$_2$)$_2$}$_4$Co(SCN)$_4$] or C$_8$H$_{16}$N$_{12}$CoS$_8$Hg. This compound melts at 183°C (Czakis-Sulikowska and Sołoniewicz 1963) and crystallizes in the tetragonal structure with the lattice parameters $a = 1727 \pm 2$ and $c = 427 \pm 1$ pm and the experimental and calculated densities 2.09 and 1.91 g·cm$^{-3}$, respectively (Korczyński 1963a, Korczyński and Porai-Koshits 1965). It was prepared by the addition at 20°C saturated aqueous solution of thiourea to 100 mL of 0.5 M aqueous solution of (NH$_4$)$_2$[Hg(SCN)$_4$] containing 2 mL of concentrated HNO$_3$ up to dissolution of formed precipitate (Czakis-Sulikowska and Sołoniewicz 1963). Then, 100 mL of the solution containing 0.5 M ZnSO$_4$ and 20 g NH$_4$SCN was added by stirring. The formed precipitate was crystallized from the aqueous solution.

[Hg$_2$Co(SCN)$_6$·C$_6$H$_6$] or C$_{12}$H$_6$N$_6$CoS$_6$Hg$_2$. This compound crystallizes in the triclinic structure with the lattice parameters $a = 759$, $b = 837$, and $c = 887$ pm and $\alpha = 98.31°$, $\beta = 101.01°$, and $\gamma = 95.50°$ and the experimental and calculated densities 2.65 and 2.71 g·cm$^{-3}$, respectively (Grønbaek and Dunitz 1964) ($a = 749$, $b = 860$, and $c = 873$ pm and $\alpha = 98°53'$, $\beta = 99°06'$, and $\gamma = 95°35'$ [Baur et al. 1962]). From a solution containing a Hg$^{2+}$ salt and thiocyanate (molar ratio of about 1:4), a solution of Co$^{2+}$ salt precipitates, in the presence of benzene, bright-pink microcrystalline title compound, which can also be obtained in well-developed crystals (Baur et al. 1962). The obtained compound is thermodynamically unstable.

[Hg$_2$Co(SCN)$_6$·C$_7$H$_8$] or C$_{13}$H$_8$N$_6$CoS$_6$Hg$_2$. From a solution containing a Hg$^{2+}$ salt and thiocyanate (molar ratio of about 1:4), a solution of Co$^{2+}$ salt precipitates, in the presence of toluene, red microcrystalline title compound, which can also be obtained in well-developed crystals (Baur et al. 1962). The obtained compound is thermodynamically unstable.

[HgCo(SCN)$_4$(Py)$_2$] or C$_{14}$H$_{10}$N$_6$CoS$_4$Hg. This compound crystallizes in the monoclinic structure with the lattice parameters $a = 1145.9 \pm 0.6$, $b = 1329.3 \pm 0.7$, and $c = 1449.8 \pm 0.7$ pm and $\beta = 105.65° \pm 0.03°$ and the experimental and calculated densities $2.02 \pm 0.01$ and 2.030 g·cm$^{-3}$, respectively (Beauchamp et al. 1976). It was prepared by the following procedure. An ethanolic solution of Co(SCN)$_2$ and excess of Hg(SCN)$_2$ were mixed, and the mixture was filtered to obtain a solution of the pink compound HgCo(SCN)$_4$·2EtOH. Crystals of the title compound were grown by vapor diffusion of Py into the pink solution.

## 7.51  Hg–H–C–N–Ni–S

Hg$_2$Ni(SCN)$_6$·C$_6$H$_6$ or C$_{12}$H$_6$N$_6$NiS$_6$Hg$_2$. This multinary compound is formed in this system. From a solution containing a Hg$^{2+}$ salt and thiocyanate (molar ratio of about 1:4), a solution of Ni$^{2+}$ salt precipitates, in the presence of benzene, light-blue microcrystalline title compound (Baur et al. 1962). The obtained compound is thermodynamically unstable.

## 7.52  Hg–H–C–P–As–S

(Ph$_4$P)$_2$[Hg$_2$As$_4$S$_9$] or C$_{48}$H$_{40}$P$_2$As$_4$S$_9$Hg$_2$. This multinary compound is formed in this system. It crystallizes in the monoclinic structure with the lattice parameters $a = 1011.9 \pm 0.2$, $b = 1801.0 \pm 0.4$, and $c = 1493.2 \pm 0.3$ pm and $\beta = 103.98° \pm 0.02°$, a calculated density of 2.098 g·cm$^{-3}$ and an energy gap of 2.9 eV (Chou and Kanatzidis 1995). To obtain it, a Pyrex tube (ca. 4 mL) containing HgCl$_2$ (0.5 mM), K$_3$AsS$_3$ (0.5 mM), Ph$_4$PBr (1 mM), and 0.3 mL of H$_2$O was sealed under vacuum and kept at 130°C for 1 week. Large pale-yellow rod-like crystals were obtained by washing with water, MeOH, and anhydrous ether.

## 7.53  Hg–H–C–P–O–S

Some compounds are formed in this system.

[Hg{(Pr$^i$O)$_2$PS$_2$}$_2$] or C$_{12}$H$_{28}$P$_2$O$_4$S$_4$Hg. This compound crystallizes in the monoclinic structure with the lattice parameters $a = 1180.0 \pm 0.1$, $b = 892.5 \pm 0.2$, and $c = 2216.7 \pm 0.2$ pm and $\beta = 94.988° \pm 0.007°$ and a calculated density of 1.791 g·cm$^{-3}$ (Casas et al. 1999) ($a = 3170 \pm 2$, $b = 865 \pm 1$, and $c = 2234 \pm 1$ pm and $\beta = 129.34° \pm 0.02°$ and the experimental and calculated densities $1.74 \pm 0.02$ and $1.758 \pm 0.004$ g·cm$^{-3}$, respectively [Lawton 1971]). It was prepared by dissolving P$_4$S$_{10}$ (1.34 g) in excess of Pr$^i$OH to give 50 mL of the acid HS(S)P(OPr$^i$)$_2$, and the mixture was treated with a solution of PhHgOAc (3.3 g) in Pr$^i$OH. A yellowish solution was formed, which soon decomposed to give a black precipitate. This was filtered, and the clear colorless filtrate was evaporated to produce colorless crystals of the title compound (Casas et al. 1999). Soft, light-gray needles were obtained by recrystallization of Hg[(Pr$^i$O)$_2$PS$_2$]$_2$ from warm absolute EtOH (Lawton 1971).

[HgMe{S(O)PPh$_2$}] or C$_{13}$H$_{13}$POSHg. This compound melts at 137°C–138°C and crystallizes in the monoclinic structure with the lattice parameters $a = 2633.9 \pm 0.2$, $b = 617.7 \pm 0.1$, and $c = 1770.2 \pm 0.5$ pm and $\beta = 96.118° \pm 0.001°$ and a calculated density of 2.082 g·cm$^{-3}$ (Casas et al. 1997). To obtain it, diphenylphosphinothioic acid (1 mM) dissolved in EtOH (20 mL) was mixed with a solution of MeHgOH (1 mM) in $H_2O$ at room temperature and stirred for 2 h. A white crystalline substance deposited after concentrating the solution. The solid was filtered, washed with heptane, and dried.

**[HgPh{S(O)PPh$_2$}] or C$_{18}$H$_{15}$POSHg**. This compound melts at 181°C–183°C (Casas et al. 1997). It was prepared by the next procedure. A solution containing PhHgOAc (2 mM) 30 mL of EtOH was mixed with a solution of diphenylphosphinothioic acid (2 mM) dissolved in the same solvent (25 mL) allowed to stand without stirring. After ca. 10 min, a white crystalline precipitate (long needles) deposited, which was filtered, washed with heptane, and dried.

**[Hg(S$_2$COPr$^i$)$_2$PPh$_3$] or C$_{26}$H$_{29}$PO$_2$S$_4$Hg**. This compound crystallizes in the triclinic structure with the lattice parameters $a = 1180.7 \pm 0.3$, $b = 1321.2 \pm 0.2$, and $c = 1021.5 \pm 0.2$ pm and $\alpha = 107.84° \pm 0.01°$, $\beta = 108.98° \pm 0.02°$, and $\gamma = 83.10° \pm 0.02°$ and the experimental and calculated densities 1.67 and 1.698 g·cm$^{-3}$, respectively (Abrahams et al. 1986).

**[Hg(S$_2$COPr$^i$)$_2$P($c$-C$_6$H$_{11}$)$_3$] or C$_{26}$H$_{47}$PO$_2$S$_4$Hg**. This compound crystallizes in the monoclinic structure with the lattice parameters $a = 1019.2 \pm 0.2$, $b = 2009.8 \pm 0.2$, and $c = 1572.8 \pm 0.3$ pm and $\beta = 102.80° \pm 0.01°$ and the experimental and calculated densities 1.58 and 1.589 g·cm$^{-3}$, respectively (Abrahams et al. 1986).

**[Ph$_4$P][Hg(SC{O}Me)$_3$] or C$_{30}$H$_{29}$PO$_3$S$_3$Hg**. This compound was obtained by the following procedure (Sampanthar et al. 1999). Thioacetic acid (14 mM) was added dropwise to a stirred aqueous solution of Et$_3$N (14 mM) to obtain a solution of Et$_3$NHMe{O}CS. To this HgCl$_2$ (4.5 mM) in 11 mL of MeOH/H$_2$O (1:10 volume ratio) was added. The resulting solution was stirred for 10 min. When a solution of Ph$_4$PCl (4.5 mM) in H$_2$O was added, a cream-colored precipitate was formed. The mixture was stirred for 1 h, after which CH$_2$Cl$_2$ (4 mL) was added. The precipitate in the aqueous layer was extracted into the CH$_2$Cl$_2$ layer, which became yellow. CH$_2$Cl$_2$ was separated and washed with 15 mL portions of deionized water three or four times to remove the side product, Et$_3$NHCl. Et$_2$O was added to the CH$_2$Cl$_2$ solution until turbidity was noticed, when the mixture was set aside in a refrigerator overnight, to obtain yellow crystals. They were decanted, washed with Et$_2$O, and dried in vacuo.

**(Ph$_4$P)[Hg(SOCPh)$_3$] or C$_{45}$H$_{35}$PO$_3$S$_3$Hg**. This compound crystallizes in the monoclinic structure with the lattice parameters $a = 1320.2 \pm 0.2$, $b = 1427.6 \pm 0.2$, and $c = 2110.8 \pm 0.2$ pm and $\beta = 90.92° \pm 0.01°$ and a calculated density of 1.589 g·cm$^{-3}$ for pink crystals and $a = 1322.7 \pm 0.1$, $b = 1429.7 \pm 0.2$, and $c = 2114.2 \pm 0.2$ pm and $\beta = 90.82° \pm 0.01°$ for yellow crystals (Vittal and Dean 1996). To obtain this compound, a portion of liquid PhCOSH (15.0 mM) was added to Et$_3$N (14.2 mM) in 15 mL of MeOH. To this stirred mixture, a solution of HgCl$_2$ (4.75 mM) in 10 mL of MeOH by syringe very close to the surface was added to produce a clear yellow solution. The addition of HgCl$_2$ in this way is necessary to avoid any formation of black particles. Upon the addition of Ph$_4$PBr (4.75 mM) in 10 mL of MeOH, a bright-lemon precipitate was formed, which on continued stirring slowly underwent partial dissolution. After the addition of MeCN (10 mL) and CH$_2$Cl$_2$ (10 mL), the mixture was heated to ca. 50°C for ca. 10 min, giving a clear yellow solution. This was allowed to cool to room temperature (ca. 20 min), then refrigerated at 5°C overnight, producing long yellow needles. The crystalline product was isolated by decantation, then washed, and dried. The washings were combined with the decantate and kept at room temperature. On partial evaporation of the solvent, many light-pink block-like crystals were formed, contaminated with a small amount of a black powder. These crystals were separated, washed with MeOH and $Et_2O$, and dried in Ar. All preparations were carried out under an Ar atmosphere.

$[Ph_4P]_2[Hg(1,2-S_2C_6H_{10})_2]\cdot4H_2O$ or $C_{60}H_{68}P_2O_4S_4Hg$. This compound crystallizes in the triclinic structure with the lattice parameters $a = 1368.5 \pm 0.2$, $b = 1829.0 \pm 0.2$, and $c = 1330.7 \pm 0.1$ pm and $\alpha = 110.445° \pm 0.008°$, $\beta = 111.292° \pm 0.008°$, and $\gamma = 69.157 \pm 0.009$ (Govindaswamy et al. 1992). The reaction of $HgCl_2$ with three equivalents of the lithium salt of 1,2-trans-cyclohexanedithiolate and two equivalents of $Ph_4PBr$ in $H_2O$ gives crystals of $[Ph_4P]_2[Hg(1,2-C_6H_{10})_2]\cdot4H_2O$.

## 7.54   Hg–H–C–P–O–Se–Cl–S

$[Hg_4(SePh)_6\{SP(c-C_6H_{11})_3\}_3$ or $C_{90}H_{129}P_3O_8Se_6Cl_2S_3Hg_4$. This multinary compound is formed in this system. To obtain it, into a solution of $Hg(SPPh_3)_3(ClO_4)_2$ (0.15 mM) in 3 mL of $CH_2Cl_2$ at room temperature solid $SP(c-C_6H_{11})_3$ (0.15 mM) was stirred first, then solid $Hg(SePh)_2$ (0.45 mM) (Dean and Manivannan 1990a). The mixture was stirred for 5 min, then filtered to remove traces of Hg. Onto the yellow filtrate was layered $Et_2O$ (4 mL). Crystallization occurred upon cooling to 5°C in the refrigerator. The yellow crystalline product was separated by decantation of the mother liquor, washed with $Et_2O$, and dried in vacuo.

## 7.55   Hg–H–C–P–O–Te–Cl–S

Some compounds are formed in this system.

$[Hg_4(TePh)_6\{SP(c-C_6H_{11})_3\}_3]$ or $C_{90}H_{129}P_3O_8Te_6Cl_2S_3Hg_4$. To obtain this compound, a solution of $Hg(SPPh_3)_3(ClO_4)_2$ (0.15 mM) in 3 mL of $CH_2Cl_2$ at room temperature was stirred first solid $SP(c-C_6H_{11})_3$ (0.15 mM), then solid $Hg(TePh)_2$ (0.45 mM) (Dean and Manivannan 1990a). The mixture was stirred for 5 min, then filtered to remove traces of Hg. Onto the yellow filtrate was layered $Et_2O$ (4 mL). Crystallization occurred upon cooling to 5°C in the refrigerator. The orange crystalline product was separated by decantation of the mother liquor, washed with $Et_2O$, and dried in vacuo.

$[(\mu-TePh)_6(HgSPPh_3)_4](ClO_4)_2$ or $C_{108}H_{90}P_4O_8Te_6Cl_2S_4Hg_4$. To obtain this compound, a solution of $Hg(SPPh_3)_3(ClO_4)_2$ (0.15 mM) in 3 mL of $CH_2Cl_2$ at room temperature was stirred first solid $SPPh_3$ (0.15 mM), then solid $Hg(TePh)_2$ (0.45 mM) (Dean and Manivannan 1990a). The mixture was stirred for 5 min, then filtered to remove traces of Hg. Onto the yellow filtrate was layered $Et_2O$ (4 mL). Crystallization occurred upon cooling to 5°C in the refrigerator. The yellow crystalline product was separated by decantation of the mother liquor, washed with $Et_2O$, and dried in vacuo.

## 7.56   Hg–H–C–P–O–Cl–S

Some compounds are formed in this system.

**[Hg{SP(c-C₆H₁₁)₃}₂](ClO₄)₂ or C₃₆H₆₆P₂O₈Cl₂S₂Hg.** This compound was prepared by the next procedure (Dean and Manivannan 1990a). To a solution of HgCl₂ (2.46 mM) in 20 mL of Me₂CO a solution of SP(c-C₆H₁₁)₃ (4.95 mM) in 80 mL of Me₂CO was added. The mixture was reduced in volume and upon cooling gave a white solid, which was separated by filtration and dried in vacuo. The chloride complex was converted to the ClO₄⁻ salt using AgClO₄. After recrystallization from a CH₂Cl₂/hexane mixture, the title compound was obtained.

**[Hg(SPPh₃)₃](ClO₄)₂ or C₅₄H₄₅P₃O₈Cl₂S₃Hg.** This compound was synthesized by the direct reaction of Hg(ClO₄)₂ and Ph₃PS (Dean and Manivannan 1990a). A mixture of Hg(ClO₄)₂·3H₂O (1.2 mM) and Ph₃PS (3.6 mM) in CH₂Cl₂ (60 mL) was stirred at room temperature and then warmed carefully on in a steam bath until dissolution of the ClO₄⁻ salt was complete. The solution was filtered and then concentrated, at which point the addition of hexane gave the crude material. After recrystallization from a CH₂Cl₂/hexane mixture, the title compound was obtained.

**[Hg₄(SPh)₆(PEt₃)₄](ClO₄)₂ or C₆₀H₉₀P₄O₈Cl₂S₆Hg₄.** To obtain this compound, a 1.5 mM portion of solid Hg(SPh)₂ was stirred into a mixture of 0.50 mM of Hg(PEt₃)₂(ClO₄)₂ and 1.0 mM of PEt₃ in 10 mL of acetone, producing a clear colorless solution (Dean et al. 1987). This was saturated with diethyl ether and cooled to 5°C for crystallization. The colorless crystals were separated by decantation, washed with cyclohexane, and dried in vacuo. All syntheses were performed under an Ar atmosphere.

**[Hg₄(SMe)₆(PPh₃)₄](ClO₄)₂ or C₇₈H₇₈P₄O₈Cl₂S₆Hg₄.** To obtain this compound, a slurry of 0.90 mM of Hg(SMe)₂ in 10 mL of CHCl₃ was added with stirring to a solid mixture containing 0.30 mM of Hg(PPh₃)₂(ClO₄)₂ and 0.65 mM of PPh₃ (Dean et al. 1987). After 10 min, most of the solids had dissolved to give a colorless solution. This was filtered and left to crystallize at 5°C. The colorless transparent crystals obtained in ca. 12 h were separated by decantation, washed with diethyl ether, and dried in vacuo. All syntheses were performed under an Ar atmosphere.

**[Hg₄(SEt)₆(PPh₃)₄](ClO₄)₂·CH₂Cl₂ or C₈₅H₉₂P₄O₈Cl₄S₆Hg₄.** To obtain this compound, a solution of 1.5 mM of Hg(SEt)₂ in 5 mL of CH₂Cl₂ was added to a solution containing 0.50 mM of Hg(PPh₃)₂(ClO₄)₂ and 1.0 mM of PPh₃ in 10 mL of CH₂Cl₂ (Dean et al. 1987). The mixture was stirred for 10 min and then filtered, saturated with diethyl ether, and left for crystallization at 5°C. The white crystals were collected by filtration, washed with ether, and dried in vacuo. All syntheses were performed under an Ar atmosphere.

**[Hg₄(SPh)₆(PPh₃)₄](ClO₄)₂·1.5CHCl₃ or C₁₀₉.₅H₉₁.₅P₄O₈Cl₆.₅S₆Hg₄.** To obtain this compound, to a solution of Hg(PPh₃)₂(ClO₄)₂ (0.30 mM) in 10 mL of CHCl₃ was added 0.60 mM of solid PPh₃, producing a clear solution (Dean et al. 1987). Into this was stirred 0.90 mM of Hg(SPh)₂ to give a yellow solution with traces of suspended fine solid. The mixture was stirred for 10 min and then filtered. The filtrate was covered with a layer of diethyl ether and left undisturbed at 0°C–5°C overnight. The resulting light-yellow needlelike crystals were separated by decantation of the mother liquor, washed with diethyl ether, and dried in vacuo. All syntheses were performed under an Ar atmosphere.

## 7.57   Hg–H–C–P–S

Some compounds are formed in this system.

**[HgMe(S$_2$PEt$_2$)] or C$_5$H$_{13}$PS$_2$Hg**. This compound melts at 40°C (Casas et al. 1994). It was obtained as follows. To a suspension of MeHgOAc (3.75 mM) in EtOH (20 mL) at 0°C the stoichiometric quantity of NaS$_2$PEt$_2$·2H$_2$O (0.79 g) dissolved in the same solvent (20 mL) was added. The mixture was stirred for 3 h at 0°C, and the solution obtained was vacuum concentrated and stored for several days at 4°C. The white crystals formed were filtered off and vacuum dried.

**[HgPh(S$_2$PEt$_2$)] or C$_{10}$H$_{15}$PS$_2$Hg**. This compound melts at 92°C–93°C and crystallizes in the monoclinic structure with the lattice parameters $a = 623.8 \pm 0.1$, $b = 2016.2 \pm 0.2$, and $c = 1128.1 \pm 0.1$ pm and $\beta = 104.63° \pm 0.01°$ and a calculated density of 2.085 g·cm$^{-3}$ (Casas et al. 1994). It was obtained by the next procedure. To a suspension of PhHgOAc (2.97 mM) in EtOH (25 mL) at 0°C the stoichiometric quantity of NaS$_2$PEt$_2$·2H$_2$O (0.63 g) dissolved in the same solvent (20 mL) was added. After stirring for 3 h at 0°C, the solution was filtered and stored at 4°C for several days until crystals suitable for XRD formed.

**[MeHgS$_2$PPh$_2$] or C$_{13}$H$_{13}$PS$_2$Hg**. This compound melts at 85°C and crystallizes in the triclinic structure with the lattice parameters $a = 918.9 \pm 0.1$, $b = 1194.5 \pm 0.3$, and $c = 699.30 \pm 0.09$ pm and $\alpha = 105.79° \pm 0.01°$, $\beta = 96.03° \pm 0.01°$, and $\gamma = 78.92° \pm 0.02°$ and a calculated density of 2.133 g·cm$^{-3}$ (Zukerman-Schpector et al. 1991). It was prepared as follows. To a methanolic suspension of MeHgAcO in an ice-water bath was added one equivalent of ammonium diphenyldithiophosphinate in the same solvent. The mixture was stirred for 3 h, and the white precipitate of MeHgS$_2$PPh was then filtered off.

**[HgMe{S$_2$P(C$_6$H$_{11}$)$_2$}] or C$_{13}$H$_{25}$PS$_2$Hg**. This compound melts at 145°C–148°C (Casas et al. 1994). It was obtained as follows. To a suspension of MeHgOAc (3.64 mM) in chloroform (25 mL) at 0°C the stoichiometric quantity of HS$_2$P(C$_6$H$_5$)$_2$ (0.96 g) dissolved in the same solvent (25 mL) was added. The solution obtained was stirred for 3 h at 0°C and vacuum concentrated. The white precipitate obtained by the addition of cold MeOH was filtered off and vacuum dried.

**[PhHgS$_2$PPh$_2$] or C$_{18}$H$_{15}$PS$_2$Hg**. This compound melts at 170°C (Zukerman-Schpector et al. 1991). It was prepared by the next procedure. To a methanolic suspension of PhHgAcO in an ice-water bath one equivalent of ammonium diphenyldithiophosphinate in the same solvent was added. The mixture was stirred for 3 h, and the white precipitate of the title compound was then filtered off.

**[HgPh{S$_2$P(C$_6$H$_{11}$)$_2$}] or C$_{18}$H$_{27}$PS$_2$Hg**. This compound melts at 177°C (Casas et al. 1994). It was obtained by the next procedure. To a suspension of PhHgOAc

(3.00 mM) in chloroform (25 mL) at 0°C the stoichiometric quantity of $HS_2P(C_6H_5)_2$ (0.79 g) dissolved in the same solvent (25 mL) was added. The solution obtained was stirred for 3 h at 0°C and the solvent was removed under vacuum. The solid obtained was dissolved in $CH_2Cl_2$ and the solution was cooled to –30°C for several days. A white precipitate formed, and was filtered off and vacuum dried.

**$[(C_4H_3S)_2Hg]_2[(Ph_2P)_2CH_2]$ or $C_{42}H_{34}P_2S_4Hg_2$.** This compound melts at 110°C–112°C (Bell et al. 2004). The reaction of 2-thienylmercury(II) chloride with bis(diphenylphosphino)methane in ethanolic solution produced the title compound as white powder.

**$[(Ph_4P)(PhS)_3Hg]$ or $C_{42}H_{35}PS_3Hg$.** This compound melts at 112°C with decomposition (Liesic von and Klar 1977) and crystallizes in the monoclinic structure with the lattice parameters $a = 1318$, $b = 1358$, and $c = 2005$ pm and $\beta = 106°$ and the experimental and calculated densities 1.89 and 1.91 g·cm$^{-3}$, respectively (Behrens et al. 1977). It was prepared from the reaction of $Hg(SPh)_2$, NaSPh, and $Ph_4PCl$ in the liquid $NH_3$ (Behrens et al. 1977, Liesic von and Klar 1977). Its single crystals were grown by recrystallization from a THF/EtOH solution of the title compound.

**$[Ph_4P]_2[Hg_2(SCH_2CH_2S)_3]$ or $C_{54}H_{52}P_2S_6Hg_2$.** This compound crystallizes in the monoclinic structure with the lattice parameters $a = 1039.3 \pm 0.2$, $b = 1432.9 \pm 0.2$, and $c = 1710.3 \pm 0.4$ pm and $\beta = 96.91° \pm 0.02°$ and a calculated density of 1.782 g·cm$^{-3}$ (Henkel et al. 1985). It was obtained when an excess of ethane-1,2-dithiolate was added to a solution of $HgCl_2$ in MeOH in the presence of $Na_2S_2$.

**$[Ph_4P]_2[Hg_3(SCH_2CH_2S)_4]$ or $C_{56}H_{56}P_2S_8Hg_3$.** This compound crystallizes in the monoclinic structure with the lattice parameters $a = 2271.3 \pm 0.7$, $b = 950.1 \pm 0.2$, and $c = 2932.3 \pm 0.7$ pm and $\beta = 119.42° \pm 0.02°$ and a calculated density of 1.988 g·cm$^{-3}$ (Henkel et al. 1985). This compound was obtained when an excess of ethane-1,2-dithiolate was added to a solution of $HgCl_2$ in MeOH.

**$[Ph_4P]_2[Hg(SPh)_4]$ or $C_{72}H_{60}P_2S_4Hg$.** This compound melts at 136°C with decomposition (Liesic von and Klar 1977). It was prepared from the reaction of $Hg(SPh)_2$, NaSPh, and $Ph_4PCl$ in the liquid $NH_3$.

## 7.58 Hg–H–C–P–F–S

**$[Hg(SCF_3)_2\cdot(2PPh_3)]$ or $C_{38}H_{30}P_2F_6S_2Hg$.** This multinary compound is formed in this system. It melts at 166°C–168°C with decomposition (Man et al. 1959).

## 7.59 Hg–H–C–P–Cl–S

Some compounds are formed in this system.

**$[(CH_3)_3PS]_2[HgCl_2]$ or $C_6H_{18}P_2Cl_2S_2Hg$.** This quinary compound is formed in this system. It was obtained by the direct reaction of trimethylphosphine sulfide, $(CH_3)_3PS$, with $HgCl_2$ in absolute ethanol (Meek and Nicpon 1965).

**$[C_4H_3SHgCl\cdot(Ph_2PCH_2)_2]$ or $C_{30}H_{27}P_2ClSHg$.** This compound melts at 165°C–167°C (Bell et al. 2004). The reaction of 2-thienylmercury(II) chloride with bis(diphenylphosphino)ethane in ethanolic solution produced the title compound as white powder.

**[HgCl₂·Ph₃PS] or C₁₈H₁₅PCl₂SHg**. This compound melts at 225°C with decomposition (at 230°C [King and McQuillan 1967]) and crystallizes in the monoclinic structure with the lattice parameters $a = 1030$, $b = 1284$, and $c = 1774$ pm and $\beta = 53°0'$ and the experimental and calculated densities 2.00 and 2.01 g·cm⁻³, respectively (Dalziel et al. 1967). It was prepared by heating HgCl₂ (10 mM) and Ph₃PS (3.2 g, excess) under reflux in absolute EtOH and was recrystallized by Soxhlet extraction with the same solvent as diamond-shaped plates.

**[(C₄H₃SHgCl)₂·{(Ph₂P)₂CH₂}] or C₃₃H₂₈P₂Cl₂S₂Hg₂**. This compound melts at 210°C–212°C (Bell et al. 2004). The reaction of 2-thienylmercury(II) chloride with bis(diphenylphosphino)methane in ethanolic solution produced the title compound as white powder.

## 7.60   Hg–H–C–P–Br–S

Some compounds are formed in this system.

**[HgBr₂·Ph₃PS] or C₁₈H₁₅PBr₂SHg**. This compound melts at 234°C with decomposition (at 237°C [King and McQuillan 1967]) and crystallizes in the monoclinic structure with the lattice parameters $a = 1027$, $b = 1308$, and $c = 1720$ pm and $\beta = 55°30'$ and the experimental and calculated densities 2.32 and 2.28 g·cm⁻³, respectively (Dalziel et al. 1967). It was prepared by heating HgBr₂ (10 mM) and Ph₃PS (3.2 g, excess) under reflux in absolute EtOH and was recrystallized by Soxhlet extraction with the same solvent as diamond-shaped plates. Reaction with a threefold excess of Ph₃PS under the same conditions gave only the title compound.

**(Ph₄P)₄[Hg(S₄)₂]Br₂ or C₉₆H₈₀P₄Br₂S₈Hg**. This compound melts at ca. 250°C and crystallizes in the tetragonal structure with the lattice parameters $a = 1956.3 \pm 0.4$ and $c = 1153.1 \pm 0.2$ pm and the experimental and calculated densities 1.48 g·cm⁻³ (Bailey et al. 1991). To obtain it, Na₂S·9H₂O (1 mM) was partially dehydrated, and HgS (1 mM) and S (6 mM) were added, followed by DMF (5 mL). On ultrasonication, the solids quickly dissolved to yield a dark-green solution. After removal of a small amount of white solid, a solution of Ph₄PBr (2 mM) in MeCN (6.5 mL) was added. Red crystals began to grow within 4 h and were separated after 2 days, cleaned with filter paper and vacuum dried. (Ph₄P)₄[Hg(S₄)₂]Br₂ is hygroscopic and decomposes quickly to HgS in a moist atmosphere. All preparations were performed under N₂ as inert gas.

## 7.61   Hg–H–C–P–I–S

**HgI₂·Ph₃PS or C₁₈H₁₅PI₂SHg**. This multinary compound is formed in this system. It melts at 166°C with decomposition (at 173°C [King and McQuillan 1967]) and crystallizes in the orthorhombic structure with the lattice parameters $a = 2052$, $b = 908$, and $c = 2170$ pm and the experimental and calculated densities 2.44 and 2.46 g·cm⁻³, respectively (Dalziel et al. 1967). It was prepared by heating finely divided $HgI_2$ (10 mM) and $Ph_3PS$ (3.5 g, excess) under reflux for ca. 4 h in benzene and was recrystallized by Soxhlet extraction with absolute EtOH as yellow needles.

## 7.62 Hg–H–C–As–O–S

$(Ph_4As)[Hg(SOCPh)_3]$ or $C_{45}H_{35}AsO_3S_3Hg$. This multinary compound is formed in this system. It crystallizes in the monoclinic structure with the lattice parameters $a = 1325.2 \pm 0.7$, $b = 1430.2 \pm 0.6$, and $c = 2116.4 \pm 1.1$ pm and $\beta = 91.42° \pm 0.04°$ (yellow crystals) and $a = 1327.7 \pm 0.9$, $b = 1432.3 \pm 0.7$, and $c = 2114.4 \pm 0.6$ pm and $\beta = 91.37° \pm 0.04°$ (red crystals) (Vittal and Dean 1996). To obtain this compound, a portion of liquid PhCOSH (10.6 mM) was added to $Et_3N$ (9.6 mM) in 15 mL of MeOH. To this stirred mixture, a solution of $HgCl_2$ (2.4 mM) in 10 mL of MeOH was added by a syringe very close to the surface to produce a clear yellow solution. Addition of $HgCl_2$ in this way is necessary to avoid any formation of black particles. Upon the addition of $Ph_4AsCl \cdot H_2O$ (4.76 mM) in 10 mL of MeOH, a precipitate was formed, which on continued stirring slowly underwent partial dissolution. After the addition of MeCN (10 mL) and $CH_2Cl_2$ (20 mL), the mixture was heated to ca. 50°C for ca. 10 min, giving a clear solution. This was allowed to cool to room temperature (ca. 20 min), then refrigerated at 5°C overnight, producing very light, creamy, needlelike crystals. The crystalline product was isolated by decantation, then washed, and dried. The washings were combined with the decantate and kept at room temperature. On partial evaporation of the solvent, there are many light-red and block-shaped crystals. These crystals were separated, washed with MeOH and $Et_2O$, and dried in Ar. All preparations were carried out under an Ar atmosphere.

## 7.63 Hg–H–C–As–O–Cl–S

$[Hg_4(SPh)_6(AsPh_3)_4](ClO_4)_2 \cdot 1.5CHCl_3$ or $C_{109.5}H_{91.5}As_4O_8Cl_{6.5}S_6Hg_4$. This multinary compound is formed in this system. To obtain it, to a solution of $Hg(AsPh_3)_2(ClO_4)_2$ (0.30 mM) in 10 mL of $CHCl_3$ 0.60 mM of solid $AsPh_3$ was added, producing a clear solution (Dean et al. 1987). Into this 0.90 mM of $Hg(SPh)_2$ was stirred to give a yellow solution with traces of suspended fine solid. The mixture was stirred for 10 min and then filtered. The filtrate was covered with a layer of diethyl ether and left undisturbed at 0°C–5°C overnight. The resulting light-yellow needlelike crystals were separated by decantation of the mother liquor, washed with diethyl ether, and dried in vacuo. All syntheses were performed under an Ar atmosphere.

## 7.64 Hg–H–C–As–Cl–S

Some compounds are formed in this system.

$[HgCl_2 \cdot Ph_3AsS]$ or $C_{18}H_{15}AsCl_2SHg$. This compound melts at 228°C (King and McQuillan 1967).

$[Hg(SC_6Cl_5)_4][(C_6H_5)_4As]$ or $C_{48}H_{20}AsCl_{20}S_4Hg$. This multinary compound, which melts with decomposition at 153°C, is formed in this system (Lucas and Peach 1969). It has been formed using penthachlorophenylthio anion as a ligand and was prepared by adding an aqueous solution of $Hg(NO_3)_2 \cdot H_2O$ to an aqueous solution of $C_6Cl_5SH$. After filtration, an aqueous solution of tetraphenylarsenium chloride was added to the filtrate to precipitate $[Hg(SC_6Cl_5)_4][(C_6H_5)_4As]$.

## 7.65   Hg–H–C–As–Br–S

$[HgBr_2 \cdot Ph_3AsS]$ or $C_{18}H_{15}AsBr_2SHg$. This multinary compound is formed in this system. It melts at 238°C (King and McQuillan 1967).

## 7.66   Hg–H–C–As–I–S

$[HgI_2 \cdot Ph_3AsS]$ or $C_{18}H_{15}AsI_2SHg$. This multinary compound is formed in this system. It melts at 198°C (King and McQuillan 1967).

## 7.67   Hg–H–C–Sb–I–S

$[HgI_2 \cdot Ph_3SbS]$ or $C_{18}H_{15}SbI_2SHg$. This multinary compound is formed in this system. It melts at 162°C (King and McQuillan 1967).

## 7.68   Hg–H–C–O–S

Some compounds are formed in this system.

**$[MeHgSAc]$ or $C_3H_6OSHg$, methylmercury thioacetate**. This compound melts at 138°C–140°C (Bach and Weibel 1976). To obtain it, MeHgCl (10 mM), HSAc (10 mM), 50 mL of MeOH, and 10 mL of 1 M aqueous NaOH (10 mM) were combined. Evaporation of solvent led to grayish solid, which was purified by recrystallization from $CH_2Cl_2$ to give a white solid.

**$[MeHgS_2Ac]$ or $C_3H_6OS_2Hg$**. This compound melts at 64°C–65°C (Casas et al. 2002) and crystallizes in the orthorhombic structure with the lattice parameters $a = 710.2 \pm 0.1$, $b = 687.5 \pm 0.3$, and $c = 1473.9 \pm 0.2$ pm and a calculated density of 2.979 $g \cdot cm^{-3}$ (Tiekink 1986b). It was prepared by the following procedure (Casas et al. 2002). MeHgCl (1.26 g) was dissolved in 50 mL of $CH_2Cl_2$, which was treated with 0.90 g (excess) of solid, finely powdered $NaS_2COMe$, and the mixture was stirred at room temperature for 1 h. The solution was filtered and left to evaporate in open air. A slightly decomposed solid residue (gray color) was obtained. The solid was redissolved in 5 mL $CH_2Cl_2$ and filtered, and the filtrate was covered with petroleum ether and left to crystallize in the refrigerator. The needlelike, colorless crystals were filtered and dried in vacuo.

The title compound could also be prepared from the reaction of MeHgCL (0.4 g) in $CH_2Cl_2$ (25 mL) and equimolar amount of $KS_2COMe$ in $H_2O$ (Tiekink 1986b). The mixture was stirred for 30 min, separated, and the nonaqueous layer dried over anhydrous $Na_2SO_4$. Petroleum ether (40°C–60°C) was added and the solution was allowed to stand until pale-yellow crystals were deposited.

**$[Hg(C_3H_6OS_2)]$ or $C_3H_6OS_2Hg$**. This compound melts at 160°C with decomposition (Canty and Kishimoto 1977). To obtain it, 2,3-dimercapto-1-propanol ($C_3H_8OS_2$)

(2.14 mM) dissolved in $H_2O$ (5 mL) was added to $HgCl_2$ (2.14 mM) in $H_2O$ (15 mL). A white precipitate that formed immediately was collected and washed with $H_2O$ and EtOH. A similar preparation using $Hg(OAc)_2$ gave an identical product.

**[Hg(C₃H₆OS₂)] or C₃H₆OS₂Hg.** This compound melts at 97°C with decomposition (Canty and Kishimoto 1977). To obtain it, 1,3-dimercapto-2-propanol ($C_3H_8OS_2$) (2.74 mM) dissolved in $H_2O$ (2 mL) was added to $HgCl_2$ (2.60 mM) in $H_2O$ (18 mL). A white precipitate that formed immediately was collected and washed with $H_2O$ and EtOH.

$Hg(C_3H_6OS_2)$ (1.16 mM) in Py (20 mL) deposited colorless crystals over 14 h (Canty and Kishimoto 1977). The crystals were collected but not washed or dried. These crystals contain 1.5 M of Py for 1 M of the title compound. On loss of Py, the white powder formed has analysis consistent with $Hg(C_3H_6OS_2)$ and melts at 202°C.

**[MeSHgOAc] or C₄H₆O₂SHg, acetato(methanethiolato)mercury(II).** This compound melts with decomposition at 100°C–104°C (Canty et al. 1978a) and crystallizes in the monoclinic structure with the lattice parameters $a = 529.2 \pm 0.2$, $b = 1591.2 \pm 0.7$, and $c = 708.6 \pm 0.2$ pm and $\beta = 92.96° \pm 0.03°$ and the experimental and calculated densities more than 2.9 and 3.42 g·cm⁻³, respectively (Canty et al. 1978b). It was prepared by stirring a suspension of $(MeS)_2$ in an aqueous solution of $Hg(OAc)_2$ for ca. 24 h, and the insoluble products were collected, washed with $H_2O$, and dried.

**[Hg(S₂COMe)₂] or C₄H₆O₂S₄Hg.** This compound crystallizes in the orthorhombic structure with the lattice parameters $a = 1060.5 \pm 0.1$, $b = 800.3 \pm 0.2$, and $c = 2347.8 \pm 0.2$ pm and a calculated density of 2.766 g·cm⁻³ (Tiekink 1987a). Colorless needles were grown from the solution of the title compound in acetone.

**[MeHg(S₂COEt)] or C₄H₈OS₂Hg.** This compound melts at 68°C–69°C and crystallizes in the monoclinic structure with the lattice parameters $a = 724.21 \pm 0.05$, $b = 689.9 \pm 0.3$, $c = 834.2 \pm 0.1$ pm $\beta = 106.457° \pm 0.008°$ and a calculated density of 2.798 g·cm⁻³ (Casas et al. 2002). To obtain it, MeHgCl (1.26 g) dissolved in 50 mL of $CH_2Cl_2$ was treated with 0.90 g (excess) of solid, finely powdered $NaS_2COEt$, and the mixture was stirred at room temperature for 1 h. The solution was filtered and left to evaporate in open air. A slightly decomposed solid residue (gray color) was obtained. The solid was redissolved in 5 mL $CH_2Cl_2$ filtered, and the filtrate was covered with petroleum ether and left to crystallize in the refrigerator. The needle-like, colorless crystals were filtered and dried in vacuo. The title compound was also prepared from the reaction of MeHgCl (0.4 g) in $CH_2Cl_2$ (25 mL) and equimolar amount of $KS_2COEt$ in $H_2O$ (Tiekink 1986b). The mixture was stirred for 30 min and separated, and the nonaqueous layer dried over anhydrous $Na_2SO_4$. Petroleum ether (40°C–60°C) was added and the solution was allowed to stand until pale-yellow crystals were deposited.

**[MeHgSCH₂OAc] or C₄H₈O₂SHg, methylmercury methylmercaptoacetate.** To obtain this compound, to a stirred solution of 5 mM of MeHgCl and 1.7 mL of 3N NaOH (5.1 mM) in 25 mL of MeOH 0.53 g $HSCH_2OAc$ in 10 mL of MeOH was added (Bach and Weibel 1976). After 5 min, the product was extracted with ether from a saturated NaCl solution. Removal of the ether afforded a pale-yellow oil.

**[EtSHgOAc] or C₄H₈O₂SHg, acetato(ethanethiolato)mercury(II).** This compound melts with decomposition at 133°C–136°C (Canty et al. 1978a). It was prepared by stirring a suspension of EtS₂ in an aqueous solution of Hg(OAc)₂ for ca. 24 h, and the colorless crystals were collected, washed with H₂O, and dried.

**[Hg(C₂H₅OS)₂] or C₄H₁₀O₂S₂Hg.** This compound melts at 122°C (Canty and Kishimoto 1977). On addition of 1-mercapto-2-ethanol (9.51 mM) to Hg(SCN)₂ (4.26 mM) in EtOH (10 mL), a white precipitate formed immediately. It was collected and washed with EtOH.

**[MeHgS₂COPr$^i$] or C₅H₁₀OS₂Hg.** This compound melts at 52°C and crystallizes in the monoclinic structure with the lattice parameters $a = 688.4 \pm 0.1$, $b = 705.9 \pm 0.2$, and $c = 1927.3 \pm 0.4$ pm and $\beta = 97.00° \pm 0.02°$ and a calculated density of 2.507 g·cm⁻³ (Casas et al. 2002). It was prepared by the next procedure. MeHgCl (1.26 g) dissolved in 50 mL of CH₂Cl₂ was treated with 0.90 g (excess) of solid, finely powdered NaS₂COPr$^i$, and the mixture was stirred at room temperature for 1 h. The solution was filtered and left to evaporate in open air. A slightly decomposed solid residue (gray color) was obtained. The solid was redissolved in 5 mL CH₂Cl₂ and filtered, and the filtrate was covered with petroleum ether and left to crystallize in the refrigerator. The needlelike, colorless crystals were filtered and dried in vacuo.

The title compound could also be prepared from the reaction of MeHgCl (0.4 g) in CH₂Cl₂ (25 mL) and equimolar amount of KS₂COPr$^i$ in H₂O (Tiekink 1986b). The mixture was stirred for 30 min and separated and the nonaqueous layer dried over anhydrous Na₂SO₄. Petroleum ether (40°C–60°C) was added and the solution was allowed to stand until pale-yellow crystals were deposited.

**[(MeHg)₂(C₃H₆OS₂)] or C₅H₁₂OS₂Hg₂.** To obtain this compound, 2,3-dimercapto-1-propanol (C₃H₈OS₂) (1.30 mM) in benzene (5 mL) was added dropwise to a solution of methylmercuric acetate (2.61 mM) in the same solvent (20 mL) (Canty and Kishimoto 1977). After 15 min, a "gluey" solid was collected and washed with benzene (10 mL) twice by decanting off the solvent.

**[(MeHg)₂(C₃H₆OS₂)] or C₅H₁₂OS₂Hg₂.** This compound melts at 103°C–104°C (Canty and Kishimoto 1977). To obtain it, 1,3-dimercapto-2-propanol (C₃H₈OS₂) (1.11 mM) in EtOH (5 mL) was added to a solution of methylmercuric acetate (2.11 mM) in the same solvent (30 mL). After 24 h, a colorless crystalline precipitate was collected and washed with EtOH.

**[Hg(S₂COEt)] or C₆H₁₀O₂S₄Hg, mercury ethylxanthate.** This compound crystallizes in the monoclinic structure with the lattice parameters $a = 990.4 \pm 0.2$, $b = 690.3 \pm 0.7$, and $c = 910.4 \pm 0.8$ pm and $\beta = 100.21° \pm 0.03°$ and the experimental and calculated densities 2.49 and 2.40 g·cm⁻³, respectively (Watanabe 1977). Its crystals were grown as thin plates or rectangular tablets from the solution of the title compound in CCl₄.

**[OAcHgC₄H₂SCOOH] or C₇H₆O₄SHg, 5-acetoxymercuriothiophen-2-carboxylic acid.** It was prepared by slow addition of an aqueous solution (ca. 50 mL) of Hg(OAc)₂ (1.57 mM; with some drops of 2 M acetic acid) to an aqueous solution (ca. 30 mL) of thiophene-2-carboxylic acid (1.57 mM) (Popović et al. 2000b). The crude mercurated product was treated with DMSO, and insoluble solid was separated by filtration. Acetone was then added very slowly to the filtrate. The precipitated monomercurated thiophen-2-carboxylic acid left overnight, then filtered off, washed with acetone, and dried.

**[MeHgPhCOS] or $C_8H_8OSHg$.** This compound melts at 61°C (Sytsma and Kline 1973). To obtain it, MeHgOAc was dissolved in water and added to thiobenzoic acid dissolved in MeOH or methanolic KOH solution. The crude products were purified by the recrystallization in $H_2O$/acetone.

**[PhHg($S_2$COMe)] or $C_8H_8OS_2Hg$.** This compound melts at 120°C–121°C and crystallizes in the monoclinic structure with the lattice parameters $a = 3733 \pm 2$, $b = 482.5 \pm 0.1$, and $c = 1268.6 \pm 0.1$ pm and $\beta = 101.21° \pm 0.02°$ and a calculated density of 2.542 g·cm$^{-3}$ (Tiekink 1987b). It was prepared by the following procedure. To a stirred solution of PhHgCl (0.5 g, 50 mL of $CH_2Cl_2$) the stoichiometric quantity of $KS_2$COMe (20 mL of $H_2O$) was added. After 1 h of stirring, the organic layer was separated and dried over $Na_2SO_4$. Crystals were deposited as the volume was reduced. The bulk product was recrystallized from a solution in $CH_2Cl_2$.

**[MeHg$C_7H_5O_2S$] or $C_8H_8O_2SHg$.** This compound melts at 168°C (Sytsma and Kline 1973). To obtain the title compound, it is necessary to use an aqueous solution of MeHgOH prepared from MeHgI and excess amount of freshly precipitated $Ag_2O$. MeHgOH solution, from which AgI and excess $Ag_2O$ have been removed by filtration, was added to a methanolic solution of $o$-mercaptobenzoic acid. The crude product was purified by the recrystallization in hexane/$CH_2Cl_2$.

**[PhHgS$(CH_2)_2$OH] or $C_8H_{10}OSHg$.** This compound melts at 80°C (Canty and Kishimoto 1977). It was prepared by the interaction of 2-mercapto-1-ethanol with phenylmercuric acetate and recrystallized from chloroform.

**[Hg($S_2$COPr$^n$)$_2$] or $C_8H_{14}O_2S_4Hg$.** This compound crystallizes in the orthorhombic structure with the lattice parameters $a = 737.1 \pm 0.3$, $b = 853.4 \pm 0.4$, and $c = 1161.8 \pm 0.4$ pm and a calculated density of 2.140 g·cm$^{-3}$ (Hounslow and Tiekink 1991). It was prepared from the reaction of $HgCl_2$ and potassium xanthate (1:2 molar ratio) in aqueous solution.

**[Hg($S_2$COPr$^i$)$_2$] or $C_8H_{14}O_2S_4Hg$.** This compound crystallizes in the monoclinic structure with the lattice parameters $a = 1422.6 \pm 0.8$, $b = 980.6 \pm 0.8$, and $c = 2141.0 \pm 1.0$ pm and $\beta = 100.13° \pm 0.10°$ and the experimental and calculated densities 2.00 and 2.08 g·cm$^{-3}$, respectively (Watanabe 1981). It was precipitated from an aqueous solution of potassium isopropylxanthate and $HgCl_2$. Yellow needlelike crystals were grown from acetone by rapid evaporation.

**[(OAcHg)$_2C_4$HSCOOH] or $C_9H_8O_6SHg_2$, 4,5-bis(acetoxymercurio)thiophen-2-carboxylic acid.** It was prepared by slow addition of an aqueous solution (ca. 50 mL) of Hg(OAc)$_2$ (1.57 mM; with some drops of 2 M acetic acid) to an aqueous solution (ca. 30 mL) of thiophene-2-carboxylic acid (0.79 mM) (Popović et al. 2000b). On mixing the solutions, the title compound was obtained immediately. It was then filtered off, washed with $H_2O$, and dried in vacuo.

**[MeHgS$_2$COBz] or C$_9$H$_{10}$OS$_2$Hg.** This compound melts at 98°C–99°C and crystallizes in the monoclinic structure with the lattice parameters $a = 589.64 \pm 0.07$, $b = 872.20 \pm 0.08$, and $c = 2201.7 \pm 0.2$ pm and $\beta = 93.095° \pm 0.008°$ and a calculated density of 2.343 g·cm$^{-3}$ (Casas et al. 2002). To prepare it, MeHgCl (1.26 g) dissolved in 50 mL of CH$_2$Cl$_2$ was treated with 0.90 g (excess) of solid, finely powdered NaS$_2$COBz, and the mixture was stirred at room temperature for 1 h. The solution was filtered and left to evaporate in open air. A slightly decomposed solid residue (gray color) was obtained. The solid was redissolved in 5 mL CH$_2$Cl$_2$ and filtered, and the filtrate was covered with petroleum ether and left to crystallize in the refrigerator. The needlelike, colorless crystals were filtered and dried in vacuo.

**[PhHg(EtCOS$_2$)] or C$_9$H$_{10}$OS$_2$Hg.** This compound melts at 120°C–122°C and crystallizes in the orthorhombic structure with the lattice parameters $a = 1114 \pm 1$, $b = 2709.2 \pm 0.3$, and $c = 717.9 \pm 0.4$ pm and a calculated density of 2.446 g·cm$^{-3}$ (Tiekink 1987b, 1994). Its crystals suitable for XRD were grown by vapor diffusion of diethyl ether into an acetonitrile solution of the title compound.

**[Hg(C$_3$H$_5$OS$_2$)$_2$] or C$_{10}$H$_{10}$O$_2$S$_4$Hg.** This compound crystallizes in the monoclinic structure with the lattice parameters $a = 930.0 \pm 0.2$, $b = 669.3 \pm 0.2$, and $c = 1958.5 \pm 0.9$ pm and $\beta = 100.94° \pm 0.03°$ and the experimental and calculated densities 2.44 ± 0.01 and 2.456 g·cm$^{-3}$, respectively (Chieh and Moynihan 1980). When 0.5 g of solid HgCl$_2$ was added to an aqueous solution containing 0.575 g of potassium ethylxanthate (Hg$^{2+}$/ethylxanthate—1:2 molar ratio), a gray precipitate was obtained. The product was dissolved in acetone and filtered to remove the black suspended material, leading to a clear solution that gave colorless prismatic crystals of the title compound when the solvent had evaporated at 25°C.

**[(OAcHg)$_2$C$_4$HSCH$_2$COOH] or C$_{10}$H$_{10}$O$_6$SHg$_2$, 4,5-bis(acetoxymercurio)-2-thienylethanoic acid.** It was prepared by slow addition of an aqueous solution (ca. 50 mL) of Hg(OAc)$_2$ (1.57 mM; with a few drops of 2 M acetic acid) to an aqueous solution (ca. 30 mL) of 2-thienylethanoic acid (0.79 mM) (Popović et al. 2000b). Immediately after mixing the solutions of the reactants, the mercurated product was obtained. It was filtered off, washed with water and dried in vacuo.

**[PhHg(S$_2$COPr$^i$)] or C$_{10}$H$_{12}$OS$_2$Hg.** This compound melts at 76°C–77°C and crystallizes in the monoclinic structure with the lattice parameters $a = 1367.8 \pm 0.5$, $b = 2134.7 \pm 0.7$, and $c = 1457.0 \pm 0.6$ pm and $\beta = 114.99° \pm 0.02°$ and a calculated density of 2.134 g·cm$^{-3}$ (Tiekink 1987b). It was prepared by the following procedure. To a stirred solution of PhHgCl (0.5 g, 50 mL of CH$_2$Cl$_2$), the stoichiometric quantity of KS$_2$COPr$^i$ (20 mL of H$_2$O) was added. After 1 h of stirring, the organic layer was separated and dried over Na$_2$SO$_4$. Crystals were deposited as the volume was reduced. The bulk product was recrystallized from a solution in CH$_2$Cl$_2$.

**[Hg(S$_2$COBu$^n$)$_2$] or C$_{10}$H$_{18}$O$_2$S$_4$Hg.** This compound crystallizes in the orthorhombic structure with the lattice parameters $a = 741.8 \pm 0.5$, $b = 2566 \pm 5$, and $c = 861 \pm 3$ pm and a calculated density of 2.022 g·cm$^{-3}$ (Cox and Tiekink 1999b).

**[Hg(C$_{12}$H$_8$OS$_2$)] or C$_{12}$H$_8$OS$_2$Hg.** This compound was prepared using the next procedure (Block et al. 1990). A solution containing HgCl$_2$ (1.0 mM), Et$_3$N (2.0 mM), and bis(2-mercaptophenyl) ether (1.0 mM) in MeOH (15 mL) was stirred at room temperature for 1 h. The solid product, which formed immediately upon mixing of the reagents, was collected by filtration and washed with MeOH and distilled water. After solvent removal in vacuo, a white powder of the title compound was obtained.

**[Hg{S₂CO(CH₂)₂C(H)Me₂}₂] or C₁₂H₂₂O₂S₄Hg**. This compound crystallizes in the monoclinic structure with the lattice parameters $a = 2872 \pm 2$, $b = 582.8 \pm 0.9$, and $c = 1141 \pm 2$ pm and $\beta = 99.7° \pm 0.1°$ and a calculated density of 1.859 g·cm⁻³ (Cox and Tiekink 1999b).

**[Hg(S₂COC₅H₁₁-$n$)₂] or C₁₂H₂₂O₂S₄Hg**. This compound crystallizes in the monoclinic structure with the lattice parameters $a = 2732 \pm 1$, $b = 585.4 \pm 0.7$, and $c = 1138.4 \pm 0.7$ pm and $\beta = 95.35° \pm 0.05°$ and a calculated density of 1.931 g·cm⁻³ (Cox and Tiekink 1999b).

**[Hg{S₂CO(CH₂)₂CMe₃}₂] or C₁₄H₂₆O₂S₄Hg**. This compound crystallizes in the monoclinic structure with the lattice parameters $a = 3004.5 \pm 0.4$, $b = 595.5 \pm 0.8$, and $c = 1171.2 \pm 0.2$ pm and $\beta = 97.07° \pm 0.1°$ and a calculated density of 1.773 g·cm⁻³ (Cox and Tiekink 1999b).

**[(PhHg)₂(C₃H₆OS₂)] or C₁₅H₁₆OS₂Hg₂**. This compound melts at 66°C (Canty and Kishimoto 1977). To obtain it, 2,3-dimercapto-1-propanol (C₃H₈OS₂) (1.49 mM) was added to a solution of phenylmercuric acetate (2.98 mM) in H₂O (200 mL) with stirring. A white precipitate that formed immediately and after several hours of stirring was collected and washed with H₂O.

**[(PhHg)₂(C₃H₆OS₂)] or C₁₅H₁₆OS₂Hg₂**. This compound melts at 62°C–66°C (Canty and Kishimoto 1977). To obtain it, 1,3-dimercapto-2-propanol (C₃H₈OS₂) (1.85 mM) was added to a suspension of phenylmercuric acetate (3.71 mM) in H₂O (200 mL) with stirring. A white precipitate formed immediately and after several hours of stirring was collected and washed with H₂O.

## 7.69 Hg–H–C–O–F–S

Some compounds are formed in this system.

**[CF₃SHgOAc] or C₃H₃O₂F₃SHg, trifluoromethylthiomercuric acetate**. This compound melts at 80°C–85°C and decomposes at 170°C (Downs et al. 1962). The reaction of trifluoromethylthiomercuric chloride (CF₃SHgCl) (0.324 g) with AgOAc (0.161 g) gave a solid residue after the filtrate had been freeze dried (Emeléus and Pugh 1960). When this was sublimed in vacuum a small amount of unchanged chloride sublimed at 40°C. The sublimate at 70°C was CF₃SHgOAc, which was free from chloride. The title compound could also be obtained by evaporating a solution of Hg(OAc)₂ (1.51 g) and bis(trifluoromethylthio)mercury (1.91 g) in MeOH with the next vacuum sublimation (Downs et al. 1962).

**[CF₃COOHgC₄H₂SCH₂COOH] or C₈H₅O₄F₃SHg, 5-trifluoroacetoxymercurio-2-thienylethanoic acid**. It was prepared by slow addition of an aqueous solution (ca. 50 mL) of Hg(CF₃COO)₂ (1.17 mM; with a few drops of 2 M acetic acid) to an aqueous solution (ca. 30 mL) of 2-thienylethanoic acid (1.17 mM) (Popović et al. 2000b). The reaction mixture was left overnight, and then the precipitated product was filtered off, washed with water, and dried in vacuo.

## 7.70   Hg–H–C–O–Cl–S

Some compounds are formed in this system.

**[HgCl$_2$·(DMSO)] or C$_2$H$_6$OCl$_2$SHg**. This compound melts at 125°C–126°C (Selbin et al. 1961). It was obtained if HgCl$_2$ was dissolved in DMSO at about 60°C; cooling produced crystals that were recrystallized from acetone. HgCl$_2$·(DMSO) could also be prepared if a saturated solution of HgCl$_2$ was treated with DMSO; white crystals formed immediately. These were washed with ether and air-dried.

**[2HgCl$_2$·(DMSO)] or C$_2$H$_6$OCl$_4$SHg$_2$**. This compound melts at 126°C–128°C (Biscarini et al. 1971). It was prepared by the next procedure. To a boiling ethanol solution (500 mL) of HgCl$_2$ (0.4 M), a solution of DMSO (0.1 M) in 95% EtOH was added. The solvent was evaporated under reduced pressure, and the residue was crystallized from chloroform to constant melting point.

**[Hg(Me$_3$Hg$_3$S)ClO$_4$] or C$_3$H$_9$O$_4$ClSHg$_4$**. This compound melts at 142°C (Clarke and Woodward 1967). It was prepared in MeOH by the reaction between stoichiometric amounts of anhydrous AgClO$_4$ and (MeHg)$_2$S. After filtering off the precipitate of Ag$_2$S, the solvent was removed from the filtrate by evaporation under reduced pressure, and the remaining white solid was recrystallized from MeOH.

**[HgCl$_2$·(Et$_2$SO) or C$_4$H$_{10}$OCl$_2$SHg**. This compound melts at 84°C–86°C (Biscarini et al. 1971). It was prepared by the next procedure. To a boiling ethanol solution (500 mL) of HgCl$_2$ (0.4 M), a solution of Et$_2$SO (0.1 M) in 95% EtOH was added. The solvent was evaporated under reduced pressure, and the residue was crystallized from benzene to constant melting point.

**[2HgCl$_2$·(Et$_2$SO)] or C$_4$H$_{10}$OCl$_4$SHg$_2$**. This compound melts at 87°C–90°C (Biscarini et al. 1971). It was prepared by the next procedure. To a boiling ethanol solution (500 mL) of HgCl$_2$ (0.4 M), a solution of Et$_2$SO (0.1 M) in 95% EtOH was added. The solvent was evaporated under reduced pressure, and the residue was crystallized from benzene to constant melting point.

The title compound could also be prepared as follows (Biscarini et al. 1971). To a saturated solution of the addition compound of Et$_2$S with HgCl$_2$ (0.11 M) in EtOH, a solution of H$_2$O$_2$ (0.1 M) in the same solvent was added, and the extent of reaction was followed by iodometric titration. When 90% of H$_2$O$_2$ was reacted (ca. 100–200 h), the solvent was evaporated under reduced pressure without heating. The residue was crystallized.

**[HgCl$_2$·2(DMSO)] or C$_4$H$_{12}$O$_2$Cl$_2$S$_2$Hg**. This compound melts at 112°C–113°C (Jain and Rivest 1969b). It was prepared by dissolving HgCl$_2$ in a minimum amount of warm DMSO forming a clear solution. The title compound crystallized out on adding about 5 mL of absolute EtOH to the reaction mixture. It was filtered out under suction and the excess DMSO was removed under vacuum.

**[3HgCl₂·2(DMSO)]** or **C₄H₁₂O₂Cl₆S₂Hg₃**. This compound melts at 123°C–125°C (Biscarini et al. 1971) and crystallizes in the triclinic structure with the lattice parameters $a = 667.2 \pm 0.7$, $b = 928.6 \pm 0.8$, and $c = 876.4 \pm 0.8$ pm and $\alpha = 60.0° \pm 0.1°$, $\beta = 95.5° \pm 0.1°$, and $\gamma = 90.1° \pm 0.2°$ and the experimental and calculated densities 3.25 and 3.45 g·cm⁻³, respectively (Biscarini et al. 1974). It was prepared by the next procedure (Biscarini et al. 1971). To a boiling ethanol solution (500 mL) of HgCl₂ (0.4 M), a solution of DMSO (0.1 M) in 95% EtOH was added. The solvent was evaporated under reduced pressure, and the residue was crystallized from benzene to constant melting point.

**[(ClHg)₂C₄HSCOOH]** or **C₅H₂O₂Cl₂SHg₂**, **4,5-bis(chloromercurio)thiophen-2-carboxylic acid**. It was prepared by slow addition of an aqueous solution (ca. 50 mL) of HgCl₂ (5.52 mM) and NaOAc·3H₂O (11.04 mM) to an aqueous solution (ca. 30 mL) of thiophene-2-carboxylic acid (2.76 mM) (Popović et al. 2000b). The reaction mixture was heated for 3 h on a water bath and left overnight, and then the precipitated product was filtered off, washed with water, and dried in vacuo.

**[ClHgC₄H₂SCOOH]** or **C₅H₃O₂ClSHg**, **5-chloromercuriothiophen-2-carboxylic acid**. It was prepared by slow addition of an aqueous solution (ca. 50 mL) of HgCl₂ (1.84 mM) and NaOAc·3H₂O (3.68 mM) to an aqueous solution (ca. 30 mL) of thiophene-2-carboxylic acid (1.84 mM) (Popović et al. 2000b). The reaction mixture was left overnight, and then the precipitated product was filtered off, washed with water, and dried in vacuo.

**[(ClHg)₂C₄HSCH₂COOH]** or **C₆H₄O₂Cl₂SHg₂**, **4,5-bis(chloromercurio)-2-thienylethanoic acid**. It was prepared by slow addition of an aqueous solution (ca. 50 mL) of HgCl₂ (5.52 mM) and NaOAc·3H₂O (22.08 mM) to an aqueous solution (ca. 30 mL) of 2-thienylethanoic acid (2.76 mM) (Popović et al. 2000b). The reaction mixture was left to stand for a week, and then the precipitated product was filtered off, washed with water, and dried in vacuo.

**[ClHgC₄H₂SCH₂COOH]** or **C₆H₅O₂ClSHg**, **5-chloromercurio-2-thienylethanoic acid**. It was prepared by slow addition of an aqueous solution (ca. 50 mL) of HgCl₂ (1.84 mM) and NaOAc·3H₂O (3.68 mM) to an aqueous solution (ca. 30 mL) of 2-thienylethanoic acid (1.84 mM) (Popović et al. 2000b). The reaction mixture was left overnight, and then the precipitated product was filtered off, washed with water, and dried in vacuo.

**[HgCl₂·(Prⁿ₂SO)]** or **C₆H₁₄OCl₂SHg**. This compound melts at 75°C–83°C (Biscarini et al. 1971). It was prepared by the next procedure. To a boiling ethanol solution (500 mL) of HgCl₂ (0.4 M), a solution of Prⁿ₂SO (0.1 M) in 95% EtOH was added. The solvent was evaporated under reduced pressure, and the residue was crystallized from benzene/light petroleum mixture to constant melting point.

The title compound could also be prepared as follows (Biscarini et al. 1971). To a saturated solution of the addition compound of Prⁿ₂S with HgCl₂ (0.11 M) in EtOH, a solution of H₂O₂ (0.1 M) in the same solvent was added, and the extent of reaction was followed by iodometric titration. When 90% of H₂O₂ was reacted (ca. 100–200 h), the solvent was evaporated under reduced pressure without heating. The residue was crystallized.

**[2HgCl$_2$·(Pr$^i_2$SO)]** or **C$_6$H$_{14}$OCl$_4$SHg$_2$.** This compound melts at 97°C–100°C (Biscarini et al. 1971). It was prepared by the next procedure. To a boiling ethanol solution (500 mL) of HgCl$_2$ (0.4 M), a solution of Pr$^i_2$SO (0.1 M) in 95% EtOH was added. The solvent was evaporated under reduced pressure, and the residue was crystallized from benzene/light petroleum mixture to constant melting point.

The following procedure could also be used for the preparation of the title compound (Biscarini et al. 1971). The reaction in solution, pH = 1 for sulfuric acid, was conducted as described in the succeeding text for 3HgCl$_2$·2(MeEtSO) (second procedure) and reaches completion (90%) in ca. 1–2 h. Then the reaction was stopped by neutralization with solid NaHCO$_3$, which also causes formation of a brown precipitate of basic mercury carbonate, and partial decomposition of the addition compound present in solution. After the elimination of the solvent under reduced pressure at room temperature, the residue of the title compound was crystallized.

**[3HgCl$_2$·2(MeEtSO)]** or **C$_6$H$_{16}$O$_2$Cl$_6$S$_2$Hg$_3$.** This compound melts at 86°C–87°C (Biscarini et al. 1971). It was obtained by the next procedure. To a boiling ethanol solution (500 mL) of HgCl$_2$ (0.4 M), a solution of MeEtSO (0.1 M) in 95% EtOH was added. The solvent was evaporated under reduced pressure, and the residue wascrystallized from benzene/light petroleum mixture, CCl$_4$, or benzene to constant melting point.

The title compound could also be prepared as follows (Biscarini et al. 1971). To a saturated solution of the addition compound of MeSEt with HgCl$_2$ (0.11 M) in EtOH, a solution of H$_2$O$_2$ (0.1 M) in the same solvent was added, and the extent of reaction was followed by iodometric titration. When 90% of H$_2$O$_2$ was reacted (ca. 100–200 h), the solvent was evaporated under reduced pressure without heating. The residue was crystallized.

**[(2-MeSC$_7$H$_5$O)·(HgCl$_2$)$_2$]** or **C$_8$H$_8$OCl$_4$SHg$_2$** (C$_7$H$_6$O = tropone). This compound crystallizes as pale-yellow prisms and melts at 135°C–136°C (Asao and Kikuchi 1972).

**C$_8$H$_{16}$O$_2$Cl$_2$S$_2$Hg, dichlorobis-(1,4-thioxan)mercury(II).** This compound crystallizes in the orthorhombic structure with the lattice parameters $a$ = 1531, $b$ = 1969, and $c$ = 433 pm and a calculated density of 2.44 g·cm$^{-3}$ (McEwen and Sim 1967b).

**[Bu$^n_2$SO·HgCl$_2$]** or **C$_8$H$_{18}$OCl$_2$SHg.** This compound melts at 63°C–65°C (Biscarini et al. 1971) and crystallizes in the monoclinic structure with the lattice parameters $a$ = 2071.0 ± 1.0, $b$ = 811.3 ± 0.3, and $c$ = 809.0 ± 0.4 pm and $\beta$ = 94.74° ± 0.05° and the experimental and calculated densities 2.15 and 2.127 g·cm$^{-3}$, respectively (Biscarini et al. 1981). It was obtained by the next procedure. To a boiling ethanol solution (500 mL) of HgCl$_2$ (0.4 M), a solution of Bu$^n_2$SO (0.1 M) in 95% EtOH was added. The solvent was evaporated under reduced pressure, and the residue was crystallized from light petroleum to constant melting point (Biscarini et al. 1971).

The title compound could also be prepared as follows (Biscarini et al. 1971, 1981). To a saturated solution of the addition compound of Bu$^n_2$S·2HgCl$_2$ (8.8 mM) in EtOH, a solution of H$_2$O$_2$ (8.8 mM) in the same solvent was added, and the extent of reaction was followed by iodometric titration. When 90% of H$_2$O$_2$ was reacted (ca. 100–200 h), the solvent was evaporated under reduced pressure without heating. The residue was crystallized.

The following procedure could also be used for the preparation of the title compound (Biscarini et al. 1971). The reaction in solution, pH = 1 for sulfuric acid, was conducted as described in the preceding text and reaches completion (90%) in ca. 1–2 h. Then, the reaction was stopped by neutralization with solid NaHCO$_3$, which also causes the formation of a brown precipitate of basic mercury carbonate, and partial decomposition of the addition compound present in solution. After elimination of the solvent under reduced pressure at room temperature, the residue of the title compound was crystallized.

**[Bu$^i_2$SO·HgCl$_2$] or C$_8$H$_{18}$OCl$_2$SHg.** This compound melts at 54°C–59°C (Biscarini et al. 1971). It was obtained by the next procedure. To a boiling ethanol solution (500 mL) of HgCl$_2$ (0.4 M), a solution of Bu$^i_2$SO (0.1 M) in 95% EtOH was added. The solvent was evaporated under reduced pressure, and the residue was crystallized from ligroin to constant melting point (Biscarini et al. 1971).

The title compound could also be prepared as follows (Biscarini et al. 1971). To a saturated solution of the addition compound of Bu$^i_2$S·2HgCl$_2$ (0.11 mM) in EtOH, a solution of H$_2$O$_2$ (0.1 mM) in the same solvent was added, and the extent of reaction was followed by iodometric titration. When 90% of H$_2$O$_2$ was reacted (ca. 100–200 h), the solvent was evaporated under reduced pressure without heating. The residue was crystallized.

The following procedure could also be used for the preparation of the title compound (Biscarini et al. 1971). The reaction in solution, pH = 1 for sulfuric acid, was conducted as described in the preceding text and reaches completion (90%) in ca. 1–2 h. Then the reaction was stopped by neutralization with solid NaHCO$_3$, which also causes the formation of a brown precipitate of basic mercury carbonate, and partial decomposition of the addition compound present in solution. After elimination of the solvent under reduced pressure at room temperature, the residue of the title compound was crystallized.

**[3HgCl$_2$·2(Et$_2$SO)] or C$_8$H$_{20}$O$_2$Cl$_6$S$_2$Hg$_3$.** This compound melts at 82°C–89°C (Biscarini et al. 1971). It was prepared by the next procedure. To a boiling ethanol solution (500 mL) of HgCl$_2$ (0.4 M), a solution of Et$_2$SO (0.1 M) in 95% EtOH was added. The solvent was evaporated under reduced pressure, and the residue the crystallized from benzene to constant melting point.

**[Hg(ClO$_4$)$_2$·(4DMSO)] or C$_8$H$_{24}$O$_{12}$Cl$_2$S$_4$Hg.** This compound crystallizes in the monoclinic structure with the lattice parameters $a = 1249.0 \pm 0.6$, $b = 2288 \pm 1$, and $c = 1727 \pm 1$ pm and $\beta = 105.67° \pm 0.04°$ and the experimental and calculated densities $1.96 \pm 0.02$ and $1.99$ g·cm$^{-3}$, respectively (Sandström 1978). Its recrystallization was performed from methanol solutions at about 0°C.

**[{Hg(C$_{10}$H$_{20}$S$_4$)(H$_2$O)}(ClO$_4$)$_2$] or C$_{10}$H$_{22}$O$_9$Cl$_2$S$_4$Hg.** This compound crystallizes in the orthorhombic structure with the lattice parameters $a = 1658.3 \pm 0.2$, $b = 1398.6 \pm 0.3$, and $c = 893.2 \pm 0.2$ pm and the experimental and calculated densities $2.24 \pm 0.05$ and $2.195$ g·cm$^{-3}$, respectively (Alcock et al. 1978). It was prepared by mixing solutions of Hg(ClO$_4$)$_2$·$x$H$_2$O (excess) and 1,4,8,11-tetrathiacyclotetradecan (C$_{10}$H$_{20}$S$_4$) (0.134 g) in aqueous MeOH (10 mL). The white powder that immediately precipitated was filtered off and washed with H$_2$O and chloroform.

Recrystallization was accomplished by slow evaporation of an extremely dilute solution (0.1 g in 500 mL) of the powder in 80% aqueous MeOH. Small poorly formed fragments of crystalline title compound were obtained.

**[C$_6$H$_4$(HgS$_2$COMe)$_2$]·(CH$_2$Cl$_2$) or C$_{11}$H$_{12}$O$_2$Cl$_2$S$_4$Hg$_2$.** This compound crystallizes in the triclinic structure with the lattice parameters $a = 1063.7 \pm 0.2$, $b = 1202.2 \pm 0.2$, and $c = 769.2 \pm 0.1$ pm and $\alpha = 101.60° \pm 0.01°$, $\beta = 98.75° \pm 0.01°$, and $\gamma = 89.96° \pm 0.01°$ and a calculated density of 2.709 g·cm$^{-3}$ (Tiekink 1986a). It was obtained from the reaction (1:1 molar ratio) of PhHgCl in CH$_2$Cl$_2$ and KS$_2$COMe in H$_2$O.

**[Ph$_2$SO·HgCl$_2$] or C$_{12}$H$_{10}$OCl$_2$SHg.** This compound melts at 100°C–102°C (Biscarini et al. 1971) and crystallizes in the triclinic structure with the lattice parameters $a = 835.4 \pm 0.8$, $b = 830.4 \pm 0.7$, and $c = 1111.5 \pm 0.8$ pm and $\alpha = 108.2° \pm 0.1°$, $\beta = 100.3° \pm 0.1°$, and $\gamma = 102.3° \pm 0.1°$ and the experimental and calculated densities 2.4 and 2.28 g·cm$^{-3}$, respectively (Biscarini et al. 1973). It was obtained by the next procedure. Hydrated HgCl$_2$ and Ph$_2$SO were each dissolved in EtOH, and the obtained solutions were mixed slowly with stirring (Gopalakrishnan and Patel 1967). The mixture, upon slow evaporation by passing dry air at room temperature, yielded crystals of the title compound in about a day. These were separated, washed several times with petroleum ether to remove any adhering Ph$_2$SO, and recrystallized either from acetone or ethanol. It could also be prepared as follows (Biscarini et al. 1971). To a boiling ethanol solution (500 mL) of HgCl$_2$ (0.4 M), a solution of Ph$_2$SO (0.1 M) in 95% EtOH was added. The solvent was evaporated under reduced pressure, and the residue was crystallized from a light ligroin to a constant melting point. Its crystals are instable in air (Biscarini et al. 1973).

**[Ph$_2$SO·2HgCl$_2$] or C$_{12}$H$_{10}$OCl$_4$SHg$_2$.** This compound melts at 115°C–117°C (Biscarini et al. 1971). It was obtained by recrystallization of Ph$_2$SO·HgCl$_2$ from benzene/light petroleum mixture.

**[Hg(C$_{12}$H$_{24}$O$_4$S$_2$)Cl$_2$]·HgCl$_2$ or C$_{12}$H$_{24}$O$_4$Cl$_4$S$_2$Hg$_2$.** This compound crystallizes in the triclinic structure with the lattice parameters $a = 841.0 \pm 0.2$, $b = 1135.1 \pm 0.9$, and $c = 1234.7 \pm 0.6$ pm and $\alpha = 92.48° \pm 0.05°$, $\beta = 94.52° \pm 0.03°$, and $\gamma = 105.66° \pm 0.05°$ pm and the experimental and calculated densities 2.43 and 2.46 g·cm$^{-3}$, respectively (Dalley and Larson 1981). It was prepared by mixing an aqueous solution of 1,4,7,10-tetraoxa-13,16-dithiacyclooctadecane with a large excess of HgO dissolved in concentrated HCl.

**[Hg(C$_{12}$H$_{24}$S$_4$)](ClO$_4$)$_2$ or C$_{12}$H$_{24}$O$_8$Cl$_2$S$_4$Hg, 1,5,9,13-tetrathiocyclohexadecane mercury(II) perchlorate.** This compound crystallizes in the monoclinic structure with the lattice parameters $a = 1003.3 \pm 0.3$, $b = 1342.1 \pm 0.4$, and $c = 1596.0 \pm 0.4$ pm and $\beta = 96.48° \pm 0.02°$ and a calculated density of 2.165 g·cm$^{-3}$ (Setzer et al. 1991) ($a = 1344.8 \pm 0.5$, $b = 953.8 \pm 0.4$, and $c = 868.0 \pm 0.2$ pm and $\beta = 95.10° \pm 0.03°$ [Jones et al. 1979]). To obtain it, a solution of 1,5,9,13-tetrathiacyclohexadecane (0.136 mM) in 8 mL of anhydrous nitromethane was added to a solution of Hg(ClO$_4$)$_2$·3H$_2$O (0.0937 mM) in 2 mL of anhydrous nitromethane and 4 drops of acetic anhydride. Colorless crystals were grown from the reaction mixture by solvent diffusion with anhydrous diethyl ether. The crystalline solid was dried overnight under high vacuum to give the title compound as a colorless crystalline solid (Setzer et al. 1991). The crystals of the title compound could also be grown from a MeOH/H$_2$O solvent mixture (Jones et al. 1979).

**[Hg(ClO$_4$)$_2$(6DMSO)]** or **C$_{12}$H$_{36}$O$_{14}$Cl$_2$S$_6$Hg**. This compound melts at 125°C–128°C (Ahrland and Björk 1974) and crystallizes in the triclinic structure with the lattice parameters $a = 741.8 \pm 0.2$, $b = 1083.1 \pm 0.6$, and $c = 1124.1 \pm 0.4$ pm and $\alpha = 118.37° \pm 0.04°$, $\beta = 93.10° \pm 0.03°$, and $\gamma = 92.60° \pm 0.03°$ pm and the experimental and calculated densities 1.76 ± 0.03 and 1.82 g·cm$^{-3}$, respectively (Sandström and Persson 1978). To obtain it, Hg(ClO$_4$)$_2$·3H$_2$O (10 mM) was dissolved in EtOH (17 mM). Sometimes, it was necessary to decant the solution in order to remove a slight residue of insoluble matter. When DMSO (60 mM) was added, precipitate occurred immediately. The crystals were filtered in a dry N$_2$ atmosphere and dried in vacuum over night. On standing, Hg(I) was formed in the ethanol solution with the production of acetaldehyde. DMSO should be therefore added without delay (Carlin et al. 1962, Ahrland and Björk 1974).

Hg(ClO$_4$)$_2$(DMSO)$_6$ could also be obtained using the following methodological procedure. Saturated DMSO solution, 0.94 M at 25°C, was prepared by dissolving Hg(ClO$_4$)$_2$(DMSO)$_4$ in pure DMSO. By slow evaporation or by cooling, colorless prismatic crystals of this compound were formed. They were filtered off and washed with cold acetone. The crystals are very hygroscopic and decomposed rapidly in air (Sandström and Persson 1978).

**[HgCl$_2$·C$_{17}$H$_{24}$O$_5$S$_2$]** or **C$_{17}$H$_{24}$O$_5$Cl$_2$S$_2$Hg**. This compound melts at 208°C–209°C and crystallizes in the monoclinic structure with the lattice parameters $a = 1874.9 \pm 0.4$, $b = 802.5 \pm 0.2$, and $c = 1511.5 \pm 0.3$ pm and $\beta = 95.84° \pm 0.02°$ and a calculated density of 1.891 g·cm$^{-3}$ (Kubo et al. 1996). Its single crystals were obtained by crystallization of equimolar mixture of HgCl$_2$ and 5,8,11,14-tetraoxa-2,17-dithiobicyclo[16.4.1] tricosa-1(22),18,20-triene-23-one (C$_{17}$H$_{24}$O$_5$S$_2$) from MeCN.

**C$_{24}$H$_{16}$O$_2$Cl$_2$S$_2$Hg, dichlorobis(phenoxathiin)mercury(II)**. This compound crystallizes in the monoclinic structure with the lattice parameters $a = 3100$, $b = 1946$, and $c = 395$ pm and $\beta = 111°18'$ (Cheung et al. 1965). It readily dissociates and the crystals at room temperature and exposed to the atmosphere lose their phenoxathiin content in a few days.

**[Hg(C$_{24}$H$_{31}$O$_4$S)Cl]** or **C$_{24}$H$_{31}$O$_4$ClSHg**. This compound melts at 180°C with decomposition (Pouskouleli et al. 1977). To prepare it, spironolactone (208.2 mg) was dissolved in aqueous EtOH (less than 45% in H$_2$O) and HgCl$_2$ in equimolar amount (135.8 mg) dissolved in the same solvent was added to the aforementioned solution. After refluxing for 3 h, the solution became cloudy. A white precipitate started to be formed after heating for 1 h at 50°C; it was left overnight, collected, and washed with H$_2$O and EtOH. After filtration, the color of the precipitate started to change to yellow. Using EtOH for recrystallization, the white precipitate gave a turbid solution from which a yellow precipitate of the title compound was obtained.

The same yellow precipitate gave the aforementioned reaction when MeOH, acetone, and Pr$^n$OH were used as solvents. The white precipitate became yellow even at low temperatures and in the absence of light. This compound is extremely unstable and was purified using a mixture of ethyl acetate/chloroform (1:1 volume ratio).

[Hg(ClO$_4$)$_2${(CH$_2$CH$_2$)$_2$SO}$_6$] or C$_{24}$H$_{48}$O$_{14}$Cl$_2$S$_6$Hg. Interaction of either MeOH or EtOH solution of tetrahydrothiophene oxide [(CH$_2$CH$_2$)$_2$SO] with a similar solution of hydrated mercuric(II) perchlorate results in the precipitation of white crystals of this compound (Carlin et al. 1962). It can be recrystallized from the same solvent. The obtained compound was dried in vacuo at 60°C.

[Hg(ClO$_4$)$_2${O(CH$_2$CH$_2$)$_2$SO}$_6$] or C$_{24}$H$_{48}$O$_{20}$Cl$_2$S$_6$Hg. Interaction of either MeOH or EtOH solution of thioxane oxide [O(CH$_2$CH$_2$)$_2$SO] with a similar solution of hydrated mercuric(II) perchlorate results in the precipitation of white crystals of this compound (Carlin et al. 1962). It can be recrystallized from the same solvent. The obtained compound was dried in vacuo at 60°C.

## 7.71   Hg–H–C–O–Br–S

Some compounds are formed in this system.

C$_{22}$H$_{31}$O$_2$BrSHg. This compound crystallizes in the tetragonal structure with the lattice parameters $a = 1295.3 \pm 0.4$ and $c = 1272.7 \pm 0.4$ pm and the experimental and calculated densities $1.97 \pm 0.02$ and 1.989 g·cm$^{-3}$, respectively (Terzis et al. 1980). Reaction of the steroid 7α-(thioacetyl)-(17R)-spiro[androst-4-ene-17,2(3H)-furane (spiroxazone) with HgBr$_2$ produced a crystalline solid C$_{22}$H$_{31}$O$_2$BrSHg. Its crystals were obtained from an ethanol solution.

[Hg(C$_{24}$H$_{31}$O$_4$S)Br] or C$_{24}$H$_{31}$O$_4$BrSHg. This compound melts at 125°C with decomposition (Pouskouleli et al. 1977). To prepare it, spironolactone (208.3 mg) and HgBr$_2$ (108.2 mg) were dissolved in aqueous EtOH (30%) and refluxed for about 4 h. The cloudy solution after one night gave a grayish precipitate, which was collected and washed with EtOH and H$_2$O.

## 7.72   Hg–H–C–F–S

Some compounds are formed in this system.

[MeHgC$_6$F$_5$S] or C$_7$H$_3$F$_5$SHg. This compound melts at 102°C (Sytsma and Kline 1973). To obtain it, MeHgOAc was dissolved in water and added to pentafluorothiophenol dissolved in MeOH or methanolic KOH solution. The crude products were purified by the recrystallization in H$_2$O/acetone.

[MeHgSC$_6$H$_4$F-p] or C$_7$H$_7$FSHg. This compound melts at 81°C (Sytsma and Kline 1973). To obtain it, MeHgOAc was dissolved in water and added to p-fluorothiophenol dissolved in MeOH or methanolic KOH solution. The crude products were purified by the recrystallization in hexane.

## 7.73   Hg–H–C–Cl–S

Some compounds are formed in this system.

**[MeClHgS] or CH₃ClSHg**. This compound crystallizes in the monoclinic structure with the lattice parameters $a = 749.0 \pm 0.4$, $b = 739.5 \pm 0.2$, and $c = 781.5 \pm 0.3$ pm and $\beta = 92.93° \pm 0.04°$ and a calculated density of 4.350 g·cm⁻³ (Canty et al. 1979a).

**[EtSHgCl] or C₂H₅ClSHg**. This compound melts at 213°C (Canty et al. 1978a). It was prepared by the addition of liquid ethanethiol to stirred ethanol solution of HgCl₂ in 1:1 molar ratio. The precipitate was collected, washed with EtOH, and dried over P₂O₅ in a vacuum.

**C₃H₆Cl₂S₃Hg, dichloro-(1,3,5-trithian)mercury**. This compound crystallizes in the orthorhombic structure with the lattice parameters $a = 429$, $b = 1433$, and $c = 1362$ pm and a calculated density of 3.25 g·cm⁻³ (Costello et al. 1966). It crystallizes from acetone solution of 1,3,5-trithian and HgCl₂ in 1:1 molar ratio.

**[MeEtS·HgCl₂] or C₃H₈Cl₂SHg**. This compound melts at 104°C (McAllan et al. 1951).

**[{Me₃S}{HgCl₃}] or C₃H₉Cl₃SHg, trimethylsulfonium mercurate(II)**. This compound melts at 191°C–192°C and crystallizes in the monoclinic structure with the lattice parameters $a = 889.4 \pm 0.6$, $b = 1460.3 \pm 0.8$, and $c = 747.6 \pm 0.6$ pm and $\beta = 97.85° \pm 0.05°$ and the experimental and calculated densities 2.70 and 2.65 g·cm⁻³, respectively (Biscarini et al. 1977). To obtain it, DMSO and HgCl₂ (1:1 molar ratio) were warmed until white fumes appeared in the reaction flask (ca. 150°C). The solid residue was extracted with MeOH from which white crystals of [Me₃S][HgCl₃] were obtained.

**[C₄H₃SHgCl] or C₄H₃ClSHg, 2-thienylmercury(II) chloride**. This compound melts at 185°C–187°C (Bell et al. 2004). It was prepared by the mercuration of thiophene.

**[HgCl₂·C₄H₈S] or C₄H₈Cl₂SHg**. This compound crystallizes in the triclinic structure with the lattice parameters $a = 938 \pm 5$, $b = 761 \pm 4$, and $c = 583 \pm 4$ pm and $\alpha = 98.8° \pm 0.5°$, $\beta = 104.7° \pm 0.5°$, and $\gamma = 90.2° \pm 0.5°$ pm and the experimental and calculated densities 3.0 g·cm⁻³ (Bränden 1964b). It was prepared by mixing tetrahydrothiophene with HgCl₂ in a molar ratio of 1:1 using EtOH as a solvent. Needle-shaped single crystals were obtained by repeated recrystallization from acetone.

**[BuᵗS·HgCl] or C₄H₉ClSHg**. This compound melts at 147°C with decomposition (Canty et al. 1978a). To obtain it, Buᵗ₂S·HgCl₂ (6.6 g) was suspended in EtOH (60 mL), brought to the boil, and quickly filtered (Biscarini et al. 1972). The residual solid was treated several times with new portions of EtOH. From the filtered solution, a white product separates slowly, which does not melt below 260°C. Decomposition of this compound leads to the formation of Hg₃S₂Cl₂ (Biscarini et al. 1972). It was also prepared by the addition of liquid t-butanethiol to a stirred ethanol solution of HgCl₂ in 1:1 molar ratio. The precipitate was collected, washed with EtOH, and dried over P₂O₅ in a vacuum.

**[MePr$^i$S·HgCl$_2$] or C$_4$H$_{10}$Cl$_2$SHg.** This compound melts at 144.5°C (McAllan et al. 1951).

**[Et$_2$S·HgCl$_2$] or C$_4$H$_{10}$Cl$_2$SHg.** This compound melts at 76.5°C–77°C (Faragher et al. 1929). It was isolated as white needles by the interaction of HgCl$_2$ and Et$_2$S in EtOH.

**[MePr$^n$S·2HgCl$_2$] or C$_4$H$_{10}$Cl$_4$SHg$_2$.** This compound melts at 165°C (McAllan et al. 1951).

**[Et$_2$S·2HgCl$_2$] or C$_4$H$_{10}$Cl$_4$SHg$_2$.** This compound melts at 119°C–119.5°C (Faragher et al. 1929, McAllan et al. 1951) and crystallizes in the triclinic structure with the lattice parameters $a = 743 \pm 1$, $b = 928 \pm 1$, and $c = 922 \pm 1$ pm and $\alpha = 108.7° \pm 0.1°$, $\beta = 99.0° \pm 0.1°$, and $\gamma = 103.1° \pm 0.1°$ pm and the experimental and calculated densities 3.4 and 3.35 g·cm$^{-3}$, respectively (Bränden 1964a). It was prepared by mixing Et$_2$S with excess of HgCl$_2$ using EtOH as a solvent (Bränden 1964a). It could also be isolated by the interaction of HgCl$_2$ dissolved in EtOH and Et$_2$S dissolved in benzene or acetone (Faragher et al. 1929). Its single crystals were obtained as white plates by repeated recrystallization from acetone (Bränden 1964a).

**[2Me$_2$S·3HgCl$_2$] or C$_4$H$_{12}$Cl$_6$S$_2$Hg$_3$.** This compound melts at 158°C (McAllan et al. 1951) (at 150°C–151°C [Phillips 1901]). Upon adding methyl sulfide to a solution of HgCl$_2$, a bulky precipitate of a white color was produced. This compound could be a mixture of Me$_2$S·HgCl$_2$ and Me$_2$S·2HgCl$_2$, as it has no definite melting (Faragher et al. 1929). By heating rapidly, a clear liquid is obtained at 151°C.

**[Hg(C$_5$H$_4$S$_5$)Cl$_2$]$_2$ or C$_5$H$_4$Cl$_2$S$_5$Hg.** This compound crystallizes in the triclinic structure with the lattice parameters $a = 855.0 \pm 0.2$, $b = 984.76 \pm 0.08$, and $c = 829.8 \pm 0.2$ pm and $\alpha = 100.69° \pm 0.01°$, $\beta = 118.07° \pm 0.02°$, and $\gamma = 102.79° \pm 0.01°$ pm and a calculated density of 2.874 g·cm$^{-3}$ (Dai et al. 1998) ($a = 837.3 \pm 0.3$, $b = 850.9 \pm 0.3$, and $c = 980.0 \pm 0.7$ pm and $\alpha = 102.71° \pm 0.04°$, $\beta = 100.67° \pm 0.04°$, and $\gamma = 118.07° \pm 0.03°$ pm [Wang et al. 1998]). It was prepared by the following procedure. To an acetone solution (4 mL) of 4,5-ethylenedithio-1,3-dithiole-2-thione (C$_5$H$_4$S$_5$) (0.05 mM) HgCl$_2$ (0.1 mM) in acetone (2 mL) was added (a solution of C$_5$H$_4$S$_5$ [0.5 mM] in CH$_2$Cl$_2$ [3 mL] was added to the solution of HgCl$_2$ [0.5 mM] in acetonitrile [20 mL] [Wang et al. 1998]). The mixture was stirred for 20 min at room temperature under an Ar atmosphere. A red precipitate was formed, and it was filtered and dried in argon. Suitable crystals for XRD were grown at –5°C from the filtrate, which were sealed in a glass tube (Dai et al. 1998).

**[EtPr$^i$S·HgCl$_2$] or C$_5$H$_{12}$Cl$_2$SHg.** This compound melts at 82.5°C (McAllan et al. 1951).

**[EtPr$^n$S·2HgCl$_2$] or C$_5$H$_{12}$Cl$_4$SHg$_2$.** This compound melts at 106°C (McAllan et al. 1951).

**[PhSHgCl] or C$_6$H$_5$ClSHg.** This compound that was prepared for the first time by Lecher (1920) melts at 189°C–192°C (Gregg et al. 1951). To obtain it, a solution containing 12 mM of HgCl$_2$ in 10 mL of EtOH was added at room temperature to 8 mM of PhSH in 100–150 mL of EtOH either warm or at room temperature.

MeOH was also used as the reaction solvent. PhSHgCl could also be prepared if a solution containing 4 mM HgCl$_2$ in 10 mL EtOH was added to 2.7 mM of triphenyl-methyl phenyl sulfide in 100–150 mL of EtOH either warm or at room temperature. Ether, acetone, or ethyl acetate could be substituted for EtOH and MeOH as reaction solvents. However, yields were lower and appeared to vary in proportion to the volume of solvent used.

**[EtBu$^n$S·HgCl$_2$] or C$_6$H$_{14}$Cl$_2$SHg**. This compound melts at 47.5°C (McAllan et al. 1951).

**[EtBu$^i$S·HgCl$_2$] or C$_6$H$_{14}$Cl$_2$SHg**. This compound melts at 79°C (McAllan et al. 1951).

**[EtBu$^s$S·HgCl$_2$] or C$_6$H$_{14}$Cl$_2$SHg**. This compound melts at 50°C (McAllan et al. 1951).

**[EtBu$^t$S·HgCl$_2$] or C$_6$H$_{14}$Cl$_2$SHg**. This compound melts at 83°C (McAllan et al. 1951).

**[Pr$^n_2$S·HgCl$_2$] or C$_6$H$_{14}$Cl$_2$SHg**. This compound melts at 87.5°C–88°C (Faragher et al. 1929, McAllan et al. 1951). It was isolated as long white needles by the interaction of HgCl$_2$ and Pr$^n_2$S in EtOH.

**[Pr$^n$Pr$^i$S·HgCl$_2$] or C$_6$H$_{14}$Cl$_2$SHg**. This compound melts at 75.5°C (McAllan et al. 1951).

**[Pr$^i_2$S·HgCl$_2$] or C$_6$H$_{14}$Cl$_2$SHg**. This compound melts at 148.5°C (McAllan et al. 1951).

**[Hg$_2$(SPr$^i$)$_2$Cl$_2$] or C$_6$H$_{14}$Cl$_2$S$_2$Hg$_2$, (isopropylthio)mercury(II) chloride**. This compound crystallizes in the monoclinic structure with the lattice parameters $a = 2142.4 \pm 0.6$, $b = 466.8 \pm 0.2$, and $c = 673.4 \pm 0.3$ pm and $\beta = 90.45° \pm 0.05°$ and a calculated density of 3.07 g·cm$^{-3}$ (Biscarini et al. 1984).

**[Pr$^n_2$S·2HgCl$_2$] or C$_6$H$_{14}$Cl$_4$SHg$_2$**. This compound melts at 121°C–122°C (Faragher et al. 1929). It was isolated as white plates by the interaction of HgCl$_2$ dissolved in EtOH and Pr$^n_2$S dissolved in benzene.

**[(MeEtS)$_2$(HgCl$_2$)$_3$] or C$_6$H$_{16}$Cl$_6$S$_2$Hg$_3$**. This compound melts at 106°C–107°C (Biscarini and Nivellini 1969).

**[(C$_7$H$_6$S)·(HgCl$_2$)] or C$_7$H$_6$Cl$_2$SHg (C$_7$H$_6$S = thiotropone)**. Yellow crystals of this compound melt at 145°C with decomposition (Asao and Kikuchi 1972).

**[MeHgSC$_6$H$_4$Cl] or C$_7$H$_7$ClSHg**. This compound that was obtained for the first time by Märcker (1865) melts at 63°C–65°C (Sytsma and Kline 1973, Bach and Weibel 1976, Castiñeiras et al. 1986). To obtain it, to a solution of 2.0 mM of MeHgCl in 20 mL of hot MeOH a warm solution of 0.29 g of p-chlorothiophenol (2.0 mM) in 10 mL of MeOH and 1 mL of 5 N aqueous NaOH was added. An excess NaOH solution was added until a white precipitate formed and remained. The solution was then heated to boiling and a crystalline precipitate formed upon cooling. Filtration gave a crude product. Recrystallization from CH$_2$Cl$_2$ gave white needles of [MeHgSC$_6$H$_4$Cl] (Bach and Weibel 1976, Castiñeiras et al. 1986).

It could also be obtained if MeHgOAc was dissolved in water and added to
*p*-chlorothiophenol dissolved in MeOH or methanolic KOH solution (Sytsma and
Kline 1973). The crude products were purified by the recrystallization in hexane.

**[*m*-TolSHgCl], [*p*-TolSHgCl], and [*o*-TolSHgCl] or $C_7H_7ClSHg$**. These compounds
melt at 187°C–190°C, 225°C–227°C, and 177°C–180°C, respectively (Gregg et al.
1951). To obtain them, a solution containing 12 mM of $HgCl_2$ in 10 mL of EtOH was
added at room temperature to 8 mM of *m*-TolSH (*p*-TolSH, *o*-TolSH) in 100–150 mL
of EtOH either warm or at room temperature. MeOH was also used as the reaction
solvent. They could also be prepared if a solution containing 4 mM $HgCl_2$ in 10 mL
EtOH was added to 2.7 mM of triphenylmethyl *m*-tolyl (*p*-tolyl, *o*-tolyl) sulfide in
100–150 mL of EtOH either warm or at room temperature. Ether, acetone, or ethyl
acetate could be substituted for EtOH and MeOH as reaction solvents. However,
yields were lower and appeared to vary in proportion to the volume of solvent used.

**$C_8H_{12}Cl_4S_2Hg_2$, 1,6-dithiacyclodeca-*cis*-3,*cis*-8-dienebis(mercuric chloride)**.
This compound crystallizes in the monoclinic structure with the lattice parameters
$a = 729$, $b = 1701$, and $c = 620$ pm and $\beta = 92°34'$ and a calculated density of 3.08
$g \cdot cm^{-3}$ (Cheung et al. 1965, Cheung and Sim 1965).

**[$Bu^i_2S \cdot HgCl_2$] or $C_8H_{18}Cl_2SHg$**. This compound melts at 116°C (Faragher et al. 1929).
It was isolated as long white needles by the interaction of $HgCl_2$ and $Bu^i_2S$ in EtOH.

**[$Bu^i_2S \cdot 2HgCl_2$] or $C_8H_{18}Cl_4SHg_2$**. This compound melts at 131°C (Faragher et al.
1929). It was isolated as white plates by the interaction of $HgCl_2$ dissolved in EtOH
and $Bu^i_2S$ dissolved in benzene.

**[$Bu^n_2S \cdot 2HgCl_2$] or $C_8H_{18}Cl_4SHg_2$**. This compound melts at 112°C–113°C (Faragher
et al. 1929). It was isolated as white plates by the interaction of $HgCl_2$ dissolved in
EtOH or $H_2O$ and $Bu^n_2S$ dissolved in benzene.

**[$(Et_2S)_2(HgCl_2)_3$] or $C_8H_{20}Cl_6S_2Hg_3$**. This compound melts at 81°C–82°C (Biscarini
and Nivellini 1969).

**[$Hg_2(C_{10}H_{20}S_4)]Cl_4$ or $C_{10}H_{20}Cl_4S_4Hg_2$**. This compound crystallizes in the monoclinic structure with the lattice parameters $a = 808.0 \pm 0.2$, $b = 1138.9 \pm 0.2$, and
$c = 1070.6 \pm 0.2$ pm and $\beta = 92.15° \pm 0.01°$ and the experimental and calculated densities $2.80 \pm 0.05$ and $2.736$ $g \cdot cm^{-3}$, respectively (Alcock et al. 1976, 1978). Its crystals
were prepared by mixing 5 mL solution of $HgCl_2$ (0.272 g) and 1,4,8,11-tetrathiacy-
clotetradecan ($C_{10}H_{20}S_4$) (0.134 g) in boiling $MeNO_2$. The precipitate was collected,
washed with chloroform and $H_2O$, dried and recrystallized from $MeNO_2$.

**[$(Am^n_2S)(HgCl_2)_2$] or $C_{10}H_{22}Cl_4SHg_2$**. This compound melts at 86°C–87°C
(Biscarini and Nivellini 1969).

**[$(MeBu^tS)(HgCl_2)]_2$ or $C_{10}H_{24}Cl_4S_2Hg_2$**. This compound melts at 132°C–133°C
(Biscarini and Nivellini 1969).

**[$(C_{12}H_5)S \cdot HgS \cdot HgCl$] or $C_{12}H_5ClS_2Hg_2$**. The white crystals of this compound
were obtained by the interaction of $(C_{12}H_5)S \cdot HgS$ and $HgCl_2$ in the solution of
EtOH (Vogt 1861).

**[Hg$_2$(C$_{12}$H$_{24}$S$_4$)]Cl$_4$ or C$_{12}$H$_{24}$Cl$_4$S$_4$Hg$_2$.** To obtain this compound, a solution of 1,5,9,13-tetrathiacyclohexadecane (0.141 mM) in 8 mL of anhydrous nitromethane was added to a solution of HgCl$_2$ (0.104 mM) in a mixture of anhydrous nitromethane (1 mL) and absolute EtOH (1 mL) (Setzer et al. 1991). A colorless precipitate was immediately formed. The precipitate was washed with 5 mL of Et$_2$O and vacuum dried to give the title compound as a colorless powder.

**[{Hg(C$_{12}$H$_{18}$S$_2$)Cl$_2$}$_2$] or C$_{24}$H$_{36}$Cl$_4$S$_4$Hg$_2$.** This compound crystallizes in the triclinic structure with the lattice parameters $a = 755.2 \pm 0.4$, $b = 949.3 \pm 0.4$, and $c = 1220.1 \pm 0.4$ pm and $\alpha = 70.98° \pm 0.04°$, $\beta = 77.99° \pm 0.04°$, and $\gamma = 88.07° \pm 0.03°$ pm and a calculated density of 2.04 g·cm$^{-3}$ (Romero et al. 1996). Synthesis of the title compound was carried out using standard Schlenk techniques under dry N$_2$. HgCl$_2$ (0.44 mM) and concentrated HCl (0.5 mL) were added to a solution of 1,3-bis(ethylthiomethyl)benzene (0.44 mM) in MeOH (25 mL). The resulting mixture was refluxed for 2 h, then filtered, and allowed to stand at room temperature for 4 days. A white crystalline precipitate formed, which was filtered off and vacuum dried.

## 7.74   Hg–H–C–Br–S

Some compounds are formed in this system.

**[MeSHgBr] or CH$_3$BrSHg.** This compound crystallizes in the monoclinic structure with the lattice parameters $a = 777.0 \pm 0.9$, $b = 750.0 \pm 0.5$, and $c = 794.5 \pm 1.0$ pm and $\beta = 91.71° \pm 0.03°$ and a calculated density of 4.70 g·cm$^{-3}$ (Canty et al. 1979b). It precipitated as a white powder on addition of methanethiol (6.04 mM) to a solution of HgBr$_2$ (6.04 mM) in EtOH (20 mL). Crystals of the title compound formed upon slow evaporation of its solution in pyridine.

**[EtSHgBr] or C$_2$H$_5$BrSHg.** This compound melts at 198°C (Canty et al. 1978a). It was prepared by the addition of liquid ethanethiol to stirred ethanol solution of HgBr$_2$ in 1:1 molar ratio. The precipitate was collected, washed with EtOH, and dried over P$_2$O$_5$ in a vacuum.

**[Bu$^t$SHgBr] or C$_4$H$_9$BrSHg.** This compound melts at the temperature higher than 230°C (Canty et al. 1978a). It was prepared by the addition of liquid $t$-butanethiol to stirred ethanol solution of HgBr$_2$ in 1:1 molar ratio. The precipitate was collected, washed with EtOH, and dried over P$_2$O$_5$ in a vacuum.

**[Hg(C$_5$H$_4$S$_5$)Br$_2$]$_2$ or C$_5$H$_4$Br$_2$S$_5$Hg.** This compound was prepared by the following procedure (Dai et al. 1998). To an acetone solution (4 mL) of 4,5-ethylenedithio-1,3-dithiole-2-thione (C$_5$H$_4$S$_5$) (0.05 mM) Hg(ClO$_4$)·H$_2$O and Bu$^n_4$NBr (1:2 molar ratio, 0.1 mM) in acetone (2 mL) was added. The mixture was stirred for 20 min at room temperature under an Ar atmosphere. A precipitate was formed, and it was filtered and dried in argon.

**[MeHgC$_6$H$_4$BrS] or C$_7$H$_7$BrSHg.** This compound melts at 71°C (Sytsma and Kline 1973). To obtain the title compound, it is necessary to use an aqueous solution of MeHgOH prepared from MeHgI and excess amount of freshly precipitated Ag$_2$O. MeHgOH solution, from which AgI and excess Ag$_2$O have been removed by filtration, was added to a methanolic solution of *p*-bromothiophenol. The crude product was purified by the recrystallization in hexane/CH$_2$Cl$_2$.

**[HgBr$_2$(C$_{10}$H$_{20}$S$_4$)] or C$_{10}$H$_{20}$Br$_2$S$_4$Hg, dibromo(1,4,8,11-tetrathiacyclotetradecane)mercury(II)**. This compound crystallizes in the orthorhombic structure with the lattice parameters $a = 1035 \pm 2$, $b = 837 \pm 1$, and $c = 1989 \pm 3$ pm (Galešić et al. 1986).

**[(HgBr$_2$)$_2$(C$_{12}$H$_{24}$S$_6$)] or C$_{12}$H$_{24}$Br$_4$S$_6$Hg$_2$, tetrabromo(1,4,7,10,13,16-hexathiacyclooctadecane)dimercury(II)**. This compound crystallizes in the monoclinic structure with the lattice parameters $a = 1175.6 \pm 0.3$, $b = 1268.3 \pm 0.2$, and $c = 844.8 \pm 0.2$ pm and $\beta = 100.42° \pm 0.01°$ and a calculated density of 2.899 g·cm$^{-3}$ (Herceg and Matković-Čalogović 1992). Its colorless crystals were obtained by the crystallization from nitromethane.

**[Hg$_7$(SC$_6$H$_{11}$)$_{12}$Br$_2$] or C$_{72}$H$_{132}$Br$_2$S$_{12}$Hg$_7$**. This compound crystallizes in the rhombohedral structure with the lattice parameters $a = 2286.0 \pm 0.4$ and $c = 1482.3 \pm 0.2$ pm (hexagonal settings) at 200 K (Alsina et al. 1992a). It was prepared by mixing methanolic solutions of Na[SC$_6$H$_{11}$] and HgBr$_2$ with thiolate: mercury molar ratios in the range 1–1.75:1 in several different experiments. It precipitated slowly and was recrystallized from pyridine by slow evaporation to give colorless crystals of approximately cubic shape.

## 7.75  Hg–H–C–I–S

Some compounds are formed in this system.

**[Me$_3$S·HgI$_3$] or C$_3$H$_9$I$_3$SHg**. This compound crystallizes in the monoclinic structure with the lattice parameters $a = 851 \pm 3$, $b = 1557 \pm 5$, and $c = 1169 \pm 12$ pm and $\beta = 128.0° \pm 0.4°$ and the experimental and calculated densities 3.5 and 3.58 $\pm$ 0.4 g·cm$^{-3}$, respectively (Fenn 1966b) ($a = 847$, $b = 1551$, and $c = 1158$ pm and $\beta = 127.8°$ [Fenn et al. 1963]). It was prepared from Me$_2$S, MeI, and HgI$_2$, and needle-shaped crystals were grown from a solution of the title compound in acetone (Fenn et al. 1963, Fenn 1966b).

**[(Me$_3$S)$_2$·HgI$_4$] or C$_6$H$_{18}$I$_4$S$_2$Hg**. This compound crystallizes in the orthorhombic structure with the lattice parameters $a = 1322 \pm 5$, $b = 1669 \pm 4$, and $c = 913 \pm 5$ pm and the experimental and calculated densities 2.81 $\pm$ 0.05 and 2.85 $\pm$ 0.02 g·cm$^{-3}$, respectively (Fenn 1966a) ($a = 1306$, $b = 1681$, and $c = 927$ pm [Fenn et al. 1963]). Its crystals, which have the form of thick plates, were grown by slow evaporation of a solution of the title compound in acetone.

**[HgI$_2$(C$_{10}$H$_{20}$S$_4$)] or C$_{10}$H$_{20}$I$_2$S$_4$Hg, diiodo(1,4,8,11-tetrathiacyclotetradecane) mercury(II)**. This compound crystallizes in the orthorhombic structure with the lattice parameters $a = 1064.1 \pm 0.3$, $b = 850.7 \pm 0.3$, and $c = 2008.7 \pm 0.8$ pm and the experimental and calculated densities 2.62 and 2.640 g·cm$^{-3}$, respectively (Galešić et al. 1986). It was prepared by the reaction of C$_{10}$H$_{20}$S$_4$ and HgI$_2$ in acetone. The light-yellow crystalline product formed was filtered off, washed with acetone, and dried. Crystals were recrystallized from benzene.

## 7.76   Hg–H–As–O–S

**HgS–HgO–As$_2$O$_5$.** The phase diagram is not constructed. **Hg$_3$(AsO$_3$S)$_2$·2H$_2$O** quinary compound is formed in this system, which crystallizes in the orthorhombic structure and exists with two molecules of hydration water (Gigauri et al. 1999). Thermal decomposition of this compound can be represented by the next scheme: Hg$_3$(AsO$_3$S)$_2$·2H$_2$O (50°C–130°C) → Hg$_3$(AsO$_3$S)$_2$ (130°C–330°C) → Hg$_3$As$_2$O$_4$S (330°C–400°C) → HgS·Hg(AsO$_2$)$_2$ (400°C–460°C) → HgS. This compound was obtained by the interaction of Hg(NO$_3$)$_2$ with sodium monothioarsenate at room temperature: 3Hg(NO$_3$)$_2$ + 2Na$_3$AsO$_3$S = Hg$_3$(AsO$_3$S)$_2$·2H$_2$O + 6NaNO$_3$.

## 7.77   Hg–K–Rb–Ge–S

**HgS–K$_2$S–Rb$_2$S–GeS$_2$.** The phase diagram is not constructed. **K$_{1.6}$Rb$_{0.4}$Hg$_3$Ge$_2$S$_8$** quinary compound is formed in this system. It crystallizes in the monoclinic structure with the lattice parameters $a = 964.9 \pm 0.5$, $b = 839.3 \pm 0.5$, and $c = 972.0 \pm 0.2$ pm and $\beta = 95.08° \pm 0.03°$ and a calculated density of 4.658 g·cm$^{-3}$ (Kanatzidis et al. 1997a,b; Liao et al. 2003). This compound was prepared by the heating of mixture HgS + K$_2$S + Rb$_2$S + Ge + S at 400°C for 72 h with the next cooling to 385°C over 120 h and then to 165°C at a rate of 2°C/h.

## 7.78   Hg–K–C–N–S

**Hg(SCN)$_2$·KSCN.** This quinary compound is formed in this system. It crystallizes in the monoclinic structure with the lattice parameters $a = 1108.1$, $b = 407$, and $c = 1091.5$ pm and $\beta = 114°55'$ and the experimental and calculated densities 3.00 and 2.86 g·cm$^{-3}$ (Zhdanov and Sanadze 1952). It was prepared by the slow mixing of a diluted aqueous solution of HgSO$_4$ and concentrated solution of KSCN to obtain insoluble precipitate by heating this mixture. After cooling the solution, colorless needlelike crystals were obtained.

## 7.79   Hg–K–C–N–Br–S

**K$_2$[HgBr$_2$(SCN)$_2$].** This multinary compound, which crystallizes in the hexagonal structure with the lattice parameters $a = 1289.6 \pm 0.1$ and $c = 1129.3 \pm 0.1$ pm, is formed in this system (Gacemi et al. 2005). It was prepared from 1 mM of Hg(SCN)$_2$, which was dissolved in 50 mL of H$_2$O with stoichiometric amount of KBr. The mixture was refluxed during 1 h, and the resulting clear solution was slowly evaporated at room temperature, yielding transparent prisms of the title compound. Crystals were filtered off, washed with a small amount of MeOH, and dried at room temperature.

## 7.80   Hg–Rb–C–N–S

**RbHg(SCN)$_3$.** This quinary compound is formed in this system. It crystallizes in the monoclinic structure with the lattice parameters $a = 1140.67 \pm 0.50$, $b = 416.70 \pm 0.08$, and $c = 1110.26 \pm 0.36$ pm and $\beta = 115.20° \pm 0.02°$ and the experimental and calculated densities 3.17 and 3.20 g·cm$^{-3}$, respectively (Thiele and Messer 1976). RbHg(SCN)$_3$ was obtained by the interaction of RbSCN and Hg(SCN)$_2$ at stoichiometric ratio in a EtOH/H$_2$O solution.

## 7.81  Hg–Rb–O–Cl–S

**Rb$_3$[Hg(SO$_4$)$_2$][HgSO$_4$Cl]**. This quinary compound is formed in this system. It crystallizes in the monoclinic structure with the lattice parameters $a = 786.2 \pm 0.2$, $b = 979.0 \pm 0.4$, and $c = 1000.2 \pm 0.1$ pm and $\beta = 111.12° \pm 0.02°$ and the experimental and calculated densities 4.45 and 4.48 g·cm$^{-3}$, respectively (Bosson 1976). Its single crystals were prepared by melting RbCl, Rb$_2$SO$_4$, and anhydrous HgSO$_4$ in a molar ratio 1:1:2 in a sealed gold tube at 700°C. The structure of this compound may be considered as ionic made up of Hg(SO$_4$)$_3$$^{2-}$, HgSO$_4$Cl$^-$ and Rb$^+$ ions.

## 7.82  Hg–Cs–C–N–S

**CsHg(SCN)$_3$**. This quinary compound is formed in this system. It crystallizes in the monoclinic structure with the lattice parameters $a = 954.31 \pm 0.22$, $b = 1089.86 \pm 0.20$, and $c = 1996.49 \pm 0.36$ pm and $\beta = 108.39° \pm 0.04°$ and the experimental and calculated densities 3.38 and 3.42 g·cm$^{-3}$, respectively (Thiele et al. 1974, Thiele and Messer 1976). CsHg(SCN)$_3$ was obtained by the interaction of CsSCN and Hg(SCN)$_2$ at stoichiometric ratio in a EtOH/H$_2$O solution.

## 7.83  Hg–Cu–Tl–As–S

**HgS–Cu$_2$S–Tl$_2$S–As$_2$S$_3$**. The phase diagram is not constructed. **CuHg$_2$TlAs$_2$S$_6$** quinary compound (mineral routhierite) is formed in this system. Routhierite was defined originally as having a formula of the type MHgAsS$_3$, in which M is (Tl, Cu, Ag) with Tl dominant (Jambor and Grew 1990). It crystallizes in the tetragonal structure with the lattice parameters $a = 998.21 \pm 0.11$ and $c = 1131.22 \pm 0.12$ pm and a calculated density of 5.83 g·cm$^{-3}$ (Bindi 2008) ($a = 998.6 \pm 0.5$ and $c = 1134.8 \pm 0.8$ pm [Jambor and Grew 1990]; $a = 997.7 \pm 0.2$ and $c = 1129.0 \pm 0.3$ pm and a calculated density of 5.83 g·cm$^{-3}$ [Fleischer and Mandarino 1975]). This mineral contains Ag, Zn, and Sb as impurities.

## 7.84  Hg–Cu–C–N–S

**CuHg(SCN)$_4$**. This quinary compound exists in this system. It crystallizes in the orthorhombic structure with the lattice parameters $a = 765 \pm 1$, $b = 902 \pm 1$, and $c = 1517 \pm 2$ pm (Korczyński 1961, 1962) ($a = 903$, $b = 768$, and $c = 1515$ pm [Porai-Koshits 1963]) and the experimental and calculated densities 3.18 and 3.20 g·cm$^{-3}$, respectively. Its crystals were prepared by the crystallization from the solution of Cu(NO$_3$)$_2$ and K$_2$[Hg(SCN)$_4$] as the dark-green rhombic bipyramids.

## 7.85   Hg–Cu–As–Sb–S

$Cu_6(Cu_{5.26}Hg_{0.75})(As_{2.83}Sb_{1.17})S_{13}$. This quinary compound (mineral schwazite) is formed in this system (Pervukhina et al. 2010b). It crystallizes in the cubic structure with the lattice parameter $a = 1028.90 \pm 0.01$ pm and a calculated density of 4.942 g·cm$^{-3}$ for $Cu_{11.28}Hg_{0.72}As_{3.08}Sb_{0.92}S_{13}$ composition.

## 7.86   Hg–Ag–C–N–O–F–S

$[(CF_3S)_2Hg]\cdot[AgNO_3]$ or $C_2AgNO_3F_6S_2Hg$. This multinary compound is formed in this system. To prepare it, bis(trifluoromethylthio)mercury (2.18 g) was heated with a solution of $AgNO_3$ (1.84 g; 10 mL) (Downs et al. 1962). The precipitate was filtered off, washed with $H_2O$ and benzene, and dried in vacuo over $P_2O_5$. The title compound melted with decomposition at about 170°C and did not have appreciable dissociation pressure below 50°C. At 100°C, however, given after 10–12 h in vacuo, a sublimate of bis(trifluoromethylthio)mercury and an involatile residue of $AgNO_3$.

## 7.87   Hg–Ag–As–Sb–S

$AgHg(As,Sb)S_3$ quinary compound (mineral laffittite) is formed in this system (Pervukhina et al. 2010a). It crystallizes in the monoclinic structure with the lattice parameters $a = 775.60 \pm 0.03$, $b = 1133.40 \pm 0.04$, and $c = 666.50 \pm 0.03$ pm and $\beta = 115.233° \pm 0.004°$ and a calculated density of 6.078 g·cm$^{-3}$ for the $AgHgAs_{0.88}Sb_{0.12}S_3$ composition.

## 7.88   Hg–Ag–Cl–Br–I–S

$HgS–AgCl–AgBr–AgI$. The phase diagram is not constructed. The crystal structure analysis supports evidence from EPMA that a range of composition for mineral perroudite exists with the ideal structural formula of $Hg_5Ag_4S_5(Cl, Br, I)_4$ (Mumme and Nickel 1987). This range may be expressed as $Hg_{5-x}Ag_{4+x}S_{5-x}(Cl, Br, I)_{4+x}$ at $-1.4 < x < 1.4$. This mineral crystallizes in the orthorhombic structure with the lattice parameters $a = 1743 \pm 2$, $b = 1224 \pm 2$, and $c = 435 \pm 1$ pm for $Hg_{4.6}Ag_{4.4}S_{4.6}Cl_{2.0}Br_{0.8}I_{1.6}$ (Mumme and Nickel 1987) (for $Hg_{5.18}Ag_{4.75}S_{6.0}Cl_{2.40}Br_{0.98}I_{1.81}$ [Sarp et al. 1987]); $a = 1747 \pm 3$, $b = 1223 \pm 2$, and $c = 429 \pm 2$ pm for $Hg_{5.04}Ag_{4.03}S_{3.96}Cl_{1.55}Br_{0.87}I_{1.55}$; and $a = 1746 \pm 5$, $b = 1222 \pm 4$, and $c = 429 \pm 3$ pm for $Hg_{5.00}Ag_{4.2}S_{5.45}Cl_{1.55}Br_{0.25}I_{1.6}$ (Sarp et al. 1987). Average calculated density of perroudite is 6.92 g·cm$^{-3}$.

According to the data of Mason et al. (1992), $HgAg(Cl, Br, I)S$ multinary compound (mineral capgaronnite) exists in this system. It crystallizes in the orthorhombic structure with the lattice parameters $a = 680.3 \pm 0.8$, $b = 1287 \pm 1$, and $c = 452.8 \pm 0.7$ pm and a calculated density of 6.43 $\pm$ 0.01 g·cm$^{-3}$.

## 7.89   Hg–Tl–As–Sb–S

$HgS–Tl_2S–As_2S_3–Sb_2S_3$. The phase diagram is not constructed. $Hg_3Tl_4As_8Sb_2S_{20}$ quinary compound (mineral vrbaite) is formed in this system. It crystallizes in the orthorhombic structure with the lattice parameters $a = 1339.9$, $b = 2338.9$, and $c = 1128.7$ pm and the experimental and calculated densities $5.30 \pm 0.03$ and $5.40$ g·cm$^{-3}$, respectively (Cayé et al. 1967, Ohmasa and Nowacki 1971) ($a = 1335 \pm 5$, $b = 2332 \pm 5$, and $c = 1123 \pm 5$ pm and a calculated density of $5.29$ g·cm$^{-3}$ [Frondel 1941]).

## 7.90  Hg–Tl–O–Cl–S

**Tl$_3$[Hg(SO$_4$)$_2$][HgSO$_4$Cl]**. This quinary compound is formed in this system. It crystallizes in the monoclinic structure with the lattice parameters $a = 787.0 \pm 0.1$, $b = 975.2 \pm 0.1$, and $c = 994.5 \pm 0.1$ pm and $\beta = 110.72° \pm 0.01°$ and the experimental and calculated densities $6.22$ and $6.10$ g·cm$^{-3}$, respectively (Bosson 1976). Its single crystals were prepared by melting TlCl, Tl$_2$SO$_4$, and anhydrous HgSO$_4$ in a molar ratio 1:1:2 in a sealed gold tube at 700°C. The structure of this compound may be considered as ionic made up of Hg(SO$_4$)$_3$$^{2-}$, HgSO$_4$Cl$^-$ and Tl$^+$ ions.

## 7.91  Hg–C–Sn–N–Cl–S

**Hg(SCN)$_2$·SnCl$_4$**. This multinary compound, which melts at the temperature higher than 400°C, is formed in this system (Jain and Rivest 1969a). It was prepared by the interaction of the Hg(SCN)$_2$ suspension with SnCl$_4$ during 4–6 days with magnetic stirring at room temperature. A cream-colored solid powdery compound was obtained, which was filtered off, washed with the solvent, and dried in vacuo at room temperature. All experimental manipulations were carried out in a dry box continuously flushed with dry N$_2$.

## 7.92  Hg–C–Sn–N–Br–S

**2Hg(SCN)$_2$·SnBr$_4$**. This multinary compound, which melts at a temperature higher than 400°C, is formed in this system (Jain and Rivest 1969a). It was prepared by the interaction of the Hg(SCN)$_2$ suspension with SnBr$_4$ during 4–6 days with magnetic stirring at room temperature. A cream solid powdery compound was obtained, which was filtered off, washed with the solvent, and dried in vacuo at room temperature. All experimental manipulations were carried out in a dry box continuously flushed with dry N$_2$.

## 7.93  Hg–C–Ti–N–Cl–S

**Hg(SCN)$_2$·TiCl$_4$**. This multinary compound, which melts at 170°C–172°C with decomposition, is formed in this system (Jain and Rivest 1969a). It was prepared by the interaction of the Hg(SCN)$_2$ suspension with TiCl$_4$ during 4–6 days with magnetic stirring at room temperature. An orange solid powdery compound was obtained, which was filtered off, washed with the solvent, and dried in vacuo at room temperature. All experimental manipulations were carried out in a dry box continuously flushed with dry N$_2$.

## 7.94　Hg–C–Ti–N–Br–S

**Hg(SCN)₂·TiBr₄**. This multinary compound, which melts at the 168°C–170°C with decomposition, is formed in this system (Jain and Rivest 1969a). It was prepared by the interaction of the $Hg(SCN)_2$ suspension with $TiBr_4$ during 4–6 days with magnetic stirring at room temperature. A cream-colored solid powdery compound was obtained, which was filtered off, washed with the solvent, and dried in vacuo at room temperature. All experimental manipulations were carried out in a dry box and continuously flushed with dry $N_2$.

## 7.95　Hg–C–N–O–F–S

**CF₃SHgNO₃ or C₃NO₃F₃SHg, trifluoromethylthiomercuric nitrate**. This multinary compound is formed in this system. To obtain it, trifluoromethylthiomercuric chloride (CF₃SHgCl) (0.300 g) and AgNO₃ (0.150 g) were dissolved separately in $H_2O$ (10 mL) acidified with $HNO_3$ and the solution was mixed (Emeléus and Pugh 1960). AgCl was filtered off, and the filtrate was freeze-dried. The residue was sublimed in vacuum at 100°C and gave CF₃SHgNO₃.

## 7.96　Hg–C–N–F–S

**Hg[N(C₂F₃S)₄]₂ or C₁₆N₂F₂₄S₈Hg, [2,3,4,5-tetrakis(trifluoromethylthio)pyrrolyl] mercury**. This quinary compound is formed in this system. It crystallizes in the monoclinic structure with the lattice parameters $a = 1263.0 \pm 0.2$, $b = 1485.9 \pm 0.2$, and $c = 1755.3 \pm 0.1$ pm and $\beta = 102.917° \pm 0.007°$ and the experimental and calculated densities $2.33 \pm 0.01$ and 2.34 g·cm⁻³, respectively (Brauer 1979).

## 7.97　Hg–C–N–Cl–S

**HgClSCN**. This quinary compound is formed in this system. It crystallizes in the orthorhombic structure with the lattice parameters $a = 1040.18 \pm 0.05$, $b = 418.43 \pm 0.04$, and $c = 1020.21 \pm 0.05$ pm (Mosset and Bagieu-Beucher 2002) ($a = 1014$, $b = 422$, and $c = 1038$ pm and a calculated density of 4.30 g·cm⁻³ [Zvonkova and Zhdanov 1952]). To obtain it, a mixture of $HgCl_2$ and $Hg(SCN)_2$ was stirred in boiling water for 15 min. Upon cooling, crystallization of the prepared compound starts immediately. Transparent needles suitable for XRD are obtained after 2 days.

## 7.98　Hg–C–N–Br–S

**HgBrSCN**. This quinary compound is formed in this system. It crystallizes in the monoclinic structure with the lattice parameters $a = 623$, $b = 427$, and $c = 872$ pm and the experimental and calculated densities 4.74 and 4.77 g·cm⁻³, respectively (Zvonkova and Zhdanov 1952).

## 7.99　Hg–C–N–Mn–S

**HgMn(SCN)₄**. This quinary compound is formed in this system. It crystallizes in the tetragonal structure with the lattice parameters $a = 1130.99$ and $c = 425.37$ pm

(Joseph et al. 2006) ($a = 1132.16$ and $c = 427.19$ pm [Wang et al. 2002b]). $HgMn(SCN)_4$ was obtained by the interaction of $NH_4SCN$, $HgCl_2$, and $MnCl_2$ (4:1:1 molar ratio). It was prepared also by the following procedure. Heated and dissolved in 300 mL $H_2O$ with stirring were 304.5 g $NH_4SCN$ and 342.6 g $Hg(NO_3)_2 \cdot 2H_2O$ (Wang et al. 2001). To the colorless solution, 358 g $Mn(NO_3)_2$ aqueous solution (50 mass.%) was added. After the pale-red solution was left standing at room temperature for a while, precipitated $HgMn(SCN)_4$ (pale-greenish-yellow in color) was separated. To obtain relatively large and transparent single crystals, $NH_4SCN$ (1 M) was dissolved in hot $H_2O$ (500 mL); $Hg(NO_3)_2 \cdot H_2O$ (1 M) was added in the solution and heated to 70°C. To the pale-yellow solution, 198 g $MnCl_2$ was added and acidified with 0.1 M $HNO_3$ or HCl to pH 5 3. The resulting aqueous solution was allowed to slowly cool to 40°C. After 1 day, pale-greenish-yellow, transparent, and tetrahedral long-bar-shaped single crystals were crystallized out. Slow solvent evaporation method at a constant temperature is suitable for growing high-quality single crystals of the title compound (Wang et al. 2001, 2002b, Joseph et al. 2006).

## 7.100   Hg–C–N–Co–S

$HgCo(SCN)_4$. This quinary compound is formed in this system. It crystallizes in the tetragonal structure with the lattice parameters $a = 1110.9 \pm 0.2$ and $c = 437.9 \pm 0.2$ pm and the experimental and calculated densities $3.020 \pm 0.005$ and $3.018 \pm 0.003$ g·cm$^{-3}$, respectively (Jeffery 1947, Jeffery and Rose 1968). $HgCo(SCN)_4$ was obtained by the interaction of KSCN, $HgCl_2$, and $CoCl_2$ (4:1:1 molar ratio) (Potheher et al. 2008). Its crystals were produced by precipitation during crystallization, and they were deep-purple, square prisms (Jeffery 1947, Jeffery and Rose 1968).

## 7.101   Hg–C–N–Ni–S

$Hg_2Ni(SCN)_6$. This quinary compound is formed in this system. It crystallizes in the orthorhombic structure with the lattice parameters $a = 736 \pm 2$, $b = 1898 \pm 2$, and $c = 1285 \pm 2$ pm and the experimental and calculated densities $2.979 \pm 0.003$ and 2.990 g·cm$^{-3}$, respectively (Iizuka 1978). Its single crystals were obtained by slow cooling of the solution of the title compound from ca. 40°C to room temperature in a thermostat for about 1 week.

## 7.102   Hg–C–F–O–S

$CF_3SHgCF_3COO$ or $C_3O_2F_6SHg$, trifluoromethylthiomercuric acetate. This quinary compound is formed in this system. It melts at 100°C–101°C with some decomposition (Downs et al. 1962). It could be obtained by evaporating a solution of $Hg(CF_3COO)_2$ (2.23 g) and bis(trifluoromethylthio)mercury (2.10 g) in MeOH with the next vacuum sublimation.

## 7.103   Hg–C–F–Cl–S

$HgCl(SCF_3)$ or $CF_3ClSHg$, trifluoromethylthiomercuric chloride. This quinary compound is formed in this system. It crystallizes in the orthorhombic structure with the lattice parameters $a = 601.5 \pm 0.2$, $b = 816.6 \pm 0.4$, and $c = 2219 \pm 1$ pm and the experimental and calculated densities 4.11 and 4.05 g·cm$^{-3}$, respectively (Borrajo et al. 1976). It was prepared by the evaporation of the solution, which was obtained by the interaction of $Hg(SCF_3)_2$ (0.301 g) and $HgCl_2$ (0.204 g) (or $CF_3SCl$); each, in ether (10 mL), was mixed and shaken at room temperature (Haszeldine and Kidd 1953, Eмеléus and Pugh 1960, Downs et al. 1961). The product, after removal of ether, was sublimed in a vacuum at 40°C–50°C and was $HgCl(SCF_3)$. It crystallizes from chloroform as colorless, thin, rectangular plates (Borrajo et al. 1976). This compound could also be obtained by the interaction of $PCl_3$ (3.09 mM) and $Hg(SCF_3)_2$ (3.10 mM) in a sealed tube at room temperature (Eméleus and Pugh 1960).

## 7.104   Hg–Cl–Br–I–S

$HgS–HgCl_2–HgBr_2–HgI_2$. The phase diagram is not constructed. $Hg_3S_2(Cl, Br, I)_2$ quinary compound (mineral grechishchevite) exists in this system. It crystallizes in the orthorhombic structure with the lattice parameters $a = 1324.9 \pm 0.3$, $b = 1325.9 \pm 0.3$, and $c = 871.0 \pm 0.2$ pm and a calculated density of 7.180 g·cm$^{-3}$ (in the hexagonal structure with the lattice parameters $a = 1322.5 \pm 0.5$ and $c = 868.5 \pm 0.5$ pm and the calculated and experimental densities of 7.23 and 7.16 g·cm$^{-3}$, respectively [Vasilyev et al. 1989]) for $Hg_3S_2Cl_{0.5}BrI_{0.5}$ (Pervukhina et al. 2003, 2004) and $a = 1323.3 \pm 0.3$, $b = 1322.5 \pm 0.4$, and $c = 871.8 \pm 0.2$ pm and a calculated density of 7.236 g·cm$^{-3}$ for $Hg_3S_2Cl_{0.4}BrI_{0.6}$ (Borisov et al. 1999).

## REFERENCES

Abrahams B.F., Corbett M., Dakternieks D., Gable R.W., Hoskins B.F., Tiekink E.R.T., Winter G. N.M.R. studies of phosphine adducts of mercury and cadmium xanthates: Crystal and molecular structures of $Cd(S_2COPr^i)_2PPh_3$, $Hg(S_2COPr^i)_2PPh_3$ and $Hg(S_2COPr^i)_2P(c-C_6H_{11})_3$, *Aust. J. Chem.*, 39(12), 1993–2001 (1986).
Ahmad S., Sadaf H., Akkurt M., Sharif S., Khan I.U. Bis(1,3-dibuthylthiourea) dicyanidomercury(II), *Acta Crystallogr.*, E65(10), m1191–m1192 (2009).
Ahrland S., Björk N.-O. Metal halide and pseudohalide complexes in dimethylsulfoxide solution. I. Dimethyl sulfoxide solvates of silver(I), zinc(II), cadmium(II), and mercury(II), *Acta Chem. Scand.*, 28A(8), 823–828 (1974).
Ahuja I.S., Garg A. Complexes of some pyridines and anilines with zinc(II), cadmium(II) and mercury(II) thiocyanates, *J. Inorg. Nucl. Chem.*, 34(6), 1929–1935 (1972).
Alcock N.W., Herron N., Moore P. Comparison of the different modes of bonding of the macrocycle in μ-(1,4,8,11-tetrathiacyclotetradecane-$S^1S^4;S^8S^{11}$)-bis[dichloromercury-(II)] and aqua(1,4,8,11-tetrathiacyclotetradecane)mercury(II) perchlorate by x-ray structural analysis, *J. Chem. Soc. Dalton Trans.*, (5), 394–399 (1978).
Alcock N.W., Herron N., Moor P. New mode of bonding for 1,4,8,11-tetrathiacyclotetradecane (ttp): X-ray crystal structure of [($Cl_2Hg)_2$(ttp)], *J. Chem. Soc. Chem. Commun.*, (21), 886–887 (1976).
Alsina T., Clegg W., Fraser K.A., Sola J. [$Hg_7(SC_6H_{11})_{12}Br_2$], a novel mercury cage molecule with bridging and terminal thiolate ligands and with terminal and μ$_6$ bromide, *J. Chem. Soc. Chem. Commun.*, (14), 1010–1011 (1992a).
Alsina T., Clegg W., Fraser K.A., Sola J. Homoleptic cyclohexanthiolato complexes of mercury(II), *J. Chem. Soc. Dalton Trans.*, (8), 1393–1399 (1992b).

Altaf M., Stoeckli-Evans H., Ahmad S., Isab A.A., Al-Arfaj A.R., Malik M.R., Ali S. Crystal structure of trinuclear mercury(II) cyanide complex of tetramethylthiourea, [{(tetram ethylthiourea)$_2$Hg(CN)$_2$}$_2$·Hg(CN)$_2$], *J. Chem. Crystallogr.*, 40(12), 1175–1179 (2010a).

Altaf M., Stoeckli-Evans H., Batool S.S., Isab A.A., Ahmad S., Saleem M., Awan S.A., Shaheen M.A. Mercury(II) complexes of pyrrolidinedithiocarbamate, crystal structure of bis{[μ$^2$-(pyrrolidinedithiocarbamato-*S,S'*)(pyrrolidinedithiocarbamato-*S,S'*) mercury(II), *J. Coord. Chem.*, 63(7), 1176–1185 (2010b).

Alvarez H.M., Tran T.B., Richter M.A., Alyounes D.M., Rabinovich D., Tanski J.M., Krawiec M. Homoleptic Group 12 metal bis(mercaptoimidazolyl)borate complexes M(Bm$^R$)$_2$ (M = Zn, Cd, Hg), *Inorg. Chem.*, 42(6), 2149–2156 (2003).

Alyea E.C., Ferguson G., Restivo R.J. Structural studies of steric effects in phosphine complexes. The crystal and molecular structure of dithiocyanato(tricyclohexylphosphine) mercury(II), *J. Chem. Soc. Dalton Trans.*, (19), 1845–1848 (1977).

Amoedo-Portela A., Carballo R., Casas J.S., García-Martínez E., Gómez-Alonso C., Sánchez-González A., Sordo J., Vázquez-Lopez E.M. The coordination chemistry of the versatile ligand bis(2-pyridylthio)methane, *Z. anorg. und allg. Chem.*, 628(5), 939–950 (2002).

Anjali K.S., Vittal J.J., Dean P.A.W. Syntheses, characterization and thermal properties of [M(Spy)$_2$(SpyH)$_2$] (M = Cd and Hg; Spy$^-$ = pyridine-4-thiolate; SpyH = pyridinium-4-thiolate) and [M(SpyH)$_4$](ClO$_4$)$_2$ (M = Zn, Cd and Hg), *Inorg. Chim. Acta*, 351, 79–88 (2003).

Asao T., Kikuchi (née Ninomiya) Y. The preparation of some tropone complex, *Chem. Lett.*, 1(5), 413–416 (1972).

Aucken I., Drago R.S. Mercury(II) chloride thiourea complexes, *Inorg. Synth.*, 6, 26–30 (1960).

Bach R.D., Weibel A.T. Nuclear magnetic resonance studies on anion-exchange reactions of alkylmercury mercaptides, *J. Am. Chem. Soc.*, 98(20), 6241–6249 (1976).

Baggio R., Garland M.T., Perec M. Synthesis and solid-state structural characterization of *N,N'*-dicyclohexyldithiooxamide complexes of HgX$_2$ (X = SCN or Cl), *J. Chem. Soc. Dalton Trans.*, (6), 987–992 (1995).

Bailey T.D., Banda R.M.H., Craig D.C., Dance I.G., Ma I.N.L., Scudder M.L. Mercury polysulfide complexes, [Hg(S$_x$)(S$_y$)]$^{2-}$: Syntheses, $^{199}$Hg NMR studies in solution, and crystal structure of (Ph$_4$P)$_4$[Hg(S$_4$)$_2$]Br$_2$, *Inorg. Chem.*, 30(2), 187–191 (1991).

Banda R.M.H., Craig D., Dance I.G., Scudder M. Tetrahedral HgS$_4$ and linear HgS$_2$ coordination in the crystal structure of Na$_2$Hg$_3$S$_4$(H$_2$O)$_2$, *Polyhedron*, 10(1), 41–45 (1991).

Bandoli G., Clemente D.A., Sindellari L., Tondello E. Preparation, properties, and crystal structure of dichlorobis-(*O*-ethyl thiocarbamate) mercury(II), *J. Chem. Soc. Dalton Trans.*, (5), 449–452 (1975).

Baraldi M., Malavasi W., Grandi R. Mercury(II) dibromo-bis-(3-amino-5-mercapto-1,2,4-triazole): Synthesis, crystal structure, and infrared characterization, *J. Chem. Thermodyn.*, 26(1), 63–66 (1996).

Barrera H., Bayon J.C., Gonzalez-Duarte P., Sola J., Viñas J.M., Briansó J.L., Briansó M.C., Solans X. Synthesis and vibrational spectra of [RSHg$^{II}$]$^{2+}$(ClO$_4$)$_2$ complexes. Crystal and molecular structure of bis(*N*-methylpiperidinium-4-thiolato)-mercury(II) perchlorate, *Polyhedron*, 1(7–8), 647–654 (1982).

Baur R., Schellenberg M., Schwarzenbach G. Aromatenkomplexe von Quecksilber, *Helv. Chim. Acta*, 45(3), 775–783 (1962).

Bayon J.C., Casals I., Gaete W., Gonzalez-Duarte P., Ros J. Complexes of 1-methyl-4-mercaptopiperidine with zinc(II), cadmium(II) and mercury(II) halides, *Polyhedron*, 1(2), 157–161 (1982).

Beauchamp A., Pazdernik L., Rivest R. Bispyridine adduct of cobalt(II) mercury(II) tetrathiocyanate, *Acta Crystallogr.*, B32(2), 650–652 (1976).

Beauchamp A.L., Saperas B., Rivest R. Crystal structure of *cis*-dithiocyanatobis(1,10-phen-anthroline)mercury(II), a six-coordinate mercury complex, *Can. J. Chem.*, 49(21), 3579–3580 (1971).

Beauchamp A.L., Saperas B., Rivest R. Structure cristalline et moléculaire du *cis*-dithiocya natobis(phénanthroline-1,10)mercure(II), Hg(SCN)$_2$(C$_{12}$H$_8$N$_2$)$_2$, *Can. J. Chem.*, 52(16), 2923–2927 (1974).

Beck W., Stetter K.H., Tadros S., Schwarzhans K.E. Darstellung, IR- und [19]F-KMR-Spektren von Pentafluorphenylmercapto-Metallkomplexen, *Chem. Ber.*, 100(12), 3944–3954 (1967).

Behrens U., Hofmann K., Klar G. Chalkogenolat-Ionen und ihre Derivate, III. Kristall- und Molekülstruktur von Tetraphenylphosphonium-tris(tellurophenolato)mercurat(II), *Chem. Ber.*, 110(11), 3672–3677 (1977).

Bell N.A., Branston N.N., Clegg W., Creighton J.R., Cucurull-Sánchez L., Elsegood M.R.J., Raper E.S. Complexes of heterocyclic thiones and Group 12 metals. Part 3. Preparation and characterisation of 1:2 complexes of mercury(II) halides with 1-methylimidazo-line-2(3*H*)-thione: The crystal structures of [(HgX$_2$)(1-methylimidazoline-2(3*H*)-thi-one)$_2$] (X = Cl, Br, I) at 160 K, *Inorg. Chim. Acta*, 303(2), 220–227 (2000a).

Bell N.A., Branston N.N., Clegg W., Parker L., Raper E.S., Sammon C., Constable C.P. Complexes of heterocyclic thiones and Group 12 metals. Part four. Preparation and characterisation of 1:1 complexes of mercury(II) halides with 1,3-thiazolidine-2-thi-one and 1,3-benzothiazoline-2-thione. Crystal structure of the discrete trans dimer [(μ-dibromo)bis(trans{(bromo)(1,3-thiazolidine-2-thione)}mercury(II))], *Inorg. Chim. Acta*, 319(1–2), 130–136 (2001a).

Bell N.A., Clegg W., Creighton J.R., Raper E.S. Complexes of heterocyclic thiones and Group 12 metals. Part 2: The chemical and electrochemical synthesis of mercury(II) complexes of 1-methylimidazoline-2(3*H*)-thionate. The crystal structure of *trans*-[bis-{(η¹-*S*-1-methylimidazoline-2(3*H*)-thione)(η¹-*S*-1-methylimidazoline-2(3*H*)-thionate)-(μ$_2$-*S*,*N*-1-methylimidazoline-2-thionate)mercury(II)}] at 160 K, *Inorg. Chim. Acta*, 303(1), 12–16 (2000b).

Bell N.A., Coles S.J., Constable C.P., Hibbs D.E., Hursthouse M.B., Mansor R., Raper E.S., Sammon C. Complexes of heterocyclic thiones and Group 12 metals. Part 5. Reactions of 1,3-thiazolidine-2-thione and benzo-1,3-thiazoline-2-thione with mercury(II) halides in a 2:1 ratio. Crystal structures of bis(1,3-thiazolidine-2-thione) mercury(II) bromide and bis(benzo-1,3-thiazolinato)mercury(II), *Inorg. Chim. Acta*, 323(1–2), 69–77 (2001b).

Bell N.A., Crouch D.J., Jaffer N.E. Coordination complexes of 2-thienyl- and 2-furyl-mercu-rials, *Appl. Organometal. Chem.*, 18(3), 135–138 (2004).

Bellon P.L., Demartin F., Devillanova F.A., Isaia F., Verani G. Reactivity and mechanisms in the reactions between mercury(II) halides and 5,5-dimethylimidazolidine-2-thione-4-one. Crystal structure of a mercury bromide complex, *J. Coord. Chem.*, 18(4), 253–261 (1988).

Benedetti A., Fabretti A.C., Preti C. Structure, IR, and NMR spectra of tetrakis(4-methylpi-peridinedithiocarbamate) dimercury(II), *J. Cryst. Spect. Res.*, 18(6), 685–692 (1988).

Bermejo E., Carballo R., Castiñeiras A., Domínguez R., Liberta A.E., Maichle-Mössmer C., Salberg M.M., West D.X. Synthesis, structural characteristics and biological activities of complexes of Zn$^{II}$, Cd$^{II}$, Hg$^{II}$, Pd$^{II}$, and Pt$^{II}$ with 2-acetylpyridine 4-methylthiosemi-carbazone, *Eur. J. Inorg. Chem.*, (6), 965–973 (1999).

Bermejo E., Castiñeiras A., Garcia I., West D.X. Spectral and structural studies of mercury(II) complexes of 2-pyridineformamide *N*(4)-dimethylthiosemicarbazone, *Polyhedron*, 22(8), 1147–1154 (2003).

Bermejo E., Castiñeiras A., García-Santos I., West D.X. Structural and coordinative variability in zinc(II), cadmium(II), and mercury(II) complexes of 2-pyridineformamide 3-hexameth-yleneiminylthiosemicarbazone, *Z. anorg. und allg. Chem.*, 630(7), 1096–1109 (2004).

Beurskens P.T., Cras J.A, Noordik J.H., Spruijt A.M. Crystal and molecular structure of diiodo-*N,N,N',N'*-tetramethylthiuramdisulphidemercury(II), *J. Cryst. Mol. Struct.*, 1(1), 93–98 (1971).

Bindi L. Routhierite, Tl(Cu,Ag)(Hg,Zn)$_2$(As,Sb)$_2$S$_6$, *Acta Crystallogr.*, C64(12), i95–i96 (2008).

Biscarini P., Foresti E., Pradella G. Organothiometallic compounds. Crystal structure and spectroscopic properties of (isopropylthio)mercury(II) chloride, *J. Chem. Soc. Dalton Trans.*, (5), 953–957 (1984).

Biscarini P., Fusina L., Nivellini G.D. Oxidation of addition compounds between organic sulphides and mercury(II) chloride, *J. Chem. Soc. A: Inorg. Phys. Theor.* 1128–1131 (1971).

Biscarini P., Fusina L., Nivellini G., Mangia A., Pelizzi G. Crystal and molecular structure of the 1:1 adduct between diphenyl sulphoxide and mercury(II) chloride, *J. Chem. Soc. Dalton Trans.*, (2), 159–161 (1973).

Biscarini P., Fusina L., Nivellini G., Mangia A., Pelizzi G. Crystal structure and spectroscopic properties of mercury(II) halide complexes. Part II. The dimethyl sulphoxide–mercury(II) chloride (2/3) adduct, *J. Chem. Soc. Dalton Trans.*, (17), 1846–1849 (1974).

Biscarini P., Fusina L., Nivellini G., Pelizzi G. Three-co-ordinate mercury: Crystal structure and spectroscopic properties of trimethylsulphonium mercurate(II), *J. Chem. Soc. Dalton Trans.*, (7), 664–668 (1977).

Biscarini P., Fusina L., Nivellini G., Pelizzi G. Crystal structure and spectroscopic properties of mercury(II) halide complexes. Part 3. The di-*n*-butyl sulphoxide–mercury(II) chloride (1/2) adduct, *J. Chem. Soc. Dalton Trans.*, (4), 1024–1027 (1981).

Biscarini P., Nivellini G.D. Addition compounds of organic sulphides with mercuric chloride. Infrared spectra in the solid state, *J. Chem. Soc. A: Inorg. Phys. Theor.*, 2206–2210 (1969).

Block E., Brito M., Gernon M., McGowty D., Kang H., Zubieta J. Mercury(II) and methylmercury(II) complexes of novel sterically hindered thiolates: [13]C and [199]Hg NMR studies and the crystal and molecular structures of [MeHg(SC$_6$H$_2$-2,4,6-Pr$^i_3$)], [Hg(SC$_6$H$_4$-2-SiMe$_3$)$_2$], [Hg(2-SC$_5$H$_3$N-3-SiMe$_3$)$_2$], and [Hg{(2-SC$_6$H$_4$)$_2$SiMe$_2$}]$_2$, *Inorg. Chem.*, 29(17), 3172–3181 (1990).

Bochmann M, Webb K.J., Powell A.K. Synthesis and structure of [Hg(SC$_6$H$_2$Bu$^t_3$)$_2$(py)]: A T-shaped complex of mercury, *Polyhedron*, 11(5), 513–516 (1992).

Bond A.M., Colton R., Hollenkamp A.F., Hoskins B.F., McGregor K. Voltammetric, coulometric, mercury-199 NMR, and other studies characterizing new and unusual mercury complexes produced by electrochemical oxidation of mercury(II) diethyldithiocarbamate. Crystal and molecular structure of octakis(*N,N*-diethyldithiocarbamato)pentamercury(II) perchlorate, *J. Am. Chem. Soc.*, 109(7), 1969–1980 (1987).

Book L., Chieh C. The structure of *catena*-{di-μ-chloro-tetrakis[μ-(*N,N*-diethyldithiocarbamato-*S,S'*)]trimercury(II)}, [Hg$_3$(C$_5$H$_{10}$NS$_2$)$_4$Cl$_2$], *Acta Crystallogr.*, B36(2), 300–303 (1980).

Borisov S.V., Magarill S.A., Romanenko G.V., Pervukhina N.V. Features of the crystal chemistry of the rare mercury minerals of a supergene origin—Possible sources of mercury in surface water and air basins [in Russian], *Khim. v interesah ustoych. razvit*, 7(5), 497–503 (1999).

Borrajo J., Varetti E.L., Aymonino P.J., Silva A.M. Crystallographic data for mercuric chloride perfluoromethanethiolate, HgCl(SCF$_3$), *Z. Kristallogr.*, 144(1–6), 110–115 (1976).

Bosson B. The crystal structures of Tl$_3$[Hg(SO$_4$)$_2$][HgSO$_4$Cl] and Rb$_3$[Hg(SO$_4$)$_2$][HgSO$_4$Cl], *Acta Chem. Scand. A*, 30(4), 241–248 (1976).

Bowmaker G.A., Dance I.G., Dobson B.C., Rogers D.A. Syntheses and vibrational spectra of some tris(alkanethiolato)mercurate(II) complexes, and crystal structure of the hexakis(methanethiolato)dimercurate(II) dianion, *Aust. J. Chem.*, 37(8), 1607–1618 (1984).

Bowmaker G.A., Dance I.G., Harris R.K., Henderson W., Laban I., Scudder M.L., Oh S.-W. Crystallographic, vibrational and nuclear magnetic resonance spectroscopic characterization of the [(PhS)$_2$Hg(μ-SPh)$_2$Hg(SPh)$_2$]$^{2-}$ ion, *J. Chem. Soc. Dalton Trans.*, (11), 2381–2388 (1996).

Bränden C.I. The crystal structure of 2HgCl$_2$(C$_2$H$_5$)$_2$S, *Ark. Kemi*, 22(1), 83–91 (1964a).

Bränden C.I. The crystal structure of HgCl$_2$·C$_4$H$_8$S, *Ark. Kemi*, 22(6), 495–500 (1964b).

Brauer D.J. Structure of bis[2,3,4,5-tetrakis(trifluoromethylthio)pyrrolyl]mercury, Hg[N(CSCF$_3$)$_4$]$_2$, *Acta Crystallogr.*, B35(8), 1770–1773 (1979).

Bravo J., Casas J.S., Castaño M.V., Gayoso M., Mascarenhas Y.P., Sanchez A., Santos de C.O.P., Sordo J. Methyl- and phenylmercury(II) derivatives of 2-mercaptobenzothiazole. Crystal structure of (2-benzothiazolylthio)methylmercury(II), *Inorg. Chem.*, 24(21), 3435–3438 (1985).

Brinkhoff H.C., Cras J.A., Steggerda J.J., Willemse J. The oxidation of dithiocarbamato complexes of nickel, copper and zinc, *Rec. trav. chim. Pays-Bas*, 88(6), 633–640 (1969).

Brinkhoff H.C., Dautzenberg J.M.A. Complexes of *N,N*-dialkyldithiocarbamate esters with mercury(II) dihalides, *Rec. trav. chim. Pays-Bas*, 91(1), 117–125 (1972).

Brinkhoff H.C., Grotens A.M., Steggerda J.J. A NMR study of *N,N,N′,N′*-tetra-alkylthiuram disulfides and their complexes with Zn, Cd and Hg, *Rec. trav. chim. Pays-Bas*, 89(1), 11–17 (1970).

Brodersen von K., Humnel H.-U. Die Kristallstruktur von MgHg(SCN)$_4$·2H$_2$O, *Z. anorg. und allg. Chem.*, 491(1), 34–38 (1982).

Brotherton P.D., Epstein J.M., White A.H., Willis A.C. Crystal structure of di-μ-chloro-dichlorobis(methyl pyrrolidine-1-carbodithioate)dimercury (II), *J. Chem. Soc. Dalton Trans.*, (21), 2341–2343 (1974).

Brotherton P.D., Healy P.C., Raston C.L., White A.H. Crystal structure of chlorobis(thiourea) mercury(II) chloride, *J. Chem. Soc. Dalton Trans.*, (3), 334–336 (1973).

Brotherton P.D., White A.H. Crystal structure of a new form (β) of tetrakis(thiourea) mercury(II) chloride, *J. Chem. Soc. Dalton Trans.*, (23), 2696–2698 (1973a).

Brotherton P.D., White A.H. Crystal structure of a new mercury(II) complex dichloromercury-2/3 thiourea, *J. Chem. Soc. Dalton Trans.*, (23), 2698–2700 (1973b).

Cannas M.,. Devillanova F.A., Marongiu G., Verani G. X-ray and infrared analysis of Hg(II) halide complexes with *N*-mono and *N,N′*-disubstituted imidazolidine-2-thione, *J. Inorg. Nucl. Chem.*, 43(10), 2383–2388 (1981).

Canty A.J., Kishimoto R. Mercury(II) and organomercury(II) complexes of thiols and dithiols, including british anti-lewisite, *Inorg. Chim. Acta*, 24, 109–122 (1977).

Canty A.J., Kishimoto R., Tyson R.K. Synthesis and spectroscopic studies of RSHg$^{II}$ complexes, including complexes of Bu$^t$SHgX (X=Cl,Br) and RSHgO$_2$CMe (R=Me,Et) with pyridine, *Aust. J. Chem.*, 31(3), 671–676 (1978a).

Canty A.J., Marker A. Methylmercury(II) derivatives of pyridine-2(1H)-thione, *Aust. J. Chem.*, 29(6), 1383–1387 (1976).

Canty A.J.,Raston C.L.,White A.H.Crystal and molecular structure of polymeric MeSHgO$_2$CMe and MeSHgO$_2$CMe·C$_5$H$_5$N and the tetranuclear complex (Bu$^t$S)$_4$Cl$_4$Hg$_4$(C$_5$H$_5$N)$_2$, *Aust. J. Chem.*, 31(3), 677–684 (1978b).

Canty A.J., Raston C.L., White A.H. Crystal structure of methanethiolatomercury(II) chloride, *Aust. J. Chem.*, 32(5), 1165–1168 (1979a).

Canty A.J., Raston C.L., White A.H. Structural studies of RSHg$^{II}$ complexes containing (–Hg–SR–)$_n$ rings and chains, *Aust. J. Chem.*, 32(2), 311–320 (1979b).

Canty A.J., Tyson R.K. Interaction of MeSHgO$_2$CMe with imidazole and related ligands, *Inorg. Chim. Acta*, 29, 227–230 (1978).

Carlin R.L. Roitman J., Dankleff M., Edwards J.O. Six-coordinate mercury(II), *Inorg. Chem.*, 1(1), 182–184 (1962).

Carson G.K., Dean P.A.W. The metal NMR spectra of thiolate and phenylselenolate complexes of zinc(II) and mercury(II), *Inorg. Chim. Acta*, 66, 157–161 (1982).

Carty A.J., Malone S.F., Taylor N.J. The selenium-mercury interaction: Synthesis, spectroscopic and x-ray structural studies of methylmercury-selenourea complexes, *J. Organomet. Chem.*, 172(2), 201–211 (1979).

Casals I., González-Duarte P., Clegg W., Foces-Foces C., Cano F.H., Martínez-Ripoll M., Gómez M., Solans X. Synthesis and structures of tetranuclear 2-(dimethylamino)ethanethiolato complexes of zinc, cadmium and mercury involving both primary and secondary metal-halogen bonding, *J. Chem. Soc. Dalton Trans.*, (10), 2511–2518 (1991a).

Casals I., González-Duarte P., Clegg W. Preparation of complexes of the 3-trimethylammonio-1-propanethiolato ligand, $[M\{S(CH_2)_3NMe_3\}_2](PF_6)_2$, M = Zn, Cd, Hg. Crystal structures of monomeric and polymeric forms of the mercury complex, *Inorg. Chim. Acta*, 184(2), 167–175 (1991b).

Casals I., González-Duarte P., Sola J., Font-Bardia M., Solans L., Solans X. Polymeric thiolate complexes of Group 12 metals. Crystal and molecular structures of *catena*-[μ-(3-dimethylammonio-1-propanethiolate)]-dichlorocadmium(II) and bis[3-(dimethylammonio)propyl] disulphide tetrabromocadmate(II), *J. Chem. Soc. Dalton Trans.*, (10), 2391–2395 (1987).

Casals I., González-Duarte P., Sola J., Miraitlles C., Molins E. Mercury–sulphur stretching frequencies, synthesis and x-ray crystal structure of *catena*-μ-(3-dimethylammonio-1-propanthiolato)dichloromercury(II), *Polyhedron*, 7(24), 2509–2514 (1988).

Casas J.S., Castellano E.E., Ellena J., Haiduc I., Sánchez A., Semeniuc R.F., Sordo J. Supramolecular self-assembly in the crystal structures of methylmercury xanthates, MeHgS(S)COR, R = Et, *i*Pr and CH$_2$Ph, *Inorg. Chim. Acta*, 329(1), 71–78 (2002).

Casas J.S., Castellano E.E., Ellena J., Haiduc I., Sánchez A., Sordo J. The crystal and molecular structure of mercury(II) bis(isopropyl)dithiophosphate, $Hg[S_2P(OPr^i)_2]_2$, revisited: New comments about its supramolecular self-organization, *J. Chem. Crystallogr.*, 29(7), 831–836 (1999).

Casas J.S., Castiñeiras A., Haiduc I., Sánchez A., Sordo J., Vázquez-López E.M. Supramolecular self-organization in methyl(diphenylphosphinothioato)mercury(II), $[HgMe\{S(O)PPh_2\}]$, a polymer containing $Hg_2O_2P_2S_2$ eight-membered rings interconnected through secondary Hg⋯O bonds, *Polyhedron*, 16(5), 781–787 (1997).

Casas J.S., Castiñeiras A., Sánchez A., Sordo J., Vázquez-López E.M. Dicyclohexyldithiophosphinates and diethyldithiophosphinates of methylmercury(II) and phenylmercury(II): Crystal and molecular structure of $[HgPh(S_2PEt_2)]$, *J. Organomet. Chem.*, 468(1–2), 1–6 (1994).

Cassidy I., Garner M., Kennedy A.R., Potts G.B.S., Reglinski R., Slavin P.A., Spicer M.D. The preparation and structures of Group 12 (Zn, Cd, Hg) complexes of the soft tripodal ligand hydrotris(methimazolyl)borate (Tm), *Eur. J. Inorg. Chem.*, (5), 1235–1239 (2002).

Castiñeiras A., García I., Bermejo E., Ketcham K.A., West D.X., El-Sawaf A.K. Coordination of Zn$^{II}$, Cd$^{II}$, and Hg$^{II}$ by 2-pyridineformamide-3-piperidyl-thiosemicarbazone, *Z. anorg. und allg. Chem.*, 628(2), 492–504 (2002).

Castiñeiras A., Hiller W., Strahle J., Bravo J., Casas J.S., Gayoso M., Sordo J. Methyl- and phenyl-mercury(II) derivatives of 2-mercaptopyridine. Crystal and molecular structure of methyl(pyridine-2-thiolato)mercury(II), *J. Chem. Soc. Dalton Trans.*, (9), 1945–1948 (1986).

Cayé, R., Picot P., Pierrot R., Permingeat F. Nouvelles données sur la vrbaïte, sa teneur en mercure, *Bull. Soc. Fr. Minéral.*, 90, 185–191(1967).

Cecconi F., Ghilardi C.A., Midollini S., Orlandini A. Synthesis and x-ray structure of the thioethereal mercury(II) complex $[N(CH_2CH_2SCHMe_2)_3HgCl]_2(Hg_2Cl_6)$, *Inorg. Chim Acta*, 269(2), 274–278 (1998).

Cheung K.K., McEwen R.S., Sim G.A. Structures of some mercury(II) complexes, *Nature*, 205(4969), 383–384 (1965).

Cheung K.K., Sim G.A. Complexes of mercury. Part I. X-ray analysis of 1,6-dithiacyclodeca-*cis*-3,*cis*-8-dienebis(mercuric chloride), *J. Chem. Soc.*, 5988–6004 (1965).

Chieh C. A three-coordinate complex of mercury: Crystal structure of bis(iodo-*N*,*N*-diethyldithiocarbamatomercury(II)), *Can. J. Chem.*, 55(1), 65–69 (1977a).

Chieh C. Synthesis and structure of dichlorobis(thiosemicarbazide)mercury(II), *Can. J. Chem.*, 55(9), 1583–1587 (1977b).

Chieh C. Synthesis and crystal structure of diiodotetraethylthiuramdisulfidemercury(II), *Can. J. Chem.*, 55(7), 1115–1119 (1977c).

Chieh C. Crystal structure of dibromotetraethylthiuramdisulfidemercury(II), *Can. J. Chem.*, 56(7), 974–975 (1978a).

Chieh C. Crystal structure of methyl,2-mercaptopyrimidinatomercury(II), *Can. J. Chem.*, 56(4), 560–563 (1978c).

Chieh C. Crystal structure of mercury(II) bromide dithiocarbamate, $[HgBr_2(S_2CNEt_2)Hg(S_2CNEt_2)]_n$, an inorganic polymer, *Can. J. Chem.*, 56(4), 564–566 (1978b).

Chieh C., Cheung S.K. A crystallographic and spectroscopic study of mercury(II) dithiocarbamate, *Can. J. Chem.*, 59(18), 2746–2749 (1981).

Chieh C., Cowell D.H. Synthesis and crystal structure of thiosemicarbazidedichloromercury (II), *Can. J. Chem.*, 55(22), 3898–3900 (1977).

Chieh C., Leung L.P.C. Crystal structure and vibrational spectra of methyl *N*,*N*-diethyldithiocarbamato mercury(II), *Can. J. Chem.*, 54(19), 3077–3084 (1976).

Chieh C., Moynihan K.J. Xanthate and dithiocarbamate complexes of group IIb elements, and an interesting relationship between two mercury(II) ethylxanthate phases, *Acta Crystallogr.*, B36(6), 1367–1371 (1980).

Choudhury S., Dance I.G., Guerney P.J., Rae A.D. Tetramethylammonium tetrakis(4-chloro-benzenethiolato)mercurate(II). Synthesis and molecular structure, *Inorg. Chim. Acta*, 70, 227–230 (1983).

Chou J.-H., Kanatzidis M.G. Hydrothermal synthesis of $(Ph_4P)_2[Hg_2As_4S_9]$ and $(Me_4N)$ $[HgAs_3S_6]$. Extended chains and layers based on the condensation of $[AsS_3]^{3-}$ units, *Chem. Mater.*, 7(1), 5–8 (1995).

Christou G., Folting K., Huffman J.C. Mononuclear, three-coordinate metal thiolates: Preparation and crystal structures of $[NBu^n_4][Hg(SPh)_3]$ and $[NPr^n_4][Pb(SPh)_3]$, *Polyhedron*, 3(11), 1247–1253 (1984).

Clarke J.H.R., Woodward L.A. Vibrational spectra of the tris(methylmercuric) sulphonium and tris(methylmercuric)oxonium ions, *Spectrochimica Acta, Part A*, 23(7), 2077–2087 (1967).

Colombini G., Preti C. Complexes of thiomorpholin-3-one and thiazolidine-2-thione with zinc(II), cadmium(II) and mercury(II) thiocyanates, acetates and fluoborates, *J. Inorg. Nucl. Chem.*, 37(5), 1159–1165 (1975).

Costello W.R., McPhail A.T., Sim G.A. Complexes of mercury. Part II. X-ray analysis of dichloro-(1,3,5-tri-thian)mercury(II), *J. Chem. Soc. A: Inorg. Phys. Theor.*, 1190–1193 (1966).

Cox M.J., Tiekink E.R.T. Structural variations in the mercury(II) bis(1,1-dithiolate)s: The crystal and molecular structure of $[Hg(S_2CNMe_2)_2]$, *Z. Kristallogr.*, 212(7), 542–544 (1997).

Cox M.J., Tiekink E.R.T. Structural diversity in the mercury(II) bis(*O*-dialkyldithiocarbonate) compounds, *Z. Kristallogr.*, 214(8), 486–491 (1999b).

Cox M.J., Tiekink E.R.T. Structural diversity in the mercury(II) bis(*N*,*N*-dialkyldithiocarbamate) compounds: An example of the importance of considering crystal structure when rationalising molecular structure, *Z. Kristallogr.*, 214(9), 571–579 (1999a).

Cox M.J., Tiekink E.R.T. The crystal structure of monomeric bis(dicyclohexyldithiocarbamato) mercury(II), *Main Group Met. Chem.*, 23(12), 793–794 (2000).

Cras J.A., Willemse J., Gal A.W., Hummelink-Peters B.G.M.C. Preparation, structure and properties of compounds containing the dipositive tri-copper hexa($N$,$N$-di-$n$-butyldithiocarbamato) ion, compounds with copper oxidation states II and III, *Rec. trav. chim. Pays-Bas*, 92(6), 641–650 (1973).

Criado A., Conde A., Moreno E., Márquez R. The crystal and molecular structure of Cl(SH) Hg[SC(NH$_2$)$_2$], *Z. Kristallogr.*, 141(3–4), 193–202 (1975).

Cristiani F., Demartin F., Devillanova F.A., Diaz A., Isaia F., Verani G. Reactivity of mercury(II) perchlorate towards 5,5-dimethylimidazolidine-2-thione-4-one. Structure of bis(5,5-dimethylimidazolidine-2-thione-4-one)mercury(II) perchlorato triaquo, *J. Coord. Chem.*, 21(2), 137–146 (1990).

Czakis-Sulikowska M., Sołoniewicz R. The crystalline complex compounds containing thiourea and thiocyanates ions. I. Systems Hg$^{++}$–CS(NH$_2$)$_2$–Zn$^{++}$–SCN$^-$ and Hg$^{++}$–CS(NH$_2$)$_2$–Co$^{++}$–SCN$^-$ [in Polish], *Rocz. Chem.*, 37(11), 1405–1410 (1963).

Dai J., Munakata M., Bian G.-Q., Xu Q.-F. Kuroda-Sowa T., Maekawa M. Mercury(II) complexes of 4,5-ethylenedithio-1,3-dithiole-2-thione, having S⋯S and S⋯Cl contact assembled ribbon structure, *Polyhedron*, 17(13–14), 2267–2270 (1998).

Dalley N.K., Larson S.B. Structure of dichloro(1,4,7,10-tetraoxa-13,16-dithiacyclooctadecane-$S$,$S'$)mercury(II)-mercury dichloride, *Acta Crystallogr.*, B37(12), 2225–2227 (1981).

Dalziel J.A.W., Holding A.F. le C., Watts B.E. Addition compounds of triphenylphosphine sulphide with mercury(II) and copper(I) halides, *J. Chem. Soc. A: Inorg. Phys. Theor.*, 358–361 (1967).

Das A.K., Seth S. The crystal and molecular structure of (HgL$_2$)$_n$ (L=4,6-dimethylpyrimidine-2-thiolate)—An unusual helical supramolecular assembly in solid phase: In search of a new antidote to mercury poisoning, *J. Inorg. Biochem.*, 65(3), 207–218 (1997).

Davidović N., Matković-Čalogović D., Popović Z., Vedrina-Dragojević I. Di-μ-chloro-bis[(2-ammoniobenzenethiolato-$S$)chloromercury(II)], *Acta Crystallogr.*, C54(5), 574–576 (1998).

Davies A.G., Peddle G.J.D. $N$-mercuricarbamates, *Chem. Commun.*, (5), 960 (1965).

Davis A.R., Murphy C.J., Plane R.A. Triphenylphosphine and triphenylarsine complexes of mercury(II) thiocyanate, nitrate, and perchlorate, *Inorg. Chem.*, 9(2), 423–425 (1970).

Dean P.A.W., Manivannan V. Synthesis and multinuclear ($^{13}$C, $^{77}$Se, $^{125}$Te, $^{199}$Hg) magnetic resonance spectra of adamantane-like anions of mercury(II), [(μ-ER)$_6$(HgX)$_4$]$^{2-}$ (E=S, Se, Te; X=Cl, Br, I), *Inorg. Chem.*, 29(16), 2997–3002 (1990b).

Dean P.A.W., Manivannan V. A multinuclear ($^{13}$C, $^{31}$P, $^{77}$Se, $^{125}$Te, $^{199}$Hg) magnetic resonance study of the adamantane-like clusters [(μ-ER)$_6$(HgE'PR'$_3$)$_4$]$^{2+}$ (E=S, Se, Te; E'=S, Se; R'=Ph, c-C$_6$H$_{11}$) and the precursors [Hg(E'PR'$_3$)$_n$]$^{2+}$, *Can. J. Chem.*, 68(2), 214–222 (1990a).

Dean P.A.W., Vittal J.J., Trattner M.H. New adamantane-like mercury–chalcogen cages. Synthetic and multinuclear ($^{31}$P, $^{77}$Se, $^{199}$Hg) NMR study of [(μ-ER)$_6$(HgL)$_4$]$^{2+}$ (E=S or Se; L=tertiary phosphine or arsine) and related species with mixed ligands, *Inorg. Chem.*, 26(25), 4245–4251 (1987).

De Filippo D., Devillanova F., Preti C., Verani G. Thiomorpholin-3-one and thiazolidine-2-thione complexes with group IIB metals, *J. Chem. Soc. A: Inorg. Phys. Theor.*, 1465–1468 (1971).

Dehnicke K. Das Schwingungsspektrum von Phenylquecksilberthiocyanat, *J. Organomet. Chem.*, 9(1), 11–17 (1967).

Devillanova F.A., Verani G. Reactions of 1,3-oxazolidine-2-thione with zinc(II), cadmium(II) and mercury(II) halides, *J. Coord. Chem.*, 7(3), 177–180 (1978).

Dillen J., Lenstra A.T.H., Haasnoot J.G., Reeddijk J. Transition metal thiocyanates of 5,7-dimethyl[1,2,4]triazolo[1,5-a]-pyrimidine studied by spectroscopic methods. The crystal structures of diaquabis (dimethyltriazolopyrimidine-$N^3$)bis(thiocyanato-$N$) cadmium(II) and bis(dimethyltriazolopyrimidine-$N^3$)bis(thiocyanato-$S$)-mercury(II), *Polyhedron*, 2(3), 195–201 (1983).

Downs A.J., Ebsworth E.A.V., Emeléus H.J. The Raman spectra of bis(trifluoromethylthio) mercury and derived compounds, *J. Chem. Soc.*, 3187–3192 (1961).

Downs A.J., Ebsworth E.A.V., Emeléus H.J. Spectroscopic studies of the reaction of bis(trifluoromethylthio)mercury with salts of oxyacids, *J. Chem. Soc.*, 1254–1260 (1962).

Dupont L., Dideberg O., Rulmont A., Nyssen Ph. Structure du dichloro(thiocyanato-$S$) mercurate(II) d'ammonium, *Acta Crystallogr.*, C39(3), 323–326 (1983).

Du Z.-X., Li J.-X. Crystal structure of *catena*-poly{[chloromercury(II)]-μ-[2-(2-pyridylmethylamino)ethanesulfonato-$\kappa^4N,N',O{:}O'$]}, [Hg(C$_8$H$_{11}$N$_2$O$_3$S) (Cl)]$_n$, *Z. Kristallogr. New Cryst. Struct.*, 223(3), 211–212 (2008)

Emeléus H.J., Pugh H. Reactions of bistrifluoromethylthiomercury and trifluoromethylthio-mercuric chloride, *J. Chem. Soc.*, 1108–1112 (1960).

Esmhosseini M., Safari N., Amani V. Crystal structure of [hydroxy-methoxy-di(2-pyridyl) methane-$\kappa^2N,N'$](μ$_2$-thiocyanato-$N{:}S$)(thiocyanato-$\kappa S$)mercury(II), Hg(C$_{12}$H$_{12}$N$_2$O$_2$) (SCN)$_2$, *Z. Kristallogr. New Cryst. Struct.*, 226(4), 541–542 (2011).

Fan J., Yin X., Zhang W.-G., Zhang Q.-J., Lai C.-S., Tiekink E.R.T., Fan Y., Nuang M.Y. Synthesis, structure and thermal stability of metal complexes with $N,N'$-dibenzyl dithiocarbamate [in Chinese], *Acta Chim. Sin.*, 62(17), 1626–1634 (2004).

Faragher W.F., Morrell J.C., Comay S. Interaction of alkyl sulfides and salts of mercury, *J. Am. Chem. Soc.*, 51(9), 2774–2781 (1929).

Faraglia G., Graziani R., Sindellari L., Forsellini E., Casellato U. Mercury(II) halide complexes with $N,N$-dimethyl $O$-ethylthiocarbamate. The crystal structure of Hg(DMTC) Cl$_2$, *Inorg. Chim. Acta*, 58, 167–172 (1982).

Faraglia G., Sindellari L., Zarli B. Complexes of mercury(II) halides with $N$-methyl $O$-ethylthiocarbamate, *Inorg. Chim. Acta*, 53, L245–L247 (1981).

Fenn R.H., Oldham J.W.H., Phillips D.C. Crystal structure of (CH$_3$)$_3$S·HgI$_3$ and the configuration of the HgI$_3^-$-ion, *Nature*, 198(4878), 381–382 (1963).

Fenn R.H. The structure of mercuri-iodide ions. II. Bistrimethylsulphonium mercuritetraiodide [(CH$_3$)$_3$S]$_2$HgI$_4$, *Acta Crystallogr.*, 20(1), 24–27 (1966a)

Fenn R.H. The structure of mercuri-iodide ions. I. Trimethylsulphonium mercuritriiodide (CH$_3$)$_3$SHgI$_3$, *Acta Crystallogr.*, 20(1), 20–23 (1966b).

Fleischer M., Mandarino J.A. New mineral names, *Am. Mineral.*, 60(9–10), 945–947 (1975).

Frondel C. Unit cell and space group of vrbaite (Tl(As,Sb)$_3$S$_5$), seligmannite (CuPbAsS$_5$) and samsonite (Ag$_4$MnSb$_2$S$_6$), *Am. Mineral.*, 26(1), 25–28 (1941).

Gacemi A., Benbertal D., Bagieu-Beuchera M., Lecchi A., Mosset A. Two mercury chloro thiocyanate complexes: NH$_4$[HgCl$_2$(SCN)] and NH$_4$[HgCl(SCN)$_2$], *Z. anorg. und allg. Chem.*, 629(14), 2516–2520 (2003).

Gacemi A., Benbertal D., Gautier-Luneau I., Mosset A. Crystal structure of dipotassium dibromo-bis(thiocyanato)mercurate(II), K$_2$[HgBr$_2$(SCN)$_2$], *Z. Kristallogr. New Cryst. Struct.*, 220(3), 311–312 (2005).

Gagnon C., Beauchamp A.L. Structure du bis(thiocyanato)(triphénylphosphine)mercure(II), *Acta Crystallogr.*, B35(1), 166–169 (1979).

Gal A.W., Beurskens G., Cras J.A., Beurskens P.T., Wilklemse J. Preparation, properties and crystal structure of hexakis($N,N$-di-$n$-butyldithiocarbamato)(copper(III), -copper(II), -gold(III)) tetrabromodi-μ-bromodimercurate(II), [(Cu$_2$Au)(Bu$_2$dtc)$_6$][Hg$_2$Br$_6$], *Rec. trav. chim. Pays-Bas*, 95(7–8), 157–159 (1976).

Galešić N., Herceg M., Sevdić D. Structure of diiodo(1,4,8,11-tetrathiacyclotetradecane) mercury(II), *Acta Crystallogr.*, C42(5), 565–568 (1986).

Gallagher J.F., Alyea E.C., Ferguson G. Structural studies of steric effects in phosphine complexes: Dimers of bis-{dithiocyanato(1-phenyl-dibenzophosphole)mercury(II)} associate with significant Hg–N and Hg–S interdimer interactions, *Croat. Chem. Acta*, 72(2–3), 243–250 (1999).

Gigauri R.D., Machhoshvili R.I., Helashvili G.K., Gigauri R.I., Indzhiya M.A. Ag, Zn, Cd, Hg(II) monothioarsenates [in Russian], *Zhurn. neorgan. khim.*, 44(1), 26–29 (1999).

Gopalakrishnan J., Patel C.C. Diphenyl sulphoxide complexes of some divalent metal ions, *Inorg. Chim. Acta*, 1, 165–168 (1967).

Govindaswamy N., Moy J., Millar M., Koch S.A. A distorted $[Hg(SR)_4]^{2-}$ complex with alkanethiolate ligands: The fictile coordination sphere of monomeric $[Hg(SR)_x]$ complexes, *Inorg. Chem.*, 31(26), 5343–5344 (1992).

Green M., Prince P., Gardener M., Steed J. Mercury(II) *N,N'*-methyl-phenylethyl-dithiocarbamate and its use as a precursor for the room-temperature solution deposition of β-HgS thin films, *Adv. Mater.*, 16(12), 994–996 (2004).

Gregg D.C., Iddles H.A., Stearns, Jr. P.W. Triphenylmethyl aryl sulfides. I. Hydrogenolysis with Raney nickel. Reactions with mercuric chloride, *J. Org. Chem.*, 16(2), 246–252 (1951).

Grønbaek R., Dunitz J.D. Struktur des Hexarhodano-kobalt(II)-di-Quecksilber(II)-Benzol-Komplexes, *Helv. Chim. Acta*, 47(7), 1889–1897 (1964).

Gruff E.S., Koch S.A. Trigonal-planar $[M(SR)_3]^{1-}$ complexes of cadmium and mercury. Structural similarities between mercury–cysteine and cadmium–cysteine coordination centers, *J. Am. Chem. Soc.*, 112(3), 1245–1247 (1990).

Harding (née Aitken) M.M. The crystal structure of the mercury dithizone complex, *J. Chem. Soc.*, 4136–4143 (1958).

Haszeldine R.N., Kidd J.M. Reactions of fluorocarbon radicals. Part XI. Synthesis and some reactions of trifluoromethanethiol and trifluoromethanesulphenyl chloride, *J. Chem. Soc.*, 3219–3225 (1953).

Healy P.C., White A.H. Crystal structure of bis(*NN*-diethyldithiocarbamato)mercury(II), *J. Chem. Soc. Dalton Trans.*, (3), 284–287 (1973).

Henkel G., Betz P., Krebs B. $[Hg_3(SCH_2CH_2S)_4]^{2-}$ and $\{[Hg_2(SCH_2CH_2S)_3]^{2-}\}_n$: Examples of trinuclear and quasi-isolated binuclear polymeric mercury thiolate anions, *J. Chem. Soc. Chem. Commun.*, (21), 1498–1499 (1985).

Herceg M., Matković-Čalogović D. Tetrabromo(1,4,7,10,13,16-hexathiacyclooctadecane) dimercury(II), *Acta Crystallogr.*, C48(10), 1779–1782 (1992).

Hounslow A.M., Tiekink E.R.T. Correlations between nuclear magnetic resonance spectra and crystal structure. III. A $^{13}C$ nuclear magnetic resonance study in the solid state of bis(xanthato) complexes of mercury(II); The crystal and molecular structure of bis(*n*-propyldithiocarbonato)mercury(II), *J. Crystallogr. Spectrosc. Res.*, 21(2), 133–137 (1991).

Hou W., Yuan D., Xu D., Zhang N., Yu W., Liu M., Tao X., Sun S., Jiang M. A new organometallic nonlinear optical material—Triallylthiourea mercury bromide (ATMB) crystal: Growth and structure, *J. Cryst. Growth*, 133(1–2), 71–74 (1993).

Howie R.A., Tiekink E.R.T., Wardell J.L., Wardell S.M.S.V. Complementary supramolecular aggregation via O–H···O hydrogen-bonding and Hg···S interactions in bis[*N,N'*-di(2-hydroxyethyl)-dithiocarbamato-*S,S'*]mercury(II): $Hg[S_2CN(CH_2CH_2OH)_2]_2$, *J. Chem. Crystallogr.*, 39(4), 293–298 (2009).

Hubert J., Beauchamp A.L., Rivest R. Crystal and molecular structure of dithiocyanato(triphenylarsine)mercury(II), *Can. J. Chem.*, 53(22), 3383–3387 (1975).

Iizuka M. Crystal structure of nickel mercuric thiocyanate, $NiHg_2(SCN)_6$, Annual Report of Bunkyo University, No. 11, Bunkyo University, Koshigaya, Saitama Prefecture, Japan, pp. 65–68 (1978).

Isab A.A., Fettouhi M., Malik M.R., Ali S., Fazal A., Ahmad S. Mercury (II) cyanide of thioureas and the crystal structure of $[Hg(N\text{-methylthiourea})_2(CN)_2]$, *Korrd. khim.* 37(3), 181–186 (2011).

Isab A.A., Perzanowski H.P. $^1H$, $^{13}C$ and $^{199}Hg$ NMR studies of the complexation of $HgCl_2$ by imidazolidine-2-thione and its derivatives, *J. Coord. Chem.*, 21(3), 247–252 (1990).

Isab A.A., Perzanowski H.P. $^1H$, $^{13}C$ and $^{199}Hg$ NMR studies of the –NHCS-containing ligands with mercuric halides, *Polyhedron*, 15(14), 2397–2401 (1996).

Isab A.A., Wazeer M.I.M. Complexation of Zn(II), Cd(II) and Hg(II) with thiourea and selenourea: A $^1H$, $^{13}C$, $^{15}N$, $^{77}Se$ and $^{113}Cd$ solution and solid-state NMR study, *J. Coord. Chem.*, 58(6), 529–537 (2005).

Ito M., Iwasaki H. The structure of the monomeric form of mercury(II) N,N-diisopropyldithiocarbamate [bis(N,N-diisopropyldithiocarbamato)mercury(II)], *Acta Crystallogr.*, B35(11), 2720–2721 (1979).

Ivanov A.V., Korneeva E.V., Bukvetskii B.V., Goryan A.S., Antzutkin O.N., Forshling W. Structural organization of mercury(II) and copper(II) dithiocarbamates from EPR and $^{13}C$ and $^{15}N$ MAS NMR spectra and x-ray diffraction, *Russ. J. Coord. Chem.*, 34(1), 59–69 (2008).

Iwasaki H. Crystal structure of dichloro-tetrakis(diethyldithiocarbamato)-trimercury, $Hg_3Cl_2(S_2CNEt_2)_4$, *Chem. Lett.*, 1(11), 1105–1106 (1972).

Iwasaki H., Ito M., Kobayashi K. Coexistence of monomeric and dimeric complex molecules in a crystal: The crystal structure of the β-form of mercury(II) N,N-diisopropyldithiocarbamate, *Chem. Lett.*, 7(12), 1399–1402 (1978).

Iwasaki H. The crystal structure of dimeric and monomeric forms of mercury(II) N,N-diethyldithiocarbamate, $Hg_2(S_2CNEt_2)_4$ and $Hg(S_2CNEt_2)_2$, *Acta Crystallogr.*, B29(10), 2115–2124 (1973).

Jain S.C., Rivest R. Complexes of Group IV halides with mercuric thiocyanate and mercuric cyanide as ligands, *Can. J. Chem.*, 47(12), 2209–2214 (1969a).

Jain S.C., Rivest R. Coordination complexes of mercury(II) pseudo-halides. Preparation and infrared spectra studies of $Hg^{II}$ pseudo-halides with N,N,N',N'-tetramethylethylenediamine, 3-cyanopyridine and dimethylsulfoxide as ligands, *Inorg. Chim. Acta*, 3, 552–558 (1969b).

Jain S.C., Rivest R. Coordination complexes of mercury(II) pseudo-halides. Preparation and infrared spectral studies of the complexes of $Hg^{II}$ pseudo-halides with 1,10-phenanthroline, 2,2'-bipyridine and triphenylphosphine as ligands, *Inorg. Chim. Acta*, 4, 291–295 (1970).

Jambor J.L., Grew E.S. New mineral names, *Am. Mineral.*, 75(7–8), 931–937 (1990).

Jeffery J.W. Crystal structure of $Co[Hg(CNS)_4]$, *Nature*, 159(4044), 610 (1947).

Jeffery J.W., Rose K.M. The structure of cobalt mercury thiocyanate, $Co(NCS)_4Hg$, *Acta Crystallogr.*, B24(5), 653–662 (1968).

Jellinek F., Lagowski J.J. Some addition compounds of bis(trifluoromethylthio) mercury, *J. Chem. Soc.*, 810–814 (1960).

Jones T.E., Sokol L.S.W.L., Rorabacher D.B., Glick M.D. Encircling of mercury(II) by a macrocyclic ligand: X-ray crystal structure of 1,5,9,13-tetrathiacyclohexadecanemercury (II) perchlorate, *J. Chem. Soc. Chem. Commun.*, (3), 140–141 (1979).

Joseph G.P., Philip J., Rajarajan K., Rajasekar S.A., Pragasam A.J.A., Thamizharasan K., Kumar S.M.R., Sagayaraj P. Growth and characterization of an organometallic nonlinear optical crystal of manganese mercury thiocyanate (MMTC), *J. Cryst. Growth*, 296(1), 51–57 (2006).

Kanatzidis M.G., Liao J.H., Marking G.A. Alkali metal quaternary chalcogenides and process for the preparation of thereof, USA Patent 5, 614, 128. Appl. No. 606565. Filed February 26, 1996; Data of Patent March 25, 1997 (1997a).

Kanatzidis M.G., Liao J.H., Marking G.A. Alkali metal quaternary chalcogenides and process for the preparation of thereof, USA Patent 5, 618, 471. Appl. No. 606886. Filed February 26, 1996; Data of Patent April 08, 1997 (1997b).

Kennedy B.P., Lever A.B.P. Studies of the metal–sulfur bond. Complexes of the pyridine thiols, *Can. J. Chem.*, 50(21), 3488–3507 (1972).

King M.G., McQuillan G.P. Metal halide complexes of triphenylphosphine sulphide and some related ligands, *J. Chem. Soc. A: Inorg. Phys. Theor.*, 898–901 (1967).

Kinoshita H., Ouchi A. Synthesis, and crystal and molecular structure of bis(*N*-methylformamide)tetrakis(thiocyanato)cobalt(II)mercury(II), $CoHg(SCN)_4(HCONHCH_3)_2$, *Bull. Chem. Soc. Jpn.*, 59(11), 3495–3499 (1986).

Kinoshita H., Ouchi A. The synthesis and the crystal and molecular structure of tris(*N*-ethyl-2-pyrrolidone)tetrakis(thiocyanato)cobalt(II)mercury(II), $[CoHg(SCN)_4(C_6H_{11}NO)_3]$, *Bull. Chem. Soc. Jpn.*, 61(9), 3350–3352 (1988).

Kinoshita H., Shimoi M., Ouchi A. The crystal and molecular structure of tetrakisthiocyanatobis(triphenylphosphine)cobalt(II)mercury(II) in dimer form, $[CoHg(SCN)_4\{(C_6H_5)_3P\}_2]_2$, *Bull. Chem. Soc. Jpn.*, 58(4), 1304–1307 (1985b).

Kinoshita H., Shimoi M., Ouchi A., Sato S. The synthesis, and the crystal and molecular structure of [diaquatetrakis(thiocyanato)cobalt(II)mercury(II)]-*N*,*N*-dimethylacetamide (1/2), $CoHg(SCN)_4(H_2O)_2 \cdot 2\{CH_3CON(CH_3)_2\}$, *Bull. Chem. Soc. Jpn.*, 58(10), 2782–2785 (1985a).

Kinoshita H., Shimoi M., Ouchi A. The structure of diaquatetrakis(thiocyanato)cobalt(II)mercury(II)-*N*-methyl-2-pyrrolidine(1/2), $CoHg(SCN)_4(H_2O)_2 \cdot 2(C_3H_6CONCH_3)$, *Bull. Chem. Soc. Jpn.*, 59(4), 1253–1254 (1986).

Kim Y.-Y., Lee H.-J., Nam H.J., Noh D.-Y. Mononuclear three-coordinate mercury(II)-thiolate complex: Synthesis and x-ray crystal structure, *Bull. Korean Chem. Soc.*, 22(1), 17–18 (2001).

Klevtsova R.F., Glinskaya L.A., Zemskova S.M., Larionov S.V. Crystal structure of mixed ligand compound $[HgPhen\{(C2H_5)_2NCS_2\}_2]$ and nature of intermolecular interaction in the structures of $[MPhen\{(C_2H_5)_2NCS_2\}_2]$ (M = Zn, Cd, Hg) complexes [in Russian], *Zh. Strukt. Khim.*, 43(2), 346–353 (2002).

Korczyński A. The crystal structure of copper tetrathiocyanatomercurate $Cu[Hg(SCN)_4]$ [in Polish], *Rocz. Chem.*, 35(4), 1173–1174 (1961).

Korczyński A. The crystal structure of $Cu[Hg(SCN)_4]$ [in Polish], *Rocz. Chem.*, 36(10), 1539–1541 (1962).

Korczyński A. An X-ray study of crystalline $Hg[SC(NH_2)_2]_4Co(SCN)_4$ [in Polish], *Rocz. Chem.*, 37(12), 1645 (1963a).

Korczyński A. The crystal structure of $Hg[SC(NH_2)_2]_2(SCN)_2$. Part I [in Polish], *Rocz. Chem.*, 37(12), 1647–1648 (1963b).

Korczyński A. The crystal structure of $Hg(H_2NCSNH_2)_2(SCN)_2$ [in Polish], *Rocz. Chem.*, 40(4), 547–565 (1966).

Korczyński A. The crystal structure of $Hg[H_2NSCNH_2)_3Cl_2$ [in Polish], *Rocz. Chem.*, 42(7–8), 1207–1220 (1968).

Korczyński A., Porai-Koshits M.A. The crystal structure of $Hg[SC(NH_2)_2]_4Co(SCN)_4$, *Rocz. Chem.*, 39(11), 1567–1584 (1965).

Kozarek W.J., Fernando Q. X-ray crystal and molecular structure of the adduct of mercury(II) chloride with dehydrodithizone, *J. Chem. Soc. Chem. Commun.*, (10), 604–605 (1972).

Kozarek W.J., Fernando Q. Crystal and molecular structure of $C_{13}H_{10}N_4S \cdot HgCl_2$, the adduct of anhydro-5-mercapto-2,3-dyphenyltetrazolium hydroxide and mercury(II) chloride, *Inorg. Chem.*, 12(9), 2129–2131 (1973).

Kubo K., Kato N., Mori A., Takeshita H. 20-Dicyanomethylene-5,8,11,14-tetraoxa-2,17-dithiabicyclo[16.4.1]-tricosa-1(23),18,21-triene and its mercury(II) dichloride complex, *Acta Crystallogr.*, C56(6), 644–646 (2000).

Kubo K., Mori A., Kato N., Takeshita H. Preferential coordination of mercury(II) to oxygen over sulfur atoms in a tropone-attached dithiocrown ether, *Acta Crystallogr.*, C52(7), 1656–1658 (1996).

Lai C.S., Lim Y.X., Yal T.C., Tiekink E.R.T. Molecular paving with zinc thiolates, *CrystEngChem*, 4(99), 596–600 (2002).

Lai C.S., Tiekink E.R.T. Refinement of the crystal structure of bis[bis(*N,N*-diethyldithio-carbamato)mercury(II)], [Hg(S₂CNEt₂)₂]₂, *Z. Kristallogr. New Cryst. Struct.*, 217(4), 593–594 (2002).

Lai C.S., Tiekink E.R.T. Bis(pyrrolinedithiocarbamato)mercury(II), *Appl. Organometal. Chem.*, 17(2), 143 (2003a).

Lai C.S., Tiekink E.R.T. Chloro(*N,N*-diethyldithiocarbamato)(4,7-dimethyl-1,10-phenanth-roline)mercury(II)hemi-chloroformsolvate, *Appl. Organometal. Chem.*, 17(2), 141–142 (2003b).

Lai C.S., Tiekink E.R.T. Phenyl(N,N-di-n-propyldithiocarbamato)mercury(II), *Appl. Organometal. Chem.*, 17(3), 194 (2003c).

Lai C.S., Tiekink E.R.T. Supramolecular association in organomercury(II) 1,1-dithio-lates. Complementarity between Hg···S and hydrogen bonding interactions in organomercury(II) 2-amino-cyclopent-1-ene-1-carbodithioates, *CrystEngComm*, 5(44), 253–261 (2003d).

Lai C.S., Tiekink E.R.T. Bis(*N,N*-dibenzyldithiocarbamato)mercury(II), *Appl. Organometal. Chem.*, 18(2), 104 (2004).

Larionov S.V., Kirichenko V.N., Zemskova C.M., Oglezneva I.M. Synthesis of complexes of diethyldithiocarbamates of zinc(II), cadmium(II), mercury(II) with a nitrogen-contain-ing ligands, and the study of their sublimation [in Russian], *Koord. khim.*, 16(1), 79–84 (1990).

Lavertue P., Hubert J., Beauchamp A.L. Crystal structure of dichlorobis(6-mercaptopurine) mercury(II), *Inorg. Chem.*, 15(2), 322–325 (1976).

Lawton S.L. Crystal and molecular structure of the polymeric complex mercuric O,O′-diisopropylphosphorodithioate, Hg[*i*-C₃H₇O)₂PS₂]₂, *Inorg. Chem.*, 10(2), 328–335 (1971).

Leßmann F., Beyer L., Sieler J. Synthesis and x-ray structure of the first chloro-bridged thiourea mercury(II) complex [C₆H₅C(OCH₃)NC(S)N(C₂H₅)₂HgCl₂]₂, *Inorg. Chem. Commun.*, 3(2), 62–67 (2000).

Lecher H. Beiträge zum Valenzproblem des Schwefels. II. Über das Thiophenol-quecksilber, *Ber. dtsch. chem. Gesl.*, 48(2), 1425–1432 (1915).

Lecher H. Über das Phenylmercapto-quecksilberchlorid, *Ber. dtsch. chem. Ges.*, 53(4), 568–577 (1920).

Lee H.-J., Nam H.J., Noh D.-Y. Dimeric mercury(II) chloride complex of sulfur-rich ligand: Synthesis and x-ray crystal structure of *trans*-[{Hg(μ-Cl)Cl(dPhEDT-DTT)}₂]·(CH₃CN)₂, *Bull. Korean Chem. Soc.*, 20(11), 1368–1370 (1999).

Liao J.-H., Marking G.M., Hsu K.F., Matsushita Y., Ewbank M.D., Borwick R., Cunningham P., Rosker M.J., Kanatzidis M.G. α- and β-A₂Hg₃M₂S₈ (A = K, Rb; M = Ge, Sn): Polar quaternary chalcogenides with strong nonlinear optical response, *J. Am. Chem. Soc.*, 125(31), 9484–9493 (2003).

Liesic von J., Klar G. Chalkogenolat-Ionen und ihre Derivate. II. Komplexe von Chalko-genophenolat-Ionen mit Quecksilber(II), *Z. anorg. und allg. Chem.*, 435(1), 103–112 (1977).

Lobana T.S., Sharma R., Sharma R., Sultana R., Butcher R.J. Metal derivatives of 1,3-imid-azolidine-2-thione with divalent d¹⁰ metal ions. (Zn–Hg): Synthesis and structures, *Z. anorg. und allg. Chem.*, 634(4), 718–723 (2008).

Lopez-Garzon R., Gutierrez-Valero M.D., Godino-Salido M.L., Keppler B.K., Nuber B. Spectroscopic studies of metal-pyrimidine complexes. Crystal structures of 4,6-dimethyl-2-thiopyrimidine complexes with Zn(II) and Cd(II), *J. Coord. Chem.*, 30(2), 111–123 (1993).

López-Torres E., Mendiola M.A., Pastor C.J. Mercury and methylmercury complexes with a triazine-3-thione ligand, *Polyhedron*, 25(6), 1464–1470 (2006).

López-Torres E., Mendiola M.A., Rodríguez-Procopio J., Sevilla M.T., Colacio E., Moreno J.M., Sobrados I. Synthesis and characterisation of zinc, cadmium and mercury complexes of benzilbisthiosemicarbazone. Structure of cadmium derivative, *Inorg. Chim Acta*, 323(1–2), 130–138 (2001).

Lou S.-F., Wang Q., Ding J. Crystal structure of bis{[4,4,5,5-tetramethyl-2-(1-methyl-1*H*-benzimidazol-2-yl)-4,5-dihydroimidazole-1-oxyl-3-oxide-$\kappa^2 N,O$]-($\mu_2$-thiocyanato-*N*:*S*)(thiocyanato-$\kappa S$)mercury(II)}, $Hg_2(C_{15}H_{19}N_4O_2)_2(SCN)_4$, *Z. Kristallogr. New Cryst. Str.*, 227(2), 105–106 (2012).

Lucas C.R., Peach M.E. Metal derivatives of penthachlorothiophenol, *Inorg. Nucl. Chem. Lett.*, 5(2), 73–76 (1969).

Mahadevappa D.S., Murthy A.S.A. Some complexes of zinc(II), cadmium(II), and mercury(II) with thiosemicarbazide, *Aust. J. Chem.*, 25(7), 1565–1568 (1972).

Mahjoub A.R., Morsali A. Hg(II), Tl(III), Cu(I), and Pd(II) complexes with 2,2'-diphenyl-4,4'-bithiazole (DPBTZ), syntheses and x-ray crystal structure of [Hg(DPBTZ)(SCN)$_2$], *J. Coord. Chem.*, 56(9), 779–785 (2003).

Mahjoub A.R., Ramazani A., Morsali A. Crystal structure of 2,2'-bis(4,5-dimethylimidazole) (dithiocyanato)mercury(II), $Hg(C_{10}H_{14}N_4)(CNS)_2$, *Z. Kristallogr. New Cryst. Struct.*, 218(4), 435–436 (2003).

Makhija R.C., Beauchamp A.L., Rivest R. The crystal structure of $Hg(SCN)_2 \cdot As(C_6H_5)_3$, a three-co-ordinate mercury complex, *J. Chem. Soc., Chem. Commun.*, (18), 1043–1044 (1972).

Makhija R.C., Beauchamp A.L., Rivest R. Crystal and molecular structure of dithiocyanatobis (triphenylphosphine)-mercury(II), *J. Chem. Soc. Dalton Trans.*, (22), 2447–2450 (1973).

Makhija R.C., Rivest R., Beauchamp A.L. Structure of a new crystalline modification of dit hiocyanato(triphenylphosphine)mercury(II), *Can. J. Chem.*, 57(19), 2555–2559 (1979).

Malik M.R., Ali S., Ahmad S., Altaf M., Stoeckli-Evans H. Dicyanidobis(*N*,*N*-dimethylthiourea-$\kappa S$)mercury(II), *Acta Crystallogr.*, E66(9), m1060–m1061 (2010).

Man E.H., Coffman D.D., Muetterties E.L. Synthesis and properties of bis(trifluoromethylthio)-mercury, *J. Am. Chem. Soc.*, 81(14), 3575–3577 (1959).

Märcker C. Ueber einige schwefelhaltige Derivate des Toluols, *Justus Liebigs Ann. Chem.*, 136(1), 75–95 (1865).

Mason B., Mumme W.G., Sarp H. Capgaronnite, $HgS \cdot Ag(Cl,Br,I)$, a new sulfide-halide mineral from Var, France, *Am. Mineral.*, 77(1–2), 197–200 (1992).

Matković-Čalogović D., Popović Z., Pavlović G., Soldin Ž., Giester G. Diiodo(3,4,5,6-tetrahydropyrimidinium-2-thiolato-S)mercury(II), *Acta Crystallogr.*, C57(4), 409–411 (2001).

McAllan D.T., Cullum T.V., Dean R.A., Fidler F.A. The preparation and properties of sulfur compounds related to petroleum. I. The dialkyl sulfides and disulfides, *J. Am. Chem. Soc.*, 73(8), 3627–3632 (1951).

McCormick B.J., Greene D.L. Thiocarbamate complexes. II. Derivatives of zinc(II), cadmium(II) and mercury(II), *Inorg. Nucl. Chem. Lett.*, 8(7), 599–703 (1972).

McEwen R.S., Sim G.A. Complexes of mercury. Part IV. Crystal and molecular structure of the adduct of mercury dimethoxide and phenyl isothiocyanate, *J. Chem. Soc. A: Inorg. Phys. Theor.*, 1552–1558 (1967a).

McEwen R.S., Sim G.A. Complexes of mercury. Part III. X-ray analysis of dichlorobis-(1,4-thioxan)mercury(II), *J. Chem. Soc. A: Inorg. Phys. Theor.*, 271–275 (1967b).

Meek D.W., Nicpon P. Metal complexes of tertiary phosphine sulfides, *J. Am. Chem. Soc.*, 87(21), 4951–4952 (1965).

Mendoza C.S., Kamata S., Kawamina M. Crystal structure of bis($N$,$N$-diethyldithiocarbamate)mercury(II), *Anal. Sci.*, 13(3), 517–518 (1997).

Mosset A., Bagieu-Beucher M. Redetermination of the crystal structure of mercury chlorothiocyanate, HgClSCN, *Z. Kristallogr. New Cryst. Struct.*, 217(1), 1–2 (2002).

Mufakkar M., Tahir M.N., Sadaf H., Ahmad S., Waheed A. Dichloridobis($N$,$N'$-diethylthiourea-$\kappa S$)-mercury(II), *Acta Crystallogr.*, E66(8), m1001–m1002 (2010).

Mumme W.G., Nickel E.H. Crystal structure and crystal chemistry of perroudite: A mineral from Coppin Pool, Western Australia, *Am. Mineral.*, 72(11–12), 1257–1262 (1987).

Nam H.J., Lee H.-J., Noh D.-Y. Novel mercury(II) complexes of 1,3-dithiole-2-thiones containing the 2-pyridyl moiety: Syntheses, x-ray crystal structures and solution behavior, *Polyhedron*, 23(1), 115–123 (2004).

Nardelli M., Chierici I. Complessi di cloruri metallici con tiosemicarbazide, *Ric. Sci.*, 30(2), 276–279 (1960).

Nawaz S., Sadaf S., Fettouhi M., Fazal A., Ahmad S. Dichloridobis($N$,$N$,$N'$,$N'$-tetramethylthiourea-$\kappa S$)mercury(II), *Acta Crystallogr.*, E66(8), m952 (2010).

Noh D.-Y., Underhill A.E., Hursthouse M.B. Synthesis and crystal structure of an unusual bimetallic mercury–dithiolene complex, *Chem. Commun.*, (22), 2211–2212 (1997).

Norris A.R., Palmer A., Beauchamp A.L. Crystal structures of three methylmercury(II) complexes with thiazolidine-2-thione, *J. Crystallogr. Spectrosc. Res.*, 20(1), 23–30 (1990).

Norris A.R., Taylor S.E., Buncel E., Bélanger-Gariépy F., Beauchamp A. Crystal structures of two methylmercury complexes with 1-methylimidazoline-2-thione, *Can. J. Chem.*, 61(7), 1536–1541 (1983).

Ohmasa M., Nowacki W. The crystal structure of vrbaite $Hg_3Tl_4As_8Sb_2S_{20}$, *Z. Kristallogr.*, 134(5-6), 360–380 (1971).

Onwudiwe D.C., Ajibade P.A. Synthesis, characterization and thermal studies of Zn(II), Cd(II) and Hg(II) complexes of $N$-methyl-$N$-phenyldithiocarbamate: The single crystal structure of $[(C_6H_5)(CH_3)NCS_2]_4Hg_2$, *Int. J. Mol. Sci.*, 12(3), 1964–1978 (2011).

Pavlović G., Popović Z., Soldin Ž., Matković-Čalogović D. A 1:2 complex of mercury(II) chloride with 1,3-imidazole-2-thione, *Acta Crystallogr.*, C56(1), 61–63 (2000a).

Pavlović G., Popović Z., Soldin Ž., Matković-Čalogović D. Diiodobis(1-methyl-1,3-imidazolium-2-thiolato-$S$)mercury(II), *Acta Crystallogr.*, C56(7), 801–803 (2000b).

Perchard C., Zuppiroli G., Gouzerh P., Jeannin Y., Robert F. Etude spectroscopique de derives mercuriques d'amides-thiols: Partie II. Structure cristalline et spectres de vibration du bis(méthylcarbamoyléthanethiolato)mercure(II), *J. Mol. Struct.*, 72, 119–129 (1981).

Pervukhina N.V., Borisov S.V., Magarill S.A., Naumov D.Yu., Vasil'ev V.I., Nenashev B.G. Crystal structure of synthetic analogue of mineral grechishchevite and topology of the Hg–X (X = S, Se, Te) bonds in the mercury chalcogenides structures [in Russian], *Zhurn. strukt. khim.*, 45(3), 462–470 (2004).

Pervukhina N.V., Borisov S.V., Magarill S.A., Vasil'ev V.I., Kurat'eva N.V., Kozlova S.G. Refinement of the crystal structure of As-schwazite $(Cu_{5.26}Hg_{0.75})(As_{2.83}Sb_{1.17})S_{13}$ (Aktash, Mountain Altai) [in Russian], *Zhurn. strukt. khim.*, 51(5), 934–939 (2010a).

Pervukhina N.V., Borisov S.V., Magarill S.A., Vasil'ev V.I., Kurat'eva N.V., Kozlova S.G. Refinement and crystallographic analysis of the structure of the Sb-containing laffittite AgHg(As,Sb)S_3 from Chauvai (Kyrgyzstan) [in Russian], *Zhurn. strukt. khim.*, 51(4), 712–716 (2010b).

Pervukhina N.V., Vasil'ev V.I., Borisov S.V., Magarill S.A., Naumov D.Yu. The crystal structure of a polymorph of $Hg^{2+}_3S_2Br_{1.0}Cl_{0.5}I_{0.5}$, *Can. Mineral.*, 41(6), 1445–1453 (2003).

Phillips F.C. Compounds of methyl sulphide with halides of metals, *J. Am. Chem. Soc.*, 23(4), 250–258 (1901).

Pickardt J., Dechert S. Kristallstrukturen "supramolekularer" (Benzo-18-krone-6)kaliumtetrathiocyanatometallate: Ein dimerer Komplex {[K(Benzo-18-krone-6)]$_2$[Hg(SCN)$_4$]}$_2$ und zwei isomere Komplexe [K(Benzo-18-krone-6)][Cd(SCN)$_3$] mit kettenförmigen Trithiocyanatocadmat-Anionen, *Z. anorg. und allg. Chem.*, 625(1), 153–159 (1999).

Pickardt J., Gong G.-T., Hoffmeister I. Kristallstruktur des Quecksilber(II)-Cryptats [Hg(cryptand 222)][{Hg(SCN)$_3$}$_2$] mit einem kettenförmigen Tri(thiocyanato) mercurat(II)-Anion, *Z. Naturforsch.*, 50B(6), 993–996 (1995a).

Pickardt J., Shen J., Gong G.-T. Kristallstrukturen der "supramolekularen"1,4-Diazabicyclo[2.2.2]octan-Quecksilber(II)-Komplexe [Hg(dabco)I$_2$] und [Hg(dabco)(SCN)$_2$], *Z. Naturforsch.*, 50B(4), 833–836 (1995b).

Popović Z., Matković-Čalogović D., Hasić J., Vikić-Topić D. Preparation and spectroscopic properties of the complexes of mercuric thiocyanate with pyridine-2-thione and pyridine-2-carboxylic acid. Crystal and molecular structure of two polymorphs of Hg(SCN)$_2$(C$_5$H$_5$NS)$_2$, *Inorg. Chim. Acta*, 285(2), 208–216 (1999a).

Popović Z., Matković-Čalogović D., Pavlović G., Soldin Ž., Giester G., Rajić M., Vikić-Topić D. Preparation, thermal analysis and spectral characterization of the 1:1 complexes of mercury(II) halides and pseudohalides with 3,4,5,6-tetrahydropyrimidine-2-thione. Crystal structures of bis(3,4,5,6-tetrahydropyrimidine-2-thione-*S*)mercury(II) tetrachloro and tetrabromomercurate(II), *Croat. Chem. Acta*, 74(2), 359–380 (2001).

Popović Z., Matković-Čalogović D., Soldin Ž., Pavlović G., Davidović N., Vikić-Topić D. Mercury(II) compounds with 1,3-imidazole-2-thione and its 1-methyl analogue. Preparative and NMR spectroscopic studies. The crystal structures of di-μ-iodo-bis[iodo(1,3-imidazolium-2-thiolato-*S*)mercury(II)], bis[bromo(1,3-imidazolium-2-thiolato-S)]mercury(II) and bis[μ-(1-*N*-methyl-1,3-imidazole-2-thiolato-*S*)] mercury(II), *Inorg. Chim. Acta*, 294(1), 35–46 (1999b).

Popović Z., Pavlović G., Matković-Čalogović D., Soldin Ž., Rajić M., Vikić-Topić D., Kovaček D. Mercury(II) complexes of heterocyclic thiones. Part 1. Preparation of 1:2 complexes of mercury(II) halides and pseudohalides with 3,4,5,6-tetrahydropyrimidine-2-thione. X-ray, thermal analysis and NMR studies, *Inorg. Chim. Acta*, 306(2), 142–152 (2000a).

Popović Z., Soldin Ž., Matković-Čalogović D., Pavlović G., Rajić M., Giester G. Mercury(II) complexes with heterocyclic thiones—Preparation and characterization of the 1:1 and 1:2 mercury(II) complexes with benzo-1,3-imidazole-2-thione, *Eur. J. Inorg. Chem.*, (1), 171–180 (2002a).

Popović Z., Soldin Ž., Pavlović G., Matković-Čalogović D., Mrvoš-Sermek D., Rajić M. Mercury(II) compounds with 1,3-benzothiazole-2-thione. Spectral, thermal, and crystal structure studies, *Struct. Chem.*, 13(5-6), 425-436 (2002b).

Popović Z., Soldin Ž., Plavec J., Vikić-Topić D. Mercuration of thiophene-2-carboxylic acid, 2-thienylethanoic acid and 3-(2-thienyl)alanine: Preparation and spectral characterization, *Appl. Organomet. Chem.*, 14(10), 598–603 (2000b).

Porai-Koshits M.A. The structural motifs of the crystals of some thiocyanato compounds of divalent nickel and copper, *J. Struct. Chem.*, 4(4), 531–539 (1963).

Potheher I.V., Madhavan J., Rajarajan K., Nagaraja K.S., Sagayaraj P. Growth and characterization of diaquatetrakis (thiocyanato) cobalt(II) mercury(II) *N*-methyl-2-pyrolidone (CMTWMP) single crystals, *J. Cryst. Growth.*, 310(1), 124–130 (2008).

Pouskouleli G., Kourounakis P., Theophanides T. Mercury and cadmium thiosteroid complexes, *Inorg. Chim. Acta*, 24, 45–51 (1977).

Preti C., Tosi G., De Filippo D., Verani G. Group IIB metal complexes with thiazolidine-2-selenone and thiazolidine-2-one as ligands, *J. Inorg. Nucl. Chem.*, 36(12), 3725–3729 (1974).

Preti C., Tosi G. Tautomeric equilibrium study of thiazolidine-2-thione. Transition metal complexes of the deprotonated ligand, *Can. J. Chem.*, 54(10), 1558–1562 (1976).

Ramazani A., Morsali A., Haji-Abolfath A. Crystal structure of (2,2'-biquinolyl-*N*,*N*) bis(thiocyanato)mercury(II) benzene hemisolvate, [Hg(C$_{18}$H$_{12}$N$_2$)(SCN)$_2$]·0.5C$_6$H$_6$, *Z. Kristallogr. New Cryst. Struct.*, 219(3), 245–246 (2004).

Raper E.S., Creighton J.R., Bell N.A., Clegg W., Cucurull-Sánchez L. Complexes of heterocyclic thiones and group twelve metals. Part 1. Preparation and characterisation of 1:1 complexes of mercury(II) halides with 1-methylimidazoline-2(3*H*)-thione: The crystal structure of [(μ$_2$-dibromo)bis(*trans*{(bromo)(1-methyl-imidazoline-2(3*H*)-thione)} mercury(II))] at 160 K, *Inorg. Chim. Acta*, 277(1), 14–20 (1998).

Relf J., Cooney R.P., Henneike H.F. A laser Raman, infrared and nuclear magnetic resonance spectral study of complexation of methylmercury(II) thiocyanate by ionic thiocyanate, *J. Organomet. Chem.*, 39(1), 75–86 (1972).

Rodríguez A., Sousa-Pedrares A., García-Vázquez J.A., Romero J., Sousa A. Electrochemical synthesis and structural characterisation of cadmium and mercury complexes containing pyrimidine-2-thionate ligands, *Eur. J. Inorg. Chem.*, (11), 2242–2254 (2005).

Romero I., Sánchez-Castelló G., Teixidor F., Whitaker C.R., Rius J., Miravitlles C., Flor T., Escriche L., Casabo J. Silver(I), mercury(II) and copper(I) complexes of acyclic and macrocyclic dithioether, *meta*-xylyl based ligands, *Polyhedron*, 15(12), 2057–2065 (1996).

Sadaf H., Ahmad S., Sharif S., Khan I.U., Akkurt M., Ng S.W., Khan M.I., Bashir S.A., Mufakkar M. Crystal structures of [Hg(*N*-ethylthiourea)$_2$(CN)$_2$] and [Hg(*N*-propylthiourea)$_2$(CN)$_2$], *Zhurn. strukt. khimii*, 53(1), 155–159 (2012).

Sakhri A., Beauchamp A.L. Crystal structure of tetraphenylphosphonium trithiocyanato-mercurate(II), *Inorg. Chem.*, 14(4), 740–743 (1975a).

Sakhri A., Beauchamp A.L. Structure cristalline du tétrathiocyanatomercurate(II) de tétraphénylphosphonium, [(C$_6$H$_5$)$_4$P]$_2$[Hg(SCN)$_4$], *Acta Crystallogr.*, B31(2), 409–413 (1975b).

Sampanthar J.T., Deivaraj T.C., Vittal J.J., Philip A.W, Dean P.A.W. Thioacetate complexes of Group 12 metals. Structures of [Ph$_4$P][Zn(SC{O}Me)$_3$(H$_2$O)] and [Ph$_4$P][Cd(SC{O}Me)$_3$], *J. Chem. Soc. Dalton Trans.*, (24), 4419–4423 (1999).

Sandström M. Crystal and molecular structure of Hg(ClO$_4$)$_2$·4(CH$_3$)$_2$SO, *Acta Chem. Scand.*, 32A(6), 527–532 (1978).

Sandström M., Persson I. Crystal and molecular structure of hexakis(dimethylsulfoxide) mercury(II) perchlorate, [Hg((CH$_3$)$_2$SO)$_6$](ClO$_4$)$_2$, *Acta Chem. Scand.*, 32A(2), 95–100 (1978).

Sarp H., Birch W.D., Hlava P.F., Pring A., Sewell D.K.B., Nickel E.H. Perroudite, a new sulfide-halide of Hg and Ag from Cap-Garonne, Var, France, and from Broken Hill, New South Wales, and Coppin Pool, Western Australia, *Am. Mineral.*, 72(11–12), 1251–1256 (1987).

Scouloudi H. The crystal structure of mercury tetrathiocyanate–copper diethylenediamine, [Hg(SCN)$_4$][Cu(en)$_2$], *Acta Crystallogr.*, 6(7), 651–657 (1953).

Secondo P.M., Land J.M., Baughman R.G., Collier H.L. Polymeric octahedral and monomeric tetrahedral Group 12 pseudohalogeno (NCX$^-$: X = O, S, Se) complexes of 4-(*N*,*N*-dimethylamino)pyridine, *Inorg. Chim. Acta*, 309(1–2), 13–22 (2000).

Selbin J., Bull W.E., Holmes, Jr. L.H. Metallic complexes of dimethylsulphoxide, *J. Inorg. Nucl. Chem.*, 16(3–4), 219–224 (1961).

Setzer W.N., Tang Y., Grant G.J., VanDerveer D.G. Synthesis and x-ray crystal structures of heavy-metal complexes of 1,5,9,13-tetrathiacyclohexadecane, *Inorg. Chem.*, 30(19), 3652–3656 (1991).

Shunmugam R., Sathyanarayana D.N. Complexes of imidazoline-2-thione and its 1-methyl analogue with Cu(II), Zn(II), Cd(II) and Hg(II) salts, *J. Coord. Chem.*, 12(3), 151–156 (1983).

Silver A., Koch S.A., Millar M. X-ray crystal structures of a series of $[M^{II}(SR)_4]^{2-}$ complexes (M = Mn, Fe, Co, Ni, Zn, Cd and Hg) with $S_4$ crystallographic symmetry, *Inorg. Chim. Acta*, 205(1), 9–14 (1993).

Singh P.P., Singh M. 1,1'-Bis(thiocyanatomercurio)ferrocene as ligand towards $M(NCS)_2$ [M = Mn(II), Co(II), Ni(II), Cu(II), Zn(II)], *Bull. Chem. Soc. Jpn.*, 59(4), 1229–1233 (1986).

Skelton B.W., White A.H. Crystal structures of $N,N,N',N'$-tetramethylthiuram monosulphide and diiodo($N,N,N',N'$-tetramethylthiuram monosulphide)mercury(II), *Aust. J. Chem.*, 30(8), 1693–1699 (1977).

Sousa-Pedrares A., Romero J., Garcia-Vázquez J., Durán M.L., Casanova I., Sousa A. Electrochemical synthesis and structural characterization of zinc, cadmium and mercury complexes of heterocyclic bidentate ligands (N, S), *Dalton Trans.* (7), 1379–1388 (2003).

Stålhandske C.M.V., Persson I., Sandström M., Åberg M. Crystal structure of bis($N,N$-dimethylthioformamide)mercury(II) iodide, bromide, and chloride and solvation of the neutral mercury(II) halide complexes in $N,N$-dimethylthioformamide, a resonance-induced hydrogen-bonding solvent and ligand, *Inorg. Chem.*, 36(22), 4945–4953 (1997).

Stuart D.A., Nassimbeni L.R., Hutton A.T., Koch K.R. Methylmercury derivatives of mercaptopyrimidines: Structures of (4-amino-5-methyl-2-pyrimidinethiolato) methylmercury(II) and (4-amino-2-mercapto-6-pyrimidinonato)methylmercury(II) monohydrate, *Acta Crystallogr.*, B36(10), 2227–2230 (1980).

Sytsma L.F., Kline R.J. Mercury-proton spin-spin coupling constants of some methylmercury compounds,, *J. Organomet. Chem.*, 54, 15–21 (1973).

Tallon J., Garcia-Vazquez J.A., Romero J., Louro M.S., Sousa A., Chen Q., Chang Y., Zubieta J. Electrochemical synthesis of mercury(II) complexes with some heterocyclic thiones: The crystal and molecular structures of bis(6-tertbutyldimethylsilyl)2-pyridyl disulphide (6-Bu$^t$SidmepyS-Spy-6-Bu$^t$Sidme) and Hg(6-Bu$^t$SidmepyS)$_2$, *Polyhedron*, 14(17–18), 2309–2317 (1995).

Taylor N.J., Carty A.J. Nature of $Hg^{2+}$-L-cysteine complexes implicated in mercury biochemistry, *J. Am. Chem. Soc.*, 99(18), 6143–6145 (1977).

Taylor N.J., Wong Y.S., Chieh P.C., Carty A.J. Syntheses, x-ray crystal structure, and vibrational spectra of L-cysteinato(methyl)mercury(II) monohydrate, *J. Chem. Soc. Dalton Trans.*, (5), 438–442 (1975).

Terzis A., Faught J.B., Pouskoulelis G. Crystal and molecular structure of a mercurythiosteroid complex, *Inorg. Chem.*, 19(4), 1060–1063 (1980).

Thiele G., Bauer R., Messer D. Die Kristallstrukturen von $CsHg(CN)_3$ und $CsHg(SCN)_3$, *Naturwissenschaften*, 61(5), 215–216 (1974).

Thiele G., Messer D. Die Kristallstrukturen der Trithiocyanatomercurate $RbHg(SCN)_3$ und $CsHg(SCN)_3$, *Z. anorg. und allg. Chem.*, 421(1), 24–36 (1976).

Thirumaran S., Venkatachalam V., Manohar A., Ramalingam K., Bocelli G., Cantoni A. Synthesis and characterization of bis($N$-methyl-$N$-ethanol-dithiocarbamato)M(II) (M = Zn, Cd, Hg) and bis($N,N$-(iminodiethylene)-bisphthalimidedithiocarbamato)M(II) (M = Zn, Cd, Hg) complexes. Single crystal x-ray structure of bis(di(2-hydroxyethyl)-dithiocarbamato)zinc(II), *J. Coord. Chem.*, 44(3–4), 281–288 (1998).

Tiekink E.R.T. Crystal structure of a 1,2-phenylenedimercury dixanthate, *J. Organomet. Chem.*, 303(3), C53–C55 (1986a).

Tiekink E.R.T. Methylmercury xanthates, *Inorg. Chim Acta*, 112(1), L1–L2 (1986b).

Tiekink E.R.T. Bis($O$-methyldithiocarbonato)mercury(II), *Acta Crystallogr.*, 43(3), 448–450 (1987a).

Tiekink E.R.T. Phenylmercury dithiolates. The crystal and molecular structures of ($C_6H_5$) $Hg(S_2COR)$ (R = Me; $^i$Pr) and ($C_6H_5$)$Hg(S_2CNEt_2)$, *J. Organomet. Chem.*, 322(1), 1–10 (1987b).

Tiekink E.R.T. (*O*-Ethyl dithiocarbonato-κ*S*)phenylmercury, [Hg(C₃H₅OS₂)(C₆H₅)], *Acta Crystallogr.*, C50(6), 861–862 (1994).

Tiekink E.R.T. Crystal structure of the 1,10-phenanthroline adduct of chloro-(di-n-butyl dithiocarbamato)mercury(II), C₂₁H₂₆ClHgN₃S₂, *Z. Kristallogr. New Cryst. Struct.*, 216(3), 439–440 (2001).

Udupa M.R., Krebs B. Crystal and molecular structure of mercury(II) tetrathiocyanatobis(di methylformamide)cobaltate(II), *Inorg. Chim. Acta*, 42, 37–41 (1980).

Vasilyev V.I., Usova L.V., Pal'chik N.A. Grechishchevite—Hg₃S₂(Br,Cl,I)₂—A new supergene sulfohalogenide of mercury [in Russian], *Geol. geofiz.*, (7), 61–69 (1989).

Vittal J.J., Dean P.A.W. Chemistry of thiobenzoates: Syntheses, structures, and NMR spectra of salts of [M(SOCPh)₃]⁻ (M = Zn, Cd, Hg), *Inorg. Chem.*, 35(11), 3089–3093 (1996).

Vittal J.J., Dean P.A.W. Tetramethylammonium tris(thiobenzoato-*O*,*S*)mercury(II), *Acta Crystallogr.*, C53(4), 409–410 (1997).

Vogt C. XVII. Ueber Benzylmercaptan und Zweifach–Schwefel–Benzyl, *Justus Liebigs Ann. Chem.*, 119(2), 142–153 (1861).

Wachhold M., Rangan K.K., Billinge S.J.L., Petkov V., Heising J., Kanatzidis M.G. Mesostructured non-oxidic solids with adjustable worm-hole shaped pores: M–Ge–Q (Q = S, Se) frameworks based on tetrahedral [Ge₄Q₁₀]⁴⁻ clusters, *Adv. Mater.*, 12(2), 85–91 (2000a).

Wachhold M., Rangan K.K., Lei M., Thorpe M.F., Billinge S.J.L., Petkov V., Heising J., Kanatzidis M.G. Mesostructured metal germanium sulfide and selenide materials based on the tetrahedral [Ge₄S₁₀]⁴⁻ and [Ge₄Se₁₀]⁴⁻ units: Surfactant templated three-dimensional disordered frameworks perforated with worm holes, *J. Solid State Chem.*, 152(1), 21–36 (2000b).

Wang Q.-H., Long D.-L., Huang J.-S. Self-assembled one-dimensional zigzag structure of mercury(II) complex with sulphur rich ligand 4,5-ethylenedithio-1,3-dithiole-2-thione, *Polyhedron*, 17(21), 3665–3669 (1998).

Wang S., Fackler, Jr. J.P. Mercury complexes with one-dimensional chain structures. Syntheses and crystal structures of [Hg(C₅H₄NS)(CH₃CO₂)]ₙ, Hg(C₅H₄NS)₂, and Hg(CH₂P(S)Ph₂)₂, *Inorg. Chem.*, 28(13), 2615–2619 (1989).

Wang X.Q., Xu D., Cheng X.F., Lü M.K., Yuan D.R., Huang J., Lu G.W., Xie T.X., Wang S.L., Lü G.T., Zhao J.Q., Yang Z.H. Preparation and properties of a novel nonlinear optical crystal: MnHg(SCN)₄(H₂O)₂·2(C₃H₆CONCH₃), *Cryst. Res. Technol.*, 37(6), 551–563 (2002a).

Wang X., Xu D., Lü M., Yuan D., Cheng X., Li S., Huang J., Wang S., Liu H. Crystal growth, spectral and thermal properties of nonlinear optical crystal: MnHg(SCN)₄, *J. Cryst. Growth*, 245(1–2), 126–133 (2002b).

Wang X.Q., Xu D., Lu M.K., Yuan D.R., Xu S.X. Crystal growth and characterization of the organometallic nonlinear optical crystal: Manganese mercury thiocyanate (MMTC), *Mater. Res. Bull.*, 36(5–6), 879–887 (2001).

Wang X.-Q., Yu W.-T., Xu D., Li T.-B., Wang Y.-L., Zhang G.-H., Xue G., Yang H.-L., Ren Q. Crystal structure of bis(tetra-*n*-butylammonium) bis(1,3-dithiole-2-thione-4,5-dithiolato)mercurate(II), [(C₄H₉)₄N]₂[Hg(C₃S₅)₂], *Z. Kristallogr. New Cryst. Struct.*, 220(2), 177–179 (2005a).

Wang X.-Q., Yu W.-T., Xu D., Zhang G.-H. Polymeric bis(*N*-methylformamide) tetrathiocyanatocadmium(II)mercury(II), *Acta Crystallogr.*, E61(6), m1147–m1149 (2005b).

Wang X.-Q., Yu W.-T., Xu D., Lu M.-K., Yuan D.-R., Liu J.-R., Lu G.-T. Polymeric diaquatetra-μ-thiocyanato-manganese(II)mercury(II) bis(*N*,*N*-di-methylacetamide) solvate, *Acta Crystallogr.*, C56(11), 1305–1307 (2000a).

Wang X.-Q., Yu W.-T., Xu D., Lu M.-K., Yuan D.-R., Lu G.-T. Manganese mercury thiocyanate (MMTC) glycol monomethyl ether, *Acta Crystallogr.*, C56(6), 647–648 (2000b).

Wang X.-Q., Yu W.-T., Xu D., Lu M.-K., Yuan D.-R. Poly[[bis(dimethyl sulfoxide-κO)tris-(thiocyanato-κN)manganese(II)]-μ-thiocyanato-κ²N:S-mercury(II)], *Acta Crystallogr.*, C56(4), 418–420 (2000c).

Wang Y.-L., Yu W.-T., Xu D., Wang X.-Q., Guo W.-F., Zhang G.-H. Crystal structure of bis(N-ethylpyridinium) bis(2-thioxo-1,3-dithiole-4,5-dithiolato)mercurate(II), (C₇H₁₀N)₂[Hg(C₃S₅)₂], *Z. Kristallogr. New Cryst. Struct.*, 221(1), 27–28 (2006).

Watanabe Y. Mercury ethylxanthate, *Acta Crystallogr.*, B33(11), 3566–3568 (1977).

Watanabe Y. The structure of mercury(II) isopropylxanthate, *Acta Crystallogr.*, 37(3), 553–556 (1981).

Watton S.P., Wright J.G., MacDonnell F.M., Bryson J.W., Sabat M., O'Halloran T.V. Trigonal mercuric complex of an aliphatic thiolate: A spectroscopic and structural model for the receptor site in the Hg(II) biosensor MerR, *J. Am. Chem. Soc.*, 112(7), 2824–2826 (1990).

White J.L., Tanski J.M., Churchill D.G., Rheingold A.L., Rabinovich D. Synthesis and structural characterization of 2-mercapto-1-*tert*-butylimidazole and its Group 12 metal derivatives (Hmim$^{tBu}$)₂MBr₂ (M = Zn, Cd, Hg), *J. Chem. Crystallogr.*, 33(5–6), 437–445 (2003).

White J.L., Tanski J.M., Rabinovich D. Bulky tris(mercaptoimidazolyl)borates: Synthesis and molecular structures of the Group 12 metal complexes [Tm$^{tBu}$]MBr (M = Zn, Cd, Hg), *J. Chem. Soc. Dalton Trans.*, (15), 2987–2991 (2002).

Wojnowski W., Wojnowski M., Schnering von H.G., Noltemeyer M. Beiträge zur Chemie der Silicium-Schwefel-Verbindungen. XXXVI. Das Quecksilber(II)-bis-tri-*tert*-butoxysilanthiolat, *Z. anorg. und allg. Chem.*, 531(12), 153–157 (1985).

Wong Y.S., Carty A.J., Chieh C. X-ray crystallographic and spectroscopic studies of the binding of methylmercury by the sulphur amino-acid DL-penicillamine: Models for methylmercury poisoning of proteins, *J. Chem. Soc. Dalton Trans.*, (19), 1801–1808 (1977).

Wong Y.S., Chieh P.C., Carty A.J. Binding of methylmercury by amino-acids: X-ray structure of DL-penicillaminatomethylmercury(II), *J. Chem. Soc. Chem. Commun.*, (19), 741–742 (1973a).

Wong Y.S., Chieh P.C., Carty A.J. The interaction of organomercury pollutants with biologically important sites: An x-ray study of the 2:1 complex between methylmercury and penicillamine, *Can. J. Chem.*, 51(15), 2597–2599 (1973b).

Wu Z.-Y., Xu D.-J., Hung C.-H. Synthesis and crystal structure of diiodobis(thiourea) mercury(II)-bis(diazafluoren-9-one), *J. Coord. Chem.*, 57(9), 791–796 (2004).

Yu W.-T., Wang X.-Q., Xu D., Lu M.-K., Yuan D.-R. Tetrathiourea mercury(II) tetrathiocyanatomanganate(II), *Acta Crystallogr.*, C57(2), 145–146 (2001).

Yusof M.S.M., Yamin B.M., Kassim M.B. Bis(o-chlorophenylbenzoylthiourea-κS) diiodomercury(II), *Acta Crystallogr.*, E60(1), m98–m99 (2004).

Zemskova S.M., Gromilov S.A., Larionov S.V. Synthesis and X-ray investigation of diethyl-dithiocarbamato-iodide complexes of zinc(II), cadmium(II), mercury(II) [in Russian], *Sib. khim. zhurn.*, (4), 74–77 (1991).

Zhang X.-M., Li P., Wang H., Chen H.-Y. Crystal structure of bis(tetrabutylammonium) bis(1,2,3-trioxocyclopent-4-ene-4,5-dithiolato-S,S')mercury(II)—water (1:0.28), [N(C₄H₉)₄]₂[Hg(C₅O₃S₂)₂]·0.28H₂O, *Z. Kristallogr. New Cryst. Struct.*, 227(2), 230–232 (2012).

Zhang Y.-M., Yang L.-Z., Lin Q., Wei T.-B. Synthesis and crystal structure of bis{(μ-chloro)-chloro[N-benzoyl-N'-(2-hydroxyethyl)thiourea]mercury(II), *J. Coord. Chem.*, 58(18), 1675–1679 (2005).

Zhdanov G.S., Sanadze V.V. Crystal structure of thiocyanates. IV. X-ray analysis of the crystal structure of Hg(SCN)₂·ASCN; A – K, NH₄ [in Russian], *Zhurn. fiz. khim.*, 26(4), 467–478 (1952).

Zvonkova Z.V., Zhdanov G.S. Crystal structure of thiocyanates. V. Crystal structures of mercury halogenothiocyanates [in Russian], *Zhurn. fiz. khimii*, 26(4), 586–591 (1952).

Zukerman-Schpector J., Vázquez-López E.M., Sánchez A., Casas J.S., Sordo J. Synthesis and structural characterization of diphenyldithiophosphinates of methyl- and phenylmercury(II). Crystal structure of MeHgS$_2$PPh$_2$, *J. Organomet. Chem.*, 405(1), 67–74 (1991).

# 8 Systems Based on HgSe

## 8.1 Hg–H–Na–C–N–O–Se

$[Na(C_8H_{16}O_4)_2]_2[Hg(Se_4)_2]\cdot1.5DMF$ or $C_{36.5}H_{74.5}Na_2N_{1.5}O_{17.5}Se_8Hg$. This multinary compound is formed in this system (Ahle et al. 1993). It crystallizes in the monoclinic structure with the lattice parameters $a = 2884 \pm 2$, $b = 1407.7 \pm 0.7$, and $c = 2843 \pm 2$ pm and $\beta = 93.93° \pm 0.05°$ and a calculated density of 1.94 g·cm$^{-3}$. It has been prepared by the reaction of $Na_2Se_4$ with $Hg(OAc)_2$ in DMF solution in the presence of 12-crown-4 ($C_8H_{16}O_4$), forming dark-red needlelike crystals.

## 8.2 Hg–H–Na–C–O–Se

$[Na(C_{10}H_{20}O_5)]_2[Hg(Se_4)_2]$ or $C_{20}H_{40}Na_2O_{10}Se_8Hg$. This multinary compound is formed in this system (Adel et al. 1988). It crystallizes in the monoclinic structure with the lattice parameters $a = 2094.9 \pm 0.6$, $b = 1048.4 \pm 0.1$, and $c = 1745.5 \pm 0.7$ pm and $\beta = 98.38° \pm 0.03°$ and a calculated density of 2.31 g·cm$^{-3}$. To prepare it, a suspension containing 4.93 mM of $Hg(OAc)_2$ and an equimolar quantity of $[Na(C_{10}H_{20}O_5)]_2Se_6$ was stirred at room temperature for 12 h. Then the mixture was heated at reflux for 12 h, filtered hot, and allowed to cool slowly to room temperature. After 3 days dark-red single crystals were obtained. These were filtered off, washed with cold EtOH, and dried in vacuo.

## 8.3 Hg–H–K–C–Sn–O–Se

Some compounds are formed in this system.

$[K_{10}(H_2O)_{16}(MeOH)_{0.5}][Hg_4(\mu_4\text{-}Se)(SnSe_4)_4]$ or $C_{0.5}H_{34}K_{10}Sn_4O_{16.5}Se_{17}Hg_4$. This compound crystallizes in the tetragonal structure with the lattice parameters $a = 1562 \pm 2$ and $c = 2544.0 \pm 0.5$ at 150 K, a calculated density of 3.546 g·cm$^{-3}$, and an energy gap of 2.36 eV (Brandmayer et al. 2004). To obtain it, $K_4[SnSe_4]\cdot1.5MeOH$ (0.175 mM) was suspended in MeOH (5 mL) and added to a solution of $Hg(OAc)_2$ (0.140 mM) in $H_2O$ (5 mL). After stirring for 24 h, the precipitate was removed and the filtrate was layered with THF (10 mL). Small, orange truncated octahedra of this compound was formed, besides small amounts of $[K_4(MeOH)_4][Sn_2Se_6]$ after 1 week.

$[K_6(H_2O)_3][Hg_4(\mu_4\text{-}Se)(SnSe_4)_3]\cdot MeOH$ or $CH_{10}K_6Sn_3O_4Se_{13}Hg_4$. This compound crystallizes in the rhombohedral structure with the lattice parameters (in hexagonal setting) $a = 1512.5 \pm 0.4$ and $c = 1624.7 \pm 0.3$ at 203 K and a calculated density of 3.878 g·cm$^{-3}$ (Brandmayer et al. 2004). To obtain it, $K_4[SnSe_4]\cdot1.5MeOH$ (0.15 mM) was suspended in MeOH (5 mL) and added to a solution of $Hg(OAc)_2$ (0.150 mM) in $H_2O$ (5 mL) with immediate formation of a black precipitate. After stirring for 24 h, the precipitate was removed and the filtrate was layered with

THF (10 mL). Over 2 weeks, small orange cuboids of this compound crystallized together with small quantities of $[K_4(MeOH)_4][Sn_2Se_6]$.

## 8.4   Hg–H–K–C–O–Se

$[\{K(C_{12}H_{24}O_6)\}_2[Hg(Se_4)_2\}]_2$ or $C_{24}H_{48}K_2O_{12}Se_8Hg$. This multinary compound is formed in this system (Kräuter et al. 1989). It crystallizes in the triclinic structure with the lattice parameters $a = 1057.4 \pm 0.2$, $b = 1130.6 \pm 0.2$, and $c = 2000.1 \pm 0.2$ pm and $\alpha = 88.52° \pm 0.01°$, $\beta = 85.86° \pm 0.01°$, and $\gamma = 66.22° \pm 0.01°$ and a calculated density of 2.19 g·cm$^{-3}$. The dark-red crystals of the title compound have been prepared by the reaction of $[K(C_{12}H_{24}O_6)]_2Se_6$ with $Hg(OAc)_2$ in ethanolic solution.

## 8.5   Hg–H–Cs–C–O–Se

$[Cs(C_{12}H_{24}O_6)_2][Hg_2(Se_4)_3]$ or $C_{24}H_{48}Cs_2O_{12}Se_{12}Hg_2$. This compound crystallizes in the monoclinic structure with the lattice parameters $a = 2018 \pm 1$, $b = 1097.0 \pm 0.2$, and $c = 2317.1 \pm 0.5$ pm and $\beta = 99.40° \pm 0.03°$ and a calculated density of 2.81 g·cm$^{-3}$ (Magul et al. 1991). It has been prepared as dark-red crystals by the reaction of a lithium polyselenide solution in DMF with $Hg(OAc)_2$ in the presence of CsBr and 18-crown-6 ($C_{12}H_{24}O_6$).

## 8.6   Hg–H–C–Ge–N–Se

Some compounds are formed in this system.

$(C_{12}H_{25}NMe_3)_2[HgGe_4Se_{10}]$ or $C_{30}H_{68}Ge_4N_2Se_{10}Hg$. To obtain this compound, $(C_{12}H_{25}NMe_3)_4Ge_4Se_{10}$ was dissolved in a warm (ca. 60°C) EtOH/H$_2$O mixture (ca. 20 mL; 1:1 volume ratio) (Wachhold et al. 2000a,b). A solution of HgCl$_2$ in EtOH/H$_2$O was then added dropwise to the stirred $(C_{12}H_{25}NMe_3)_4Ge_4Se_{10}$ solution to give instantly a yellow-brownish precipitate. The resulting suspension was stirred for another hour, followed by product isolation by centrifugation and subsequent decantation of the solvent. The product was washed with water and acetone to remove all residues of $(C_{12}H_{25}NMe_3)_4Ge_4Se_{10}$ and dried with ether.

This compound could also be obtained if $K_4Ge_4Se_{10}$ and two equivalents of $C_{12}H_{25}NMe_3Br$ were dissolved in H$_2$O/MeOH (ca. 20 mL; 1:1 volume ratio) and heated to 50°C–60°C to give a yellow solution (Wachhold et al. 2000a,b). The solution of HgCl$_2$ in H$_2$O (100 mL) was slowly added to this mixture under vigorous stirring, which led to the immediate precipitation of the product. The isolation and purification were the same as for the first method. $(C_{12}H_{25}NMe_3)_2[HgGe_4Se_{10}]$ is a wide band gap semiconductor.

$(C_{14}H_{29}NMe_3)_2[HgGe_4Se_{10}]$ or $C_{34}H_{76}Ge_4N_2Se_{10}Hg$. This compound was prepared by the same procedure as $(C_{12}H_{25}NMe_3)_2[HgGe_4Se_{10}]$ was obtained using $(C_{14}H_{29}NMe_3)_4Ge_4Se_{10}$ instead of $(C_{12}H_{25}NMe_3)_4Ge_4Se_{10}$ (first method) and $C_{14}H_{29}NMe_3Br$ instead of $C_{12}H_{25}NMe_3Br$ (second method) (Wachhold et al. 2000a,b). It is a wide band gap semiconductor with an energy gap of 2.37 eV.

$(C_{16}H_{33}NMe_3)_2[HgGe_4Se_{10}]$ **or** $C_{38}H_{84}Ge_4N_2Se_{10}Hg$. This compound was prepared by the same procedure as $(C_{12}H_{25}NMe_3)_2[HgGe_4Se_{10}]$ was obtained using $(C_{16}H_{33}NMe_3)_4Ge_4Se_{10}$ instead of $(C_{12}H_{25}NMe_3)_4Ge_4Se_{10}$ (first method) and $C_{16}H_{33}NMe_3Br$ instead of $C_{12}H_{25}NMe_3Br$ (second method) (Wachhold et al. 2000a,b). It is a wide band gap semiconductor.

$(C_{18}H_{37}NMe_3)_2[HgGe_4Se_{10}]$ **or** $C_{42}H_{92}Ge_4N_2Se_{10}Hg$. This compound was prepared by the same procedure as $(C_{12}H_{25}NMe_3)_2[HgGe_4Se_{10}]$ was obtained using $(C_{18}H_{37}NMe_3)_4Ge_4Se_{10}$ instead of $(C_{12}H_{25}NMe_3)_4Ge_4Se_{10}$ (first method) and $C_{18}H_{37}NMe_3Br$ instead of $C_{12}H_{25}NMe_3Br$ (second method) (Wachhold et al. 2000a,b). It is a wide band gap semiconductor.

## 8.7 Hg–H–C–Sn–N–Se

Some compounds are formed in this system.

**[dbnH][Hg$_{0.5}$Sn$_{0.5}$Se$_2$]** or $C_7H_{13}Sn_{0.5}N_2Se_2Hg_{0.5}$ (dbn = 1,5-diazabicyclo[4.3.0]non-5-ene, $C_7H_{12}N_2$). This compound crystallizes in the tetragonal structure with the lattice parameters $a = 1286.05 \pm 0.02$ and $c = 685.560 \pm 0.010$ pm, a calculated density of 2.594 g·cm$^{-3}$, and an energy gap of 2.46 eV (Xiong et al. 2013b). To obtain it, a mixture of Sn powder (0.24 mM), HgCl$_2$ (0.50 mM), Se (1.48 mM), $C_7H_{12}N_2$ (1 mL), and H$_2$O (0.5 mL) was stirred under ambient conditions. The resulting mixture was sealed into an autoclave equipped with a Teflon liner (23 mL) and heated at 190°C for 8 days. After cooling to room temperature, light-yellow rod crystals of [dbnH][Hg$_{0.5}$Sn$_{0.5}$Se$_2$] were obtained by filtration, washed several times with EtOH, and selected by hand.

**[Hg$_2$Sn$_2$Se$_7$][dbuH]$_2$** or $C_{18}H_{34}Sn_2N_4Se_7Hg_2$ (dbuH = 1,5-diazabicyclo[5.4.0]undec-7-ene, $C_9H_{16}N_2$). This compound crystallizes in the triclinic structure with the lattice parameters $a = 1039.0 \pm 0.2$, $b = 1224.4 \pm 0.2$, and $c = 1397.7 \pm 0.3$ pm, and $\alpha = 86.58° \pm 0.03°$, $\beta = 79.51° \pm 0.03°$, and $\gamma = 70.14° \pm 0.03°$, a calculated density of 3.025 g·cm$^{-3}$, and an energy gap of 2.43 eV (Xiong et al. 2013a). Yellow block crystals of this compound were synthesized by heating a mixture of HgCl$_2$, Sn, Se, $C_9H_{16}N_2$, H$_2$O, and polyethylene glycol 400 (surfactant) at 160°C for 20 days.

**[Hg$_2$Sn$_2$Se$_6$(Se$_2$)][dbuH]$_2$** or $C_{18}H_{34}Sn_2N_4Se_8Hg_2$ (dbuH = 1,5-diazabicyclo[5.4.0]undec-7-ene, $C_9H_{16}N_2$). This compound crystallizes in the monoclinic structure with the lattice parameters $a = 1472.89 \pm 0.02$, $b = 1860.450 \pm 0.010$, and $c = 1352.000 \pm 0.010$ pm and $\beta = 111.5710° \pm 0.0010°$, a calculated density of 3.040 g·cm$^{-3}$, and an energy gap of 2.17 eV (Xiong et al. 2013a). Red rod crystals of this compound were synthesized by heating a mixture of HgCl$_2$, Sn, Se, $C_9H_{16}N_2$, H$_2$O, and polyethylene glycol 400 (surfactant) at 160°C for 4 days. If surfactant was removed and the reaction was performed under the same conditions, only [Hg$_2$Sn$_2$Se$_7$][dbuH]$_2$ was obtained.

**[(C$_{21}$H$_{38}$N)$_{4-2x}$Hg$_x$SnSe$_4$] (1.0 < $x$ < 1.3).** This compound is a medium band gap semiconductor with $E_g = 2.2$ eV (Trikalitis et al. 2001). It was prepared by the following procedure. A solution of 9.94 mM of surfactant (cetylpyridinium bromide, $C_{21}H_{38}NBr·H_2O$) in 20 mL of formamide was heated at 75°C for a few minutes forming a clear solution. To this solution, 1.00 mM of $K_4SnSe_4$ was added and the mixture was stirred for 30 min, forming a clear deep-red solution. To this, a solution of 1.00 mM of the mercury salt in 10 mL of formamide was added slowly, using a pipet. A precipitate formed immediately, and the mixture was stirred while warm for 20 h. The products were isolated by suction, filtered, and washed with large amount of warm formamide and $H_2O$. The solids were dried under vacuum overnight. A dark-orange product was formed. All reactions were carried out under $N_2$ atmosphere.

## 8.8   Hg–H–C–N–P–I–Se

$[Ph_3P)_2N[[Hg_4(SePh)_6I_3]$ or $C_{72}H_{60}NP_2I_3Se_6Hg_4$. This multinary compound is formed in this system (Dean and Manivannan 1990a). It was prepared by the following procedure. A mixture of $HgI_2$ (0.11 mM), $[(Ph_3P)_2N]I$ (0.22 mM), and $Hg(SePh)_2$ (0.32 mM) in $CH_2Cl_2$ (3 mL) gave a clear yellow solution. This solution was layered with a mixture of $Me_2CO$ (2 mL) and $Et_2O$ (2 mL), and the mixture was left at 0°C–5°C overnight for crystallization to occur. The yellow crystalline product was separated by decantation, washed with $Et_2O$, and dried in vacuo. All synthesis and preparation of samples were carried out under an Ar or $N_2$ atmosphere.

## 8.9   Hg–H–C–N–As–Se

Some compounds are formed in this system.

$(Me_4N)[HgAsSe_3]$ or $C_4H_{12}NAsSe_3Hg$. This compound crystallizes in the monoclinic structure with the lattice parameters $a = 711.5 \pm 0.1$, $b = 1746.4 \pm 0.5$, and $c = 935.6 \pm 0.2$ pm, and $\beta = 91.34° \pm 0.01°$ at 173 K, a calculated density of 3.35 g·cm$^{-3}$, and an energy gap of 2.4 eV (Chou and Kanatzidis 1996). To obtain it, a mixture of $HgCl_2$, $K_3AsSe_3$, and $Me_4NCl$ was thoroughly mixed and sealed in a Pyrex tube with some water. The reaction was run at 110°C for 2 days. Above 650°C $(Me_4N)[HgAsSe_3]$ completely evaporates. All syntheses were carried out under a dry $N_2$ atmosphere.

$(Et_4N)[HgAsSe_3]$ or $C_8H_{20}NAsSe_3Hg$. This compound crystallizes in the monoclinic structure with the lattice parameters $a = 717.5 \pm 0.2$, $b = 1890.7 \pm 0.4$, and $c = 1089.7 \pm 0.3$ pm and $\beta = 99.56° \pm 0.02°$ at 173 K, a calculated density of 2.93 g·cm$^{-3}$, and an energy gap of 2.4 eV (Chou and Kanatzidis 1996). To obtain it, a mixture of $HgCl_2$, $K_3AsSe_3$ and $Et_4NCl$ was thoroughly mixed and sealed in a Pyrex tube with some water. The reaction was run at 110°C for 2 days. Above 650°C $(Et_4N)[HgAsSe_3]$ completely evaporates. All syntheses were carried out under a dry $N_2$ atmosphere.

## 8.10   Hg–H–C–N–O–Se

Some compounds are formed in this system.

$[MeHg\{SeC(NH_2)_2\}]NO_3$ or $C_2H_7N_3O_3SeHg$. This compound crystallizes in the orthorhombic structure with the lattice parameters $a = 752.4 \pm 0.1$, $b = 1120.4 \pm 0.2$, and $c = 973.8 \pm 0.2$ pm and the experimental and calculated densities 3.22 and 3.241 g·cm$^{-3}$, respectively (Carty et al. 1979). To obtain this compound, a hot deoxygenated aqueous solution of MeHgNO$_3$ (5.0 mM) was added to a refluxing solution of a stoichiometric amount of selenourea under N$_2$. After ca. 30 min, the solution was filtered and slowly evaporated to low volume. Failure to adequately protect the solution from air leads to extensive decomposition. Colorless light-sensitive crystals of the title compound were filtered off, dried in a desiccator, and stored in the dark.

**[Se(HgMe$_3$)$_3$]NO$_3$ or C$_3$H$_9$NO$_3$SeHg$_3$.** This compound melts at 120°C with decomposition (Breitinger and Morell 1974). It is formed on the reaction of H$_2$Se with MeHgNO$_3$ in 50 mL of diethyl ether under an N$_2$ atmosphere. The colorless precipitate that is formed immediately is filtered by suction and recrystallized from Pr$^i$OH.

**[(Ph$_2$Se)$_2$Hg$_2$(NO$_3$)$_2$] or C$_{24}$H$_{20}$N$_2$O$_6$Se$_2$Hg$_2$.** This compound melts at 108°C and decomposes at 160°C (Brodersen et al. 1977). It was obtained by the following procedure. To 50 mL of 0.01 M solution of diphenylselenide in methanol 25 mL of 0.01 M solution Hg$_2$(NO$_3$)$_2$ in the same solvent was added with stirring. Colorless needlelike crystals that formed were filtered off, washed quickly with a small amount of MeOH, and dried in air. The title compound was recrystallized from methanol.

## 8.11   Hg–H–C–N–O–Cl–Se

**[MeHg{SeC(NH$_2$)$_2$}]ClO$_4$ or C$_2$H$_7$N$_2$O$_4$ClSeHg.** This multinary compound is formed in this system (Carty et al. 1979). To obtain it, hot deoxygenated aqueous solution of MeHgClO$_4$ (5.0 mM) was added to a refluxing solution of a stoichiometric amount of selenourea under N$_2$. After ca. 30 min, the solution was filtered and slowly evaporated to low volume. Failure to adequately protect the solution from air leads to extensive decomposition. Colorless light-sensitive crystals of the title compound were filtered off, dried in a desiccator, and stored in the dark.

## 8.12   Hg–H–C–N–Se

Some compounds are formed in this system.

**[MeSeHgCN] or C$_2$H$_3$NSeHg.** When dimethyl diselenide was added to an excess of 5% aqueous mercuric cyanide, a yellow solid separated (Bird and Challenger 1942). It decomposes at 101°C with frothing and turns black. When the fission was repeated in aqueous alcoholic solution, a yellow precipitate, presumably MeSeHgCN, was obtained.

**[(Me$_3$EtN)$_2${Hg(Se$_4$)$_2$}] or C$_{10}$H$_{28}$N$_2$Se$_8$Hg.** This compound crystallizes in the monoclinic structure with the lattice parameters $a = 1251.03 \pm 0.02$, $b = 1129.42 \pm 0.02$, and $c = 1669.40 \pm 0.03$ pm, and $\beta = 94.980° \pm 0.001°$ at 173 K (Kim and Kim 2014). All experiments were performed under an atmosphere of dry Ar or N$_2$ using either a glovebox or a Schlenk line. To a 50 mL DMF solution of 1.5 mM K$_2$Se$_4$ and 1.5 mM of Me$_3$EtNI, a 10 mL DMF solution of 0.75 mM HgCl$_2$ was added dropwise over a 20 min period. After removing undissolved precipitates by filtration, 60 mL of ether was slowly layered over the filtrate solution. Upon standing at room temperature for 3 days, dark-purple crystals were yielded. These crystals were isolated and washed with ether several times. More crystals were obtained upon layering an additional 50 mL of ether over the solution after the isolation of the first crop of crystals.

**[(Et$_4$N)$_2${Hg(Se$_4$)$_2$}] or C$_{16}$H$_{40}$N$_2$Se$_8$Hg.** To obtain this compound, a mixture of Li$_2$Se (2 mM), black Se (6 mM), and Et$_4$NCl (2 mM) in MeCN (30 mL) and Et$_3$N (5 mL) was stirred for 1 h. A solution of mercury ethylxanthate, Hg(EtCOS$_2$)$_2$, (1 mM) in DMF (10 mL) was added to that mixture. The resulting solution was stirred for 1 h and then filtered. The volume of the filtrate was reduced to 20 mL, and Pr$^i$OH (20 mL) was layered on it. Crystals were produced in 2 days. All the manipulations were carried out under an N$_2$ atmosphere (Ansari et al. 1990).

**[(Et$_4$N)$_4${Hg$_7$Se$_9$}]] or C$_{32}$H$_{80}$N$_4$Se$_9$Hg$_7$.** This compound crystallizes in the orthorhombic structure with the lattice parameters $a = 1606.6 \pm 0.5$, $b = 1459.4 \pm 0.3$, and $c = 2475.4 \pm 0.7$ pm at 166 K and a calculated density of 3.02 g·cm$^{-3}$ (Kim and Kanatzidis 1991). The reaction of HgCl$_2$ with Na$_2$Se in DMF solution in the presence of Et$_4$NBr followed by filtration and careful dilution with ether produces (Et$_4$N)$_4$[Hg$_7$Se$_9$].

**[(Et$_4$N)$_4${Hg$_7$Se$_{10}$}]] or C$_{32}$H$_{80}$N$_4$Se$_{10}$Hg$_7$.** This compound crystallizes in the triclinic structure with the lattice parameters $a = 1228.1 \pm 0.4$, $b = 1255.5 \pm 0.6$, and $c = 2115.5 \pm 0.6$ pm and $\alpha = 96.15° \pm 0.03°$, $\beta = 98.48° \pm 0.02°$, and $\gamma = 107.84° \pm 0.03°$ and a calculated density of 2.97 g·cm$^{-3}$ (Kim and Kanatzidis 1991). It was obtained by the reaction between a DMF solution of HgCl$_2$ and Na$_2$Se$_2$ in a 2:3 molar ratio in the presence of Et$_4$NBr.

## 8.13   Hg–H–C–N–Cl–Se

Some compounds are formed in this system.

**[MeHg{SeC(NH$_2$)$_2$}]Cl or C$_2$H$_7$N$_2$ClSeHg.** To obtain this compound, hot deoxygenated aqueous solution of MeHgCl (5.0 mM) was added to a refluxing solution of a stoichiometric amount of selenourea under N$_2$ (Carty et al. 1979). After ca. 30 min, the solution was filtered and slowly evaporated to low volume. Failure to adequately protect the solution from air leads to extensive decomposition. Colorless light-sensitive crystals of the title compound were filtered off, dried in a desiccator, and stored in the dark.

**[Hg{SeC(NH$_2$)$_2$}$_2$Cl$_2$] or C$_2$H$_8$N$_4$Cl$_2$Se$_2$Hg.** This compound melts at 302°C–303°C with decomposition (Isab and Wazeer 2005). It was prepared at room temperature by mixing solutions of selenourea with HgCl$_2$ in a minimum amount of distilled H$_2$O (2–10 mL) in a 2:1 molar ratio and by stirring for 1 or 2 h. The isolated product was dried thoroughly.

**[Hg(C$_3$H$_6$N$_2$Se)$_2$Cl$_2$] or C$_6$H$_{12}$N$_4$Cl$_2$Se$_2$Hg.** This compound melts at 229°C–230°C with decomposition (Devillanova and Verani 1977). It was obtained by boiling an absolute EtOH solution of HgCl$_2$ under reflux with ethyleneselenourea (1:2 molar ratio).

**[{HgCl(C$_5$H$_5$N)$_{0.5}$(SeBu$^t$)}$_4$] or C$_{26}$H$_{46}$N$_2$Cl$_4$Se$_4$Hg$_4$.** This compound crystallizes in the monoclinic structure with the lattice parameters $a = 1215.1 \pm 0.5$, $b = 1673.8 \pm 0.7$, and $c = 1013.8 \pm 0.6$ pm and $\beta = 90.93° \pm 0.04°$ and a calculated density of 2.65 g·cm$^{-3}$ (Arnold and Canty 1982). A pyridine solution containing equimolar amounts of $Hg(SeEt)_2$ and $HgCl_2$ gave colorless crystals of $[\{HgCl(C_5H_5N)_{0.5}(SeBu^t)\}_4]$.

**$[\{HgCl(C_5H_5N)(SeEt)\}_4]$ or $C_{28}H_{40}N_4Cl_4Se_4Hg_4$.** This compound crystallizes in the monoclinic structure with the lattice parameters $a = 804.4 \pm 0.5$, $b = 1738.7 \pm 1.4$, and $c = 1558.5 \pm 2.1$ pm and $\beta = 101.75° \pm 0.02°$ and a calculated density of 2.63 g·cm$^{-3}$ (Arnold and Canty 1982). A pyridine solution containing equimolar amounts of $Hg(SeEt)_2$ and $HgCl_2$ gave crystals of $[\{HgCl(C_5H_5N)(SeEt)\}_4]$ that were recrystallized from 50% pyridine in EtOH.

**$[Et_4N]_2[(\mu\text{-}SePh)_6(HgCl)_4]$ or $C_{52}H_{70}N_2Cl_4Se_6Hg_4$.** To obtain this compound, into a mixture of $HgCl_2$ (0.14 mM) and $Et_4NCl·H_2O$ (0.27 mM) in $CH_2Cl_2$ (5 mL) at room temperature solid $Hg(SePh)_2$ (0.40 mM) was stirred, producing a yellow solution (Dean and Manivannan 1990a). The solution was filtered, then the filtrate was layered with $Et_2O$ (1 mL), and the mixture was left in the refrigerator overnight for crystallization to occur. The yellow crystalline product was separated by decantation, washed with $Et_2O$, and dried in vacuo. All synthesis and preparation of samples were carried out under an Ar or $N_2$ atmosphere.

## 8.14   Hg–H–C–N–Br–Se

Some compounds are formed in this system.

**$[MeHg\{SeC(NH_2)_2\}]Br$ or $C_2H_7N_2BrSeHg$.** To obtain this compound, a hot deoxygenated aqueous solution of MeHgBr (5.0 mM) was added to a refluxing solution of a stoichiometric amount of selenourea under $N_2$ (Carty et al. 1979). After ca. 30 min, the solution was filtered and slowly evaporated to low volume. Failure to adequately protect the solution from air leads to extensive decomposition. Colorless light-sensitive crystals of the title compound were filtered off, dried in a desiccator, and stored in the dark.

**$[MeHg\{SeC(NH_2)_2\}(NMe_2)]Br$ or $C_4H_{13}N_3BrSeHg$.** This compound crystallizes in the triclinic structure with the lattice parameters $a = 625.1 \pm 0.1$, $b = 834.8 \pm 0.1$, and $c = 957.0 \pm 0.2$ pm and $\alpha = 104.39° \pm 0.01°$, $\beta = 78.29° \pm 0.01°$, and $\gamma = 97.06° \pm 0.01°$ and the experimental and calculated densities 3.14 and 3.140 g·cm$^{-3}$, respectively (Carty et al. 1979). To obtain this compound, hot deoxygenated aqueous solution of $MeHg(NMe_2)Br$ (5.0 mM) was added to a refluxing solution of a stoichiometric amount of selenourea under $N_2$. After ca. 30 min, the solution was filtered and slowly evaporated to low volume. Failure to adequately protect the solution from air leads to extensive decomposition. Colorless light-sensitive crystals of the title compound were filtered off, dried in a desiccator, and stored in the dark.

**$[Hg(C_3H_6N_2Se)_2Br_2]$ or $C_6H_{12}N_4Br_2Se_2Hg$.** This compound melts at 220°C–221°C with decomposition (Devillanova and Verani 1977). It was obtained by boiling an absolute EtOH solution of $HgBr_2$ under reflux with ethyleneselenourea (1:2 molar ratio).

## 8.15   Hg–H–C–N–I–Se

[Hg(C$_3$H$_6$N$_2$Se)$_2$I$_2$] or C$_6$H$_{12}$N$_4$I$_2$Se$_2$Hg. This multinary compound is formed in this system (Devillanova and Verani 1977). It melts at 216°C–217°C with decomposition and was obtained by boiling an absolute MeOH solution of HgI$_2$ under reflux with ethyleneselenourea (1:2 molar ratio).

## 8.16   Hg–H–C–P–As–Se

(Ph$_4$P)$_2$[Hg$_2$As$_4$Se$_{11}$] or C$_{48}$H$_{40}$P$_2$As$_4$Se$_{11}$Hg$_2$. This multinary compound is formed in this system (Chou and Kanatzidis 1996). It crystallizes in the triclinic structure with the lattice parameters $a = 1032.9 \pm 0.2$, $b = 1701.7 \pm 0.3$, and $c = 1748.5 \pm 0.3$ pm and $\alpha = 92.70° \pm 0.01°$, $\beta = 99.56° \pm 0.02°$, and $\gamma = 103.71° \pm 0.01°$ at 173 K, a calculated density of 2.62 g·cm$^{-3}$, and an energy gap of 2.0 eV. To obtain it, a mixture of HgCl$_2$, K$_3$AsSe$_3$, and Ph$_4$PBr was thoroughly mixed and sealed in a Pyrex tube with some water. The reaction was run at 110°C for 1 week. Above 650°C (Ph$_4$P)$_2$[Hg$_2$As$_4$Se$_{11}$]] completely evaporates. All syntheses were carried out under a dry N$_2$ atmosphere.

## 8.17   Hg–H–C–P–O–Cl–Se

Some compounds are formed in this system.

[Hg{SeP(c-C$_6$H$_{11}$)$_3$}$_2$](ClO$_4$)$_2$ or C$_{36}$H$_{66}$P$_2$O$_8$Cl$_2$Se$_2$Hg. This compound was prepared by the next procedure (Dean and Manivannan 1990b). To a solution of HgCl$_2$ (2.46 mM) in 20 mL of Me$_2$CO a solution of SeP(c-C$_6$H$_{11}$)$_3$ (4.95 mM) in 80 mL of Me$_2$CO was added. The mixture was reduced in volume and upon cooling gave a white solid, which was separated by filtration and dried in vacuo. The chloride complex was converted to the ClO$_4^-$ salt using AgClO$_4$. After recrystallization from a CH$_2$Cl$_2$/hexane mixture, the title compound was obtained.

[Hg(SePPh$_3$)$_3$](ClO$_4$)$_2$ or C$_{54}$H$_{45}$P$_3$O$_8$Cl$_2$Se$_3$Hg. This compound was synthesized by direct reaction of Hg(ClO$_4$)$_2$ and Ph$_3$PSe (Dean and Manivannan 1990b). A mixture of Hg(ClO$_4$)$_2$·3H$_2$O (1.2 mM) and Ph$_3$PSe (3.6 mM) in CH$_2$Cl$_2$ (60 mL) was stirred at room temperature and then warmed carefully on a steam bath until dissolution of the ClO$_4^-$ salt was complete. The solution was filtered and then concentrated, at which point the addition of hexane gave the crude material. After its recrystallization from a CH$_2$Cl$_2$/hexane mixture, the title compound as a yellow crystalline product was obtained.

[Hg$_4$(SePh)$_6$(PEt$_3$)$_4$](ClO$_4$)$_2$ or C$_{60}$H$_{90}$P$_4$O$_8$Cl$_2$Se$_6$Hg$_4$. To obtain this compound, to a solution of Hg(PEt$_3$)$_2$(ClO$_4$)$_2$ (0.30 mM) and PEt$_3$ (0.62 mM) in 5 mL of CHCl$_3$ 0.90 mM of solid Hg(SePh)$_2$ was added with stirring (Dean et al. 1987). A clear yellow solution was obtained, from which the product separated as thick yellow oil at the bottom. After 10 min, the solvent was removed under a flow of Ar. The viscous yellow product was dissolved in 5 mL of acetone. The acetone solution was filtered and cooled at 0°C before the addition of 10 mL of diethyl ether to form a layer over the top. The mixture was left overnight at 0°C for crystallization to occur. The yellow crystalline solid product was separated by decantation of the mother liquor, washed with diethyl ether, and dried in vacuo. All syntheses were performed under an Ar atmosphere.

**[Hg$_4$(SePh)$_6$(PPh$_3$)$_4$](ClO$_4$)$_2$** or **C$_{108}$H$_{90}$P$_4$O$_8$Cl$_2$Se$_6$Hg$_4$**. To obtain this compound, solid PPh$_3$ (0.46 mM) was added to a solution, containing 0.23 mM of Hg(PPh$_3$)$_2$(ClO$_4$)$_2$ in 10 mL of acetone precooled to 0°C (Dean et al. 1987). The solid dissolved to give a clear solution. Into this was stirred 0.35 g of Hg(SePh)$_2$ to produce, within a few minutes, a clear yellow solution. Cyclohexane was added carefully to form a layer. After the solution was kept at 5°C for 24 h, the yellow flaky crystalline product was collected by filtration, washed with cyclohexane, and dried in vacuo. All syntheses were performed under an Ar atmosphere.

## 8.18   Hg–H–C–P–Se

Some compounds are formed in this system.

**[(Ph$_4$P)(PhSe)$_3$Hg]** or **C$_{42}$H$_{35}$PSe$_3$Hg**. This compound melts at 102°C with decomposition (Liesic von and Klar 1977) and crystallizes in the monoclinic structure with the lattice parameters $a = 1095$, $b = 1358$, and $c = 2457$ pm and $\beta = 98°$ (Behrens et al. 1977). It was prepared from the reaction of Hg(SePh)$_2$, NaSePh, and Ph$_4$PCl in the liquid NH$_3$ (Behrens et al. 1977, Liesic von and Klar 1977). Its single crystals were grown by recrystallization from a THF/EtOH solution of the title compound.

**[(Ph$_4$P)$_2${Hg(Se$_4$)$_2$}]** or **C$_{48}$H$_{40}$P$_2$Se$_8$Hg**. This compound crystallizes in the monoclinic structure with the lattice parameters $a = 1046.0 \pm 0.3$, $b = 2129.4 \pm 0.2$, and $c = 2213.1 \pm 0.3$ pm and $\beta = 97.62° \pm 0.02°$ and a calculated density of 2.07 g·cm$^{-3}$ (Magul et al. 1991) ($a = 2312.8 \pm 0.7$, $b = 2031.1 \pm 0.5$, $c = 2395.6 \pm 0.8$ pm and $\beta = 119.43° \pm 0.01°$ [Banda et al. 1989]). Bis-tetraselenide complex [Hg(Se$_4$)$_2$]$^{2-}$ could be readily prepared by the reaction of Se, Na, and HgCl$_2$ in DMF and crystallized with Ph$_4$P$^+$ (Banda et al. 1989). The general preparative procedure involves HgCl$_2$, Se, and Na in a molar ratio of 1:8:4 under N$_2$ atmosphere, with Ph$_4$PBr slowly diffused into the filtered reaction solution.

The title compound could also be prepared as dark-red crystals by the reaction of [(Ph$_4$P)$_2${Sn(Se$_4$)$_3$}] with Hg(OAc)$_2$ in DMF solution (Magul et al. 1991). It is possible for this compound to have two polymorphic modifications.

**[Ph$_4$P]$_2$[Hg(SePh)$_4$]** or **C$_{72}$H$_{60}$P$_2$Se$_4$Hg**. This compound melts at 96°C with decomposition (Liesic von and Klar 1977). It was prepared from the reaction of Hg(SePh)$_2$, NaSePh, and Ph$_4$PCl in the liquid NH$_3$.

**Hg$_4$[Ph$_3$P]$_2$(C$_6$H$_5$Se)$_8$** or **C$_{84}$H$_{70}$P$_2$Se$_8$Hg$_4$**. This compound melts at 134.5°C–135.5°C and crystallizes in the triclinic structure with the lattice parameters $a = 1373.2 \pm 0.1$, $b = 1478.8 \pm 0.1$, and $c = 2264.1 \pm 0.2$ pm and $\alpha = 73.095° \pm 0.002°$, $\beta = 73.579° \pm 0.002°$, and $\gamma = 65.783° \pm 0.001°$ at 100 K (Wang et al. 2005). The reactions were carried out under a nitrogen atmosphere using standard Schlenk techniques. At room temperature, a solution of HgBr$_2$ (0.5 mM) and Ph$_3$P (0.5 mM) in acetonitrile (15 mL) was added to a mixture of benzeneselenol (1.5 mM) and $Et_3N$ (1.5 mM) in acetonitrile (5.0 mL). Three days later, crystals were obtained from slow evaporation of an acetonitrile and methanol solution of the title compound.

$[(C_6H_5)_4P]_3[Hg_2(SeC_6H_5)_7]$ or $C_{114}H_{95}P_3Se_7Hg_2$. This compound melts at 96°C with decomposition. It was prepared from the reaction of $Hg(SePh)_2$, NaSePh, and $Ph_4PCl$ in the liquid $NH_3$ (Liesic von and Klar 1977).

## 8.19   Hg–H–C–P–Te–Se

Some compounds are formed in this system.

$[Ph_4P]_2[HgTe_{0.58}Se_{7.42}]$ or $C_{48}H_{40}P_2Te_{0.58}Se_{7.42}Hg$. This compound crystallizes in the monoclinic structure with the lattice parameters $a = 2291.1 \pm 0.3$, $b = 2016.1 \pm 0.4$, and $c = 2368.5 \pm 0.2$ pm and $\beta = 118.65° \pm 0.01°$ at 108 K and a calculated density of 2.131 g·cm$^{-3}$ (Bollinger and Ibers 1995). To prepare it, $[Ph_4P]_2[Hg(Se_4)_2]$ (0.25 mM) and $TePEt_3$ (0.25 mM) were added to separate Schlenk flasks and reacted as described for $[Ph_4P]_2[Hg(Te_2Se_2)_2]$. The reaction mixture darkened slightly with each drop of $TePEt_3$ solution added, but the final solution was not much darker than the original. This mixture was filtered into a fresh flask, and $Et_2O$ (5.0 mL) was layered on top. Several very large, red, block-shaped crystals grew overnight.

$[Ph_4P]_2[HgTe_{1.52}Se_{6.48}]$ or $C_{48}H_{40}P_2Te_{1.52}Se_{6.48}Hg$. This compound crystallizes in the monoclinic structure with the lattice parameters $a = 2294.0 \pm 0.3$, $b = 2022.0 \pm 0.2$, and $c = 2379.7 \pm 0.2$ pm and $\beta = 118.50° \pm 0.01°$ at 108 K and a calculated density of 2.173 g·cm$^{-3}$ (Bollinger and Ibers 1995). To prepare it, $[Ph_4P]_2[Hg(Se_4)_2]$ (0.25 mM) and $TePEt_3$ (0.50 mM) were added to separate Schlenk flasks and reacted as described for $[Ph_4P]_2[Hg(Te_2Se_2)_2]$. The reaction mixture darkened slightly with each drop of $TePEt_3$ solution added, and the final solution was somewhat darker than the original. This mixture was filtered into a fresh flask, and $Et_2O$ (5.0 mL) was layered on top. A large number of red, block-shaped crystals grew overnight.

$[Ph_4P]_2[Hg(Te_2Se_2)_2]$ or $C_{48}H_{40}P_2Te_4Se_4Hg$. This compound crystallizes in the tetragonal structure with the lattice parameters $a = 2117.4 \pm 0.5$ and $c = 1097.6 \pm 0.3$ pm at 108 K and a calculated density of 2.302 g·cm$^{-3}$ (Bollinger and Ibers 1995). To obtain it, $[Ph_4P]_2[Hg(Se_4)_2]$ (0.25 mM) and $TePEt_3$ (1.00 mM) were placed in separate Schlenk flasks. Because $TePEt_3$ decomposes under vacuum by the loss of $PEt_3$, the flask containing $TePEt_3$ was filled with dry $N_2$ by performing 10 quick cycles of evacuation and $N_2$ refill. The flask containing $[Ph_4P]_2[Hg(Se_4)_2]$ was filled with $N_2$ by the more normal process of two 15 min cycles of evacuation followed by $N_2$ refills. $[Ph_4P]_2[Hg(Se_4)_2]$ was dissolved in DMF (10 mL) to afford a bright-red solution, and $TePEt_3$ was dissolved in DMF (1 mL) to afford a bright-yellow solution. The $TePEt_3$ solution was added dropwise via cannula to the stirred $[Ph_4P]_2[Hg(Se_4)_2]$ solution. The reaction mixture darkened with each drop of $TePEt_3$ solution added, until the final solution became dark red–brown. A 1.0 M $PEt_3$ solution in THF (0.2 mL) was added to the solution, and the mixture was then filtered into a fresh flask. Toluene (5 mL) was layered on top of the solution, and a large number of dark, needle-shaped crystals grew overnight. Additional crystals were obtained by layering more toluene on the mother liquor.

## 8.20 Hg–H–C–P–F–Se

Some compounds are formed in this system.

[(CF$_3$Se)$_2$Hg·Ph$_3$P] or C$_{20}$H$_{15}$PF$_6$SeHg. This compound was prepared by mixing strong benzene solutions of the reactants in stoichiometric proportions and recrystallized from benzene (Clase and Ebsworth 1965).

[(CF$_3$Se)$_2$Hg·2Ph$_3$P] or C$_{38}$H$_{30}$P$_2$F$_6$SeHg. This compound was prepared by mixing strong benzene solutions of the reactants in stoichiometric proportions and recrystallized from benzene (Clase and Ebsworth 1965).

## 8.21 Hg–H–C–P–Cl–Se

[HgCl$_2$·Ph$_3$PSe] or C$_{18}$H$_{15}$PCl$_2$SeHg. This multinary compound is formed in this system. It melts at 231°C (King and McQuillan 1967) and crystallizes in the monoclinic structure with the lattice parameters $a = 1051 \pm 7$, $b = 1305 \pm 7$, $c = 1783 \pm 5$ pm and $\beta = 128°30' \pm 20'$ and the experimental and calculated densities 2.10 and 2.11 g·cm$^{-3}$, respectively (Glasser et al. 1969). Flat, slightly elongated, diamond-shaped crystals were precipitated upon mixing solutions of Ph$_3$PSe and HgCl$_2$ in acetone.

## 8.22 Hg–H–C–P–Br–Se

[HgBr$_2$·Ph$_3$PSe] or C$_{18}$H$_{15}$PBr$_2$SeHg. This multinary compound is formed in this system (King and McQuillan 1967). It sublimes at 260°C.

## 8.23 Hg–H–C–P–I–Se

Some compounds are formed in this system.

[HgI$_2$·Ph$_3$PSe] or C$_{18}$H$_{15}$PI$_2$SeHg. This compound melts at 221°C (King and McQuillan 1967). Its light-yellow crystals were prepared by the interaction of Ph$_3$PSe and HgI$_2$ in EtOH or MeOH (Nicpon and Meek 1966).

[HgI$_2$·2Bu$^n_3$PSe] or C$_{24}$H$_{54}$P$_2$I$_2$Se$_2$Hg. This compound was obtained by the interaction of Bu$^n_3$PSe and HgI$_2$ in EtOH or MeOH (Nicpon and Meek 1966).

## 8.24 Hg–H–C–As–I–Se

[HgI$_2$·Ph$_3$AsSe] or C$_{18}$H$_{15}$AsI$_2$SeHg. This multinary compound is formed in this system (Nicpon and Meek 1966). It was prepared by the interaction of Ph$_3$AsSe and HgI$_2$ in EtOH or MeOH.

## 8.25   Hg–H–C–O–Se

**[MeSeHgO₂CMe] or C₃H₆O₂SeHg, acetato(methaneselenolato)mercury(II).**
This quinary compound is formed in this system (Arnold and Canty 1981). To obtain
it, Hg(SeMe)₂ (0.34 mM) was added to a solution of Hg(OAc)₂ (0.34 mM) in H₂O
(5 mL) and EtOH (3 mL). After 15 min, the colorless solution was filtered to remove
a small amount of unreacted Hg(SeMe)₂, and the filtrate deposited colorless crystals
upon slow evaporation. They were collected and dried over P₂O₅.

## 8.26   Hg–H–C–O–Cl–Se

**[(Ph₂Se)₄Hg₂(ClO₄)₂] or C₄₈H₄₀O₈Cl₂Se₄Hg₂.** This multinary compound is formed
in this system. It exists in two polymorphic modifications. The first one crystallizes in
the monoclinic structure with the lattice parameters $a = 1391.0 \pm 0.9$, $b = 1457.5 \pm 0.9$,
and $c = 1210.9 \pm 0.6$ pm and $\beta = 92.6°$ and the experimental and calculated densities
2.08 and 2.27 g·cm⁻³, respectively (Brodersen et al. 1977). The second modifications
also crystallize in the monoclinic structure with the lattice parameters $a = 1160.9$,
$b = 1446.2$, and $c = 1439.1$ pm and $\beta = 98.2°$ and the experimental and calculated
densities 2.27 and 2.13 g·cm⁻³, respectively (Brodersen et al. 1978). To obtain the
title compound, 0.01 M solution of diphenylselenide in MeOH (40 mL) was added to
0.01 M solution (10 mL) of Hg₂(ClO₄)₂ and 2 mL of concentrated HCl in methanolic
solution (Brodersen et al. 1977). Incubation overnight in the refrigerator resulted
in the formation of red and yellow cylinder-like crystals in the ratio of 4:1 that was
filtered, washed with a small amount of MeOH, and dried in air. Red cylinder-like
crystals begin to melt at 185°C and yellow at 195°C. The clear melt is formed only
at 270°C.

## 8.27   Hg–H–C–Cl–Se

Some compounds are formed in this system.

**[MeSeClHg·HgCl₂] or CH₃Cl₃SeHg₂.** This compound melts at 123°C (Bird
and Challenger 1942). Adding Me₂Se₂ alone, or with alcohol, to the mixture of
HgCl₂ (10 g), H₂O (80 mL), and HCl (20 mL) gave yellow insoluble needles of
MeSeClHg·HgCl₂.

**[Me₂Se·HgCl₂] or C₂H₆Cl₂SeHg.** This compound melts at 153°C with decomposition (Bird and Challenger 1942).

**[MeEtSe·HgCl₂] or C₃H₈Cl₂SeHg.** This compound melts at 98°C and decomposes
at 140°C (Bird and Challenger 1942). To obtain it, Me₂Se₂ was dissolved in EtOH and
treated with just over two atomic proportions of sodium, giving an orange solution
and a black precipitate. Upon the addition of ethyl iodide (2 mol), reaction occurred
at once. The clear yellow solution was refluxed for 30 min, water added, and the
mixture distilled in steam. Volatile matter was removed in an air stream through
mercuric cyanide into the mixture of HgCl₂ (10 g), H₂O (80 mL), and HCl (20 mL),
where white needles formed.

**[MeSeEt·2HgCl₂]** or **C₃H₈Cl₄SeHg₂**. This compound melts at 141.5°C (Bird and Challenger 1942).

**[Et₂Se·HgCl₂] or C₄H₁₀Cl₂SeHg**. This compound melts at 93°C (Bird and Challenger 1942).

**[MePrⁿSe·HgCl₂]** or **C₄H₁₀Cl₂SeHg**. This compound melts at 88°C (Bird and Challenger 1942).

**[Prⁿ₂Se·HgCl₂]** or **C₆H₁₄Cl₂SeHg**. This compound melts at 93°C–94°C (Bird and Challenger 1942).

## 8.28  Hg–C–O–F–Se

**[(CF₃O₂Se)₂Hg]** or **C₂O₄F₆Se₂Hg, mercuric trifluoromethylselenonate**. This quinary compound is formed in this system (Clase and Ebsworth 1965). It was obtained if (CF₃Se)₂Hg·Ph₃P or (CF₃Se)₂Hg·2Ph₃P was dissolved in concentrated HNO₃ with the evolution of brown fumes; the reaction occurs more rapidly upon heating. A white crystalline solid of the title compound was separated as the solution cooled.

## 8.29  Hg–C–F–Cl–Se

**[CF₃Se·HgCl]** or **CF₃ClSeHg**. This quinary compound is formed in this system. It melts at 175°C–177°C (Clase and Ebsworth 1965) [at 185°C–190°C (Dale et al. 1958)] and was obtained by evaporation of the solvent from a methanolic solution of HgCl₂ and Hg(SeCF₃)₂ (Clase and Ebsworth 1965) or by evaporation of the solvent from a solution of Hg(SeC₃F₇)₂ (0.212 g) and HgCl₂ (0.116 g) in ether (Dale et al. 1958).

## 8.30  Hg–C–F–Br–Se

**[CF₃Se·HgBr]** or **CF₃BrSeHg**. This quinary compound is formed in this system (Clase and Ebsworth 1965). It melts at 160°C–162°C with decomposition and was obtained as pale-yellow crystalline solid from methanolic solution containing Hg(SeCF₃)₂ and HgBr₂ and was purified by recrystallization from benzene.

## 8.31  Hg–C–F–I–Se

**[CF₃Se·HgI]** or **CF₃ISeHg**. This quinary compound is formed in this system (Clase and Ebsworth 1965). It was prepared by fractional sublimation in an evacuated glass tube surrounded by a copper cylinder, the base of which was maintained at 45°C (4–7 days), of a fused mixture (approximately equimolar) of HgI₂ and Hg(SeCF₃)₂. The compound was obtained as lemon-yellow crystals, softening with decomposition at 110°C and stable in the dark at room temperature but darkening in light and decomposing to give HgI₂ at 45°C in 12 h.

# REFERENCES

Adel J., Weller F., Dehnicke K. Synthese und Kristallstrukturen der spirocyclischen Polyselenido-Komplexe [Na-15-Krone-5]$_2$[M(Se$_4$)$_2$] mit M = Zink, Cadmium und Quecksilber, *Z. Naturforsch.*, 43B(9), 1094–1100 (1988).

Ahle A., Dehnicke K., Maczek K., Fenske D. Synthese und Kristallstruktur von [Na(12-Krone-4)$_2$]$_2$[Hg(Se$_4$)$_2$·1.5 DMF, *Z. anorg. und allg. Chem.*, 619(10), 1699–1703 (1993).

Ansari M.A., Mahler C.H., Chorghade G.S., Lu Y.-J., Ibers J.A. Synthesis, structure, spectroscopy, and reactivity of M(Se$_4$)$_2$$^{2-}$ anions, M = Ni, Pd, Zn, Cd, Hg, and Mn, *Inorg. Chem.*, 29(19), 3832–3839 (1990).

Arnold A.P., Canty A.J. Synthesis, structure and spectroscopic studies of mercury(II) selenolates and MeHgSeBu$^t$, *Inorg. Chim. Acta*, 55, 171–176 (1981).

Arnold A.P., Canty A.J. Mercury(II) selenolates. Crystal structures of polymeric Hg(SeMe)$_2$ and the tetramericpyridinates[{HgCl(py)(SeEt)}$_4$ and [{HgCl(py)$_{0.5}$(SeBu$^t$)}$_4$], *J. Chem. Soc., Dalton Trans.*,(3), 607–613 (1982).

Banda R.M.H., Cusick J., Scudder M.L., Craig D.C., Dance I.G. Syntheses and x-ray structures of molecular metal polyselenide complexes [M(Se$_4$)$_2$]$^{2-}$ M = Zn, Cd, Hg, Ni, Pb, *Polyhedron*, 8(15), 1995–1998 (1989).

Behrens U., Hofmann K., Klar G. Chalkogenolat-Ionen und ihre Derivate, III. Kristall- und Molekülstruktur von Tetraphenylphosphonium-tris(tellurophenolato)mercurat(II), *Chem. Ber.*, 110(11), 3672–3677 (1977).

Bird M.L., Challenger F. Potassium alkaneselenonates and other alkyl derivatives of selenium, *J. Chem. Soc.*, 570–574 (1942).

Bollinger J.C., Ibers J.A. Reactions of [M(Se$_4$)$_2$]$^{2-}$ anions with TePEt$_3$: $^{77}$Se and $^{125}$Te spectra of [MTe$_n$Se$_{8-n}$]$^{2-}$ (M = Zn, Cd, Hg; $n$ = 0–4) and preparation and crystal structure of [PPh$_4$]$_2$[Hg(Te$_2$Se$_2$)$_2$], *Inorg. Chem.*, 34(7), 1859–1867 (1995).

Brandmayer M.K., Clérac R., Weigend F., Dehnen S. Ortho-chalcogenostannates as ligands: Syntheses, crystal structures, electronic properties, and magnetism of novel compounds containing ternary anionic substructures [M$_4$(μ$_4$-Se)(SnSe$_4$)$_4$]$^{10-}$ (M = Mn, Zn, Cd, Hg), {[Hg$_4$(μ$_4$-Se)(SnSe$_4$)$_3$]$^{6-}$}, or {[HgSnSe$_4$]$^{2-}$}, *Chem. Eur. J.*, 10(20), 5147–5157 (2004).

Breitinger D., Morell W. Tris(methylmercurio) selenonium salts [Se(HgCH$_3$)$_3$]X and bis(methylmercuric) chalcogenides Ch(HgCH$_3$)$_2$ (Ch = Se and Te), *Inorg. Nucl. Chem. Lett.*, 10(5), 409–411 (1974).

Brodersen K., Liehr G., Rosenthal M. Stabile Quecksilber(I)–Selen–Verbindungen, 1. Synthese und Struktur des Tetrakis(diphenylseleno)-diquecksilber(I)diperchlorats, Synthese von Bis(diphenylseleno)-diquecksilber(I)dinitrat, *Chem. Ber.*, 110(10), 3291–3296 (1977).

Brodersen K., Liehr G., Rosenthal M., Thiele G. Stabile Quecksilber(I)-Selen-Verbindungen, 2. Mitteilung. Die Struktur der roten Modifikation des Tetrakis(diphenylseleno)-diquecksilber(I)-bis(chlorat(VII)), *Z. Naturforsch.*, 33B(11), 1227–1230 (1978).

Carty A.J., Malone S.F., Taylor N.J. The selenium-mercury interaction: Synthesis, spectroscopic and x-ray structural studies of methyl mercury-selenourea complexes, *J. Organomet. Chem.*, 172(2), 201–211 (1979).

Chou J.-H., Kanatzidis M.G. Hydrothermal synthesis and characterization of (Me$_4$N) [HgAsSe$_3$], (Et$_4$N)[HgAsSe$_3$], and (Ph$_4$P)$_2$[Hg$_2$As$_4$Se$_{11}$]: Novel 1-D mercury selenoarsenates, *J. Solid State Chem.*, 123(1), 115–122 (1996).

Clase H.J., Ebsworth E.A.V. Some reactions and derivatives of bis(trifluoromethylseleno)mercury, *J. Chem. Soc.*, 940–947 (1965).

Dale J.W., Emeléus H.J., Haszeldine R.N. Organometallic and organometalloidal fluorine compounds. Part XIV. Trifluoromethyl derivatives of selenium, *J. Chem. Soc.*, 2939–2945 (1958).

Dean P.A.W., Manivannan V. A multinuclear ($^{13}$C, $^{31}$P, $^{77}$Se, $^{125}$Te, $^{199}$Hg) magnetic resonance study of the adamantane-like clusters [($\mu$-ER)$_6$(HgE′PR′$_3$)$_4$]$^{2+}$ (E = S, Se, Te; E′ = S, Se; R′ = Ph, c-C$_6$H$_{11}$) and the precursors [Hg(E′PR′$_3$)$_n$]$^{2+}$, *Can. J. Chem.*, 68(2), 214–222 (1990a).

Dean P.A.W., Manivannan V. Synthesis and multinuclear ($^{13}$C, $^{77}$Se, $^{125}$Te, $^{199}$Hg) magnetic resonance spectra of adamantane-like anions of mercury(II), [($\mu$-ER)$_6$(HgX)$_4$]$^{2-}$ (E = S, Se, Te; X = Cl, Br, I), *Inorg. Chem.*, 29(16), 2997–3002 (1990b).

Dean P.A.W., Vittal J.J., Trattner M.H. New adamantane-like mercury–chalcogen cages. Synthetic and multinuclear ($^{31}$P, $^{77}$Se, $^{199}$Hg) NMR study of [($\mu$-ER)$_6$(HgL)$_4$]$^{2+}$ (E = S or Se; L = tertiary phosphine or arsine) and related species with mixed ligands, *Inorg. Chem.*, 26(25), 4245–4251 (1987).

Devillanova F.A., Verani G. Complexes of znc(II), cadmium(II) and mercury(II) halides with ethylenselenourea, *Trans. Met. Chem.*, 2(1), 9–11 (1977).

Glasser L.S.D., Ingram L., King M.G., McQuillan G.P. Crystal and molecular structure of the dimeric 1:1 addition compound of mercury(II) chloride and triphenylphosphine selenide, *J. Chem. Soc. A: Inorg. Phys. Theor.*, 2501–2504 (1969).

Isab A.A., Wazeer M.I.M. Complexation of Zn(II), Cd(II) and Hg(II) with thiourea and selenourea: A $^1$H, $^{13}$C, $^{15}$N, $^{77}$Se and $^{113}$Cd solution and solid-state NMR study, *J. Coord. Chem.*, 58(6), 529–537 (2005).

Kim K.-W., Kanatzidis M.G. Synthesis and structural characterization of [Hg$_7$Se$_{10}$]$^{4-}$ and [Hg$_7$Se$_9$]$_n^{4n-}$: Novel metal-chalcogenide frameworks, *Inorg. Chem.*, 30(9), 1966–1969 (1991).

Kim K.-W., Kim J. Crystal structure of bis(N,N, N-trimethylethanaminium)bis(tetraselenido-$\kappa^2$Se$^1$,Se$^4$))mercurate(II), C$_{10}$H$_{28}$HgN$_2$Se$_8$, *Z. Kristallogr. New Cryst. Struct.*, 229(1), 35–36 (2014).

King M.G., McQuillan G.P. Metal halide complexes of triphenylphosphine sulphide and some related ligands, *J. Chem. Soc. A: Inorg. Phys. Theor.*, 898–901 (1967).

Kräuter G., Weller F., Dehnicke K. Synthesen und Kristallstrukturen der Kronenether-Komplexe [Li$_3$(12-Krone-4)$_3$O$_2$CCH$_3$][Cd(Se$_4$)$_2$], {[K(18-Krone-6)]$_2$[Hg(Se$_4$)$_2$]}$_2$ und [Na(15-Krone-5)]NO$_3$, *Z. Naturforsch.*, 44B(4), 444–454 (1989).

Liesic von J., Klar G. Chalkogenolat-Ionen und ihre Derivate. II. Komplexe von Chalkogenophenolat-Ionen mit Quecksilber(II), *Z. anorg. und allg. Chem.*, 435(1), 103–112 (1977).

Magul S., Neumüller B., Dehnicke K. Synthese und Kristallstrukturen der Polyselenido-Mercurate (PPh$_4$)$_2$[Hg(Se$_4$)$_2$] und [Cs(18-Krone-6)]$_2$[Hg$_2$(Se$_4$)$_3$], *Z. Naturforsch.*, 46B(8), 985–991 (1991).

Nicpon P., Meek D.W. Metal complexes of tertiary arsine sulphides and phosphine selenides, *Chem. Commun.*, (13), 398–399 (1966).

Trikalitis P.N., Rangan K.K., Bakas T., Kanatzidis M.G. Varied pore organization in meso-structured semiconductors based on the [SnSe$_4$]$^{4-}$ anion, *Nature*, 410(6829), 671–675 (2001).

Wachhold M., Rangan K.K., Billinge S.J.L., Petkov V., Heising J., Kanatzidis M.G. Mesostructured non-oxidic solids with adjustable worm-hole shaped pores: M–Ge–Q (Q = S, Se) frameworks based on tetrahedral [Ge$_4$Q$_{10}$]$^{4-}$ clusters, *Adv. Mater.*, 12(2), 85–91 (2000a).

Wachhold M., Rangan K.K., Lei M., Thorpe M.F., Billinge S.J.L., Petkov V., Heising J., Kanatzidis M.G. Mesostructured metal germanium sulfide and selenide materials based on the tetrahedral [Ge$_4$S$_{10}$]$^{4-}$ and [Ge$_4$Se$_{10}$]$^{4-}$ units: Surfactant templated three-dimensional disordered frameworks perforated with worm holes, *J. Solid State Chem.*, 152(1), 21–36 (2000b).

Wang R.-F., Zhang W.-G., Fan J., Wang S.-L. Crystal structure of bis(triphenylphosphine) hexakis(μ-phenylselenido)bis(phenylselenido)tetramercury(II), $Hg_4[P(C_6H_5)_3]_2$ $(C_6H_5Se)_8$, Z. Kristallogr. New Cryst. Struct., 220(1), 61–64 (2005).

Xiong W.-W., Li P.-Z., Zhou T.-H., Tok A.I.Y., Xu R., Zhao Y., Zhang Q. Kinetically controlling phase transformations of crystalline mercury selenidostannates through surfactant media, Inorg. Chem., 52(8), 4148–4150 (2013a).

Xiong W.-W., Li P.-Z., Zhou T.-H., Zhao Y., Xu R., Zhang Q. Solvothermal syntheses of three new one-dimensional ternary selenidostannates: $[DBNH][M_{1/2}Sn_{1/2}Se_2]$ (M=Mn, Zn, Hg), J. Solid State Chem., 204, 86–90 (2013b).

# 9 Systems Based on HgTe

## 9.1 Hg–H–K–C–N–O–Te

$[(C_{18}H_{36}N_2O_6K)_2HgTe_2 \cdot (C_2H_8N_2)]$ or $C_{38}H_{80}K_2N_6O_{12}Te_2Hg$. This multinary compound is formed in this system. It crystallizes in the triclinic structure with the lattice parameters $a = 1126.3 \pm 0.1$, $b = 1217.1 \pm 0.2$, and $c = 1033.5 \pm 0.2$ pm and $\alpha = 107.00° \pm 0.02°$, $\beta = 91.18° \pm 0.02°$, and $\gamma = 88.36° \pm 0.02°$ and a calculated density of 1.65 g·cm$^{-3}$ (Burns and Corbett 1981). To obtain it, 0.266 mM of the alloy composition KHgTe was allowed to react with 0.266 mM of 2,2,2-crypt (4,7,13,16,21,24-hexaoxa-1,10-diazabicyclo[8.8.8]hexacosane, $C_{18}H_{36}N_2O_6$) in ca. 8 mL of dry ethylenediamine ($C_2H_8N_2$). This initially gave a red solution around the alloy, which upon gentle shaking or diffusion quickly produced a yellow solution. There was also evidence for an additional blue intermediate. After ca. 3 days, the red color ceased to form around the alloy, and the solution was intensely yellow. Slow evaporation of the $C_2H_8N_2$ from the decanted solution gave large, block-shaped crystals, which were washed with ethylamine to remove any excess 2,2,2-crypt. The crystals instantly turn black upon exposure to moist air.

## 9.2 Hg–H–C–Si–Te

$[Hg\{SiTe(SiMe_3)_3\}_2]_2$ or $C_{18}H_{54}Si_8Te_2Hg$. This quinary compound is formed in this system. It melts at 166°C–172°C (Bonasia and Arnold 1992). To obtain it, a solution of $Hg[N(SiMe_3)_2]_2$ (1.0 mM) in 25 mL of hexane was combined with $HTeSi(SiMe_3)_3$ (2.0 mM) dissolved in 30 mL of hexane to give a yellow-brown solution. After 30 min of stirring, no further darkening was observed, and the solvent was removed under reduced pressure. Extraction with hexamethyldisiloxane (45 mL) gave a yellow solution, which was filtered, concentrated, and cooled to −40°C. After 24 h, filtration afforded the title compound as transparent green crystals.

It could be prepared also by another procedure (Bonasia and Arnold 1992). Diethyl ether (70 mL) was added to a stirred mixture of 2.4 mM of $(THF)_2LiTeSi(SiMe_3)_3$ and 1.20 mM of $HgCl_2$. The brown mixture was stirred for 12 h, after which no further darkening was observed. The solvent was removed under reduced pressure, and the residue was extracted with pentane (50 mL). The yellow solution was filtered, concentrated to 10 mL, and cooled to −40°C. After 12 h, 0.763 g clear green-yellow crystals were isolated by filtration.

## 9.3 Hg–H–C–N–Te

Some compounds are formed in this system.

$[(Me_4N)_3Hg_3Te_7 \cdot 0.5(C_2H_8N_2)]$ or $C_{13}H_{40}N_4Te_7Hg_3$. This compound crystallizes in the triclinic structure with the lattice parameters $a = 1208.0 \pm 0.3$, $b = 1726.1 \pm 0.7$, and $c = 990.5 \pm 0.1$ pm and $\alpha = 94.22° \pm 0.03°$, $\beta = 92.26° \pm 0.01°$, and $\gamma = 80.12° \pm 0.03°$ (Dhingra et al. 1994). It was prepared by the constant current cathodic dissolution of a $Hg_2Te_3$ electrode into an ethylenediamine solution of $Me_4NI$. Flat, black needles were obtained.

**$[(Et_4N)_2Hg_2Te_4]$ or $C_{16}H_{40}N_2Te_4Hg_2$.** This compound crystallizes in the monoclinic structure with the lattice parameters $a = 1486.9 \pm 0.8$, $b = 1226.3 \pm 0.4$, and $c = 1529.7 \pm 0.6$ pm, and $\beta = 97.43° \pm 0.04°$ and a calculated density of 2.815 g·cm$^{-3}$ (Dhingra et al. 1994). Treatment of an ethylenediamine extract of an alloy of nominal composition $K_2Hg_2Te_3$ with $Et_4NI$ gives $(Et_4N)_2Hg_2Te_4$.

**$[(Me_4N)_4\{Hg_4Te_{12}\}]$ or $C_{16}H_{48}N_4Te_{12}Hg_4$.** This compound crystallizes in the monoclinic structure with the lattice parameters $a = 1032.4 \pm 0.3$, $b = 2185.1 \pm 0.6$, $c = 1112.5 \pm 0.4$ pm, and $\beta = 110.63° \pm 0.03°$ and a calculated density of 3.719 g·cm$^{-3}$ (Kim and Kanatzidis 1994). It was obtained from the reaction of $HgCl_2$ and $K_2Te$ in a 2:5 molar ratio in DMF. To a solution of 1.1 mM $K_2Te$ in 20 mL DMF, a 15 mL DMF solution of 0.44 mM HgCl was added dropwise over 15 min. $K_2Te$ is not soluble in DMF but dissolved to give a dark-brownish solution upon the addition of $HgCl_2$ solution. Following filtration to remove a precipitate, an 80 mL MeOH solution of 0.46 mM $Me_4NCl$ was layered over the filtrate solution. Black rectangular platelet crystals grew allowing the solution to stand for a week. These crystals were isolated and washed with ether several times.

**$[(Bu^n_4N)_4Hg_4Te_{12}]$ or $C_{64}H_{144}N_4Te_{12}Hg_4$.** This compound crystallizes in the monoclinic structure with the lattice parameters $a = 1665.3 \pm 0.6$, $b = 2688.8 \pm 0.6$, $c = 2706.8 \pm 0.6$ pm, and $\beta = 90.72° \pm 0.03°$ (Haushalter 1985). To obtain it, $K_2Hg_2Te_3$ dissolves in ethylenediamine to give very dark-brown-orange solution. Treatment of these dark extracts with a solution of $Bu^n_4NBr$ in MeOH furnishes dark-brown $[Bu^n_4N]_4Hg_4Te_{12}$.

## 9.4 Hg–H–C–N–Cl–Te

**$[(Et_4N)_2(\mu\text{-}TePh)_6(HgCl)_4]$ or $C_{52}H_{70}N_2Cl_4Te_6Hg_4$.** This multinary compound is formed in this system. It was prepared from $HgCl_2$ (0.19 mM) and $Et_4NCl \cdot H_2O$ (0.38 mM) and $Hg(TePh)_2$ (0.56 mM) in $CH_2Cl_2$ (4 mL) at room temperature (Dean and Manivannan 1990b). The mixture was stirred, but a clear solution was never obtained. Over about a 30 min period, the red color of the solid $Hg(TePh)_2$ was replaced by the yellow color of the solid product, and the completion of the reaction was detected visually. The crystalline product was separated by decantation, washed with $Et_2O$, and dried in vacuo. All syntheses and preparation of samples were carried out under an Ar or $N_2$ atmosphere.

## 9.5 Hg–H–C–N–Cl–Mn–Te

**$[\{Mn(C_2H_8N_2)_3\}_2Cl_2(Hg_2Te_4)]$ or $C_{12}H_{48}N_{12}Cl_2Mn_2Te_4Hg_2$.** This multinary compound is formed in this system. It crystallizes in the monoclinic structure with the lattice parameters $a = 1572.0 \pm 0.3$, $b = 1681.2 \pm 0.3$, $c = 1522.4 \pm 0.3$ pm, and

$\beta = 114.31° \pm 0.01°$ (Li et al. 1996). It has been synthesized at 180°C by a solvothermal reaction using ethylenediamine as a solvent.

## 9.6  Hg–H–C–N–Mn–Te

$[\{Mn(C_2H_8N_2)_3\}_2Hg_2Te_9]$ or $C_{12}H_{48}N_{12}Mn_2Te_9Hg$. This multinary compound is formed in this system. It crystallizes in the monoclinic structure with the lattice parameters $a = 1201.9 \pm 0.1$, $b = 2873.1 \pm 0.4$, $c = 1826.7 \pm 0.4$ pm, and $\beta = 98.04° \pm 0.01°$ (Li et al. 1996). It has been synthesized at 180°C by a solvothermal reaction using ethylenediamine as a solvent.

## 9.7  Hg–H–C–N–Fe–Te

$[\{Fe(C_2H_8N_2)_3\}_2Hg_2Te_9]$ or $C_{12}H_{48}N_{12}Fe_2Te_9Hg_2$. This multinary compound is formed in this system. It crystallizes in the monoclinic structure with the lattice parameters $a = 1191.5 \pm 0.2$, $b = 1910.5 \pm 0.5$, $c = 1818.0 \pm 0.5$ pm, and $\beta = 98.64° \pm 0.01°$ and a calculated density of 3.282 g·cm$^{-3}$ (Li et al. 1995). It was synthesized by the reaction of HgCl (0.25 mM), FeCl$_2$ (0.25 mM), K$_2$Te (0.25 mM), and Te (0.75 mM) in ethylenediamine at 160°C. The starting material was mixed in a glovebox and loaded in a Pyrex tube containing ca. 3.7 mL of ethylenediamine. The tube was then evacuated and sealed with a torch. The sample was heated at 160°C for 7 days and then cooled to room temperature. The product was washed with EtOH followed by diethyl ether. The black, column-like crystals were obtained.

## 9.8  Hg–H–C–P–O–Cl–Te

Some compounds are formed in this system.

$[Hg_4(TePh)_6(PEt_3)_4](ClO_4)_2$ or $C_{60}H_{90}P_4O_8Cl_2Te_6Hg_4$. To obtain this compound, solid Hg(TePh)$_2$ (1.5 mM) was added with stirring to a mixture of Hg(PEt$_3$)$_2$(ClO$_4$)$_2$ (0.51 mM) and PEt$_3$ (1.0 mM) in 7 mL of acetone (Dean et al. 1989). After ca. 30 min, most of the Hg(TePh)$_2$ was dissolved to give a reddish-orange solution. The small amount of solid that remained was removed by filtration. After Et$_2$O (6 mL) was layered onto the filtrate, crystallization occurred on keeping the mixture at 0°C overnight. The yellowish-orange product was separated by decantation of the mother liquor, washed with Et$_2$O, and dried in vacuo. All syntheses were carried out under an Ar atmosphere.

$[Hg_4(TeMe)_6(PPh_3)_4](ClO_4)_2$ or $C_{78}H_{78}P_4O_8Cl_2Te_6Hg_4$. To obtain this compound, the crude Hg(TeMe)$_2$ was added incrementally with stirring to a mixture of Hg(PPh$_3$)$_2$(ClO$_4$)$_2$ (0.111 mM) and PPh$_3$ (0.22 mM) in 6 mL of CHCl$_3$ at room temperature, until the $^{31}$P NMR spectrum of the supernatant liquid at −60°C showed no evidence for Hg(PPh$_3$)$_4]^{2+}$ (Dean et al. 1989). At this point, the heterogeneous mixture was filtered. The yellow filtrate was covered with a layer of Et$_2$O (1 mL) and left undisturbed at 0°C–5°C overnight. The resultant yellow crystalline solid was separated by decantation of the mother liquor, washed with Et$_2$O, and dried in vacuo to yield a yellow powder of $[Hg_4(TeMe)_6(PPh_3)_3](ClO_4)_2$. All syntheses were carried out under an Ar atmosphere.

**$[Hg_4(TePh)_6(P\{4\text{-}C_6H_4Cl\}_3)_3](ClO_4)_2$ or $C_{90}H_{66}P_3O_8Cl_{11}Te_6Hg_4$.** This compound was prepared in the same manner as $[Hg_4(TePh)_6(PPh_3)_4](ClO_4)_2$, except that it was unnecessary to add $Et_2O$ to induce crystallization (Dean et al. 1989). The yellow product that formed when all the $Hg(TePh)_2$ had reacted was separated from the mother liquor by decantation, washed with $Et_2O$, and dried in vacuo. This same compound was formed as an immediate yellow precipitate when $Et_2O$ was layered onto a solution of $[Hg_4(TePh)_6(P\{C_6H_4Cl\}_3)_3)_4(ClO_4)_2$ in $CH_2Cl_2$.

**$[Hg_4(TePh)_6(PPh_3)_3](ClO_4)_2$ or $C_{90}H_{75}P_3O_8Cl_2Te_6Hg_4$.** To obtain this compound, a solution of $Hg(PPh_3)_2(ClO_4)_2$ (0.13 mM) in 5 mL of $CH_2Cl_2$ at room temperature was stirred solid $PPh_3$ (0.13 mM) and then solid $Hg(TePh)_2$ (0.39 mM) (Dean et al. 1989). After 10 min, the mixture was filtered and $Et_2O$ was layered onto the yellow filtrate. Crystallization began within 10 min at room temperature and was completed by cooling to 0°C–5°C in the refrigerator for ca. 18 h. The yellow crystals were separated by decantation of the mother liquor, washed with $Et_2O$, and dried in vacuo to yield $[Hg_4(TePh)_6(PPh_3)_3](ClO_4)_2$. Crystallinity was lost during the drying process. A sample of this compound, as a fine yellow powder, was obtained without the addition of an inert diluent when a solution of $[Hg_4(TePh)_6(PPh_3)_3](ClO_4)_2$ in $CHCl_3$ was allowed to stand for several weeks at room temperature in a capped vessel. All syntheses were carried out under an Ar atmosphere.

**$[Hg_4(TePh)_6(P\{4\text{-}C_6H_4Me\}_3)_3](ClO_4)_2$ or $C_{99}H_{93}P_3O_8Cl_2Te_6Hg_4$.** This compound was prepared in the same manner as $[Hg_4(TePh)_6(PPh_3)_3](ClO_4)_2$, but by using $CHCl_3$ as the initial solvent and with crystallization induced by cooling the filtrate to 0°C–5°C overnight without the addition of $Et_2O$ (Dean et al. 1989).

**$[Hg_4(\mu\text{-}TePh)_6(PPh_3)_4](ClO_4)_2$ or $C_{108}H_{90}P_4O_8Cl_2Te_6Hg_4$.** Triphenylphosphine (10 mM) and $(PPh_3)_2Hg(ClO_4)_2$ (5 mM) were stirred together in 50 mL of acetone until the solution became clear (~10 min) (Singh et al. 1990). Solid $Hg(TePh)_2$ (15 mM) was added to this solution with continued stirring. Within 30 min, a clear yellow solution was obtained. Cyclohexane (20 mL) was slowly added to this solution so that it formed a separate layer. The mixture was kept at 5°C for 48 h. The resulting yellow crystals were filtered, washed thoroughly with cyclohexane, recrystallized from chloroform, and dried in vacuo. All operations were performed under an atmosphere of argon.

**$[Hg_4(\mu\text{-}TeC_6H_4\text{-}p\text{-}OMe)_6(PPh_3)_4](ClO_4)_2$ or $C_{114}H_{102}P_4O_{14}Cl_2Te_6Hg_4$.** This compound was prepared in the manner described for $[Hg_4(\mu\text{-}TePh)_6(PPh_3)_4](ClO_4)_2$, except that the reaction medium was chloroform (Singh et al. 1990).

## 9.9   Hg–H–C–P–Te

Some compounds are formed in this system.

**$[Hg(TePh)_2\cdot depe]$ or $C_{22}H_{34}P_2Te_2Hg$ [depe = 1,2-bis(diethylphosphino)ethane].** This compound melts at 95°C, decomposes at 128°C–129°C, and crystallizes in the orthorhombic structure with the lattice parameters $a = 1668.3 \pm 0.3$, $b = 1711.4 \pm 0.3$, and $c = 1918.7 \pm 0.3$ pm and a calculated density of 1.896 g·cm$^{-3}$ (Brennan et al. 1990). To obtain it, depe (1.94 mM) was added to Hg(TePh)$_2$ (0.67 mM) suspended in toluene (35 mL). After stirring for 3 h, the solution was filtered, concentrated to ca. 15 mL, and cooled ($-10°$C) to give yellow crystals. All reactions were performed in an inert atmosphere.

**[{Hg(TePh)$_2$}$_2$·dmpe] or C$_{30}$H$_{36}$P$_2$Te$_4$Hg$_2$ [dmpe = 1,2-bis(dimethylphosphino) ethane].** This compound melts at 163.9°C–165.5°C (Brennan et al. 1990). To obtain it, dmpe (0.53 mM) was added to Hg(TePh)$_2$ (0.66 mM) in pyridine (20 mL). The solution was saturated with heptane (5 mL) and cooled ($-10°$C) to give yellow needles. All reactions were performed in an inert atmosphere.

**[(Ph$_4$P)(PhTe)$_3$Hg] or C$_{42}$H$_{35}$PTe$_3$Hg.** This compound melts at 130°C with decomposition (Liesic von and Klar 1977) and crystallizes in the monoclinic structure with the lattice parameters $a = 1553.4 \pm 1.9$, $b = 1448.1 \pm 1.6$, and $c = 1793.2 \pm 2.1$ pm and $\beta = 96.74° \pm 0.05°$ (Behrens et al. 1977). It was prepared from the reaction of Hg(TePh)$_2$, NaTePh, and Ph$_4$PCl in the liquid NH$_3$. Its single crystals were grown by recrystallization from a THF/EtOH solution of the title compound.

**[(Ph$_4$P)$_2$Hg$_2$Te$_5$] or C$_{48}$H$_{40}$P$_2$Te$_5$Hg$_2$.** This compound crystallizes in the monoclinic structure with the lattice parameters $a = 1040.0 \pm 0.2$, $b = 1732.4 \pm 0.4$, $c = 1374.2 \pm 0.3$ pm, and $\beta = 94.02°$ (Haushalter 1985). It forms black, spear-shaped needles with curved crystal faces, when the solution of K$_2$Hg$_2$Te$_3$ in ethylenediamine and an ethylenediamine solution of Ph$_4$PBr are allowed to diffuse slowly (3 weeks) into each other in an *H*-tube.

**[(Ph$_4$P)$_2$HgTe$_7$] or C$_{48}$H$_{40}$P$_2$Te$_7$Hg.** This compound crystallizes in the monoclinic structure with the lattice parameters $a = 1275.5 \pm 0.3$, $b = 1432.7 \pm 0.3$, $c = 1393.2 \pm 0.3$ pm, and $\beta = 105.58° \pm 0.03°$ at 110 K (McConnachie et al. 1993). To obtain it, a polytelluride solution, which was prepared by dissolving Li$_2$Te powder and 3 equivalents of Te powder in DMF, was added to a DMF solution of Hg(EtCOS$_2$) and Ph$_4$PCl with stirring. Et$_3$P was added, and the mixture was heated for 1 h at 90°C, followed by filtration and layering with ether.

**[(Ph$_4$P)$_2$Hg(Te$_4$)$_2$] or C$_{48}$H$_{40}$P$_2$Te$_8$Hg.** This compound crystallizes in the tetragonal structure with the lattice parameters $a = 2136.5 \pm 0.3$ and $c = 1126.8 \pm 0.2$ and a calculated density of 2.454 g·cm$^{-3}$ (Bollinger et al. 1995). It was prepared by the following procedure. To a solution of Li$_2$Te$_4$, (2 mM) in DMF (15 mL) was added at room temperature a solution of HgCl$_2$ (1 mM) dissolved in DMF (5 mL). To the resulting red-brown reaction mixture was added Ph$_4$PBr (2.0 mM) dissolved in DMF (10 mL). The solution was filtered and the filtrate was layered with Et$_2$O (50 mL). Red-brown crystals of the title compound were grown, while the solution was stirred overnight at room temperature. A second bath of product was obtained by layering additional Et$_2$O on the mother liquor.

**[{(C$_6$H$_5$)$_4$P}$_5${Hg$_3$(TeC$_6$H$_5$)$_{11}$}] or C$_{186}$H$_{155}$P$_5$Te$_{11}$Hg$_3$.** This compound melts at 96°C with decomposition. It was prepared from the reaction of Hg(TePh)$_2$, NaTePh, and Ph$_4$PCl in liquid NH$_3$ (Liesic von and Klar 1977).

## 9.10   Hg–H–C–O–Te

**[(*p*-EtOC₆H₄Te)₂Hg]** or **C₁₆H₁₈O₂Te₂Hg**. This multinary compound is formed in this system. It melts at 110°C–112°C (Dance and Jones 1978). To obtain it, (*p*-EtOC₆H₄)₂Te₂ (0.5 mM) and metallic Hg (25 mM) were shaken with 20 mL of dry benzene for 48 h. The resulting deep-red solid was ground to a powder and the obtained suspension in benzene was decanted from unreacted Hg. Filtration produced a red powder.

## 9.11   Hg–H–C–O–Cl–Te

Some compounds are formed in this system.

**[*p*-MeOC₆H₄TeHgCl₂]** or **C₇H₇OCl₂TeHg**. This compound melts at 114°C–115°C (Dance and Jones 1978). To obtain it, (*p*-MeOC₆H₄)₂Te₂ (1 mM) and HgCl₂ (2 mM) were dissolved in the minimum quantities of hot EtOH and mixed. Upon cooling, the yellow precipitate was filtered and washed with cold EtOH.

**[*p*-EtOC₆H₄TeHgCl]** or **C₈H₉OClTeHg**. This compound melts at 160°C–161°C (Dance and Jones 1978). To obtain it, equimolar quantities of (*p*-EtOC₆H₄Te)₂Hg and HgCl₂ were refluxed for 30 min in a small volume of EtOH. The deep-red solid was rapidly replaced by a fine yellow powder which was filtered and washed with EtOH.

**[*p*-EtOC₆H₄TeHgCl₂]** or **C₈H₉OCl₂TeHg**. This compound melts at 110°C–112°C (Dance and Jones 1978). To obtain it, (*p*-EtOC₆H₄)₂Te₂ (1 mM) and HgCl₂ (2 mM) were dissolved in the minimum quantities of hot EtOH and mixed. Upon cooling, the yellow precipitate was filtered and washed with cold EtOH.

**[(*p*-EtOC₆H₄)₂TeHgCl₂]** or **C₁₆H₁₈O₂Cl₂TeHg**. This compound melts at 157°C–158°C (Dance and Jones 1978).

## 9.12   Hg–H–C–O–Br–Te

Some compounds are formed in this system.

**[(*p*-EtOC₆H₄)₂TeHgBr₂]** or **C₁₆H₁₈O₂Br₂TeHg**. This compound melts at 157°C–158°C (Dance and Jones 1978).

**[(*p*-EtOC₆H₄)₂Te₂HgBr₂]** or **C₁₆H₁₈O₂Br₂Te₂Hg**. This compound melts at 120°C–122°C (Dance and Jones 1978). To obtain it, (*p*-EtOC₆H₄)₂Te₂ and HgBr₂ were dissolved in the minimum quantities of hot EtOH and mixed. Upon cooling, the precipitate was filtered, washed with EtOH, and dried under vacuum. The stoichiometry of the compound formed was dependent on the ratio of reactants and on the rate and order of mixing. The stoichiometry was in all cases intermediate between [*p*-EtOC₆H₄TeHgBr₂] and [(*p*-EtOC₆H₄)₂Te₂HgBr₂]. A sample of a brown complex with stoichiometry (*p*-EtOC₆H₄)₂Te₂HgBr₂ has a melting point of 109°C–110°C.

## 9.13 Hg–H–C–O–I–Te

Some compounds are formed in this system.

**[($p$-EtOC$_6$H$_4$)$_2$TeHgI$_2$] or C$_{16}$H$_{18}$O$_2$I$_2$TeHg**. This compound melts at 129°C–130°C (Dance and Jones 1978).

**[($p$-EtOC$_6$H$_4$)$_2$Te$_2$HgI$_2$] or C$_{16}$H$_{18}$O$_2$I$_2$Te$_2$Hg**. This compound melts at 125°C–126°C (Dance and Jones 1978). To obtain it, equimolar quantities of ($p$-EtOC$_6$H$_4$)Te$_2$ and HgI$_2$ were mixed and a brown flocculent precipitate formed. This was filtered, washed with EtOH, and dried.

**[(EtOC$_6$H$_4$)$_2$Te$_2$HgI$_2$] or C$_{16}$H$_{18}$O$_2$I$_2$Te$_2$Hg**. This compound melts at 202°C–204°C (Dance and Jones 1978). When (EtOC$_6$H$_4$)Te$_2$ (1 mM) and HgI$_2$ (2 mM) were mixed in hot EtOH, a yellow precipitate was formed upon cooling. This was washed with cold EtOH and dried.

## 9.14 Hg–H–C–Cl–Te

**[(Ph$_2$Te)HgCl$_2$] or C$_{12}$H$_{10}$Cl$_2$TeHg**. This multinary compound is formed in this system. It melts at 158°C–160°C (Dance and Jones 1978) and was prepared by mixing equimolar quantities of HgCl$_2$ and Ph$_2$Te solutions in EtOH (Einstein et al. 1983).

## 9.15 Hg–H–C–Br–Te

**[(Ph$_2$Te)HgBr$_2$] or C$_{12}$H$_{10}$Br$_2$TeHg**. This multinary compound is formed in this system. It melts at 150°C–152°C (Dance and Jones 1978) and was prepared by mixing equimolar quantities of HgBr$_2$ and Ph$_2$Te solutions in EtOH (Einstein et al. 1983).

## 9.16 Hg–H–C–Br–I–Te

**[Me$_2$TeI$_2$·HgBr$_2$] or C$_2$H$_6$Br$_2$I$_2$TeHg**. This multinary compound is formed in this system. It melts at 127°C–128°C (Dance and Jones 1978). To obtain it, equimolar quantities of α-Me$_2$TeI$_2$ and HgBr$_2$ were dissolved in the minimum volume of absolute EtOH and mixed. The yellow α-Me$_2$TeI$_2$ became clear and, in standing in stream of N$_2$, produced yellow needles of Me$_2$TeI$_2$·HgBr$_2$. It was also formed, following the same procedure, from α-Me$_2$TeBr$_2$ and HgI$_2$.

## 9.17 Hg–H–C–I–Te

Some compounds are formed in this system.

**[(Ph$_2$Te)HgI$_2$] or C$_{12}$H$_{10}$I$_2$TeHg**. This compound melts at 150°C–151°C (Dance and Jones 1978) and crystallizes in the monoclinic structure with the lattice parameters $a = 1482.7 \pm 0.2$, $b = 1343.8 \pm 0.3$, and $c = 1524.5 \pm 0.2$ pm, and $\beta = 90.32° \pm 0.1°$ and the experimental and calculated densities 3.19 and 3.220 g·cm$^{-3}$, respectively (Einstein et al. 1983). It was prepared by mixing the solution of HgI$_2$ in EtOH with the solution of PhTe$_2$ in some excess in the same solvent.

**[Ph₂Hg·2Me₂TeI₂]** or **C₁₆H₂₂I₄Te₂Hg**. This compound melts at 127°C–128°C (Dance and Jones 1978).

**[(Et₂TeI₂)₂·HgPh₂]** or **C₂₀H₃₀I₄Te₂Hg**. Upon mixing equimolar solutions in chloroform of HgPh₂ and of the α-Et₂TeI₂, the brown color of the solution became yellow, and after 5 s, a yellow precipitate was suddenly formed (Gilbert and Lowry 1928). This precipitate was a heavy, yellowish-green oil; when cooled to −8°C, it solidified to a pale-yellow crystalline mass and then remained solid up to 94°C.

## 9.18 Hg–H–Sn–C–N–Te

**(Et₄N)₂[HgSnTe₄]** or **C₁₆H₄₀N₂SnTe₄Hg**. This multinary compound is formed in this system. It crystallizes in the tetragonal structure with the lattice parameters $a = 1461.4 \pm 0.2$ and $c = 758.5 \pm 0.1$ pm and a calculated density of 2.235 g·cm⁻³ (Dhingra and Haushalter 1994). It was prepared by ethylenediamine extraction of K₂HgSnTe₄. The alloy was crushed to a fine black powder and stored under an He atmosphere. The red-brown extract obtained by the addition of 10 mL of ethylenediamine to 1.00 g of K₂HgSnTe₄ was filtered and carefully layered with a solution of 0.566 g of Et₄NI in 10 mL of ethylenediamine, which after 4 days gave large crystals of (Et₄N)₂[HgSnTe₄].

## 9.19 Hg–Cu–In–Mn–Te

**2HgTe–CuInTe₂–2MnTe**. The $T(x)$ diagram for Hg₂ₓ(Cu,In)ᵧMn₂zTe₂ at $z = 0$ and the $T(z)$ diagrams at $y = 3x$, $x = y$, and $x = 3y$ are given in Grima et al. (1992). In the case of $z = 0$ section, there is an effectively single-phase solid solution across the complete composition range, with the order–disorder chalcopyrite α to zincblende β transformation being shown as a single line, varying from 670°C at $x = 0$ to ~ 220°C at $x = 0.25$. The sections $y = 3x$, $x = y$, and $x = 3y$ show very similar behavior to one another. However, in contrast to the $x = 0$ section, only the $y = 3x$ section shows the presence of the α chalcopyrite phase and only at temperatures below ~200°C. In all three sections, transitions are observed, which can be attributed to Mn ordering (β–β′). For this system, the transition occurs at temperatures up to 430°C. In all three sections also, the transition at 325°C is observed for all compositions investigated. In the case of the $x = 3y$ section, the two-phase field L + β is reduced to the line $z = 0$, which is the limit of the three-phase field L + β + γ. In all of these sections, the β-phase disappears at ~800°C. Each sample was annealed after preparing for 1 month at 450°C.

## 9.20 Hg–Ag–In–Mn–Te

**2HgTe–AgInTe₂–2MnTe**. The $T(z)$ diagram for Hg₂ₓ(Ag,In)ᵧMn₂zTe₂ at $x = 0$, $y = 3x$, $x = y$, and $x = 3y$ are given in Grima et al. (1992). Below 325°C, a two-phase β + η field is shown, but the amount of the η-phase is very small and is not observed by XRD. Such segregation effect could be attributed to Te segregation from the AgInTe₂ phase. Each sample was annealed after preparing for 1 month at 450°C.

# REFERENCES

Behrens U., Hofmann K., Klar G. Chalkogenolat-Ionen und ihre Derivate, III. Kristall- und Molekülstruktur von Tetraphenylphosphonium-tris(tellurophenolato)mercurat(II), *Chem. Ber.*, 110(11), 3672–3677 (1977).

Bollinger J.C., Roof L.C., Smith D.M., McConnachie J.M., Ibers J.A. Synthesis, x-ray crystal structures, and NMR spectroscopy of [PPh$_4$]$_2$[M(Te$_4$)$_2$], M = Hg, Cd, Zn, *Inorg. Chem.*, 34(6), 1430–1434 (1995).

Bonasia P.J., Arnold J. Zinc, cadmium, and mercury tellurolates: Hydrocarbon solubility and low coordination numbers enforced by sterically encumbered silyltellurolate ligands, *Inorg. Chem.*, 31(12), 2508–2514 (1992).

Brennan J.G., Siegrist T., Carroll P.J., Stuczynski S.M., Reynders P., Brus L.E., Steigerwald M.L. Bulk and nanostructure group II-VI compounds from molecular organometallic precursors, *Chem. Mater.*, 2(4), 403–409 (1990).

Burns R.C., Corbett J.D. Heteroatomic polyanions of post transition metals. Synthesis and crystal structure of a salt of ditelluromercurate(II), HgTe$_2$$^{2-}$, *Inorg. Chem.*, 20(12), 4433–4434 (1981).

Dance N.S., Jones C.H.W. A spectroscopic investigation of some organotellurium-mercury(II) complexes, *J. Organomet. Chem.*, 152(2), 175–185 (1978).

Dean P.A.W., Manivannan V. Synthesis and multinuclear ($^{13}$C, $^{77}$Se, $^{125}$Te, $^{199}$Hg) magnetic resonance spectra of adamantane-like anions of mercury(II), [(μ-ER)$_6$(HgX)$_4$]$^{2-}$ (E = S, Se, Te; X = Cl, Br, I), *Inorg. Chem.*, 29(16), 2997–3002 (1990b).

Dean P.A.W., Manivannan V., Vittal J.J. New adamantane-like mercury–chalcogen cages. 2. Synthetic and multinuclear ($^{31}$P, $^{77}$Se, $^{125}$Te, $^{199}$Hg) magnetic resonance study of tellurolate-bridged mercury(II) clusters [(μ-TeR)$_6$(HgPR′$_3$)$_4$]$^{2+}$ and [(μ-TeR)$_6$(Hg)(HgPR′$_3$)$_3$]$^{2+}$ and related species with mixed-bridging chalcogenates, *Inorg. Chem.*, 28(12), 2360–2368 (1989).

Dhingra S.S., Haushalter R.C. One-dimensional inorganic polymers: Synthesis and structural characterization of the main group metal polymers K$_2$HgSnTe$_4$, (Et$_4$N)$_2$HgSnTe$_4$, (Ph$_4$P) GeInTe$_4$, and RbInTe$_2$, *Chem. Mater.*, 6(12), 2376–2381 (1994).

Dhingra S.S., Warren C.J., Haushalter R.C., Bocarsly A.B. One-dimensional mercury telluride polymers: Synthesis and structure of (Et$_4$N)$_2$Hg$_2$Te$_4$ and (Me$_4$N)$_4$Hg$_3$Te$_7$(0.5en), *Chem. Mater.*, 6(12), 2382–2385 (1994).

Einstein F.W.B., Jones C.H.W., Jones N., Sharma R.D. Crystal structure and vibrational spectra of (TePh$_2$)HgI$_2$, *Inorg. Chem.*, 22(26), 3924–3928 (1983).

Gilbert F.L., Lowry T.M. Studies of valency. Part XII. Isomeric derivatives of diethyl telluride, *J. Chem. Soc.*, 3179–3189 (1928).

Grima P., Quintero M., Woolley J.C. T(z) diagrams and lattice parameter values for the Hg$_{2x}$(CuIn)$_y$Mn$_{2z}$Te$_2$ and Hg$_{2x}$(AgIn)$_y$Mn$_{2z}$Te$_2$ systems, *J. Cryst. Growth*, 119(3–4), 381–390 (1992).

Haushalter R.C. Synthesis and structure of new Hg–Te polyanions: [Hg$_4$Te$_{12}$]$^{4-}$, a cluster containing Te$^{2-}$, Te$_2$$^{2-}$, and Te$_3$$^{2-}$ ligands and [Hg$_2$Te$_5$]$^{2-}$, a new one-dimensional inorganic polymer, *Angew. Chem. Int. Ed. Engl.*, 24(5), 433–435 (1985).

Kim K.-W., Kanatzidis M.G. Synthesis, structure, and properties of the polychalcogenides [M$_4$Te$_{12}$]$^{4-}$ (M = Cd, Hg), *Inorg. Chim. Acta*, 224(1–2), 163–169 (1994).

Liesic von J., Klar G. Chalkogenolat-Ionen und ihre Derivate. II. Komplexe von Chalkogenophenolat-Ionen mit Quecksilber(II), *Z. anorg. und allg. Chem.*, 435(1), 103–112 (1977).

Li J., Chen Z., Kelley J.L., Proserpio D.M. Crystal growth at low temperatures—Solvothermal synthesis of two mercury tellurides, *Mater. Res. Soc. Symp. Proc.*, 453, 29–34 (1996).

Li J., Rafferty B.G., Mulley S., Proserpio D.M. [Fe(en)$_3$]$_2$(Hg$_2$Te$_9$): A novel tellurometalate containing one-dimensional chains of weakly bound Zintl anions (Hg$_2$Te$_9$)$^{4-}$, *Inorg. Chem.*, 34(26), 6417–6418 (1995).

McConnachie J.M., Ansari M.A., Bollinger J.C., Salm R.J., Ibers J.A. Synthesis and structural characterization of the telluroargentate [PPh$_4$]$_2$[NEt$_4$][AgTe$_7$] and telluromercurate [PPh$_4$]$_2$[HgTe$_7$] compounds containing the unprecedented $\eta^3$-Te$_7^{4-}$polytelluride anion, *Inorg. Chem.*, 32(15), 3201–3202 (1993).

Singh A.K., Khandelwal B.L., Srivastava V. Synthesis of first aryltellurolate-mercury(II) clusters having Hg$_4$Te$_6$ core, *Phosphorus, Sulfur, Silicon*, 48(1–4), 169–171 (1990).

# Index